Conceptual Aircraft Design

Aerospace Series

Conceptual Aircraft Design: An Industrial Approach	Kundu, Price & Riordan	January 2019
Helicopter Flight Dynamics: Including a Treatment of Tiltrotor Aircraft, 3rd Edition	Gareth Padfield	October 2018
Space Flight Mechanics	Craig A. Kluever	March 2018
Introduction to Nonlinear Aeroelasticity	Grigorios Dimitriadis	April 2017
Introduction to Aerospace Engineering with a Flight Test Perspective	Stephen Corda	March 2017
Adaptive Aeroservoelastic Control	Tewari	March 2016
Theory and Practice of Aircraft Performance	Kundu, Price and Riordan	November 2015
The Global Airline Industry, Second Edition	Belobaba, Odoni and Barnhart	July 2015
Modeling the Effect of Damage in Composite Structures: Simplified Approaches	Kassapoglou	March 2015
Introduction to Aircraft Aeroelasticity and Loads, 2nd Edition	Wright and Cooper	December 2014
Aircraft Aerodynamic Design: Geometry and Optimization	Sóbester and Forrester	October 2014
Theoretical and Computational Aerodynamics	Sengupta	September 2014
Aerospace Propulsion	Lee	October 2013
Aircraft Flight Dynamics and Control	Durham	August 2013
Civil Avionics Systems, 2nd Edition	Moir, Seabridge and Jukes	August 2013
Modelling and Managing Airport Performance	Zografos, Andreatta and Odoni	July 2013
Advanced Aircraft Design: Conceptual Design, Analysis and Optimization of Subsonic Civil Airplanes	Torenbeek	June 2013
Design and Analysis of Composite Structures: With Applications to Aerospace Structures, 2nd Edition	Kassapoglou	April 2013
Aircraft Systems Integration of Air-Launched Weapons	Rigby	April 2013
Design and Development of Aircraft Systems, 2nd Edition	Moir and Seabridge	November 2012
Understanding Aerodynamics: Arguing from the Real Physics	McLean	November 2012
Aircraft Design: A Systems Engineering Approach	Sadraey	October 2012
Introduction to UAV Systems 4e	Fahlstrom and Gleason	August 2012
Theory of Lift: Introductory Computational Aerodynamics with MATLAB and Octave	McBain	August 2012
Sense and Avoid in UAS: Research and Applications	Angelov	April 2012
Morphing Aerospace Vehicles and Structures	Valasek	April 2012
Gas Turbine Propulsion Systems	MacIsaac and Langton	July 2011
Basic Helicopter Aerodynamics, 3rd Edition	Seddon and Newman	July 2011
Advanced Control of Aircraft, Spacecraft and Rockets	Tewari	July 2011
Cooperative Path Planning of Unmanned Aerial Vehicles	Tsourdos et al	November 2010
Principles of Flight for Pilots	Swatton	October 2010
Air Travel and Health: A Systems Perspective	Seabridge et al	September 2010
Design and Analysis of Composite Structures: With applications to aerospace Structures	Kassapoglou	September 2010
Unmanned Aircraft Systems: UAVS Design, Development and Deployment	Austin	April 2010
Introduction to Antenna Placement & Installations	Macnamara	April 2010

Principles of Flight Simulation	Allerton	October 2009
Aircraft Fuel Systems	Langton et al	May 2009
The Global Airline Industry	Belobaba	April 2009
Computational Modelling and Simulation of Aircraft and the Environment: Volume 1 - Platform Kinematics and Synthetic Environment	Diston	April 2009
Handbook of Space Technology	Ley, Wittmann Hallmann	April 2009
Aircraft Performance Theory and Practice for Pilots	Swatton	August 2008
Aircraft Systems, 3rd Edition	Moir & Seabridge	March 2008
Introduction to Aircraft Aeroelasticity and Loads	Wright & Cooper	December 2007
Stability and Control of Aircraft Systems	Langton	September 2006
Military Avionics Systems	Moir & Seabridge	February 2006
Design and Development of Aircraft Systems	Moir & Seabridge	June 2004
Aircraft Loading and Structural Layout	Howe	May 2004
Aircraft Display Systems	Jukes	December 2003
Civil Avionics Systems	Moir & Seabridge	December 2002

Conceptual Aircraft Design

An Industrial Approach

Ajoy Kumar Kundu
Queen's University Belfast
UK

Mark A. Price
Queen's University Belfast
UK

David Riordan
Bombardier Aerospace
UK

Registered Offices
John Wiley & Sons, Inc., 111 River Street, Hoboken, NJ 07030, USA
John Wiley & Sons Ltd, The Atrium, Southern Gate, Chichester, West Sussex, PO19 8SQ, UK

Editorial Office
The Atrium, Southern Gate, Chichester, West Sussex, PO19 8SQ, UK

For details of our global editorial offices, customer services, and more information about Wiley products visit us at www.wiley.com.

Wiley also publishes its books in a variety of electronic formats and by print-on-demand. Some content that appears in standard print versions of this book may not be available in other formats.

Library of Congress Cataloging-in-Publication Data

Names: Kundu, Ajoy Kumar, 1934- author. | Price, Mark A. (Aerospace engineer)
    author. | Riordan, David (Aerospace engineer) author.
Title: Conceptual aircraft design : an industrial approach / Professor Ajoy
    Kumar Kundu, Queen's University Belfast, UK, Professor Mark A. Price,
    Queen's University Belfast, UK, Mr David Riordan, Bombardier Aerospace.
Description: First edition. | Hoboken, NJ : John Wiley & Sons, 2019. |
    Includes bibliographical references and index. |
Identifiers: LCCN 2018032342 (print) | LCCN 2018033454 (ebook) | ISBN
    9781119500278 (Adobe PDF) | ISBN 9781119500261 (ePub) | ISBN 9781119500285
    (hardcover)
Subjects: LCSH: Airplanes–Design and construction.
Classification: LCC TL671.2 (ebook) | LCC TL671.2 .K763 2019 (print) | DDC
    629.134/1–dc23
LC record available at https://lccn.loc.gov/2018032342

Cover Design: Wiley
Cover Image: Courtesy of Bombardier Inc.

Set in 10/12pt WarnockPro by SPi Global, Chennai, India

We dedicate this book to all those who will enjoy studying our book.

Contents

Series Preface *xxxvii*

Preface *xxxix*

Individual Acknowledgements By Ajoy Kumar Kundu *xli*

By Mark A. Price *xlv*

By David Riordan *xlvii*

List of Symbols and Abbreviations *xlix*

Road Map of the Book *lvii*

Part I   Prerequisites *1*

1   **Introduction** *3*

1.1   Overview *3*

1.2   Brief Historical Background *4*

1.2.1   Flight in Mythology *4*

1.2.2   Fifteenth to Nineteenth Centuries *4*

1.2.3   From 1900 to World War I (1914) *5*

1.2.4   World War I (1914–1918) *6*

1.2.5   Period between World War I and World War II – Inter War Period, the Golden Age (1918–1939) *9*

1.2.6   World War II (1939–1945) *9*

1.2.7   Post-World War II *10*

1.3   Aircraft Evolution *10*

1.3.1   Aircraft Classifications and their Operational Environments *11*

1.4   Current Aircraft Design Trends for both Civil and Military Aircraft (the 1980s Onwards) *13*

1.4.1   Current Civil Aircraft Trends *15*

1.4.2   Current Military Aircraft Trends *15*

1.5   Future Trends *16*

1.5.1   Civil Aircraft Design *17*

1.5.2   Military Aircraft Design *19*

1.5.3   UAVs/UASs *20*

1.5.4   Military Applications *21*

1.5.5   Civil and Commercial Applications *21*

1.5.6   Recreational Applications *21*

1.5.7   Research and Development Applications *21*

1.5.8   Rocket Applications in Future Aircraft Design Trends *22*

1.6   Forces and Drivers *23*

1.7	Airworthiness Requirements	*23*
1.8	Current Aircraft Performance Analyses Levels	*25*
1.8.1	By the Designers	*25*
1.9	Aircraft Classification	*26*
1.9.1	Comparison between Civil and Military Design Requirements	*26*
1.10	Topics of Current Research Interest Related to Aircraft Design (Supersonic/Subsonic)	*27*
1.11	Cost Implications	*30*
1.12	The Classroom Learning Process	*30*
1.12.1	Classroom Learning Process versus Industrial Practices	*31*
1.12.2	Use of Computer-Assisted Engineering (CAE)	*32*
1.12.3	What is Not Dealt With in Depth in this Book?	*33*
1.13	Units and Dimensions	*34*
1.14	Use of Semi-Empirical Relations and Datasheets	*34*
1.14.1	Semi-Empirical Relations Compared to Statistical Graphs	*35*
1.14.2	Use of Semi-Empirical Weights Relations versus use of Weight Fractions	*35*
1.14.3	Aircraft Sizing	*36*
1.15	The Atmosphere	*36*
1.15.1	Hydrostatic Equations and Standard Atmosphere	*36*
1.15.2	Non-Standard Atmosphere/Off-Standard Atmosphere	*43*
1.15.3	Altitude Definitions – Density Altitude (Off Standard)	*43*
1.15.4	Humidity Effects	*45*
1.15.5	Greenhouse Effect	*45*
	References	*45*
2	**Aircraft Familiarity, Aircraft Design Process, Market Study**	*46*
2.1	Overview	*46*
2.2	Introduction	*47*
2.3	Aircraft Familiarisation	*48*
2.3.1	Civil Aircraft and Its Component Configurations	*48*
2.3.1.1	Subsonic Jet Aircraft	*48*
2.3.2	Turboprop Aircraft	*49*
2.3.3	Military Aircraft and Its Component Configurations	*50*
2.4	Typical Aircraft Design Process	*53*
2.4.1	Input	*53*
2.5	Market Survey – Project Identification	*53*
2.5.1	Civil Aircraft Market – Product Identification	*56*
2.6	Four Phases of Aircraft Design	*57*
2.6.1	Understanding Optimisation	*58*
2.6.2	Typical Resources Deployment	*60*
2.6.3	Typical Cost Frame	*61*
2.6.4	Typical Time Frame	*61*
2.7	Typical Task Breakdown in Each Phase	*62*
2.7.1	Functional Tasks During Conceptual Study (Phase I)	*63*
2.7.2	Project Activities for Small Aircraft Design	*64*
2.8	Aircraft Specifications for Three Civil Aircraft Case Studies	*67*
2.9	Military Market – Some Typical Military Aircraft Design Specifications	*70*
2.9.1	Aircraft Specifications/Requirements for Military Aircraft Case Studies	*71*
2.10	Airworthiness Requirements	*73*
2.10.1	Code of Federal Regulations (CFR) – Title 14	*73*
2.10.1.1	FAR23	*73*

2.10.1.2   Standard Category (Volume 1, Section 23-3: Aeroplane Categories Dated 2011-01-0)   *73*
2.10.1.3   Special Category (FAR Part 1.1)   *74*
2.10.1.4   FAR25   *74*
2.10.1.5   The Parts of FAR Requirements   *74*
2.11   Coursework Procedures – Market Survey   *75*
        References   *76*

3   Aerodynamic Fundamentals, Definitions and Aerofoils   *78*
3.1   Overview   *78*
3.2   Introduction   *79*
3.3   Airflow Behaviour – Laminar and Turbulent   *80*
3.3.1   Aerofoils   *84*
3.4   Flow Past an Aerofoil   *84*
3.5   Generation of Lift   *85*
3.6   Aircraft Motion, Forces and Moments   *86*
3.6.1   Motion   *87*
3.6.2   Forces   *88*
3.6.3   Moments   *88*
3.6.4   Basic Control Deflections – Sign Convention   *89*
3.6.5   Aircraft Loads   *89*
3.7   Definitions of Aerodynamic Parameters   *91*
3.8   Aerofoils   *91*
3.8.1   Subsonic Aerofoils   *92*
3.8.2   Aerofoil Lift Characteristics   *92*
3.8.3   Groupings of Subsonic Aerofoils – NACA/NASA   *93*
3.8.3.1   NACA Four-Digit Aerofoil   *93*
3.8.3.2   NACA Five-Digit Aerofoil   *94*
3.8.3.3   NACA Six-Digit Aerofoil   *95*
3.8.3.4   NACA Seven-Digit Aerofoil   *96*
3.8.3.5   NACA Eight-Digit Aerofoil   *97*
3.8.3.6   Peaky-Section Aerofoil   *97*
3.8.3.7   NASA Supercritical Aerofoil   *98*
3.8.3.8   Natural Laminar Flow (NLF) Aerofoil   *98*
3.8.3.9   NACA GAW Aerofoil   *99*
3.8.3.10   Supersonic Aerofoils   *99*
3.8.3.11   Other Types of Subsonic Aerofoil   *100*
3.9   Reynolds Number and Surface Condition Effects on Aerofoils – Using NACA Aerofoil Test Data   *101*
3.9.1   Camber and Thickness Effects   *102*
3.9.2   Comparison of Three NACA Aerofoils   *105*
3.10   Centre of Pressure and Aerodynamic Centre   *105*
3.10.1   Relation Between Centre of Pressure and Aerodynamic Centre   *107*
3.10.1.1   Estimating the Position of the Aerodynamic Centre, *a.c.*   *107*
3.10.1.2   Estimating the Position of the Centre of Pressure, *c.p.*   *108*
3.11   Types of Stall   *109*
3.11.1   Gradual Stall   *109*
3.11.2   Abrupt Stall   *109*
3.12   High-Lift Devices   *110*
3.13   Flow Regimes   *112*
3.13.1   Compressibility Correction   *114*

3.13.2   Transonic Effects   *115*
3.13.3   Supersonic Effects   *115*
3.14   Summary   *117*
3.14.1   Simplified Aerofoil Selection Methodology   *120*
3.15   Aerofoil Design and Manufacture   *123*
3.15.1   Direct Aerofoil Design Method   *123*
3.15.2   Inverse Aerofoil Design Method   *124*
3.15.2.1 Current Practice   *124*
3.15.2.2 Manufacture   *124*
3.16   Aircraft Centre of Gravity, Centre of Pressure and Neutral Point   *125*
3.16.1   Aircraft Centre of Gravity (CG)   *125*
3.16.2   Aircraft Neutral Point   *125*
3.16.3   Summary   *125*
          References   *125*

4   **Wings**   *127*
4.1   Overview   *127*
4.2   Introduction   *128*
4.3   Generic Wing Planform Shapes   *128*
4.3.1   Unswept Wing Planform   *128*
4.3.2   Swept Wing Planform   *128*
4.3.3   Civil Aircraft   *129*
4.3.4   Military Aircraft (Supersonic Wings)   *131*
4.4   Wing Position Relative to Fuselage   *132*
4.4.1   Advantages and Disadvantages – Civil and Military Aircraft   *133*
4.4.1.1   High-Wing   *133*
4.4.1.2   Low Wing   *133*
4.4.1.3   Mid-Wing   *134*
4.4.2   Definitions and Description of Anhedral/Dihedral   *134*
4.4.2.1   Outer-Wing Dihedral   *135*
4.4.2.2   Gull Wing   *135*
4.4.2.3   Inverted Gull Wing   *135*
4.4.2.4   Parasol   *136*
4.5   Structural Considerations   *136*
4.6   Wing Parameter Definitions   *137*
4.6.1   Wing Reference (Planform) Area, S_W *137*
4.6.2   Wing Sweep Angle, $\Lambda_{1/4}$ *138*
4.6.3   Wing Aspect Ratio, AR *138*
4.6.4   Wing Root Chord (c_{root}) and Tip Chord (c_{tip})   *139*
4.6.5   Wing Taper Ratio, λ *139*
4.6.6   Wing Twist   *139*
4.7   Spanwise Variation of Aerofoil t/c and Incidence   *139*
4.7.1   Wing Stall Pattern and Wing Twist   *139*
4.8   Mean Aerodynamic Chord (MAC)   *140*
4.8.1   Linearly Tapered Trapezoidal Wing   *141*
4.8.2   Kinked Wing – Double Linearly Tapered Trapezoidal Wing   *142*
4.9   Wing Aerodynamics   *145*
4.9.1   Downwash and Effective Angle of Attack, α_{eff} *145*
4.9.2   Induced Drag of the Elliptical Wing   *147*
4.9.3   Induced Drag of the Non-Elliptical Wing – Oswald's Efficiency Factor   *147*

4.9.3.1   Downwash   *148*
4.9.4   AR Correction of 2D Aerofoil Characteristics for a 3D Finite Wing   *149*
4.9.5   Wing Moment Curve Slope   *150*
4.10   Wing Load   *153*
4.10.1   Schrenk's Method – An Approximate Method to Compute Wing Load   *155*
4.10.2   Wing Planform Load Distribution   *159*
4.11   Compressibility Effect: Wing Sweep   *160*
4.11.1   Compressibility Drag/Wave Drag   *160*
4.11.2   Wing Sweep   *160*
4.11.2.1   Sweep Wing Effects   *162*
4.11.3   Relationship Between Wing Sweep, Mach Number and Aerofoil t/c   *162*
4.11.3.1   Torenbeek'S Method   *163*
4.11.3.2   Shevell's Method   *165*
4.11.4   Variable Sweep Wings – Reconfiguring in Flight   *166*
4.12   Transonic Wings   *167*
4.13   Supersonic Wings   *167*
4.13.1   Supersonic Wing Planform   *170*
4.14   Additional Vortex Lift – LE Suction   *170*
4.15   High-Lift Devices on the Wing – Flaps and Slats   *170*
4.15.1   High-Lift Device Evolution and Mechanism   *171*
4.15.2   High-Lift Device C_{Lmax} *172*
4.15.3   Aileron Size   *174*
4.16   Additional Surfaces on the Wing   *175*
4.17   The Square-Cube Law   *176*
4.18   Influence of Wing Area and Span on Aerodynamics   *177*
4.18.1   Aircraft Wetted Area (A_W) versus Wing Planform Area (Sw)   *177*
4.19   Summary of Wing Design   *179*
4.19.1   Simplified Wing Design Methodology   *180*
            References   *183*

5   Bodies – Fuselages, Nacelle Pods, Intakes and the Associated Systems   *184*
5.1   Overview   *184*
5.2   Introduction   *185*
5.2.1   Generic Fuselages   *185*
5.2.1.1   Civil Aircraft   *185*
5.2.1.2   Transport Category   *186*
5.2.1.3   Small Aircraft Category up to Eight Seats   *186*
5.2.1.4   Military Category Fuselage   *186*
5.2.1.5   Military Aircraft Fuselage Types   *186*
5.2.2   Generic Nacelles Pods/Intakes/Auxiliary Bodies   *187*
            CIVIL AIRCRAFT   *188*
5.3   Fuselage Geometry – Civil Aircraft   *188*
5.3.1   Aircraft Zero-Reference Plane/Fuselage Axis   *188*
5.3.2   Fuselage Length, l_{fus} *189*
5.3.3   Front-Fuselage Closure Length, l_f *189*
5.3.4   Aft-Fuselage Closure Length, l_a *189*
5.3.5   Mid-Fuselage Constant Cross-Section Length, l_m *189*
5.4   Fuselage Closures – Civil Aircraft   *189*
5.4.1   Front (Nose Cone) and Aft-End Closure   *190*
5.4.2   Fuselage Upsweep Angle   *192*

5.4.3   Fuselage Plan View Closure Angle   *192*
5.5   Fuselage Fineness Ratio (FR)   *192*
5.6   Fuselage Cross-Sectional Geometry – Civil Aircraft   *194*
5.6.1   Fuselage Height, H *194*
5.6.2   Fuselage Width, W *194*
5.6.3   Average Diameter, D_{ave} *194*
5.6.4   Cabin Height, H_{cab} *194*
5.6.5   Cabin Width, W_{cab} *194*
5.6.6   Pilot Cockpit/Flight Deck   *195*
5.6.7   Cabin Interior Details   *195*
5.7   Fuselage Abreast Seating – Civil Aircraft   *195*
5.7.1   Narrow Body   *196*
5.7.2   Wide Body   *196*
5.7.3   More than Two Aisles - Blended Wing Body   *196*
5.8   Cabin Seat Layout   *197*
5.8.1   Narrow-Body, Single-Aisle Aircraft   *200*
5.8.1.1   Two Abreast (4–24 Passengers)   *200*
5.8.1.2   Three Abreast (24–50 Passengers)   *201*
5.8.1.3   Four Abreast (Around 44–80 Passengers)   *201*
5.8.1.4   Five Abreast (80–150 Passengers)   *202*
5.8.1.5   Six Abreast (120–230 Passengers)   *202*
5.8.2   Wide-Body, Double-Aisle Aircraft   *203*
5.8.2.1   Seven Abreast (160–260 Passengers)   *203*
5.8.2.2   Eight Abreast (250–380 Passengers, Wide-Body Aircraft)   *203*
5.8.2.3   Nine to Ten Abreast (350–480 Passengers, Wide-Body Aircraft)   *204*
5.8.2.4   Ten Abreast and More (More than 400–Almost 800 Passenger Capacity, Wide-Body and
             Double-Decked)   *205*
5.8.2.5   To summarise   *205*
5.9   Fuselage Layout   *205*
5.9.1   Fuselage Configuration – Summary   *206*
5.10   Fuselage Aerodynamic Considerations   *206*
5.10.1   Fuselage Drag   *207*
5.10.2   Transonic Effects – Area Rule   *207*
5.10.3   Supersonic Effects   *208*
5.11   Fuselage Pitching Moment   *208*
5.12   Nacelle Pod – Civil Aircraft   *213*
5.12.1   Nacelle/Intake External Flow Aerodynamic Considerations   *214*
5.12.2   Positioning Aircraft Nacelle Pod   *215*
5.12.3   Wing-Mounted Nacelle Position   *216*
5.12.3.1 Summary – Lateral Nacelle Position   *216*
5.12.3.2 Vertical Position   *216*
5.12.3.3 Summary – Vertical Nacelle Position   *217*
5.12.3.4 Over-Wing Nacelle   *217*
5.12.4   Fuselage-Mounted Nacelle Position   *217*
5.12.4.1 Summary – Fuselage Mounted Nacelle Position   *218*
5.12.5   Trijet Centre Engine   *219*
5.12.6   Some Structural Considerations   *219*
5.13   Exhaust Nozzles – Civil Aircraft   *220*
5.13.1   Civil Aircraft Thrust Reverser (TR)   *220*
             MILITARY AIRCRAFT   *222*

5.14   Fuselage Geometry – Military Aircraft   *222*
5.14.1   Fuselage Axis/Zero-Reference Plane – Military Aircraft   *222*
5.14.2   Fuselage Length, l_{fus} *223*
5.14.3   Fuselage Nose Cone, l_{cone} *223*
5.14.4   Front-Fuselage Length, l_{front} *223*
5.14.5   Mid-Fuselage Length, l_{mid} *223*
5.14.6   Aft-Fuselage Closure Length, l_{aft} *223*
5.14.7   Fuselage Height, H_{fus} *223*
5.14.8   Fuselage Width, W_{fus} *223*
5.14.9   Fuselage Cross-Sectional Area, A_{fus} *223*
5.15   Pilot Cockpit/Flight Deck – Military Aircraft   *224*
5.15.1   Flight Deck Height, H_{fd} *224*
5.15.2   Flight Deck Width, W_{fd} *224*
5.15.3   Egress   *224*
5.16   Engine Installation – Military Aircraft   *224*
5.16.1   Military Aircraft Intake   *225*
           References   *228*

6   Empennage and Other Planar Surfaces   *229*
6.1   Overview   *229*
6.2   Introduction   *230*
6.2.1   The Role of the Empennage – Stabilising and Controlling   *230*
6.2.2   The Role of Ailerons – Stabilising and Controlling   *230*
6.3   Terminologies and Definitions of Empennage   *231*
6.3.1   Horizontal Tail (H-Tail)   *231*
6.3.2   Vertical Tail (V-Tail)   *232*
6.4   Empennage Mount and Types   *232*
6.4.1   Empennage Positional Configuration – Single Boom   *232*
6.4.1.1   Conventional type   *232*
6.4.1.2   *High tee-tail (t-tail)*   *233*
6.4.1.3   *Mid-tail (cruciform tail)*   *234*
6.4.1.4   Unconventional Empennage types   *234*
6.4.2   Empennage Positional Configuration – Multi- (Twin) Boom   *235*
6.4.2.1   Conventional Empennage types   *235*
6.5   Different Kinds of Empennage Design   *235*
6.5.1   Ruddervator   *235*
6.5.2   Elevon   *236*
6.5.3   Stabilator/Taileron   *237*
6.5.4   Canard Aircraft   *237*
6.6   Empennage Tail Arm   *237*
6.6.1   Canard Configuration   *238*
6.6.2   Tail Volume Coefficients   *239*
6.6.2.1   H-tail volume coefficient, C_{HT} *239*
6.6.2.2   V-tail volume coefficient, C_{VT} *239*
6.6.2.3   Canard volume coefficient, C_{CT} *240*
6.7   Empennage Aerodynamics   *240*
6.7.1   Wing Downwash on the H-tail   *240*
6.7.2   H-tail – Longitudinal Static Stability: Stick-Fixed and Power-Off   *241*
6.7.3   V-tail – Directional Static Stability: Rudder-Fixed   *244*
6.7.3.1   Rudder lock   *247*

6.7.4 Aircraft Component Stability *252*
6.7.5 Post Stall Behaviour *253*
6.7.6 Aerodynamic Shielding of the V-tail *254*
6.7.7 Wing Wake on the H-tail *255*
6.8 Aircraft Control System *256*
6.8.1 Civil Aircraft Control Sub-System *256*
6.8.2 Military Aircraft Control Sub-System *256*
6.8.3 Reversible and Irreversible Control *258*
6.9 Aircraft Control Surfaces and Trim Tabs *259*
6.9.1 Aerodynamic Balancing of Control Forces *260*
6.9.2 Power Assisted Managing of Control Forces *262*
6.9.3 Active Control Systems *262*
6.10 Empennage Design *262*
6.10.1 Simplified Empennage Design Methodology *263*
6.10.2 Control Surfaces Design *263*
6.10.3 Effect of Power in Moment Contribution *264*
6.11 Other Planar Surfaces *264*
6.11.1 Ventral Fins *264*
6.11.2 Dorsal Fins (LERX on the V-tail) *265*
6.11.3 LERX on the H-tail *266*
6.11.4 Pylons *266*
6.11.5 Speed Brakes and Dive Brakes *267*
References *267*

7 **Aircraft Statistics, Configuration Choices and Layout** *268*
7.1 Overview *268*
7.2 Introduction *269*
CIVIL AIRCRAFT *270*
7.3 Civil Aircraft Mission (Payload Range) *270*
7.3.1 Civil Aircraft Economic Considerations *271*
7.4 Civil Subsonic Jet Aircraft Statistics (Sizing Parameters) *271*
7.4.1 Civil Aircraft Regression Analysis *272*
7.4.2 Maximum Takeoff Mass versus Number of Passengers *272*
7.4.3 MTOM versus Operational Empty Mass *274*
7.4.4 MTOM versus Fuel Load *275*
7.4.5 MTOM versus Wing Area *276*
7.4.6 Wing Geometry – Area versus Loading, Span and Aspect Ratio *278*
7.4.7 Empennage Area and Tail Volume Coefficients versus Wing Area *278*
7.4.8 Aircraft Lift to Drag Ratio (L/D) versus Wing Aspect Ratio (AR) *280*
7.4.9 MTOM versus Engine Power *280*
7.5 Internal Arrangements of Fuselage – Civil Aircraft *282*
7.5.1 Flight Crew (Flight Deck) Compartment Layout *282*
7.5.2 Cabin Crew and Passenger Facilities *283*
7.5.3 Seat Arrangement, Pitch and Posture (95th Percentile) Facilities *283*
7.5.4 Passenger Facilities *285*
7.5.5 Doors – Emergency Exits *285*
7.5.6 Cargo Container Sizes *288*
7.6 Some Interesting Aircraft Configurations – Civil Aircraft *288*
7.7 Summary of Civil Aircraft Design Choices *291*
MILITARY AIRCRAFT *292*

7.8   Military Aircraft: Detailed Classification, Evolutionary Pattern and Mission Profile   *292*
7.8.1   Fighter Aircraft Generations   *297*
7.9   Military Aircraft Mission   *299*
7.10   Military Aircraft Statistics (Regression Analysis)   *299*
7.10.1   Military Aircraft MTOM versus Payload   *301*
7.10.2   Military MTOM versus OEM   *301*
7.10.3   Military MTOM versus Fuel Load, M_f *301*
7.10.4   Military MTOM versus Wing Area   *301*
7.10.5   Military MTOM versus Engine Thrust   *302*
7.10.6   Military Empennage Area versus Wing Area   *304*
7.11   Military Aircraft Component Geometries   *304*
7.11.1   Fuselage Group (Military)   *304*
7.11.2   Wing Group (Military)   *305*
7.11.3   Empennage Group (Military)   *308*
7.11.4   Intake/Nacelle Group (Military)   *309*
7.12   Miscellaneous Comments   *310*
7.13   Summary of Military Aircraft Design Choices   *310*
          References   *311*

Part II   Aircraft Design   *313*

8   Configuring Aircraft – Concept Definition   *315*
8.1   Overview   *315*
8.2   Introduction   *317*
8.2.1   Starting Up Aircraft Conceptual Design   *318*
8.2.1.1   Civil Aircraft   *318*
8.2.1.2   Military Aircraft   *319*
8.2.1.3   Military Trainer Aircraft   *319*
8.2.2   Aircraft CG and Neutral Point   *320*
          CIVIL AIRCRAFT   *321*
8.3   Prerequisites to Initiate Conceptual Design of Civil Aircraft   *321*
8.3.1   Starting-Up (Conceptual Definition – Phase I)   *322*
8.3.2   Methodology for Shaping and Laying out of Civil Aircraft Configuration   *323*
8.3.3   Shaping and Laying out of Civil Aircraft Components   *324*
8.4   Fuselage Design   *325*
8.4.1   Considerations Needed to Configure the Fuselage   *325*
8.4.1.1   Methodology for Fuselage Design   *326*
8.5   Wing Design   *327*
8.5.1   Considerations in Configuring the Wing   *328*
8.5.2   Methodology for Wing Design   *329*
8.5.3   Structural Consideration for a Wing Attachment Along a Fuselage Layout   *330*
8.6   Empennage Design   *330*
8.6.1   Horizontal Tail   *331*
8.6.2   Vertical Tail   *331*
8.6.2.1   Typical Values of Tail Volume Coefficients   *332*
8.6.3   Considerations in Configuring the Empennage   *333*
8.6.4   Methodology for Empennage Design – Positioning and Layout   *333*
8.6.4.1   Check Aircraft Variant Empennage   *334*
8.7   Nacelle and Pylon Design   *334*

8.7.1   Considerations in Configuring the Nacelle   *335*

8.7.2   Methodology for Civil Aircraft Nacelle Pod and Pylon Design   *335*

8.8   Undercarriage   *337*

8.9   Worked-Out Example: Configuring a Bizjet Class Aircraft   *337*

8.9.1.1   Aircraft Specifications/Requirements   *337*

8.9.1.2   Technology Level Adopted   *338*

8.9.1.3   Statistics of Existing Design with the Class of Aircraft   *338*

8.9.2   Bizjet Aircraft Fuselage: Typical Shaping and Layout   *338*

8.9.3   Bizjet Aircraft Wing   *342*

8.9.4   Bizjet Empennage Design – Positioning and Layout   *344*

8.9.4.1   Checking Variant Designs   *346*

8.9.5   Bizjet Aircraft Nacelle – Positioning and Layout of an Engine   *347*

8.9.6   Bizjet Aircraft Undercarriage – Positioning and Layout   *348*

8.9.7   Finalising the Preliminary Bizjet Aircraft Configuration   *348*

8.9.7.1   Consolidate Summary of Bizjet and Its Variants at this Stage of 'Concept Definition'   *349*

8.9.7.2   Baseline Aircraft External Dimensions   *349*

8.9.8   Miscellaneous Considerations in the Bizjet Aircraft   *349*

    MILITARY AIRCRAFT   *350*

8.10   Prerequisite to Initiate Military (Combat/Trainer) Aircraft Design   *350*

8.10.1   Starting-Up (Conceptual Definition – Phase I)   *352*

8.10.2   Methodology for Shaping and Laying Out of Conventional Two-Surface Military Aircraft   *353*

8.10.3   Shaping and Laying out of Military Aircraft Components   *353*

8.11   Fuselage Design (Military – Combat/Trainer Aircraft)   *354*

8.11.1   Considerations Needed to Configure a Fuselage   *354*

8.11.2   Methodology for Fuselage Design   *355*

8.11.3   Structural Considerations for the Fuselage Layout   *356*

8.12   Wing Design (Military – Combat/Trainer Aircraft)   *356*

8.12.1   Considerations in Configuring the Wing   *356*

8.12.2   Methodology for Wing Design   *357*

8.13   Empennage Design (Military – Combat/Trainer Aircraft)   *358*

8.13.1   Considerations in Configuring the Empennage   *358*

8.13.1.1   Typical Values of Tail Volume Coefficients   *359*

8.13.2   Methodology for Empennage Design   *359*

8.14   Engine/Intake/Nozzle (Military – Combat/Trainer Aircraft)   *360*

8.14.1   Considerations in Configuring the Intake/Nozzle   *360*

8.14.2   Methodology for Configuring the Intake/Nozzle   *361*

8.15   Undercarriage (Military – Combat/Trainer Aircraft)   *361*

8.16   Worked-Out Example – Configuring Military AJT Class Aircraft   *361*

8.16.1   Use of Statistics in the Class of Military Trainer Aircraft   *363*

8.16.2   AJT – Fuselage   *365*

8.16.3   AJT – Wing   *367*

8.16.4   AJT – Empennage   *369*

8.16.5   AJT – Engine/Intake/Nozzle   *370*

8.16.6   Undercarriage Positioning   *371*

8.16.7   Miscellaneous Considerations – Military Design   *372*

8.16.8   Variant CAS Design   *372*

8.16.9   Summary of Worked-Out Military Aircraft Preliminary Details   *373*

8.17   Turboprop Trainer Aircraft (TPT)   *374*

8.17.1   Use of Statistics in the Class of Turboprop Trainer Aircraft (TPT)   *376*

8.17.2   TPT – Fuselage   *377*

8.17.3   TPT – Wing   *378*
8.17.4   TPT – Empennage   *380*
8.17.5   TPT – Intake/Exhaust   *382*
          References   *383*

9   **Undercarriage**   *384*
9.1   Overview   *384*
9.2   Introduction   *385*
9.3   Types of Undercarriage   *387*
9.4   Undercarriage Description   *388*
9.4.1   Tyre (or Tire)   *390*
9.4.2   Brakes   *390*
9.4.3   Wheel Gears   *390*
9.5   Undercarriage Nomenclature and Definitions   *391*
9.5.1   Wheel Stability   *392*
9.5.2   Alignment   *392*
9.6   Undercarriage Retraction and Stowage   *393*
9.6.1   Stowage Space Clearances   *394*
9.7   Undercarriage Design Drivers and Considerations   *394*
9.7.1   Turning of an Aircraft   *396*
9.8   Tyre Friction with the Ground: Rolling and Braking Friction Coefficient   *396*
9.9   Load on Wheels and Shock Absorbers   *397*
9.9.1   Load on Wheel Gears   *398*
9.9.1.1   Static Condition   *399*
9.9.1.2   Dynamic Condition – Brake Application – Nose Wheel Load   *400*
9.10   Energy Absorbed   *400*
9.10.1   Energy Absorption by Strut   *401*
9.10.2   Energy Absorption by Tyre   *401*
9.10.3   Deflection under Load   *402*
9.11   Equivalent Single Wheel Load (ESWL)   *402*
9.11.1   Floatation   *402*
9.12   Runway Pavement   *403*
9.13   Airfield/Runway Strength and Aircraft Operating Compatibility   *404*
9.13.1   California Bearing Ratio (CBR)   *404*
9.13.2   Pavement Classification Number (PCN)   *404*
9.13.3   Aircraft Classification Number (ACN)   *405*
9.13.4   ACN/PCN Method   *405*
9.13.5   Load Classification Number (LCN) and Load Classification Group (LCG)   *405*
9.14   Wheels and Tyres   *407*
9.14.1   Wheels   *407*
9.14.1.1 Divided Split Wheel Assembly   *407*
9.14.1.2 Demountable Wheel Flange Assembly   *408*
9.14.2   Tyres   *408*
9.14.3   Tyre Construction   *409*
9.14.3.1 Tyre Material   *410*
9.14.3.2 Bias Ply Aircraft Tyres   *410*
9.14.3.3 Radial Ply Aircraft Tyre   *410*
9.14.4   Common Construction Layout Arrangements for Both Bias and Radial Ply Tyres   *410*
9.14.4.1 Difference in Layout Arrangement for both Bias and Radial Ply Tyres   *411*
9.15   Tyre Nomenclature, Classification, Loading and Selection   *411*

9.15.1   Tyre Type Classification   *412*
9.15.2   Tyre Deflection   *413*
9.15.2.1   Inflation   *414*
9.15.2.2   Aircraft Tyre Pressure and Size   *414*
9.15.2.3   Wheel Spacing   *414*
9.16   Configuring Undercarriage Layout and Positioning   *414*
9.16.1   Undercarriage Layout Methodology   *416*
9.17   Worked-Out Examples   *417*
9.17.1   Civil Aircraft: Bizjet   *418*
9.17.1.1   Undercarriage Layout and Positioning   *418*
9.17.1.2   Step-by-Step Approach   *418*
9.17.1.3   Wheel load: *420*
9.17.1.4   Tyre selection   *420*
9.17.1.5   Deflection   *420*
9.17.1.6   Aircraft LCN   *421*
9.17.2   Advanced Jet Trainer (Military): AJT   *421*
9.17.2.1   Undercarriage Layout and Positioning   *422*
9.17.2.2   Step-by-step approach – CAS variant   *422*
9.17.2.3   Wheel load   *424*
9.17.2.4   Aircraft LCN   *425*
9.17.2.5   Tyre Selection (CAS)   *425*
9.17.2.6   Deflection   *425*
9.17.2.7   CAS Variant of AJT   *425*
9.17.3   Turboprop Trainer: TPT   *425*
9.18   Discussion and Miscellaneous Considerations   *426*
9.18.1   Undercarriage and Tyre Data   *426*
References   *427*

10   **Aircraft Weight and Centre of Gravity Estimation**   *428*
10.1   Overview   *428*
10.1.1   Coursework Content   *429*
10.2   Introduction   *429*
10.2.1   From 'Concept Definition' to 'Concept Finalisation'   *431*
10.3   The Weight Drivers   *431*
10.4   Aircraft Mass (Weight) Breakdown   *432*
10.5   Aircraft CG and Neutral Point Positions   *433*
10.5.1   Neutral Point, NP   *435*
10.5.2   Aircraft CG Travel   *435*
10.6   Aircraft Component Groups   *436*
10.6.1   Aircraft Components   *436*
10.7   Aircraft Component Mass Estimation   *438*
10.7.1   Use of Semi-Empirical Weight Relations versus Use of Weight Fractions   *439*
10.7.1.1   Weight Fraction Method   *439*
10.7.1.2   Semi-Empirical Method   *439*
10.7.2   Limitations in Use of Semi-Empirical Formulae   *441*
CIVIL AIRCRAFT   *443*
10.8   Mass Fraction Method – Civil Aircraft   *443*
10.8.1   Mass Fraction Analyses   *444*
10.9   Graphical Method – Civil Aircraft   *445*
10.10   Semi-Empirical Equation Method (Statistical)   *446*

10.10.1 Fuselage Group (M_F) – Civil Aircraft *447*
10.10.2 Wing Group – Civil Aircraft *449*
10.10.3 Empennage Group – Civil Aircraft *450*
10.10.4 Nacelle Group – Civil Aircraft *450*
10.10.5 Undercarriage Group – Civil Aircraft *452*
10.10.6 Miscellaneous Group – Civil Aircraft *452*
10.10.7 Power Plant Group – Civil Aircraft *452*
10.10.8 Systems Group – Civil Aircraft *453*
10.10.9 Furnishing Group – Civil Aircraft *454*
10.10.10 Contingency and Miscellaneous – Civil Aircraft *454*
10.10.11 Crew – Civil Aircraft *454*
10.10.12 Payload – Civil Aircraft *454*
10.10.13 Fuel – Civil Aircraft *455*
10.11 Centre of Gravity Determination *455*
10.12 Worked-Out Example – Bizjet Aircraft *456*
10.12.1 Fuselage Group Mass *457*
10.12.2 Wing Group Mass *457*
10.12.3 Empennage Group Mass *458*
10.12.4 Nacelle Group Mass *458*
10.12.5 Undercarriage Group Mass *458*
10.12.6 Miscellaneous Group Mass *458*
10.12.7 Power Plant Group Mass *459*
10.12.8 Systems Group Mass *459*
10.12.9 Furnishing Group Mass *459*
10.12.10 Contingency and Miscellaneous Group Mass *459*
10.12.11 Crew Mass *459*
10.12.12 Payload Mass *459*
10.12.13 Fuel Mass *459*
10.12.14 Weight Summary *459*
10.12.15 Bizjet Aircraft Mass and CG Location Example *460*
10.12.16 First Iteration to Fine-Tune CG Position Relative to an Aircraft and Components *460*
MILITARY AIRCRAFT *461*
10.13 Mass Fraction Method – Military Aircraft *461*
10.14 Graphical Method to Predict Aircraft Component Weight – Military Aircraft *463*
10.15 Semi-Empirical Equations Method (Statistical) – Military Aircraft *463*
10.15.1 Military Aircraft Fuselage Group (SI System) *464*
10.15.2 Military Aircraft Wing Mass (SI System) *465*
10.15.3 Aircraft Empennage *465*
10.15.4 Nacelle Mass Example – Military Aircraft *465*
10.15.5 Power Plant Group Mass Example – Military Aircraft *465*
10.15.6 Undercarriage Mass Example – Military Aircraft *466*
10.15.7 System Mass – Military Aircraft (Higher Mass Fraction for Lighter Aircraft) *467*
10.15.8 Aircraft Furnishing (Ejection Seat) – Military Aircraft *467*
10.15.9 Miscellaneous Group (M_{MISC}) – Military Aircraft *467*
10.15.10 Contingency (M_{CONT}) – Military Aircraft *467*
10.15.11 Crew Mass *467*
10.15.12 Fuel (M_{FUEL}) *467*
10.15.13 Payload (M_{PL}) *467*
10.16 CG Determination – Military Aircraft *468*
10.17 Classroom Example of Military AJT/CAS Aircraft Mass Estimation *468*

10.17.1   AJT Fuselage (Based on CAS Variant)   *468*
10.17.2   AJT Wing (Based on CAS Variant)   *469*
10.17.3   AJT Empennage (Based on CAS Variant)   *469*
10.17.4   AJT Nacelle Mass (Based on CAS Variant)   *469*
10.17.5   AJT Power Plant Group Mass (Based on AJT Variant)   *470*
10.17.6   AJT Undercarriage Mass (Based on CAS Variant)   *470*
10.17.7   AJT Systems Group Mass (Based on AJT Variant)   *470*
10.17.8   AJT Furnishing Group Mass   *470*
10.17.9   AJT Contingency Group Mass   *470*
10.17.10   AJT Crew Mass   *470*
10.17.11   AJT Fuel Mass (M_{FUEL})   *470*
10.17.12   AJT Payload Mass (M_{PL})   *470*
10.18   AJT Mass Estimation and CG Location   *471*
10.19   Classroom Example of a Turboprop Trainer (TPT) Aircraft and COIN Variant Weight Estimation   *472*
10.19.1   TPT Fuselage Example   *474*
10.19.2   TPT Wing Example   *474*
10.19.3   TPT Empennage Example   *474*
10.19.4   TPT Nacelle Mass Example   *475*
10.19.5   TPT Power Plant Group Mass Example   *475*
10.19.6   TPT Undercarriage Mass Example (Based on CAS Variant)   *475*
10.19.7   TPT Systems Group Mass Example   *475*
10.19.8   TPT Furnishing Group Mass Example   *475*
10.19.9   TPT Contingency Group Mass Example   *475*
10.19.10   TPT Crew Mass Example   *476*
10.19.11   TPT Fuel (M_{FUEL})   *476*
10.19.12   TPT Payload (M_{PL})   *476*
10.20   Classroom Worked-Out TPT Mass Estimation and CG Location   *476*
10.20.1   COIN Variant   *476*
10.21   Summary of Concept Definition   *478*
           References   *478*

11   **Aircraft Drag**   *479*
11.1   Overview   *479*
11.2   Introduction   *480*
11.3   Parasite Drag Definition   *481*
11.4   Aircraft Drag Breakdown (Subsonic)   *482*
11.4.1   Discussion   *483*
11.5   Understanding Drag Polar   *483*
11.5.1   Actual Drag Polar   *484*
11.5.2   Parabolic Drag Polar   *485*
11.5.3   Comparison Between Actual and Parabolic Drag Polar   *485*
11.6   Aircraft Drag Formulation   *487*
11.7   Aircraft Drag Estimation Methodology (Subsonic)   *488*
11.8   Minimum Parasite Drag Estimation Methodology   *489*
11.8.1   Geometric Parameters, Reynolds Number and Basic C_F Determination   *489*
11.8.2   Computation of Wetted Areas   *490*
11.8.2.1   Lifting Surfaces   *490*
11.8.2.2   Fuselage   *490*
11.8.2.3   Nacelle   *490*

11.8.3   Stepwise Approach to Compute Minimum Parasite Drag   *491*
11.9   Semi-Empirical Relations to Estimate Aircraft-Component Parasite Drag   *491*
11.9.1   Fuselage   *491*
11.9.2   Wing, Empennage, Pylons and Winglets   *494*
11.9.3   Nacelle Drag   *495*
11.9.3.1   Intake Drag   *496*
11.9.3.2   Base Drag   *497*
11.9.3.3   Boat Tail Drag   *497*
11.9.3.4   Nacelle 3D Effects   *498*
11.9.3.5   Total Nacelle Drag   *498*
11.9.4   Recovery Factor   *499*
11.9.5   Excrescence Drag   *499*
11.9.6   Miscellaneous Parasite Drags   *499*
11.9.6.1   Air-Conditioning Drag   *500*
11.9.6.2   Trim Drag   *500*
11.9.6.3   Aerials   *500*
11.10   Notes on Excrescence Drag Resulting from Surface Imperfections   *500*
11.11   Minimum Parasite Drag   *501*
11.12 ΔC_{Dp} Estimation   *501*
11.13   Subsonic Wave Drag   *502*
11.14   Total Aircraft Drag   *503*
11.15   Low-Speed Aircraft Drag at Takeoff and Landing   *503*
11.15.1   High-Lift Device Drag   *504*
11.15.2   Dive Brakes and Spoiler Drag   *506*
11.15.3   Undercarriage Drag   *506*
11.15.4   One-Engine Inoperative Drag   *508*
11.16   Propeller-Driven Aircraft Drag   *508*
11.17   Military Aircraft Drag   *509*
11.18   Supersonic Drag   *509*
11.19   Coursework Example – Civil Bizjet Aircraft   *511*
11.19.1   Geometric and Performance Data   *511*
11.19.2   Computation of Wetted Areas, *Re* and Basic C_F *513*
11.19.3   Computation of 3D and Other Effects to Estimate Component C_{Dpmin} *514*
11.19.4   Summary of Parasite Drag   *518*
11.19.5 ΔC_{Dp} Estimation   *518*
11.19.6   Induced Drag   *518*
11.19.7   Total Aircraft Drag at LRC   *519*
11.20   Classroom Example – Subsonic Military Aircraft (Advanced Jet Trainer – AJT)   *519*
11.20.1   AJT Details   *522*
11.21   Classroom Example – Turboprop Trainer (TPT)   *522*
11.21.1   TPT Details   *524*
11.22   Classroom Example – Supersonic Military Aircraft   *527*
11.22.1   Geometric and Performance Data of Vigilante – RA C5 Aircraft   *527*
11.22.2   Computation of Wetted Areas, *Re* and Basic C_F *528*
11.22.3   Computation of 3D and Other Effects to Estimate Component C_{Dpmin} *529*
11.22.4   Summary of Parasite Drag (ISA Day, 36 182 ft Altitude and Mach 0.9)   *532*
11.22.5 ΔC_{Dp} Estimation   *532*
11.22.6   Induced Drag   *533*
11.22.7   Supersonic Drag Estimation   *533*
11.22.8   Total Aircraft Drag   *536*

11.23	Drag Comparison *537*	
11.24	Some Concluding Remarks *538*	
	References *538*	

12	**Aircraft Power Plant and Integration** *540*	
12.1	Overview *540*	
12.2	Background *540*	
12.3	Definitions *543*	
12.3.1	Recovery Factor, RF *545*	
12.4	Introduction – Air-Breathing Aircraft Engine Types *546*	
12.4.1	Simple Straight-Through Turbojets *546*	
12.4.2	Turbofan – Bypass Engine (Two Flow – Primary and Secondary) *546*	
12.4.3	Three Flow Bypass Engine *548*	
12.4.4	Afterburner Engines *548*	
12.4.5	Turboprop Engines *548*	
12.4.6	Piston Engines *549*	
12.5	Simplified Representation of a Gas Turbine (Brayton/Joule) Cycle *551*	
12.6	Formulation/Theory – Isentropic Case (Trend Analysis) *551*	
12.6.1	Simple Straight-Through Turbojet Engine – Formulation *552*	
12.6.2	Bypass Turbofan Engines – Formulation *553*	
12.6.3	Afterburner Engines – Formulation *555*	
12.6.4	Turboprop Engines – Formulation *556*	
12.6.4.1	Summary *556*	
12.7	Engine Integration to Aircraft – Installation Effects *556*	
12.7.1	Subsonic Civil Aircraft Nacelle and Engine Installation *557*	
12.7.2	Turboprop Integration to Aircraft *558*	
12.7.3	Combat Aircraft Engine Installation *559*	
12.8	Intake/Nozzle Design *560*	
12.8.1	Civil Aircraft Subsonic Intake Design *560*	
12.8.2	Military Aircraft Supersonic Intake Design *561*	
12.8.2.1	Intake at Fuselage Nose with a Centre Body *561*	
12.8.2.2	Intake at the Fuselage Side *562*	
12.9	Exhaust Nozzle and Thrust Reverser (TR) *563*	
12.9.1	Civil Aircraft Exhaust Nozzles *565*	
12.9.2	Military Aircraft TR Application and Exhaust Nozzles *565*	
12.10	Propeller *566*	
12.10.1	Propeller-Related Definitions *566*	
12.11	Propeller Theory *568*	
12.11.1	Momentum Theory – Actuator Disc *568*	
12.11.2	Blade Element Theory *571*	
12.12	Propeller Performance – Use of Charts, Practical Engineering Applications *572*	
12.12.1	Propeller Performance – Blade Numbers $3 \leq N \geq 4$ *575*	
	References *575*	

13	**Aircraft Power Plant Performance** *577*	
13.1	Overview *577*	
13.2	Introduction *578*	
13.2.1	Engine Performance Ratings *578*	
13.2.1.1	Takeoff Rating *578*	
13.2.1.2	Flat-Rated Takeoff Rating *578*	

13.2.1.3   Maximum Continuous Rating   *579*
13.2.1.4   Maximum Climb Rating   *579*
13.2.1.5   Maximum Cruise Rating   *579*
13.2.1.6   Idle Rating   *579*
13.2.1.7   Recovery Factor, RF   *580*
13.2.2   Turbofan Engine Parameters   *580*
13.3   Uninstalled Turbofan Engine Performance Data – Civil Aircraft   *581*
13.3.1   Performance with BPR ≈ 4 ± 1 (Smaller Engines, e.g. Bizjets)   *582*
13.3.1.1   Takeoff Rating   *582*
13.3.1.2   Maximum Climb Rating   *583*
13.3.1.3   Maximum Cruise Rating   *584*
13.3.1.4   Idle Rating   *584*
13.3.2   BPR Around ≈6 ± 1 (Larger Engines)   *585*
13.3.2.1   Turbofan performance   *585*
13.3.2.2   Takeoff Rating   *585*
13.3.2.3   Maximum Climb Rating   *585*
13.3.2.4   Maximum Cruise Rating   *585*
13.3.3   BPR Around ≈10 ± 2 (Larger Engines – Big Jets, Wide Body Aircraft)   *585*
13.3.4   Uninstalled Turbofan Engine Performance Data – Military (BPR < 1)   *585*
13.3.5   Uninstalled Turboprop Engine Performance Data (All Types up to 100 Passenger Class)   *587*
13.3.5.1   Takeoff Rating   *589*
13.3.5.2   Maximum Climb Rating   *589*
13.3.5.3   Maximum Cruise Rating   *589*
13.4   Installed Engine Performance Data of Matched Engines to Coursework Aircraft   *590*
13.4.1   Turbofan Engines (Smaller Engines for Bizjets – BPR ≈ 4)   *590*
13.4.1.1   Takeoff Rating (Bizjet) – STD Day   *590*
13.4.1.2   Maximum Climb Rating (Bizjet) – STD Day   *590*
13.4.1.3   Maximum Cruise Rating (Bizjet) – STD Day   *592*
13.4.1.4   Idle Rating (Bizjet) – STD Day   *592*
13.4.2   Turbofans with BPR Around 6 ± 1 (Larger Engines – Regional Jets and Above)   *592*
13.4.3   Military Turbofan (Advanced Jet Trainer/CAS Role – BPR < 1) – STD Day   *593*
13.5   Installed Turboprop Performance Data   *594*
13.5.1   Propeller Performance at STD Day – Worked-Out Example   *594*
13.5.2   Turboprop Performance at STD Day   *596*
13.5.2.1   Maximum Takeoff Rating (Turboprop) – STD Day   *596*
13.5.2.2   Maximum Climb Rating – STD Day   *597*
13.5.2.3   Maximum Cruise Rating (Turboprop) – STD Day   *597*
13.6   Piston Engine   *598*
13.7   Engine Performance Grid   *602*
13.7.1   Installed Maximum Climb Rating (TFE731-20 Class Turbofan)   *603*
13.7.2   Maximum Cruise Rating (TFE731-20 Class Turbofan)   *604*
13.8   Some Turbofan Data (OPR = Overall Pressure Ratio)   *606*
             References   *607*

14   **Aircraft Sizing, Engine Matching and Variant Derivatives**   *608*
14.1   Overview   *608*
14.1.1   Summary   *609*
14.2   Introduction   *609*
14.2.1   Civil Aircraft   *609*
14.2.2   Military Aircraft   *610*

14.3 Theory *610*
14.3.1 Sizing for Takeoff Field Length (TOFL) – Two Engines *611*
14.3.1.1 Civil Aircraft Design: Takeoff *613*
14.3.1.2 Military Aircraft Design: Takeoff *613*
14.3.2 Sizing for the Initial Rate of Climb (All Engines Operating) *614*
14.3.3 Sizing to Meet Initial Cruise *614*
14.3.4 Sizing for Landing Distance *615*
14.4 Coursework Exercise – Civil Aircraft Design (Bizjet) *615*
14.4.1 Takeoff *615*
14.4.2 Initial Climb *616*
14.4.3 Cruise *616*
14.4.4 Landing *617*
14.5 Sizing Analysis – Civil Aircraft (Bizjet) *617*
14.5.1 Variants in the Family of Aircraft Designs *618*
14.5.2 Bizjet Family *619*
14.6 Coursework Exercise – Military Aircraft (AJT) *619*
14.6.1 Takeoff – Military Aircraft *620*
14.6.2 Initial Climb – Military Aircraft *620*
14.6.3 Cruise – Military Aircraft *620*
14.6.4 Landing – Military Aircraft *621*
14.6.5 Sizing for the Turn Requirement of 4g at Sea Level *621*
14.7 Sizing Analysis – Military Aircraft (AJT) *623*
14.7.1 Single Seat Variants in the Family of Aircraft Designs *624*
14.7.1.1 Configuration *624*
14.7.1.2 Thrust *625*
14.7.1.3 Drag *625*
14.8 Aircraft Sizing Studies and Sensitivity Analyses *625*
14.8.1 Civil Aircraft Sizing Studies *625*
14.8.1.1 Military Aircraft Sizing Studies *625*
14.9 Discussion *626*
14.9.1 The AJT *629*
References *630*

15 Aircraft Performance *631*
15.1 Overview *631*
15.1.1 Section 15.13: Summarised Discussion of the DesignClasswork Content *631*
15.2 Introduction *632*
15.2.1 Aircraft Speed *633*
15.2.2 Some Prerequisite Information *633*
15.2.3 Cabin Pressurisation *634*
15.3 Takeoff Performance *635*
15.3.1 Civil Transport Aircraft Takeoff [FAR (14CFR) 25.103/107/109/149] *636*
15.3.2 Balanced Field Length (BFL) – Civil Aircraft *637*
15.3.2.1 Normal All-Engine Operating (Takeoff) – Civil Aircraft *637*
15.3.2.2 Unbalanced Field Length (UBFL) – Civil Aircraft *638*
15.3.3 Civil Aircraft Takeoff Segments [FAR (14CFR) 25.107 – Subpart B] *638*
15.3.4 Derivation of Takeoff Equations *640*
15.3.4.1 Acceleration *640*
15.3.4.2 Takeoff Field Length (TOFL) Estimation – Distance Covered from zero to V_2 *640*
15.4 Landing Performance *642*

15.4.1   Approach Climb and Landing Climb and Baulked Landing *643*
15.4.2   Derivation of Landing Performance Equations *643*
15.4.2.1 Ground Distance During Glide, S_{glide} *643*
15.5   Climb Performance *644*
15.5.1   Derivation of Climb Performance Equations *645*
15.5.2   Quasi-Steady State Climb *645*
15.5.2.1 Constant EAS Climb *645*
15.5.2.2 Constant Mach Climb *646*
15.6   Descent Performance *648*
15.6.1   Derivation of Descent Performance Equations *648*
15.7   Checking of the Initial Maximum Cruise Speed Capability *649*
15.8   Payload-Range Capability – Derivation of Range Equations *649*
15.9   In Horizontal Plane (Yaw Plane) – Sustained Coordinated Turn *651*
15.9.1   Kinetics of a Coordinated Turn in Steady (Equilibrium) Flight *651*
15.9.2   Maximum Conditions for a Turn in the Horizontal Plane *653*
15.10   Aircraft Performance Substantiation – Worked-Out Classroom Examples – Bizjet *653*
15.10.1   Checking TOFL (Bizjet) – Specification Requirement 4400 ft *654*
15.10.1.1 All-Engine Takeoff – 20° Flap *655*
15.10.1.2 One Engine Inoperative – Balanced Field Takeoff (BFL) Inoperative *656*
15.10.2   Checking Landing Field Length (Bizjet) – Specification Requirement 4400 ft *660*
15.10.3   Checking Takeoff Climb Performance Requirements (Bizjet) *661*
15.10.4   Checking Initial En Route Rate of Climb – Specification Requirement is 2600 ft min^{-1} *661*
15.10.5   Integrated Climb Performance (Bizjet) *663*
15.10.6   Checking Initial High-Speed Cruise (Bizjet) – Specification Requirement of High-Speed Cruise Mach 0.75 at 41 000 ft Altitude *663*
15.10.7   Specific Range (Bizjet) *664*
15.10.8   Descent Performance (Bizjet) – Limitation Maximum Descent Rate of 1800 ft min^{-1} *664*
15.10.9   Checking out the Payload-Range Capability – Requirement of 2000 nm *667*
15.11   Aircraft Performance Substantiation – Military AJT *668*
15.11.1.1 AJT data *669*
15.11.1.2 Mission Profile *669*
15.11.2   Checking TOFL (AJT) – Specification Requirement 3600 ft (1100 m) *671*
15.11.2.1 Takeoff with 8° Flap *671*
15.11.3   Checking the Second Segment Climb Gradient at 8° Flap *674*
15.11.4   Checking Landing Field Length (AJT) – Specification Requirement 3600 ft *674*
15.11.5   Checking the Initial Climb Performance – Requirement 50 m s^{-1} (10 000 ft min^{-1}) at Normal Training Configuration (NTC) *675*
15.11.6   Checking the Maximum Speed – Requirement Mach 0.85 at 30 000 ft Altitude at NTC *675*
15.11.7   Compute the Fuel Requirement (AJT) *676*
15.11.8   Turn Capability – Check n_{max} at the Turn (AJT) *677*
15.12   Propeller-Driven Aircraft – TPT (Parabolic Drag Polar) *677*
15.13   Summarised Discussion of the Design *678*
15.13.1   The Bizjet *679*
15.13.2   The AJT *680*
      References *681*

16   Aircraft Cost Considerations 682
16.1   Overview *682*
16.2   Introduction *683*
16.3   Aircraft Cost and Operational Cost *686*

16.3.1   Operating Cost (OC)   *687*
16.4   Rapid Cost Modelling   *690*
16.4.1   Nacelle Cost Drivers   *692*
16.4.2   Nose Cowl Parts and Subassemblies   *694*
16.4.3   Methodology (Nose Cowl Only)   *694*
16.4.4   Cost Formulae and Results   *697*
16.5   Aircraft Direct Operating Cost (DOC)   *701*
16.5.1   Formulation to Estimate DOC   *703*
16.5.2   Worked-Out Example of DOC – Bizjet   *705*
16.6   Aircraft Performance Management   *707*
16.6.1   Methodology   *709*
16.6.2   Discussion – The Broader Issues   *710*
          References   *710*

Part III   Further Design Considerations   *713*

17   Aircraft Load   *715*
17.1   Overview   *715*
17.2   Introduction   *715*
17.2.1   Buffet   *716*
17.2.1.1 Q-Corner – (Coffin Corner)   *716*
17.2.2   Flutter   *717*
17.3   Flight Manoeuvres   *718*
17.3.1   Pitch-Plane (x–z-Plane) Manoeuvre: (Elevator/Canard Induced)   *718*
17.3.2   Roll-Plane (y–z-Plane) Manoeuvre: (Aileron Induced)   *718*
17.3.3   Yaw-Plane (y–x-Plane) Manoeuvre: (Rudder Induced)   *718*
17.4   Aircraft Loads   *718*
17.4.1.1 On-Ground   *718*
17.4.1.2 In Flight   *719*
17.5   Theory and Definitions   *719*
17.5.1   Load Factor, n *719*
17.6   Limits – Load and Speeds   *720*
17.6.1   Maximum Limit of Load Factor, n *720*
17.7 V-n Diagram   *721*
17.7.1   Speed Limits   *722*
17.7.2   Extreme Points of the V-n Diagram   *723*
17.7.2.1 Positive Loads   *723*
17.7.2.2 Negative Loads   *723*
17.7.3   Low-Speed Limit   *724*
17.7.4   Manoeuvre Envelope Construction   *725*
17.7.5   High-Speed Limit   *725*
17.8   Gust Envelope   *726*
17.8.1   Gust Load Equations   *726*
17.8.2   Gust Envelope Construction   *728*
          References   *729*

18   Stability Considerations Affecting Aircraft Design   *730*
18.1   Overview   *730*
18.2   Introduction   *730*
18.3   Static and Dynamic Stability   *731*

18.3.1   Longitudinal Stability – Pitch-Plane (Pitch Moment, M)   *733*
18.3.2   Directional Stability – Yaw-Plane (Yaw Moment, N)   *733*
18.3.3   Lateral Stability – Roll-Plane (Roll Moment, L)   *734*
18.3.4   Summary of Forces, Moments and their Sign Conventions   *736*
18.4   Theory   *736*
18.4.1   Pitch-Plane   *736*
18.4.2   Yaw-Plane   *739*
18.4.3   Roll-Plane   *740*
18.5   Current Statistical Trends for Horizontal and Vertical Tail Coefficients   *741*
18.6   Stick Force – Aircraft Control Surfaces and Trim Tabs   *741*
18.6.1   Stick Force   *741*
18.7   Inherent Aircraft Motions as Characteristics of Design   *743*
18.7.1   Short-Period Oscillation and Phugoid Motion (Long-Period Oscillation)   *743*
18.7.2   Directional/Lateral Modes of Motion   *744*
18.7.3   Spinning   *747*
18.8   Design Considerations for Stability – Civil Aircraft   *747*
18.9   Military Aircraft – Non-Linear Effects   *750*
18.10   Active Control Technology (ACT) – Fly-by-Wire (FBW)   *752*
18.11   Summary of Design Considerations for Stability   *754*
18.11.1   Civil Aircraft   *754*
18.11.2   Military Aircraft – Non-Linear Effects   *755*
          References   *755*

19   **Materials and Structures**   *756*
19.1   Overview   *756*
19.2   Introduction   *756*
19.3   Function of Structure – Loading   *759*
19.3.1   Wing   *759*
19.3.2   Empennage/Tail   *759*
19.3.3   Fuselage   *760*
19.3.4   Undercarriage   *760*
19.3.5   Torsion   *761*
19.4   Basic Definitions – Structures   *761*
19.4.1   Structure   *761*
19.4.2   Load   *762*
19.4.3   Limit Load   *762*
19.4.4   Proof Load   *762*
19.4.5   Ultimate Load   *762*
19.4.6   Structural Stiffness   *762*
19.5   From Structure to Material   *762*
19.6   Basic Definitions – Materials   *763*
19.6.1   Stress   *763*
19.6.2   Strain   *764*
19.6.3   Hooke's Law   *765*
19.7   Material Properties   *765*
19.7.1   Stiffness   *765*
19.7.2   Yield Stress   *765*
19.7.3   Fracture Toughness   *765*
19.7.4   Ductility   *766*
19.7.5   Durability   *766*
19.8   Considerations with Respect to Design   *766*

19.8.1	Material Selection	*766*
19.8.2	Ashby Scatter Plots	*771*
19.8.3	Material Cost Considerations	*771*
19.8.4	Manufacture	*774*
19.8.5	Integrated Decisions	*776*
19.8.6	Design Exercise	*776*
19.9	Structural Configuration	*776*
19.9.1	Bending	*776*
19.9.2	Torsion	*778*
19.9.3	Buckling	*781*
19.9.3.1	Column/Beam Buckling	*782*
19.9.3.2	Plate Buckling	*783*
19.10	Materials – General Considerations	*784*
19.11	Metals	*786*
19.12	Wood and Fabric	*788*
19.13	Composite Materials	*788*
19.13.1	Composite Materials, Fibre and Fabric	*789*
19.13.1.1	Plain Weave	*790*
19.13.1.2	Twill Weave	*790*
19.13.1.3	Satin Weave	*790*
19.13.2	Types of Synthetic Composite Material Fibre	*790*
19.13.3	Matrix Bond for Composite Material Fibre	*790*
19.13.4	Strength Characteristics of Composite Materials	*791*
19.13.5	Honeycomb Structural Panels	*791*
19.14	Structural Configurations	*793*
19.14.1	Skin	*793*
19.14.2	Fuselage	*794*
19.14.3	Wings	*794*
19.14.4	Spars	*796*
19.14.5	Ribs and Stringers	*797*
19.14.6	Fuselage Structural Considerations	*797*
19.14.7	Assembly and Wing Box	*798*
19.14.8	Empennage	*799*
19.15	Rules of Thumb and Concept Checks	*800*
19.16	Finite Element Analysis (FEA)/Finite Element Method (FEM)	*804*
	References	*805*
20	**Aircraft Manufacturing Considerations**	*806*
20.1	Overview	*806*
20.2	Introduction	*808*
20.3	Design for Manufacture and Assembly (DFM/A)	*808*
20.4	Manufacturing Practices	*809*
20.5	Six-Sigma Concept	*811*
20.6	Tolerance Relaxation at the Wetted Surface	*812*
20.6.1	Cost Versus Tolerance Relationship	*813*
20.7	Reliability and Maintainability (R&M)	*814*
20.8	The Design Considerations	*814*
20.9	'Design for Customer' (A Figure of Merit)	*817*
20.9.1	Index for the 'Design for Customer'	*818*
20.9.2	Worked-Out Example	*818*

20.9.3   Discussion   *819*
20.10   Digital Manufacturing Process Management   *821*
20.10.1   The Product, Process and Resource (PPR) Hub   *822*
20.10.2   Integration of CAD/CAM, Manufacturing, Operations and In-Service Domains   *822*
20.10.3   Shop-Floor Interface   *823*
20.10.4   Design for Maintainability and 3D-Based Technical Publication Generation   *824*
            References   *824*

21   Miscellaneous Design Considerations   *825*
21.1   Overview   *825*
21.2   Introduction   *826*
21.3   History of FAA – the Role of Regulation   *827*
21.3.1   The Role of Regulation   *830*
21.4   Flight Test   *831*
21.5   Contribution by the Ground Effect on Takeoff   *832*
21.6   Aircraft Environmental Issues   *833*
21.6.1   Noise Emissions   *833*
21.6.2   Engine Exhaust Emissions   *838*
21.7   Flying in Adverse Environments   *838*
21.7.1   Group 1 – Adverse Environments due to Loss of Visibility   *839*
21.7.2   Group 2 – Adverse Environments due to Aerodynamic and Stability/Control Degradation   *839*
21.7.3   Bird Strikes   *841*
21.8   Military Aircraft Flying Hazards   *842*
21.8.1   Aircraft Combat Survivability   *842*
21.9   End-of-Life Disposal   *842*
21.10   Extended Range Twin-Engine Operation (ETOP)   *843*
21.11   Flight and Human Physiology   *843*
21.11.1   Aircraft Design Considerations for Human Factors   *844*
21.11.2   Automation – Unmanned Aircraft Vehicle (UAV)/Unmanned Aircraft System (UAS)   *844*
21.12   Some Emerging Scenarios   *845*
21.12.1   Counter-Terrorism Design Implementation   *845*
21.12.2   Health Issues   *845*
21.12.3   Damage From Runway Debris (an Old Problem Needs a New Look)   *845*
            References   *846*

22   Aircraft Systems   *847*
22.1   Overview   *847*
22.2   Introduction   *848*
22.3   Environmental Issues (Noise and Engine Emission)   *849*
22.3.1   Noise Emissions   *849*
22.4   Safety Issues   *851*
22.4.1   Doors – Emergency Egress   *851*
22.4.2   Escape Slide/Chute – Emergency Egress   *852*
22.4.2.1   Classroom Exercise   *852*
22.5   Aircraft Flight Deck (Cockpit) Layout   *853*
22.5.1   Air Data Instruments and Flight Deck   *854*
22.5.2   Altitude Measurement – Altimeter   *855*
22.5.3   Airspeed Measuring Instrument – Pitot-Static Tube   *855*
22.5.4   Angle of Attack Probe   *856*
22.5.5   Vertical Speed Indicator (VSI)   *857*

22.5.6   Temperature Measurement   *857*
22.5.7   Turn/Side Slip Indicator   *858*
22.5.8   Multi-Functional Display (MFD)/Electronic Flight Instrument System (EFIS)   *858*
22.5.9   Civil Aircraft Flight Deck   *860*
22.5.10   Combat Aircraft Flight Deck   *861*
22.5.11   Heads-Up Display (HUD)   *861*
22.5.12   Hands on Throttle and Stick (HOTAS) and Side Stick Controller   *862*
22.5.13   Helmet Mounted Display (HMD)   *862*
22.5.14   Voice Operated Control   *862*
22.6   Aircraft Systems   *862*
22.6.1   Aircraft Control Subsystems   *863*
22.6.2   Engine and Fuel Control Subsystems   *865*
22.6.2.1   Piston Engine/Fuel Control System (Total System Weight Around 1–1.5% of MTOW)   *865*
22.6.2.2   Turbofan Engine/Fuel Control System (Total System Weight Around 1.5–2% of MTOW)   *866*
22.6.2.3   Fuel Storage and Flow Management   *867*
22.6.3   Emergency Power Supply   *867*
22.6.4   Avionics Subsystems   *868*
22.6.4.1   Military Application   *868*
22.6.4.2   Civil Application   *869*
22.6.5   Electrical Subsystems   *869*
22.6.6   Hydraulic Subsystem   *870*
22.6.7   Pneumatic System   *871*
22.6.7.1   Environment Control System (ECS) – Cabin Pressurisation/Air-Conditioning   *872*
22.6.8   Oxygen System   *874*
22.6.8.1   Oxygen Supply   *874*
22.7   Flying in Adverse Environments and Passenger Utility   *874*
22.7.1   Anti-Icing/De-Icing Systems   *874*
22.7.1.1   Use of Hot Air Blown Through Ducts   *874*
22.7.1.2   Use of Electrical Impulses   *875*
22.7.1.3   Use of Chemicals   *875*
22.7.1.4   Use of Boots (Pneumatic/Electric)   *875*
22.7.2   De-Fogging and Rain Removal System   *875*
22.7.3   Lightning and Fire Hazards   *876*
22.7.4   Utility Subsystems   *877*
22.7.5   Passenger Services and Utility Usage   *878*
22.7.6   Aircraft Sound Horn   *878*
22.8   Military Aircraft Survivability   *878*
22.8.1   Military Emergency Escape – Egress   *878*
22.8.2   Aircraft Combat Survivability   *882*
22.8.3   Returning to Home Base   *882*
22.8.4   Military Aircraft Stealth Considerations   *882*
22.8.5   Low Observable (LO) Aircraft Configuration   *883*
22.8.5.1   Heat Signature   *883*
22.8.5.2   Radar Signature   *884*
            References   *885*

23   Computational Fluid Dynamics   *886*
23.1   Overview   *886*
23.2   Introduction   *887*
23.3   Current Status   *888*

23.4   Approach Road to CFD Analyses   *889*
23.5   Some Case Studies   *892*
23.6   Hierarchy of CFD Simulation Methods   *893*
23.7   Summary of Discussions   *896*
23.7.1   CFD Analyses   *896*
23.7.2   Wind Tunnel Tests   *897*
23.7.3   Flight Tests   *897*
         References   *897*

24   Electric Aircraft   *899*
24.1   Overview   *899*
24.2   Introduction   *900*
24.3   Energy Storage   *902*
24.3.1   Lithium-Ion Battery   *904*
24.3.2   Fuel Cells   *904*
24.3.3   Solar Cell   *904*
24.4   Prime Mover – Motors   *905*
24.4.1   Controller   *905*
24.5   Electric Powered Aircraft Power Train   *906*
24.5.1   Battery Powered Aircraft Power Train   *906*
24.5.2   Fuel Cell Powered Aircraft Power Train   *906*
24.5.3   Solar Energy (Photovoltaic Cell) Powered Aircraft Power Train   *907*
24.6   Hybrid Electric Aircraft (HEA)   *908*
24.7   Distributed Electric Propulsion (DEP)   *910*
24.8   Electric Aircraft Related Theory/Analyses   *911*
24.8.1   Battery and Motor   *911*
24.8.2   Conventional Aircraft Performance   *911*
24.8.3   Battery Powered Electric Aircraft Range Equation   *912*
24.8.4   Fuel Cell Powered Electric Aircraft Range Equation   *913*
24.9   Electric Powered Aircraft Sizing   *914*
24.10   Discussion   *916*
24.10.1 Overall Aircraft Performance with Battery Powered Aircraft   *917*
24.10.2 Operating Cost of Battery Powered Aircraft   *917*
24.10.3 Battery Powered Unmanned Aircraft Vehicle (UAV)/Unmanned Aircraft System (UAS)   *918*
24.11   Worked-Out Example   *918*
         References   *919*

Appendix A   Conversions and Important Equations   *920*

Appendix B   International Standard Atmosphere Table Data from Hydrostatic Equations   *923*

Appendix C   Fundamental Equations (See Table of Contents for Symbols and Nomenclature.)   *926*
C.1   Kinetics   *926*
C.2   Thermodynamics   *926*
C.3   Supersonic Aerodynamics   *927*
C.4   Normal Shock   *928*
C.5   Oblique Shock   *929*
C.6   Supersonic Flow Past a 2D Wedge   *929*
C.7   Supersonic Flow Past 3D Cone   *930*
C.8   Incompressible Low Speed Wind Tunnel (Open Circuit)   *931*

Appendix D   Some Case Studies – Aircraft Data   *932*
D.1   Airbus320 Class Aircraft   *932*
D.1.1   Dimensions (to Scale the Drawing for Detailed Dimensions)   *932*
D.2   Drag Computation   *932*
D.2.1   Fuselage   *932*
D.2.2   Wing   *933*
D.2.3   Vertical Tail   *934*
D.2.4   Horizotal Tail   *935*
D.2.5   Nacelle, C_{Fn} *935*
D.2.6   Pylon   *936*
D.2.7   Roughness Effect   *936*
D.2.8   Trim Drag   *936*
D.2.9   Aerial and Other Protrusions   *936*
D.2.10   Air Conditioning   *936*
D.2.11   Aircraft Parasite Drag Build-Up Summary and C_{Dpmin} *937*
D.2.12 ΔC_{Dp} Estimation   *937*
D.2.13   Induced Drag, C_{Di} *937*
D.2.14   Total Aircraft Drag   *938*
D.2.15   Engine Rating   *938*
D.2.16   Weights Breakdown (There Could be Some Variation)   *938*
D.2.17   Payload Range (150 Passengers)   *939*
D.2.18   Cost Calculations (US$ – Year 2000)   *941*
D.3   The Belfast (B100) – A Fokker F100 Class Aircraft   *942*
D.3.1   Customer Specification and Geometric Data   *942*
D.3.1.1   Customer Specification   *942*
D.3.1.2   Family Variants   *942*
D.3.1.3   Fuselage (Circular Constant Section)   *942*
D.3.1.4   Wing   *943*
D.3.1.5   V-tail (Aerofoil 64-010)   *943*
D.3.1.6   H-tail (Tee Tail, Aerofoil 64-210 – Installed with Negative Camber)   *943*
D.3.1.7   Nacelle (Each – 2 Required)   *943*
D.3.2   The B100 Aircraft Weight Summary   *943*
D.3.2.1   Weight Summary   *943*
D.3.2.2   Engine (Use Figures 13.5 to 13.7 – Uninstalled Values)   *943*
D.3.2.3   Other Pertinent Data   *943*
D.3.2.4   High Lift Devices (Flaps and Slats)   *944*
D.3.2.5   B100 Drag   *944*
D.4   The AK4 (4-Place Utility Aircraft) – Retractable Undercarriage   *944*
D.4.1   Customer Specification and Geometric Data   *944*
D.4.1.1   Customer Specification   *944*
D.4.1.2   Fuselage   *944*
D.4.1.3   Wing   *944*
D.4.1.4   V-tail (Aerofoil 64-010)   *945*
D.4.1.5   H-tail (Tee Tail, Aerofoil 64-210 – Installed with Negative Camber)   *945*
D.4.1.6   High Lift Devices (Slotted Flaps)   *945*
D.4.1.7   Engine Details – STD Day Performance (Figure D.5)   *945*
D.4.1.8   Engine Data   *946*
D.4.1.9   Propeller   *946*
D.4.2   The AK4 Aircraft Its Component Weights (kg)   *947*

Appendix E   Aerofoil Data   *948*

Data Courtesy of I. R. Abbott and A. E. Von Doenhoff, *Theory of Wing Sections.*   *948*

Appendix F   Wheels and Tyres   *959*

F.1   Glossary – Bias Tyres   *959*
F.2   Glossary – Radial Bias Tyres   *960*
F.3   Tyre Terminology   *961*
F.4   Typical Tyre Data   *962*

Index   *965*

Series Preface

The field of aerospace is multidisciplinary and wide ranging, covering a large variety of disciplines and domains, not only in engineering but in many related supporting activities. These combine to enable the aerospace industry to produce innovative and technologically advanced products. The wealth of knowledge and experience that has been gained by expert practitioners in the aerospace field needs to be passed on to others working in the industry and to researchers, teachers, and the student body in universities.

The *Aerospace Series* aims to be a practical, topical, and relevant series of books for people working in the aerospace industry, including engineering professionals and operators, academics, and allied professions such as commercial and legal executives. The range of topics is intended to be wide ranging, covering design and development, manufacture, operation, and support of aircraft, as well as topics such as infrastructure operations and advances in research and technology.

Peter Belobaba
Jonathan Cooper
Allan Seabridge

Preface

This new book is a natural follow up, retaining some DNA, of the book titled *Aircraft Design* (ISBN-10: 0521885167 and ISBN-13: 978-0521885164) by Ajoy Kumar Kundu, published by Cambridge University Press in April 2010. This new book is a collaborative effort by the original author with his colleagues Professor Mark A. Price of Queens University Belfast (QUB) and Mr. David Riordan of Bombardier Aerospace-Belfast.

Aircraft design fundamentals have remained virtually unchanged over most of the last two decades. Therefore, some amount of material from the first author's earlier publication is retained in this book. Based on the feedback, this new book takes the opportunity for a complete update and overhaul, by eliminating the deficiencies of the earlier edition, adding better clarification, incorporating improved diagrams in all chapters, and moving forwards with additional treatises, for example, on issues with aircraft performance monitoring, electric aircraft, future aircraft configuration and structural matters, which are included in an introductory manner. The benefits of using semi-empirical methods are explained in Section 1.14. CFD, as a design tool, has advanced to the point where it can be considered a 'digital wind tunnel', almost replacing the older practices of using datasheets as explained in Chapter 23. Electric aircraft design is dealt in Chapter 24. We have omitted vertical takeoff and landing/short takeoff and landing (VTOL/STOL), as well as helicopters: these subjects require their own extensive treatment.

We have recast the arrangement of the book into three parts: 'Part I – Prerequisites' deals with introductory information, definitions, gaining familiarity with aircraft and their configuration; examining options available and ultimately leading into suggesting candidate configurations specified by market requirements. Part I serves as the necessary prerequisites, but it is in Part II where aircraft design is dealt with. 'Part II – Aircraft Design' concentrates on the theory and practice of conceptual aircraft design involving aerodynamic, propulsive, structural and system considerations to fine-tune candidate configurations to arrive at the final design. Next, it is sized, and then its performance capabilities are substantiated. We are fortunate to offer industry standard procedures with worked-out examples of aircraft designs close to some existing aircraft, for example, similar to the Learjet45, Tucano turboprop trainer and a Hawk advanced jet trainer. 'Part III – Further Design Considerations' covers additional information designers must have.

The authors are grateful to access Bombardier data on the Learjet45, and the Tucano aircraft to corroborate the worked example data. To maintain commercial confidence, the Bombardier data is not given in this book but used only to check that the worked-out results are satisfactory. Part III serves as additional information that designers must know. The last chapter is on electric aircraft.

'The Road Map of the Book' at the beginning suggests a pathway to progress, but the book can be read in any order and serves equally well as a reference manual to reinforce elements of theory or to check examples for comparison. Our book titled *Theory and Practice of Aircraft Performance* (ISBN: 987-1-119-07417-5) published by John Wiley & Sons, Ltd in 2016 serves as a complementary book giving full performance estimation of the three worked example aircraft provided in this book.

The classical aerodynamic, structural and propulsive treatises of military aircraft design have the same fundamentals; the difference is in having the configurations and analyses to suit the mission role. Therefore, some discussions on military aircraft design have been included to allow for some progress in the study of this domain. Unlike the older book, we intend to keep the treatment on both the civil and military aircraft designs in one volume for convenience. It has to be borne in mind that the complexities of military aircraft are significantly more difficult. Military combat aircraft invariably have supersonic flight and high manoeuvre capabilities. The complex problems arising from the effects of a shock wave present a difficult challenge in designing a supersonic wing. There is not enough design data available in the public domain to undertake such in an undergraduate course. In addition, the integration of complex microprocessor-based electronics and systems engineering are beyond of the scope of this book. We doubt there is any undergraduate textbook specialising in advance fifth generation combat aircraft design of the class F35, Su37 and so on. Yet, we thought of including some basic concepts of supersonic aircraft design. Also included is the very important topic of supersonic aircraft drag estimation using Lockheed's well established 'Delta Method' and also the uninstalled low by-pass engine performance, should any instructor venture to offer such coursework. It is hoped that someday a supersonic aircraft designer from industry will write a book along the lines of the prevailing subsonic aircraft design textbooks for academic usage.

We have not included any futuristic advance design details as these are yet to be constructed. These are still under research; some types have scaled remote controlled test configurations showing promise with future potential. However, we have discussed some interesting futuristic aircraft configuration with a list of reference literature for those who wish to take up such studies. Moreover, such advanced aircraft cannot be well understood without the fundamentals in this book first being mastered.

The use of CAD for 2D and 3D drawings is assumed in academic work and students are always encouraged to use these tools. But, the use of manual drawing and any associated topics are not considered here as such detailed technical drawing is no longer practiced in industry. However, the use of three-view hand drawn sketches may prove helpful, if required, and this skill is strongly encouraged as it is a major help in understanding a design and in communicating the concept within the team and to customers.

For undergraduate classroom project work, we consider it best to take up an aircraft design project that can be compared with the existing types, as given in this book for both civil and military aircraft design study. This will help the student master the fundamentals before progressing to advanced studies and configurations.

We gratefully acknowledge full support of Queens University Belfast (QUB) and Bombardier Aerospace-Belfast, to bring out this book. Contributions made by our QUB colleagues Dr. Danielle Soban, Dr. Damian Quinn and Dr. Simao Marques are thankfully acknowledged. We also offer our thanks to our ex-colleague Professor Emmanuel Benard, now of Institut Supérieur de l'Aéronautique et de l'Espace, Toulouse, France, for his help.

We thank Professor Michael Niu, Professor Jan Roskam (DARcorp), Professor Egbert Torenbeek, Dr Bill Gunston, the late Dr John McMasters (Boeing Aircraft Company) and the late Dr L. Pazmany, Mr Richard Ferrier, Yevgeny Pashnin and Pablo Quispe Avellaneda who allowed us to use their figures.

We are indebted to *Jane's All the World Aircraft Manual*, Flightglobal, BAE Systems, Europa Aircraft Company, Airbus, NASA, MIT, Boeing, Defense Advanced Research Projects Agency, Hamilton Standards, Virginia Tech Aerospace and Ocean Engineering, and General Atomics to use their figures free of cost. All names are duly credited in their figures.

We are grateful to Cambridge University Press and their senior editor Mr Peter Gordon for allowing us the images used in the book titled *Aircraft Design* (ISBN 978-0-521-88516-4) by Ajoy Kumar Kundu.

We offer our sincere thanks to Mrs Anne Hunt, Associate Commissioning Editor, Mechanical Engineering of John Wiley & Sons, Ltd, the publisher of this book. Her clear, efficient and prompt support proved vital to reach the goal. Our thanks to the Reviewers and the Wiley Aerospace Series Editors for their observations, which helped the authors to shape the book. We thank Mustaq Ahamed Noorullah of SPi-Global for managing the publication process so patiently, courteously and efficiently, contributing substantially to the improvements to our book.

As ever, reader feedback is the best way to improve future editions – this is very important for the authors who have the obligation to disseminate good knowledge to the next generation. Readers may feel free to send their remarks to 'm.price@qub.ac.uk' or 'david.riordan@aero.bombardier.com'.

Ajoy Kumar Kundu
Mark A. Price
David Riordan

Individual Acknowledgements By Ajoy Kumar Kundu

No work on aircraft design can be done alone. Over time, throughout my passionate engagement in this fascinating subject, I have been advised, taught, supported, shared, allowed to see rare documents and even taught to become a private pilot, which exposed me to aeronautics in a fortunate manner. One-third of my career has been spent in academia and two-thirds in industry. I owe a lot to many. Now, after my retirement with time in hand. I am now in an age group that bridges more than half a century of aerospace progress. My aerospace career started in 1960 and now this is the twenty-first century. I have had the good fortune of witnessing many aerospace achievements, especially putting man on the Moon.

There is a clear gap between academic pursuits and what industry expects from fresh graduates as finished university 'products'. Both the USA and the UK are aware of this problem and both make periodic recommendations. One of the reasons could be that there appears to be a misconception within certain sectors of academics/research that how engineers generate good designs is a 'mystery'. All aircraft books outline methodologies devoid of any mystery; the differences are in the content. Keeping an undergraduate curriculum simple may ease teaching but could mask the rigours of the industrial practice when the gap appears. In fact, engineering design is a process and today's practices are so mature that they demonstrate systematic patterns. The freshly inducted students must be given good exposure to industrial practices at an early stage, to have a feel for what to expect when they join industry to start their professional lives.

The role of scientists and engineers are defined (Von Karman said 'A scientist discovers what already exists. An engineer creates that never was'). Converting ideas into reality for customer usage proves more difficult than adding any number of publications to a list (an exception being those papers that break new ground or advance a cause that is being adapted to wealth generation). Maybe the measure by which scientists are judged should be like that of engineers, namely, how much wealth the work has generated (where wealth must be defined in broad terms to include all that encompasses the common wealth). It is to be clearly understood that both scientists and engineers have to work together and not in a fallacious hierarchy in which advanced degrees stand above significant experience. Just consider the names Johnson, Mitchell, Dassault and so on.

Today's engineers will have to be strong in both analytical and applied abilities to convert ideas into profitable products. We hope that this book will serve the cause by combining analytical methods and engineering practices adapted to aircraft design. Therefore, the prerequisites would be second year (UK)/junior (USA) level mathematics and aerodynamics. It is not very difficult for those who are rusty or have missed out in acquiring these prerequisites – just an academic term of effort in what is offered on any university syllabus. This book offers more than just analysis by including 'experience'. Aircraft design must be practised.

I thank my teachers, heads of establishments/supervisors, colleagues, students, shop-floor workers and all those who taught and supported me during my career. I remember the following (in no particular order) who have influenced me – the list is compacted for the sake of brevity; there are many more individuals to whom I owe my thanks.

Teachers/Academic Supervisor/Instructors:

The late Professor Triguna Sen of Jadavpur University.
Professor James Palmer of Cranfield University.
Professor Arthur Messiter and Professor Martin Sichel of University of Michigan.
The late Professor Holt Ashley and Professor Samuel McIntosh of Stanford University.
Reverend Dr John Watterson, Ex-Senior Lecturer, Queens University Belfast, UK.
The late Squadron Leader Ron Campbell, RAF (Retrd.), Chief Flying Instructor, Cranfield University.

Heads of Establishments/Supervisors:

The late Dr Vikram Sarabhai, the Indian Space Research Organisation.
James Fletcher, Short Brothers and Harland, Belfast.
Robin Edwards, Canadair Limited, Montreal.
Kenneth Hoefs, Head of the New Airplane Project group, Boeing Company, Renton, USA.
Wing Commander Baljit Kapur, Chairman of Hindustan Aeronautics Limited, Bangalore.
The late Mr Raj Mahindra, MD (D&D), Hindustan Aeronautics Limited (HAL), Bangalore.
Tom Johnston, Director and Chief Engineer, Bombardier Aerospace-Belfast.
Dr Tom Cummings, Chief Aerodynamicist, Bombardier Aerospace-Belfast.

My Ex-Students and Most of My Ex-Colleagues (Both Shop-Floor and Office)

Without them, I do not have my profession.

I offer my special thanks to HAL shop-floor workers who stood by me and offered unconditional support in many difficult situations.

I am grateful to all the establishments associated with names given here. I learned a lot from them. Many of my examples are based on my work in these companies. I started my aeronautical career with Bombardier Aerospace, Belfast (then Short Brothers and Harland Ltd) and after a long break rejoined and then retired from the company, the first aerospace company to celebrate its centenary.

Personal Observation:

If I might deviate slightly, I have found from personal experience that one of the major hindrances to progress in some of the developing world comes from the inability to administrate technological goals even when there is no dearth of technical manpower – who perform better when working in the advanced world. People know about political asylum but professional asylum, also known as the 'brain drain', is a real issue. While design is not done by democratic process, the design culture should encourage free sharing of knowledge and a liberal distribution of due recognition to subordinates. Lack of accountability in higher offices is one of the root causes; the failure to exploit the full potential of natural and human resources. Technology can be purchased, but progress has to be earned. I hope to prepare the readers to contribute to progress. Progress and awareness come together to change management culture, something is bound to happen, I am hoping that it catches up fast to exploit the true potential of the nation.

My grandfather, the late Dr Kunja Behari Kundu; my father, the late Dr Kamakhya Prosad Kundu and my cousin-brother, the late Dr Gora Chand Kundu are long gone but they kept me inspired and motivated to remain studious. In a similar way, today, I must mention my mother, the late Mrs Haimabati Kundu and the three aunties, the late Mrs Kalyani Kundu, the late Mrs Ishani Kundu and the late Mrs Chandra Kundu, who contributed to my upbringing in different ways to become what I am today. My wife Gouri's tireless support saved me from becoming a hunter-gatherer, sparing the time to write this book. I am grateful to these people – thanks are not enough.

From the finest experience we enjoyed together to publish our book titled *Theory and Practice of Aircraft Performance* (ISBN: 9871119074175), it is natural to join hands again under this task of publication. Indeed, I am fortunate again to join with my long-standing colleagues Professor Mark A. Price, Pro-Vice-Chancellor for the Faculty of Engineering and Physical Sciences at Queen's University Belfast (formerly, he was Head of the School of Mechanical and Aerospace Engineering) and Mr David Riordan, Fellow, Nacelle Design and Power-Plant Integration (Bombardier Aerospace-Belfast) as co-authors. With more than five years added since we started our first adventure, this older person now needed more support than the last time. I am certain a second edition will follow in the capable hands of my coauthors' hands, but I do not think I will do anything more than remain in a supporting role.

I had my aeronautical education in the United Kingdom and in the United States of America; I worked in India, the United Kingdom and in North America. In today's world of cooperative ventures among countries, especially in the defence sector, the methodologies adopted in this book should apply. These organisations gave me the best education, their best jobs and their best homes.

Ajoy Kumar Kundu

By Mark A. Price

I have had the honour and privilege of working with Ajoy for some 20 years now since he came to Queen's after working in Bombardier. I have always learned much from his vast experience and wisdom, and have always thoroughly enjoyed partnering with him. Most recently, this was in our venture writing the book titled *Theory and Practice of Aircraft Performance* (ISBN: 9871119074175), along with David, whom I have also been fortunate to know from a variety of research programmes and his engagement in our degree programmes. He has added much value to these with his industrial practitioner's experience and, of course, many years of technical knowledge.

When it comes to the subject of aircraft design there are few who really have the depth of perspective and breadth of knowledge to be able to convey the concepts to those wishing to learn. Although there are many aircraft flying today, those who have actually designed an aircraft that has successfully flown and become operational are few. Ajoy is one of those few. My own experience lies far more in the details of design and manufacturing, focusing more on how those concepts can be realised in a manufacturing environment. This is still the business of aircraft design, but there is a particular ability required of the concept designers. They require a knowledge of many aspects of the aeroplane. Understanding of aerodynamics, performance, structures, materials, controls, costs, manufacturing systems, and so on is needed, and to this must be added the skill of understanding how these complex concepts and theories integrate and come together in the completed aeroplane. This is not easy.

As someone who has taught a range of technical subjects at undergraduate and post-graduate level I had developed a deep understanding of many engineering science disciplines and of pedagogical approaches becoming a successful educator. An essential skill in the world of academia. However, the most challenging course I taught was aircraft design, for the reasons I mentioned above. But, moreover, due to the scarcity of experienced individuals who had designed successful aircraft, the range of text books available was small, and the necessary industrial knowledge and exemplars for students to learn from was absent. The frustration for me was that I could not find the resources to fill the gaps in my knowledge and so help the students move to the next level.

The opportunity, therefore, to work with Ajoy in developing a textbook that can provide students with the breadth and depth needed to develop their skills, and especially that brings a practical industrial perspective, with many real examples to follow, was something I could not pass up. Moreover, as we now see a transport revolution coming, with concepts in personal air vehicles and the appearance of large sophisticated drones, there is a real need to have many more capable aircraft designers in industry. The need is strong and therefore I am very pleased to join with Ajoy and David to help a new generation of aspiring designers learn their trade.

I have many people to thank for being in the right place and right time to have this opportunity. I am thankful to Queen's University Belfast for providing an environment supportive of educational development, and in particular the noble aim to provide graduates valuable to industry. Together with Ajoy and David, my contribution in this effort has been in shaping this book to offer course material in line with industrial standard approaches.

There are many who have supported me in my career and my life thus far. My mentors, Mr Sam Sterling, Professor Raghu Raghunathan and Professor Cecil Armstrong, who have all provided much in the way of guidance, wise words and sharp wit, in addition to standing as exemplars of their profession, providing excellence in education and research. My colleague, Dr Adrian Murphy, who has worked alongside me from the start and shared many of the risks we took in developing new ideas to bring to a sometimes sceptical world. We have learned much together as we trod the path of mistakes and blind alleys. I thank the outstanding team of academics and support staff in Mechanical and Aerospace Engineering and across our whole faculty of Engineering and Physical Sciences who make work such an enjoyable part of my life. I cannot thank enough my family, my wife, Denise, and my daughter, Rachel, who have shown patience beyond their calling in allowing me space and time to fulfil my dreams. And lastly, my parents. My mother, Ann, who gave me the gift of words, and my late father, Matt, who instilled in me the virtue of delivering to the customer what they actually need, and hence my enthusiasm for this book; to fulfil a need for industry and the graduates they require.

Mark A. Price

By David Riordan

Aircraft have transformed the way in which we live; opening up for an ever-increasing number of people the opportunity to explore new places and cultures, vacation in different climates and travel long distances on transport vehicles that are fast, efficient and comfortable. The refinement of aircraft design to improve the experience for all involved and decrease associated costs so that more and more people can enjoy it relies on a wide spectrum of subject matter experts, with few in the industry receiving exposure to the how a clean-sheet aircraft design evolves. The principal author of this book (Ajoy) is one of those few, having accumulated a wealth of experience and seemingly very much enjoyed doing so along the way.

Since the first time I met Ajoy Kundu during the early 1980s, when we worked together in the Aerodynamics Department at Short Brothers PLC, Belfast, I have admired how he has mastered the fundamentals of many aspects of aircraft design and, with enthusiasm, wants to share this knowledge and experience with others. I find it sobering that, Ajoy, even though long retired from working at Shorts, invests long hours each day researching the subject matter that he loves. The effort he has made in writing this book has been a true labour of love, which I trust will be to the benefit for those who might care to study any aspect of the subject matter that is within this book.

I was very pleased when asked by Ajoy to join him and Mark in pulling this book together and since then have increasingly been proud to be associated with its content. The aerospace industry is most enjoyable to work in and continues to offer many challenges and experiences that I would encourage anyone with even a little interest to pursue. I am immensely grateful to those who have facilitated my career development within Bombardier and look forward to unabated enjoyment of the same in the years to come. My wife, Hazel, and two sons, Matthew and Jack, have, with gracious patience, permitted me to indulge in the job that that I love. I am truly appreciative of the love and encouragement shown to me by all three.

So in summary, for those who may be reading this book with a whole lifetime of opportunity to look forward to, make time to learn new things; ensure to enjoy what you do and never fail to recognise how much we rely on investments others make on our behalf.

David Riordan

List of Symbols and Abbreviations

Symbols

A	area
A_1	intake high light area
APR	Augmented power rating
A_{th}	throat area
A_W	wetted area
AR	aspect ratio
a	speed of sound, acceleration
$\bar{a}$	average acceleration at $0.7\ V_2$
ac	aerodynamic centre
a_{0_w}	wing zero lift line relative to fuselage reference line
α_{ht}	angle of attack at the H-tail
b	span
C_R, C_B	root chord
C_D	drag coefficient
C_{Di}	induced drag coefficient
C_{Dp}	parasite drag coefficient
C_{Dpmin}	minimum parasite drag coefficient
C_{Dw}	wave drag coefficient
C_v	specific heat at constant volume
C_F	overall skin friction coefficient, force coefficient
C_f	local skin friction coefficient, coefficient of friction
C_L	lift coefficient
C_l	sectional lift coefficient, rolling moment coefficient
C_{Li}	integrated design lift coefficient
$C_{L\alpha}$	lift curve slope
$C_{L\beta}$	side slip curve slope
C_m	pitching moment coefficient
C_n	yawing moment coefficient
C_p	pressure coefficient, power coefficient, specific heat at constant pressure
C_T	thrust coefficient
C_{HT}	horizontal tail volume coefficient
C_{VT}	vertical tail volume coefficient
C_{xxxx}	cost with subscript identifying parts assembly
C'_{xxxx}	cost heading for the type
CC	combustion chamber
CG	centre of gravity

c	chord
c_{root}	root chord
c_{tip}	tip chord
cp	centre of pressure
D	drag, diameter
D_{skin}	skin friction drag
D_{press}	pressure drag
d	diameter
E	modulus of elasticity, stored energy
E^*	mass (gravimetric) specific energy,
E_V^*	volumetric specific energy density
e	Oswald's factor
ε_u	deflected upwash angle
ε_d	deflected downwash angle
F	force
f	flat plate equivalent of drag, wing span
f_c	ratio of speed of sound (altitude to sea level)
F_{ca}	aft fuselage closure angle
F_{cf}	front fuselage closure angle
F_B	body axis
F_I	inertia axis
F_W	wind axis
F_{xxx}	component mass fraction, subscript identifies the item (Section 8.8)
F/m_a	specific thrust
FR	fineness ratio
G	gearing ratio
g	acceleration due to gravity
H	height
h	vertical distance, height
i_f	incidence of the fuselage camber line relative to fuselage reference line
i_t	angle between H-tail MAC_{HT} and fuselage reference line
i_w	wing and fuselage setting angle.
J	advance ratio
k	constant, sometimes with subscript for each application
L	length, lift
L_{FB}	nacelle fore-body length
L_N	nacelle length
L_{VT}, l_{VT}	vertical tail arm
L_{HT}, l_{HT}	horizontal tail arm
L	length
M	mass, moment
M_f	fuel mass
M_i	component group mass, subscript identifies the item (Section 8.6)
M_{xxx}	component item mass, subscript identifies the item (Section 8.6)
m	mass
$\dot{m}_a$	air mass flow rate
$\dot{m}_f$	fuel mass flow rate
$\dot{m}_p$	primary (hot) air mass flow rate (turbofan)
$\dot{m}_s$	secondary (cold) air mass flow rate (turbofan)
N	revolution per minute, number of blades, normal force

N_e	number of engine
n	load factor
ng	load factor times acceleration due to gravity
P, p	static pressure, power
P^*	mass (gravimetric) specific power
P_{bat}	electric power of the battery
P_t, p_t	total pressure
p'	angular velocity about Y-axis
p_e	exit plane static pressure
p_∞	atmospheric (ambient) pressure
Q	heat energy per of the system
q	dynamic head, heat energy per unit mass
q'	angular velocity about z-axis
R	gas constant, reaction
Re	Reynolds number
Re_{crit}	critical Reynolds number
r	radius, angular velocity
r'	angular velocity about the x-axis
S	area, most of the time, with a subscript identifying the component
S_{Bs}	project side area of the fuselage
S_H, S_{HT}	horizontal tail reference area
S_n	maximum cross sectional area
S_W	wing reference area
S_V, S_{VT}	vertical tail reference area
sfc	specific fuel consumption
T	temperature, thrust, time
T_C	non dimensional thrust
T_F	non dimensional force (for torque)
T_{SLS}	sea level static thrust at takeoff rating
T/W	thrust loading
t/c	thickness to chord ratio
tf	turbofan
U_g	vertical gust velocity
U_∞	free stream velocity
u	local velocity along the x-axis
V	free stream velocity
V_A	aircraft stall speed at limit load
V_B	aircraft speed at upward gust
V_C	aircraft maximum design speed
V_D	aircraft maximum dive speed
V_S	aircraft stall speed
V_e	exit plane velocity (turbofan)
V_{ep}	primary (hot) exit plane velocity (turbofan)
V_{es}	secondary (cold) exit plane velocity (turbofan)
W	weight, width, rate of energy usage (per hour).
W_A	useful work done on aircraft
W_E	mechanical work produced by engine
W/S_w	wing loading
w_u	upwash deflected velocity
w_d	downwash deflected velocity

x	distance along the x-axis
y	distance along the y-axis
z	vertical distance

Greek Symbols

α	angle of attack
α_{eff}	effective angle of incidence
β	CG angle with vertical at main wheel, blade pitch angle, side slip angle
Γ	dihedral angle, circulation
γ	ratio of specific heat, fuselage clearance angle
Δ	increment measure
δ	boundary layer thickness
ε	downwash angle
η_t	thermal efficiency
η_p	propulsive efficiency
η_o	overall efficiency
Λ	wing sweep, subscript indicates the chord line.
λ	taper ratio
μ	friction coefficient, wing mass
ρ	density
θ	elevation angle, flight path angle, fuselage upsweep angle
π	constant = 3.14
σ	atmospheric density ratio
τ	thickness parameter, torque
υ	velocity
φ	roll angle, bank angle
ψ	azimuth angle, yaw angle
ω	angular velocity

Subscripts (in many cases the subscripts are spelled out and not listed here)

a	aft
ave	average
ep	primary exit plane
es	secondary exit plane
f	front, fuselage
f_b	blockage factor for drag
f_h	drag factor for nacelle profile drag (propeller driven)
fus	fuselage
HT	horizontal tail
M	middle
N, nac	nacelle
o	free stream condition
p	primary (hot) flow
s	stall, secondary (cold) flow
t, tot	total
w	wing
VT	vertical tail
∞	free stream condition

Abbreviations

AB	After burner
ACAS	Advanced Close Air Support
ACT	Activate Control Technology
AEA	Association of European Airlines
AEW	Airborne Early Warning
AF	Activity Factor
AFM	Aircraft Flight-track Monitoring
AGARD	Advisory Group for Aerospace Research and Development
AHM	Aircraft Health Monitoring
AIAA	American Institute for Aeronautics and Astronautics
AJT	Advanced Jet Trainer
AMPR	Aeronautical Manufacturer's Planning Report
APM	Aircraft Performance Monitoring
APU	Auxiliary Power Unit
ARINC	Aircraft Radio Inc.
ASDA	Acceleration Stop Distance Available
AST	Air Staff Target
ATA	Aircraft Transport Association
ATF	Advanced Tactical Support
AVGAS	Aviation GASoline (petrol)
AVTUR	Aviation Turbine Fuel
BAS	Bombardier Aerospace–Shorts
BFL	Balanced Field Length
BLDC	Brushless DC Motor
BMS	Battery Management System
BOM	Bill Of Material
BPR	By-Pass Ratio
BRM	Brake Release Mass
BWB	Blended Wing Body
BVR	Beyond Visual Range
CAD	Computer Aided Design
CAE	Computer Aided Engineering
CAM	Computer Aided Manufacture
CAS	Close Air Support, Calibrated Air Speed
CCV	Control Configured Vehicle
CFD	Computational Fluid Dynamics
CFL	Critical Field Length
CG	Centre of Gravity
CV	Control Volume
CWY	Clearway
DCPR	Design Controller's Planning Report
DBT	Design Build Team
DEP	Distributed Electric Propulsion
DFSS	Design For Six Sigma
DFM/A	Design for Manufacture/Assembly
DOC	Direct Operating Cost
DoD	Department of Defence
DOD	Depth of Discharge

DOT	Department Of Transport
EAS	Equivalent Air Speed
EASA	European Aviation Safety Agency
ECS	Environment Control System
EFIS	Electronic Flight Information System
FS	Factor of Safety
EPNL	Effective Perceived Noise Level
EPR	Exhaust Pressure Ratio
ESHP	Equivalent SHP
ETOPS	Extended Twin Operations
EW	Electronic Warfare
FAA	Federal Aviation Administration
FADEC	Full Authority Digital Electronic Control
FBW	Fly-By-Wire
FEM	Finite Element Method
FPS	Foot/Pound System
HEA	Hybrid Electric Aircraft
HMD	Helmet Mounted Display
HOTAS	Hands On Throttle And Stick
HR	Hybrid Ratio
HUD	Heads-Up Display
HP	Horse Power, High Pressure
H-tail	Horizontal Tail
IATA	International Air Transport Association
IC	Internal Combustion
ICAO	International Civil Aviation Organisation
IOC	Indirect Operational Cost
INCOSE	International Council Of Systems Engineering
IPPD	Integrated Product and Process Development
IATA	International Air Transport Association
ICAO	International Civil Aviation Organization
ISA	International Standard Atmosphere
JAA	Joint Aviation Authority
JAR	Joint Airworthiness Regulation
JUCAS	Joint Unmanned Combat Air System
KE	Kinetic Energy
KEAS	Knots Equivalent Air Speed
L/D	Lift-to-Drag (ratio)
LE	Leading Edge
LF	Load Factor
LFL	Landing Field Length
LP	Low Pressure
LCA	Light Combat Aircraft
LCC	Life Cycle Cost
LCN	Load Classification Number
LCR	Lip Contraction Ratio
LDA	Landing Distance Available
LERX	Leading Edge EXtension (strake)
LOH	Liquid Hydrogen
LPO	Long Period Oscillation

MAC	Mean Aerodynamic Chord
MDA	Multi-Disciplinary Analysis
MDO	Multi-Disciplinary Optimisation
MEM (W)	Manufacturer's Empty Mass (Weight)
MFD	MultiFunctional Display
MFR	Mass Flow Rate
MoD	Ministry of Defence
MOGAS	MOtor GASoline (petrol)
MPM	Manufacturer Process Management
MTM	Maximum Taxi Mass
MTOM (W)	Maximum Take-Off Mass (Weight)
NACA	National Advisory Committee for Aeronautics
NASA	National Aeronautics and Space Administration
NBAA	National Business Aviation Association
NHA	Negative High Angle Of Attack
NIA	Negative Intermediate Angle of attack
NiCad	Nickel-Cadmium
NiMH	Nickel-Metal Hybrid
NLA	Negative Low Angle of attack
NRC	Non-Recurring Cost
NTC (M)	Normal Training Configuration (mass)
OEM (W)	Operating Empty Mass (Weight)
OEMF	Operating Empty Mass Fraction
PAX	Passenger
PCN	Pavement Classification Number
PCU	Power Control Unit
PE	Potential Energy
PHA	Positive High Angle of attack
PIA	Positive Intermediate Angle of attack
PLA	Positive Low Angle of attack
PLM	Product Life cycle Management
PNdB	Perceived Noise Decibel
PNL	Perceived Noise Level
PPR	Product, Process and Resource
PV	Photovoltaic
RAeS	Royal Aeronautical Society
RAT	Ram Air Turbine
RC	Rate of Climb, Recurring Cost
RCS	Radar Cross-section Signature
RD&D	Research Design and Development
RDDMC	Research Design, Development, Manufacture and Cost
RDD&T	Research Design, Development and Test
R&M	Reliability and Maintainability
RF	Range Factor
RFP	Request for Proposal
RJ	Regional Jet
RLD	Required Landing Distance
RPM	Revolution per Minute, Revenue Passenger Mile
RPV	Remotely Piloted Vehicle
RSA	Runway Safety Area

SATS	Small Aircraft Transportation System
SEP	Specific Excess Power
sfc	Specific Fuel Consumption
SHP	Shaft Horse Power
SI	System International
SPL	Sound Pressure Level
SPO	Short Period Oscillation
STOL	Short TakeOff and Landing
SWY	StopWaY
TAF	Total Activity Factor
TET	Turbine Entry Temperature
TGT	Turbine Guide-vane Temperature
TOFL	Takeoff Field Length
TOC	Total Operating Cost
TPT	TurboProp Trainer
TQM	Total Quality Management
TR	Thrust Reverser
TTOM	Typical TakeOff Mass (military)
T&E	Training and Evaluation
UAV	Unmanned Air Vehicle
UAS	Unmanned Aircraft System
UCA	Unmanned Combat Aircraft
UHBPR	Ultra High BPR
ULD	Unit Load Device
V-tail	Vertical Tail
ZFM (W)	Zero Fuel Mass (Weight)

Road Map of the Book

1 Objectives and Aims

The objectives and aims of this book are to give an understanding of aircraft design fundamentals as practised during the conceptual phase of a new aircraft project. Attempts are made to adhere, as best as possible, to the typical industrial methodologies, simplified for academic usage. The book is geared towards understanding the physics involved in shaping aircraft components and then assembling into a finished aircraft configuration. Many of the details of research findings are avoided unless the information can be made useful by applying in a meaningful manner; wherever appropriate, demonstrated through worked-out examples. Sufficient references are provided on the related research.

Piecemeal topical design examples are avoided. Aircraft design has to be practised in totality. From the given aircraft design specifications, the work flow of the worked-out examples starts from scratch, progressing linearly chapter-by-chapter, going through the full process, ending in a final frozen aircraft configuration and their performance substantiation for management-review to consider for the project 'go-ahead' or not. Worked-out examples of three different classes of aircraft (Bizjet, AJT and TPT) are configured from scratch to completeness. The three different aircraft bear close similarities with aircraft currently in operation for the readers to compare the results. To suit academic curriculum, only one configuration from each class of aircraft is worked out, adhering to the methodology adapted in industries, in a simplified manner. Industry takes a thorough and extended study with different types of aerofoil and wing planform shapes on each of the several candidate aircraft configurations (e.g., high or low wing, fuselage shape, nacelle location, and empennage arrangement – see Figure 8.2). For example, say, an industry finds three candidate configurations and studies each configuration with various wing designs (say three or more), totalling to at least nine aircraft to generate points for parametric optimisation (using graphs). The choice for the objective function(s) of the parametric optimisation depends on the aircraft mission requirements. For commercial transport aircraft, the objective function is mainly to minimise the Direct Operating Cost (DOC). To avoid a lengthy repeat computations of each deign up to DOC estimates, parametric optimisation has not been carried out here. It may prove useful if the readers can do parametric optimisation for their configuration choices. The issues arising from Multi-disciplinary Analysis (MDA) and Optimisation (MDO) are discussed in Section 2.6.1.

Computational fluid dynamics (CFD), as a design tool, has advanced to the point where it can be considered a 'digital wind tunnel', almost replacing the older practices of using datasheets (DATCOM/ESDU) as explained in Section 1.14. This book uses semi-empirical relations, those well adopted in industries, some of them are embedded in the datasheets. Analyses of CFD output require thorough understanding of the underlying theory and practices of the aircraft design process. Generally, introduction to CFD comes late in the senior years of undergraduate study, by that time students are likely to gain some exposure to aircraft design. We recommend the use of computer-aided drawing (CAD) in generating configurations, which facilitates any subsequent CFD work (see Section 1.12.2). Using CFD during conceptual study is now a routine industrial practice to determine the aerodynamic characteristics of aircraft and its components. However, airplane design methodology cover wider topics beyond the aerodynamic analyses. After the project 'go-ahead' is obtained, wind-tunnel tests are carried out to verify analyses and finally substantiated through flight tests. It is suggested that use of software for conceptual aircraft design may be postponed till the design fundamental are mastered by manual computation to get a feel for numbers and assess the labour content involved.

2 The Arrangement

In a step-by-step manner in this book, an approach road to industry-standard aircraft design methodology at the conceptual stage is presented that can be followed in classrooms, from the initial stages of finding a market to the final stages of freezing the aircraft configuration. In the aircraft industry, after the go-ahead is obtained, the development programme moves to the next phase (i.e. the Project [or Product] Definition Phase), which is not within the scope of this book. The book covers two semesters of work as outlined in Section 3.

Aircraft design is not generated from any amateurish wish-list or school-day fantasies of exotic *Star Wars* shapes. Aircraft design is a rigorous discipline with a conservative approach, requiring thorough understanding of the related sciences embracing almost all branches of technology. Therefore, it is essential to learn the basics through conventional designs and then move on to innovations after mastering these basics. Course-work methodology should be in harmony with industrial practices; otherwise, the gap between academia and industry would interfere.

The book is split in three parts as described next. The chapters are arranged linearly; however, a classroom course will require the instructor to choose any topic as intended for the course. Section 2 suggests a typical curriculum for course work. We attempt to keep the treatise interesting by citing historical cases. The main driver for readers is the motivation to learn. The road map of the book is described as follows.

Part 1 – Prerequisites (Chapters 1–7)

These seven chapters cover the historical past, related introductory information, definitions, gain familiarity with an aircraft and its configuration and examine options available leading into suggesting candidate configurations specified by the market requirements. These chapters form the basis to start design study with industry-standard worked-out examples close to current operating aircraft. The readers must know what is covered in these seven chapters to undertake aircraft design study. Practically no coursework is included in these seven chapters except carrying out a mock market study.

Chapter 1 – Introduction
This is an introductory chapter serving as a 'starter package' for this book. We recommend it for leisure reading for all. It begins with a brief historical outline intended to inspire readers' interest in our aerospace heritage (one of the few areas in which reality can be more interesting than fiction). The fascinating stories of human achievement are motivational, students are urged to read books and peruse Internet web sites dedicated to aerospace history. They cover the full range of human emotions: from disappointment due to failures and fatalities to the joy of successes; from light-hearted circus flying to flying in spectacular displays that defy imagination. Chapter 1 continues with a description of typical current designs and associated market drivers. It is followed by looking into the future possibilities and topics of research interest to advance technologies. Sections 1.12–1.14 give some interesting points for the intended coursework, for example, the learning process, justifying the use of both the SI and the FPS unit systems, the pros and cons of using datasheets/semi-empirical formulae, 'what is not covered in this book' and so on. Finally, the chapter ends with the understanding the atmosphere through which aircraft have to fly; that is, introducing the atmosphere comprehensively.

Chapter 2 – Design Process
This chapter begins with introducing aircraft (both civil and military) to gain familiarity with the product lines the readers are to study. It is followed by giving some idea on how aircraft industry works in a generic manner, that is, phases of new aircraft design project from the conceptual stage to the finished product and continuing with product support to end of life disposal, in short from 'cradle to grave'. It continues with a discussion of the

importance of market information. Marketing and airworthiness are the two most important requirements that give the product viability. Students are encouraged to conduct a short mock market study to generate a specification for which experienced guidance is required. For commercial aircraft, the specification is primarily the mission profile for the payload range capability. The differences between military and civil aircraft specifications and the associated financial outlay are significant. Military specifications are substantially more complex, depending on the specific combat role. They vary widely and complexity spirals when multirole capabilities are required. Substantiation of airworthiness regulations is mandatory in the industry and also is discussed. The U.S. Federal Aviation Regulations (FARs) are now in widespread use and adhered to. The recently established European Aviation Safety Agency (EASA) standards are similar to FAR and therefore are not discussed here.

Chapter 3 – Aerodynamics

This chapter introduces the associated aeronautical fundaments, for example, classification of aerofoil (Appendix C includes the aerofoil details used in this book), various definitions used and aerodynamic considerations that are central to shaping a streamlined aircraft configuration. Extensive treatment of aerodynamics is provided separately in all aeronautical schools; here, only the necessary aerodynamic information has been compiled for reference as the aircraft design coursework progresses. The readers are required to know the facts and to refer to and apply them as and when required.

Chapters 4, 5 and 6 – Aircraft Components

These three chapters deal with isolated aircraft components, in sequence, starting with the wing followed by bodies of revolution such as the fuselage nacelle and other lifting surfaces, for example, empennage, etc. A wide variety of options is given in these chapters. The chapters discuss various possible aircraft component configurations currently in use to assist in rational selection. These serve as the building blocks (in a Lego/Mechano concept) to assemble in an appropriate manner from the choices available. However, there is always scope to introduce some proven innovative concepts from creative minds. Readers are reminded that there is no scope to introduce any unproven concept that may jeopardise safety, any aspect cannot be compromised.

Chapter 7 – Statistics

This is the last chapter in Part I. It presents aircraft statistics to observe a strong statistical correlation within various designs of the classes of aircraft. Products from different origins show similarities that indicate a strong statistical pattern, which provides an idea of what is to be expected in a new design. A new design, with commercial considerations, must be a cautious progression, advancing through the introduction of the latest proven technologies to stay ahead of competition. Military aircraft designs necessarily must be bolder and make bigger leaps to stay decisively ahead of potential adversaries, regardless of the cost. Eventually, older, declassified military technology trickles down to commercial use. One example is fly-by-wire (FBW) technology.

Jane's All the World's Aircraft Manual (published annually) has served many generations of aeronautical engineers around the world for more than half a century. This is an indispensable source for vital aircraft statistics.

Part II – Aircraft Design (Chapters 8–16)

Chapters 8–15 concentrate on the theory and practices of aircraft design involving aerodynamics, propulsive, structural and systems considerations to select a suitable one from the candidate configurations and fine-tune to a preliminary aircraft configuration in the *concept definition*. Subsequently, it is sized and then its performance capabilities are substantiated to *concept finalisation*. There are iteration process involved and should progress quickly by using spreadsheets for repetitive calculations.

We are fortunate to offer industry-standard procedures with worked-out aircraft designs close to some of the existing kinds, for example, similar to the Learjet45, Tucano turboprop trainer (TPT) and a

Hawk advanced jet trainer (AJT). Authors are grateful to have been able to access Bombardier data on the Learjet45 and the Tucano aircraft to corroborate the worked-out example data. To keep commercial in confidence, the Bombardier data is not given in this book but was used only to check the worked-out data to be very satisfactory.

Chapter 8 (Configuring Aircraft – Concept Definition)

In this long Chapter 8, configuring a new aircraft with given specification (market study) starts in a formal manner, serving as the backbone of this book. Readers are to retrieve information from Chapters 4–6 to generate preliminary aircraft. Statistics of current aircraft (Chapter 7) within the class of study offer the starting point with guessed maximum takeoff weight (MTOW), engine thrust (T_{SLS}/engine) to give definite geometric dimensions that meet market requirements. The objective is to generate aircraft components, piece by piece and mate them in the way as shown in the middle diagram of Figure 2.1. These serve as *boilerplate* template geometries of aircraft components assembled to generate candidate aircraft. One of them will be selected as the preliminary configuration seen as *concept finalisation*, which will be sized with matched engines to *concept finalisation* (Chapter 14), thereby freezing the configuration for management to consider for a go-ahead decision. The aircraft conceptual-design study must consider offering a family of variants to cover a wider market at low cost by retaining significant component commonalities. This point is emphasised throughout the book. Consideration of family of variants must begin at the initial stage to make products right the first time (i.e. the Six-Sigma approach). Civil and military aircraft configurations are discussed separately because they are quite different in their mission role.

Chapter 9 – Undercarriage

With the preliminary aircraft configuration arrived at in Chapter 8, the next step is to layout the undercarriage with sized wheel base and wheel track in Chapter 9. The role of aircraft centre of gravity (CG) travel range and neutral point locations (guessed) play an important role in layout undercarriage. Some trial and error is required to satisfy the constraints of tipping on any side and the wheel load constraints. This may involve some reposition of components. This gives the opportunity to make the first iteration to refine the preliminary aircraft configuration. Undercarriage load, deflection and selection of tyre sizes are dealt with here.

Chapter 10 Weights and CG

Semi-empirical relations are used to estimate aircraft component weights to arrive at aircraft weights at different stages of loading conditions, for example, to maximum permissible load, empty load and so on. This allows location of the aircraft CG and range of travel for the extreme loading conditions. The computed weights are more accurate and likely to differ from the guessed weights to arrive at the preliminary aircraft configuration as worked out in Chapter 8. This will require iteration of the preliminary aircraft configuration making slight changes to the components' geometries.

Chapter 11 – Drag

This chapter addresses the difficult aspect of drag estimation for both military and civil aircraft. Successful understanding of aircraft drag is of paramount importance for the readers. This book is presenting the industry-standard methodology to estimate aircraft and its component drag.

Chapters 12 and 13 – Powerplant

Relevant information on aircraft power plants is integral to aircraft design. Although this book does not focus on aircraft engine design, aircraft designers should thoroughly understand the propulsion system as the 'heart' of the aircraft. Chapter 12 introduces various aircraft propulsion system as well as related topics concerning engine and aircraft integration. This information is necessary for shaping nacelles and estimating their installed drag. It is followed by Chapter 13 giving typical engine performances.

Chapter 14 (Sizing and Engine Matching – Concept Finalisation)

With the drag polar (Chapter 11) and engine performance (Chapter 13), in this chapter the preliminary aircraft configuration (the baseline aircraft), thus obtained in Chapter 10 after the necessary iterations, is sized with matched engines simultaneously satisfying the specification requirements of field performances, initial climb and cruise capabilities. This phase closely conforms to industry practices. The sized baseline aircraft is likely to have some variance from the preliminary configuration. It will require reiteration with revised MTOW and T_{SLS}/engine to achieve *concept finalisation*, with a configuration that may now be frozen, from which variant designs in a family are derived, if required, with minor changes. The procedure offers a 'satisfying' solution for the most important sizing parameters, complying with constraints imposed by market specifications. Parametric sensitivity studies are required, which eventually prove to be the key to success through balancing comfort with cost in a fiercely competitive market. Safety is never compromised. The go-ahead decision waits until the aircraft operating costs are evaluated (Chapter 16) to examine whether the new design offers better economics than the current operational aircraft.

Note: The authors avoided early sizing and first presented a concept definition and the sized concept finalisation through a formal industry-standard aircraft sizing and engine matching procedure. By taking aircraft sizing in early chapters using a mass-fraction method without knowing how the aircraft looks can mask the merits of new aircraft design superiority. This is explained in Section 10.8.1.

Chapter 15 – Performance Analyses

The sized baseline aircraft performance now need be substantiated to check whether it can meet the mission profile while at the same time establishing the operating cost for competition analysis. This is the proof that the product demonstrates compliance with the customer's requirements as listed in the specifications. In case the performance capabilities are unable to meet the specified requirements, then the project has to recycle through another series of iterations: an expensive outcome. Fortunately, with current methodology using CAE this does not happen to the extent that it occurred in the past. The variant aircraft performances starts as the fall-out of the design but can be tweaked with some improvements. In this chapter, the derivation of aircraft performance equations is kept to a minimum. Our book titled *Theory and Practice of Aircraft Performance*, ISBN: 9871119074175, published by John Wiley & Sons, Ltd, 2016 [8] serves as a complementary book, giving full performance estimations of the three worked-out examples of aircraft given in this book.

Chapter 16 – Cost Analyses

The success and failure of the new aircraft, capable of meeting the performance requirements, hinges on its ability to compete in the operational area at a better operating cost than current operational aircraft offer. This chapter computes the aircraft Direct Operation Cost (DOC) to examine competitiveness. Since about a third of DOC is contributed to by the unit aircraft cost (price), it is important to know the aircraft manufacturing cost that determines the unit selling price. This chapter also gives an overview on aircraft cost methodology (aircraft selling price can vary from time to time depending on what market bears during the negotiating period with potential customers: this book only considers the standard declared price). When the go-ahead decision is given, the baseline aircraft with its variants enter into the next phase of the new aircraft project, beyond the scope of this book.

Part III – Further Design Considerations (Chapters 17–24)

The new aircraft design study is not yet over. Chapters 17–23 provide a considerable amount of additional information related to aircraft design. The designers must know about these to have a good grasp of aircraft and its functional capabilities.

Chapter 17 – Loads

This chapter addresses aircraft limitations to retain structural integrity as well human physical limitations under manoeuvre loads defining the operational envelope (i.e. the *V-n* diagram).

Chapter 18 – Stability

This chapter covers aircraft static stability, an essential criteria for safe operation. Aircraft CG location and empennage sizes contribute to aircraft stability and control capabilities. The equations governing aircraft static stability are derived here. Fortunately, aircraft dynamic behaviour and control responses are not addressed in the conceptual phase – they are considered after the configuration is finalised. If required later, the control geometries are tailored or adjusted, possibly requiring another iteration to update the configuration. To save time in the classroom, the iterations of control surface tailoring are avoided. The design configuration is now complete but still requires fine tuning of the aircraft mass and CG location.

Chapter 19 – Materials and Structures

Material and manufacturing considerations can significantly affect aircraft cost and weight. Introductory coverage of the types of aircraft materials and their properties is presented. Then the types used to fabricate load-bearing structural parts in the main aircraft components are shown. Finally, a worked-out example demonstrates how to size a load-bearing part. The core objective is to select a material that gives superior strength-to-weight ratio at an affordable cost.

Chapter 20 – Manufacturing Considerations

Some background on manufacturing considerations is given here for the readers to be aware that the aircraft thus far configured could be manufactured at the lowest cost to keep DOC low (Chapter 16). Therefore, manufacturing cost control is of utmost importance to remain competitive in the fiercely competitive market. Design for Manufacture and Assembly (DFM/A) methodology, design and cost implications are addressed in this chapter. These considerations are also part of Phase I study carried out by separate costing departments.

Chapter 21 – Miscellaneous Design Considerations

There are many other issues – for example, concerning environment, safety, operational matters, 'end of life' valorisation, some emerging new scenarios and so on – that affect design considerations at the very conceptual stage of the project and these are covered here.

Chapter 22 – Aircraft Systems

This chapter continues with those issues dealing with bought-out items, hardware considerations and selection and human interface items, for example, flight deck instruments, that would affect aircraft weight and cost. It gives a brief overview of some of the main aircraft systems that play essential roles in an aircraft becoming airworthy and must be known by the designers engaged in conceptual aircraft design to generate aircraft configurations.

Chapter 23 – CFD

CFD is now an indispensable tool in aircraft design. This is seen as digital wind tunnel. Very quickly and at low cost, the aerodynamic properties of a moving body through air can be obtained to some satisfactory accuracy, which will be eventually verified in wind-tunnel tests. This chapter gives an overview of the role of CFD in the conceptual design study phase. This book is not about CFD, which is an exhaustive subject in itself to which scientists and engineers can devote their entire careers. Today, almost all undergraduate aeronautical engineering courses introduce CFD early on so that students can gain proficiency in application software and apply it.

Chapters 24 – Electric Aircraft

The scope of electrically powered aircraft is given. It describes the types of electricity storage devices and harvesting technologies available in the context of today's design obligations. It is followed by the brief description of the prime mover, that is, the electric motor. Then some generalised theories applicable to both conventional and electrical fixed wing aircraft are shown, followed by specific theories for meant for electric aircraft performance analyses.

Appendices

Appendices A and C gives the usual support materials on conversion factors, atmospheric tables and some useful equations used in this book. Appendix D gives a case study of the Airbus320 type aircraft. The design study starts from scratch using the methodologies presented in the book. Only the Airbus supplied three-view diagram was used to begin with. It is for the readers to judge what can be achieved when compared with actual Airbus data available in the public domain. It is for this reason the authors recommend that a class should be of the type close to current production aircraft for substantiation. Working on arbitrary aircraft piecemeal is unlikely to withstand the test of accuracy. Appendix E gives seven sets of aerofoil data with the sources acknowledged. Lastly, Appendix F gives Type III tyre data taken from the Goodyear Tyre Data.

2.1 Discussion

Each chapter of the book starts with an overview, a summary of what is to be learned, and the coursework content. There are no exercises at the end of the chapters as the classroom aircraft design project is an essential part of the coursework. Each of the chapters is arranged in sequence in continuation of project progression. Extensive industry-standard worked-out examples cover from start to finish. Associated industry-standard examples in the book are of the class of Learjet45, the Royal Air Force (RAF) Hawk AJT aircraft and the RAF Tucano TPT. Case studies are indispensable to the coursework and classroom exercises must be close to actual aircraft that have been modified to maintain 'commercial in confidence'. Additional examples in Appendix D are based on actual designs worked out by the authors. The results are not from the industry but have been compared with available performance data. The industry is not liable for what is presented herein. Many categories of aircraft have been designed; this book covers a wide range for coursework exercises and provides adequate exposure to important categories. After students become proficient, they can then undertake less conventional aircraft designs.

Developing a configuration within a family concept so that variants can be designed at low cost and cover a wider market area is emphasised. One might even say, 'design one and get the second at half the development cost'. The jet transport aircraft is recommended as the most suitable for coursework projects. Chapter 2 lists a few projects of interest to students. Other projects could be extracted from the competitions held by RAeS in the United Kingdom and organisations such as NASA, the Federal Aviation Administration (FAA) and AIAA in the USA.

Designing an F22 class of aircraft is beyond the scope of this book – we question whether any textbook can be used for undergraduate coursework without first offering an exercise on simpler designs. Nevertheless, advanced work on military designs is possible only when the basics have been mastered; the aim of this book.

For classroom practice, using manual computation is recommended with spreadsheets developed by students because the repetitive aspect is part of the learning process. It is essential for students to develop a sense for numbers and to understand the labour content of design (it is expensive to make midcourse changes). We decided not to provide CDs with companion software. We do not follow this practice because the software for handling repetitive tasks constrains students from interacting more with the governing equations and is part of the learning experience.

If students elect to use off-the-shelf software, then it must be reputable. For US readers, well-circulated NASA programmes are available. However, these are more meaningful after the subject of aircraft design is well understood, that is, after completing the coursework using manual computations. This leads to an appreciation of how realistic the computer output is, as well as how changes in input to improve results are made. It is better to postpone using conceptual-design software until entering the industry or doing postgraduate work. In academia, students can use CFD and finite element method (FEM) analyses to complement the aircraft design learning process.

Finally, we recommend that aircraft designers should have some flying experience – it should prove very helpful to understand flying the vehicle one is trying to design. Getting a licence requires some effort and financial resources, but getting a few hours of planned flight experience would still be instructive. One may plan and discuss with the instructor what needs to be demonstrated, for example, aircraft characteristics in response to control input, stalling, '*g*' force in steep manoeuvre, stick forces and so on. Some universities/schools offer a few hours of flight tests as an integral part of their aeronautical engineering courses, but we are suggesting a little more than that: hands-on experience under the supervision of an instructor. A driver with good knowledge of car design features appreciates the car better.

Flying radio-controlled model aircraft may be interesting to students, but we do not think it is relevant because it is not an industrial practice unless the project concerns radio-controlled aircraft such as remotely piloted vehicles (RPVs) and unmanned air vehicles (UAVs). Some combat aircraft have an unpowered, accurately scaled, radio-controlled model dropped from the mother aircraft to test stability behaviour. However, if there is interest, instructors and students can take up model aircraft flying for demonstrations/hobbies.

2.2 Note

The authors avoided early sizing and first presented a concept definition and the sized to concept finalisation through a formal industry-standard aircraft sizing and engine matching.

Many aircraft design books have presented aircraft sizing in early chapters using a mass-fraction method without knowing how the aircraft looks. The disadvantage of such an approach is that the aircraft weight estimation may not match the early sizing when the geometry is established through the formal method to concept finalisation. There is a lack of any worked-out example progressed linearly to concept finalisation to substantiate early sizing results. Section 10.8.1 discusses this point.

The authors left aircraft sizing to Chapter 14: the first task is to generate the proposed new aircraft configuration (with a three-view diagram), which is the concept definition done in Chapter 8. Only then does the project progress towards concept finalisation through a formal method of aircraft sizing with matched engines. Chapters 9–13 cover the necessary prerequisites of undercarriage placement, drag estimation and 'rubberised' engine performance data leading to concept finalisation. Once the aircraft is sized with matched engines, Chapter 15 concludes the conceptual aircraft design by substantiating (aircraft performance) compliance with the FAA requirements and design specifications. Complete aircraft performance estimation of the worked-out examples (Bizjet, AJT and TPT) is carried out in [8].

The worked-out examples in this book start from a given market specification that includes requirements for payload range, field performance, climb and cruise performance and so on. There is then a linear progression from concept definition to concept finalisation in the steps as described before. The formal aircraft sizing and engine matching tasks are to find the baseline aircraft in a family concept of derivatives with variant aircraft. The object is to find a 'satisfying' design to retain component commonality, none of them are optimum designs. Other aircraft design books do not consider the design of aircraft that is the basis for subsequent derivative designs. It is not easy to apply the mass-fraction method of sizing to such a task. This is an observed weakness in academic studies of arbitrary aircraft.

Mass-fraction analysis, using equations such as Equation 10.13, does offer good insight for trend analyses, optimisation studies and rapid sensitivity studies.

3 Suggested Route for Coursework

The authors suggest the following path for the two-semester coursework. Each semester entails 36 hours of lecture and coursework: specifically, 20–24 hours of lectures by the instructor, followed by 16–12 hours of computational work in tutorial classes. Any unguided work may be left for routine computation to complete the assignment of the chapter. The final week of coursework is reserved for report writing. An outline of the final-report requirements may be given to students at the beginning of the course when the instructor may cover any suitable additional topics. Students are required to submit brief preliminary reports at the completion of each chapter so that the instructor can offer improvement guidelines. This reduces student workload at the end of the semester and enables them to complete their report without loss of quality. The coursework progresses sequentially following the chapters of this book.

Use of CAD for 2D diagrams and 3D modelling is recommended. Worked-out examples used throughout in this book are the following.

1. A 10-passenger, 2000 nm Bizjet in the Bombardier Learjet 45 class
2. AJT in the BAE Hawk class
3. TPT in the Bombardier Tucano class

First Semester 2 (12 weeks – 36 contact hours)

Lecture hours (24 hours)

1. Establish the project specification with a mock market study as described in Chapter 2 to identify the project to be undertaken.
2. Select aerofoil and establish wing characteristics (Chapters 3 and 4).
3. Configure the aircraft (Chapter 8 with input from Chapters 4–7).
4. Complete undercarriage layout and tyre sizing (Chapter 9).
5. Estimate component and aircraft weight and determine the CG location (Chapter 10).
6. Estimate aircraft drag (Chapter 11).
7. Establish engine data (Chapters 12 and 13).
8. Size the aircraft and find a matched engine and determine the family of variant designs (Chapter 14).
9. Conduct a performance evaluation to check whether the market specification is met (Chapter 15).
10. Evaluate operating costs for transport aircraft project (Chapter 16).
11. Cover loads, stability, systems and structural considerations (Chapters 17–23)
12. **Electric aircraft (Chapter 24).

(Students may use the last few weeks for writing up their project.)
Classroom work hours with the instructor: Total 24 hours
Classroom tutorial hours for the project (one per week) 12 hours
**A project on electric aircraft may prove topical if the instructor can have battery power and storage capacity with matched motor data available in the market, then specific energy consumption, aircraft speed and range/endurance and other performance capabilities can be computed.

Classroom management and requirements for submission of work in report form is determined by the instructor.

Second Semester (Optional)

This book is meant for one semester of work as suggested before. However, time permitting, for more detailed studies a second semester of work can continue to refine work done in the first semester by undertaking CFD analyses to validate aerodynamics considerations.

Set-up CFD analyses and continue with it. While students are progressing with CFD analyses, some of the following design studies may be carried out:

1. Make preliminary structural layout. Discuss material and structural considerations.
2. Establish system and instrument requirements (e.g. electrical, mechanical, control and communication navigation).
3. Review airworthiness requirements, safety and environmental issues in detail.
4. Study manufacturing and cost considerations. Prepare a Gantt chart (see Table 2.9).
5. Outline the scope of ground and flight test schedules.
6. Return to analyses CFD work and assess the progress made.

Classroom contact hours as designed by the instructor
**Alternatively, if an electric aircraft project is taken in the first semester, then a rig test (iron bird?) may be fabricated to establish the motor and battery characteristics to estimate aircraft performance. Actual data

for the electric prime mover chain are not easily available. An in-house prototype electric aircraft for flight testing is a possibility.

4 Project Assignment

It is recommended that a project assignment as a course requirement should be an aircraft of some existing kind so that the results can be verified. There is a wide variety to choose from. The public domain has sufficient geometric and performance data that can be used for comparison. Still better is acquire pilot manuals, which could be obtained from manufacturer or operator for educational purposes to be retained by academic institutes.

5 Suggestions for the Class

Coursework starts with a mock market survey to get a sense of how an aircraft design is conceived (its importance is highlighted in Section 2.3). Inexperienced students depend on instruction; therefore, a teacher's role is important at the beginning. Here, we offer some of my experiences in the hope that they may be helpful.

The teacher divides the class into groups, say about four to five students (class strength of total 20), which then work as teams. After introducing the course content and expectations, the teacher assigns (with student participation) the type of aircraft to be undertaken in the coursework (the example of the Learjet45 class of aircraft is used in this book). The teacher gives the students the payload and range for the aircraft and asks them to list what they think are the requirements from the operator's (customer's) perspective and directs them to produce a scaled three-view sketch. We recommend that students consult the referenced sources [1–7] to study similar designs and tabulate the statistical data to arrive at their proposition. (Relevant web sites also provide substantial information.) Understandably, in most cases, the specifications and concept configuration designs may not be realistic; however, some students could arrive at surprisingly advanced concepts.

It is unrealistic to assume full understanding by students at the start of the design exercise, but we found that comprehension of task obligations improves rapidly. The teacher explains the merits and demerits of each team's proposition, retaining only the best cases. Finally, the teacher selects one configuration (after pooling ideas from the groups) but allows the students to retain configuration differences (e.g. high or low wing, or tail position) that have been tailored to a realistic shape and will be systematically fine-tuned as the class progresses to the final design. When specifications have been standardized and the configurations decided, the class assumes a smooth routine. We recommend that the teacher encourage differences among configurations to compare the designs at the end of the semester. The comparison of the final design with their initial propositions, as the evidence of the learning process, will provide students with satisfaction.

This type of project work does not have closed-book final examination. Grades are based on project documents submitted by students. Grading is at the discretion of the teacher, as it should be, but peer review contributes. Working in teams requires honest feedback among students because the teacher cannot track individuals working on their own. Leadership qualities of individual students should be recognised but should not overshadow a quieter student's performance. The students will soon be experiencing the reality of the industry, and a spirit of teamwork must be understood in the classroom. This spirit is not only about cooperation with others; it also is about being an effective contributing member working in harmony within a team. By this time, the teacher would have adequate feedback on individual work quality and capability.

A note of caution: What is accomplished in 36 hours of classroom lectures takes approximately 36 weeks in industry, not including the work put in by about the same number of experienced engineers engaged full-time in the work. The undergraduate coursework must stay on schedule to conclude on time. Therefore, to maintain the schedule, the teacher must remain in close contact with the students.

References

Some useful references on periodicals and web site URLs are given next:

1 Jane's All the World's Aircraft Manual (yearly publication). [The first author's name is in 1983–1984 and 1984–1985 editions – under India.]

2 Flight International (weekly publication).

3 *Aviation Week and Space Technology (AW&ST) Journal* (weekly publication).

4 *Interavia* (no longer in circulation but older copies contain some useful information).

5 Aerospace America. An AIAA publication.

6 Mason, W.H. Aircraft Design Information Sources published online at the VPI web site www.dept.aoe.vt .edu/~mason/Mason/ACinfoTOC.html (accessed June 2018).

7 Kundu, A.K., Price, M.A., and Riordan, D. (2016). *Theory and Practice of Aircraft Performance*. Wiley. ISBN: 9871119074175.

8 Web sites: Some useful references on periodicals and web site URLs are given here. Please note that URLs are subject to change, hence web-links can get lost. Readers are recommended to search for alternative sites on related topics. There are many available and these can provide useful information. www .flightglobal.com/cutaways www.globemaster.de fighterjets.milavia.net/ www.aerosite.net/caravelle.htm www.commonswikimedia.org www.soton.ac.uk/~jps7/AircraftDesignResources/LloydJenkinson%20data www .the-blueprints.com http://aerospace.illinois.edu/m-selig/ads/coord_database.html#N http://airfoiltools.com/ airfoil/details?airfoil=n64212mb-il

Part I

Prerequisites

1

Introduction

1.1 Overview

This book covers design studies on fixed winged aircraft and the conceptual stage of a project in detail, covering both civil and military aircraft types except for the V/STOL. Chapter 1 begins with a brief historical introduction, surveying our aeronautical legacy to motivate readers by describing the remarkable progress made from mythical conceptions of flight to high-performance aircraft with capabilities unimagined by early aeronautical pioneers. This is followed by presenting the issues involved with units and dimensions as well other pertinent topics; for example, cost implications and ending with a study of the atmosphere.

This chapter covers the following topics:

Section 1.2: A Brief Historical Background
Section 1.3: Aircraft Evolution
Section 1.4: Current Design Trends for Civil and Military Aircraft
Section 1.5: Future Design Trends for Civil and Military Aircraft
Section 1.6: Forces and Drivers
Section 1.7: Airworthiness Requirements
Section 1.8: Current Aircraft Performance Analyses Levels
Section 1.9: Comparison Between Civil and Military Design Requirements
Section 1.10: Topics of Current Research Interest Related to Aircraft Design
Section 1.11: Cost Implications
Section 1.12: The Classroom Learning Process
Section 1.13: Units and Dimensions
Section 1.14: Use of Semi-Empirical Relations and Graphs
Section 1.15: Atmosphere

Current trends indicate maturation in the technology of classical aeronautical sciences with diminishing returns on investment, making the industry cost conscious. To sustain the industry, newer avenues are being researched through better manufacturing philosophies and exploiting the trends in price and weight reduction of newer bought-out avionic items and of composite materials. Future trends indicate globalisation, with multinational efforts to advance technology making it better, faster and less expensive than the existing limits.

Some of the discussions in this chapter are based on personal experiences and are shared by the authors' many colleagues in several countries. Aerospace is not only multidisciplinary but also multidimensional – it may look different from varying points of view. The final product converges to similarity for the comparable designs.

The readers must go through Section 1.15 in detail. The examples given here should cover the needs of this book and readers must be able work through them.

Conceptual Aircraft Design: An Industrial Approach, First Edition. Ajoy Kumar Kundu, Mark A. Price and David Riordan.
© 2019 John Wiley & Sons Ltd. Published 2019 by John Wiley & Sons Ltd.

1.2 Brief Historical Background

This section provides a very brief history of flight, designed to motivate individuals to explore human aerial achievements further. Many books cover the broad sweep of aeronautical history while others discuss specific accomplishments and famous people's achievements in aeronautics. The references [1–4] are good places to start your exploration. Innumerable web sites on historical topics and technological achievements exist; simply enter keywords such as *Airbus*, *Boeing* or anything that piques your curiosity and you will find a wealth of information.

1.2.1 Flight in Mythology

People's desire to fly is ancient – every civilisation has their early imaginations embedded in mythologies. There are the well-known examples, for example, in Greek mythology Daedalus and his son Icarus flew on wings made from wax and feathers, flying chariots in Indian mythologies and their successors vimanas, flying-carpets in many Middle Eastern folk traditions and so on. Flying creatures also populate our early myths including the bird-man (Garuda), flying horses (Pegasus/Sleipnir) and flying dragons. The imagination of flight is universal. Readers may explore these and other mythical conceptions of flight through Internet searches.

History is unfortunately more 'down-to-earth' than mythology with stories about early pioneers who leapt from towers and cliffs, only to leave the Earth in a different but predictable manner because they did not respect natural laws. Human dreams and imagination became reality only a little over 100 years ago on 17December, 1903, when the Wright brothers succeeded with the first powered heavier-than-air flight. It only took 65 years from that date to land a man on the Moon.

1.2.2 Fifteenth to Nineteenth Centuries

Our chronicle of known flight begins with tethered kites flown in China as long ago as 600 B.C. The first scientific attempts to design a mechanism for aerial navigation are credited to Leonardo da Vinci (1452–1519). He was the grandfather of modern aviation, even if his machines never defied gravity. He sketched many contraptions (Figure 1.1a) in his attempt to make a mechanical bird. Birds, who provided his inspiration, possess such refined design features that the initial human path into the skies could not take that route. Today's micro-air devices are increasingly exploring natural designs. After da Vinci, there was an apparent scientific lull for more than a century until Sir Isaac Newton (1642–1727) computed the power required to make sustained flight. The idle period may simply be the lack the documentary evidence for it is inconceivable that human fascination with flight abated in this period. Flight is essentially a practical matter, so real progress paralleled other industrial developments (e.g. isolating gas for buoyancy).

While it appears that Bartolomeu de Gusmao may have demonstrated balloon flight in 1709 [4] in Portugal; information on this event is still lean. So, we credit Jean-François Pilâtre de Rozier and François Laurent d'Arlandes as the first people to effectively defy gravity, using a balloon designed by the Montgolfier brothers (France) (Figure 1.1b) in 1783. For the first time, it was possible to sustain with limited control in air flight. These balloon pioneers were subject to the prevailing winds and were thus limited in their navigational options. In 1784, Jean-Pierre Blanchard (France) with Dr John Jeffries (USA) added a hand-powered propeller to a balloon and made the first aerial crossing of the English Channel on 7 January, 1785. (Jules Verne's fictional balloon trip around the world in 80 days became a reality when Steve Fossett circumnavigated the globe in fewer than 15 days in 2002.) In 1855, Joseph Pline was the first to use the word *aeroplane* in a paper he wrote proposing a gas-filled dirigible with a propeller.

It was not until 1804 that the first recorded controllable heavier-than-air machine to stay freely airborne was recorded when Englishman Sir George Cayley constructed and flew a kite-like glider (Figure 1.2a) with movable control surfaces. In 1842, an English engineer Samuel Henson secured a patent on an aircraft design that was driven by a steam engine.

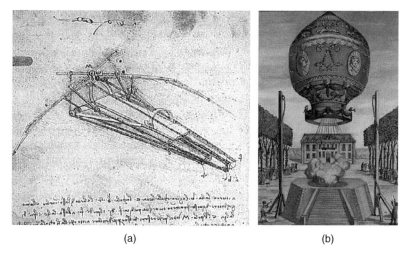

(a)   (b)

Figure 1.1 Early concepts and reality of flying. (a) Leonardo's flying machine (idea) and (b) Montgolfier Balloon (reality). Source: reproduced with permission of NASA.

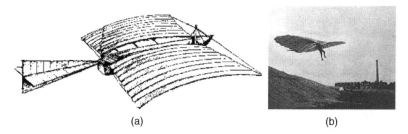

(a)   (b)

Figure 1.2 Heavier-than-air unpowered aircraft. (a) Cayley's kite plane and (b) One of Lilienthal's gliders. Source: reproduced with permission of NASA.

With his brother Gustav, Otto Lilienthal was successfully flying gliders (Figure 1.2b) in Berlin more than a decade (1890) before the Wright brothers' first experiments. His flights were controlled but not sustained. Sadly, Lilienthal's aerial developments ended abruptly and his experience was lost when he died in a crash in 1896.

The early flight machine designs were hampered by an overestimation of the power requirement needed for sustained flight. This mistake (based in part on Newton and others calculations) may have discouraged attempts of the best German engine makers of the time to build aircraft engines because they would have been too heavy.

1.2.3   From 1900 to World War I (1914)

The question of who was first in flight is an important event to remember. The Wright brothers (United States) are recognised as the first to achieve sustained, controlled flight in a heavier-than-air manned flying machine (*Wright Flyer*, Figure 1.3a). Before discussing their achievement, some 'also-rans' deserve mention. John Stringfellow accomplished the first powered flight of an unmanned heavier-than-air machine in 1848 in England. In France, Clement Ader also made a successful flight in his 'Eole'. Gustav Weisskopf (Whitehead), a Bavarian who immigrated to the United States, claimed to have made a sustained, powered flight [3] on August 14, 1901, in Bridgeport, Connecticut. Karl Jatho of Germany made a 200-ft hop (longer than the Wright brothers' first flight) powered (10-HP Buchet engine) flight on August 18, 1903. At what distance a 'hop' becomes a 'flight' could be debated.

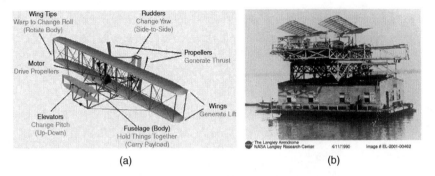

(a) (b)

Figure 1.3 Heavier-than-air powered aircraft. (a) The *Wright Flyer* and (b) Langley's Aerodrome catapult launch. Source: reproduced with permission of NASA.

Perhaps most significant are the efforts of Samuel P. Langley, who made three attempts to get his designs ('Aerodrome') airborne with a pilot at the controls (Figure 1.3b). His designs were aerodynamically superior to the Wright Flyer, but the strategy to ensure pilot safety resulted in structural failure while catapulting from a ramp towards water. His model aircraft were flying successfully in 1902. (To prove the capability, subsequently, in 1914 Curtiss could make a short flight with a modified Aerodrome.) The failure of his aircraft also broke Professor Langley – a short time afterward, he died of a heart attack. Professor Langley, a highly qualified scientist, had substantial government funding whereas the Wright brothers were mere bicycle mechanics without any external funding.

The Wright brothers' aircraft was an inherently unstable aircraft but, good bicycle mechanics that they were, they understood that stability could be sacrificed if sufficient control authority was maintained. They employed a foreplane (canard) for pitch control, which also served as a stall-prevention device. Modern designs have reprised this solution as seen in the Burt Rutan-designed aircraft. Exactly a century later, a flying replica model of the Wright Flyer failed to lift-off on its first flight (more details in Internet). A full-scale non-flying replica of the Wright Flyer is on display at the Smithsonian Museum in Washington, D.C. This exhibit along with those at other similar museums are well worth a trip to visit.

Having shown that sustained and controlled flight was possible, the Wright brothers' commercial success was limited as they were outpaced by a new generation of aerial entrepreneurs. Inventions followed in rapid succession from pioneers such as Alberto Santos Dumas, Louis Bleriot and Glenn Curtiss to name a few. Each inventor presented a new contraption, some of which demonstrated genuine design improvements. Fame, adventure and 'Gefühl' (feelings) were major drivers of innovation since the early years saw little financial gain from selling 'joy rides' and air shows – spectacles never seen before then and still appealing to the public today.

It did not take long to demonstrate the advantages of aircraft for mail delivery and military applications. At approximately 100 miles per hour (mph), on average, aircraft were travelling three times faster than any surface vehicle – and in straight lines. Mail was delivered in less than half the time. The potential for military applications was dramatic and well demonstrated during World War I. About a decade after the first flight in 1903, aircraft manufacturing had become a lucrative business. The first author started his aeronautical-engineering career with Short Brothers and Harland (now part of the Bombardier Aerospace group), a company that started aircraft manufacturing by contracting to fabricate the Wright designs. One of the co-authors is also currently employed, in a senior position, in the company. The company is now the oldest surviving aircraft manufacturer still in operation. In 2008, it celebrated its centenary, the first aircraft company to do so.

1.2.4 World War I (1914–1918)

Balloons were the earliest (the second half of nineteenth century) airborne military vehicle but controlled aircraft replaced their role as soon their effectiveness were demonstrated just before WWI. Their initial role started as observation platform and soon their military offensive capabilities (bombing, dogfights etc.)

were established. Their combat effectiveness became a decisive factor for military strategy. This attracted entrepreneur both in private and public sectors to grow in a rapid rate. In both the sides of Atlantic the number of aircraft and engine design and manufacturing establishments could exceed more than 100 organisations. With the growing recognition of the potential for aircraft as military aircraft application, the actual demand was in the European scenarios. Serious military aircraft design activities began not until WWI started. German aeronautical science and technologies made rapid advances.

This section shows how fast aircraft industry grew within a decade of first flight at first, driven by military application. This is the period which lay the foundation of what is to come subsequently; developed to the extent as it stands today. The section is kept brief only few aircraft examples are shown. The readers are recommended to explore web sites and obtain details of aircraft data.

In the USA: In 1908, the US Army accepted tender for military aircraft and after extensive tests the Signal Corps accepted Wright Model A powered by 35 hp. in 1909. In 1912 the Wright Model B was used for the first time to demonstrate the firing of a machine gun from a aeroplane. Soon after, Glenn Curtiss designed aircraft became the dominant US aircraft designer. Curtiss aircraft introduced naval carrier-based flying during 1910–1911. The company became early pioneers of producing military flying boats, planes that could takeoff and land in water. One of the earlier designs was Curtiss F4 (Figure 1.4 and Table 1.1). The Boeing Company was started around this time. Among the famous names of early aviation are Martin, Packard, Vaught and so on; possibly in excess of two dozen aircraft and engine design and manufacturing companies emerged in the USA during the period. Despite this, America introduced arguably superior European designed military aircraft in their armed forces.

In the UK: Upon the recommendation of British Defence Ministry in 1911, the Royal Flying Corps (RFC) was formed in 1912. In 1918 it merged with the Royal Naval Air Service to form the Royal Air Force (RAF). The Royal Aircraft Factory B.E.2 was a single-engine two-seat biplane, in service with the RFC in 1912. They were used as fighters, interceptors, light bombers, trainers and reconnaissance aircraft. A more successful design

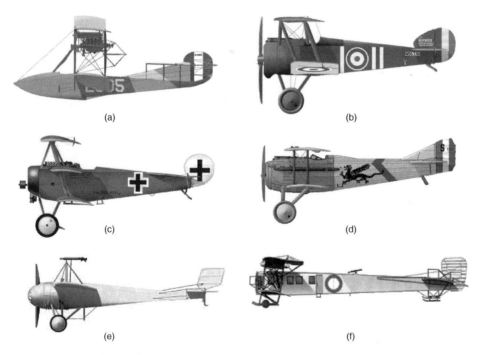

Figure 1.4 Very early aircraft (World War I). (a) Curtiss F4 (USA) Source: Courtesy of: www.wp.scn.ru, (b) Fokker Dr1 (Germany) Source: Courtesy of: www.fokkerdr1.com, (c) Caproni Ca.20 (Italy), Source: Courtesy of : www.airlinepicture.blogspot.com, (d) Sopwith Camel (UK) Source: Courtesy of: www.worldac.de, (e) SPAD S VII (France) Source: Courtesy of: www .greatwarflyingmuseum.com, (f) Sikorsky Ilya Muromets (Bomber) Source: Courtesy of: www.aviastar.org/air/russia.

Table 1.1 Performance summary of the aircraft in Figure 1.4.

	Curtiss	Sopwith	Fokker	SPAD	Caproni	Ilya
	Flying boat	Camel	DR1	SVII	Ca20	Mouromet
Engine-HP	2×275	130	110	150	110	4×148
Wing area-ft2	1216	231	201	192	144	1350
MTOM-lb	10 650	1455	1292	1632	≈ 1290	12 000
Max. Speed-knot	85	115	185	119	100	110

with better capabilities was the single-seat Sopwith Pup. It entered service in the autumn of 1916. The Avro 504 (100–130 hp) and Sopwith Camel (1913; 110 hp, see Figure 1.4) are some of the well-known aircraft of their time. Some other famous UK aircraft of the time bore the names of Armstrong-Whitworth, A.V. Roe, Blackburn, Bristol, Boulton/Paul, De Havilland, Fairey, Handley Page, Short Brothers, Supermarine, Vickers, Westland and so on.

In Germany: Die Fliegertruppen des deutschen Kaiserreiches (the Flier Troops of the German Kaiser Empire), of the Imperial German Army Air Service was formed in 1910 and changed the name to the *Luftstretkräfte* in 1916 (this became the *Luftwaffe* in the mid-1930s). Advances made by the German aeronautical science and technologies produced many types of relatively high-performance aircraft of the time. These saw action during World War I. The triplane Fokker Dr1 (Figure 1.4) was perhaps the most famous fighter of the period. The triplane was flown by the famous 'Red Baron', Rittmeister Manfred Freiherr von Richthofen, the top-scoring ace of World War 1 with 80 confirmed kills. Another successful German military aeroplane, the Albatross III, served on the Western Front until the end of 1917. The Junkers D.I was the first ever cantilever monoplane design to enter production. It utilised corrugated metal wings and front fuselage with fabric covering only being used on the rear fuselage. The Friedrichshafen FF.33 was one of the earliest German single-engine amphibious reconnaissance biplanes (1914). Some of the other famous German aircraft of the time bore the names of AEG, Aviatik, DFW, Fokker, Gotha (Gothaer Waggonfabrik), Halberstadt, Hannoversche, Junkers, Kondor, Roland, LVG, Zeppelin and so on.

In France: The French Air Force (*Armée de l'Air*, ALA) is the air force of the French Armed Forces. It was formed in 1909 as the *Service Aéronautique*, as part of the French Army, and was made an independent military branch in 1933. The first Bleriot XIs entered military service in France in 1910. Other famous French military aircraft are Nieuport 10 (1914–80 HP France) and their subsequent designs, The SPAD S VII (Figure 1.4) was a successful French fighter aircraft of the First World War used by many countries. The Caudron G.4 series was the first French built twin-engine bomber biplane platform introduced in the early years of World War 1. Some of the other famous French aircraft of the time bore the names of Hanriot, Maurice Farman, Moraine-Saulnier, Salmson and so on. Many countries, for example the UK, the USA, Italy and Russia, bought French military aircraft for their Air Force.

Other European Countries: Aircraft design and manufacturing activities in other European countries – for example, Italy, Russia, Scandinavian countries, the countries of the Iberian Peninsular and so on – were also vigorously pursued. Only Italian and Russian designs are briefly given next.

Italy could claim to be among the earliest to experiment with military aviation. As early as 1884 before powered heavier-than-air vehicles, the *Regio Esercito* (Italian Royal Army) operated balloons as observation platforms. During early World War I period Caproni developed a series of successful heavy bombers. The Caproni Ca. 20 (1914) was one of the first real fighter planes (Figure 1.4). It is a monoplane that integrated a movable, forward-firing drum-fed Lewis machine gun 2 ft above the pilot's head, firing over the propeller arc. Some of the other famous Italian aircraft of the time bore the names of Società Italiana Aviazione, Ansaldo and so on.

The Russian Empire under Czar had the Imperial Russian Air Force possibly before 1910. Russian aeronautical sciences had advanced research of the time with the famous names like Tsiolkovsky and Zhukovsky. The history of military aircraft in Imperial Russia is closely associated with the name of Igor Sikorsky. He immigrated to the USA in 1919; aircraft bearing his name are still produced. In 1913–14 Sikorsky built the first four-engine biplane, the Russky Vityaz. His famous bomber aircraft, the Ilya Muromets is shown in Figure 1.4. Other famous aircraft of Russian origin of the time had the names Anade, Antara, Anadwa, Grigorvich and so on.

1.2.5 Period between World War I and World War II – Inter War Period, the Golden Age (1918–1939)

Urgent necessity for military activities during World War I advanced the aeronautical science and technology to the point when post World War I could exploit its potential by presenting attractive proposition for business growth. The aeronautical activities in the peace period were deployed to expand industrial and national growth. The enhanced understanding of aerodynamic, aircraft control laws, thermodynamics, metallurgy, structural and system analyses made aircraft and engine size and performance to grow in rapid strides. A wide variety of innovative new designs emerged to cover wide grounds of application both in military and civil operations. Records for speed, altitude and payload capabilities kept updated in frequent intervals. This period is seen as the Golden Age of aeronautics and reinforced the foundation of the growth led to the present advancement.

With enhanced aeronautical knowledge to increase aircraft capabilities, availability of experienced pilots and public awareness offered ideal environment to make commercial aviation a reality. Surplus post war experienced pilots was available who could easily adapt to newer designs. They kept them engaged with performing in air shows and offers joy rides. In this period, aircraft industries geared up in defence applications and in civil aviation, with financial gain as the clear driver. The free market economy of the West contributed much to aviation progress; its downside, possibly reflecting greed, was under-regulation. The proliferation showed signs of compromise with safety issues, and national regulatory agencies quickly stepped in, legislating for mandatory compliance with airworthiness requirements (USA – 1926). Today, every nation has its own regulatory agency.

One of the earliest application of commercial operation with passenger flying was done on the modified Sikorsky Ilya Muromets (Bomber – Figure 1.4). It had an insulated cabin with heating and lighting, comfortable seats, lounge and a toilet. Fokker was a Dutch aircraft manufacturer named after its founder, Anthony Fokker. The company operated under several different names, starting out in 1912 in Schwerin, Germany, moving to the Netherlands in 1919. In the 1920s, Fokker entered its glory years, becoming the world's largest aircraft manufacturer. Its greatest success was the F.VIIa/3m trimotor passenger aircraft, which was used by 54 airline companies worldwide. It shared the European market with the Junkers all-metal aircraft but dominated the American market until the arrival of the Ford Trimotor that copied the aerodynamic features of the Fokker F.VII, and Junkers structural concepts. In May 1927, Charles Lindberg won the Ortega prize for the first individual non-stop transatlantic flight.

Early aircraft design was centred on available engines, and the size of the aircraft depended on the use of multiple engines. The combination of engines, materials, and aerodynamic technology enabled aircraft speeds of approximately 200 mph; altitude was limited by human physiology. In the 1930s, Durener Metallwerke of Germany introduced *duralumin*, with higher strength-to-weight ratios of isotropic material properties, and dramatic increases in speed and altitude resulted.

1.2.6 World War II (1939–1945)

The introduction of duralumin brought a new dimension to manufacturing technology. Structure, aerodynamics and engine development paved the way for substantial gains in speed, altitude and manoeuvring capabilities. These improvements were seen pre-eminently in World War II designs such as the Supermarine Spitfire, the North American P-51, the Focke-Wolfe 190 and the Mitsubishi Jeero-Sen. Multiengine aircraft also grew to sizes never before seen.

The invention of the jet engine (independently by Whittle of the United Kingdom and von Ohain of Germany) realised the potential for unheard-of leaps in speed and altitude, resulting in parallel improvements in aerodynamics, materials, structures and systems engineering. Heinkel He 178 is the first jet powered aircraft (27 August, 1939), followed by the Gloster E.28 on 15 May, 1941.

1.2.7 Post-World War II

A better understanding of supersonic flow and a suitable rocket engine made it possible for Chuck Yeager to break the sound barrier in a Bell X1 in 1949 (the aircraft is on exhibit at the Smithsonian Air and Space Museum in Washington, D.C). Tens of thousands of the Douglas C-47 Dakota and Boeing B17 Flying Fortress were produced. Post-war, the De Havilland Comet was the first commercial jet aircraft in service; however, plagued by several tragic crashes, it failed to become the financial success it promised.

The 1960s and 1970s saw rapid progress with many new commercial and military aircraft designs boasting ever-increasing speed, altitude and payload capabilities. Scientists made considerable gains in understanding the relevant branches of nature: in aerodynamic [4] issues concerning high lift and transonic drag; in materials and metallurgy, improving the structural integrity; and in significant discoveries in solid-state physics. Some of the outstanding designs of those decades emerged from the Lockheed Company, including the F104 Starfighter, the U2 high-altitude reconnaissance aircraft and the SR71 Blackbird. These three aircraft, each holding a world record of some type, were designed in Lockheed's Skunk Works, under the supervision of Clarence (Kelly) Johnson. I recommend that readers study the design of the nearly half-century-old SR71, which still holds the speed–altitude record for aircraft powered by air-breathing engines.

During the late 1960s, the modular approach to gas-turbine technology gave aircraft designers the opportunity to match aircraft requirements (i.e. mission specifications and economic considerations) with 'rubberised' engines (vide Section 12.2). This was an important departure from the 1920s and 1930s, when aircraft sizing was based around multiples of fixed-size engines. Chapter 12 describes the benefits of modular engine design. This advancement resulted in the development of families of shorter and longer aircraft designs. Plugging the fuselage sections and, if necessary, allowing wing growth covered a wider market area at a lower development cost because considerable component commonality could be retained in a family: a cost-reduction design strategy. Capitalistic objectives render designers quite conservative, forcing them to devote considerably more time to analysis. Military designs emerge from more extensive analysis – for example, the strange-looking Lockheed F117 is configured using stealth features to minimise radar signature. Now, more matured stealth designs look conventional (e.g. the Lockheed F22).

1.3 Aircraft Evolution

Figure 1.5 shows the history of progress in speed and altitude capabilities. The impressive growth in one century is astounding – leaving the Earth's surface in a heavier-than-air vehicle and returning from the Moon in fewer than 66 years!

It is interesting that for air-breathing engine powered aircraft, the speed–altitude record is still held by the more than 40-year-old design, the SR71 (Blackbird), capable of operating at around Mach 3.0 and a 100 000-ft altitude. Aluminium-alloy properties would allow a flight speed up to Mach 2.5. Above Mach 2.5, a change in material and/or cooling would be required because the stagnation temperature would approach 600 K, exceeding the strength limit of aluminium alloys. Aircraft speed–altitude capabilities have remained stagnant since the 1960s. A recent breakthrough was the success of *Spaceship One* that took aircraft to the atmosphere edge to 100 km altitude. In civil aviation, the supersonic transport (SST) aircraft *Concorde* was designed nearly four decades ago and has not yet been supplanted. Concorde's speed–altitude capability is Mach 2.2 at around 60 000 ft.

In military aircraft scenarios, gone (almost) are the days of 'dogfights' that demanded a high-speed chase to bring an adversary within machine-gun firing range (i.e. low projectile speed, low impact energy, and no homing); if the target was missed, the hunter became the hunted. In the post-World War II period, around the late 1960s, air-superiority combat required fast acceleration and speed (e.g. the Lockheed F104

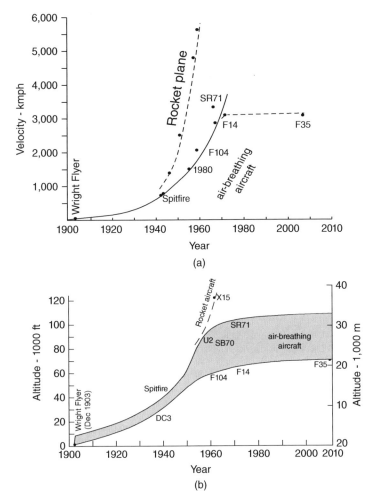

Figure 1.5 Aircraft operational envelope. (a) Speed and (b) altitude.

Starfighter) to engage with infrared homing missiles firing at a relatively short distance from the target. As missile capabilities advanced, the current combat aircraft design trend showed a decrease in speed capabilities. Instead, high turning rates and acceleration, integrated with superior missile capabilities (i.e. guided, high speed and high impact even when detonated in proximity of the target), comprise the current trend. Target acquisition beyond visual range (BVR) – using an advance warning system from a separate platform – and rapid aiming comprise the combat rules for mission accomplishment and survivability. Current military aircraft operate below Mach 2.5; hypersonic aircraft are in the offing.

1.3.1 Aircraft Classifications and their Operational Environments

There are many types of aircraft in production serving different sectors of mission requirements; civil and military missions differ substantially. It is important to classify aircraft category to isolate and identify strong trends existing within a class. An aircraft can be classified based on its role, use, mission, power plants and so forth, as shown in Table 1.2. Here, the first level of classification is based on operational role (i.e. civil or military discussion on military aircraft is given online) and this chapter is divided into these two classes. In the second level, the classification is based on the generic mission role, which also would indicate size. The third level proceeds with classification based on the type of power plant used and so on. The examples worked

Table 1.2 Aircraft classification.

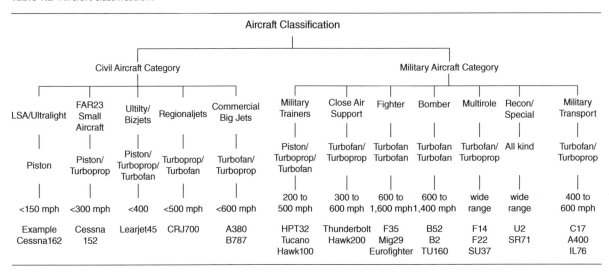

						Aircraft Classification						
		Civil Aircraft Category						Military Aircraft Category				
LSA/Ultralight	FAR23 Small Aircraft	Ultilty/ Bizjets	Regionaljets	Commercial Big Jets		Military Trainers	Close Air Support	Fighter	Bomber	Multirole	Recon/ Special	Military Transport
Piston	Piston/ Turboprop	Piston/ Turboprop/ Turbofan	Turboprop/ Turbofan	Turbofan/ Turboprop		Piston/ Turboprop/ Turbofan	Turbofan/ Turboprop	Turbofan Turbofan	Turbofan Turbofan	Turbofan/ Turboprop	All kind	Turbofan/ Turboprop
<150 mph	<300 mph	<400	<500 mph	<600 mph		200 to 500 mph	300 to 600 mph	600 to 1,600 mph	600 to 1,400 mph	wide range	wide range	400 to 600 mph
Example Cessna162	Cessna 152	Learjet45	CRJ700	A380 B787		HPT32 Tucano Hawk100	Thunderbolt Hawk200	F35 Mig29 Eurofighter	B52 B2 TU160	F14 F22 SU37	U2 SR71	C17 A400 IL76

out in this book are the types that cover a wide range of aircraft design, which provides an adequate selection for an aircraft design course.

Readers are suggested to examine what could be the emerging design trends within the class of aircraft. In general, new commercial aircraft designs are extensions of the existing designs incorporating proven newer technologies (some are fallouts from declassified military applications) in a very conservative manner.

Figure 1.6 indicates the speed–altitude regimes for the type of power plant used. Currently, low-speed–low-altitude aircraft are small and invariably powered by piston engines of no more than 500 horsepower (HP) per engine (turboprop engines start to compete with piston engines above 400 HP). World War II had the

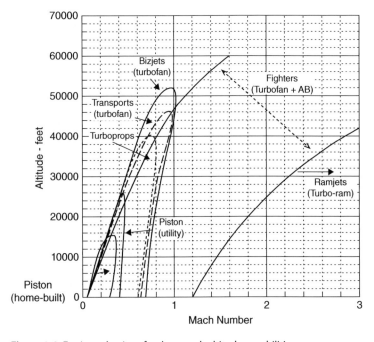

Figure 1.6 Engine selections for the speed–altitude capabilities.

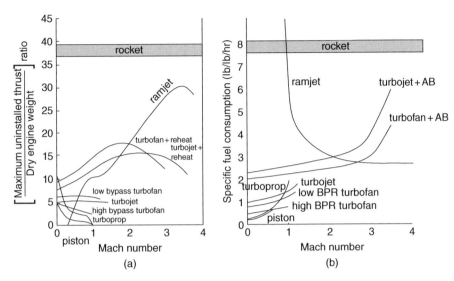

Figure 1.7 Engine performance. (a) Thrust to weight ratio and (b) specific fuel consumption (modified from the diagrams of Harned, M., Aero Digest, Vol, 69, No. 1, July, 1954.

Spitfire aircraft powered by Rolls Royce Merlin piston engines (later by Griffon piston engines) that exceeded 1000 HP; these are nearly extinct, surviving only in museum collections. Moreover, aviation gasoline (AVGAS) for piston engines is expensive and in short supply.

The next level in speed–altitude is by turboprops operating at shorter ranges (i.e. civil aircraft application) and not critical to time due to a slower speed (i.e. propeller limitation). Turboprop fuel economy is best in the gas-turbine family of engines. Subsonic cargo aircraft and military transport aircraft may be more economical to run using turboprops because the question of time is less critical, unlike passenger operation that is more time critical with regard to reaching their destinations.

The next level is turbofans operating at higher subsonic speeds. Turbofans (i.e. bypass turbojets) begin to compete with turboprops at ranges of more than 1000 nm due to time saved as a consequence of higher flight speed. Fuel is not the only factor contributing to cost – time is also money. A combat aircraft power plant uses lower bypass turbofans; in earlier days, there were straight-through (i.e. no bypass) turbojets. Engines are discussed in more detail in Chapters 12 and 13.

Figure 1.7a illustrates the thrust-to-weight ratio of various types of engines. Figure 1.7b illustrates the specific fuel consumption (*sfc*) at sea level static takeoff thrust (T_{SLS}) rating in an International Standard Atmosphere (ISA) day for various classes of current engines. At cruise speed, the *sfc* would be higher.

1.4 Current Aircraft Design Trends for both Civil and Military Aircraft (the 1980s Onwards)

Aircraft design is affected by a number of issues not the least is the economic considerations for the product whether civilian or military. The current aircraft design trends for civil and military aircraft are discussed separately in the following two sub-sections.

A major concern that emerged in the commercial aircraft industry from the market trend and forecast analysis of the 1990s was the effect of inflation on aircraft manufacturing costs. Since then, all the major manufacturers and the subcontracting industries have implemented cost-cutting measures. As many more factors have become more important, the conceptual phase of aircraft design is now conducted using a multidisciplinary approach (i.e. concurrent engineering), which must include manufacturing engineering and an appreciation for the cost implications of early decisions; the buzzword is integrated product and process development (IPPD). Margins of error have shrunk to the so-called zero tolerance so that tasks are done right the first time. Two of the management tools to achieve this end are (i) Design for Manufacture and Assembly (DFM/A), an engineering approach with the objectivity to minimise cost of production without sacrificing design integrity and (ii) Design for Six-Sigma (DFSS); an integrated approach to design with the key issue to reduce any scope for mistakes/inefficiencies, that is, make the product right the first time to prevent waste of company resources. The importance of environmental issues emerged, forcing regulatory authorities to impose limits on noise and engine emission levels. Recent terrorist activities are forcing the industry and operators to consider preventive design features. Also, the consideration for better on-board medical attention is always there.

With rising fuel prices, air travellers have become cost-sensitive. In commercial aircraft operations, the direct operating cost (DOC) depends more on the acquisition cost (i.e. unit price) than on the fuel cost (2000 prices) consumed for the mission profile. Today, for the majority of mission profiles, fuel consumption constitutes between 15 and 25% of the DOC (DOC, see Section 16.3), whereas the aircraft unit price contributes between three and four times as much, depending on the payload range [5]. For this reason, manufacturing considerations that can lower the cost of aircraft production should receive as much attention as the aerodynamic saving of drag counts. The situation would change if the cost of fuel exceeds the current airfare sustainability limit (see Section 1.7 and Chapter 16). The price of fuel in 2008 was approaching the limit when drag-reduction efforts were regaining ground.

The last three decades witnessed a 56% average annual growth in air travel, exceeding 2×10^9 revenue passenger miles (RPM) per year. Publications by the International Civil Aviation Organization (ICAO), National Business Aviation Association (NBAA) and other journals provide overviews of civil aviation economics and management. The potential market for commercial aircraft sales is on the order of billions of dollars per year. However, the demand for air travel is cyclical and – given that it takes about 4 years from the introduction of a new aircraft design to market – operators must be cautious in their approach to new acquisitions: They do not want new aircraft to join their fleet during a downturn in the air travel market. Needless to say, market analysis is important for both the manufacturers and operators in planning new projects and purchases.

Deregulation of airfares has made airlines compete more fiercely in their quest for survival. The growth of budget airlines compared to the decline of established airlines is another challenge for operators.

It became clear that the current design trends for civil aircraft is customer-driven strategy as the best approach for survival in a fiercely competitive marketplace. The new paradigm of 'better, farther and cheaper to market' replaced, in a way, the old mantra of 'higher, faster and farther' [5].

Readers are suggested to examine what could be the emerging design trends within the class of aircraft. In general, new commercial aircraft designs are extensions of existing designs incorporating proven newer technologies (some are fallouts from declassified military applications) in a very conservative manner. It is seen as working in *verified design space*, that is, incorporating proven but advanced technologies.

The military aircraft has priority of national defence interest over commercial considerations. To stay ahead of any potential adversary, the research and design organisations constantly search for new technologies never tried before. Although these new technologies are required to be demonstrated satisfactorily before implementation, these are yet to be proven in operational arena. On account of these unknown factors, it is seen as working in an *aspirational design space*.

While commercial aircraft can earn self-sustaining revenue, military operations depend totally on government expenditures with limited potential for continuing revenue beyond the occasional cash from approved export sales. The cost of developing and building a new design has risen sufficiently to strain the economy of most single nations. Not surprisingly, the number of new designs proposals has drastically reduced, with military designs moving towards multinational collaborations among allied nations. The retention of confidentiality in defence matters is an additional complicating factor.

Combat roles are classified as interdiction, air-superiority, air defence and, when missions overlap, multi-role (see Section 10.4 for details). Action in hostile environments calls for special attention to: design for survivability; systems integration for target acquisition and weapons management; and design considerations for reliable navigation and communication. All told, it is a complex system – mostly operated by a single pilot – an inhuman task unless the workload were relieved by microprocessor-based decision making. Fighter pilots are special breed of aircraft operators demanding the best emotional and physical conditioning to cope with the stresses involved. Aircraft designers have deep obligation to ensure combat pilot survival. Unmanned Aerial Vehicle (UAV) technology is in the offing – the Middle-East conflict saw the successful use of UAVs both for surveillance and military operations.

1.4.1 Current Civil Aircraft Trends

When Boeing introduced its 737 twinjet aircraft (derived from their then best-selling three-engine B727,), the DouglasDC-9 and BAe 111 were the most popular two-engine commercial transport aircraft. The Boeing737 series, spanning nearly four decades of production to this day, has become the best-selling aeroplane in the history of the commercial aircraft market with more than 10 000 shipped and ordered. Of course, in that time, considerable technological advancements have been incorporated, improving the B737's economic performance at least by about 50%.

Current commercial transport aircraft in the 100–300-passenger classes all have a single slender fuselage, backward-swept low-mounted wings, two under-slung wing-mounted engines, and a conventional *empennage* (i.e. a horizontal and a vertical tail); this conservative approach is revealed in the similarity of configuration. The similarity in larger aircraft is the two additional engines; there have been three-engine designs but only on a few aircraft because the configuration was rendered redundant by variant engine sizes that cover the in-between sizes and extended twin operations (ETOPS). The largest commercial-jet-transport aircraft, the Airbus380 (Figure 1.8a) made its first flight on 27 April, 2005 and is currently in service. Boeing87 Dreamliner (Figure 1.8b) is the replacement for its successful Boeing 767 and 777 series, entered service on January, 2011 aiming at competitive economic performance. The new addition is the Airbus350, slightly larger than B787, entered service on January, 2015.

The gas-turbine turboprop engine offers better fuel economy than to current turbofan jet engines. However, because of propeller limitations, the turboprop-powered aircraft's cruise speed is limited to about two-thirds of the high-speed subsonic turbofan-powered aircraft. Operators have to consider that at lower operational ranges (e.g. distances less than 1000 nautical miles [nm]) the difference in sortie time would generally be less than a half hour, with an approximate 20% saving in fuel cost. If a long-range time delay can be tolerated (e.g. for cargo or military heavy-lift logistics), then large turboprop aircraft operating over longer ranges become more economical. Advances in propeller technology are pushing turboprop-powered aircraft cruise speeds close to the high subsonic cruising speed of turbofan-powered aircraft high subsonic cruise speeds.

1.4.2 Current Military Aircraft Trends

There are differences between civil and military design requirements (see Section 1.9.1). However, there are some similarities in their design processes up to the point when a new breakthrough is introduced – one thinks

(a)   (b)

Figure 1.8 Current wide-body large commercial transport aircraft. (a) Airbus380 (Courtesy of Airbus) and (b) Boeing787 Dreamliner. Source: Copyright: Boeing.

(a)

(b)

(c)

Figure 1.9 Current combat aircraft. (a) F117 Nighthawk. Source: Courtesy of the U.S Airforce/Sgt Aaron Allmon, (b) X-35 (F35 experimental) Source: courtesy of the U.S Airforce/Dana Russo and (c) B2 Bomber. Source: courtesy of U.S Airforce/Sgt Jeremy Wilson.

instinctively of how the jet engine changed designs in the 1940s. Consider the F117 Nighthawk (Figure 1.9): in order to incorporate stealth technology appeared as an aerodynamicist's nightmare; but it too is now conforming to something familiar in the shape of the F35 lightening II; its prototype X35 is given in Figure 1.9. We must not forget that military roles are more than just combat: they extend to transportation and surveillance (reconnaissance, intelligence gathering and electronic warfare). The F35, Eurofighter, Rafale, Gripen, Sukhoi PAK-FA and so on, are the current frontline fighter aircraft. In strategic bombing B52 served for four decades and is to continue for another two decades – some design! The latest B2 (Figure 1.9) bomber looks like an advanced flying wing without the vertical tail.

1.5 Future Trends

It is no exception from past trends that speed, altitude and payload will be expanded in both civilian and military capabilities. Coverage on the aircraft design process in the next few decades is given in [1, 3]. In the near-future trends the vehicle-capability boundaries will be pushed to the extent permitted by economic and defence factors and infrastructure requirements (e.g. navigation, ground handling, support etc.). Section 1.7 gives a short list of new technologies currently under investigation. In technology, smart material (e.g. adaptive structure) will gain ground, microprocessor-based systems will advance to reduce weight and improve functionality and manufacturing methodology will become digital. However, unless the price of fuel increases beyond affordability, investment in aerodynamic improvement will be next in priority. Small flyers and unmanned aircraft are emerging as new markets. Readers are advised to make periodic search on various web sites and journals for updated information on this topic.

Future designs consist of the Smarter Skies vision studying concepts to reduce waste in the system (waste in time, waste in fuel and reduction of CO_2). The European Union's NACRE (New Aircraft Concepts Research) project is studying potential radical overhauls to aircraft design with the goal of improving eco-efficiency, optimising performance and reducing costs. Within this effort, the NACRE Pro Green aircraft specifically aims at the reduction of an airliner's operational environmental footprint.

Compared to current-generation aircraft, the design for ultra-low emissions will cause some limitations in operational aspects. The slow cruise speed would reduce the number of flights per day and thus affect its economic productivity, with a passenger's overall travelling time increasing by between 5 and 10%. The high precision surfaces for the laminar flow wing could require special treatment and care in the manufacturing process. Likewise, the engine position on top of the fuselage is expected to require specific attention and increased effort during maintenance, compared to under-wing installations.

In the military scenario, it is the electronics that would play the main role, although aerodynamic challenges on stealth, manoeuvre and improved capability/efficiency would be as much as in demand as would be for structural/material considerations. Engine development would also be in parallel development with all of these discoveries/inventions.

Fighter aircraft systems will have enhanced capabilities in areas such as reach, persistence, survivability, net-centricity, situational awareness, human-system integration and weapons effects. The future system will have to counter adversaries equipped with next-generation advanced electronic attack, sophisticated integrated air defence systems, passive detection, integrated self-protection, directed energy weapons and cyber-attack capabilities. In the immediate future, fighters are expected to use advanced engines such as Adaptive Versatile Engine Technology to allow longer ranges and higher performance.

Both operators and manufacturers will be alarmed if the price of fuel continues to rise to a point where the air-transportation business finds it difficult to maintain profitable operations. It is likely that demand for power plants using alternative fuels, biofuel, liquid hydrogen (LOH) and possibly nuclear power for large transport aircraft covering long ranges. Aircraft fuelled by LOH have been used in experimental flight for some time, and fossil fuel mixed with biofuel is currently being flight tested especially with the US military's biofuel initiative.

Long-distance hypersonic attack aircraft represent a strong candidate for short-time deployment strike aircraft. Again, it is the electronics that plays the main role, although aerodynamic challenges of stealth manoeuvre, and improved capability/efficiency are also in as much demand for structural/material considerations. Engine development is also in parallel development with all of these discoveries/inventions.

This book does not deal with these futuristic designs. One must first master the fundamentals presented in this book to carry out such futuristic designs. If enough information is available, then these futuristic military aircraft could be more suited material for postgraduate teamwork on aircraft design, undertaken by those who already have some proficiency in aeronautical engineering and have the time for longer project work. Without systems integration and matched weapon integration for the mission role, mere aerodynamic shaping and a list of weaponry wish list exercises would prove meaningless. Representative details of systems architecture and their capabilities affecting aircraft performance are still not fully available in the public domain. Working on such an important aspect based on piecemeal information is not the best procedure to attempt in the undergraduate curriculum when there is so much to learn from conventional designs. Chapter 21 briefly covers miscellaneous design considerations.

1.5.1 Civil Aircraft Design

Any extension in payload capacity will remain subsonic for the foreseeable future until gains made by higher speed vehicles demonstrate operational success. High-capacity operations will likely remain around the size of the Airbus 380.

Some carefully modelled and researched futuristic designs seen in Figure 1.10 have the potential to remarkably increase capacity. A blended-wing body (BWB-B2) can use the benefits of the wing-root thickness sufficiently large to permit merging (Figure 1.10a, top – NASA green aircraft goals) with the fuselage, thereby benefiting from the fuselage's contribution to lift and additional cabin volume. Another alternative seen in Figure 1.10a, bottom would be that of the joined-wing concept. Studies of twin-fuselage and joined-fuselage large transport aircraft are given in Figure 1.10b and show additional potential. In the conventional design, although yet to be realised, there is no reason why an over-wing nacelle pod (Figure 1.10c) mounted large capacity commercial aircraft configuration not to succeed in operation. The concept is not new. The VFW 614 aircraft with over-wing pods were produced in the 1970s. The Hondajet

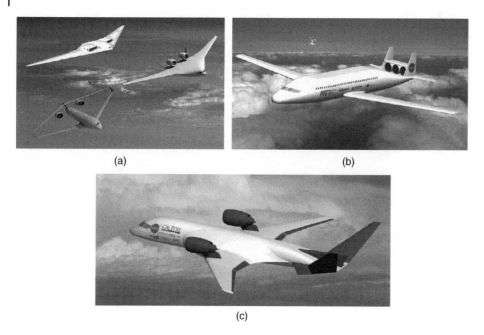

(a)

(b)

(c)

Figure 1.10 Some well-studied futuristic subsonic aircraft designs. (a) Blended-wing aircraft (top) and joined wing (bottom), (b) joined fuselage and (c) over-wing nacelle. Credit: Massachusetts Institute of Technology. Source: all photos courtesy of Courtesy of NASA.

with an over-wing nacelle is currently going through the Federal Aviation Administration (FAA) certification process.

The speed–altitude extension will progress initially through SSTs and then hypersonic transport (HST) vehicles. The SST technology is well understood after three decades of studying the performance of the Anglo–French-designed Concorde, which was designed to operate above Mach 2 at a 50 000-ft altitude carrying 128 passengers.

Contemporary planning for the next-generation SST suggests similar speed–altitude capability but with sizes varying from as few as 10 business passengers to approximately 300 passengers in at least transatlantic and transcontinental operations. Transcontinental operations will demand sonic-shock-strength reduction through aerodynamic gains because operating at anything less than Mach 1.6 has less to offer in terms of time savings.

The next-generation SST will have about the same speed–altitude capability (possibly less in speed capability, around Mach 1.8), but the size will vary from as few as 10 business passengers (Figure 1.11a) to approximately 300 passengers (Figure 1.11b) to cover at least transatlantic and transcontinental operations. Transcontinental operations would demand sonic-shock-strength reduction through aerodynamic gains rather than speed reduction; anything less than Mach 1.6 has less to offer in terms of time savings.

The real challenge will be to develop an hypersonic transport aircraft (HST) (Figure 1.11c) operating at approximately Mach 6 that would require operational altitudes above 100 000 ft. Speed above Mach 6 offers diminishing returns in time saved because the longest distance necessary is only about 12 000 nm (i.e. about 3 hours of flight time).

Smaller Bizjets and regional jets will morph, and unfamiliar shapes may appear on the horizon, but small air-craft in personal ownership used for utility and pleasure flying are likely to revolutionise the concept of flying through their popularity, similar to how the automobile sector grew. The revolution will occur in short-field capabilities, as well as vertical takeoffs, and safety issues in both design and operation. Smaller aircraft used for business purposes will see more private ownership to stay independent of the more cumbersome airline operations. There is a good potential for airparks to grow. Various 'roadable' aircraft (flying car) have been

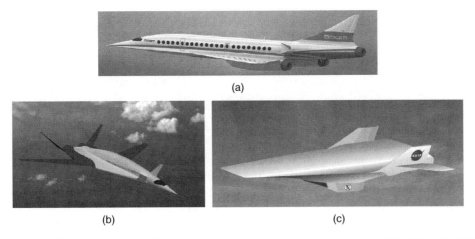

(a)

(b)   (c)

Figure 1.11 Some well-studied futuristic supersonic/hypersonic aircraft designs. (a) Small supersonic transport aircraft. Source: Credit: Boom Technology, USA. (b) Large supersonic transport aircraft. (c) Hypersonic aircraft. Source: Credit: NASA, USA.

designed. The major changes would be in system architecture through miniaturisation, automation and safety issues for all types of aircraft.

The NASA, the US Department of Transportation (USDOT), FAA, industry stakeholders and academia have joined forces to pursue a National General Aviation Roadmap leading to a Small Aircraft Transportation System (SATS). This strategic undertaking has a 25-year goal to bring the next generation of technologies to and improve travel between remote communities and transportation centres in urban areas by utilising the nation's 5400 public use general aviation airports (United States). The density of these airfields in Europe is much higher. The major changes would be in system architecture through miniaturisation, automation and safety issues for all types of aircraft.

A new type of vehicle known as a ground-effect vehicle (GEV) is a strong candidate for carrying a large payload (e.g. twice that of the Boeing 747) and flying close to the surface, almost exclusively over water. (A GEV is not really new: The Russians built a similar vehicle (Figure 1.12) called the 'Ekranoplan', but it did not appear in the free market economy.). It operated around 350 mph with undeniable future potential.

1.5.2 Military Aircraft Design

Progress in military would defy all imaginations (readers may also research this on the Internet). Size and shape could be as small as an insect for surveillance to as large as any existing kind. Vehicles as small as 15 cm and of a 1 kg mass have been successfully built for operation. Prototypes much smaller are already successfully flown. Reliance on in-built intelligence would certainly lead to more remotely piloted vehicles (RPV) in operation.

Figure 1.12 Lun class Ekranoplan (https://en.wikipedia.org/wiki/Lun-class_ekranoplan).

Other terminologies include unmanned, unoccupied, pilotless – it is better that I settle with one term – I call this RPV. These are piloted remotely or autonomously. However, the Unmanned or Unoccupied Air Vehicle (UAV) is also a prevalent terminology. I saw in literature 'unoccupied micro-vehicle' – wondering who would be the possible occupant, otherwise?

It is the electronics that would play the main role, although aerodynamic challenges on stealth, manoeuvre and improved capability/efficiency would be as much in demand as would structural/material considerations. Engine development would also be in parallel to all of these discoveries/inventions.

Long-distance hypersonic attack aircraft represent strong candidates for short-time deployment strike aircraft. Again, it is the electronics that plays the main role, although aerodynamic as mentioned before are also in as much demand and engine development is parallel development to these. Military applications for HST vehicles are likely to precede civilian applications. Small-scale HST vehicles have been flown recently.

In the military scenario, the USAF seeks Next-Generation Tactical Aircraft/Next Gen TACAIR. In the immediate future, the sixth-generation fighters are expected to use advanced engines. The long-range strike systems are seen in the form of the Long-Range Strike Bomber (LRS-B), currently under study in the USA. Russia is also developing a new long-range bomber called the PAK-DA, with a BWB that has an outer wing section with swing wing capability.

Once again it is electronics, aerodynamics, structural/material considerations and engine development would also develop further.

1.5.3 UAVs/UASs

The UAVs are powered, aerial vehicles flown without a human pilot on board. They have had many names, previously known as drones and subsequently have had names, for example, RPVs, unpiloted aerial vehicles (UAVs), remotely piloted aircraft (RPAs) and so on. In 2005, a designated name unmanned aircraft system (UAS) was introduced that has been now accepted for use by the US DOD, FAA, CAA and ICAO. This book uses the synonymous terms of UAV and UAS interchangeably.

The concept of a drone is not new; as early as World War I (WWI), they were deployed. WWII saw their extensive usage. Since the 1950s and the advent of transistors, by using hand held controllers, recreation model aircraft flying became popular. Subsequently, with microprocessor-based on-board instrumentation, military usage of UASs has been successfully demonstrated. Nations who can afford the technology have already entered the race to develop UAVs/UASs.

While this book does not deal with UAV designs, there exits considerable similarity in aerodynamic and propulsive design considerations. The difference is in the type of electronic payload carried on board. Hence, UAVs are briefly introduced here so that the readers may take an interest if they wish to specialise; UAVs are normally offered in separate courses involving systems integration to specific types of UAV applications. UAVs are often preferred for missions that are too dangerous or in emergencies for manned aircraft.

The general approach to configure the external geometry during conceptual design phase (Phase I) follows the same route as dealt with in this book, bearing in mind the need to cater for the laid down mission specifications, for example, payload and range/endurance and manoeuvre limits. The aerodynamic, structure and propulsive design considerations adopt a similar methodology to satisfy the requirements. Being without any occupant, the certification standards are different and in certain areas less stringent. Cost factors also allows to search for cost effective, material, systems and power plant specifically meant for UAV/UAS.

UASs fall under two categories as follows:

i) *Autonomous* – Aircraft with the ability to make decisions without human intervention. These are mostly used in military applications as missiles, as target for training aircraft gunners and so on. These are not suitable for civilian usage and seen unsuitable for regulation due to legal and liability issues.

ii) *Remotely piloted* with a human interface for decision making. These can be both of military and civil types. With proliferation in civil applications, for safety reasons, the UAS are now subjected to government regulations of the respective countries.

There are many types of usages, broadly falling under the following categories.

1.5.4 Military Applications

Pilot survivability constraints are being taken out of the design process as system-processing power grows. In the future, remote-controlled or pilotless aircraft will have the advanced computer capability to make weapon delivery decisions and advanced accuracy that could eliminate an on-board human interface thus permitting the aircraft to operate with greater loads and improved combat capability. More RPVs are anticipated to come into operation. Military UAV/UAS have the following applications:

- Combat – To providing attack capability for high-risk missions. Capable of very high-g rapid turning for tactical gains, very low altitude flying for stealth and others specific to mission requirements that are beyond the capabilities of manned aircraft.
- Reconnaissance – To gather intelligence for defence usage.
- Logistic – Deliver military cargo in high-risk and accessible areas.
- Target and decoy – To simulate enemy aircraft or missiles for countermeasures and aiming practise.

Progress in military aircraft may defy all our contemporary imaginations. Flyers sized as small as insects (micro aircraft – dragonfly drones 15 cm and 1 kg mass) for surveillance [10]) have been successfully built for operation. Much smaller prototypes are being successfully flown.

1.5.5 Civil and Commercial Applications

These cover a wide of applications to a fast-growing industry. Some of the application areas by governments include non-military security work, policing, traffic and crowd monitoring. Reconnaissance operations, border patrol missions, forest fire detection, surveillance, coordinating humanitarian aid, search and rescue missions, detection of illegal human trafficking, illegal hunting, land surveying, fire and large-accident investigation, landslide measurement, illegal landfill detection, firefighting, such as inspection of power or pipelines and so on.

Usage by private business sector and individuals include aerial surveying of crop search and rescue operations, crop spraying, inspecting power lines and pipelines, counting wildlife, delivering parcel and medical supplies to remote or otherwise inaccessible regions, media, surveillance, recreational (see next), land assessment, filming and so on.

The downsides are the proliferations, unnecessarily invading privacy, accidents causing damage and injuries and so on, for which steps to regulate are gradually evolving in this relatively new application domain.

1.5.6 Recreational Applications

Recreational model flying is allowed to operate below 400 ft in designated areas.

1.5.7 Research and Development Applications

These aim to further develop UAV/UAS technologies. Today, non-recreational UAV/UAS usage must require certification of authorisation and must be flown by qualified ground-based pilots. Recreational model flying is allowed to operate below 400 ft in designated areas.

UAS designs have explored helicopter like UAVs with vertical takeoff and landing capabilities. They come in many configurations that include multirotors, for example, quadcopters; hexacopters, octocopters and so on. They also come in many shapes, from small to large heavy lifters. The term Vertical Takeoff and Landing Tactical Unmanned Aerial Vehicle (VTUAV) is used in military applications.

Figure 1.13 shows an operational UAV, the Ikhana, used for imaging. Figure 1.13b shows military UAS configuration types.

(a)   (b)

Figure 1.13 Future unmanned aircraft system (UAS). (a) Ikhana Source: Courtesy of General Aeronautical Systems, Inc. and (b) X47B Northrop Grumman (https://news.usni.org).

1.5.8 Rocket Applications in Future Aircraft Design Trends

Von Braun [7] mentions that he took the idea from Tippu's success for his V2 rocket, paving the way for today's achievement in space flight as an expanded envelope beyond winged flight vehicles and originally deemed likely to begin commercial operation by 2012. Rocketry first entered the Western European experience when Tippu Sultan used rockets against the British-led Indian army at the Battle of Srirangapatna in 1792. The propellants were based on a Chinese formula nearly 1000 years old. The experience of Tippu's rockets led the British to develop missiles at the Royal Laboratory of Woolwich Arsenal, under the supervision of Sir William Congreve, in the late eighteenth century.

A new type of speed–altitude capability will come with suborbital space flight (tourism) using rocket-powered aircraft, as demonstrated (2004) by designer Burt Rutan's *Space Ship One* hitch-hiked to the jet engine powered White Knight aeroplane to suborbital space. Interest in this aircraft has continued to grow – the prize of $10 million offered could be compared with that of a transatlantic prize followed by commercial success. The advent and success of the Rutan-designed *Space Ship One* will certainly bring about the large market potential of rocket-powered aeroplanes.

A larger *Space Ship Two* is currently being developed. A new type of speed–altitude capability will come from suborbital space flight (tourism) using rocket-powered aircraft, as demonstrated by Rutan's *Space Ship Two* that hitchhikes with the White Knight to altitude (Figure 1.14), from where it makes the ascent. Interest in this aircraft has continued to grow – particularly in view of the prize and potential for success described before.

There are other contenders to exploit the growing potential of this kind of space tourism.

Figure 1.14 White Knight carrying *Space Ship Two*. Reproduced with permission of Virgin Atlantic.

1.6 Forces and Drivers

The current aircraft design strategy is linked to industrial growth, which in turn depends on national infrastructure, governmental policies, workforce capabilities and natural resources; these are generally related to global economic-political circumstances. More than any other industry, the aerospace sector is linked to global trends. A survey of any newspaper provides examples of how civil aviation is affected by recession, fuel price increases, spread of infectious diseases and international terrorism. In addition to its importance for national security, the military aircraft sector is a key element in several of the world's largest economies. Indeed, aerospace activities must consider the national infrastructure as an entire system. A skilled labour force is an insufficient condition for success if there is no harmonisation of activity with national policies; the elements of the system must progress in tandem. Because large companies affect regional health, they must share socio-economic responsibility for the region in which they are located.

The current status stems from the 1980s when returns on investment in classical aeronautical technologies such as aerodynamics, propulsion and structures began to diminish. Around this time, however, advances in microprocessors enabled the miniaturisation of control systems and the development of microprocessor-based automatic controls (e.g. fly-by-wire: FBW), which also had on additional weight-saving benefit. Dramatic but less ostensive radical changes in aircraft management began to be embedded in design. At the same time, global political issues raised new concerns as economic inflation drove man-hour rates to a point at which cost-cutting measures became paramount. In the last three decades of the twentieth century, man-hour rates in the West rose four to six times (depending on the country), resulting in aircraft price hikes (e.g. typically by about six times for the Boeing 737 – of course, accompanied by improvements in design and operational capabilities.) Lack of economic viability resulted in the collapse or merger/takeover of many well-known aircraft manufacturers. The number of aircraft companies in Europe and North America shrunk by nearly three-quarters; currently, only two aircraft companies (Boeing and Airbus, i.e. in the West) are producing large commercial transport aircraft. Bombardier Aerospace and Embraer of Brazil have recently entered the large-aircraft market with capacity of more than 100 passengers. Also, in the are the Russians, the Chinese and the Japanese. Over time, aircraft operating-cost terminologies have evolved and, currently, the following are used in this book.

Over time, aircraft operating-cost terminologies have evolved and currently, the following are used in this book (Section 16.5 gives details).

IOC – Indirect Operating Cost: Comprises Costs not directly involved with the sortie (trip).
COC – Cash Operating Cost: Comprises the trip (sortie) Cost elements.
FOC – Fixed Operating Cost: Comprises Cost elements even when not flying but related to trip cost.
DOC – Direct Operating Cost: = COC + FOC.
TOC – Total Operating Cost: = IOC + DOC.
Because there are variances in definitions, this book uses the standardised definitions.

The importance of environmental issues emerged, forcing regulatory authorities to impose limits on noise and engine emission levels. Of late terrorist activities have made the industry and operators think hard about preventive design considerations.

1.7 Airworthiness Requirements

From the days of barnstorming and stunt flying in the 1910s, it became obvious that commercial interest had the potential to short-circuit safety considerations. Government agencies quickly stepped in to safeguard people's security and safety without deliberately harming commercial interest. Western countries developed and published thorough systematic rules – these are in the public domain (see relevant websites). In civil applications, they are Federal Aviation Regulation (FAR) (US) and JAR (the newly formed designation is European

Aviation Safety Agency – EASA); both are quite close. The author prefers to work with the established FAR at this point. In military applications, the standards are Milspecs (US) and Defence Standard 970 (earlier AvP 970 – UK); they do differ in places.

The Government of United States of America have 50 titles of Code of Federal Regulations (CFRs) published in the Federal Register by the Federal Government covering wide areas subjected to federal regulations. The FARs, are rules prescribed by the FAA governing all aviation activities in the USA under title 14 of the CFRs, which covers wide varieties of aircraft related activities in many parts, of which this book concerns mainly with Parts 23, 25, 33 and 35. However, another set of regulations in Title 48 of CFRs is the 'Federal Acquisitions Regulations', and this has led to confusion with the use of the acronym 'FAR'. Therefore, the FAA began to refer to aerospace specific regulations by the term '14 CFR part XX' instead of FAR. There is a growing tendency in the industry to adapt to using 14 CFR part XX. However, retaining the use of FAR meaning FAR is still acceptable, and in this book the authors continue with the use of the older practice of the term.

Safety standards were developed through multilateral discussions between manufacturers, operators and government agencies, which continue even today. These are the minimum standards comes as regulations and are mandatory requirements to comply. The regulatory aspects have two kinds of standards, as follows.

1. *Airworthiness standards.* These are concerned with aircraft design by manufacturers complying with regulatory requirements to ensure design integrity for limiting performance. These are outlined in FAR 25/JAR25 in extensive detail in a formal manner and are revised when required. After substantiating the requirements through extensive testing, Aircraft Flight Manuals (AFMs) are issued by the manufacturers for each type of aircraft designed.
2. *Operating standards.* These are concerned with the technical operating rules to be adhered by operators and are outlined in FAR 121/JAR-OPS-1 in extensive detail in a formal manner, revised when required. Aircraft operational capabilities are substantiated by the manufacturer through extensive flight testing and are certified government certification agencies, for example, the FAA/JAA. The contents of the AFM are recast in the Flight Crew Operating Manual (FCOM) that outlines the aircraft's limitations and procedures, along with the full envelope of aircraft performance data. Today, with the integration of computers in aircraft operation, it is possible to perform aircraft performance monitoring (APM) for optimum operations. Today, the operational aspects require full understanding of operating microprocessor-based aircraft design.

In civil aviation, every country requires safety standards to integrate with their national infrastructure and climatic conditions for aircraft operation, as well as to relate with their indigenous aircraft designs. Therefore, each country had their own design and operations regulations. As aircraft started to cross international borders, the standards to the foreign designed aircraft have had to be re-examined and possibly re-certified to be allowed to operate safely with their country. To harmonise the diverse nature of the various demands, the ICAO was formed in 1948, to recommend the international minimum recommended standards. It has now become legal for the International practices. However, within their own country there could their own operational regulations; while countries in North America and some European countries adopt FAR 121 while some other European countries follow JAR-OPS-1.

Aircraft operation is prone to litigations as mishaps do occur. To avoid ambiguity as well to ensure clarity to design, FAA documentations are written in a very elaborate, in-depth and articulate manner demanding intense study to understand and apply. It is for this reason that this book does not exactly copy the FAR lines and instead quotes the relevant part number along with outlining the requirements with explanations and supported by worked-out examples. The authors recommend that readers to access the latest FAA publication, their web site (www.faa.gov/regulations_policies/faa_regulations) should prove useful. The readers may have to contact FAA to make relevant the documents available. Most academic/aeronautical institute libraries necessarily keep FAR documents. For those who are in industry, these documents will be available there.

Aeronautical engineering does not progress without these documents to guarantee a minimum safety in design and operation.

The FAR (14CFT) Part 25 has the most stringent airworthy compliance requirements. The FAR 23 (general aviation aircraft – currently under revision) and the FAR Part 103 (ultra-light aircraft) have considerable lower level of requirements and use the same performance equations for performance analyses. This book deals only with the FAR (14CFR) Part 25.

1.8   Current Aircraft Performance Analyses Levels

Aircraft performance analysis is required to be carried at the very early stages of conceptual design phase and continued in every phases of programme updating capabilities as more accurate data are available till it is substantiated through flight tests. At the conceptual stage the performance prediction has to be sufficiently accurate to obtain management 'go-ahead' for a programme that bears promise of eventual profit making or successfully act against adversary. In the next phase, the performance figures are fine-tuned to promise guarantee to potential operators. Industry must be able to make aircraft performance analysis to a high degree of accuracy.

The analyses of aircraft performance cascade down from preliminary study to final refinement by design engineers followed by flight test substantiation and to the engineers preparing the AFM and the FCOM for operational usage. The various levels where aircraft performances are evaluated are briefly given next.

1.8.1   By the Designers

1. *At research level (feasibility study).* In this stage, engineers examine new technologies and their capabilities to advance new aircraft designs, examine possible modifications to improve existing design and so on. At this level researchers explore newer aircraft performance capabilities, optimise operational procedures using close-form equations that yield quick results for comparison and selection.
2. *At conceptual design level (Phase I of a project).* This is an outcome of feasibility study serious enough to progress towards market launching. In this phase, the study need to be done in a specific manner to freeze configurations in a family concept by sizing the aircraft with matched engines. In this phase, full aircraft analysis is not required. It only covers what is required in dialogue with potential customers with promising performance specifications to make comparative study to eliminate competition. If successful, go-ahead for the programme is given at the end of the study.
3. *Detailed design level (Phases II and III).* These are post go-ahead phase analyses to give guarantee to the potential customers. By now, more aerodynamic information is available through wind-tunnel test and computational fluid dynamics (*CFD*) analyses. More detailed and accurate aircraft performance estimation is now possible.
4. *Final level (Phases IV).* This is the final design phase and aircraft performances are carried out for substantiation (flight test) purpose to obtain certification of airworthiness. All technical/engineering and ground/flight tested substantiation data are the passed to a dedicated group to prepare the AFM and the FCOM for operational usage.

The format of presenting aircraft performance in the AFM and the FCOM is different from the format of aircraft performance documents used by the designers; the former is derived from the latter. The performance documents prepared during Phases I, II and III are used by engineers and contain predicted data that are substantiated through ground/flight tests. The full set of engineering data are given to experienced performance engineers at the dedicated customer support group who prepare the AFM and FCOM manuals for airline operators. Typically, the design office uses the prevailing terminologies, but the AFM and the FCOM must incorporate standard formal terminologies specified by the airworthiness agencies to avoid any ambiguity.

While preparing operational manuals does not involve extensive computation, it requires articulated presentation as errors or lack of clarity is not acceptable. This book follows the typical terminologies used by engineers along with introducing the synonymous formal terminologies to keep the readers informed.

1.9 Aircraft Classification

There are many types of aircraft in production, serving different sector of mission requirements – the civil and military missions differ substantially. It is important to classify the aircraft category to isolate and identify strong trends existing within the class. Readers are recommended to examine what could be the emerging design trends within the class of aircraft. In general, new commercial aircraft designs are extensions of the existing designs incorporating proven newer technologies (some are fallouts from declassified military applications) in a very conservative manner.

Table 1.2 in Section 1.3.1 gives aircraft classification based on its mission role and usage, power plants, configuration type, size and so on. Here, the first level of classification is based on operational role, that is, civil or military – this book is split into these two classes. In the next level, the classification branches into their generic mission role that would also indicate size. The next level proceeds with classification based on the type of power plant used and so on as the reader would notice. The examples worked out in this book are convenient for a classroom course. It should be noted that there is a lot to choose from without stretching too far into studying exotic types.

1.9.1 Comparison between Civil and Military Design Requirements

Design lessons learned so far on the current trend are summarised as follows:

- *Civil aircraft design.* For the foreseeable future, aircraft will remain subsonic and operating below 60 000 ft (large subsonic jets <45 000 ft). However, aircraft size could grow even larger if the ground infrastructure can handle the volume of passenger movement. Lower acquisition costs, lower operational costs and improved safety and environmental issues would act as design drivers. The SST would attempt an entry and HST operations still could be several decades away.
- *Military aircraft design.* Very agile aircraft incorporating extensive microprocessor-based control and systems management operating below Mach 2.5, high altitude (>60 000 ft), and BVR capabilities would be the performance demand. The issue of survivability is paramount – if required, aircraft could be operated unmanned. The military version of hypersonic combat aircraft could arrive sooner, paving the way to advance civil aircraft operations. Armament-and-missile-development activities would continue at a high level and would act as one of the drivers for vehicle design.

The military aircraft picture is not much different even when national interest has priority over commercial considerations. Whereas commercial aircraft can earn self-sustaining revenue, military operations depend totally on taxpayers' money, with no cash flow coming in, other than export sales that carry the risk of disclosure of tactical advantages. The cost frame of a new design has risen sufficiently to strain the economy of single nations. The typical project cost of a new high-technology combat aircraft is approximately $200 billion, an amount that exceeds the total cost incurred by all Western aircraft companies half a century ago. At approximately $100–200 million apiece, the price of a new combat aircraft is equivalent to nearly 1000 World War II Spitfires. Not surprisingly, the number of new designs has drastically dropped, and military designs are moving towards multinational collaborations among allied nations, where the retention of confidentiality in defence matters is possible.

Although derived from the same scientific basis, the aircraft design philosophy for the military and civil aircraft differs on account of their mission role. Table 1.3 compares the two classes of aircraft design, the civil and military.

It can now be seen how much different military aircraft design is as compared to civil design.

Table 1.3 Comparison between civil and military design requirements.

Issue	Civil aircraft	Military aircraft
Design space	Varied	Aspirational
Certification standards	Civil (FAR – US)	Military (Milspecs – US)
Operational environmental safety issues	Friendly Uncompromised, no ejection	Hostile Survivability requires ejection
Mission profile	Routine and monitored by air traffic control (ATC)	As situation demands and could be unmonitored
Flight performance	Near-steady-state operation and scheduled; gentle manoeuvres	Large variation in speed and altitudes; pilot is free to change briefing schedule; extreme manoeuvres
Flight speed	Subsonic and scheduled (not addressing SST here)	Have supersonic segments; in combat, unscheduled
Engine performance	Set throttle dependency, no afterburner (subsonic)	Varied throttle usage, with afterburner
Field performance	Mostly metal runways, generous in length, with ATC support	Different surfaces with restricted lengths; marginal ATC
Systems architecture	Moderately complex, high redundancies, no threat analysis	Very complex, lower redundancies, threat acquisition
Environmental issues	Strictly regulated; legal minimum standards	Relaxed; peacetime operation in restricted zones
Maintainability	High reliability with low maintenance cost	High reliability but at a considerably higher cost
Ground handling	Extensive ground-handling support with standard equipment	Specialised and complex ground-support equipment
Economics	Minimise DOC; cash flow back through revenue earned	Minimise LCC; no cash flow back
Training	Routine	Specialised and more complexed

1.10 Topics of Current Research Interest Related to Aircraft Design (Supersonic/Subsonic)

Research and development efforts continue in all branches of science and technology to incorporate meaningful design improvements to stay ahead of competition/threats. Funding has to be carefully appropriated with priorities in areas where there are the highest returns on investment for the objectives to be achieved. There is a tendency for diminishing return on investment on the conventional areas of aerodynamics, structures, propulsion and systems. The current interest on research and development appears to be in the following areas. While the external appearance of long-range jet aircraft has not changed much, advances in technology have actually transformed the entire design and manufacturing process through advances in computer aided engineering (CAE), for example, computer aided drawing (CAD), CFD, computational structural mechanics (CSM/finite element method (FEM)), multidisciplinary analyses (MDA) and multidisciplinary optimisation (MDO) in IPPD.

The aim is to enable future aircraft to carry more payload (20% for commercial transport aircraft) for the same amount of fuel and to takeoff and land quieter as compared to current capabilities. Military aircraft design challenges are considerably higher and remains in aspirational design space with new technologies to stay ahead of any potential adversary whose capabilities are kept in secret.

Aerodynamics (analytical study, simulation and wind-tunnel testing)

1. Aerodynamic performance of aerofoil/wing at large angle of incidence (steady and unsteady properties). The unsteady aerodynamics of slender wings and aircraft undergoing large amplitude manoeuvres.
2. Aero-acoustic studies on high-lift wing slat track and cut-out system to reduce noise. Aero-acoustic effects of high-lift wing slat track and cut-out system.
3. Studies on sophisticated aerodynamic concepts of new aircraft configurations with a high level of integration and functionality.
4. Transonic aircraft design – aerodynamic shape optimisation.
5. Effects of ice accretions on aircraft aerodynamics.
6. Experimental and design studies of stalls, spins and methods of the spin recovery including simulation of the automated handling improvement system operation.
7. Development, research and aerodynamic design of high-performance aircrafts configuration including aircrafts with thrust-vectoring module.
8. Natural laminar flow and hybrid laminar flow control and separation control.
9. Laminar flow requires surfaces that are smoother, with fewer steps, gaps or contamination than the surfaces that can typically be manufactured currently. To get the full benefits of laminar flow aerodynamic improvements research is carried out to give improved build tolerances and new joining technologies.

Materials and structures (analytical study, simulation and testing)

1. New materials (metal and non-metals) and manufacturing processes that do not exist today.
2. Develop smart composites structural concept with the ability to arrest cracks and self-repair the damage.
3. Develop materials for adaptive structures.
4. The development of composites that can store electrical energy as well as acting as a structural material.
5. Nanotechnology including carbon nanotubes, graphene and other new materials to be developed.
6. Sophisticated structural concepts with a high level of integration and functionality for new aircraft design.
7. Structural concepts for composites to tailored structures that place each fibre of the carbon material to optimise properties to reduce weight and hence fuel burn and improve aero-elastic properties.
8. Studies on multi-material structures (MMSs) aero structures. These MMSs include continuous fibre reinforced laminates, textile laminates, textile composites for high temperature applications, layered materials, sandwich structures with a variety of cores (honeycombs, foams, truss grids, functionally graded materials) and nanoparticle reinforced polymers, require advanced analysis tools for characterising their mechanical, thermal and electrical behaviour. Also, studies on metal matrix composites.
9. Studies on resin modifications for composite structures.

Propulsion (study, simulation, and testing) and integration

1. To develop lighter, quieter, more efficient engines (reduce fuel burn and environmental impact) and at competitive cost. Reduce noise and emission by 20% for commercial transports.
2. Aerodynamic analyses of unsteady flow and aero-acoustic prediction capabilities for airframe and propulsion noise sources associated with future concepts.
3. Introduction of a second generation composite fan.
4. Introduce revolutionary engine technologies with a radically different manufacturing philosophy.
5. Develop the next-generation engine core demonstrator engine incorporating high efficiency, high temperature compressor and turbine, low emissions combustor, low loss air system and high temperature materials.
6. It is also a key enabler to meeting the requirements of future engines, where changes in architecture will drive towards larger, lower pressure ratio fans and smaller higher pressure ratio core components at a given thrust size with geared gas-turbine engines.
7. Turbo-machinery flow and turbulence

8. Propellers: Turboprops offer up to 30% savings in fuel burn compared to an equivalent turbofan-powered aircraft. Increases in the cost of fuel and increasing environmental pressures have highlighted the need for more efficient, cleaner and quieter aircraft. The search is on for a number of aerodynamic and acoustic innovations to improve noise and further improve the efficiency of turboprop aircraft, thereby increasing market potential.
9. Supersonic and hypersonic intake aerodynamics.

Systems (avionics/electrical/mechanical)

1. Highly efficient operation to reduce fuel burn by innovative thermal and electrical management and advanced avionics systems.
2. Aircraft systems that are more connected, more interactive and increasingly wireless, like many household applications of today.
3. New materials to allow equipment to operate robustly in the harshest environments. Harsh environment electronics to support the more-electric aircraft and ultimately all-electric-aircraft concepts foreseen for the next-generation single aisle platform
4. Highly efficient and reliable electrical actuation
5. Advanced electrical power generation and distribution for greener aircraft
6. Advanced systems integration through the adoption of digital FBW, advanced navigational techniques and Full Authority Digital Engine Control (FADEC).
7. Advanced health management and intelligent systems
8. To close current technological gaps in motors and control electronics, which will support the introduction of all-electric aircraft.
9. To develop the highly complex and automated cockpit to handle abnormal situations, loss-of-control accident.

Manufacturing philosophy

1. Green manufacturing technologies to be developed include out-of-autoclave technologies such as microwave curing that will dramatically reduce energy usage and improve competitiveness
2. Advanced high-rate airframe manufacturing systems that optimise and integrate all the individual specialties together. Deployment of microprocessor-based machine tools, auto-riveting and so on.
3. Market projections for the next-generation single aisle sector indicate worldwide demand of up to 100 aircraft per month, requiring production rates for complex new technology aircraft parts well in excess of rates seen today. New manufacturing technologies will be needed to integrate with supply chain management and parts logistics technologies that can locate any component part in any part of the worldwide manufacturing supply chain at any time of the day or night
4. Integrated manufacturing and supply systems to link all enabling technologies and supply chains.
5. Integrated and multifunctional structural technologies to produce de-risking demonstrators.
6. Environmentally efficient manufacturing technologies to industrialise new manufacturing capability.
7. Advanced manufacturing techniques suited to design for manufacture, assembly and for quality.
8. Studying the best-practices manufacturing techniques, for example, jigless assembly, flyaway tooling, gaugeless tooling, inline assembly, automatic riveting, Six-Sigma and supporting methodologies.
9. Design and manufacture to integrate new product concepts and enabling technologies, study of digital manufacturing process management, simulation and so on.

Designers must keep themselves informed on technological progress and recognise what can be incorporated to advance new aircraft design to stay ahead of competition. To convert a new proven technology in a product line and obtain airworthiness approval is a challenge that designers have to undertake with passion.

1.11   Cost Implications

Aircraft design strategy is constantly changing. Initially driven by the classical subjects of aerodynamics, structures and propulsion, the industry is now customer-driven and design strategies consider the problems for manufacture and assembly that lead the way in reducing manufacturing costs. Chapter 16 addresses cost considerations in detail. In summary, an aircraft designer must be cost-conscious now and even more so in future projects.

It is therefore important that a basic exercise on cost estimation (i.e. second semester class-work) be included in the curriculum. A word of caution: academic pursuit on cost analysis to find newer tools may not amenable to industrial use. Manufacturers must rely on their own costing methodologies, which are not likely to appear in the public domain. How industry determines cost is sensitive information used to stay ahead in free market competition.

It is emphasised here that there is a significant difference between civil and military programmes in predicting costs related to aircraft unit-price costing. The civil aircraft design has an international market with cash flowing back from revenues earned from fare-paying customers (i.e. passengers and freight) – a regenerative process that returns funds for growth and sustainability to enhance the national economy. Conversely, military aircraft design originates from a single customer demand for national defence and cannot depend on export potential – it does not have cash flowing back and it strains the national economy out of necessity. Civil aircraft designs share common support equipment and facilities, which appear as IOCs and do not, significantly, load aircraft pricing. The driving cost parameter for civil aircraft design is the DOC, omitting the IOC component. Therefore, using a generic term of life cycle cost (LCC) = (DOC + IOC) in civil applications, it may be appropriate in context but would prove to be off the track for aircraft design engineers. Military design and operations incorporating discreet advances in technology necessarily have exclusive special support systems, equipment and facilities. The vehicles must be maintained for operation-readiness around the clock. Part of the supply costs and support costs for aircraft maintenance must be borne by manufacturers that know best and are in a position to keep confidential the high-tech defence equipment. The role of a manufacturer is defined in the contractual agreement to support its product 'from cradle to grave' – that is, the entire life cycle of the aircraft. Here, LCC is meaningful for aircraft designers in minimising costs for the support system integral to the specific aircraft design. Commercial transports would have nearly five times more operating hours than military vehicles in peacetime (i.e. hope for the life of the aircraft). Military aircraft have relatively high operating costs even when they sit idle on the ground. Academic literature has not been able to address clearly the LCC issues in order to arrive at an applicable standardised costing methodology.

Aircraft design and manufacture are not driven by cost estimators and accountants; they are still driven by engineers. Unlike classical engineering sciences, costing is not based on natural laws; it is derived to some extent from manmade policies, which are rather volatile, being influenced by both national and international origins. The academic pursuit to arrest costing in knowledge-based algorithms may not prove readily amenable to industrial applications. However, the industry could benefit from the academic research to improve in-house tools based on actual data. We are pleased to present in this book a relevant, basic cost-modelling methodology [11] from an engineer's perspective reflecting the industrial perspective so engineers may be aware of the labour content to minimise cost without sacrificing design integrity. The sooner engineers include costing as an integral part of design; the better the competitive edge will be.

1.12   The Classroom Learning Process

To meet the objectives of offering close-to-industrial practice in this book, it is appropriate to reiterate and expand on remarks made in the Preface about the recognised gap between academia and the industry. A teacher can considerably assist the newly initiated by streamlining course material with problem sets pertaining directly to industrial methodology – it has to be intelligently set without any 'fluff'. The course material should also inculcate passion for design to the young minds. A practitioner must enjoy his work.

The level of mathematics in this book is not an advanced one but contains much technological information. The purpose of the book is to provide close-to-industry standard computations in the coursework and engineering approaches sciences necessary for analysis. This is an indispensable way to expose new comers to work as a team; an essential skill required in the industry.

1.12.1 Classroom Learning Process versus Industrial Practices

Academic practices are very different from industrial practices. Both the USA and the UK are aware of this problem (0.16–0.21) and make periodic recommendations (see Preface).

Traditionally, universities develop analytical abilities by offering the fundamentals of engineering science. Courses are structured with all the material available in textbooks or notes; problem assignments are straightforward with unique answers. This may be termed a 'closed-form' education. Closed-form problems are easy to grade and a teacher's knowledge is not challenged (relatively). Conversely, industry requires the tackling of 'open-form' problems for which there is no single answer. The best solution is the result of interdisciplinary interaction of concurrent engineering within design built teams (DBTs), in which Total Quality Management (TQM) is needed to introduce 'customer-driven' products at the best value. Offering open-ended courses in design education that cover industrial requirements is more difficult and will challenge a teacher, especially when industrial experience is lacking. The associative features of closed and open-form education are shown in Figure 1.15 modified from [9].

Academics made studies on how to conduct design studies, especially suggesting efficient algorithms at the starting phases. On the other hand, industries have evolved from the legacies of their past practices, constantly updating to adopt newer technologies to the extent investment allows. Industries do interact with universities to explore new ideas but must remain conservative in implementing any attractive proposition. Industries cannot afford to take chances with unproven schemes, no matter how impressive it may be. The changes come in a gradual step-by-step manner, unless it is a case of taking over of a run-down company. The aim of this book is to adhere, as close as possible, to industrial practices as experienced by the authors. Therefore, the good academic research publications suggesting various methodologies on conceptual aircraft design are not included here.

To meet industry's needs, newly graduated engineers need a brief transition time before they can become productive, in line with the specialised tasks assigned to them. They must have a good grasp of the mathematics and engineering sciences necessary for analysis and sufficient experience for decision making. They must be capable of working under minimal supervision with the creative synthesis that comes from experience that academia cannot offer. The industrial environment will require new recruits to work in a team, with an appreciation of time, cost, and quality under TQM – which is quite different from classroom experience. Today's conceptual aircraft designers must master many trades and specialise in at least one, not ignoring the state-of-the-art 'rules of thumb' gained from past experiences; there is no substitute. They need to be good 'number-crunchers' with good analytical ability. They also need assistance from an equally good support team to encompass wider areas. This is the purpose of my book to provide in the coursework close-to-industry standard computations and engineering approaches necessary for analysis and enough experience to work on a team.

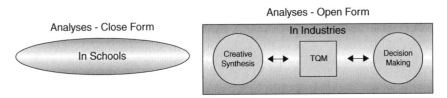

Figure 1.15 Associative features of 'close' and 'open' form education (modified from Nicolai [9]).

It is clear that unless an engineer has sufficient analytical ability, it will be impossible for him or her to convert creative ideas to a profitable product. Today's innovators who have no analytical and practical skills must depend on engineers to accomplish routine tasks under professional investigation and analysis and to make necessary decisions to develop a marketable product.

Here, it can be compared with what can be achieved in about 36 hours of classroom lectures plus 60 hours by each of about 30 inexperienced students to what is accomplished by 20 experienced engineers each contributing 800 hours ($\approx$6 months). Once the task was clearly defined shadowing industrial procedures, leaving out multiple iterations, It is found that a reduced workload is possible in a classroom environment. It cuts down man-hour content, especially when iterations are minimised to an acceptable level. The authors' goal is to offer inexperienced students a powerful analytical capability without underestimating the importance of innovation and decision making.

For this reason, it is emphasised that introductory class-work projects should be familiar to students so that they can relate to the examples and subsequently substantiate their work with an existing type. Working on an unfamiliar nonexistent design does not enhance the learning process at the introductory level.

Today, the use of CAD to generate 2D drawings and 3D modelling is the practice, especially in aerospace engineering. All modern aircraft manufacturers have adopted CAE, discarding the old articulated drawing boards for which there was practically no takers as even small general companies have changed to CAD. The developing countries have invested in CAD process with falling computer and software prices. Even for private usage, around $600 a satisfactory PC with low cost CAD programmes with sufficient capability for academic classroom work offers satisfactory results. Academic establishments without such facilities is unthinkable as without such facility competitive curriculum is difficult to sustain and will soon have to change to modern methods, at least in the minimalistic way if finance is a constraint.

1.12.2 Use of Computer-Assisted Engineering (CAE)

There are indispensable advantages in using CAD. The old method of lofting on large drawing boards is completely eliminated – the CAD algorithm takes of it with accuracy manually not possibly. Modifications can be stored and retrieved when required for comparison. The output can be directly fed to numerically controlled machine when metal can be cut; the total time to do the complete task in a fraction of time with practically no rejection.

There are considerably more benefits from 3D CAD solid modelling: It can be uploaded directly into CFD analysis to continue with aerodynamic estimations, as one of the first tasks is to estimate loading (CFD) for structural analysis using the FEM. The solid model offers accurate surface constraints for generating internal structural parts. CAD drawings can be uploaded directly to computer aided manufacture (CAM) operations, ultimately leading to paperless design and manufacture offices (see Chapter 20). Vastly increased computer power has reached the desktop with parallel processing. CAE (e.g. CAD, CAM, CFD, FEM and systems analyses) is the accepted practice in the industry. Those who can afford supercomputers will have the capability to conduct research in areas hitherto not explored or facing limitations (e.g. high-end CFD, FEM and MDO). This book is not about CAE; rather, it provides readers with the basics of aircraft design that are in practice in the industry and that would prepare them to use CAD/CAE (Figure 1.16 shows a CAD drawing example). Three-dimensional modelling provides fuller, more accurate shapes that are easy to modify, and it facilitates maintenance of sequential configurations – benefits that become evident as one starts to configure.

Finally, it is recommended that aircraft designers have some flying experience, which is most helpful in understanding the flying qualities of aircraft they are trying to design. Obtaining a licence requires effort and financial resources, but even a few hours of planned flight experience would be instructive. One may plan and discuss with the flight instructor what needs to be demonstrated – that is, aircraft characteristics in response to control input, stalling, g force in steep manoeuvres, stick forces and so forth. Some universities offer a few hours of flight tests as an integral part of aeronautical-engineering courses; however, it is suggested that even some hands-on experience under the supervision of a flight instructor. A driver with a good knowledge of the design features has more appreciation for the automobile.

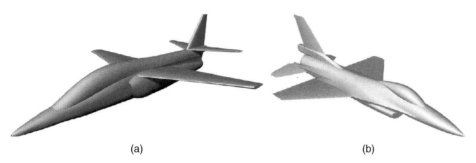

(a)   (b)

Figure 1.16 CAD drawing examples. (a) Advanced Jet Trainer (AJT) and (b) F16 (Courtesy of Pablo Quispe Avellaneda, Naval Engineer, Peru).

1.12.3 What is Not Dealt With in Depth in this Book?

Aircraft design involves with large number of disciplines of science and technology. The conceptual stage of a new aircraft project is the starting phase to assess whether a marketable product can be taken up as part of any manufacturer's ambition to stay in business. At this stage, the management allocates marginal budget to examine the feasibility of a new aircraft. It is for this reason, the methodology adopted during conceptual design stage engages only with small areas of science and technology what is necessary to bring out a design capable of satisfying the market requirement and stay ahead of competition. If a go-ahead is obtained, the proposed design is systematically improved to production standard through subsequent design phases (Section 2.4), not surprisingly, the improvements can be as high as 1015% better that what was proposed by the conceptual study.

Therefore, the aim of this book is to develop the geometric shape, that is, the external mould-lines to propose a realistic new aircraft configuration in the conceptual design phase to 'concept definition' and subsequently formally sized with matched engine to 'concept finalisation'. This book deals only with those aspects of technology associated to develop the final aircraft configuration that necessitates to start with some 'guesstimates' the statistics of existing aircraft. However, those influencing topics associated with conceptual study will be touched upon for information. Some of the associated topics that are not covered are the following.

1. Location of the aircraft neutral point (aircraft aerodynamic centre – Section 3.10) and the extent of aircraft centre of gravity (CG) travel (fore and aft limits to cater for aircraft weight variations – Section 10.4) are guessed from the statistics of past designs. However, details of aircraft, component weights and the CG location for the maximum takeoff weight will be will be estimated.
2. Aircraft structural design philosophy and choice of material affect aircraft weight, which in turn affects aircraft performance hence the aircraft size (see Chapter 19). Only the necessary information on material choices and general structural layout philosophy, sufficient to be aware of what is required during the conceptual study are presented here. It may also be noted that without the accurate structural component geometries, moment of inertia cannot be credibly estimated. Aircraft structural design courses are separately offered in academies.
3. Aircraft dynamic stability requires the knowledge of aircraft moment of inertia. Since aircraft dynamic stability is not analysed, assessment of moment of inertia is left out. Moreover, the structural geometries are not known during the conceptual design phases of project. These are done in the next phase of aircraft detailed design. Separate courses are routinely offered on aircraft static and dynamic stabilities in order to assess aircraft control characteristics. Only the static stability of aircraft is explained in this book in order to be aware of the role of aircraft CG location and its travel limits.
4. In this book, the planform areas of the control areas taken from statistics are positioned to the geometric locations. Two examples of using the DATCOM method to size empennage are shown in Chapters 5 and 6. Typically, as cost saving measures, accurate control areas sizing are postponed until project go-ahead is obtained and carried out in the Phase II of the project. In the past, the datasheet served to get the control areas, this practice is now supplanted by CFD analyses followed by a wind-tunnel test to substantiate

the findings by analytical tools. Control area sizing is notorious for requiring flight testing in qualitative assessments by several pilots, which requires some tailoring to harmonise pilot exerted efforts.

5. Use of datasheets is kept to the minimum – see Section 1.14.
6. Only the pertinent FAA regulations required for conceptual design study are dealt with in the book without copying the regulation texts from the FAR documents.
7. The authors avoided early aircraft sizing by the mass-fraction method (see Section 0.2.2).

1.13 Units and Dimensions

The aeronautical industry is yet to fully convert to the system international (SI) measurement system. Aeronautical industry extensively uses graphs/statistical data and semi-empirical relations. The user must be aware of implications of their usages, which type and where. These two issues are addressed in this section.

The post-war dominance of British and American aeronautics has kept the use of the foot–pound–second (FPS, also known as the Imperial System) system current, despite the use of non-decimal fractions and the ambiguity of the word *pound* in referring to both mass and weight. The benefits of the SI are undeniable: a decimal system and a distinction between mass and weight. However, there being 'nowt so queer as folk', the authors are presented with an interesting situation in which both FPS and SI systems are used. Operational users prefer FPS (i.e. altitudes are 'measured' in feet); however, scientists and engineers find SI more convenient. This is not a problem if one can become accustomed to the conversion factors. Appendix A provides an exhaustive conversion table that adequately covers the information in this book. However, readers will be relieved to know that in most cases, the text follows current international standards in notation units and the atmospheric table. Aircraft design and performance are conducted at the ISA (see Section 3.3). References are given when design considerations must cater to performance degradation in a non-standard day.

1.14 Use of Semi-Empirical Relations and Datasheets

The necessity of semi-empirical formulae arises because of exact mathematical analyses of nature arresting true complexities are difficult to solve and sometime not possible on account non-linearity in the physical phenomena. DATCOM (USA) and RAE DATA sheets (UK, recently replaced by Engineering Sciences Data Unit (ESDU)) served many generations of engineers for more than a half century and are still in use. These are well substantiated compilation of data both in graphical representation along with their associated semi-empirical relations. ESDU constantly updates with revisions to include latest data. Inclusion many of DATCOM/ESDU semi-empirical relations and graphs proves meaningless unless their use is shown in worked out examples. Use of datasheets requires cross referencing, sometimes in an involved manner. In text book usage, it needs to be explained with worked-out examples. While the authors included two examples of the DATCOM method in Chapters 5 and 6, they think that the usage of such datasheet methods is better suited to industries than in classroom undergraduate work. Examples of DATCOM methods are useful to obtain stability data. Since this book deals with the preliminary stage of conceptual design, usage of datasheets is kept in minimal.

Over time, as technology advanced, new tools using CAE have somewhat replaced earlier methods. The CAE has helped considerably but at a price of having some inherent errors arising from discretising the complex non-linear mathematical expressions into algebraic form. Depending on the areas of application, in certain cases, for example in CFD analyses, the results can remain under question. While drag analyses of complete aircraft with a guaranteed result is yet to be achieved, other areas of analyses have advanced to the point to consider CFD as a 'Digital Wind Tunnel'. But in most other applications, CFD is now an integral part of the design process and applied extensively during the conceptual design phase.

This book extensively uses semi-empirical relations in drag estimation and weight of aircraft and its components as the standard industrial practices. Data and semi-empirical methods give close results as both of them are derived from test data. This section explains the merit of using semi-empirical formulae. Semi-empirical relations and their associated graphs cannot guarantee exact results; at best it is coincidental to be error-free. A user of semi-empirical relations and graphs must be aware of the extent of error that can incur. Even when providers of semi-empirical relations and graphs give the extent of error range, it is difficult to substantiate any errors in a particular application.

If test results are available, they should be used in conjunction with the semi-empirical relations and graphs. Tests (e.g. aerodynamics, structures and systems) are expensive to conduct but they are indispensable to the process. Certifying agencies impose mandatory requirements on manufacturers to substantiate their designs with test results. These test results are archived as a databank to ensure that in-house semi-empirical relations are kept 'commercial in confidence' as proprietary information. CFD and FEM are the next in priority. It has to be proven conclusively about the consistency of CFD in predicting drag. At this stage semi-empirical relations and graphs are used extensively in drag prediction. This also is true for weight prediction. It is important for instructors to compile as many test data as possible in their resources.

1.14.1 Semi-Empirical Relations Compared to Statistical Graphs

Graphs plotted from statistical data can only capture the trend. They will require theoretical considerations to refine into developing semi-empirical formulae. These formulae will must be substantiation by recognised bodies, for example, DATCOM/ESDU, for wide acceptance. Typically, graphs plotted from statistical data in text books are limited to extracting from what is available in public domain, mostly of older designs are with error-band much higher than well-established semi-empirical formulae/graphs and not suitable for accurate design but useful to study the associated trends. Therefore, this book relies on the widely used semi-empirical relations, especially in the most critical requirement to establish aircraft drag. Following subsection discusses the case of use of semi-empirical formulae for aircraft and its component weight estimation.

Data reading from graphs is normal engineering practices since graphs are readily available and data can be quickly obtained. Accuracy in reading data from graphs depends on the resolution of graphs. The graphs given in this book have their limitations as described. The plot sizes are small and do not have an adequate resolution, therefore the readings are unlikely to be accurate. It is recommended that the readers plot graphs of latest aircraft in the category of their study with consistent accurate data in high resolution using high-end graph plotting software.

1.14.2 Use of Semi-Empirical Weights Relations versus use of Weight Fractions

Aircraft and its component weight estimation using weight fraction offers trends but are not as accurate as well-substantiated semi-empirical formulae. The weight fractions can quickly give some idea of what to expect. Also, the weight fractions are expressed in simple mathematical expressions; those can be easily manipulated to examine the interrelation between different parameters and also can be used for research; for example, optimisation. These kind researches are not undertaken in this book. Since weight estimates require iterations to become refined as more information is obtained, this book uses semi-empirical formulae as will be dealt with in relevant chapters.

This subsection discusses the merit of using the semi-empirical weights formulae. Unless the weight fraction data and related graphs are generated from the statistics of the latest aircraft in operation in the class, the use of any text book data is discouraged. Statistical-based weight fractions data offers a good check to examine if the results obtained using the semi-empirical weights formulae are in agreement or not. Industries maintain data base of existing aircraft and possible competition aircraft to give some idea of what is expected and serve to initiate the conceptual design study and make comparison. After that, the typical trend in industry is to use their in-house semi-empirical weights formulae for aircraft component weight estimation with substantiated accuracy within 5% to begin with and gets refined to less than to 2% before the final assembly of the first aircraft is completed. Once manufactured, accurate weights data for components and the whole aircraft can be quickly weighed to obtain the exact value.

CAD modelling can give component volume and its weight can be determined from the density of the material used. However, keeping account of thousands of components requires considerable attention. Although weight prediction using CAD is improving, as of today, use of semi-empirical formulae is the standard industrial practice. Over the aircraft production span its weights keep changing, it always grows with modifications and additional requirements. Any major weight reduction will require some form of re-engineering with a product to come out as a new variant.

To estimate aircraft and its components drag, there is no option but to use semi-empirical formulae as the industry standard practice (Section 1.11).

1.14.3 Aircraft Sizing

A section in the Roadmap of this Book (in the front matter) explains the reasons for the formal aircraft sizing carried out later in Chapter 14, instead of early sizing using the *mass-fraction method*.

1.15 The Atmosphere

Since interaction with air is essential for flight, aircraft engineers must have full understanding of the environment (the atmosphere) and the properties of air. Knowledge of the atmosphere is an integral part of aerospace engineering. The science of atmosphere is quite complex; however, the aircraft performance engineers only use standardised tables. This section gives the background of important fundamentals, perhaps a little more than what the book uses.

The atmosphere around the Earth is never uniform nor is in steady state – nature exhibits variance in properties at all the time around at all the places. The scientific community needed some standardisation to compare aircraft performance. After substantial data generation, an international consensus was reached to obtain the ISA [8], which follows hydrostatic relations at static, dust free and zero humidity condition. ISA Data (Appendix B) given up to 32 miles in altitude is nearly identical to US Standard Atmosphere Data.

The atmosphere, in the classical definition up to a 40-km altitude, is dense (continuum): its homogeneous constituent gases are nitrogen (78%), oxygen (21%) and others (1%). At sea level (with subscript 0), the ISA condition gives the following properties:

Pressure, $p_0 = 101\,325\,\mathrm{N\,m^{-2}}$ (14.7 lb/in.²); Density, $\rho_0 = 1.225\,\mathrm{kg\,m^{-3}}$ (0.02378 slugs/ft³); Temperature, $T_0 = 288.16\,\mathrm{K}$ (518.69°R); Viscosity, $\mu_0 = 1.789 \times 10^{-5}\,\mathrm{N\,(s\,m^2)^{-1}}$ (3.62×10^{-7} lb –s ft⁻²)

Acceleration due to gravity, $g_0 = 9.807\,\mathrm{m\,s^{-2}}$ (32.2 ft s⁻²).

1.15.1 Hydrostatic Equations and Standard Atmosphere

The hydrostatic equation of a vertical column of air of density ρ in the elemental height dh standing on one unit area is derived as follows (Figure 1.17a). At any cross-section of column, the force is the pressure. Force on height h_1 (lower cross-section) = Weight of cube + Force on height h_2

that is $p_2 + \rho g \mathrm{d}h = p_1$

that is $p_2 - p_1 = -\rho g \mathrm{d}h$ (pressure decreases with altitude)

or

$$\mathrm{d}p = -\rho g \mathrm{d}h \tag{1.1}$$

In the Equation (1.1), ρ and g varies with altitude, h. To simplify integration using perfect gas laws, the variable ρ is replaced by $p/(RT)$ and g is held constant at sea level value of g_0. The justification of holding g constant is given next. Then Equation (1.1) can be written as follows.

$$\mathrm{d}p/p = -g_0 \mathrm{d}h/(RT) \tag{1.2}$$

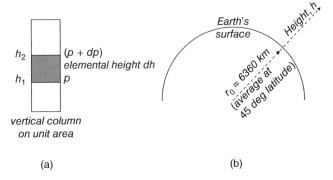

Figure 1.17 Hydrostatic equation and altitude effect. (a) Force diagram for the hydrostatic equation and (b) aircraft distance from the centre of the Earth.

on integrating

$$\int_0^h \frac{dp}{p} = -g_0 \int_0^h \frac{dh}{RT} \tag{1.3}$$

Relations in Equation (1.1) show dependency on acceleration due gravity, g, which is inversely proportional to the square of the radius (average radius $r_0 = 6360$ km) from the centre of the Earth. If h is the altitude of an aircraft from the Earth's surface, then it is at a distance $(r_0 + h)$ from the centre of the Earth. Figure 1.17b shows schematically the aircraft distance from the centre of the Earth.

Acceleration due to gravity, g, at height h is expressed as:

$$g = g_0 \left(\frac{r_0}{r_0 + h} \right)^2 \tag{1.4}$$

where $g_0 = 9.807$ m s^{-2} is the acceleration due to gravity at sea level (surface).

Substituting the values: $g = 9.807 \times \left(\frac{6,360 \times 10^3}{6,360 \times 10^3 + h} \right)^2$

Example 1.1   Find g at 30 km: $9.807 \times \left(\frac{6,360}{6,390} \right)^2 = 0.9906 \times 9.807$ (less than 1% change in g)

For terrestrial flight, h is much less than r_0; there is less than a 1% change in g up to 30 km. To simplify integration of Equation (1.2), variation in acceleration due to gravity value is held constant at sea level value up to 30 km altitude. For keeping g_0 constant, the potential energy at a slightly lower altitude equals the potential energy of pressure altitude, h with actual varying g. The small error arising from keeping g_0 constant results in geopotential altitude, h_p, which is slightly lower than the pressure (geometric) altitude, h. This book uses the ISA table that gives geopotential pressure altitude, h_p. The relation between h and h_p can be derived as follows.

For the same potential energy, PE, following relationship can be set up.

$$PE = m \int_0^h g dh = mg_0 h_p \quad \text{or} \quad \int_0^h g dh = g_0 h_p$$

Substituting the value of g from Equation (1.4), into the above equation can be written as

$$g_0 \int_0^h \left(\frac{r_0}{r_0 + h} \right)^2 dh = g_0 \left(\frac{r_0 h}{r_0 + h} \right) = g_0 h_p \text{ this shows } h > h_p \tag{1.5}$$

that is geopotential altitude that is slightly lower than the pressure (geometric) altitude. At 25 km altitude, the difference between the two is less than half of a percent. This book uses the geopotential altitude from the ISA table.

The centripetal force due to Earth's rotation is collinear with gravity vector, ***g***. The centripetal force is very small compared to gravitational force and is ignored.

Since it is impossible to measure tapeline geometric altitude, pressure is used to compute altitude. In ideal situation in a ISA day properties of geometric altitudes in which gravity varies are identical with hydrostatic equation. The hydrostatic Equation (1.3) shows that with altitude gain, the pressure decreases. Temperature profile with altitude variation is known from observation and is used to obtain other properties using hydrostatic equation. In reality, an ISA day is difficult to find; nevertheless, it is used to standardise aircraft performance to a reference condition for assessment and comparison.

Atmospheric temperature variation behaves strangely with altitude – it is influenced by the natural phenomenon: it decreases linearly up to 11 km at a lapse rate (λ) of $-$ 6.5 K km^{-1} reaching 216.66 K (in FPS, $\lambda = 0.00357$ R/ft. up to 36 089 ft). Temperature gradient breaks at 11 km altitude known as *tropopause*; below the tropopause is the *troposphere* and above it is the *stratosphere* extending up to 54 km. The initial temperature profile in stratosphere up to 20 km remains constant at 216.66 K. From 20 km upwards temperature increases linearly at a rate of 1 K/km up to 32 km. This books interest is within 32 km. Figure 1.18a shows the typical variation of atmospheric properties with altitude up to 100 km (Figure 1.18b is enlarged by reducing the *y*-axis scale). From 32 to 76 km, the atmospheric data are currently considered tentative. Above a 76 km altitude, variations in atmospheric data are considered speculative. It is found that up to 90 km mean molecular mass is nearly homogenous and termed there homosphere, above it is termed the hetrosphere. Nearly half the air-mass is within 18 000 ft. altitude.

Currently, no atmospheric dependent aircraft flying above 32 km and therefore the atmospheric properties above this altitude are beyond the scope of this book. Typical atmospheric stratification (based primarily on temperature variation) is as follows (the applications in this book do not exceed 65 000 ft [$\approx$20 km].

Troposphere	up to 11 km (36 089 ft)
Tropopause	11 km (36 089 ft)
Lower stratosphere (isothermal)	from 11 to 20 km (36 089–65 600 ft)
Upper stratosphere	from 20 to 47 km (65 600 ft–154 300 ft) – break at 32 km
Stratopause	47 km (154 300 ft)
Upper stratosphere (isothermal)	from 47 km (154 300 ft) to 54 km
Mesosphere	from 54 to 90 km
Mesopause	90 km
Thermosphere	from 100 to 550 km (can extend and overlap with ionosphere)
Ionosphere	from 550 to 10 000 km
Exosphere	above 10 000 km

Enough data are collected up to stratopause to have an ISA day. Above that altitude, data are being collected using sounding rockets and are treated as tentative.

Based on temperature profile the atmosphere up to 32 km there are three distinct zones as follows.

Zone 1 (from sea level to 11 km altitude): ***Troposphere*** in negative temperature gradient with lapse rate of -0.0065 K m^{-1} reaching 216.66°C at tropopause.

Zone 2 (from 11 to 20 km altitude): ***Stratosphere (lower)*** at constant isothermal temperature of 216.66°C.

Zone 3 (from 20 to 32 km altitude): ***Stratosphere (upper)*** in positive temperature gradient with lapse rate of 0.001 K m^{-1}.

From the temperature distribution at distinct altitude zones as shown in Figure 1.18, the following hydrostatics equations are obtained that give the related properties for the given altitude, *h*, in metres. Computation of each zone uses the properties at the base of the zone and need to be computed out at the beginning. The ISA table is obtained from these equations that can be used if the ISA table is not available.

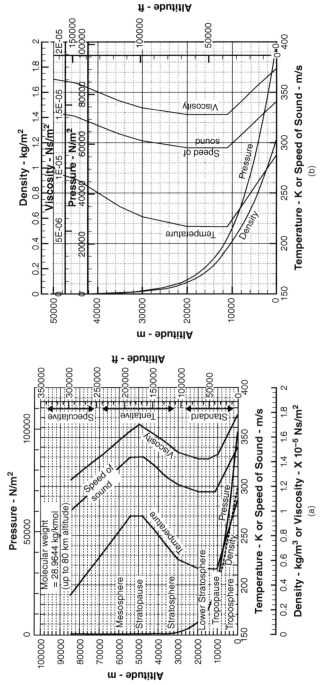

Figure 1.18 Standard atmosphere [14] (use Appendix B for accurate values). (a) Up to 100 000 m and (b) Magnified up to 50 000 m.

Using hydrostatic equations up to 32 km, the properties of air in SI unit are derived next.(for simplicity, only these set of equations the symbol for geopotential altitude have h instead of h_p, that is subscript 'p' is dropped).

The following two equations will prove derivation of the pressure, p, temperature, T, density, ρ, and viscosity, μ.

The perfect gas law:

$$\rho = \frac{p}{(T \times R)} \tag{1.6a}$$

and **Sutherland's** equation for viscosity is given as

$$\mu = \mu_0 \frac{(T_0 + C)}{(T + C)}\left(\frac{T}{T_0}\right)^{1.5} = \frac{\lambda T^{1.5}}{(T + C)} = \frac{1.458 \times T^{1.5}}{(T + 110.4)} \tag{1.6b}$$

in SI units, the constants $\lambda = 1.458 \times 10^{-6}$ (kg/(m s $K^{0.5}$) and $C = 114.4$.

In FPS

$$\mu = \mu_0 \frac{(T_0 + 198.72)}{(T + 198.72)}\left(\frac{T}{T_0}\right)^{1.5} \text{ lb s ft}^{-2} \tag{1.6c}$$

Following three non-dimensional parameters, δ, σ and θ are useful in aircraft performance analyses. All atmospheric properties in the three zones can be expressed in terms of these three ratios.

Relative pressure, $\delta = p/p_0$, relative density, $\sigma = \rho/\rho_0$, and relative temperature, $\theta = T/T_0$,

Troposphere (from sea level to 11 km altitude)

Nature exhibits temperature drop with altitude gain, the rate of decay is known as 'lapse rate', λ. An international standard of lapse rate in SI is taken as, $\lambda = -0.0065$ K m^{-1} (in FPS, $\lambda = 0.00357$ R/ft) reaching 216.66 K at 11 km. In SI, the constants $g_0 = 9.807$ m s^{-2} and R $= 287.053$ J kg K^{-1}. At sea level standard day $p_0 = 101\,325$ N m^{-2} and $T_0 = 288.16$ K.

Temperature, T, in K $= T_0 - (0.0065 \times h)$ up to 11 000 m in tropopause.

or

$$h = (1/0.0065)\,(T_0 - T) = \frac{T_0}{0.0065}\left(1 - \frac{T}{T_0}\right), \text{ where } T_0 = 288.16 \text{ K} \tag{1.7a}$$

That gives

$$dh = T/(-0.0065) \tag{1.7b}$$

Pressure

Substituting Equation (1.7b) in Equation (1.3) and integrating with the limits of sea level ($p_0 = 101\,325$ N m^{-2} and $T_0 = 288.16$ K) and h (p and T), the following is obtained.

$$\int_0^h \frac{dp}{p} = -g_0 \int_0^h \frac{dh}{RT} = g_0 \int_0^h \frac{dT}{0.0065RT} \tag{1.8a}$$

Or

$$\frac{(p)}{(p_0)} = \left(\frac{T}{T_0}\right)^{\frac{g_0}{0.0065R}} \text{ or } = \left(\frac{T}{T_0}\right) = \left(\frac{p}{p_0}\right)^{\frac{0.0065R}{g_0}} \tag{1.8b}$$

In SI, Pressure

$$p \text{ in N/m}^2 = 101325 \times \left(\frac{T}{288.16}\right)^{\frac{g_0}{0.0065R}} = 101325 \times \left(\frac{T}{288.16}\right)^{5.2562} \tag{1.8c}$$

And Equation (1.7b and 1.8b) gives $h = \dfrac{T_0}{0.0065}\left[1 - \left(\dfrac{p}{p_0}\right)^{\frac{0.0065R}{g_0}}\right] = \dfrac{T_0}{0.0065}\left[1 - \left(\dfrac{p}{p_0}\right)^{0.1906}\right]$ (1.8d)

Density

Using gas laws (Equation (1.5)), Equation (1.8b) becomes

$$\frac{(\rho)}{(\rho_0)} = \frac{pT_0}{p_0 T} = \left(\frac{T}{T_0}\right)^{\frac{g_0}{0.0065R}} \left(\frac{T_0}{T}\right) = \left(\frac{T}{T_0}\right)^{\frac{g_0}{0.0065R}-1} = \left(\frac{T}{T_0}\right)^{\frac{g_0-0.0065R}{0.0065R}} \tag{1.8e}$$

Or

$$\left(\frac{T}{T_0}\right) = \left(\frac{\rho}{\rho_0}\right)^{\frac{0\ 0.0065R}{g_0-0.0065R}}$$

In SI units

$$\rho = 1.225 \times \left(\frac{T}{288.16}\right)^{\frac{9.807-0.0065\times287.053}{0.0065\times287.053}} = 1.225 \times \left(\frac{T}{288.16}\right)^{4.256}$$

And from Equation (1.7a),

$$h = \frac{T_0}{0.0065}\left(1 - \left(\frac{\rho}{\rho_0}\right)^{\frac{0.0065R}{g_0-0.0065R}}\right) = \frac{T_0}{0.0065}\left[1 - \left(\frac{\rho}{\rho_0}\right)^{0.235}\right] \tag{1.8f}$$

Kinematic viscosity
v in m^2 s^{-1} = μ/ρ in m^2 s^{-1}

Example 1.2   Find the atmospheric properties at geopotential height of 8 km on an ISA day.
Temperature at 8 km is $T = 288.16 - (0.0065 \times 8000) = 288.16 - 52 = 236.16$ K

$$Pressure, p = 101325 \times \left(\frac{236.16}{288.16}\right)^{\frac{9.807}{0.0065\times287.053}} = 101325 \times (0.8195)^{5.2562} = 35\ 598.98 \text{ N m}^{-2}$$

From Equation (1.8c)

$$Density, \rho = 1.225 \times \left(\frac{236.16}{288.16}\right)^{4.256} = 1.225 \times 0.4286 = 0.525 \text{ in kg m}^{-3}$$

Sutherland's Equation (1.6b) for viscosity in SI units, the constants b and C become as follows.
$\lambda = 1.458 \times 10^{-6}$ (kg/(m s K$^{0.5}$)) and $C = 114.4$
From Equation (1.6b),

$$\mu = \frac{1.458 \times 10^{-6} \times 236.16^{1.5}}{236.16 + 110.4} = 1.458 \times 1.0471 \times 10^{-5-} = 0.000015267 \text{ kg (m s)}^{-1}$$

Stratosphere (lower – isothermal) – from 11 to 20 km

Temperature, $T = 216.66$ K stays constant up to 20 km. Equation (1.3) gives
Pressure – lower stratosphere
Equation 1.3 it gives

$$\int_{hb}^{h} \frac{dp}{p} = -g_0 \int_{hb}^{h} \frac{dh}{RT} = -(g_0/RT) \int_{hb}^{h} dh$$

Integrating within the limits of the base pressure at $h_p = 11$ km and altitude h:
$p_b = 22\ 633$ N m^{-2} and $T_b = 216.66$ K to altitude h,

$$Pressure, p \text{ N m}^{-2} = 22{,}633 \times e^{\frac{-g_0}{216.66R}(h-11000)}$$

Density–lower stratosphere

$$Density, \rho \text{ kg/m}^3 = \rho_1 e^{-\left[\frac{g_0}{RT}\right](h-h_1)} \text{ or use gas law } \rho = \frac{p}{(216.66 \times R)} \tag{1.8g}$$

where p is obtained from the this equation.

Example 1.3   Find the atmospheric properties at geopotential height of 15 km.
Temperature at 15 km is constant at $T = 216.65$ K.

Pressure, p in

$$\text{N m}^{-2} = 22\ 633 \times e^{\frac{-9.807}{216.66 \times 287.053}(15000-11000)} = 22\ 633 \times e^{-0.6307} = 22\ 633 \times 0.5322$$

$$= 12\ 045.7 \text{ N m}^{-2}$$

$$Density, \rho \text{ kg m}^{-3} = \frac{p}{(216.66R)} = \frac{12045.7}{(216.66 \times 287.053)} = \frac{12045.7}{62192.9} = 0.1937 \text{ kg m}^{-3}$$

or

$$\rho = 0.36392 \times e^{-\left[\frac{9.807}{287.053 \times 216.66}\right](15000-11000)} = 0.36392 \times e^{-0.6307} = 0.36392 \times 0.5322 = 0.1937 \text{ kg m}^{-3}$$

Stratosphere (upper) – from 20 to 32 km

It is in a positive temperature gradient in upper stratosphere from 20 to 32 km.
Here lapse rate in SI is, $\lambda = 0.001$ K m^{-1} (in FPS, $\lambda = 0.000549$ R/ft)
Temperature, $T = 216.65 + (0.001 \times h)$ up to 32 000 m altitude
That gives d$h = T/(0.001)$ and substituting in Equation (1.3), the following is obtained.
(*Kinematic viscosity* v in m^2/s $= \mu/\rho$ in m^2/s)

Pressure – upper stratosphere

Equation (1.3) gives $\int_0^h \frac{dp}{p} = -g_0 \int_0^h \frac{dh}{RT} = -g_0 \int_0^h \frac{dT}{0.001RT}$

or

$$\frac{(p)}{(p_0)} = \left(\frac{T}{T_0}\right)^{\frac{-g_0}{0.001R}} \text{ or } = \left(\frac{T}{T_0}\right) = \left(\frac{p}{p_0}\right)^{\frac{-0.001R}{g_0}}$$

Integrating within the limits of the base pressure at $h_p = 20$ km and altitude h:
$p_b = 5475.21$ N m^{-2} and $T_b = 216.66$ K to altitude h,

$$Pressure, p \text{ in N/m}^2 = 5475.21 \times \left(\frac{T}{216.65}\right)^{\frac{-g_0}{0.001R}} \tag{1.8h}$$

Density – upper stratosphere

$$Density, \rho \text{ in kg/m}^3 = 0.08803 \times \left(\frac{T}{216.66}\right)^{\frac{-g_0}{0.001R}-1} \tag{1.8i}$$

$$Viscosity, \mu \text{ kg (m s)}^{-1} = \frac{1.458 \times 10^{-6}T^{1.5}}{T + 110.4} \text{ (from Equation (1.6b))}$$

Kinematic viscosity v in m^2 s^{-1} $= \mu/\rho$ in m^2 s^{-1}

Example 1.4   Find the atmospheric properties at geopotential height of 25 km.

$$Temperature, T, \text{K} = 216.66 + [0.0047 \times (h - 20\ 000)] = 216.66 + (0.001 \times 5\ 000) = 221.66 \text{ K}.$$

$$Pressure, p \text{ in N m}^{-2} = 5475.21 \times \left(\frac{221.66}{216.66}\right)^{\frac{-9.807}{0.001 \times 287.053}}$$

$$= 5475.21 \times (1.0231)^{-34.164} = 5475.21/2.1819 = 2510.34 \text{ N m}^{-2}$$

$$Density, \rho \text{ in kg m}^{-3} = \frac{p}{(T \times R)} = \frac{2510.34}{(221.66 \times 287.053)} = \frac{2510.34}{63628.17} = 0.0395 \text{ kg m}^{-3}$$

or

$$\rho = 0.08803 \times \left(\frac{221.66}{216.66}\right)^{\frac{-9.807}{0.001 \times 287.053}-1} = 0.08803 \times (1.0231)^{-34.1644-1} = 0.08803 \times 0.448 = 0.0394 \text{ kg m}^{-3}$$

$$Viscosity, \mu \text{ kg/m s} = \frac{1.458 \times 10^{-6}T^{1.5}}{T + 110.4} \text{ (from Equation (1.6b))}$$

Appendix B gives the ISA table up to a 25 km altitude, which is sufficient for this book because all aircraft (except rocket-powered special-purpose aircraft – e.g. the space plane), described fly below 25 km. Linear interpolation of properties may be carried out between altitudes.

1.15.2   Non-Standard Atmosphere/Off-Standard Atmosphere

While there is an internationally accepted standard atmosphere (ISA), there are no accepted standards for atmospheric conditions different from an ISA day, which is the typical situation. Typically, the hydrostatic equations are used to obtain the off-standard day atmospheric properties. The difference between non- and off-standard atmosphere is that the former has no standards and the latter takes only the incremental temperature ΔT effects on ISA atmosphere. This book deals with the off-standard atmosphere expressed as $(T_{ISA} \pm \Delta T)$.

As mentioned previously, an aircraft rarely encounters the ISA day. Wind circulation over the globe is always occurring. Surface wind current such as *doldrums* (i.e. slow winds in equatorial regions), *trade winds* (i.e. predictable wind currents blowing from subtropical to tropical zones), *westerlies* (i.e. winds blowing in the temperate zone) and *polar easterlies* (i.e. year-round cold winds blowing in the Polar Regions) are well-known. In addition, there are characteristic winds in typical zones – for example, *monsoon* storms; wind-tunnelling effects of strong winds blowing in valleys and ravines in mountainous and hilly regions; steady up and down drafts at hill slopes and daily coastal breezes. At higher altitudes, these winds have an effect. Storms, twisters and cyclones are hazardous winds that must be avoided. There are more complex wind phenomena such as wind-shear, high-altitude jet streams and vertical gusts. Some of the disturbances are not easily detectable, such as clear air turbulence (CAT). Humidity in the atmosphere is also a factor to be considered. The air-route safety standards have been improved systematically through a round-the-clock surveillance and reporting. In addition, modern aircraft are fitted with weather radars to avoid flight paths through disturbed areas. Flight has never been safer apart from manmade hazards. This book deals primarily with ISA day, with the exception of *gust load*, which is addressed in Chapter 17, for structural integrity affecting aircraft weights. On aircraft performance the critical issues related to non-standard atmosphere is discussed with worked out examples. Aircraft performance engineers must ensure aircraft capabilities in a hot or cold day (off-standard day) operations; hot day can be critical on aircraft performance, especially at takeoff. More details of flying in adverse environment is given in Section 18.6.

Unlike ISA, in off-standard atmosphere, ambient pressure in geopotential altitude is not the same as pressure altitude. Appendix B gives an off-standard chart in which only the change in temperature is given as an input. The results from the hydrostatic equations for the geopotential altitude and its corresponding pressure show the changes in density, speed of sound, viscosity and coefficient of friction as a consequence of change in ambient temperature.

On a hot day, the density of air decreases resulting in degradation of aircraft and engine performances. Certification authorities (i.e. FAA and CAA) require that aircraft demonstrate the ability to perform as predicted in hot and cold weather and in gusty wind. The certification process also includes checks on the ability of the environmental control system (ECS, e.g. anti-icing/de-icing and air-conditioning) to cope with extreme temperatures. In this book, performance degradation on an off-standard day is addressed to an extent. The procedure to address off-standard atmospheres is identical to computation using the ISA conditions, except that the atmospheric data are different. The atmosphere is not fully dry; there is always some humidity present affecting ISA conditions. This book does not deal with humidity effects but discusses them briefly in Section 1.15.4.

1.15.3   Altitude Definitions – Density Altitude (Off Standard)

It is not possible to measure the geometric (tapeline) altitude of aircraft. Aerospace application has to depend on the hydrostatic relation between altitude and pressure (Equation (1.2)). Density Altitude is formally defined as 'geopotential pressure altitude corrected for off-standard temperature variations'. Only temperature variation is considered for computing density altitude.

The ISA day is hard to find and therefore off-standard day atmospheric behaviour also needs to be looked into from the point of view aircraft performance, which depends primarily on air density. Aircraft lift and engine performance depends on ambient air density. For example, on a hot day the sea level air density will be thinner and have lower than standard day pressure. The flight deck altimeter (see Section 3.3.1) will then indicate higher altitude corresponding to the lower pressure. At the same time engine will produce less thrust and aircraft generates less lift at the same true air speed (see Section 3.2). Therefore, density altitude is what is used for off-standard atmosphere. Density altitude is the pressure altitude corrected for off-standard conditions. For example, if the top level of a hilly area is 3000 ft and if the altimeter indicates 4000 ft on a hot off-standard day when actually he is below 3000 ft, then a pilot on a foggy day may suddenly see they are flying at the hill and may hit it due to not having enough reaction time to pull up.

With an off-standard day the atmospheric properties are denoted by the subscript '$_{off}$'. For off-standard atmospheric conditions let

$$T_{off} = T_{ISA} \pm \Delta T$$

Gas laws give $\rho_{off} = \frac{p_{off}}{RT_{off}}$, noting $p = p_{off}$. The off-standard density can be obtained as

$\rho_{off} = \frac{\rho_0 p T_0}{p_0 (T \pm \Delta T)}$, Noting that density altitude is only the function of density, the density altitude h (geopotential) can be computed using Equation (1.8f),

$$h = \frac{T_0}{0.0065}\left(1 - \left(\frac{\rho_{off}}{\rho_0}\right)^{\frac{0.0065R}{g_0 - 0.0065R}}\right) = \frac{T_0}{0.0065}\left[1 - \left(\frac{pT_0}{p_0(T \pm \Delta T)}\right)^{0.235}\right] \tag{1.9a}$$

A density altitude chart can be generated using Equation (1.9a) and is given in Appendix B.

However, if only off-standard temperature at an altitude is known then the density ρ_{off} can be computed without the ISA table to get the pressure. However, its application in practice is low as it is not in demand.

$$p = p_0\left(\frac{T}{T_0}\right)^{\frac{g_0}{0.0065R}} = p_0\left(\frac{T_0 - 0.0065h}{T_0}\right)^{5.2562} = p_0\left(1 - \frac{0.0065h}{288.16}\right)^{5.2562} \tag{1.9b}$$

and substituting the T_{off}, it becomes

$$\rho_{off} = \frac{101325(1 - 0.0000256h)^{5.2562}}{287.053(T_{ISA} \pm \Delta T)} = \frac{353.3(1 - 0.0000256h)^{5.2562}}{(T_{ISA} \pm \Delta T)} \tag{1.9c}$$

Exercise 1.5   In Problem 1.2, the atmospheric properties at 8 km in STD have been computed. In this problem compute the geopotential density altitude of Problem 1.2 in a $T_{off} = ISA + 10°C$ and $T_{off} = ISA + 40°C$ day.

$T_{off} = ISA \pm 10°C$

Example 1.2 gives at 8 km STD day, $p = 35\,598.98\,\text{N m}^{-2}$ and $\rho = 0.525$ in kg m^{-3} and $T = 236.16\,\text{K}$. $T_{off} = ISA + 10°C = 246.16\,\text{K}$. Use Equation (1.9a), the following is obtained.

$$h = \frac{T_0}{0.0065}\left[1 - \left(\frac{pT_0}{p_0(T \pm \Delta T)}\right)^{0.235}\right] = \frac{288.16}{0.0065}\left[1 - \left(\frac{35598.98 \times 288.16}{101325 \times 246.16)}\right)^{0.235}\right]$$

or

$h = 44332.3 \times [1 - 0.4112^{0.235}] = 8335\,m$ that is, indicating 335 m higher altitude.

$T_{off} = ISA \pm 40°C$

Example 1.2 gives at a 8 km STD day, $p = 35\,598.98\,\text{N m}^{-2}$ and $\rho = 0.525$ in kg m^{-3} and $T = 236.16\,\text{K}$. $T_{off} = ISA + 10\,C = 276.16\,\text{K}$. Use Equation (1.9a), the following is obtained.

$$h = \frac{T_0}{0.0065}\left[1 - \left(\frac{pT_0}{p_0(T \pm \Delta T)}\right)^{0.235}\right] = \frac{288.16}{0.0065}\left[1 - \left(\frac{35598.98 \times 288.16}{101325 \times 276.16)}\right)^{0.235}\right]$$

or

$h = 44332.3 \times [1 - 0.3666^{0.235}] = 44332.3 \times [1 - 0.79] = 9310\,m$; that is, indicating 1310 m higher altitude.

The same procedure is adopted for zones 2 and 3 in the stratosphere using relevant equations.
In the stratosphere from 11 to 20 km, density altitude,

$$h_p = 11000 - \frac{(RT_S)}{g_0} \ln \frac{\rho}{\rho_s} \tag{1.9d}$$

On a hot day, a pilot could hit the ground while the altimeter shows altitude and vice versa on a cold day. Serious errors can occur if the altimeter is not adjusted to the condition of the territory. Density altitude is used more by pilots than engineers. Pilots communicate with the ground station to adjust the altimeter accordingly.

1.15.4 Humidity Effects

ISA is modelled for a dry atmosphere, but there is always some humidity present, in a monsoon climate it can be very high. The molecular mass ratio of water vapour to dry air is 0.62:1. Humidity affects the density of air. Presence of water vapour decreases the air density to a small extent, that is, an increase of the relative humidity in the air decreases the air density. Therefore, it is considered when computing density altitude. Relative humidity also affects aircraft lift and engine performances. The readers are recommended to look at humidity charts to obtain air density to compute altitude. The humidity effect is not used in this book.

1.15.5 Greenhouse Effect

The greenhouse effect concerns aerospace engineers. Atmospheric gas compositions greatly influence climatic temperature, affecting environment. About 99% of atmosphere is composed of diatomic gases, for example, N_2 and O_2 that are transparent to radiation. But the very small percentages of triatomic and higher gases, for example, carbon dioxide (CO_2), methane (CH_4), ozone (O_3) and chlorofluorocarbons (CFCs – industrial gases like aerosols, refrigeration gases etc.) are responsible for atmospheric warming and are known as greenhouse gases. These are seen as pollutants and exist mostly within 32 km. An increase of CO_2, CH_4 and CFCs acts to create greenhouse blanket effects that increase ambient temperature. On the other hand, depletion of O_3 thins out the shield for solar ultraviolet radiation that harms the population. The silent group of aerospace engineers is no less concerned than vociferous environmentalists and acts to minimise pollution working directly to resolve industrial challenges. Currently, aerospace produces less than 2% of total manmade pollution. Airworthiness authorities have imposed strict limits to contain pollution form gas emission. Aeronautical infrastructure deserves credit for it.

References

1 Anderson, J. (2002). *The Airplane: A History of its Technology*. AIAA Publications.
2 (2001). *Jane's Fighting Aircraft of World War I*. Random House Group.
3 Taylor, M.J.H. (1983). *Milestones of Flight*. Chancellor Press.
4 Anderson, J.D. (1998). *History of Aerodynamics*. Cambridge University Press.
5 Murman, M.W. and Rebentisch, E. (2000). Challenges in the better, faster, cheaper era of aeronautical design, engineering and manufacturing. *The Aeronautical Journal* 104 (1040): 481–489.
6 Kundu, A., Curran, R., Crosby, S. et al. (2002). *Rapid Cost Modeling at the Conceptual Stage of Aircraft Design*. Los Angeles, USA: AIAA Aircraft Technology, Integration and Operations Forum.
7 Von Braun, W. and Ordway, F.I. (1969). *History of Rocketry and Space Travel*. Thomas Y. Crowell Co.
8 ICAO (1993). *Manual of the ICAO Standard Atmosphere (extended to 80 kilometers (262 500 feet))*, Doc 7488- CD, 3e. ISBN: ISBN 92-9194-004-6.
9 Nicolai, L.M. (1992). Designing a Better Engineer. *Aerospace America AIAA*, 46: 30–33.

2

Aircraft Familiarity, Aircraft Design Process, Market Study

2.1 Overview

 The ve ry first thing when studying aircraft design, as the topic of study in this book, is to get familiar with the product hardware as an introduction for the newly initiated reader. The road to a successful aircraft design has a formal step-by-step approach through phases of activity and must be managed. It is appropriate that readers should have some idea of how aircraft design projects are managed in a company and this is covered at an early stage of study. It is recommended that newly initiated readers read through this chapter because it tackles important parts of the work. To start a project requires generation of customer specifications, based on which the configured aircraft has the potential to succeed. A small part of the coursework starts in this chapter.

The go-ahead for a programme comes after careful assessment of the design with a finalised aircraft configuration that evolved during the conceptual study (i.e. Phase I). The prediction accuracy at the end of Phase I must be within at least ±5%. In Phase II of the project, when more financing is available after obtaining the go-ahead, the aircraft design is fine-tuned through testing and more refined analysis. This is a time and cost consuming effort, with prediction accuracy now at less than ±2 to ±3%, offering guarantees to potential buyers. This book does not address project definition activities (i.e. Phase II); these are in-depth studies conducted by specialists.

This book is concerned with the task involved in the conceptual design phase but without rigorous optimisation. Civil aircraft design lies within a verified design space; that is, it is a study within an achievable level of proven but leading-edge technology involving routine development efforts. Conversely, military aircraft design lies within an aspirational design space; that is, it is a study of unproven advanced technology requiring extensive development efforts. Obviously, the latter is technologically more complex, challenging and difficult. Generally, the go-ahead for a military combat aircraft project is preceded by a scaled down technology demonstrator aircraft to prove the technology concept.

This chapter covers the following topics:

Section 2.2: Introduction
Section 2.3: Aircraft Familiarisation (Civil/Military)
Section 2.4: Aircraft Design Process; Systems Approach to Design and Project Phases
Section 2.5: Four Phases of Aircraft Project
Section 2.6: Task Breakdown in Each Phase and Functional Activities
Section 2.7: Market Survey (Civil and Military)
Section 2.8: Three Civil Aircraft Case Studies – Aircraft Specifications and Requirements for
Section 2.9: Some Typical Military Aircraft Design Specifications
Section 2.10: Airworthiness Requirements
Section 2.11: Coursework Procedures

The coursework activity begins in Section 2.5 with a mock market survey to generate aircraft specifications and requirements and helps students understand its importance in the success or failure of a product. With

Conceptual Aircraft Design: An Industrial Approach, First Edition. Ajoy Kumar Kundu, Mark A. Price and David Riordan.
© 2019 John Wiley & Sons Ltd. Published 2019 by John Wiley & Sons Ltd.

guidance from the instructor, students conduct a mock market survey. Students generate a bar chart (i.e. Gantt chart) to monitor progress during the semester. The remainder of the chapter is recommended easy reading. The following would prove useful resources.

Jane's all the World's aircraft manual. This is an indispensable source of aircraft statistics vital for any aircraft design work.

Flight international. This was a weekly publication from the UK, now comes as an online publication. It is a newsletter-type journal, providing the latest brief coverage of aerospace activities around the world.

Flight global. This is an on-line news and information website related to the aviation and aerospace industries. It also offers aircraft diagrams and data.

Aviation week and space technology. This is a weekly publication from the United States that provides more in-depth analysis of aerospace developments. It thoroughly covers the US scenario as well as worldwide coverage.

Interavia. A bimonthly publication from Switzerland that covered aerospace news and commercial airline business. It has now ceased to be published.

A list of aircraft design books is given in the Reference Section [1–24].

2.2   Introduction

Existing aircraft indicate how the market is served and should indicate what is needed for the future. Various aircraft have been designed, and new designs should perform better than any existing designs. Designers are obligated to search for proven advanced technologies that emerge. There could be more than one option so the design team must conduct trade-off studies to arrive at a 'satisfying' design that will satisfy the customer. Balancing between 'optimised' and 'satisfied' aircraft design is the key to succeed. Economy and safety are possibly the strongest drivers in commercial transport. Aircraft design drivers for combat are performance capability and survivability (i.e. safety).

Despite organisational differences that exist among countries, one thing is common to all: namely, the constraint that the product must be 'fit for purpose'. It is interesting to observe that organisational structures in the East and the West are beginning to converge in their approach to aircraft design. The West is replacing its vertically integrated setup with a major investor master company in the integrating role along with risk-sharing partners. Since the fall of communism in Eastern Europe, the socialist bloc is also moving away from specialist activities to an integrated environment with risk-sharing partners. Stringent accountability has led the West to move away from vertical integration – in which the design and manufacture of every component were done under one roof – to outsourcing design packages to specialist companies. The change was inevitable and it has resulted in better products and profitability, despite increased logistical activities.

The aircraft design process is now set in rigorous methodology and there is considerable caution in the approach due to the high level of investment required. The process is substantially front-loaded, even before the project go-ahead is given. In this chapter, generic and typical aircraft design phases are described as practised in the industry, which includes market surveys and airworthiness requirements. A product must comply with regulatory requirements, whether in civil or military applications. New designers must realise from the beginning the importance of meeting mandatory design requirements imposed by the certifying authorities.

Exceeding budgetary provisions is not uncommon. Military aircraft projects undergo significant technical challenges to meet time and cost frames; in addition, there could be other constraints. (The 'gestation' period of the Eurofighter project has taken nearly two decades. An even more extreme example is the Indian Light Combat Aircraft, which spanned nearly three decades; the original specifications likely to become obsolete by then.) Some fighter aircraft projects have been cancelled after the prototype aircraft was built (e.g. the Northrop F20 Tiger shark and the BAC TSR2). A good design organisation must have the courage to abandon concepts that are outdated and mediocre. The design of combat aircraft cannot be compromised because of national pride; rather, a nation can learn from mistakes and then progress step-by-step to a better future.

2.3   Aircraft Familiarisation

This section introduces generic civil and military aircraft. Geometric definitions relevant to aerodynamic considerations are addressed in Chapter 3 and detailed descriptions of various types of aircraft and their classification are provided in Chapter 4. A diagram of aircraft with major subassemblies as components is provided herein. Indeed, aircraft design has become highly modular in the interests of the 'family' concept, which facilitates low development cost by maintaining a high degree of parts commonality.

Aircraft span, length, and height are currently restricted by the International Civil Aviation Organisation (ICAO) to 80 m, 80 m and 80 ft, respectively, for ground handling and storage considerations. The height is in feet but the span and length are in metres; this restriction may change. Section 1.6 highlights the mix of SI and foot–pound–second (FPS) units in aerospace engineering.

2.3.1   Civil Aircraft and Its Component Configurations

In general, the civil aircraft category includes five types: (i) small club trainers, (ii) utility aircraft, (iii) business aircraft, (iv) single aisle narrow-body commercial transporters (regional aircraft to midsize) and (v) double aisle wide-body large transporters. The various types of available configuration options are described in Chapters 4–6.

2.3.1.1   Subsonic Jet Aircraft

The typical subsonic commercial jet aircraft structural components subsections are shown in Figure 2.1 (Lockheed L1011). These consist of typically, wing, fuselage, nacelle and empennage; others (e.g. winglets, strakes and auxiliary control surfaces) are less obvious but play vital roles – otherwise, they would not be included. Because there are many options, components are associated in groups for convenience, as described in the following subsections.

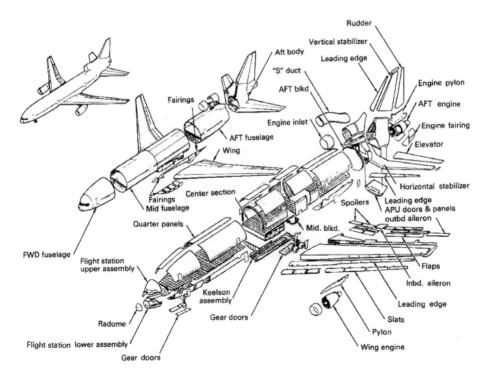

Figure 2.1 Lockheed 1011 blowout diagram. Source: Courtesy of Michael Niu – [24].

1. *Fuselage group.* This group includes the nose cone, the constant midsection fuselage, the tapered aft fuselage and the tail cone. The fuselage belly fairing (shown in Figure 2.1 as several subassembly components below the fuselage) may be used to house equipment at the wing-fuselage junction, such as the undercarriage wheels.
2. *Wing group.* This group consists of the main wing, high-lift devices, spoilers, control surfaces and tip devices. The example of Lockheed L1011 has low wing configuration with the structural wing box passing through the fuselage belly. High-lift devices include leading-edge slats or trailing edge flaps. The leading-edge slats are shown attached to the main wing and the trailing edge flaps and spoilers are shown detached from the port wing. Spoilers are used to decelerate aircraft on descent; as the name suggests, they 'spoil' lift over the wing and are useful as 'lift dumpers' on touchdown. This allows the undercarriage to more rapidly absorb the aircraft's weight, enabling a more effective application of the brakes. In some aircraft, a small differential deflection of spoilers with or without the use of ailerons is used to stabilise an aircraft's rolling tendencies during disturbances. Winglets (not in the figure) are one of a set of tip treatments that can reduce the induced drag of an aircraft.
3. *Empennage group.* The empennage is the set of stability and control surfaces at the back of an aircraft. The Lockheed example, as shown in Figure 2.1, has vertical tail (known as the *V-tail*) split into a fin in the front and a rudder at trailing edge, with an end cap on the top. The horizontal tail (known as the *H-tail*) is attached to the fuselage as low tail configuration consisting of the fixed stabiliser and the movable elevator at the trailing edge.
4. *Nacelle group.* Podded nacelles are slung under the wings and one is mounted on the aft fuselage; pylons affect the attachment. Engines can be mounted on each side of the fuselage. The nacelle design is discussed in detail in Chapter 12. Turbofans are preferred for higher subsonic speed.
5. *Undercarriage group.* The undercarriage, or landing gear, usually consists of a nose-wheel assembly and two sets of main wheels that form a tricycle configuration. Tail-dragging, bicycle and even quad configurations are possible, depending on the application of an aircraft. Wheels are usually retracted in flight, and the retraction mechanism and stowage bay comprise part of the undercarriage group. Undercarriage design is discussed in Chapter 9.

Not shown in Figure 2.1 are the trimming surfaces used to reduce control forces experienced by the pilot. During the conceptual phase, these surfaces generally are shown schematically, with size based on past experience. The sizing of trim surfaces is more appropriate once the aircraft configuration is frozen (i.e. a Phase II activity). Trim-surface sizing is accomplished by using semi-empirical relations/computational fluid dynamics (CFD) analyses and is fine-tuned by tailoring the surfaces and areas or adjusting the mechanism during flight trials. In this book, trim surfaces are treated schematically – the main task is to size the aircraft and finalise the configuration in Phase I. On larger aircraft, powered controls are used; pitch trimmings in conjunction with moving tail planes.

2.3.2 Turboprop Aircraft

A cutaway diagram of a propeller-driven Bombardier Dash 8-300 turboprop aircraft is shown in Figure 2.2. Propeller-driven aircraft speeds are below Mach 0.5. Some modern aircraft with advanced propeller design have the capability to cruise above Mach 0.6. Turboprop structural components subsections groups are similar to subsonic jet aircraft shown in Figure 2.1 and are associated in groups for convenience, as described in the following subsections.

1. *Fuselage group.* These have a near circular cross-section fuselage. This group includes the nose cone, the constant midsection fuselage, the tapered aft fuselage and the tail cone. It is also assembled from components in the similar manner to that shown in Figure 2.1.
2. *Wing group.* Figure 2.2 shows the high wing configuration with the structural wing box passing over fuselage and the external surface is faired to make it streamlined. This group consists of the main wing, high-lift devices, spoilers, control surfaces and tip devices.

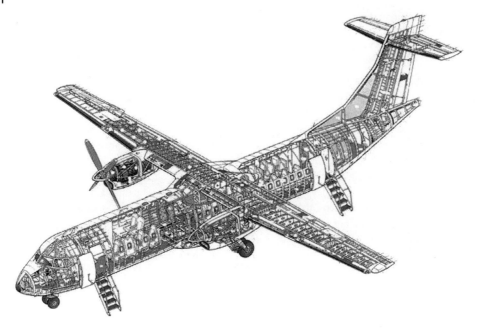

Figure 2.2 ATR 72 turboprop aircraft cutaway diagram.

3. *Empennage group.* Figure 2.2 shows the empennage has the horizontal tail as a T-tail (see Section 6.4) set at the top of the V-tail, and consists of the stabiliser and the elevator.
4. *Nacelle group.* Podded nacelles are slung under the wings without a pylon. The nacelle houses the undercarriage.
5. *Undercarriage group.* These have a long undercarriage retracted into the turboprop nacelle pod. Undercarriage design is discussed in Chapter 9.

2.3.3 Military Aircraft and Its Component Configurations

Table 1.3 in Section 1.9.1 compared the difference between civil and military aircraft design and operation. On account of its mission requirements military configurations are more diverse than civil designs. Figures 2.3 depicts a blowout diagram of the General Dynamics (now Boeing) F16 showing the internal structural layout.

Due to design differences on account of the mission role (Table 1.3), the combat aircraft structural subcomponent groups differ from the civil aircraft structural subgroups as described earlier. This is mainly evident in fuselage structure, which invariably houses the power plant (except large bombers and transport aircraft), unlike commercial civil aircraft designs. Combat aircraft structural components subsections are grouped as follows. The densely packed fuselage design typically has three subgroups as follows.

1. *Front-fuselage sub-group* (*Military*). The front fuselage starts with the nose cone, which has to be pointed for supersonic capability, and then houses a radar that could be of around 1 m diameter. The nose cone can be swung open to access for radar maintenance. The cone section is attached to flight deck module to house the flight crew, instruments, system black boxes and so on, all requiring relatively more frequent inspection and maintenance. The nose undercarriage is attached to this section. It may or may not have the intake, depending on the design.

   The military aircraft pilot seat has more freedom to recline so as to shorten carotid artery height to reduce blood starvation to the brain at high *g* manoeuvres that causes pilot blackouts.
2. *Mid-fuselage group* (*Military*). This is a complicated structural subgroup that bears the wing loads, air-brake loads and the main undercarriage loads. Wing are attached to this section and it houses the

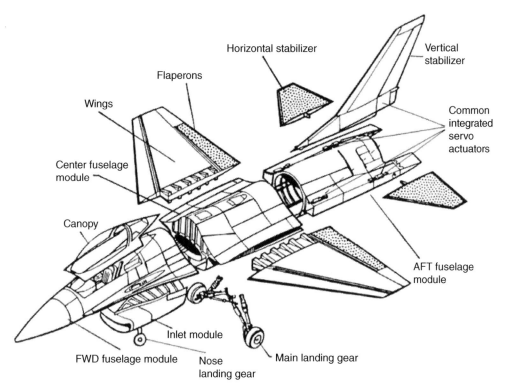

Figure 2.3 Schematic blown-out diagram of Boeing F16 showing the internal structural layout. Source: Courtesy of Michael Niu – [24].

main undercarriage as well as having a hollow intake duct for the engines. Practically all the cable, fuel and oil pipes, pneumatic ducts and linkages go through this section. Narrowing of fuselage section for area-ruling in some designs is done. This makes military aircraft mid-fuselage design considerably more complex compared to hollow constant cross-section transport aircraft mid-fuselage design considerations.

The fuselage belly fairing would house accessories; in most cases, the undercarriage. Current tendencies for the wing-body fairing with considerable blending for superior aerodynamic considerations, for example, to improve lift to drag ratio and fly at higher angle of attack. In a blended fuselage it is hard to isolate the fuselage, possibly a convenient choice would be where the wing root is attached.

3. *Aft-fuselage sub-group*. The fighter aircraft fuselage would invariably house the power. Fuselage aft ends up as engine exhaust system and therefore will not have closure as in civil design. In a case where the engine dangles below the fuselage spine (F4 Phantom), then a pointed aft end closure follows. Power plants require periodic maintenance for which the aft fuselage needs to be spilt or the mid-fuselage to be opened to access the engine, especially if it has to be taken out.

4. *Lifting surfaces group* (*Military*). In military aircraft design wing, stability and control surfaces group are dealt with together as some designs have both empennage and canard all working together for extreme manoeuvres during combat, unlike civil aircraft designs **Wing**. The evolution of fighter aircraft shows the dominant delta or short trapezoidal wing planform. This is for the obvious reasons of having high leading-edge sweep; a low aspect ratio to negotiate high g manoeuvres would generate a high wing root bending moment. It could restrict span growth but encourage a large wing root chord of delta or trapezoid shape with strake planform as in the cases of the F16, F18 and so on. For control reasons, it could have additional surfaces. The following are the configuration choices (strakes are taken as part of the wing). Section 4.17 describes in detail with illustrations the following types of configurations.

i) *One-surface configuration.* Pure delta planform or its variation – the trailing edge of the delta like wing can be made to work like H-tail as an integral part of the wing (Mirage 2000).

ii) *Two-surface configuration.* Two-surface configuration has two possibilities – tail in back or tail in front. Delta like wing or trapezoidal wing with conventional H-tail for pitch control (MIG 21). In some designs the H-tail is replaced by a canard surface for pitch control in relaxed stability (has destabilising effect). Two-surface configuration with strake is shown in. Variants are double delta (SAAB Viggen).

iii) *Three-surface configuration.* The ultimate kind is of three-surface configuration. It has wing, H-tail in aft end and canard in the front end (Sukhoi Su37).

5. *Empennage group.* Combat aircraft empennage shaping and sizing is a complex procedure (Section 6.15.6), primarily on account of short tail arm and the need to fly in relaxed stability to execute fast and hard manoeuvres. The B2 apparently appears to be without a tail. The F22 has a large canted V-tail. Options for control surface configuration are be shown along with wing options. The delta wing has an H-tail integrated with it. This book will adhere to the conventional configuration of H-tail and V-tail for the trainer aircraft example. Modern designs deploy tailerons (a stabilator – see Section 16.9.1) to initiate pitch and roll control by the H-tail.

6. *Intake group.* Instead of pod mounted engines, military aircraft have engines embedded into the fuselage with integrated intake. Military aircraft intake design is a complex procedure as the power plant is kept within the fuselage unlike the simpler pod mounted configuration of civil aircraft. Therefore, instead of having a nacelle, air-breathing intakes become an integral part of the fuselage.

These six groups of aircraft components offer the preliminary shape of candidate combat aircraft configurations. Eventually, after the wing sizing and engine matching exercise, the choice for configuration has to be narrowed down to one that would offer the best choice for the mission. The family derivatives of military aircraft are quite different, again depending on the mission role, for example, use of additional crew, trainer version, carrier borne version, longer range version, improved variant version and so on. Undercarriage information is dealt with separately in Chapter 9.

Military configuration study would also require some iterations to position the empennage and undercarriage with respect to the wing, as initially the centre of gravity (CG) position is not known. Weights are estimated from a provisional positioning and then the positions fine-tuned through iterations when the CG is known. In a classroom exercise, one iteration suffices.

The role of the canard in military applications is quite different from that of the role in civil aircraft designs. It has been found that strakes can also provide additional vortex lift and fast responses to pitch control with a conventional tail. The choice for strake or canard is still not properly researched in the public domain. It is interesting to note that US designs have strakes while the European ones have a canard. A detailed study of aircraft control laws and fly-by-wire (FBW) system architecture is required to make the choice. Up until the 1990s, flaws in FBW software caused several serious accidents.

Wing attachment to fuselage varies from case to case. The leading edge can have slats and the trailing edge would invariably have flaps. Centrally mounted large air brakes to decelerate have practically eliminated the role of spoilers. Landing in a shorter airfield may require deployment of a brake parachute.

Since military aircraft are expected to encounter transonic flight, aircraft cross-sectional area distribution becomes an important consideration. A seamless smooth distribution of cross-section (area-rule) is explained in Section 3.13.

Again, it is emphasised that this book is introductory in nature. Due to not having enough information on modern fighter design considerations, the author restricts military aircraft design exercises to a trainer class of aircraft. In this book, a military trainer in the class of Royal Air Force (RAF) Hawk is dealt with. An example of an Advanced Jet Trainer (AJT) with a Close Air Support (CAS) variant is described and worked on as a military trainer aircraft design, greatly simplifying the objective for military aircraft design. The readers will find that there is a lot to learn from this class of aircraft to have a feel for military aircraft design considerations.

2.4 Typical Aircraft Design Process

The typical aircraft design process follows the classical systems approach pattern. The official definition of a *system*, adopted by the International Council of Systems Engineering (INCOSE) [25] is: 'A system is an interacting combination of elements, viewed in relation to function'. The design system has an input (i.e. a specification or requirement) that undergoes a process (i.e. phases of design) to obtain an output (i.e. certified design through substantiated aircraft performance), as shown in Figure 2.4.

In the following are the definitions of the terms used in Figure 2.4.

2.4.1 Input

The *project identification* process through year-round market study (in consultation with potential operators) and research for the extent of advanced technologies to be adopted) to establish a new aircraft project within a company's capability to successfully compete in the market in order to grow, if not sustain. Market study presents the requirements/specification of the new aircraft presented to management to consider.

As for subsystems, the components of an aircraft are interdependent in a multidisciplinary environment, even if they have the ability to function on their own (e.g. wing-flap deployment on the ground is inert, whereas in flight it affects vehicle motion). Individual components such as the wings, nacelle, undercarriage, fuel system and air-conditioning also can be viewed as subsystems. Components are supplied for structural and system testing in conformance with airworthiness requirements in practice. Close contact is maintained with the planning engineering department to ensure that production costs are minimised, the schedule is maintained and build tolerances are consistent with design requirements.

Chart 2.1 suggests a generalised functional envelope of aircraft design architecture as a system, which is in line with the Aircraft Transport Association (ATA) index [26] for commercial transport aircraft. Further descriptions of subsystems are provided in subsequent chapters.

Extensive wind tunnel, structure and aircraft systems testing are required early in the design cycle to ensure that safe flight tests result in airworthiness certification approval. The multidisciplinary systems approach to aircraft design is carried out within the context of integrated product and process development (IPPD). Four phases comprise the generic methodology (discussed in the next section) for a new aircraft to be conceived, designed, built and certified.

Civil aircraft projects usually proceed to preproduction aircraft that will be flight-tested and sold, whereas military aircraft projects proceed with technical demonstrations of prototypes (technology demonstrator) before the go-ahead is given. The technologies demonstrators are typically scaled down aircraft meant to substantiate cutting-edge technologies and are not sold for operational use.

2.5 Market Survey – Project Identification

A very important part of manufacturers is connecting their capabilities to customers, more so for capital intensive product like aircraft. Market survey plays a central role as a communication tool.

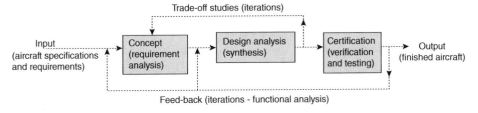

Figure 2.4 Aircraft design process (see 2.1).

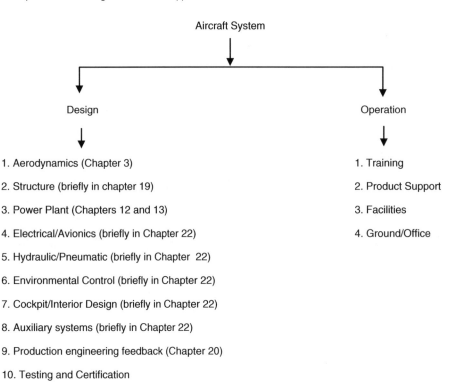

Chart 2.1 Aircraft system.

Typically, market survey and analyses are carried out by a separate marketing group, which may be outside the aircraft design bureau but stays in close interaction to arrive at a product line to be considered by management as a candidate project to be undertaken. The marketing group stays in touch with potential customers (e.g. airline operators) and tries to harmonise diverse requirements to consolidate a product in a family concept of designs catering for a wide range of demand. This chapter attempts to introduce to the readers what is expected from the design engineers. It is recommended that class may conduct a mock market survey to arrive at a project to be undertaken as a classroom exercise.

In a free market economy, an industry cannot survive unless it grows. In a cost intensive civil-market economy, governmental sustenance is only a temporary relief. The starting point to initiate a new aircraft design project is to establish the key drivers – that is, the requirements and objectives based on market, technical, certification and organisational requirements. These key drivers are systematically analysed and then documented by aircraft manufacturers (Chart 2.2).

Documents, in several volumes, describing details of the next layer of design specifications (requirements), are issued to those organisations involved with the project. Market survey is a way to determine customer requirements; user feedback guides the product. In parallel, the manufacturers incorporate the latest but proven technologies to improve design to stay ahead of the competition, always confined within the financial viability of what the market can afford. Dialogue between manufacturers and operators continues all the time to bring out the best in the design.

Military product development has a similar approach but would require some modifications to Chart 2.2. Here, the government is both the single customer and the regulatory body. Therefore, competition is only between the bidding manufacturers. The market is replaced by operational requirements arising from perceived threats from potential adversaries. Column 1 of Chart 2.2 then becomes 'operational drivers' that include weapons management, counter intelligence and so on. Hence, this section on market survey is divided into civil and military customers as shown in Chart 2.3. 'Customer' is a broad based term and is defined in this book in the manner given in the chart.

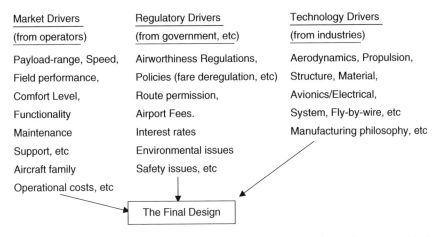

Chart 2.2 The design drivers (in the free market economy, a design has to face competition).

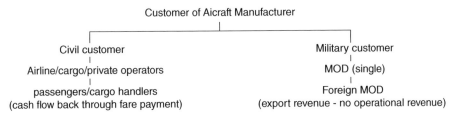

Chart 2.3 Customers of an aircraft manufacturer.

In the UK military, the Ministry of Defence (MoD), as the single customer, searches for a product and circulates a Request for Proposal (RFP) to the national infrastructure, where most manufacturing is run privately. It is similar in the United States, but using a different terminology. The product search is a complex process – the MoD must know a potential adversary's existing and future capabilities and administrate national research, design and development (RD&D) infrastructures to be ready with discoveries and innovations to supersede an adversary's capabilities.

The air staff target (AST) is an elaborate aircraft specification as a customer requirement. A military project is of national interest and, in today's practice, capable companies are invited to first produce a technology demonstrator as proof of concept. The loser in the competition is paid by the government for the technology demonstrator aircraft and learns about advanced technology for the next RFP or civilian design. Therefore, in a sense, there is no loser and the nation hones its technical manpower.

Although it is used, the authors do not think an 'RFP' is appropriate terminology in civilian applications: Who is making the request? It is important for aircraft manufacturers to know the requirements of many operators and supply a product that meets the market's demands in performance, cost and time frame. Airline, cargo and private operators are direct customers of aircraft manufacturers, which do not have direct contact with the next level of customers (i.e. passengers and cargo handlers – see Chart 2.3). Airlines do their market surveys of passenger and freight requirements and relay the information to manufacturers. The surveys often are established by extensive studies of target-city pairs, current market coverage, growth trends and passenger input. Inherent in the feedback are diverse requirements that must be coalesced into a marketable product. A major order from a single operator could start a project, but manufacturers must cater to many operators to enlarge and stabilise their market share. The civilian market is searched through a multitude of queries to various operators (i.e. airlines), both nationally and internationally. In civil aviation, the development of the national infrastructure must be coordinated with aircraft manufacturers and operators to ensure national growth. Airlines generate revenue by carrying passengers and freight, which provide the cash flow that supports the maintenance and development of the civil aviation infrastructure. Cargo generates important revenues for airlines and airports and the market for it should not be underestimated – even if it means modifying older aeroplanes. Manufacturers and operators are in continual contact to develop product lines with new and/or modified aircraft. Aircraft manufacturers must harmonise the diversity in requirements such that management decides to undertake a conceptual study to obtain the go-ahead. There is nothing comparable to the process taken by the MoD to initiate an RFP with a single customer demand.

The private or executive aircraft market is driven by operators that are closely connected to business interests and cover a wide spectrum of types, varying from four passengers to specially modified midsized jets.

In its lifespan, military aircraft utilisation in peacetime is approximately 7500 hours, about one-tenth that of commercial transport aircraft (i.e. $\approx$75 000 hours). Annual peacetime military aircraft utilisation is low (i.e. $\approx$600 h yr^{-1}) compared to annual civil aircraft utilisation, which can exceed 3000 h yr^{-1}. Although derived from the same scientific basis, the aircraft design philosophy for the military and civil aircraft differs on account of their mission role (see Section 1.9.1, Table 1.3).

2.5.1 Civil Aircraft Market – Product Identification

Following up on the review in Chapter 1 about the current status of the civil aircraft market, this section describes how to generate aircraft specifications that will help to sell the product and generate a profit. The coursework starts here with a mock (i.e. representative) market survey leading to what must be designed – that is, the conception of the aircraft, the Phase I obligations.

Input from operators to manufacturers is significant and varied. The manufacturer needs to group the requirements intelligently in a family of aircraft sizes and capabilities. It is necessary to cover as much ground as the market demands, yet maintain component commonalities in order to lower development costs of the derivative aircraft in the family. This book lists only those market parameters that affect aircraft aerodynamic design, the most important being the payload-range capability of the aircraft, which has the greatest influence in shaping the aircraft. Details of other requirements (e.g. systems requirements, maintenance and passenger services) are not discussed here but are briefly introduced.

For the mock market studies, students may be asked to table aircraft requirements as if they are representing the airlines' interest. It is understandable that they may list requirements that are not practicable. It is therefore the instructor's responsibility to provide reasons for discarding each impracticable point and then coalescing the remainder into a starting point. Section 2.8 suggests interesting cases for coursework experience. There is a wide variety of civil aircraft in operation; in the following are the requirements for the three classes addressed in the scope of this book.

There are various models of new aircraft design process published in public domain, most of them suggested by researchers/academics. Unfortunately, the methodologies adopted in industry are not readily available. While the published materials offer some insight, the question remains on how effective the suggested research publications are in the aircraft industries.

During the early 1970s, the Japanese innovative management practices known as the Quality Function Deployment (QFD) [27] showed remarkable success in manufacturing process to become world leaders in bringing out quality products that made enormous gains in their export market. During the 1980s, the USA made extensive studies on the Japanese approach and presented an approach road with a tool known as the House of Quality (HOQ) [28] that can be applied in the western environment.

The HOQ is a management tool applied at the conceptual stages of product development that maps attributes of the parameters, cascading down the phases of design. The method brings out the perceived importance of various parameters to set targets for design. The HOQ presents a series of matrices in diagram form called a House of Quality on account of having a roof-like structure in its top. This house can be divided in 'rooms', representing the attributes to consider, for example, design requirements, their relative importance, customer feedback and so on. The method demonstrated improvements for the organisations those who are in the consumer market, where customers are the general public and the manufacturers bring out products perceived as having customer appeal backed up by aggressive advertisement.

However, the aerospace market differs from the consumer market. There are limitations in applying the QFD/HOQ method in high-technology with very high investment and with low production volume of high cost aerospace products. The main customers in aerospace industries are the (airline) operators and general public feedback to manufacturing industries comes through operators. The (airline) operators have in-house engineers, some with design experience from industries and are knowledgeable about their requirements. Aircraft manufacturers are in constant dialogue with the operators to explore what kind of new advanced technologies can be introduced to stay ahead of competition, these serving as the drivers to lay down the new aircraft specifications meeting the mission requirements, not overlooking public opinion. Here, the HOQ analyses may prove useful by the marketing department of aircraft manufacturers to incorporate some of the user requests. Military aircraft design office starts with the specifications laid by the MoD in their RFP. A technology demonstrator will refine the advanced technology to be incorporated in the new design.

The aircraft design office is supplied by the new aircraft project requirements/specifications. It is suggested that a mock survey in a classroom exercise may avoid HOQ analyses. Once specifications are finalised, the search is constrained with little scope for the aircraft design bureau to apply the QFD/HOQ tool. Each requirement is of importance in the competitive high investment endeavours, setting relative importance may be counterproductive in a fiercely competitive environment. Industry management may take an interest in QFD/HOQ, primarily aimed at production planning but the design office have their set procedure developed over time. It is for this reason that this book bypasses dealing with QFD/HOQ in new aircraft project design phases as described in Section 2.6. (Applying QFD/HOQ in aircraft design phases may have limitations but it is by no means a flawed system as the consumer market has shown uniform gains by adopting it. Readers are recommended to explore related publications in the public domain [27, 28].)

After the market demand and manufacturer's capabilities are established, the new aircraft design starts to progress with the following considerations.

1. Candidates' aircraft configurations are presented for management review to cater for the diverse range customer requirements that can be combined into a family of designs retaining component commonality as a cost reduction measure, each variant catering for the needs of the customer for their mission requirements. The technology level to be adopted in all areas of the project has to be established, both for design and manufacture.
2. The most suited configuration from the candidates' designs is taken up with a view towards obtaining the go-ahead for the project. After the go-ahead, it then goes through detailed analyses to ensure aircraft performance, handling criteria, structural and systems integrity complying with airworthiness standards. A family of variant designs are also offered to cover the wider market at a lower cost.
3. Establish the human interface for both the crew and customer for ease of operation of the bought-out items offering best value for money as well as being low weight.
4. Management strategies to keep cost low and remain competitive. Efforts are made to make the product right first time at every stage of progress. Extensive risk analyses are made.
5. Establish the manufacturing philosophy and ensure quality.
6. Ensure maintainability, training, support for customers and disposal at the end of aircraft life.

2.6 Four Phases of Aircraft Design

Given next are the generic details of the new aircraft design process, in four phases. At the end of the standalone Phase I of the new aircraft project devoted to conceptual study, after the approval to go-ahead is obtained, the project enters the 'point of no return'. The next three phases run continuously without any marked delineation, the beginning of next stage overlapping the end of the previous. Industrial practices may vary from company to company but the underlying theme is nearly the same, as given in the next section.

Concept design phase (*Phase I*). This is the *project definition* and *project finalisation* phase, conducted by a small number of experienced engineers in a dedicated group to quickly present a viable aircraft configuration and performance capabilities at a credible accuracy level, sufficient to guarantee the operators what to expect, to the management to decide to give a go-ahead or not. Naturally, budget allocation at this stage is kept low due to the uncertainty in advancing the project. This is the subject matter of this book.

Design analyses. This consists of two phases (Phases II and III) as described next.

i) Project Development Phase (Preliminary Design) – Phase II

This starts after go-ahead for the new aircraft project is obtained when management commits with extended budgetary provision to make detailed analyses. The project is now committed as a point of no return with an expectation of an *investment return*. This has two stages as follows. In this phase, the prediction of aircraft capability is fine-tuned to high accuracy with detailed computer-aided engineering (CAE) analyses and extensive aerodynamic, structure and system tests. Detailed structural design starts now (details in Sections 2.6 and 2.7).

ii) Detailed Design Phase (Full-Scale Product Development) – Phase III

In this phase, for civil aviation, the preproduction or prototype, and for combat aircraft scaled down technology demonstrator building, is accelerated (details in Sections 2.6 and 2.7).

Certification/verification (*Phase IV*). This is the final phase to get preproduction fight tested and certified for operation (details in Sections 2.6 and 2.7).

Aircraft manufacturers conduct year-round exploratory work on research, design and technology development as well as market analysis to search for a product. A new project is formally initiated in the four phases shown in Chart 2.4, which is applicable for both civil and military projects.

Among organisations, the terminology of the phases varies. Chart 2.4 offers a typical, generic pattern prevailing in the industry. The differences among terminologies are trivial because the task breakdown covered in various phases is approximately the same. For example, some may see the market study, specifications and requirements as Phase I and the conceptual study as Phase II; others may define the project definition phase (Phase II) and detailed design phase (Phase III) as the preliminary design and full-scale development phases, respectively. Some prefer to invest early in the risk analysis in Phase I to get more information and delay go-ahead; however, the general trend is to make risk analyses in Phase II when the design is better defined, thereby saving the Phase I budgetary provisions in case the project fails to obtain the go-ahead. A military programme may require early risk analysis because it would be incorporating technologies not yet proven in operation. Some may define disposal of aircraft at the end as a design phase of a project.

Adhering to the general pattern, companies conduct the design phases as evolved over time and became ingrained to their practices. It is for this reason, the topic is not elaborated on any further. A new employee should be able to sense the pulse of organisational strategies and practices as soon joining a company.

Company management establishes a dedicated Design-Build-Team (DBT) to meet at regular intervals to conduct design reviews and make decisions on the best compromises through multidisciplinary analyses (MDA) and multidisciplinary optimisation (MDO), as shown in Chart 2.5; this is what is meant by an IPPD (i.e. concurrent engineering) environment.

The conceptual phase of aircraft design is now conducted using a multidisciplinary approach (i.e. concurrent engineering), which must include manufacturing engineering and an appreciation for the cost implications of early decisions in IPPD environment. As mentioned in Chapter 1, the chief designer's role has changed from *telling* to *listening*; he or she synthesises information and takes full command if and when differences of opinion arise. Margins of error have shrunk to the so-called zero tolerance so that tasks are done right the first time; the Six-Sigma approach is one management tool used to achieve this end.

2.6.1 Understanding Optimisation

Specialist areas may optimise design goals, but in an IPPD environment, compromise must be sought. It is emphasised frequently that optimisation of individual goals through separate design considerations may prove counterproductive and usually prevents the overall (i.e. global) optimisation of ownership cost. MDO offers

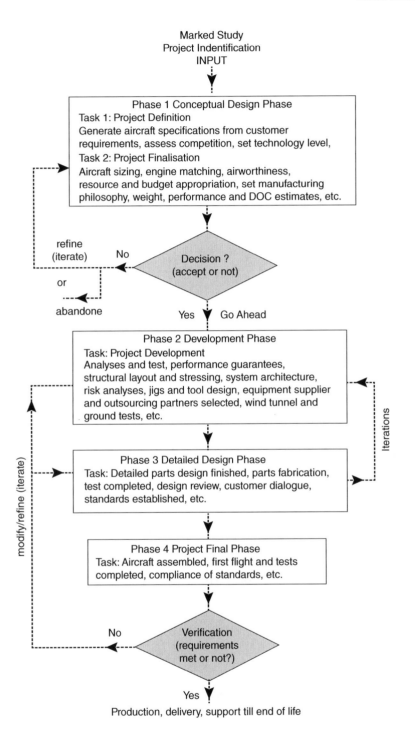

Chart 2.4 Four phases of the aircraft design and development process.

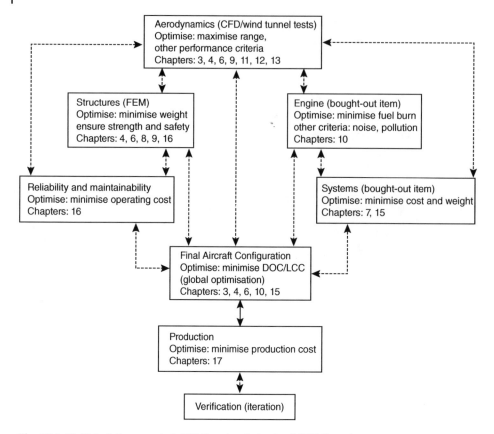

Chart 2.5 Multidisciplinary analysis (MDA) and optimisation (MDO) flow chart.

good potential but it is not easy to obtain global optimisation; it is still evolving. In a way, global MDO involving many variables is still an academic pursuit. Industries are in a position to use sophisticated algorithms in some proven areas. An example is reducing manufacturing costs by reshaping component geometry as a compromise – such as minimising complex component curvature. The compromises are evident in offering a family of variant aircraft because none of the individuals in the family is optimised, whereas together, they offer the best value. In other words the aim is to have a 'satisfied' rather than an 'optimised' design.

Industry is aware of the importance of optimisation but in the design practice, simple parametric optimisation taking one variable at a time yields a satisfactory result. It is for this reason this book does not deal with any optimisation process unless it is supported by worked-out examples substantiated with existing designs.

2.6.2 Typical Resources Deployment

All phases do not work under uniform manpower loading; naturally, Phase I starts with light manpower during the conceptual study and reaches peak manpower (100%) at Phase III; it decreases again when flight testing starts, by then the design work is virtually done and support work continues.

Figure 2.5 is a typical distribution of cost and manpower loading (an average percentage is shown); the manpower loading forecast must be finalised during the Phase I study. The figure also shows the cumulative deployment. At the end of a project, it is expected that the actual figure should be close to the projected figure. Project costs consist primarily of salaries (most of the cost), bought-out items, and relatively smaller miscellaneous amounts (e.g. advertising, travel and logistics). Chain lines in Figure 2.5 illustrate the cost-frame outlay.

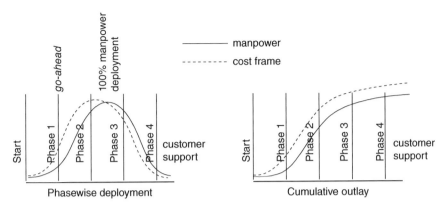

Figure 2.5 Resource deployments (manpower and finance).

2.6.3 Typical Cost Frame

A crude development cost, up to certification (in the year 2015 in US Dollars), is shown in Table 2.1. Typical unit aircraft costs by class are also given (there is variation among companies). A substantial part of the budget is committed to Phase I.

2.6.4 Typical Time Frame

Typical time frames for the phases of different types of projects are shown in Chart 2.6. All figures are the approximate number of months. Exploratory work continues year-round to examine the viability of incorporating new technologies and to push the boundaries of company capabilities – which is implied rather than explicit in Chart 2.6.

When an aircraft has been delivered to the operators (i.e. customers), a manufacturer is not free from obligation. Manufacturers continue to provide support with maintenance, design improvements and attention to operational queries until the end of an aircraft's life. Modern designs are expected to last for three to four decades of operation. Manufacturers may even face litigation if customers find cause to sue. Compensation payments have crippled some well-known general aviation (GA) companies. Fortunately, the 1990s saw a relaxation of litigation laws in GA for a certain period after a design is established, a manufacturer's liabilities are reduced; this resulted in a revitalisation of the GA market. Military programmes involve support from 'cradle to grave'.

This emphasises that the product must be done right the first time. Midcourse changes add unnecessary costs that could be detrimental to a project – a major change may not prove sustainable. Procedural methodologies such as the Six-Sigma approach have been devised to ensure that changes are minimised.

Table 2.1 Development costs up to certification included (cost at 2015 prices).

Aircraft class (turbofan)	Development cost (US$[a])	Unit cost (US$*)
6-passenger general aviation aircraft	7–12 million	$\approx \pm 2$ million
10-passenger business aircraft	25–40 million	5–10 million
50-passenger regional aircraft	60–100 million	25–30 million
150-passenger midsized aircraft	200–500 million	50–70 million
500+ - passenger large aircraft	2–10 billion?	200–300 million
Military combat aircraft (high end)	5–15 billion?	100+ million?

a) Does not include production launch cost.

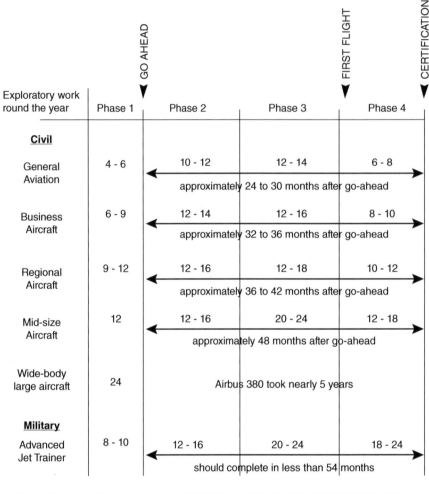

Chart 2.6 Typical project time frame.

2.7 Typical Task Breakdown in Each Phase

Typical task obligations in each phase of civil aircraft design are defined in this section. Military aircraft designs follow the same pattern but more rigorously. Military aircraft must deal with new technologies, which could still require operational proving; therefore, there is uncertainty involved in military aircraft projects.

Phase I: Conceptual Study Phase (Feasibility Study)

Much of the work in the conceptual study phase can be streamlined through a good market study to identify a product line within a company's capabilities. In this phase, findings of the market study are developed with candidate configurations; the technology to be adopted is firmed up and the economic viability is finalised. This is accomplished through aircraft sizing, engine matching, preliminary weight estimation, and evolution of a family of aircraft with payload and range combinations (i.e. aircraft performance) for all configurations.

Planning portfolios with budgetary provisions, manpower requirements, progress milestones, potential subcontract/risk-sharing partners' inputs and so forth are included as the starting point of the design process. In general, at the end of this phase, management decision for a go-ahead is expected with a final configuration selected from the candidate configurations offered. Continuous interaction with potential customers (i.e. operators and subcontractors) occurs during this phase, with the objective of arriving at a family of aircraft as the most 'satisfying' design with compromises rather than an 'optimum' solution. Management may request a level of detail (e.g. risk analysis) that could extend the study phase or flow into the next phase, thereby delaying the go-ahead decision to the early part of Phase II. This is likely if the candidate aircraft configurations are short-listed instead of finalised. For those designers who have planned ahead, Phase I should finish early – especially if they are well versed in the product type and have other successful designs in their experience.

Phase II: Project Definition Phase (Preliminary Design)

This phase begins after the go-ahead has been given to a project, and a 'point of no return' is reached during this phase. Project definition sometimes may overlap with the detailed design phase (i.e. Phase III). During the advanced design phase, the project moves towards a finer definition, with a guarantee that the aircraft capabilities will meet if not exceed the specifications. Some iteration invariably takes place to fine-tune the product. Details of the technology level to be used and manufacturing planning are essential, and partnership outsourcing is initiated in this phase. Procurement cost reviews and updates also are ongoing to ensure that project viability is maintained. Many fine aircraft projects have been stalled for lack of proper planning and financial risk management. The beginning of metal cutting and parts fabrication as well as deliveries of bought-out items (e.g. engine and avionics) must be completed in Phase II. In this phase, extensive wind tunnel testing, CFD analysis, detailed weights estimation, detailed structural layout and finite element method (FEM) analysis, system definitions, production planning and so forth are carried out. (Readers may study recent case histories of products such as the Swearingen SJ30 [now certified and under production] and the Fairchild–Dornier 928.)

Phase III: Detailed Design Phase (Full-Scale Product Development)

In this phase, manufacturers push towards completion – when peak manpower is deployed for the project. Normally, projects cannot sustain delay – time is money. All aspects of detailed design and systems architecture testing are completed in this phase. (The test rig is called an 'iron bird' – it simulates full-scale control and system performance.) At the end of Phase III, the aircraft assembly should near, if not achieve, completion.

Phase IV: Final Phase (Certification)

Phase IV must start with completion of the aircraft assembly for ground-testing of installed systems and other mandatory structural strength-testing to prepare for flight testing. In general, two to four aircraft are needed to complete nearly 200–800 flight testing sorties (depending on the type of aircraft) towards substantiation for certification of the airworthiness standard. At this stage, there should be no major setbacks because the engineers have learned and practised aircraft design well with minimal errors.

Each project has a characteristic timeline; this book uses a four-year project time. Remember, however, that some projects have taken more or less time. Section 2.4.2 is a detailed breakdown of a small aircraft project for a small or medium company. The authors recommend that similar detailed milestone charts be drawn for coursework projects to give an idea of the manpower requirements.

2.7.1 Functional Tasks During Conceptual Study (Phase I)

Because this book is concerned only with Phase I, it is important to delineate functional task obligations assigned to individual designers – also known as top-level definition. Market specifications should first be delineated to develop task content, as shown in Chart 2.5, for the top-level study of civil aircraft during Phase

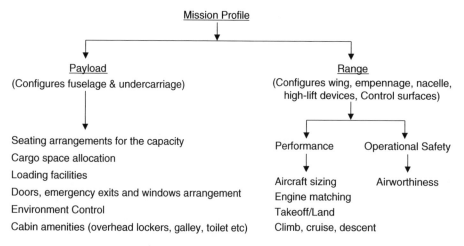

Chart 2.7 Top level definition (Phase I, conceptual study).

I. Payload determines the fuselage size and shape and leads into undercarriage design, depending on wing and engine positioning. Wing design largely determines the range, operational envelope and field-performance objectives. Considering all requirements together, the aircraft configuration evolves: there can be more than one candidate configuration (e.g. high or low wing, nacelle location and empennage arrangement). Chart 2.7 can be divided into functional work group activity to focus attention on specific areas – necessarily in an IPPD environment for MDA.

The military aircraft design approach is not significantly different except that the payload is the armament, which is generally under-slung or kept inside the fuselage bay. The mission profile is the sortie profile for the military role meeting the target range with the capabilities to return to base.

2.7.2 Project Activities for Small Aircraft Design

Progress monitoring is essential to keep an aircraft design project on schedule and within the budgetary provisions. To monitor progress, milestones are laid down in as much detailed work breakdown as possible as soon as the go-ahead is obtained. There are several ways to depict the work breakdown showing completion dates, all are nearly similar and the choice depends on the preferences of the manufacturer. By far the most popular and widely practised depiction is in the form bar chart known as Gantt chart. This lists activities and check points in convenient details corresponding to the planned time frame. The activities are listed next, phase-wise.

The project DBT meets regularly in an IPPD environment to monitor progress and streamline the way if any bottle-neck is envisaged. For any critical obstacle, the DBT meets as and when required to remove it. Cash flow and bought-out supply do not show in a Gantt chart, and finance and purchase departments are kept in the monitoring loop and have their own progress monitoring chart made to their convenience, which may not be in the form of a Gantt chart. In fact, all departments work together to achieve the goal to make the project a success.

Table 2.2 gives a schematic bar chart (Gantt chart), in short, of these tasks just to give an example. The readers are required to make a detailed Gantt chart, expanding all tasks to segmented detail to keep track on progress. Any bottlenecks must be foreseen by the respective line managers and corrective measures taken. Once again it is stressed that all drawings are made in CAD.

Typical work content and milestones for a small aircraft project are given here in blocks of time; readers need to expand this in bar chart form (the coursework involved in drawing the Gantt chart may alter the contents of the table, as required). Larger aircraft design follows similar activities in an expanded scale suited to task obligations.

Table 2.2 Gantt chart for the activities of the small aircraft project. Figures in parentheses are the time frames in months.

Months (readers are to prepare in details)		1 – 6 **Phase-1** (6 months)	7 – 18 **Phase-2** (15 months)	19 – 36 **Phase-3** (12 months)	37 – 42 **Phase-4** (9 months)
Conceptual studies, Phase-1 (6 months)					
CAD surface of aircraft	(6)				
Preliminary CFD/Wind tunnel	(4)				
Layout of internal structures	(6)				
Preliminary weights analysis	(3)				
Mock-up construction–prefer	(6)				
CAD layouts of:	(6)				
-Control system concept					
-Elecl/Avionics systems concept					
-Mechanical systems concept					
-Powerplant installation concept					
Data base for materials and parts	(4)				
Manufacturing/tooling philosophy and CAM	(6)				
Initial aircraft sizing/engine matching	(4)				
Freeze designs for management	(3)				
Logistics planning	(3)				
Budget planning	(6)				
GO-AHEAD					
Project Definition studies, Phase-2 (15 months)					
Detailed component drawings	(15)				
Stress analyses (FEM)	(15)				
Extensive CFD/Wind tunnel analyses	(12)				
Flutter analyses	(6)				
Detailed analyses of the following	(15)				
-Engine performance					
-Aircraft performance					
-Stability/Control analyses					
Cost analyses	(8)				
Standards requirements	(8)				
Issuance of production drawings	(12)				
Jigs/Tools design and set-up	(8)				
Start of component manufacture	(12)				
Place orders for bought-out items	(6)				
Performance guarantees	(6)				
Detailed Design, Phase-3 (12 months)					
(Some work runs in conjunction with Phase 2)					
Completion of all drawings(CAD)	(12)				
Completion of all aircraft components	(15)				
Receive all bought-out items	(6)				
Detailed weight/CGanalyses	(8)				
Fine tune aircraft/engine/stability	(12)				
Schedules and Check-lists	(6)				
Start Ground Tests	(12)				
Flight tests planning	(6)				
Prototype build status schedules	(6)				
Detailed cost analyses	(8)				
Start component sub-assemblies	(8)				
Certification, Final Phase-4 (9 months)					
(Some work runs in conjunction with Phase 3)					
Final prototype assembly and equipping	(6)				
Complete all ground tests(structure/systems)	(6)				
Design reviews	(3)				
Complete Flight tests	(6)				
Review analyses	(3)				
Certification	(6)				
Ready for customer delivery					
Man power loading			(maximum manpower)		

Phase 1 Conceptual Design – 6 months

1. Market Survey – establish aircraft specification from customer requirements. Information is extracted from year-round exploratory work.
2. Layout candidate aircraft configurations with the fuselage, then wing, empennage (canard if considered), undercarriage, power plant and so on.
3. Establish wing parameters as the most important component in synthesising aircraft. Select the aerofoil profile, its thickness to chord ratio, wing reference and planform areas, aspect ratio, wing sweep, taper ratio, wing twist, spar location, flap area, flight control and wing location with respect to the fuselage. The same considerations apply to all other lifting areas.
4. Initiation of computer-aided drawing (CAD) 3D surface modelling.
5. Preliminary CFD analysis to establish pressure distribution and loads on aircraft.
6. May conduct some preliminary wind tunnel tests.
7. Preliminary weights and CG estimates.
8. Aircraft preliminary drag estimate.
9. Size aircraft and match engine.
10. Establish engine data.
11. Preliminary aircraft and engine performance.
12. Freeze configuration to one aircraft along with a family of aircraft if considered.
13. Layout of internal structures and arrangement of fuselage interior.
14. Mock-up drawings, construction and initial evaluation.
15. Control system concept layout in CAD.
16. Electrical/avionics systems concept layout in CAD.
17. Mechanical systems concept layout in CAD.
18. Power plant installation concept in CAD.
19. Data base for materials and parts.
20. Plan for bought-out items and delivery schedule.
21. Planning for outsourcing, if any.
22. Preliminary cost projection.

Obtain management go-ahead

Phase II: Project Definition – 9 months

1. Integrated and component drawings in CAD.
2. FEM stress analysis of all components – wing, fuselage and so on.
3. Completion of mock-up and final assessment.
4. Advanced CFD analysis.
5. Wind tunnel model testing and CFD substantiation.
6. Flutter analysis.
7. Extensive and final aircraft and engine performance.
8. Detailed part design and issuance of manufacturing/production drawings in CAD.
9. Aircraft stability and control analysis and control surface sizing.
10. Finalise the control system design in CAD.
11. Finalise electrical/avionics systems design in CAD.
12. Finalise mechanical systems design in CAD.
13. Finalise power plant installation design in CAD.
14. Jigs and tool design.
15. Subcontracting, if any.
16. Place order for bought-out items and start receiving items.
17. Cost analysis.
18. Design review.
19. Continue with customer dialogue and updating (no change of specification).

Phase III: Detailed Design (Product Development) Phase – 12 months

1. Completion of detailed component design in CAD.
2. Completion of stress analysis.
3. Completion of CFD analysis.
4. Revise to final weights analysis.
5. Completion and issuance of all production drawings in CAD/computer aided manufacture (CAM).
6. Completion of production jigs and tools.
7. Parts manufacture and start of aircraft component subassembly.
8. All bought-out items received.
9. Standards, schedules and check lists.
10. Finalise ground/flight test schedules.
11. Prototype shop status schedules.
12. Revise cost analysis.
13. Start of ground tests.
14. Design review.
15. Continue with customer dialogue and updating (no change of specification).

Phase IV (Testing and Certification Phase) – 9 months

1. Final assembly and prototype equipping.
2. Completion of ground and flight tests and their analysis.
3. Review of analysis and, if required, modify design (aim to get it right first time).
4. Overall design review.
5. Review cost estimate.
6. Customer dialogue and sales arrangement.
7. Continue design review and support.

Note that production launch costs are normally kept separate from design and development costs. Total time to complete the project is three years (2½ from go-ahead). The time is tight but possible. It can be stretched to a four-year development programme for companies with limited resources.

2.8 Aircraft Specifications for Three Civil Aircraft Case Studies

It is recommended that the introductory coursework exercise uses one of the three specifications provided as a starting point. Accordingly, the initial follow-up activity is limited to work on the Learjet 45 class aircraft (see the second design specification).

Design Specifications of a Four-Seater Piston Engine Aircraft as a Baseline (Federal Aviation Regulation (FAR) 23)

Payload	4 passengers (including pilot) + baggage (e.g. two golf bags) = 4×85 (averaged) + 60 = 400 kg
Range	800 miles + reserve
Maximum cruise speed	Above 200 mph
Cruise altitude	Unpressurised cabin; approximately 10 000 ft (ceiling could be higher)
Takeoff distance	500 m @ sea level to 35 ft
Landing distance	500 m (at takeoff weight) @ sea level from 50 ft
Initial rate of climb	8 metres per second (m s^{-1})
Undercarriage	Retractable
Cabin comfort	Cabin heating, side-by-side seating, cabin interior width = 50 in.
Technology level	Conventional
Power plant	Piston engine

Derivative Version: A Lighter Two-Seater Light Club Trainer/Usage Aircraft (FAR 23).

(Derivatives are more difficult to develop for smaller aircraft because there is less room with which to work. Fuselage unplugging is difficult unless the baseline design made provision for it. There are considerable savings in certification cost.)

Payload	2 passengers plus light baggage = 200 kg
Range	400 miles + reserve
Maximum cruise speed	140 mph
Takeoff distance	300 m @ sea level to 35 ft
Landing distance	300 m (at takeoff weight) @ sea level from 50 ft
Initial rate of climb	$5\,\mathrm{m\,s^{-1}}$
Undercarriage	Fixed
Cabin comfort	Cabin heating, side-by-side seating, cabin interior width = 46 in.

(The other specifications are the same as in the baseline four-seater design.)

Derivative versions are achieved by shortening the wing root and empennage tips, unplugging the fuselage section (which is difficult if it is not a continuous section but is possible if the design of the baseline four-passenger aircraft considers this in advance), lightening the structural members, re-engineering to lower the power and so forth.

Design Specifications of a Baseline 8–10-Passenger (Learjet 45 Class) Aircraft (FAR 25).

Payload	8–10 passengers and 2 pilots + baggage
High comfort level	8×100 (averaged) + 300 = 1100 kg
Medium comfort level	10×80 (averaged) + 300 = 1100 kg
Range	2000 miles + reserve
Maximum cruise speed	Mach 0.7
Cruise altitude	Above 40 000 ft (ceiling over 50 000 ft)
Takeoff distance	1000 m @ sea level to 15 m
Landing distance	1000 m (at takeoff weight) @ sea level from 15 m
Initial rate of climb	$16\,\mathrm{m\,s^{-1}}$
Undercarriage	Retractable
Cabin Comfort	Pressurised cabin with air-conditioning and oxygen supply
Cabin interior width	=58 in.
Technology level	Advanced
Power plant	Turbofan engine

Shortened Derivative Version: Four To Six Passengers In a Baseline Aircraft Family (Far 25).

(This derivative works by unplugging continuous-section fuselage barrel on both sides of the wing.)

Payload	4–6 passengers and 2 pilots + baggage
High comfort level	4×100 (averaged) + 200 = 600 kg
Medium comfort level	$6 \times$ (averaged) + 120 = 600 kg
Range	2000 miles + reserve
Maximum cruise speed	Mach 0.7
Cruise altitude	Above 40 000 ft (ceiling over 50 000 ft)
Takeoff distance	800 m @ sea level to 15 m
Landing distance	800 m (at takeoff weight) @ sea level from 15 m

(The other specifications are the same as in the baseline design.)

Lengthened Derivative Version: 12–14 Passengers In the Baseline Aircraft Family (Far 25). (The longer derivative works in the same way by inserting continuous-section fuselage plugs on both sides of the wing.)

Payload	12–14 passengers and 2 pilots + baggage
High comfort level	12×100 (averaged) + 300 = 1500 kg
Medium comfort level	14×80 (averaged) + 380 = 1500 kg
Range	2000 miles + reserve
Takeoff distance	1200 m @ sea level to 15 m
Landing distance	1200 m (at takeoff weight) @ sea level from 15 m

(The other specifications are the same as in the baseline design.)

Design Specifications of a Baseline 150-Passenger (Airbus 320 Class) Aircraft (FAR 25)

Payload	150 passengers = 90×150 = 14 500 kg
Range	2800 nm (nautical miles) + reserve
Crew	2 pilots + 5 attendants
Maximum cruise speed	0.75 Mach
Cruise altitude	Above 30 000 ft (ceiling over 40 000 ft)
Takeoff distance	2000 m @ sea level to 15 m
Landing distance	2000 m (at 95% takeoff weight) @ sea level from 15 m
Initial rate of climb	$14 \, \text{m s}^{-1}$
Undercarriage	Retractable
Cabin comfort	Pressurised cabin with air-conditioning and oxygen supply,
Cabin interior diameter	=144 in.
Technology level	Advanced
Power plant	Turbofan engine

Derivative version in the aircraft family (typically Airbus 319 and Airbus 321 class on both sides of the baseline Airbus 320 aircraft). This is accomplished by plugging and unplugging the fuselage as in a Bizjet design. Readers are referred to *Jane's All the World's Aircraft* for derivative details and Appendix D for an example. Wide-body aircraft design follows the methodology.

(Note: The author encourages readers to explore market surveys for other classes of aircraft. To diversify, in the following are brief specifications for two interesting examples [17]).

A. Agriculture Applications Aircraft
    1. Airframe must be highly corrosion resistant.
    2. Airframe must be easily cleaned (i.e. removable side panels).
    3. Airframe must be flushed with water after last flight.
    4. Airframe must be easily inspected.
    5. Airframe must be easily repaired.
    6. Airframe must be highly damage tolerant.
    7. Dry and wet chemicals must be loaded easily and quickly.
    8. Cockpit must have excellent pilot crash protection.
    9. Pilot must have excellent visibility (i.e. flagman, ground crew and obstacles).
    10. The stall speed must be 60 knots or less.
    11. The service ceiling is 15 000 ft.
    12. Takeoff performance: 20 000-ft field length (rough field) with 50-ft obstacles.
    13. Hopper capacity: 400 US gallons/3200 lbs.

It is suggested that the design be approached through use of FAR Parts 137, 135 and 123. Readers may review current designs from *Jane's All the World's Aircraft*. Key considerations include choice of materials, configuration and structural layout and system design. In every other respect, the design should follow the standard approach described herein.

B. Airport Adaptive Regional Transport with Secondary Role to Support US Homeland Security (Abridged from [17])

Payload	49 passengers + flight and cabin crew
Range	1500 miles with reserve
Takeoff and landing Field length	2500 ft
Maximum speed	400 knots
Mission profile	Multiple takeoffs and landings without refuelling

For the airport adaptive role, the aircraft can simultaneously approach a major airport in non-interfering adverse weather and takeoff and land from shorter, largely unused runways, sub-runways and taxiways. The aircraft will be evaluated for an automatic spiral-descending, decelerating approach in instrument meteorological conditions (IMCs) (Category 3C) conditions and be able to continue with one engine inoperative. The aircraft also has the following secondary roles:

- Serve the civil reserve fleet and be available during a homeland-security crisis
- Serve as an ambulance
- Serve as transport firefighters to remote wilderness areas
- Serve as an emergency response vehicle for urban terrorism or a natural disaster by changing passenger-accommodation fitment

The aircraft will have half of the payload and a 750-mile range into makeshift landing zones of at least 1000 ft. More information is required for the specifications, but the level of technology is not within the scope of this book.

Other than drag estimation and certification regulations (e.g. noise), the supersonic transport (SST) design is similar to subsonic transport, aircraft design methodology. Supersonic drag estimation is addressed in Chapter 9.

In contrast to the civil market, the military aviation market starts with meeting the national defence requirements. The MoD organisations constantly review perceived threats and endeavour to stay ahead of the adversary. MoD floats a RFPs with ASTs. Many uncertainties are embedded in the road to an operational product. The development cost for these hi-tech machines is high and in many cases exceeds projected appropriations. The dominant certification standards are the Milspecs (US) and Defence Standards 970 (earlier, AvP 970 – UK). These certification standards are not as similar as are the FAR and JAR (Joint Airworthiness Regulation) requirements.

2.9 Military Market – Some Typical Military Aircraft Design Specifications

In contrast to the civil market, the military aviation market starts with meeting national defence requirements. The MoD organisations constantly review perceived threats and endeavour to stay ahead of the adversary. MoD floats RFPs with AST. A lot of uncertainties are embedded in the road to an operational product. Development cost for these hi-tech machines is high and in many cases exceed projected appropriations. The dominant certification standards are Milspecs (US) and Defence Standards 970 as before.

2.9.1 Aircraft Specifications/Requirements for Military Aircraft Case Studies

The author recommends that the introductory classroom work on military aircraft design may start with the Turboprop Trainer (TPT) aircraft followed by an AJT design specifications as given next.

Design Specifications of an TPT – UK Def Standards

Technology level	metal frame and with glass cockpit (EFIS)
Certification standard	UK Def Standards 970
Basic mission	intermediate level pilot training up to jet type
Mission profile	small, agile for sortie profile
Payload	2 80-kg pilots and 1000-kg armament
Seating	tandem
Normal training configuration (NTC)	clean configuration with four pylons
Engine	one turboprop
Maximum speed	500 kmph at 15 000 ft
Maximum sustained speed	400 kmph
Stalling speed	130 kmph (flaps extended)
Service ceiling	10 000 m
Initial rate of climb	$1200\,\mathrm{m\,s^{-1}}$ at NTC
Time to climb to 7 km	10 min at NTC
Turn performance	4 g at sea level (@mean weight)
Manoeuvre	+8 g to −4 g (fully aerobatic)
Roll rate	$75°\,\mathrm{s^{-1}}$ at 250 knot
Range	1500 km at cruise
Takeoff distance	500 m at NTC to clear 15 m (sea level)
Landing distance	550 m at NTC (no brake parachute – sea level)
Undercarriage	retractable
Flight deck	ejection seat, pressurised with oxygen supply
Cabin interior width	28 in

Design Specifications (Requirements) of a Baseline AJT

Technology level	advanced, multi-functional display (EFIS)
Certification standard	UK Def Standards 970
Basic mission	training in jet aircraft up to operational conversion type
Mission profile	small, agile (see Chapter 13) for sortie profile
Payload	over 1500-kg armament (prefer 1800 kg)
Number of pylons	5
Crew	two 90-kg pilots in tandem seating
Normal training configuration (NTC)	clean configuration with pylons only
Engine	one turbofan with low bypass ratio – no afterburning
Maximum speed	Mach 0.75 (920 kmph)
Maximum sustained speed	Mach 0.7 (860) kmph
Stalling speed	220/180 kmph (no flaps/flaps, respectively)
Service ceiling	14 000 m
Initial rate of climb	$40\,\mathrm{m\,s^{-1}}$

Time to climb to 12 km	12 minutes
Turn performance	4 g at sea level (@mean weight)
Manoeuvre	+8 g to −4 g (fully aerobatic)
Roll rate	200° s^{-1}
Range	700 km at sea level and 1200 km at 9-km cruise altitude
Endurance	2.5 hours with reserve
Takeoff distance	1100 m to clear 15 m (sea level)
Landing distance	1000 m (no brake parachute – sea level)
Undercarriage	retractable
Flight deck	ejection seat, pressurised with oxygen supply
Cabin interior width	30 in
Technology level	advanced multifunctional display
Structure	primary structure of metal frame; secondary structure in composite

Derivative version in the family of a Baseline AJT – a single-seat CAS aircraft (all performance figures at NTC)

Technology level	advanced, multi-functional display (EFIS)
Certification standard	UK Def Standards 970
Mission profile	small, agile (see Chapter 13 for sortie profile)
Payload	2500-kg armament
Crew	single 90-kg pilot
Number of pylons	5
Engine	one turbofan with low bypass ratio
Maximum speed	maximum Mach 0.75 (910 kmph) level flight
Maximum sustained speed	0.7 (850 kmph)
Stalling speed	240/200 kmph (no flaps/flaps, respectively)
Service ceiling	14 000 m
Initial rate of climb	50 m s^{-1}
Time to climb to 12 km	8 min.
Turn performance	5 g at sea level (@mean weight)
Manoeuvre	+8 g to −4 g (fully aerobatic)
Roll rate	200° s^{-1}
Range	700 km at sea level (no drop tanks);
	1500 km at 9-km cruise (with drop tank)
Sortie duration	1.5 hours (no drop tanks) with reserve
Takeoff distance	1400 m to clear 15 m (sea level)
Landing distance	1200 m (at landing weight, sea level – no brake parachute)
Undercarriage	retractable
Flight deck	ejection seat, pressurised with oxygen supply
Cabin interior width	30 in.
Structure	primary structure of metal frame; secondary structures in composite

2.10   Airworthiness Requirements

From the days of barnstorming and stunt-flying in the 1910s, it became obvious that commercial interests had the potential to short-circuit safety considerations. Government agencies quickly stepped in to safeguard people's security and safety without deliberately harming commercial interests. Safety standards were developed through multilateral discussions, which continue even today. Western countries developed and published thorough and systematic rules – these are in the public domain (see relevant web sites).

All aircraft design must satisfy the minimum safety standards of the countries of operation governed by their certification agencies (authorities). In the USA the certification agency is the Federal Aviation Administration (FAA) and in Europe it is the European Aviation Safety Agency (EASA), both standards have almost similar requirements and working together to harmonise to equivalent standards. This book follows the FAA standards, which is more prevalent internationally. Civil aviation is one of two major categories of flying, representing all non-military aviation, both private and commercial. All civil aircraft must comply with FAA and satisfy their regulations (FAR) as defined for each category of aircraft design.

The authors' preference is to work with the established FAR; the pertinent FAR is cited when used in the text and examples. This book is only concerned with FAR Parts 23 (General Aviation) and 25 (scheduled airline operations and non-scheduled air transport operations for remuneration or hire – commercial transport category) – in short, FAR23 and FAR25, respectively. FAR documentation for certification has branched out into many specialist categories. Appendix G gives more on FARs. The FAR documentations are done in an elaborate and articulate manner and for any new reader it may appear quite challenging to grasp them in a logical manner. The basic FAR requirements, as applicable to this book, are treated here. This will give some idea on how industries deal with FAR substantiation. Industries are the places where fresh engineers will be learning about the intricacies of FAR in a formal manner.

ICAO and work together with member countries to establish common standards and recommended practices for civil aviation.

GA, a sector of civil aviation, is defined as non-military and non-airlines sector of aviation and broadly classified into two categories as the (i) Standard Category and (ii) Special Category. Small aircraft within GA are separated into Special Category discussed in Chapter 16. GA covers a large range of activities, both commercial and non-commercial, including flying clubs, flight training, agricultural aviation, light aircraft manufacturing and maintenance.

The Transport Category (scheduled airline operations and non-scheduled air transport operations for remuneration or hire) has a higher level of requirements under FAR25.

2.10.1   Code of Federal Regulations (CFR) – Title 14

CFR classifies all civil aircraft into 11 categories, 10 of them under FAR23 all below 12 500 lb and one under FAR25 in a higher specification in which all transport aircraft 12 500 lb are to be certified.

2.10.1.1   FAR23
This has two main categories, (i) Standard (four types) and (ii) Special (six types) as follows.

2.10.1.2   Standard Category (Volume 1, Section 23-3: Aeroplane Categories Dated 2011-01-0)

(i) *Normal category.* Aeroplanes that have a seating configuration, excluding pilot seats, of nine or less, a maximum certificated takeoff weight of 12 500 pounds or less and are intended for non-aerobatic operation.

(ii) *Utility category.* Limited to aeroplanes that have a seating configuration, excluding pilot seats, of nine or less, a maximum certificated takeoff weight of 12 500 pounds or less and are intended for limited acrobatic operation. Aeroplanes certificated in the utility category may be used in any of the operations covered under paragraph (a) of this section and in limited acrobatic operations. Limited acrobatic operation includes: (i) spins (if approved for the particular type of aeroplane) and (ii) lazy eights, chandelles, steep turns or similar manoeuvres, in which the angle of bank is more than 60° but not more than 90°.

(iii) *Acrobatic category.* Limited to aeroplanes that have a seating configuration, excluding pilot seats, of nine or less, a maximum certificated takeoff weight of 12 500 pounds or less and are intended for use without restrictions, other than those shown to be necessary as a result of required flight tests.

(iv) *Commuter category.* Limited to propeller-driven, multiengine aeroplanes that have a seating configuration, excluding pilot seats, of 19 or less, and a maximum certificated takeoff weight of 19 000 pounds or less. The commuter category operation is limited to any manoeuvre incident to normal flying, stalls (except whip stalls) and steep turns, in which the angle of bank is not more than 60°.

Except for the commuter category, aeroplanes may be type certificated in more than one category if the requirements of each requested category are met.

2.10.1.3   Special Category (FAR Part 1.1)

This consists of the following six categories not dealt with in the book. These are *mostly* small aircraft (see Chapter 16).

(v) Light Sports Aircraft (LSA) have two sub-divisions, Special Light-Sport Aircraft (S-LSA) and Experimental Light-Sport Aircraft (E-LSA). FAR Part 1.1.

(vi) *Experimental.* These are Experimental Amateur-Built (E-AB) aircraft, or experimental exhibition) – also known as home built/kit built aircraft. If complied with, these can also be certified under the LSA category. The various applications in this class are as follows:

    a) An aircraft whose purpose is to test new design concept and demonstrate its airworthiness, equipment or operating techniques. It is a research and development activity. It can be used for market survey.

    b) Exhibition and air racing.

(vii) *Limited aircraft.* Surplus military aircraft converted to civilian use (Experimental Exhibition category).

(viii) *Primary aircraft.* Aircraft in this category are of a simple design and intended exclusively for pleasure and personal use. Although these aircraft may be available for rental and flight instruction under certain conditions, the carrying of persons or property for hire is prohibited.

(ix) *Provisional aircraft.* Aircraft with a 'provisional' category type certificate for special operations and operating limitations.

(x) *Restricted aircraft.* These include agriculture, forest/wildlife conservation, aerial surveying, patrolling, weather observation, aerial advertising and other operations specified by the administrator.

2.10.1.4   FAR25

The FAR 25 applies to Transport Category Aircraft (scheduled airline operations and non-scheduled air transport operations for remuneration or hire.) and has higher level of requirements compared to FAR23. This maximum takeoff weight (MTOW) is over 12 500 lb for the FAR25 category of transport aircraft.

2.10.1.5   The Parts of FAR Requirements

For convenience, the airworthiness requirements to ensure the safety standards are classified by their specialised areas along with their associated allied topics into several parts as shown in the Table 2.3. As this book only deals only with FAR23 and FAR25, the table lists only the standards for these two. Each part may have several documents describing meticulously the certification requirements in legal frame to clear any

Table 2.3 FAR categories of airworthiness standards.

Aircraft types	General aviation	Normal	Transport
Aircraft	FAR Part 23	FAR Part 23	FAR Part 25
Engine	FAR Part 33	FAR Part 33	FAR Part 33
Propeller	FAR Part 35	FAR Part 35	FAR Part 35
Noise	FAR Part 36	FAR Part 36	FAR Part 36
General operations	FAR Part 91	FAR Part 91	FAR Part 91
Agriculture	FAR Part 137		
Large commercial transport	Not applicable		FAR Part 121

Table 2.4 Aircraft categories.

Aircraft types	General aviation	Normal	Transport
MTOW – lb	less than 12 500	less than 12 500	Over 12 500
No. of engine	zero or more	more than one	more than one
Type of engine	All types	Propeller only	All types
Flight crew	One	Two	Two
Cabin crew	None	None up to 19 PAX	None up to 19 PAX
Maximum no. of occupants	10	23	Unrestricted
Maximum operating altitude	25 000 ft	25 000 ft	Unrestricted

ambiguity. For design substantiation, various parts may be cross-referenced to meet the standards required. The Appendix lists the pertinent requirements used in this book.

Table 2.4 gives the definitions for the GA, Normal Category and Transport of Aviation.

In military applications, the standards are Milspecs (US) and Defence Standard 970 (previously AvP 970) (UK); they are different in some places.

Since 2004, in the United States, new sets of airworthiness requirements came into force for light aircraft in the category LSA class (UK – Very Light Aircraft; CS-VLA – Certification Specifications for Very Light Aeroplanes) designs and have eased certification procedures and litigation laws, rejuvenating the industry in the sector. Europe also has a similar approach but its regulations differ to an extent. Small/light aircraft and microlight types have different certification standards not discussed in this book.

With unmanned aerial vehicles (UAVs) gaining applications in civil aviation, there will be growing demand to involve the FAA to ensure safety and regulate activities. There are other classes, for example, ultralights, trikes, hang gliders, parasols, lighter than air dirigibles, helicopters and so on, which are beyond the scope of this book.

2.11 Coursework Procedures – Market Survey

The coursework task is to conduct a mock market study. The instructor divides the class into groups of four or five students who will work as a team (see the Road Map of the Book, which gives the typical allotted time). However, how the class is conducted is at the instructor's discretion. Designing a conventional civil or a military trainer aircraft is appropriate for undergraduate introductory work. In this book, a Bizjet and an AJT aircraft design are used.

Step 1. The instructor decides which class of aircraft will be used for the design project. Each group is required to submit their ideas of the aircraft configured with assigned design considerations. Students will have input but the instructor ultimately explains why a certain aircraft is chosen.

Step 2. The instructor discusses each suggestion, discarding the impractical and coalescing the feasible. The instructor will add anything that is missing, with explanations.

Step 3. Each team must then submits a scaled, three-view sketch of the proposed design. There will be differences in the various configurations. CAD is recommended.

Step 4. The instructor discusses each configuration, tailoring the shape, with explanation, to a workable shape. Each team works on its revised configuration; preferably, the class will work with just one design.

References

1 Wood, K.D. (1966). *Aircraft Design*. Boulder, CO: Johnson Publishing Co.

2 Corning, G. (1960). *Supersonic and Subsonic Airplane Design*. Gerald Corning, published by the author.

3 Nicolai, L.M. (1975). *Fundamentals of Aircraft Design*, 95120. San Jose, CA: METS, Inc.

4 Torenbeek, E. (1982). *Synthesis of Subsonic Airplane Design*. Delft University Press.

5 Roskam, J. *Aircraft Design*, Published by the author as an 8 volume set, 1985–1990.

6 Stinton, D. (1983). *The Design of the Airplane*. New York: van Nostrand Reinhold.

7 Brandt, S.A., Stiles, R., Bertin, J., and Whitford, R. (2015). *Introduction of Aeronautics – A Design Perspective*, 3e. Reston, Virginia: AIAA.

8 Raymer, D. (1992). *Aircraft Design – A Conceptual Approach*. Reston, VA: AIAA.

9 Thurston, D. (1995). *Design for Flying*, 2e. Tab Books.

10 Fielding, J. (1999). *Introduction to Aircraft Design*. Cambridge University. Press.

11 Jenkinson, L.R., Simpson, P., and Rhodes, D. (1999). *Civil Jet Aircraft Design*. Arnold Publications.

12 Jenkinson, L.R. and Marchman, J.F. (2003). *Aircraft Design Project Studies*. Butterworth.

13 Howe, D. (2000). *Aircraft Conceptual Design Synthesis*. London: Professional. Engineering Ltd.

14 Whitford, R. (2000). *Fundamentals of Fighter Design*. Shrewsbury, UK: Airlife.

15 Schaufele, R.D. (2000). *The Elements of Aircraft Preliminary Design*. Santa Ana, CA: ARIES Publications.

16 Kuchemann, D. (1978). *Aerodynamic Design of Aircraft*. Pergamon International Library.

17 Kundu, A.K. (2010). *Aircraft Design*. Cambridge University Press.

18 Huenecke, K. (1987). Flugzeugentwerf Umdruck zur Vorlesung (Aircraft Design Handout for Reading). In: *Modern Combat Aircrarft Design*. Shrewsbury, UK: Airlife.

19 Schmitt, D. Technische Universitat Munchen, 2000.

20 McMasters, J. H. (1994). Boeing Commercial Aircraft General Engineering Division, Summer Intern Training Program.

21 Talay, T.A. NASA SP-367;(1975). *Introduction to the Aerodynamics of Flight*. NASA.

22 Gudmundsson, S. (2013). General aviation aircraft design. In: *Applied Methods and Procedures*, 1e. Butterworth Heinemann.

23 Torenbeek, E. (2013). *Advanced Aircraft Design: Conceptual Design, Technology and Optimization of Subsonic Civil Airplanes*. Wiley.

24 Niu, M. (1999). *Airframe Structural Design*. Hong Kong: Commlit Press Ltd.

25 Jackson, S. (1997). *Systems Engineering for Commercial Aircraft*. Routledge. ISBN: 0-291-39846-4.

26 Aircraft Transport Association of America (ATA)., (1989). Specifications for Manufacturers' Technical Data. Specification 100. ATA.

27 Berna, L., Dornberger, U., Suvelza, A., and Byrnes, T. (2009). *Quality Ffunction Deployment (QFD) for Sservices Handbook*. Universitat Leipzig.

28 Olewnik, A. and Lewis, K. (2008). Limitations of the house of quality to provide quantitative design information. *International Journal of Quality & Reliability Management* 25 (2): 125–146.

3

Aerodynamic Fundamentals, Definitions and Aerofoils

3.1 Overview

Any object moving through air interacts with the medium at each point of the wetted (i.e. exposed) surface, creating a pressure field around the aircraft body. An important part of aircraft design is to exploit this pressure field by shaping its geometry to arrive at the desired performance of the vehicle, for example, shaping to generate lifting surfaces, to accommodate payload, to house a suitable engine in the nacelle and to tailor control surfaces. This chapter is concerned with the aerodynamic information required at the conceptual design stage of a new aircraft design project. Also, this chapter covers the necessary details of one of the most important geometries concerning aircraft design, the aerofoil.

Aeronautical engineering schools offer a series of aerodynamic courses, starting with the fundamentals and progressing towards the cutting edge. It is assumed that readers of this book have been exposed to aerodynamic fundamentals. Presented herein is a brief compilation of applied aerodynamics without detailed theory beyond what is necessary. Many excellent textbooks are available in the public domain for reference. Because the subject is so mature, some almost half-century-old introductory aerodynamics books still serve the purpose of this course; however, more recent books relate better to current examples. The information in this chapter and carried to the next chapters is essential for designers and must be understood. It is recommended that the readers go through Chapters 3–7 thoroughly.

This chapter covers the following topics:

Section 3.2: Introduction to Aerodynamics
Section 3.3: Airflow Behaviour: Laminar and Turbulent
Section 3.4: Flow Past the Aerofoil
Section 3.5: Aerofoil and Lift Generation
Section 3.6: Reynolds Number and Surface Condition Effect
Section 3.7: Aircraft Motion, Forces and Moments
Section 3.8: Definitions of Aerodynamic Parameters
Section 3.9: Centre of pressure and aerodynamic centre
Section 3.10: Types of Stall
Section 3.11: High-Lift Devices
Section 3.12: Flow Regimes (area rule), Compressibility Correction
Section 3.13: Transonic and Supersonic Effects
Section 3.14: Summary of the Chapter
Section 3.15: Aerofoil Design and Manufacture
Section 3.16: Aircraft Centre of Gravity, Centre of Pressure and Neutral Point

Classwork is postponed until Chapter 8, except for the mock market survey in Chapter 2. The readers should return to Chapters 3–7 to extract the information necessary to configure the aircraft.

Conceptual Aircraft Design: An Industrial Approach, First Edition. Ajoy Kumar Kundu, Mark A. Price and David Riordan.
© 2019 John Wiley & Sons Ltd. Published 2019 by John Wiley & Sons Ltd.

3.2   Introduction

Aircraft conceptual design starts with shaping an aircraft, finalising geometric details through aerodynamic considerations in a multidisciplinary manner (see Section 2.3) to arrive at the technology level to be adopted. In the early days, aerodynamic considerations dictated aircraft design; gradually, other branches of science and engineering gained equal importance.

All fluids have some form of viscosity. Air has a relatively low viscosity, but it is sufficiently high to account for its effects. Mathematical modelling of viscosity is considerably more difficult than if the flow is idealised to have no viscosity (i.e. inviscid); then, simplification can obtain rapid results for important information. For scientific and technological convenience, all matter can be classified as shown in Chart 3.1.

This book is concerned with air (gas) flow. Air is compressible and its effect is realised when aircraft flies through it. Aircraft design requires an understanding of both incompressible and compressible fluids. Nature is conservative (other than nuclear physics) in which mass, momentum and energy are conserved. Aerodynamic forces of lift and drag (see Section 3.5) are the resultant components of the pressure field around an aircraft. Aircraft designers seek to obtain the maximum possible lift-to-drag ratio (i.e. a measure of minimum fuel burn) for an efficient design (this simple statement is complex enough to configure, as will be observed throughout the coursework). Aircraft stability and control are the result of harnessing these aerodynamic forces. Aircraft control is applied through the use of aerodynamic forces modulated by the control surfaces (e.g. elevator, rudder and aileron). In fact, the sizing of all aerodynamic surfaces should lead to meeting the requirements for the full flight envelope without sacrificing safety.

To continue with sustained flight, an aircraft requires a lifting surface in the form of a plane – hence, *aeroplane* (the term *aircraft* is used synonymously in this book). The secret of lift generation is in the sectional characteristics (i.e. *aerofoil*) of the lifting surface that serve as wings, similar to birds. This chapter explains how the differential pressure between the upper and lower surfaces of the wing is the lift that sustains the aircraft weight. Details of aerofoil characteristics and the role of the empennage that comprise the lifting surfaces are explained as well. The stability and control of an aircraft are aerodynamic-dependent and discussed in Chapter 18.

Minimising the drag of an aircraft is one of the main obligations of aerodynamicists. Viscosity contributes to approximately two-thirds of the total subsonic aircraft drag. The effect of viscosity is apparent in the wake of an aircraft as disturbed airflow behind the body. Its thickness and intensity are indications of the extent of

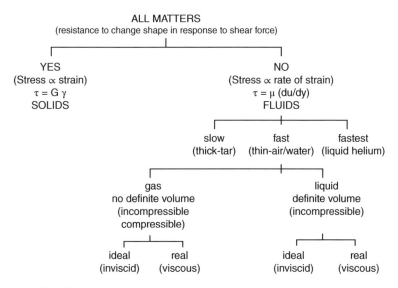

Chart 3.1 All matters.

drag and can be measured. One way to reduce aircraft drag is to shape the body such that it will result in a thinner wake. The general approach is to make the body a teardrop shape with the aft end closing gradually compared to the blunter front-end shape for subsonic flow. Behaviour in a supersonic flow is different but it is still preferable for the aft end to close gradually. The smooth contouring of teardrop shaping is called *streamlining*, which follows the natural airflow lines around the aircraft body – it is for this reason that aircraft have attractive smooth contour lines. Streamlining is synonymous with speed and its aerodynamic influence in shaping is revealed in any object in a relative moving airflow (e.g. boats and automobiles).

Streamlines: In a flowing fluid, an identifiable physical boundary defined as control volume (CV – see Section 12.6.1) can be chosen to describe mathematically the flow characteristics. A CV can be of any shape but the suitable CVs confine several streamlines like well-arranged 'spaghetti in a box' in which the ends continue along the streamline, crossing both cover ends but not the four sides. Streamline is a useful concept to deal with one strand of fluid that can be related with the domain of fluid in the CV, especially for inviscid fluid consideration.

The conservative laws within the CV for steady flow (independent of time, t) that are valid for both inviscid incompressible and compressible flow are provided herein. These can be equated between two stations (e.g. Stations 1 and 2) of a streamline. Inviscid (i.e. ideal) flow undergoing a process without any heat transfer is called the *isentropic process*. During the conceptual study phase, all external flow processes related to aircraft aerodynamics are considered isentropic, making the mathematics simpler. (Air flow through engines is an internal process.)

Some useful equations are given in Appendix C.

3.3   Airflow Behaviour – Laminar and Turbulent

Understanding the role of the viscosity of air is important to aircraft designers. The simplification of considering air as inviscid may simplify mathematics, but it does not represent the reality of design. Inviscid fluid does not exist, yet it provides much useful information rather quickly. Subsequently, the inviscid results are improvised. To incorporate the real effects of viscosity, designs must be tested to substantiate theoretical results.

The fact that airflow can offer resistance due to viscosity has been understood for a long time. Navier in France and Stokes in England independently arrived at the same mathematical formulation; their equation for momentum conservation embedding the viscous effect is known as the Navier–Stokes equation. It is a nonlinear partial differential equation still unsolved analytically, except for some simple body shapes. In 1904, Ludwig Prandtl presented a flow model that made the solution of viscous flow problems easier [1, 2]. He demonstrated by experiment that the viscous effect of flow is realised only within a small thickness layer over the contact surface boundary; the rest of the flow remains unaffected. This small thickness layer is called the boundary layer. Today, numerical methods (i.e. computational fluid dynamics, CFD) can address viscous problems to a great extent.

The best way to model a continuum (i.e. densely packed) airflow is to consider the medium to be composed of very fine spheres of the molecular scale (i.e. diameter 3×10^{-8} cm and intermolecular space 3×10^{-6} cm). Like sand, these spheres flow one over another, offering friction in between while colliding with one another. Air flowing over a rigid surface (i.e. acting as a flow boundary) will adhere to it, losing velocity; that is, there is a depletion of kinetic energy of the air molecules as they are trapped on the surface, regardless of how polished it may be. On a molecular scale, the surface looks like the crevices shown in Figure 3.1a, with air molecules trapped within to stagnation. The contact air layer with the surface adheres and it is known as the 'no-slip' condition. The next layer above the stagnated no-slip layer slips over it – and, of course, as it moves away from the surface, it will gradually reach the airflow velocity. The pattern within the boundary layer flow depends on how fast it is flowing.

Here is a good place to define the parameter called the *Reynolds Number* (Re) given in Equation (3.1). Re is a useful and powerful parameter – it provides information on the flow status with the interacting body involved.

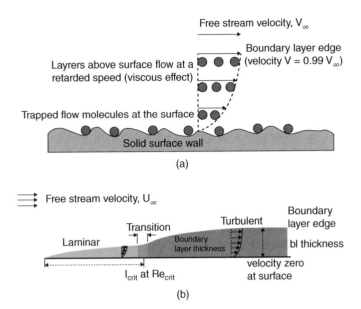

Figure 3.1 Boundary layer. (a) Magnified view of airflow over a rigid surface (boundary). (b) Boundary layer over a flat plate with a sharp leading edge.

Re increases with length.

$$\mathrm{Re} = (\rho_\infty U\infty l)/\mu_\infty \tag{3.1}$$

$$= (\text{density} \times \text{velocity} \times \text{length})/\text{coefficient of viscosity} = (\text{inertia force})/(\text{viscous force})$$

where μ_∞ = coefficient of viscosity.

It represents the degree of skin friction depending on the property of the fluid. The subscript infinity, ∞, indicates the condition, that is, undisturbed infinite distance ahead of the object. Re is a grouped parameter, which reflects the effect of each constituent variable, whether they vary alone or together. Therefore, for a given flow, characteristic length, l, is the only variable in Re. Re increases along the length. In an ideal flow (i.e. inviscid approximation), Re becomes infinity – not much information is conveyed beyond that. However, in real flow with viscosity, it provides vital information: for example, on the nature of flow (turbulent or laminar), on separation and on many other characteristics.

Figure 3.1b describes a boundary layer of airflow over a flat surface (i.e. plate) with sharp leading edge (LE) aligned to the flow direction (i.e. x-axis). Initially, when the flow encounters the flat plate at the LE, it develops a boundary layer that keeps growing thicker until it arrives at a critical length, when flow characteristics then make a transition and the profile thickness suddenly increases. The friction effect starts at the LE and flows downstream in an orderly manner, maintaining the velocity increments of each layer as it moves away from the surface – much like a sliding deck of cards (in lamina). This type of flow is called a *laminar flow*. Surface skin friction depletes the flow energy transmitted through the layers until at a certain distance (i.e. the critical point) from the LE, flow can no longer hold an orderly pattern in lamina, breaking down and creating turbulence. The boundary layer thickness is shown as δ at a height where 99% of the free streamflow velocity is attained.

The region where the transition occurs is called the *critical point*. It occurs at a predictable distance l_{crit} from the LE, having a critical Re of R_{ecrit} at that point. At this distance along the plate, the nature of the flow makes the transition from laminar to *turbulent flow*, when eddies of the fluid mass randomly cross the layers. Through mixing between the layers, the higher energy of the upper layers energises the lower layers. The physics of turbulence that can be exploited to improve performance (e.g. dents on a golf ball forces a laminar flow to a turbulent flow) is explained later.

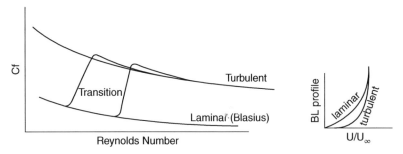

Figure 3.2 Viscous effect of air on flat plate (C_f is defined next).

With turbulent mixing, the boundary layer profile changes to a steeper velocity gradient and there is a sudden increase in thickness, as shown in Figure 3.2. For each kind of flow situation, there is a R_{ecrit}. As it progresses downstream of l_{crit}, the turbulent flow in the boundary layer is steadily losing its kinetic energy to overcome resistance offered by the sticky surface. If the plate is long enough, then a point may be reached where further loss of flow energy would fail to negotiate the surface constraint and would leave the surface as a *separated flow*. Separation also can occur early in the laminar flow.

The extent of velocity gradient, du/dy, at the boundary surface indicates the tangential nature of the frictional force; hence, it is shear force. At the surface where $u = 0$, du/dy is the velocity gradient of the flow at that point. If F is the shear force on the surface area, A, is due to friction in fluid, then shear stress is expressed as follows:

$$F/A = \tau = \mu(du/dy), \text{ where } \mu \text{ is the coefficient of viscosity} \tag{3.2}$$

$\mu = 1.789 \times 10^{-5}$ kg m^{-1} s^{-1} or Ns m^{-2} (1 g m^{-1} s^{-1} = 1 P) for air at sea level ISA.

Kinematic viscosity, $v = \mu/\rho$m^2/s (1 m^2 s^{-1} = 104 St), where ρ is density of fluid.

The measure of the frictional shear stress is expressed as a *coefficient of friction*, C_f, at the point:

$$\text{coefficient of friction, } C_f = \tau/q_\infty, = (\text{shear stress/dynamic head}) \text{ at the point.} \tag{3.3}$$

Friction drag for the length l, $D_f = \int_0^l \tau dx$ (τ varies along the length).

Average coefficient of friction, C_F is computed for a flat plate of unit length for the length l, that is, over the area $(1 \times l)$ is expressed as

$$C_F = \frac{1}{1}\int_0^l \frac{\tau dx}{0.5\rho U_\infty^2} = \frac{D_f}{0.5\rho U_\infty^2 A} \tag{3.4}$$

The difference of du/dy between laminar and turbulent flow is shown in Figure 3.2; the latter has a steeper gradient – hence, it has a higher C_f. The up arrow indicates increase and vice versa; for incompressible flow, *temp*↑ makes μ↓, which reads as viscosity decreases with a rise in temperature, and for compressible flow, *temp*↑ makes μ↑.

The pressure gradient along flat plate gives $dp/dx = 0$. Airflow over curved surfaces accelerates or decelerates depending on which side of the curve the flow is negotiating $(dp/dx \neq 0)$. Extensive experimental investigations on the local skin friction coefficient, C_f, and I_F on a 2D flat plate are available for a wide range of Re (Figure 11.22). The overall coefficient of skin friction over a surface is expressed as C_F and is higher than the 2D flat plate. C_f increases from laminar to turbulent flow, as can be seen from the increased boundary layer thickness. Correction is applied to flat plate C_f, and C_F to get curved surface C_f and C_F. In general, C_F is computed semi-empirically from the flat plate C_f (see Chapter 11).

For flat plate in zero pressure gradient the boundary layer thickness, δ and local skin friction C_f can be expressed as given in Table 3.1 [2].

It is recommended to use test values as presented in graphic form in Figure 11.24. These graphs are readily available in many publications.

Table 3.1 Boundary layer thickness, δ, and local skin friction, Cf.

	Laminar	Turbulent	
Boundary layer thickness, δ	$\dfrac{5.2x}{\sqrt{Re_x}}$	$\dfrac{0.37x}{Re^{0.2}}$	(3.5a)
Local skin friction C_f	$\dfrac{1.328}{\sqrt{Re_l}}$	$\dfrac{0.074}{R_l^{0.2}}$	(3.5b)

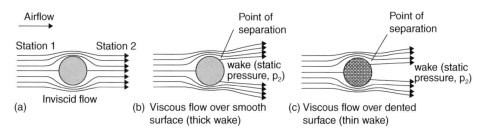

Figure 3.3 Air flow past a sphere. (a) Inviscid flow ($p_1 = p_2$). (b) Viscous flow ($p_1 > p_2$). (c) Dented surface ($p_1 > p_2$).

Scientists have been able to model the random pattern of turbulent flow using statistical methods. However, at the edges of the boundary layer, the physics is unpredictable. This makes accurate statistical modelling difficult, with eddy patterns at the edge extremely unsteady and the flow pattern varying significantly. It is clear why the subject needs extensive treatment.

To explain the physics of drag of 3D body, the classical example of flow past a sphere is shown in Figure 3.3. A sphere in inviscid flow will have no drag (Figure 3.3 - left) because it has no skin friction and there is no pressure difference between the front and aft ends there is nothing to prevent the flow from negotiating the surface curvature. Diametrically opposite to the front stagnation point is a rear stagnation point, equating forces on the opposite sides. This ideal situation does not exist in nature but can provide important information.

In the case of a real fluid with viscosity, the physics changes nature offering drag as a combination of skin friction and the pressure difference between fore and aft of the sphere. At low Re, the low-energy laminar flow near the surface of the smooth sphere (Figure 3.3 - centre) separates early, creating a large wake in which the static pressure, p, at the aft (subscript '2') cannot recover to its initial value at the front of the sphere (subscript '1'). The pressure at the front is now higher at the stagnation area, resulting in a pressure difference that appears as pressure drag. It would be beneficial if the flow was made turbulent by denting the sphere surface (Figure 3.3 - right). In this case, high-energy flow from the upper layers mixes randomly with flow near the surface, reenergising it. This enables the flow to overcome the spheres curvature and adhere to a greater extent, thereby reducing the wake. Therefore, a reduction of pressure drag compensates for the increase in skin friction drag (i.e. C_f increases from laminar to turbulent flow). This concept is applied to golf ball design (i.e. low Re, low velocity and small physical dimension). The dented golf ball would go farther than an equivalent smooth golf ball due to reduced drag.

$$\text{Therefore: } \text{drag} = \text{skin friction drag} + \text{pressure drag} \tag{3.6}$$

The situation changes drastically for a body at high Re (i.e. high velocity and/or large physical dimension; for example, an aircraft wing or even a golf ball hit at a very high-speed that would require more than any human effort), when flow is turbulent almost from the LE. A streamlined aerofoil shape does not have the highly steep surface curvature over large area like that of a dent less golf ball; therefore, separation occurs very late, resulting in a thin wake. Therefore, pressure drag is low. The dominant contribution to drag comes from skin friction, which can be reduced if the flow retains laminarisation over more surface area (although it is not applicable to a golf ball). Laminar aerofoils have been developed to retain laminar flow characteristics over a relatively large part of the aerofoil. These aerofoils are more suitable for low-speed operation (but Re higher than the golf ball application) such as gliders and have the added benefit of a very smooth surface made of composite materials.

Clearly, the drag of a body depends on its profile that is, how much wake it creates. The blunter the body, the greater the wake size will be; it is for this reason that aircraft components are streamlined. This type of drag is purely viscous-dependent and is termed *profile drag*. In general, in aircraft applications, it is also called *parasite drag*, as explained in Chapter 11.

3.3.1 Aerofoils

Aerofoils (Figure 3.4) are the 2D bread-slice like sections of the plane wing with a streamlined geometrical shape to generate the least drag for the desired lift generated. Compared with a cylinder of a diameter equal to the maximum thickness of a well-designed symmetrical aerofoil, the former generates 10 times more drag. A 2D flat plate of the same height placed normal to the same airflow generates 16.7 times more drag than the aerofoil. Even 3D bodies like the fuselage, nacelle and so on, are made to streamlined shapes as drag reduction measures.

3.4 Flow Past an Aerofoil

A typical airflow past an aerofoil is shown in Figure 3.4 later.

In Figure 3.4, the front curvature of the aerofoil causes the flow to accelerate with the associated drop in pressure, until it reaches the point of inflection on the upper surface of the aerofoil. This is known as a *region of favourable pressure gradient* because the lower pressure downstream favours airflow. Past the inflection point, airflow starts to decelerate, recovering the pressure (i.e. flow in an adverse pressure gradient) that was lost while accelerating. For inviscid flow, it would reach the trailing edge (TE), regaining the original free stream flow velocity and pressure conditions. In reality, the viscous effect depletes flow energy, preventing it from regaining the original level of pressure. Along the aerofoil surface, airflow depletes its energy due to friction (i.e. the viscous effect) of the aerofoil surface. The viscous effect appears as a wake behind the body.

The result of a loss of energy while flowing past the aerofoil surface is apparent in adverse pressure gradient; it is like climbing uphill. A point may be reached where there is not enough flow energy left to encounter the adverse nature of the downstream pressure rise – the flow then leaves the surface to adjust to what nature allows. Where the flow leaves the surface is called the *point of separation*, and it is critical information for aircraft design. When separation happens over a large part of the aerofoil, it is said that the aerofoil has *stalled* because it has lost the intended pressure field. Generally, it happens on the upper surface; in a stalled condition, there is a loss of low-pressure distribution and, therefore, a loss of lift. This is an undesirable situation for an aircraft in flight. There is a minimum speed below which stalling will occur in every winged aircraft. The speed

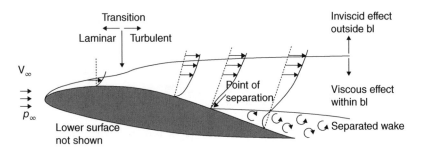

Figure 3.4 Air flow past an aerofoil.

at which an aircraft stalls is known as the *stalling speed*, V_{stall}. At stall, an aircraft cannot maintain altitude and can even become dangerous to fly; obviously, stalling should be avoided.

For a typical surface finish, the magnitude of skin friction drag depends on the nature of the airflow. Below Re_{crit}, laminar flow has a lower skin friction coefficient, C_f, and, therefore, a lower friction (i.e. lower drag). The aerofoil LE starts with a low Re and rapidly reaches Re_{crit} to become turbulent. Aerofoil designers must shape the aerofoil LE to maintain laminar flow as much as possible.

Aircraft surface contamination is an inescapable operational problem that degrades surface smoothness, making it more difficult to maintain laminar flow. As a result, Re_{crit} advances closer to the LE. For high-subsonic flight speed (high Re), the laminar flow region is so small that flow is considered fully turbulent.

This section points out that designers should maintain laminar flow as much as possible over the wetted surface, especially at the wing LE. As mentioned previously, gliders – which operate at a lower Re – offer a better possibility to deploy an aerofoil with laminar flow characteristics. The low annual utilisation in private usage favours the use of composite material, which provides the finest surface finish. However, although the commercial transport wing may show the promise of partial laminar flow. With six degrees of freedom in body axes F_B laminar flow at the LE, the reality of an operational environment at high utilisation does not guarantee adherence to the laminar flow. For safety reasons, it would be appropriate for the governmental certifying agencies to examine conservatively the benefits of partial laminar flow. This book considers the fully turbulent flow as starting from the LE of any surface of a high-subsonic aircraft.

3.5 Generation of Lift

Figure 3.5a is a qualitative description of the flow field and its resultant forces on the aerofoil. The result of skin friction is the drag force, shown in Figure 3.5b. The lift is normal to the flow.

Section 3.8.1 explains that a typical aerofoil has an upper surface more curved than the lower surface, which is represented by the camber of the aerofoil. Even for a symmetrical aerofoil, the increase in the angle of attack increases the velocity at the upper surface and the aerofoil approaches stall, a phenomenon described in Section 3.4.

Figure 3.6a shows the pressure field around the aerofoil. The pressure at every point is given as the pressure coefficient distribution, as shown in Figure 3.6b. The upper surface has lower pressure, which can be seen as a negative distribution. In addition, cambered aerofoils have moments that are not shown in the figure.

Figure 3.7a shows flow physics around the aerofoil. At the LE, the streamlines move apart: One side negotiates the higher camber of the upper surface and the other side negotiates the lower surface. The higher curvature at the upper surface generates a faster flow than the lower surface. They have different velocities when they meet at the TE, creating a vortex sheet along the span. The phenomenon can be decomposed into a set of straight streamlines representing the free streamflow condition and a set of circulatory streamlines of a strength that matches the flow around the aerofoil. The circulatory flow is known as the *circulation*, Γ, of the aerofoil (Figure 3.7b). The concept of circulation provides a useful mathematical formulation to represent lift.

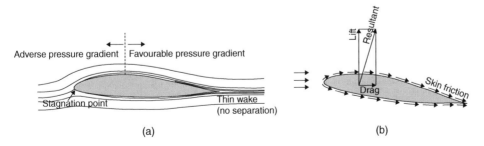

(a)   (b)

Figure 3.5 Flow field around an aerofoil. (a) Streamline pattern over aerofoil. (b) Resultant force on an aerofoil.

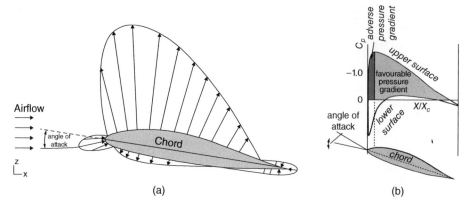

Figure 3.6 Pressure field representations around an aerofoil. (a) Pressure field distribution. (b) Cp distribution over an aerofoil.

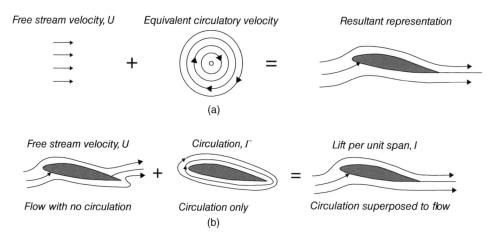

Figure 3.7 Lift generation on an aerofoil. (a) Mathematical representation by superimposing free vortex flow over parallel flow. (b) Inviscid flow representation of flow around an aerofoil.

Circular flow is generated by the effect of the aerofoil camber, which gives higher velocity over the upper wing surface. The directions of the circles show the increase in velocity at the top and the decrease at the bottom, simulating velocity distribution over the aerofoil.

The flow over an aerofoil develops a lift per unit span of $l = \rho U \Gamma$ (see other textbooks [2–7] for the derivation). Computation of circulation is not easy. This book uses accurate experimental results to obtain the lift.

3.6 Aircraft Motion, Forces and Moments

An aircraft is a vehicle in motion; in fact, it must maintain a minimum speed above the stall speed. The resultant pressure field around the aircraft body (i.e. wetted surface) is conveniently decomposed into a usable form for designers and analysts. The pressure field alters with changes in speed, altitude and orientation (i.e. attitude). This book primarily addresses a steady level flight pressure field; the unsteady situation is considered transient in manoeuvres. Chapter 17 addresses certain unsteady cases (e.g. gusty winds) and references are made to these design considerations when circumstances demand it. This section provides information on the parameters concerning motion (i.e. kinematics) and force (i.e. kinetics) used in this book.

Figure 3.8 Six degrees of freedom in body axes.

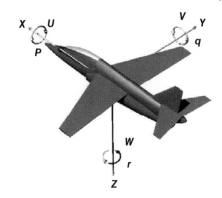

3.6.1 Motion

Unlike an automobile, which is constrained by the road surface, an aircraft is the least constrained vehicle, having all the degrees of freedom (Figure 3.8): three linear and three rotational motions along and about the three axes. These can be represented in any coordinate system; however, in this book, the right-handed Cartesian coordinate system is used. Controlling motion in six degrees of freedom is a complex matter. Careful aerodynamic shaping of all components of an aircraft is paramount, but the wing takes top priority. Aircraft attitude is measured using Eulerian angles – ψ (azimuth), θ (elevation) and Φ (bank) – and these are in demand for aircraft control; however, this is beyond the scope of this book. In classical flight mechanics, many types of Cartesian coordinate systems are in use. The three most important are as follows:

1. *Body-fixed axes*, F_B, is a system with the origin at the aircraft centre of Gravity (CG) and the x-axis pointing forward (in the plane of symmetry), the y-axis going over the right wing, and the z-axis pointing downward.
2. *Wind-axes system*, F_W, also has the origin (gimballed) at the CG and the x-axis aligned with the relative direction of airflow to the aircraft and points forward. The y- and z-axes follow the right-handed system. Wind axes vary, corresponding to the airflow velocity vector relative to the aircraft.
3. *Inertial axes*, F_I, fixed on the Earth. For speed and altitudes below Mach 3 and 100 000 ft., respectively, the Earth can be considered flat and not rotating, with little error, so the origin of the inertial axes is pegged to the ground. Conveniently, the x-axis points north and the y-axis east, making the z-axis point vertically downwards in a right-handed system.

In a body-fixed coordinate system, F_B, the components of six degrees of freedom are as follows:

Linear velocities	U along x-axis (+ve forward)
	V along y-axis (+ve right)
	W about z-axis (+ve down)
Angular velocities	p about x-axis known as roll (+ve) – aileron activated
	q about y-axis known as pitch (+ve nose up) – elevator activated
	r about z-axis known as yaw (+ve) – rudder activated
Angular acceleration	$\dot{p}$ about x-axis known as roll rate (+ve)
	$\dot{q}$ about y-axis known as pitch rate (+ve nose up)
	$\dot{r}$ about z-axis known as yaw rate (+ve)

In the wind-axes system, F_W, the components of six degrees of freedom are as follows:

Linear velocities	V along x-axis (+ve forward)
Linear accelerations	$\dot{V}$ along x-axis (+ve forward)
and so on.	

If the parameters of one coordinate system are known then parameters in other coordinate system can be found out through transformation relationship.

3.6.2 Forces

In a steady-state level flight, an aircraft is in equilibrium under the applied forces (i.e. lift, weight, thrust and drag) as shown in Figure 3.9. Lift is measured perpendicular to aircraft velocity (i.e. free stream flow) and drag is opposite to the direction of aircraft velocity (naturally, the wind axes, F_W, are suited to analyse these parameters). In a steady level flight, lift and weight are opposite one another; opposite forces may not be collinear.

In steady level flight (equilibrium, un-accelerated flight), the aircraft weight is exactly balanced by the lift produced by the wing (fuselage and other bodies could share a small part of lift – to be discussed later). Thrust provided by the engine is required to overcome drag.

It gives $\sum$ Force $= 0$,

That is, in vertical direction, lift (L) = weight (W) and in horizontal direction, thrust (T) = drag (D).

in short:

$$L = W \text{ and } T = D \tag{3.7}$$

that is, in equilibrium the vertical direction, lift = weight, and in the horizontal direction, thrust = drag. The aircraft weight is exactly balanced by the lift produced by the wing (the fuselage and other bodies could share a part of the lift – discussed later). Thrust provided by the engine is required to overcome drag.

3.6.3 Moments

The forces are not necessarily collinear and also the components of an aircraft have their inherent moments. Moments arising from various aircraft components are summed to zero to keep in straight flight, that is, in steady level flight $\sum$ Moment $= 0$.

Any force/moment imbalance would show up in the aircraft flight profile. This is how an aircraft is manoeuvred – through force and/or moment imbalance – even for the simple actions of climb and descent.

When not in equilibrium the accelerating forces are taken into account at the instant of computation to find its resultant affecting aircraft flight condition. If there were any force/moment imbalance, it would show up in the aircraft flight profile. That is how an aircraft is manoeuvred – through force and/or moment imbalance – even for simple actions of climb and descent.

A summary of forces and moments in body axes F_B is tabulated in Table 3.2.

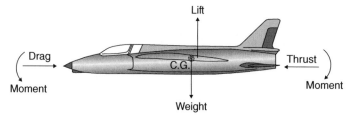

Figure 3.9 Equilibrium flight (C_G at ⊕) (Folland Gnat: the 1960s, United Kingdom – world's smallest fighter aircraft. Fuselage length = 9.68 m, span = 7.32 m, height = 2.93 m).

Table 3.2 Aircraft forces and moments in body frame, FB.

Axis	Force	Moment	Velocity Component	Angle	Angular Velocity Component
x	X	L	U	φ	p
y	Y	M	V	θ	q
z	Z	N	W	ψ	r

In wind axes, F_W, the forces and moments are transformed to different magnitude and direction where the force components X, Y and Z are resolved to lift (L), drag (D) and side force (C).

3.6.4 Basic Control Deflections – Sign Convention

Conventional aircraft have four basic controls: the elevators, ailerons, rudder and throttle. The elevator and the throttle are longitudinal controls, that is, they affect the two longitudinal degrees of freedom: changes of speed along the Ox-axis and pitch about the Oy-axis. Likewise, the ailerons and the rudder are called lateral controls because they affect the three lateral degrees of freedom: sideslip along the Oy-axis; roll about the Ox-axis and yaw about the Oz-axis.

(i) *Elevator*. The elevator is used to control the angle of attack of the aircraft, that is, pitch change and therefore its airspeed at constant power. This will be demonstrated when we come to the consideration of Longitudinal Static Stability. Note that positive elevator angle, η, generates negative pitching moment, M, and that this is achieved by pushing the control column forward (Figure 3.10).

(ii) *Aileron*. The ailerons are used to control the aircraft in roll. More specifically, the ailerons apply a rolling moment to the aircraft and so are used to demand *roll rate*. Roll rate is used to achieve bank angle, and bank angle is used to initiate a turn. Note that positive aileron angle (right hand side down), ξ, generates negative rolling moment, L (Figure 3.10).

(iii) *Rudder*. The rudder is used to control the side slip angle of the aircraft. This might be of turn. Note that positive rudder angle, ζ, generates negative yawing moment, N (Figure 3.10).

Heaving up and down results mainly from encountering sudden gust or arise as transient state of a manoeuvre.

The symbols employed to label the forces, moments and linear and angular velocities in this coordinate system are given in Table 3.1. This notation is used in flight dynamics (there are other conventions notably that are used in USA). (Note that L is used to denote the lift and V to denote the true airspeed of the aircraft. This will not cause confusion because the context will always make it clear whether, for example, L refers to lift or rolling moment.)

Heaving (whole aircraft vertical movement) in the Oz direction can occur as a result of encountering a vertical gust or as a transient state on account of control application.

3.6.5 Aircraft Loads

Chapter 17 deals with aircraft load in detail. In this section, only a preliminary definition is given to get an idea on what aircraft load means and its basic definition.

An aircraft is subject to load at any time. In flight, aircraft load varies with manoeuvres and/or when gusts are encountered. Early designs resulted in many structural failures in operation. Figure 3.11 gives the force diagram of an aircraft in accelerated flight (pull-up).

Newton's law states that change from an equilibrium state requires an additional applied force, that would associate with some form of acceleration, a. When applied in the pitch plane to increase the angle of incidence,

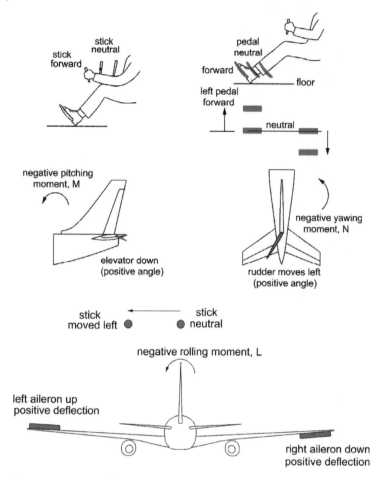

Figure 3.10 Control deflection and sign convention.

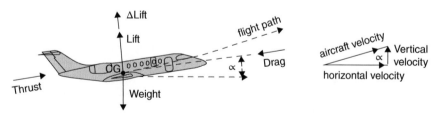

Figure 3.11 Equilibrium flight.

α, initiated by rotation of the aircraft, the additional force appears as an increment in lift, ΔL, resulting in gain in height (Figure 3.11).

From Newton's law, $\Delta L = $ acceleration $\times$ mass $= a \times W/g$ 　　　　　(3.8a)

The resultant force equilibrium gives $L' = L + \Delta L = W + a \times W/g = W(1 + a/g)$ 　　　(3.8b)

Load factor, n, is defined as $n = (1 + a/g) = L/W + \Delta L/W = 1 + \Delta L/W$ 　　　(3.9)

The load factor, *n*, indicates the increase in force contributed by the centrifugal acceleration, *a*. The load factor, $n = 2$, indicates a two-fold increase in weight; that is, a 90-kg person would experience a 180-kg weight. The load factor, *n*, is loosely termed as the *g load*; in this example, it is the 2-*g* load.

3.7 Definitions of Aerodynamic Parameters

Section 3.3 has already given the definition of *Re* and described the physics of the laminar/turbulent boundary layer. This section gives other useful non-dimensional coefficients and derived parameters frequently used in this book. The most common nomenclature, without any conflicts between both the sides of Atlantic are listed here – these are internationally understood.

$$\text{Let } q_\infty = \tfrac{1}{2}\rho V_\infty^2 = \text{dynamic head} \tag{3.10}$$

(subscript ∞ represents free stream condition and is sometimes dropped)

'*q*' is a parameter extensively used to non-dimensionalise lumped parameters.

The coefficients of 2D aerofoil and 3D wing differ as shown next (lower case subscripts represent the 2D aerofoil and the capital letters are for the 3D wing).

$$\text{2D aerofoil section (subscripts with lower case letters):} \tag{3.11}$$

C_l = sectional aerofoil lift coefficient = *section lift/qc*
C_d = sectional aerofoil drag coefficient = *section drag/qc*
C_m = aerofoil pitching moment coefficient = *section pitching moment/qc²* (+ nose up)

$$\text{3D wing (subscripts with upper case letters), replace chord, } c, \text{ by wing area, } S_W: \tag{3.12}$$

C_L = lift coefficient = *lift/qS_W*
C_D = drag coefficient = *drag/qS_W*
C_M = pitching moment coefficient = *lift/qS_W²* (+ nose up)

Chapter 4 deals with 3D wings, where correction to 2D results necessarily arrives at 3D values. The next section deals with lift generation over an aerofoil and pressure distribution in at any point over the surface in terms of pressure coefficient, C_p, which is defined as:

Equation (1.4) defined pressure coefficient at a point as,

$$C_p = (p_{local} - p_\infty)/q \tag{3.13}$$

Then sectional lift coefficient C_l of an aerofoil can be expressed as

$$C_l = \frac{1}{c}\int_0^c (C_{pl} - C_{pu})dx = \int_0^1 (C_{pl} - C_{pu})d(x/c) \tag{3.14}$$

3.8 Aerofoils

The cross-sectional shape of a wing (i.e. the bread-slice-like sections of a wing comprising the aerofoil) is the crux of aerodynamic considerations. The wing is a 3D surface (i.e. span, chord and thickness). An aerofoil represents 2D geometry (i.e. chord and thickness). Aerofoil characteristics are evaluated at mid-wing to eliminate the finite 3D wing tip effects. Typically, its characteristics are expressed over unit chord permitting scaling to fit any suitable size.

Aircraft performance depends on the type of aerofoil incorporated in its wing design. Aerofoil design is a specialised topic not dealt with in this book. There are many well proven existing aerofoils catering for a wide range of performance demands. NACA/NASA have systematically developed aerofoils in a family of designs and these are widely used by many aircraft. The majority of general aviation (GA) aircraft have NACA aerofoils. However, to stay ahead of competition, all major aircraft manufacturers develop their own aerofoils that are kept 'commercial in confidence' as proprietary information. These are not available in public domain. Moreover, unless these aerofoils can be compared it is not recommended to use them in undergraduate studies. The NACA series aerofoil will prove sufficient to make a good GA aircraft and this book only depends on this family.

Nowadays, there are specialised application software packages for designing aerofoil sections and these can bring out credible results. The only way to apply them is to verify the analytical results by comparing with wind tunnel test results. Certification of any new aerofoil is an additional task.

This section introduces various types of aerofoil and their geometrical characteristics. Subsequent chapters deal with aerofoil characteristic and how to make the selection. The 3D effects of a wing are discussed in Section 4.7.

3.8.1 Subsonic Aerofoils

To standardise aerofoil geometry, Figure 3.12 provides the universally accepted definitions that should be well understood [6].

Chord length is the maximum straight-line distance from the LE to the TE. The *mean line* represents the mid-locus between the upper and lower surfaces. The *camber* of an aerofoil is expressed as the percent deviation of the mean line from the chord line. The mean line is known as the *camber line*. Coordinates of the upper and lower surfaces are denoted by y_U and x_L for the distance measured from the chord line.

The chord line is kept aligned to the x-axis of the Cartesian coordinate system and measured from the LE. The *thickness* (t) of an aerofoil is the distance between the upper and the lower contour lines at the distance along the chord, measured perpendicular to the mean line and expressed in percentage of the full chord length. Conventionally, it is expressed as the *thickness to chord* (t/c) ratio as a percentage. A small radius at the LE is necessary to smooth out the aerofoil contour. It is convenient to present aerofoil data with the chord length non-dimensionalised to unity so that the data can be applied to any aerofoil size.

Aerofoil pressure distribution is measured in a wind tunnel (also in CFD) to establish its characteristics, as shown in [6]. Wind tunnel tests are conducted at mid-span of the wing model so that results are as close as possible to 2D characteristics. These tests are conducted at several Re values. A higher Re indicates a higher velocity; that is, it has more kinetic energy to overcome the skin friction on the surface, thereby increasing the pressure difference between the upper and lower surfaces and, hence, more lift.

3.8.2 Aerofoil Lift Characteristics

Figure 3.13 shows the typical test results of an aerofoil as plotted against a variation of the angle of attack, α. These graphs show aerofoil characteristics.

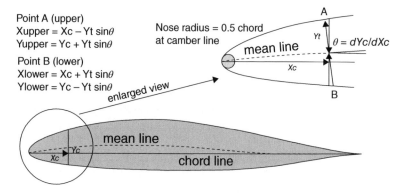

Figure 3.12 Aerofoil section and definitions – NACA family.

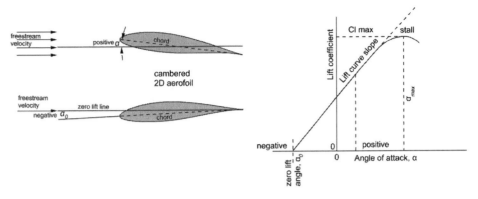

Figure 3.13 Typical aerofoil characteristics.

Angle between free stream velocity vector and aerofoil is the angle of attack, α. As α increases, the upper curvature effect increases over the aerofoil airflow velocity, thereby increasing circulation that results in generating more lift up to a point when viscous effects deplete flow energy to make the aerofoil stall as explained in Section 3.5. This is shown in Figure 3.13: that until the incipient stage of stall the lift increases, it varies linearly with α increase. This is the operating range of aircraft at α well below the incipient stage of stall. A cambered aerofoil develops some lift at $\alpha = 0$. In this case, the zero lift angle α_0 is negative. Initially, the variation is linear; then, at about 8° to 10° α, it starts to deviate reaching maximum C_{lmax} at α_{max}. Past α_{max}, the C_l drops rapidly – if not drastically – when stall is reached. Stalling starts at reaching α_{max}. The lift curve slope $dC_l/d\alpha$ is slope of the lift curve within its linear range of variation, typically measured between α from about $-2°$ to 6°.

3.8.3 Groupings of Subsonic Aerofoils – NACA/NASA

From the early days, European countries and the United States undertook intensive research to generate better aerofoils to advance aircraft performance. By the 1920s, a wide variety of aerofoils appeared and consolidation was needed. Since the 1930s, NACA generated families of aerofoils benefiting from what was available in the market and beyond by further testing. It presented the aerofoil geometries and test results in a systematic manner, grouping them into family series. The generic pattern of the NACA aerofoil family is listed in [6] with well-calibrated wind tunnel results. The book was published in 1949 and has served aircraft designers (civil and military) for more than half a century and is still useful. The NACA was subsequently reorganised to the NASA and continued with aerofoil development, concentrating on having better laminar flow characteristics over an aerofoil. Also, the industry undertook research and development to generate better aerofoils for specific purposes, but they are kept 'commercial in confidence'.

Designations of the NACA series of aerofoils are as follows: four-, five- and six-digit, given here are the ones of the interest to this book. Many fine aircraft have used the NACA aerofoil. However, brief comments on other types of aerofoil are also included.

The NACA four- and five-digit aerofoils were created by superimposing a simple camber-line shape with a thickness distribution that was obtained by fitting with the following polynomial [6]:

$$y = \pm (t/0.2) \times \left(0.2969 \times x^{0.5} - 0.126 \times x - 0.3537 \times x^2 + 0.2843 \times x^3 - 0.1015 \times x^4\right) \tag{3.15}$$

3.8.3.1 NACA Four-Digit Aerofoil

Each of the four digits of the nomenclature represents a geometrical property, as explained here using the example of the NACA 2315 aerofoil, is shown in Figure 3.14.

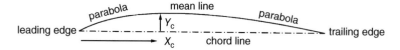

Figure 3.14 Camber-line distribution of the NACA 2315.

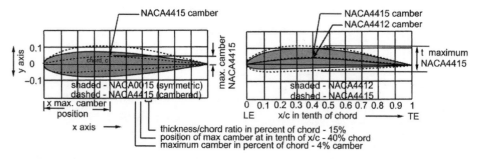

Figure 3.15 Comparing NACA 0015, NACA 4412 and NACA 4415 geometries.

2	3	15
Maximum camber, y_c as a percentage of chord	Location of $y_{c\text{-max}}$ in 1/10 of chord from the LE ratio as a percentage of chord	The last two digits give the maximum thickness to the chord

The camber line of four-digit aerofoil sections is defined by a parabola from the LE to the position of maximum camber followed by another parabola to the TE (Figure 3.14). This constraint did not allow the aerofoil design to be adaptive. For example, it prevented generation of aerofoil with more curvature towards the LE in order to provide better pressure distribution. Figure 3.15 compares NACA 0015, NACA 4412 and NACA 4415 geometries.

3.8.3.2 NACA Five-Digit Aerofoil

After the four-digit sections came the five-digit sections. The first two and last two digits represent the same definitions as the four-digit NACA aerofoil and are still applicable except the camber shape differs as defined by the middle digit. The middle digit stands for the aft position of the mean line bringing the change in defining camber-line curvature. The middle digit has only two options of zero for a straight (i.e. standard) and one for an inverted cube. These are explained next for the NACA 23012 with NACA 23112 aerofoil.

2	3	0 or 1	12
Maximum camber as a percentage of chord, Y_C The design-lift coefficient is 3/2 of it, in tenths	Max. thickness of max camber in 1/20 of chord X_C	0 – straight/standard 1 – inverted cube	The last two digits give the maximum thickness to chord ratio as a percentage of chord

Explanation for NACA 23012

First digit, 2. It has 0.02c maximum amount of camber with design-lift coefficient $= 2 \times (3/20) = 0.3$.
Second digit, 3. Position of maximum camber at $3 \times 2/200 = 15\%$ chord length from LE.
Third (middle) digit 0. Aft camber shape is straight (standard).
Last two digits, 12. It has maximum thickness to chord ratio $= 0.12$, i.e. 12% of the chord length.

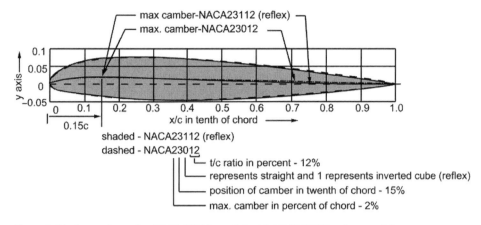

Figure 3.16 Comparison of an NACA 23012 aerofoil with NACA 23112 reflex aerofoils.

NACA 23112: Same as before, except that its aft camber shape has a inverted cube shape; that is, curved (NACA Report 537]. Figure 3.16 compares the NACA 23012 with the NACA 23112 aerofoil.

The NACA five series have been well used in many GA/utility category aerofoil. They have higher C_{lmax}, low $C_{m\text{-}ac}$ with good C_{dmin} and $C_{l\alpha}$. However, these aerofoils are not suitable for ab nitio trainer aircraft due to abrupt stall characteristics that are not forgiving for student pilots.

3.8.3.3 NACA Six-Digit Aerofoil

The five-digit family was an improvement over the four-digit *NACA* series aerofoil; however, researchers subsequently found better geometric definitions to represent a new family of a six-digit aerofoil. The state-of-the-art for a good aerofoil often follows reverse engineering – that is, it attempts to fit a cross-sectional shape to a given pressure distribution. The NACA six-digit series aerofoil came much later (it was first used for the P51 Mustang design in the late 1930s) from the need to generate a desired pressure distribution instead of being restricted to what the relatively simplistic four and five-digit NACA series could offer. The six-digit series aerofoils were generated from a more or less prescribed pressure distribution and were designed to achieve some laminar flow. This was achieved by placing the maximum thickness far back from the LE. Their low-speed characteristics behave like the four- and five-digit series but show much better high-speed characteristics. However, the drag bucket seen in wind tunnel test results may not show up in actual flight. Some six-digit aerofoils are more tolerant to production variation compared to typical five-digit aerofoils.

NACA six-digit aerofoils are possibly those most popular, widely used in various classes of aircraft. Their success was followed by increased effort to develop an aerofoil with laminar flow characterises over a wide speed regime. The definition for the NACA six-digit aerofoil example using NACA 63_2-212 is as follows.

An NACA six-digit aerofoil example 63_2-212 definition is:

6	3	Subscript $_2-$	2	12
Six Series	Location of min Cp in 1/10 chord	Half width of low drag bucket in 1/10 of C_l	Ideal C_l in tenths (design)	Max thickness as a percentage of chord

The six-digit aerofoil nomenclature follows the following sequence.

first number. The number '6' indicating the series.

*second number. One digit describing the distance of the minimum c_p (pressure) area in tens of percentage points of the chord.

***third number**. The subscript digit gives the range of lift coefficient in tenths above and below the design-lift coefficient in which favourable pressure gradients exist on both surfaces.

A hyphen in between.

fourth number. Ideal aerofoil design-lift coefficient C_1 in tenths.

fifth and sixth numbers. The last two digits is the maximum thickness as a percentage of the chord.

*It does not refer to camber. In the complete form, the six series mean line type is indicated by an associated letter 'a' (p. 121 of [6]), not given here.

**The subscript can also be expressed in parentheses, for example 63_2-212 as 63(2)-212. Within the parentheses there could be more numbers as explained in [6]. A modified aerofoil can carry the letter 'A' (e.g. 64_1A212 given in p. 122 of [6].

The nomenclature of NACA six series is an involved one. The readers are recommended to refer to [1] (section 3.8 and p. 119–122) for the exact definition. This book only covers how to use the graphs of the extent applied in worked-out examples.

In the example of the six-series aerofoil NACA 65_2-415 given in Figure 3.17, the minimum pressure position is at the 50% chord location indicated by the second digit. This is the point of inflection at the upper surface. The subscript 2 indicates that the minimum drag coefficient (drag bucket) is near its minimum value over a range of C_1 of 0.2 above and below the design-lift coefficient. The next digit indicates the design-lift coefficient of 0.4, and the last two digits indicate the maximum thickness in percentage of the chord of 15%.

Three NACA six-series aerofoils are compared with location of minimum C_p from 0.3c to 0.5c from the LE.

3.8.3.4 NACA Seven-Digit Aerofoil

The seven-digit family of aerofoil followed to further maximising laminar flow achieved by separately identifying the low-pressure zones on upper and lower surfaces of the aerofoil. The aerofoil is described by seven digits in the following sequence (Figure 3.18).

first number. The number '7' indicating the series.

second number. One digit describing the distance of the minimum pressure (C_p) area on the upper surface in tens of percent of the chord.

third number. One digit describing the distance of the minimum pressure area on the lower surface in tens of percent of the chord.

fourth is one letter referring to a standard profile from the earlier NACA series.

fifth number. Single digit describing the lift coefficient in tenths.

sixth and seventh number. The last two digits give the maximum thickness as a percentage of the chord.

Figure 3.18 is an example of the NACA 747A315 of the seven-digit aerofoil series. The first digit '7' indicates the series number. The second digit '4' signifies that it has favourable pressure gradient on the upper surface to the extent of 40% of chord peaking to minimum c_p and thereafter starts the adverse pressure gradient. The third digit '7' says the same for the lower surface and in this case up to 70%. The last three digits are same

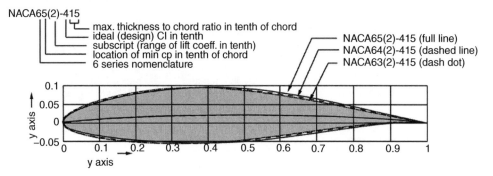

Figure 3.17 Comparison of NACA65_2 415, NACA64_2 415 and NACA63_2 415 aerofoils.

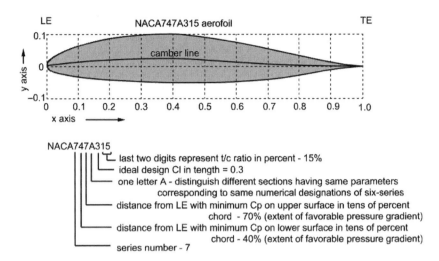

Figure 3.18 The Seven Series NASA747A315 aerofoil.

nomenclature as for the six-digit NACA aerofoil, indicating it has a design-lift coefficient of $C_l = 0.3$ and a maximum thickness of 15% of the chord, preceded by a letter 'A' to distinguish that it has a class of different sections.

3.8.3.5 NACA Eight-Digit Aerofoil

The NACA eight-digit aerofoil series are the *supercritical aerofoils* designed to independently maximise airflow above and below the wing. These are a variation of the six-series and seven-series and the eight-digit series has the same numbering system as the seven-series aerofoils except that the first digit is an '8' to identify the series. Application of NACA eight-digit aerofoils to aircraft is yet to be proven and not discussed here.

3.8.3.6 Peaky-Section Aerofoil

Peaky-section aerofoils were developed during the early 1960s by the large commercial aircraft manufacturers to fly at higher M_{crit} (20 count drag rise – see Section 3.14). It was done by tweaking their well-proven high performing aerofoil of the time by slightly drooping down the aerofoil nose section and re-contouring. This allowed a rise of local airspeed near the LE on account well designed nose droop, causing negative C_p to peak up, hence the name (Figure 3.19). Distributed weak local shocks were allowed to form at that area that do not cause flow separation until it may happen further downstream when higher M_{crit} is reached.

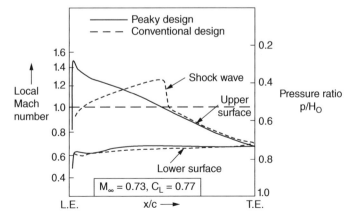

Figure 3.19 Comparison between a peaky-section aerofoil with a conventional aerofoil.

The NASA supercritical aerofoil appeared during the mid-1960s and gradually replaced the peaky-section aerofoil.

3.8.3.7 NASA Supercritical Aerofoil

In an effort to develop aerofoil to operate at a higher subsonic speed yet retaining good low-speed characteristics better than what existed, Richard T. Whitcomb of the Langley Research Center developed the supercritical aerofoil during the early 1960s. The goal was to increase the drag divergence Mach number (Figure 3.13), thereby reducing drag and allowing for more efficient flight in the transonic regime. It has the characteristic shape of a flat top following a large LE radius and curved tail with thickness at the TE.

This distinctive aerofoil shape helps the local supersonic flow with isentropic recompression on account of reduced curvature over the middle region of the upper surface and substantial aft camber.

The National Aeronautics and Space Administration (NASA) made systematic studies in three phases during the 1970s to develop this family of aerofoils with thicknesses from 2 to 18% and design-lift coefficients from 0 to 1.0. These were called the 'supercritical aerofoils'.

Three phases of development are as follows:

Phase 1. Supercritical Aerofoils: Early Supercritical Aerofoils that increased the drag divergence Mach number beyond the six-series NACA family.

Phase 2. Supercritical Aerofoils: The extension of Phase 1 Supercritical Aerofoils with target pressure distributions.

Phase 3. Supercritical Aerofoils: aerofoils developed for studies to reduce Phase 2 LE radii. The study was eventually abandoned.

The supercritical aerofoil number designation is in the form in the example of SC(2)-0412 as shown in Figure 3.20. The aerofoil designation is broken down into two segments – the first segment of three characters starts with SC, indicating it is a supercritical aerofoil and the bracketed number shows the development phase – in this case, Phase II of three phases of development as described next. The last segment starts with its first two digits as the design-lift coefficient in tenths and the last two are the thickness in percent chord, in the example $C_l = 0.4$ and $t/c = 12\%$.

3.8.3.8 Natural Laminar Flow (NLF) Aerofoil

The NLF class of aerofoil was designed by NASA during the early 1980s as a follow-up of the successful NACA six-series aerofoil for low subsonic speed GA aircraft operation offering low profile drag at high Re. NLF(1)-0213 and NLF(1)-0414 exhibited good laminar flow up to 70% of chord length at Mach 0.4 and $Re = 10 \times 10^6$. NLF type aerofoils suit the composite wing, that can have smooth polished surface, better than a metal wing.

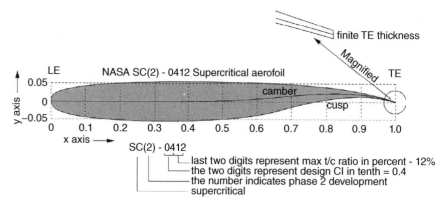

Figure 3.20 Supercritical aerofoil NASA SC(2)-0412.

Figure 3.21 Natural laminar flow aerofoil.

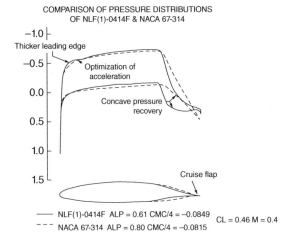

COMPARISON OF PRESSURE DISTRIBUTIONS
OF NLF(1)-0414F & NACA 67-314

Typically, NLF aerofoil shows 'flat roof-top' pressure distribution. NLF(1)-0414 has achieved $C_{lmax} = 1.83$ at $\alpha = 18°$ operating at Re $= 10 \times 10^6$. Figure 3.21 compares the NLF(1)-0414 with NACA 67-314.

Later for high-speed applications, for example, business jets, HSNLF(1) – 0213 was designed to suit applications in compressible flow (HSNLF = high speed natural laminar flow aerofoil).

3.8.3.9 NACA GAW Aerofoil

The NASA General Aviation Wing (GAW) series evolved later for low-speed applications and use by GA (Figure 3.22). Although the series showed better lift-to-drag characteristics, their performance with flap deployment, tolerance to production variation and other issues are still in question. As a result, the GAW aerofoil has yet to compete with some of the older NACA aerofoil designs. However, a modified GAW aerofoil has appeared with improved characteristics.

The numbering system is similar to the supercritical aerofoil.

3.8.3.10 Supersonic Aerofoils

A supersonic aerofoil is a cross-section geometry designed to generate lift efficiently at supersonic speeds (Figure 3.23). The need for such a design arises when an aircraft is required to operate consistently in the supersonic flight regime. Supersonic aerofoils are necessarily thin in the range of $\approx 0.04 < (t/c) < \approx 0.07$.

Supersonic aerofoils generally have a thin section formed of either angled planes (called 'double wedge aerofoil') or opposed arcs (called 'biconvex aerofoil') with very sharp leading and TEs. The sharp edges prevent the formation of a detached bow shock in front of the aerofoil as it moves through the air. This shape is in contrast to subsonic aerofoils, which often have rounded LEs to reduce flow separation over a wide range of angle of attack A rounded edge would behave as a blunt body in supersonic flight and thus would form a bow shock, which greatly increases wave drag. The aerofoils' thickness, camber and angle of attack are varied to achieve a design that will cause a slight deviation in the direction of the surrounding airflow.

However, since a round LE decreases an aerofoil's susceptibility to flow separation, a sharp LE implies that the aerofoil will be more sensitive to changes in angle of attack. Therefore, to increase lift at lower speeds,

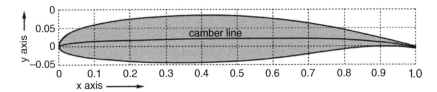

Figure 3.22 NASA/Langley/Whitcomb LS(2)-0413 (GA(W)-2) general aviation aerofoil.

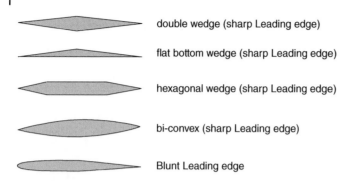

double wedge (sharp Leading edge)

flat bottom wedge (sharp Leading edge)

hexagonal wedge (sharp Leading edge)

bi-convex (sharp Leading edge)

Blunt Leading edge

Figure 3.23 Supersonic aerofoil.

aircraft that employ supersonic aerofoils also use high-lift devices such as LE and TE flaps. The thin six and seven series aerofoils have been used in combat aircraft design.

The supersonic aerofoil designation is as follows.

$$NACA\ NS - (X1)(Y1) - (X2)(Y2)$$

The letter 'N' is replaced by the series number, the number '1' being used for wedge-shape profiles and the number '2' being used for circular arc profiles. The letter 'S' denotes it is supersonic. The letter 'X1' represents the distance along the chord from the LE to the point of maximum thickness 'Y1' for the upper surface. The letters 'X2' and 'Y2' represent the corresponding values for the lower surface. 'X' and 'Y' are the percentage aerofoil cord length. In the following are some examples of 6% t/c aerofoil.

Subsonic	
NACA 66-006	blunt LE aerofoil
Supersonic	
NACA 1S-(30)(03)-(30)(03)	wedge shaped aerofoil (double wedge, max. thickness at 30% chord)
NACA 1S-(70)(03)-(70)(03)	
NACA 2S-(30)(03)-(30)(03)	
NACA 2S-(50)(03)-(50)(03)	circular arc aerofoil (in this case biconvex, max. thickness in middle)
NACA·2S-(70)(03)-(70)(03)	

Typically, the sharp LE thin supersonic aerofoil, at its clean basic configuration, has low C_{lmax} (in the order of 0.8 to 0.9). Its LE droop increases the aerofoil camber giving ΔC_{lmax} in the order of 0.4 to 0.5. A 20° TE deflection can give nearly twice the C_{lmax} of the basic aerofoil C_{lmax}.

3.8.3.11 Other Types of Subsonic Aerofoil

NACA's earliest attempt (in the 1930s) to make a systemic generic type was the NACA 1-Series (or 16 series) [6]. This new approach to aerofoil design had its shape mathematically derived from the desired lift characteristics. Prior to this, aerofoil shapes were first created and then had their characteristics measured in a wind tunnel. The 1-series aerofoils are described by five digits. Since this type is no longer used, it is not discussed here.

Subsequently, after the six series sections, aerofoil design became more specialised with aerofoils designed for their particular application. In the mid-1960s, Whitcomb's 'supercritical' aerofoil allowed flight with high critical Mach numbers (operating with compressibility effects, producing in wave drag) in the transonic region.

The NACA seven and eight series were designed to improve some aerodynamic characteristics. In addition to the NACA aerofoil series, there are many other types of aerofoil in use.

To remain competitive, the major industrial companies generate their own aerofoil. One example is the peaky-section aerofoil that were popular during the 1960s and 1970s for the high-subsonic flight regime. Aerofoil designers generate their own purpose-built aerofoil with good transonic performance, good maximum lift capability, thick sections, low drag and so on – some are in the public domain but most are held commercial in confidence for strategic reasons of the organisations. Subsequently, more transonic supercritical aerofoils were developed, by both research organisations and academic institutions. One such baseline design in the United Kingdom is the RAE 2822 aerofoil section, whereas the CAST 7 evolved in Germany. It is suggested that readers examine various aerofoil designs.

There are many other types of aerofoil, for example, Eppler, Liebeck (used in gliders) and many older types, for example, the Wortmann, Gottingen, Clark Y, Royal Air Force (RAF) aerofoils and so on, not discussed here. There are large number of other aerofoil developed by many scientists, in addition to proprietary aerofoil developed by industry. However at this stage, the well-used and established NACA series aerofoils will serve adequately until the readers join industry to use their data and analyses methods, today using CFD. While, NACA series aerofoil test data are still prevalent, the use of DATCOM (the short name for the *USAF Data Compendium for Stability and Control*)/ Engineering Sciences Data Unit (ESDU) for aerofoil analyses is gradually receding. URLs [8–11] may prove useful to get some information on various types of aerofoil.

Discussion In earlier days, drawing the full-scale aerofoils of a large wing and their manufacture was not easy and great effort was required to maintain accuracy to an acceptable level; their manufacture was also not easy. Today, computer-aided drawing/computer-aided manufacture (CAD/CAM) and microprocessor-based numerically controlled lofters have made things simple and very accurate. In December 1996, NASA published a report outlining the theory behind the U.S. National Advisory Committee for Aeronautics (NACA) (predecessor of the present-day NASA) aerofoil sections and computer programs to generate the NACA aerofoil.

Aerofoil characteristics are sensitive to geometry and require hard tooling with tight manufacturing tolerances to manufacture to adhere closely to the profile.

Often, a wing design has several aerofoil sections varying along the wing span. Appendix F provides six types of aerofoil for use in this book. Readers should note that the 2D aerofoil wind tunnel test is conducted in restricted conditions and will need corrections for use in real aircraft. Section 3.14.1 gives a simplified aerofoil selection method.

3.9 Reynolds Number and Surface Condition Effects on Aerofoils – Using NACA Aerofoil Test Data

This section explains the NACA aerofoil test result that shows the Re effect. NACA aerofoil test data is compiled in a specific format [6]. Figure 3.24 gives the traced NACA 65-410 aerofoil test data [6]. A higher Re implies higher airspeed that increases C_{lmax} (shown magnified in Figure 3.24), assisted by increased energy of airflow.

A small aircraft with 4 ft chord flying at 100 mph (146.7 ft s^{-1}) at sea level standard has Re $= 3.85 \times 10^6$ ($\mu_0 = 3.62 \times 10^{-7}$ lb-sec ft^{-2}). This will require a large wind tunnel to do the tests. To keep cost low, tests are conducted on smooth aerofoil model at three Re numbers. Tests are also done with rough surface models to represent realistic operational situation, because wing skin gets degraded/contaminated during usage. With a rough surface, there is considerable all-round degradation of aerofoil capability with the loss of the drag bucket, resulting in drag rise. Corrections are applied to the full scale by extrapolating test data.

The designers examine the test data focusing on the following five aerofoil properties. The test results of the NACA 65-410 aerofoil with a rough surface at Re $= 6 \times 10^6$ are given next.

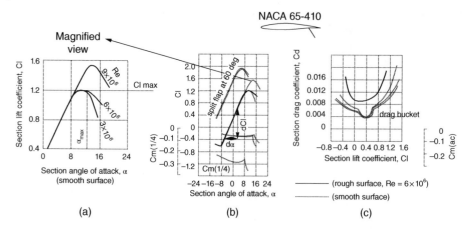

Figure 3.24 Traced NACA 65-410 aerofoil test data at three Re values [1]. (a) Magnified graph. (b) C_l and $C_{m1/4}$ result. (c) C_d and $C_{m\text{-ac}}$ result.

1. High C_{lmax} at high angle of attack, α	$C_{lmax} \approx 1.2$ at $\alpha \approx 12°$
2. High-lift curve slope ($dC_l/d\alpha$)	$dC_l/d\alpha \approx 0.01^{a)}$
3. Low drag (have drag bucket)	$C_d \approx 0.009$
4. Low moment C_m	$C_{m(1/4)} \approx -0.07$
5. Zero lift angle, $C_{l\alpha = 0}$	$-2°$
6. Stall characteristics	It has moderately gentle stall

a) $dC_l/d\alpha$ is to be evaluated at the linear range of the C_l graph in the operating range of the aircraft, in this case from $-4°$ to $6°$ angle of attack.
Two-dimensional aerofoil results are to be corrected for 3D wing application (treated in Section 4.7.2).

3.9.1   Camber and Thickness Effects

Section 3.8.1 gave the definition of camber line (Figure 3.12). It represents the curvature of the aerofoil mean line. For a symmetric aerofoil, the camber is zero. Camber assists top surface aft end gradient to assist the favourable pressure gradient to delay flow separation (Figure 3.25). Zero lift angle of cambered aerofoil becomes negative. Test results/CFD analyses give the extent of the zero lift angle associated with camber. Positive camber gives a negative nose-down moment and vice versa. Therefore, an aircraft with a wing that

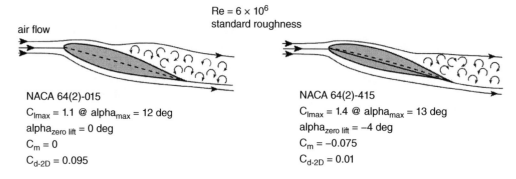

Figure 3.25 Camber effect comparison.

has a positive camber will require a horizontal tail with a negative camber to balance the flight to stable equilibrium.

When compared with same aerofoil without camber (i.e. symmetric aerofoil), its introduction is necessary to delay separation with an increase in angle of attack, hence higher C_{lmax}, at the cost of some drag and moment rise. Figure 3.25 shows that NACA 64(2) 415 with camber offers higher C_{lmax} associated with higher α_{max} and negative zero lift angle at the expense of higher drag and nose-down moment. A compromise is sought aiming to get the optimum camber that gives the maximum (L/D) with the least amount of associate C_m.

Figure 3.26 explains the physics by comparing test data at Re = 6×10^6 of three aerofoil (NACA 65_2-015, NACA 65_2-015 and NACA 65_2-415), one symmetric and the other two with camber.

Increase of camber shifts the lift graph to the left ($C_{l\alpha=0}$ more negative), that is, increases sectional lift coefficient, C_l at the same α at the expense of increased drag and moment coefficients.

The amount of camber is linked with the thickness to chord level and its distribution. Typically, the camber is 2–8% for subsonic applications and 0–2% for supersonic applications. The lower the aircraft design speed, the more the camber may be.

There are special aerofoils suited to horizontal tailless aircraft such as all wing (blended-wing body, BWB) aircraft. These wings have mixed camber, positive in the front and a small amount of negative camber at the TE to balance. These are known as camber with reflex. The NACA 23112 aerofoil (Figure 3.9) has small amount of reflex.

Figure 3.27 studies the NACA four-digit aerofoil camber (NACA 1412, 2412 and 4412) and thickness (NACA 2412, 2415 and 2418) effects.

The lift characteristics degrades after certain increase in thickness, in this case past a 15% thickness to chord (t/c) ratio. Since structural engineers prefer thicker aerofoils to attain structural integrity and increased wing internal volume for fuel accommodation, typically, for subsonic aircraft applications, the selection remains between 10 and 18% t/c. For low-speed aircraft below Mach 0.3 the aerofoil thickness is on the high side; many successful aircraft have 15–18% t/c. But for aircraft operating above Mach 0.7, aerofoil thickness will depend on the wing sweep. Higher speed application with less wing sweep has lower t/c (see Figure 4.23 – Section 4.11.1).

Figure 3.28 shows typical trends in thickness and camber variations comparing NACA six-digit aerofoil. Such trends also show in other NACA series aerofoil. The exact value must be read from the test data given in [6]. The camber effect of the other NACA series confirms similar characteristics as in the NACA six series

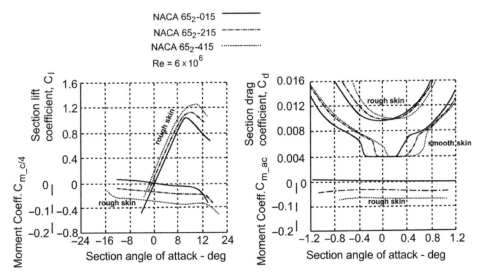

Figure 3.26 Comparing three test aerofoils (NACA 65_2-015, NACA 65_2-215 and NACA 65_2-415).

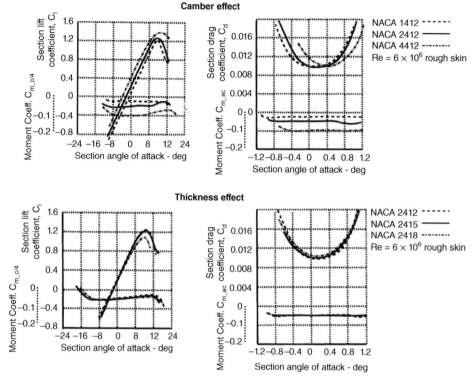

Figure 3.27 Comparing NACA four-digit aerofoils (NACA 1412, 2412, 2415, 2418 and 4412),

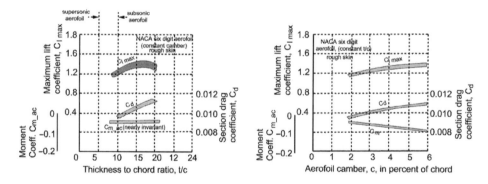

Figure 3.28 Typical trends in thickness and camber effects – comparing NACA six-digit aerofoils.

shown in Figure 3.28 except that the NACA four-digit does not have a drag bucket. The thickness effect shows practically no change in moment characteristics. Thickness effect can be seen as thicker the aerofoil more is the pressure drag on account higher blockage are and of small increase in contour length increasing the skin friction drag (at the same C_L). However, thicker aerofoils offer benefits with better structural integrity, lower weight and increased volume that can accommodate more fuel, if required.

Current research is being conducted on aerofoils with a self-adaptive camber using smart materials to suit the full range of operational flight speed for the best L/D. Smart materials can be customised to generate a specific response to a combination of inputs. These materials include piezo-electrics and electrostrictives, as well as shape memory alloy (SMA), which is lightweight and produces high force and large deflection.

NACA Four, Five and Six series aerofoil

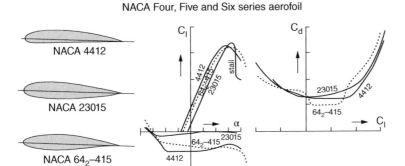

Figure 3.29 Comparison of three NACA aerofoils [6].

3.9.2 Comparison of Three NACA Aerofoils

The NACA 4412, NACA 23015 and NACA 642-415 are three commonly used aerofoils – there are many different types of aircraft that use one of these. Figure 3.29 compares these three popular NACA aerofoils (source NACA) to examine what is needed to make the choice.

The NACA 23015 has sharp stalling characteristics; however, it can give a higher sectional lift, C_l and lower sectional moment, C_m, than others. Drag-wise, the NACA 64_2-415 has a bucket to give the lowest sectional drag. The NACA 4412 is the oldest and, for its time, was the favourite. Of these three examples, the NACA 64_2-415 is the best for gentle stall characteristics and low sectional drag, offsetting the small drag due to the relatively not so high moment coefficient.

The five-digit NACA 23015 gives the highest C_{lmax} at higher angle of attack, α, but has abrupt stall not suitable for the ab initio club trainer but serves the utility category of aircraft. The benefits of the newer six-digit NACA 64_2-415 offers all round benefits with gentler stall characteristics including a lower C_m and a drag bucket to offer lower drag at low C_L compared to four-digit NACA 4414. However, the four-digit NACA series is tolerant to production variation, hence it is popular in small aircraft designs as it is cheaper to manufacture.

The advanced NACA six-digit series aerofoil came after NACA five-digit series aerofoil making this successful aerofoil to fall behind.

An aerofoil designer must produce a suitable aerofoil that encompasses the best of all five qualities – a difficult compromise to make. Flaps are also an integral part of the design. Flap deflection effectively increases the aerofoil camber to generate more lift. Therefore, a designer also must examine all five qualities at all possible flap and slat deflections.

3.10 Centre of Pressure and Aerodynamic Centre

As stated earlier, an aerofoil is 2D bread-slice like wing section is considered weightless. The pressure field around aerofoil develops forces and moment.

The *centre of pressure, c.p.,* is defined as a point where moment vanishes and the resultant force of the pressure field acts. In other words, *c.p.* is the point where the resultant force represents the equivalent of what the pressure field generates on the aerofoil; the resultant force vector acting at the centre of pressure is the value of the integrated pressure field where there is no moment. The resultant force is resolved into lift and drag. Since the pressure field changes with angle of attack, the position *c.p.* changes accordingly. This makes *c.p.* a difficult parameter to deal with.

As the angle of attack, α, is increased, the aerofoil gets front loaded and the *c.p.* moves forward. With decrease of α the *c.p.* moves back. When α is reduced to a low value, for cambered aerofoil the inherent

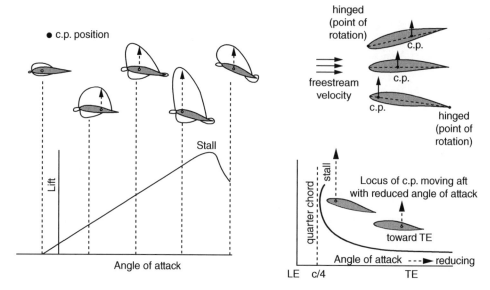

Figure 3.30 Movement of centre of pressure with change in lift.

nose-down moment may move the *c.p.* rearward outside of the aerofoil. Low resultant force at low angle of attack requires a large moment arm to balance, pushing the *c.p.* further aft. At zero lift, the *c.p.* approaches infinity. For a typical aerofoil, the *c.p.* stays within the aerofoil for the operating range. Figure 3.30 depicts the movement of centre of pressure, *c.p.*, with a change in lift. The figure shows that, past α_{max}, the C_l drops rapidly – if not drastically – when stall is reached. Stalling starts at reaching α_{max}. It will be shown in Section 3.10.1 that *c.p.* is always aft of the quarter chord. This makes for a realistic aerofoil, as the aft is loaded with a nose-down (−ve) moment. The extent of aft loading depends on how far behind is the minimum *c.p.* point. The second digit of the NACA six series gives the minimum *c.p.* point in tenth of the chord from the LE.

The aerodynamic centre, *ac*, is concerned with moments about a point, typically on the chord line (Figures 3.31). It does not deal with force by itself. However, it is noticed that the moment about the quarter chord of the aerofoil is invariant to the angle of attack ($dC_m/d\alpha = 0$) until stall occurs. This point is invariant to α change and is known as the *a.c.*, which is a natural reference point through which all forces and moments are defined to act. The '*a.c.*' is close to the quarter chord point. There could be minor variations of the position of the *a.c.* around the quarter chord among aerofoils. To standardise, the fixed point at the quarter chord from the LE is also measured. In term of coefficient, these are C_{m_ac} and $C_{m1/4}$ are measured in wind tunnel tests. A symmetric aerofoil has $C_{m\_ac} = 0$ but there is small variation of $C_{m1/4}$ with α variation. The relation between C_{m_ac} and $C_{m1/4}$ can expressed in mathematical relations as dealt with in Section 3.10.1. Aerofoil characteristics given in Appendix F show gives both the C_{m_ac} and the $C_{m1/4}$.

The *a.c.* offers much useful information. At the *a.c.*, although the $dC_m/d\alpha = 0$, it has some moment (except symmetrical aerofoil) and is not the *c.p.*, which is always aft of the *a.c.* The *a.c.* is an useful parameter in stability analyses when aircraft CG has to be taken into account. The higher the positive camber, the more lift is generated for a given angle of attack; however, this leads to a greater nose-down moment. To counter this nose-down moment, conventional aircraft have a horizontal tail with the negative camber supported by an elevator. For tailless aircraft (e.g. delta wing designs in which the horizontal tail merges with wing), the TE is given a negative camber as a 'reflex'. This balancing is known as *trimming* and it is associated with the type of drag known as *trim drag*. Aerofoil selection is then a compromise between having good lift characteristics and a low moment.

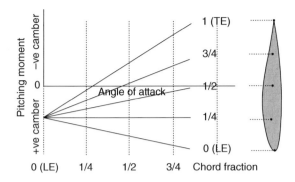

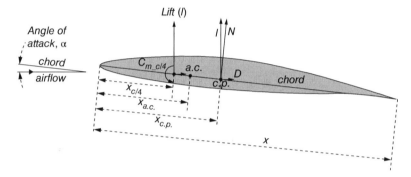

Figure 3.31 Aerodynamic centre invariant near the quarter chord (fractional chord position about which the moment is taken).

When series of aerofoil are stacked up side by side to form a wing that is to be integrated with the aircraft, then contribution of isolated wing weight has to be considered its moment characteristics. Typically, wing Mean Aerodynamic Chord (MAC) (Section 4.8) represents its reference geometry to compute wing aerodynamic characteristics. The point of CG location is about which the wing moment is considered. The CG can be ahead or aft of the quarter chord point of the wing MAC. Chapter 18 explains that static stability requires to simultaneously satisfy (i) that aerofoil C_m has to be positive at $\alpha = 0$ and (ii) $dC_m/d\alpha$ has to be negative as well. A typical cambered aerofoil has a negative C_m at $\alpha = 0$. Therefore, the wing will require an additional surface behind it (a small wing like horizontal tail at aft fuselage or a reflex wing TE) to keep nose up to retain static stability. The wing and empennage are dealt with in Chapters 4 and 6, respectively.

3.10.1 Relation Between Centre of Pressure and Aerodynamic Centre

Typically, the a.c. is within 22–27% of the aerofoil chord from the *LE*. Since the point varies from aerofoil to aerofoil, therefore, for standardisation, test results are given about a fixed point of quarter chord, *c*/4, from the *LE*. From the test result of a aerofoil moment around the quarter chord, the aerodynamic centre can be accurately determined and also given in a graph (Appendix F).

Figure 3.31 shows a typical aerofoil (NACA six series) with quarter chord *c*/4 and the aerodynamic centre, *a.c.*, shown at a distance $x_{c/4}$ and $x_{a.c.}$ from the *LE*, respectively. At the $x_{c/4}$, the sectional lift, L, and moment $M_{c/4}$ are shown in the diagram (drag is small and its moment contribution is negligibly small).

3.10.1.1 Estimating the Position of the Aerodynamic Centre, *a.c.*

As mentioned before, the quarter chord $x_{c/4}$ is at a fixed position and test results of $C_{m\_c/4}$ are available [6]. Also, C_{m_ac} is invariant with respect to angle of attack, α (i.e. C_l). Taking the moment about the *a.c.* the moment equilibrium gives (moment contribution by drag is small as the moment arm is negligible). The aerofoil is considered weightless and the definitions of the aerodynamic coefficients are given in Equation (1.2).

$$\sum M = 0 \text{ that is, } M_{ac} = l \times (x_{a.c.} - x_{c/4}) + M_{c/4}$$

In coefficient form, $C_{m\_ac} = (C_l/c) \times (x_{a.c.} - x_{c/4}) + C_{m\_c/4}$ (3.16)

Differentiating with respect to angle of attack α, it becomes (note that C_{m_ac} is constant),

$$0 = (dC_l/d\alpha) \times (x_{a.c.}/c - x_{c/4}/c) + dC_{m\_c/4}/d\alpha$$

Transposing $(x_{a.c.}/c - x_{c/4}/c) = -(dC_{M\_c/4}/d\alpha)/(dC_l/d\alpha) = -(m_0/a_0)$

where, m_0 = slope of the $C_{M\_c/4}$ curve and a_0 = is the lift curve slope.

substituting in Equation (3.16), $C_{m\_c/4} = C_{m\_ac} + \dfrac{(dC_{m\_c/4}/d\alpha)}{(dC_l/d\alpha)}(C_l) = C_{m\_ac} - \dfrac{(m_0)}{(a_0)}(C_l)$ (3.17)

m_0 and a_0 can be evaluated from the test results (Appendix F).

or $x_{a.c.}/c = x_{c/4}/c - m_0/a_0 = 0.25 - m_0/a_0$ (in terms of percentage of chord) (3.18)

3.10.1.2   Estimating the Position of the Centre of Pressure, *c.p.*

Estimating *c.p.* is lot more difficult as it moves (Figure 3.30), with the operating range from mid-chord, $c/2$, to *a.c.* at aircraft stall.

Let the aerofoil section lift l act at the aerofoil *c.p.* Taking moment about the quarter chord, $c/4$.

$$\sum M = 0, \text{ that is, } 0 = l \times (x_{c.p.} - x_{c/4}) + M_{c/4.}$$

In coefficient form, $C_l \times (x_{c.p.} - x_{c/4}) + C_{m\_c/4} = 0$
Substituting the value of $C_{m\_c/4}$ from Equation (3.18) into this equation

$$C_l \times (x_{c.p.}/c - x_{c/4}) + C_{m\_ac} - \frac{(m_0)}{(a_0)}(C_l) = 0$$

or $(x_{c.p.}/c - x_{c/4}/c) = -(C_{m\_ac})/C_l + \dfrac{(m_0)}{(a_0)}$ (3.19)

C_{m_ac} is negative for conventional aerofoil, that makes $x_{c.p.}/c > x_{c/4}$, always.

Replacing $m_0/a_0 = (x_{a.c.}/c - x_{c/4}/c)$ from the relation given below Equation (3.16) in this equation it becomes:

$$(x_{c.p.}/c - x_{c/4}/c) = -(C_{m\_ac})/C_l + (x_{a.c.}/c - x_{c/4}/c)$$

or $x_{c.p.}/c = (x_{a.c.}/c) - (C_{m\_ac})/C_l$ (3.20)

It can be seen that the *c.p.* is aft of the *a.c.* At the *a.c.*, although the $dC_m/d\alpha = 0$, it has some moment (except symmetrical aerofoil).

Example 3.1   Find the aerodynamic centre, *a.c.* and centre of pressure, *c.p.*, at 6° of angle of incidence, α, for the NACA 65-410 aerofoil from the test results given in Appendix F.

Solution

From the test result graph NACA 65-410 gives $dC_l/d\alpha = a_0 = (1.05-0.25)/8 = 0.1$/degree
and $m_0 = (-0.0755 - (-0.062))/[8 - (-4)] = -0.0135/12 = -0.00125$/degree
Equation (3.17) gives, $x_{a.c.}/c = x_{c/4}/c + m_0/a_0 = 0.25 - (-0.00125/0.1) \approx 0.2625$ of the chord from the LE.
Page 186 of [6] gives ≈ 0.262. (The ideal C_l for NACA 65-410 is 0.3.)
Then at 6° angle, $C_l = 0.9$ and $C_{m\_c/4} = -0.075$ from the graph in Appendix F.
Then at angle of zero lift $C_{m\_ac} = C_{m\_c/4} + C_l \times (x_{a.c.}/c - x_{c/4}/c) = -0.075 - 0.9 \times (0.012) = -0.0846$.
Equation (3.20) gives, $x_{c.p.}/c = (x_{a.c.}/c) - (C_{m\_ac})/C_l = 0.25 + (0.0846/0.9) = 0.25 + 0.094 = 0.344$.

The centre of pressure and aerodynamic centre can be analytically determined if the aerofoil properties can be analytically expressed. Within in the operating range, the aerofoil lift characteristics can be expressed as a straight line and the moment characteristic is not exactly a straight line but can be tolerated as straight line; however the drag characteristics may not be amenable to fit as an equation of parabola resulting in inaccuracies. To obtain industry standard results, it is recommended that test data may be used [6].

It is stressed that the readers must understand the role of *a.c.* and *c.p.* This gives a good insight into how an aerofoil behaves.

3.11 Types of Stall

Section 3.3 describes the physics of stall phenomena over an aerofoil. It is essential that designers understand stalling characteristics because wing stall is an undesirable state for an aircraft to enter. Figure 3.32 shows the general types of stall that can occur. This section describes how these different types of stall affect aircraft design.

3.11.1 Gradual Stall

This is a desirable pattern and occurs when separation is initiated at the TE of the aerofoil; the remainder maintains the pressure differential. As the separation moves slowly towards the LE, the aircraft approaches stall gradually, giving the pilot enough time to take corrective action. The forgiving and gentle nature of this stall is ideal for an ab initio trainee pilot. The type of aerofoil that experiences this type of stall has a generously rounded LE, providing smooth flow negotiation but not necessarily other desirable performance characteristics.

3.11.2 Abrupt Stall

This type of stall invariably starts with separation at the LE, initially as a small bubble. Then, the bubble either progresses downstream or bursts quickly and catastrophically (i.e. abruptly). Aerofoils with a sharper LE, such as those found on higher-performance aircraft, tend to exhibit this type of behaviour. At stall, C_m degrades to a sudden increase of nose-down moment. Aircraft stall is affected by wing stall, which depends on aerofoil characteristics. Stall characteristics with high-lift devices deployed have to be examined at the takeoff and landing operations. Thin aerofoil have abrupt stall. Aircraft with thin aerofoil are flown by experienced pilots trained to recognise the incipient stages of stall to take corrective actions in time.

A thin aerofoil with a sharp LE exhibits an abrupt stall.

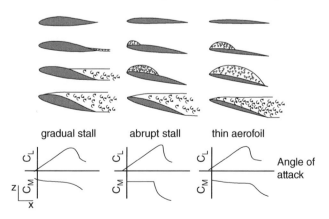

Figure 3.32 Stall pattern.

3.12 High-Lift Devices

For pilot ease and safety, takeoff and landing are done at the lowest achievable speed as conceived by the specification requirements. A large commercial transport aircraft cruising at about 500 kn at 39 000 ft altitude, lifts off at about 150 kn and approaches to land at about 130 kn. Flying less than a third of cruise speed is at high C_L will require some sort of lift augmentation by incorporating some form of high-lift devices that changes the aerofoil configuration from its retracted position (loosely termed 'clean') to a more cambered and extended chord aerofoil (loosely termed a 'dirty' configuration).

High-lift devices are small aerofoil-like elements that are fitted at the TE of the wing as a *flap* and/or at the LE. A typical high performance high-lift device is shown in Figures 3.33. It is a double slotted Fowler flap with a slot at the LE of the wing. In this example, the Fowler flap system has a 'vane' to give the first gap for the slot followed by a flap that creates the gap for the second slot.

In typical cruise conditions, the flaps and slats are retracted within the contour of the aerofoil. Flaps and slats can be used independently or in combination. At low speed, they are deflected, rendering the aerofoil more curved as if it had more camber. The Fowler flap extends the chord, which increases the wing area. A typical flow field around the flaps and slats is shown in Figure 3.33. The entrainment effect through the gap between the wing and the flap allows mixing of boundary layer flow to remain attached in order to provide the best possible lift. Gaps and overlaps are to be optimised to obtain the best high-lift.

In order to augment lift to achieve low-speed capabilities, the deployment of high-lift devices penalise by increase in drag and nose-down moment, as well as the stall margin gets reduced associated with unfavourable abrupt stall. While it achieves the goal of lift-off at shorter field length at takeoff, the drag increase degrades the climb rate and may prove critical to clear obstacles and changes the pitch attitude requiring high degree of trim to ease stick force. Approach to land occurs at steeper angles. These aspects are dealt with in detail in subsequent chapters.

In order to augment lift to achieve low-speed capabilities, the deployment of high-lift devices penalise by increase in drag and nose-down moment, as well as the stall margin gets reduced and have unfavourable abrupt

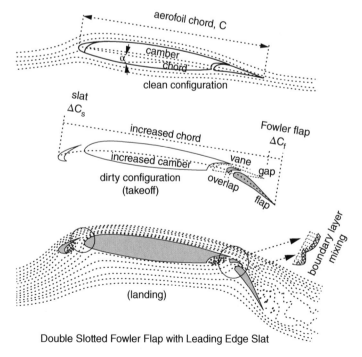

Figure 3.33 High-lift device.

stall. While it achieves the goal of lift-off at shorter field length at takeoff, the drag increase degrades the climb rate and prove critical to clear obstacles and changes the pitch attitude requiring high degree of trim to ease stick force. Approach to land occurs at steeper angles. These aspects are dealt with in detail in subsequent chapters.

A qualitative lift augmentation by using high-lift devices is shown in Figure 3.34. While the lift curve slope $dC_l/d\alpha$ is not greatly affected, the deployment of flaps gives a higher C_l at the same angle of attack, α, on account of an increased camber. Note that LE slot itself does change the camber holding on to the same α at zero lift, $\alpha_{Cl=0}$, but all other high-lift devices reconfigure the basic aerofoil to a higher camber when $\alpha_{Cl=0}$ takes a more negative value.

There are various kind of high-lift devices as shown in Figure 3.35, the choice depends on the category of aircraft as listed in Section 4.15. LE and TE devices are shown separately. A high-lift system selectively

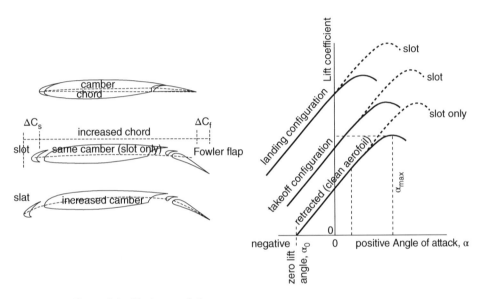

Figure 3.34 Flap and slat lift characteristics.

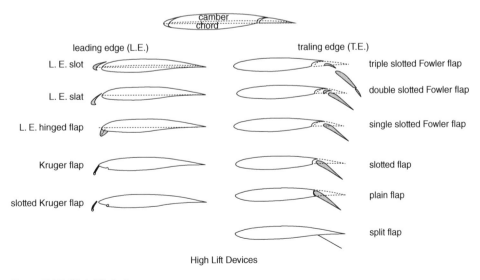

Figure 3.35 High-lift devices.

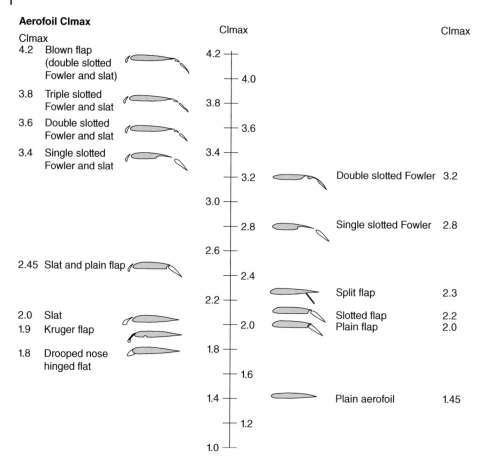

Figure 3.36 High-lift devices and typical values of 2D C_{lmax}.

combines as required for the design. Section 4.13 continues with high-lift design and this is discussed in detail there.

Figure 3.36 lists the experimental values of the lift coefficients of aerofoil (lift augmentation by *blown flaps* is not dealt with here) at the landing configuration. These values are representative of types of NACA aerofoils and may be used if actual data are not available. When high-lift devices are installed on a wing, its finite span 3D effects reduces the C_{lmax} to a lower value of C_{Lmax}, as shown in Figure 4.31. This topic extends to wing flap design in Section 4.13.

Higher-performance, high-lift devices are complex in construction and therefore heavier and more expensive. Selection of the type is based on cost-versus performance trade-off studies – in practice, past experience is helpful in making selections.

Figure 3.37 depicts the corresponding distribution of the pressure coefficient C_p at an angle of attack of 15°.

Deflection of either the control surface or a change in the angle of attack will alter the pressure distribution. The positive y-direction has a negative pressure on the upper surface. The area between the graphs of the upper and lower surface C_p distribution is the lift generated for the unit span of this aerofoil.

3.13 Flow Regimes

Disturbance in air creates pressure wave and if within the audible frequency range then it manifests as a sound wave. In other words, disturbance created by a moving object (a body, say an aircraft) travels through air at

Figure 3.37 Aerofoil C_p distribution characteristics.

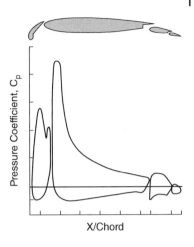

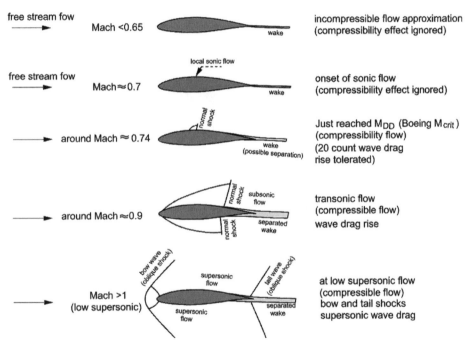

Figure 3.38 Flow regimes.

the speed of sound. As the speed of the object increases, the difference between the object speed and speed of sound decreases. Figure 3.38 shows the various flow regimes over an aerofoil (2D) section.

Approaching Mach 0.5, air molecules between the source and the limiting speed of sound start to coalesce and the effect is felt as *compressibility effects* offering some resistance to the moving object and is termed *wave drag*. Airflow over a curved surface accelerates higher than free stream velocity known as *super velocity*. Section 3.13.2 explains the compressibility effect on an aerofoil resulting from the onset of sonic flow at a point on account of the thickness distribution along the chord. The onset of wave drag is taken as a critical Mach number, M_{crit}. Below Mach 0.5, the compressibility effects are weak and can be ignored. The body speed approaching around Mach 0.75, an arbitrary limit is imposed accepting a 20 count ($\Delta C_D = 0.002$) drag rise at the speed known as the *drag divergence Mach number*, M_{DD}. Between M_{DD} and speed of sound, the compressibility effect is strong with a complex mix of subsonic and supersonic flow regimes over the body as *transonic flow*. Also, the local coalescing of pressure appears as a local shock wave on the surface, which interacts with the local boundary layer with the potential to rupture the flow and to separate leading to the undesirable result of loss of lift.

To repeat, *drag divergence* is a sudden increase in drag. A 20-count drag rise ($C_D = 0.002$) at the Mach number is known as the *drag divergence Mach* (M_{DD}), shown in Figure 3.38. The critical Mach (M_{crit}) is the onset of the transonic flow field and is lower than the M_{DD}. In the past, the NACA six-digit series of aerofoils were used in aircraft flying at transonic speeds at relatively low M_{DD}. Special attention is necessary when generating the aerofoil section designs that have a flat upper surface.

There several other ways to define drag divergence Mach. Airbus and Boeing take M_{crit} when drag rises to $\Delta C_D = 0.002$ due to local shock forming on the wing at a critical point when transonic effects need to be considered. The onset of local shock formation is yet to reach a critical point as the highest Mach number for aircraft to cruise is the M_{CR} when the compressibility effects are small enough to be ignored. The Douglas Aircraft Company (since taken over by Boeing) had another definition: when the rate of change of parasite drag, C_{DP} with respect to aircraft Mach number, M, shows a 10% gain, that is, $d(C_{DP})/dM = 0.1$. These are arbitrarily accepted. It appears that civil aircraft design prefers $\Delta C_D = 0.002$ and military aircraft design prefers $d(C_{DP})/dM = 0.1$ to define the M_{DD}. In this book, henceforth, the Boeing definition will be used based on the first author's work experience at the Boeing Aircraft Company, that is M_{CR} and M_{crit} and instead of M_{crit} and M_{DD}, respectively (Figure 3.38).

When the body speed reaches the speed of sound (Mach = 1), the pressure wave coalesces to a wave front known as a *shock wave*, it starts as normal shock ahead of the body with some thickness and subsonic flow behind the normal shock. As the body speed increases above that of sound, the shock wave gradually turns to oblique shock and gets attached to the body if it has a sharp edge. In 3D, the oblique shock is in the shape of a cone/wedge whose subtended angle reduces with increase in the body speed. Since the disturbance created by the body cannot travel faster than speed of sound, the flow field ahead of the shock cone/wedge remains undisturbed and is seen as a *zone of silence*.

This section is meant to give an overview of the physics for the readers to review and apply. Wing M_{DD} and M_{crit} are dealt in Section 4.11.

3.13.1 Compressibility Correction

Compressibility effects require application of correction factors (Prandtl–Glauert Rule); high-subsonic and supersonic corrections differ slightly. In the following are some of the corrections of interest.

Parameter	High-subsonic	Supersonic

2D lift curve slope, a_0
$$a_{0\_comp} = \frac{a_{0\_incomp}}{\sqrt{(1 - M_\infty^2)}} = \frac{a_{0\_incomp}}{\sqrt{(M_\infty^2 - 1)}} \qquad (3.21)$$

pressure coefficient, C_p
$$C_{p\_comp} = \frac{C_{p\_incomp}}{\sqrt{(1 - M_\infty^2)}} = \frac{C_{p\_incomp}}{\sqrt{(M_\infty^2 - 1)}} \qquad (3.22)$$

lift coefficient, C_l
$$C_{l\_comp} = \frac{C_{L\_incomp}}{\sqrt{(1 - M_\infty^2)}} \qquad (3.23)$$

lift coefficient, $C_{l\alpha}$
$$C_{l\_\alpha\_comp} = \frac{C_{L\_\alpha\_incomp}}{\sqrt{(1 - M_\infty^2)}} \qquad (3.24)$$

For low subsonic speeds below ≈0.3 Mach, the correction factors are small and can be neglected. For high-subsonic speeds, the correction factor can be applied up to the onset of drag rise (M_{cr}, Boeing definition) ≈ 0.7 Mach as above this speed close to $M_\infty = 1.0$, the correction diverges to unrealistic values. The same is true for the supersonic speed, the correction factor is applicable above Mach 1.3. Fortunately, supersonic aircraft flying below 1.4 Mach do not offer aerodynamic and economic gains.

In a way, theoretical analyses of supersonic flow is simple compared to incompressible flow analyses. For a low aspect ratio (AR) wing, further correction is required to obtain a lift curve slope. This book uses experimental data and the Prandtl–Glauert correction factor is not used.

3.13.2 Transonic Effects

This section explains the transonic effects resulting from the thickness distribution along an aerofoil chord. At high-subsonic speeds, the local velocity along a curved surface (e.g. on aerofoil surface) can exceed the speed of sound, whereas flow over the rest of the surface stays subsonic. In this case, the aerofoil is said to be in *transonic flow*. At higher angles of attack, transonic effects can appear at lower flight speeds. Aerofoil-thickness distribution along the chord length is the parameter that affects the induction of transonic flow. A local shock on account of supersonic flow interacting with the boundary layer can trigger early separation, resulting in unsteady vibration.

Transonic characteristics exhibit an increase in wave drag on account of the compressibility effect. There is a rapid drag increase. Military aircraft in hard manoeuvre can enter into such an undesirable situation, even at a lower speed. As much as possible, designers try to avoid, delay or minimise the onset of flow separation over the wing due to local shocks. These effects are undesirable but unavoidable; however, aircraft designers keep the transonic effect as low as possible.

The patented Whitcomb aerofoil section appeared later with a flatter top that delayed the M_{crit} (Airbus/Boeing definition) to a higher Mach number. This was called the *supercritical aerofoil section*. It allowed the aircraft to fly faster with a 20 count wave drag rise. The geometrical characteristics exhibit a round LE, followed by a flat upper surface and rear-loading with camber; the lower surface at the TE shows the cusp. All modern high-subsonic aircraft have the supercritical aerofoil section characteristics. Figures 3.39 and 3.40 compare a typical transonic aerofoil (i.e. the Whitcomb section) and its characteristics with a NACA six series aerofoil. Manufacturers develop their own section or use any data available to them.

Structural engineers prefer aerofoil sections to be as thick as possible, which favours structural integrity at lower weights and allows the storage of more fuel onboard. However, aerodynamicists prefer the aerofoil to be as thin as possible to minimise the transonic flow regime in order to keep the wave drag rise lower. Wing planform shape influences the choice of aerofoil thickness. Wing planform is dealt with in Chapter 4.

The choice decides the extent of wing sweep required to lower the t/c ratio to achieve the desired result (i.e. to minimise the compressible drag increase for the cruise Mach number) while also satisfying the structural requirements. To standardise drag rise characteristics, the flow behaviour is considered to be nearly incompressible up to M_{cr} and can tolerate up to M_{crit}, allowing a 20-count drag increase ($\Delta C_D = 0.002$).

3.13.3 Supersonic Effects

As a result of the presence of shock wave, the flow physics of supersonic flow differs from subsonic flow. In case of supersonic flow, the disturbance generated by the body cannot travel ahead of the body and stays within the shock cone. Subsonic disturbances travels ahead of the object at the speed of sound.

Figure 3.39 Wing sweep versus aerofoil thickness to chord ratio.

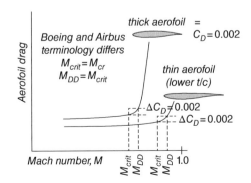

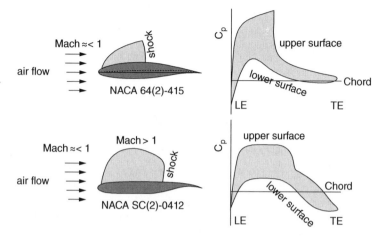

Figure 3.40 Transonic flow comparison (supercritical Whitcomb aerofoil).

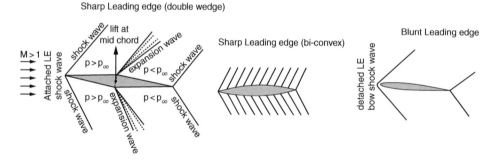

Figure 3.41 Shock patterns over three types of supersonic aerofoil.

Figure 3.41 shows the shock pattern over the three well-used types of supersonic aerofoil of six supersonic aerofoils in Figure 3.16.

The double wedge aerofoil has a sharp LE and the oblique shocks stay attached. Reaching a point of inflection, expansion waves develop to negotiate the favourable pressure gradient to accelerate flow reaching the TE again in a shock wave.

The biconvex aerofoil also has sharp LE but on smooth curvature the oblique shocks are distributed until the inflection is from where the distributed expansion continues with flow accelerating ending in tail oblique shock.

The blunt LE thin aerofoil has a small normal shock with a small zone of subsonic flow. As the aerofoil slope changes rapidly, the shock turns to oblique shock with an angle corresponding to the free stream Mach number (Figures 3.40 and 3.41). Downstream, the shock pattern is like this too. The NACA 66 206 is an example of this kind. Due to manufacturing limitation and operational maintenance, a blunt LE with a very small LE radius is widely used, especially if the wing is within the shock cone (see Section 4.13).

A thin supersonic aerofoil with a sharp LE in fully immersed supersonic flow gives (linear thin aerofoil theory) the following.

$$2\text{D lift coefficient, } C_1 = \frac{4\alpha}{\sqrt{M^2 - 1)}} \text{ where } \alpha = \text{angle of attack within the linear range}$$

$$\text{or lift curve slope, } \frac{dC_l}{d\alpha} = \frac{4}{\sqrt{M^2 - 1)}}$$

Figure 3.42 Flow over a supersonic aerofoil.

Supersonic aerofoils are very thin (4–% t/c) and have little camber. The cruise C_L values are low, that is, it operates at low α. These conditions allow the governing nonlinear equations to be linearised, yielding good results within the operating range. To obtain pressure distribution around the aerofoil, the linearised supersonic equations can then be split into three components of: (i) straight line at an angle of attack, α, (ii) aerofoil camber at zero angle of attack and (iii) thickness distribution of the aerofoil as shown in Figure 3.42 and superimpose the solutions by adding to get the results.

The low thickness to chord ratio (t/c), low camber and low angle of attack, α, indicate that the shock wave is very weak and its angle, β, can be approximated to a Mach wave angle, μ (i.e. $\beta \approx \mu$, refer to Section 3.3). This simplifies the wing design considerations discussed in Section 4.13.

3.14 Summary

Aerofoil design is a subject on its own involving theoretical CFD and extensive wind tunnel tests beyond the scope of this book. Since the 1980s, with the advent of CFD, the approach to aerofoil design started changing. As CFD capabilities became more refined, it is now possible to producing credible 2D aerofoil characteristics (lift, drag, moment and stall characteristics) in high fidelity, minimising wind tunnel testing to the point of making it nearly redundant, saving cost and time. CFD is now accepted as a *digital wind tunnel* to obtain aerofoil characteristics. This book only deals with the NACA series aerofoil. It is recommended to use aerofoil test data [6]. Six sets of suitable aerofoil experimental data are provided in Appendix F. This section gives a coarse guide to making an aerofoil selection.

This chapter covers a considerable amount of information, many in graphic representations to facilitate the newly initiated to appreciate the underlying technologies in choosing an aerofoil. The heart of aircraft design is in the selection of an appropriate aerofoil to arrive at the wing planform geometry fit for the purpose, that is, mainly to offer high-lift/drag ratio (L/D), low drag and low moment for the wing. The primary parameters of interest in selecting aerofoils are C_{lmax}, $C_{l\alpha}$, α_{stall}, $\alpha_{Cl=0}$, C_{m-ac} and C_{dmin}. There could be several candidate aerofoils with similarities in characteristics to satisfy performance requirements. It is also important to consider how tolerant the aerofoil is to manufacturing deviation. It requires experience to select an aerofoil relying on past experience and currently operating aircraft performances.

It is stressed here that aerofoil selection is not a standalone exercise. The degradation of aerodynamic properties of a 2D aerofoil as finite wing sections (see Section 4.7) has to be considered at this stage. Also the C_{lmax}, α_{stall}, $\alpha_{zero-lift}$, C_{m-ac} and C_{dmin} characteristics of high-lift devices deployed, as well stall characteristics, need to be examined in choosing a suitable aerofoil.

The compressibility effect of airflow influences the shaping of aircraft. Airflow below Mach 0.3 behaves as nearly incompressible – in such a regime all aircraft are propeller (piston engine) driven. From Mach 0.3 to 0.6 compressibility effects gradually build up but propellers are still effective up to Mach 0.6, however, its magnitude is small enough to be ignored. Above Mach 0.6, aircraft component geometry has to cater for compressibility effects. Jet propulsion becomes more suitable above Mach 0.6. Therefore, the aircraft component configuration is divided in two classes, one suited to flying below Mach 0.6 and one above Mach 0.6. The latest propeller technology is pushing flight speed towards Mach 0.7.

Aerofoil selection is mission specific. Aircraft speed requirements influence the aerofoil choice, of course, the other considerations, for example low C_m, low C_d, high t/c and so on, in a best compromise have to be taken into account. Based on the specified aircraft design speed, aircraft are grouped in the following category. This gives the Re of the wing.

Low subsonic (below Mach 0.5) – propeller driven – ($\Lambda_{1/4}$ wing sweep $\approx 0 \pm 5\%$ degree)

Gliders/Microlights operating around 80 mph in a very low Re No. Typical suitable aerofoils are chosen from the NACA four-digit, Liebeck, Lissaman, Eppler aerofoils series. Some of these offer laminar flow characteristics. Typical (t/c) ratio is around 0.14–0.18. This class of aircraft is not dealt with in this book.

Home-builds (Light Sports Aircraft – LSA) operating at around 120 mph with a low Re. A well proven NACA four-digit serves well. The LSA certification category of aircraft is not dealt with in this book. The design approach for this class is similar to the club trainer/recreational category of aircraft given next. Typical (t/c) ratio is around 0.12–0.18.

Club trainers/recreational aircraft (Normal Category, FAR 23) operating at around 150 mph in a low Re. Ab initio trainer aircraft must have gentle and gradual stall characteristics and be of a forgiving type, providing safety. Typical suitable aerofoils for trainer aircraft are chosen from the NACA four-digit series. The NACA five-digit series may be considered for recreational aircraft. With abrupt stalling characteristics, the NACA five-digit series are not suitable for trainer aircraft. Of late, the GAW aerofoils have been used in this category of aircraft. Typical (t/c) ratio is around 0.12–0.18.

Aerobatic Category (FAR 23) operating around 200 mph. On account of requiring better performance and handling qualities during inverted flying, a symmetrical aerofoil may also be considered. The NACA four- and six-digit aerofoils suit well. Typical (t/c) ratio is around 0.12–0.16.

Utility/Regional aircraft operating around 300 mph. This class of aircraft are flown by experienced pilots in fare-earning sorties at higher speeds but are well below compressibility effects to be considered. NACA five- and NACA six-digit aerofoils are now more commonly used.

High-subsonic (Mach 0.7 to Mach 0.95) – jet propulsion – ($\Lambda_{1/4}$ wing sweep ≈ 10–$35°$)

Incorporating wing sweep influences the choice for aerofoil t/c as shown in Figure 4.27 (forward sweep aircraft are uncommon but possible). In general, in this category of aircraft design, industries develop their own aerofoils and these are kept 'commercial in confidence'.

The *Bizjet/regional jet* operating between Mach 0.7 and 0.85 requires a swept wing. The use of the NACA six-digit aerofoil dominates high-subsonic aircraft design. Thickness to chord ratio t/c is around 0.09–0.15, depending on sweep (see Section 4.11.2).

Commercial jets operating between Mach 0.75 and 0.92. Aircraft with longer range missions have higher speed requiring relative wing $\Lambda_{1/4}$ sweep of 20–35° that influences the choice of aerofoil t/c. These high-subsonic aircraft operating in transonic regimes require careful consideration. Industry accepts a wave drag rise (compressibility effect) of 20 counts (see Section 4.11.1) for high-speed cruise at M_{DD}. The higher the M_{DD}, the thinner the t/c for the given wing sweep. The current practice is to adopt a supercritical aerofoil. Supercritical aerofoils have a good stretch of laminar flow from the LE. Laminar flow characteristics are sought after, but the extent of laminar flow during operation is not guaranteed on account of natural surface degradation during usage. There are excellent supercritical aerofoils in the NASA SC series. Six-digit aerofoils have also been used. The t/c values are around 0.12–0.15.

Supersonic category [$\Lambda_{1/4}$ wing sweep ≈ 0–$60°$ (delta wing)]

Combat aircraft operating between Mach 1.8 and 2.4. *Concorde* was the only commercial aircraft in operation; a new generation of supersonic aircraft is now under consideration. On account of the presence of shock waves, the supersonic aerofoil sections are of quite a different class for the reasons dealt with in Section 3.8.3.10. These are necessarily very thin aerofoils, typically with a t/c around 0.04–0.07.

Table 3.3 gives a few examples of NACA aerofoils that are widely used in industry. Note the absence of aerofoils used by Airbus and Boeing, as they develop their own.

The trends in the relationship of various parameters are shown in Figures 3.22, 3.31, 3.32 and 3.33. These show the trends for readers to compare and establish what are the desirable characteristics for the aerofoil to be chosen. In short, these graphs indicate some important aerodynamic properties that are summarised next. Table 3.4 gives the qualitative lists on the advantages and the disadvantages of the NACA series aerofoil.

Figure 3.43 gives some of the well-used aerofoil parameters. Surface roughness reduces the C_{lmax} and increases C_{dmin} from the smooth (clean) surface aerofoil data. Rough surfaces represent an operational environment. Suggestions given next are from well-proven NACA aerofoils widely in use.

Table 3.3 Aircraft using an NACA aerofoil [8–11].

Conventional Aircraft:	Wing Root Aerofoil	Wing Tip Aerofoil
Propeller Driven Aircraft		
Aero Commander 200	NACA 23015	NACA 4412
Aero Commander 500 Shrike	NACA 23012	NACA 23012
Aeronca 11-AC Chief	NACA 4412	NACA 4412
Auster J1B	NACA 23012	NACA 23012
Avro 748	NACA 23018	NACA 4412
Beagle B.121 Bulldog	NACA 63-615	NACA 63-615
Beech 300 Super King Air	NACA 23018	NACA 23012
Beech 77 Skipper	NASA GA(W)-1	NASA GA(W)-1
Britten-Norman BN-2A Islander	NACA 23012	NACA 23012
Canadair CL-215	NACA 4417 mod	NACA 4417 mod
Cessna 150	NACA 2412	NACA 0012
Cessna 170	NACA 2412	NACA 2412
Shorts S312 Tucano T.Mk1	NACA 63A415	NACA 63A212
RANS S-16 Shekari	NASA GA(W)-1	NASA GA(W)-1
Supermarine 378 Spitfire IX	NACA 2213	NACA 2209.4
Shorts SC.7 Skyvan	NACA 63A414	NACA 63A414
Shorts SD-330 (C-23)	NACA 63A418	NACA 63A414
Shorts SD-360	NACA 63A418	NACA 63A41
SIAI-Marchetti SF.600	NASA GA(W)-1	NASA GA(W)-1
Jet Propulsion Aircraft		
Cessna 560 Citation V Excel	NACA 23014 mod	NACA 23012
Gulfstream Aerospace GIV	NACA 0012 Mod	NACA 64A008.5 Mod
Learjet 55	NACA 64A109	NACA 64A109 mod
Learjet 60	NACA 64A109 mod	NACA 64A109 mod
Supersonic Combat Aircraft		
Boeing F-15 Eagle	NACA 64A006.6	NACA 64A203
Lockheed F-104	Biconvex 3.36%	Biconvex 3.36%
Lockheed Martin F-16	NACA 64A204	NACA 64A204
McDonnell Douglas F-15 Eagle	NACA 64A006.6	NACA 64A203
McDonnell Douglas F-18 Hornet	NACA 65A005 mod	NACA 65A003.5 mod
McDonnell Douglas F-4 Phantom II	NACA 0006.4-64 mod	NACA 0003-64 mod
North American Aviation RA-5C	NACA 65A005	NACA 65A005
Northrop F-5 Tiger	NACA 65A004.8	NACA 65A004

Figure 3.43 shows that the five-digit NACA 23015 gives the highest C_{lmax} at a higher angle of attack, α, but has abrupt stall characteristics not suitable for the ab initio club trainer, however, it serves utility category aircraft. The benefits of the newer six-digit NACA 64_2-415 are that it offers all round gains with gentler stall characteristics, lower C_m and a drag bucket to offer lower drag at the design C_l. The four-digit NACA series aerofoils are tolerant to production variation, hence they are popular in small aircraft designs being cheaper to manufacture.

Table 3.4 Generalised summary (see Section 3.15 at the end of Chapter 4).

Aerofoil	Desirable	Undesirable	Application
NACA four-digit	good gentle stall	moderate C_{lmax}	Low-speed application
(Manufacturing deviation)	very tolerant	high $C_{m\text{-}ac}$	ab initio trainer aircraft
Surface roughness	tolerant	high profile drag	small aircraft, club usage, empennage design
NACA five-digit	high C_{lmax}	abrupt stall	moderate speed
	low $C_{m\text{-}ac}$	high profile drag	utility aircraft
(Manufacturing deviation)		intolerant	
Surface roughness	tolerant		
NACA six-digit	acceptable stall		wide range of designs
	good C_{lmax}		(all category, extending
	moderate $C_{m\text{-}ac}$		to high-subsonic operation
	low profile drag (drag bucket)		thin sections for supersonic designs)
(Manufacturing deviation)	somewhat tolerant		
Surface roughness		intolerant	
GAW aerofoil	very high C_{lmax}		not many aircraft using it
	high α_{stall}		
z	low profile drag in cruise (drag bucket)		tin section to supersonic designs)
(Manufacturing deviation)	somewhat tolerant		
Surface roughness		intolerant	
Supersonic aerofoil	These are different class of aerofoil with considerable low values of C_{lmax} and α_{stall}. Post stall state shave C_{lmax} extend to around 20°, allowing high α manoeuvres. With additional vortex lift with wing strake, it can extend to as high as around 40° manoeuvres.		

Wing planform shape (dealt with in Chapter 4) influences the choice of aerofoil thickness as shown in Figure 4.20. The figure gives the relation between wing aerofoil t/c and aircraft design speed within the band of wing sweep. Increase of aircraft speed decreases aerofoil t/c.

3.14.1 Simplified Aerofoil Selection Methodology

A simplified step-by-step generalised approach to making an aerofoil selection from available test data (NACA/NASA series) is outlined in a generalised manner so as to give a feel for the process: in industry it is done in a more rigorous manner (this is not the only way to select an aerofoil). In most large aircraft manufacturing companies, there is a separate group of researchers who develop advanced technologies. The role of aircraft designer is to interact with the researchers to develop a suitable advanced aerofoil to perform better than the competition. Aerofoil design is now very mature; to get a better one is not easy. Similar classes of NACA/NASA aerofoils have characteristics close enough to serve many smaller aircraft.

Aerofoil selection must match the requirements of the wing capability, hence it is finalised along with the considerations of wing design as summarised in Section 4.19 to meet the mission specifications. In civil aircraft design, aerofoils with good L/D capability are sought. Aerofoils show best L/D at their design C_l. Test results have to be studied to obtain the design C_l for the NACA four- and five-digit aerofoils. For combat aircraft, C_l has to cover a wider range catering for manoeuvres as well. The step-by-step approach is given next.

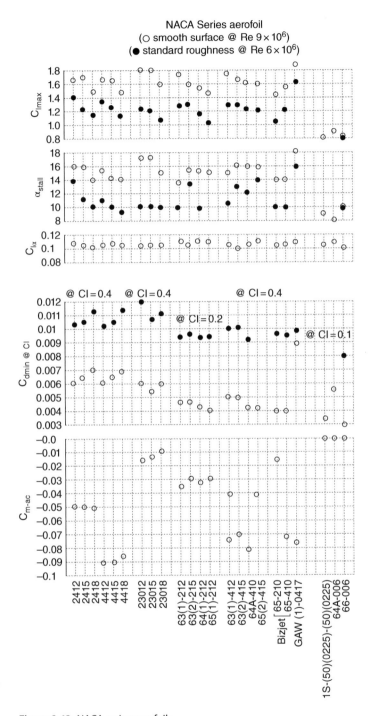

Figure 3.43 NACA series aerofoils.

Step 1

The first task is the find the design C_L for the new aircraft project. For civil aircraft, this is where the aircraft operates for the longest time. In transport aircraft, this is at the mid-cruise weight flying at a Long Range Cruise (LRC) Mach number.

Given here are some typical design C_L values for some classes of aircraft (more details in Section 4.19). Higher wing loads (W/S_W) give higher design C_L, operating at same-speed altitudes.

Aircraft category	Aircraft design C_L	Typical preferred aerofoil[a]
Large transport aircraft	≈0.25–0.35	NACA six-digit
Regional jets/Bizjets	≈0.3–0.4	NACA five- and six-digit aerofoil
Small club class aircraft	≈0.3–0.4	NACA four-digit aerofoil
Jet trainer	≈0.15–0.2	NACA six-digit aerofoil
Combat aircraft	≈0.1–0.2	Supersonic aerofoil

a) These are only typical NACA type suggestions. There are many other kinds that can be considered.

Step 2

Convert (see Section 4.9.4) the finite span wing design C_L to a 2D aerofoil C_l. Given next are some 2D aerofoil design C_l values that may be used at this stage. These coarse factors depend on the wing AR. This is not an industrial practice; that is considerably more rigorous using CFD, but it can start until the CFD generated aerofoil design C_l value is available. Unless the aircraft configuration has some wing body blending, the contribution to lift by the rest of aircraft is less than 3% and can be neglected. Oswald's efficiency factor, e (Section 4.9.3), gives the suitable factor taking AR and other effects into account.

Aircraft category	AR	Factor	Aerofoil design C_l
Large transport aircraft	6–8	≈0.8–0.85	≈0.3–0.4
Large transport aircraft	8–10	≈0.9–0.95	≈0.3–0.45
Regional jets/Bizjets	6–8	≈0.8–0.85	≈0.3–0.4
Small club class aircraft	5–6	≈0.7–0.8	≈0.4–0.5
Jet trainer	4–6	≈0.7–0.8	≈0.2–0.3
Combat aircraft	Case to case based study is required. A coarse way to initiate		
	combat aircraft study is to take the wing design C_L as the aerofoil C_l		

Step 3

The higher the camber, the higher the C_l at the same angle of attack, α. It is a desirable property but Figure 3.27 shows that a higher camber gives a higher $C_{m\text{-ac}}$, an undesirable property. Camber shifts the zero lift angle $\alpha_{Cl=0}$ and the C_d.

Lowest values of C_d and $C_{m\text{-ac}}$ are preferred but both at the lowest may not be possible. This requires seeking the extent of camber required that can compromise between C_d and $C_{m\text{-ac}}$, the lower the better. The NACA four-digit aerofoil camber is given by the first number (only camber extents 2 and 4% of chord length are considered). The NACA five-digit aerofoil camber is also given by the first number (camber of 2% of chord length is considered) However, the NACA six-digit series does not give the camber but the second digit indicates the location of minimum C_p: important information as an indication for C_d and $C_{m\text{-ac}}$ that requires examining the graphs. The second digit of a combat aircraft extends from 3 to 6.

Step 4

Decide the aerofoil thickness to chord ratio, t/c. Structural engineers prefer a thicker t/c that gives better rigidity lowering wing weight and offers provision between the spars to accommodate bigger volume for fuel storage. On the other hand, aerodynamicists desire thinner aerofoils to keep drag, C_d, low. Wing sweep (Section 4.8) permits an increase in the t/c. Typically, low subsonic aerofoils have a t/c between 12 and 18% and high-subsonic aerofoils have t/c between 9 and 15%. The thickness to chord ratio, t/c, may vary along the wing span, being lowest at the tip section. In the case of a supersonic aerofoil, $t/c \approx 0.05$ and camber is nearly zero.

Step 5

Lastly, high $C_{l\alpha}$ is sought. Figure 3.43 indicates that most subsonic aerofoils have $C_{l\alpha}$ around $0.105 \pm 5\%$ per degree. Interestingly, the choice of C_{lmax} comes later. Ab initio trainers prefer high C_{lmax} and high α_{stall}.

Step 6

Examine the aerofoil characteristics with a high-lift device. The NACA test graphs give data with a split flap. Absence of full details of aerofoil characteristics with contemporary flaps restrict the study in this book.

Other import aspects are to examine is the post-stall behaviour of the aerofoil with and without high-lift devices and how tolerant the aerofoil characteristics will be under production variations. Ab initio trainers require gentle and gradual stall that offers safety.

Although operational usage cannot guarantee laminar flow, aerofoil selection with a good laminar flow characteristic is desirable.

Start with picking up several suitable aerofoil sections (say 3–5) that have the design C_l. Choose the higher t/c ratio permitted by the wing sweep for the category of aircraft as given before. Next, to plot and superimpose the characteristics to compare and make the final choice.

Aerofoil selection for the worked out examples of *Bizjet*, *Advanced Jet Trainer* (AJT) and *Turboprop Trainer Aircraft* (TPT) are carried out in Chapter 8.

Since no single aerofoil can offer all of the desired properties at its maximum, compromise is required.

3.15 Aerofoil Design and Manufacture

Aerofoil design and manufacture are very important processes in any new aircraft project. Success depends on the fidelity to adhere to accuracy so that it performs as predicted. These are involved topics for which special study/training is required and are beyond the scope of this book.

This section is meant to make newcomers aware of the situation and give what is expected, should they be interested in pursuing the subject for research or production. Aerofoil design is theory intensive and special courses are offered by all aeronautical institutes. There are two ways to design aerofoils:

i. the direct aerofoil design method
ii. the inverse aerofoil design method

After obtaining the suitable aerofoil, the aircraft designers progress with the aircraft design (the objective of this book). The finished product is drawn and then passed on to the production engineers to be manufactured. The role of aircraft designer is to work in an integrated product and process development (IPPD) environment as part of a team, bridging all sides, to make the product right first time. The Chief Engineer must have a good grasp of all aspects of aircraft design to integrate various components as a system.

3.15.1 Direct Aerofoil Design Method

This method involves theoretical analyses of a given aerofoil geometry drawn from designers' past experiences and data banks. The analyses starts with the inviscid flow considerations that simpler to solve, which gives a good estimates of $dC_l/d\alpha$. For other parameters, for example, C_{lmax}, C_m and so on, viscous flow considerations involved with nonlinear differential Navier–Stokes equations are to be analysed; these are hard to solve if not sometimes impossible. Basically, the analytical procedure deals with tackling the two influencing parameters, the camber-line geometry and the thickness distribution along it, conducted at several angles of attack. In earlier days without computers, this was a time consuming process.

The results obtained from theoretical analyses necessitate substantiation by wind tunnel tests of several aerofoil geometries in a parameter manner to choose the best from the lot, not really an efficient way to optimise. The NACA four- and five-digit family of aerofoils were developed in this way. This is a lengthy, difficult and expensive process that could give a better aerofoil but the researchers want to achieve more.

3.15.2 Inverse Aerofoil Design Method

To overcome the limitations faced by the direct method, researchers succeeded by attempting an inverse method to analyse. The process starts with establishing an achievable superior C_p distribution and then making complex theoretical analyses requiring solution of difficult equations to find the matching geometry, if required iterating the C_p distribution to bring out what is achievable close to the specified one. This process delivers better aerofoils and has delivered new aerofoil identification designations.

3.15.2.1 Current Practice

Numerical methods for solving differential equations have been prevailing for some time. Solving the Navier–Stokes equations in compressible viscous flow was difficult, if not impossible in certain situations. Mathematicians devised methods to discretise the difficult nonlinear partial differential equations into algebraic forms, which are solvable. During the early 1970s, CFD results of simple 2D bodies in inviscid flow were demonstrated showing that certain parameters were comparable with wind tunnel test results and analytical solutions.

With more refinement in the numerical algorithm and with progress in computer processing power, the aerofoil design process has changed. Today, optimised aerofoil geometry can be found at a fraction of cost and time compared to the older methods. The confidence in the accuracy of CFD results for 2D aerofoils reached points where wind-tunnel tests for substantiation could be reduced to a few tests. The 2D aerofoil CFD results are now acceptable to the certification agencies. CFD analyses is now seen as a digital wind tunnel.

There are excellent free aerofoil design software packages developed by well-known researchers/academics available in the market. Some of them are extensively used in industry, large or small.

3.15.2.2 Manufacture

Manufacturing an accurate wing profile with aerofoil shaped ribs was possibly harder than designing them. It was difficult to produce large components in exact dimensions with the desired low tolerance that incurred a large number of rejections. The production process requires specified tolerances; the aerodynamicists want them as low as possible and the production engineers want them as high as possible. The compromise will associate with a corresponding loss of quality resulting in drag rise, moment change and degradation in C_{lmax}. The full-scale drawing of large aerofoils (say in the order exceeding 10 ft) had to be marked around the technician sitting inside it with a multitude of heavy weight location pins, which not amenable to maintaining tight tolerance. The lofting of wing skin required trial and error methods with some wastage. It was a painful process. The manual riveting had dimples, there were mismatches at the joints, waviness in the wing skin and so on. These methods could not avoid some costly rejections of the finished components.

All changed with the adaptation to the computer-aided engineering (CAE) (CFD/Finite Element Method (FEM)/CAD/CAM etc.) method; a computer-based design and manufacturing process. Today, accurate CAD modelling/drawings have dispensed with manual lofting and tighter tolerances can be specified as the CAD files can be directly sent to microprocessor-based Numerically Controlled (NC) machines that can machine/cut/rivet automatically at higher accuracy and at higher speeds that were hitherto impossible. The rejection rate has come down, man-hours have come down and machine hours have reduced.

It is strongly recommended that the readers may compare the finished surfaces of some older aircraft designs, say the class of the Mustang P, Douglas DC3 and so on, with some newer aircraft designs, say the class of F35, Airbus 320 and so on, to get a feel for the difference.

3.16 Aircraft Centre of Gravity, Centre of Pressure and Neutral Point

So far, mainly aerofoil aerodynamics has been dealt with. Two important aircraft parameters: (i) the aircraft CG and (ii) the aircraft Neutral Point (NP), need to be defined and explained at this early stage as they will be used in subsequent chapters for the whole aircraft. Shaping and positioning of aircraft components, for example, wing planform, fuselage shape, nacelle/intake shapes and empennage are established by satisfying the position of aircraft CG with respect to the location of the aircraft NP.

3.16.1 Aircraft Centre of Gravity (CG)

CG is the point where the resultant weight of all the assembled aircraft components and equipment acts vertically downward. Aircraft CG has to be computed, shown in the worked-out examples in Chapter 10. Assembling the aircraft components by placing them in relation to each other is to be made in a way that will keep the aircraft CG location in a desirable place. If requires, internal equipment can be repositioned to bring the CG to the desired location. Centre of pressure, *c.p.*, and aerodynamic centre, *a.c.*, are discussed in Section 3.11.1.

3.16.2 Aircraft Neutral Point

Aircraft NP (see Section 10.5) is associated with the location of the aircraft CG to assess aircraft stability. If aircraft CG is at the NP then the aircraft becomes neutrally stable, hence it is called the NP; the aircraft $dM/d\alpha = 0$.

The differences between aircraft centre of pressure, wing aerodynamic centre and aircraft NP are summarised in Section 3.16.3.

3.16.3 Summary

i) CG is the point where the resultant weight of all the assembled aircraft components and equipment acts vertically downward.
ii) Centre of pressure, *c.p.*, of an aircraft is the point where the lift acts. It has nothing to do with the aircraft CG.
iii) Aerodynamic centre, *a.c.*, is the point in the weightless aerofoil/wing where the pitching moments are invariant to the changes in angle of attack. It also has nothing to do with the aircraft CG.
iv) NP is a fixed point for the aircraft configuration and is associated with the location of aircraft CG, which can vary depending how the aircraft is loaded with payload and fuel. Aircraft CG located at the NP makes aircraft neutrally stable, hence the name.

References

1 Shapiro, A. (1953 and 1954). *The Dynamics and Thermodynamics of Compressible Fluid Flow*, vol. 1 and II. New York: Ronald (Wiley) Press.
2 Kuethe, A.M. and Chow, C.-Y. (1986). *Foundations of Aerodynamics*. Wiley.
3 Bertin, J.J. and Cummings, R.M. (2009). *Aerodynamics for Engineers*. Prentice Hall: Pearson.

4 Anderson, J. (2011). *Fundamentals of Aerodynamics SI*. McGraw Hill.

5 Schlichting, H. and Gersten, H. (2003). *Boundary Layer Theory*, 8e.

6 Abbott, I.H. and von Doenhoff, A.E. (1949). *Theory of Wing Sections, Including a Summary of Aerofoil Data*. New York: McGraw Hill.

7 Kuchemann, D. (2012). *The Aerodynamic Design of Aircraft*. AIAA.

8 ESDU. Website. Available online at: www.esdu.com/cgi-bin/ps.pl?sess=unlicensed_1180626100613ksk&t=gen&p=home (accessed June 2018).

9 Lednicer, D. Airfoils of US and Canadian Aircraft. Available online at: www.aerofiles.com/aerofoils.html (accessed June 2018).

10 Lednicer, D. (2010). The Incomplete Guide to Airfoil Usage. Available online at: http://m-selig.ae.illinois.edu/ads/aircraft.html (accessed June 2018).

11 USAF DATCOM. (2017) Public domain aeronautical software. Available online at /www.pdas.com/datcom.html (accessed June 2018).

4

Wings

4.1 Overview

The last chapter dealt with the pertinent aspects of aerodynamic considerations of 2D geometries as the basis for the aerodynamic design of aircraft. This and the next two chapters deal with the aerodynamic considerations of 3D geometries, as well as aircraft components. These are studied at the conceptual stage of new aircraft design project.

There are basically two types of geometrical shapes that constitute aircraft components, for example (i) planar bodies meant to deal with force and moments for aircraft to perform, for example, wings, empennage and control surfaces, dealt with in Chapters 4 and 6 and (ii) bodies of revolution, for example, fuselages, nacelles and so on, dealt with in Chapter 5.

These three Chapters 4–6 outline the components of aircraft and serve as the building blocks to arrive at aircraft configuration. The objective is to generate aircraft components piece by piece in a Lego/Mechano style and mate them in the way shown in the middle diagram of Figure 2.1. This information serves to develop 'boilerplate' template geometries of aircraft components to be put together to generate candidate aircraft configurations as a 'concept definition'. During the conceptual design phase, these will eventually be sized to 'concept finalisation' (typically, planar planform to zoom in/out) with matched engines (Chapter 14) to finalise the design.

This chapter begins with standard definitions of the various parameters of planar surfaces that will be used in this book and covers the following topics:

Section 4.2: Introduction to Aerodynamics
Section 4.3: Generic Wing Planform Shapes
Section 4.4: Wing Positional Relative to Fuselage
Section 4.5: Structural Considerations
Section 4.6: Wing Parameter Definitions
Section 4.7: Spanwise Variation of Aerofoil of t/c and Incidence
Section 4.8: Mean Aerodynamic Chord
Section 4.9: Wing Aerodynamics
Section 4.10: Aspect Ratio Correction
Section 4.11: Compressibility/Transonic Effects
Section 4.12: Transonic Wing
Section 4.13: Supersonic Wing
Section 4.14: Additional Vortex Lift
Section 4.15: Additional Surfaces on Wing
Section 4.16: High-Lift Devices
Section 4.17: Square-Cube Law
Section 4.18: Influence of Wing Area and Span
Section 4.19: Summary of Wing Design

Conceptual Aircraft Design: An Industrial Approach, First Edition. Ajoy Kumar Kundu, Mark A. Price and David Riordan.
© 2019 John Wiley & Sons Ltd. Published 2019 by John Wiley & Sons Ltd.

Section 4.19 summarises this long chapter containing considerable information on what aerodynamic considerations affect wing design.

4.2 Introduction

Wings as lift generation surfaces have relatively small thicknesses compared to their characteristic lengths. The quest is to design a wing in a streamline shape in order to get high-lift, low drag and a low moment. In other words, the aim is to maximise the aircraft lift-to-drag ratio for the mission role; for transport aircraft this is about 40–50% higher than fighter aircraft.

The wing contributes to lift generation and the characteristics are based on the chosen aerofoil section that fits the purpose with the aim to be better than the existing designs. Aircraft design cannot progress without selecting suitable aerofoil for wing and empennage with desirable parameters, for example, C_l, C_m and C_d values versus angle of attack showing post stall characteristics. The National Advisory Committee for Aeronautics (NACA) aerofoil 2D test data are widely available and have served aircraft designers for decades, still continuing today. These 2D aerofoil test data are corrected for 3D wing applications as shown in Section 4.9.4. Although datasheets (DATCOM: the short name for the *USAF Data Compendium for Stability and Control*/ESDU: Engineering Sciences Data Unit) [1, 2] give empirical relations to obtain $dC_L/d\alpha$, the preferred method is to work it out from the test data. Today, computational fluid dynamics (CFD) has made datasheets almost redundant and, in this book, the use of datasheets is minimised.

Wing planform shape and its associated geometries are established from the aircraft mission role, primarily from their speed and manoeuvre capabilities. Wing planform shape is subsequently sized to requirements with matched engines (e.g. zoom in/out).

International Civil Aviation Organization (ICAO) aeroplane design codes (or groups, in the case of the Federal Aviation Administration: FAA) are based primarily on wingspan. The legacy 747 family has been categorised under ICAO Code E, which has a span limit of up to but not including 65 m. Its equivalent, the FAA Group V limit is up to but not including 214 ft. The 747-8 wingspan is about 224.4 ft (68.4 m), making it the first Boeing commercial aeroplane to be categorised as ICAO Code F (or FAA Group VI limit is 262 ft). However, the 747-8's wingspan is much shorter than the maximum ICAO Code F wingspan of 80 m. Both the ICAO and FAA share the same concept of designing airports based on critical aeroplane dimensions and grouping of airport sizes.

4.3 Generic Wing Planform Shapes

There are basically two types of generic wing planform shapes: (i) unswept wing and (ii) swept wing as described in the Subsections 4.3.1 and 4.3.2 (Figure 4.1). (The elliptical wing planform is a special case beyond the scope of generic shapes.)

4.3.1 Unswept Wing Planform

An unswept rectangular or tapered wing planform has its quarter-chord line normal to the fuselage axis, represented by $\Lambda_{1/4} = 0$ (definition of sweep is given in Section 4.6.2). Rectangular wings are cheaper to manufacture due to having ribs of identical geometry. The trapezoidal planform has variable ribs along the span and costs more to manufacture. These kinds of wing planform geometries are used in low speed aircraft below Mach 0.3 (some supersonic wings can be designed with $\Lambda_{1/4} \approx 0$ – see Section 4.13).

4.3.2 Swept Wing Planform

By far the majority of aircraft operating above Mach 0.5 have a wing planform with some degree of sweep, that is, $\Lambda_{1/4} \neq 0$, backwards or forwards. Military aircraft planforms have more variety to offer. These can be

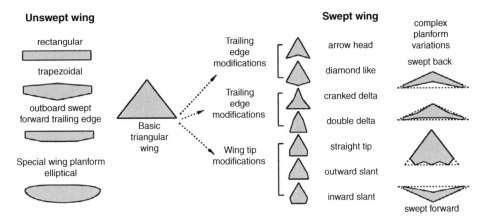

Figure 4.1 Generic wing planform shapes.

presented in a unified manner, that is from delta shape to trapezoidal shape as shown in Figure 4.1. This starts from a basic triangular (delta) planform. Combination of the basic types of modifications can bring out any planform types currently in use. For example, a swept trapezoid wing is a cropped arrowhead delta with a large span. The B2 Sirit wing planform is a good example (as shown later in Figure 4.3i).

Post-World War II (WWII) aircraft designs need to be separated between civil and military types as the mission roles moved away from some commonality as was there before the War. Civil aircraft operation remained subsonic (except for *Concorde* now not in service) while combat aircraft are dominantly supersonic. For subsonic speeds exceeding Mach 0.6, sweeping the wing backward (Figure 4.2d) or forward (Figure 4.3f) is necessary to delay the compressibility effects on the wing. Except for a very few thin wings, all wings carry a fuel load. Larger multi-engine transport category aircraft have wing pod-mounted engines. Fuel and engine loads serve as a wing bending moment relief to counter bending due to wing lift forces. Military supersonic aircraft wing require special considerations as shown in Section 4.10. Supersonic wings are very thin and have a lower aspect ratio (AR). Combat aircraft do not have wing mounted engines. Figures 4.2 and 4.3 illustrate the fundamental types of planform shape for civil and military classes of design with brief discussions.

The blended-wing-body (BWB) configuration can be considered for all wing aircraft derived from the basic triangular shape delta planform. Large subsonic aircraft suit the BWB configuration when the thicker wing root can blend into fuselage height. A new terminology has now appeared, Hybrid-Wing-Body (HWB), which does not look like a fully all-wing aircraft but has the wing blended to the fuselage with part of the fuselage at the aft end having an empennage. Studies exhibit that both the BWB and HWB have considerably better fuel economy compared to the current large subsonic transport jet aircraft.

4.3.3 Civil Aircraft

1. *Rectangular planform.* The rectangular wing planform (Figure 4.2a) is used for low speed (i.e. incompressible flow) aircraft below Mach 0.3. It is the most elementary shape with constant rib sections along the wingspan. Therefore, the cost of manufacture is lower because only one set of tooling for ribs is needed for the entire wing. However, this planform has the least efficient spanwise loading. This type of planform is well-suited to small aircraft, typically for private ownership and homebuilt types. There are larger aircraft that have the rectangular wing (e.g. Shorts SD360 series and Britten-Norman (BN) Islander).

2. *Tapered (trapezoidal) planform.* This is the most common planform shape in use because it offers good aerodynamics with a good spanwise load distribution. The taper ratio can vary. In Figure 4.2b, the leading edge (LE) has a small backward sweep; other designs have a straight LE and the SAAB Safir has a forward sweep, which provides pilot visibility in a high-wing aircraft. With almost no sweep, this type of wing can be designed for a maximum speed of Mach 0.5. If it must go faster, then wing sweep is required. The production

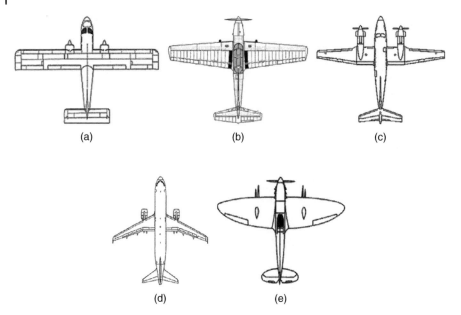

(a)   (b)   (c)

(d)   (e)

Figure 4.2 Examples of aircraft with different wing planform shapes. (a) Rectangular (BN-2B). (b) Taper (DH Chipmunk). (c) Cranked wing (Beech King Air). (d) Sweep back (A320). (e) Elliptical (Spitfire-military).

costs of a tapered wing are higher than for a rectangular wing because the ribs are different spanwise. However, the tapered wing maintains straight lines at the leading and trailing edges, which provides some ease in tool (jig and fixture) designs.

3. *Cranked (kinked)-wing planforms* are a combination of planform types with a break in LE. A tapered wing can be modified with a crank (kinked) incorporated (i.e. two tapered wings blended into one – Figure 4.2c). The Beech 200 is a good example of combining the available options. In this case, the centre section is rectangular and the outboard wings are tapered suiting wing mounted engines. Other combinations are possible.

4. *Swept wing (backward/forward) planform.* High-subsonic benefits from sweeping (Figure 4.2d). Section 4.8 discusses the aerodynamic benefits of wing sweep.

5. *Elliptical planform.* The Spitfire aircraft shown in Figure 4.2e is a fine example of an elliptical wing, which offers the best aerodynamic efficiency due to the best spanwise load distribution (Section 4.9). However, it is the most expensive to manufacture and designers should avoid it because a good tapered planform approximates elliptical load distribution, yet its manufacturing cost is substantially lower. Curved-wing leading and trailing edges require relatively more expensive tooling. The Spitfire aircraft reached very high speeds for its time.

6. *Blended Wing Body (BWB/HWB).* BWB configured military bomber aircraft designs have been in operation for several decades (see the next section). Gaining from experience, BWB/HWB configurations for larger multi-aisle seating in civil aircraft designs are currently under study (see Figure 1.10a). Some small experimental aircraft BWB configurations have been designed and flown successfully. While the technology has advanced to the point to ascertain economic benefits, its application is yet to be readied for civil operational usage. A dedicated cargo aircraft with large payload-range capability may come out earlier than its passenger version.

Currently, there are no delta wing civil aircraft except some small experimental types and prototype supersonic transport. With advances in technology, several companies are now considering the manufacture of supersonic passenger carrying transport aircraft from small to large sizes with viable economics, promising growth (see Figure 1.11).

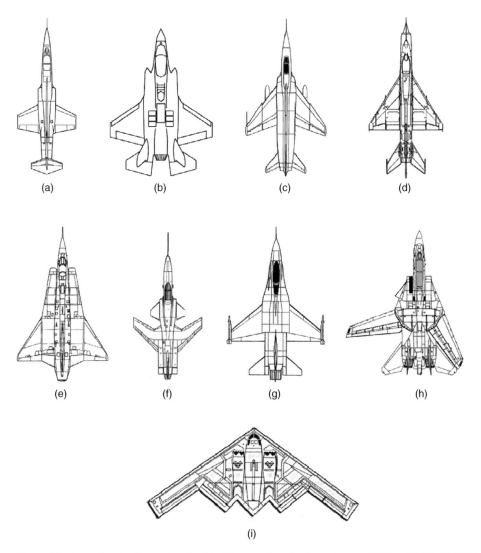

Figure 4.3 Examples of military aircraft with different wing planform shapes. (a) F104. (b) F35. (c) HF24. (d) MIG21. (e) SAAB Draken. (f) X-29. (g) F16. (h) F14 Tomcat. (i) B2 Spirit. Source: Reproduced with permission from Ferrier.

4.3.4 Military Aircraft (Supersonic Wings)

1. *Unswept trapezoidal (tapered) planform.* A short AR trapezoidal wing can be designed with low, almost zero sweep ($\Lambda\,{}^{1}\!/_{4} \approx 0$) shown in Figure 4.3a,b. It offers good aerodynamics with good spanwise load distribution.
2. *Swept wing (trapezoidal) planform.* Along with being trapezoidal, swept wings have been the dominant wing planform shape. High speed operation benefits from sweeping. By far the majority of swept wings are swept backwards on account of aerodynamic and structural considerations (Figure 4.3c). However, with the promise of better aerodynamics, the forward sweep technology demonstrator X-29 was built (Figure 4.3f). Forward swept wings can have problems with aeroelastic divergence. Newer materials have the potential to reduce the aeroelastic divergence problems.
3. *Delta wing planform.* Supersonic aircraft requires a high degree of sweep. The taper ratio of trapezoidal wing with an extreme value of zero becomes a delta wing. Military aircraft with low AR and high sweep

make the wing planform approach a delta shape or some very similar shapes with a cranked delta as shown in (Figure 4.3e).

4. *Strake wing planform.* The leading edge root extension (LERX), also known as a strake, assists vortex generation to provides additional lift at high angles of attack, α, discussed in Section 4.14. Some modern combat aircraft incorporate strakes, for example, the F16 (Figure 4.3g).

5. *Variable sweep wing planform.* Long range supersonic aircraft have to operate in a wide range of speed schedules (see Section 4.11.4). A variable sweep aircraft (Figure 4.3h) offers attractive design considerations to make wing sweep transition as required to suit the aircraft speed. The wing stays unswept at takeoff/landing, is partially swept at subsonic cruise close to the combat zone and moves to fully swept at supersonic dash in the combat arena to engage and rapidly disengage (variable sweep wing design for commercial supersonic aircraft is an attractive proposition).

6. *BWB/HWB.* Large bomber class aircraft with small fuselage volumes are meant primarily to accommodate a flight deck and power plant, which do not require a large diameter. This suits the BWB configuration when the thicker wing root can blend into a relatively smaller fuselage height (Figure 4.3i). An earlier successful design was the Avro Vulcan type aircraft that had a V-tail; the H-tail merged in the wing as a reflex trailing edge. With the advent of Fly-by-Wire (FBW), a V-tail can be dispensed of; for example, in the subsonic B2 Spirit bomber. More such designs are in the offing, for example, the possibly of the futuristic US Long Range Strike-B (LRS-B) heavy bomber with supersonic capability.

4.4 Wing Position Relative to Fuselage

Aerodynamic and structural considerations influence wing configuration and its position with respect to the fuselage. Depending on the design drivers, basically, the wing has three main positions with respect to a fuselage. The wing can be placed anywhere from (i) the top (i.e. high-wing) to (ii) the bottom (i.e. low wing) of the fuselage or (iii) in between (i.e. midwing), as shown in Figure 4.4.

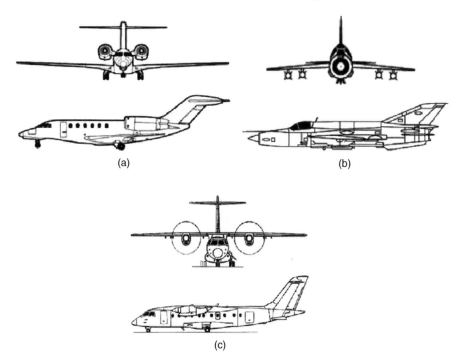

Figure 4.4 Positioning of the wing with respect to the fuselage (all T-tail configurations). (a) Low-wing, Cessna Citation X. (b) Mid-wing (T-tail), MIG21. (c) High-wing, Dornier 328. Source: Reproduced with permission from Ferrier.

The dominant configuration for civil transport aircraft has been a low wing, which provides a wider main-undercarriage wheel track allowing better ground manoeuvring. A low wing also offers a better crashworthy safety feature in the extremely rare emergency situation of a belly landing.

Aircraft with a high-wing allow better ground clearance and also allow the fuselage to be closer to the ground, which makes loading easier – especially with a rear-fuselage cargo door. Large propeller-driven aircraft favour a high-wing configuration to allow sufficient ground clearance for the propeller. The main undercarriage is mounted on the fuselage sides with a bulbous fairing causing some additional drag. However, this configuration provides better wing aerodynamics (e.g. the BAe RJ100 and Dornier 328 are successful high-wing designs). However, the authors believe more high-winged, large commercial-transport aircraft could be considered in future. The high-winged commercial-transport aircraft/Bizjets can gain a competitive edge as these offer better aerodynamics. Design trends show that military transport aircraft have predominantly high wings with large rear-mounted cargo doors.

4.4.1 Advantages and Disadvantages – Civil and Military Aircraft

The advantages and the disadvantages of each kind are given in the following subsections.

4.4.1.1 High-Wing
Advantages

Better lift generation – clean upper surface.
Lower interference drag – only at the under-wing fuselage corner.
Have a stabilising rolling moment.
Better ground clearance – engine mounted high on the wing.
Fuselage mounted undercarriage are smaller and have simpler articulation, hence they are lighter.
Built-in boarding ladder – some designs have an articulated ladder with the main doors.
Under wing maintenance/support crew movement (in general, passengers are not allowed).
Capable of having rear loading door – military aircraft are invariable with a rear loading door.
Suits flying boats that need to stay well above water and avoids wave interaction.

Disadvantages

High refuelling points.
Requires heavy reinforced structural attachments to suspend the heavy fuselage.
Less protection at belly landing on the ground or water – there is no wing to absorb shock or to stay afloat.
Accessibility for maintenance/cleaning may be cumbersome.
Narrower wheel track.
Under wing turboprop nacelles can house an undercarriage with long struts and are heavy.

Remarks

Suits military transport and civil cargo aircraft designs (C17, C130, A400 etc.).
Suits flying boat designs (Shorts, Dornier, Beriev etc., flying boats).
Large civil aircraft shied away from this configuration but the authors believe that there is no reason why we may not see more high-wing commercial-transport aircraft appear on the market (BAe 146, DO328 etc., are successful designs).

4.4.1.2 Low Wing
Advantages

Lower interference drag – only at the upper wing-fuselage corner.
Has a strong ground effect cushion that can benefit takeoff, but will take a longer landing length.
Easier access wing refuelling points.

Wider wheel track – steady ground manoeuvre.

Good protection at belly landing on the ground or water – a wing to absorb shock or help to stay afloat for some time.

Disadvantages

Lift generation interfered with by the presence of a fuselage.

Have a destabilising rolling moment.

Low ground clearance. For small aircraft, the engines must be mounted over the wing or on the fuselage.

All ground logistic movements must go round a large wing span.

Difficult for small aircraft under-wing maintenance/support crew movement.

Long undercarriage struts with rather complex articulation, hence they are relatively heavy.

Requires an airport boarding ladder. Some rare designs have a belly opening rear door built-in articulated ladder – these have not been popular.

Remarks

Popular civil aircraft design (typical narrow body Airbus/Boeing aircraft in operation).

4.4.1.3 Mid-Wing

Advantages

May have a neutral rolling moment.

Easier manufacturing philosophy – through the fuselage wing box.

Less wing root fairing – lighter structure.

For large aircraft, under-wing maintenance/support crew movement (passengers are not allowed).

Disadvantage

High interference drag both at the upper and lower wing-fuselage corner.

Lift generation interfered with by the presence of a fuselage.

Longer undercarriage struts with rather complex articulation, hence they are relatively heavy.

Remarks

A compromise between a low tail and a T-tail.

Popular civil aircraft design (typical wide body Airbus/Boeing aircraft in operation).

4.4.2 Definitions and Description of Anhedral/Dihedral

Aircraft in a yaw/roll motion have a cross-flow over the wing affecting the aircraft roll stability (Chapter 18), the extent depends on the position of the wing with respect to the fuselage. The wing profile configuration seen from the aircraft front view shows the dihedral/anhedral angle (Figure 4.5). The desirable roll stability is achieved by setting the wing inclined to the fuselage. Definitions of 'dihedral' and 'anhedral' are given next. The dihedral or anhedral angle also can be applied to the horizontal tail.

The *dihedral angle,* Γ, (i.e. the wing tip chord raised above the wing root chord) assists roll stability. A typical dihedral angle is between 2° and 4°; high-subsonic large aircraft can have dihedral angle as high as 6–7°.

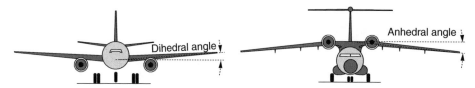

Figure 4.5 Wing dihedral and anhedral angles.

Figure 4.5a shows the dihedral angle with a low-wing configuration also allows more ground clearance for the wing tip.

The opposite of a dihedral angle is an *anhedral angle,* which lowers the wing tip with respect to the wing root and is typically associated with high-wing aircraft (Figure 4.5b). An anhedral angle is between 0 and −2°; A high-wing aircraft has inherent roll stability acting opposite to low-wing design and if it has too much stability, then an anhedral angle (−ve dihedral) is required to compensate it.

Figure 4.6 shows various design considerations for wing setting as seen from the front. There are variations in dihedral/anhedral angles arising out of other broader special purpose applications that are worth studying.

4.4.2.1 Outer-Wing Dihedral

The Jo Del aircraft series started with their first aircraft D9 with a configuration of outer-wing dihedral (Figure 4.6a). This became a standard feature (like a trademark) of many successful subsequent Jo Del models including the changeover of the name to Robin. The D9 had a one piece straight inner wing section with a constant chord and no dihedral. About a third of the wing semi-span with tapered outer section was a joined 14° dihedral that gave stability and tip clearance from ground. There are clear constructional advantages that also serve for ease of mounting into the cockpit by stepping on a horizontal part of the inner wing.

4.4.2.2 Gull Wing

Flying boat wings are invariably high and possibly gull wing to stay above the water line. The gull wing design found its way into flying boats starting from pre-WWII designs. The gull wing allowed designers to ensure adequate propeller tip clearance over the water by placing the engines on the highest point of the wing. A gull wing raises the engine further above to keep propeller tips clear from water. High roll stability on account of high-wing and inboard dihedral is compensated for by anhedral giving a 'gull-shape' wing. The Beriev 12 is a good example of a gull wing shape as shown in Figure 4.6b. Belfast based Shorts large flying boats were the first to introduce the gull wing designs.

4.4.2.3 Inverted Gull Wing

The inverted gull wing was developed at the same time having the same design driver to keep propeller tip clearance from ground. Large propeller diameter is associated with high power engines on smaller aircraft, for example, fighter aircraft of WWII. Instead of making a long heavy undercarriage to maintain propeller clearance, an innovative inverted gull wing design is a good proposition. The inverted gull wing consists of a centre section with large anhedral suited to an undercarriage design with short struts and compensated for by an adequate dihedral for stability that also gives wing tip clearance. Some successful designs are the Vought F4U Corsair (aircraft carrier based) and JU 87 Stuka (Figure 4.6c) fighter aircraft of WWII. The short undercarriage has a low weight penalty, also offers low drag and may dispense with retraction, as in the case of the Stuka, or can be retracted, as in the case of the Corsair.

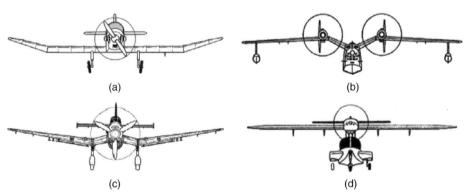

(a)   (b)

(c)   (d)

Figure 4.6 Wing profiles associated with dihedral and anhedral angles. (a) Jo Del Bebe D9. (b) Beriev B-12B. (c) JU87-Stuka. (d) Dornier S-7400. Source: Reproduced with permission from Ferrier.

4.4.2.4 Parasol

Parasol wings (little or no dihedral/anhedral), placed on struts high above the fuselage, provide another alternative to flying boat design. These also provide other advantages of an unrestricted view below and good lateral stability; therefore, the configuration can be applied to land based aircraft. In a parasol wing design, the wing is a separate sub-assembly component without a direct fuselage assembly joint. Instead, the wing is secured over the fuselage on a pylon or with a cabane strut (a box-frame structure with set wing incidence angle) centre wing support, with or without additional wire bracing. Many successful parasol aircraft with cabane struts were built in the past. Since drag associated with a cabane strut is high, the newer designs prefer a single pylon support, for example, the Dornier Do S7-400 flying boat. Another successful parasol design aircraft of USA origin was the WWII famed PBY (Petrol Bomber, code *Y*) Catalina flying boat (Figure 4.6d – pylon and struts).

4.5 Structural Considerations

Position of wing with respect to fuselage (e.g. high, low or middle) is a strong design driver for wing-body structural considerations that affect aircraft external mouldlines (Figure 4.7). The wing centre section should not interfere with the cabin passage-height clearance – this is especially critical for smaller aircraft.

The wing box arrangement for smaller aircraft should pass over (e.g. high-wing DO328) or under (e.g. Learjet 45, Cessna Citation X) the fuselage. Both the cases have a generous fairing (Figure 4.4) that conceals the fuselage mouldline kink (i.e. drag-reduction measure), which would otherwise be visible. Midwing (or near-midwing) designs are more appropriate for larger aircraft with a passenger cabin floor-board high enough to allow the wing box positioned underneath it.

Unpressurised propeller-driven aircraft operating at lower altitudes can have rectangular cross-sections to reduce manufacturing costs, as well as offer more space (e.g. Shorts 360 aircraft). A pressurised fuselage cross-section would invariably be circular or nearly circular to minimise weight from the point of hoop-stress considerations.

High-subsonic aircraft with Truss Braced Wing (TBW) research is showing some benefits. This has to undergo considerable development effort before it can appear on the market for operational usage.

Attaching the wing to the fuselage could have a local effect on fuselage external shape. The following are the basic types of attachments:

1. *Carry-through wing box.* For larger aircraft, this is separately constructed and attached to the fuselage recess. Subsequently, wings are mated at each side in accurate assembly jigs. For smaller aircraft, this could

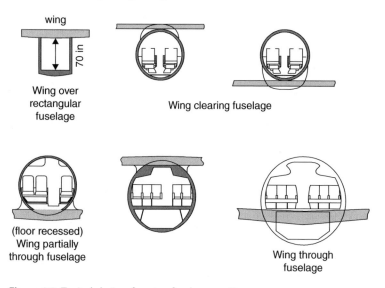

Figure 4.7 Typical choices for wing-fuselage positions.

be integral to the wing and then attached to the fuselage recess. In that case, the wing box is built into the wing, either in two halves or as a tip-to-tip assembly. A fairing at the junction reduces the interference drag. A central wing box is a part of the wing structure that integrates with the fuselage and is positioned high, low or at a convenient mid-location.

2. *Central beam and wing root attachments.* These have a simpler construction and therefore are less costly, suited to smaller aircraft.

3. *Wing roots (with multispar) joined to a series of fuselage frames.* These are mostly suited to military aircraft designs. They are heavier and can be tailored to varying fuselage contours. The wing root is then secured to the fuselage structure, sometimes outside the shell, with attachments.

4. *Strut/braced wing support.* Thus far, this structural arrangement is suited to smaller, low-speed, high-wing aircraft. Some low-wing agricultural aircraft have braced wings. Struts add to drag but for a low-speed operation the increment can be tolerated when it is less costly to build and lighter in construction.

   For smaller aircraft, the wing must not pass through the fuselage interior, which would obstruct passenger movement. If the wing is placed outside the fuselage (i.e. top or bottom), then a large streamlined fairing on the fuselage would accommodate the wing box. The example of the Cessna Excel shows a low-wing design; the DO328 includes a fairing for the high-wing design. The Dornier 328 conceals the fairing that merges with the fuselage mould lines. The extra volume could be beneficial; however, to arrive at such a configuration, a proper Direct Operating Cost (DOC) analysis must demonstrate its merits. High-wing aircraft must house the undercarriage in a fuselage fairing, although some turboprop aircraft have the undercarriage tucked inside the wing mounted engine nacelles (see Figure 12.18).

   Of late, NASA studies indicate that strut/braced wing support promises use of high $AR \approx 15$ cantilevered wing to reduce induced drag, yet keeping aeroelastic issues under operational limits.

5. *Swing wing.* Attachment of a swing wing is conveniently outside the fuselage such that the pivots have space around them to allow wing rotation.

4.6   Wing Parameter Definitions

This section defines the parameters used in wing design and explains their roles. The parameters are: the wing reference area, S_W; wing-sweep angle, Λ and wing taper ratio, λ. Dihedral and twist angles are given after the reference area is established. Reference areas are concerned with the projected rectangular/trapezoidal area of the wing.

4.6.1   Wing Reference (Planform) Area, S_W

In the aircraft design exercise, one of the most important parameters that can be modified after being manufactured in order to improve capabilities is the aircraft wing area. Therefore, instead of using a new modified wing each time, a reference area may be defined at the conceptual stage that can be used for the modified wing with adjustments made where required. Wing reference area, S_W, is the nominal projected area of an isolated wing (i.e. portion buried in the fuselage is included) and not the total surface area of both the upper and lower surface areas. S_W is used for computational purposes throughout the design process, as will be shown in subsequent chapters. For a linearly trapezoidal wing, S_W is exactly the same as the geometrical projected area of an isolated wing.

Being a nominal area, the definition of S_W differs from design bureau to design bureau of various aircraft manufacturers. In general commercial-transport aircraft design, there are primarily two types of definitions used on the either side of the Atlantic. The differences in the definitions of S_W are irrelevant as long as the type is known and adhered to. This book uses the type prevalent in the US that has straight edges extending to the fuselage centre line buried into fuselage (Figure 4.8a). Some European definitions show the part buried inside the fuselage as a rectangle (Figure 4.8b); that is, the edges are not extended up to the centreline unless

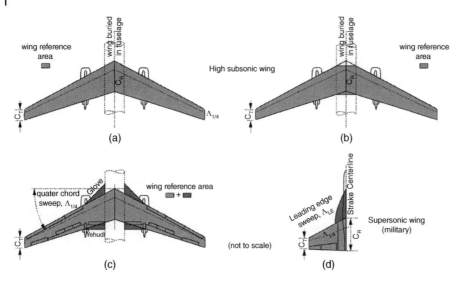

Figure 4.8 Wing reference area definition. (a) Fully tapered (USA). (b) Tapered up to fuselage side (Europe). (c) Tapered wing with glove and yehudi. (d) Supersonic wing.

the wing itself is rectangular wing normal to the centre line. Each industry benefits from their own definition in usages.

A typical subsonic commercial-transport type (Boeing 737) is shown in Figure 4.8c. An extension at the LE of the wing root is called a *glove* and an extension at the trailing edge is called a *yehudi* (this is a Boeing terminology). The yehudi's low-sweep trailing edge offers better flap characteristics. These extensions can originate in the baseline design or as an extension added to the existing platform to accommodate a larger wing area. A glove and/or a yehudi can be added later as modifications; however, this is not easy because the aerofoil geometry would be affected. The areas of 'glove' and 'yehudi' are not included in S_W. (The Boeing aircraft company has two deifications of S_W. Earlier they had S_W, as shown in Figure 4.8a. Subsequently, possibly since the 1970s, the definition was modified to include the glove and yehudi (plus a fraction of them) as S_W. This book uses the definition of the wing reference area, S_W, as shown in Figure 4.8a. Therefore, other definitions are not dealt with here.)

In military aircraft application, the LERXs, also known as strakes (Figure 4.8d), assist the vortex lift generation discussed in Section 4.14. The associated definition of a typical supersonic wing (F18 combat aircraft) is shown in the figure.

4.6.2 Wing Sweep Angle, $\Lambda_{1/4}$

The wing quarter-chord line is the spanwise locus of a quarter of the chord of the reference wing planform area measured from the LE, as shown in Figure 4.8. The wing sweep is measured by the angle, $\Lambda_{1/4}$, of the quarter-chord line extended with the line perpendicular to the centre line. LE sweep, Λ_{LE}, and trailing edge sweep, Λ_{TE}, of trapezoidal wing have different angles.

4.6.3 Wing Aspect Ratio, *AR*

For the rectangular wing planform area, the aspect ratio is defined as, $AR = (\text{span}, b)/(\text{chord}, c)$. For a generalised trapezoidal wing planform area:

$$\text{Aspect ratio}, AR = (b \times b)/(b \times c) = (b^2)/(S_W) \tag{4.1}$$

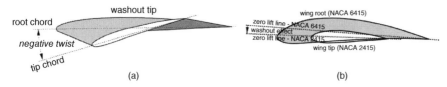

Figure 4.9 Wing twist. (a) Geometric wing twist. (b) Aerodynamic wing twist.

4.6.4 Wing Root Chord (c_{root}) and Tip Chord (c_{tip})

These are the aerofoil chords parallel to the aircraft centreline, the wing root chord (c_{root}) is at the centreline and the wing tip chord (c_{tip}) is at the tip of S_W.

4.6.5 Wing Taper Ratio, λ

This is defined as the ratio of the wing tip chord to the wing root chord (c_{tip}/c_{root}). The best taper ratio is in the range of 0.3–0.5. The taper ratio improves wing efficiency by providing a higher Oswald's efficiency factor (Section 4.9.3).

4.6.6 Wing Twist

The wing can be twisted geometrically by making the wing tip nose down (i.e. *washout*) relative to the wing root (Figure 4.9a), which causes the wing root to stall earlier (i.e. retain aileron effectiveness at the wing tips). Typical twist-angle values are 1–2° and rarely exceed 3°. Twisting the wing tip upward is known as *washin*.

Aileron effectiveness can also be retained by reducing the camber along the span from wing root to tip without a geometric twist and is known as an aerodynamic twist (Figure 4.9b).

4.7   Spanwise Variation of Aerofoil *t/c* and Incidence

High-subsonic transport aircraft wings are fine-tuned for the best spanwise load distribution to extract high L/D, as well good low speed handling qualities, by incorporating a spanwise aerofoil *t/c* variation gradually to the thinnest at the tip and introducing a downwash twist to prevent wing tip stall as shown in Figure 4.10. A thicker inboard wing offers better structural integrity and increased fuel storage volume. Camber distribution along the semi-span also varies, the extent depends mainly on the compromise between geometric and aerodynamic twist as well as lift capability. Typically, camber reduces from root to tip. The supersonic *Concorde* aircraft wing design exhibits complex distribution of *t/c*, α and camber distribution along the wing span.

Wing aerofoil *t/c* is an important parameter in aircraft design exercise, for example, in deciding wing sweep, wing parasite drag and so on. Since many commercial aircraft designs have *t/c* varying spanwise, a simplistic approach may be taken at the conceptual design stage to take a representative value of *t/c* that can give satisfactory results. There are several empirical formulae available to compute a representative *t/c*. For classroom usage, this book suggests using *t/c* at the Mean Aerodynamic Chord (MAC) as a safe value. In the case of example given in Figure 4.10, the $(t/c)_{MAC} = 11.8\%$. Readers may compare a *t/c* computed with empirical relations available in the public domain with the *t/c* value at MAC.

4.7.1 Wing Stall Pattern and Wing Twist

The lower the speed at landing, the easier it is for the pilot to handle the aircraft safely. An aircraft landing occurs near the wing stall condition when the aileron effectiveness should be retained to avoid possible asymmetric wing drop, a tip hitting the ground. In other words, when approaching the stall condition, its gradual

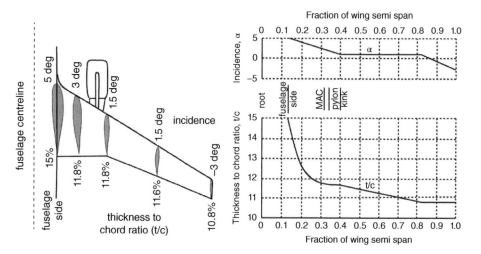

Figure 4.10 Typical wing (A310) spanwise variation of aerofoil *t/c* and incidence. Source: Credit Airbus

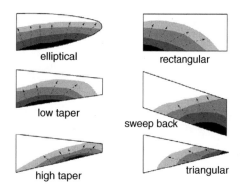

Figure 4.11 Stall progressing from the trailing edge as angle of attack is increased.

development should start from the wing root, which allows the aileron at the wing tip to retain its ability to maintain level flight.

Figure 4.11 (also see Figure 3.32) shows typical wing stall propagation patterns on various types of wing planform. Because a swept-back wing tends to stall at the tip first, twisting of the wing tip nose downward (i.e. washout) is necessary to force the root section to stall first, thereby retaining roll control during the landing. A good way to ensure the delay of the wing tip stall is to twist the wing about the *y*-axis so that the tip LE is lower than the wing root LE.

4.8 Mean Aerodynamic Chord (*MAC*)

One of the most important parameters is the *MAC*, which is the chord-weighted average chord length of the wing, defined as follows:

$$MAC = \frac{2}{S_W} \int_0^{b/2} c(y)^2 dy \tag{4.2}$$

Position of *MAC* from the root chord

$$y_{MAC} = \frac{2}{S_W} \int_0^{b/2} c(y)y\,dy \tag{4.3}$$

where *c* is local wing chord as function of *y* and S_W is the wing reference area.

The *MAC* is preferred for computation instead of the simpler mean geometric chord for the aerodynamic quantities whose values are weighted more strongly by local chord that is reflected by their contribution to the area. The *MAC* is often used in defining the non-dimensional pitching moments and is also used to compute the reference length for calculation of Reynolds number (*Re*) as part of the wing drag estimation.

4.8.1 Linearly Tapered Trapezoidal Wing

Trapezoidal wing reference area, S_W (with sweep) – see Figure 4.12 (note $C_R = C_{root}$ and $C_T = C_{tip}$)
For full wing when span, *b*, wing area,

$$S_W = \frac{1}{2}(C_R + C_T) \times b \tag{4.4}$$

In the diagram: $AC' = C_R + C_T = B'D$
From the proportional properties of triangle, from at any distance '*y*' from C_R the chord length '*c*' can be written as

$$c = C_R - 2(C_R - C_T)y/b. \tag{4.5}$$

or $c^2 = [C_R - 2(C_R - C_T)y/b]^2$,
On substituting in the integral, Eq. (4.2) becomes

$$
\begin{aligned}
MAC &= \frac{2}{S_W} \int_0^{b/2} c^2 dy = \frac{2}{S_W} \int_0^{b/2} \left[C_R - 2(C_R - C_T)(y/b)\right]^2 dy \\
&= \frac{2}{S_W} \left[(y C_R^2)_0^{b/2} - \left\{ \frac{2y^2 C_R}{b}(C_R - C_T) \right\}_0^{b/2} - \left\{ 4(C_R - C_T)^2 \frac{y^3}{3b^2} \right\}_0^{b/2} \right] \\
&= \frac{2}{S_W} \left[\frac{b}{2} C_R^2 - \frac{b C_R}{2}(C_R - C_T) - \frac{b}{6}(C_R - C_T)^2 \right]
\end{aligned}
$$

Using Eq. (4.4) for $S_W = \frac{1}{2}(C_R + C_T) \times b$

$$MAC = [2/(C_R + C_T)][C_R^2 - C_R^2 + C_R C_T + C_R/3 + C_T^2/3 - 2C_R C_T/3] \tag{4.6a}$$

$$= [2/(C_R + C_T)][C_R^2/3 + C_T^2/3 - C_R C_T/3] = \frac{2}{3}\left[\frac{(C_R + C_T)^2}{(C_R + C_T)} - \frac{C_R C_T}{(C_R + C_T)} \right] \tag{4.6b}$$

$$= 2/3 \times [C_R + C_T - C_R \times C_T/(C_R + C_T)] \tag{4.7}$$

The spanwise location y_{MAC} of the *MAC* can be obtained from Eqs. (4.3) and (4.5).

$$y_{MAC} = \frac{2}{S_W} \int_0^{b/2} c(y) y \, dy = \frac{2}{S_W} \int_0^{b/2} \left[C_R - 2(C_R - C_T)(y/b)\right] y \, dy$$

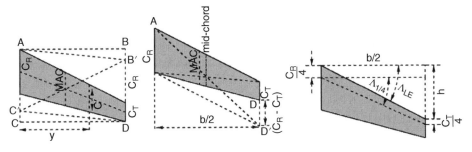

Figure 4.12 Trapezoidal wing planform – MAC.

$$\text{or } y_{\text{MAC}} = \frac{2}{S_W} \left[\frac{C_R b^2}{8} - \frac{2}{b}\left[\left(\frac{C_R b^3}{3 \times 8}\right)\right] - \left[\frac{C_T b^3}{3 \times 8}\right]\right] = \frac{b}{6}\left(\frac{C_R + 2C_T}{C_R + C_T}\right) \tag{4.8}$$

The relation between $\Lambda_{1/4}$ and Λ_{LE} can be derived by eliminating '*h*' from the last diagram of Figure 4.12, as follows.

$$(2/b)\tan\Lambda_{1/4} + C_R - C_T = (2/b)\tan\Lambda_{\text{LE}}$$

$$\text{or } \tan\Lambda_{1/4} + C_R - C_T = \tan\Lambda_{\text{LE}} - (b/2)C_R(\lambda - 1) \tag{4.9}$$

For wings with chord extensions (glove/yehudi), the *MAC* may be computed by evaluating the *MAC* of each linearly tapered portion then taking an average, weighted by the area of each portion. In many cases, however, the *MAC* of the reference trapezoidal wing is used.

4.8.2 Kinked Wing – Double Linearly Tapered Trapezoidal Wing

A kinked wing may not be seen as modifications with glove and yehudi, but for a new wing that has two trapezoidal wings with reference areas (both side of the centreline), the inboard reference is $S_{\text{W-in}}$ and outboard reference $S_{\text{W-out}}$. Together, $S_W = (S_{\text{W-in}} + S_{\text{W-out}})$. The kink line separating the two areas is placed at y_K from the centreline as shown in Figure 4.13. Chord length at y_K is C_K. Wing span is *b*.

Equation (4.4) gives

Inboard wing area, $S_{\text{W-in}} = \frac{1}{2}(C_R + C_K) \times 2Y_K =$
Outboard wing area, $S_{\text{W-out}} = \frac{1}{2}(C_K + C_T) \times (b - 2Y_K) =$
Therefore, the total wing area, $S_W = \frac{1}{2}(C_R + C_K) \times 2Y_K + \frac{1}{2}(C_R + C_K)(b - 2Y_K) =$

$$\text{or } S_W = (C_R - C_T)Y_K + \frac{1}{2}(C_K + C_T)b = (bC_R/2)[(1-\lambda)(Y_K/b) + (\lambda_{\text{in}} + \lambda)]$$

$$\text{Wing aspect ratio, } AR = (b^2/S_W) = \left(\frac{2b}{C_R(1-\lambda)\left(2Y_K/b\right) + (\lambda_{\text{in}} - \lambda)}\right) \tag{4.10}$$

Wing *MAC* is defined as the area-weighted average of inboard and outboard wing *MACs*, which can be computed as follows.

$$MAC = \left(\frac{MAC_{in} \times S_{W-in} + MAC_{out} \times S_{W-out}}{S_W}\right) \tag{4.11}$$

In the same manner y_{MAC} can be found by first computing $y_{\text{MAC-in}}$ and $y_{\text{MAC-out}}$ and then take the weighted average as follows.

$$y_{\text{MAC}} = \left(\frac{y_{MAC-in} \times S_{W-in} + y_{MAC-out} \times S_{W-out}}{S_W}\right) \tag{4.12}$$

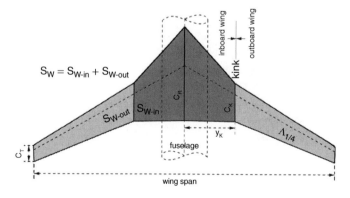

Figure 4.13 Kinked wing – double tapezoidal wing. Source: Reproduced with permission from Cambridge University Press.

Derivation becomes more complex if the LE kink point and the trailing edge kink points are not in the same chord line as in Figure 4.12. In that case, the wing is divided into three surface, the inner, the outer and a narrow strip in between stemming from the two kink points. The computation can then progress as before with three surfaces.

Example 4.1   For a small aircraft, find the root chord (C_R), tip chord (C_R), *MAC* and the spanwise location (y_{MAC}) of the *MAC* of a 100ft^2 linearly tapered (trapezoidal) wing area with *AR* 6.25 and taper ratio, $\lambda = 0.33$.

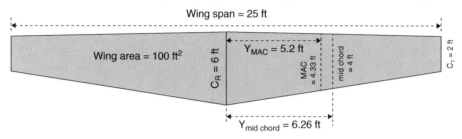

Solution:
Equation (4.1) gives, $b = \sqrt{(AR \times S_W)} = \sqrt{625} \approx 25$ ft
   For a trapezoidal wing, Eq. (4.4) gives $S_W = \frac{1}{2}(C_R + C_T) \times b$
   or $(C_R + C_T) = (2 \times 100)/25 = 8$, solving for $\lambda = C_T/C_R = 0.33$, it gives $C_R \approx 6$ ft an $C_T \approx 2$ ft
   Equation (4.5a) gives $MAC = 2/3 \times [C_R + C_T - C_R \times C_T/(C_R + C_T)] = 2/3 \times [6 + 2 - 6 \times 2/(6 + 2)] = 4.33$ ft
from Eq. (4.8) $y_{MAC} = \frac{b}{6}\left(\frac{C_R + 2C_T}{C_R + C_T}\right) = (25/6)[(6 + 2 \times 2)/(6 + 2)] = 5.2$ ft from the wing centreline.
   The geometric mean chord (mid-chord) is 4 ft located at 6.25 ft from wing centreline. (Note: a wing with sweep will have the same dimensions, except that the sweep will move C_T and *MAC* further backwards.)

Example 4.2   A commercial aircraft wing, as shown here, has a wing span = 144 ft, quarter-chord sweep, $\Lambda = 28°$, root chord, C_R, = 21 ft and tip chord, C_R, = 7.4 ft. Find the wing area, S_W, *AR*, taper ratio, λ, *MAC* and the spanwise location of *MAC*.)

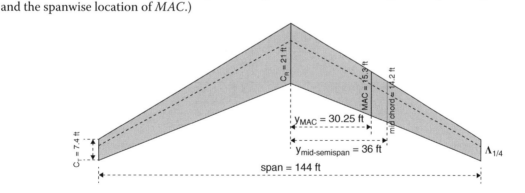

Solution:
Equation (4.4) gives the wing area, $S_W = \frac{1}{2}(C_R + C_T) \times b = \frac{1}{2}(21 + 7.4) \times 144 = 2044.8$ ft$^2 \approx 2045$ ft^2
   Equation (4.1) gives $AR = b^2/S_W = 144^2/2045 = 10.14$
   Taper ratio, $\lambda = C_T/C_R = 7.4/21 = 0.35$
   Equation (4.7) gives

$$MAC = 2/3 \times [C_R + C_T - (C_R \times C_T)/(C_R + C_T)] = 2/3 \times [21 + 7.4 - 21 \times 7.4/(21 + 7.4)]$$

or $MAC = 2/3 \times [28.4 - 155.4/(28.4)] = 2/3 \times [28.4 - 5.472] = 15.283$ ft

Equation (4.8)

$$y_{MAC} = \frac{b}{6}\left(\frac{C_R + 2C_T}{C_R + C_T}\right) = (144/6)[(21 + 2 \times 7.4)/(21 + 7.4)] = 24 \times [(35.8)/(28.4)] = 30.25 \text{ ft}$$

Mid-chord = 14.2 ft, position at a mid-semi span of 36 ft.

Example 4.3   A commercial aircraft kinked wing, as shown in Figure 4.12 has the following. Span $b = 144$ ft, $C_R = 43$ ft, $C_T = 8$ ft and $C_K = 17.5$ ft and $Y_K = 24$ ft. Find wing area, S_W, taper ratio, λ, *MAC* and y_{MAC}.

Solution:
Wing taper ratio, $\lambda = 8/43 = 0.186$
  Inboard wing taper ratio, $\lambda_{in} = 17.5/43 = 0.407$
  Outboard wing taper ratio, $\lambda_{out} = 8/17.5 = 0.457$

$$S_W = (bC_R/2) \times [(1 - \lambda)(2Y_K/b) + (\lambda_{in} + \lambda)] = (144 \times 43/2) \times [(1 - 0.186)(2 \times 24/144) + (0.407 + 0.186)]$$

$$= (3096) \times [(0.814)(0.33) + 0.593] = (3096)(0.862) = 2668 \text{ ft}^2 \approx 2670 \text{ ft}^2$$

$$S_{W-in} = \tfrac{1}{2}(C_R + C_K) \times 2Y_K = \tfrac{1}{2}(43 + 17.5) \times 2 \times 24 = 1452 \text{ ft}^2 \approx 1450 \text{ ft}^2$$

$$S_{W-out} = \tfrac{1}{2}(C_K + C_T) \times 2(b/2 - Y_K) = \tfrac{1}{2}(17.5 + 8) \times 2(72 - 24) = 1224 \text{ ft}^2 \approx 1220 \text{ ft}^2$$

$$S_W \approx 2670 \text{ ft}^2$$

From Eq. (4.10),

$$AR = \left(\frac{2b}{C_R\left[(1 - \lambda)(2Y_K/b) - (\lambda_{in}\,\lambda)\right]}\right) = \left(\frac{2 \times 144}{43 \times \left[(1 - 0.186) \times \left[\frac{2 \times 24}{144}\right] + (0.407 + 0.186)\right]}\right) = 7.8$$

From Eq. (4.7),
$C_R = 43$ ft, $C_T = 8$ ft and $C_K = 17.5$ ft, $Y_K = 24$ ft

$$MAC_{in} = 2/3 \times [C_R + C_K - C_R \times C_K/(C_R + C_K)] = 2/3 \times [43 + 24 - 43 \times 24/(43 + 24)]$$

$$= 2/3 \times [67 - 1032/67] = 2/3 \times [67 - 1032/67] = 2/3 \times 51.6 = 34.4$$

$$MAC_{out} = 2/3 \times [C_K + C_T - C_K \times C_T/(C_K + C_T)] = 2/3 \times [17.5 + 8 - 17.543 \times 28/(17.5 + 8)]$$

$$= 2/3 \times [25.5 - 490/25.5] = 2/3 \times [25.5 - 19.22] = 2/3 \times 6.28 = 4.2$$

From Eq. (4.11),

$$MAC = \left(\frac{MAC_{in} \times S_{W-in} + MAC_{out} \times S_{W-out}}{S_W}\right) = \left(\frac{34.4 \times 1450 + 4.2 \times 1220}{2670}\right) = 20.84 \text{ ft}$$

$$y_{MAC-in} = \frac{2 \times Y_K}{6}\left(\frac{C_R + 2C_K}{C_R + C_K}\right) = \frac{2 \times 24}{6}\left(\frac{43 + 2 \times 24}{43 + 24}\right) = 8 \times \left(\frac{43 + 48}{67}\right) = 10.87 \text{ ft}$$

$$y_{MAC-out} = \frac{2 \times (b - Y_K)}{6}\left(\frac{C_K + 2C_T}{C_K + C_T}\right) = \frac{2 \times (144 - 24)}{6}\left(\frac{24 + 2 \times 8}{24 + 8}\right) = 40 \times \left(\frac{40}{32}\right) = 50 \text{ ft}$$

$$y_{MAC} = \left(\frac{y_{MAC-in} \times S_{W-in} + y_{MAC-out} \times S_{W-out}}{S_W}\right) = \left(\frac{10.87 \times 1450 + 50 \times 1220}{2670}\right) = 28.8 \text{ ft}$$

4.9   Wing Aerodynamics

Similar to a bird's wing, an aircraft's wing is the lifting surface to sustain the weight of the aircraft that makes flight possible. Proper wing planform shape and size are crucial to improving aircraft aerodynamic efficiency and performance; however, aerofoil parameters are often compromised.

A 3D finite wing produces vortex flow as a result of tip effects, as shown in Figure 4.14. The high pressure from the lower surface rolls up at the free end of the finite wing, creating the tip vortex. Also shown in the figure is the downwash distribution for a typical non-elliptical wing. The direction of vortex flow is such that it generates downwash, which is distributed spanwise at varying strengths. A reaction force of this downwash is the lift generated by the wing. Energy consumed by the downwash appears as lift-dependent induced drag, D_i, and its minimisation is a goal for aircraft designers.

In the case of an elliptical wing, as shown in Figure 4.15, the downwash is distributed uniformly along the wing span as an efficient distribution (see Section 4.9.1).

4.9.1   Downwash and Effective Angle of Attack, α_{eff}

An elliptical wing planform (e.g. the Spitfire fighter of WWII) creates ideal uniform spanwise downwash at its lowest magnitude and leads to minimum induced drag. This is used in formulations based on Prandtl lifting-line theory. It postulates a bound vortex sheet, the leading vortex line with constant circulation Γ with downwash uniformly distributed along the wing aerodynamic centre, *a.c.* (here, taken at the quarter chord), bound by two tip vortices and a trailing vortex.

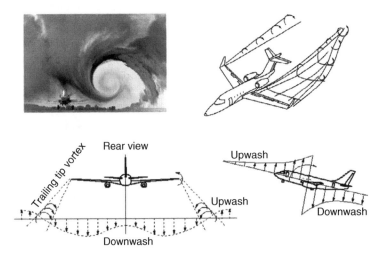

Figure 4.14 Wing tip vortex. Source: Photo: Courtesy NASA. Reproduced with permission from Cambridge University Press.

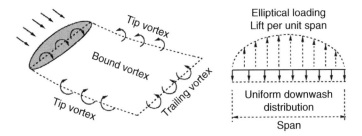

Figure 4.15 Pressure, flow pattern and down-wash effect of finite 3D wing.

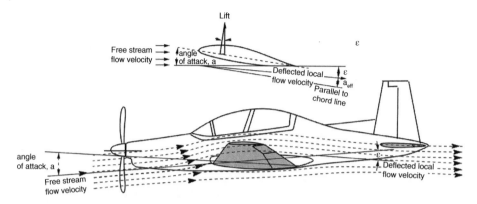

Figure 4.16 Downwash angle and its distribution on elliptical wing platform.

The LE vortex line creates strong upwash just ahead of wing LE, which decreases sharply and gives the theoretical downwash velocity 'w' at the wing aerodynamic centre, *a.c.*, subtending downwash angle, ε, as shown in Figure 4.16. In summary, the downwash effect of a 3D elliptical wing deflects free streamflow, V_∞, by an angle, ε, to V_{local}. It can be interpreted as the section of a 3D wing behaving as a 2D infinite wing with:

$$\textit{effective angle of incidence, } \alpha_{\text{eff}} = (\alpha - \varepsilon), \tag{4.13}$$

where α is the angle of attack at the aerofoil section, by $V\infty$.

Behind the wing aerodynamic centre, *a.c.* (quarter chord), the downwash continues but attenuates to deflected velocity to another *w* totalling 2*w*, which is twice the downwash velocity *w* as defined previously.

The velocities and angles associated with the lifting-line theory applied an *elliptical* planform wing are summarised next.

w: By definition of the lifting-line theory is the uniform downwash velocity at the aerodynamic centre. It is of a constant value for the geometry and attitude under study.

ε: The downwash angle as $\varepsilon = w/V$, uniform span wise.

α_{eff}: The downwash, *w*, gives the local deflection of the velocity that reduces the free stream angle attack, α, to an effective angle of attack α_{eff}.

Other associated parameters may be defined as follows.

w_u: The varying strength of upwash deflected velocity, highest just ahead of the wing LE.

ε_u: The varying strength of the local angle of the deflected upwash caused by *wu*.

w_d: The varying strength of downwash deflected velocity, highest just aft of the wing LE.

ε_d: The varying strength of the local angle of the deflected downwash caused by *wd*.

The total deflected downward velocity at the H-tail is 2*w*, which is not the downwash, *w*, defined at the wing aerodynamic centre but twice that value. Far downstream, the deflected downward velocity of the vortex sheet dissipates to zero on account of viscosity. The deflected downward velocity can be very strong emanating from large aircraft. Accidents have happened to small aircraft trying to land immediately after larger aircraft take off. In fact, the FAA has a regulation regarding keeping a certain distance from the aircraft in front to allow the vortex effects to dissipate.

Flow modelling of the physics relating the downwash angle, ε, is complex. Prandtl's lifting-line theory is a well-accepted method of obtaining the expression for ε. Aerodynamics textbooks [3–5] may be consulted to theoretically derive the downwash angle for elliptical wings as follows:

$$\varepsilon = C_L/\pi AR \text{ (in radians)} = 57.3 C_L/\pi AR \text{ (in deg)} \tag{4.14}$$

Equation (4.14) indicates that downwash angle, ε, is *AR* dependent.

The magnitude of downwash angle increases with angle of attack, α. Within the linear range of lift curve, downwash angle, ε, can be expressed as follows.

$$\varepsilon = \varepsilon_0 + \frac{d\varepsilon}{d\alpha} \times \alpha \tag{4.15}$$

where ε_0 = the downwash angle at $\alpha = 0$ on account of camber effect. For symmetric aerofoil, $\varepsilon_0 = 0$.

4.9.2 Induced Drag of the Elliptical Wing

Average downwash angle, $\varepsilon = C_L/e\pi AR$ (in radians) $= 57.3 C_L/e\pi AR$ (in deg) $\tag{4.16}$

Local lift, L_{local}, produced by a 3D wing, is resolved into components perpendicular and parallel to free streamflow, V, that is, aircraft velocity. In coefficient form, the integral of these forces over the span gives the following:

$$C_L = L\cos\varepsilon/qS_W \text{ and } C_{Di} = L\sin\varepsilon/qS_W \text{ (the induced-drag coefficient)}$$

For small angles, ε, it reduces to:

$$C_L = L/qS_W \text{ and } C_{Di} = L\varepsilon/qS_W = C_L\varepsilon, \text{ where } q = 0.5\,\rho V^2 \tag{4.17}$$

C_{Di} is the drag generated from the downwash angle, ε, and is lift-dependent (i.e. induced); hence, it is called the *induced-drag coefficient*. For the elliptical wing planform using Eq (4.14), Eq. (4.17) becomes:

$$C_{Di} = C_L\varepsilon = C_L \times C_L/\pi AR = C_L^2/\pi AR \tag{4.18}$$

[The same formulation of in Eq. (4.18) can be derived by applying Newton's second law, that is, the applied lift force equals the rate of change of momentum emanating from the deflected downwash caused by $2w$. Take a circular stream-tube, of diameter equal to wing span, b, outside of which the wing contribution is neglected. The stream-tube velocity is V and cross-section area is $A = \pi b^2/4$. Past the wing, the stream-tube gets deflected by the downwash velocity $2w$.

Lift force, $L = \dot{m}\,2w$, where $\dot{m} = \rho AV$, gives the rate of change of momentum

Therefore, $L = (\rho AV) \times 2w = [\rho(\pi b^2/4)V] \times 2w$, where $A = \pi b^2/4$
By definition, $L = 0.5\rho V^2 S_W C_L$, where S_W is the wing area
Equating, the following is obtained.
$0.5\rho V^2 S_W C_L = [\rho(\pi b^2/2)V] \times w$
Simplifying, $VS_W C_L = [(\pi b^2)] \times w$
Noting that $AR = b^2/S_W$, it becomes
$w = (VC_L)/(\pi AR)$ is the downwash velocity at the *a.c.*
Again from the velocity components, down angle, $\varepsilon = w/V$
The $\varepsilon = C_L/\pi AR$ (in radians)]

4.9.3 Induced Drag of the Non-Elliptical Wing – Oswald's Efficiency Factor

For a non-elliptical wing planform, the distributed spanwise downwash varies in strength. The distribution of downwash varies from wing to wing depending on their planform shape and AR. For the same AR and at the same C_L, the average downwash over a non-elliptical wing is higher than the ideal distribution over elliptical wing will be higher and semi-empirical. Based on past test data, industries use a semi-empirically established correction factor, e, called *Oswald's efficiency factor* (always ≤ 1) that is applied as follows:

Average downwash angle, $\varepsilon = C_L/e\pi AR$ (in radians) $= 57.3 C_L/e\pi AR$ (in deg) $\tag{4.19}$

4.9.3.1 Downwash

Equation (4.19) following can be rewritten for trapezoidal wing

$$\varepsilon = 57.3 \times (C_{L0} + C_{L\alpha} \times \alpha)/(e\pi AR) \tag{4.20}$$

Therefore, $\frac{\partial \varepsilon}{\partial \alpha}$ can be obtained as follows.

$$\frac{\partial \varepsilon}{\partial \alpha} = (57.3 \times C_{L\alpha})/(e\pi AR) \tag{4.21}$$

Local lift, L_{local}, produced by a 3D wing, is resolved into components perpendicular and parallel to free streamflow, V; that is, aircraft velocity. In coefficient form, the integral of these forces over the span gives the following. For a non-elliptical wing planform, Eq. (4.18) becomes:

$$C_{\text{Di}} = C_L \varepsilon = C_L \times C_L/e\pi AR = C_L{}^2/e\pi AR \tag{4.22}$$

For the same AR and C_L, the induced drag, C_{Di}, of non-elliptical is higher than that of elliptical wing. The factor 'e' indicates the degree of the non-elliptical working inefficiently.

Equation (4.19) show that the downwash decreases with an increase in the AR. When the AR reaches infinity, there is no downwash and the wing becomes a 2D infinite wing (i.e. no tip effects) and its sectional characteristics are represented by aerofoil characteristics. The downwash angle, ε, is small; in general, less than 5° for aircraft with a small AR. The aerofoil section of the 3D wing apparently could produce less lift than the equivalent 2D aerofoil. Therefore, 2D aerofoil test results would require correction for a 3D wing application, as explained in the following section.

Oswald's efficiency factor, e, is a measure of how effective the wing is in having reducing the finite span contributed induced drag, C_{Di}, as indicated in Eq. (4.19) for a given AR. The elliptical has the least C_{Di} with $e = 1$. The higher the AR, the lower the C_{Di} but AR is restricted to what the structural considerations permit. Wing planform shape also affects the induced drag. A well designed trapezoidal wing planform can bring spanwise wing loading close to elliptical span loading. $AR \approx 10$ and above with a taper ratio $\lambda \approx 0.33$ can make 'e' approach close to unity at C_L giving C_{Dpmin} (Figure 11.2).

In summary, induced drag is lowest for an elliptical wing planform, when $e = 1$, however, it is costly to manufacture. In general for wings with $AR > 6$, the industry uses a trapezoidal planform with a taper ratio, $\lambda \approx 0.3$–0.5, resulting in an e value ranging from 0.85 to 0.98 (an optimal design approaches 1.0). A rectangular wing has a ratio of $\lambda = 1.0$ and a delta wing has a ratio of $\lambda = 0$ (typically $AR < 6$), which result in an average e below 0.8. A rectangular wing with its constant chord is the least expensive planform to manufacture for having the same-sized ribs along the span.

Also note that spanwise loading of an isolated wing will get altered with the interference of an integrated fuselage. In that case, the wing capability to reduce C_{Di} will also change, that is, will have a slightly different 'e'. The difference being small, at this stage of Phase I design activities, Oswald's efficiency factor, e, is kept invariant.

Table 4.1 gives the values of Oswald's efficiency factor, e, for classwork.

The values in Table 4.1 are typical averages taken from the statistics of existing designs and may be used at the initial stages of Phase I of a new project when not much information is available. However, industries will have to progress to get as much as possible accurate values through CFD analyses and subsequently substantiate through wind-tunnel test after 'go-ahead' is obtained when more funds are available.

Table 4.1 Oswald's efficiency, e, (aspect ratio correction).

	AR ≈ 5–7.5	AR ≈ 6–9	AR ≈ 9–12	AR > 12
Rectangular wing	≈0.7 to ≈0.78	≈0.75 to ≈0.8		
Taper wing ($\lambda \approx 0.33$)	≈0.75 to ≈0.8	≈0.8 to ≈0.9	≈0.9 to ≈0.95	≈> 0.95

The extent of downwash is lift-dependent; that is, it increases with an increase in C_L. Strictly speaking, Oswald's efficiency factor, e, varies with wing incidence; however, the values used are considered an average of those found in the cruise segment and remain constant. In that case, for a particular aircraft design, the average downwash angle, ε, is treated as a constant taken at the mid-cruise condition. Advanced wings of commercial-transport aircraft can be designed in such a way that at the design point, $e \approx 1.0$.

4.9.4 AR Correction of 2D Aerofoil Characteristics for a 3D Finite Wing

To incorporate the tip effects of a 3D wing, 2D test data need to be corrected for Re and AR. Eq. (4.19) indicates that a 3D wing will produce α_{eff} at an attitude when the aerofoil is at the angle of attack, α. Because α_{eff} is always less than α, the wing produces less C_L corresponding to aerofoil C_l (see Figure 4.17). This section describes how to correct the 2D aerofoil data to obtain the 3D wing lift coefficient, C_L, versus the angle of attack, α, relationship. Within the linear variation, $dC_L/d\alpha$ needs to be evaluated at angles, typically, from $-2°$ to $8°$.

The 2D aerofoil lift-curve slope $a_0 = (dC_L/d\alpha)$, where α = angle of attack (incidence).

The 2D aerofoil will generate the same lift at a lower angle of attack, α_{eff} than what the wing will generate at α ($\alpha_{3D} > \alpha_{2D}$) to get $C_L = C_l$. Therefore, using the 2D aerofoil data, the wing lift coefficient C_L can be worked at the angle of attack, α, as shown here (all angles are in degrees). The wing lift at an angle of attack, α, is as follows (refer to Eq. (4.13)):

$$C_L = a_0 \times \alpha_{eff} + \text{constant} = a_0 \times (\alpha - \varepsilon) + \text{constant},$$

where constant $= C_{L@\alpha=0}$ and a_0 is the aerofoil lift-curve slope, $dC_l/d\alpha$. Using equation 4.14 for e, following is obtained

$$C_L = a_0 \times (\alpha - 57.3\, C_L/e\pi/AR) + C_{L@\alpha=0}$$

$$or\ C_L + (57.3\, C_L \times a_0/e\pi AR) = a_0 \times \alpha + C_{L@\alpha=0}$$

$$or\ C_L = (a_0 \times \alpha)/[1 + (57.3 \times a_0/e\pi AR)] + C_{L@\alpha=0}/\left[1 + (57.3 \times a_0/e\pi AR\right] \tag{4.23}$$

For a given wing, the last term of Eq. (4.23) is constant.

Differentiating Eq. (4.23) with respect to α, it becomes:

$$dC_L/d\alpha = a_0/[1 + (57.3a_0/e\pi AR)] = a = \text{lift} - \text{curve slope of the wing} \tag{4.24}$$

Therefore,

$$C_L = C_{L0} + (dC_L/d\alpha)\alpha, \tag{4.25}$$

where C_{L0} is C_L at $\alpha = 0$.

Evidently, the wing-lift-curve slope, $dC_L/d\alpha = a$, is less than the 2D aerofoil lift-curve slope, $dC_l/d\alpha = a_0$. Figure 4.19 shows the degradation of the wing-lift-curve slope, $dC_L/d\alpha$, from its 2D aerofoil value.

The correction is applied to the linear part of the lift-curve slope but not the stall angle in the non-linear part of the C_L versus α graph. CFD analyses/wind-tunnel tests are the best methods to get the α_{max} for the

Figure 4.17 Lift curve slope correction for aspect ratio.

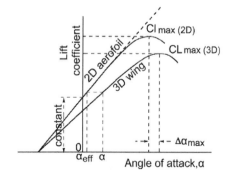

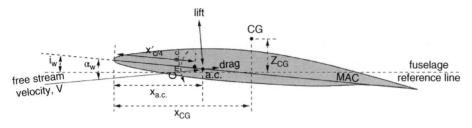

Figure 4.18 Wing pitching moment.

C_{Lmax}. Wing integration with fuselage will affect the C_L versus α graph characteristics to a small extent, which is ignored at this of stage of the study in this book.

The wing tip effect delays the stall by a few degrees because the outer-wing flow distortion reduces the local angle of attack; shown as α_{max}. Note that $\Delta\alpha_{max}$ is the shift of C_{Lmax}; this value $\Delta\alpha_{max}$ is determined experimentally. In this book, the empirical relationship of $\Delta\alpha_{max} = 1$–$2°$, for $AR > 5$–12, $\Delta\alpha_{max} = 1°$, for $AR > 12$–20, and $\Delta\alpha_{max} = 0°$, for $AR > 20$. Extend the 3D lift-curve slope to maximum C_{Lmax} line and fair the graph.

4.9.5   Wing Moment Curve Slope

Section 3.10.1 developed the moment equations for aerofoils. In this section, the moment equation for wing is derived. Moment has be about a point, in this derivation it is the Centre of Gravity (CG). Aircraft CG moves corresponding to how the aircraft is loaded, primarily amount of fuel and payload placement position. Wing aerodynamic centre, $a.c._w$, is a fixed point and also wing moment about $a.c._w$ is invariant to changes in angle of attack, α_W (angle of the incidence free stream velocity V to chord). Since, span wise, the wing chord length can vary, its weighted average as MAC is taken as representing the wing. Figure 4.18 the force and moment diagram of the MAC. Typically, $a.c._w$, is very close to the quarter-chord. Therefore, to simply x_{ac} is taken as $\frac{1}{4}MAC$ without much loss in accuracy in establishing M_{cg}.

Moment equilibrium gives $\sum\text{Moment} = 0$.

Resolving the forces and moments as shown in Figure 4.18, the following relation is derived. Moment arm is along the fuselage reference line, which has free stream velocity angle to as $(\alpha_w - i_w)$, where i_w, is the wing-body incidence Subscript 'w' represents wing only.

$$M_{CG\_W} = L_w \cos(\alpha_w - i_w)(x_{CG} - x_{ac}) + D_w \sin(\alpha_w - i_w)(x_{CG} - x_{ac})$$
$$+ L_w \sin(\alpha_w - i_w)(z_{CG}) + D_w \cos(\alpha_w\, i_w)(z_{CG}) + M_{ac\_W}$$

To note that the sign of M_{ac_W} is based aerofoil characteristics. Example 4.5 demonstrates the use of sign associated with each term.

Note that angle of attack, $(\alpha_W - i_w)$, is small enough to consider $\sin(\alpha_W - i_w) \approx (\alpha_W - i_w)$ and $\cos(\alpha_W - i_w) \approx 1$. Also, consider that $C_L \gg C_D$ and the height of CG from fuselage reference line is sufficiently small to justify to neglect its contribution.

Writing directly in coefficient form by deriving $(\frac{1}{2}\rho V^2 S_W MAC)$, the following is obtained.

$$C_{M\_CG\_W} = C_{m\_ac\_w} + C_{LW}\left(\frac{x_{CG}}{MAC} - \frac{x_{ac}}{MAC}\right) \tag{4.26}$$

C_{LW} can be replaced by Eq. (4.23)

$$C_{M\_CG\_W} = C_{m\_ac\_w} + (C_{L0W} + C_{L\alpha W})\alpha_W\left(\frac{x_{CG}}{MAC} - \frac{x_{ac}}{MAC}\right)$$

where C_{L0W} is C_{LW} at $\alpha = 0$.

It can written as

$$C_{M\_CGW=} C_{M\_0W} + (C_{M\_\alpha W})\alpha_W \tag{4.27}$$

where

$$C_{M\_0W} = C_{m\_ac\_w} + (C_{L0W})\left(\frac{x_{CG}}{MAC} - \frac{x_{ac}}{MAC}\right) \tag{4.28}$$

and

$$C_{M\_\alpha W} = C_{L\alpha W}\left(\frac{x_{CG}}{MAC} - \frac{x_{ac}}{MAC}\right) \tag{4.29}$$

is the moment curve slope.

Example 4.4   Given the NACA 2412 aerofoil data (see Appendix D), construct a wing C_L versus α graph for a rectangular wing planform of $AR = 7$, with an Oswald's efficiency factor, $e = 0.75$, at a flight $Re = 1.5 \times 10^6$.

From the 2D aerofoil test data at $Re = 6 \times 10^6$, find $dC_l/d\alpha = a_0 = 0.095$ per degree (evaluate within the linear range: -2– $8°$). $C_{l@\alpha = 0 = 4.0}$ and $\alpha_{@Cl = 0} = -4°$, C_{lmax} is at $\alpha = 16°$.
   Use Eq. (4.24) to obtain the 3D wing-lift-curve slope:

$$dC_L/d\alpha = a = a_0/[1 + (57.3a_0/e\pi AR)] = 0.095/[1 + (57.3 \times 0.095)/(0.75 \times 3.14 \times 7)]$$
$$= 0.095/[1 + 5.444/16.485] = 0.095/[1 + 0.33] = 0.0714$$

C_L at $\alpha = 0$ is a $\alpha = 4 \times 0.0714 = 0.2856$. Therefore, the equation for the 3D lift-curve slope in the linear range (up to $\approx 8°$ deg) is as follows.

$$C_L = 0.2856 + 0.0714\alpha$$

From the 2D test data, C_{lmax} for three Res for smooth aerofoils and one for a rough surface, interpolation results in a wing $C_{lmax} = 1.25$ at flight Re 1.5×10^6. Finally, for $AR = 7$, the $\Delta\alpha_{max}$ increment is $1°$, which means that the wing stalls at $(16 + 1) = 17°$.
   The wing has lost some lift-curve slope (i.e. less lift for the same angle of attack) and stalls at a slightly higher angle of attack compared to the 2D test data. Draw a vertical line from the 2D stall $\alpha_{max} + 1°$ (the point where the wing maximum lift is reached). Then, draw a horizontal line with $C_{Lmax} = 1.25$.
   Finally, translate the 2D stalling characteristic to the 3D wing-lift-curve slope joining the portion to the C_{Lmax} point following the test-data pattern. This demonstrates that the wing C_L versus the angle of attack, α, can be constructed, as shown in Figure 4.17.

Example 4.5   The NACA 65-410 aerofoil (see Appendix D) and wing data are as follows. The fight Reynolds number $= Re = 8.7 \times 10^6$.

Aerofoil	Wing
$C_{lmax} = 1.3$ (rough surface)	$S_W = 323\ \text{ft}^2$ (trapezoid planform)
$dC_l/d\alpha = a_0 = 0.11$ per degree	$MAC = 7\ \text{ft}$
$C_{m\_ac} = -0.07$	$AR = 7.5$
$\alpha_{stall} = 12°$ (rough surface)	Oswald's efficiency factor, $e = 0.85$
$\alpha_{L = 0} = -2°$ (zero lift angle)	$x_{ac} \approx 0.25\ MAC$

Construct a wing C_L versus α graph for the wing and find the lift-curve slope, $dC_L/d\alpha$. Establish the C_L equation for the linear range. Determine the downwash of the wing at $C_L = 0.45$. Find the $\frac{d\varepsilon}{d\alpha}$ of the downwash. Also find the expression for the wing moment coefficient, C_M (i) with CG ahead of ac at 20% MAC and (ii) with CG aft of $a.c.$ at 30% of MAC.

Solution:
From the 2D aerofoil test data at $Re = 6 \times 10^6$, find $dC_l/d\alpha = a_0 = 0.11$ per degree (evaluate within the linear range: -2–$8°$). C_{lmax} is at $\alpha_{stall} = 12°$ (rough surface).

Use Eq. (4.24) to obtain the 3D wing-lift-curve slope:

$$dC_L/d\alpha = a = a_0/[1 + (57.3a_0/e\pi AR)] = 0.11/[1 + (57.3 \times 0.11)/(0.9 \times 3.14 \times 7.5)]$$

$$= 0.11/[1 + 6.3/22.2] == 0.11/[1 + 0.297] = 0.0848 \text{ per deg} = 4.86 \text{ per rad}$$

$$C_L = (a_0 \times \alpha)/[1 + (57.3a_0/e\pi AR)]$$

C_{L_0W} at $\alpha = 0$ can be computed as

$$C_{L\_0W} == (0.11 \times 2)/\left[1 + (57.3 \times 0.11/(0.9 \times 3.14 \times 7.5)\right] = 0.22/1.29 = 0.17$$

Then the C_L equation for the linear range is as follows.

$$C_L = 0.17 + 0.0848 \times \alpha$$

From the 2D test data, C_{lmax} for three *Re*s for smooth aerofoils and one for a rough surface, Interpolation results in a wing $C_{lmax} = 1.3$ at flight $Re = 8.7 \times 10^6$. Finally, for $AR = 7.5$, the $\Delta\alpha_{max}$ increment is 1°, which means that the wing stalls at $(12 + 1) = 13°$.

Equation (4.19) gives $\dfrac{\partial\varepsilon}{\partial\alpha} = (57.3 \times C_{L\alpha})/(e\pi AR) = (57.3 \times 0.0848)/(0.85 \times 3.14 \times 7.5) = 4.86/20.2 = 0.24$

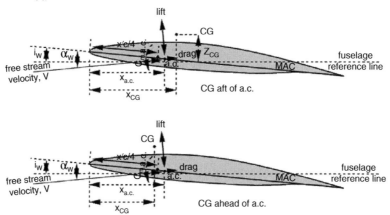

(i) *Aerodynamic centre ahead of CG*
Given $(x_{CG}/MAC) = 0.3$ and $(x_{ac}/MAC) \approx 0.25$
Use Eq. (4.28), for CG behind of *a.c.*

$$C_{M\_0W} = C_{m\_ac\_w} + (C_{L0W})\left(\frac{x_{CG}}{MAC} - \frac{x_{ac}}{MAC}\right) = -0.07 + (0.17)(0.3 - 0.25)$$

$$= -0.07 + (0.17)(0.3 - 0.25) = -0.07 + 0.0085 = -0.0615$$

Use Eq. (4.29),

$$C_{M\_\alpha W} = C_{L\alpha W}\left(\frac{x_{CG}}{MAC} - \frac{x_{ac}}{MAC}\right) = 0.0848 \times (0.3 - 0.25) = 0.00424 \text{ per deg} = 0.243 \text{ per rad}$$

Moment Eq. (4.27)

$$C_{M\_CGW} = C_{M\_0W} + (C_{M\_\alpha W})\alpha_W = -0.0615 + (0.00424)\alpha, \text{ where } \alpha \text{ is in deg.}$$

(ii) *Aerodynamic centre aft of CG*
Given that $(x_{CG}/MAC) = 0.15$ and $(x_{ac}/MAC) \approx 0.25$
Use Eq. (4.28),

$$C_{M\_0W} = C_{m\_ac\_w} + (C_{L0W})\left(\frac{x_{CG}}{MAC} - \frac{x_{ac}}{MAC}\right) = -0.07 + (0.17)(0.20 - 0.25) = -0.0785$$

Use Eq. (4.29),

$$C_{M\_\alpha W} = C_{L\alpha W} \left(\frac{x_{CG}}{MAC} - \frac{x_{ac}}{MAC} \right) = 0.0848(0.2 - 0.25) = -0.00424 \text{ per deg} = -0.243 \text{ per rad}$$

Moment Eq. (4.27)

$$C_{M\_CGW=} C_{M\_0W} + (C_{M\_\alpha W})\alpha_W = -0.0785 - (0.00424)\alpha$$

4.10 Wing Load

Theories involved in predicting spanwise wing load distribution are complex. Relying on Helmholtz's vortex theorems, Ludwig Prandtl (Göttingen, Germany) and Frederick Lancaster (UK) were the first scientists to formulate theories to predict spanwise lift distribution on isolated 3D (finite) wings. The mathematical formulation is known as Prandtl lifting-line theory (also known Lancaster–Prandtl wing theory).

The theory suggests that the wing can be represented by a vortex filament line (Figure 4.19) obeying Helmhotlz's three theorems. Recall that a vortex is a circulatory flow with circulation strength, Γ, that matches to generate the lift (Section 3.6). The circulatory flow of a vortex has a downward velocity at the trailing edge as downwash, discussed in Section 4.9, to derive the expression for the induced drag, C_{Di}. For inviscid airflow, the vortex has infinite core velocity *bound* inside the wing and can be excluded in computation but the circulatory externally around the wing complies with the necessary conditions to progress with the solution. At the tips of the wing, when lift becomes zero, the vortices, one at each tip, become *free*. Aircraft in motion continuously shed trailing vortices downstream to infinity closing to the start line when the aircraft first started to move. Sufficiently downstream, the vortices dissipates and takes a U-shape, termed a *horse-shoe* vortex. The flow velocity of the circulation can be computed using the Biot–Savart law. (Refer to [2–6] for details of derivations).

The formulation of the theory was based on Fourier functions and is difficult to solve. With the difficulties in tackling a vortex filament line of constant circulation, Prandtl refined his analyses, replacing a single vortex filament by large number of horse-shoe vortices with varying strength nestled to a line produced better representation of spanwise lift distribution, resulting in a vortex sheet shedding from the wing trailing edge (Figure 4.19). In this case, instead of having one wing tip vortex there is an infinite series of horse-shoe vortices with varying strength emanating from the wing trailing edge. The finding of lift distribution over an elliptical shaped wing planform laid the foundation for the advances made in aircraft design.

His solution for spanwise lift distribution of an isolated planar (zero twist) elliptical wing planform and with invariant aerofoil section profile also exhibits an elliptical shape given by the following equation (Figure 4.20).

$$l = l_0 \sqrt{1 - \left(\frac{2y}{b} \right)^2} \tag{4.30}$$

where l = sectional lift per at any station, y, from wing centreline, with a y-axis along the wing span. l_0 is the lift (z-axis) at the wing centre line and b is the wing span. Wing lift, L, is the area of the semi-ellipse. He

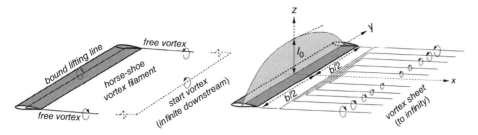

Figure 4.19 Spanwise lift distribution.

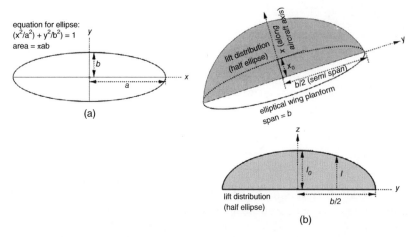

Figure 4.20 Properties of an ellipse. (a) Ellipse. (b) In terms of wing-based symbols.

demonstrated that elliptical spanwise lift distribution has constant downwash offering the least induced drag, C_{Di}. Constant spanwise downwash generates a constant sectional lift coefficient, C_l, across the wing span.

Geometric Properties of an Ellipse

In the conventional Cartesian coordinate system, the generalised equation for an ellipse is given by: $(x^2/a^2) + (y^2/b^2) = 1$, it is a horizontally elongated ellipse with $a =$ semi-major axis (semi-wing span) and $b =$ semi-minor axis. (An ellipse is a special type of circle with a radius 'b' elongated by a factor of a/b.) The area of ellipse is $S_{ellip} = \pi ab$.

Rewriting the equation of the ellipse in terms of an isolated elliptical wing geometry with associated symbols (Figure 4.20b), the wing elliptical wing planform equation becomes as follows (x- and y-axes have interchanged).

$$(y^2)/(b/2)^2 + (x^2/X_0^2) = 1$$

Wing area, $S_W = \pi(b/2)X_0$ that gives, $X_0 = 2S_W/(\pi b)$
Then this equation becomes $(y^2)/(b/2)^2 + (b\pi x^2/2S_W)^2 = 1$
or $[(x^2/2S_W)/b\pi]^2 = 1 - (y^2)/(b/2)^2$

$$\text{or } x = \left(\frac{2S_W}{\pi b}\right)\sqrt{1 - \left(\frac{2y}{b}\right)^2}, \tag{4.31}$$

where x is half the local chord, 'c' at station 'y' from the wing centreline.
Therefore,

$$\text{local wing chord}, c = 2x = \left(\frac{4S_W}{\pi b}\right)\sqrt{1 - \left(\frac{2y}{b}\right)^2} \tag{4.32}$$

Aerodynamic Properties of an Elliptic Wing

Lift Distribution

Consider the spanwise life distribution (z-axis) that is also to be semi-elliptical, as discovered by Prandtl.
Its elliptical equation is $(y^2/(b/2)^2) + (z^2/l_0^2) = 1$
The semi-elliptical lift distribution area represents wing lift $S_{L\_semi-ellip} = \pi(b/4)l_0 = L$.
Then,

$$l_0 = (4S_{L\_semi-ellip})/(\pi b) = (4L)/(\pi b) \tag{4.33}$$

It gives an elliptical lift distribution equation as, $(y^2/(b/2)^2) + (\pi b)^2 \, z^2/(4S_{L\_ellip})^2 = 1$

or $z^2 = [(4S_{L\_ellip})^2/(\pi b)^2] \times [1 - (y^2/(b/2)^2)]$. Replacing z with the section lift, l, the following is obtained.

$$\text{or } l = \left(\frac{4L}{\pi b}\right)\sqrt{1 - \left(\frac{2y}{b}\right)^2}, \tag{4.34}$$

where $l_0 = \left(\frac{4L}{\pi b}\right)$.

At any distance along the span measured from wing centreline, y, the wing chord is c along the x-axis (Eq. (4.31)) and the magnitude of local lift l represented by z. For a given aircraft wing with span b, it shows that only the local chord, c, at y contributes to the lift.

Circulation, $\Gamma(y)$ Distribution – Elliptic Wing

The concept of circulation, Γ, was briefly explained in Section 3.5. In this section, its role in lift generation is explained.

Lift of wing in term of circulation, Γ, is given by the following equation (for the derivation, consult [2–5]).

$$L = \Gamma\rho v b = 0.5\rho v^2 S_W C_L \tag{4.35}$$

where wing area, $S_W = $ (wing span, b) $\times$ (standard mean chord, $\bar{c}$) $= b\bar{c}$

Lift 'L' is constant for a particular aircraft flying at a steady level speed.

By definition, standard mean chord (SMC) is not the MAC. This is the chord of a rectangular wing of the same span and area of any wing under study (also known as the geometric mean chord). SMC is a geometric relation, for example, for a trapezoidal wing it is $\frac{1}{2}(C_R + C_T)$.

$$\text{Lift per unit span}, (L/b) = \Gamma\rho v = 0.5\rho v^2 \bar{c} C_L = q\bar{c} C_L, \tag{4.36}$$

where $q = \frac{1}{2}\rho v^2$.

Again Eq. (4.35) can be rewritten as

$$\Gamma = \frac{1}{2}v\bar{c} C_L \text{ m}^2/\text{s} \tag{4.37}$$

Equations (4.35) and (4.36) brings out the section role of $(\bar{c}C_L)$. For a particular aircraft in steady level flight, ρ and v remain constant. Therefore, lift per unit span (L/b) and circulation (Γ) depends only on the parameter $(\bar{c}C_L)$ and not just the C_L.

From the elliptic lift distribution, the circulation, Γ, follows the elliptic spanwise circulation $\Gamma(y)$ distribution as follows (not derived – refer to [4]).

$$\Gamma(y) = \Gamma_0\sqrt{1 - \left(\frac{2y}{b}\right)^2}, \tag{4.38}$$

where $\Gamma_0 = \dfrac{4L}{\rho v b \pi}$.

Using Eq. (4.35), substitute $L = \Gamma\rho v b$,

This gives $\Gamma_0 = \dfrac{4\Gamma}{\pi}$.

4.10.1 Schrenk's Method – An Approximate Method to Compute Wing Load

Following Prandtl's lifting theory [6] to get some idea of load distribution on isolated wings, scientists continued to work on finding better methods to compute spanwise wing load distribution. Important steps forward in the quest to predict wing load on subsonic lifting surface theory were presented by Hans Multhopp [7] (Göttingen, Germany). His findings include fuselage contribution on spanwise wing load distribution. Multhopp's analytical method is relatively difficult to solve. Subsequently, followed by other researchers, for example, Oskar Schrenk [8], Franklin Diederich [9] and so on, offered some easier solutions with some approximations. Recently in industry, modern methods have been using CFD analyses to obtain considerably better results. Wing loading data are used for wing stress analyses and aeroelastic studies, which are beyond the scope of this book. CFD analysis to obtain realistic wing load distribution of aircraft configuration has replaced the older methods.

Schrenk's approximate method is the easier one to obtain spanwise wing load distribution of isolated wings. The results are close enough to what CFD analyses can predict. The method was widely used industry to get preliminary wing load at the conceptual design stage to proceed with the structural design of wings. Schrenk's method is still in use to obtain quick results. Given in this section is a simplified Schrenk's method to expose this to the readers. Schrenk's method does not use circulation, Γ.

The basis of Schrenk's method is the elliptical wing sectional with constant spanwise C_l distribution. Schrenk observed that spanwise lift distribution between elliptical and non-elliptical wings follows similar pattern; only the local cC_l Eq.s (4.35) and (4.36) parameter influences the lift distribution. Schrenk's approximation suggests that the lift distribution of non-elliptical wing is the mean chord distribution of the elliptical and the non-elliptical wings.

The first task is to construct elliptical distribution to a scalable standard wing at $C_L = 1$ (close to C_L stall). Lift per unit span of the wing is as follows.

$l = qcC_l = qc$ (since $C_L = 1$ and is uniform across elliptic wingstrip area $= 1 \times c$)

or, $l/q = c$ spanwise lift distribution depends only on the local chord, c. (4.39)

Given next is Schrenk's method to obtain spanwise lift distribution step-by-step.

Example 4.6 Worked out here is the Bizjet example undertaken throughout this book (Section 8.9) with the following information to compute wing load distribution using Schrenk's method.

Trapezoidal planar wing area, $S_W = 323\,\text{ft}^2$ (for simplification downwash twist angle is ignored), span, $b = 49.2\,\text{ft}$, taper ratio, $\lambda = 0.4$, $C_R = 9.38\,\text{ft}$, $C_T = 3.75\,\text{ft}$ and $AR = 7.5$.

Solution: (results obtained are plotted here)
Step 1
Elliptical wing
Construct its equivalent elliptical wing with same span, $b = 49.2\,\text{ft}$ and wing area, $S_{W\_ell} = 323\,\text{ft}^2$ using the relation of Eq. (4.32).

Elliptic wing chord distribution, $c(y) = \left(\dfrac{4S_w}{\pi b}\right)\sqrt{1 - \left(\dfrac{2y}{b}\right)^2}$

At the centreline, the maximum chord, $c_0 = (4 \times 323)/(3.14 \times 49.2) = 1.274 \times 6.565 = 8.363\,\text{ft}$
Rewrite this equation in terms of the equivalent elliptic wing:

$$c(y) = 8.363\sqrt{1 - \left(\dfrac{2y}{49.2}\right)^2}$$

Equation (4.39) gives the wing planform spanwise lift distribution as cC_l. Standardising to $C_l = 1$, the lift distribution is the chord, c, distribution of the elliptic wing is given in the table here and plotted in Figure 4.21.
Semi-spanwise distance from elliptic wing centreline – Spanwise chord, c, distribution ($C_l = 1$).

Station	0	5	10	15	20	24.6
$\sqrt{[1 - (2y/49.2)^2]}$	1	0.979	0.914	0.793	0.582	0
$c(y)$ – ft	8.363	8.19	7.64	6.628	4.869	0

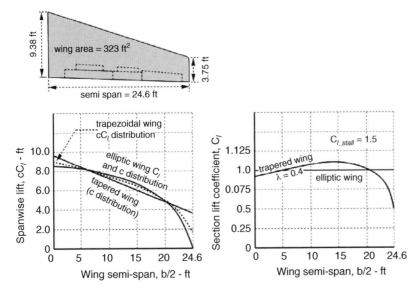

Figure 4.21 Bizjet spanwise wing loading.

Step 2

Trapezoid wing (wing area, $S_W = 323\,\text{ft}^2$)

Trapezoid wing $\lambda = 0.4$, wing chord reduces by 0.2288 ft for 1 ft run along the span.

Schrenk's approximation suggests that spanwise lift distribution $l/q = cC_l$ (Eq. (4.39)) of the trapezoidal wing is the mean and between the two is the spanwise lift distribution for the rectangular wing as given in the following.

Spanwise chord, c, and C_l distribution

Station	0	5	10	15	20	24.6
$c(y)$	9.38	8.24	7.09	5.95	4.8	3.75
$cC_l = \frac{1}{2} \times (c_{\text{ellip}} + c_{\text{trap}})$	8.87	8.21	7.37	6.29	4.84	1.875
C_l	0.946	0.997	1.0387	1.0573	1.007	0.5

The plotted results given here show that the tapered wing ($\lambda = 0.4$) sectional C_l approaches a stall value at about 61% of the span from the wing centreline. Decreasing the taper ratio, λ, moves the sectional C_l approaching stall move towards the wing tips, which is not desirable as the aircraft loses lateral controllability, hence a downwash twist (geometric or aerodynamic) given to the wing to make its root section stall first.

Step 3

Load on trapezoidal wing

Load on trapezoidal wing ($S_W = 323\,\text{ft}^2$). Divide the half wing span into five sections (higher division increases accuracy, a minimum of 5 is suggested: see Figure 4.22).

Aircraft cruising at sea level at $C_L = 1.0$ ($\rho = 0.002378\,\text{slugs/ft}^3$) and $v = 231.6\,\text{ft s}^{-1}$).

$q = \frac{1}{2}\rho v^2 = 63.82\,\text{slugs/ft}^2$

Load on wing section $= 0.5\rho v^2 \times (\text{section area}) = 63.82 \times (\text{section area})$

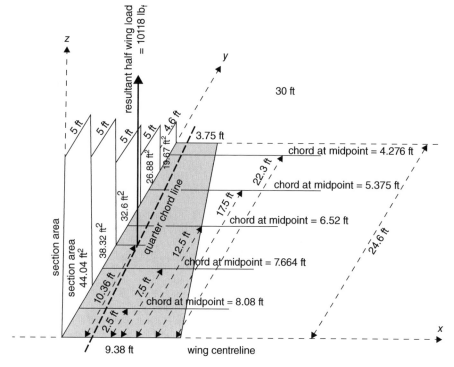

Figure 4.22 Bizjet wing - elliptical.

Sections	0–5	5–10		10–15	15–20	20–24.6
Midpoint	2.5	7.5		12.5	17.5	22.3
Chord at midpoint, C_{ave}	8.808	7.664		6.52	5.375	4.276
Δ section width	5	5		5	5	4.6
Area of the section, ft^2	44.04	38.32		32.60	26.88	19.67
Midpoint C_l	0.972	1.018		1.048	1.032	0.754
Lift on section, lb	2730.8	2489.7		2180.5	1770.5	946.0
Load on half wing, lb	10 117.57					
Load on full wing, lb	20 235.13	1.2% variation (a good value)[a]				
Wing load, lb ft^{-2}	62.64747					

a) Bizjet maximum takeoff mass (MTOM) is computed to be 20 680 lb (Table 10.7). By the time it takes off to 1500 ft and makes a sea level test run at the given condition for this problem, it may be assumed to have consumed, say, 200 lb of fuel to arrive at a weight of 20 480 lb. Accuracy can be increased by increasing the number of span division. Also incorporating the effect of wing twist improves accuracy (not done here).

Resultant force in half wing

Moment of section load	6827.0	18 672.7	27 256.4	30 983.6	21 097.5
Total moment = 104 837.2					
Resultant at from centreline	10.362 ft				

Having standardised to $C_L = 1$, for any other C_L, multiply this table by the new C_L. For gust load of the load factor, n, multiply L in Eq. (4.35) by n.

With this information, wing bending moment and shear force can computed for wing structural design, which is beyond the scope of this book (Chapter 19 briefly introduces aircraft structural design considerations).

4.10.2 Wing Planform Load Distribution

Figure 4.23 compares spanwise load distribution on a wing with a taper ratio, $\lambda = 0–0.1.0$. It was mentioned earlier that the induced drag is lowest for an elliptical wing planform when $e = 1$. It can be observed that with taper ratio, $\lambda \approx 0.3–0.4$ the load distribution is close to an elliptical wing. In general, the industry uses a trapezoidal planform with a taper ratio, $\lambda \approx 0.3–0.5$, resulting in an e value ranging from 0.85 to 0.98 (at optimal design it approaches 1.0). A rectangular wing has a ratio of $\lambda = 1.0$ with $e \approx 0.75–0.85$ depending on the wing AR. The delta wing has a ratio of $\lambda = 0$, which results in an average e below 0.75.

Most aircraft wings have some amount of twist when the shape will change as shown in Figure 4.24 for an $AR = 4.0$ tapered wing. The curves in Figure 4.24 indicate that an increase in sweepback angle from 20° to 60° results in a large increase in the value of the loading parameter near the tip relative to that at the root for wings of AR 4.0 and taper ratio 0.4. Reducing the taper ratio from 0.6 to 0.25 on wings of AR 4.0 and 40 of sweepback causes a corresponding increase in the relative amount of load carried near the wingtip, as shown

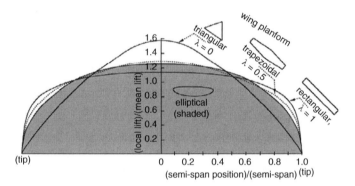

Figure 4.23 Spanwise load distribution (no twist).

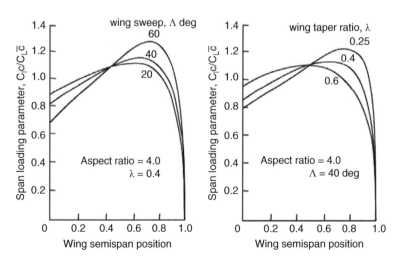

Figure 4.24 Spanwise load distribution (aspect ratio = 4).

in the figure. Variations in the AR for a given sweepback angle and taper ratio also have an important influence on the shape of the span loading curve. A tweaked tapered wing with $\lambda \approx 0.33$–0.35 is the choice.

4.11 Compressibility Effect: Wing Sweep

Section 3.13 explains the compressibility effect on aerofoil resulting from the onset of sonic flow at a point, known as the *super velocity* effect, on the upper surface on account of the thickness distribution along its chord. On the wing, the same phenomenon can occur along the wing chord. Evidently, the local super velocity over the wing is greater than the free stream Mach, M_∞. A typical consequence of higher coverage of sonic flow over the wing, there is the rapid drag increase due to compressibility effects (example on aerofoil is given in Figure 3.38).

4.11.1 Compressibility Drag/Wave Drag

Section 3.14 indicated that the compressibility effect contributes to drag rise. This is defined as wave drag C_{Dw}. The critical Mach number, M_{crit}, (Boeing definition), arbitrarily defined at the point when $C_{Dw} = 0.002$. This has to be added to the aircraft drag as defined by Eq. 11.2 for incompressible flow as follows.

Aircraft Drag = Parasite Drag + Induced drag + Compressibility drag

In coefficient form,

$$C_D = C_{DP} + C_{Di} + C_{Dw}. \tag{4.40}$$

In supersonic flow, the shock wave drag has to be added. For convenience in the book keeping of various types of aircraft drag components, supersonic drag at zero lift is added to Eq. (4.18). This topic is dealt with in detail in Section 11.18.

At this juncture, it is necessary to introduce the two kinds of aircraft cruise speed schedules (see Chapter 15 for details) in operational practice. These are (i) Long Range Cruise (LRC) Mach number, M_{CR}, at the airspeed when the wing indicates the initiation of just forming sonic speed at the upper surface, hence it is considered free from compressible wave drag and (ii) High Speed Cruise (HSC) Mach number, M_{crit}, is at the airspeed when local shock wave is tolerated to allow 20 counts of drag rise ($\Delta C_D = 0.002$ as wave drag), in order to save cruise time at the expense of a small loss of range for the same amount of onboard fuel load. At a constant cruise speed, the aircraft lift coefficient C_L changes from a high to a low value as the aircraft becomes lighter with fuel burn. The design lift coefficient, C_{L_des}, is taken at the mid-cruise weight at HSC speed schedule. Depending on the mission role, design lift coefficient, C_{L_des}, can be at a different condition as will be shown in military aircraft design.

Using Airbus/Boeing definition (see Section 3.13), the onset of the sonic flow field is M_{cr}, a suitable speed to cruise without incurring wave drag (compressibility effect – see Section 11.13) penalty. A 20-count drag rise ($C_D = 0.002$), the extent of drag divergence above which it becomes critical and hence termed the *critical Mach number* (M_{crit}). This drag rise on account of wave drag is tolerated for HSC operation, typically, about $\Delta 0.4$ Mach to $\Delta 0.8$ Mach higher than the M_{cr}.

4.11.2 Wing Sweep

One way to delay the M_{crit} is to sweep the wing either backward or forward (Figure 4.3). Wing sweep encourages the undesirable spanwise flow distortion, which can be reduced by having a 'glove' (Figure 4.8) in front of wing root or installing a fence (Figure 4.37, later) on the wing. Wing sweep is measured by the quarter-chord sweep angle, $\Lambda_{1/4}$, to make the aerofoil thinner along free stream flow, as shown in Figure 4.25. If the trailing edge remains unswept (with yehudi), then flap effectiveness is less degraded due to a quarter-chord sweep.

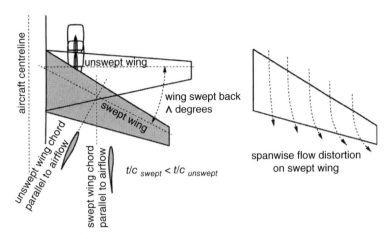

Figure 4.25 Sweep of wing.

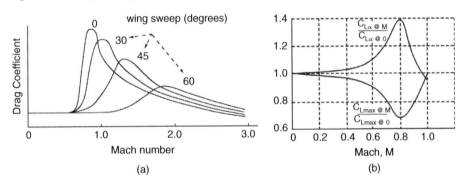

Figure 4.26 Sweep of wing. (a) Drag comparison (aerofoil specific). (b) Mach number effect.

Note that 'wing slide' (i.e. in which the chord length remains the same) is different from wing sweep, in which the chord length is longer by the secant of the sweep angle (Figure 4.25).

Wing sweep thins the aerofoil t/c ratio and delays the sudden drag rise (Figure 4.26a). The swept-back wing is by far more prevalent because of structural considerations. Another effect of speed gain is a change in C_{Lmax}. For a particular wing, the ratio of $(C_{Lmax\,comp}/C_{Lmax\,incom})$ decreases approximately to 0.7, as shown in Figure 4.26b.

Shown here is the relationship between the sweep angle and wing geometries (Figure 4.19a). From the geometry, the chord length of a swept wing increases, resulting in a decrease in the t/c ratio as expressed by the relations given here:

$$chord_{swept} = (chord_{unswept})/(\cos \Lambda_{1/4})$$

This results in:

$$(thickness/chord_{swept}) < (thickness/chord_{unswept}) \tag{4.41}$$

Wing sweep directly increases the drag divergence Mach number, M_{DD}, by divided by the cosine of the sweep angle as given in Eq. (4.42).

$$(M_{crit@\Lambda1/4\,with\,sweep}) = (M_{crit@\Lambda1/4=0})/(Cos\Lambda_{1/4\,with\,sweep}) \tag{4.42}$$

The sweep also degrades the C_{Lmax}. Since C_{Lmax} is not a geometric parameter, the cosine of $\Lambda_{1/4}$ is modified to the following relations, which gives a reasonable value matching the statistics of several existing aircraft.

$$C_{Lmax-swept} = C_{Lmax\_unswept} \times \sqrt{Cos\Lambda_{1/4}} \tag{4.43}$$

4.11.2.1 Sweep Wing Effects

The choice of M_{crit} that gives the design lift coefficient, $C_{L\text{-des}}$, and aerofoil t/c decides the extent of wing sweep required to achieve the desired result (i.e. to maximise M_{crit}) while also satisfying structural requirements. Wing sweep will bring the aircraft aerodynamic centre more aft. There is 14–16% loss of C_{Lmax} of account of wing sweep at the landing configuration.

Structural engineers prefer aerofoil sections to be as thick as possible, which favours structural integrity at lower weights and increases wing volume allowing more onboard fuel storage. However, aerodynamicists prefer the aerofoil to be as thin as possible to minimise the transonic flow regime in order to keep the wave drag rise lower. A compromise is sought.

4.11.3 Relationship Between Wing Sweep, Mach Number and Aerofoil t/c

Wing design is the heart of aircraft design. The relationship between wing sweep, $\Lambda_{1/4}$, wing aerofoil thickness to chord ratio, t/c, and the cruise Mach number are important considerations for wing design. For a given wing area and AR, a relationship exists between aircraft wing sweep, $\Lambda_{1/4}$, design Mach number (i.e. the design $C_{L\text{-des}}$) and aerofoil average thickness to chord ratio, t/c_{ave}.

Aircraft with smaller range could fly at a lower speed compared to an aircraft mission with a long range when time saving on account of higher speed is appreciable. An aircraft with a lower cruise speed is cheaper to design and manufacture offering lower operational costs attractive for operators. Market requirements are the starting point to specify what should be achievable cruise to compete in the market place. Aircraft manufacturers' researchers continually search for improved aerofoils. Drag reduction at the same operating speed is the main quest for researchers. The lower the aircraft cruise speed, the lower the sweep can be, allowing a higher AR for the same size of wing reference area, S_W, that has lower induced drag C_{Di}. It is not easy to push Mach capabilities much higher when high-subsonic transport aircraft are already cruising at transonic speed close to sonic speed. Next, is to consider how to minimise the cost of manufacture. With the same t/c, a wing with smaller sweep is slightly cheaper to manufacture. Therefore, it is important to find out the correct wing geometry.

The following values are typical aircraft speed schedules adopted for mission range that are used in current practice in the majority of operating aircraft with the exception of a few special ones. The values given in Table 4.2 are used in this book.

Wing planform shape influences the choice of aerofoil thickness as shown in Figure 4.27. The figure gives the relation between wing aerofoil t/c and aircraft design speed within the band of wing sweep. Increase of aircraft speed decreases aerofoil t/c.

At the initial conceptual design stages of a new aircraft project when not much information is known, considerable experience is required to generate these basic wing parameters to progress towards a sized wing. Many researcher have proposed semi-empirical relations correlating the three parameters, for example, $\Lambda_{1/4}$, t/c and $C_{L\text{-des}}$, matched to the aircraft speed schedule. These relationships are aerofoil specific. At the initial stage of a new aircraft project, predictions based on semi-empirical relations develop from the statistics of past designs and backed-up theory can offer a good starting point. Subsequently, when more information (e.g. CFD/wind-tunnel tests) is available as the project progresses, the wing geometry gets refined in an iterative process. The authors have selected the findings of Torenbeek [11] and Shevell [12] as suitable for academic studies. These are briefly described in the following.

Table 4.2 Typical cruise Mach number for the mission range (Boeing terminology).

Aircraft category	Range, R (nm)	LRC @ M_{CR}	HSC @ M_{crit}	ΔM
Bizjet/Regional jets	R < 2000	≈0.7–0.75	≈0.75–0.78	≈0.04–0.07
Mid-range jets	2000 < R < 4000	≈0.75–0.8	≈0.8–0.85	≈0.04–0.08
Long-range jets	R > 4000	≈0.85–0.88	≈0.9–0.95	≈0.04–0.08

Figure 4.27 Thickness to chord ratio.

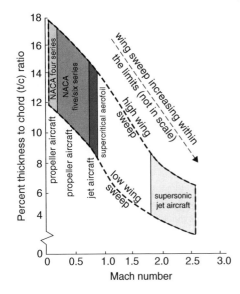

4.11.3.1 Torenbeek'S Method

Torenbeek gives the semi-empirical relation based on theoretical considerations to get some idea on the relation between M_{CR}, $\Lambda_{1/4}$ and $(t/c)_{ave}$. Brief details on the topic are given here. For full details, readers are recommended to refer to [11, pp. 246–249]. To maintain uniformity with the Boeing symbols used in this book, the following are used:

The onset of shock, M_{CR} (Boeing) = M_{cr} (Torenbeek).
The drag divergence Mach number (20 count drag rise), M_{crit} (Boeing) = M_{crD} (Torenbeek).

From the experimental data on symmetrical aerofoil, the relation between thickness to chord ratio (t/c) and $(C_{p\_incomp})_{min}$ can be approximately represented by the following equation:

$$(C_{p\_incomp})_{min} = \text{constant} \times (t/c)^{1.5}$$

Substituting the Prandtl–Glauert compressibility correction (Eq. (3.52)), $C_{p\_comp} = \frac{C_{p\_incomp}}{\sqrt{(1-M_\infty^2)}}$ into the above equation, the following is obtained for compressible flow:

$$(t/c) = k \times \left\{ C_{p\_comp} \sqrt{1 - M_\infty^2} \right\}^{2/3} \tag{4.44}$$

where k is a constant, a function of t/c distribution along the aerofoil chord.

At the critical speed M_∞ becomes M_{CR} (see Section 3.14, Boeing definition) when local speed at the crest (point of inflexion) on the upper surface of the aerofoil hits sonic speed. Then $(C_{p\_comp})_{min}$ becomes C_{p_CR} (has negative sign). Typically, aircraft LRC cruise speed is kept at the onset of sonic speed (Figure 3.39). The expression for C_{p_CR} is given below:

$$C_{p\_CR} = \left[\left(\frac{2}{\gamma M_{CR}^2} \left\{ \left[1 - \frac{2 + (\gamma - 1)M_{CR}^2}{\gamma + 1} \right]^{3.5} - 1 \right\} \right) \right] \quad \text{where for air,} \quad \frac{\gamma}{(\gamma - 1)} = 3.5$$

By substituting the above into Eq. (4.44), the relation between t/c and M_{cr} can be obtained for unswept wing with conventional aerofoil section as given below:

$$(t/c) = k \times \left[\left[\left(\frac{2}{\gamma M_{CR}^2} \right) \left\{ \left[1 - \frac{2 + (\gamma - 1)M_{CR}^2}{\gamma + 1} \right]^{3.5} \right\} \sqrt{1 - M_{CR}^2} \right] \right]^{2/3} \tag{4.45}$$

where $k = 0.24$ for conventional symmetrical aerofoil, e.g. NACA four-digit/NACA five-digit types.

For more advanced aerofoil, e.g. NACA six-digit aerofoil and NASA supercritical aerofoil type, Eq. (4.45) is semi-empirically modified to Eq. (4.46) meant for symmetrical aerofoil at zero lift:

$$(t/c) = 0.3 \times \left\{ \left[1 - \left(\frac{5 + M_{CR}^2}{5 + (M_*)^2} \right)^{3.5} \right] \left(\frac{\sqrt{1 - M_{CR}^2}}{M_{CR}^2} \right) \right\}^{2/3} \tag{4.46}$$

where M^* is a constant and nothing to do with Mach number. M^* is a user-defined aerofoil specific semi-empirical factor for unswept wing. M^* represents aerodynamic capability for the type of the advanced supercritical aerofoil (capable of operating at M_{crit}, above M_{CR}).

Torenbeek gives the value of M^* for unswept wing, as follows:

Conventional aerofoil, NACA six-digit series (low subsonic, <0.75 Mach), $M^* = 1.0$.
  *NACA six-digit series (high subsonic), peaky section aerofoil (high subsonic, $0.75 <$ Mach < 0.85), $M^* = 1.03–1.05$.
  *Slightly modified by the authors.
Supercritical aerofoil, (high subsonic, Mach > 0.85) $M^* = 1.12–1.15$.

Applying to high subsonic flight speed requires wing sweeping back expressed by its quarter-chord sweep angle, $\Lambda_{1/4}$, to suit M_{CR}. Swept wing aerofoil encounters component of the free stream component of M_{CR} that makes airspeed $M_{CR} \times \cos\Lambda_{1/4}$ normal to the wing. Torenbeek modified Eq. (4.46) to the semi-empirical Eq. (4.47) given below. Unlike (t/c) for 2D aerofoil, 3D wing may have varied (t/c) along its span. Therefore, the average thickness to chord ratio $(t/c)_{ave}$ has to be considered. The modified equation is given below.

$$(t/c)_{ave} = \left(\frac{0.3}{M_{CR}} \right) \left[(M_{CR} \cos \Lambda_{1/4})^{-1} - (M_{CR} \cos \Lambda_{1/4}) \right]^{1/3} \times \left[1 - \left\{ \frac{5 + (M_{CR} \cos \Lambda_{1/4})^2}{5 + (M^*)^2} \right\}^{3.5} \right]^{2/3}$$

$$\tag{4.47}$$

The user factor M^* is modified for the swept wing and the camber effects at C_{L_des} at M_{crit} is given below. M^* is reduced by the factor $\frac{(0.25 \times C_{L\_des})}{(\cos \Lambda_{1/4})^2}$.

  Conventional aerofoil, $M^* = \left(1.0 - \frac{(0.25 \times C_{L\_des})}{(\cos \Lambda_{1/4})^2} \right)$.
  Peaky section aerofoil (high subsonic), $M^* = \left(1.03 - \frac{(0.25 \times C_{L\_des})}{(\cos \Lambda_{1/4})^2} \right)$ to $\left(1.05 - \frac{(0.25 \times C_{L\_des})}{(\cos \Lambda_{1/4})^2} \right)$.
  Supercritical aerofoil, $M^* = \left(1.12 - \frac{(0.25 \times C_{L\_des})}{(\cos \Lambda_{1/4})^2} \right)$ to $\left(1.15 - \frac{(0.25 \times C_{L\_des})}{(\cos \Lambda_{1/4})^2} \right)$.

The equation is tried out for some supercritical aerofoil, yielding satisfactory results. While NACA six-series aerofoil are not as modern as the peaky section aerofoil, they can compete satisfactorily; many high subsonic aircraft has been built with thinner NACA six series; some of them can have $(t/c)_{max}$ past $0.30c$.

There are two unknowns, e.g. $\Lambda_{1/4}$ and $(t/c)_{ave}$. But finding $\Lambda_{1/4}$ for the $(t/c)_{ave}$ is more difficult. A good approach is to substitute the desired $\Lambda_{1/4}$ and find the $(t/c)_{ave}$.

The $(t/c)_{ave}$ of the wing aerofoil is represented by the relations, as follows:

$(t/c)_{ave}$ of trapezoidal wing $= (t_{root} + t_{tip})/2$, for t/c is uniformly varying along the span.

$(t/c)_{ave}$ (trapezoidal wing with kink – Figure 4.10) = $(t/c)_{@MAC}$.

There are several other definitions, one such given by Shevell (Eq. (4.48)).

Industries use CFD and data bank. Torenbeek's method may be used when not much information is available. Equation (4.45) gives a good insight to the relation between M_{CR}, $\Lambda_{1/4}$ and $(t/c)_{ave}$. Torenbeek's Eq. (4.47) is workable and reliable, being backed by theory. The equation being semi-empirical nature can only give an approximate ('ball park') result. Torenbeek gives an example with C_{L_des} at M_{crit}, allowing for 20 counts drag rise.

Example 4.7

i. Find the $(t/c)_{ave}$ for a wing with $C_{L\_des} \approx 0.44$ flying at Mach 0.74 (M_{DD}) and $\Lambda_{1/4} = 14°$.
   Solution: This gives $M^* = [1.0 - (0.25 \times 0.44)/0.941] = 0.846$, resulting in $(t/c)_{ave} = 0.14$.
ii. Find the $(t/c)_{ave}$ for a wing with $C_{L\_des} \approx 0.2$ flying at Mach 0.85 (M_{DD}) and $\Lambda_{1/4} = 20°$.
   Solution: This gives $M^* = [1.03 - (0.25 \times 0.2)/0.64] = 0.977$, $(t/c)_{ave} = 0.125$.

As can be appreciated, the complex aerodynamics involving many variables are not easily amenable to being derived in a semi-empirical manner that will not guarantee an exact solution. Designers like to have the thinnest aerofoil without compromising wing integration considerations, e.g. wing volume, strength, etc. For this reason, it is imperative to use CFD results, subsequently substantiating the results with wind tunnel tests. Typical relation between M_{CR} and M_{crit} for the types of aircraft is given in Table 4.2.

4.11.3.2 Shevell's Method

Shevell's [12] method is aerofoil specific, generated from industrial data and is semi-empirical in nature. It gives an approximate relationship between M_{cr} and M_{crit}. To maintain uniformity with the Boeing symbols used in this book, the following are used:

The onset of shock, M_{CR} (Boeing) = M_{cc} (Shevell).

The drag divergence Mach number, M_{crit} (Boeing – 20 count drag rise) = M_{DIV} (Shevell – drag rise based on dC_D/(dM), as explained in Section 3.14).

Shevell presented the graph as shown in Figure 4.28a presenting the relation between with wing t/c, cruise Mach number, M_{CR}, at various C_{L_des} for wing with no sweep ($\Lambda_{1/4} = 0$). This C_L is a nominal value to be used in the graph without reference to wing area. Given any two variables, the third can be obtained from the graph.

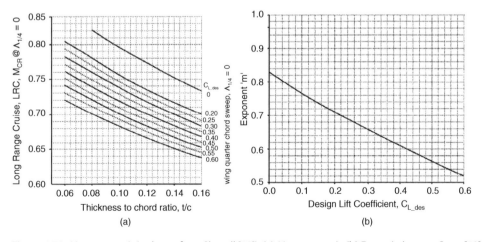

Figure 4.28 M_{CR} versus t/c (redrawn from Shevell [12]). (a) M_{CR} versus t/c. (b) Factor 'm' versus C_{L_des} [12].

Aerofoil t/c can vary along the span and Shevell suggested the following relation as the weighted average of the aerofoil t/c to be taken for the use in context.

$$(t/c)_{\text{ave}} \text{ weighted} = (t_{\text{root}} + t_{\text{tip}})/(c_{\text{root}} + c_{\text{tip}}) \tag{4.48}$$

The graph then gives the $M_{\text{CR}\_\Lambda=0}$ for an unswept wing that needs to be corrected to obtain the quarter-chord sweep, $\Lambda_{1/4}$, suiting the specified Mach number, M_{CR}, which has to be higher than the $M_{\text{CR }\Lambda1/4=0}$.

Apply the following relation to obtain the quarter-chord sweep, $\Lambda_{1/4}$.

$$M_{\text{CR}} = (M_{\text{CR }\Lambda1/4=0}/\cos\Lambda_{1/4}) \tag{4.49}$$

Lastly, one more correction is applied to suit M_{CR} to match the aerofoil t/c aligned to the free stream on account of wing sweep, which is thinner than the aerofoil t/c. The following semi-empirical relation is used to obtain the index 'm' corresponding to $C_{\text{L\_des}}$ from Figure 4.28b to be applied in the equation:

$$M_{\text{CR}} = (M_{\text{CR }\Lambda1/4=0}/\cos^m\Lambda_{1/4}) \tag{4.50}$$

If this matches the specified M_{CR}, then the wing quarter-chord sweep, $\Lambda_{1/4}$, is established for the $(t/c)_{\text{ave}}$. Otherwise, it has to be iterated by varying the wing sweep.

Shevell gives the relation between M_{CR} and M_{crit} as follows. This definition is not used in this book. $M_{\text{crit}} = M_{\text{CR}} \times [1.02 + (1 - \cos\Lambda_{1/4}) \times (0.08)]$.

Shevell's method gives a good insight to the relation between M_{cr} and M_{crit}. In industries, this older method is now replaced by CFD and data bank. At this stage, the statistical values given in Table 4.2 will be used in this book.

4.11.4 Variable Sweep Wings – Reconfiguring in Flight

The previous sections explain the benefits of having wing sweep to increase M_{crit}, that is, aircraft can operate at higher speed as a result of delaying compressible drag (wave drag rise, ΔC_{Dw}) rise. However, the downside is that, for the same area and taper ratio, the wing aspect area is reduced, increasing the induced drag, C_{Di}. Since delaying the C_{Dw} at M_{crit} saves wave drag more than increase in C_{Di}, this makes higher speed operation attractive. Also Eq. (4.43) shows that wing sweep reduces C_{Lmax}.

However, if an aircraft with an unswept wing is operated at a lower speed below C_{Dw} rise, then there is distinct advantage of fuel saving on account of lower engine throttle settings and flying at a higher L/D to get a higher AR, that is, less C_{Di}. An unswept wing with higher C_{Lmax} offers better safety for field performance with lower-speed schedules for takeoff/landing and higher payload capabilities. In summary, a swept wing is more suitable for high speed operations, while an unswept wing suits operations at lower speeds.

Long range supersonic aircraft have to operate in a wide range of speed schedules. Currently, there are no civil aircraft operating at supersonic speed. For military aircraft in interdiction roles, the supersonic operation primarily starts close to the combat area, otherwise there is no point flying supersonic for long with a very high rate of fuel consumption. A well planned operational sortie can have a longer radius of action by flying subsonic until it reaches the operational zone to accelerate to supersonic speed. A variable sweep aircraft offers attractive design considerations to make wing sweep transition as required to suit the aircraft speed. The pilot keeps wing unswept at takeoff/landing, uses partial sweep at subsonic cruise close to the combat zone and then supersonic dash at the combat arena to engage and rapidly disengage. Many of the past generation of aircraft, for example, the F111, F14, B1, Tornado, MIG23, Tu160 and so on, have variable swept wings.

Note that the latest generation combat aircraft are not pursuing variable sweep aircraft design on account of advances in technologies, for example, stealth, beyond visual range (BVR) missile homing, counter measures, surveillance and so on. The swing mechanism is complex and heavy, adding to maintenance and reliability issues.

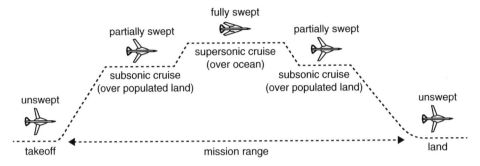

Figure 4.29 Variable sweep commercial-transport aircraft operation.

Variable sweep design for commercial supersonic aircraft (>Mach 2 operation) is an attractive proposition. Heavy aircraft with over 100 passengers operating above Mach 2.0 develop a strong sonic boom noise. Therefore, over populated landmasses, operations are required to be carried out at high-subsonic speed or with a recent development at speeds below 1.6 Mach. But for long stretches over oceans, continuous high supersonic speed flight can be carried out. A pilot operated variable sweep transition facilitates to suit the aircraft speed as required (Figure 4.29).

Some supersonic Bizjets have reduced design speeds around Mach 1.6: with the advancement of sonic boom noise reduction, the configuration does need not to consider variable sweep aircraft.

4.12 Transonic Wings

To obtain a civil aircraft transonic wing planform, this follows the same procedure adopted for high-subsonic wing planform designs in a higher wing quarter-chord wing sweep of around 30° and with the additional consideration of the area ruling of the fuselage (Section 5.10.2). The basis of the transonic wing is in super-critical aerofoil selection as presented in Section 3.13.2. The wing planform selection and aerofoil selection goes together as aerofoil-planform integration. The objective is to minimise compressibility drag by delaying M_{crit} to a higher value.

NASA's test aircraft SCW F8 (Super Critical Wing, Figure 4.30) is a good example of a transonic wing mounted on a Crusader fighter aircraft. Boeing and Lockheed proposed transonic aircraft based on this wing concept during the early 1970s but the lean period of time did not suit the market. Much later, in 2001, the Boeing Sonic cruiser was proposed but not pursued due to lean demand of the time.

Military aircraft are supersonic and the strake/canard (Section 4.13) is used for transonic hard manoeuvres.

4.13 Supersonic Wings

Since any disturbance travelling through air cannot cross beyond the barrier of the shock wave, air ahead of the shock wave front remains undisturbed. Unlike subsonic aircraft, noise emanating from supersonic aircraft cannot be heard ahead of the aircraft, in a *zone of silence*. The typical aerofoil used in a supersonic wing is shown in Figure 3.23.

Section 3.13.3 outlined that using linearised theory, the pressure distribution around the aerofoil can be obtained by spitting the problem into three components of (i) straight line at an angle of attack, α, (ii) aerofoil camber at zero angle of attack and (iii) thickness distribution of the aerofoil as shown in Figure 3.42 and superimpose the solutions by addition to get the results. The thin wing problem can also be tackled in the similar manner by considering it as flat plate at small angle of incidence, α. This allows approximation of the shock wave angle, β, close enough to be taken as the Mach wave angle, μ, that is, $\beta \approx \mu$, permitting development of a wing planform template that will be followed up with CFD analyses, wind-tunnel tests and final sizing.

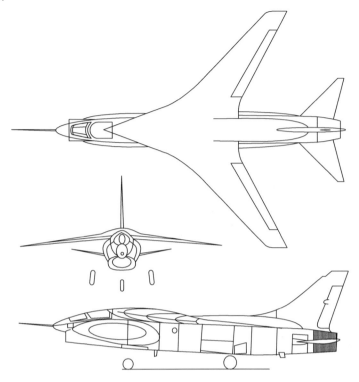

Figure 4.30 Transonic wing – NASA SCW F8 test aircraft. Source: Credit NASA.

Equation (3.21) gives $\sin\mu = (1/M_\infty)$ or $\mu = \sin^{-1}(1/M_\infty)$.
or $M_\infty \sin\mu = 1 = (U_\infty \sin\mu/a)$, where U_∞ is the free stream velocity

$$U_\infty \sin\mu = a = \text{speed of sound} \tag{4.51}$$

In other words, the Mach wave front travels at the speed of sound in the direction normal to it. For this reason, the Mach wave angle, μ, converges as flight Mach number increases.

The supersonic wing has many possibilities and hence can be quite challenging and complex. This section briefly outlines the various design considerations required for wing design. Only two basic types of supersonic wing planform are discussed briefly in this book. Figure 4.31a shows a wing with higher sweep within the Mach cone and Figure 4.31b shows a wing with lower sweep partly outside the Mach cone.

Figure 4.31 (top figure) shows an isolated flat wing in supersonic free stream of Mach 1.8, which gives the Mach angle, $\mu = 37.5°$ measured from line aligned to free stream flow. The entire isolated wing LE $\Lambda_{LE} = 35°$ is in supersonic flow. The break points of the wing are at the tips and at the centre line where the Mach angle covers only part of the wing shown as shaded. This area within the shock cone (angle) requires 3D analyses. The unshaded area is outside the Mach cone and can be dealt with by 2D analyses. Define $\varepsilon = (90 - \Lambda_{LE})$ to measure from the same line from where Mach angle, μ, is measured.

The two basic supersonic wings shown in the bottom sketches of Figure 4.31 are as follows.

i. The left hand delta wing LE $\varepsilon = 90-60 = 30°$, that is, the wing span is within the Mach wave angle $\mu = 37.5°$. The airspeed normal to the wing LE is subsonic. Therefore, a suitable thin supersonic aerofoil can have a rounded LE.

ii. The right hand trapezoidal wing LE $\varepsilon = 90-35 = 55°$; that is, more than $\mu = 37.5°$ making the entire wing LE in supersonic flow. Here, the wing span is greater than what the Mach wave angle, μ, can cover. The airspeed normal to the wing LE is supersonic. Therefore, a suitable thin supersonic aerofoil has a sharp LE.

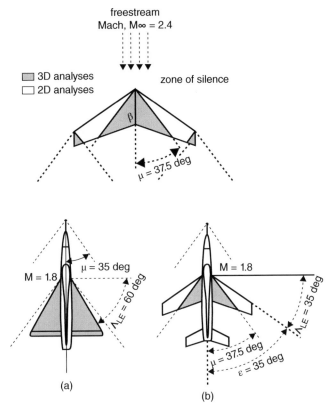

Figure 4.31 Supersonic wing. (top) Isolated wing. (a) Wing with $\varepsilon < \mu$. (b) Wing with $\varepsilon > \mu$.

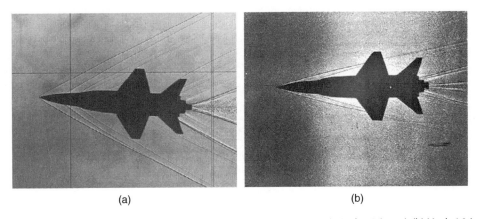

Figure 4.32 X-15 Rocket powered aircraft. (Source: Courtesy NASA). (a) Mach 3.3 ($\varepsilon < \mu$). (b) Mach 6.0 ($\varepsilon > \mu$).

Figure 4.32 shows a Schlieren photograph of the high supersonic/low hypersonic shock wave pattern on an X-15 aircraft in a wind-tunnel test at Mach 3.3 and Mach 6.0. At Mach 3.0 the wing span is within the shock wave making the LE subsonic. At Mach 6.0, part of the outboard wing is outside the shock wake with its LE exposed to supersonic flow.

Most supersonic aircraft wing spans are designed to be within the Mach wave angle, μ, as in the case of the delta wing example, but they can also be at any other geometry with a wing span within the Mach wave angle, μ.

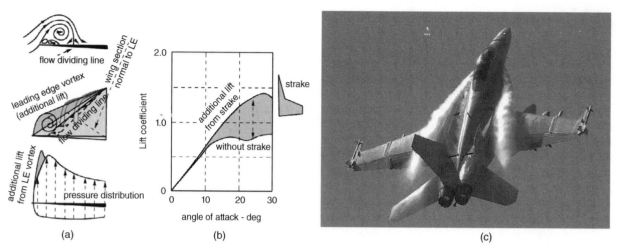

Figure 4.33 Additional vortex lift. (a) Additional vortex lift (half wing shown). (b) Additional lift by strake. (c) Strake vortex over an F18. Source: (c) Credit NASA.

4.13.1 Supersonic Wing Planform

There are basically two types, for example (i) with $\Lambda_{1/4} \approx 0$ and (ii) $\Lambda_{1/4} > 0$. For the same wing area, sweep allows an increase in t/c. Also, for a low taper ratio ($\lambda < 0.3$), in the extreme case for a delta wing, $\lambda = 0$, the long wing root chord and thicker t/c provides increased fuel volume.

4.14 Additional Vortex Lift – LE Suction

Stalling of conventional wings, such as those configured for high-subsonic civil aircraft, occurs around the angle of attack, α, anywhere from 14° to 18°. Difficult manoeuvring demanded by military aircraft requires a much higher stall angle (i.e. 30–40°). This can be achieved by having a carefully placed additional low AR lifting surface; for example, having a LERX (leading edge strake, e.g. F16 and F18) or a canard (e.g. Eurofighter and Su37).

At high angles of attack, the LE of these surfaces produces a strong vortex tube, as shown in Figure 4.33. The vortex flow sweeping past the main wing reenergises the streamlines, delaying flow separation at a higher angle of attack. Vortex flow has low pressure at its core, where the velocity is high (refer to aerodynamic textbooks for more information). This phenomena produces additional vortex lift as shown in Figure 4.33b. There is a limit of high angle of attack, α, beyond which the vortex flow breakdowns and separation takes place resulting in sharp stall. Estimating the additional vortex lift uses LE suction analogy [13].

At airshows during the early 1990s, MIG-29s demonstrated flight at very high angles of attack (i.e. above 60°); their transient 'cobra' movement had never been seen before by the public.

4.15 High-Lift Devices on the Wing – Flaps and Slats

Chapter 3 dealt with 2D sections of aerofoil with flaps and slats. The C_{lmax} values associated with various types and combinations are given in Figure 3.36. This section discusses their application on wings. An excellent review on high-lift devices is given in [16, 17].

Essentially, the role of high-lift devices is to increase the camber of the aerofoil section in a spanwise distribution, executed by the operator as and when required in stages for the mission segment. Deployment of high-lift devices offers the requisite lift at lower aircraft speed but at the cost of undesirable increase in drag and nose down moment.

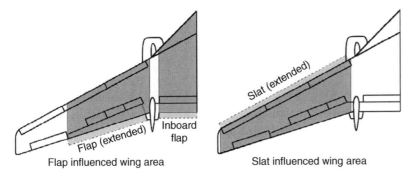

Figure 4.34 Flaps and slats on a wing.

Low speed operation near round offers safety by easing pilot workload. Airfield operations (takeoff and landing) require low speed aircraft operation to ease pilot workload. This is achieved by deploying high-lift devices resulting in drag rise. During takeoff, the aircraft accelerates to climb preferring a high climb gradient, but at landing the aircraft decelerates to stop at the shortest distance. Drag increase will reduce the excess thrust available required to climb after lift-off at takeoff resulting into degradation of climb gradient. Therefore, at takeoff the flap/slat deployment is at a low setting (typically between a 10° and 20° flap setting) only to meet the takeoff requirements stipulated by the governmental regulatory bodies. On the other hand, at landing, the requirement is different as the aircraft is decelerating down to stop. It is therefore desirable to touch down at minimum speed using maximum possible flap/slat deployment (typically extended to 40°). A serious situation may arise if there is a baulked landing when the aircraft is required to go around for the circuit to try to land again. This time there is severe loss of climb gradient on account full flap/slat deployment.

Flap/slat deployment does not necessarily cover the full wing span. In Figure 4.34 the shaded area shows the influenced wing area by flap/slat deployment. In this case, together the flap/slat influenced wing area cover over about 90% of the exposed wing planform area.

As in the case of aerofoil, the deflection effects of the wing mounted high-lift devices increase drag and nose down moment, as well as reduce the stall margin with unfavourable abrupt stall. High-lift devices yield to shorter field length at takeoff, but on account of drag increase the climb rate degrades that may prove critical to clear obstacles. High-lift devices change the pitch attitude requiring high degree of trim to ease stick force. Approach to land occurs at steeper angles with higher deflection.

4.15.1 High-Lift Device Evolution and Mechanism

Figure 4.35 gives a brief evolution of high-lift devices in the four decades at the second half of the twentieth century. Earlier, with a rather complex three-slot Fowler flap, a very high order of C_{Lmax} was achieved advancing high-subsonic transport aircraft to grow larger, which was associated with high-wing loading. These were complex, heavy and expensive components that drew attention for needing improvement by making them simpler and lighter, yet retaining comparable C_{Lmax} values,

First, the Fowler mechanism makes the flap translate backward to extend the wing chord and then rotates to increase the camber. There are four main kinds of Fowler flap mechanism as follows ([16, 17]):

  i. *Hinged flap.* This is the simplest type, effective but does not extend much to give the best value C_{Lmax}.
 ii. *Linked flap.* This allows to translate further, then rotate to give better C_{Lmax}.
iii. *Tracked flap.* The mechanism moves on a laid guided track in a predetermined translation and rotation to give an alternate way to get a high C_{Lmax}. But the restricted movement can be improved by the mechanism described next.
 iv. *Hybrid link-tracked.* A compromise to give the best design.

Figure 4.36 shows a hinged and link flap mechanism as well as with a spoiler. To reduce drag, the under-slung mechanism is enclosed in a streamlined fairing.

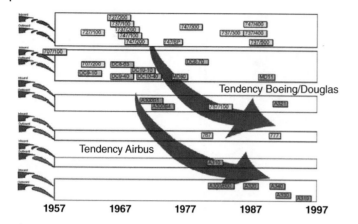

Figure 4.35 High-lift device evolution [17]. Source: Ref. [16].

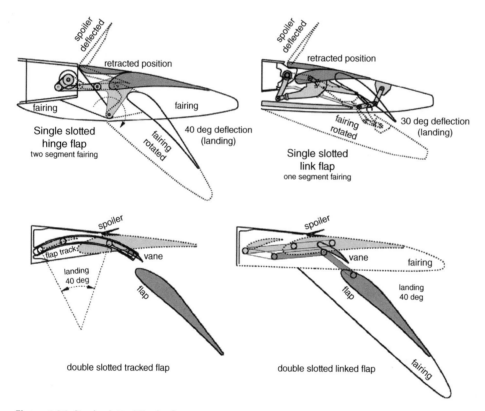

Figure 4.36 Single-slotted Fowler flap.

4.15.2 High-Lift Device C_{Lmax}

It can now be appreciated how complex designing the high-lift devices can be. Since the end of WWII, for nearly for four decades, DATCOM and RAe Datasheets were the only way to obtain preliminary high-lift configurations to suit aircraft performance requirements. It was a relatively lengthy procedure, cross-referencing a multitude of graphs/empirical relations not satisfactorily enough. It required a large number of wind-tunnel tests in a series of several possible configurations, taking nearly a year to complete in an expensive way to get the best suited one in a parametric optimisation. As CFD advanced to the point to get a reliable 2D aerofoil,

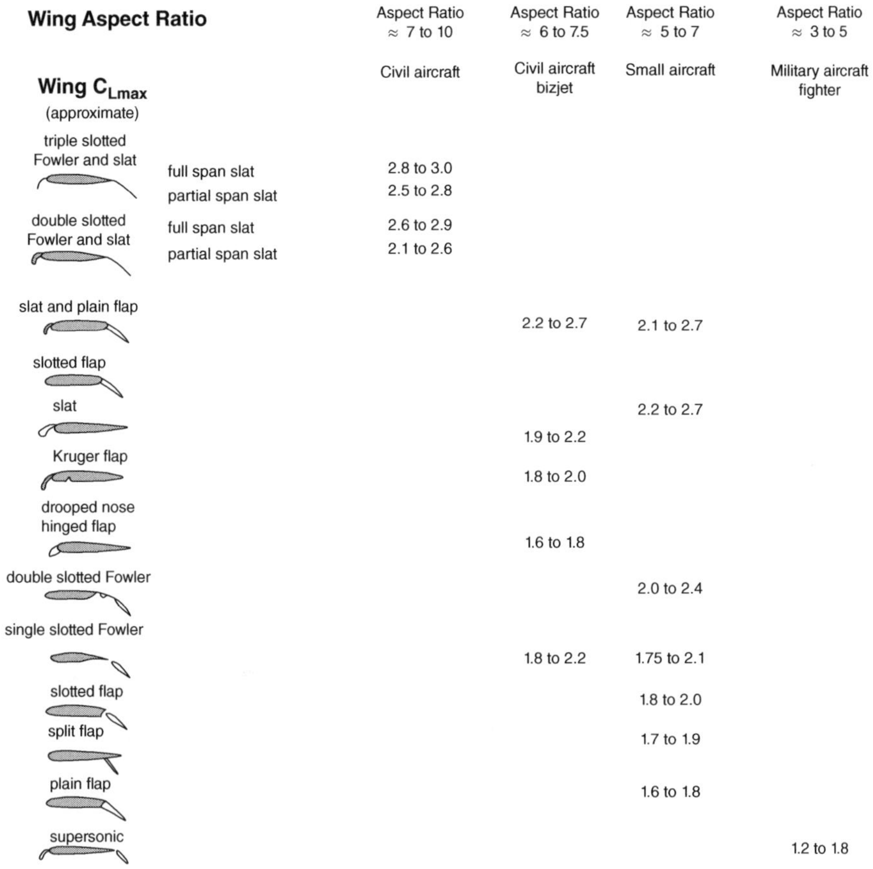

Wing Aspect Ratio / Wing C_{Lmax} (approximate)		Aspect Ratio ≈ 7 to 10 — Civil aircraft	Aspect Ratio ≈ 6 to 7.5 — Civil aircraft bizjet	Aspect Ratio ≈ 5 to 7 — Small aircraft	Aspect Ratio ≈ 3 to 5 — Military aircraft fighter
triple slotted Fowler and slat	full span slat	2.8 to 3.0			
	partial span slat	2.5 to 2.8			
double slotted Fowler and slat	full span slat	2.6 to 2.9			
	partial span slat	2.1 to 2.6			
slat and plain flap			2.2 to 2.7	2.1 to 2.7	
slotted flap				2.2 to 2.7	
slat			1.9 to 2.2		
Kruger flap			1.8 to 2.0		
drooped nose hinged flap			1.6 to 1.8		
double slotted Fowler				2.0 to 2.4	
single slotted Fowler			1.8 to 2.2	1.75 to 2.1	
slotted flap				1.8 to 2.0	
split flap				1.7 to 1.9	
plain flap				1.6 to 1.8	
supersonic					1.2 to 1.8

Figure 4.37 Wing C_{Lmax} with flaps and slats.

the last two depended entirely on CFD making the datasheet method redundant. Today, a complex 3D wing with a complex type high-lift device can be analysed by CFD to get a better optimised configuration. For this reason, this book leaves out usage of datasheets as this will not serve any useful purpose to the fresh graduates. Instead, for classroom usage, Figure 4.37 gives useful wing C_{Lmax} with flaps and slats, corrected for wing sweep and AR, for the use in this book. Flap chord length also affects C_{Lmax}. References [14, 15] give lift coefficient values for different types of high-lift configurations.

For its complexities, design of high-lift devices are carried by specialist groups, the extent required is beyond the scope of this book.

Also presented is an empirical method to obtain wing C_{Lmax} from the aerofoil C_{lmax} given in Figure 3.36. Given next is a simple and quick method to get a preliminary C_{Lmax} for a wing with the chosen high-lift device. This is not exactly an industrial practice but follows the logic to get a value to initiate starting to set up the wing for CFD analyses. A 2D aerofoil with a high-lift device should be corrected for 3D wing geometry for its AR using Table 4.1 and sweep effects by $\sqrt{\cos \Lambda_{1/4}}$. The correction factors are summarised here.

Aspect ratio correction	k_{AR}	Use Table 4.1
$\Lambda_{1/4}$ sweep correction	k_Λ	Use $\sqrt{\cos \Lambda_{1/4}}$
(Flap chord/MAC) correction	k_c	Use 1.0 for 0.3 flap chord/MAC
	k_c	Use 1.05 for 0.32 flap chord/MAC
	k_c	Use 1.08 for 0.35 flap chord/MAC

This gives

$$C_{\text{Lmax}} = k_{\text{AR}} \times k_{\Lambda} \times k_{\text{c}} \times C_{\text{lmax}} \tag{4.52}$$

High-lift devices on the same aircraft can be a mixture of several types and span coverage which contribute to the overall aircraft C_{Lmax}. The following table gives some typical C_{Lmax} based on the simplified assumption that the high-lift type is same for the wing and covers the wing to the extent as shown in Figure 4.33. It is recommended to use test values, if these are not available then use Figure 4.33.

Examples that follow are of the typical aircraft types with the assumption of having a flap chord/MAC of 0.3. The correction factor applied is (k_{Λ} and k_{AR}), k_{c} being 1.0. These are not manufacturers' values. Readers may check.

Aircraft type	$\Lambda_{1/4}$ deg	AR	High-lift type	$C_{\text{lmax}@\Lambda1/4 = 0}$	Correction	$C_{\text{Lmax}@\Lambda1/4}$
B727	32	7.0	Triple slotted Fowler + slat	3.8	0.8×0.92	2.65
B737-900	25	9.45	Triple slotted Fowler + slat	3.8	0.9×0.952	3.1
B747	37.5	7.0	Triple slotted Fowler + slat	3.8	0.8×0.89	2.6
DC9	25	8.7	Double slotted Fowler + slat	3.6	0.85×0.952	2.9
DC10	35	7.0	Double slotted Fowler + slat	3.6	0.8×0.905	2.6
A320	25	10.5	Single slotted Fowler + slat	3.4	0.92×0.952	3.15
A330	30	10	Single slotted Fowler + slat	3.4	0.92×0.93	2.9
Bizjet	14	7.5	Single slotted Fowler	2.6	0.8×0.98	2.0
AJT	25	5.3	Double slotted Fowler	3.2	0.75×0.952	2.28

Figure 4.38a shows a double slotted flap on an aircraft on the ground with spoiler deflected and Figure 4.38b shows a triple slotted flap extended on an aircraft in flight.

4.15.3 Aileron Size

Aileron size is the product of its span and average chord that can be related with wing span and chord. Statistical data/empirical equations can be used to earmark the aileron dimensional provision on the wing, but this is not aileron design. Aileron design relates to control response characteristics beyond the scope of this book.

At the conceptual design stage of conventional aircraft design, a good way to start is to keep the aileron span a 30–40% of wing span and the local aileron chord 40–30% of the local wing chord covering about 8–12% of wing area. A tapered wing will result in a tapered aileron. Holding the base area, readers may trade with aileron span and chord, that is, span increase needs to adjust with decrease in local chord and vice versa. It is suggested that readers should generate more appropriate statistical data in the same class of aircraft under study instead of taking generalised statics of all classes of aircraft.

(a) (b)

Figure 4.38 Fowler flap extended. (a) Double slotted – spoiler deflected. (b) Triple slotted.

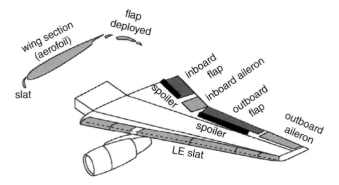

Figure 4.39 Wing high-lift devices.

Large high-subsonic commercial-transport aircraft use also use an inboard aileron (Figure 4.39) for finer roll control acting as dampers integrated in an FBW system. A full span flaperon acts both as a flap and with differential movement as an aileron. In this case, the flaperon chord is reduced to around 20–25% of the chord.

In the past, datasheets were used to arrive at a preliminary aileron size. Today, aileron sizing is carried out though CFD analyses in less time and more accurately to minimise wind-tunnel substantiation, thereby saving cost and time. Final aileron sizing invariably requires tailoring through subjective flight tests (control parameters) by several test pilots.

4.16   Additional Surfaces on the Wing

Flaps and slats on a 2D aerofoil are described in Section 3.12. This section describes their installation on a 3D wing.

Flaps comprise about two-thirds of an inboard wing at the trailing edge and are hinged on the rear spar (positioned at 60–66%; the remaining third by the aileron) of the wing chord, which acts as a support. Slats run nearly the full length of the LE. The deployment mechanism of these high-lift devices can be quite complex. The associated lift-characteristic variation with incidence for 2D aerofoil is shown in Figure 3.37. The lift-characteristic variation with incidence for 3D wing has similar pattern but with degraded $dC_L/d\alpha$ and C_{Lmax} at a slightly higher α_{stall}.

The aileron acts as the roll-control device and is installed at the extremities of the wing for about a third of the span at the trailing edge, extending beyond the flap. The aileron can be deflected on both sides of the wing to initiate roll on the desired side. In addition, ailerons can have trim surfaces to alleviate pilot loads. A variety of other devices are associated with the wing (e.g. spoiler, vortex generator and wing fence).

Spoilers (or *lift dumpers*) (Figure 4.39) are flat plates that can be deployed nearly perpendicular to the airflow over the wing. They are positioned close to the CG (i.e. *x*-axis) at the MAC to minimise the pitching moment and they also act as air brakes to decrease the aircraft speed. They can be deployed after touchdown at landing when they would 'spoil' the flow on the upper wing surface, which destroys the lift generated (the US terminology is *lift dumper*). This increases the ground reaction for more effective use of wheel brakes.

There other control devices, for example, the flaperon and spoileron, which combine functions of aileron with flap or spoiler.

Winglets (also known as *sharklets*, a new term) are wing tip devices to reduce induced drag by reducing the intensity of the wing tip vortex. Figure 4.40 shows the prevalent type of winglets, which modify the tip vortex to reduce induced drag. At low speed, the extent of drag reduction is minimal and many aircraft do not have winglets. At higher speeds, it is now recognised that there is some drag reduction, no matter how small ($\approx 5\% \pm 1\%$), and this has begun to appear in almost all high-subsonic jet aircraft – even as a styling trademark on some. The blended winglet shows the best gain and is seen in newer high-subsonic aircraft. The Hoerner type and sharp-raked winglets are used in lower-speed aircraft.

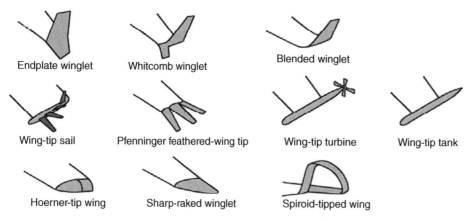

Figure 4.40 Wing flow modifier and vortex generators: types of winglets (from NASA).

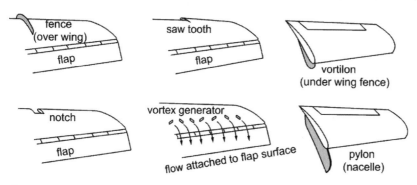

Figure 4.41 Leading edge flow modifier and vortex generators.

Extensive analyses and tests indicate that approximately 2–5% of induced-drag reduction may be possible with a carefully designed winglet. Until the 1980s, the winglet was not prominent in aircraft. In this book, no credit is taken for its use. Coursework can incorporate the winglet in project work.

Wing flow modifier devices (Figure 4.41) are intended to improve the flow quality over the wing. In the figure, a *fence* is positioned at about half the distance of the wingspan. The devices are carefully aligned to prevent airflow that tends to move spanwise (i.e. outward) on swept wings.

Figure 4.41 also shows examples of *vortex generators*, which are stub wings carefully placed in a row to generate vortex tubes that energise flow at the aft wing. This enables the flow to remain attached; however, additional drag increase due to vortex generators must be tolerated to gain this benefit. Vortex generators and/or a fence also can be installed on a nacelle to prevent separation.

4.17 The Square-Cube Law

For an example, increase the linear dimensions of a solid cube from 1 to 2 units. From the following example, it can be seen that the increase in weight is faster than the increase in area (the subscript 1 represents the small cube and the subscript ^{2} represents the larger cube):

$\text{area}_2 = \text{area}_1 \times (\text{length}_2/\text{length}_1)^2$, a four-fold increases from 6 to 24 square units
$\text{volume}_2 = \text{volume}_1 \times (\text{length}_2/\text{length}_1)^3$, an eight-fold increase from 1 to 8 cube units

Applying this concept to a wing, increasing its span (i.e. linear dimension, b, maintaining geometric similarity) would increase its volume faster than the increase in surface area, although not at the same rate as for a cube. Volume increase is associated with weight increase, which in turn would require stiffening of the structure, thereby further increasing the weight in a cyclical manner. This is known as the *square-cube law* in aircraft design terminology. The following logic was presented a half a century ago by those who could not envisage very large aircraft.

weight, $W \propto \text{span}^3$
wing planform area, $S_w \propto \text{span}^2$

This indicates that, for the given material used, because of excessive weight growth, there should be a size limit beyond which aircraft design may not be feasible. If the fuselage is considered, then it would be even worse with the additional weight.

Yet, aircraft size keeps growing – the size of the Airbus A380 would have been inconceivable to earlier designers. In fact, a bigger aircraft provides better structural efficiency, as shown in Figure 7.3, for operating empty mass fraction (OEMF) reduction with maximum takeoff weight (MTOW) gain. Researchers have found that advancing technology with newer materials – with considerably better strength-to-weight ratio, weight reduction by the miniaturisation of systems, better high-lift devices to accommodate higher wing-loadings, better fuel economy and so forth – has defied the square-cube law. Strictly speaking, there is no apparent limit for further growth (up to a point) using the current technology.

The authors believe that the square-cube law needs better analysis to define it as a law. Currently, it indicates a trend and is more applicable to weight growth with an increase in AR. What happens if the AR does not change? Section 4.18 provides an excellent example of how a low AR can compete with a high AR design.

4.18 Influence of Wing Area and Span on Aerodynamics

For a given wing loading (W/S_W), aerodynamicists prefer a large wingspan to improve the AR in order to reduce induced drag at the cost of a large wing root bending moment. Structural engineers prefer to see a lower span resulting in a lower AR. The BWB design for larger aircraft has proven merits over conventional designs but awaits technological and market readiness. Interesting deductions are made in the following sections.

Toreenbeek [4] made a fine comparison to reveal the relation between aircraft wetted area, A_W, and wing planform area, S_w. Later, Roskam [18] also presented his findings to reinforce the point. Toreenbeek's result is shown in Figure 4.42.

4.18.1 Aircraft Wetted Area (A_W) versus Wing Planform Area (S_W)

Section 4.17 raised an interesting point on aircraft size, especially related to wing geometry. This section discusses another consideration on how the aircraft wing planform area and the entire AW surface areas can be related. Again, the wing planform area, S_W, serves as the reference area and does not account for other wing parameters (e.g. dihedral and twist).

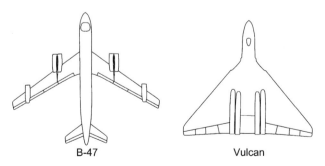

B-47   Vulcan

Figure 4.42 Torenbeek's comparison between the B47 and Vulcan.

The conflicting interests between aerodynamicists and stress engineers on the wing AR presents a challenge for aircraft designers engaged in conceptual design studies (this is an example of the need for concurrent engineering). Both seek to give the aircraft the highest possible *lift-to-drag* (*L/D*) ratio as a measure of efficient design. Using Eq. (4.24), the following can be shown (i.e. incompressible flow):

$$\text{drag}, D = qS_W C_D = qS_W(C_{DP} + C_{Di})$$

$$\text{or } C_D = (C_{DP} + C_{Di}) \tag{4.53}$$

where Eq. (4.18) gives $C_{Di} = C_L^2/\pi AR$

Clearly, $C_{DP} \propto$ wetted area, A_W, where A_W = aircraft wetted area and C_{Di} is $\propto (1/AR) = S_W/b^2$.

For a conventional two-surface aircraft, it can be said that higher the AR higher is the L/D ratio. Chapter 7 provides statistical data for various designs. It is evident from Table 4.3 that BWB can offer comparable L/D at lower AR as discussed next.

Define the wetted-area *AR* as follows:

$$AR_{wet} = b^2/A_W = AR/(A_W/S_W) \tag{4.54}$$

This is an informative parameter to show how close the configuration is to the wing-body configuration.

Torenbeek compared an all-wing aircraft (i.e. the Avro Vulcan bomber) to a conventional design (i.e. Boeing B47B bomber) with a similar weight of approximately 90 000 kg and a similar wing span of about 35 m. It was shown that these designs can have a similar L/D ratio despite the fact that the all-wing design has an AR less than one-third of the former. This was possible because the all-wing aircraft precludes the need for a separate fuselage, which adds extra surface area and thereby generates more skin-friction drag. Lowering the skin-friction drag by having a reduced wetted area of the all-wing aircraft compensates for the increase in induced drag for having the lower AR. All-wing aircraft provide the potential to counterbalance the low AR by having a lower wetted area. Again, the concept of BWB gains credence.

Table 4.3 provides statistical information to demonstrate that a BWB is a good design concept to satisfy both aerodynamicists and stress engineers with a good L/D ratio and a low AR wing, respectively. In the table, a new parameter – wetted AR, $b^2/A_W = AR/(A_W/S_W)$ – is introduced. The table provides the relationship among the AR, wing area and wetted area and how it affects the aircraft aerodynamic efficiency in terms of the ratio. Within the same class of wing planform shape, the trend shows that a higher AR provides a better L/D ratio. All-wing aircraft (e.g. BWB) provide an interesting perspective, as discussed in this section.

Table 4.3 Wing area and span comparison.

Aircraft	B47	Vulcan
Gross wing area (sq. ft)	1430	3446
Total wetted area (sq. ft)	11 300	9500
Span (ft)	116	99
Max. wing loading (W/S_W)	140	43.5
Max. span loading (W/b)	1750	1520
AR	9.43	2.84
C_{D0}	0.0198	0.0069
$L/D_{max}/C_{Lopt}$	17.25/0.682	17.0/0.235
C_L at maximum cruise speed	0.48	0.167

4.19   Summary of Wing Design

Section 3.14 summarised a methodology to select a bread-slice aerofoil section from a 2D infinite span wing. In this section, a methodology for wing design is outlined. This requires careful consideration to decide the planform shape of the wing, that is, wing sweep, $\Lambda_{1/4}$, taper, λ, span, b, and so on, for the required wing area that gets formally sized to meet the aircraft requirement. In addition, the wing has to be tweaked to satisfy the handling and stability issues.

Aircraft speed requirement influences the wing planform geometry. The aircraft weight, along with its field performance, determine its area, of course, the other considerations have to be taken into account. In this chapter, only the planform geometry, based on the specified aircraft design speed, aircraft are grouped in the following category. This gives the *Re* of the wing. As the aerofoil and wing design are interrelated, the grouping outlined in Section 3.14 also suits to configure wing planform geometry. Given next are the possible planform shapes for the classes of aircraft.

Low subsonic (below Mach 0.5) – propeller driven – ($\Lambda_{1/4}$ wing sweep $\approx 0 \pm 5°$)
   Gliders/Microlights operating around 80 mph with a very low Re number, these have rectangular wings and are cheap to manufacture. This class of aircraft is not dealt with in this book.
   Home-builds (Light Sports Aircraft – LSA) operating around 120 mph at a low Reynolds number invariably have rectangular wings and are cheap to manufacture, a few have tapered wings. The LSA certification categories of aircraft are not dealt with in this book. The design approaches for this class have similarity with the club trainer/recreational category of aircraft given next.
   Club trainers/recreational aircraft (Normal Category, Federal Aviation Regulation (FAR) 23) operating around 150 mph at a low Reynolds number. Ab initio trainer aircraft must have gentle and gradual stall characteristics of a forgiving type providing safety. Although the rectangular wing planform dominates, there are designs with tapered wing planforms to extract induced drag reduction. The market can absorb some cost increase if the customer opts for better performance.
   Aerobatic Category (FAR 23) operating at around 200 mph. Generally with a tapered wing planform, but this can be of a rectangular type.
   Utility/Regional aircraft operating around 300 mph. The tapered wing planform dominates although there are aircraft with rectangular planforms as a cost effective measure.
High-subsonic (Mach 0.7–Mach 0.95) – jet propulsion – ($\Lambda_{1/4}$ wing sweep $\approx 10–35°$)
   Incorporate wing sweep influence the choice for aerofoil *t/c* as shown in Figure 4.27 (forward sweep aircraft are uncommon but possible). In general, in this category of aircraft design, industries develop their own aerofoils, but those are kept commercial in confidence.
   Bizjet/regional jet operating between Mach 0.7 and Mach 0.85 requires a swept wing of 10° to 30°. Invariably with a tapered wing planform.
   Commercial jets operating between Mach 0.75 and Mach 0.92. Aircraft with a longer range mission have higher speeds requiring relative wing $\Lambda_{1/4}$ sweep of 20°–35° that influences the choice for aerofoil *t/c*. Invariably has a tapered wing planform.
Supersonic category ($\Lambda_{1/4}$ wing sweep $\approx 0–60°$ [delta wing])
   Combat aircraft operating between Mach 1.8 and Mach 2.4. The *Concorde* was the only commercial aircraft in operation; a new generation of supersonic aircraft are now under consideration. On account of the presence of shock waves, the supersonic wings are a different class dealt with in Section 4.13.
   Table 4.4 gives a few examples of jet propelled aircraft wings that fly above Mach 0.7 and require wing sweep, which are widely used in industry.
   Given in Section 4.19.1 is a simplified wing design methodology (this is not the only way to select aerofoil). A simplified step-by-step approach to make an aerofoil section from available test data (NACA/NASA series) is outlined in a generalised manner to provide a feel for the process; in industry it is done in a more rigorous manner. This is not the only way to select an aerofoil. To stay ahead of the competition, the bigger aircraft manufacturers search for better aerofoils than those in current practice. CFD analyses are extensively used.

Table 4.4 Aircraft wing types (@ maximum design speed).

Conventional aircraft	Type*	Wing area, S_W—m²/ft²	Span b—m/ft	Aspect ratio ($b^2/S_W = AR$)	Wing sweep LE/$\Lambda_{1/4}$ deg
Jet Propulsion Aircraft					
Cessna Citation XLS+ (0.68 Mach)		34.35/370	17.17/56.4	8.6	0
Gulfstream Aerospace GIV (0.88 Mach)		88.3/950.39	23.72/77'10″	6.4	
Learjet 60 (0.74 Mach)		24.57/264.5	13.35/43'10″	7.25	
Airbus 320 (@ 0.78/0.82 Mach)		122.4/1318	35.8/117'5″	10.5	25
Airbus 330 (@ 0.82/0.86 Mach) ($t/c = 15.25\%$ at root, 11.27% at kink and 10.6% at tip)		361.6/3892	60.3/197.83	10.06	30
Airbus 380 (@ /0.89 Mach)		845/9100	79.75/261.6	7.53	33.5
Boeing 737-200 (@ 0.72/0.82 Mach)		102/1098	28.32/93	7.85	25
Boeing 747-200B (@ 0.86/0.92 Mach) ($t/c = 9.4\%$ average)		511/5500	59.5/195'8″	7.0	37.5
Boeing 767-200 (@ 0.71/0.76 Mach) ($t/c = 15.1\%$ at root and 10.3% at tip)		283.3/3050	47.57/156	8.0	31.5
Boeing 777-200 (@ 0.84/0.89 Mach)		428.8/4605	60.93/199'11″	8.65	31.6
Boeing 777-200 (@ 0.84/0.89 Mach)		428.8/4605	60.93/199'11″	8.65	31.6
Boeing 787 (@ 0.85/0.89 Mach)		360.5/3800	60.12/197'3″	10	32.2
Supersonic Combat Aircraft					
Boeing F-15 Eagle(@ 2.4 Mach) ($t/c = $ NACA64A006.6 at root and NACA64A203 at tip)		56.5/608	13.05/42'10″	3.01	
Lockheed Martin F-16 ($t/c = $ NACA64A204 at root and tip)		27.87/300	9.96/32'8″	3.2	40/
McDonnell Douglas F-18 Hornet					
North American Aviation RA-5C ($t/c = 5\%$)		65.03/700	19.2/53.14	3.73	43/39.5
Northrop F-5 Tiger ($t/c = $ NACA65A004.8 at root and NACA64A004.8 at tip)		17.28/186	8.13/26.8	3.82	

4.19.1 Simplified Wing Design Methodology

Typical choices for the wing design are suggested here. This concerns the planform shape, its sweep, taper ratio, twist and distribution of the types of aerofoil used as they are not necessarily uniform along the wing span. It was suggested earlier to choose the highest permissible thickness to chord ratio for wing structural integrity and internal volume for fuel storage/accessibility. Cost consideration in wing design is important. However, the rectangular wing planform is the cheapest to manufacture, Section 4.10.2 shows the benefits of a trapezoidal wing planform.

The starting point is the market specification that defines the kind of aircraft to be designed. There are three main categories of aircraft in broad classification as described before, for example (i) low subsonic (below Mach 0.6) – propeller driven, (ii) high-subsonic (Mach 0.6–Mach 0.95) – jet propulsion and (iii) the supersonic category. Taking each category separately, a step-by-step methodology is outlined here for classroom work, staying close to industrial practices.

The wing is built up by stacking up the selected aerofoil to a suitable spanwise arrangement to satisfy the aircraft requirements. Aerofoil selection and wing shaping is mission specific. Given next is a step-by step methodology (this is not the only way to design wing). Since the aerofoil and wing design are interrelated, Figure 4.27 may be referred to examine the choice.

The aircraft wing design plays the most important part, along with all design considerations, to make the new aircraft project a success. The starting point are the aircraft specifications laid by the market study or Request for Proposal (RFP) by the Ministry of Defence (MoD).

The following specifications are the basis for a wing design.

i. Aircraft mission profile, for example payload, range, radius of action and so on. This gives the MTOM.
ii. Aircraft design speed, which gives the C_{L_des}.
iii. The preferred position of wing with respect to fuselage is generally given, that is, high, middle or low wing (see Section 4.4.1 for their advantages and disadvantages).

Strictly speaking, wing design can only start when there is some idea of its general configuration is available from guesstimates (statistics – Chapter 7), evidently this involves some iterative process. This gives the preliminary C_{L_des}. The wing design progresses systematically, first arriving at a preliminary configuration as a *concept definition* worked out in Chapter 8 and finally sized to *concept finalisation* in Chapter 15). (Even if the aerofoil choice offers laminar flow characteristics, no credit for drag benefits are claimed for certification purposes.)

(i) Low subsonic (below Mach 0.6) – incompressible flow:

Step 1
1. Find the design C_{L_des} mid-operational weight. Aircraft operating speed along with C_{L_des} will allow selection of the aerofoil, a methodology given in Section 3.14.1. Desired aerofoil qualities are high $dC_l/d\alpha$, low C_d and low C_m. Within these constraints, the highest possible aerofoil thickness to chord ratio, t/c, is sought for structural integrity at lower wing weight and provides adequate volume for fuel storage. Ab-initio trainers and small recreational club flying aircraft need aerofoils with gentle gradual stalling characteristics that offer safety.

Step 2
1. This category of aircraft with low subsonic speed (negligible compressibility effect) will invariable have a wing sweep $\Lambda_{1/4} \approx 0$. Any small amount sweep ($<\pm 5°$) is not on account of aerodynamic considerations, but arises from geometric considerations, for example, CG position, pilot view, structural, manufacturing considerations and so on.

Step 3
1. This type of high-lift is decided by the required C_{Lmax} for takeoff/landing for the available runway length and its load bearing capability of the intended airfield runway of operation. Small aircraft can get away with the simplest types of flaps, for example, plain, split or slotted. These are cheap to manufacture and serve well for low-wing loading aircraft. However, for bigger and heavier utility category aircraft, some have faster operating speeds with a turboprop and will have higher wing loading when a single slotted Fowler flap will prove helpful.

Step 4
1. High AR is desired for induced drag, C_{Di}, reduction. Small aircraft operating below 150 mph are typically rectangular and have an AR between 5 and 7 and with a taper ratio, $\lambda = 0$. Some, like the turboprop powered Shorts SD360, have a high rectangular AR and have served well. Bigger and heavier utility category aircraft operating above 200 mph benefit from having tapered wings at an additional cost. In any case, the best value in this category of aircraft has a taper ratio $0.35 < \lambda < 0.5$.

Step 5
1. Finally, it is necessary to settle the dihedral and wing twist. In this category of aircraft, dihedral angle for low-wing aircraft is between 2° and 4°. High-wing anhedral angle is generally between 0° and −2°. Wing twist with a washout property is the prevalent type. Use statistics to check the values at this stage project study.

The overall aim is to get a high L/D at low cost (may need some compromise). Winglets are unnecessary for this class of low-speed aircraft.

(ii) High-subsonic (Mach 0.7 – Mach 0.95 – jet propulsion) – compressible flow

On account of considering the compressible effects, this category of aircraft wing design is lot more complex that the low subsonic type aircraft wing design. Desired operating speed is specified. For short ranges below 3000 mile, these operate at HSC at around 0.75–0.8 Mach. Missions in excess of 3000 miles benefit with a higher HSC at around 0.80–0.85 Mach. Very long range aircraft operate around 0.85–0.9 Mach.

Step 1

1. Find the design C_{L_des} mid-operational weight. Aircraft operating speed along with C_{L_des} will allow selection of an aerofoil, a methodology given in Section 3.14.1. Desired aerofoil qualities are high $dC_l/d\alpha$, low C_d and low C_m. Within these constraints, the highest possible aerofoil t/c is sought for structural integrity at a lower wing weight and provides adequate volume for fuel storage.

Step 2

1. The dominant type is the low-wing type, but a large diameter passenger carrying fuselage has the wing passing through the floor-boards, say at a third of fuselage height from its keel mouldline. There are few passenger carrying aircraft that have a high-wing configuration, for example, the BAe 145 aircraft (see Section 4.4.1 for their advantages and disadvantages). Rear loading aircraft invariably have a high wing.

2. The rapid drag rise due to compressibility can be delayed by sweeping the wing dominantly backward. Selecting the aerofoil with a suitable t/c, use Torenbeek's Eq. (4.47) to get what extent of $\Lambda_{1/4}$ is required for the specified operating speed at the design C_{L_des}. The relationship between wing sweep, $\Lambda_{1/4}$, wing aerofoil thickness to chord ratio, t/c, and the cruise Mach number is summarised in Section 4.11.3.

Step 3

1. The type of high-lift is decided by the required C_{Lmax} for takeoff/landing for the available runway length and its load bearing capability of the intended airfield runway of operation. The smaller Bizjet/Regional aircraft usually has a single slotted Fowler flap. As a heavier aircraft weight increases (higher wing loading), a more complex form of high-lift device, for example, double or triple slotted Fowler flaps, are needed to get a high enough C_{Lmax} to operate safely for the operating airfield runway length. Wings with very high-wing loadings will have both flaps and slats. Complexities add to cost and weight increase.

Step 4

1. Wings in this category of aircraft will invariably have a tapered planform, if required with a kink (with glove and/or yehudi – Figure 4.8). The best value in this category of aircraft has $0.275 < \lambda < 0.375$. High AR is desired for induced drag C_{Di} reduction. The smaller Bizjet/Regional AR varies from 7–8. Bigger ones have an AR of around 8–10; structures permitting it can be still slightly higher. The future concept of aircraft with a braced wing will be capable of much higher AR (≈ 15).

Step 5

1. Finally, settle the dihedral angle and wing twist. For a low wing, the smaller Bizjet/Regional dihedral angle for low-wing aircraft is between 2° and 4°. The dihedral angle for large commercial jets reaches as high as 6–7°. High-wing anhedral angle is generally between 0° and −2°. Wing twist with a washout property is the prevalent type. Use statistics to check the values at this stage project study. Winglets (Figure 4.40) are now becoming standard features that can reduce induced drag, C_{Di}, up to 5%.

The overall aim is to get high L/D at low cost (may need some compromise). Preliminary wing designs for the types of aircraft carried out in this book are worked out in Chapter 8 and sized to final geometry in Chapter 15. To repeat, the current industrial practices are gradually becoming more reliant on CFD/test data – these offer better accuracy and are reliable. Therefore, this book suggests using test data as sufficient for classroom practice.

(iii) Supersonic – with shock waves (supersonic wing design exercise has not been undertaken here)

Step 1

1. Find the design C_{L_des} mid-operational weight and aircraft operating speed along with C_{L_des}. Supersonic aerofoils are thin and have no camber (Figure 3.23). They have has low C_d and low C_m as well as a low $dC_l/d\alpha$. Supersonic wings carry little or no fuel.

Step 2

1. It is preferable to keep the wing within the Mach cone at its maximum permissible speed. Typical wing planform shapes are delta or trapezoidal with a taper ratio $0 < \lambda < 0.25$.

Step 3

1. The type of high-lift is decided by the required C_{Lmax}. Being very thin, supersonic wings do not have as many choices as those available for subsonic wings for LE and TE high-lift devices. High manoeuvre aircraft have some form of LERX.

References

1 Fink, R. D., USAF Stability and Control, DATCOM. Report Number AFWAL-TR-83-3048, Flight Dynamic Laboratory, Wright-Patterson Air Force Base, Ohio, 1978.

2 Engineering Sciences Data Unit (ESDU), IHS ESDU, Website.

3 Anderson, J. (2017). *Fundamentals of Aerodynamics*, 6e. New York: McGraw-Hill.

4 Kuethe, A.M. and Chow, C.Y. (1986). *Foundations of Aerodynamics*. Wiley.

5 Bertin, J.J. and Cummings, R.M. (2009). *Aerodynamics for Engineers*, 5e. Pearson.

6 Prandtl, L., Theory of Lifting Surfaces, Part II, NACA TN 10, 1920.

7 Multhopp, H. (1938). The calculation of the lift distribution of serofoils. *Luftfahrtforschung* 15: 153, RTP translation No: 2392, ARC 8516.

8 Schrenk, O., A simple approximate method for obtaining the spanwise lift distribution, NACA TM 948. Luftwissen, Vol. 4, No. 7, 1940.

9 Diederich, F., A simple approximate method for calculating spanwise lift distributions and aerodynamic influence coefficients at subsonic speed., NACA TN 2751, 1952.

10 Loftin, Jr., L. K., Quest for Performance - The Evolution of Modern Aircraft, NASA SP-468, 2004.

11 Torenbeek, E. (1982). *Synthesis of Subsonic Airplane Design*. Delft University Press.

12 Shevell, R.S. (1989). *Fundamentals of Flight*. Prentice Hall.

13 Polhamus, E.C. (1966). *Concept of the Vortex Lift of a Sharp Edge Delta Wing Based on a Leading Edge Suction Analogy*, NASA TN D-3767. Hampton: Langley Research Centre.

14 Perkins, C.D. and Hage, R.E. (1949). *Aircraft Performance Stability and Control*. Wiley.

15 Furlong, G. C. and McHugh, J. G., A summary and analysis of the low-speed longitudinal characteristics of swept wings at high Reynolds number. NACA TR 1339, 1957.

16 Reckzeh, D., Aerodynamic Design of Airbus High-Lift Wings in a Multidisciplinary Environment, European Congress on Computational Methods in Applied Sciences and Engineering, 24–28 July, 2004, Finland.

17 Rudolph, P.K.C. (1996). *High-Lift System on Commercial Subsonic Airliners*, NASA Contractor Report 4746. Ames, CA, USA: Ames Research Centre.

18 Roskam, J. (2017). *Aircraft Design*, 8 volume set. Lawrence, KA, USA: DARcorporation.

19 Jane's All the World's Aircraft Manual (yearly publication) Jane's Information Group.

5

Bodies – Fuselages, Nacelle Pods, Intakes and the Associated Systems

5.1 Overview

The last chapter dealt with the pertinent aspects of the planar 3D geometries of wing design considerations. This chapter follows up with the design considerations of 3D geometries for bodies, for example, the fuselage, nacelle pods and items integral to them. The aim is to shape objects in a teardrop-like streamlined geometry that will minimise drag generation. The purpose of such bodies is to accommodate payload, consumables, equipment and so on, and produce very little lift. Some of the bodies house engines as nacelle pods with integrated engine intake and exhaust ducts, hence they are included in this chapter. The aim is still the same; that is, to shape object in a teardrop streamlined geometry that will minimise drag generation. Bodies of revolution offer a destabilising moment. Care must be taken to keep it at an acceptable level. Making an aircraft streamlined also makes it look elegant. With engines inside the fuselage, combat aircraft have their air intake as part of the fuselage and this is dealt with in this chapter.

Some dominant geometries of fuselage, nacelles, and other bodies along with the design data are presented in order to suggest possible choices available to configure new aircraft designs and arrive at a *concept definition*. No analytical optimisation is carried out here as these are beyond the scope of this book. In industry, Cockpit/Flight Decks (CFDs) are used throughout to fine-tune the external aircraft geometry. It is to be noted that, along with aerodynamic considerations to shape the component geometries, their structural considerations maintenance, repair, and overhaul (MRO) aspects will also have to be taken into account.

The chapter begins with standard definitions of the various parameters of bodies that will be used in this book. Civil and military missions differ and they are explained in detail separately. This chapter covers the following topics:

Section 5.2: Introduction
Section 5.3: Fuselage Geometry – Civil Aircraft
Section 5.4: Fuselage Closures – Civil aircraft
Section 5.5: Fuselage Fineness Ratio – Civil Aircraft
Section 5.6: Fuselage Cross-Section Geometry Civil Aircraft
Section 5.7: Fuselage Abreast Seating
Section 5.8: Fuselage Cabin Layout
Section 5.9: Fuselage Structural Considerations
Section 5.10: Fuselage Aerodynamics Considerations
Section 5.11: Fuselage Pitching Moment
Section 5.12: Nacelle Pod – Civil Aircraft
Section 5.13: Exhaust Nozzle – Civil Aircraft
Section 5.14: Fuselage Geometry – Military Aircraft
Section 5.15: Pilot Cockpit/Flight Deck – Military Aircraft
Section 5.16: Engine Installation – Military Aircraft

Conceptual Aircraft Design: An Industrial Approach, First Edition. Ajoy Kumar Kundu, Mark A. Price and David Riordan.
© 2019 John Wiley & Sons Ltd. Published 2019 by John Wiley & Sons Ltd.

5.2 Introduction

The reason for making the fuselage streamlined is to minimise drag, which in turn gives a high aircraft lift to drag (lift/drag) ratio as the objective of the design. The aim is to maximise the aircraft lift-to-drag ratio for the mission role; for transport aircraft this is about, typically, 50% higher than fighter aircraft. It is possible to blend the wing and body where the fuselage is fused into the wing, known as Blended-Wing-Body (BWB) aircraft. Such aircraft have been constructed and are currently in operation, one such example is the B2 bomber (Spirit). It is meaningful to also make bodies in streamlined shape to minimise drag and if possible, extract as much lift as they can offer, no matter how small it may be. Rear-loading fuselage shaping requires compromise to ensure that it does not adversely affect aircraft stability.

The body shapes are basically of two types; (i) fuselages and (ii) nacelles/pods/auxiliary attachments and so on. The design considerations for the fuselage are dealt with first followed by the considerations for nacelles and other bodies.

5.2.1 Generic Fuselages

The term 'fuselage' is derived from the French word *fuselage* meaning 'spindle' shape. All fuselages have a flight deck to serve as crew station. Fuselages of small utility, transport and military category aircraft differ according to their design specifications and mission roles. Typical differences between them are outlined in the following.

5.2.1.1 Civil Aircraft

Large transport aircraft have a long constant cross-section, circular or close to circular in shape. However, small civil aircraft do not require a constant cross-section. Single engine aircraft have an engine mounted at the extremities of the fuselage, jet engines are mounted at the rear. The dominant parameters in fuselage aerodynamic design are its maximum cross-sectional area, and its front and aft-end closure shapes. Since the cross-section may not be exactly circular (Figure 5.1), for the main types of civil aircraft, the following definitions represent diameter (Figure 5.2 shows military fuselage cross-sections for comparison).

The equivalent cross-sectional area of the fuselage is defined as follows.

$$D_{eq-fus} = \sqrt{((4 \times A_{cross-section})/\pi)} \text{ where } A_{cross-section} \tag{5.1}$$

For an elliptical cross-section the effective diameter, D_{eff}, is used that reduces Eq. (5.1) as follows.

$$D_{eff} = \sqrt{(H + W)}, \text{ where, } H = \text{fuselage height and } W = \text{fuselage width} \tag{5.2}$$

Fuselages with close to circular constant cross-sections can use the definition of average diameter, $D_{ave-fus}$, as given here.

$$D_{ave-fus} = (H + W)/2, \text{ where, } H = \text{fuselage height and } W = \text{fuselage width} \tag{5.3}$$

The BWB aircraft configuration has a fuselage merged with the wing, yet it can be delineated to be dealt with as required.

Figure 5.1 Transport aircraft fuselage cross-section.

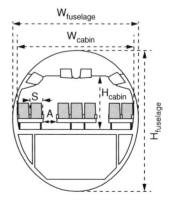

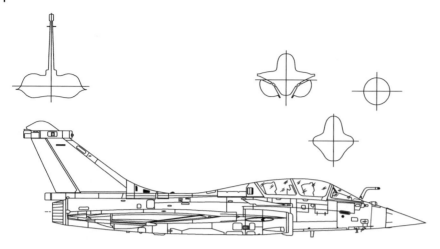

Figure 5.2 Military aircraft fuselage cross-sections.

5.2.1.2 Transport Category

All transport type aircraft fuselages have a circular or near circular cross-section with a constant midsection to house the payload (passenger/cargo) and equipment. In some cases they may accommodate part of the fuel load. These fuselages have closures at both ends, for subsonic types the front end is blunter than the gradual closure at the aft, that is, a tear-drop streamline shape (Figure 5.3a). Aft-loading fuselages with loading ramps have a blunt aft end (Figure 5.3b).

5.2.1.3 Small Aircraft Category up to Eight Seats

Small aircraft operating at low subsonic speeds do not have a long enough constant fuselage cross-section (Figure 5.3c). They are contoured to accommodate 2–8 persons and the luggage they carry. However, the fuselage design process has the same approach as that of high subsonic aircraft to house the occupants, consumables and systems that may or may not include the engine, depending on whether there is a single or multi-engine configuration.

5.2.1.4 Military Category Fuselage

The role of military aircraft fuselage is very different (Figures 5.2 and 5.4) and configurations differ from design to design. The supersonic aircraft front end has to have a necessarily sharp nose cone to minimise shock wave drag. This is a tightly packed housing for engines (located at the aft end), fuel and most of the aircraft systems hardware.

5.2.1.5 Military Aircraft Fuselage Types

Shaping of military aircraft fuselages must have the freedom to generate variable cross-sections to comply with the best aerodynamic contour, tightly hugging the arranged layout to house densely packed equipment

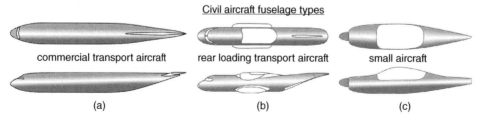

Figure 5.3 Generic fuselage shapes of civil aircraft. (a) Gradual closure. (b) Blunt end and high upsweep. (c) Varying cross-section.

Figure 5.4 Military aircraft fuselage types. (a) Varying cross-section (boat-tail aft end). (b) Varying complex cross-section. (c) Military aircraft fuselage interior Chengdu J-10 (Source: https://www.sinodefenceforum.com).

(Figure 5.4). Fuel tanks and engines placed at the aft end, themselves, do not have constant cross-sections. The their aft end closes in a boat-tail shape with the engine nozzle exit plane at the end. In other words, military fuselages have variable crossing sections and are densely packed with power plants and their intake ducts, exhaust nozzles, electronic black boxes, radar, system equipment, undercarriage, fuel and so on. None of the definitions given in Eqs. (5.1)–(5.3) serve as useful other than as figures of merit for comparison.

5.2.2 Generic Nacelles Pods/Intakes/Auxiliary Bodies

Engines and their accessories/systems not buried inside fuselage need to be housed in specifically designed nacelle pods. The role of nacelle is solely meant to house the engine and its accessories and is dealt with separately. In this chapter only the external geometries are considered. The internal geometries depending on the aerodynamic considerations of air inhalation is discussed in Chapter 12 covering with aircraft propulsion.

This chapter deals with the various configuration options to consider for the choice of nacelle/intake. Military aircraft engines are fuselage mounted and hence do not have nacelles, unlike some in earlier bomber aircraft designs. Today's bomber designs, like the B2 Spirit, have engines buried into the BWB configuration.

There are other types of closed bodies of revolution, for example, drop tanks, armaments and so on that are not dealt with in this book. Their drag estimation is relatively simple. Enough information is given in the Chapter 11 on aircraft drag to evaluate auxiliary body drag.

CIVIL AIRCRAFT

5.3 Fuselage Geometry – Civil Aircraft

A civil aircraft fuselage is designed to carry revenue-generating payloads, primarily passengers but the cargo version can also carry containers or suitably packaged cargo. It is symmetrical to a vertical plane and maintains a constant cross-section with front and aft-end closures in a streamlined shape. The aft fuselage is subjected to adverse pressure gradients and therefore is prone to separation. This requires a shallow closure of the aft end so that the low-energy boundary layer adheres to the fuselage, minimising pressure drag. The fuselage can also produce a small amount of lift, but this is typically neglected in the conceptual stages of a configuration study. The following definitions are associated with fuselage geometry.

5.3.1 Aircraft Zero-Reference Plane/Fuselage Axis

The aircraft zero-reference and the fuselage axis are used to position and locate aircraft components to facilitate computation and manufacturing processes. They are orthogonal to each other. The aircraft zero-reference plane is a near vertical plane, typically passes through the farthest point of the nose cone, as shown in Figures 5.5a and 5.6, but designers can choose any station for their convenience, within or outside of the fuselage.

Given here are several ways to define fuselage axis as desired by the design bureau.

  (i) Fuselage axis as a mean line of the constant cross-sectional part of the mid-fuselage (Figure 5.6).
 (ii) Fuselage axis passing through the nose cone if it is close to the line as defined before (Figure 5.5a).
(iii) Fuselage axis is close to the principal inertia axis of the aircraft.

Fuselage axis may not pass through the aft-end closure point. The fuselage axis may or may not be parallel to the ground (tail dragger aircraft is an example). If the fuselage axis is parallel to the ground then the zero-reference plane is vertical.

The overall fuselage length, l_{fus} consists of the (i) front-fuselage nose cone (l_f), (ii) constant cross-section midsection barrel (l_m), and (iii) aft-end closure (l_a). The constant cross-section mid-fuselage length is established from the passenger seating capacity. The following geometrical definitions are extensively used in this book (see Figure 5.5).

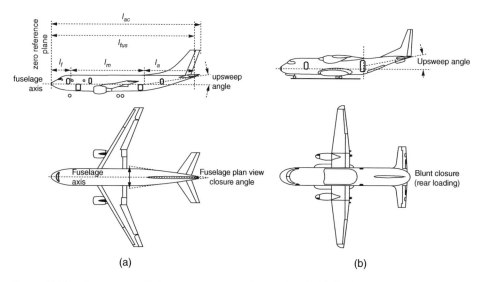

(a)   (b)

Figure 5.5 Fuselage geometrical parameters – Lengths associated with fuselages. (a) Conventional aft end. (b) Rear-loading aft end – blunt closure.

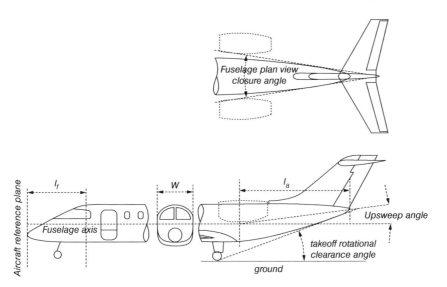

Figure 5.6 Front and aft-end closure.

5.3.2 Fuselage Length, l_{fus}

Fuselage length, l_{fus}, is the length along the fuselage axis, taken by measuring the length of the fuselage from the tip of the nose cone to the tip of the tail cone. This is not the same as the aircraft length, l_{ac}, as shown in Figure 5.5a. Aircraft length, l_{ac}, may not be equal to fuselage length, l_{fus}, if any other part of the aircraft extends beyond the fuselage extremities (e.g. the tail sweep may go beyond the tail cone of the fuselage). Depending on the rated passenger capacity, the fuselage length changes in discrete steps of rows and width changes in increments of one seat pitch and width at a time.

5.3.3 Front-Fuselage Closure Length, l_f

This is the length of the front-fuselage from the tip of the nose cone to the onset of the constant cross-sectional barrel of the mid-fuselage (Figure 5.5a). It encloses the pilot CFD and the windscreen, followed by the mid-fuselage constant-section barrel.

5.3.4 Aft-Fuselage Closure Length, l_a

This starts from the end of the constant cross-sectional barrel of the mid-fuselage up to the tip of the tail cone (Figure 5.5a). It may enclose the last few rows of passenger seating, rear exit door, toilet, and – for a pressurised cabin – the aft pressure bulkhead, which is an important component from a structural design perspective ($l_a > l_f$).

5.3.5 Mid-Fuselage Constant Cross-Section Length, l_m

This is the constant cross-sectional mid-barrel of the fuselage, where passenger seating and other facilities are accommodated (including windows and emergency exit doors etc.).

5.4 Fuselage Closures – Civil Aircraft

The spindle-shaped closure of the fuselage at both ends of the constant midsection keeps the nose cone blunter than the gradually tapered aft cone, as shown in Figure 5.6. It illustrates the front-fuselage closure (i.e. nose cone) length, l_f. Being in a favourable pressure gradient of flow, it is blunter than the aft closure. The aft-fuselage closure (tail cone) length, l_a, encloses the rear pressure bulkhead with a gradual closure in an adverse pressure gradient and has some degree of upsweep. In the centre, the cross-sectional view of the fuselage is shown.

The shaping of front and aft closures are carefully developed by analysing the pressure distribution pattern to make it as streamlined as possible for the full flight envelop. Front-fuselage upper curvature provides pilot visibility needed to ensure that local shock formation at M_{crit} is at its minimum. The aft gradual closure (as well plan view closure – Figure 5.6) needs to ensure minimum separation, having a flow energy of the thick boundary layer. The aft-fuselage upsweep angle needs to clear the fuselage rotation angle at takeoff. Some aft fuselages may require curved upsweep.

In the past, empirical relations were used, but today's CFD analyses have made the empirical relations obsolete. For simplicity, this book uses the statistics of past designs as shown in Figure 5.7 and Tables 5.1 and 5.2.

5.4.1 Front (Nose Cone) and Aft-End Closure

The front-end closure of bigger aircraft appears to be blunter than on smaller aircraft because the nose cone is sufficiently spacious to accommodate pilot positioning and instrumentation. A kink appears in the windscreen mouldlines of the fuselage to fit flat glasses on a curved fuselage body; flat surfaces permit wiper installation and are less costly to manufacture. Some small aircraft have curved windscreens that permit smooth fuselage mouldlines.

The aft-end closure is shallower to minimise airflow separation when the boundary layer becomes thicker. All fuselages have some upsweep for aircraft rotational clearances at takeoff. Designers must configure a satisfactory geometry with attention to all operation and structural requirements (e.g. pilot vision polar, pressure-bulkhead positions and various doors).

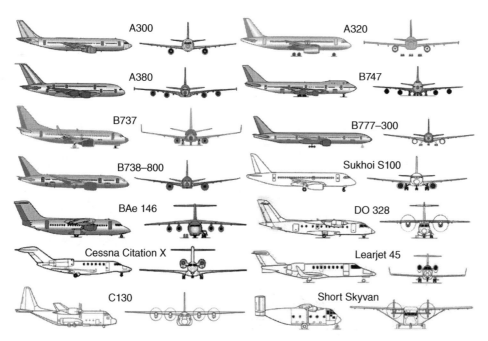

Figure 5.7 Front (nose cone) and aft-end closure (not to scale – various sources).

Table 5.1 Fuselage closure parameter (see Figure 4.16 – nomenclature at the bottom).

Aircraft	L (m)	D (m)	H – (m)	W – (m)	H/W	L_f/D	L_a/D	UA	CA
A300-600 (TA, TF, LW)	53.62	5.64	5.64	5.64	1	1.6	3.103	5	9
A310-300 (TA, TF, LW)	46.66	5.64	5.64	5.64	1	1.6	3.4	5	11
A320-200 (TA, TF, LW)	37.57	3.96	3.96	3.96	1	1.5	3.4	4	8
A330-300 (TA, TF, LW)	59	5.64	5.64	5.64	1	1.82	3.64	8	11
A340-600 (TA, TF, LW)	59.39	5.64	5.64	5.64	1	1.6	3.32	8	9
A380 (TA, TF, LW)	70.4	7.78	8.41	7.14		1.5	3.91	5	11
Boeing737 (TA, TF, LW)	31.28	3.95	4.11	3.79		1.10	2.80	7	15
Boeing747 (TA, TF, LW)	68.63	7.3	8.1	6.5		1.35	3.31	5	11
Boeing757 (TA, TF, LW)	45.96	4.05	4.0	4.10		1.64	2.91	6	13
Boeing767 (TA, TF, LW)	47.24	5.03	5.03	5.03	1	1.17	2.67	7	15
Boeing777 (TA, TF, LW)	63.73	6.2	6.2	6.2	1	1.23	2.85	7	13
MD11 (TA, TF, LW)	58.65	6.02	6.02	6.02	1	1.45	2.82	5	13
Tupolev 204 (TA, TF, LW)	46.1	3.95	3.8	4.1		1.46	2.96	5	9
Fokker 100 (TA, TF, LW)	32.5	3.3	3.05	3.49		1.42	3.42	2	10
Dornier 728 (TA, TF, LW)	27.03	2.56	2.05	3.25		1.34	2.6	5	13
Dornier 328 (RA, TF, LW)	20.92	2.42	2.425	2.415		1.27	2.64	5	10
Dash8 Q400 (RA, TP, HW)	25.68	2.07	2.03	2.11		1.71	3.22	4	9
Bae RJ85 (RA, TP, HW)	28.55	3.56	3.56	3.56	1	1.46	2.62	4	12
Skyvan (RA, TP, HW)	12.22	square	2.2	2.2		0.95	2.0	9	0
Cessna 560 (BJ, TF, LW)	15.79	5.64	5.64	5.64	1	2.05	2.91	2	8
Learjet 31A (BJ, TF, LW)	x	5.64	1.63	1.63		2.17	3.64	2	5
Cessna 750 (BJ, TF, LW)	21	1.8	1.8	1.8	1	2.00	3.00	7	15
Cessna 525 (BJ, TF, LW)	14	1.6	1.6	1.6	1	2.00	2.56	7	13
Learjet 45 (BJ, TF, LW)		5.64	1.75	1.72		1.91	2.86	8	4
Learjet 60 (BJ, TF, LW)	17.02	3.96	1.96	1.96	1	1.91	2.82	2	5
CRJ 700 (RA, TF, LW)		2.69	2.69	2.69	1	1.60	3.15	5	12
ERJ 140 (RA, TF, LW)	26.58					2.00	2.89	3	14
ERJ 170 (RA, TF, LW)	29.9	3.15	3.35	2.95		1.56	2.67	3	13
C17 (MT, TF, HW)	49.5	6.85	6.85	6.85	1	0.85	3.41	10	12
C130 (MT, TF, HW)	34.37	4.33	4.34	4.32		0.95	2.56	9	12

TA – Transport Aircraft; LW – Low Wing; H – Fuselage Height; RA – Regional Aircraft; HW – High Wing; W – Fuselage Width; BJ – Business Jet; L – Fuselage Length; L_f – Front closure length; MT – Military Transport; D – Fuselage Diameter; L_f – Aft closure length; TF – Turbofan; UA – Upsweep angle, deg; TP – Turboprop; CA – Closure angle, degree.

Table 5.2 Fuselage front and aft closure ratios (no rear door).

Seating abreast	Front-fuselage closure ratio, F_{cf}	Aft-fuselage closure ratio, F_{ca}	Aft closure angle – degrees
≤3	≈1.7–2	≈2.6–3.5	≈5–10
4–6	≈1.5–1.8	≈2.5–3.75	≈8–14
≥7	≈1.5–1.7	≈2.5–3.75	≈10–15

5.4.2 Fuselage Upsweep Angle

In general, the fuselage aft end incorporates an upsweep (Figure 5.6) for ground clearance at rotation on take-off. The upsweep angle is measured from the fuselage axis to the mean line of aft-fuselage height. It may not be a straight line if the upsweep is curved like a banana; in that case, it is segmented to smaller straight lines. The rotation clearance angle is kept to 12–16°; however, the slope of the bottom mouldline depends on the undercarriage position and height. Rear-loading aircraft have a high wing with the undercarriage located close to the fuselage belly. Therefore, the upsweep angle for this type of design is high. The upsweep angle can be seen in the elevation plane of a three-view drawing. There is significant variation in the upsweep angle among designs. A higher upsweep angle leads to more separation and, hence, more drag.

5.4.3 Fuselage Plan View Closure Angle

The closure angle is the aft-fuselage closure seen in a plan view of the three-view drawing and it varies among designs. The higher the closure angle, the greater the pressure drag component offered by the fuselage. In rear-loading aircraft, the fuselage closes at a blunt angle; combined with a large upsweep, this leads to a high degree of separation and, hence, increased pressure drag. A finer aft closure angle is desired, but for larger aircraft the angle increases and attempts are made to keep length L_f to an acceptable level to save weight and cost.

A finer aft closure angle is desired; however, for larger aircraft, the angle increases to keep the length (L_f) to an acceptable level to reduce weight and cost.

Figure 5.7 shows several examples of current types of commercial transport aircraft designs [1]. Statistical values for the front- and aft-fuselage closure are summarised in Table 5.1. There are special designs that may not fall in this generalised table. Designers may exercise their own judgement in making a suitable streamline shape to allow for an upsweep to clear for aircraft rotation at takeoff.

Table 5.1 lists the front and aft-fuselage closure statistics. The front-fuselage closure ratio is $F_{cf} \approx 1\text{--}2$ and aft-fuselage closure ratio is $F_{ca} \approx 2\text{--}4$.

Define:

$$\text{Front-fuselage closure ratio, } F_{cf} = l_f / D_{ave} \tag{5.4}$$

$$\text{Aft-fuselage closure ratio, } F_{ca} = l_a / D_{ave}$$

Table 5.2 gives typical guidelines for the fuselage front and aft-end closure ratios (Eq. (5.4)) – the range represents the current statistical values.

5.5 Fuselage Fineness Ratio (FR)

Fuselage configuration is dictated by the aircraft mission specifications. Fuselage geometry is deterministic as its volume requirement is established from the aircraft mission (payload-range) specification, as will be shown in Chapter 8. The aerodynamic task for fuselage shaping is to minimise drag and pitching moment for the volume required to accommodate the required items, depending whether it is a civil or military aircraft. A useful fuselage (also applied to a nacelle) design parameter is the FR (also known as the Slenderness Ratio) and is defined next.

Using Eq. (5.3), the fuselage FR is defined as,

$$FR = (\text{fuselage length/Average fuselage diameter}) = l_f / D_{ave\text{-}fus} \tag{5.5}$$

From test data, empirical relations give aerodynamically optimum values for a realistic FR as above 16. Yet, no aircraft has been built with that long and slender a fuselage, so as to avoid adverse structural issues that may creep in. (Readers are recommended to refer to Section 2.5.1 on the role of optimisation that has to be

Table 5.3 Number of passenger versus number of abreast seating and fineness ratio.

Baseline aircraft	Passenger capacity	Abreast seating	Fuselage Dia_{ave} (m)	Length (m)	Fineness	Cross-section
Learjet 45	6 (4–8)	2	1.75	17.2	≈10	circular
Dornier 228	18	2		≈		rectangular
Dornier 328	24	3	2.2	20.92	≈	circular
ERJ135	37	3	2.28	24.39	≈10.7	circular
ERJ145	50	3	2.28	27.93	≈12.25	circular
Canadair CL600	19	4	2.69	18.77	≈7	circular
Canadair RJ200	50	4	2.69	24.38	≈9.06	circular
Canadair RJ900	86	4	2.69	36.16	≈13.44	circular
Boeing717	117	5	3.34	35.34	≈10.28	non-circular
BAe145 (RJ100)	100	5	3.56	30	≈8.43	circular
Airbus318	107	6	3.96	30.5	≈7.7	circular
Airbus321	185	6	3.96	44	≈11.1	circular
Boeing737–100	200	6	3.66	28	≈7.65	non-circular
Boeing737–900	200	6	3.66	42.11	≈11.5	non-circular
Boeing757–300	230	6	3.66	54	≈14.7	highest ratio
Boeing767–300	260	7	5.03	53.67	≈10.7	circular
Airbus330–300	250	8	5.64	63	≈11.2	circular
Airbus340–600	380	8	5.64	75.3	≈13.35	circular
Boeing777–300	400	9	6.2	73.86	≈11.9	circular
Boeing747–400	500	10	≈6.5	68.63	≈10.55	partial double deck
Airbus380	600	10	≈6.7	72.75	≈10.8	fully double deck

understood for applying the process in fuselage design.) Statistics of existing designs give FR to be within 8–15 for subsonic transport aircraft (see Table 5.3 and Figure 5.8).

The high subsonic commercial transport aircraft fuselage section is basically a circular or near circular constant cross-sectional tube with a blunt front end and tapered aft end closure. Table 5.3 lists fuselage FR values between ≈8 and ≈15. Making it shorter or longer is associated with problems arising from aerodynamic and structural issues. Aft-loading blunt end fuselages invariably associate with some of these problems when ventral fins are installed to improve flow instability.

It must be stressed that the transport aircraft product line should be offered in a family of variants to cover a wide market demand, at lower unit cost, by maintaining component commonalities. The fuselage length is extended or shortened to arrive at variant designs in the family, that is, having different FRs in the family of variants. The baseline aircraft FR may be kept around 10.

Supersonic commercial transport aircraft necessarily have high FR (above 20) to deal with supersonic wave drag. Supersonic fuselage aircraft weight per passenger is considerably higher compared to subsonic fuselage with the same passenger capacity.

Table 5.3 lists the FRs of some of the existing designs [1]. A good value for commercial transport aircraft design is 10 ± 2. The B757-300 has the highest FR at 15.7.

Figure 5.8 summarises the statistics of fuselage FR for the abreast seating arrangement. It is to be noted that current International Civil Aviation Organization (ICAO) limit on fuselage length is 80 m. This limit is an artificial one based on current airport infrastructure size and handling limitations. To get better resolution, readers are recommended to plot a similar graph, collecting data for as many aircraft by category of aircraft in the project under study. This will serve to check aircraft configured in project work, as worked out in Chapter 8.

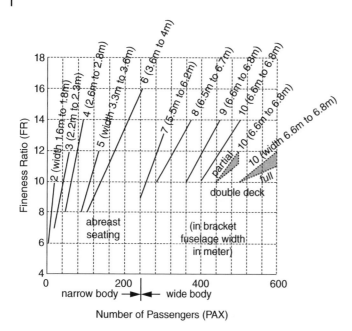

Figure 5.8 Abreast seating versus number of passengers (standard configuration) and fineness ratio.

5.6   Fuselage Cross-Sectional Geometry – Civil Aircraft

Fuselage cross-sectional geometry is not only concerned with aerodynamic considerations but also internal arrangement design considerations. This section only deals with the transport category of aircraft.

5.6.1   Fuselage Height, *H*

This is the maximum distance of the fuselage from its underside (not from the ground) to the top in the vertical plane (Figure 5.9).

5.6.2   Fuselage Width, *W*

This is the widest part of the fuselage in the horizontal plane. For a circular cross-section, this is the diameter shown in Figure 5.9.

5.6.3   Average Diameter, D_{ave}

For a non-circular cross-section, this is the average of the fuselage height and width at the constant cross-section barrel part ($D_{ave} = (H + W)/2$). Sometimes this is defined as $D_{effective} = \sqrt{(H*W)}$; another suitable definition is $D_{equivalent} = \text{perimeter}/2\pi$. This book uses the first definition.

5.6.4   Cabin Height, H_{cab}

This is the internal cabin height from the floor, as shown in Figure 5.9.

5.6.5   Cabin Width, W_{cab}

This is a the internal cabin width, as shown in Figure 5.9.

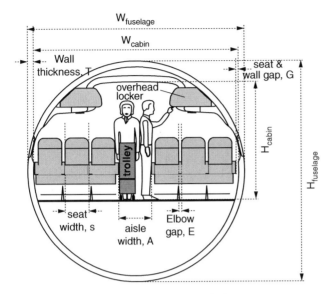

Figure 5.9 Fuselage cross-section geometrical parameters.

5.6.6 Pilot Cockpit/Flight Deck

This is a term used for the enclosed space for the flight crew in the front-fuselage. Chapter 15 describes the flight deck in more detail.

5.6.7 Cabin Interior Details

Details of cabin interior details, such as the seat details/pitch, passenger amenities and so on.

5.7 Fuselage Abreast Seating – Civil Aircraft

The minimum number of seats abreast is one row, which is not a practical design – one would have to crawl into the cabin space. There must be at least two-abreast seating (e.g. Beech 200 and Learjet 45); the largest to-date is the 10-abreast seating arrangement with two aisles in the wide-bodied Boeing747 and Airbus380. The two-aisle arrangement is convenient for wider body six-abreast seating.

Currently, transport aircraft abreast seating is of two kinds as follows. All fuselage cross-sections are symmetrical to the vertical plane.

1. Narrow body – up to six abreast seating with one aisle. The abreast seating is designated as an X-X arrangement (Figure 5.10).
2. Wide body – more than six abreast seating with two aisles. The abreast seating is designated as an X-X-X arrangement. BWB layout arrangement has the possibility of having more than two aisles (Figure 5.11).

The hyphens represent aisles and Xs are the clusters of seats. In general, aircraft with four-abreast seating and more have space below the cabin floor for baggage and cargo.

As passenger capacity exceeds six hundred (if not in a double-deck arrangement), the fuselage depth allows an attractive design with BWB when more than two aisles are possible. A BWB military combat aircraft has been successfully designed but its high-capacity civil aircraft version awaits development, delayed primarily by the technology-development and airport-infrastructure limitations; the market has yet to evolve as well.

5.7.1 Narrow Body

Single-aisle passenger seating arrangements are called narrow-bodied transport aircraft. A '3–3' arrangement indicates that it is a narrow-bodied aircraft, has one aisle and has total of six seats in a cluster of three seats at each window. Figure 5.10 shows the various options for an aircraft fuselage cross-section to accommodate different cabin seating arrangements from two- to six-abreast seating in a row.

5.7.2 Wide Body

When the seating number is increased to more than six abreast, the number of aisles is increased to two to alleviate congestion in passenger movement. More than one aisle (currently two, but possible may grow to three) is regarded as wide-bodied transport aircraft. For example a '3-4-3' arrangement indicates that it is a wide-bodied aircraft has a total of 10 seats with two aisles and a cluster of three seats at each window side with a cluster of four seats in the centre flanked by two aisles.

Because of the current fuselage-length limitation of 80 m, larger-capacity aircraft have a double-deck arrangement (e.g. the B747 and the A380). Figure 5.11 shows options for typical wide-body and double-deck aircraft fuselage cross-sections to accommodate different cabin seating arrangements from seven- to 10-abreast seating in a row.

It is interesting to study a six-abreast wide-body seating in 2-2-2 arrangement with only a 19 in. increase in fuselage diameter by having two 20 in. width aisles (Figure 5.11). This eliminates a centre seat, offering better comfort, easy movement, rapid evacuation and so on, possibly with a roughly 5% (not computed) increase in Direct Operating Cost (DOC): this should appeal to customers, both the operators and passengers.

5.7.3 More than Two Aisles - Blended Wing Body

A three-aisle arrangement with 10-abreast seating would eliminate the cluster of four seats together. A BWB would have more than two aisles; there is no reason to not consider a triple-deck arrangement. Figure 5.12 shows various options for futuristic aircraft fuselage cross-sections to accommodate different cabin seating arrangements.

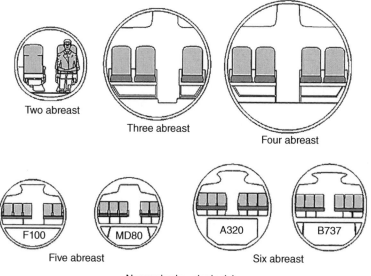

Figure 5.10 Typical narrow body single-aisle civil aircraft fuselage cross-sections (not to scale).

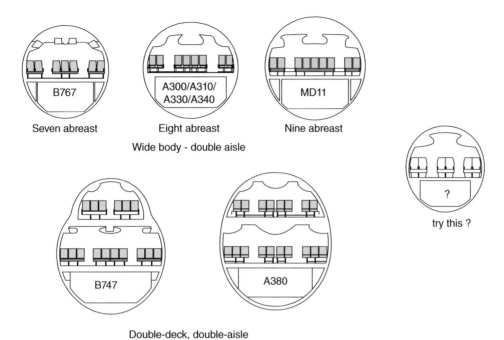

Figure 5.11 Typical wide-body double-aisle civil aircraft fuselage cross-sections (not to scale).

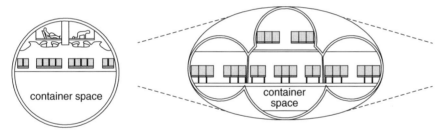

Figure 5.12 Various options for futuristic aircraft fuselage cross-sections.

5.8 Cabin Seat Layout

There are two parameters of size: fuselage width W and fuselage length l_f, which determine the constant-section fuselage-barrel length. In turn, this depends on the seat pitch and width for the desired passenger comfort level. Fuselage geometry is determined from the designed passenger capacity (see Figure 5.8 and Table 5.3). This is a typical relationship between the number of passengers and the number of abreast seating – a new design would be similar. The width and length of the fuselage must be determined simultaneously, bearing in mind that the maximum growth potential in the family of variants cannot be too long or short and keeping the FR at 7–14 (a good baseline value is around 10).

The first task is to determine the abreast seating for passenger capacity. The standard practice for seat dimensions is to cater to the 95th percentile of European men. Dimensions of seat pitch and width at various comfort level is given in Table 7.2. Elbowroom is needed on both sides of a seat; in the middle seats, it is shared. Typical elbowroom is from 1.5 to 2 in. for economy class and double that for first class. In addition, there is a small space between the window elbowrest and the fuselage wall, larger for more curved smaller aircraft – typically about an inch. A wider cabin provides more space for passenger comfort at an additional cost and drag. A longer seat pitch and wider seats offer better comfort, especially for oversized people. Fuselage width is the result of adding the thickness of the fuselage structural shell and soft wall furnishings to the cabin width. During Phase II (i.e. the project-definition stage), when sufficient structural details emerge, the interior-cabin geometric dimensions are defined with better resolution; the external geometry remains unaffected. The number of abreast seating and total passenger capacity determine the number of rows.

The first parameter to determine for the fuselage average diameter is the number of abreast seating for passenger capacity. There is an overlap on choice for the midrange capacity in the family of design; for example, an A330 with 240–280 passengers has seven-abreast seating whereas the same passenger capacity in a B767 has eight-abreast seating. When seating number is increased to more than six abreast, the number of aisles is increased to two to alleviate congestion in passenger movement. Because of the current fuselage-length limitation of 80 m, larger-capacity aircraft have a double-deck arrangement (e.g. the B747 and the A380). It would be interesting to try a two-aisle arrangement with six-abreast seating that would eliminate a middle seat.

Typical geometric and interior details for aircraft with two- to 10-abreast seating accommodating from 4 to 600 passengers with possible cabin width, fuselage length, and seating arrangements are described in this subsection and shown in Figures 5.13 and 5.14. The public domain has many statistics for seating and aisle dimensions relative to passenger number, cabin volume and so forth. The diagrams in this section reflect current trends. Figures 5.13 and 5.14 show the spaces for toilets, galleys, wardrobes, attendant seating and so forth, but these are not indicated as such. There are considerable internal dimensional adjustments required for the compromise between comfort and cost.

Dimensions listed in Tables 5.4 and 5.5 are estimates for narrow- and wide-body aircraft. The figures of seat pitch, seat width and aisle width are provided as examples of what exists on the market. The dimensions in the tables can vary to a small extent, depending on customer requirements.

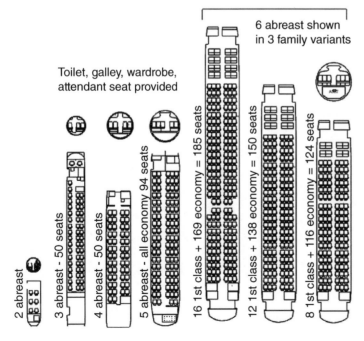

Figure 5.13 Narrow body, single-aisle fuselage layout (not to scale).

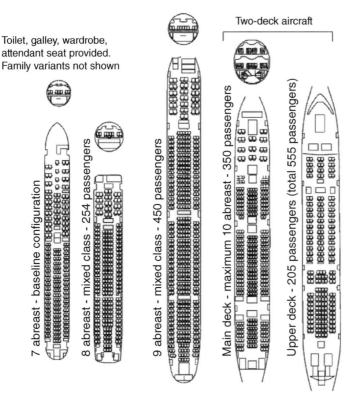

Figure 5.14 Wide-body, double-aisle fuselage layout (not to scale). Source: Reproduced with permission from Cambridge University Press.

Table 5.4 Fuselage seating dimensions – narrow body – all dimensions are in inches. Medium comfort level. Refer to Figure 5.10 for the symbols used.

	Two-abreast (1–1)	Three-Abreast (1–2)	Four-abreast (2–2)	Five-abreast (2–3)	Six-abreast (3–3)
Seat width, B (LHS)	19	19	2×18	2×18	3×18
Aisle width, A	17	18	19	20	21
Seat width, B (RHS)	19	2×19	2×18	3×18	3×18
Total elbow room	4×1.5	5×1.5	6×1.5	7×2	8×2
Gap between wall and seat, G	2×1.5	2×1	2×1	2×0.5	2×0.5
Total cabin width, W_{cabin}	64	85	102	126	141
Total wall thickness, T	2×2.5	2×4	2×4.5	2×5	2×8.5
Total fuselage width, $W_{fuselage}$	69	93	111	136	151
Cabin height, H_{cabin}	60[a]	72[a]	75	82	84
Typical fuselage height, H_{fus}	70	85	114	136	151

a) Recessed floor.

Table 5.5 Fuselage seating dimensions – wide body. Medium comfort level. Refer to Figure 5.9 for the symbols used.

	Seven-abreast (2-3-2)	Eight-abreast (2-4-2)	Nine-abreast (2-5-2)	Ten-abreast (3-4-3)
Seat width, B (LHS)	2×19	2×19	2×19	3×19
Aisle width, A	22	22	22	22
Seat width, B (centre)	3×19	4×19	5×19	4×19
Aisle width, A (RHS)	22	22	22	22
Seat width, B (RHS)	2×19	2×19	2×18	3×19
Total elbow room	9×1.5	10×1.5	11×1.5	12×1.5
Gap between wall and seat, G	2×0.5	2×0.5	2×0.5	2×0.5
Total cabin width, W_{cabin}	192	212	232	253
Total wall thickness, T	2×6	2×8.5	2×7	2×8.5
Total fuselage width, $W_{fuselage}$	204	225	246	268
Cabin height, H_{cabin}	84	84	84–86	84–86
Typical fuselage height, H_{fus}	204	225	246	268

5.8.1 Narrow-Body, Single-Aisle Aircraft

Figure 5.13 shows a typical seating arrangement for a single-aisle, narrow-body aircraft with two to six passengers abreast and seating carrying up to about 220 passengers (all economy class). Sections 5.4 and 5.5 give the general considerations of closure angles and FR. Table 5.4 provides typical dimensions for establishing narrow-body fuselage widths. Details of seat, internal facilities and doors, and so on for each type are given subsequently.

5.8.1.1 Two Abreast (4–24 Passengers)

Two-abreast seating is the lowest arrangement. The 10-passenger capacity extends from 4 to 19 (e.g. Beech 1900D) and could expand to 24 passengers in an extreme derivative version. Current regulations do not require a cabin crew for up to 19 passengers, but some operators prefer to have one crew member, who uses a folding seat secured in a suitable location. An expanded variant of two-abreast seating can exceed 19 passengers, but a new high-capacity design should move into three-abreast seating, described next. The baggage area is at the rear, which is the preferred location in smaller aircraft. Also, it is preferred to have a toilet at the rear.

Summary   Typical two-abreast seating (1-1 abreast seating arrangement) could be as follows:

Cabin width. This consists of one seat on each side of the centre aisle. To avoid tightness of space in a smaller aircraft, seats could be slightly wider, sacrificing aisle width where there is little traffic. Typically, cabin width is between 64 and 70 in.

Cross-section. Fuselage cross-section is typically circular or near circular (overall width > height). Designers will have to make compromise the choice to maximise sale. The bottom half could open up for better leg room. There is no payload space below the floor but it can be used for aircraft equipment/fuel storage. Luggage space is at the aft fuselage.

Front/aft Closure. See Table 5.1 for the desirable range of dimensions.

Fuselage length. This depends on the number of passengers and facilities provided.

Seat pitch. See Table 7.2 gives the dimensions for comfort level. Medium comfort level is at 30 in. This determines the fuselage midsection length. Add front and aft closures to the midsection.

Family variants. Addition/subtraction of fuselage plugs to a maximum of four rows conveniently distributed on each side of the wing is possible. The worked-out example baseline version starts with 10 passengers (Section 8.9).

5.8.1.2 Three Abreast (24–50 Passengers)

A typical three-abreast (2-1) seating arrangement accommodates 24–45 passengers, but variant designs change that from 20 to 50 passengers (e.g. ERJ145). Full standing headroom is possible; for smaller designs, a floorboard recess may be required (see Figure 5.9). A floorboard recess could trip passengers when they are getting to their seat. Space below the floorboards is still not adequate for accommodating any type of payload. Generally, space for luggage in the fuselage is located in a separate compartment at the rear but in front of the aft pressure bulkhead (the luggage compartment door is sealed). A toilet is provided at the aft end.

Note: At least one cabin crew is required for up to 30 passengers. Above this number, two cabin crew members are required for up to 50 passengers. A new design with potential to grow to more than 50 passengers should start with four-abreast seating described in the next section.

Summary

Cabin width. This comprises of two in a cluster and one seat arrangement on each side of the aisle. The aisle width could also be increased to ease cabin crew access. Cabin width is kept within 82–88 in. depending upon customer demand for comfort level.

Cross-section. Fuselage cross-section is typically circular but will follow cabin-section contour with added wall thickness. There is no payload space below the floor but can be used for aircraft equipment/fuel storage.

Front/aft Closure. See Table 5.1 for the desirable range of dimensions.

Fuselage length. This depends on the number of passengers and facilities provided.

Seat pitch. See Table 7.2 gives the dimensions for comfort level. Medium comfort level is at 30 in. This determines the fuselage midsection length. Add front and aft closures to the midsection.

Family variants. Addition/subtraction of fuselage plugs to a maximum of five rows, conveniently distributed on each side of wing. The baseline version could start with 36 passengers with the least being 24 and the most 50 passengers.

5.8.1.3 Four Abreast (Around 44–80 Passengers)

A typical four abreast (2-2) seating arrangement accommodates 44–80 passengers but variant designs could extend from 40 to 96 passengers (Canadair CRJ1000 and the Canadair CL-600 is an executive version that can take 19 passengers – another example of a derivative). Cabin crew increases to at least three for higher passenger loads. Increase in fuselage diameter can offer below-floor space for payload usage, but it is still on the tight side. To maximise below-floor space, fuselage height could be in a slightly oval shape, the upper lobe kept semicircular and the bottom half elongated to suit smaller container sizes. Note the facilities and luggage compartment arrangement. As fuselage radius is increased, the gap between the elbowrest and fuselage wall can be reduced to 1 in. (2.54 cm) each side and seat width increased.

Summary

Cabin width. Four abreast is arranged as two seats in a cluster on both the sides of centre aisle. Cabin width is kept within 100 to 106 in. depending upon customer demand for the comfort level. The aisle width could also be increased to ease cabin crew access and passenger traffic.

Cross-section. The fuselage cross-section is typically circular but can be elongated. It follows the cabin-section contour with added wall thickness (see Table 8.1). Full standing headroom is easily achievable. There is aft-fuselage luggage space.

Front/aft closure. See Table 5.1 for the range of dimensions.

Fuselage length. This depends on number of passengers and facilities provided.

Seat pitch. See Table 7.2 gives the dimensions for the comfort level. Medium comfort level is at 30 in. This determines the fuselage midsection length. Add front and aft closures to the midsection.

Family variants. Addition or subtraction of fuselage plugs, to a maximum of seven rows, conveniently distributed on each side of the wing, is possible. The baseline version could start with 60 passengers and range from 40 to 96 passengers.

5.8.1.4 Five Abreast (80–150 Passengers)

A typical five-abreast (3-2) seating arrangement can accommodate 85–130 passengers, but variant designs could extend that number somewhat on both sides. The number of cabin crew increases with passenger capacity. There are not many aircraft with five-abreast seating because the increase from four abreast to six abreast better suited market demand. A prominent five-abreast design is the MD-9 series (now the Boeing717).

The fuselage diameter widens to provide more generous space. Space below the floorboards is conspicuous to accommodate standard containers (see Section 7.7.6). The fuselage aft closure could affect seating – that is, the last row could be reduced to four abreast. To ease cabin access, the aisle width widens to at least 20 in. plus the armrest at each side. To maximise the below-floor space, the fuselage could be slightly elongated, with the bottom half stretched to accommodate container sizes. A separate cargo space exists at the rear fuselage in the closure area.

Summary

Cabin width. Five abreast is seating arranged as three in a cluster on one side of the single aisle and two in a cluster on the other side. Very little gap is required between the armrest and the cabin wall because the fuselage radius is adequate. Cabin width is from 122 to 130 in. depending on customer demand for comfort level. The aisle width could be increased to facilitate passenger and crew traffic.

Cross-section. The fuselage cross-section is typically circular but can be elongated. It follows the cabin-section contour with added wall thickness. Full standing headroom is easily achievable. There is potential for aft-fuselage luggage space.

Front/aft closure. See Table 5.1 for the range of dimensions.

Fuselage length. This depends on number of passengers and facilities provided.

Seat pitch. See Table 7.2 gives the dimensions for the comfort level. Medium comfort level is at 30 in. This determines the fuselage midsection length. Add front and aft closures to the midsection.

Family variants. Addition or subtraction of fuselage plugs, to a maximum of eight rows, conveniently distributed on each side of the wing, is possible. The baseline version could start with 100 passengers and range from 85 to 150 passengers.

5.8.1.5 Six Abreast (120–230 Passengers)

A typical six-abreast (3-3) seating arrangement can accommodate 120–230 passengers, but variant designs could extend that number somewhat on both sides. This class of passenger capacity has the most commercial transport aircraft in operation (more than 8000), including the Airbus320 family and the Boeing737 and 757 families. The Boeing757–300 has the largest passenger capacity of 230 and the highest FR of 14.8. There is considerable flexibility in the seating arrangement to accommodate a wide range of customer demands.

Figure 5.13 shows an aircraft family of variant designs to accommodate three different passenger-loading capacities in mixed classes. A typical six-abreast seating arrangement accommodates 120–200 passengers, but variant designs could change that number from 100 to 230 passengers. The number of cabin crew increases accordingly. The fuselage diameter is wider to provide generous space. Space below the floorboard can accommodate standard containers (see Section 7.8.8). To maximise the below floorboard space, the fuselage height could be slightly elongated, with the bottom half suitable for container sizes. A separate cargo space is located at the rear fuselage.

Summary

Cabin width. Six-abreast seating is arranged as three in a cluster on both sides of the single centre aisle. Very little gap is required between the armrest and the cabin wall because the fuselage radius is adequate. Cabin width is from 138 to 145 in., depending on customer demand for comfort level. The aisle width is increased to facilitate passenger and crew traffic.

Cross-section. The fuselage cross-section is typically circular but can be elongated. It follows the cabin-section contour with added wall thickness (see Table 5.4). Full standing headroom is adequate. There is potential for aft-fuselage luggage space.

Front/aft closure. See Table 5.1 for the range of dimensions of seat pitch and width.

Fuselage length. This depends on number of passengers and facilities provided.

Seat pitch. See Table 7.2 gives the dimensions for the comfort level. Medium comfort level is at 30 in. This determines the fuselage midsection length. Add front and aft closures to the midsection.

Family variants. Addition or subtraction of fuselage plugs, to a maximum of 10 rows, conveniently distributed on each side of the wing, is possible. The baseline version could start with 150 passengers and range from 85 to 210 passengers. The Boeing757 baseline starts with a higher passenger load, enabling the variant to reach 230 passengers.

The dimensions in the tables can vary to a small extent, depending on customer requirements. The seat arrangement is shown by numbers in clusters of seats, as a total for the full row with a dash for the aisle.

5.8.2 Wide-Body, Double-Aisle Aircraft

These aircraft are also known as wide-bodied aircraft. Figure 5.14 shows a typical seating arrangement for a double-aisle, wide-body aircraft with 7–10 passengers abreast carrying up to 555 passengers; however, high-density seating of all economy-class passengers can exceed 800 (e.g. A380). These large passenger numbers require special attention to manage comfort, amenities and movement. Sections 5.4, and 5.5 give the general considerations for closure angles and FR. Table 5.5 provides typical dimensions for establishing wide-body fuselage widths. Details of seats, internal facilities, doors and so on of each type are given subsequently.

5.8.2.1 Seven Abreast (160–260 Passengers)

A typical seven-abreast fuselage (with better comfort) would have the following features:

the Boeing767 appears to be the only aircraft with seven-abreast seating (with better comfort) and it can reconfigure to eight-abreast seating. Typical seven-abreast seating accommodates 160–260 passengers, but variant designs could change that number on either side. The number of cabin crew increases accordingly. The fuselage diameter is wider to provide generous space. Space below the floorboards can accommodate cargo containers. To maximise the below floorboard space, the fuselage height could be slightly elongated, with the bottom half suitable for container sizes. A separate cargo space is located at the rear fuselage.

Summary

Cabin width. Seven-abreast seating is arranged as 2-3-2 in a cluster of two at the window sides and a cluster of three at the centre between the two aisles. Very little gap is required between the armrest and the cabin wall because the fuselage radius is adequate. The cabin width is from 190 to 196 in., depending on customer demand for comfort level. The aisle width could be increased to facilitate cabin crew access and passenger movement.

Cross-section. The fuselage cross-section is typically circular but can be oval.

It follows the cabin-section contour with added wall thickness (see Table 5.2). Full standing headroom is no longer an issue. There is potential for aft-fuselage luggage space.

Front/aft closure. See Table 5.1 for the range of dimensions.

Fuselage length. This depends on the number of passengers and facilities provided.

Seat pitch. See Table 7.2 gives the dimensions for comfort level. Medium comfort level is at 30 in. This determines the fuselage midsection length. Add front and aft closures to the midsection.

Family variants. Addition or subtraction of fuselage plugs, to a maximum of 10 rows, conveniently distributed on each side of the wing, is possible. The baseline version could start with 200 passengers and range from 160 to 260 passengers.

5.8.2.2 Eight Abreast (250–380 Passengers, Wide-Body Aircraft)

The Airbus300/310/330/340 series have all been configured for eight-abreast seating. Figure 5.11 shows an example of an eight-abreast seating arrangement for a total of 254 passengers (in mixed classes; for all economy-class, 380 passengers in a variant design is possible). Space below the floorboards can accommodate larger containers. Seat width, pitch, and layout with two aisles result in considerable flexibility to cater to a wide range of customer demands. The cross-section is typically circular, but to maximise below floorboard space it could be slightly elongated, with the bottom half suitable for cargo container sizes. There is potential for a separate cargo space at the rear of the fuselage.

Summary

Cabin width. Eight-abreast seating is arranged as 2-4-2 in a cluster of two at the window sides and a cluster of four in the centre between the two aisles. Very little gap is required between the armrest and the cabin wall because the fuselage radius is adequate. The cabin width is from 210 to 216 in., depending on customer demand for comfort level. The aisle width is nearly the same as for a wide-bodied layout to facilitate cabin crew and passenger movement.

Cross-section. The fuselage cross-section is typically circular but can be oval. It follows the cabin-section contour with added wall thickness (see Table 5.2). Full standing headroom is adequate. There is potential for aft-fuselage luggage space.

Front/aft closure. See Table 5.1 for the range of dimensions.

Fuselage length. This depends on number of passengers and facilities provided.

Seat pitch. See Table 7.2 gives the dimensions for the comfort level. Medium comfort level is at 30 in. This determines the fuselage midsection length. Add front and aft closures to the midsection.

Family variants. Addition or subtraction of fuselage plugs, to a maximum of 11 rows, conveniently distributed on each side of the wing, is possible. The baseline version could start with 300 passengers and range from 250 to 380 passengers.

5.8.2.3 Nine to Ten Abreast (350–480 Passengers, Wide-Body Aircraft)

The current ICAO restriction for fuselage length is 80 m. The associated passenger capacity for a single-deck aircraft is possibly the longest currently in production. It appears that only the Boeing777 has been configured to nine or ten-abreast seating in a single deck. Figure 5.11 is an example of a nine-abreast seating layout for a total of 450 passengers. Seat width, pitch and a layout with two aisles has a similar approach to the earlier seven-abreast seating designs, which embeds considerable flexibility for catering to a wide range of customer demands. Cabin crew numbers can be as many as 12. Space below the floorboards can carry larger containers (i.e. LD3). The cross-section is typically circular, but to maximise below floorboard space it could be slightly elongated, with the bottom half suitable for container sizes. There is potential for a separate cargo space at the rear fuselage.

Summary

Cabin width. Nine-abreast seating is arranged as 2-5-2 in a cluster of two at the window sides and a cluster of five in the centre between the two aisles. A 3-3-3 arrangement is also possible but not shown. Very little gap is required between the armrest and the cabin wall because the fuselage radius is adequate. The cabin width is from 230 to 236 in., depending on customer demand for comfort level. The aisle width is nearly the same as for the wide-bodied layout to facilitate cabin crew access and passenger movement.

Cross-section. The fuselage cross-section is typically circular but can be oval. It follows the cabin-section contour with added wall thickness (see Table 8.2). Full standing headroom is no longer an issue. There is potential for an aft-fuselage luggage space.

Front/aft closure. See Table 5.1 for the range of dimensions.

Fuselage length. This depends on number of passengers and facilities provided.

Seat pitch. See Table 7.2 gives the dimensions for the comfort level. Medium comfort level is at 30 in. This determines the fuselage midsection length. Add front and aft closures to the midsection.

Family variants. Addition or subtraction of fuselage plugs, to a maximum of 12 rows, conveniently distributed on each side of the wing, is possible. The baseline version could start with 400 passengers and range from 300 to 480 passengers.

5.8.2.4 Ten Abreast and More (More than 400–Almost 800 Passenger Capacity, Wide-Body and Double-Decked)

A more than 450-passenger capacity provides the largest class of aircraft with variants exceeding an 800-passenger capacity. This would invariably become a double-decked configuration to keep fuselage length below the current ICAO restriction of 80 m. Double-decking could be partial (e.g. Boeing747) or full (e.g. Airbus380), depending on the passenger capacity; currently, there are only two double-decked aircraft in production.

With a double-decked arrangement, there is significant departure from the routine adopted for a single-decked arrangement. Passenger numbers of such large capacity would raise many issues (e.g. emergency escape compliances servicing and terminal handling), which could prove inadequate compared to current practice. Reference [2] may be consulted for a double-decked aircraft design. The double-decked arrangement produces a vertically elongated cross-section. Possible and futuristic double-decked arrangements are shown in Figure 5.12. The number of cabin crew increases accordingly. The space below the floorboards is sufficient to accommodate larger containers (i.e. LD3).

5.8.2.5 To summarise

Cabin width. The lower deck of a double-decked aircraft has at most 10 abreast, arranged as 3-4-3 in a cluster of three at the window sides and a cluster of four in the centre between the two aisles. Very little gap is required between the armrest and the cabin wall because the fuselage radius is adequate. The cabin width is from 250 to 260 in., depending on the customer's demand for the comfort level. The aisle width is nearly the same as for a wide-bodied layout to facilitate cabin crew and passenger movement.

Cross-section. A double-decked fuselage cross-section is elongated at this design stage. It follows the cabin-section contour with added wall thickness (see Table 8.2). Full standing headroom is no longer an issue. There is potential for aft-fuselage luggage space.

Front/aft closure. Add front and aft closures to the fuselage midsection. See Table 5.1 for the range of dimensions.

Fuselage length. This depends on number of passengers and facilities provided.

Seat pitch. See Table 7.2 gives the dimensions for the comfort level. Medium comfort level is at 30 in. This determines the fuselage midsection length. Add front and aft closures to the midsection.

Family variants. This depends on the number of passengers and facilities. Addition or subtraction of fuselage plugs, to a maximum of 10 rows, conveniently distributed on each side of the wing, is possible. Fuselage length is less than 80 m.

Table 5.5 provides typical dimensions to establish a wide-body fuselage width. All dimensions are in inches, and decimals are rounded up. More fuselage-interior details are given in Table 5.5. Designers are free to adjust the dimensions.

The dimensions in the tables can vary to a small extent, depending on customer requirements. The seat arrangement is shown by numbers in clusters of seats, as a total for the full row with a dash for the aisle.

5.9 Fuselage Layout

When the interior arrangement is determined, the constant cross-section mid-fuselage needs to be closed at the front and aft ends. The midsection fuselage could exhibit closure trends at both the front and aft ends, with diminishing interior arrangements at the extremities. The front-end fuselage mouldlines have a favourable pressure gradient and are therefore blunter with large curvatures for rapid front-end closure. Basically, a designer must consider the space for the flight crew at the front end and ensure that the pilot's view polar is adequate. Conversely, the aft end is immersed in an adverse pressure gradient with low energy and a thick boundary layer therefore, a gradual closure is required to minimise airflow separation (i.e. minimise pressure drag). The aft end also contains the rear pressure-bulkhead structure. The longer aft-end space could be used for payload (i.e. cargo) and has the scope to introduce artistic aesthetics without incurring cost and performance penalties.

An important current trend is a higher level of passenger comfort (with the exception of low cost airlines). Specifications vary among operators. Designers should conduct tradeoff studies on cost versus performance in consultation with the operators to satisfy as many potential buyers as possible and to maximise sales. This is implied at every stage of aircraft component sizing, especially for the fuselage.

Variants in the family of aircraft are configured by using a constant cross-section fuselage plug in/out units of one row of pitch. For larger variant, fuselage plugs with seat row are inserted, distributed in front and aft of the wing. When in odd numbers, their distribution is dictated by the aircraft Centre of Gravity (CG) position. In most cases, the front of the wing has the extra row. Conversely, a decrease in passenger numbers is accomplished by removing the fuselage plug using the same logic. For example, a 50-passenger increase at 10-abreast seating of a wide-body aircraft variant requires five seat rows to be inserted; plugs distributed with three rows subassembly in front of the wing and a subassembly of two rows aft of the wing. Conversely, a 50-passenger decrease is accomplished by removing three rows from the rear and two from the front. For smaller aircraft with smaller reductions, unplugging may have to be entirely from the front of the wing.

5.9.1   Fuselage Configuration – Summary

Configuration for fuselage for civil aircraft is deterministic and does not rely on empirical relations. It is mainly decided by the number of passengers that has to be accommodated and at a specified comfort level (seat width and pitch – Section 7.5.2).

Given next is a generic step-by-step approach to configure a fuselage. It should be noted that for the same abreast seating there is about 5% variation fuselage width variation in different designs. Also, the cross-section varies from a circle to a somewhat elongated height. Therefore, the readers may use their own discretion to make a design with difference that is better than the existing. This is a challenging task for all designers.

Step 1. Decide the seat abreast arrangement for passenger capacity (Figure 5.8).

Step 2. Determine fuselage width based on the abreast seating arrangement (Tables 5.4 and 5.5). Minimum aisle width is a certification requirement that is taken in account in the tables.

Step 3. Develop a fuselage cross-section based on fuselage width and under-floor space provision (see Section 19.13). Container sizes are given in (Section 7.5.6).

Step 4. Establish the constant cross-section mid-fuselage length, 'l_m' (Figures 5.5, 5.13 and 5.14) by computing the number of seat rows required. In case of a wide-body layout, the last few rows may use the space in the converging section of the aft fuselage (Figures 5.14–5.16).

Step 5. Generate the front-fuselage length, 'l_f' and aft fuselage length, 'l_a' using the ratios given in Tables 5.1 and 5.2. Plan view closure angle is in line with the ratios used.

Step 6. Fuselage upsweep angle has to clear the fuselage rotation angle at takeoff. At this stage, use the statistical values given in Table 5.1. Subsequently after configuring and positioning the undercarriage (Chapter 10), the upsweep angle may be revised.

Step 7. Make sure that internal passenger/crew facilities (Section 7.5.4) are adequately catered for in Steps 4 and 5.

Step 8. Finally, provide doors and windows as required, bearing in mind that the door requirements have to satisfy Federal Aviation Regulation (FAR) requirements (Section 7.5.3).

5.10   Fuselage Aerodynamic Considerations

Unlike planer surfaces (wings, empennage etc., required to generate forces and moments, the role of bodies of revolutions – Chapter 4), fuselages, nacelles and so on serve as containers meant primarily to house crew, payload, consumables and host of many types of system equipment; in the case of combat aircraft, to house the engine. This makes aerodynamic considerations to develop relevant geometries less complex.

The static stability of a spindle-shaped body, for example the fuselage, is inherently destabilising. Therefore, their aerodynamic design must not only minimise parasitic drag and extract as much as lift possible as an added bonus, but also simultaneously reduce pitching moment. From a stress loading point of view, a pressurised fuselage with 192 circular cross-section is the lightest. A circular cross-section fuselage with streamlined front and aft closures offers minimum drag and is favoured by aerodynamicists. However, operational specifications to accommodate standard under floorboard space in cargo containers may force the fuselage cross-section to elongate to a near circular shape. Double-decker fuselage cross-sections have a necessarily elongated shape to accommodate two decks (see Figure 5.11). The separation line between the wing and fuselage of BWB configurations is not clearly defined, but may conveniently be separated as shown in Figure 5.4b, typically along the wing assembly joining line with the fuselage assembly in production planning that gives a good delineation.

Some gliders and light aircraft operating at low Re with composite material construction have a bulbous front-fuselage to accommodate occupants and the power plant. Aft of the cabin, volume the fuselage narrows to a near tube-like aft end. This favours extension of laminar flow over a smooth composite surface giving a high L/D ratio, that is, increases the glide ratio. Although the fundament design principles are the same, these kinds of aircraft are not dealt with in this book.

5.10.1 Fuselage Drag

Spindle-like 3D fuselage shapes generate very little or no induced drag, C_{Di}. They mainly develop parasite drag, C_{Dp} (friction and pressure drag, see Chapter 9). Friction drag depends on surface skin friction (in coefficient form, C_F, at the Re) and the wetted surface area S_{fus} and pressure drag depends on the aft-end closure shape. The minimum surface area for a given volume is a sphere, an impracticable shape that associates with high pressure drag (Section 3.4) as compared to a teardrop streamline shape with lower pressure drag. The optimum streamline shape offering minimum parasite drag has FR around 3.0, which is also not practical for fuselage design. Fuselage aft end closure shape also plays a role in pitching moment contribution. CFD optimisation can offer some solution to configure fuselage aft end closure complying with constraints of upsweep angle permitting the required takeoff rotation. This is a good example of how and where to apply optimisation process (see Section 2.5.1), an aspect beyond the scope of this book. This book examines aft-end shapes of existing design to make the choice. Various types of fuselage design considerations are given next to assist to make the choice.

The characteristic length of fuselage is its length, l_{fus}. It should be long enough to have $Re > R_{crit}$, hence separation occurs at the aft end of fuselage, its severity depends on the fuselage length and its aft end closure shape. Separation contributes to drag increase. In addition, if there is any lateral instability then the aircraft may enter into certain kinds of unsteady harmonic oscillations. The aerodynamic quest is to shape the fuselage to force separation as far aft as possible, if required, use vortex generators such as those in wing or ventral fins to reduce if not eliminate these undesirable characteristics.

For high subsonic transport aircraft flying above Mach 0.75, the front-fuselage closure ratio, F_{cf} is kept around 2.0. At lower speed, it can get blunter, rarely goes below 1.5. Aft-end closure ratio, F_{ca} is kept between 2.5–3.75. To facilitate takeoff rotation, fuselage upsweep angle varies from 2 to 10° depending on the mission requirements/specifications. Table 5.1 lists the data for some existing designs that means the mission requirement can be compared to a particular aircraft in question.

5.10.2 Transonic Effects – Area Rule

For an aircraft configuration, it has been shown that the cross-sectional area distribution along the body axis affects the wave drag associated with transonic flow. The bulk of this area distribution along the aircraft axis comes from the fuselage and the wing. The best cross-sectional area distribution that minimises wave drag is a cigar-like smooth distribution (i.e. uniform contour curvature; lowest wave drag) known as the *Sears–Haack ideal body* (Figure 5.15). The fuselage shape approximates it; however, when the wing is attached, there is a

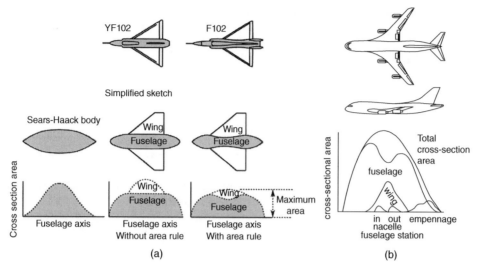

Figure 5.15 Area rule. (a) Transonic area rule. (b) Boeing transonic aircraft.

sudden jump in volume distribution (Figure 5.15). In the late 1950s, Whitcomb demonstrated through experiments that 'waisting' of the fuselage in a 'coke-bottle' shape could accommodate wing volume, as shown in the last part of Figure 5.15b. This type of procedure for wing-body shaping is known as the *area rule*. A smoother distribution of the cross-sectional area reduces wave drag.

Whitcomb's finding was deployed on F102 Delta Dragger fighter aircraft (Figure 5.15a). The modified version with area ruling showed considerably reduced transonic drag. For current designs with wing-body blending, it is less visible, but designers still study the volume distribution to make it as smooth as possible. Even the hump of a Boeing747 flying close to transonic speed helps with the area ruling. The following subsection considers wing (i.e. 3D body) aerodynamics.

In addition, the latest generation combat aircraft have leading edge root extension (LERX) (strakes), mostly above the engine intakes, where delineation between wings and body becomes difficult; normally taken at wing-fuselage assembly joint line. These designs require developing fuselage mouldlines hugging the distributed internal contents, satisfying the CG location for a full flight envelope. Fuselage cross-sections vary for the different types of combat aircraft configuration evolved by incorporating advanced technologies and mission requirement to encounter the potential adversaries with capabilities that can only be guessed. A BWB configuration like the B1 Spirit bomber aircraft design is seen as a flying wing, the fuselage merged and serves as part of the wing that also considers smooth changes in aircraft sectional area along its length.

5.10.3 Supersonic Effects

As a result of the presence of a shock wave, the flow physics of supersonic flow differs from subsonic flow. Nose cone of fuselage spearheads through air and develops a shock cone. Unlike thin aerofoils, the fuselage has conical angle of 2β (see Section 3.1) to displace air flow to produce a shock wave with and angle of shock $\beta > \mu$.

5.11 Fuselage Pitching Moment

Fuselage pitching moment characterises contribute to the horizontal tail sizing (Chapter 6). Spindle-shaped fuselage bodies are inherently unstable in the pitch plane.

Researchers and academics have proposed various semi-empirical relations, extracted from test data and backed-up by theories, to obtain pitching moment coefficients, C_M. The results can only give approximate values. Some of them are included in DATCOM, which progressed further to be unified and refined in an extensive manner for industry usage, serving many generations. DATCOM also shows the extent of discrepancies that can arise by comparing with test results. This evolved from Munk's original research [3] on dirigible like circular cross-section bodies in inviscid flow. Such bodies do not produce lift, only pure moment. Therefore, the centre of pressure is considered in the limiting sense infinitely ahead of the body. Weight of the body is not an aerodynamic force, hence the position of CG has no role in developing an aerodynamic moment. He stated that the rate of change of moment developed with change in angle of attack is a function of body volume and the dynamic head 'q' ($0.5\rho V^2$) of the relative airflow over it as given next.

$$dM/d\alpha = f(\text{body volume}, q)$$

Later, Multhopp [4] refined with the generalisation to apply to non-circular cross-section spindle as in fuselage bodies. In real viscous flow fuselage lift generation is insignificantly small, hence it is neglected and hence the role of CG does not enter to have a lift contributed moment. Perkins and Hege [2] give the explanation of Multhopp's method. Nelson [5] and Pamadi [6] present DATCOM method [7] examples that can be used in academic courses.

Although the authors consider the DATCOM methods are complex and should be kept to minimum usage in undergraduate courses, two examples of the DATCOM method are presented in this section and in Section 6.7.3.

Fuselage moment at zero incidence, C_{Mf_0}

Equation (5.6) gives the algebraic from of the relation to estimate fuselage moment coefficient, C_{mf_0} (moment at zero angle of attack), for engineering computation.

$$C_{Mf\_0} = \frac{(k_2 - k_1)}{36.5 S_W \times MAC} \sum_0^{l_f} w_f^2 \left(a_{0\_w} + i_f\right) \Delta x \tag{5.6}$$

where

S_W is the wing reference area.
MAC is the Mean Aerodynamic Chord of the wing
Δx is the segment length
$(k_2 - k_1)$ is the correction factor for fuselage fineness ratio – use Figure 5.16a.
w_f is the average width of the fuselage cross-sections
a_{0_w} wing zero lift line relative to fuselage reference line.

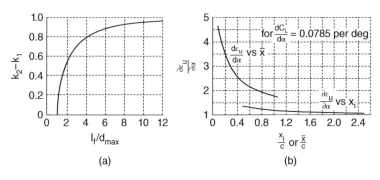

(a)   (b)

Figure 5.16 Fuselage moments factors. (a) Fuselage fineness correction factor. (b) Variation.

i_f is the incidence of the fuselage camber line relative to fuselage reference line at the centre of the corresponding Δx_i. Nose droop and aft-sweep are each positive values.

Figure 5.16a gives the values of $(k_2 - k_1)$ and is reproduced from [7] (it is the same as given in NACA TM 1036, figure 2).

Fuselage moment variation with angle of attack, CMf_α

Equation (5.7) shows the variation with angle of attack, α, given algebraic form for computational ease as shown in the worked-out example that follows. The fuselage is inherently unstable, the front-fuselage contributes much to the instability.

$$\text{Fuselage moment curve slope, } C_{\text{Mf}\_\alpha} = \frac{1}{36.5 S_W \times MAC} \sum_0^{l_f} w_f^2 \frac{\partial \varepsilon_u}{\partial \alpha} \Delta x \qquad (5.7)$$

where $\frac{\partial \varepsilon_u}{\partial \alpha}$ is explained next. Other symbols are as before.

Airflow ahead of the wing has upwash and downstream aft of the wing has downwash (Figure 5.17, Example 5.1). Local angle of attack, α, changes along the flow for the fuselage length. Therefore, each segment in the plan view has a different α. The term $\frac{\partial \varepsilon_u}{\partial \alpha}$ is the rate of upwash/downwash varying with α change, on account of contribution by the wing and the width of the fuselage contributes. The shaded fuselage section (darker shaded area by width, C) within the wing in the plan view does not contribute to C_{maf}.

Front and with upwash deflected flow aft fuselage with downwash deflected flow $\frac{\partial \varepsilon_u}{\partial \alpha}$ values are computed differently.

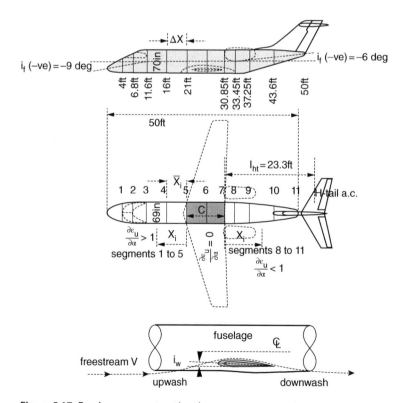

Figure 5.17 Fuselage moments estimation.

For a considerably higher local α, the upwash ahead of wing contributes to the C_{maf} more with $\frac{\partial \varepsilon_u}{\partial \alpha} > 1$ than at the downstream, as can be seen directly from Figure 5.16b, Only one section next to the wing uses the top graph given in Figure 5.16b, all other sections ahead of it use the bottom graph: the strength attenuates to lowest value at the nose cone.

Figure 5.16b for $\frac{\partial \varepsilon_u}{\partial \alpha}$ is taken from [7] (slightly modified). The top graph is used for one section just ahead for upwash values and all other sections use the bottom graph. These graphs are based on a wing $AR = 8$ with $C_{L\alpha\_AR=8} = 4.5$/rad (0.07 853/degree). For any other wing it has to adjusted by its $C_{L\alpha}$ as follows.

$$\frac{\partial \varepsilon_u}{\partial \alpha} = (C_{L\alpha})/0.07\,853) \times \left[\frac{\partial \varepsilon_u}{\partial \alpha} \text{ as computed} \right] - \text{see Example 5.1.}$$

Downwash flow deflection gradually gets reduced as the downwash moves away from the wing to a value of $2w$ (not the downwash 'w' at the wing a.c.). Downwash deflection at the H-tail $= 2w$ and the deflected angle ε_{def} (not the same ε to get the α_{eff} but twice more).

Using Eq. (4.19), $\varepsilon_{def} = 2C_L/(\pi AR)$.

$$\text{Therefore, } \left(\frac{\partial \varepsilon_{def}}{\partial \alpha} \right) = 2C_{L\alpha}/(\pi AR) \tag{5.8}$$

$\frac{\partial \varepsilon_d}{\partial \alpha}$ estimation for downwash flow deflection does not use the graphs. The following relation is used

$$\left(\frac{\partial \varepsilon_d}{\partial \alpha} \right) = \frac{x_i}{l_h} \left(1 - \frac{\partial \varepsilon_{def}}{\partial \alpha} \right) \tag{5.9}$$

Together with Eqs. (5.6) and (5.7) the moment characteristics are obtained.

$$\text{Fuselage moment coefficient } C_{Mf} = C_{Mf\_0} + C_{Mf\_\alpha} \tag{5.10}$$

Example 5.1, given in a step-by-step manner, expands the procedure of computation in better details].

Example 5.1

The wing-body configuration of a Bizjet type aircraft is shown in Figure 5.17 and has the following data. (This problem is in line with one of the two methods shown in Section 4.2.2.1 of DATCOM [7]. This is a relatively simple example and shows the extent of work involved. For no test data, the value of $(C_{L\alpha})_{WB} = 0.05$ is assumed, close to the value used in DATCOM sample problem of a smaller aircraft.)

Aerofoil: NACA 65-410

Aerofoil zero lift angle, $\alpha_{0L} = -2°$ Wing

Wing area, $S_W = 323$ ft^2	Fuselage length $= 50$ ft
$AR = 7.5$ and $MAC = 7$ ft	Fuselage height $= 70$ in. (5.83 ft)
Wing $dC_L/d\alpha = 0.0848$/degree (from Problem 4.5)	Fuselage width $= 69$ in. (5.75 ft)
Length, $l_{ht} = 23.3$ ft	Fuselage fineness ratio $= 8.6$
Wing incidence with respect to fuselage reference line, $i_w = 2°$	
Wing zero lift angle relative to fuselage reference line, $\alpha_{0w} = 0°$	

Step 1

Typical fuselages have a drooped nose and aft end upsweep. Draw the fuselage camber line joining the mid-points of the fuselage elevation view. Draw the fuselage reference line. This is convenient in this exercise as the straight line goes through the mid-points of the mid-fuselage. This gives the reference line coinciding with part of the mid-camber line.

Step 2

Divide both the elevation (to evaluate C_{Mf_0}) and plan (to evaluate C_{Mf_α}) views into segments, the higher the better. It is a good idea to take at least five segments in front of the wing and segments at back of the wing as shown in the figure. The shaded fuselage section within the wing in the plan view does not contribute to C_{Mf_α}, therefore the distance x_i is the mid-point of the section measured from the wing edges. The sections in the two views need not be identical; in this example they are identical.

Step 3

Measure the angle 'i_f' at both the ends of the fuselage camber line. Nose droop aft end upsweep are negative values here.

Step 4 – To compute C_{Mf_0}.

Tabulate the algebraic equations given in Eq. (5.6) as computed here.

Fuselage pitching moment at zero incidence, C_{Mf_0}

Station	Δx-ft	w_f – ft	$a_{0\_w} + i_f$	$w_f{}^2 \times (a_{0\_w} + i_f) \times \Delta x$
1	4.0	2.46	9.0	−217.86
2	2.8	4.3	9.0	−465.95
3	4.8	5.7	0	0
4	4.4	5.83	0	0
5	5.0	5.83	0	0
6	5.0	5.83	0	0
7	4.85	5.83	0	0
8	2.6	5.83	0	0
9	3.8	5.76	6.0	−756.45
10	6.35	4.5	6.0	−771.53
11	6.4	2.54	6.0	−247.74
Total				−2459.52

Substituting the values in the above table in Eq. (5.6), the following is obtained. Figure 5.16a gives the correction factor $(k_2 - k_1) = 0.86$.

$$C_{Mf\_0} = \frac{0.86}{36.5 \times 323 \times 7}(-2459.52) = -0.0256$$

Step 5 – To compute C_{Mf_α}.

Since the flow field ahead of wing have upwash airflow deflection and downwash deflection aft of wing, both sides in varying strength along the fuselage, the computational work is dealt with separately. Note that the downwash stabilised deflection aft of wing around H-tail is $2w$ and not the downwash 'w' at the *a.c.* defined in Section 4.7.1. The fuselage length block by the wing is 9.85 ft.

Front-fuselage ahead of wing

Upwash deflection ahead of wing contributes a great deal to the C_{Mf_α}. The graphs in Figure 5.16b are obtained. These graphs are based on wing $AR = 8$ with wing-body lift curve slope of $(C_{L\alpha\_AR = 8})_{WB} = 4.5/\text{rad}$ (0.07853/degree). The value of $(C_{L\alpha})_{WB\_current} = 0.05/\text{degree}$ for the current configuration is assumed. A correction factor of $((C_{L\alpha})_{WB\_current})/[(C_{L\alpha\_AR = 8})_{WB}] = 0.05/0.07853 = 0.59$.

Fuselage length covered by wing, $C = 9.85$ ft

Use the top graph of Figure 5.16b for Section 5 and Sections 1–4 the bottom of Figure 5.16b. Apply the correction factor of 0.59 in the last column.

Station	Δx-ft	$\bar{x}/C$	x_i – ft	x_i/C	w_f – ft	$\left(1+\frac{\partial \varepsilon_u}{\partial \alpha}\right)$	$w_f^2 \times \frac{\partial \varepsilon_u}{\partial \alpha} \times \Delta x$	Corrected
1	4.0		19	1.93	3.0	1.03	37.8	22.3
2	2.8		15.6	1.58	4.72	1.03	65.36	38.56
3	4.8		11.8	1.2	5.23	1.12	145.0	8.54
4	4.4		7.2	0.73	5.75	1.25	181.84	107.28
5	5.0	0.51				2.3	380.22	224.32
6	5.0			0				0
7	4.85			0				0
	Total							478.0 ft^3

Aft fuselage behind the wing

Length from aft fuselage to H-tail ¼MAC, l_h = 19.73 ft

Downwash deflection at the H-tail = $2w$ and the deflected angle ε_{def} (not the same ε to get the α_{eff} but twice more). Using Eq. (5.8), $\varepsilon_{def} = 2C_L/(\pi AR)$.

Therefore, $\left(\frac{\partial \varepsilon_{def}}{\partial \alpha}\right) = 2C_{L\alpha}/(\pi AR) = (2 \times 57.3 \times 0.0848)/(3.14 \times 7.5) = 0.412$/rad.

Use Eq. (5.9) to compute $\left(\frac{\partial \varepsilon_d}{\partial \alpha}\right) = \frac{x_i}{l_h}\left(1 - \frac{\partial \varepsilon_{def}}{\partial \alpha}\right)$

Station	Δx-ft	x_i – ft	x_i/C	w_f – ft	x_i/l_h	$\left(\frac{\partial \varepsilon_d}{\partial \alpha}\right)$	$w_f^2 \times \frac{\partial \varepsilon_u}{\partial \alpha} \times \Delta x$
8	2.6	1.3	0.132	5.75	0.066	0.0388	3.66
9	3.8	4.2	0.467	5.23	0.228	0.134	15.34
10	6.35	9.575	0.972	4.35	0.485	0.286	37.72
11	6.4	15.95	1.62	2.84	0.808	0.474	27.0
Total							76.22 ft^3

Adding front and aft fuselage, $w_f^2 \times \frac{\partial \varepsilon_u}{\partial \alpha} \times \underline{\Delta x} = (478 + 76.22) = 554.24$ ft^3.

Substituting the values in Eq. 5.7, the following is obtained.

$C_{Mf\_\alpha} = \frac{554.24}{36.5 \times 323 \times 7} = 0.0067$ per degree = 0.384 per rad (AR correction is applied).

Using Eq. (5.10), fuselage moment coefficient $C_{Mf} = -0.0256 + 0.0067\alpha$ (where α is in degrees).

The result is not verified but is representative.

DATCOM methods worked well for many generations, even when the results were approximate they were close enough to work with. But today, once the pre-processing in CFD is set-up, results can be obtained in minutes with a higher degree of accuracy.

5.12   Nacelle Pod – Civil Aircraft

Nacelles are the structural housing for aircraft engine. The first commercial transport aircraft was the De Havilland Comet with four engines buried into the wing (Figure 5.18a). The configuration was found out to have high intake drag. The proven pod-mounted nacelle, was there and the configuration with improved aerodynamic and structural features became the standard design (Figure 5.18b). Today, all multi-engine civil aircraft nacelles are invariably externally pod-mounted, either slung under or mounted over the wing or attached to the aft fuselage. The front part of the nacelle is the intake and the aft end is the nozzle for the hot

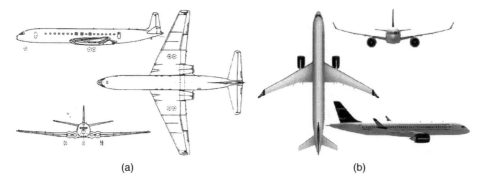

(a)   (b)

Figure 5.18 Aircraft nacelle. (a) De Havilland Comet. (b) Bombardier CS-100. Credit Bombardier.

engine exhaust flow, which should not impinge on any aircraft surface (hence fuselage mounted engines are positioned at the aft end).

In a way, the external geometry of civil aircraft nacelle pod is a body of revolution with a near circular cross-section varying along its length. Equation (5.1) is valid to obtain nacelle average diameter for a cross-section. A pod casing may be seen as wrapped wing around an engine with aerofoil shaped casing sections facilitating a large diameter intake with short-duct length for the air-breathing engines. The crown-cut section is thinner than the keel-cut section as can be seen in Figure 12.5. The keel-cut section is thicker to house accessories and its fuller lip contour helps to avoid separation at high angle of attack. In principle, it is desirable to have circular cross-sectional areas for the intake throat area, but it may not always be possible, say for ground clearance. It may be noted that the Boeing737 nacelle has a flat keel line to gain some ground clearance. In this book, the intake areas would be considered circular.

Propeller driven aircraft generally operates at low subsonic speed. For the same power output, the air-mass flow demand for turboprops is considerably lower as compared to turbofans. Their nacelle pod serves the same function to house the power plant but design consideration differs. They invariably have pilot intake but the internal airflow ducts are short and may have bends, still acting moderately as diffusers (Figure 5.19).

For subsonic turbofans, the intake acts as a diffuser with an acoustic lining to abate noise generation. The inhaled air-mass flow demanded by an engine varies considerably: Intake design must also cater for flight at high incidence and yaw attitudes for the full flight envelope so that the flow distortion, separation and turbulence within the duct are at an acceptable level for engine to operate without compressor blade stall or engine flame out.

In principle, the external contour lines of good nacelle designs are not necessarily symmetrical to its vertical plane. But to keep cost down by maintaining commonality, some nacelle designs are made symmetrical to the vertical plane. This would allow manufacturing jigs to produce interchangeable nacelles between port and starboard sides and be able to minimise the essential difference at the finishing end.

5.12.1 Nacelle/Intake External Flow Aerodynamic Considerations

Nacelle pod is also like the fuselage 3D bodies of revolution as it has air flowing over it, but it differs as it also has internal airflow flowing through its intake duct facilitating air-breathing into the engine. Nacelle aerodynamics need to be considered for both the external and internal flow characteristics. This section gives a brief introductory description of nacelle external aerodynamic considerations. Nacelle external flow aerodynamics have similar considerations as those discussed in Section 5.10.

In addition to housing the engine, the main purpose of the nacelle is to facilitate the internal airflow reaching the engine face (or the fan of gas turbines) with minimum distortion over a wide range of aircraft speeds and attitudes. Aircraft designers need to position the nacelle at an orientation to receive the incoming free stream tube into the intake duct with minimum distortion for the full flight envelope (see Section 5.8.1).

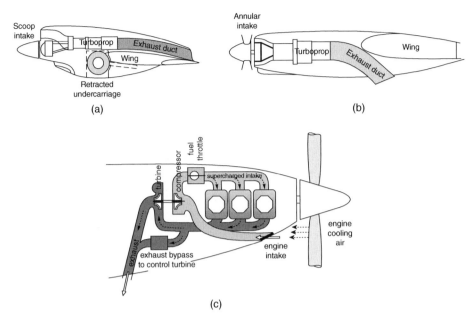

Figure 5.19 Sketch of a propeller driven engine nacelle pods. (a) Scoop intake TP nacelle. (b) annular intake TP nacelle. (c) Piston engine nacelle.

The nacelle aerodynamic design (external mouldline shaping and internal contouring) has progressed to a point of diminishing return on the efforts made and approaching to a generic shape. The characteristics length of nacelle is its length, l_{nac}. It is long enough to have $Re > R_{crit}$, hence separation can occur at the aft end of nacelle, its severity depends on its aft end closure shape. Separation contributes to drag increase. If required, vortex generators such as a few vanes are positioned at the critical positions. The entrainment effects of high energy engine exhaust flow helps to minimise separation of external flow over the aft end of fuselage. Nacelle pods are isolated bodies attached to aircraft with a pylon and have much lower length to diameter ratio (FR) compared to the fuselage.

Military aircraft with engines buried into the fuselage do not have nacelle pods; here, airflow to engine passes through a long intake duct. These are dealt with in better detail in Chapter 12.

5.12.2   Positioning Aircraft Nacelle Pod

In the vicinity of wing and/or fuselage, airflow past nacelle pod will be interfered with and that will give rise to interference drag. Positioning the nacelle pod with respect to the aircraft is an important design consideration.

The best position to place a nacelle pod is on the wing. The engine weight relieves some of the wing bending moment developed from wing lift force in flight. This also favours the aircraft CG to stay forward. However, low-wing smaller aircraft do not have enough ground clearance for under-slung nacelles, therefore they have to be mounted on the fuselage, always at the aft of it. Interestingly, Hondajet has defied the convention and positioned the nacelles over the wing. The authors believe that lessons from Hondajet's operational experiences will see more aircraft designs with over wing-nacelle configuration. The considerations for a wing-mounted nacelle (e.g. A320) and fuselage mounted nacelle (e.g. B727) differ as described in the following sub-sections.

To minimise yaw moment at one-engine inoperative situations, the engine thrust line should be as near to fuselage centre line as possible but at the same time not too close to cause high interference drag. As a compromise, typically, wing-mounted nacelle pods are placed at about twice the maximum nacelle diameter $\approx [\pm (2 \times D_{nac\_max})]$ away from fuselage and aft-fuselage nacelles placed at $\approx [\pm (1.5 \times D_{nac\_max})]$.

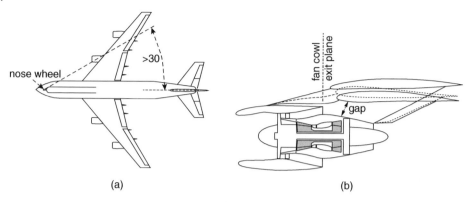

(a)   (b)

Figure 5.20 Typical nacelle positions. (a) Nose wheel spray. (b) Position of nacelle with respect to the wing [8].

The closest an inboard engine should be kept to is at least 30° away from the nose wheel spray angle as shown in Figure 5.20 (B747 is a little widely spaced). It is found that the most inboard engines placed 30° from the nose wheel give the wing-mounted nacelle position $>[(2 \times D_{nac\_max})]$.

Aerodynamic, structural and CG locations influence the position of the nacelle with respect to the aircraft. This section discusses location of nacelle positions based on past experiences; those that are extracted from existing designs and those that are used during the conceptual design phase. Subsequently, in the next project design phase, the configuration is refined through CFD analyses.

5.12.3   Wing-Mounted Nacelle Position

Structurally, outboard nacelle locations are desirable to reduce wing bending moments in flight but flutter requirements are complex and may show more inboard locations to be more favourable. The latter also favours directional control after engine failure. Finally, the lateral position of the engines affects ground clearance, an issue of special importance for large, four-engine aircraft. Given next are some guidelines that may be used in classroom exercises. The following should be considered when positioning nacelles.

5.12.3.1   Summary – Lateral Nacelle Position
Table 5.5 Wing-mounted lateral nacelle position

1. To keep nacelle positioned $>[\pm(2 \times D_{nac\_max})]$.
2. To stay clear of nose wheel spray during takeoff (Figure 5.20a).
3. To keep exhaust flow not to impinge on flaps.
4. To keep engine at a position to satisfy CG position with respect to the aircraft.

As a guide, aircraft designers will have to make the best compromises for where to position engine on wing. In the classroom, the following may be adopted to position wing-mounted nacelles.

two engines 0.3–0.32 of semi-wing span from the aircraft centre line.
three engines same as above – the third one at aircraft centre line.
four engines inboard at 0.29–0.32 and outboard at 0.62–0.66 of the wing semi-span.

Ideally, nacelles at either side should be identical to keep product commonality as a measure of cost reduction. However, they are not identical. A mirror image of slightly banana-shaped nacelles gives the best flow alignment on account of being affected by the presence of the fuselage and wing.

5.12.3.2   Vertical Position
Typically, wing-mounted nacelles are under slung below the wing supported by a pylon. The dominating design consideration is to minimise wing-nacelle interference drag. All nacelles are hung well over (Figure 5.20b)

and ahead of the wing to keep interference drag low, almost to zero. Most high bypass ratio turbofans have short-duct nacelles (Section 12.7.1), the cold secondary fan exhaust flow ejects over the primary duct case. The nacelle pod should be forward enough with a short fan cowl exit plane that stays ahead of the wing leading edge. A typical gap between the nacelle and wing may be taken as ≈15% of the local wing chord length.

The current design practice is to keep the nacelle position as far ahead of the wing with an orientation to capture the three stream tube diffused through a short intake duct length, normal to the fan face. A forward nacelle position favours adjusting the aircraft CG position in the conceptual design stage. There is no quick answer for the degree of incidence (Figure 5.20b), which is design specific and varies for the type of installation. It depends on the engine position with respect to the wing, for example, how much inboard on the wing, the flexure of the wing during flight and so on. However, the considerations for fuselage mounted nacelles, as given next, offer some idea of wing-mounted nacelle positions. Post conceptual design studies using CFD, wind tunnel and flight tests would fine-tune nacelle geometry and positional geometry to production standards.

Three typical relative positions of the nacelle with respect to wing are shown in Figure 5.21 [1] in detail: the top wing represents a B747, the middle represents an A300 and the bottom wing represents a DC10.

Other considerations are to keep adequate clearance for 15° roll during takeoff and landing (Figure 5.23).

5.12.3.3   Summary – Vertical Nacelle Position

1. Short fan cowl exit plane stays ahead of the wing leading edge.
2. Take the gap between nacelle and the wing as ≈15% of the local wing chord length.
3. Keep adequate clearance for 15° roll during takeoff and landing.

5.12.3.4   Over-Wing Nacelle

One of the earliest over-wing podded jet engine nacelles was the experimental VWF-Fokker 614 and now is the Hondajet. Over-wing nacelle design (Figure 5.21) has not caught up yet. The authors believe that over-wing nacelle design shows good potential. During the 1970s, the BoeingYC-14 experimental STOL aircraft was built. Later, the Antanov 72 and 74, using a similar design concept, were produced.

5.12.4   Fuselage-Mounted Nacelle Position

The wing-mounted nacelle position, as discussed in Section 5.12.3, also offers some ideas for considerations of fuselage mounted nacelle positions (Figures 5.22). Fuselage mounted nacelle contours have similarity in design, but because there are no constraints of the wing, the interference drag is lower and can be brought closer to the fuselage, thus clearing the thicker boundary growth at the aft fuselage. A gap of roughly at least half the nacelle diameter may be left between fuselage and the nacelle. The vertical position is where it offers the least pitching moment developed by engine thrust about the CG for the whole aircraft, taking into account of contribution by the H-tail. The position can be anywhere close to the fuselage centreline to high up on fuselage.

With an aft engine installations, the nacelles must be placed to be free of interference from wing wakes. The DC-9 was investigated thoroughly for wing and spoiler wakes and the effects of yaw angles, which might cause the fuselage boundary layer to be ingested. Here, efficiency is not the concern because little flight time is spent yawed, with spoilers deflected or at a high angle of attack. However, the engine cannot tolerate excessive fan face flow distortion.

An aft fuselage mounted nacelle has many special problems. The pylons should be as short as possible to minimise drag but long enough to avoid aerodynamic interference between fuselage, pylon and nacelle. To minimise this interference without excessive pylon length, the nacelle cowl should be designed to minimise local velocities on the inboard size of the nacelle. On a DC-9 a wind tunnel study compared cambered and symmetrical, long and short cowls and found the short banana shaped cambered cowl to be best and lightest in weight. The nacelles are cambered in both the plan and elevation views to compensate for the angle of attack at the nacelle (Figure 5.22).

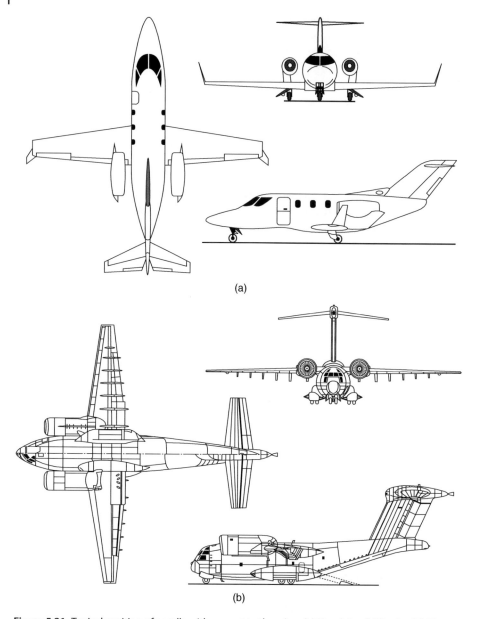

(a)

(b)

Figure 5.21 Typical position of nacelle with respect to the wing. (a) Hondajet. (b) BoeingYC-12.

5.12.4.1 Summary – Fuselage Mounted Nacelle Position

Aircraft designers will have to make the best compromises for where to position an engine on a wing. For a classroom exercise, the following may be considered when positioning a nacelle on a fuselage.

1. To stay clear of wing wake.
2. To keep nacelle positioned $>[\pm(1.5 \times D_{\text{nac\_max}})]$, clear of fuselage boundary layer ingestion.
3. To ensure exhaust flow does not interfere with empennage.
4. To keep the engine forward enough to satisfy CG position with respect to the aircraft.
5. Attitude with respect to the fuselage centre line may have a 3° offset as shown in Figure 5.22.

Figure 5.22 Fuselage mounted nacelle position.

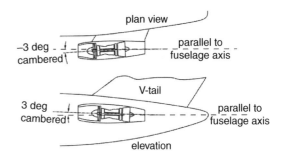

5.12.5 Trijet Centre Engine

A trijet aircraft configuration has its odd engine at the centreline placed at the aft of the aircraft fuselage (Figure 5.23). The dominant choices are to have an S-duct (B727) with the engine at the tail cone of the fuselage or straight through a duct (DC10) clear above the aft fuselage with a V-tail on its top. This large nacelle pod over the fuselage also serves as part of a fin stabiliser for the aircraft.

The S-duct configuration is lighter in construction but has lower recovery factor (RF – Section 13.2.1) with intake duct loss. The straight duct configuration is heavier but has a high RF in the order of a conventional nacelle pod. A tradeoff study (DOC comparison) is required to make the choice.

The S-bend has a lower engine location and uses the engine exhaust to replace part of the fuselage boat-tail (saves drag). It has more inlet loss, a distortion risk and higher inlet drag from the S-duct bend. The straight-through inlet with the engine mounted on the fin has an ideal aerodynamic inlet free of distortion, but does have a small inlet loss due to the length of the inlet and an increase in fin structural weight to support the engine.

5.12.6 Some Structural Considerations

Structurally, outboard nacelle locations are desirable to reduce wing bending moments in flight but flutter requirements are complex and may show more inboard locations to be more favourable. The latter also favours directional control after engine failure. Finally, the lateral position of the engines affects ground clearance, an issue of special importance for large, four-engine aircraft.

Nacelle pod clearance from ground must ensure avoidance of debris ingestion. Also, crashworthiness of the pod in case of nose wheel collapse has to be catered for as shown in Figure 5.24.

Although rare, high heavy turbine discs at high rpm have disintegrated under high stress of centrifugal force operating at elevated temperature. The fragments have enough energy to cut through the fuselage shell causing catastrophic accidents. To contain fragments, the inner cowls around the rotor are reinforced by Kevlar, a material is also used as tank armour plates.

Another influence of wing-mounted nacelles is the effect on flaps. The high temperature, high 'q' exhaust impinging on the flap increases flap loads and weight and may require a titanium structure (more expensive).

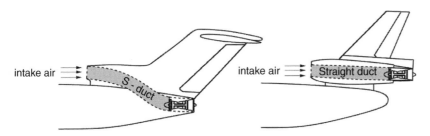

Figure 5.23 Trijet centre engine.

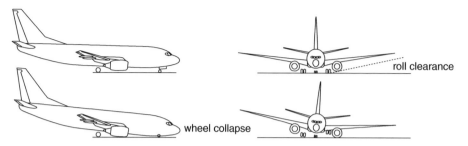

Figure 5.24 Nacelle pod clearance from ground. Source: Reproduced with permission from Cambridge University Press.

The impingement also increases drag, a significant factor in takeoff climb performance after engine failure. Eliminating the flap behind the engine reduces C_{Lmax}.

In principle, the external contour lines of good nacelle designs are not necessarily symmetrical to its vertical plane. But to keep cost down by maintaining commonality, most nacelle designs are made symmetrical to the vertical plane. This would allow manufacturing jigs to produce interchangeable nacelles between port and starboard sides and be able to minimise the essential difference at the finishing end. The nacelle aerodynamic design (external mould-line shaping and internal contouring) has progressed to a point of diminishing return on the efforts made and approaching to a generic shape.

Post-conceptual design studies using CFD, wind tunnel and flight tests would fine-tune the nacelle geometry and its positional geometry to production standards.

5.13 Exhaust Nozzles – Civil Aircraft

Civil aircraft nozzles are conical in which a thrust reverser (TR) is integrated. Small turbofan aircraft may not need a TR but aircraft of the RJ size and above use a TR. Inclusion of a TR may slightly elongate nozzle length – this will be ignored in this book.

In general, nozzle exit area is sized as a perfectly expanded nozzle ($p_e = p_\infty$) at long range cruise (LRC) conditions. At higher engine ratings, $p_e > p_\infty$. The exit nozzle of a long duct turbofan does not run choked at cruise ratings. At takeoff ratings, the back pressure is high at lower altitude, therefore a long duct turbofan could escape from running choked (low pressure secondary flow mixes within the exhaust duct. The exhaust nozzle runs in a favourable pressure gradient, hence its shaping is relatively simple to establish geometrical dimensions. However, it is not a simple engineering task to suppress noise level and withstand elevated temperature.

Nozzle exit plane is at the end of engine. Its length from the turbine exit plane is about 0.8–1.5 of the fan face diameter. Nozzle exit area diameter may be taken coarsely as half to three-quarters of intake throat diameter in this study.

Each possibility entails compromises of weight, inlet loss, inlet distortion, drag, reverser effectiveness and maintenance accessibility.

5.13.1 Civil Aircraft Thrust Reverser (TR)

TRs are not required by the regulatory authorities (Federal Aviation Administration, FAA/Civil Aviation Authority, CAA). They are expensive components, heavy and only applied on ground, yet their impact on aircraft operation is significant on account of having additional safety through better control, reduced time to stop and so on, especially at aborted takeoffs and other related emergencies. Airlines want to have TRs even at the cost of increased DOC.

Broadly speaking, there are two types of TRs: (i) those operating on a mixed fan and core flow and (ii) those operating on fan flow only. Their choice depends on BPR, nacelle location and customer specification.

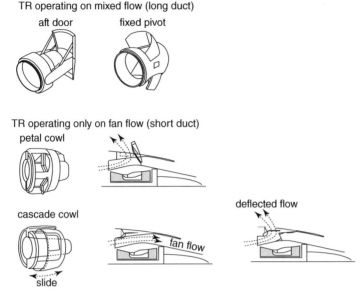

Figure 5.25 Type of thrust reversers.

The first type operating on the total flow (both fan and core) is shown in the top of Figure 5.25. There are two types: (i) sliding port aft door type when the doors slides to the aft end as it opens up to deflect the exhaust flow and (ii) fixed pivot type when the doors rotates to position to deflect the exhaust flow.

The second type of TR operates on the fan flow only. There are two types: (i) the petal cowl type – its mechanism is the middle of Figure 5.25 – and (ii) the cascade cowl type shown at the bottom of Figure 5.25. There are two cascade types: the conventional type and the natural blockage type. Bombardier CRJ700/900 aircraft use their own patented cascade TR of the natural blockage type. The external cowls translate back blocking the fan flow when it escapes through fixed cascades that reverse the flow. This is an attractive design with low parts count, scalable, easier to maintain and offers relatively higher retarding energy. The petal type operating on fan flow is suitable to short-duct nacelles as shown in the figure. The petal doors open on a hinge to block the secondary flow of the fan when it deflects to develop reverse thrust.

TRs are applied below 150 kts and are retracted back at around 50 kts when wheel brakes are effective. The choice for the type would depend on designer's comprise from the available technology at his/her disposal. Aircraft designers must ensure TR efflux is well controlled – there should be no adverse impingement of on aircraft surface nor there be re-ingestion in the engine. Figure 5.26 gives a typical satisfactory TR efflux pattern.

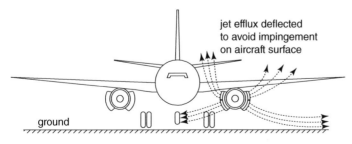

Figure 5.26 TR efflux pattern (from [9]).

MILITARY AIRCRAFT

5.14 Fuselage Geometry – Military Aircraft

Unlike the transport aircraft fuselage geometries, military fuselages have varied cross-sections as shown in Figure 5.27. The military geometries are generated based on mission depended design specifications stemming from the technology level adopted. Military mission roles are listed in Section 5.2. Choices of military aircraft fuselages are also given subsequently in Chapter 7. The following geometrical definitions of civil aircraft design are extensively used in this book.

A combat aircraft with a blunt engine exit plane does not have a pointed aft fuselage closure. A single engine configuration may have boat-tail like aft end (Figure 5.4a) closure but twin side-by-side engine configurations do not require boat tailing (Figure 5.4b). With a shorter wheel base and longer wheel strut there is little or no fuselage upsweep for the takeoff rotation clearance.

The mission requirements for combat aircraft are different from civil aircraft requirements (Section 1.9.1). In a way, there is no payload except to carry ammunitions for internal mounted gun as consumables. The avionics pods, drop fuel tanks and expendable armaments are carried externally (except for bombers). Combat aircraft have their engine installed at the aft end of fuselage, facilitating the hot exhaust to flow out. These factors act as driver to have combat aircraft with varied cross-sections along the fuselage as shown Figure 5.27.

5.14.1 Fuselage Axis/Zero-Reference Plane – Military Aircraft

Fuselage axis is a line parallel to the centreline of the constant cross-section part of the fuselage barrel. It typically passes through the farthest point of the nose cone, facilitating the start of reference planes normal to it. The fuselage axis may or may not be parallel to the ground. The principal inertia axis of the aircraft can be close to the fuselage axis. In general, the zero-reference plane is at the nose cone, but designers can choose any station for their convenience, within or outside of the fuselage. This book considers the fuselage zero-reference plane to be at the nose cone, as shown in Figure 5.27.

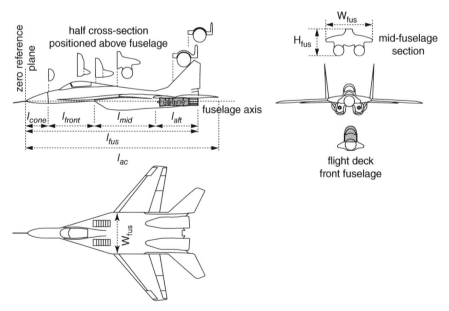

Figure 5.27 MIG29 type of combat aircraft (extracted from various sources – not to scale). Source: Reproduced with permission from Cambridge University Press.

5.14.2 Fuselage Length, l_{fus}

This is along the fuselage axis, measuring the length of the fuselage from the tip of the nose cone to engine exit plane. Fuselages are further sectioned at the joints of the fabricated subassembly sections built in their dedicated jigs. As the manufacturing philosophies may differ, the fuselage sections subassemblies differ from design to design. Figure 5.27 gives a typical example.

5.14.3 Fuselage Nose Cone, l_{cone}

The pointed nose cone is the front closure is the aerodynamic requirement for supersonic flight and also serves as the housing for the radar, which is mounted at the front of the fuselage front section, l_{front}, bulkhead.

5.14.4 Front-Fuselage Length, l_{front}

Following the nose cone is the front-fuselage sub-assembly with the length, l_{front}. This is the dedicated section to house the flight deck (crew station) for the crew, mostly single seated but can be dual (as required for the role), either in tandem or side-by-side arrangement. A canopy with streamlined windscreen mould lines covers pilot CFD.

5.14.5 Mid-Fuselage Length, l_{mid}

The mid-fuselage length, l_{mid}, of the subassembly follows the l_{front} up to the engine face carrying the intake ducts, cables, linkages and, depending on the design the fuel tanks, filling up the section cavity.

5.14.6 Aft-Fuselage Closure Length, l_{aft}

Aft-fuselage closure length, l_{aft}, is the last subassembly section following the l_{mid} housing the power plants. As mentioned earlier, combat aircraft with a blunt engine exhaust plane have no transport aircraft like pointed closure. Typically, this does not require an upsweep angle.

5.14.7 Fuselage Height, H_{fus}

Fuselage height, H_{fus}, is the maximum distance of the fuselage from its underside (not from the ground) to the top in the vertical plane (Figure 5.27). Combat aircraft fuselage cross-section varies along the fuselage length, hence each cross-section along the fuselage length has to be considered separately.

5.14.8 Fuselage Width, W_{fus}

Fuselage width, W_{fus}, is the widest part of the fuselage in the horizontal plane. Combat aircraft fuselage cross-section varies along the fuselage length, hence each cross-section along the fuselage length has to be considered separately.

5.14.9 Fuselage Cross-Sectional Area, A_{fus}

Combat aircraft cross-section area along fuselage stations is an important parameter to establish fuselage volume distribution required to estimate aircraft supersonic drag (Section 11.18). Being non-circular, the concept of average or equivalent cross-section diameter is not relevant. The term effective diameter, D_{eff}, as given in Eq. (5.2), exists for a non-circular cross-section as a reference value. It is not used in this book.

Because of complex and varying fuselage cross-sections along the length, there is difficulty in interpreting military aircraft FR. The definition of FR with average or equivalent diameter serves little purpose, except as a reference number.

5.15 Pilot Cockpit/Flight Deck – Military Aircraft

The 'crew station' for the operator, in most design a single pilot but in some designs there may be second crew as weapons operator to alleviate pilot work-load. The operator can have a side-by-side or tandem seating arrangement. This is the space allocated in the front-fuselage to accommodate pilot/crew. A canopy seals the space not only as protection but also facilitates environmental control. Current canopy design has moved away from raised flat-shield design to bubble canopy offering generous vision polar. The older practice of raised canopy are now applied to bomber like bigger aircraft.

A single-piece bird-proof polycarbonate bubble canopy give the pilot 360° all-round unobstructed vision, greatly improved vision over the side and to the rear. The upward vision is unrestricted, the side vision typically is 40° look-down and forward vision is typically 15° looking down over the nose cone.

Flight deck design is an integral part of fuselage layout as shown in Figure 5.28. A raised bubble canopy offers unrestricted view for the pilot in the upper hemisphere. The nose cone cavity houses the forward-looking radar and other black boxes.

The reclining angle pilot seat is also moved away from the rather upright older angle of around 12° to tilting backward at 30°. A more inclined position of pilot seating reduces the height of the carotid artery to the brain, giving an additional margin on high-*g* manoeuvres to avoid causing black out.

5.15.1 Flight Deck Height, H_{fd}

This is the internal cabin height from the flight deck floorboard to canopy top, as shown in Figure 5.28.

5.15.2 Flight Deck Width, W_{fd}

Flight deck width, W_{fd} (Figure 5.28), should be generous for accommodate the 99% percentile crew size. The W_{fd} should be more than 38 in. width. Fuselage width at the canopy interface should be adequate for ejection seat egress with its width close to 38 in. (trainer aircraft = 0.35–0.37 in.). Details of ejection process is given in Chapter 17.

5.15.3 Egress

See Section 22.8.1.

5.16 Engine Installation – Military Aircraft

Combat aircraft have engines integral with the fuselage buried inside. Therefore, pods do not feature unless there is a requirement for more than two engines on a large aircraft.

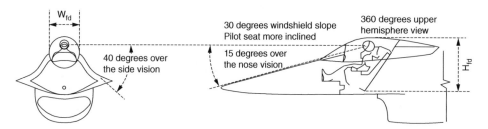

Figure 5.28 Flight deck (cockpit) layout – military aircraft. Source: Reproduced with permission from Cambridge University Press.

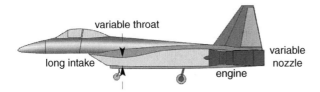

Figure 5.29 Installed engine in a combat aircraft.

Figure 5.29 shows a turbofan installed onto a supersonic combat aircraft. Power plants are placed at the aft end, hence they have a longer duct acting as a diffuser incurring higher flow energy loss. The external contour of the engine housing is integral to fuselage mouldlines. More details are given in Section 5.8.2; internal flow through the intake duct is discussed in Chapter 9. Military aircraft internal contours design of the intake and exit nozzle are the obligation of aircraft designers done in consultation with the engine manufacturer.

5.16.1   Military Aircraft Intake

While military aircraft do not have pods, their intake geometries have many configuration possibilities. Chart 5.1 gives the dominant types of military aircraft intake configurations: all have an engine buried in the fuselage.

Figures 5.30 and 5.31 give examples of the intake configurations given in the Chart 5.1. Earlier designs had an intake at the nose of aircraft for both subsonic (e.g. F86 with pitot intake) and supersonic (e.g. MIG21 with a movable centre body) operations. A centre body is required for aircraft speed capability above Mach 1.8, otherwise it can be pitot intake. The intake ducts were long and had bends to go below the pilot and then up again to the engine face, incurring flow energy loss resulting in low RF.

Later, with side intakes, bifurcated to both sides of fuselage considerably improved the RF by shortening the duct length. The loss from the bends on account of bifurcation of the intake duct is minimised through careful contouring, nowadays with CFD analyses. Other possibilities with a short intake duct is to have chin intake (e.g. F16) under the fuselage or intake over the fuselage (e.g. F107). Shorter intake ducts have the benefits of weight saving.

Intake design must also cater for flight at high incidence and yaw attitudes for the full flight envelope, so the flow distortion, separation, turbulence within the duct are at acceptable levels for engine to operate with compressor blade stall or engine flame outs.

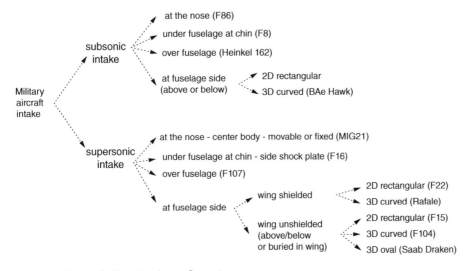

Chart 5.1 Types of military intake configuration.

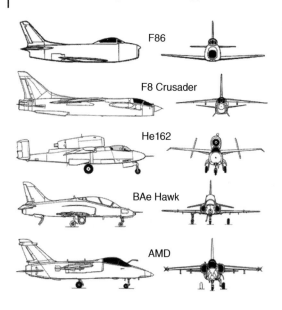

Figure 5.30 Subsonic combat aircraft intakes (various sources).

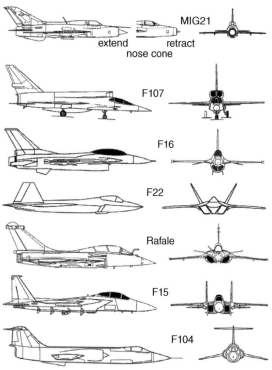

Figure 5.31 Supersonic combat aircraft intakes (various sources).

Unlike the aircraft front intake, the side intakes are subjected to a breath of air flowing past the front section of the fuselage with the consequence of having boundary layer adhered to the fuselage surface in a favourable pressure gradient of the increasing fuselage cross-section. Ingestion of this boundary layer through the intake will degrade engine performance. This problem is overcome by installing a splitter plate (Figure 5.32) acting as diverter to bleed the boundary layer to spill out from both sides the splitter/diverter plate. An adjustable ramp (F5) on splitter plate acts as shock diffuser. Front-fuselage mounted gun is to be positioned to avoid ingestion of hot gas efflux from firing rounds.

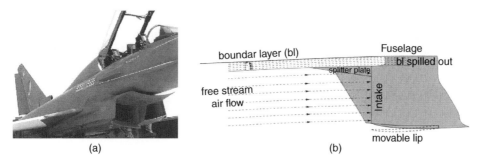

Figure 5.32 Splitter plate for boundary layer bleed (Eurofighter Typhoon). (a) Typhoon intake. (b) Splitter plate.

Figure 5.33 F16 modified with diverterless supersonic intake (DSI) for flight testing. (a) F16 Diverterless supersonic intake. (b) DSI side view. (Source: http://globalmilitaryreview.blogspot.co.uk/.)

Subsequently, further advancement have been achieved by using Diverterless Supersonic Inlet (DSI) is a type of jet engine air intake used by some modern combat aircraft to control air flow into their engines. It consists of a 'bump' (Figure 5.33) and a forward-swept inlet cowl, which work together to divert boundary layer airflow away from the aircraft's engine. This eliminates the need for a splitter plate, while compressing the air to slow it down from supersonic to subsonic speeds. The DSI can be used to replace conventional methods of controlling supersonic and boundary-layer airflow. This is a good example of using CFD to design DSI intake. It is practically impossible to find the shape of the bump for the DSI without using CFD.

The DSI bump functions as a compression surface and creates a pressure distribution that prevents the majority of the boundary layer air from entering the inlet at speeds up to Mach 2. There is no need for variable geometry supersonic intake for shock diffusion. There are considerable advantages with the DSI, as follows. Diverterless inlets lighten the aircraft weight.

1. Eliminates complex and heavy mechanical systems. There is a low part count.
2. Eliminates drag producing splitter plate through the boundary layer bleed passage.
3. No moving parts; for example, travelling centre body for shock diffusion.

Powerful gas turbine air intake at maximum power creates a strong suction vortex as shown in Figure 5.34a with ground water sucked in at the high-wing mounted military transport aircraft. This is capable of ingesting solid debris damaging the engine that may lead in to engine failures. Foreign object debris (FOD) ingestion by engine is a real harmful issue and must be prevented. Airfields for civil operation keeps the metalled cleaned several times a day, also as and when required as observed from frequent inspections. Boeing737 pods are low slung too close to the ground, hence they have vortex dissipaters to reduce the airflow suction at the bottom part of the intake to reduce the chances of FOD ingestion. Personnel in front of operating engines is strictly

(a)   (b)

Figure 5.34 Intake vortex suction and protection. (a) Ground water suck. (b) MIG 29 auxiliary air intake.

forbidden – fatalities have occurred. Some turbofans do not have stationary fan guide vanes in the front of the engine.

High powered supersonic turbofan intakes in military aircraft are positioned much lower and may need to be operated from a rougher macadam airfield when gravel ingestion is certain to occur. As an adaptation to rough-field operations, some designs, for example, the MIG 29 (Figure 5.34b), have the provision to close the front main air inlet, supported by properly sized louvre like the auxiliary air inlet on the upper side of the wing strakes under which the main inlet is positioned. The pilot operated auxiliary louvre intake is engaged during in field performance during takeoff and landing only. One of the other possibilities to prevent FOD ingestion is to have mesh screens in the main intake. For an S-shaped duct, the side intake has a spring loaded flapped door suitably positioned at the end of the S-bend so that the harmful heavier FOD failing to negotiate the bend hits a flap door to be thrown outside. S-duct design incurs flow energy loss and is not desirable.

Last but not the least in nacelle/intake design considerations, is retaining engine accessibility for MRO activities.

References

1 Jane's All the World's Aircraft Manual (yearly publication). Jane's Information Group.
2 Perkins, C.D. and Hege, R.E. (1988). *Airplane Performance, Stability and Control*. New York: Wiley.
3 Munk, M. M., The Aerodynamic Forces on Airship Hull, NACA TR 184, 1924.
4 Multhopp, H. (1942). Aerodynamics of Fuselage, NACA TM 1036.
5 Nelson, R.C. (1989). *Flight Stability and Automatic Control*. New York: McGraw-Hill.
6 Pamadi, B.N. (2015). *Performance, Stability, Dynamics and Control of Airplanes*, 3e. USA: AIAA, Reston, VI.
7 Fink, R. D. (1978). USAF Stability and Control, DATCOM. Report Number AFWAL-TR-83-3048, Flight Dynamic Laboratory, Wright-Patterson Air force Base, Ohio, April.
8 Cohen, R.H. and Saravamutto, H.I.H. (2001). *Gas Turbine Theory*, 5e. Harlow: Pearson.
9 Gunston, W. (2006). *The Development of Jet and Turbine Aeroengines*, 4e. Patrick Stephens Ltd.

6

Empennage and Other Planar Surfaces

6.1 Overview

Chapters 4 and 5 dealt with the pertinent aspects regarding the aerodynamic considerations of the wing and fuselage. The wing and body are found to be inherently unstable. Therefore, an empennage is required to balance this combined wing-body instability as well as offer the requisite control to execute manoeuvres. This chapter is devoted, primarily, to describing the role of empennage, the physics involved, aerodynamic considerations, possible choices and the advantages and disadvantages of the options available. In this book, while the horizontal tail (H-tail) and the vertical tail (V-tail) are sized for new aircraft configuration, the control surfaces are not sized. Control surfaces are earmarked into their positions with statistically obtained typical sizes.

This chapter deals also discusses other types of lifting surfaces, for example, ailerons to execute roll control, LERXs (strakes) for additional vortex lift and other planar surfaces, as aircraft components, these are studied at the conceptual stage of a new aircraft design project. These planer surfaces have an aerofoil like cross-section with chord length many times longer than the thickness. They are meant to generate forces and moments, some as modifiers to improve flow around bodies (e.g. dorsal fins, speed brakes serving as drag generators for deceleration) and others are mere support structures (e.g. a pylon).

This chapter covers the following topics:

Section 6.2: Introduction
Section 6.3: Terminologies and Definitions of the Empennage
Section 6.4: Empennage Mount and Types
Section 6.5: Different Kinds of Empennage Designs
Section 6.6: Empennage Tail Arm
Section 6.7: Empennage Aerodynamics
Section 6.8: Aircraft Trim Tab and Control Surfaces types
Section 6.9: Empennage Structural Considerations
Section 6.10: Empennage Design
Section 6.11: Other Planar Surfaces

Designing the empennage is not easy; it requires additional knowledge of aircraft stability and control; these subjects are offered in all academic institutes as separate courses requiring extensive treatment. The basics of aircraft stability and control required for aircraft design study are covered in Chapter 18. Preliminary configuration relies on the statistics of past designs, a practice widely used in the industry at the early stage of conceptual design study. Together with the wing and body, the empennage, engine and undercarriage present a complete new aircraft configuration. As mentioned earlier, this information serves to develop 'boilerplate' template geometries of aircraft components to be assembled to generate candidate aircraft configurations during the conceptual design phase (Chapter 8). Eventually, the chosen one will be formally sized as shown in Chapter 14.

Conceptual Aircraft Design: An Industrial Approach, First Edition. Ajoy Kumar Kundu, Mark A. Price and David Riordan.
© 2019 John Wiley & Sons Ltd. Published 2019 by John Wiley & Sons Ltd.

Designers have to ensure that there is adequate trim authority available (trim should not run-out) at any condition. This is normally done in Phase II after the configuration is frozen. In this book, the trim surfaces are earmarked and are not sized.

6.2   Introduction

In order to maintain steady flight, the aircraft requires empennage (or canard) surfaces to maintain stability. Section 3.8 introduced the idea that an aircraft has six degrees of motion in 3D space. The three angular motions in pitch, yaw and roll are controlled by the control surfaces as described in the following subsections.

6.2.1   The Role of the Empennage – Stabilising and Controlling

The role of the empennage is to offer stability to keep the aircraft flightworthy as well provide controllability to manoeuvre, as required in its mission role. The two angular motions, pitch and yaw, in the six degrees of freedom are controlled by two lift generating planer surfaces as an empennage mounted at the aft end of the fuselage in conventional designs at the aft of aircraft. These are achieved as follows (refer to Figure 3.11 for the body axes, F_B, system). Figure 3.13 in Section 3.7.3 gives the sign convention associated with control deflection.

i. Pitch stability and control (nose-up/down in the x–z-plane of symmetry – the pitch plane) by the lift generating planer surface as the H-tail. Its fixed part (stabiliser) offers longitudinal stability and the movable part (elevator) provides the control (pitch moment, C_M).
ii. Directional stability and control (yaw turns the x–z-plane) by the symmetrical V-tail. Its fixed part (fin) offers directional stability and its movable part (rudder) provides directional control (lateral moment, C_N). Having no constraints, the centrifugal force developed in a pure turn in a level wing will be associated with slide slip. It has to be a coupled motion with some degree of roll (bank) using wing mounted ailerons to accomplish a no sideslip coordinated turn.

6.2.2   The Role of Ailerons – Stabilising and Controlling

The third angle of motion is rolling. Lateral roll control in the y–z-plane can be performed with ailerons at the wing tips' trailing edges. The opposite deflection of ailerons at each side gives the rolling moment, C_L (not to be confused with the lift coefficient, C_L). Ailerons conform to the aerofoil profile of the wing and are not part of the empennage. Roll has coupled motion with yaw.

Like wings, empennage planform shapes are also basically of two types: (a) unswept and (b) swept. H-tails can also have dihedral/anhedral angles for the same reason as they are applied to the wing configuration (see Section 4.4.1). The empennage configuration follows similar considerations as wing design to establish the aspect ratio (AR), sweep, taper ratio, dihedral angle and twist. There are basically three types of empennage configurations as shown in Figure 6.1.

1. The most common method, as the conventional practice, is to install a horizontal planar surface (termed as H-tail) at the far aft end for pitch stability and control of the aircraft (Figure 6.1a). Lateral stability and control is achieved via a vertical planar surface (termed the V-tail) at the far aft end in the plane of symmetry. Since it is at the end, it is seen as the tail of an aircraft as a part of the empennage.
2. Described in Section 1.2, the Wright brothers' 'Flyer' configuration had a canard configuration (tail in front). This kind of design has steadily gained ground in the last three decades, especially in combat aircraft design (Figure 6.1b).

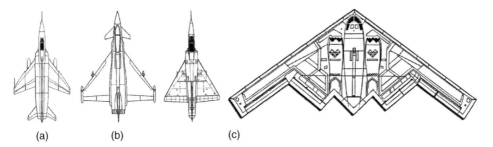

Figure 6.1 Typical empennage designs. (a) Conventional, (b) canard configuration and (c) H-tail merged with the wing (tailless).

3. For a single surface delta wing aircraft (Mirage 5) and all-wing configuration (B2), part of the tail offering a nose-up moment is merged with the wing as its reflex movable trailing edge (Figure 6.1c). The differential movements of these trailing edges also act as ailerons, known as elevons (see Section 6.5). Depending on the technology adopted, blended-wing-body (BWB) aircraft may or may not have a V-tail.

Except for very small number of designs, all aircraft are symmetrical to their vertical plane, that is, from its centreline, the port side and starboard side are the mirror images of each other. Therefore, the aircraft is without any lateral moments in straight flight along its plane of symmetry. It becomes essential to provide a vertical surface (V-tail) that can be manipulated to generated lateral moment (C_N) to the extent required to execute a yaw turn. This is known as the V-tail part of the empennage. It has a symmetric aerofoil.

Although two examples of DATCOM methods are given in this chapter to understand the physics of the problem, this book relies on the statistics of past designs as given in Chapter 7.

6.3   Terminologies and Definitions of Empennage

Empennage shaping observes the rules of wing design with wings being lifting surfaces. Modern aircraft with *Fly-by-Wire* (FBW) technology can operate with more relaxed stability margins, especially for military designs, and hence require smaller empennage areas compared with older conventional designs (see Figure 18.18). The terminologies associated with the empennage are given in the following subsections (Figure 6.2).

6.3.1   Horizontal Tail (H-Tail)

A H-tail is like a small wing at the tail (i.e. the aft end of the fuselage). It consists of the fixed or moving stabiliser and the moving elevator for controlling the pitch degree of freedom (Figure 6.2a). In general, the H-tail has a negative camber to give a nose-up moment to counterbalance the inherent nose-down moment of a tailless wing-fuselage combination (see Section 6.7.4). The H-tail can be positioned low at the fuselage side, in the middle cutting through the V-tail or at the top of the V-tail to form a T-tail (see Figure 6.3). Like the wing

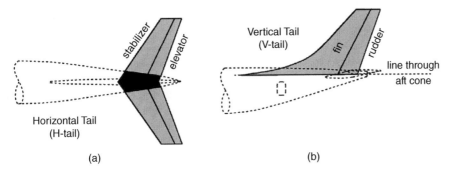

Figure 6.2 (a) Horizontal tail and (b) vertical tail.

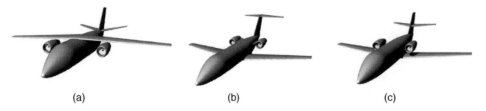

(a)   (b)   (c)

Figure 6.3 The dominant options for the wing, tail and nacelle positions. (a) Low tail, under-wing pods, (b) high T-tail, fuselage-mounted pods and (c) mid-tail (cruciform), over wing pods.

planform definition, the H-tail reference area, S_{HT}. For a low tail, part of S_{HT}, is buried inside the fuselage. The T-tail at the top of the V-tail has a fully exposed planform. The mid-tail has a small part of S_{HT} buried within the V-tail.

Typically, for civil aircraft, H-tail planform area is anywhere from one-third to a quarter of the wing planform size, AR ≈ 3.0, sweep is about same as the wing and there is no twist, but it can have dihedral/anhedral angles. The T-tail on a swept back V-tail permits a longer tail arm, L_{HT} (Section 6.8).

Military aircraft empennage sizing requires special attention as they require more control authority for greater manoeuvrability. Modern combat aircraft have lower wing AR values and have shorter tail arms requiring larger H-tail areas.

6.3.2   Vertical Tail (V-Tail)

The V-tails have a symmetrical aerofoil section unbiased to either side to turn. The V-tail consists of a fixed fin and a moving rudder to control primarily the yaw degrees of freedom but has coupled motion with roll. The V-tail is positioned at the plane of aircraft symmetry and has a symmetrical aerofoil. The V-tail of a fuselage-mounted propeller driven aircraft may be given skew (small amount of incidence angle) to counter yaw developed by its swirling slipstream.

Figure 6.2b shows the geometrical definition of a conventional V-tail surface reference area, S_{VT}. Depending on the closure angle of the aft fuselage, the root end of the V-tail is fixed arbitrarily through a line drawn parallel to the fuselage reference line, passing through tail cone, hence it can be partly buried into the fuselage. The projected area up to its root end is considered the reference area, S_{VT} of the V-tail. Other definitions takes the V-tail root at the fuselage reference line, which may not pass through the tail cone.

Typically, for civil aircraft, the V-tail planform area is about 12–20% of wing reference area. V-tail design is critical to takeoff (to satisfy V_{mc} – see Chapter 15) – especially to tackle yawed ground speed arising from cross-winds and/or asymmetric power of multi-engine aircraft. A large V-tail can cause snaking of the flight path at low speed, which can be easily cured by introducing a 'yaw-damper'. At cruise, a relatively large V-tail is not a problem but ensures safer operation. V-tail is half span, its AR is based on full span.

6.4   Empennage Mount and Types

Chart 6.1 tabulates the variety of empennage configuration options available in a systematic way. Note that other kinds of empennage configurations are possible (canard design not included).

6.4.1   Empennage Positional Configuration – Single Boom

6.4.1.1   Conventional type
By far, the majority of civil aircraft designs have two surfaces almost orthogonal to each other as the H-tail and V-tail. The H-tail can be positioned low at the fuselage (Figure 6.3a – has a dihedral angle) or at the top as a T-tail (Figure 6.3b – has an anhedral angle) or anywhere in between (Figure 6.3c) as a mid-tail (also known as

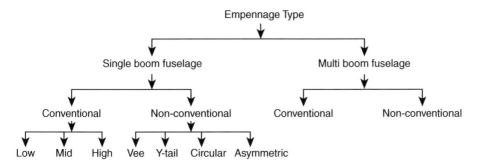

Chart 6.1 Types of empennage configurations.

a cruciform tail) configuration. Any combination of the scheme is feasible but it is decided on from the various aerodynamic, stability and control considerations. The V-tail is kept symmetric to the aircraft centreline and can be swept (there are exceptions). Twin V-tails (military aircraft) can be straight or inclined. The advantages and the disadvantages of the three setups are given next.

Low tail

This is the well-proven configuration, adopted by the majority of aircraft designs, used especially in large commercial aircraft applications.

Advantages

i. Robust structural integrity for being attached to the girth of an aft fuselage offering best resistance to torsion under empennage load.
ii. Wing downwash on the H-tail can be well utilised, especially in a wing flap extended position.
iii. Careful positioning of the H-tail can avoid wing wake interfering with the H-tail as shown in Figure 6.16 later.

Disadvantages

i. Can shield the V-tail in spin (see Figure 6.15).
ii. High interference drag – both at the upper and lower surfaces of the H-tail joint.

Examples: All Airbus and Boeing commercial transport aircraft currently in production.

6.4.1.2 *High tee-tail (t-tail)*
The other extreme of H-tail position is placed at the top of the V-tail suiting an aircraft with aft fuselage-mounted engines to stay away from the engine exhaust.

Advantages

i. On account of V-tails being made swept back, the H-tail arm is longer for the same size fuselage. Longer tail arms reduce H-tail size for the same tail volume coefficients (Section 6.10).
ii. Does not shield the V-tail (see Figure 6.15).
iii. Low interference drag – only at the lower surfaces of the H-tail joint.

Disadvantages

i. The V-tail becomes too heavy to support a T-tail and provide torsion rigidity.
ii. In deep stall, wing wake may affect the T-tail elevator authority that may cause aircraft to enter into flat spin when recovery may become difficult. For example, the BAe 111 flight test crash in October 1963 and

similar crashes that occurred with many other aircraft. The stabilator (see Section 6.5.3) type of T-tail may become a necessity.

iii. Prone to flutter.

Examples: Learjet series, Boeing727, Beech King aircraft, C17 military transport aircraft and so on.

6.4.1.3 *Mid-tail (cruciform tail)*

This configuration is in between the two extremes as described here, meant for aft-mounted engines. This is a good compromise by raising the H-tail just above engine exhaust interference.

Advantages

i. May have a neutral rolling moment.
ii. Lower structural weight as compared to the T-tail of the same size.
iii. Partial rudder available for spin recovery.

Disadvantage

i. High interference drag – both at the upper and lower surfaces of the H-tail joint.
ii. Does not have a tail arm as long as the T-tail can get for the same fuselage length.

Examples: Sud Aviation Caravelle, Jet Stream, Dassault Falcon and so on.

6.4.1.4 Unconventional Empennage types

Figure 6.4 shows the dominant unconventional empennage design options (this figure can also be used to illustrate the wing and nacelle position options). The merits of the unconventional empennage have its applicability but by far the majority of designs have one horizontal and one vertical surfaces. This book deals only with the conventional designs.

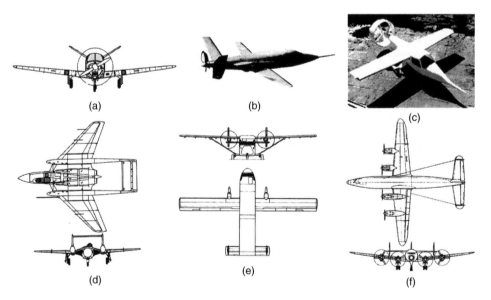

Figure 6.4 Other types of civil aircraft empennage design option. (All figures except the LearFan and Marvel aircraft). (a) Vee-tail Beech Bonanza. (b) Y-tail (Lear Fan). (c) Circular tail (Courtesy: www.aviastar.org/ Courtesy: Mississippi air/usa/learavia_learfan.php State University helicopters@inbox.ru Marvel. (d) Twin V-tail on twin boom: De Havilland Sea Vixen. (e) Twin V-tail on H-tail: Short Sky van. (f) Three V-tail (Lockheed Constellation). Source: Reproduced with permission of Ferrier.

Vee-tail/Y-tail (Figure 6.4)

The Beechcraft Bonanza 35 has a *vee-shaped* tail (not be confused with a V-tail) and in some designs it can be an inverted vee-tail (ruddervator – see Section 6.5.1). One of the early Lear designs had Y-shaped empennage – a vee-tail along with a vertical fin extending below the fuselage. Vee-tail/Y-tail designs are more complex but follow the same routine as the conventional ones, that is, resolving the forces on the surfaces into vertical and horizontal directions. The MacDonnell Douglas F4 Phantom H-tail has a high degree of anhedral angle, almost like an inverted vee-tail, like a wide open inverted Y-tail. A proper Y-tail configuration can be seen in the LearFan 2100 manufactured at their Belfast based factory.

Twin tail/triple tail (Figure 6.4)

If the V-tail size is large on account of a short tail arm, then the area could to be split into two (Short Skyvan) or three V-tails (Lockheed Constellation) arising from structural and aerodynamic considerations. Multiple V-tails are placed symmetrically about the aircraft plane of symmetry. The short 330 aircraft is a good example of a twin-tail configuration with rear loading facility. The end plate effect of a V-tail on H-tail increases its effectivity as well as decreases H-tail induced drag.

Circular empennage (Figure 6.4)

An experimental aircraft with a circular tail was designed by the Mississippi State University in the 1960s as a research platform. It does not appear to have shown definite advantages over conventional empennage design to supplant it.

6.4.2 Empennage Positional Configuration – Multi- (Twin) Boom

6.4.2.1 Conventional Empennage types

Fuselage mounted

Many successful aircraft and Unmanned Aerial Vehicles (UAVs) have been designed with a twin boom accommodating an empennage; for example, the Lockheed P38, North American F-82 Twin Mustang, De Havilland Fairchild C119, Vampire/Sea Vixen, Cessna Skymaster and so on all have a V-tail each on the fuselage twin booms and the H-tail joining them (or an inverted vee-tail). It suits rear mounted pusher aircraft designs, centrally placed jet engines and, in the case of the P38, the booms extends from piston engine nacelles. The propeller slipstream from each of the boom mounts assists the empennage effectivity.

6.5 Different Kinds of Empennage Design

Apart from the conventional type of empennage, there are other kinds described in the next subsections.

6.5.1 Ruddervator

The ruddervator is a novel concept first suggested in the early 1930s by combining the actions of H-tail and V-tail into a '*vee*'-shaped tail, which can be seen as a dihedral of an H-tail to about half way at $\Gamma \approx 45°$ and eliminating the V-tail. Its inverted shape is also possible with $\Gamma \approx -45°$.

The lift forces normal to the inclined ruddervator are resolved along the pitch plane and yaw plane. When the pilot control stick is pulled or pushed, then the movable part at both sides of the ruddervator move together up or down to give nose-up or nose-down, respectively; here yaw-plane force components cancel each other. For turning, the rudder pedals are used when differential movement at each side occurs. This time, the force components in the yaw plane are in the same direction as shown in the last diagram of Figure 6.5 for a left turn. However, force components in pitch are equal and opposite but produce a couple of forces to give some rolling moment to the left, assisting a coordinated turn with bank angle.

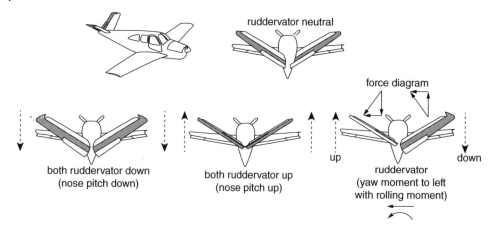

Figure 6.5 Ruddervator (Vee-tail).

While there are successful models are in operation, for example, the Beechcraft Bonanza 35 (small propeller driven utility aircraft), Fouga Magister C170 and variants (military jet trainer of French origin), and few others.

A properly sized vee-tail design has some advantages as follows:

Control surfaces are not shielded and control authority is available in spin.
No-slip coordinated turn is possible without or needing little application of ailerons.
There is sufficient ground clearance if the aircraft tips at one side and so on.

However, there are some unfavourable issues as follows:

It requires a complex control mechanism and is relatively heavy.
It needs a larger area, which also makes it heavy.
It has coupled control and controlling aircraft in gusty weather during takeoff and landing can prove to be more difficult compared to using a conventional empennage with an H-tail, V-tail and so on.

An inverted vee-tail is a possibility offering similar design considerations. This is a rare application and only came as proposition (Blohm Voss VM 213) i some microlites (with a twin boom joined by an inverted vee-tail), but is used noticeably in UAV designs. Studying the UAV configurations, it is clear that the aft-mounted propeller pusher UAV design configurations have long undercarriages for ground clearances allowing an inverted vee-tail (see the Ikhana in Figure 1.17a).

It is for these reasons that the vee-tail/Y-tail and their inverted version concepts did not catch on. Practising engineers may refer to NACA-ACR-L5A03 [1] (also known as NACA TR-823) for detailed study.

6.5.2 Elevon

Elevons are aircraft control surfaces that combine the functions of the elevator (used for pitch control) and the aileron (used for roll control), hence the name: a portmanteau (blended word) of *elev*ator and ailer*on*. (Figure 6.1b). They are frequently used on tailless aircraft such as flying wings.

Elevons are installed on each side of the aircraft at the trailing edge of the wing. When moved in the same direction (up or down) they will cause a pitching force (nose-up or nose-down) to be applied to the airframe. When moved differentially (one up, one down), they will cause a rolling force to be applied. These forces may be applied simultaneously by appropriate positioning of the elevons, for example, one wing's elevon completely down and the other wing's partly down.

An aircraft with an elevon is controlled as though the pilot still has separate aileron and elevator surfaces at his disposal, controlled by the yoke or stick. The inputs of the two controls are mixed either mechanically or electronically to provide the appropriate position for each elevon. Delta wings with reflex trailing edges have elevons, for example, on the F-102A Delta Dagger, Mirage 5 and so on.

6.5.3 Stabilator/Taileron

A stabilator (a blended word of 'stabiliser-elevator'), also known as a taileron, is a fully movable aircraft stabiliser. It has several other terminologies, for example (i) all-moving tailplane, (ii) all-movable tail (plane), (iii) all-moving stabiliser, (iv) all flying tail (plane), (v) full-flying stabiliser, (vi) flying tail, (vii) slab tailplane and (viii) taileron. It serves the usual functions of longitudinal stability, control and stick force requirements otherwise performed by the separate fixed and movable parts of a conventional horizontal stabiliser. Apart from a higher efficiency at a high Mach number, it is a useful device for changing the aircraft balance within wide limits and for mastering the stick forces. Stabilators may not prove best suited to subsonic aircraft design.

Military aircraft can have all-moving H-tails with emergency splitting to act as elevators in case there is failure of the stabilator and there are several choices for positioning it for splitting, typically at 40% of the mean aerodynamic chord (MAC).

6.5.4 Canard Aircraft

Canard is French for duck, which in flight stretches out its long neck with its bulbous head in front. When a horizontal surface is placed in front of the aircraft, it presents a similar configuration; hence, this surface is sometimes called a canard. A canard is an aerodynamic arrangement wherein a small forewing or foreplane is placed forward of the main wing of a fixed-wing aircraft. The term 'canard' may be used to describe the aircraft itself, the wing configuration or the foreplane (see Section 6.6.1).

Despite the use of a canard surface on the first powered aeroplane, the Wright Flyer of 1903, canard designs were not built in quantity until the appearance of the Saab Viggen jet fighter in 1967. The aerodynamics of the canard configuration are complex and require careful analysis.

Rather than use the conventional tailplane configuration found on most aircraft, an aircraft designer may adopt the canard configuration to reduce the main wing loading, to better control the main wing airflow, or to increase the aircraft's manoeuvrability, especially at high angles of attack or during a stall. Canard foreplanes, whether used in a canard or three-surface configuration, have important consequences on the aircraft's longitudinal equilibrium, static and dynamic stability characteristics. A military canard is meant for high manoeuvrability and a civil canard is meant for static stability and weight reduction.

6.6 Empennage Tail Arm

The destabilising moments from wing, fuselage and nacelle are balanced by the empennage moments about the aircraft centre of gravity (CG) [2]. The *tail arms* are the length of the moment arms of the lift forces developed by the H-tail and V-tail. The respective tail arms are denoted l_{HT} for the H-tail and l_{VT} for the V-tail both acting at the respective aerodynamic centre close enough to ¼MAC. These are important parameters for sizing the empennage for aircraft stability and control, measured from the CG; its position changes according to the payload and fuel load distribution change.

Therefore, it is convenient if fixed moment arms can be considered if they are close to the moment arm measured from the CG (Figure 6.6). To generate statistics from past designs, a fixed position between wing and empennage proves convenient. In this case, for conventional two-surface aircraft configuration, a *tail arm* is defined as the length between the quarter point of wing $(MAC)_{¼\_wing}$ and quarter point of tail $(MAC)_{¼\_H\_tail}$ denoted l'_{HT} and $(MAC)_{¼\_V\_tail}$ denoted l'_{VT}. Typically, CG variation in civil aircraft is less than 10–15% of the MAC and for military aircraft around 5–10% of the MAC keeping its quarter-chord point within the margin of movement. Typically, conventional aircraft have their CG close to $(MAC)_{¼\_wing}$, making $l_{HT} \approx l'_{HT}$ and $l_{VT} \approx l'_{VT}$, removing ambiguity as they are treated as the same.

The terms l_{HT} and l_{VT} are important parameters to size empennage areas obtained from the non-dimensional tail volume coefficients defined in Section 6.6.2. Strictly speaking, statistics of tail volume coefficients from both CG and $MAC_{¼}$ should be separately considered. However, the differences between them are very small,

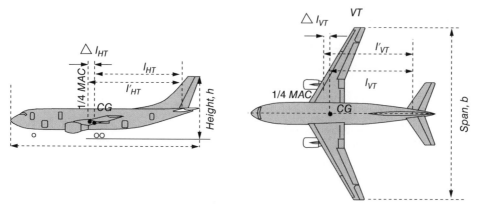

Figure 6.6 Geometric parameters for the tail volume coefficient.

contained well within the bands of statistical dispersion. Therefore, from now onwards, only the terms l_{HT} and l_{VT} are used to denote the respective tail arms. Whichever method is used, the final result will be the same so long as appropriate interpretation is followed in the computational process.

At the conceptual design stages, when the aircraft wing position relative to the empennage is yet to be settled, it is more appropriate to initially guess a CG position to compute tail arms and subsequently fine-tune the tail arm lengths as the CG changes. It is for this reason in this book the tail arms are measured from the aircraft CG.

6.6.1 Canard Configuration

The Wright Brothers' Flyer had a control surface at the front (with a destabilising effect), which resulted in a sensitive control surface. The last two decades have seen the return of aerodynamic surfaces placed in front of the wing. Figure 6.7a shows a three-surface canard Piaggio P180 Avanti and Figure 6.7b shows a two-surface canard Rutan Long-EZ aircraft. The canard surface can share some lift (in civil aircraft designs) with the wing.

In general, the inherent nose-down moment (unless a reflex trailing edge is employed) of a wing requires a downward force by the H-tail to maintain level flight. This is known as *trimming force*, which contributes to trim drag. For an extreme CG location (which can happen as fuel is consumed), high trim drag can exist in a large portion of the cruise sector. The incorporation of a canard surface can reduce trim drag as well as the H-tail area, S_{HT}; if not, eliminate the aft H-tail as in the case of the Long-EZ. Until recently, the benefit from the canard application in large transport aircraft has not been marketable.

A successful Bizjet design is the Piaggio P180 Avanti has achieved a very high speed for its class of aircraft through careful design considerations embracing not only superior aerodynamics but also the use

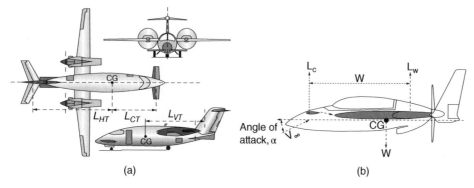

Figure 6.7 Canard configurations. (a) Three-surface canard (Piaggio P180 Avanti) and (b) two-surface canard.

of composite materials to reduce weight. Military aircraft use a canard to enhance pitch control. Canards in military aircraft have a special role to execute extreme manoeuvres.

6.6.2 Tail Volume Coefficients

The CG position is shown in Figure 6.6. The distances from the CG to the quarter-chord of the V-tail and H-tail MAC (i.e. $MAC_{VT/4}$ and $MAC_{HT/4}$) are designated l_{HT} and l_{VT}, respectively. The aircraft *a.c.* is taken at the quarter-chord of the wing $MAC_{W/4}$.

The tail volume coefficients are non-dimensional parameters: this lumps together all the influencing variables that arise in stability analyses (Eqs. (6.1) and (6.2)). Figure 7.8 gives statistical values of tail volume coefficients and tail areas with respect to wing area.

6.6.2.1 H-tail volume coefficient, C_{HT}

Define H-tail volume coefficient is derived in terms of the H-tail plane reference area, S_{HT}, as follows.

$$C_{HT} = \frac{S_{HT} \times l_{HT}}{S_W \times MAC} \tag{6.1}$$

or

$$S_{HT} = \frac{C_{HT} \times S_W \times MAC}{l_{HT}}$$

where C_{HT} is the H-tail volume coefficient, $0.5 < C_{HT} < 1.2$; a good value is between 0.6 and 0.9. l_{HT} is the H-tail arm = distance between the aircraft CG to the $MAC_{HT/4}$. In general, the area ratio $S_{HT}/S_W \approx 0.20$–0.35. H-tail span is about a third of the wing span. Figure 7.8 gives a cluster of V-tail designs with tail volume coefficients A good starting point is to take $C_{VT} \approx 0.07$, $AR \approx 1.0$–1.75, sweep is about same as the wing and no twist.

6.6.2.2 V-tail volume coefficient, C_{VT}

Define V-tail volume coefficient is derived in terms of the H-tail plane reference area, S_{VT}, as follows.

$$C_{VT} = \frac{S_{VT} \times l_{VT}}{S_W \times b} \quad \text{where } 'b' \text{ is the wing span} \tag{6.2}$$

or

$$S_{VT} = \frac{C_{VT} \times S_W \times b}{l_{VT}}$$

where C_{VT} is the V-tail volume coefficient, $0.05 < C_{VT} < 0.1$; a good value is between 0.06 and 0.09. l_{VT} is the H-tail arm = distance between the aircraft CG to the a.c. of MAC_{VT}. In general, the area ratio $S_{VT}/S_W \approx 0.15$–0.26.

Equations (6.1) and (6.2) give a good indication of what kind of empennage size is to be expected. Having the tail arm and wing area established from the configuration obtained in the preliminary study (Chapter 8 gives the worked-out examples), the empennage sizes depend on the values of C_{HT} and C_{VT} chosen for the design from the existing statistics for the H- and V-tail, respectively. It is recommended to take higher values of C_{HT} and C_{VT} to have a generous sized empennage to ensure safety with better control authority. The weight penalty is small; if required, the empennage can be fine-tuned after prototype flight testing by tailoring to smaller than the initial sizes.

Figure 7.8 gives a cluster of V-tail designs with tail volume coefficients A good starting point is to take $C_{VT} \approx 0.07$, $AR \approx 1.0$–1.75, sweep about same as the wing and no twist. The T-tail acts as an end plate at the tip of V-tail when, for the T-tail configuration, tail volume, C_{VT}, could be reduced. Like wing design, the V-tail can have sweep, but dihedral/anhedral and twist is meaningless, as the V-tail needs to be symmetric about the fuselage centreline. For propeller driven aircraft, the V-tail could be kept slightly skewed ($\approx 1°$) to offset swirled slipstream effects and gyroscopic torque of rotating engines and propellers. Sweeping of the V-tail would effectively increase the tail arm L_{VT}, an important dimension to sizing the V-tail. It is important to ensure that the V-tail, especially the rudder, is not shielded by the H-tail to retain effectiveness, especially during spin recovery. With a T-tail, there is no shielding.

Military aircraft empennage sizing requires special attention as these require more control authority for greater manoeuvrability. Modern combat aircraft with FBW system architecture have a lower wing AR and shorter tail arms, requiring relatively larger tail areas. With FBW system architecture, this reduces the V-tail size with $C_{VT} \approx 0.07–0.09$, AR ≈ 3.0, sweep is about the same as the wing. The V-tail area varies from 15 to 25% of the wing reference area. Fighter aircraft have an all-movable tail for control. If the V-tail is too large then it is split into two.

6.6.2.3   Canard volume coefficient, C_{CT}

This also is derived from the pitching-moment equation for steady-state (i.e. equilibrium) level flight. The canard reference area, SC, has the same logic for its definition as that of the H-tail. Its tail arm is l_{CT}. The canard reference area is given as:

$$C_{CT} = \frac{S_{CT} \times l_{CT}}{S_W \times MAC} \tag{6.3}$$

or

$$S_{CT} = \frac{C_{CT} \times S_W \times MAC}{l_{CT}}$$

where C_{CT} is the H-tail volume coefficient, $0.5 < C_{HT} < 1.2$; a good value is 0.6–0.9, depending on whether it is a conventional H-tail. The l_{CT} is the H-tail arm = distance between the aircraft CG to the ac of MAC_{CT}. In general, $S_{CT}/S_W \approx 0.2–0.6$.

6.7   Empennage Aerodynamics

Empennage design involves detailed stability and control analyses, which are beyond the scope of this book. Aircraft static stability is dealt with in a generalised manner in Chapter 18. However, to understand empennage design, the pertinent topics on aircraft stability and the associated aerodynamics are briefly described in this section. Only conventional empennages are dealt with here, both having fixed and a movable surfaces. The H-tail is discussed first. Readers may refer to References [2–9] dedicated to aircraft stability and control for further reading.

6.7.1   Wing Downwash on the H-tail

Section 4.9.1 (Figure 4.16) dealt with wing downwash, ε, which flows over the H-tail (typically with an inverted aerofoil) at an angle of incidence, α_{ht}, with respect to the stabiliser as shown in Figure 6.8. (It is convenient to express all the relevant angles associated with downwash measured with respect to a fixed fuselage reference line established by the designer; typically, from aircraft nose to tail tips.)

From Figure 6.8, the following angles are observed.

α = angle of attack, the angle between free stream velocity, V, and the wing chord line. Depending on the aircraft attitude, α, can vary.

i_w = Wing and fuselage setting angle. This is the fixed geometric angle measured between MAC and fuselage reference line. Being invariant, it offers convenience to measure all other angles that can vary. The angle is meant to offer minimum drag at long range cruise (LRC) speed. Typically, $i_w \approx 2 \pm 1°$.

α_{ht} = This is at angle of attack at the H-tail, the angle between wing downwash deflection velocity of $2w$ with the H-tail MAC_{HT}. The angle of attack, α_{ht}, can vary, depending on aircraft velocity and pitch attitude. α_{ht} has to be computed.

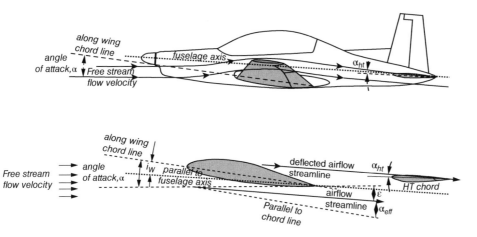

Figure 6.8 H-tail aerodynamics.

i_t = This is angle between the H-tail MAC_{HT} and fuselage reference line. It can be either fixed or can be rigged to a different setting angle on the ground, or can be set by the pilot in flight.

ε_d = This is the downwash deflected angle and is twice the downwash angle, ε, at the aerodynamic centre, *a.c.*, of the wing MAC.

From this information,

$$\alpha_{ht} = \alpha - \varepsilon_d + i_t \tag{6.4}$$

The intention is to develop the C_{L_HT} required to produce downward force to balance the destabilised nose-down moment of the tailless aircraft.

For small aircraft, the stabiliser part of the H-tail is fixed to fuselage after satisfactory completion of flight testing prototype aircraft before the final version is produced. Some smaller utility type aircraft can get H-tail rigged on ground before a sortie to suit the loading condition. However, for larger aircraft, to expand the elevator control authority, the stabiliser can be adjusted in flight by the pilot. To achieve fast response to control, combat aircraft have an all-moving H-tail as shown in Figure 6.19, later.

6.7.2   H-tail – Longitudinal Static Stability: Stick-Fixed and Power-Off

Aircraft pitch stability analyses is in its pitch plane with rotation about the *y*-axis. The H-tail, with a fixed stabiliser and a movable elevator, requires the stabiliser part to provide adequate pitch stability. In normal operation, the control stick is held by the pilot, keeping the elevator in a fixed position is known as the *stick-fixed* case. When the elevator fails (or the pilot leaves the control stick free) that will make the elevator to flap freely about its hinge line contributing nothing, rendering it useless. This situation of the elevator moving freely is termed the *stick-free* case, a design consideration to ensure safety in this undesirable situation. To relieve pilot fatigue by continuously applying force to hold the stick in place, trimming devices (see Section 6.8) are installed, when the pilot can ease off but should not control stick-free. This section deals only with the *stick-fixed* and power-off case. The most critical situation for H-tail design is at the permissible aft-most aircraft CG when aircraft stability and control authority are in high demand.

Figure 6.9 gives the geometric and kinetic information about a conventional aircraft to explain the physics behind aircraft longitudinal static stability. In this simplification, it is considered that the vertical height of aircraft CG is very close to the fuselage reference line, thus making the wing and nacelle close enough to make small contributions to moments in comparison with other contributions, hence they are neglected. (The rigorous method follows the same procedure without ignoring any contribution to pitching moment generation.)

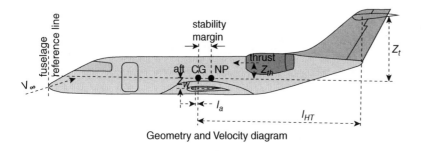

Geometry and Velocity diagram

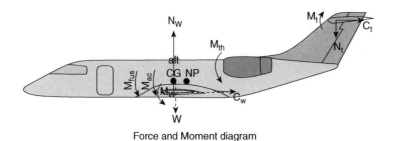

Force and Moment diagram

Figure 6.9 Aircraft forces and moments (power-off).

In summary, the vertical distances (z) of each component could be above or below the CG, depending on configuration as described next.

i. Drag of low wing below CG (z_a) will have a nose-down moment and vice-versa for high wing. For mid-wing positions, it has to be noted on which side of CG it lies and could have a z_a small enough to be ignored for a preliminary analysis.

ii. The H-tail area is considerably smaller than the wing, so is its drag that can be neglected even when positioned as a T-tail at this stage of study.

iii. Under-wing slung engines have a z_{th} below the CG generating a nose-up moment. For fuselage-mounted engines, z_{th} is likely to be above the aircraft CG and its thrust would generate a nose-down moment offering stability. For most military aircraft, the thrust line is very close to the CG and hence for a preliminary analysis, the z_{th} term can be ignored; that is, no moment generated with thrust (unless vectored). Since, the section considers a power-off situation as the critical case, moment contribution by thrust is not considered. Nacelle height, being closer to CG, means its drag is also not taken into account under this simplification.

In equilibrium flight, the role of the H-tail is to provide a sufficient opposite moment to balance the moments developed by the rest of the aircraft for the full flight envelope, as equated next.

$$C_{M\_cg} = \text{(Pitching moment by wing)} + \text{(Pitching moment by fuselage)} + \text{(Pitching moment by H} - \text{tail)}$$

The governing generalised equation for aircraft stability in coefficient form, derived in Equation (18.8), is shown in Eq. (6.5). In equilibrium, $C_{M\_cg} = 0$, that makes,

$$C_{M\_cg} = - \left[C_{Lw} \left(l_W/c \right) + C_{M\_w} \right] - C_{M\_fus} + \left[C_{L\_HT} \left(S_{HT}/S_w \right)\left(l_{HT}/c \right) \eta_t + C_{M\_t} \left(S_{HT}/S_w \right) \eta_t \right] = 0 \qquad (6.5)$$

where, η_t ($=q_t/q_\infty$) represent the wing wake effect on the H-tail, producing downwash, q_t is the incident dynamic head and q_∞ is the free stream dynamic head. The value of η_t varies from 0.95–1.0. Note that for conventional aircraft with aft-most CG, the C_{LHT} requires downward force to keep nose-up, hence it carries the opposite sign. The worked-out example, Example 6.1, assigns the appropriate signs.

By transposing, Eq. (6.5) can written as follows.

$$\left[C_{L\_HT}\,(S_{HT}/S_w)(l_{HT}/c)\,\eta_t + C_{M\_ht}\,(S_{HT}/S_w)\,\eta_t\right] = \left[C_{Lw}\,(l_W/c) + C_{M\_w}\right] + C_{M\_fus} \tag{6.6}$$

Given next is the simplified version to establish the methodology to size the H-tail. Further simplification can be made by studying the order of magnitude of the moment contributed by each term. Typically, an H-tail has a low negative camber, if not without camber, in the case of small recreational types of aircraft, one can get away with a flat plate. The C_{M_t} of the H-tail stays low as well the ratio of $(S_{HT})/(S_w)$ is in the order of 0.25–0.3: on the other hand, the long tail arm, l_t, is the main contributor to the nose-up moment. In other words, $[C_{M\_ht} \times (S_{HT}/S_w) \times \eta_t] < < [C_{Lt} \times (S_{HT}/S_w)(l_{HT}/c) \times \eta_t]$, hence it can be neglected, yet a realistic sized H-tail area, S_{HT}, can be obtained.

Equation (6.6) can now be written as

$$\left[C_{L\_HT}\,(S_{HT}/S_w)(l_{HT}/c)\,\eta_t\right] = \left[C_{Lw}\,(l_W/c) + C_{M\_w}\right] + C_{M\_fus}$$

Using Eq. (6.1),

$$C_{LHT} \times \left[(l_{HT}/S_{HT})/(S_w c)\right]\,\eta_{HT.} = C_{HT} \times \eta_{HT} C_{LHT}$$

where $C_{HT} = \frac{S_{HT} \times l_{HT}}{S_w \times MAC}$ (for a T-tail above the wing wake, $\eta_t = 1.0$ and here $c =$ wing MAC).

Equation (6.6) reduces to a simple usable expression for easy computation as:

$$\left[C_{L\_HT}\,(S_{HT}/S_w)(l_{HT}/c)\,\eta_t\right] = \left[C_{Lw}\,(l_W/c) + C_{M\_w}\right] + C_{M\_fus} \tag{6.7}$$

$$C_{HT} \times C_{LHT} = \left[C_{Lw}\,(l_W/c) + C_{Mw}\right] + C_{Mfus}$$

For a given wing, Figure 6.10 gives the limits of H-tail size. H-tail growth moves the neutral point (NP) aft and the CG can be allowed to move aft retaining the safe static margin for stability. H-tail growth also permits the CG limit to move forward as more elevator power is available at the cost of increased stick force. The H-tail size is at the allowable stick force limit at the forward-most CG position that gives the sized CG travel within the acceptable limit. The shaded area shown in the figure gives a choice to for H-tail size within the prescribed CG travel limits and the stick force limit.

Stability requires the CG to be ahead of NP offering a sufficient *stability margin*, defined as follows.

$$\text{Stability Margin} = \left(\frac{x_{NP} - x_{CG}}{MAC_{1/4}}\right) \tag{6.8}$$

expressed as a percentage of wing *MAC*.

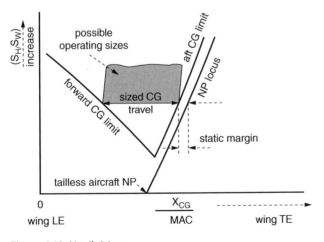

Figure 6.10 H-tail sizing.

6.7.3 V-tail – Directional Static Stability: Rudder-Fixed

Aircraft directional stability analyses follows the same routine as of the longitudinal stability analyses, this time about z-axis rotation in the yaw plane instead of about y-axis rotation in the pitch plane. The aircraft, being symmetrical about the vertical (pitch) plane, has directional stability that behaves like a weathercock. Therefore, in principle, aircraft flying straight along the plane of symmetry do not produce a yaw moment. In reality, even when aircraft flystraight in along the plane of symmetry there will be yawing moments on account of (i) spiralling propeller slipstream flow, (ii) asymmetric power setting for multi-engine aircraft, (iii) any cross-wind gusts and so on. Unlike aircraft pitching stability, the physics of directional stability can be complex as the direction (yaw) and lateral (roll) motions are cross-coupled. V-tail rudder actuation offers both yaw and roll moments. While an aircraft is meant to rotate about the z-axis to get yaw, it also develops significant roll about the x-axis. Ailerons, not part of the empennage, also have cross-coupled motions. Activation of roll about the x-axis with aileron deflections is also coupled with yaw. Uncoupled yaw without roll will make an aircraft side slip (Figure 6.11).

To encounter yaw moment, aircraft in yaw attitude must have static directional stability to return to its equilibrium state. Also, an aircraft requires deliberate yaw moment by rudder deflection to turn in a yaw plane (it is not a coupled coordinated turn but associates with side slip). In this book, only the *rudder-fixed* directional stability is discussed. It is analogous to the *stick-fixed* case in the study of pitch stability.

Figure 6.11 gives the geometric and kinematic details considered in analyses. The aircraft is shown flying at a velocity V_∞ at a positive side slip to starboard side, $\beta = \tan^{-1}(v/u)$. Note that yaw angle, ψ, is different from sideslip angle, β. Yaw angle, ψ, is a geometric (Euler) angle measured from a reference line, say aircraft turned 90 has $\psi = 90°$. Sideslip angle, β, is the angle between the aircraft velocity vector and aircraft x-axis in the plane of symmetry. In a no sideslip ($\beta = 0$) coordinate turn, ψ indicates the extent of turn that has a value. When the aircraft is in a straight side slip, it has the aircraft velocity vector as the reference line. In this case with the sign convention, $\beta = -\psi$. This analysis deals only with the sideslip angle, β.

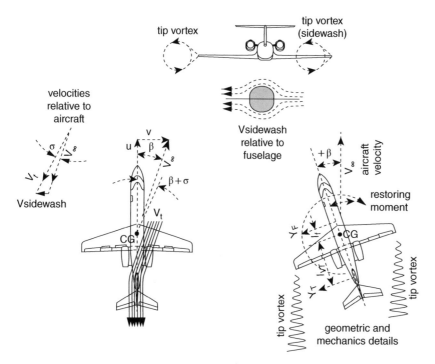

Figure 6.11 Direction stability (also see Section 18.3.2 for more explanation).

At yaw, the starboard wing tip vortex is inclined towards the V-tail creating side-wash and the fuselage yawed adding to the side-wash, making side-wash velocity $V_{sidewash}$ = (tip vortex contribution + fuselage contribution) when added to the free stream velocity, V_∞, the resultant relative, V_t is the increases the side slip angle to $(\beta + \sigma)$ on the V-tail. Y_T is V-tail lift, L_T, component and fuselage side force along the y-axis.

In equilibrium, the yaw moment of an aircraft about its CG is given next:

Aircraft yawing moment = (yawing moment by tail less aircraft) + (yawing moment by V − tail) = 0

(its mathematical form is given in Eq. (6.9a).)

Quoting DATCOM [10] (Section 5.2.3.1 Wing-body sideslip derivative, $C_{N\beta}$, in the linear angle of attack range):

> *The wing-body yawing moment due to sideslip can be considered as the sum of the yawing moments of the body, the wing and the wing-body interference. The wing contribution is important only at large incidences and is deleted from the solution of wing-body yawing moment due to sideslip. The method provided herein gives the total wing-body sideslip derivative $C_{N\beta}$ as the sum of the body and wing-body interference contributions.*

> *The unstable directional stability contribution of the fuselage is dominant in this analysis, and since slender-body theory predicts the body contribution to be essentially independent of Mach number, the method presented is considered to be valid for all speeds*
> *Experimental investigations have shown that the wing-body interference contribution to the yawing-moment derivative is essentially independent of sweep, taper ratio and Mach number. Furthermore, the evidence from Refs [1–3] is that the effect of wing vertical position on wing-body yawing moment is small.*

In the simplest form, the main contributors to yaw are from the fuselage and the V-tail. In order to understand the physics of the directional stability following simplifications considered in this section.

i. Thin surface-like wing and H-tail in parallel to the yaw plane contribute very little to the yawing moment, hence they not taken into account to keep analyses simple.
ii. Position of the wing and H-tail with respect to the fuselage has a small effect on the contribution to yaw moment [10] and is not considered here.
iii. Wing sweep and dihedral effects are not considered.
iv. Note that the role of CG is practically irrelevant when an aircraft is horizontal in the yaw plane because the weight vector has no component in the horizontal yaw plane. However, CG is useful reference point about which yaw moments are evaluated.

Therefore, neglecting lower order of yawing-moment contributions, this aircraft yawing-moment relation about its CG in equilibrium can be written in coefficient terms of an equation, as follows.

$$C_{N\_cg} = C_{NF} - C_{NVT} = 0 \ (V - \text{tail balancing fuselage yaw moment in side slip}) \tag{6.9a}$$

(the yaw force, Y_T, developed by the V-tail is the lift, L_{VT}, in coefficient form, C_{NVT})

The aircraft being symmetric about vertical plane all $C_N = 0$ at zero sideslip $(\beta = 0)$. Then the alternative form of Eq. (6.9a) can also be expressed in terms of rate of β change. This is a more useful relation to get C_{N_cg} at any β.

$$dC_{N\_cg}/d\beta = C_{NF\beta} - C_{NVT\beta} \tag{6.9b}$$

At equilibrium flight, $C_{N\_cg} = 0$; that is, $C_{NVT} = C_{NF}$ \hfill (6.10)

C_{NF} can be evaluated by Multhopp's method as was done in Section 5.11 to evaluate the fuselage pitching moment coefficient, C_{Mf}.

In coefficient form, the fuselage contribution can be written as

$$C_{NF} = -k_N k_{Rl} \left[(S_{Bs} l_b)/(S_w b) \right] \beta \tag{6.11}$$

where

k_N = an empirical wing-body interference factor
k_{Rl} = an empirical correction factor
S_{Bs} = the project side area of the fuselage
l_b = the fuselage length
b = wing span

In coefficient form, the V-tail contribution can be written as (see Equation (18.12), the lift force on V-tail, L_{VT}, is in coefficient form C_{LVT}):

$$C_{NVT\beta} = \left[(l_{VT}/S_{VT})/(S_w \times b) \right] \eta_{VT} (dC_{LVT}/d\beta) = C_{VT} \eta_{VT} (dC_{LVT}/d\beta) \tag{6.12}$$

where Eq. (6.2) gives, C_{VT} = Vertical Tail Volume Coefficient = $\frac{S_{VT} \times l_{VT}}{S_W \times b}$

or $S_{VT} = \dfrac{C_{VT} \times S_W \times b}{l_{VT}}$

Equation (6.9b) as the aircraft yawing-moment relation about its CG becomes:

$$C_{N\_cg} = -k_N k_{Rl} \left[(S_{Bs} l_b)/(S_w b) \right] \beta + C_{VT} \times \eta_{VT} C_{LVT} \tag{6.13}$$

K_n and K_{Rl} can be obtained from Figure 6.12a,b. (Reproduced from [10].)

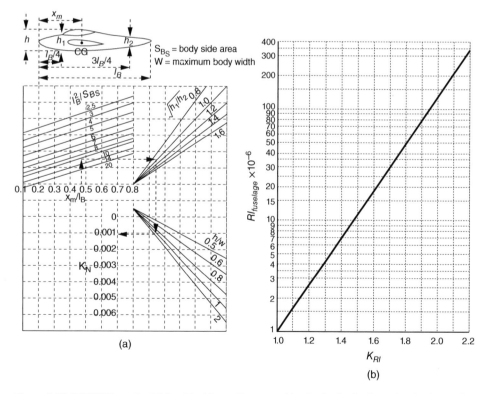

Figure 6.12 DATCOM graphs. (a) Empirical factor K_N versus side slip $C_{N\beta}$ for body + wing-body interference and (b) effect of fuselage Re on wing-body $C_{N\beta}$.

Noting that C_{LVT}, is subjected to the side-wash effect with angle $(\beta + \sigma)$
Therefore,

$$C_{\text{LVT}} = C_{\text{LVT}\beta} \times (\beta + \sigma)$$

or

$$(dC_{\text{LVT}}/d\beta) = C_{\text{LVT}\beta} \times (1 + d\sigma/d\beta)$$

Using Eq. (6.9a,b) it can be written as follows.

$$dC_{\text{N\_cg}}/d\beta = -k_{\text{N}}k_{\text{Rl}}\left[(S_{\text{Bs}}l_{\text{b}})/(S_{\text{w}}b)\right] + C_{\text{VT}}\,\eta_{\text{VT}} \times C_{\text{LVT}\beta} \times (1 + d\sigma/d\beta) \tag{6.14}$$

DATCOM [10] gives the algebraic relation as follows.

$$\eta_{\text{VT}}\,(1 + d\sigma/d\beta) = 0.724 + \frac{3.06(S_{VT}/S_W)}{(1 + \cos\Lambda_w)} + 0.4(z_w/d) + 0.009 \times AR_{\text{W}} \tag{6.15}$$

In this book, examples of z_w are small and neglected. Equation (6.15) reduces to

$$\eta_{\text{VT}}\,(1 + d\sigma/d\beta) = 0.724 + \frac{3.06(S_{VT}/S_W)}{(1 + \cos\Lambda_w)} + 0.009 \times AR_{\text{W}} \tag{6.16}$$

With the conclusion arrived at in evaluated $C_{\text{M\_HT}}$, the V-tail will be sized by using the statistical data as worked out in Example 6.3.

6.7.3.1 Rudder lock

Rudder lock is an undesirable situation that can happen if the V-tail is not properly designed. If an aircraft flies at an high sideslip angle, β, then the V-tail can stall and lose its control authority. The powerless rudder flicks aligning with the local flow. If the fin size is relatively small compared to the rudder, then the rudder gets locked into that flicked position and the pilot will then have to apply force to reposition to regain authority. If the extent of force required is in excess of what an average pilot can exert, then control recovery will not take place endangering safety. Accidents and incidents have happened with rudder lock, some ended in fatality. It is one of the examples of the complexities involved in empennage design.

Example 6.1 Find the H-tail (it is a T-tail) size of the Bizjet example that has the following data (α is in degrees) see the figure. Take the wing MAC aerodynamic centre to be at its quarter-chord $c/4$.

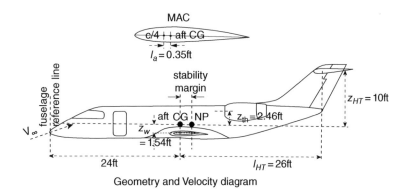

Geometry and Velocity diagram

Pertinent geometric details are as follows:

Wing	H-tail
Aerofoil: NACA 65–410	Aerofoil: NACA 65–210
MAC = 7 ft	$MAC_t = 4$ ft
$AR = 7.5$	$AR_t = 4.25$
$C_{L\_0} = -2.5$	$C_{L\_0} = -2°$
$i_w = 2°$	$i_t = 0°$ (negative camber)
$C_{mw\_1/4} = -0.7$	$C_{mt\_1/4} = -0.15$

Solution

From Eq. (6.7)

In equilibrium, with proper sign convention

$$\left[C_{L\_HT} \, (S_{HT}/S_w)(l_t/c) \, \eta_t \right] = \left[C_{Lw} \, (l_a/c) + C_{mw} \right] + C_{mfus}$$

$$V_{HT} \, C_{LHT} = \left[C_{Lw} \, (l_a/c) + C_{mw} \right] + C_{mfus}$$

where $V_{HT} = (S_{HT}/S_w)(l_t/c)$

At HSC

Wing contribution

Design $C_L = 0.4$
From Example 4.5,

$$C_{LW} = 0.17 + 0.0848 \times \alpha \text{ or } 0.17 + 0.0848 \times \alpha = 0.4$$

resulting into $\alpha = 0.23/0.0848 = 2.7°$

$$(l_a/\text{MAC}) = 0.35/7 = 0.05$$

$$C_{Lw} \, (l_a/c) = 0.40 \times 0.05 = 0.02 \text{ (CG aft of C/4, hence nose} - \text{up, +ve)}$$

From Example 4.5,

$$C_{M\_CGW} = C_{M\_0W} + (C_{M\_\alpha W})\alpha_W = -0.0615 + (0.00424)\alpha$$

or $C_{M\_CGW} = -0.0615 + (0.00424)\alpha = -0.0615 + (0.00424) \times 3.9 = -0.045$
or $[C_{Lw} \, (l_a/c) + C_{mw}] = -[0.02 + 0.045] = -0.065$

Fuselage contribution

From Example 5.1, fuselage moment coefficient

$$C_{mfus} = -0.0256 + 0.0067\alpha \text{ (where } \alpha \text{ is in degrees)}$$

$$\alpha_f = \alpha - i_w = 2.7 - 2 = 0.7$$

substituting, $C_{mfus} = -0.0256 + 0.0067 \times 0.7 = -0.0209$
substituting in Eq. (6.7),

$$\left[C_{L\_HT} \times (S_{HT}/S_w)(l_t/c) \, \eta_t \right] = \left[C_{Lw} \times (l_a/c) + C_{mw} \right] + C_{mfus}$$

$= 0.065 + 0.0209 = 0.0859$ (righting moment nose-up with a +ve sign countering the nose-down moment with −ve sign)

Tail contribution

It has a T-tail, hence downwash angle is not applied and $\eta_t = 0$.

It is has freedom to adjust.

Set $\alpha_{ht} = 0°$

C_{L_HT} (NACA 64–210) is set with inverted, that is, with −ve camber with following properties.

$$C_{l@\alpha=0} = 0.15, \alpha_{zero-lift} = (-2°) \text{ and } C_{m-cg} = -0.15$$

substituting in Eq. (6.7),

$$\left[C_{L\_HT} (S_{HT}/S_w)(l_t/c) + C_{mt} (S_{HT}/S_w) \right] = \left[C_{Lw} (l_a/c) + C_{mw} \right] + C_{mfus} = 0.065 + 0.0209 = 0.0859$$

NACA 64–210 3D wing correction is as follows.

3D correction:

$$dC_L/d\alpha = a = a_0/ \left[1 + (57.3a_0/e\pi AR) \right] = 0.105/ \left[1 + (57.3 \times 0.105)/(0.7 \times 3.14 \times 4.25) \right]$$

or $a = 0.105/(1 + 0.644) = 0.0638$

$$C_{L\_0} = (a_0 \times \alpha_{zero-lift})/ \left[1 + (57.3a_0/e\pi AR) \right] = (0.105 \times 2)/(1 + 0.644) = 0.127$$

On substituting

$$\left[0.127 \times (S_{HT}/S_w)(l_t/c) \right] + [C_{mt} (S_{HT}/S_w)] = 0.0859$$

$$(S_{HT}/S_w) \times [(0.127 \times (l_t/c) + C_{mt}))] = (S_{HT}/S_w) \times [(0.127 \times (26/7) + 0.15)] = 0.62 \times (S_{HT}/S_w) = 0.0859$$

or $S_{HT} = 0.0859 \times 323/0.62 \approx 45 \text{ ft}^2$: this is the minimum size for *stick-fixed* pitch stability.

In the case of inadvertent loss of the elevator, the aircraft becomes *stick free* reducing its static stability. This requires making the H-tail larger (typically ≈50% larger) than the minimum size estimated previously. The increased size also caters for gusts and manoeuvre for the *stick-fixed* case. At the early conceptual stages of the new aircraft project, statistical values are used to obtain H-tail size as worked in Example 6.3. This suits classroom practices.

Comments

This example is not exactly an industrial practice but meant to provide some insight to understand the physics of static stability and the role of the H-tail. The accuracy of the results are not substantiated, nor is the methodology tried with other aircraft. The results obtained cannot ensure the adequacy of the requirements. Figure 6.7 gives a typical pattern of moment characteristics (not related to this problem) contributed by the various components.

Therefore, a simplified empennage methodology using statics data is given next. Similar methods with better statistical details are practised in many industries during conceptual stages, especially for small aircraft design, yielding reliable results that will get finalised as design progresses. Example 6.3 gives an example on how to obtain empennage sizes in a simple but effective way.

History has shown tendency of empennage to grow larger. A generous empennage size ensures safety, with very small increase in weight and drag.

Example 6.2   Given the wing-body configuration, determine the V-tail size (*rudder-fixed* case only). Body characteristics (values taken from the worked-out Bizjet example in Chapter 8).

	V-tail aerofoil: NACA 65–010
Wing area $S_w = 323 \text{ ft}^2$	V-tail area, $S_V = 47.3 \text{ ft}^2$, $l_{vt} = 23.5 \text{ ft}$
Wing span $b = 49.2 \text{ ft}$	V-tail span, $b_V = 7 \text{ ft}$
Wing aspect ratio, $AR_W = 7.5$	V-tail aspect ratio, $AR_V = 1.04$ ($dC_{LVT}/d\beta$) = 0.04 (low aspect ratio)

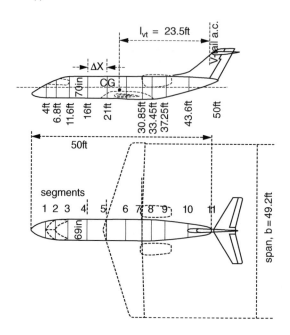

The projected side area of the body, S_{Bs}.

Station	Δx, ft	Area, ft²
1	4.0	$4 \times 2.46 = 9.84$
2	2.8	$2.8 \times 4.3 = 12$
3	4.8	$4.8 \times 5.7 = 27.4$
4	4.4	$4.4 \times 5.83 = 26.25$
5	5.0	$5 \times 5.83 = 29.15$
6	5.0	$5 \times 5.83 = 29.15$
7	4.85	$4.85 \times 5.83 = 28.3$
8	2.6	$2.6 \times 5.83 = 15.16$
9	3.8	$3.8 \times 5.76 = 21.9$
10	6.35	$6.35 \times 4.5 = 28.6$
11	6.4	$6.4 \times 2.54 = 16.25$
	Total	244 ft²

$l_b = 50$ ft	$S_{Bs} = 244$ ft²	$x_m = 23.8$ ft (CG)	$h = 5.83$ ft
w = 5.75 ft (body width)	$h_1 = 4.4$ ft @ 12.5 ft ($\frac{1}{4}l_b$)		
$h_2 = 5.6$ ft @ 37.5 ft ($\frac{3}{4}$@ 37.5 ft $(3/4l_b)l_b$)			

Wing characteristics:

$S_w = 323$ sq²	$b = 49.2$ ft

Additional characteristics:

$M = 0.74$	$R_1 = (9.2/50) \times 10^7$ per feet $= 1.84 \times 10^6$ per feet

Compute:

$X_m/l_b = 23.8/50 = 0.476$

$(l_b{}^2)/S_{Bs} = 50^2/244 = 10.25$

$\sqrt{(h_1/h_2)} = \sqrt{(4.4/5.6)} = 0.886$

$h/w = 5.83/5.75 = 1.014$

Figure 6.12a gives, $K_N = 0.0011$ per degree (Ref. [10], Figure 5.2.3.1–8)

$R_{efus} = 9.2 \times 10^7$

Figure 6.12b gives, $K_{Rl} = 1.92$ (Ref. [10], Figure 5.2.3.1–9)

Solution

$$C_{NF\_\beta} = -k_N k_{Rl}\left[(S_{BS} \times l_b)/(S_w b)\right] = -0.0011 \times 1.92 \times \left[(244 \times 50)/(323 \times 49.2)\right]$$

$$= -0.00211 \times (12\,200/15892) = -0.00211 \times 0.768 = -0.00162 \text{ per deg (based on } S_w \text{ and } b)$$

This destabilising directional moment is to be countered by the V-tail for the directional stability. Consider *rudder-fixed* case. Eq. (6.14) gives:

$$dC_{N\_cg}/d\beta = C_{NF\_\beta} + dC_{NVT}/d\beta = -k_N k_{Rl}\left[(S_{Bs}l_b)/(S_w b)\right] + C_{VT}\,\eta_{VT} \times C_{LVT\beta} \times (1 + d\sigma/d\beta)$$

where $C_{LVT\beta} = 0.035$ for a low aspect ratio V-tail of $AR_{VT} = 1.04$ (see Section 8.11.4)

From the DATCOM empirical relations, we get

$$\eta_{VT} \times (1 + d\sigma/d\beta) = 0.724 + \frac{3.06(S_{VT}/S_W)}{(1 + \cos \Lambda_w)} + 0.009 \times AR_W = 0.724 + \frac{3.06 \times (47.3/323)}{(1 + \cos 14)} + 0.009 \times 7.5$$

$$= 0.724 + (0.448/1.97) + 0.0675 = 1.019,$$

It is interesting to note that, for having a relatively low sweep, side-wash, σ, effects contribute only about a 2% increase.

Substituting

$$dC_{NVT}/d\beta = C_{VT} \times C_{LVT\beta} \times \eta_{VT}(1 + d\sigma/d\beta) = C_{VT} \times 0.035 \times 1.019 = 0.036\,C_{VT} \text{ per degree}$$

This gives in equilibrium

$$dC_{N\_cg}/d\beta = -k_n k_{Rl}[(S_{BS} \times l_b)/(S_w b)] + C_{VT}\eta_{VT}(dC_{LVT}/d\beta) = -0.00162 + 0.036\,C_{VT} = 0 \text{ or } C_{VT} = 0.045$$

This gives $S_{VT} = \frac{C_{VT} \times S_W \times b}{l_{VT}} = (0.045 \times 323 \times 49.2)/23.5 = 30.2 \text{ ft}^2$.

This is minimum V-tail size for the *rudder-fixed* consideration.

In case of the inadvertent rudder failure, the aircraft becomes unintentionally *rudder free*, when a good part of V-tail is lost reducing the static stability. Additional V-tail authority is required to make the V-tail a larger size (typically $\approx 50\%$ larger) than the minimum size estimated above, making it $1.5 \times 30.2 = 45.2 \text{ ft}^2$. The increased size caters for a cross-wind, one engine inoperative situation for multi-engine aircraft and manoeuvres for the *rudder-fixed* case. The result is not substantiated, but comes close to the figure obtained using the statistical value as worked out in Example 6.3.

Comments

This example is not exactly an industrial practice but is meant to provide some insight to understand the physics of static stability and the role of the V-tail as well as the use of DATCOM. The accuracy of the results are not substantiated, nor is the methodology tried on other aircraft. The results obtained cannot ensure the adequacy of the requirements.

Therefore, a simplified empennage methodology using statics data is given in Example 6.3. Similar methods with better statistical details are practised in many industries during conceptual stages, especially for small aircraft design, yielding reliable results that will get finalised as design progresses. Example 6.3 gives an example of how to obtain empennage sizes in a simple but effective way.

History has shown a tendency for empennages to grow larger. A generous empennage size ensures safety with a very small increase in weight and drag.

Example 6.3 Using statistical data, the following are taken to obtain empennage sizes of the aircraft given in Example 6.1. Use the information given in the diagram (from the statistic, the C_{HT} and C_{VT} values are taken to be higher than those obtained in Example 6.1 to keep empennage sizes on the generous side.

H-tail volume coefficient, $C_{HT} = 0.7$ (taken higher than what was obtained in Example 6.1 for the reason given before).

V-tail volume coefficient, $C_{VT} = 0.07$, same as in Example 6.2.

Solution
Use Eq. (6.1)

$$S_{HT} = (C_{HT})(S_W \times MAC)/L_{HT}$$

$S_{HT} = (0.7 \times 30 \times 2.132)/7.62 = 5.88\,m^2$ ($63.3\,ft^2$), which is about 20% of the wing area.
Use Eq. (6.2)

$$S_{VT} = (C_{VT})(S_W \times wing\,span)/L_{VT}$$

The $S_{VT} = (0.07 \times 30 \times 15)/7.16 = 4.4\,m^2$ ($47.3\,ft^2$), which is about 15% of the wing area.

This is a simple and effective way to obtain preliminary empennage sizes at the concept definition stage. The results compare well with the existing types.

The rest of the empennage geometric details are worked out in Section 8.11.4.

6.7.4 Aircraft Component Stability

A wing is meant to develop lift to sustain flight but is typically unstable with a destabilising nose-down moment (−ve). The cylindrical shape fuselage also has a destabilising moment contribution. Therefore, a wing-body combination is not flyable unless some restoring moment can be incorporated. One way is to add a horizontal with a positive nose-up moment to an extent that makes the total aircraft stable for the full flight envelope as shown in Figure 6.13 (reproduced from [2]). Chapter 18 is dedicated to aircraft stability and may be referred to for the sign conventions.

The degree of pitching moment curve slope, $C_{M\alpha}$ (−ve is stable), indicates the strength of stability; the lower the value, the more relaxed the pitch stability. Commercial aircraft require a high degree of pitch stability; on the other hand, the demands of fast response for combat aircraft benefits from relaxed stability (FBW aircraft are capable of making quick manoeuvres in slightly unstable stability). Utility aircraft/small aircraft are comfortable with intermediate level of pitch stability. The following values are indicative of current designs. (Requirements for specific mission role will require $C_{M\alpha}$ to suit performance capability.)

$C_{M\alpha} = -1.4 \pm 0.2$ highly stable (commercial aircraft)
$C_{M\alpha} = -0.8 \pm 0.2$ very stable (utility aircraft/small aircraft)
$C_{M\alpha} = -0.4 \pm 0.2$ relatively relaxed stability for rapid response (combat aircraft)

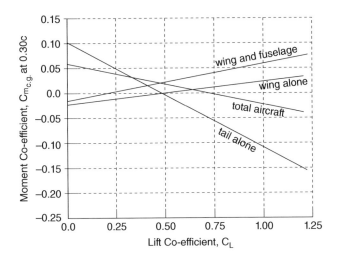

Figure 6.13 Tail contribution to stability.

To size the H-tail, it needs to produce an equal nose-up to balance the wing-body inherent nose-down moment. Therefore, the wing-body moment characteristic and the tail only moment characteristic have to be obtained.

The V-tail moment coefficient, C_N, characteristics are similar to the C_M of the H-tail, except it is free from any downwash effects. The degree of yawing-moment curve slope, $C_{N\beta}$ (+ve is stable), indicates the strength of stability; the lower the value, the more relaxed the pitch stability. In the case of the V-tail, combat aircraft require good high stability to stay locked on target without swaying and have a larger rudder for rapid control response. Modern military aircraft have an FBW system architecture to ensure the best available performance. Commercial aircraft can have less than military requirements. Modern commercial transport aircraft also have an FBW control system. Even some smaller utility aircraft have some form of yaw-damper to avoid 'snaking' in the demand to stay aligned with the runway.

Commercial aircraft: $\approx 0.08 < C_{N\beta} < 0.16$
Utility aircraft/small aircraft: $\approx 0.06 < C_{N\beta} < 0.14$
Combat aircraft: $\approx 0.15 < C_{N\beta} < 0.3$

The side force that the fuselage and V-tail could contribute to the rolling moment is discussed in Section 18.4.2.

Wind tunnel tests/computational fluid dynamics (CFD) analyses are required to establish the wing-body moment characteristic and the tail only moment characteristics. Nowadays, datasheets are rarely used in industry. This book uses statistical data that should offer comparable results.

6.7.5 Post Stall Behaviour

Figure 6.13 gives typical pitch stability characteristics of aircraft as a whole and component wise, showing the need for some stabilising force to keep in equilibrium in flight and controllable in manoeuvre. Figure 6.14 extends the characteristics into post stall phase of two typical modern civil aircraft examples. Post stall characteristics can only be determined through wind tunnel testing, or nowadays by CFD analyses and subsequently verified by flight tests. The behaviour varies from design to design and not easily ascertainable. The post region shows nose-up characteristics that contribute to aircraft behaviour requiring careful design considerations to keep in safe operation.

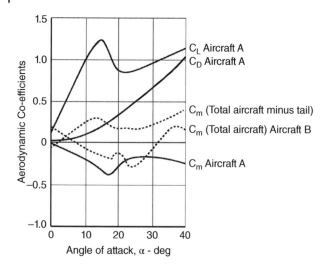

Figure 6.14 Post stall moment (two typical civil aircraft examples).

Post stall exhibits some form of nose-up tendencies on account that when a separated thick wing wake acts over the H-tail, airflow becomes unsteady. Loss of wing lift and the downwash angle of attack on H-tail results in a nose-up pitching moment. Any asymmetry of forces about the vertical can lead into an incipient spin stage that may prove dangerous during flight tests. It is more severe in the case of high AR and low sweep wings.

Aircraft A in Figure 6.14 represents typical civil aircraft design characteristics showing C_L, C_D and C_M variations with increasing angle of attack, α, starting from its operating range to well past stalled phase. Stall occurs at around $\alpha = 15°$ with rapid loss of lift, drag steadily rising and pitching moment showing pitch-up tendencies.

Aircraft A in Figure 6.14 represents typical civil aircraft design characteristics showing C_L, C_D and C_M variations with increasing angle of attack, α, starting from its operating range to well past stalled phase. Stall occurs at around $\alpha = 15°$ with a rapid loss of lift, drag steadily rising and pitching moment showing pitch-up tendencies.

In the case of the Aircraft B, in Figure 6.14 is the tail contribution in post stall region with C_M trimmed just above 30° angle of attack. In this case, the elevator loses its control authority and if enters a flat spin, recovery becomes difficult. Empennage design is rather complex and not dealt with in this book. Practising aeronautical engineers may refer to NACA TR 1339, which summarises longitudinal characteristics of a swept wing at a high Reynolds number.

6.7.6 Aerodynamic Shielding of the V-tail

The H-tail position relative to the V-tail is a significant consideration; the options available are shown in Figure 6.15. It can be from the lowest position through the fuselage to the other extreme, on the top as a T-tail. Any position in between is considered the mid-tail position.

Designers must ensure that the H-tail does not shield the V-tail. The wake (i.e. dashed lines in Figure 6.15 from the H-tail should not cover more than 50% of the V-tail surface and should also have more than 50% of the rudder area free from its wake to maintain control effectiveness, especially during spin and stall recoveries. Shifting the V-tail aft with the rudder extending below the fuselage will bring the fin and rudder adequately outside the wake. A dorsal and ventral fin can bring out more fin surface outside the wake, but the rudder must be larger to retain effectiveness. Lowering the H-tail would move the wake aft; however, if it is too low, it may hit the ground at rotation – especially if the aircraft experienced a sudden bank due to wind gusts.

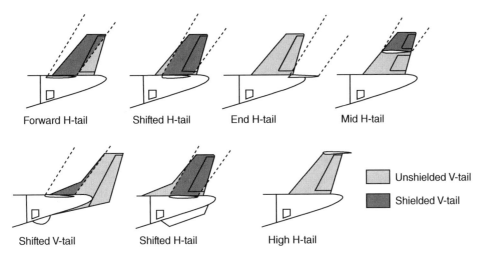

Figure 6.15 Positioning of the horizontal tail – V-tail shielding.

6.7.7 Wing Wake on the H-tail

Careful positioning of the H-tail can avoid wing wake avoid interfering with H-tail as shown in Figure 6.16. At a high angle of attack, the wing wake should avoid the H-tail in the near-stall condition so that the pitch control remains adequate. Stabiliser and elevator may be place away from the zone of stalled wing wake as shown in the diagram. Until the stall wing wake position can be established, the values of the angles in the figure may be taken to start with. CFD/wind tunnel test will show the position. An all-moving T-tail ensures higher elevator control authority to overcome any inadequacy if the tail is within the wing wake.

Care must be taken so that the H-tail is not within the entrainment effects of the jet exhaust situated at the aft end, which is typical for aft-mounted or within-fuselage jet engines. A military aircraft engine is inside the fuselage, which may require a pen-nib type extension to shield the jet-efflux effect on the H-tail. In that case, the H-tail is moved up to either the midlevel or the T-tail.

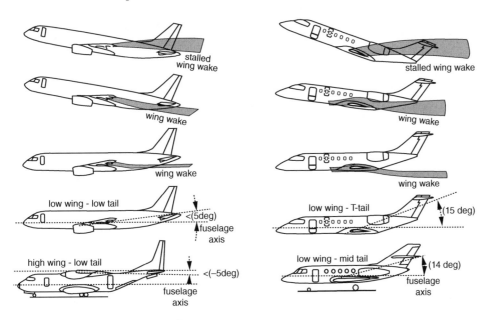

Figure 6.16 Wing wake on the H-tail.

6.8 Aircraft Control System

Chapter 18 covers the analytical considerations of aircraft motion in six degrees of freedom and its control. Figure 3.8 shows a Cartesian representation of the six degrees, comprising of three linear and three rotational motions. This section offers a brief description of associated control hardware and design considerations (for more details, refer to [2–9]). Aircraft control system weight is about 1–2% of *maximum takeoff weight* (MTOW).

6.8.1 Civil Aircraft Control Sub-System

The three-axes (pitch, yaw and roll) aircraft control has evolved considerably. Use of trim tabs and aerodynamic/mass balances alleviate hinge moments of the deflecting control surfaces easing pilot work-load. Some of the types in operation are given here.

   i. *Wire-pulley type.* This is the basic type. Two wires per axis act as tension cables, moving over low friction pulleys to pull the control surfaces in each direction. Although there are many well designed aircraft using this type of mechanism, frequent maintenance is required to check tension level and possible fraying of wire strands. If it is in improper tension then the wires can jump out of pulleys making the system inoperable. Other associated problems are dirt getting in, the rare occasion of jamming, elastic deformation of support structures leading to loss of tension and so on. Figure 6.17 shows both the wire-pulley (rudder and aileron) and push-pull rod (H-tail) types of control linkages. The push-pull rod is described next.

  ii. *Push-pull rod type.* The problems of wire-pulley type are largely overcome by the use of push-pull rods to move the control surfaces. Designers must ensure that the rods do not buckle under compressive load. In general, it is slightly heavier and a little more expensive but worth installing for ease of maintenance. Many aircraft use a combination of wire-pulley and push-pull rod arrangements (Figure 6.17).

  iii. *Mechanical control linkage boosted by a Power Control Unit* (*PCU*). As aircraft size increases, the forces required to move the control surfaces increase to a point when pilot work-load crosses the specified limit. Power assistance by PCU overcomes the problem. One of the problems of using PCU is that the natural feedback feel of control forces is obscured. Then artificial feel is incorporated for finer adjustment leading to smoother flight. PCUs can be hydraulic, linear driven electric motors or rotary actuators (there are many types). Figure 18.17 is supported by PCU.

  iv. *Electro-mechanical control system.* Large aircraft can save considerable weight by replacing mechanical linkages with electric signals to drive the actuators. Aircraft with FBW use this type of control system (Figure 18.16). Currently, many aircraft routinely use secondary controls (e.g. high-lift devices, spoilers, trim tabs etc.) driven by electrically signalled actuators.

  v. *Active control system. Feedback control systems* such as *stability augmentation systems* (SAS – e.g. a yaw-damper) were routinely deployed for some time. Modern aircraft, especially combat aircraft control systems have become very sophisticated. A FBW architecture integrated with a Full Authority Digital Engine Control (FADEC) is essential to these very complex systems when aircraft can fly under relaxed stability margins (see Section 18.10). Enhanced performance requirements and safety issues have increased the design complexities incorporating various kinds of additional control surfaces. Figure 6.18 shows typical subsonic transport aircraft control surfaces.

  vi. *Optically signalled control system.* The latest innovation uses optically signalled actuators. Advanced aircraft are already flying using fibreoptic lines to communicate with control systems.

6.8.2 Military Aircraft Control Sub-System

Figure 6.19 shows the various control surfaces/areas and system retractions required for a three-surface configuration. It can be seen that it has a lot more than modern civil aircraft have.

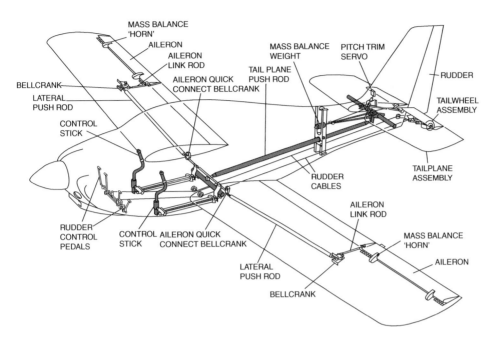

Figure 6.17 Wire-pulley and push-pull rod control system. (Source: Courtesy of Europa-Aircraft).

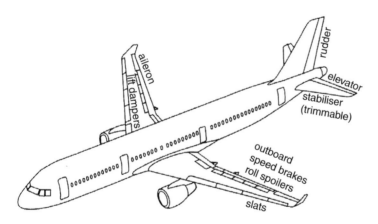

Figure 6.18 Civil aircraft control surfaces.

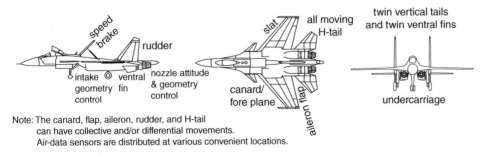

Figure 6.19 Military aircraft control surfaces.

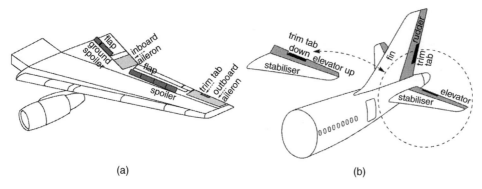

Figure 6.20 Aircraft control surfaces. (a) Wing control and (b) empennage control.

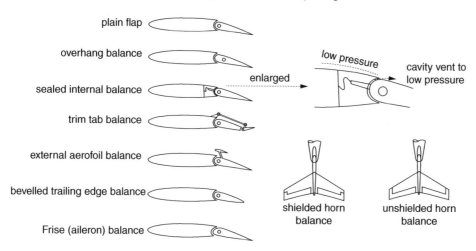

Figure 6.21 Types of aerodynamic balancing of controls.

Military aircraft control requirements are at higher level on account of the demand for hard manoeuvres. FBW assisted controls allows aircraft to fly with small amounts of negative stability margin for rapid response. The F117 is incapable of flying without FBW. Additional controls are the canard, intake scheduling and thrust vectoring devices. Fighter aircraft may use stabilators (see Section 6.5.3) in which the elevators can move differentially to improve roll capability. Stabilators are used collectively for pitch and differentially for roll control. Also aileron and rudder can be interconnecting. There could be automatic control in parallel with the basic system.

Military aircraft with short fuselage results in a short tail arm. Therefore, to make combat aircraft capable of hard and fast manoeuvres, a large tail surface is required that may give rise to some rolling moment problems and make it heavy. It is convenient to split a single large V-tail into a twin tail half as light (canted or vertical) as shown in Figure 6.19.

6.8.3 Reversible and Irreversible Control

Depending on the type of designs, based on the linkage between the control stick and control surface and their responses, the flight control systems are classified into two main types: (i) reversible control and (ii) irreversible control, as explained next.

Reversible control. A reversible flight control system results when force input from either side, that is, between control stick or control surface, can be felt by the other side. With a direct mechanical linkage (wire-pulley type or push-pull rod type – Figure 6.17) between control stick and control surface, the pilot force application is translated to control surface action to generate a hinge moment. In a similar manner, if the force is applied on the control surface, it will translate to the pilot directly leading them to experience the feel of the force on their control stick corresponding to the force on the control surface. This can be tested when the aircraft is on the ground, when pre-flight checks of the control surface movement move the flight deck controls. Mostly, small aircraft (around 1500 kg and below) have a reversible flight control system (see Section 6.8).

Irreversible control. As aircraft size increases, the forces required to move the control surfaces increase to a point when pilot work-load crosses the specified limit when power assisted controls are required. Various types power systems are described in Section 6.8. In the case of an irreversible flight control system, there is no direct mechanical linkage between the control stick/wheel and the flight control surface. A PCU is installed in between, delinking from manual input to more powerful mechanical output to supply the higher level of force required for control surface activation. The FBW technology is the most advanced type incorporating microprocessor-based power assisted controls. This makes the flight control system an irreversible system. In this case, any force input on the control will not be felt by the pilot, or intervened with by the PCU or other power assistance device installed. This desensitises the pilot, being unable to gauge the extent of input required. This problem is overcome by incorporating a calibrated artificial feel (Q-feel), simulating a reversible flight control system.

6.9   Aircraft Control Surfaces and Trim Tabs

Aerodynamic force (in this, case the lift force) on a movable control surface contributes to aircraft attitude change as an act of manoeuvre. The aerodynamic force develops a moment about the hinge of the rotating movement. Figure 6.20 shows wing and empennage control surfaces and associated trim tabs. There are two types of control surfaces, (i) primary control surfaces and (ii) secondary control surfaces. Sign conventions associated with controls deflection are given in Figure 3.11. (Section 18.6 deals with stick force.)

i. Primary control surfaces for 3° of angle of rotation:
   A central stick or yoke for pitch (elevator – stick movement as pull/push) and roll (aileron – stick movement sideways, right/left).
   Foot operated control yaw (rudder operation – right and left operated in a linked mechanism).
   Elevator and rudder constitute empennage. Various kind of empennage systems are described in Section 6.5.
   (The throttle is the engine power control system.)
ii. Secondary control surfaces
   Mainly, the secondary control surfaces supplements primary control surfaces for pilot ease by relieving pilot force, which otherwise could be high. These are trim surfaces incorporated on all primary control surfaces as required.
   There are many other kind of secondary control surfaces, for example, high-lift devices, spoilers, airbrakes and so on.

Control actuation requires pilot effort; the bigger the aircraft, the bigger are the control surfaces, requiring corresponding higher forces that can induce pilot work-load stress. This situation shows up in aircraft as small as 2000 lb in weight. The empennage sizing must guarantee to offer adequate authority in the full flight envelope, with special attention to critical manoeuvres. Control forces are dealt with in Section 18.6.

The design objective is therefore to alleviate pilot stick (pitch and roll) and pedal forces (yaw) to an qualitatively acceptable level stipulated by the certification agencies and accepted by qualitative flight tests. For bigger and/or advanced aircraft, this is achieved by incorporating power assisted controls. The methods to relieve pilot force are by balancing control surfaces, for example, with counter weights, aerodynamically with horn balance and so on. For bigger and/or advanced aircraft, this is achieved by incorporating power assisted controls.

Here, only the concept of trimabilty is introduced to establish the incorporation of trim tabs on the control surfaces as shown in Figure 18.21. To relieve the pilot work-load, especially when a large force is required to control, designers have discovered various methods to achieve that. An aerodynamic trim tab deflects in the opposite direction to the elevator deflection. This develops opposite elevator hinge moment to relieve some pilot (pull) force.

The analysis procedure adopted for elevators is also applicable to rudder and aileron hinge moment analyses and they also have trim tabs. Certification agencies have stipulated the limit of pilot force required in a harmonised manner for the three axes of controls. There are several methods to ways to relieve pilot forces, for example, aerodynamic mechanical and power assisted methods, as shown in Subsection 6.9.1. (Control surface instability, e.g. flutter, reversal, etc., should be avoided. These are specialised topics beyond the scope of this book.)

6.9.1 Aerodynamic Balancing of Control Forces

The design objective of balancing control force is to alleviate pilot stick (pitch and roll) and pedal forces (yaw) to an acceptable level stipulated by the certification agencies and accepted by qualitative flight tests. There are several methods to achieve this as described briefly in this subsection. It is desired that the empennage is sized and rigged in a manner such that a pilot will need to typically apply the least amount of force to actuate control deflection during the desired flight path. The empennage sizing must guarantee to offer adequate authority in the full flight envelope, with special attention to critical manoeuvres. (In this book, the control force balancing system is not designed but just introduces the possibilities to the readers. Determining hinge moments is notoriously difficult. Movable aircraft control design and sizing are topics of their own and beyond the scope of this book.)

In the following, are the main type of aerodynamic balance mechanisms shown in Figure 6.22.

i. *External overhang*. The hinge line is moved rearward away from its nose to have the forward overhung part move down while the active rear portion aft of the hinge line rotates up to give an aircraft nose-up

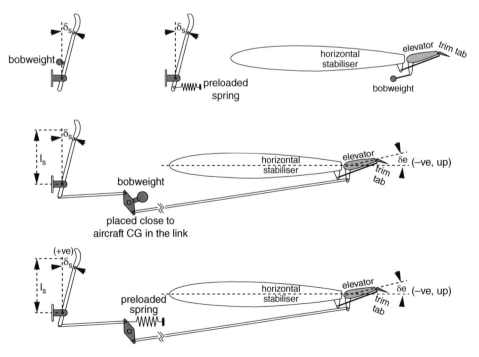

Figure 6.22 Bob-weight and spring.

pitching moment. The extent of front part from the hinge line is sized to provide an opposite force to relieve the active force on the larger part of the control surface (elevator) behind the hinge line.

ii. *Sealed internal overhang*. An innovative way to utilise the pressure field around the leading edge of the control surface. Here the overhang part is inside the wing aerofoil within a chamber and does not protrude out as in the case external overhang mentioned in point (i). This suits a thicker aerofoil of the wing with laminar flow characteristics so separation effects at the trailing edge are relatively small. Within the chamber, a sealed partition from the overhang extending from the leading edge of the control surface in enclosed with carefully placed gaps on both side, positioned to take the low pressure from the lift producing part of the curved leading edge. The differential of pressure between the upper and lower surfaces assists movement in the direction relieving pilot force.

iii. *Trim tab*. An effective way to relieve pilot force by incorporating trim tabs at the trailing edge of the control surface deflecting in the opposite direction to the control surface deflection. It is a mechanical linkage (or electrically operated) to extract the opposite moment of aerodynamic force from an additional extended surface (Figure 6.21). The linkage may be made as a fixed one or adjustable to suit pilot requirements after control is actuated (see Section 18.6).

iv. *External aerofoil balance*. Not so common in practice. The external aerofoil offers the additional force in the direction it is required.

v. *Bevel trailing edge*. Bevelling the trailing edges of elevator increases the hinge moment in the direction of rotation to relieve pilot force [11]. Boundary layer separation at the trailing edge affects the elevator hinge moment. When the elevator is deflected downward, then airflow over its upper surface separates earlier than at the lower surface causing change in elevator hinge moment. By sharp bevelling of the elevator trailing edge, the change can be minimised, assisting the pilot.

vi. *Frise (aileron)*. Ailerons are positioned at the trailing edge and at the outboard side of the wing. These are rigged in such a manner that each side deflects in opposite direction. The upward deflected side goes down on account loss of lift, while the other side moves with downward deflection: the wing goes up with the additional lift. Rolling of an aircraft by aileron activation is invariably associated with adverse yawing of aircraft in the other direction. The aerofoil of the Frise aileron is specifically shaped in such way that its lower side of the leading edge is sharper than upper side and, if required ,the hinged line is lowered. When the Frise aileron is deflected upwards, the sharp end of the overhang part ahead of hinge line protrudes outside the wing contour offering more drag than the other side of the wing, thereby reducing the yaw tendency. The overhang part also acts as balancing force, as described before. (A differential extent of deflection, that is, upward deflection more than downward is a better method to reduce adverse yaw. This also delays tip stall of the outward wing for having the aileron in a reduced angle of attack. Large aircraft incorporate a roll spoiler to increase drag on the wing, the side going down. Of course, the best method is to use a rudder for a coordinated turn without aircraft yawing.)

vii. *Horn balance*. This is like external overhang but instead of moving the hinge line aft, the free ends of the elevator tips have a 'horn' like extension ahead of the hinge line to relieve pilot force. It can be shielded or unshielded (Figure 6.21).

Mechanical balancing of control forces use a down-spring and/or a bob-weight in the linkage or on the stick itself (Figure 6.22). A properly sized moment arm to the bob-weight load can reduce the magnitude of the weight and it is desirable to place it close to aircraft CG.

Both are meant for the same usage and can installed singly or together. These are also seen as giving a 'feel' of the control in terms of control force per 'g'. For a pull up action from trimmed level flight, the down-spring/bob-weight resists with elevator down 'feel' sized to the requirement. Typically, a bob-weight provides a higher stick force per 'g'. Certification requirements are that the minimum stick force per g is 3 lbs per 'g' for a single handed stick controlled aircraft (one hand) and 4 lbs per 'g' for a two handed wheel controlled aircraft. The stick force gradient (Section 18.6) requires 1 lb force per 6 kts aircraft speed change.

But there is a difference between them. The bob-weight under the influence of inertial load pushes the stick forward. Therefore, in high 'g' manoeuvre, the pilot will experience higher stick force per 'g' when pulled. This will not be case for spring loaded control stick and stick force per 'g' remains the same what spring stiffness offers.

A bob-weight ahead of the hinge line of the control surface balances the weight of each side about its rotational axis. In this case, only the aerodynamic balancing is to be considered.

6.9.2   Power Assisted Managing of Control Forces

When aircraft becomes big enough with larger control surfaces, then a power assisted mechanical and or electric system is incorporated in the control linkage.

The *mechanical circuit*, which links the cockpit controls with the hydraulic circuits. Like the mechanical flight control system, it consists of rods, cables, pulleys and sometimes chains. The *hydraulic circuit*, which has hydraulic pumps, reservoirs, filters, pipes, valves and actuators. The actuators are powered by the hydraulic pressure generated by the pumps in the hydraulic circuit (refer to Section 6.8.1).

6.9.3   Active Control Systems

Refer to Sections 6.8.1 and 18.10.

6.10   Empennage Design

Empennage design is cost intensive, requiring generation of accurate design information; for example, the related design coefficients and derivatives required to make the analyses. Textbooks [6–9, 11] give the fundamental theories related to the role of the empennage. These detailed studies (CFD/wind tunnel) are carried out more intensively during Phase II of the project. DATCOM/Engineering Sciences Data Unit (ESDU) ([10, 12]) give empirical relations to obtain C_M, C_N, C_L and aircraft NP, to get a satisfactory preliminary empennage. This is a time consuming lengthy procedure.

Empennage design is complex. Like any other feed-back system analysis, control characteristics are rarely amenable to the exactness of theory due to lack of exact information of the system. Theoretical methods can produce a result but because of their limitations in modelling, the accuracy is difficult to substantiate; instead datasheet/CFD/test methods are used in industry. Datasheet methods served well until CFD analyses took over yielding better results much faster.

The empennage sizing must guarantee to offer adequate authority in the full flight envelope, with special attention to critical manoeuvres. It is important that the design should keep *flying qualities* within desirable levels by shaping the aircraft appropriately. Flying qualities can only be determined by actual flight tests. In a marginal situation, recorded test data could satisfy airworthiness regulations yet may not prove satisfactory to the pilots. Typically, several pilots evaluate aircraft flying qualities to iron out if there are debatable points. Flight tests start with the prototype empennage then move to fine-tuning by tailoring and riggings through the tests. Practically all modern aircraft incorporate active control technology to improve aircraft flying qualities. This is a routine design exercise and offers considerable advantages to overcome any undesirable behaviour, which is automatically corrected all the time.

Classroom analytical study can only give representative results to understand the physics involved in designing empennages. It can be questioned to apply optimisation methods in empennage design; one of the pertinent questions is, what is the objective? Surely, flying qualities are what pilot feels and are hard to interpret analytically?

At the early stages of the conceptual design phase, use of statistical data of the past empennage sizes in the class is a good starting point. Statistical data exhibit strong correlations and the methodology is used in industry at the inception of the project when not much data are available. This is adopted in this book.

Structural considerations for an empennage joint are given in Section 19.12.

H-tail. H-tail design has additional constraints over wing design, as its size depends on wing and fuselage sizes and their moment characteristics. In addition, the location of aircraft CG shifts between the permissible forward-most and aft-most limits. For safety reasons, the H-tail must provide stability for the aft-most CG position, which requires a larger H-tail size than what is required at the forward-most CG position requiring higher force to control. Power assisted controls allow the H-tail to be on the larger size for safe operation.

A larger H-tail size can easily be tailored by trimming the tips. On the other hand, enlarging a smaller sized H-tail is lot more expensive; in a difficult situation it may have to be re-designed and then fabricated for re-installation.

Wing position with respect to fuselage gives the H-tail arm, L_{HT} length, satisfying the extreme CG locations. Fuselage length is deterministic (see Section 5.9.1). The moving wing position enters into wing chasing of the CG and can become unwieldy. Reposition of the H-tail can offer some adjustment to the H-tail arm, L_{HT}. Therefore, the method used here is a simple one based on the statistics of past designs.

V-tail. The procedure for V-tail design follows the same methodology as used in H-tail design, this time in the yaw-plane instead of in the pitch plane. The V-tail, being in the plane symmetry with a symmetric aerofoil, does not produce a moment in normal flying without yaw making it relatively easier to design. It deals with C_N instead of C_M.

6.10.1 Simplified Empennage Design Methodology

Typical choices for the empennage design are suggested here. This is concerned with the planform area, its shape, sweep, taper ratio and twist. Cost consideration for empennage design is important. Positioning of a conventional tail is linked with the wing position with respect to the fuselage.

Although Examples 6.1 and 6.2 demonstrated the DATCOM methods, this book uses statistical data as worked out in Example 6.3. The reason for it is explained next in the two DATCOM worked-out examples and in Section 6.7.3. Given next is a step-by-step approach to decide empennage planform size and shape.

Step 1 (Find H-tail and V-tail areas)
  Uses tail volume coefficients from statistics to obtain the empennage areas.
Step 2 (Decide on the aerofoil)
  An H-tail can have a symmetric aerofoil or low camber aerofoil placed inverted to obtain a nose-up moment at a lower incidence to the wing wake. H-tail aerofoil t/c is comparable to wing aerofoil at MAC. Small recreational aircraft operation below 150 mph can get away with thick flat plate with rounded leading edge.
  A V-tail has a symmetric aerofoil, with similar t/c comparable to wing aerofoil at MAC. Positioning of the V-tail is always at the plane of symmetry, propeller driven aircraft may require 1–2° offset to counter the propeller wash in swirl and the torque reaction of the heavy propeller. V-tail with sweep back gives an additional tail arm, lt, for a T-tail mounted on the on the top.
Step 3 (Decide on the planform)
  Decide the planform shape. Typically, it follows the wing planform shape but at a lower AR, that is, a tapered wing inherits a tapered planform empennage.
Step 4 (Decide on the dihedral/anhedral, Λ and λ)
  Once H-tail and V-tail areas obtained, their dihedral/anhedral, Λ and λ, have to be decided. At this stage, hold empennage sweep should be no less than the wing sweep, but may increase to gain in tail arms. Keep taper ratio, λ, at around 0.25–0.5, but it may take a suitable $\lambda \approx 0.3$.

6.10.2 Control Surfaces Design

Control surfaces, for example, elevator, rudder and aileron, are subcomponents as a part of H-tail, V-tail and wing, respectively. Design of these control surfaces are not dealt with in this book due to their complexities, associated with hinge moments, gaps, balancing, trim surfaces, in addition to the usual aerodynamic and

CG considerations. Instead, the required areas, taken from the statistics of similar designs, are earmarked. However, to understand the physics of control surface design, only the elevator design concept is discussed as given next.

So far, the H-tail was designed at a stick-fixed trimmed equilibrium state of aircraft flight. The static stability characteristics of aircraft are measured in term of the slope $dC_M/d\alpha$, a more negative value it makes more stable. The higher the stability, the higher the stiffness requiring more power to control and its limit is specified for pilot ease. The equilibrium state of operation is when $C_M = 0$ at a particular 'α' for the set-up, that is, at a particular C_L, in the example case at HSC with $C_L = 0.4$. However, to manoeuvre, say, in pitch plane to climb, pilot will have to use the control surface elevator up (−ve upward deflection) that will change aircraft C_L to a higher value as climb speed is slower than the cruise speed. Higher C_L, that is, at higher α, reduces the C_M as can be seen in Figure 6.13. To maintain equilibrium during climb, the aircraft setting of H-tail needs to changed.

H-tail setting can be adjusted by either (i) by rotating the entire H-tail that can pose problems in subsonic operation or (ii) by rotating the aft part of the H-tail about a hinge line as an elevator, as the conventional accepted design that is now in practically all subsonic aircraft in operation.

Examine the case for climb when additional lift in terms of ΔC_M requires the aircraft to rotate for climb by applying the elevator up. When the rotation reaches to a steady climb, the equilibrium equation, with an H-tail contribution, can be written in line with Eq. (6.7), given here.

$$C_{mcg} = \text{(Pitching moment by wing)} + \text{(Pitching moment by fuselage)} + \text{(Pitching moment by H − tail)} = 0$$

$$(C_{L\_HT} \times C_{HT}) + [(dC_L/d\alpha)_{HT}] \times \alpha_t \times C_{HT} = -\left\{\left[C_{Lw}(l_a/c) + C_{Mw}\right] + C_{Mfus}\right\}_{@steady\,climb} \tag{6.17}$$

where $[(dC_L/d\alpha)_{HT}] \times \alpha_t \times C_{HT}$ is incremental downward moment ΔC_{m_HT} developed by the upward rotation of the elevator angle by δ_e.

Trimming. This ΔC_{M_HT} increment is achieved by the pilot applying force to pull the control stick to make elevator rotating upward. Climb takes time to reach the desired height that can strain a pilot by holding the stick applying force. This force can be relived with rotating a trim tab in the opposite direction built into the system (Figure 6.17). Fine-tuning to zero force is done by the pilot by adjusting a trim wheel/switch in the cockpit.

6.10.3 Effect of Power in Moment Contribution

This is based on the position of the thrust line with respect to CG affecting the aircraft moment. Power-on and power-off moments will be different. Induced airflow on the H-tail from jet blast and propeller slipstream over the H-tail alter H-tail moment characteristics. Power effects are not dealt with in this book.

6.11 Other Planar Surfaces

Aircraft are invariably installed with other surfaces such as the ventral fin, dorsal fin, pylon, speed brakes and so on, as described in the following subsections.

6.11.1 Ventral Fins

A good aircraft should not have a ventral fin that contributes to weight. It comes from two reasons from the inadequacy shown during flight tests, as follows.

i. If there is not enough 'weathercock' stability on account of an inadequate fin area. An additional planar area is attached at the plane of symmetry at the aft fuselage under the belly to a size required to make up for the inadequacy of the fin size, as shown in Figure 6.23. The placement must ensure ground clearance at aircraft takeoff rotation, if required, install a wheel to protect impact.

Figure 6.23 Dorsal and ventral fins.

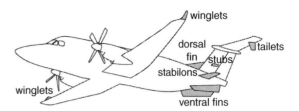

ii. If the aircraft exhibits inadequate handling characteristics such as pitching problems and/or lateral oscillations, for example, a Dutch roll/swaying. Installing a pair of ventral fins (small inverted fixed Vee-tail) in either side of the aircraft plane of symmetry, set inclined with some anhedral angle at the aft fuselage sides, as shown in Figure 6.23. The ventral chord length is considerably longer than its span and set aligned to aircraft at aircraft cruise attitude to minimise drag. Pitching problems and lateral oscillation problems have different origins, hence setting the pair of ventral fins to the fuselage has a different anhedral angle. The two cases are explained next.

When a high angle of attack, α, if the H-tail retains pitch-up tendency, then lift force on ventral fins on account of the angular attack offers nose-down moments. The size and anhedral angle of ventral fins are determined by a series of flight tests in a trial and error method, CFD analyses can reduce the number of flight tests.

iii. In case of lateral oscillation problems, a ventral fin can minimise oscillations, the size and anhedral angle are established by a series of flight tests; in a trial and error method, CFD analyses can reduce the number of flight tests.

Good examples incorporating ventral fins are the Learjet family, MIG21 and so on. Nowadays, incorporating FBW technology gives safety to keep aircraft within the safe flying envelope. All modern commercial transport has FBW technology. Advanced fifth generation combat aircraft have a vector thrust mechanism for both axes overcoming inadequacies that can arise in both pitch and yaw controls.

6.11.2   Dorsal Fins (LERX on the V-tail)

Dorsal fins act in the same way as in the case of leading edge root extension (LERX)/strakes on wings. The dorsal fin is a triangular shaped planar surface (can be flat or contoured to fin the V-tail aerofoil) installed on the fuselage in front the fin. In the case of a fin getting stalled at high angle of attack, β, making rudder ineffective and causing rudder lock (not mechanical seizure), a situation when the aircraft cannot respond to rudder deflection on account of airflow breaking away earlier at the fin surface; sometimes it may lead into a worse situation where the rudder acts in reverse. Like wing strakes (Section 4.11), the dorsal fin develops additional vortex lift covering a good part of the fin allowing an adequate part of the rudder to initiate aircraft to respond and obtain full authority eventually.

Beech 1900C presents a unique empennage design consideration. To keep cost down by retaining components commonality, the pressurised Beechcraft 1900 was derived from their earlier successful design, the Beechcraft Super King Air aircraft. In addition to having a dorsal fin and a pair of ventral fins, the Beech 1900C has another pair of stub like horizontal surfaces (known as stabilons – Figure 6.23) both in the middle of the aft fuselage plus a pair of stub like small horizontal planes in roughly the middle of the stabiliser. The T-tail also has small vertical fin like plates (called tailets – Figure 6.23) on the lower surface close to the tip. Evidently, these are installed after analysing flight tests to enhance stability. The authors do not have the manufacturers' design philosophies but stabilons and small surfaces on the fin seem to increase the static stability margin to accommodate extended CG limits. Tailets can serve to improve an aircraft's lateral stability. This is a fine example of the complexities of empennage design; a topic that stands on its own and is beyond the scope of this book as mentioned earlier.

6.11.3 LERX on the H-tail

Strakes (LERX) at the H-tail play an important role just as they do on the wing (see Section 4.6.1), to additional vortex lift at a high tail plane angle of attack, α_{ht}. This kind of strake comes as modification attachment in case flight tests show inadequacy in elevator authority (e.g. HAL Deepak, HPT32).

6.11.4 Pylons

Pylons are support structure and not intended to contribute aerodynamic forces other than their inherent drag. Pylons attached to the wing or fuselage carry a wide variety of loads, for example, engine nacelle pod, armament loads, electronic pods, drop tanks and so on. The case of a pylon carrying a nacelle pod is shown in Figure 6.24. To reduce interference drag between the nacelle and fuselage, sometimes it can be shaped into slight curvature (banana shape).

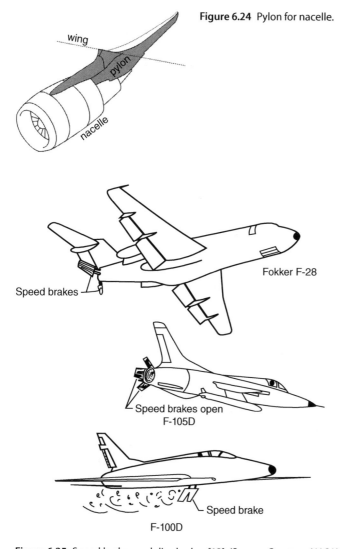

Figure 6.24 Pylon for nacelle.

Figure 6.25 Speed brakes and dive brakes [13]. (Source: Courtesy NASA).

6.11.5 Speed Brakes and Dive Brakes

Speed brakes and *dive brakes* have the same definition. They are mounted specifically on the fuselage for military aircraft and as spoilers on the wings for civil aircraft. However, there are civil aircraft that use this type of device mounted on the fuselage. Speed brakes are specifically designed to reduce speed rapidly, typically on approach and in military combat manoeuvres. Care must be taken to minimise aircraft moment change when these brakes are deployed.

Speed and dive brakes are primarily drag-producing devices positioned in those areas that will create the smallest changes in moments (i.e. kept symmetrical to the aircraft axis with the least moment arm from the CG). Figure 6.25 shows fuselage-mounted devices.

The Boeing F22 does not have a separate dive brake. It uses the two rudders of the canted V-tail deflected in opposite directions along with spoilers and flaps deflected upward and downward, respectively.

References

1 Campbell, J. P. (1945). Experimental verification of a simplified Vee-tail theory and analysis of available data on complete models with Vee tails. NACA-ACR-L5A03.

2 Perkins, C.D. and Hege, R.E. (1988). *Airplane Performance, Stability and Control*. New York: Wiley.

3 Nelson, R.C. (1998). *Flight Stability and Automatic Control*, 2e. Illinois: McGraw-Hill.

4 Etkin, B. (1972). *Dynamics of Atmospheric Flight*. Wiley.

5 Pamadi, B.N. (2015). *Performance, Stability, Dynamics and Control of Airplanes*, 3e. Reston, VA, USA: AIAA.

6 Anderson, J. (2011). *Fundamentals of Aerodynamics SI*. McGraw Hill.

7 Kuchemann, D. (2012). *The Aerodynamic Design of Aircraft*. AIAA.

8 Abzug, M.J. and Larrabee, E.E. (2002). *Aircraft Stability and Control*. Cambridge, UK: Cambridge Aerospace Press.

9 Seckel, E. (1964). *Stability and Control of Airplanes and Helicopters*. New York: Academic Press.

10 Fink, R. D. (1978). USAF Stability and Control, DATCOM. Report Number AFWAL-TR-83-3048, Flight Dynamic Laboratory, Wright-Patterson Air force Base, Ohio, April 1978.

11 Nicolosi, F., Vecchia, P.D., and Ciliberti, D. (2013). *A New Vertical Tail Design Procedure for General Aviation and Turboprop Aircraft*. Linkoping: University of Naples 'Frederico II', EWADE.

12 Engineering Sciences Data Unit (ESDU), IHS ESDU, 133 Houndsditch, London

13 Talay, T. A (1976). NASA SP-367; Introduction to the Aerodynamics of Flight.

7

Aircraft Statistics, Configuration Choices and Layout

7.1 Overview

The design considerations of aircraft components, for example, the wings, empennages, fuselages, nacelle pods, intakes and associated systems, are presented in Chapters 4–6 to generate boilerplate templates of aircraft configurations at the conceptual stage. This chapter progresses to the next phase to see aircraft as a whole, all components assembled from the available multitude of options to generate possible preliminary candidate configuration for a new aircraft project with given requirements and specifications. A large number of various types of aircraft configurations are included in this chapter for the readers to examine the options available. These large varieties of configurations should assist in some creative thinking on how to combine the various options of aircraft components. This is the last chapter in Part I, serving as the prerequisite information to carry out the aircraft design exercises dealt with in Chapter 8 (Part II). Chapter 7 gives configuration choices and supporting statistics, and details pertinent to aircraft configuration. Undercarriages and power-plants are dealt with separately in Chapters 9 and 10.

During the last century, many aircraft configurations have appeared in their time; today, most of those are not relevant to the current practices. One good example is the Douglas DC3 Dakota manufactured in its thousands. Older designs, no matter how good they were, cannot compete with today's designs. This book addresses only those well-established currently operating designs as shown in the recent *Jane's All the World's Aircraft Manual* [1]; however, references are made to some interesting and unique older aircraft configurations. These older designs may not be applicable to the current practices but they offer some interesting innovative ideas to study.

Section 1.3 showed growth patterns in the aircraft operational envelope (e.g. speed-altitude capabilities). It continued with a classification (Section 1.3.1) of generic aircraft types that show distinct patterns within the class in order to narrow down the wide variety of choices available. Statistics is a powerful tool for establishing design trends and some pertinent statistical parameters are provided herein.

The first part of this chapter presents the statistics of the past aircraft designs. This provides a tool for trend analyses of aircraft designs. The trend analyses show remarkable correlations, irrespective of their design origins from various countries. The statistical analyses give some idea of what to expect from the proposed new aircraft, that is, typical weights, wing area, thrust loading empennage sizes and so on. The current design and configuration parameters serve as the template for identifying considerations that could influence new designs with improvements.

This chapter also shows various types of aircraft and how they are configured, for example, incorporating types of wing planform, fuselage shape, intake shapes and positioning and empennage arrangements as the 'building blocks' for shaping an aircraft in many possible configurations. Artistic aesthetics are considered as long as they do not unduly penalise aircraft cost and performance. The new Boeing787 Dreamliner (see Figure 1.8) and Airbus350 aircraft shapes are good examples of the respective companies' latest high-subsonic commercial transport aircraft. It is interesting that the Dreamliner configuration transitioned to the new B787 with more conventional aero-shaping. The B787's advances in technology were not as radical a venture compared to Boeing's earlier Sonic Cruiser proposal (see Figure 7.18a), which was shelved. These

Conceptual Aircraft Design: An Industrial Approach, First Edition. Ajoy Kumar Kundu, Mark A. Price and David Riordan.
© 2019 John Wiley & Sons Ltd. Published 2019 by John Wiley & Sons Ltd.

decisions were made by one of the world's biggest and best companies; the Sonic Cruiser was not a fantasy – it simply was not timed right for market demand. This signifies the importance of conducting a market study, as emphasised in Chapter 2.

Along with the considerations to generate aircraft external mouldlines, the chapter covers the internal arrangements for the types of aircraft. These arrangements affect weight and must be studied at the conceptual design phase.

The important information given in this chapter will assist in developing a methodology of how to configure the proposed new aircraft with worked-out examples as shown in the next Chapter 8. The readers should return to Chapters 2 through 7 to extract the information necessary to configure aircraft.

This chapter covers the following topics:

Section 7.2: Chapter Introduction
Section 7.3: Civil Aircraft Mission (Payload Range)
Section 7.4: Civil Aircraft Statistics
Section 7.5: Internal Arrangements of Fuselage – Civil Aircraft
Section 7.6: Some Interesting Aircraft Configurations – Civil Aircraft
Section 7.7: Summary of Civil Aircraft Design Choices
Section 7.8: Military Aircraft Detailed Classification
Section 7.9: Military Aircraft Mission (Domain Served)
Section 7.10: Military Aircraft Statistics (Template for New Design)
Section 7.11: Military Aircraft Component Geometries (Possible Options)
Section 7.12: Miscellaneous
Section 7.13: Summary of Military Aircraft Design Choices

This chapter will be better understood after covering Chapters 8–16 (Part II), culminating in aircraft sizing and engine matching to concept finalisation by freezing the design for management to review for a possible 'go-ahead'.

7.2 Introduction

This chapter studies past aircraft designs of topical interest to demonstrate that a definite trend exists within the class of aircraft. Past designs have a strong influence on future designs; real-life experience has no substitute and is dependable. It is therefore important that past information be properly synthesised by studying statistical trends and examining all aspects of influencing parameters in shaping a new aircraft – this is one of the goals of this book. Many types of aircraft are in production serving different sector requirements – civil and military missions differ substantially. It is important to classify aircraft categories in order to identify strong trends existing within each class.

Existing patterns of correlation (through regression analysis) within a class of aircraft indicate what may be expected from a new design. There are no surprise elements until new research establishes a radical change in technology or designers introduce a new class of aircraft. In civil aircraft design, a 10–15% improvement in the operating economics above the current designs within the class is considered good; a 15–20% improvement is excellent. Of course, economic improvements must be supported by gains in safety, reliability and maintainability, which in turn add to the cost.

Readers are encouraged to examine the potential emerging design trends within the class of aircraft. In general, new commercial aircraft designs are extensions of the existing designs that conservatively incorporate newer, proven technologies (some are fall-outs from declassified military applications). Currently, the dominant aerodynamic design trends show diminishing returns on investment. Structure technologies seek suitable new materials (e.g. composites, metal alloys and smart adaptive materials) if they can reduce cost, weight and/or provide aerodynamic gains. Engine designs have scope for aerodynamic improvements to save on fuel consumption and/or weight. Chapter 1 highlights that the current challenges lie in manufacturing philosophy, better maintainability and reliability incorporating vastly improved and miniaturised systems (including microprocessor-based avionics for control, navigation, communication and monitoring). This book briefly addresses these topics, particularly from the weight-saving perspective. It is convenient to separate civil and military aircraft examples as their mission requirements are very different and cannot be compared (see Section 1.9).

CIVIL AIRCRAFT

7.3 Civil Aircraft Mission (Payload Range)

The payload-range capability constitutes the two most important parameters to represent the commercial transport aircraft operational role. These two are the basic aircraft requirements emerging out of market studies for new aircraft projects. From these two parameters, the aircraft mission profile and performance capabilities are specified.

The graph in Figure 7.1 [1–4] shows the payload-range capabilities for several subsonic-transport aircraft (i.e. turbofans and turboprops). The figure captures more than 50 different types of major current designs in operation. The points in the graph include the following aircraft: Lear 31A, Lear 45, Lear 60, Cessna 525A, Cess 650, Cess 500, Cess 550, Cess 560, Cess 560XL, ERJ 135ER, ERJ 140, ERJ 145ER, CRJ 100, CRJ 700, ERJ 170, DC-9-10, CRJ 900, ERJ 190, 737–100, 717–200, A318–100, A319–100, A320–100, Tu204, A321–100, 757–200, A310–200, 767–200, A330–200, L1011, A340–200, A300–600, A300–100, DC-10-10, MD11, 777–200, 747–100, A380, Short 330 and 360, ATR 42 and 72, Jetstream 31, SAAB 340A, Dash 7 and 8, Jetstream 41, EMB 120, EMB 120ER, Dornier 328–100 and Q400.

The trend shows that the range increases with payload increases, reflecting the market demand for the ability to fly longer distances. Long-range aircraft will have fewer sorties and will need to carry more passengers at one time. The classic comparison between the A380 versus the B747 passenger capacity is captured within the envelope shown between the two straight lines in the figure. It is interesting that there are the high-subsonic,

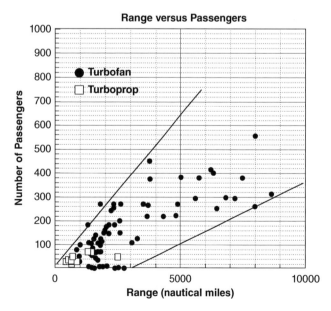

Figure 7.1 Passengers versus range – statistics of 70 aircraft (Ref. [1–4]).

long-distance executive jets, the Bombardier Global Express and Gulfstream V (not shown in the figure) carrying executives and a small number of passengers very long ranges (>6500 nm) at a considerably higher operating cost per passenger. At the other extreme, there are almost no products carrying a high passenger load for shorter ranges (i.e. <2000 nm). The authors consider that the future may show potential markets in this class of aircraft in highly populated large countries. Countries with substantial population centres could fly more passengers within their borders, such as in China, India, Indonesia, Russia and the USA.

It is obvious that turboprop aircraft flying at slower speeds cater to the shorter-range market sector as they provide better fuel economy than the jet aircraft. At short range missions, the relatively small time saving by operating faster jet aircraft is overridden by the economic gains using slower turboprop aircraft.

7.3.1 Civil Aircraft Economic Considerations

Commercial aircraft operation is singularly driven by revenue earned from the fare-paying passengers and cargo. In the operating sector, passenger *load factor* (LF_{pax}) is defined as the ratio of occupied seats to available seats, as follows.

$$Load\ factor(LF_{pax}) = \frac{Occupied\ seats}{Available\ seats}$$

Typically, for aircraft of medium sizes and larger, operational costs break even at approximately one-third full capacity (this varies among airlines; fuel costs at 2000 level fuel price with regular fares) – that is, a load factor of about 0.33. Of course, the empty seats could be filled with deregulated reduced fares, thereby contributing to the revenue earned. Until the 1960s, passenger fares were fixed under government regulation. Since the 1970s, the airfare structure has been deregulated – an airline can determine its own airfare and vary as the market demands. With the inflation and increase in fuel price, the break-even load factor, LF_{pax}, has gone up in a fluctuating manner, in some cases exceeded 50%.

It is appropriate to introduce here the definition of the dictating economic parameter, *seat-mile cost*, which represents the unit of the aircraft Direct Operating Cost (DOC) that determines airfares to meet operational costs and sustain profits. DOC is the total cost of operation for the mission sector. The US dollar is the international standard for aircraft cost estimation. Operational economics are discussed in detail in Chapter 16.

$$Seat\text{-}Mile\ Cost = \frac{Direct\ Operating\ Cost}{Number\ of\ passengers \times range\ in\ nm} = (cents/seat/nm) \qquad (7.1)$$

The seat-mile cost is the aircraft operating cost per passenger per nautical mile (nm) of the mission sector. The higher the denominator in Equation (7.1), the lower is the seat-mile cost (i.e. DOC). Therefore, the longer an aircraft flies and/or the more it carries, the lower the seat-mile cost.

To get some idea, typical Seat-Mile Costs (cents/seat/nm), are listed here. Turboprop Seat-Mile Costs are about 10–20% in the comparable class, currently not bigger than regional jets.

Small Bizjets ≈ $0.25–0.35

Regional jets ≈ $0.1–0.15

Mid-size – mid-range ≈ $0.07–0.08

Size – long range ≈ $0.05–0.06

A careful market study could fine-tune an already overcrowded marketplace for a mission profile that offers economic gains with better designs. Section 2.5.1 addresses the market study so that readers understand its importance.

7.4 Civil Subsonic Jet Aircraft Statistics (Sizing Parameters)

Section 2.3 provides familiarisation with typical civil aircraft geometry and its components. This section gives statistical relations of aircraft geometry, weight and engine thrust of large number of aircraft. To have a refined relation, readers are encouraged to generate their own statistical data for the class of aircraft to be studied. It may be possible to eliminate a parameter or two if they are not close to the regression line, as well investigate the cause being not close to the regression line. This is the starting point that culminates in Chapter 16 through formal sizing to concept finalisation of the new aircraft.

This section examines the statistics of current aircraft geometry and weight to identify aircraft sizing parameters. Regression analyses are carried out to demonstrate a pattern as proof of expectations. With available statistics, aircraft can be roughly sized to meet specifications. Sections 1.14.1 and 1.14.2 may be referred to examine the benefits of using statistics, where applicable, compared to using empirical relations.

Definitions of various types of aircraft mass (i.e. weight) are provided in Chapter 9. Some of them are required in this section, as given next (payload could be passengers and/or cargo):

MEM: Manufacturer's empty mass – the finished aircraft mass rolls out from the factory line.

OEM: Operator's empty mass = MEM + crew + consumables – it is now ready for operation.

MTOM: Maximum takeoff mass = OEM + payload + fuel, loaded to maximum design mass.

OEMF: Operational Empty Mass Fraction = ratio of OEM to MTOM.

Mass per passenger is revised to 100 kg (220 lb) from the earlier value of 90 kg (200 lb), which includes baggage allowance.

In the USA, weight unit is in lb instead of mass in kg and uses their equivalents maximum takeoff weight (MTOW), OEW (Operator's Empty Weight), MEW (Manufacturer's Empty Weight) and OEWF (Operational Empty Weight Fraction).

7.4.1 Civil Aircraft Regression Analysis

There are definite statistical relationships between aircraft weight and geometries. The choices are made to configure aircraft not arbitrary – definite reasons are associated with the choices made. Figures 7.2–7.9 (except Figure 7.7) [2] indicate strong relational trends in geometry and weight within the class of aircraft studied evolved through design considerations. Chapters 4–6 covered relevant points on aircraft component properties and status, as summarised next.

1. *Wing group.* Chapter 4 is concerned with planform shape and size of the wing. It gives various types of wing, their relation between aerofoil thickness to chord ratio and wing sweep versus Mach number. Options for high-lift devices are described in Section 4.13. This chapter gives the statistics of wing size, wing load, aspect ratio (AR) and wing span versus aircraft weight.
2. *Fuselage and nacelle group.* Chapter 5 is concerned with shaping and sizing of the fuselage. It gives statistics of fuselage; for example, cross-section types, fuselage length for the passenger capacity, fineness ratio, seating arrangement and fore and aft end closure data. It also covered nacelle geometries and locations. This chapter covers the statics of engine size versus aircraft weight, which in turns gives the nacelle size.
3. *Empennage group.* Chapter 6 is concerned with H-tail and V-tail geometries and introduces the role of tail volume coefficients. This chapter gives the statistics of H-tail and V-tail tail volume coefficients.

7.4.2 Maximum Takeoff Mass versus Number of Passengers

Figure 7.2 describes the relationship between passenger (PAX) capacity and MTOM, which also depends on the mission range for carrying more fuel for longer ranges. In conjunction with Figure 7.1, it shows that lower-capacity aircraft generally have lower ranges (Figure 7.2a) and higher-capacity aircraft are intended for higher ranges (Figure 7.2b). Understandably, at lower ranges, the effect of fuel mass on MTOM is not shown as strongly as for longer ranges that require large amounts of fuel. There is no evidence of the square-cube law, as discussed in Section 4.17. It is possible for the aircraft size to grow, provided the supporting infrastructure is sufficient.

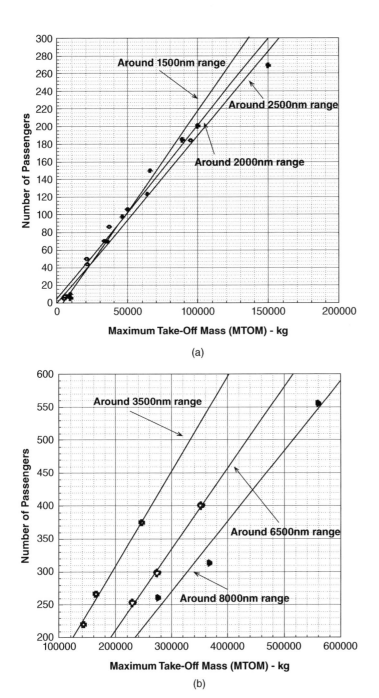

Figure 7.2 Number of passengers versus MTOM (high-subsonic jet aircraft). (a) Lower capacity and (b) higher capacity.

Figure 7.2 shows an excellent regression of the statistical data. It is unlikely that this trend will be much different in the near future. Considerable scientific/technical breakthroughs will be required to move from the existing pattern to better values (the blended-wing-body (BWB) configuration is an example when it becomes available). Light but economically viable material, superior engine fuel economy, miniaturisation of systems

Table 7.1 MTOM per passenger versus range.

Range, nm	MTOM/passenger, kg/PAX
1500	400
3500	600
6500	900
8000	1050

architecture, microprocessor-based aircraft and engine controls are some of the areas in which substantial weight reduction is possible.

In conjunction with Figure 7.2, it can be seen that longer-range aircraft have higher MTOM per passenger as shown Table 7.1 Longer-range capability requires carrying a higher proportion of fuel. Below 2500 nm, the accuracy degenerates; the weight for in-between ranges is interpolated.

Example 7.1   For a mission profile with 300 passengers and a 5000-nm range, the MTOM is estimated at $750 \times 300 = 225\,000$ kg (comparable to the Airbus 300–300 aircraft).

7.4.3   MTOM versus Operational Empty Mass

Figure 7.3 provides crucial information to establish the relationship between the MTOM and the OEM. Figure 7.3a shows the regression analysis of the MTOM versus the OEM for 26 turbofan aircraft, indicating a predictable OEM growth with MTOM being almost linear. At the lower end, aircraft with fewer than 70 passengers (i.e. Bizjet, utility and regional jet class) have a higher OEMF (around 0.6 – sharply decreasing). In the midrange (i.e. 70–200-passenger class – single-aisle, narrow-body), the OEMF is around ≈ 0.55. At the higher end (i.e. more than 200 passengers – double-aisle, wide-body), it levels out at around $\approx 0.5 \pm 0.05$; the MTOM is about twice the OEM. The decreasing trend of the weight fraction is due to better structural efficiencies achieved with larger geometries, as well allowingmore accurate design and manufacturing methods.

Aircraft load (see Chapter 17) is experienced on both the ground and in the air, which depends on the MTOM. The load in the air is a result of aircraft speed-altitude capabilities, the manoeuvrability limit and wind. A higher speed capability would increase the OEMF to retain structural integrity; however, the OEM would reflect the range capability for the design payload at the MTOM (see Figure 7.3).

Figure 7.3a is represented in higher resolution when it is plotted separately for the midsized aircraft, as shown in Figure 7.3b for midrange-size aircraft. It also provides insight to the statistical relationship between the derivative aircraft of the Boeing737 and Airbus320 families. The approaches of the two companies are different. Boeing, which pioneered the idea, had to learn the approach to the family concept of design. The Boeing737-100 was the baseline design, the smallest in the family. Its growth required corresponding growth in other aero-structures yet maintaining component commonality as much as possible. Conversely, Airbus learned from the Boeing experience: Their baseline aircraft was the A320, in the middle of the family. The elongated version became the A321 by plugging in constant cross-section fuselage sections in the front and aft of the wing, while retaining all other aero-structures. In the shortened versions, the A319 came before the even shorter A318, maintaining the philosophy of retaining component commonalities. The variants were not the optimised size, but they were manufactured at a substantially lower cost on account of low development costs from retaining component commonality, decreasing the DOC and providing a competitive edge.

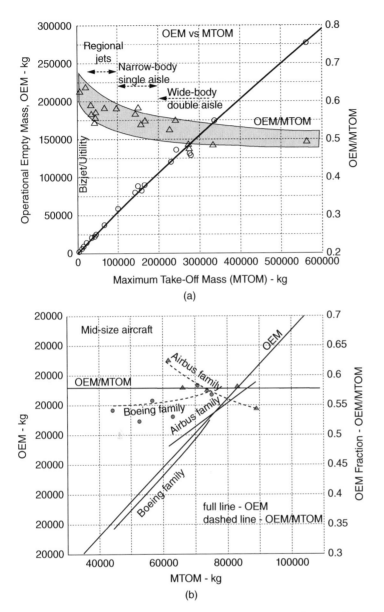

Figure 7.3 OEM versus MTOM (high-subsonic jet aircraft). (a) All types of aircraft and (b) midrange – Boeing and Airbus families.

7.4.4 MTOM versus Fuel Load

Figure 7.4 shows the relationship between fuel load, M_f and MTOM for 20 turbofan aircraft, this graph also shows the fuel fraction, M_f/MTOM. It may be examined in conjunction with Figures 7.1–7.3, which show the range increase with the increase in MTOM, fuel mass, M_f and fuel fraction, M_f/MTOM increase.

Fuel mass increases with aircraft size, reflecting today's market demand for longer ranges. The long-range aircraft fuel load, including reserves, is about half the MTOM. For the same passenger capacity, there is statistical dispersion at the low end. This indicates that, for short ranges (around 1400 nm to cross-country ranges of around 2500 nm), with a wider selection of comfort levels and choice of aerodynamic devices, the fuel content is determined by the varied market demand. At the higher end, the selection narrows, showing a linear trend.

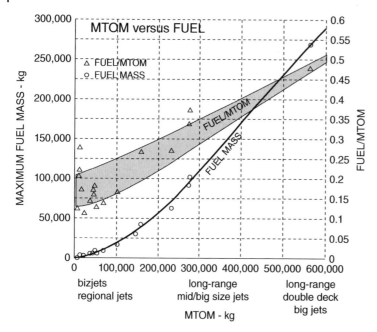

Figure 7.4 Maximum fuel load (full tank) versus MTOM (Ref. [2]).

Figure 7.3 indicates that larger aircraft have better structural efficiency and Figure 7.4 indicates that they also have a higher fuel fraction for longer ranges.

Example 7.2 An Airbus 380 aircraft with 500 passengers and 8000 miles range has the following weights breakdown.

MTOM = OEM + payload + fuel = 275 000 + 50 000 + 250 000* = 575 000 kg, which gives,

OEFM = OEM/MTOM = 275 000/575 000 = 0.478 (Figure 7.3a)

Fuel Fraction = (Fuel Load, M_f)/MTOM = 250 000/575 000 = 0.435* (Figure 7.4)

* has lower value as onboard fuel is not at the maximum capacity.

7.4.5 MTOM versus Wing Area

Whereas the fuselage size is determined from the specified passenger capacity, the wing must be sized to meet performance constraints through a matched engine (see Chapter 14). Figure 7.5 shows the relationships between the wing planform reference area, S_W and the wing loading, W/S_W, versus the MTOM.

Wing loading, W/S_w, is defined as the ratio of the MTOM to the wing planform reference area (W/S_W, kg m^{-2} or lb$_m$ ft^{-2}, if expressed in terms of weight; then, the unit becomes N m^{-2} or lb$_f$ ft^{-2}). This is a significant sizing parameter and has an important role in aircraft design.

The influence of wing loading is illustrated in the graphs in Figure 7.5. The tendency is to have lower wing loading for smaller aircraft and higher wing loading for larger aircraft operating at high-subsonic speed. High wing loading requires the assistance of better high-lift devices to operate at low speed; better high-lift devices are heavier and more expensive. On account of the cost and complexities involved in sophisticated high-lift devices, the current trend is to have lower wing load to accept simpler high-lift devices (see Figure 4.34). These graphs are useful for obtaining a starting value (i.e. preliminary sizing) for a new aircraft design that would be refined through the sizing analysis.

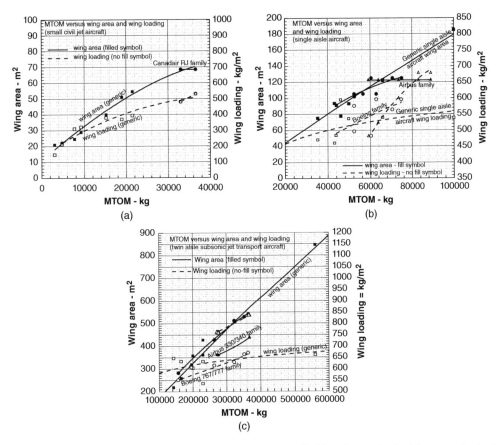

Figure 7.5 Wing area, S_W, versus MTOM (Ref. [2]). (a) Small aircraft, (b) midrange single-aisle narrow-body aircraft and (c) large twin-aisle wide-body aircraft.

The growth of the wing area with an increase in aircraft mass is necessary to sustain flight. A large wing planform area is required for better low-speed field performance, which exceeds the cruise requirement. Therefore, wing-sizing (see Chapter 14) provides the minimum wing planform area to simultaneously satisfy both the takeoff and the cruise requirements. Determination of wing loading is a result of the wing-sizing exercise.

Smaller aircraft operate in smaller airfields. To keep the weight and cost down, simpler types of high-lift devices are used in small aircraft. This results in lower wing loading 40–100 lb ft^{-2} (i.e. $\approx$200–500 kg m^{-2}), as shown in Figure 7.5a. Aircraft with a range of more than 3000 nm need more efficient high-lift devices. It was shown previously that aircraft size increases with increases in range, resulting in wing loading increases 80–150 lb ft^{-2} (i.e. $\approx$400–700 kg m^{-2} for midrange aircraft) when better high-lift devices are considered.

Here, the trends for variants in the family of aircraft variant design can be examined and given in Figure 7.5b,c. The Airbus 320 baseline aircraft is in the middle of the family. Conversely, the Boeing 737 baseline aircraft started with the smallest in the family and was forced into wing growth with increases in weight and cost; this keeps changes in wing loading at a moderate level.

Larger aircraft have longer ranges; therefore, wing loading is higher to keep the wing area low, thereby decreasing induced drag. For large wide-body twin-aisle, subsonic jet aircraft (see Figure 7.8c), the picture is similar to the midrange-sized, narrow-body single-aisle aircraft but with higher wing loadings to keep wing size relatively small (which counters the square-cube law discussed in Section 4.17).

7.4.6 Wing Geometry – Area versus Loading, Span and Aspect Ratio

The main parameters associated with the wing geometry are its area, AR, span and sweep. Figure 4.23 gives the relations of wing sweep, its aerofoil *t/c* versus maximum operating Mach number. Here, Figure 7.6 gives the relation between wing load, span and AR and wing area. It may be noted that wing AR requires consultation with the structural engineers to obtain the a high value. In the graph, the sample data are within 7–10. The AR shows a scattering trend. In the same wing-span class, the AR could be increased with advanced technology but it is restricted by the increase in wing load. Current technology requires it is in a position to have an AR up to 15.

Figure 7.6 shows that the growth of the wing span is associated with the growth in wing loading.

With steady improvements in new material properties, miniaturisation of equipment and better fuel economy, wing span is increasing with the introduction of bigger aircraft (e.g. Airbus 380). Growth in size results in a wing root thickness large enough to encompass the fuselage depth when a BWB configuration becomes an attractive proposition for large-capacity aircraft. Although technically feasible, it awaits market readiness, especially from the ground-handling perspective at airports.

Examples of some current subsonic-transport aircraft wing planforms are given in Figure 7.7.

7.4.7 Empennage Area and Tail Volume Coefficients versus Wing Area

Once the wing area is established along with fuselage length, the empennage areas (i.e. H-tail, S_{HT} and V-tail, S_{VT}) can be estimated from the static stability requirements. Section 3.22 discusses the empennage tail volume coefficients to determine empennage areas.

Statistics show large spread in the values of S_H and S_V with respect to wing area, S_W. Figure 7.8 shows growth for H-tail and V-tail surface areas with the wing area, S_W. Reference [2] is the best source to get accurate data

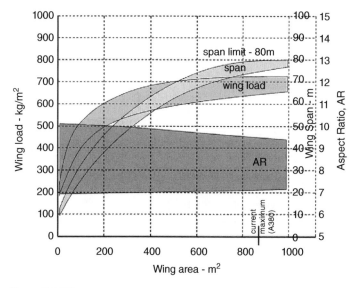

Figure 7.6 Wing geometry.

Figure 7.7 Civil aircraft wing planform shapes.

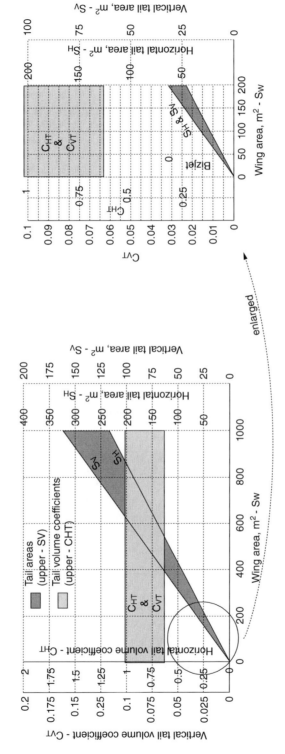

Figure 7.8 Statistics of tail volume coefficients.

on S_H and S_V, these are included in Figure 7.8. However, [2] does not give the tail volume coefficients C_{HT} and C_{VT}. These will have to be laboriously computed from the three-view diagrams given for the aircraft. This can be done with reasonable accuracy if the drawings are enlarged, the drawing scale can be determined from the linear dimensions; for example, wing span, fuselage length and so on. There is a large discrepancy in the C_{HT} and C_{VT} values as shown in the band in Figure 7.8. The same pattern can be observed in [1] of a large number of aircraft data given for older aircraft. Other good sources of aircraft statistics are found in [3, 4] and the websites listed in (Websites: (i) http://richard.ferriere.free.fr/3vues/; Specifications of F-22 (F/A-22), www.f22fighter .com/Specs.htm; Milavia, military aircraft; www.milavia.net/aircraft/; University of Southampton, Aircraft design resources; www.soton.ac.uk/~jps7/AircraftDesignResources/LloydJenkinson%20data; Elsevier, Civil Aircraft Design, book website, http://booksite.elsevier.com/9780340741528/authors/default.htm; VRR, air-craft containers, https://vrr-aviation.com/products/aircraft-containers/; Searates, air cargo ULD containers, www.searates.com/reference/uld/)

Modern aircraft with *fly-by-wire* (FBW) technology can operate with more relaxed stability margins, espe-cially for military designs, and hence require smaller empennage areas compared with older conventional designs (see Figure 12.18). It is suggested that readers make a separate plot to generate their own aircraft statistics at a better resolution for the particular class of aircraft in which they are interested in order to obtain an appropriate average value.

Typically, for civil aircraft designs, the horizontal tail, S_H, area is about 25–30% of the wing reference area, depending on the length of fuselage. For long, large aircraft, the vertical tail, S_V, area is around $\approx$12–0.15% of wing reference area, S_W. For long, large aircraft, the vertical tail area is around $\approx$12–0.15% of wing reference area, S_W. For short aircraft, the vertical tail area is around $\approx$15–0.18% of wing reference area, S_W. It is expected that readjustment of the V-tail and adjustment to S_V will be required. More detailed analyses are also expected to be carried out in the next design phase, Phase II.

In this book, the trim surfaces locations are earmarked and are not sized. Designers have to ensure that there is adequate trim authority (trim should not run-out) at any condition. This is normally done in Phase II after the configuration is frozen.

7.4.8 Aircraft Lift to Drag Ratio (L/D) versus Wing Aspect Ratio (AR)

Equation (4.18) derives the expression for aircraft induced drag, $C_{Di} = (C_L^2/\pi AR)$, which contributes a signif-icant part (as high as about a third) of aircraft total drag. The higher the AR, the lower the C_{Di}. The aim is to design aircraft to have as high as possible lift-to-drag ratio (L/D).

One of the effective way is to have as high as possible AR the structural design consideration allows. High AR wings are more expensive to design and manufacture, hence they are more meaningful for fare earning transport aircraft. Fuel saving as a result of having lower drag soon offsets the incremental cost of manufacture. The following values given in Table 7.2 may be used in this book, which are typical in the class of aircraft.

There are exceptions and readers may examine such designs.

7.4.9 MTOM versus Engine Power

The relationships between engine sizes and the MTOM are shown in Figure 7.9. Turbofan engine size is expressed as sea-level static thrust (T_{SLS}) in the International Standard Atmosphere (ISA) day at takeoff ratings, when the engine produces maximum thrust (see Chapter 11). These graphs can be used only for preliminary sizing; formal sizing and engine matching are dealt with in Chapter 14.

Thrust loading (T/W), is defined as the ratio of total thrust ($T_{SLS\,tot}$) of all engines to the weight of the aircraft. Again, a clear relationship can be established through regression analysis. Mandatory airworthiness regula-tions require that multiengine aircraft should be able to climb in a specified gradient (see Federal Aviation Administration (FAA) requirements in Chapters 14 and 15) with one engine inoperative. For a twin-engine aircraft, failure of an engine amounts to a 50% loss of power, whereas for a four-engine aircraft, it amounts to a 25% loss of power. Therefore, the T/W for a two-engine aircraft would be higher than for a four-engine aircraft.

Table 7.2 Aircraft lift to drag ratio versus wing AR.

Aircraft category	Aspect ratio range (AR)	Lift to drag ratio range (L/D)
Propeller driven		
Small aircraft ($\approx < 0.3$ Mach)	5–7 (mostly rectangular)	10–12
Large aircraft ($\approx < 0.5$ Mach)	6–8 (tapered wing dominates)	12–14
Jet propulsion (invariably with tapered wing and with some sweep)		
Smaller aircraft ($\approx 0.65 <$ Mach < 0.5)	6.5–7.5	12–15
Larger aircraft ($\approx > 0.75$ Mach)	8–12 (tapered wing dominates)	14–16
Supersonic military aircraft		
High sweep delta wing	3–4 (tapered wing dominates)	4–6
Low sweep trapezoidal wing	4–5 (tapered wing dominates)	6–12

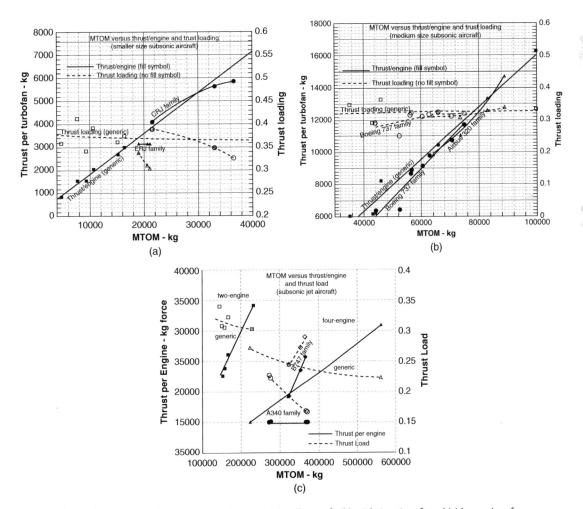

Figure 7.9 Total sea-level static thrust versus MTOM. (a) Small aircraft, (b) mid-size aircraft and (c) large aircraft.

The constraints for engine matching are that they should provide sufficient takeoff thrust to simultaneously satisfy the (i) field length specifications, (ii) initial climb requirements and (iii) initial high-speed cruise requirements satisfying the market specifications. An increase in engine thrust with increase of aircraft mass is obvious for meeting takeoff performance. Engine matching depends on wing size, number of engines and type of high-lift device used. Propeller-driven aircraft are power rated in P per kilowatt (hp or shp), which in turn provides the thrust. Turboprops use power loading, P/W, instead of T/W.

Smaller aircraft operate in smaller airfields and are generally configured with two engines and simpler flap types to keep costs down. Figure 7.9a shows thrust growth with size for small aircraft. Here, thrust loading is from 0.35 to 0.45. Figure 7.9b shows midrange statistics, mostly for two-engine aircraft. Midrange aircraft operate in better and longer airfields than smaller aircraft; hence, the thrust loading range is at a lower value, between 0.3 and 0.37. Figure 7.9c shows long-range statistics, with some two- and four-engine aircraft – the three-engine configuration is not currently in use. Long-range aircraft with superior high-lift devices and long runways ensure that thrust loading can be maintained between 0.22 and 0.34; the lower values are for four-engine aircraft.

7.5 Internal Arrangements of Fuselage – Civil Aircraft

This section covers the internal arrangements of civil aircraft fuselage, mainly on transport category. Military aircraft internal details are different and are dealt with in Section 7.8.

Internal fuselage details, as such, do not affect aircraft external geometry except the front fuselage to meet the requirements of flight deck dealt with in the next section. However, internal fuselage details affect aircraft weight, thereby requiring wing size to be resized. Therefore, it is worthwhile studying fuselage internal details in order to make arrangements to keep weight as low as possible to compete in the marketplace. In classroom project work, it is suggested to make space allocations for facilities layout in an intelligent manner complying with regulatory requirements.

7.5.1 Flight Crew (Flight Deck) Compartment Layout

The pilot flight deck, of course, is at the front-closure end of the fuselage to provide forward vision. The maximum accommodation is two side-by-side, generously spaced seats; an additional crew member for larger aircraft is seated behind the two pilots (Figure 7.10). In the past, there were two flight crews to assist two pilots; today, with improved and reliable microprocessor-based multi-functional display systems, two flight crews have become redundant. There could be provision for one. The windscreen size must allow adequate vision (see Figure 7.10), especially looking downward at high altitudes, during landing and during ground manoeuvres.

The pilot's seat is standardised as shown in Figure 7.11, with generous elbowroom to reduce physical stress. The seat can also be made to recline.

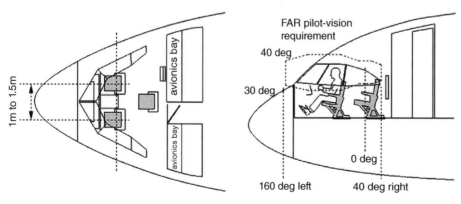

Figure 7.10 Pilot cockpit.

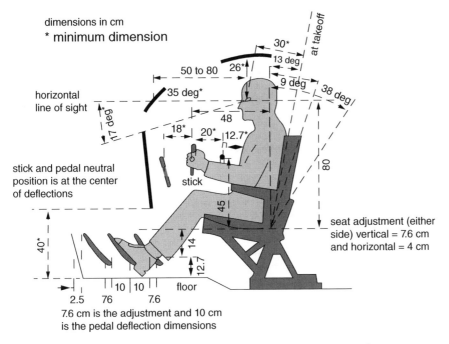

Figure 7.11 Standard pilot seat.

7.5.2 Cabin Crew and Passenger Facilities

Cabin seat arrangements are dealt with in Section 5.8. In this section, crew and passenger facilities are described. A vital fuselage design consideration is offering passenger services – the greater the passenger number, the more complex the design. This book does not cover details of interior design, a specialised state-of-the-art feature that is more than the mere functionality of safety, comfort and efficient servicing. The aesthetics also offer an appealing and welcoming friendly environment to passengers. Physiological and psychological issues such as deep vein thrombosis, claustrophobia and fear of flying can be minimised through careful design of the seat-pitch arrangement, window locations, environmental controls (i.e. pressurisation and air-conditioning) and first-aid facilities. Discussed herein are typical seat pitch, toilet and service-galley arrangements in fuselage-space management that contribute to fuselage length.

The minimum number of cabin crew is subjected to government regulations. No cabin crew is required for fewer than 19 passengers, but can be provided if an operator desires. For 19–29 passengers, at least one cabin crew is required. For 30 or more passengers, more than one cabin crew is required. The number of cabin crew increases correspondingly with the number of passengers.

7.5.3 Seat Arrangement, Pitch and Posture (95th Percentile) Facilities

Figures 7.12 and 7.13 illustrate typical passenger seating-arrangement designs, which can be more generous depending on the facilities offered by the operator. *Pitch* is the distance between two seats and varies from 28 (tight) to 36 in. (good comfort). Seat and aisle width are shown in the next figure. Typically, seat widths vary from 17 (tight) to 22 in. (good comfort). Seats are designed to meet the 16-g governmental impact regulations.

Table 7.3 gives typical seat pitch, width and aisle width (there could be a variation of dimensions from operator to operator) as currently prevalent in service. There is flexibility build in to the design to convert seating arrangements, or even remove seats, as the market demands.

Smaller aircraft with fewer passengers (i.e. up to four abreast at the lower mission range) can have a narrower aisle because there is less aisle traffic and service. For larger aircraft, the minimum aisle width should be at least 22 in. (Figure 7.13).

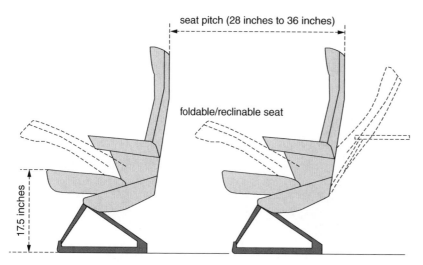

Figure 7.12 Passenger seat pitch.

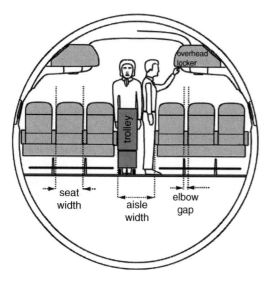

Figure 7.13 Passenger seat and aisle width.

Table 7.3 Seat/aisle pitch and width, inch (cm).

	Seat pitch	Seat width	Aisle width
Economy class			
Low comfort	28 (72)	18 (45.8)	17 (43.2)
Medium comfort	30 (76.2)	19 (48.2)	19 (48.2)
High comfort	32 (81.2)	20 (50.8)	22 (56)
Business class	34–36 (86.4–91.5)	21–22 (53.4–56)	23–25 (58.4–63.5)

Recently, some operators have offered sleeping accommodation in larger aircraft for long-range flights. This is typically accomplished by rearranging cabin space – the interior securing structure is designed with flexibility to accommodate changes.

7.5.4 Passenger Facilities

The typical layout of passenger facilities is shown in Figure 7.14 and includes toilets, service galleys, luggage compartments and wardrobes. Cabin crew are provided with folding seats, for large aircraft these are conveniently located as normal seats.

The type of service depends on the operator and ranges from almost no service for low-cost operations to the luxury of first-class service. Figure 7.15a illustrates a typical galley arrangement for a midrange passenger-carrying aircraft; other types of server trolleys are also shown. Figure 7.13 shows a trolley in the aisle being pushed by cabin crew.

Galleys are located in the passenger cabin to provide convenient and rapid service. Generally, they are installed in the cabin adjacent to the forward- and aft-galley service doors. Equipment in the galley units consists of the following:

- high-speed ovens: refrigeration
- hot-beverage containers: main storage compartments
- hot-cup receptacles

The electrical control-panel switches and circuit breakers for this equipment are conveniently located. Storage space, miscellaneous drawers and waste containers are also integrated into each galley unit.

Figure 7.15b shows a typical toilet arrangement for larger passenger-carrying aircraft. For a small Bizjet class, the toilet can be minimised (or removed – not preferred) unless there is a demand for a luxury facility.

7.5.5 Doors – Emergency Exits

FAA/Civil Aeronautics Agency (CAA) has mandatory requirements on minimum number of passenger doors, their types and corresponding sizes depending on the maximum passenger capacity the fuselage intended to accommodate. This is to ensure passenger safety – certification authorities stipulate a time limit (90 s for big jets) within which all passengers must egress if an unlikely event, for example fire, occurs. The larger the passenger capacity, the larger the number is of the minimum number of doors to be installed. Not all doors are of the same size – the emergency doors are smaller. All doors are kept armed during airborne operation.

An emergency situation (say from fire hazard, ditching on water or land etc.) would require a fast exit from the cabin to safety. FAA initially imposed a 120 s egress time but from 1967 this changed to a maximum of 90 s.

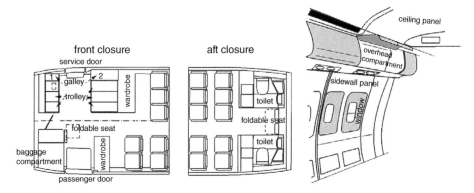

Figure 7.14 Passenger facilities.

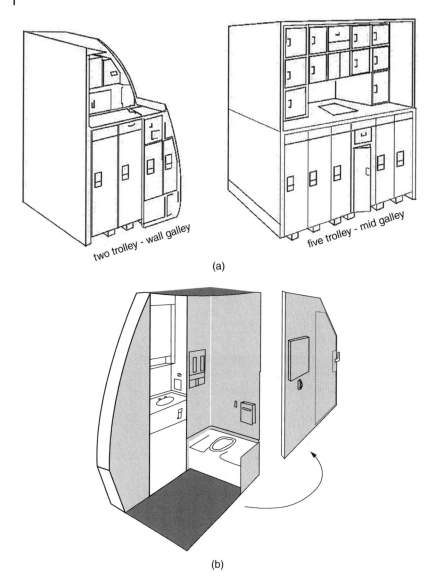

Figure 7.15 Passenger facilities. (a) Typical galleys with service trolley and (b) typical toilet arrangements.

This was possible through the advances made in slide/chute technology. To obtain airworthiness certification, aircraft manufacturers have to demonstrate by conducting simulated tests that complete egress is possible within 90 s. European Aviation Safety Agency (EASA) has similar requirements.

FAR25 Section 25.783 gives the requirements for the main cabin doors and Section 25.807 for the emergency exit doors. There are several kinds of emergency exit doors given in Table 7.4 (all measurements are in inches). All are rectangular in shape with a corner radius. The sizes are the minimum and designers can make them bigger. The oversized doors need not be rectangular so long they can inscribe the minimum rectangular size.

All doors except Type III (inside step up of 20 in. and outside step down of 27 in.) and Type IV (inside step up of 29 in. and outside step down of 36 in.) are from floor level. If a Type II is located over the wing then it can have an inside step up of 10 in. and an outside step down of 17 in. Emergency doors are placed at both the sides of the aircraft and need not be diametrically opposite, but they should be uniformly distributed

Table 7.4 Aircraft door types.

	Position	Minimum height (inch)	Minimum width (inch)	Maximum corner radii (inch)	Passenger number[a]
Type A	Floor level	72	42	7	110
Type B	Floor level	72	32	6	75
Type C	Floor level	48	30	10	55
Type I	Floor level	48	24	8	45
Type II[b]	Floor level	44	20	7	40
Type III	Over wing	36	20	7	35
Type IV	Over wing	26	19	6.3	9

a) The types of doors are related with the minimum number of passengers carried. The higher the number of passengers, the larger is the door size.

b) If Type II is located over the wing then it can have inside step up of 10 in. and outside step down of 17 in.

(no more than 60 ft away), easily accessible for even loading by passengers when it is required. Safety drill by cabin crew is an important aspect to save lives and all passengers must attend the demonstration, no matter how frequent a flier one is. From type to type there are differences.

There should be at least one easily accessible external main door. The combination of main doors and emergency doors are at the discretion of the manufacturers who will have to demonstrate evacuation within the stipulated time. Fuselage length would also decide the number of emergency doors as they should not be spaced more than 60 ft away. Table 7.5 gives the minimum number of emergency doors. It is recommended that more than minimum provision is to be made. Types A, B and C can also be used – they are deployed in larger aircraft.

There could be other types of doors, for example, a door at the tail cone, ventral doors and so on (dimensions are given in Table 7.6). Flight crew emergency exit doors are separately provided at the flight deck.

Table 7.5 Aircraft emergency door types.

Number of passengers	Minimum size emergency door type	Minimum number of emergency doors
1–9	Type IV	One in each side of the fuselage
10–19	Type III	One in each side of the fuselage[a]
20–40	Type II	Two in each side of the fuselage[a]
41–110	Type I	Two in each side of the fuselage
>110		

a) One of them could be of one size smaller.

Table 7.6 Door dimensions.

	Step height inside (outside) (inch)	Minimum height (inch)	Minimum width (inch)	Maximum corner radii
Ventral	–	≥ 48	≥ 24	8
Tail cone	24 (27)	42	72	7

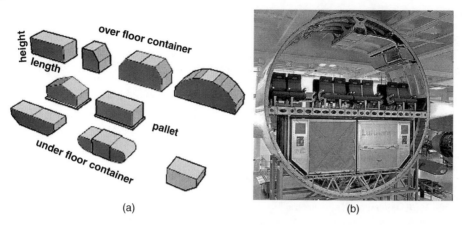

(a)   (b)

Figure 7.16 Standard containers. (a) Typical container shapes and (b) LD3 container in A300. Credit - Airbus

7.5.6 Cargo Container Sizes

As the fuselage diameter increases with passenger load, the under-floorboard space can be used for cargo and baggage transportation. With operating costs becoming more competitive, the demand for cargo shipment is increasing, to the extent that variant aircraft are being designed as cargo aircraft (e.g. no windows and a lower level of cabin pressurisation). An attractive variant is the 'combi' design, which can convert the cabin layout according to the sector payload, in which the passenger load is smaller and the cargo load is higher. The combi layout can quickly reconfigure the cabin interior for passengers in the forward part and cargo in the rear, which facilitates passenger loading and unloading through the front door.

Cargo and baggage could be handled more efficiently by keeping items in containers (Figure 7.16) and having both destination and interior-space management. At the destination, the entire container is unloaded quickly so the aircraft is free for quick turnaround utilisation. Container sizes are now standardised to fit in the fuselage and are internationally interchangeable.

The term *unit load device* (ULD) is commonly used when referring to containers, pallets and pallet nets. The purpose of the ULD is to enable individual pieces of cargo to be assembled into standardised units to ease the rapid loading and unloading of aeroplanes and to facilitate the transfer of cargo between aeroplanes with compatible handling and restraint systems.

Those containers, intended for below-floorboard placement (designated larger container, load device: LD), need to have the base smaller than the top to accommodate fuselage curvature. Those containers have rectangular cross-sections and are designated 'M'. Figure 7.16 shows typical container shapes; Table 7.7 lists standard container sizes, capacities (there are minor variations in dimensions) and designations.

7.6 Some Interesting Aircraft Configurations – Civil Aircraft

In addition to the many different types of aircraft configurations, including some old unusual types, shown in Chapters 4–6, a few more larger types of civil aircraft configuration are shown in Figure 7.17 that are worth studying. These are grouped by the number of engines installed.

(a) *Two engine.* The first one is the latest wide bodied B787 with conventional low wing configuration with two under-slung turbofans. The second one is the Beriev 200 high wing aircraft with two high fuselage mounted turbofans, a configuration that suits flying boats keeping the engine intake protected from water spray.

Table 7.7 Standard container sizes and capacity, dimensions in cm (IATA designation not given).

Type	Length, cm	Width, cm	Height, cm	Base-length, cm	Capacity, kg	Volume, m³
LD1	228	145	162.6	147	1588	5.8
LD2	156.2	153.4	162.6	119.2	1225	3.4
LD3*	200.7	153.4	162.6	157.2	1588	5.8
LD4	244	153.4	162.6	244	2450	6.1
LD6	406.4	153.4	162.6	317.5	3175	8.8
LD7	317.5	223.5	162.5	317.5	4627	9.91
LD8	317.5	153.4	162.5	243.8	2449	6.94
LD11	307	145	162.5	307	3176	7
LD 26	400	214	162.5	307	6033	12
M1	318	224	224	318	6804	17.58
PGA pallet	608	244	244	608	11 340	36.2

Note: IATA = International Air Transport Association.
The LD3 container in Airbus300 below the floorboard is shown in Figure 7.16b.

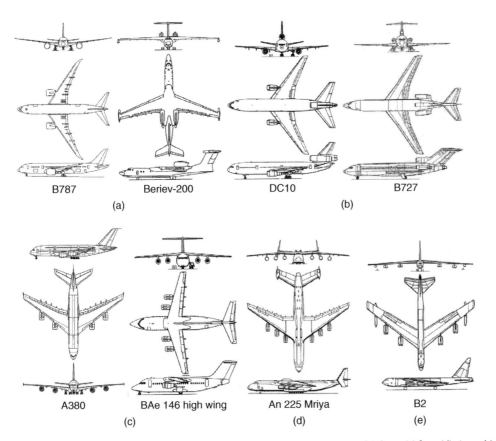

	(a)		(b)
B787	Beriev-200	DC10	B727

	(c)		(d)	(e)
A380	BAe 146 high wing	An 225 Mriya	B2

Figure 7.17 Options for conventional civil aircraft nacelle positions. (a) Two, (b) three, (c) four, (d) six and (e) eight engines. Source: Reproduced with permission from Ferrier.

(b) *Three engine*. The two well proven world's largest trijet configurations. The first one, the Douglas DC10 with a straight intake centre engine. The second one, the Boeing 727 with an S-duct intake centre engine. These are no more in production but few are still operating.

(c) *Four engine*. The first one is the world's largest passenger aircraft, the Airbus380, carrying commercial jet transport aircraft. The second one is one of the few bigger high wing commercial jet transport aircraft.

(d) *Six engine*. The world's largest cargo carrying commercial jet transport aircraft, An225, is another example of the few bigger high wing jet transport aircraft.

(e) *Eight engine*. The world's largest military aircraft, B2, serving the US Air Force for more than half a century.

Interestingly, a *seven engine* (piston engine) Russian design of the 1930s was the Kalinin K-7. Currently, with the advent of electric powered aircraft, a large number of distributed engine concepts are gaining ground. The over wing pod-mounted Honda Jet and slipper-nacelle design YC 14 Boeing STOL aircraft are shown in Figure 5.24.

The engines on single-engine aircraft are at the centreline (except on special-purpose aircraft), mostly buried into the fuselage. If propeller driven, an engine can either be a tractor (i.e. most designs) or a pusher-propeller mounted at the rear.

Some unconventional futuristic nacelle design options are shown in Figure 7.18. These are proposed designs yet to be built. Figure 7.18a shows the Boeing Super Cruiser and Figure 7.18b is the Silent aircraft BWB proposal by MIT and Cambridge Universities.

Virginia Polytechnic Institute (VPI) conducted studies on interesting aircraft configurations with potential. Through their multidisciplinary optimisation (MDO) studies of high-subsonic aircraft with engines at the tip of a strutted wing (Figure 7.18c), they found better weight and drag characteristics than in conventional cantilevered designs [5]. Although the studies have shown some merit, there are some critical issues yet to be considered; more detailed analyses are required to offer better resolution. The structural weight gain due to a truss-supported wing and the aerodynamic gain due to induced drag reduction with the possibility of having a high aspect ratio wing and from the wing-tip engines are to be studied in trade-off studies. A major concern will be to satisfy the mandatory requirement of a one-engine inoperative case. This will result in a considerably larger tail, possibly divided in half, depleting some weight benefits. Cost of production is another factor: studies must ensure there are economic gains after all these considerations. The new aircraft certification will further add to the cost. The proposed aircraft is likely to be more expensive, which may erode the DOC gains. Until more details are available, the authors do not recommend the wing-tip mounted engine installation, especially

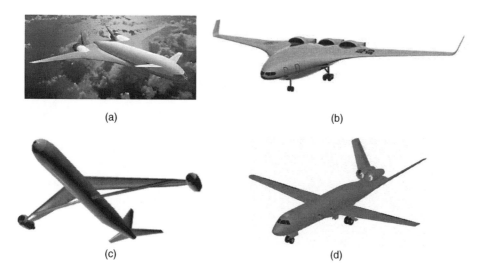

(a)   (b)

(c)   (d)

Figure 7.18 Futuristic options for nacelle position. (a) Futuristic rear engine mount, (b) BWB rear engine mount, (c) wing-tip mounted engines and (d) twin on fuselage. Source: Reproduced with permission from Cambridge University Press.

during an introductory course. Engines should be kept close to the aircraft centreline but away from any wake effects. The nose-wheel spray may require the nacelle to be at least 30°, away from the nose wheel (see Chapter 5). Detailed sensitivity studies are required for comparative analyses of this novel configuration when a simple winglet provides induced drag reduction. However, VPI's study of twin side-by-side engines between the canted vee-tail (Figure 7.18d) shows promise and concluded that it would be better with a winglet.

7.7 Summary of Civil Aircraft Design Choices

This section summarises some of the information discussed in Sections 7.5 and 7.6. The eight graphs shown in Figures 7.2–7.9 capture the actual aircraft data from *Jane's All the World's Aircraft Manual* [1] and other sources. These statistical data (with some dispersion) prove informative at the conceptual design stage for an idea of the options that can be incorporated in a new design to stay ahead of the competition with a superior product. It is amazing that with these eight graphs the reader can determine what to expect from a customer's (i.e. operator) specification on its payload range. Readers may have to wait until their project is completed to compare how close it is to the statistical data, but it will not be surprising if the coursework result falls within the statistical envelope. Civil aircraft layout methodology is summarised as follows:

1. Size the fuselage for the passenger capacity and the amenities required from the customer's specification. Next, 'guesstimate' the MTOM from Figure 7.2 (i.e. statistics) for the payload range.
2. Select the wing planform area from Figures 4.23, 7.5 and 7.7 for the MTOM. Establish the wing sweep, taper ratio and t/c for the high-speed Mach-number capability.
3. Decide whether the aircraft will be high wing, mid-wing or low wing using the customer's requirements. Decide the wing dihedral or anhedral angle based on wing position relative to the fuselage. Decide on the twist.
4. Estimate H-tail and V-tail sizes for the wing area from Figure 7.9.
5. Guesstimate the engine size for the MTOM from Figure 7.10. Decide the number of engines required. For smaller aircraft (i.e. baseline aircraft for fewer than 70 passengers), configure the engines aft-mounted; otherwise, use a wing-mounted podded nacelle.

The industry expends enormous effort to make reality align with predictions. Industries have achieved performance predictions within ± 3% for the smaller aircraft and within ±1.5% for the bigger aircraft. The generic methods adopted in this book are in line with industry; the difference is that industry makes use of more detailed and investigative analyses to improve accuracy in order to remain competitive. Industry could take 10–20 man-years (very experienced) to perform a conceptual study of midsized commercial aircraft using conventional technology. In a classroom, a team effort could take about one man-year (very inexperienced) to conduct a concise conceptual study. There may be a lower level of accuracy in coursework, yet learning to design aircraft in this way is close to industrial practices.

It is interesting to observe that no two aircraft or two engines of the same design behave identically in operation. This is primarily due to production variances within the manufacturing tolerance allocations. The difference is minor: The maximum deviation is on the order of less than ±0.2%. An older aircraft would degrade in performance. During operation, the aircraft surface becomes deformed, dented, warped and/or contaminated causing viscous drag increase and so forth. Manufacturers consider these operational problems by maintaining a record of performance of all aircraft produced. Manufacturers' performance guarantees cover average aircraft degradation only up to a point. In other words, like any product, a brand new aircraft generally would perform slightly better than indicated in the pilot's manual – and this margin serves the operators well.

If a new design fails to reach the predicted value, who is at fault? Is the shortcoming originating from the aircraft or from the engine design or from both? Is it a bad aircraft or a bad engine (if a new engine design is incorporated)? Over time, the aerospace industry has successfully approached these issues. As mentioned previously, some aerospace stories could be more exciting than fiction; readers may examine some old design cases. Today, engine and aircraft designers work cooperatively to identify the nature of and then repair shortfalls.

The compressibility effect of the airflow influences the shaping of an aircraft. Airflow below Mach 0.3 is negligible. In this speed range, all aircraft are propeller driven (mostly piston engine powered). From Mach 0.3–0.6, the compressibility effect gradually builds up but can be ignored without losing performance degradation. Turboprops are effective up to Mach 0.5. Above Mach 0.6, the aircraft component geometry caters to compressibility effects. Jet propulsion with reactionary thrust becomes more suitable above Mach 0.6. Therefore, the aircraft component configuration is divided into two classes: (i) one for flying below Mach 0.5 and (ii) one for flying above Mach 0.6. A carefully designed turboprop can operate at up to Mach 0.6, with the latest technology pushing towards Mach 0.7.

MILITARY AIRCRAFT

7.8 Military Aircraft: Detailed Classification, Evolutionary Pattern and Mission Profile

Military aircraft statistics and geometric details need to be looked into from a different angle on account of a very different mission role. Combat aircraft do not have passengers and the payloads have wide variation in expendable armament type, mostly carried externally and in some designs also internally. Their operational roles are extremely varied as listed next. The difference between civil and military aircraft design is compared in Table 1.2 (Section 1.9). A preliminary classification of military aircraft is given in Table 1.1 in Section 1.3.1, consisting of fighters, bombers, reconnaissance, transport aircraft and so on. From time to time, depending on the perceived combat, the mission requirements get closer look to arrive at specific roles, nevertheless it has to be borne in mind that there is considerable overlap in the functional capabilities between different roles. Subdivision of the fighter role has many classifications. A multi-role big combat aircraft (F14 ~ 33 000 kg) can be used in air-to-air combat as well as for interdiction precision bombing at specific targets, for example, enemy radar stations and so on. On the other hand, an air superiority role (F16 ~ 16 500 kg) calls for light agile aircraft mostly in defence roles to destroy enemy aircraft. A heavy bomber aircraft like the B52 would operate as strategic bomber with little high 'g' manoeuvre. The modern B2 bomber has stealth features to penetrate deep into enemy territories, but not much was known about its all-round capabilities up until now. In the following are the typical terminologies for various kinds of combat aircraft in use.

1. Combat Category

   Air superiority. Its role is to prevent enemy aircraft retaliation over the battlefield in enemy territory so that ground attack aircraft can carry out their tasks to disable adversary. The aircraft should be very agile in order to carry out air-to-air combat in beyond visual range (BVR) capability. As it has to fly longer distances into enemy territory and loiter in the vicinity in preparedness, it is a relatively heavier aircraft.

   Air defence. Its role is to prevent enemy aircraft gaining any superiority of the home sky. It has to out-manoeuvre the best of adversaries. Air defences are smaller, lighter and very agile and primarily meant for air-to-air combat with BVR capabilities. This requires rapid response.

   Ground attack aircraft. This caters for the tactical and other specific requirements on the battlefield. It is capable of a close air support (CAS) role (see next).

   CAS. Air to ground support (gun/missile/light bombs) on the battlefield. It is a relatively lighter aircraft, highly manoeuvrable and could be slower compared to the ground attack type. Rapid fire gunships are a variant of CAS.

   Deep air support (DAS – Interdiction). Carries heavier bombs, jdams for precision bombing in the battlefield. It has a deep penetration capability into enemy territory.

Multi-role fighter. Heavier aircraft capable of performing a variety of combat, for example, air superiority, air defence, ground attack, interdiction and so on.

Advanced tactical support (ATF). The F22 aircraft clearly illustrates the long-range air superiority mission that was envisaged for penetrating deep into enemy airspace to destroy enemy air defence aircraft and to disrupt offensive air operations. This is advanced tactics with multi-role capability – hence it is the new class.

Strategic bombing. Carpet bombing: (B52 class).

2. Surveillance Category

Maritime patrol. Has a special role to cover threats from oceans (e.g. anti-submarine role). In addition, they do surveillance and patrolling with long endurance flying.

Reconnaissance. Very high-performance aircraft beyond missile range (SR71, U2). Photographs enemy territory.

Airborne early warning (AEW). This is capable of early detection of threats with long-range sensors.

Electronic warfare (EW). Capable of electronic counter measures. Remotely piloted vehicles (RPVs) are playing an increasing role in EW.

3. Transport Category

Military transport aircraft. To serve the logistical requirements, for example, troops, equipment ferry and so on (C17 type).

Air-to-air refuelling. Larger tanker aircraft for mid-air refuelling (K135 type).

4. Training Category

Military pilot training. These are specific types of training aircraft of normally two or three types leading up to advanced combat training ready for operational conversion. Its single-seat variant can serve in a CAS role.

5. Remotely Controlled Category

Unmanned Aerial Vehicle (UAV)/ Unmanned Aircraft System (UAS)/RPV. These have no onboard pilots and are increasingly appearing on the battlefield in various roles: they could one day replace advanced manned combat aircraft.

In summary, the combat mission roles are varied as listed, but it can be compressed in to the following basic three types.

1. *Air defence.* Adequately armed (all missiles) but low range (say around 500 nm range – larger country) to keep it light for maximum agility, operating within their own (friendly) territory to defend from invading aircraft. The maritime attack role can be included in this class as aircraft carrier ships can come closer to the target zone in an enemy country.
2. *Deep penetration-multi-role.* This covers everything as listed in the combat category given before, except bomber and air defence. The longer ranges are currently limited to the order of 1000–2000 nm (crossing into enemy territory) but payload (combination of missiles and special-purpose bombs) varies according to the specific mission. All except for the CAS role have supersonic capability.
3. *Interdiction.* Bombers are slower (except the B1) carrying a large bomb load for longer distances.

For ferrying, drop tanks filled with fuel are slung at the hard points to increase range that can be in excess of twice more than the range given in Figure 7.22. Mid-air refuelling would extend range capability and it could be carried out more than once.

The reality is that capabilities are a good measure of intent, it is unrealistic to assume that any nation will expend vast sums of money to acquire specific weapons systems without seeing how that expenditure will further national interests. Long-range air superiority aircraft, such as the Phantom F4/Hornet F18/F35, have served a clearly defined role, offensive strategic air war, in their times.

A quick review of the post-World War II fighter aircraft evolutionary pattern shows rapid progress in speed-altitude and manoeuvre capabilities reflecting distinct changes that have taken place in fighter aircraft configuration. Examples of a few strikingly older designs are given in [1] and also can be found at various websites (Websites: (i) http://richard.ferriere.free.fr/3vues/; Specifications of F-22 (F/A-22), www.f22fighter .com/Specs.htm; Milavia, military aircraft; www.milavia.net/aircraft/; University of Southampton, Aircraft design resources; www.soton.ac.uk/~jps7/AircraftDesignResources/LloydJenkinson%20data; Elsevier, Civil Aircraft Design, book website, http://booksite.elsevier.com/9780340741528/authors/default.htm; VRR, aircraft containers, https://vrr-aviation.com/products/aircraft-containers/; Searates, air cargo ULD containers, www.searates.com/reference/uld/), the list is too large to include here – key words prove sufficient to locate them if they are still online.

Mission profile has a major contribution in shaping of military combat aircraft. An ultimate supersonic air superiority aircraft configuration would be quite different from the subsonic close air support type of aircraft configuration. Attempts are made here to offer a broad-based coverage for an introductory course. Configurations in Figures 7.20 and 7.21 cover the major types of aircraft in operation. These would prove sufficient examples to study for an introductory course. Abundant three-view diagrams (Figures 7.27–7.28, later) and photographs of many kinds of military aircraft are given in this book.

The US designs dominated the scene compared to the designs of other origins. There are successful European designs. The Cold War produced fine Russian designs. Some of the Russian aircraft capabilities are yet to be surpassed. Out of many, some of the outstanding US designs serving the last five decades are shown in Figure 7.21, most of them have proven their performances in various battlefields.

The F-117A Nighthawk (Figure 7.19d) is the world's first operational aircraft specifically designed to exploit low-observable stealth technology. The unique design of the single-seat subsonic F-117A provides exceptional combat capabilities. It is about the size of an F-15 Eagle and has quadruple redundant FBW flight controls. The F-117A can employ a variety of weapons and is equipped with sophisticated navigation and attack systems integrated into a state-of-the-art digital avionics suite that increases mission effectiveness and reduces pilot workload. Detailed planning for missions into highly defended target areas is accomplished by an automated mission planning system developed, specifically, to take advantage of the unique stealth capabilities. Section 22.8 briefly introduces the considerations for stealth design.

A civil aircraft operational evaluation (OP) is relatively simpler. Its DOC can be compared with the competitor to assess the viability of design. On the other hand, a military aircraft comparison is based on several criteria; for example, operation, technology, survival, cost, disposal and political. Each war has taught lessons on how factors other than purely technical and operational capability override decision for the next generation designs. The weapon capability is integral to aircraft capability and therefore the design procedure has to bear in mind the kind of weapon integration envisaged.

Typical combat aircraft design considerations include the following – they cover a lot more disciplines than the civil aircraft design.

1. *Number of crew*. Heavy workload could demand twin crew – 9-g physical limit.
2. *Sizing*. Wing loading and thrust loading, control configured sizing.
3. *Number of engines*. Survivability consideration could demand twin engines.
4. *Engine*. Selection for matching capabilities, vector trust and so on.
5. *Structure*. Choice of material, manufacturing philosophy.
6. *Operational strategy*. Air-to-Air Combat/Air to Ground Combat and so on.
7. *Configuration*. Stealth, external hard points for weapon/drop tank and so on.
8. *Performance*. Agility, speed, altitude, range, supercruise, STOL, survivability and so on.
9. *Electronics*. Weapon system, communication, navigation, data acquisition, counter measures, EW and so on.
10. *Systems*. FBW, Full Authority Digital Engine Control (FADEC), microprocessor-based management and so on.
11. *Weapon*. Type of weapon to be integrated.
12. *Life cycle*. Cost/maintenance/logistics/disposal – support from 'cradle to grave'.

Military aircraft mission profile is extremely varied and aircraft sizing depends considerably on the requirements to encounter perceived threats (there are a lot of unknowns about adversary capabilities). In addition,

Figure 7.19 Chronology of fighter aircraft design evolution (USA). (a) 1950s–1960s: Lockheed F104, Starfighter.
(b) 1960s–1970s: McDonnell F4, Phantom. (c) 1970s–1980s: Grumman F14, Tomcat (Swing wing design).
(d) 1980s–1990s: Northrop F117 Nighthawk (the first all-stealth design). (e) Twenty-first century: Lockheed F35 (stealth design).
Source: Reproduced with permission from Ferrier.

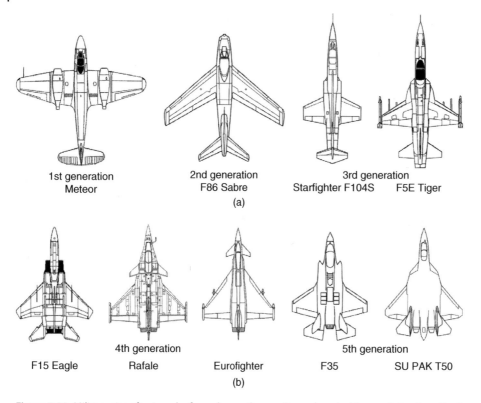

Figure 7.20 Military aircraft wing planform shapes. Source: Reproduced with permission from Ferrier.

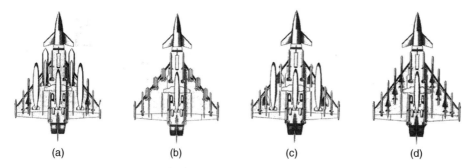

Figure 7.21 Typical multi-role missions (Ref. [1]). (a) Air interdiction, (b) close air support, (c) air defence and (d) maritime attack. Source: Reproduced with permission from Cambridge University Press.

combat and survival considerations impose severe design constraints in shaping the aircraft, for example, incorporation of stealth, manoeuvre in relaxed stability (FBW) and so on. Inclusion of stealth and FBW features requires extended studies that would substantially exceed one term work. US universities could be in a position to obtain National Aeronautics and Space Administration (NASA) software to evaluate stealth. Other nations may not be that fortunate. Control-configured FBW design would require the understanding of the control laws of relaxed stability manoeuvres that are not easy to size. A methodology to pursue these considerations in an undergraduate class could be carried out but the authors do not believe that it would do proper justice before the fundamentals are mastered. The F117 is an example of combat aircraft that incorporates stealth and FBW. It defies imagination coming closer to a 'star wars' shape – no wonder it was nicknamed aerodynamicist's nightmare. This kind of design would not prove easy for introductory classroom project work.

Typically, military aircraft structures demand extensive use of advanced materials, for example, used of composites, lithium and boron alloying with aluminium. Typically, some of the F22's external surfaces have 24% composite, 16% aluminium and some thermo-plastic material. The Eurofighter uses more than half its weight as non-metals.

A military aircraft design exercise would be incomplete without OP, which is beyond the scope of the book. In a true sense, it will require a twin dome (one flown by adversary) combat flight simulator, each flown by human pilots to assess performance capability. Here, a 100% rating means always win and 0% rating is always lose. An 80% capability can be expressed as 4 : 1 that is, in combat; one aircraft is lost against four enemy aircraft losses. Here too, the enemy aircraft and weaponry performances are based on considerable guesswork as potential adversaries are not going to declare their capabilities – it is a matter of life and death. A credible twin dome combat simulation is the nearest assessment platform designers can have – yet real life would still be different. A twin dome simulation could show significant differences in combat capability depending on the selection of weapon/system and so on. Aircraft performance capability is integral to the capabilities of the weapon system in use. In a swing role (combinations of both air-to-air and air to ground operations) the evaluation gets more complicated. Since today's combat would be BVR, a host of other external support systems (target acquisition) are required to assess the military aircraft design beyond making unusual shapes to reduce Radar Cross-section Signature (RCS – low observable). If life cycle cost (LCC) is brought into evaluation then constraints through national economy are another consideration – can these be excluded from a credible teaching exercise?

Military transport design has similarities with commercial transport design, although their operational strategies are different. This book considers military transport aircraft design to be very similar to civil design, except that its certification standards are different (Milspecs).

7.8.1 Fighter Aircraft Generations

The post-war period of development of aircraft production was marked by a scientific and technical revolution – the early era of jet aircraft. In the late 1930s, it became clear that ordinary planes with piston engines and propellers had been developed to almost its maturity and that there was not much room for further improvement unless new technologies could be implemented. Since WWII, with the introduction of the jet age, each successive generation of fighter has demonstrated increased speed and altitude compared to its predecessor, to provide a machine better adapted than that of any competitor to the needs of all potential customers.

In a way, post-World War II fighter aircraft designs evolved in discrete steps of technological advancement and are typically categorised in terms of generation of design evolution, currently in five (or six) generations as follows (Figure 7.20).

'Zeroeth' generation (1945–1955). The 'Zeroeth' generation fighter aircraft were the first military aircraft using jet engines. A few were developed during the closing days of WWII but saw very limited combat operations. These include WWII era fighters such as the Me262 and early postwar aircraft such as the F-80 and F-84.

First generation (1945–1955). The first generation can be split into two broad groups: (i) WWII era fighters such as the Me 262 (also Zeroeth generation) and (ii) the mature first-generation fighters such as the F-86 used in the Korean War. Mikoyan's design office worked on the twin-engine fighter MiG-9. Yakovlev's design office brought out the single-engine fighter Yak-15 in October 1945; it was already on the airfield for preliminary tests and for taxiing. The MiG-9 and Yak-15 promised to be lighter, easier to fly, to have better flying characteristics and be more reliable than the German planes. Examples of first generation jet fighters are the F-84, F-86 and F-100.

Second generation (1955–1960). The second-generation fighter aircraft exhibit more advanced avionics, engines and used the first guided air-to-air missiles. The period from 1950 until 1955 is marked by a dearth of significant interceptor prototypes except for the 1953 appearance of the MiG-17. Second-generation aircraft – including the MiG-19/21 and US century series fighters – were designed during the 1950s. Although they are still found in fighter inventories worldwide, older planes probably have limited combat potential when confronting more modern fighters, since they may suffer from several disadvantages. For example, they may carry less sophisticated munitions and have less capable sensors. The early MiG- and Su-series aircraft have been improved in their air-to-air role. The MiG-23 Flogger B was a second-generation fighter that had a secondary ground attack capability greater than the Fishbed or Fitter.

Third generation (1960–1970). The third-generation fighter aircraft exhibit yet more advanced avionics, engines and weapons. The changes in the fighter combat conception, new air-to-air guided missiles and the results from first- and second-generation fighter operations gave rise to the third generation, such as the later versions of the F104S and MiG-23. Third-generation aircraft may provide somewhat more military capability, especially if they have gone through extensive modifications since they were built. Designed during the 1960s through the1970s, this generation includes the MiG-27 series designed by former Soviet Union's Mikoyan Design Bureau. The third-generation fighter-bombers and tactical bombers included the Su-24 and their derivatives, the F-4s and A-7s built by the USA and the European designed Mirage 3, Mirage 5, F1 and Tornado.

Fourth generation (1970–1990). The fourth generation continued the trend towards multi-role fighters equipped with increasingly sophisticated avionics and weapon systems. These fighters also began emphasising manoeuvrability rather than speed to succeed in air-to-air combat. Fourth-generation fighters, designed during the 1960s and 1970s, include the US-designed F-14, F-15, F-16 and F/A-18; the Soviet-built SU-27 and MiG-29 and the European Mirage 2000. Fourth-generation aircraft usually have more sophisticated avionics than their predecessors, more powerful engines and are able to operate more capable missiles.

Fifth generation (1990–2010). The fifth-generation fighters use advanced integrated avionics systems to provide the pilot with a complete battlespace awareness, and use of low-observable 'stealth' technology. The F-22 and F-35 were the first fifth-generation US fighters, with Russia following with the Mikoyan Gurevich MFI prototype and the Sukhoi PAK-T50 and China with the J-20.

Making the situation even more confusing, the Chinese have their own particular order of ranking. The Chinese call the Russian fifth-generation fighter a fourth-generation machine.

The term '4.5 generation' is also sometimes seen (the Chinese call them '3.5 generation'). These are more recent fourth-generation fighters, retaining the basic characteristics of fourth-generation planes but with enhanced capabilities provided by more advanced technologies that might be seen in fifth-generation fighters. Good examples are the F/A-18E/F Super Hornet, Eurofighter Typhoon and Dassault Rafale. All make use of advanced avionics to improve mission capability and limited stealth characteristics to reduce visibility when compared to older fourth-generation aircraft.

Stealth characteristics markedly enhance the capability for survivable attack of defended targets. And that remains true, even though evolving modern air defences available on the international arms markets have increasing capability against currently deployed levels of stealth. Hence, to continue to operate effectively in the face of these defences, stealth has to be supplemented with other survivability features. Nonetheless, stealth aircraft operate at much lower levels of support than conventional aircraft and even small numbers of stealth aircraft can greatly leverage the capabilities of the remainder of the bomber force and of the tactical fighter forces.

Europe and Japan are behind in applied stealth technology as evidenced by US aircraft programmes such as the F22, B2 and F117A. Most of the applied foreign low-observable work involves basic shaping, material coating techniques and signature testing requirements. Europe has been led by France, Sweden, Germany and the UK in various types and levels of low-observable applications. Applications on fighter aircraft have generally been at fundamental applied levels, primarily using absorbent coatings, limited structural shaping and absorbent structure. Applications seem to be limited to the areas with highest signature return rather than application to an entire airframe. European firms are also working on stealth technology applications to cruise missiles and unmanned aerodynamic vehicles.

The rapid growth of a globalised economy has deepened the degree of international cooperation and expanded the variety of methods of cooperation in the international arms production industry to such an extent that a globalisation trend has also emerged in this field. The globalisation of the national defence industry refers to the change from the traditional preference for autarky to that of a globally oriented market in terms of research, production, management and sales. At present, there are mainly three ways in which globalisation exhibits itself. The first is through the purchasing of weapons from other countries and taking part in the production of these weapons (including granting of special permits, joint cooperation and development ventures and compensation trade). An example is the joint production of F16 fighter jets by USA, Holland, Denmark and Norway. The second method is through military cooperation packages covering weapons trade, production and maintenance and joint military exercises between different countries, for example, the signing of the 10-year military cooperation agreement package between India and Russia in 1999. The final means is through cross-border joint ventures and joint research and development projects between nations (including international group companies, international integration and transnational amalgamation). The four-nation joint venture for the production of the Eurofighter-2000 by the UK, Germany, Italy and Spain is an example.

7.9 Military Aircraft Mission

A typical mission profile for combat is given in Chapter 15. Figure 7.21 shows some configurations for the mission profiles to meet military aircraft mission demand, its range and armament payload are traded freely. A very high armament payload could be used for short ranges or lighter load for long-range interdiction. A relatively light armament in air defence (high g manoeuvre) role can also be just overhead that is, low range while in escort role to a relatively long range. Military aircraft has in-air refuelling capability (or use of drop tanks) to extend range. Payload mass has wide range of options – all hard points can have lighter weapons load or heavier missiles/bombs. In general, heavier aircraft will have a heavier payload. Payload being externally mounted on hard points means that aircraft drag characteristics alter substantially affecting range capability. At a design MTOM, what would be the payload would depend on mission range – here, weapon load and drop tank fuel load is traded. The B2 had to fly half the world (with mid-air refuelling) to reach the target zone. It is for these reasons, a correlation like Figure 7.1 showing passenger versus range would not offer much information for military design. Unlike civil aircraft mission profile, it clearly indicates that the same class of military aircraft can have a wide variety of payload range. It may prove convenient to assess combat aircraft with full internal fuel for the payload-range capability, quite different from what can be seen in Figure 7.1 for civil aircraft designs.

The typical multi-role armament configuration of Eurofighter is shown in Figure 7.21. In general, it consists of takeoff of heavily loaded aircraft, climb to altitude for programmed cruise that could have speed-altitude specifically tailored for the terrain releasing weapon load and perceived threat, dive down to low level high-speed dash to target zone for interdiction, then fast climb to extreme height of the lightened aircraft and return to base. For air defence, the combat would be in closer proximity to defend from attacking aircraft requiring extreme manoeuvre at high g.

7.10 Military Aircraft Statistics (Regression Analysis)

The statistics given in this chapter are from the following aircraft: B2, F14, F4, F15, F16, F18, F22, F35, F111. F117, SR71, SU37, MiG31, SU41, MiG25, Viggen, Rafale, Eurofighter, Gripen, Jaguar, Hawk, Mriage2000, Kfir, Lavi and Harrier.

In line with Section 7.5 on civil aircraft designs, this section gives the statistics of military aircraft weights and geometry. Unlike civil design progressing in an evolutionary track, the military designs tend to progress in revolutionary tracks. Military aircraft statistics are not as consistent as civil designs and require considerably more information for correlation. Military designs are operation-specific and it would dilute its specialities if they were presented in a generic fashion. The author regrets to note that not much information is available

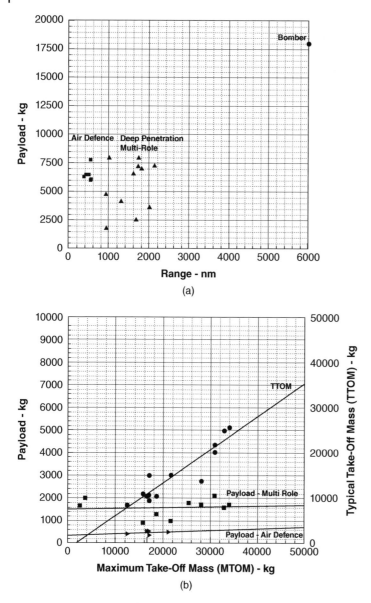

Figure 7.22 Military aircraft payload range (no drop tank or refuelling). (a) MTOM versus payload and (b) MTOM versus payload and TTOM.

in the public domain – understandably these are sensitive issues. The definition of various kinds of aircraft mass given in Figure 7.4 is applicable in this section – here payload replaces passenger capacity. To keep regression simple, linear fitment is carried out. The regression graphs given in this section can only be used for preliminary sizing.

Combat aircraft loading to MTOM would be at the sacrifice of its agility. Loading to MTOM is done when mission demands (several mission profiles are given in Chapter 13). In general, MTOMs are meant for deep penetration when considerable fuel is consumed before reaching the combat zone to make the aircraft lighter – it has option to carry the amount in drop tanks that can be jettisoned (punched out) when emptied to reduce drag. For maximum effectiveness with a balanced combat capability, military aircraft use lighter loading termed Typical Takeoff Mass (TTOM), which is typically 70% of MTOM.

7.10.1 Military Aircraft MTOM versus Payload

Figure 7.22 shows MTOM versus payload and TTOM. There is a distinct separation on armament loading capability between air defence class fighter aircraft with armament load of around 750 kg and the multi-role class aircraft carrying nearly twice the payload for a longer distance. Air Defence class fighter aircraft are lighter than the multi-role class aircraft. Understandably, in both the classes the MTOM grows with increase in payload (armament).

7.10.2 Military MTOM versus OEM

Figure 7.23 gives the relation between MTOM and OEM as well as the Operational Empty Mass Fraction (ratio of OEM to MTOM). OEM grows with growth in MTOM but there is a spread in the OEM fraction arising out of differing performance capabilities and system integration. Typically, the ratio of OEM/MTOM averages around 0.42–0.52.

7.10.3 Military MTOM versus Fuel Load, M_f

Since fuel load for combat aircraft is flexible depending on usage of drop tanks and air-to-air refuelling, onboard fuel content is taken as the standard condition to present the statistical analysis. Figure 7.24 gives the relationship between internal fuel load, M_f and fuel fraction, $M_f/MTOM$ versus $MTOM$. There is some scatter on account of diversity in requirements.

   Growth in MTOM would be associated with more fuel carrying capacity to meet range, but the fuel fraction graph shows dispersion on account of the difference in role, for example, short range air defence and longer range deep penetration types. Longer-range aircraft would be heavier due to having to carry more fuel.

7.10.4 Military MTOM versus Wing Area

The influence of wing loading shows up in the graph (Figure 7.25). Military designs could have moderate wing loadings for hard manoeuvres. Modern designs show lower wing loading – the F22 has the low order of

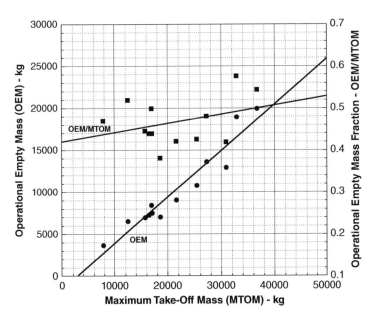

Figure 7.23 MTOM versus OEM.

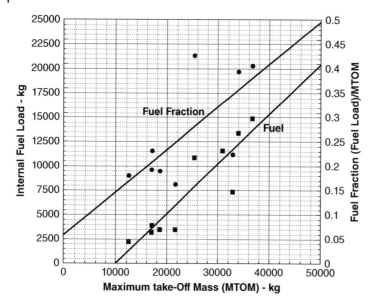

Figure 7.24 MTOM versus fuel load.

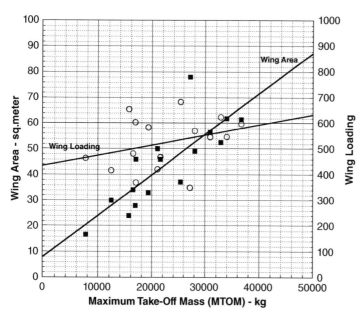

Figure 7.25 MTOM versus wing area.

350 kg m^{-2} compared to the Sepecat Jaguar with around 650 kg m^{-2}. It will be seen in the next section that the F22 also has the highest thrust loading.

7.10.5 Military MTOM versus Engine Thrust

Combat aircraft are invariably powered by jet propulsion (turboprop driven CAS aircraft are few and are excluded in the statistics). Military aircraft would require very high thrust loading, *T/W* (could exceed 1), for manoeuvres and short field performance. A high thrust requirement is of a short duration and is met by augmenting thrust by the use of afterburners (Chapter 10).

Section 7.10 explains the need for TTOM for combat effectiveness. Figure 7.26 presents the relationship between total T_{SLS} and the two types of aircraft mass, for example, MTOM and TTOM. Thrust increase is associated with aircraft mass increase. However, there is some spread in thrust loading. Typically, (total $T_{SLS}/MTOM$) averages around 0.65 but (total $T_{SLS}/TTOM$) exceeds 1. Later generations of combat aircraft have pushed thrust to weight ratio more than one would permit aircraft to accelerate in vertical climb. The F22 has the highest thrust loading as well as the lowest wing loading.

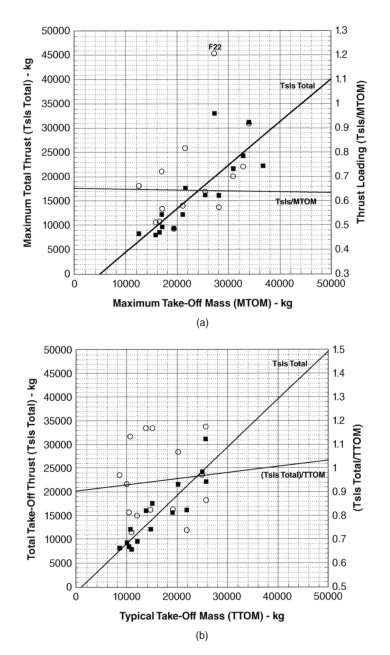

Figure 7.26 Aircraft weight versus total takeoff thrust. (a) Total thrust versus MTOM and (b) total thrust versus TTOM.

7.10.6   Military Empennage Area versus Wing Area

Military aircraft empennage configuration should be very different from civil aircraft design; many have a conventional design with an H- and V-tail. On the other hand, an extreme example of the B2 appeared to not have any tail. Examining various configurations, it can be seen that there are several options for aircraft control.

Stability and control of modern combat aircraft are invariably supported by microprocessor-based systems architecture such as FBW, when onboard computers continuously fly a slightly unstable aircraft under pilot initiated commands. This is beyond the scope of this book, so no statistical analysis could be presented here. The examples of the trainer class of aircraft would follow the conventional approach with a one H- and one V-tail design (use Figure 8.5).

Military aircraft require more control authority for greater manoeuvrability and have shorter tail arms requiring larger tail areas. Typically, the horizontal tail area is about 30–40% of the wing reference area. The vertical tail area varies from 20 to 25% of the wing reference area. Supersonic aircraft have an all movable tail for control. If the V-tail is too large then it is split into two.

7.11   Military Aircraft Component Geometries

Previous sections gave abridged statistical relations of weight and geometries for all categories of combat category aircraft. Section 2.3.3 gave some familiarisation of a typical military aircraft and its components – as mentioned earlier in a 'Lego' or 'Mechano' building block concept. Because of a large variety of combat mission profile and a large variety of technological options available, they result in wider choices to configure military aircraft. The choices are not made arbitrarily – strong reasons are associated. The following sections provide pertinent information on the (i) fuselage, (ii) wing, (iii) empennage and (iv) nacelle as military aircraft components. These four groups of aircraft component offer the preliminary shape of candidate combat aircraft configurations. Eventually, after the wing-sizing and engine-matching exercise, the choice for configuration must be narrowed down to one that would offer the best choice for the mission. Family derivatives of military aircraft are quite different, again depending on the mission role (e.g. use of additional crew, trainer version, carrier-borne version, longer-range version, improved variant version). Undercarriage information is presented separately in Chapter 9.

7.11.1   Fuselage Group (Military)

Unlike the approach to civil aircraft configuration, military design need not start with the fuselage, but it may prove convenient to do so. Fighter aircraft fuselage does not carry any internal payload – it has the singular function to accommodate the crew (or crews) and engine (or engines) along with routing of conduits of various systems (wires, pipes, linkages for the systems), fuel tanks and encasing small arms (e.g. guns). Unlike hollow civil aircraft fuselage, it is very tightly packed, minimising the fuselage volume requirement. With the engine (or engines) buried inside, air intake is an integral part of the fuselage. The large wing root of delta (or trapezoidal wing with strake) planform offers the scope to make the wing blend with the fuselage. In that case, the configuration of the wing becomes integral to configuring the fuselage, as shown in Sections 5.2.1.4 and 5.2.1.5. A variety of well-known combat aircraft images are shown in Figure 7.27.

The densely packed fuselage design starts with the nose cone, which has to be pointed for supersonic capability and then houses a radar that could be of around 1 m diameter. Fighter aircraft fuselage would invariably house the power plant and, therefore, there will not be any separate podded nacelle (some older bombers have pods). The necessity for area ruling makes for narrowing of the fuselage. Therefore, the fuselage would rarely have a constant cross-section, making fuselage shape generation quite complex.

The fuselage aft ends up as an engine exhaust system and therefore will not have closure as in civil design. In case the engine dangles below the fuselage spine (Figure 7.19b – Phantom), then a pointed aft end closure

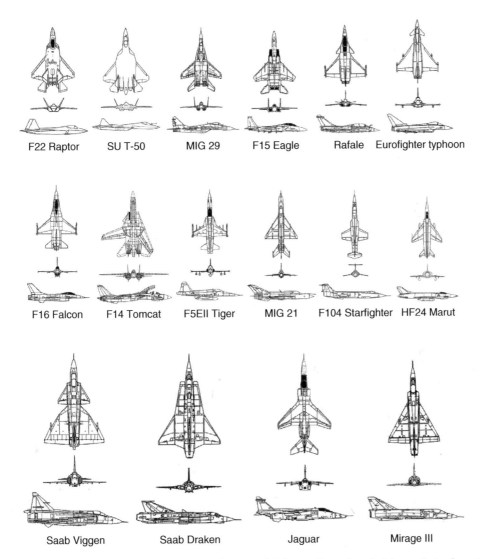

F22 Raptor   SU T-50   MIG 29   F15 Eagle   Rafale   Eurofighter typhoon

F16 Falcon   F14 Tomcat   F5EII Tiger   MIG 21   F104 Starfighter   HF24 Marut

Saab Viggen   Saab Draken   Jaguar   Mirage III

Figure 7.27 Military aircraft fuselage views (not to scale). Source: Reproduced with permission from Ferrier.

follows. The fuselage belly fairing would be housing accessories, in most cases the undercarriage. Current tendencies are for the wing-body fairing to have considerable blending for superior aerodynamic considerations, for example, to improve lift to drag ratio and fly at a higher angle of attack. In a blended fuselage it is hard to isolate it, a possible convenient choice would be where the wing root is attached.

Also, a military aircraft pilot seat has more freedom to recline to shorten carotid artery height to reduce blood starvation to the brain at high *g* manoeuvres that can cause blackouts.

7.11.2 Wing Group (Military)

The wing group is the most important component of the military aircraft. The wing planform shape needs to be established based on the operational requirements (e.g. hard manoeuvres, supersonic capabilities, short field performances). Unlike civil design, there is a large range of options for planform shape and fuel tankage space is restricted.

The evolution of fighter aircraft shows the dominant delta or short trapezoidal wing planform. This is for obvious reasons of having high leading-edge sweep and low AR to negotiate high *g* manoeuvres that would generate a high wing root bending moment. It would restrict span growth but encourage large wing root chord of delta or trapezoid with strake planform as shown in Figure 7.28g,h. For control reasons it could have additional surfaces. The following are the configuration choices (strakes are taken as part of wing):

1. *One-surface configuration*. Pure delta planform or its variation – the trailing edge of the delta like wing can be made to work like an H-tail as an integral part of the wing (Figure 7.28a).
2. *Two-surface configuration*. Delta-like wing or trapezoidal wing with a conventional H-tail for pitch control. In some designs, the H-tail is replaced by a canard surface for pitch control in relaxed stability (has a destabilising effect). A two-surface configuration has two possibilities – a tail at the back (Figure 7.28b) or

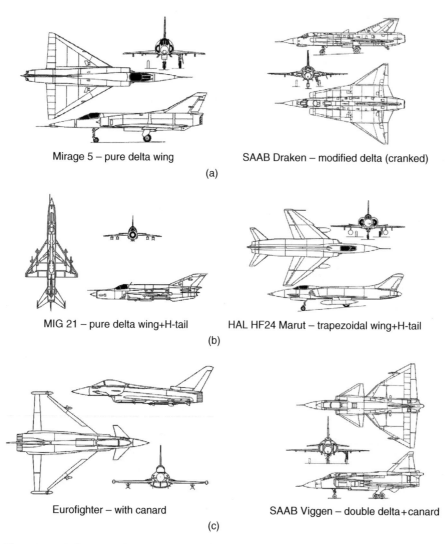

Mirage 5 – pure delta wing   SAAB Draken – modified delta (cranked)

(a)

MIG 21 – pure delta wing+H-tail   HAL HF24 Marut – trapezoidal wing+H-tail

(b)

Eurofighter – with canard   SAAB Viggen – double delta+canard

(c)

Figure 7.28 Fighter aircraft configurations. (a) One surface wing planform – trailing edge has pitch control, (b) two-surface wing planform – the conventional H-tail has pitch control, (c) two-surface wing planform – canard + wing trailing edge has pitch control, (d) two-surface wing planform – conventional layout with strakes and (e) three surface wing planform – canard + wing + H-tail. Source: Reproduced with permission from Ferrier.

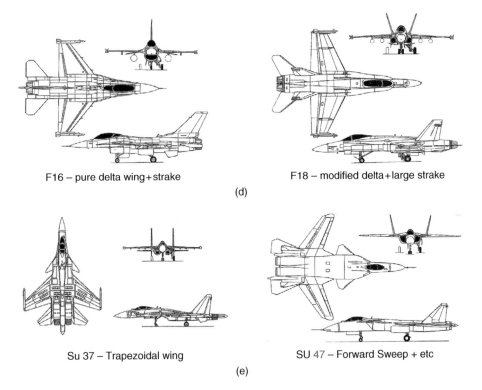

F16 – pure delta wing+strake

F18 – modified delta+large strake

(d)

Su 37 – Trapezoidal wing

SU 47 – Forward Sweep + etc

(e)

Figure 7.28 (*Continued*)

    a tail in front (canard – Figure 7.28c). A two-surface configuration with strake is shown in (Figure 7.28d). Variants include the double delta (SAAB Viggen, see Figure 7.28c).

3. *Three-surface configuration.* The ultimate kind is a three-surface configuration (Figure 7.28e). It has a wing, H-tail at the aft end and canard at the front end.

A delta wing trailing edge would have a pitch control surface integrated with it. A trapezoidal wing planform (Figure 7.20c,d) could be associated with a separate pitch control surface, typically as an H-tail. An extreme form of the three-surface arrangement exists (Sukhoi 37 – Figure 7.28e). Forward sweep has aerodynamic merits to bring the wing aerodynamic centre to move forward, which favours H-tail sizing. However, aeroelastic problems could aggravate wing twist creating instability. Carefully arranged composite material has minimised the effect of twist and there are two successful designs that have been flight tested (Su 47, see Figure 7.28e, and Grumann X29).

The role of the canard in military application is quite different from that in the role of civil aircraft designs. It has been found that strakes can also provide additional vortex lift and fast responses to pitch control with a conventional tail. The choice of strake or canard is still not properly researched in the public domain. It is interesting to note that US designs have strakes while the European kinds have canards. Detailed study of aircraft control laws and FBW system architecture is required to make the choice. Until the 1990s, flaws in FBW software have caused several serious accidents.

    Again, it is emphasised that this book is introductory in nature. For not having enough information on modern fighter design considerations, the author restricts military aircraft design exercises to a trainer class of aircraft. The readers will find that there is a lot to learn from this class of aircraft, enough to have a feel for military aircraft design considerations. Figure 7.29 shows the modern Advanced Jet Trainers that have variants for CAS roles. The Ae Hawk is a successful but relatively older design (still in production) that has a conventional configuration. EADS MAKO is one of the latest trainer aircraft designs capable of supersonic flight and light combat capability.

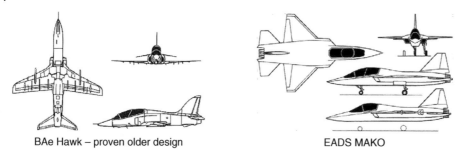

BAe Hawk – proven older design   EADS MAKO

Figure 7.29 Advanced jet trainer aircraft capable of close support combat. (a) BAe Hawk – proven older design and (b) EADS MAKO. Source: Reproduced with permission from Ferrier.

Wing attachment to fuselage varies from case to case. The leading edge can have slats and the trailing edge would invariably have flaps. Centrally mounted large air brakes to decelerate have practically eliminated the role of spoilers. Landing in shorter airfields may require deployment of a brake parachute.

Since military aircraft are expected to encounter transonic flight, aircraft cross-sectional area distribution becomes an important consideration. A seamless smooth distribution of cross-section (area-rule) is explained in Section 3.13.

7.11.3 Empennage Group (Military)

Combat aircraft empennage shaping and sizing are complex procedures, primarily on account of a short tail arm and the need to fly in relaxed stability to execute fast and hard manoeuvres (Chapter 6). The B2 apparently appears to be without a tail. The F22 has a large canted V-tail. Options for control surface configuration are shown along with wing options. The delta wing has an H-tail integrated with it. This book adheres to the conventional configuration of H- and V-tail for the trainer aircraft example. Modern designs deploy tailerons (stabilator; see Section 6.5.3) to initiate pitch and roll control by an H-tail.

Introductory comments expressed the complexity involved in control surface design, which are primarily of the empennage group. Having short tail arms (L_{HT} and L_{VT}) the stability (stabiliser/fin) and control surfaces (elevator/rudder) will necessarily have to be large in relation to the aircraft size to make fast responses. In many designs the vertical tail is split into two (could be placed slightly inclined in a shallow Vee for stealth reasons) as can be seen in the F18. The canted V-tail of the F22 is on account of stealth considerations. The horizontal tail is symmetrical to the vertical plane and has to be secured on the fuselage – some of the earlier designs had a T-tail.

Military aircraft control is through FBW technology, which processes control deflection through onboard computers ensuring safety. If the pitch control demand were high requiring flying in relaxed stability, then a canard surface in front of fuselage would prove helpful. The advent of FBW has achieved yaw control without a V-tail – it is achieved through differential use of an aileron surface that can be split to open up both in upward and downward directions, if required simultaneously. This book will not be discussing control-configured designs (control-configured vehicle: CCV) – these would require analysing the control laws, not dealt with here.

Delta wings can have H-tails integrated within with the reflex built-in at the trailing edge (Mirage 2000). The exception of an older design is shown in Figure 7.30a (YF12) with two canted vertical tails. The YF12 design paved the way for the current wing and empennage design options. An unconventional empennage exists (Figure 7.30).

The B2 in Figure 7.30 appears to be tailless. Its pitch control is at the inboard trailing edges and directional/lateral controls are carried out by controlled opening of a split aileron in both sides. Much of the future will depend on how many lifting surfaces are used. Typically, a highly manoeuvrable combat aircraft will have a large V-tail split into two and canted for stealth reasons. Trainer aircraft favour a conventional type with an H- and V-tail.

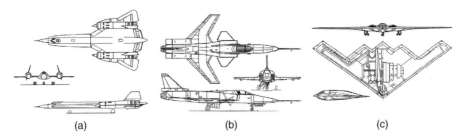

Figure 7.30 Empennage options. (a) SR71 twin tail – inclined, (b) X-29 separate H-tail with wing and (c) B2 – tailless. Source: Reproduced with permission from Ferrier.

7.11.4 Intake/Nacelle Group (Military)

Unlike a simpler pod-mounted configuration of civil aircraft, the military aircraft power plant is kept within the fuselage. Therefore, military aircraft of the combat class do not have nacelles, instead they have an (supersonic) intake that is an integral part of fuselage configuration. Supersonic intake is comparatively more complex to design. A supersonic intake has a side plate acting both as boundary layer bleeder and adjust oblique shock. Intake design requires serious considerations to cater for the full flight envelope to supply air inhalation to the engine without causing flow distortion at the engine face that can cause compressor surge and/or flame-out problems.

Combat aircraft use the aft fuselage to house the engine. Broader classification (options) of military fighter aircraft engine intake configuration is given in Table 5.1. A large number of intake configuration types are given Figure 5.34.

Single or multi-engines (so far mostly two) are kept side-by-side buried in the fuselage. Fighter aircraft intake path is longer and curvier. Older designs had a forward intake at the nose (MiG21). Aircraft with over Mach 1.8 capability have a centre body (bullet translates forward/backward to adjust bow oblique shock). These have the longest intake duct with low curvature to pass under the pilot and incur high loss, hence they are no longer pursued. Instead, chin (F8 Crusader) or side (Hawk) mounted intakes have shorter ducts and these carefully designed ducts have comparable curvature, hence less loss. A plate is kept above the fuselage boundary layer on which the intake is placed. A centre body is required for aircraft speed capability above Mach 1.8, otherwise it has a pitot intake: trapezoidal slanted side intake (F22). The B2 has over wing/fuselage intakes more suited to the BWB type of configuration.

Some older and odd type engine integration configurations with over the fuselage engine installation are shown in Figure 7.31.

In summary, intakes on fuselage have following possibilities:

1. Central forward intake (invariably circular/MiG21 – can be near circular F100). These are older designs, no longer used in practice. Considerable duct loss is associated with its long length from nose to tail and bends to pass underneath the pilot. For supersonic operation, a centre body arrangement makes diffusion through a series of oblique shocks and a normal shock. The centre intake does not have fuselage shielding at yaw and pitch attitudes.
2. Side intakes (semi-circular, rectangular, e.g. F18 and F22 etc.) – it cuts down the internal duct length by nearly half but associates with bends, which is less for two side-by-side engines. It can be over or under the wing. For supersonic operation, there is a splitter plate that bleeds the fuselage boundary layer to keep it outside the intakes. It needs to be carefully sized for flying in yawed attitude.
3. Chin mounted intake (near elliptical/kidney shaped like the Falcon F16, near rectangular like Eurofighter). These are later designed aircraft. At yaw, chin intake does not have fuselage shielding as there could be for side intakes, but being close to the ground there could be ground ingestion problems, especially during war time. This concept has proven very successful and can handle high incidence flying.
4. Central over fuselage mounted intake (this is opposed to a chin mounted intake like the Predator UCAV aircraft). This configuration is not prevalent – high incidence flying may create serious flow distortion

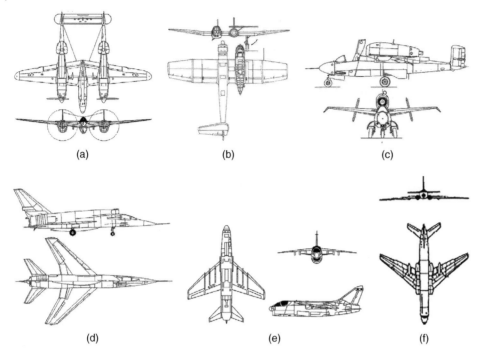

Figure 7.31 Options for engine positions of some older designs. (a) Twin fuselage engines, (b) asymmetric engine B & V141, (c) over fuselage – Heinkel 162, (d) over fuselage – F107, (e) forward intake – Corsair and (f) engines at side of fuselage. Source: Reproduced with permission from Ferrier.

    affecting engine performance. This type of configuration is gaining ground, however, on account of stealth considerations.
5. Future designs will have stealth features and F22/B2/JUCAS type nacelles will become common shapes of intake design.

7.12 Miscellaneous Comments

Sometimes stability and control necessities require additional surfaces, for example, a ventral fin, dorsal fin, delta fin and so on. Fairings between two intersecting bodies or to enclose protruding objects play an important role as flow modifiers. Vortex generators are placed wherever it is necessary – they are prominently seen on the wing's upper surface. Antennas/ducts can be seen in places as essential features to serve specific purposes. Readers are encouraged to examine real aircraft kept statically on tarmac. Every item seen on the external surface counts and contributes to drag.

    In general, during the conceptual design phase, sizing of these features is done schematically. Sizing of trim surfaces becomes more appropriate once the aircraft configuration is frozen. Initially, the control surface sizing is done empirically and by computational fluid dynamics (CFD) analysis in the second phase of study, and gets fine-tuned by tailoring the surface during flight trials. In this book, trim surfaces are kept schematically – the main task of this book is to size aircraft and freeze the configuration in the first phase of the project.

7.13 Summary of Military Aircraft Design Choices

Building blocks for aircraft configuration are presented in this chapter that will be used in Chapter 8 to configure aircraft. After establishing the specification requirements from market study (Chapter 2), the classroom work could start from Chapter 8 to layout an aircraft configuration using the typical building blocks presented in this chapter. A quick browse through this before starting aircraft layout would prove beneficial. Given next is a quick summary on how to layout military aircraft configuration.

Military Aircraft Layout (approach differs from the civil approach to configuration layout)

1. Guesstimate the *MTOM* from Figure 7.23 (statistical value) for the payload range for the class.
2. Pick a wing area from Figure 7.24 for the MTOM. Decide on a single-, two- or three-surface design (needs aircraft control analysis – in this book, kept as a two-surface conventional design). Next, decide wing geometry, for example, sweep, taper ratio and *t/c* for the high-speed Mach-number capability.
3. Decide on a high-, mid- or low wing from customer requirements. Decide whether the wing should be dihedral or anehedral based on wing position.
4. Decide number of engines required – for fighter aircraft this is unlikely to exceed two. In this book the choice is a single engine. The engine is invariably housed in the fuselage.
5. Shape the fuselage to house the engine and fit the wing and empennage. Guesstimate the H- and V-tail sizes for the wing area.

Supersonic compressibility effects dictate military aircraft design. It would invariably require reactionary thrust such as offered by jet engines. A sharp pointed nose, large wing sweep with low AR, stealth features and control configured features offer wider options. However, the designers tend to be conservative in approach with pilot survivability as the most dictating consideration, while incorporating newer concepts to stay ahead of perceived threats. As combat technology is heading towards unmanned battlefield operations, relaxed stability flying with FBW will make aircraft fly at higher *g* beyond human limits. 'Star Wars' shapes for UAV aircraft are not an impossibility.

References

1 *Jane's All the World's Aircraft Manual* (yearly publication).
2 McCanny, R (2005). Statistics and Trends in Commercial Transport Aircraft. Final Year Project, QUB.
3 Roskam, J. (2007). Airplane Design. Volumes 1 to 8.
4 Jenkinson, L.R., Simpson, P., and Rhodes, D. (1999). *Civil Jet Aircraft Design*. Arnold Publishing.
5 Yan-Yee, A. K. (2000). The Role of Constraints and Vehicle Concepts in Transport Design: A Comparison of Cantilever and Strut-Braced Wing Airplane Concepts, Master's Thesis (April 2000), Virginia Polytechnic Institute and State University.

Part II

Aircraft Design

8

Configuring Aircraft – Concept Definition

8.1 Overview

Part II of this book is devoted to configuring aircraft, as well as with design considerations of bought-out items, for example, various kinds of undercarriage, engines, onboard systems with worked-out examples wherever applicable for readers to practise. This chapter is comprehensive and serves as the basis of aircraft design, along with worked-out examples.

Note that this chapter only proposes a methodology to arrive at a tentative new aircraft configuration without the undercarriage placed (carried out in Chapter 9) because the aircraft (maximum takeoff weight (MTOW)) and its component weight and centre of gravity (CG) locations are yet to be established (carried out in Chapter 10). Therefore, at least three iterations will be required to finalise the configuration, the first one when the undercarriage is placed to suit takeoff rotational clearance and tipping back angle when a wing may need to be repositioned, the next one when the aircraft CG position is accurately estimated, which may be different from the initial guessed CG position and finally when the aircraft is sized with the matched engine to freeze the aircraft configuration, culminating in the conceptual design phase of a new aircraft project. If the initial guesses of MTOW and CG are far away from the final design, then more than three iterations may be required. Experienced engineers are capable to make good guesses (statistics of past designs help) to keep iterations to the minimum, if very lucky with no iterations required. The guessed values in worked-out examples given in this book are deliberately chosen to be close enough to avoid repetitions of iterations. In the coursework, one iteration is sufficient. Aircraft design has to be practised in completeness as classroom exercises, hence no piecemeal problem sets on any arbitrary aircraft are given in this book.

A considerable amount of background preparatory work is needed to make a product that has to be made right the first time. The prerequisite to progress with a new aircraft design is to have aircraft specifications/requirements evolved through market survey (see Section 2.7), in consultation with potential customers. The civil aircraft design exercise starts with only two vital parameters: (i) payload and (ii) range. Of course, there are other requirements (see Section 2.8) that will cover configuration into better detail. Aircraft manufacturers would like to expand the market potential, as not all operators/customers have exactly the same payload-range requirement. Therefore, typically, manufacturers offer a base line design in the middle followed by variant designs, one longer and one shorter in a family concept of aircraft configurations, retaining maximum component commonalty as a manufacturing cost saving measure. Successful designs may extend the family with variant aircraft to more than three types.

This chapter describes how to arrive at an aircraft preliminary configuration that will be best suited to market specifications and could be feasibly manufactured. Industry uses its considerable experience and imagination to propose several candidate configurations that could satisfy customer (i.e. operator) requirements and be superior to the existing designs. Finally, a design is chosen (in consultation with the operators) that would ensure the best sales prospect. In the coursework, after a quick review of possible configurations with the instructor's guidance, it is suggested that only one design be selected for classwork that would be promising in facing market competition. In a way, this chapter describes how an aircraft is conceived, first to a preliminary configuration as a *concept definition*; that is, it presents a methodology for generating a preliminary aircraft

Conceptual Aircraft Design: An Industrial Approach, First Edition. Ajoy Kumar Kundu, Mark A. Price and David Riordan.
© 2019 John Wiley & Sons Ltd. Published 2019 by John Wiley & Sons Ltd.

shape, size and weight from statistics of past designs (guesstimates). Subsequently, *concept refinement*, that is, the *finalisation* of the preliminary configuration is carried out in Chapter 14. It is a formal method of an iterative process of aircraft sizing and engine matching. It is natural to expect some differences between the preliminary and the final configurations when iteration is required.

New military aircraft design has a different approach. It is primarily meant to serve one customer, a nation's defence requirements, and with the possibilities of sale to friendly countries who can guarantee confidentiality. The appropriate government department floats a Request for Proposal (RFP) or Air Staff Target Requirements (AST) for combat aircraft based on perceived threats from potential adversaries. On account of incorporating new technologies, a new type of combat aircraft is less dependent on statistics of past designs. However, military trainer aircraft with proven existing technologies benefit from statistics of past designs. The government discusses with various aircraft manufacturers to assess capabilities and enter into a contract to develop *technology demonstrators* to assess the feasibility to give the go-ahead or not. These kinds of high-technology demonstrators are very expensive and manufacturers join hands to distribute the cost. The government also offers financial assistance.

Readers need to review Chapters 4–7 on design options and the discussion to gain insights from past experience. Statistics is a powerful tool that should be used discriminately. Current market trends show stabilised statistics as given in Chapter 7. Statistical data offer an initial guesstimate on what to expect from a new aircraft design, which has to do better than the existing ones by incorporating proven newer technologies.

This chapter covers the following topics:

Section 8.2: Introduction to Configuring Aircraft Geometry: Shaping and Layout
Section 8.3: Prerequisites to Initiate Conceptual Design of Civil Aircraft
Section 8.4: Fuselage Design – Civil Aircraft
Section 8.5: Wing Design – Civil Aircraft
Section 8.6: Empennage Design – Civil Aircraft
Section 8.7: Nacelle and Pylon Design – Civil Aircraft
Section 8.8: Undercarriage Design Considerations – Civil Aircraft
Section 8.9: Worked-Out Example – Bizjet
Section 8.10: Prerequisites to Initiate a Conceptual Design of Military Aircraft
Section 8.11: Fuselage Design – Military Aircraft
Section 8.12: Wing Design – Military Aircraft
Section 8.13: Empennage Design – Military Aircraft
Section 8.14: Engine Intake and Nozzle
Section 8.15: Undercarriage Design Considerations – Military
Section 8.16: Worked-Out Example – Advanced Jet Trainer (AJT) and Close-Air-Support (CAS) Variant Design
Section 8.17: Turboprop Trainer Aircraft and Carry Out Counter Insurgency (COIN) Variant Design

Civil and military aircraft configuration layouts are addressed separately because of the fundamental differences in their design methodologies, especially in the layout of the fuselage. A civil aircraft has 'hollow' fuselages to carry passengers/payload. Conversely, a combat aircraft fuselage is densely packed with fixed equipment and crew members. Differences between civil and military are given in Table 1.2. Future designs indicate changes in aircraft configuration that are currently under research and development (e.g. the blended-wing-body BWB). The basics of the current type of aircraft design must be understood first before any advanced designs can be undertaken. This is the aim of this book.

This is an important chapter as intensive coursework work starts from here. Readers are to begin with laying out of aircraft geometry dictated by customer specifications. There is little mathematics involved in this chapter; rather, past designs and their reasoning are important in configuring a new aircraft. Subsequent chapters enter into detailed mathematical analyses to refine the configuration. Chapter 2 presents several aircraft specifications and performance requirements of civil and military aircraft classes. From these examples, a Learjet 45 class, a Royal Air Force (RAF) Hawk class and a turboprop powered Tucano class aircraft in civil and military aircraft categories, respectively, are worked out as coursework examples.

8.2 Introduction

Section 2.6 stressed that the survival of the industries depends on finding a new product line with a competitive edge. A market study or RFP/AST is the tool to establish a product by addressing the fundamental questions of why, what and how. It is like 'gazing into a crystal-ball' to ascertain the feasibility of a (ad)venture, to assess whether the manufacturer is capable of producing such a product.

This chapter covers the essentials of the design considerations and options available to make choice to arrive at a candidate aircraft configuration, conducted early during the conceptual design phase (Phase I) of a new aircraft project. The considerable amount of information in this chapter in the form graphical representation and geometric data is meant to facilitate the newly initiated to appreciate the underlying technologies in shaping aircraft and thereby to make choices. This method is also practised in industry at the initial stages when not much other detail is available. Subsequently, intensive analyses are carried out to refine configuration, more intensely in the next phase (Detailed Design Phase II) of the project, beyond the scope of this book.

The market specification itself demands improvements. In civil aircraft design, it is primarily in economic gains but also in performance gains incorporating proven leading edge (LE) advanced technologies without compromising safety. A 10–15% all-around gain over existing designs, delivered when required by the operators, would provide market leadership for the manufacturers. Historically, aircraft designers played a more dominant role in establishing a product line; gradually, however, input by operators began to influence new designs. Major operators have engineers who are aware of the latest trends, and they competently generate realistic requirements for future operations in discussion with manufacturers.

Ideally, if cost were not an issue, an optimum design for each customer might be desirable, but that is not commercially viable. To encompass diverse demands by various operators, the manufacturers offer a family of variants to maximise the market share at lower unit cost by maintaining component commonalities. Readers can now appreciate the drivers of a new commercial aircraft project, primarily the economic viability. The first few baseline aircraft meant for testing are seen as *preproduction* aircraft, which are flight tested and subsequently sold to operators. Figure 8.1 shows how variants of the Boeing 737 family have evolved. Here, many of the fuselage, wing and empennage components are retained for the variants.

However, military aircraft designs are dictated by national requirements when superiority, safety and survivability are dominant, of course, but without ignoring economic constraints. In the case of military aircraft design, it is primarily to gain superiority over potential adversary. It leads into exploring new technologies that will have to be proven in scaled flying technology demonstrators to substantiate the viability of the new

Figure 8.1 Possible changes (shaded area) in civil aircraft family derivatives.

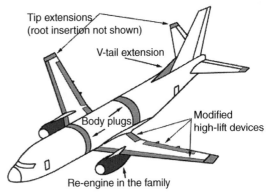

Typical modifications for derivatives in the family

concepts and associated safety issues. Today's military aircraft designs start with *technology demonstrators* to prove the advanced concepts, which are considered prototype aircraft. Production versions could be larger, incorporating the lessons learned from the demonstrator aircraft. In time, these may cascade down to civil aircraft design, one such example is fly-by-wire (FBW). The end of this chapter is devoted to some basic aspects of military aircraft design considerations pertaining to configuring simpler types of military aircraft, for example, AJTs capable of CAS and a turboprop trainer (TPT).

An interesting point to note is that no two aircraft or two engines of the same design behave identically in operation. This is primarily on account of production variance within the manufacturing tolerance allocations – there could be other reasons. The difference is of course very small – the maximum deviation would be of the order of less than ±0.2%. An old aircraft would degrade in performance being worn out and tired. During operation, the aircraft surface would get deformed, dented/warped increasing viscous drag and so on. Manufacturers take account of the real problems of operational usage by maintaining a 'status-deck' of performances of all aircraft produced. Manufacturers' quotations cover average aircraft degradation up to a point. In other words, like any engineered product, in general a brand-new aircraft would perform slightly better than what pilot's manual shows – this margin serves well for the operators.

If a new design fails to reach the predicted value then who is to blame? Is the short coming originating from aircraft or engine design or from both (readers may examine some of the old cases in context)? Is it a bad aircraft or a bad engine (if a new designed engine is installed)? Over the time, aerospace industries have addressed these issues quite successfully. Today, there is a matured approach to design with little scope to have such a blame game. As mentioned, some aerospace stories could be more exciting than fiction. Today, engine and aircraft designers work in close cooperation to identify exactly the nature of shortfalls and then repair them. In general, it is convenient if the responsibility for shaping of external nacelle mouldlines is that of airframe designer, while the internal shaping (intake duct and exhaust duct) is that of engine designer.

8.2.1   Starting Up Aircraft Conceptual Design

Aircraft size depends on its mission, which varies depending on the type role as categorised in Table 1.1 in Section 1.3.1; mainly one of two categories: (i) civil and (ii) military aircraft. For any mission the aircraft has to carry a removable mass of payload and the requisite amount of fuel, together they are seen as *useful mass* (crew are not seen as removable mass as these are integral for the mission and invariant in nature). The size of the useful mass indicates the size of the aircraft in terms of Maximum Takeoff Mass (MTOM).

When nothing other than the essential specification requirements of the users is laid down, designers have to rely on their past experiences and start with a guessed aircraft configuration based on available aircraft statistics within the class to get some idea of what to expect. The new aircraft progresses with refinements including proven innovative ideas to bring out a product that can compete in the marketplace, hoping that their product will be the best in the class. This leads into developing a concept definition with a realistic 3D model and associated three-view diagrams that gets refined as more information are generated.

The statistics of the existing designs give a good correlation between *useful mass* and MTOM. The first task, as the starting point, is to extract the aircraft MTOM from the statistics and what to expect for a new aircraft within the class. From this guessed MTOM, other aircraft-component geometries and mass can be established. Readers need to refer to Chapters 4–7 for the rationale.

Given next is the typical methodology carried out to progress with configuring a new aircraft. Because of the differences, civil and military aircraft are dealt with separately.

8.2.1.1   Civil Aircraft

The proposed new aircraft specifications (requirements) give the aircraft payload and range, good enough to start with. The aircraft needs to transport useful mass, that is, all the removal load: the payload, consumable and fuel mass – these are variable loads decided on by the operators as required for the sortie. Payload mass is specified and number of crew mass to operate the aircraft is also known. The fuel mass is proportionate to the range but yet to be established. Consumables are related to the payload (in this case, passengers). For a given range, payload itself gives an excellent statistical correlation with MTOM as shown in Figure 7.2 for different ranges of class that can be interpolated to pick the suitable point. Thereafter, from the subsequent graphs, the aircraft operator's empty mass (OEM), wing area, empennage areas, engine size and so on, can be guessed. The fuselage configuration is deterministic as shown in Section 8.6. It is followed by placing the undercarriage and determining aircraft weight and CG location leading to a *concept definition*. Next, the wing is sized to carry the aircraft weight and have a matching engine to power it to fly to *concept finalisation*.

8.2.1.2 Military Aircraft

The approach to configuring combat aircraft is different due to a very different throttle dependent mission role. Combat aircraft necessarily incorporate advanced technologies yet to be built, hence not much in the way of statistics exits within its class, but still can be generated from what exists in the close enough class. The payload is the expendable armament carried outside (except gun ammunition) the aircraft. Military does not have a range as such but a radius of combat mission that may not go as planned. However, there is a mission specific amount of fuel carried on board. This makes it hard to correlate between armament payload and radius of action as can be seen in the scatter diagram in Figure 7.22a. There are many ways to present military aircraft statistics. Figure 7.22b is one way to give the statistics. Another way would have been is to present useful load versus MTOM in one graph. In this book, Figure 7.22 is given to show the extent of scatter that exists.

Combat aircraft configuration starts with accurate computer generated sketches of possible configurations bearing close similarity with the technology demonstrator as refinements to choose one. As the combat aircraft fuselage does not have any portion with a constant cross-section as in the case of deterministic transport aircraft, its configuration is developed section by section (see Section 8.14).

National defence requirements made military designs evolve rapidly, incorporating new technologies at a considerable cost to stay ahead of a potential adversary. Whereas miniaturisations of electronic and other equipment reduces onboard mass, increased demand in combat capabilities worked in the opposite direction adds to mass. Combat aircraft size kept growing to exceed 35 000 kg for multi-role fighter aircraft. Currently, the lightest combat aircraft with proven capabilities is around 16 000 kg. The Typical Takeoff Mass (TTOM) is less than MTOM. It may look attractive to have a small lightweight fighter, but currently with less than 14 000 kg of MTOM, armament capacity and radius of action would suffer. The following points are pertinent to military aircraft-component mass estimation methodologies.

In the past, higher speed with high acceleration was sought for engagement within the visual range. Today, with advanced long-range guided missiles, it is rapid turn rate of combat aircraft that is in demand to aim quickly from beyond visual range (BVR), as far as 40–50 km away.

The authors believe that a modern combat aircraft design in the classroom of the F35 class with radar cross-section (RCS) capabilities and microprocessor-based control design, integration of systems/weapons and so on is beyond the scope of this book. A comprehensive book offering advanced combat aircraft design for classroom usage is unrealistic without first offering an exercise on simpler designs. This kind of advanced work on military designs can only be carried out when the basics are well mastered: that is aim of this book.

Therefore, this introductory book deals with military design exercises using a trainer aircraft with a CAS combat role variant as an alternative to frontline combat aircraft design. Good statistical relations exist for this class of aircraft and are used to make initial guesses for the geometric sizes.

8.2.1.3 Military Trainer Aircraft

The AJT has a specific non-combat mission profile when practise armament load can indicate the AJT class aircraft size. It requires preparing statistics within the AJT class of aircraft as given in Figures 8.11 and 8.12.

However, an intermediate level of trainer, nowadays in TPT aircraft, is primarily meant to gain proficiency in airmanship. In this case, provision is made to carry a light practice armament load in some designs without any armament practice. To get some idea on TPT size, the 'useful load' (fuel + armament, if any) is the parameter to get the possible MTOM of a new design as shown in Figure 8.19.

8.2.2   Aircraft CG and Neutral Point

Section 3.17 defined and explained aircraft CG, and Neutral Point (NP). These are required in the methodology used in this chapter to configure aircraft. Aircraft component weights and CG location are covered in Chapter 10.

Assembling the aircraft components by placing them in relation to each other is to be done in a way that will keep the aircraft CG location at a desirable place. Also, without the knowledge of CG location, which moves depending on the loading condition, the undercarriage positioning (Chapter 9) becomes unrealistic. It may require some iteration as the components' weights are not yet known. Positioning of aircraft components with respect to each other, especially the wing in relation to the fuselage, will require some iterations that may exhibit *wing chasing*.

The position of NP must be known to keep CG position and its movement within limits at a desirable position (Chapter 18).

It is difficult to establish the aircraft NP. In the past, DATCOM (the short name for the *USAF Data Compendium for Stability and Control*) methods predicted the NP and substantiation was carried out through a series of expensive wind-tunnel tests. Today, computational fluid dynamics (CFD) analysis precedes with a shorter series of wind-tunnel tests. It is for this reason that, instead of using the DATCOM method, this book takes the typical values of aircraft aft-most CG position as a percentage of mean aerodynamic chord (MAC), taken from statistics, as given in Table 8.1. Aircraft aft-most CG limit stays at least 5% ahead (conventional aircraft) of the NP.

The subsonic aerofoil '*a.c.*' lies round its quarter cord position and supersonic aerofoil *a.c.* lies round its mid-cord position (Chapter 4). But the whole aircraft NP is behind the aerofoil *a.c.* position. Here, the aircraft wing MAC plays an important role as a reference geometry in configuring aircraft.

Current supersonic aircraft fall in the combat category with a FBW fly-in relaxed stability that is close to aircraft NP, if not in a slightly negative stability position, that is, aft of the aircraft NP. These kinds of aircraft are not dealt with in this book.

The positions of aircraft CG and NP play an import role in empennage sizing, as dealt with in detail in Section 6.7. The H-tail must have pitch control capability either by having a part of the aft section hinged to move or itself all moving to provide the control authority maintaining stability. It must have adequate authority without running out at the limiting CG position (Section 10.5.1 explains).

Table 8.1   Typical aircraft aft-most CG limits.

	Wing/front mounted engines	Rear-mounted engines	
Civil aircraft	about 30–35% of MAC	about at 35–40% of MAC	
Military jet trainer aircraft		about at $\approx$45% MAC	
Military turboprop trainer aircraft		about $\approx$40% of MAC	
	Percentage of wing MAC		
	Forward-most CG as (%)	Aft-most CG (%)	Δ CG travel (%)
B767	11.0	32.0	21
DC10	8.0	18.0	10
B747	13.0	33.0	20

Source: http://www.dept.aoe.vt.edu/~mason/Mason_f/M96SC02.pdf.

CIVIL AIRCRAFT

8.3   Prerequisites to Initiate Conceptual Design of Civil Aircraft

After obtaining the new aircraft specifications/requirements, starting design work must be preceded with deciding the proven advanced technology level to be adopted to stay ahead of competition, offering all round gain in performance and economics. This is mainly concerned with aerodynamic refinements, materials selection, structure philosophy, systems architecture, choosing bought-out items (engine, avionics etc.) and establishing the manufacturing philosophy. A list of new technology considerations is given in Chapter 2. In the year round search, company-based scientists, in consultation with designers and manufacturing engineers recommend to management to decide the level of technology to be adopted after assessing the implications in the cost frame to complete in the open market. Any mid-course change of technology level could severely affect cost. It is assumed that the manufacturer has adequate funding to proceed uninterruptedly. Since cost is an important parameter to establish the value of the design, it is essential that the manufacturing philosophy must be considered during the conceptual design phase. This is a crucial prerequisite that often gets overlooked in academies to outline the starting procedure to expose the new comers. Given next is a typical summary of the prerequisites.

(1) *Aircraft specifications requirement* (generated from market study)
   (i) The mission profile – payload and range.
   (ii) Mission segment capability – takeoff and landing field lengths, climb speed, cruise speeds.
   (iii) Systems, instrumentation and engine type selections (affects weights).
   (iv) Passenger and aircraft crew facilities at the desired comfort levels.
   (v) Suggested wing position – high or low wing (to be decided during design reviews).
   (vi) Offer variant designs in family concept retaining maximum component commonality to cater for a wider market in a cost effective manner.
   (vii) Manufacturing philosophy – types of tooling and so on.
(2) *Technology level to be adopted* (should be within manufacturer's capability)
   Technology to be adopted should be cost effective and better than existing aircraft in current operation.
   (i) *Aerodynamics.* Aerofoil sections of the various lifting surfaces (profile, *t/c* ratio). Type of high-lift devices. Choice of high-lift device will affect the wing area and, as a consequence, change in MTOW. Wing and empennage planform shape. Mid-fuselage cross-section to establish passenger comfort level.
   (ii) *Structures.* Choice of material – type and extent of metals and composites used.
   (iii) *Power plant.* Type and number of power plant required. Selection from what is offered by the engine manufactures. This is a crucial selection to make a successful aircraft design.
   (iv) *Undercarriage.* Decide on the undercarriage type and its articulation kinetics.
   (v) *Systems.* Type of control and extent of automation. (electrical, electronic, mechanical, hydraulics, pneumatics).
   (vi) *Instrumentation.* Choose cost and weight effective instruments available on the market.
   (vii) *Flight deck.* Establish the layout philosophy for ease of crew operation.
   (viii) *Interior.* Layout, choice of materials, passenger interfaces and comfort level.
   (ix) *Manufacturing philosophy.* The success of new aircraft design depends considerably on the manufacturing technology adopted to minimise cost (Chapter 16). Aircraft parts and sub-assemblies must be easy to manufacture with a low parts count in a cost effective manner with associated low man-hours.

Industry makes enormous effort to make reality align with prediction; they have achieved performance predictions within $\pm 3\%$, the big aircraft within $\pm 1.5\%$. Those who are lucky could be in the spotlight. The generic methods adopted in this book are in line with industry; the difference is with industry making use of more detailed and investigative analyses to improve accuracy to remain competitive. Classwork predictions around $\pm 5\%$ compared to a similar type of operating aircraft are good.

8.3.1 Starting-Up (Conceptual Definition – Phase I)

A dedicated group (say, termed the New Aircraft Project Group, abbreviated to NAP) of very experienced designers from all areas of specialisation including manufacturing engineers, is formed to undertake the conceptual study. The NAP group conducts a conceptual phase of aircraft design in an Integrated Product and Process Development (IPPD, also known as Concurrent Engineering) environment. This multidisciplinary approach should have a good appreciation for the cost implications of early decisions to make product right the first time (see Section 2.4); aerodynamicists still play a major role in the process. Contributions by the structural engineers and production engineers are now an integral part in shaping aircraft components for making aircraft light and helping maintain easy production in order to keep both the aircraft selling price and the operating costs low thereafter.

Specialist areas may optimise their design goals but in the IPPD environment, compromise has to be sought. Optimisation of individual goals through separate design considerations may prove counterproductive and usually prevent the overall (global) optimisation of ownership cost. *Multidisciplinary optimisation* (MDO) offers good potential but to obtain global optimisation is not easy; it is still evolving. In a way, a global MDO, involving large number of variables, is still an academic pursuit. Industries are in a position to use sophisticated MDO algorithms in some proven areas of design.

Initially, the NAP relies on the statistics of existing aircraft within the class as well doing competition analysis of potential competitors. In fact, market analyses should be aware of new aircraft capabilities offered by competitors. If required, the market specifications/requirements should be revised before start-up to stay ahead of competition. Then, by incorporating proven advanced technologies, the NAP offers the best value candidate configurations.

Past designs offer good insight to move into future designs. At the start, only a few parameters will be required to propose a few candidate aircraft configurations to make the choice. Each company maintains a strong database of all operating aircraft in the class. Chapter 7 gives some statistical databases for trend analyses. To improve resolution, readers are recommended to prepare their own statistics of, say, 5–10 aircraft within the class of their project design. Experienced designers can make good guesses of what are expected, thereby reducing the number of iterations to improve accuracy, saving design man-hour costs.

Initially, the conceptual study proposes several preliminary candidate aircraft configurations to search for the best choice. Comparative studies are carried out to confirm which choice provides the best economic gains. Figure 8.2 shows six possible configurations (author-generated for coursework only). Eventually, the best configuration in the figure is selected through a series of design reviews with the customers and management (the first configuration offers the best market potential.).

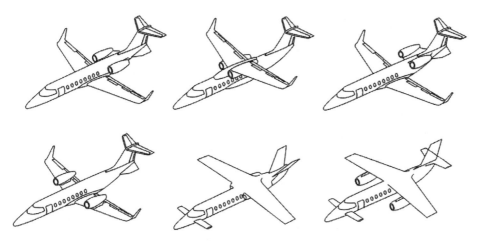

Figure 8.2 Four candidate aircraft configurations.

The first configuration offers the best market potential. The chosen configuration is sized with matched engine in an iterative process to a *satisfying* baseline design, which implies that while none of the variants is an 'optimised' design, the family of variants in the project offers a 'satisfying' design to widen the market to amortise initial investment at a slightly increased design cost frame. Retaining maximum component commonalities within the variant designs helps in reducing the aircraft price.

8.3.2 Methodology for Shaping and Laying out of Civil Aircraft Configuration

This section takes a closer look of the Phase I stage of a new aircraft project as given in Chart 2.2. The initial stages of the Phase I is the *concept definition*, as described in Chart 8.1, which gives some idea on how a project starts. This chapter deals only with Step 1 of the Chart 8.1. Thereafter, typically, the process cascades down carrying out the *concept finalisation* in a step-by-step manner.

The methodology starts with shaping the aircraft components; for example, wing planform, fuselage shape, nacelle/intake shapes and empennage and establish their external geometries. These components serve as 'building blocks' to assemble them to configure the new aircraft.

Step 1 is based on 'guesstimates' using the statistics of past designs. Designers then incorporate proven LE advanced technologies without compromising safety to stay ahead of competition. Given next are the Step 1 sequences in Chart 8.1, in a step-by-step manner,

i) From the statistics of payload-range, obtain a preliminary MTOW that will be improved through iterations.

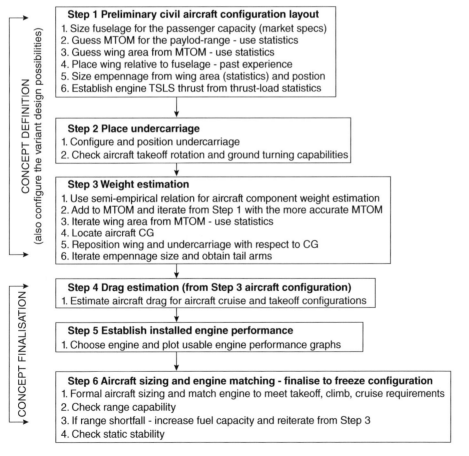

Chart 8.1 Phase I, conceptual study: methodology for finalising civil aircraft configurations.

ii) From the statistics of MTOW versus wing area, develop a preliminary wing planform with the specified technology level. Next, position the wing with respect to the fuselage in an approximate position from past experience (typically, begin with the $MAC_{1/4}$ slightly ahead of mid-fuselage). Size and position the empennage in relation to wing area (readers may study comparable existing designs). This gives the preliminary wing loading, W/S_w, which can be compared with the statistics of comparable operating aircraft. Note that moving wing position will also move the aircraft CG, therefore some iterations are required.

iii) Select the engine from what is available. The thrust loading will invariably fall within 0.32–0.35 for two-engine and 0.3–0.33 for four-engine aircraft. For propeller powered aircraft, divide the thrust by 3.5 to get the shaft horse power (SHP). Later, the engines will be properly sized to match the aircraft.

The other steps from two to six in Chart 8.1 continue in the following chapters. In the process, the preliminary geometry, weight and engine size will be revised to better accuracy leading to the final design. Finalising the aircraft configuration, as a marketable product, follows the formal methodology of aircraft sizing and engine matching (Step 6 of Chart 8.1) and is an iterative process.

8.3.3 Shaping and Laying out of Civil Aircraft Components

The objective is to generate aircraft components, piece by piece in a building-block fashion, and mate them as shown in the middle diagram of Figure 2.1. The diagram also gives a more detailed breakdown of the aircraft components in subassembly groups for a better understanding of the preliminary layout of the internal structures that facilitates preliminary cost estimates.

A lot of information has been captured within Section 7.3–7.5. This section summarises some of the points arising out of these sections. The nine graphs in Figures 7.1–7.9 capture all the real aircraft data taken from [1–9] and other sources. These statistical data (with some dispersion) prove very informative at the conceptual design stage to get an idea of what options a new design can incorporate to stay ahead of competition with a superior product. With these nine graphs one can figure out what to expect from the basic specification of *payload-range*. The readers may have to wait until the project is completed (a better appreciation will emerge after completing the sizing exercise shown in Chapter 16) to compare how close it is with the statistical data. There should be no surprise if the classroom result falls within the statistical envelope.

As it progresses with better definition, the aircraft-component weights are revised for better accuracy that may require revising the aircraft wing position to getting the aircraft CG at approximately a quarter chord of the MAC. Changing wing position changes CG location, known as *wing chasing*.

Section 6.7 dealt with empennage sizing. At this stage, it may prove cost effective to keep the control surface sizes taken from the existing statistics of comparable designs. The control surfaces can be formally sized to better accuracy in Phase II study using CFD/wind-tunnel tests (datasheets may prove useful but this practice is gradually receding). Finalisation of control area size is determined in a subjective assessment of various test pilots with a consensus of what is acceptable.

Sections 3.15 and 4.19 summarised that aircraft speed capability influences the aerofoil choice and wing shaping, respectively, implying that compressibility effects of air dictate the shaping of aircraft configuration in the following three categories.

i) *Aircraft design speed below Mach 0.6.*
   Influence of the compressibility effects of air can be neglected for aircraft maximum design speed is below Mach 0.6, when $C_{DW} \approx 0$. There are not many aircraft operating within Mach 0.6 and 0.7. Chapter 12 discusses that propeller efficiency reduces above Mach 0.6 and jet propulsion is inefficient below Mach 0.6.

ii) *Aircraft design speed above Mach 0.6 and below Mach 0.95.*
   Above Mach 0.7 speed capability, compressibility plays increasing role in shaping aircraft as speed capability increases. Shaping aircraft, for example, incorporating wing sweep, use of thinner aerofoil sections, fuselage area ruling, and so on to minimise the rise of wave drag, C_{DW}.

iii) *Aircraft design speed above Mach 0.95*

At transonic speed region, wave drag increases drastically and there no aircraft designed to fly above Mach 0.95 until it goes well past Mach 1.4. For speed above Mach 1.4 and up to 2.4, aircraft shaping requires addressing the issues arising from the effects of the presence of shock waves. The aircraft becomes slenderer with a sharper nose cone and wing leading edge, the aerofoil becomes very thin and so on. Thermal effects limit the aluminium frame not to exceed Mach 2.4. (The Lockheed SR71 Blackbird speed capability achieved Mach 3.4 with a titanium frame, a record remained unbroken for more than half a century.)

Today, industry uses computer aided drawing (CAD)-generated aircraft configurations as an integral part of the conceptual design process, which must be implemented in classwork. Having CAD 3D parametric modelling allows changes to be easily, quickly and accurately incorporated. Making 2D drawings (i.e. three-view) from 3D models is simple with few keystrokes. Sections 8.4–8.7 configure the aircraft components of civil transport category aircraft.

8.4 Fuselage Design

Chapter 5 is devoted to fuselage design. The commercial aircraft fuselage is a hollow shell accommodating the payload (passenger/cargo) and its mid-section typically has a constant cross-section. For transport aircraft design, it is convenient to start with fuselage shaping as it is determined from its specified passenger number and comfort level; that is, from the specified passenger capacity, the number of seats abreast and number of rows and facilities provided. In other words, fuselage design is not derived from empirical equations. Fuselage design parameters show strong statistical correlations as covered in Chapter 7.

Comfort level is an important parameter in free market competition. Higher comfort level with more space for passengers is preferred and results in a wider fuselage at the expense of higher drag and weight, making it more expensive to operate, thus affecting profit margins essential for industry growth and vice versa. A compromise is required to obtain the best fuselage width. In this book, medium comfort level is taken as given in Table 7.2. The aerodynamic design considerations of these types of bodies are not as stringent as wing/empennage considerations as they are not meant to generate lift to sustain flight and control aircraft performance. The main aerodynamic considerations are to reduce drag and moment of the bodies. In this book, the fore and aft closure designs of fuselage are guided by the statistics of existing designs taking care of other requirements for example, to provide takeoff rotational clearance, requirements of flight crew. Considerations for the Maintenance, Repair and Overhaul (MRO) issues and so on are to be made.

Being subsonic in operation, its front end is blunter in a favourable pressure gradient and the rear end tapers gradually to a closure in adverse pressure gradient to minimise boundary layer separation, unless it is designed for special purpose, for example, having a door for rear loading. A carefully shaped conventional fuselage can generate a very small amount lift, say about 2% of aircraft weight. The top mouldline of the fuselage contour is more curved than the bottom keel mouldline, which permits a very small amount of lift. Wing-body blended aircraft configurations are not dealt with in this book.

8.4.1 Considerations Needed to Configure the Fuselage

Section 6.7 describes typical fuselage layouts from two-abreast seating to the current widest seating of 10 abreast. The following are the general considerations for the fuselage layout.

Geometry	Aerodynamics
(1) Diameter (e.g. abreast seating, comfort level, appeal)	(1) Front and aft-end closure
	(2) Wing-body fairing
(2) Length, fineness ratio	(3) Wing and tail position

(3) Upsweep for rotation angle and rear door, if any
(4) Cross-section to suit under floor boards and headroom, cabin volume and space

(4) Surface roughness
(5) Drag

Structure (affecting weight and external geometry)	*Systems*
(1) Fuselage frame and bulkhead schemes	(1) Flight-deck design attachments
(2) Wing, empennage and undercarriage	(2) Passenger facilities
(3) Doors and windows	(3) All other systems
(4) Weight	

8.4.1.1 Methodology for Fuselage Design

Fuselage size is determined from required passenger capacity. The current International Civil Aviation Organization (ICAO) limit on fuselage length is 80 m, an artificial one based on current airport infrastructure size and handling limitations. The following are the considerations for the methodology, given in a step-by-step approach to configure a fuselage.

Step 1: Fuselage cross-section and seating arrangement.

Decide on the abreast seating arrangement corresponding to passenger number as given in Table 5.3. For the specified comfort level, find cabin and fuselage width (Figure 5.8) and height. Adding the fuselage thickness, it will give the fuselage height and width. For aircraft that are four abreast or more, under floorboard space provision needs to be considered. A pressurised fuselage would invariably be circular or near circular to minimise weight. Unpressurised cabins for aircraft operating below 4300 m (14 000 ft) need not be circular in cross-section. Smaller utility aircraft would show the benefits of having a rectangular cross-section. A box-like rectangular cross-section (Figure 4.7) would not only offer more leg room but would also be considerably cheaper to manufacture. Based on below floorboard space provision requirements and whether the cabin is pressurised or not, the fuselage cross-section shape is developed.

Step 2: Fuselage mid-section

Step 1 has established the abreast seating and fuselage width. The mid-fuselage is mostly of a constant cross-section. Determine the number of seat rows by dividing total passenger capacity by number of abreast seating. If it is not divisible then the extreme rows will have fewer seats abreast. The end extremity of the fuselage mid-section can get tapered as a start for fuselage closure. CFD analyses of the closure aerodynamics have to confirm that the pressure distribution around the closures is satisfactory with special attention to ensure to keep aft closure boundary separation to its minimum. The aft luggage space in front of the pressure bulkhead can exist, especially for small aircraft.

Decide on the passenger facilities – for example, toilets, galleys, closets, cabin crew seating and so on – and their dimensions to be added: the extent depends on number of passengers and duration of flight. Chapter 7 describes toilet and galley details. For small aircraft with shorter flight durations, it is desirable that a toilet be provided. There are small aircraft of low mission range without a toilet to keep cost down, but these could prove uncomfortable.

Step 3: Fuselage front and aft closure sections and the fuselage length

When the seating arrangement is determined in the mid-fuselage section, it must be closed at the front and aft ends for a streamlined shape, maintaining a fineness ratio within 7–14 (see Table 5.3). Typical front and aft-fuselage closure ratios are given in Table 5.1. There is a wide choice as can be observed from the past designs within the class of aircraft. While benefiting from past experience, designers develop their own configuration to improved pilot vision, drag considerations, space for storage, rotation for takeoff and so on. Adding front and aft closure to the fuselage mid-section gives the fuselage length. Front and aft fuselages have their respective bulkheads to give a sealed cabin for pressurisation.

The fuselage upsweep angle of the aft-end closure depends on the type of aircraft. If it has a rear-loading ramp as in a cargo version, then the upsweep angle is higher. The fuselage takeoff rotation clearance angle, θ (see Figure 5.6 and Figure 9.9), depends on the main-wheel position of the undercarriage relative to the

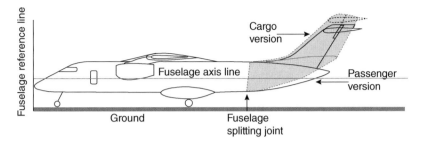

Figure 8.3 Convertible fuselage.

aircraft CG position (see Chapter 7). The typical angle for θ is between 12 and 16° to approach C_{Lmax} at aircraft rotation.

Step 4: Locate the position and sizes of aircraft doors, windows, hatchets

Federal Aviation Administration (FAA) and Civil Aeronautics Agency (CAA) have mandatory requirements on minimum number of passenger doors, their types and corresponding sizes, depending on the maximum passenger capacity the fuselage is intended to accommodate. This is to ensure passenger safety – certification authorities stipulate a time limit (90 s for big jets) within which all passengers must egress if an unlikely event, for example, a fire occurs. The larger the passenger capacity, the higher the minimum number of doors is to be installed. Not all doors are of same size – the emergency doors are smaller. Passenger doors have several categories and are dealt with in Section 5.5.5. All doors are kept armed during airborne operation.

Other Points

1. Two-abreast seating cabin height is typically less than 6 ft and has constrained leg space. Three-abreast seating, if required, with floorboard recess, will allow standing room and better leg room (Figure 5.10). For a fuselage with four-abreast seating or more, the cross-section could use below floorboard space. Full standing headroom is easily achievable for a fuselage with four-abreast seating or more.
2. Two aisles are provided for a fuselage with seven-abreast seating and more (current maximum is 10 abreast). In future, if wider cabin appears (say with BWB), then more than two aisles would be necessary.
3. The minimum number of cabin crew is dependent on the maximum number of passenger capacity the airframe can take. Cabin crew is not required up to 19 passengers but some operators provide one.
4. It is necessary to set the zero reference plane (ZRP) and assign a fuselage axis line (see Section 5.3.1). It assists in tracking the aircraft-component positions along the fuselage axis. For smaller civil aircraft, there is no constant fuselage section, and the aircraft centreline must be conveniently chosen; it is the designer's choice as long as the reference lines are clearly defined and adhered to for the entire life cycle of an aircraft that could encounter design modifications in its service life.
5. *Fuselage splitting plane.* An interesting concept is to make variants of a modular fuselage – that is, with two types of aft ends easily interchangeable. One type is for the conventional passenger version with a pointed aft-end closure, the other is for the cargo version with an increased upsweep to accommodate a rear-loading ramp. Figure 8.3 shows the concept of a 'quick-change' convertible fuselage. The aft fuselage can be split to accept either a passenger version or a rear-loading cargo version, as the mission demands. The changeover splitting joint is located behind the main undercarriage.

8.5 Wing Design

Chapter 4 is devoted to wing design. The first task for wing design is to select an aerofoil suitable for the desired aircraft performance characteristics. This book does not undertake aerofoil design; rather, it uses established

2D aerofoil data from the public domain (the NACA aerofoil data [1] in Appendix C are sufficient for this book). Industry takes an arduous route to extract as much benefit as possible from its in-house research that is kept commercial in confidence. It is an established technology in which there is a diminishing return on investment. However, the differences between the best designs and those in the public domain are enough to encourage industrial competition.

The next task is to configure a wing planform with a reference area typically for the class of aircraft. It is not determined by the passenger number as in the fuselage. Initially, corresponding to the guessed MTOM, the wing reference area is estimated from statistics. Some iteration is required because component weights are revised at the stages of progresses. Subsequently, the preliminary wing reference area must be sized using the methodology described in Chapter 14.

At the conceptual stage of the project study, typical values of wing twist and other refinements are also taken from the past experience of a designer. The values must be substantiated through CFD analysis and wind-tunnel testing to a point when the flight-test may require final local refinements (e.g. flap and aileron rigging). Initially, an isolated wing is analysed to quickly arrive at a suitable geometry and then studied with the fuselage integrated.

A generous wing root fairing is used to reduce interference drag as well as vortex intensity at the aft-fuselage flow. There is no analytical expression to specify the fairing curvature – a designer should judge the geometry from past experience and CFD analysis, considering the associated weight growth. In principle, a trade-off study between weight growth and drag reduction is needed to establish the fairing curvature. At this stage, visual approximation from past experience is sufficient. Observe the current designs and make decisions.

8.5.1 Considerations in Configuring the Wing

The following are general important considerations when designing the wing:

Geometry	Aerodynamics
(1) Aerofoil section, t/c ratio	(1) Lift and moment
(2) Wing reference area, S_W	(2) Drag
(3) Span and aspect ratio	(3) Stall, critical Mach
(4) Sweep, twist, dihedral, taper	(4) High-lift devices
(5) Glove/yehudi, if any	(5) Control surfaces
(6) Position with respect to fuselage (stability/CG)	(6) Surface roughness

Structure (affecting weight and external geometry)	*Systems*
(1) Spar, rib positions and attachment	(1) Control linkage
(2) Stiffness, aeroelasticity and torsion stability	(2) Fuel system
(3) Fuel volume	(3) Electrical
(4) Undercarriage and nacelle, if any	(4) Anti-icing
(5) Weight	

Chapter 4 is devoted to wing design, dealing with the roles of wing sweep, wing twist and wing dihedral/anhedral angle. Figure 4.27 summarised the role of aircraft speed capability influencing wing sweep and its aerofoil thickness to chord ratio. A major requirement is to make the wing root stall earlier to retain aileron effectiveness at a high angle of attack (low speed) – especially during landings. A wing twist with washout would favour such behaviour (and is the prevailing practice). Generally, the dihedral is associated with low-wing design and the anhedral with high-wing design; however, there are designs that are the reverse; a high wing can accommodate a dihedral.

8.5.2 Methodology for Wing Design

Section 4.16 may be revisited to obtain a summary of wing design. Given next is a stepwise approach to wing design.

Step 1: Decide on the aerofoil selection

Chapter 3 is devoted to aerofoil selection. Aerofoil selection is one of the most important aspects of aircraft design. Aircraft performance depends considerably on the type of aerofoil adopted. Aerofoil design is a protracted and complex process that is beyond the scope of this book. Section 3.7 outlines the strategy to search for an aerofoil that would provide a high C_{Lmax} as well as a high-lift-curve slope (dC_L/d_α), a high L/D ratio for the prescribed cruise speed, a low pitching moment and gentle stalling characteristics. Aerodynamicists prefer aerofoils to be as thin as possible but structural engineers prefer them as thick as possible. A compromise is reached based on aircraft design Mach number and the chosen wing sweep (Figure 4.27). Aerofoils can vary spanwise.

Step 2: Establish wing reference (planform) area

Initially, the wing referenced area has to be guessed from past statistics. The first estimate comes from aircraft MTOW in the payload-range capability (Figure 7.1). Next is to estimate the wing reference (planform) area, S_W, from the estimated MTOW (Figure 7.5). This will give the wing loading. Both the S_W and MTOW will be accurately sized in Chapter 14.

Step 3: Establish wing geometric shape and the associated details

Establish: (i) aspect ratio, (ii) wing sweep, (iii) taper ratio, (iv) twist and (v) dihedral (Section 3.16). The choice for wing aspect ratio, wing sweep and taper ratio are interlinked with the aircraft maximum speed to keep the compressibility drag rise within 20 drag counts at the high speed design specification (Section 4.19.1). In general, the wing planform is of a trapezoidal shape but not necessarily restricted to this; it can be modified with a glove and/or yehudi. Given next are the pertinent points associated with these five parameters.

Aspect ratio

Determine the aspect ratio, this should be the highest the structural integrity will permit (in consultation with stress-engineers) for the aerofoil thickness to chord ratio and the wing root chord based on the taper ratio. This minimises induced drag (see Eq. (4.18)). The *V-n* diagram (see Chapter 17) determines the strength requirement in pitching manoeuvres creating maximum stress from the bending moment at the wing root. Civil aircraft do not have high roll rates (unless it is a small aerobatic aircraft). Choice of material and aerofoil *t/c* ratio contributes to structural integrity.

Wing sweep

Determine the wing sweep, which is dependent on maximum cruise speed (Figure 4.23).

Taper ratio

For civil aircraft, a trapezoidal wing planform would be the dominant choice with taper ratio $\approx (0.3 < \lambda < 04)$.

Twist

At this stage, wing twist is empirically determined to improve stalling effects. Determine the wing twist; a typical value is 1–2°, mostly as washout.

Dihedral

Determine the wing dihedral/anhedral angle; initially, this is from the statistical data (Section 4.4.2). Typical values are dihedral $\approx$3–5° and anhedral $\approx$0––3°.

These five parameters are eventually substantiated through CFD analysis and wind-tunnel testing hoping that flight test results will not require further tweaking.

Step 4: Position wing Relative to the fuselage

Positioning of the wing relative to the fuselage requires the location of CG and its range of movement with weight variation (i.e. fuel and payload). The positioning of the wing should be such that the aircraft stability margin is not jeopardised by extremes of the operational CG position. The positioning of the wing relative to the fuselage is an iterative process dictated by the location of the aircraft CG at a desired position, expressed in terms of the percentage of the wing MAC. The aircraft weight distribution and CG location are yet to be established, it is initially estimated based on experience and past statistics in the aircraft class. If nothing is known, then a designer may position the wing MAC around the middle of the fuselage for rear-mounted engines or slightly ahead of the mid-fuselage for wing-mounted engines.

Subsequently, the wing position gets iterated as aircraft and its components weights are known (Chapter 10). This may not be easy because moving the wing will alter the CG position – an inexperienced engineer could encounter wing chasing. For newcomers to aircraft design, this offers an interesting exercise: However, this is not a major concern as very quickly, a 'feel' for locating the wing can be developed. The wing positions are tracked along the fuselage from the aircraft ZRP in the attempt to arrive at a desired position. Experienced designers minimise the number of iterations that could occur with wing chasing,

Step 5: Establish wing high-lift devices

Section 4.15 is devoted to high-lift device aerodynamics and their configuration types. Flaps and slats are wing components, the selection of the type depends on the field performance demands to generate high lift. In general, the more demanding the requirements, the more sophisticated the high-lift devices, which gets progressively more complex and therefore more expensive and heavier. Associated incremental lift gains by each type are shown in Figure 4.37. In general, single- or double-slotted Fowler action flaps suffice for the majority of smaller civil transport aircraft (Bizjets/Regional jets). Fowler action designs increase wing planform area and lift as well as an increase in nose-down pitching moment.

The first task is to decide the type of high-lift device required to meet the maximum C_{Lmax} to satisfy the specified field performances (takeoff and landing). Once the type of high-lift device selected, their area and other geometrical parameters are initially earmarked from statistics/semi-empirical data. Flaps are positioned behind the wing rear spar (about 60–66% of the chord) and typically run straight or piecewise. Flaps take up about two-thirds of the inner wing span. It is apparent that designers must have a good knowledge of the internal structural layout to configure an aircraft. In this book high-lift devices are not sized, but positions are earmarked.

Step 6: Wing-mounted control surfaces areas and their locations

Section 4.16 is introduces wing a host of wing-mounted control surfaces (e.g. aileron, flap, slat, spoilers and trim tabs), none of which are sized in this book. Initially, their geometries are extracted from the statistics of current designs or determined using semi-empirical relations. At this stage, their placement and positing are earmarked in this book. Control surface sizing is accomplished after the wing is sized and is addressed in subsequent design phases.

The aileron span is about a third of the wing span at the extremities. Ailerons and flaps are hinged aft of the rear spar for up and down movements; provision for them should be made during the conceptual design phase. A *flaperon* serves as both a flap and an aileron.

Not all aircraft have wing spoilers; however, aircraft with speed over Mach 0.7 generally have spoilers. These are installed close to the aircraft CG line to minimise pitch change. Spoilers act as air brakes and as *lift dumpers* during landing. The differential use of spoilers is for lateral control and they are referred to as *spoilerons*.

8.5.3 Structural Consideration for a Wing Attachment Along a Fuselage Layout

Structural considerations for attaching the wing to the fuselage are discussed in detail in Section 4.16. In summary, wing design has to consider the wing fuselage attachment option, which can affect the local fuselage external shape.

8.6 Empennage Design

Chapter 6 covered empennage design. This book deals only with conventional empennage design, for example, a V-tail and a H-tail orthogonal (near) to each other. Empennage design is quite complex as it depends on fuselage and wing sizes, those have to be configured first. The fuselage length, wing reference area (S_W), and tail arms L_{HT} and L_{VT} are the main parameters governing the empennage sizes, S_{HT} and S_{VT}. The two important parameters interlinking the relationship are defined in Eqs. (6.1) and (6.2) as follows.

H-tail area, $S_{HT} = (C_{HT})(S_W \times MAC)/L_{HT}$,

V-tail area, $S_{VT} = (C_{VT})(S_W \times wing\,span)/L_{VT}$,

Taking the C_{HT} and C_{VT} values from statistics, the respective empennage areas S_{HT} and S_{VT} can be computed from these equations. The empennage size needs to be checked out for the extreme limits of forward and aft positions of the aircraft CG when the aircraft and its component weights are known. Graphs in Figure 7.8 give the statistics and there is a wide spread in the data. The current design tendency indicates a little higher tail volume coefficient compared to the historical design trend. (Examples 6.1 and 6.2 give the DATCOM method, but at this stage it is justified to use statistics to get empennage areas as shown in Example 6.3.)

The H-tail is placed as a T-tail on a swept-back V-tail that would provide an increased tail arm, L_{HT} and L_{VT}, which would save weight by not having a longer fuselage. Smaller aircraft would benefit from a T-tail; however, to support the T-tail load, the V-tail must be made stronger with a small increase in its weight. Care must be taken to ensure that the T-tail does not enter the wing wake at a high angle of attack. This can be achieved by positioning it high above the wing wake at near stall or having a larger H-tail and/or an all-moving H-tail acting as an elevator. Also, the positioning of the H-tail has to consider its relative placement with respect to V-tail to minimise shielding. H-tail and V-tail designs are discussed separately in the following subsections.

8.6.1 Horizontal Tail

Typically, for civil aircraft, the H-tail planform area is from one-fifth to one-quarter of the wing planform size, aspect ratio, AR $\approx$ 3.0–3.5. As in wing design, the H-tail can have a sweep and a dihedral (a twist is not required). Sweeping of the H-tail would effectively increase the tail arm L_{HT}, which is an important consideration when sizing the H-tail. For a T-tail configuration, the tail arm further increases. The H-tail camber is influenced by the aircraft's CG position. In general, negative camber is used to counter a nose-down moment of the wing. At a high angle of attack, the H-tail should not remain within the wing wake; otherwise, it must be enlarged to be effective.

8.6.2 Vertical Tail

The V-tail can have a sweep, but the dihedral and anhedral angles and the twist are meaningless because the V-tail needs to be symmetric about the fuselage centreline. Typically, for civil aircraft, the V-tail planform area is about 12–20% of the wing reference area, aspect ratio, AR $\approx$ 1.0–1.5. From the statistics given in Figure 7.8, it can be seen that there is a cluster of V-tail designs with a tail volume coefficient of 0.07. For the T-tail configuration, the tail volume coefficient could be reduced to 0.06 because the T-tail acts as an endplate at the tip of the V-tail. Sweeping of the V-tail would effectively increase the tail arm L_{VT}, an important dimension in sizing the V-tail.

It is important that the V-tail remains effective for the full flight envelope. The V-tail, especially the rudder, must not be shielded by the H-tail to retain effectiveness, especially during spin recovery. Shielding of the V-tail, especially the control areas, may prove to be dangerous. A designer must ensure that at least 50% of the rudder stays unshielded at a high angle of attack. With a T-tail, there is no shielding of rudder.

The V-tail design is critical to takeoff – especially in tackling yawed ground speed resulting from a crosswind and/or asymmetric power of a multiengine aircraft. A large V-tail can cause snaking of the flight path at low speed, which can be resolved easily by introducing a 'yaw-damper' (a matter of aircraft control analysis). At cruise, a relatively large V-tail is not a major concern. For propeller-driven aircraft, the V-tail could be kept slightly skewed (less than 1°) to offset a swirled-slipstream effect and gyroscopic torque of rotating engines and propellers.

8.6.2.1 Typical Values of Tail Volume Coefficients

Typically, C_{HT} and C_{VT} depend on the class of aircraft under study. The following values of C_{HT} and C_{VT} may be used for the category of aircraft under consideration.

Category	Typical speed in knots (kt) or Mach
Home-builds – Light Sports Aircraft (LSA)	Maximum never exceed speed $\leq$120 kt
Club trainers – Normal Category (FAR 23)	Around 150 kt
Utility – Normal Category (FAR 23)	Around 200–250 kt
Aerobatic – Normal Category (FAR 23)	Around 200–300 kt
Bizjets – FAR 25	Around Mach 0.65–0.75
Commercial jets	High subsonic around Mach 0.75–0.95

Table 8.2 may be compacted in Table 8.3 based on aircraft MTOM, that is, its size. Larger civil aircraft have higher tail volume coefficients. A large V-tail is required to retain control authority during high level cross wind field performances.

Aircraft with FBW system architecture ensures that aircraft operates safely with the operating envelope, hence the empennage can be optimised to a smaller size. If a FBW control system is incorporated, the empennage sizes can be reduced because the aircraft would be able to fly safely under relaxed stabilities. However, this book is not concerned with control laws as design input in an introductory course. The FBW concept is introduced in Chapter 18 but not analysed. It will not be long until tailless aircraft, such as the B2 bomber, appear in civil aircraft designs, especially for BWB aircraft.

Table 8.2 Civil aircraft tail volume coefficients.

Aircraft category	$\approx C_{HT}$	$\approx C_{VT}$	Remarks
Home-builds/club trainers	0.6 ± 0.1	0.06 ± 0.01	To save cost, home-builds can get with flat plates as empennage.
Aerobatic category	0.6 ± 0.1	0.06 ± 0.01	Generally with tapered wing planform. Aerobatic with higher control authority.
Utility category	0.7 ± 0.1	0.07 ± 0.01	Generally with tapered wing planform (there are also rectangular planform).
Bizjet category	0.7 ± 0.1	0.07 ± 0.01	Invariably with tapered wing planform.
Commercial jets	0.9 ± 0.2	0.09 ± 0.01	Invariably with tapered wing planform.

Empennage planform geometry follows the pattern of the wing geometry. Empennage aerofoil thickness to chord ratio, t/c, is in the order of wing t/c, but type can differ.

Table 8.3 Civil aircraft tail volume coefficients.

	H-tail coefficient (C_{HT})	V-tail coefficient (C_{VT})
Small aircraft ($\approx$ < 25 000 lb)	0.6–0.8	0.05–0.08
Medium aircraft ($\approx$25 000–250 000 lb)	0.8–1.0	0.08–0.1
Large aircraft ($\approx$ > 250 000 lb)	1.0–1.2	0.1–0.12

8.6.3 Considerations in Configuring the Empennage

The following are general considerations important when configuring the empennage:

Geometry	Aerodynamics
(1) Empennage aerofoil section, t/c ratio	(1) Drag, lift, moment
(2) H-tail and V-tail reference area, S_H and S_V	(2) Tail volume coefficient
(3) Span and aspect ratio	(3) Stall, spin and yaw recovery
(4) Sweep, twist or dihedral	(4) Control and trim surfaces
(5) Position relative to wing (tail arms)	(5) Spin recovery
(6) Position H-tail to avoid shielding of V-tail	(6) Balancing
(7) H-tail position (high $\alpha_{\text{clearance}}$, T-tail)	(7) Engine-out cases

Structure (affecting weight and external geometry)	*Systems*
(1) Spar and rib positions	(1) Control linkage type
(2) Stiffness, aeroelasticity and torsion stability	(2) Control actuation type
(3) Fuel volume (if any – large aircraft)	(3) Electrical (if any)
(4) Weight	(4) FBW (if any)

8.6.4 Methodology for Empennage Design – Positioning and Layout

The empennage design has considerable similarity to the wing design. The following is a stepwise approach to empennage design: It requires some iteration as many parameters are still not known requiring estimated planform geometries taken from existing designs within the class of aircraft. Subsequently, a realistic planform will emerge that will eventually be fine-tuned through flight testing. Given next is a stepwise approach to empennage design.

Step 1: Decide the aerofoil section

In general, the V-tail aerofoil section is symmetrical but the H-tail has an inverted section with some (negative) camber. The t/c ratio of the empennage is close to the wing-aerofoil considerations. A compromise is selected based on the aircraft design Mach number and the wing sweep chosen.

Step 2: Position wing with respect to fuselage

From statistics, obtain the tentative empennage areas from Figure 7.8, which shows spread and likely to read higher areas than what is required. This preliminary area is needed to position the empennage with respect to fuselage to compute tail arms as stated in Step 3. Find the empennage areas geometry using typical aspect ratio and taper ratio and span. For aircraft with a T-tail arrangement, position the V-tail as far aft in a position suiting the structural arrangement of its main spar with a 'banjo frame' or any other arrangement going through the fuselage (Figure 19.31). Place the horizontal T-tail at the top of the V-tail. Fuselage mounted low H-tail positioning is done in conjunction with positioning the V-tail to avoid rudder shielding and wing wake consideration (Figures 6.14 and 6.15). The empennage areas will be revised in Step 4, based on empennage areas computed using a tail volume coefficient obtained from stability Eqs. (6.1) and (6.2).

Step 3: Compute tail arms and choose tail volume coefficients, C_{HT} and C_{VT}

Measure the tail arms L_{HT} from the wing $\text{MAC}_{1/4\text{-chord}}$ aircraft to the H-tail $\text{MAC}_{1/4\text{-chord}}$ and L_{VT} from the wing $\text{MAC}_{1/4\text{-chord}}$ aircraft to the V-tail $\text{MAC}_{1/4\text{-chord}}$. Empennage geometries are still not known. Take tail volume coefficients, C_{HT} and C_{VT}, respectively using Table 8.1. C_{HT} and C_{VT} also have wide spread in the statistics as shown in Figure 7.8. The spread of the tail volume coefficients have been narrowed down in Table 8.2 to usable values for this book. Industries also use the tail volume coefficients, C_{HT} and C_{VT}, from their own data bank statistics, substantiated by flight testing: these are more appropriate than the values given in Table 8.2.

Step 4: Estimate empennage reference (planform) area

From the tail volume coefficients, C_{HT} and C_{VT}, obtain in Step 3 compute the H- and V-tail planform areas using Eq. (6.1) and (6.2). Discard the earlier values empennage areas taken from statistics taken from Figure 7.8 as done in Step 2.

Step 5: Configure the geometric details of the empennage areas

Configure H- and V-tail planform geometries, that is, the aspect ratio, sweep, taper ratio and dihedral angle. These will give the empennage root chord and tip chord. Next, settle for the H-tail incidence α_{HT}, if riggable, the value can be fine-tuned after flight tests.

The empennage planform is generally but not restricted to a trapezoidal shape. A strake-like surface could be extended to serve the same aerodynamic gains as for the wing. The choices for the empennage aspect ratio, wing sweep and taper ratio are interlinked and follow the same approach as for the wing design. The empennage aspect ratio is considerably lower than that of the wing.

Step 6: Establish control surfaces

Initially, the control areas and dimensions of the elevator and the fin are earmarked from statistics and semi-empirical data. At this stage of study, the control surfaces can be postponed until more details are available to accurately size the control areas. In this book, the control surfaces are not sized. Subsequently, in the next design phase, when the finalised aircraft geometry is available, the empennage dimensions are established by formal aircraft control analysis.

All these parameters are decided from stability considerations and eventually fine-tuned through CFD analysis and wind-tunnel testing, with the hope that flight test results will not require further tweaking. Subsequently, the static stability to be computed about aircraft the forward-most and aftward most aircraft CGs.

8.6.4.1 Check Aircraft Variant Empennage

To retain component commonality, the same empennage is used in all variants. It is therefore necessary to check whether the empennage falls within the statistical range. For aircraft with a family of variant designs, the empennage design should be done for the variant most suited to retain component commonality, that is, to use the same empennage for all variants. In this case, the empennage requires a small change in planform area, it is easier to reduce the size by chopping off the tips than extending them. Manufacturing jigs should have this provision.

8.7 Nacelle and Pylon Design

Section 5.12 is devoted to nacelle design. Civil aircraft designs are invariably externally pod-mounted engines on either the wing or the aft fuselage. Most turboprop engines are mounted on the wing, except very few like the P.180 Avanti. The demonstration of high engine reliability enables extended twin operations (ETOPS) clearance by the FAA for a two-engine configuration. Three-engine designs (e.g. B727, DC10 and Lockheed Tristar) are no longer pursued except for a few designs.

Nacelles should have their thrust lines positioned close to the aircraft CG to minimise associated pitching moments. In general, the nacelle aft end is slightly inclined (i.e. 1–1.5°) downward, which also assists in takeoff. Wing-mounted engines are preferred as the engine weight gives relief to the wing bending moment in flight. An underwing-mounted nacelle should remain clear of the ground in the event of a nose wheel collapse. In case of under-wing mounted nacelle pods, a minimum of 30° of separation is necessary to avoid wheel-spray ingestion.

Because of the lack of ground clearance for smaller aircraft, engines are mounted on the fuselage aft end, forcing the H-tail to be placed higher. Aft-mounted engines are less desirable than wing-mounted engines. It is for this reason that the designers of smaller aircraft are currently considering mounting the engine over the wing, as in the Honda small-jet-aircraft design.

The keel cut is typically thicker than the crown cut to house accessories. Numerous engine accessories are part of the engine power plant. They are located externally around the casing of the engine (i.e. turbofan or turboprop). In general, these accessories are located below the engine; some are distributed at the sides (if the engine is under-wing mounted with less ground clearance). Therefore, the nacelle pods are not purely axisymmetric and show faired bulges where the accessories are located. In this book, the nacelle is symmetrical to the vertical plane but it is not a requirement. This book deals only with the long-duct nacelles for the reasons given Section 12.12.1.

Pylons are the supporting structures (i.e. a cross-section streamlined to the aerofoil shape) of the nacelle attaching to the aircraft and carrying all the linkages for engine operation. Aft-fuselage-mounted pylons are generally horizontal but can be inclined if the nacelle inlet must be raised. For wing-mounted nacelles, the pylon is invariably vertical (for aerodynamic reason to reduce interference drag take slightly curve shape like banana). The depth of the pylon is about half of the engine-face diameter; the pylon length depends on the engine position. For an aft-fuselage-mounted installation, the pylon is nearly as long as the nacelle. For a wing-mounted installation, the nacelle is positioned ahead of the wing LE to minimise wing interference.

The nacelle position depends on the aircraft size, wing position and stability considerations (see Section 4.10). Subsequently, CFD analysis and wind-tunnel testing will fine-tune the nacelle size, shape and position.

8.7.1 Considerations in Configuring the Nacelle

The following are general important considerations for configuring the nacelle (see also Section 8.7):

Geometry	**Aerodynamics**
(1) Diameter (engine dependent)	(1) Drag
(2) Length, fineness ratio and pylon geometry	(2) Interference
(3) Intake geometry and lip section (Chapter 12)	(3) Surface roughness
(4) Wing or fuselage mount position	(4) Noise/emission
(5) Cross-section to house accessories	(5) Thrust and bypass ratio level
(6) Ground clearance	

Structure (affecting weight and external geometry)	*Systems*
(1) Engine burst considerations	(1) Control linkage
(2) Foreign-object ingestion problems	(2) Fuel system
(3) Fuel volume	(3) Electrical
(4) Weight	(4) Thrust reverser
(5) Nose gear collapse	(5) Fire prevention
(6) Access	(6) Anti-icing
(7) Vibration	
(8) Pylon as nacelle support	

8.7.2 Methodology for Civil Aircraft Nacelle Pod and Pylon Design

This section provides an example for configuring the nacelle based on an engine supplied by an engine manufacturer. At this stage, the size of an engine is guessed using the statistical data given in Figure 7.9 as uninstalled T_{SLS} per engine based on the number engine installed versus the MTOM of the aircraft. Formal engine sizing and matching is done in Chapter 14. The authors recommend that the readers produce graphs at a higher resolution for the aircraft class under consideration.

It is best to obtain the engine size from the manufacturer as a bought-out item. Continuous dialogue with engine manufacturers continues with 'rubberised' engines (engines scalable and finely tuned to match the aircraft performance requirements for all variants). Unlike aircraft, in general, the external dimensions of variant engines in a family do not change – the thrust variation is accomplished through internal changes of the engine. The same nacelle geometry can be used in all variants. For major variations, the engine size changes slightly, with minimal changes affecting the nacelle mouldlines.

Given next is a stepwise approach to nacelle design.

Step 1: Find engine size

Guess T_{SLS} per engine corresponding to the MTOM from statistics (Figure 7.9). Select the engine offered by engine manufacturers. This also allows the possibility of variant engines to match the variant aircraft to be designed. Obtain the engine geometry, for example, engine fan face diameter, bare engine length and maximum diameter.

Step 2: Decide nacelle type

Deicide on the nacelle type: whether it is of a short or long-duct type, considering the reasons as given Section 12.12.1. Long-duct nacelles appear to produce higher thrust to offset the weight increase of the nacelle, while at the same time addressing environmental issues including substantial noise reduction. Also, long-duct designs could prove more suitable to certain types of thrust-reverser design. This book will only consider long-duct design but it does not prevent the choice of short-duct nacelles.

Step 3: Configure the nacelle pod

For the type of nacelle pod, its geometry is based on the chosen engine fan face diameter, bare engine length and maximum diameter. Given next is a preliminary guide line to developing long-duct nacelle geometry (short-duct nacelles follow the same line but the extent of fan cowl depends on the decisions obtained through discussion with aircraft and engine manufacturers for the best option.

$$\text{Maximum nacelle diameter} \approx< 1.5 \times \text{engine-face diameter} \tag{8.1}$$

$$\text{Intake length in front of the engine face} \approx< 1.0 \times \text{engine-face diameter} \tag{8.2}$$

(for high bypass ratio (BPR) engines, lower values – can go down as low as 0.6)

$$\text{Exhaust jet-pipe length aft of the turbine section} \approx< 1.0 \text{ to } 1.5 \times \text{engine-face diameter} \tag{8.3}$$

Note: (engine fan face diameter) > (last stage turbine diameter).

$$\text{The total nacelle length} \approx (\text{engine length}) + (k \times \text{engine-face diameter}) \tag{8.4}$$

where $1.5 < k < 2.8$, turbofans with a higher BPR have a lower value of k.

For long-duct nacelles, the fineness ratio (i.e. length/maximum diameter) is between 2 and 3.

The keel cut is typically thicker than the crown cut to house accessories. In this book, the nacelle is symmetrical to the vertical plane but it is not a requirement.

Subsonic nacelle throat area and the highlight (forward-most point of the intake) areas are sized based on rated air mass flow rate inhaled by the engine. At this stage take inlet highlight area ≈ 0.9 to 0.95 of the engine fan face area, with a diameter, D_{fan}. Taking the factor 0.95, the intake highlight area,

$$A_{highlight} = 0.95 \times \pi(D_{fan}/2)^2 = 0.746 \times (D_{fan})^2 \tag{8.5}$$

Step 4: Positioning the nacelle with respect to aircraft

Decide where to place the nacelle. For small aircraft that do not have enough ground clearance, they are positioned at the aft end (raised, medium or low position with respect to the fuselage – see Figure 8.9, later) in such a way that the exhaust plane does not interfere with empennage and the thrust line favours aircraft stability in both the pitch and yaw plane.

Step 5: Configure pylons to attach the nacelle to the aircraft

The depth of the pylon is about half of the engine-face diameter; the pylon length depends on the engine position. For an aft-fuselage-mounted installation, the pylon is nearly as long as the nacelle. For a wing-mounted installation, the nacelle is positioned ahead of the wing LE to minimise wing interference. In general, the t/c ratio of the pylon is between 8 and 10%.

8.8 Undercarriage

Chapter 10 is devoted to undercarriage design considerations. Undercarriage positioning is CG dependent. At this design stage, the CG position is not established because aircraft-component weights are not known. From experience, the undercarriage may be placed after establishing the CG position and the rotational tail clearances. Ensure that the aircraft does not tip in any direction for all possible weight distributions. (Tipping occurs in some homebuilt designs – especially the canards – when the pilot steps out of the aircraft.) This book addresses only the tricycle type – that is, a forward nose wheel followed by two main wheels behind the aft-most CG.

8.9 Worked-Out Example: Configuring a Bizjet Class Aircraft

The purpose of the worked-out example is only to substantiate the methodology outlined – intensive course-work begins now. The readers should be aware that the worked-out examples demonstrate only the proposed methodology. The readers are free to configure aircraft with their own choices and can decide their own dimensions of the class of aircraft on which they are working. Readers need not be confined to this class work example and may explore freely; simplicity can be an asset. Readers are required to work out dimensions using the information provided in the following subsections.

The worked example undertaken here is to configure a Learjet45 class Bizjet that offers variants in a family of designs. It is suggested that readers compare this with competition aircraft – this design has more room than is typical of the class. The specifications/requirements and the technology level adopted are given next.

8.9.1.1 Aircraft Specifications/Requirements

All variants have the following:

- Low-wing configuration and stabiliser as a T-tail.
- The high sweep of the V-tail to increase tail arm of the T-tail.
- Passenger facilities include a toilet (39.36 in. deep), a galley (30 in. deep), luggage space (18 in. deep) and a wardrobe (12 in. deep).
- Although cabin crew is not required up to 19 passengers, a folded seat for one cabin crew is offered.

Baseline Version (8–10 passengers) – 10 passengers at standard (medium) comfort level.

Payload: 1100 kg.
Medium (standard) comfort level: 10×100 (baggage included) $+ 100 = 1100$ kg.
(High comfort level: $8 \times 100 + 300 = 1100$ kg)
Range: 2000 miles + reserve
Cruise altitude = 40 000 ft.,
Long-range cruise (LRC) = 0.7 Mach (at high altitude)
High speed cruise (HSC) = 0.75 Mach
g-level: 3.8 to -2

Long Variant (12–14 Passengers)

Payload: 1500 kg
Standard comfort level: 14×100 (baggage included) $+ 100 = 1500$ kg
(High comfort level: $12 \times 100 + 300 = 1500$ kg)
Range: 2000 nm + reserve

Short Variant (4–6 passengers)

Payload: 600 kg

Standard comfort level: 6×100 (baggage included) = 600 kg.
(High comfort level: $4 \times 100 + 200 = 600$ kg)
Range: 2000 nm + reserve

8.9.1.2 Technology Level Adopted

Wing aerofoil: NACA 65-410.
H-Tail aerofoil: NACA 64-210 (negative camber).
V-Tail aerofoil: NACA 64-010 (symmetric).

High-lift device:

Single-slotted Fowler flap at wing trailing edge ($\approx$0% of wing), no slats.
From the test data, the following maximum lift coefficients are given:

Flap deflection (°)	0	8	20	40
C_{Lmax}	1.5	1.7	1.9	2.1

Power plant

Two rear fuselage mounted engines.
Engine: Garrett TFE731 class turbofan (thrust range in the scalable engine family: 3000–3800 lb per engine) with a bypass ratio of 4. Fan face diameter = 0.716 m (28.2 in.), Engine maximum diameter = 1 m (39.28 ft), and bare engine length of 1.547 m (60.9 in.), and a dry weight of 379 kg (836 lb), static air mass flow = 14.6 lb s^{-1}. No thrust reverser. Long-duct aft-mounted podded nacelle is specified.
Electronic Flight Information System (EFIS) type pilot display.

Material (simplified statement):

Primary structure is all metal. Some secondary structure with composite. Weight saving is taken as 5%.

8.9.1.3 Statistics of Existing Design with the Class of Aircraft

Readers are suggested to find as many aircraft in the class with a similar payload range and adapt to an approximately similar technology level. Change in technology may not fit to the pattern as shown in the graph (Figure 8.4).

To note that the empennage area size depends on tail arm length, this is not compared in the graphs. A coursework example would have a slightly smaller tail area than is shown in Figure 8.4b for a relatively larger tail arm of a T-tail, with the high sweep of the V-tail adding to the tail arm. This is an example of a designer's choice for weight reduction). It is the tail volume coefficients that decide the tail areas.

8.9.2 Bizjet Aircraft Fuselage: Typical Shaping and Layout

The following is the stepwise approach as suggested in Section 8.4.1.1; the following geometries for the baseline variant are obtained.

Step 1: Fuselage cross-section and seating arrangement
The specified comfort level has the seat pitch = 30 in. at standard (medium) comfort level), seat width = 19 in., aisle width = 19 in. and adding the elbow room of 1.5 in. at the fuselage wall side of a seat and gap of 3 in. between fuselage wall and seat. Adding together gives the cabin with as computed here.
Cabin width = 19 (seat) + 17 (aisle) + 19 (seat) + 2 × 1.5 (elbow room) + 2 × 3 (gap) = 64 in.

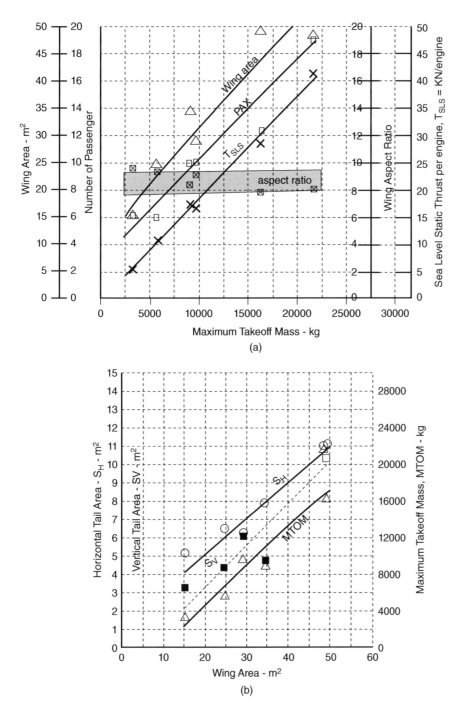

Figure 8.4 Statistics of the Bizjet class of aircraft. (a) PAX versus MTOM and wing area. (b) Empennage (S_H and S_V) – low tail. (Century, Cessna CJ2, Cessna Excel, Cessna 650, Cessna 750 and Challenger.)

Next, adding the fuselage shell thickness of 2.5 in. (6.35 cm) at side to get the fuselage width as computed here.

Fuselage width $= 64 + 2 \times 2.5 = 69$ in. (175 cm).

Fuselage Height (Depth) $= 178$ cm (70 in.)

For a pressurised cabin, the fuselage cross-section is to be as close as possible to a circular shape. As discussed previously, two-abreast seating in the cross-section results in a widening of the bottom half for legroom, shown here in the inclined position for a man in the 95th-percentile size. The fuselage mid-section is computed next.

Step 2: Fuselage mid-section

Adding the passenger facilities with dimensions given previously, the constant cross-section mid-fuselage length is computed as follows.

Baseline Bizjet – Number of seat rows $= 10/2 = 5$.

$$\text{Mid-fuselage length} = (5 \times 30) + 39.36 \,(\text{toilet}) + 30 \,(\text{galley}) + 18 \,(\text{luggage}) + 12 \,(\text{cabinet})$$
$$\approx 250 \text{ in. } (20.8 \text{ ft}, 6.35 \text{ m}).$$

The long variant Bizjet with 14 passengers at mid-comfort level has two 30 in. fuselage plugs at each side of the wing to accommodate four additional passengers in two rows.

$$\text{Long variant of Bizjet mid-fuselage length} = (7 \times 30) + 39.36 \,(\text{toilet}) + 30 \,(\text{galley}) + 18 \,(\text{luggage})$$
$$+ 12 \,(\text{wardrobe}) \approx 310 \text{ in. } (\approx 26 \text{ ft}, 7.88 \text{ m})$$

The short variant Bizjet with six passengers at mid-comfort level has two 30 in. fuselage plugs removed from the baseline variant from each side of the wing.

$$\text{Short variant of Bizjet mid-fuselage length} = (3 \times 30) + 39.36 \,(\text{toilet}) + 30 \,(\text{galley}) + 18 \,(\text{luggage})$$
$$+ 12 \,(\text{cabinet}) \approx 190 \text{ in. } (\approx 16 \text{ ft}, 4.8 \text{ m})$$

All the three variants in the Bizjet family of aircraft cabin and seat layout are shown in Figure 8.5.

Step 3: Fuselage front and aft sections and the fuselage length

Front fuselage

Take the flight-deck (cockpit) length $= 69$ in. (175 cm) as shown in Figure 8.6, and nose cone length $= 69$ in. (≈ 175 m) ensuring adequate polar pilot vision. This totals the front fuselage length $= 138$ in. (3.5 m) with the front closures ratio, $F_{cf} = 138/69 = 2$ within the range that Table 5.2 gives (smaller aircraft have larger values).

Aft fuselage

Take aft closures length, $L_a = 210$ in. (≈ 533 cm). This gives the aft closures ratio, $F_{cf} = 210/69 = 3.0$, within the range that Table 5.2 gives. Take aft-end closure angle $= 8°$ and a takeoff rotation clearance angle of $16°$ when undercarriage installed (to be checked out later).

The overall fuselage length is reached after adding front and aft closures. The windscreen shape and size must comply with Federal Aviation Regulation (FAR). This is an opportune time to streamline the fuselage, incorporating aesthetics without incurring additional cost and performance degradation. After streamlining, the various ratios are checked out to be within the acceptable range. Use the same closure lengths for all the variants the following are computed. Choosing a suitable ratio, the following dimensions are estimated:

Fuselage length

The baseline variant Bizjet fuselage length, $L = L_f + L_m + L_a = 138 + 250 + 210 = 598$ in. (≈ 50 ft, 15.2 m). Fineness ratio $= 598/69 = 8.7$.

The long variant Bizjet fuselage length, $L = L_f + L_m + L_a = 138 + 310 + 210 = 658$ in. (≈ 55 ft, 16.73 m). Fineness ratio $= 658/69 = 9.54$.

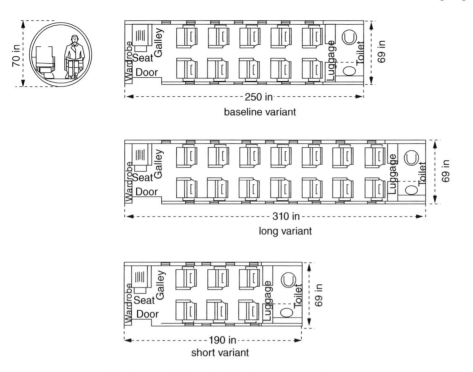

Figure 8.5 Example of configuring the fuselage for the medium comfort level.

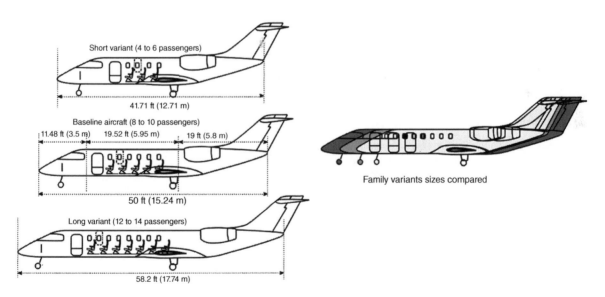

Figure 8.6 Fuselage lengths of the three variants.

The short variant *Bizjet* fuselage length, $L = L_f + L_m + L_a = 138 + 190 + 210 = 538$ in. (≈ 45 ft, 13.8 m). Fineness ratio $= 538/69 = 7.8$.

Step 4: Locate the position and sizes of aircraft doors, windows, hatchets

Passenger doors have several categories and are dealt with in Section 5.7. Based on the requirements, the positions of the doors, window and hatchets are earmarked.

The Bizjet must have the following door types:

Version	Number of passengers	Emergency door type
Baseline	10	One Type III and one Type IV
Long	14	One Type III and one Type IV
Short	6	One Type IV

It is better to keep all doors to Type III standards for component commonality, which would reduce production cost.

Discussion/Rationale

Equations/semi-empirical relations are not used in configuring fuselages. This is developed based on the specification requirements laid down for designers to consider. The fuselage cabin diameter and length is determined from the passenger number and the comfort level specified that have good statistical correlations. For subsonic aircraft, the fore-end closure is blunter (in favourable pressure gradient) than the aft-end closure with established closure angles in adverse pressure gradient.

These examples of civil aircraft family derivatives are shown in Figure 8.6 (the completed *concept definition*); the baseline aircraft is in the middle of the figure. The three variants of the family are shown with the wing positioned nearly at the middle of the fuselage. The rotation clearance is to be checked out after the undercarriage is positioned. This is not a problem because the main undercarriage length can be tailored in conjunction with the longest fuselage; this is an iterative process.

8.9.3 Bizjet Aircraft Wing

At the conceptual stage of the project study, typical values of wing twist and other refinements are taken from the past experience of a designer. The following are computed in a stepwise manner as outlined in Section 8.5.2.

Step 1: Decide the aerofoil section

For a relatively low speed cruise Mach number of 0.7 at the LRC and 0.75 at the HSC, the NACA 65-410. Design $C_L \approx 0.4$ at 0.75 Mach (Appendix C gives the NACA 65-410 aerofoil test data).

Step 2: Establish wing reference (planform) area

From Figure 8.4a, corresponding to 10 passengers, the guesstimated MTOW ≈ 9500 kg (20 900 lb) and 30 m^2 (323 ft^2). This gives a wing loading of 317.67 kg m^{-2} (3108.5 Nm$^{-2} \approx 65$ lb ft^{-2}). The selection is close to final design to avoid iteration in this book; experienced engineers can make selection to minimise iteration.

Step 3: Establish wing geometric details

In this class of subsonic Bizjet aircraft, the wing planform shape is invariably trapezoidal. As suggested in Section 4.16 that a taper ratio, $\lambda = C_T/C_R = 0.4$ is closed to having highest value of the Oswald's efficiency factor, e.

Aspect ratio and span

In consultation with structures group, the wing aspect ratio, AR = 7.5.

Wing span is worked out as, $b = \sqrt{(AR \times S_W)} = \sqrt{225} = 15$ m (49.2 ft).

Sweep

Aircraft specification given at the beginning of this section that HSC speed = 0.74 Mach at M_{DD}. At the LRC of 0.7 Mach it is at the onset of wave drag ($C_{DW} = 0$) at M_{crit}. Example 4.5 in Section 4.8.1 gives for this case wing sweep, $\Lambda_{1/4\_sweep} \approx 14°$

Root chord and tip chord

Other details of wing geometry are as follows.

Wing root and tip chord (C_R and C_T) can now be worked out from the taper ratio of $\lambda = C_T/C_R = 0.4$.

$S_W = 30 = 15 \times (C_T + C_R)/2$ solving the equations

$C_R = 2.86$ m (9.38 ft) and $C_T = 1.143$ m (3.75 ft)

Using Eq. (3.21), the wing

MAC $= 2/3 \times [2.87 + 1.148 - (2.87 \times 1.148)/(2.87 + 1.148)] = 2.132$ m (7 ft).

Twist

The wing twist is taken $-2°$, washout (see Section 3.14).

Dihedral

The wing dihedral is taken as $3°$, typical in this class of aircraft from the statistical data (Section 3.14).

Step 4: Establish wing high-lift devices

The adopted technology chooses a single-slotted Fowler flap without a LE slat that would be sufficient, saving considerably on costs. Its lift characteristics are given at the beginning of this section.

Step 5: Position wing with respect to fuselage

The wing position gets iterated as the aircraft and its components are known (Chapter 10). Positioning of the wing relative to the fuselage is an important part of configuring an aircraft. This may not be easy because moving the wing will alter the CG position – an inexperienced engineer could encounter wing chasing.

The baseline fuselage length is 15.2 m (50 ft). At this stage of design, place mid-wing MAC (2.132 m) at 7.6 m from the ZRP placed at the tip of the nose cone. There MAC$_{1/4}$ is at 7.07 m from the aircraft nose (Figure 8.8). (The CG position is still not known. The CG shown in Figure 8.8 is only a marker where to expect. The aircraft CG will be established in Chapter 10. The wing location is subsequently fine-tuned when the CG and undercarriage positions are known.)

Step 6: Control surfaces areas and location

Control areas are provisional and are sized in Phase II. Initially, a company's statistical data from previous experience serve as a good guideline. Aileron, flaps and spoilers are placed behind the wing rear spar, which typically runs straight (or piecewise straight) at about 60–66% of the chord. With a simple trapezoidal wing planform, the rear spar runs straight, which keeps manufacturing costs low and the operation simpler; therefore, it has a lower maintenance cost. With a third of the wing span exposed, the aileron area per side is about 1 m^2 (10.764 ft^2). Similarly, the flap area is 2.2 m^2 (23.68 ft^2) per side. Subsequent performance analysis would ascertain whether these assumptions satisfy field-performance specifications. If not, further iterations with improved flap design are carried out.

Figure 8.7 gives the wing geometry. To maintain component commonality, the wing should be the same for all three variants; obviously, it would be slightly larger for the smaller variant and slightly smaller for the larger variant.

For a small aircraft with limited ground clearance, the engines would be mounted on the rear fuselage. A smaller aircraft wing could be manufactured in one piece and placed under the fuselage floorboards, minimising a 'pregnant-looking' fairing. (Unlike a generous fairing as shown in Figure 4.4 to smooth the hump, this worked-out example has more streamlined fairing.)

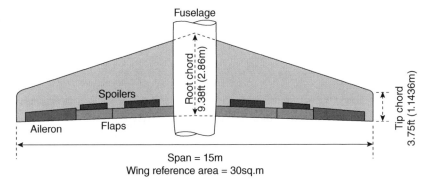

Figure 8.7 Example of wing design.

Discussion/(Rationale given in Chapters 3 and 4)

The wing along with the empennage aerodynamics is the most crucial aspect of aircraft design and requires attention. Aircraft performance and the operational economics depend on it. The wing is a lifting surface whose capability depends on the choice of aerofoil to satisfy the specified requirements.

The specified maximum operating speed for Bizjet is 0.75 Mach, a value relatively low compared to high subsonic commercial aircraft cruise speed. Based on the desirable aerofoil characterises required for the subsonic Bizjet, the NACA 65-410 is chosen that has mid-cruise design $C_L = 0.5$ at LRC of Mach 0.7 developing wave drag, $C_{DW} = 0$. The relatively thinner aerofoil thickness to chord ratio, $t/c = 10\%$ at wing sweep, $\Lambda_{1/4\_sweep} \approx 14°$ offers lower wing profile drag in addition to aim for lower induced drag mentioned before.

The wing span and aspect ratio are interrelated for the reference geometry chosen. To keep wing induced drag low, wing sweep is kept low at $\Lambda_{1/4\_sweep} \approx 14°$ allowing an aspect ratio of 7.5 for wing area of 30 m^2 resulting in wing span of 15 m which suits well for the class.

At the conceptual stage of the project study, typical values of wing twist and other refinements are taken from the past experience of a designer. The values must be substantiated and, if required, modified through CFD analysis and wind-tunnel testing to a point when the flight test may require final local refinements (e.g. flap and aileron rigging). Initially, an isolated wing is analysed to quickly arrive at a suitable geometry and then studied with the fuselage integrated. Subsequently, the wing is sized formally (see Chapter 14).

8.9.4 Bizjet Empennage Design – Positioning and Layout

The low-wing Bizjet aircraft specifications used so far to configure the fuselage and wing are sufficient for empennage design. Figure 8.4b provides empennage statistics (low tail) of the current Bizjet aircraft class. The positions of the H- and V-tail relative to the fuselage and the wing are decided by considering the aerodynamic, stability, control and structural considerations. The empennage area size depends on tail arm length. To simplify for classroom usage, Example 6.3 suggested the use of statistics instead of the DATCOM method as shown in Examples 6.1 and 6.2. The following are computed for the Bizjet empennage geometry.

Step 1: Decide the aerofoil section

H-Tail aerofoil NACA 64-210, installed with negative camber as tail. T-tail configuration is chosen to gain in tail arms.

V-Tail aerofoil NACA 64-010, symmetric.

Step 2: Position wing with respect to fuselage

From Figure 8.4b, obtain the tentatively empennage areas horizontal tail (H-tail) area, $S_{HT} = 7$ m^2 and vertical tail (V-tail) area, $S_{VT} = 5.5$ m^2. Find the empennage areas geometry using a typical aspect ratio and taper ratio and span. For aircraft with T-tail arrangements, position the V-tail as far aft a position as suits the structural arrangement. Place the horizontal T-tail at the top of the V-tail. The arrangement is shown in Figure 8.8.

Step 3: Compute tail arms and choose tail volume coefficients, C_{HT} and C_{VT}

To minimise the fuselage length, a T-tail configuration is chosen. Subsonic jet aircraft planform shape is invariably trapezoidal with each with sweep angle, $\Lambda_{1/4} = 15°$, slightly higher than a wing sweep of 14° to gain in the tail arm length. Tentatively place the H-tail and V-tail as shown in Figure 8.8. Fuselage length = 15.25 m (50 ft).

Measure the tail arms $L_{HT} = 7.62$ m (25 ft) measured from the wing MAC$_{1/4\text{-chord}}$ aircraft to the H-tail MAC$_{1/4\text{-chord}}$ and $L_{VT} = 7.16$ m (23.5 ft) measured from the wing MAC$_{1/4\text{-chord}}$ aircraft to the V-tail MAC$_{1/4\text{-chord}}$. Subsequently, the static stability is to be computed about the aircraft at the forward-most and aft-most CG. The tail arm lengths are used to compute empennage areas.

To retain component commonality between all variants, the Bizjet tail volume coefficients are taken for the baseline variant. Typical tail volumes with the class of aircraft are taken from Figure 8.4b.

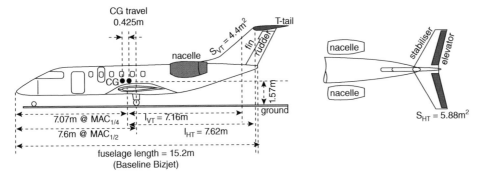

Figure 8.8 Bizjet aircraft example of empennage sizing (baseline aircraft).

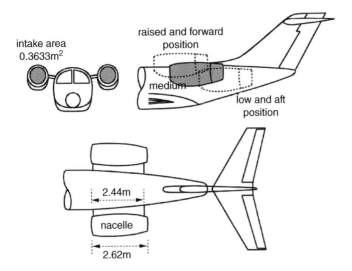

Figure 8.9 Bizjet nacelle design (baseline aircraft).

Horizontal tail:
From Figure 8.4b, take H-tail volume coefficients $C_{HT} = 0.7$.
Vertical tail:
From Figure 8.4b, take V-tail volume coefficients $C_{VT} = 0.07$.

Step 4: Estimate empennage reference (planform) area
Compute H-tail area, S_{HT}, and V-tail area, S_{VT}, using Eqs. (6.1) and (6.2), respectively. Discard the earlier values empennage areas taken from the statistics in Step 2.
Horizontal tail:
H-tail area, $S_{HT} = (C_{HT})(S_W \times MAC)/L_{HT}$
$S_{HT} = (0.7 \times 30 \times 2.132)/7.62 = 5.88$ m^2 (63.3 ft^2), which is about 20% of the wing area.
Vertical tail:
V-tail area, $S_{VT} = (C_{VT})(S_W \times \text{wing span})/L_{VT}$,
The $S_{VT} = (0.07 \times 30 \times 15)/7.16 = 4.4$ m^2 (47.3 ft^2), which is about 15% of the wing area.
Discard the earlier values empennage areas taken from Figure 7.8 as done in Step 2.

Step 5: Configure the geometric details of the empennage areas
Configure H-tail and V-tail planform geometries, that is, the aspect ratio, sweep, taper ratio and dihedral.

Horizontal tail:

Take the H-tail span to be one-third of the wing span and the taper ratio equals 0.5 (typical in the class – designers have the choice).

Span $b_{HT} = \frac{1}{3} \times 15 = 5$ m (16.4 ft)

Aspect ratio, $AR_H = b_H^2/S_H = 52/5.88 = 4.25$

Area, $S_H = \frac{1}{2}(C_R + C_T) \times b$

or $5.88 = 0.5 \times 1.5 \times C_R \times 5$

Root chord, $C_R = 1.568$ m (5.14 ft),   Tip chord, $C_T = 0.784$ m (2.57 ft)

Use Eq. (4.7) to compute the MAC.

$MAC_H = \frac{2}{3} \times [(1.568 + 0.784) - (1.568 \times 0.784)/(1.568 + 0.784)] = \frac{2}{3} \times (2.352 - 0.523) = 1.22$ m (4 ft).

Elevator area is taken as 1.21 m^2(13 ft^2)) – see Step 7 that follows.

Next, settle for the H-tail incidence α_{HT}, if riggable, the value can be fine-tuned after flight tests.

Vertical tail:

It is convenient to decide the vertical tail height from aerodynamic, structural, accessibility and in certain cases to clear obstacles. Typically, it can be about a semi-span of the H-tail. Take V-tail height (semi-span) = 2.14 m (7 ft) and the taper ratio = 0.6 to bear the load of a T-tail.

Aspect ratio, $ARV = bV2/SV = 2.142/4.4 = 1.04$.

Area, $SV = \frac{1}{2}(CR + CT) \times b$.

or $4.4 = 0.5 \times 1.5 \times CR \times 2.14$.

Root chord, $CR = 2.57$ m (8.43 ft).

Tip chord, $CT = 1.54$ m (5.05 ft).

Use Eq. (4.7) to compute the MAC.

$MAC_V = \frac{2}{3} \times [(2.57 + 1.54) - (2.57 \times 1.54)/(2.57 + 1.54)] = \frac{2}{3} \times (4.11 - 0.963) = 2.08$ m (6.8 ft).
Rudder area is taken as 0.75 m^2(8.05 ft^2) – see Step 7 that follows.

Step 6: Establish control surfaces

Initially, the control areas and dimensions of the elevator and the fin are earmarked from statistics and semi-empirical data. At this stage of study, the control surfaces can be postponed until more details are available to accurately size the control areas. In this book, the control surfaces are not sized. Subsequently, in the next design phase, when the finalised aircraft geometry is available, the empennage dimensions are established by formal stability analysis.

Horizontal tail:

H-tail area, S_{HT} is shared by the stabiliser and elevator. Typically, the elevator uses 18–25% of the H-tail area; in this case, it is 20%, which results in an elevator area of 1.21 m^2 (13 ft^2).

Vertical tail:

V-tail area, S_{VT} is shared by the fin and the rudder. Typically, the rudder encompasses 15–20% of the V-tail area – in this case, it is 17%. This gives a rudder area of 0.75 m^2 (8 ft^2).

8.9.4.1 Checking Variant Designs

To retain component commonality, the same empennage is retained, but may require some internal structural modification. It is to check the tail volume coefficients of the variant designs. The long variant with longer tail will have adequate stability. The short variant with shorter tail arm must be checked to ensure that it has tail volume coefficients.

Short Variant

H-Tail

With one seat pitch of 30 in. (0.762 m) plug removed from the aft fuselage $L_{HT\_short} = 7.62 - 0.762 = 6.86$ m (22.5 ft).

$C_{HT\_short} = (S_{HT} \times L_{HT\_short})/(S_W \times MAC_W) = (5.88 \times 6.86)/(30 \times 2.132) = 40.47/63.96 = 0.63$, it has the range as shown in Figure 8.4b.

V-Tail
With one seat pitch of 30 in. (0.762 m) plug removed from the aft fuselage $L_{VT\_short} = 7.16 - 0.762 = 6.4$ m (21 ft).

$C_{VT\_short} = (S_{VT} \times L_{VT\_short})/(S_W \times wing\,span) = (4.4 \times 6.4)/(30 \times 15) = 28.16/450 = 0.0625$, it has the range as shown in Figure 8.4b.

Long Variant
H-Tail
With adding one seat pitch of 30 in. (0.762 m) plug to the aft fuselage $L_{HT\_long} = 7.62 + 0.762 = 8.382$ m (27.5 ft).

$C_{HT\_long} = (S_{HT} \times L_{HT\_long})/(S_W \times MAC_W) = (5.88 \times 8.382)/(30 \times 2.132) = 49.45/63.96 = 0.77$, it has the range as shown in Figure 8.4b.

V-Tail
With adding one seat pitch of 30 in. (0.762 m) plug to the aft fuselage $L_{VT\_long} = 7.16 + 0.762 = 7.922$ m (26 ft).

$C_{VT\_long} = (S_{VT} \times L_{VT\_long})/(S_W \times wing\,span) = (4.4 \times 7.922)/(30 \times 15) = 34.86/450 = 0.0775$, it is with the range Figure 8.4b.

Discussion
It was mentioned earlier that the empennage aerodynamics are the most crucial to maintain stability and aircraft control without which the vehicle is inoperable. Both the H-tail and V-tail have higher than wing sweep to gain in tail arm lengths to minimise empennage surface areas. The rationale is given in Chapter 6.

8.9.5 Bizjet Aircraft Nacelle – Positioning and Layout of an Engine

Being a small aircraft, there is not enough height to accommodate under-wing nacelles, hence the aft-fuselage-mounted nacelle is chosen for the Bizjet example. It is important that a proven, reliable engine from a reputable manufacturer be chosen; of interest are the following. For variant aircraft, engine, thrust scaling of ±25% is desired.

Step 1: Find engine size and select engine
Corresponding to baseline aircraft MTOM of 9500 kg, Figure 13.10a, gives T_{SLS}/per engine = 17 000 KN ≈ 3800 lb.

There are only two engines to choose from as follows.
Honeywell (originally Garrett) TFE731 turbofan-series class.
Pratt and Whitney (Canada) PW 545 series class.

In the small engine class, Williams is coming up but is still below the required size. The Rolls Royce Viper, Adour and the Turbomeca Larzac all have a low BPR and are suited to a military application.

The Honeywell TFE731-20 turbofan is selected being in line with Learjet 45 aircraft. The following are the pertinent details of the turbofan.
Uninstalled T_{SLS} = 3800 lb. (≈17 000 N) per engine with BPR = 4 (small turbofan)
Engine Dry weight: 379 kg (836 lb)
Fan diameter: 0.716 m (28.2 in.)
Length: 1.547 m (60.9 in.)

Step 2: Decide on the nacelle type
A long-duct nacelle is chosen because it produces higher thrust to offset the weight increase of the nacelle, while also addressing environmental issues such as substantial noise reduction.

Step 3: Configure the nacelle (Figure 8.9)
Use the relationships given in Eq. (8.1):
Maximum nacelle diameter = $1.5 \times 0.716 = 1.074$ m (3.52 ft).
From Eq. (8.2), taking the factor of 0.8, compute the following.
Intake length in front of the engine face = $0.8 \times 0.716 = 0.57\,3$ m (22.6 in.)

From Eq. (8.3), take the exhaust jet-pipe length aft of the last stage turbine disc $= 1.0 \times 0.716 = 0.716$ m (2.35 ft).

Using the relation given in Eq. (8.4), the nacelle length $= 0.573 + 1.547 + 0.716 = 2.836$ m ($\approx$9.3 ft).

The nacelle fineness ratio $= 2.83/1.074 = 2.64$.

The keel cut is typically thicker than the crown cut to house accessories.

Using Eq. (8.5), intake the highlight area,

$$A_{int-\text{highlight}} = 0.95 \times \pi(D_{\text{fan}}/2)^2 = 0.95 \times 0.746 \times (0.716)^2 = 0.3633 \text{ m}^2 = 3.9 \text{ ft}^2.$$

The keel cut is thicker than the crown cut to house accessories. The nacelle is symmetrical to the vertical plane.

Step 4: Positioning the nacelle with respect to the aircraft

Being a small aircraft there is lack of ground clearance for smaller aircraft. Hence the engines are mounted on the fuselage aft end, forcing the H-tail to be placed higher. The nacelles are placed at the medium position with respect to fuselage (see Figure 8.9) at the aft end in a way the exhaust plane does not interfere with empennage and the thrust line favours aircraft stability in both the pitch and yaw plane.

Step 5: Configure pylons to attach the nacelle to the aircraft.

Being a small aircraft, the engines are aft-fuselage-mounted, one at each side. At this stage, a horizontal plate may represent the pylons that support the nacelles. The pylon length $= 2.44$ m (8 ft) has a thickness of 25 cm (9.8 in) and a symmetrical cross-section aerofoil-like structure for ease of manufacture.

Discussion

Nacelle geometry evolves around the chosen engine the Honeywell TFE731-20 turbofan. It has conventional aerodynamic design approach housing the turbofan as discussed Chapter 12.

8.9.6 Bizjet Aircraft Undercarriage – Positioning and Layout

Undercarriage positioning awaits until when the location of the aircraft CG is known. It is postponed to Chapter 10 after establishing the CG position in Chapter 9. Undercarriage positions in Figure 8.8 are shown arbitrarily in this chapter.

8.9.7 Finalising the Preliminary Bizjet Aircraft Configuration

It is interesting to observe how the aircraft is gradually taking shape – it is still based on a designer's past experience but soon will be formally sized to a satisfying rational configuration to offer the best characteristics for the design.

All aircraft components are assembled using the building-block concept to generate a preliminary aircraft configuration. To retain component commonality, the three variants maintain the same wing, empennage and nacelle geometry (some internal structures are lightened or reinforced without affecting manufacturing jigs and tools). A preliminary three-view diagram and a 3D CAD model of the Bizjet aircraft can now be drawn as shown in Figure 8.10 as the concept definition. This will undergo iterations as the project progresses to

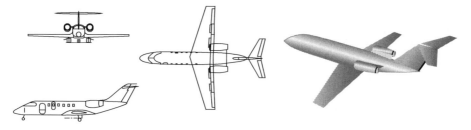

Figure 8.10 Three-view drawing and 3D model of the Bizjet aircraft.

concept finalisation. Chapter 14 sizes the aircraft to its final dimensions and finalises the configuration based on the aircraft and component mass worked out in Chapter 10.

The configuration is similar to the Learjet45 but it is not the same; there are considerable differences in configuration, component weights and performance. Readers may compare the two using Ref. [2].

The following is a summary of the worked-out civil aircraft preliminary details (from statistics).

8.9.7.1 Consolidate Summary of Bizjet and Its Variants at this Stage of 'Concept Definition'

Baseline aircraft data

MTOM: 9500 kg ($\approx$21 000 lb)

T_{SLS}/engine = 3800 lb

8.9.7.2 Baseline Aircraft External Dimensions

Fuselage (determined from passenger capacity)

Length: 15.24 m (50 ft), upsweep: 10°

Overall width: 175 cm (69 in.), overall height (depth): 178 cm (70 in.)

Average diameter: 176.5 cm (69.5 in.), fineness ratio: 8.63

Wing (aerofoil 65-410)

Planform (reference) area = 30 m², span = 15 m, AR = 7.5

Root chord, C_R = 2.87 m (9.4 ft), tip chord, C_T = 1.143 m (3.75 ft)

MAC = 2.132 (7 ft), taper ratio, λ = 0.4, $\Lambda_{1/4}$ = 14°

Dihedral = 3°, twist = 2° (washout), t/c = 10%

H-Tail (T-tail, aerofoil 64-210 – installed with a negative camber).

Planform (Reference) area: SHT = 5.88 m2 (63.3 ft2), span = 5.15 m (16.7 ft), AR = 4.22

Root chord, CR = 1.54 m (5.04 ft), tip chord, CT = 0.77 m (2.52 ft)

MACH = 1.19 m (3.9 ft), taper ratio, λ = 0.5, Λ1/4 = 15°

Dihedral = 5°, elevator = 1.21 m2 (13 ft2), t/c = 10%.

V-Tail (aerofoil 64-010) (symmetric aerofoil)

Planform (reference) area, SVT = 4.4 m2 (47.3 ft2), height: 2.14 m (7 ft), AR = 2.08

Root chord, CR = 2.57 m (8.43 ft), tip chord, CT = 1.54 m (5.05 ft)

MACV = 2.07 m (6.8 ft), t/c = 10% taper ratio, λ = 0.6, Λ1/4 = 40°

Rudder = 0.75 m2 (8 ft2), t/c = 10%

Bare Engine (each – from statistics)

Honeywell TFE731-20 turbofan

Uninstalled TSLS = 3800 lb. ($\approx$17 000 N) per engine with BPR = 4

Engine dry weight: 379 kg (836 lb)

Fan diameter: 0.716 m (28.2 in.)

Length: 1.547 m (60.9 in.)

Nacelle

Length: 2.84 m (9.3 ft), maximum diameter: 1.074 m (3.52 ft)

Short Variant (all component dimensions except the fuselage length are invariant)

Fuselage: length: 13.47 m (44.2 ft)

Long Variant (all component dimensions except the fuselage are invariant)

Fuselage: length: 16.37 m (53.7 ft).

8.9.8 Miscellaneous Considerations in the Bizjet Aircraft

The following are some additional considerations that could enhance aircraft performance but are not addressed here. At this design stage, none of the additional surfaces described need to be considered except the dorsal fin. All add to aircraft weight.

1. *Winglets*. It took some time to establish the merits of having winglets that can reduce or induce drag – some manufacturers claim a reduction as high as 5% of induced drag (i.e. approximately 1.5% in total drag reduction), which is substantial. Currently, almost all large aircraft designs incorporate winglets. Learjet has been using them for some time and they have become a symbol of its design.
2. *Dorsal fin*. A dorsal fin ahead of the V-tail could work like strakes on a wing, and they are incorporated in many aircraft – at least to a small degree. They prevent the loss of directional stability.
3. *Ventral fins*: These fins come in pairs at the aft end of the lower fuselage. Not all designs have delta fins; they are used if an aircraft shows poor stability and/or control problems. Aircraft with a flat, rear-loading, raised fuselage upsweep demonstrate these problems and delta fins are deployed to resolve them. A good design should avoid incorporating delta fins; however, on some designs, drag reduction can be achieved with their installation.

Sometimes, at the tail end, an additional vertical surface is given below the fuselage. The single surface ventral fin also serves as a skidding structure to protect the fuselage from damage at excessive early rotation, which causes tail-dragging.

Several external-surface perturbations on aircraft add to parasitic drag, including antennas, inspection-hatch covers, vent pipes and lightning dischargers. Engine and system intake and exhaust ducts and vents also increase drag.

It is suggested that readers determine whether there are any innovative requirements that should be incorporated in the conceptual design. Trends should be investigated continually for ideas to improve on aircraft design.

MILITARY AIRCRAFT

8.10 Prerequisite to Initiate Military (Combat/Trainer) Aircraft Design

The approach to conceptual design for military aircraft differs from civil aircraft study. Section 2.5 outlined that, instead of any market studies, the MoD (Ministry of Defence) floats an RFP or AST. The essential prerequisites to initiate conceptual design of a new military aircraft are (i) having the aircraft specifications requirement based on capabilities to stay ahead of potential adversary something not fully known and the (ii) new advanced technologies level to be adopted those have to be proven on scale down model technology demonstrator before they can be accepted. These prerequisites are outlined as follows. Any mid-course change of technology level could severely affect cost.

i. A combat role has a wide spectrum of activities, as outlined in Section 7.8. In general, close support/ground attack aircraft (no afterburning engines) are subsonic with quick turning capabilities, whereas air superiority aircraft have supersonic capabilities with afterburning turbofans. Predominantly, engines are buried in the fuselage; hence, there is no wing relief benefit.
ii. Modern supersonic combat aircraft configurations show a considerable amount of wing-body blending. It becomes relatively difficult to identify the delineation between the fuselage and wing. The manufacturing joint between the wing section and fuselage block is a good place to make the partition, but at the conceptual phase the manufacturing philosophy is unlikely to be finalised. For classroom purposes, one way to decouple the wing and fuselage is to see the blend as a large wing root fairing that allows the fuselage to separate when the fairing part is taken as part of the wing.
iii. Dominantly, all payload is externally mounted except for bigger designs, which could have an internal bomb load. Modern military aircraft have external load that is contoured and flushed with mould lines. Internally mounted guns are permanent fixtures. Training military aircraft pylons to carry external load are not taken as permanent fixtures. Consumables (e.g. firing rounds) are internally loaded. In general, for long-range missions, there is more than one crew and some consumables meant for the crew.
iv. Military designs are technology specific and as a result, unlike civil design, military designs exhibit large variations in statistical distribution.

For control reasons it could have additional surfaces. There are three possible choices (see Section 7.14) as follows: the (i) *one-surface configuration*, (ii) *two-surface configuration* and (iii) *three-surface configuration*.

Stealth features are integral to current combat aircraft deigns. The basic concept of stealth is to reduce the aircraft signature to enemy sensing – more details are given in Section 18.10.3. The F16 is nearly a four-decades-old design and does not have stealth features. The F117 Nighthawk is an early stealth design; its configuration shows the difficulty associated with stealth design. Designing a stealth configuration is beyond the scope of this book. Bearing this in mind, a military trainer aircraft design study is given here, exposing some major considerations for combat aircraft design.

A generous wing root fairing is used to reduce interference drag as well as vortex intensity at the aft-fuselage flow. A large aircraft BWB is an extreme example that eliminates wing root fairing problems. There is no analytical expression to specify the fairing curvature – a designer should judge the geometry from past experience and CFD analysis, considering the internal structural layout and the associated weight growth. In principle, a trade-off study between weight growth and drag reduction is needed to establish the fairing curvature. At this stage, visual approximation from past experience is sufficient: Observe the current designs and make decisions.

(1) Aircraft specifications requirement

   The following are the given specifications.

   i. Operational strategy – choice of mission profile (see Section 7.8).
   ii. Number of crew – heavy workload could demand twin crew (typically in bombers and sometime in multi-role combat aircraft).
   iii. Manoeuvre limits – positive and negative '*g*' limits – maximum 9 g.
   iv. Armament load – weapons type for the mission role.
   v. Performance – maximum speed, agility/turn capabilities, altitude, range, super-cruise (sustained supersonic flight), survivability.
   vi. Takeoff and landing field lengths.
   vii. Number of engine and type (BPR/afterburning) and with full authority digital electronic control (FADEC). To specify engine with vector thrust or not.
   viii. Wing position – high or low wing.
   ix. Systems specification – electronics, weapon system management, data acquisition, countermeasures, electronic warfare, communication, navigation, control/FBW and so on.
   x. Structural philosophy – type of material (aluminium/special alloys/composites).
   xi. Specify manufacturing philosophy.
   xii. Offer variant designs in diversified military operational role.
   xiii. Life cycle – cost/maintenance/logistics/disposal – support from cradle to grave.

(2) *Technology level to be adopted*

   i. Aerodynamics – aerofoil sections of the various lifting surfaces (profile, *t/c* ratio). Type of high-lift devices. Choice of high-lift device will affect wing area and, as a consequence, MTOW. Wing and empennage configuration concept – one-, two- or three-surface (Figure 7.28). Contour fuselage cross-sections accommodating engine with matched intake. Stealth aerodynamics.
   ii. Structures – choice of material; extent of metals used, extent of composites used. (aircraft MTOM can be made lighter by appropriate material selection).
   iii. Power plant – type and number of power plants required (low BPR).
   iv. Undercarriage – decide on the undercarriage type and its articulation kinetics.
   v. Systems – type of control and extent of automation (FBW, FADEC etc.). (electrical, electronic, mechanical, hydraulics, pneumatics).
   vi. Instrumentation – flight-deck instruments, automation to alleviate pilot work load.
   vii. Flight deck – establishes the layout philosophy for ease of crew operation, offers the crew comfort and protection.

viii. Manufacturing philosophy – the success of new aircraft design considerable on the manufacturing technology adopted to minimise cost (Chapter 16). Aircraft parts and subassemblies must be easy to manufacture with low part count in a cost-effective manner.

8.10.1 Starting-Up (Conceptual Definition – Phase I)

Unlike the civil aircraft with a New Aircraft Project Group in a company, the conglomerates for new military projects have large separate divisions to work in harmony with many different organisations. The dedicated division conducts a conceptual phase of military aircraft design progress in an IPPD environment (see Section 8.4).

The military aircraft design methodology is outlined in Chart 8.2, which differs from Chart 8.1 for civil aircraft in Step 1, after that the routine is about the same. The general approaches to military (combat) aircraft starts with guessing MTOM from statistics obtained from weapon load as payload and the radius of action as range. Combat aircraft return to the base station and therefore the definition of range is not similar to the range of civil missions, unless the aircraft is used for ferrying without weapons load.

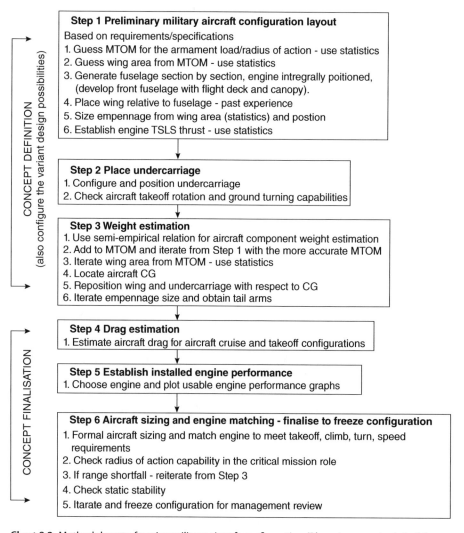

Chart 8.2 Methodology to freezing military aircraft configuration (Phase I, conceptual study).

8.10.2 Methodology for Shaping and Laying Out of Conventional Two-Surface Military Aircraft

This section takes a closer look of the Phase I stage of a new two-surface military aircraft project as given in Chart 2.2. This chapter deals only with Step 1 of the Chart 8.2. Thereafter. typically, the process cascades down from concept definition to carrying out the concept finalisation in a step-by-step manner as given next.

Typically, a concept definition of new aircraft combat configuration methodology starts with generating proportionate sketches (preferably as 3D models using CAD) of the concepts proposed by the New Combat Aircraft Project Group, incorporating new ideas hitherto not seen as operational aircraft. While relying in past experiences is essential, the new configurations present challenges to the aircraft designers on account of complexities involved. After series of 'design reviews', once a sketch of the new aircraft configuration is agreed to, the formal methodology starts.

A lot of statistical information on military aircraft has been captured within Section 7.12. These statistical data proves very informative at the conceptual design stage to get an idea what options a new design can incorporate to stay ahead of competition with a superior product. The readers may have to wait until the project is completed to compare how close it is to the statistical data.

Armament payload weight is given as per the specification. From statistics of past designs, payload requirement gives a preliminary guesstimate of MTOW. Thereafter, from other statistics of wing area, engine size are obtained. From these geometrical parameters and the specified mission range, the fuel weight can be computed. These are eventually sized to exact values (Chapter 14) through iterations, a routine procedure dealt with in this book.

The objective at the start-up is to guess preliminary aircraft geometry of the new aircraft. Step 1 is based on guesstimates using the statistics of past designs. Military aircraft layout methodology (Step 1 of Chart 8.2) can be summarised as given next and are dealt with in detail in Sections 8.14–8.6:

i. From the statistics of armament load – radius of action, obtain a preliminary MTOW that will be improved through iterations.
ii. From the statistics of MTOW versus wing area, develop a preliminary wing planform with the specified technology level. Next, position the wing with respective to the fuselage in an approximate position from the past experience (typically to begin with the $MAC_{1/4}$ close to middle of fuselage). Size and position the empennage in relation to wing area (readers may study comparable existing designs). This gives the preliminary wing loading, W/S_w, which can be compared with the statistics of comparable operating aircraft. It is to be noted that moving wing position will also move the aircraft CG, therefore some iterations are required.
iii. Select engine from what is available. The thrust loading, T/W, is considerably higher than the civil aircraft design. Combat aircraft $T/W \approx 0.8$ to 1.2 and trainer aircraft $T/W \approx 0.4$ to 0.6. For propeller powered aircraft, divide the thrust by 3.5 to get the SHP. Later, the engines will be properly sized to match the aircraft.

In the process, the preliminary geometry, weight and engine size will be revised to better accuracy leading to the final design (Step 6 of Chart 8.2), in an iterative process.

8.10.3 Shaping and Laying out of Military Aircraft Components

The objective is to generate aircraft components, piece by piece in a building-block fashion, and mate them as shown in the middle diagram of Figure 2.3. The approach to laying out civil aircraft given in Section 8.3.2 also applies to military aircraft design and hence not repeated here. The seven graphs shown in Figures 7.23 and 7.27 capture all the pertinent aircraft data taken from [2] and other sources.

However, the approach to configure military aircraft densely packed fuselage differs from civil aircraft hollowed shell fuselage design. Due to not having a constant cross-section along the length, these will have to be developed section by section. Military aircraft engines are buried into aircraft at the aft end, hence they have an air-breathing intake duct integral to the fuselage. Current design practices either have side intakes or a belly mounted chin intake. A bubble canopy is preferred.

Combat aircraft with supersonic capability have a pointed fuselage front end. Since a combat aircraft fuselage houses the power plant, its aft end is boat-tailed to the size of the exhaust nozzle. The current tendency blends the fuselage with the wing having with or without strakes.

Configuring combat aircraft is not as straightforward as making a civil aircraft *concept definition*. To stay ahead of any potential adversary, a new generation of combat aircraft incorporates new designs not seen before. As a result, configuring a new design is unlikely to follow a routine approach and, in general, there is no fixed methodology to start with.

A three-view diagram from the 3D model should show the conceptualised aircraft configuration. Making 2D drawings (i.e. three-view) from 3D models is simple with a few keystrokes. A preliminary configuration will change when it is sized. Military aircraft configuration gets revised more frequently. Having CAD 3D parametric modelling allows changes to be easily, quickly and accurately incorporated. At go-ahead, the proposed aircraft offers initial guarantee of the aircraft requirements to satisfy the MoD.

Subsequent sections give step-by-step methodology to configure aircraft components. These are then assembled and readied for the first iteration.

8.11 Fuselage Design (Military – Combat/Trainer Aircraft)

Unlike a civil aircraft that has a 'hollow' fuselage for passenger/cargo accommodation, the combat aircraft fuselage interior is densely packed with fixed equipment, contoured in a varying cross-section along its length. Note the positioning of fuel storage and engine placement. An air-breathing intake duct (no nacelle) is integral to the fuselage as engines are buried into it (Figure 5.2). Each fuselage cross-section is different, typical in today's trend of having fuselage sides that blend with the wing – the mouldlines could vary from design to design, but the considerations are about the same. The blending of fuselage with wing makes the fuselage contribute to body lift; improved area distribution reduces transonic drag and provides better wetted-area-to-volume ratio and a thicker wing root.

8.11.1 Considerations Needed to Configure a Fuselage

The following are general considerations important for the fuselage layout.

Geometry	Aerodynamics
(1) Cross-sections along fuselage	(1) Front-end closure
(2) Front fuselage flight deck with canopy configuration	(2) Surface roughness
(3) Measure the maximum width and depth	(3) Wing-body fairing
(4) Length, fineness ratio based on D_{eq}	(4) Wing and tail position
(5) Upsweep for rotation angle	(5) Stealth considerations
(6) Front fuselage closure	(6) Drag
(7) Aft fuselage end is the engine exhaust plane	
(8) Fuselage fairings where necessary	
(9) Provision for positioning intake duct	

Structure (affecting weight and external geometry)	*Systems*
(1) Structural layout scheme	(1) Flight-deck design
(2) Wing and undercarriage attachments	(2) All other systems
(3) Weapon carrying hard points	(3) Controls – FBW/FADEC
(4) Weight	(4) Undercarriage

8.11.2   Methodology for Fuselage Design

Every military aircraft fuselage design differs with its special requirements. The aircraft reference line is set at the tip of the nose cone and fuselage axis line. For manufacturing purposes, the design concept aims to make the fuselage in modular sections.

The aircraft fuselage is conveniently split up into sections (typically into four for a aft-mounted jet engine but can be into three sections for a front mounted propeller engine). Since the engine is housed in the fuselage, the intake configuration is integral to the fuselage. The pilot flight-deck section may be configured first as this part is relatively easier; the fuselage flight-deck section height width and length are have some standard sizes. The ogive/conical nose cone is installed in front of flight-deck section, accommodating the radar diameter. The load-bearing mid-fuselage is the difficult one in which to get a streamlined mouldline that will have wing box/joint, undercarriage storage bay and engine inlet duct. The aft fuselage is also relatively simple to configure; accommodate the engine (turbofan), the cylindrical jet pile and the exhaust nozzle and shape it to accept empennage integration.

There is no unique method to generate a complex combat aircraft fuselage. A typical method is to split the fuselage as shown in Figure 5.26. It shows the fuselage is split into four sections, as (i) the nose cone housing radar and other electronic boxes with length L_{cone}, followed by (ii) crew station/flight-deck section with length L_{front}, followed by (iii) the load-bearing mid-fuselage to which wing and main undercarriage are attached with length L_{mid} and, finally, (iv) aft fuselage with length L_{aft}, which is frequently split to reach the engine for it MRO needs.

The crew station/flight-deck section split is behind the rear crew seat in a vertical or inclined plane not cutting the intake if they are at the fuselage sides. However, the split goes through the under belly mounted intake.

After the engine is selected and its dimensions are known, the fuselage can then be configured to house the engine at the aft end, one/two pilots and avionics at the front end and fuel in the centre between the two intake ducts.

Using the sketch made for the new combat/trainer aircraft the fuselage is made modular. Note that although the fuselage is configured in sections, they maintain the surface tangency match between the sections as of the sketch. Given next is a step-by-step methodology to configure the fuselage with this kind of split arrangement.

Step 1: configuring the crew station/flight deck

Start with configuring the crew station/flight-deck section of length L_{front}. Decide on the crew seating arrangement if there is more than one, that is, side-by-side or tandem seating. It is important that cockpit width should be generous to have an ejection seat escape with adequate pilot space, even under restrained conditions. A minimum of 90 cm is required but a metre width of cockpit interior width is recommended. There has to be adequate space below the floorboards for cable and armour plate for protection at all sides, which increases fuselage width and height dimensions.

Step 2: front fuselage closure (configuring the nose cone)

Next, take up the nose cone to house the radar that can be of about 1 m in diameter and other electronic boxes within the length L_{cone}. The nose cones (or some kind of ogive shape) have front fuselage closure to a point for supersonic flight and some to rounded bluntness for subsonic flight. The fuselage reference plane may be positioned at the tip of the nose cone. Establish the fuselage axis.

Step 3: configuring the mid-fuselage

The load-bearing mid-fuselage to which the wings and main undercarriage are attached with a length L_{mid} and have an attachment arrangement to have a spine going through the aft end.

Step 4: Aft-fuselage closure (configuring the aft end)

Aft fuselage with length L_{aft}, which is frequently split to reach the engine for its MRO needs. Sequentially, place the four fuselage subsections along the fuselage axes. Aft-end shape depends on the type of aircraft configuration chosen. It can have plane engine exhaust plane as the aft end or may have a closure short in length. Fuselage aft-end closure should allow for aircraft rotational ground clearance. Military aircraft have rapid rotation. The fuselage clearance angle, θ, depends on the main-wheel position of the undercarriage relative to the aircraft CG position (see Chapter 7). The typical angle for θ is around 16° to approach high C_{Lmax} at aircraft rotation.

8.11.3   Structural Considerations for the Fuselage Layout

1. *The front fuselage cone.* This has the length L_{cone} and the provision to swing open for radar inspection and maintenance.
2. *The crew station/flight-deck section.* This has the length L_{front}, closed to air tightness by the front bulkhead with a radar mount ahead of it and rear bulkhead behind the pilot (rear pilot for tandem seating arrangement).
3. *The load-bearing mid-fuselage.* This has the length l_{mid} and serves as the backbone of the aircraft. Here, the wings are attached and undercarriage is installed.
4. *The aft end of mid-fuselage.* This may have provision to attach a spine going through the aft fuselage with engine attachment points and has the empennage assembled on it. The aft fuselage must have a structural arrangement for engine maintenance and removal for overhaul.

8.12   Wing Design (Military – Combat/Trainer Aircraft)

Wing design is the most important component of the military aircraft. The wing planform shape needs to be established based on the operational requirements (e.g. hard manoeuvres, supersonic capabilities, short field performances etc.). Unlike civil designs, there is a large option for planform shape. Fuel tankage space is restricted.

Subsonic military aircraft, for example, trainers and aircraft with a CAS/COIN role, follow roughly the same methodology as Section 8.5 for civil aircraft wing design. In this section, the supersonic wing design methodology is briefly outlined as it is beyond the scope of this book.

8.12.1   Considerations in Configuring the Wing

The following are general important considerations for designing the wing:

Geometry	Aerodynamics
(1) Aerofoil section, t/c ratio	(1) Drag
(2) Wing reference area, S_W	(2) Lift and moment
(3) Span and aspect ratio	(3) Stall, critical Mach
(4) Sweep, twist, dihedral, taper	(4) High-lift devices
(5) Position (for the CG)	(5) Control surfaces
(6) Wing/tail position	

Structure (affecting weight and external geometry)	*Systems*
(1) Spar and rib positions	(1) Control linkage
(2) Stiffness, aeroelasticity and torsion stability	(2) Fuel system
(3) Fuel volume	(3) Electrical
(4) Undercarriage	(4) Anti-icing
(5) Weight	

The first task for wing design is to select a suitable aerofoil suitable for the desired aircraft performance characteristics. After obtaining the MTOM from the statistics of existing aircraft (see Figure 7.23), the next step is to get the wing reference area and engine thrust requirements, also initially from statistics (see Figure 7.26).

Combat aircraft operate at much higher '*g*' loads, both in pitch and lateral manoeuvre. Military aircraft with the same flight Mach number require a higher sweep and low aspect ratio to cater for manoeuvre and other considerations. A delta and its derivative wing planform shape allows high leading edge sweep for supersonic flight operation. Section 4.18.1 gives the generic wing planform choice derived from the basic triangular delta wing for military aircraft. The dominant main wing planform shape is that of a delta or its variant; the low aspect ratio trapezoidal shape also is a candidate.

A military aircraft designer would seek low aspect ratio to have rapid manoeuvre capability. The *V-n* diagram (see Section 8.7) determines the strength requirement in pitching manoeuvres creating maximum stress from the bending moment at the wing root. A delta derivative wing planfom has a $C_T << C_R$ suiting structural designers to cater for high bending moments arising from hard manoeuvres. Choice of material and aerofoil *t/c* ratio contributes to structural integrity.

In summary, the supersonic capability of fighter aircraft demands a thin aerofoil (*t/c* 3–6%), high leading-edge sweep and low aspect ratio to negotiate high *g* manoeuvres that will minimise wing root bending moment. It would restrict span growth but encourages a large wing root chord of delta or trapezoid with a strake planform.

High sweep, small aspect ratio, thin supersonic aircraft wing-aerofoil camber variation along wing span, wing twist and dihedral/anhedral considerations are taken together to minimise drag and low speed handling capabilities. The design considerations are complex and not dealt with in this book. This book deals only with subsonic military trainer aircraft with a CAS variant.

Subsequently, the wing will be sized to the requirement (see Chapter 14). Some iteration is required because component weights are revised at the stages of the study. In coursework activity, one iteration is sufficient.

8.12.2 Methodology for Wing Design

Subsonic Wing
> The wing and empennage designs follow almost the same step-by-step procedure as for the civil aircraft design, hence they are not repeated here.

Supersonic Wing
> This is not dealt with in this book, so just an outline is given next. The readers may revisit Sections 4.13, 4.14 and 4.19.

Step 1: Decide the aerofoil section
> Decide on a suitable thin aerofoil of the type given in Figure 3.26.

Step 2: Establish wing reference (planform) area
> Guess the wing reference area, S_W, from the estimated MTOW (Figure 7.26).

Step 3: Establish wing geometric details
> Decide on a delta derivative wing planfom shape (Figure 4.1). For classroom usage, keep the wing twist and dihedral/anhedral angle at 0°. For high and mid-wing configurations, an anhedral angle of 1–2° may be given; low is kept at 0°.

Step 4: Position wing with respect to fuselage
> Positioning of the wing relative to the fuselage is an important part of configuring an aircraft. This is an iterative process dictated by the location of the aircraft CG at a desired position, expressed in terms of a percentage of the wing MAC. It has similar considerations as discussed in Section 8.5.2. It requires knowledge of the CG position and its range of movement with weight variation (i.e. fuel and armament load). Because the aircraft weight distribution is not yet established, it is initially estimated based on experience and past statistics in the aircraft class. With FBW system architecture, the static margin is small, nearly zero, and it even may be slightly negative. At this stage, the wing $MAC_{1/4}$ is placed behind the centre of fuselage. Subsequently, the wing position must be iterated after the aircraft-component weights are known and the wing is sized.

Step 5: Establish wing high-lift devices and control surfaces

In general, combat aircraft incorporate similar sorts of high-lift concepts at the wing leading and trailing edges as used in transport aircraft. Combat aircraft high-lift devices are less complex than multiple element Fowler type devices used in large commercial transport aircraft. In addition, high performance combat aircraft have vectored thrust capability for short takeoff and/or taking off from unprepared airfield. For landing, drag chutes are deployed. On an aircraft carrier landing, arrester cables are used. This book does not size these surfaces but schematically earmarks their position on the wing.

Combat aircraft with two-surface configurations have all-moving H-tails and have a stabilator/taileron (see Section 5.2.3). A flaperon serves as both a flap and an aileron. Strake A three-surface configuration aircraft have a canard serving not only as control surface but also providing vortex lift to the main wing. Combat aircraft also have airbrakes for quick deceleration from high to low speed.

8.13 Empennage Design (Military – Combat/Trainer Aircraft)

Subsonic military aircraft empennage design methodology is about the same as given in Section 8.6. Supersonic combat aircraft is considerably more complex and is beyond the scope of this book. Some general considerations regarding supersonic military aircraft empennage design are briefly outlined to assess the complexities.

The military aircraft shorter fuselages have shorter tail arm. A higher rate of manoeuvre demands relatively large empennage areas as well as operation in a smaller stability margin. For stealth consideration (radar signature), the elimination of the V-tail is desirable. If this is not possible, then reduce the area with twin-canted V-tails and position above the fuselage to get blanketed. The F22, F14, F15, F18, MIG29 and SU30 all have twin V-tails; the B2 does not. Older designs with T-tails have receded from combat aircraft design. (Use of canard and vector thrusting is a strong contribution for pitch control.)

Today's military aircraft incorporate the proven technology of FBW system architecture (MIL1553 bus); therefore, the computer is flying the aircraft within the safe envelope; hence, empennage size can be reduced to the smallest size using the contribution derived from flying with relaxed stability margins for all three axes of flight. A good knowledge of aircraft control laws is required at the conceptual design stage. However, this book does not venture into an area, which goes beyond its scope when such information is lean.

On account of a high demand for control authority, the empennage design requires special considerations. The H-tail is made as all-moving stabilator. In case of an inadvertent situation of stabilator failure, provision is kept to split into stabiliser and elevator as a measure of redundancy to ensure safety and fly back to base. It can also be made to serve as 'taileron' with differential movement at each side. A V-tail is large and, if required from structural and stealth considerations, can be split in to two. FBW system do not allow aircraft to stall. Many high performance fighter aircraft do not recover from spin and in case they flip into spin, then ejection is the routine procedure.

8.13.1 Considerations in Configuring the Empennage

The following are general considerations important for configuring the empennage (two-surface configuration):

Geometry	Aerodynamics
(1) Aerofoil section, t/c ratio	(1) Drag, lift, moment
(2) H-tail and V-tail reference area, S_H and S_V	(2) Tail volume coefficient
(3) Span and aspect ratio	(3) Stall and yaw recovery
(4) Sweep, twist or dihedral (whichever is applicable)	(4) Control and trim surfaces
(5) Position relative to wing (tail arm)	(5) Spin recovery
(6) Position H-tail to avoid shielding of V-tail	(6) Balancing
(7) H-tail position (high $\alpha_{clearance}$, T-tail)	(7) Stealth consideration

Structure	Systems
(1) Spar and rib positions	(1) Control linkage
(2) Stiffness, aeroelasticity and torsion stability	(2) Control actuation
(3) Weight	(3) Electrical

8.13.1.1 Typical Values of Tail Volume Coefficients

A coarse guideline to progress is first to establish the category of aircraft under design consideration as laid down in Table 8.4. Once the class is identified, the following may prove helpful.

Supersonic category (military combat aircraft). This category of aircraft have short tail arms and considerably lower aspect ratios with high control authority demands to execute extreme manoeuvres. Modern fighter aircraft tail volume coefficients are not readily available in the public domain. Some third- and fourth-generation aircraft tail volume coefficients (a very large spread) are given in [6]. A single surface delta wing does not have H-tail as it is merged with the wing and has L_{HT}. For the complexities involved, combat aircraft design is not dealt with in the scope of this book. While the statistics of combat aircraft tail volume coefficients are not given here, Table 8.4 gives a usable range for military trainer aircraft design.

8.13.2 Methodology for Empennage Design

Step 1: Decide on the aerofoil section
   Generally, both the H-tail and V-tail have a thin symmetrical aerofoil section. The t/c ratio of the empennage is close to the wing-aerofoil considerations.

Step 2: Position the wing with respect to the fuselage – two-surface configuration
   From statistics, obtain the tentative empennage areas. Decide if the V-tail area is required to be split into a twin V-tail configuration. This preliminary area is needed to position the empennage with respect to fuselage to compute tail arms. Find the empennage area geometry using a typical aspect ratio and taper ratio and span. Fuselage-mounted low H-tail positioning is done in conjunction with positioning the V-tail with an all-moving H-tail. The empennage areas will be revised in Step 4, based on empennage areas computed using the tail volume coefficient obtained from stability Eqs. (6.1) and (6.2).

Step 3: Compute tail arms and choose tail volume coefficients, C_{HT} and C_{VT}
   From the positions of empennage carried out in Step 2, measure the tail arms L_{HT} from the wing $MAC_{1/4\text{-chord}}$ aircraft to the H-tail $MAC_{1/4\text{-chord}}$ and L_{VT} from the wing $MAC_{1/4\text{-chord}}$ aircraft to the V-tail $MAC_{1/4\text{-chord}}$. Use tail volume coefficients, C_{HT} and C_{VT} (if not available – see [6]). Military trainer aircraft tail volume coefficients are given in Table 8.4. Industries also use the tail volume coefficients, C_{HT} and C_{VT}, from their own databanks of statistics, substantiated by flight tests.

Step 4: Estimate empennage reference (planform) area
   From the tail volume coefficients, C_{HT} and C_{VT}, obtained in Step 3, compute the H-tail and V-tail planform areas using Eqs. (6.1) and (6.2). Discard the earlier values empennage areas taken from statistics in Step 2.

Step 5: Configure the geometric details of the empennage areas
   Configure the H-tail and V-tail planform geometries, that is, the aspect ratio, sweep, taper ratio and dihedral.

Table 8.4 Military trainer aircraft tail volume coefficients.

$C_{HT} \approx 0.6 \pm 0.05$	$S_H/S_W \approx 0.3 \pm 0.05$
$C_{VT} \approx 0.07 \pm 0.01$	$S_V/S_W \approx 0.2 \pm 0.05$

Step 6: Establish control surfaces

Initially, the control areas and dimensions of the elevator and the fin are earmarked from statistics and semi-empirical data.

All these parameters are decided from stability considerations and eventually fine-tuned through CFD analysis and wind-tunnel testing, with the hope that flight test results will not require further tweaking. Subsequently, static stability is to be computed about aircraft the forward-most and aftward most aircraft CGs.

8.14 Engine/Intake/Nozzle (Military – Combat/Trainer Aircraft)

A good perspective of where the engine is installed inside the fuselage is shown in Figure 8.13. If there is more than one engine, these are kept closely coupled within the fuselage to minimise asymmetric thrust in case one engine fails. Section 6.14 may be reviewed on combat aircraft intake design considerations. Details of engine intakes and nozzles are described in Sections 12.8.

Military engines do not have large *BPR*, hence a bare engine diameter is smaller. Mission profiles are throttle dependent during training/operation. Weapons release involves serious considerations for CG shift, aerodynamic asymmetry and store separation problems. These are tackled through careful analysis using CFD and wind-tunnel testing.

Intake: As indicated earlier, combat aircraft have engines buried inside the fuselage and do not have podded nacelles. It makes the term *nacelle* redundant; instead the term *intake* is used. A single pitot type intake (e.g. MIG21) at the aircraft nose is no longer pursued on account of high inlet drag. Chin mounted intake and side intakes reduce duct length, hence the inlet drag is favoured. Weapons release involves consideration for locating the intake position. Supersonic inlets are briefly discussed in Section 12.8.2 but no design work has been undertaken.

Nozzle: As the engine is mounted at the end fuselage, the exhaust jet pipes are shorter in relation to fuselage length. It goes right up to the fuselage end. There could be significant problems with engine exhaust entrainment interfering with the low H-tail. A *pen-nib* type fuselage profile could save weight by limiting the exhaust pipe length.

8.14.1 Considerations in Configuring the Intake/Nozzle

Being housed within fuselage, there is no pod/nacelle. The following are general considerations important for configuring the intake/exhaust nozzle:

Geometry	Aerodynamics
(1) Intake position	(1) Drag
(2) Intake geometry and lip section	(2) Interference
(3) Boundary layer bleed arrangement	(3) Surface roughness
(4) Intake area	(4) Noise/emission
(5) Nozzle area	(5) Vibration
	(6) Thrust and bypass ratio level

Structure (affecting weight and external geometry)	Systems
(1) Engine burst considerations	(1) Control linkage
(2) Foreign-object ingestion problems	(2) Fuel system
(3) Fuel volume	(3) Electrical
(4) Weight and CG	(4) Thrust reverser
	(5) Fire prevention
	(6) Anti-icing

8.14.2 Methodology for Configuring the Intake/Nozzle

Supersonic and subsonic intake designs differ in their approach. Supersonic intake design is complex dealing with shock waves. Supersonic intake has to cater for a wide range of air mass demand for the full flight envelope. Doors are required to match airflow demand (see Sections 5.16.1 and 12.8.2). This book deals with subsonic intakes. Given next is a stepwise procedure for intake design.

Step 1: Decide the type of intake and the intake position
   The current dominant types are chin mounted or side mounted intakes. This book deals with side mounted intakes, placed above for low-wing configuration or below for high-wing configuration. An engine buried inside the fuselage requires fuselage side intakes with bent ducts joining the engine on the centreline. An intake duct with a gradual bend should not exceeding 6° at any point enables the engine position. The bends should be gentle to avoid separation, especially at asymmetrical flight attitudes.

Step 2: Size intake area
   From the rated engine air mass flow, size the intake throat area. At this stage take 90% of the engine fan face area. If the fan face diameter is D_{fan}, the intake area $\approx 0.95 \times \pi (D_{\text{fan}}/2)^2$. Each of the side intake areas will be half of it as follows.

$$A_{int/2} = 0.373 \times (D_{\text{fan}})^2 \tag{8.6}$$

The corners of the intake are filleted to reduce pressure loss. Its fuselage side may match the fuselage curvature. Chin mounted subsonic intake area is as follows.

$$A_{\text{int}} = 0.746 \times (D_{\text{fan}})^2 \tag{8.7}$$

It can be kidney shaped as in the case of F16, or can be kept 'boxy' with filleted corners.

Step 3: Other installation considerations associated with intake design
   Side intakes must have some form of fuselage generated boundary layer bleed arrangement. The simplest arrangement is to have splitter/diverter plate (see Figure 5.35). To keep the CG forward, the engine position should be brought as far forward as possible for the layout without creating excessive intake duct curvatures – here is where design experience counts.

Step 4: Integrate intake to the fuselage
   Being integral to fuselage, intake shaping is taken along with configuring fuselage to generate cross-sections that generates a smooth streamlined shaped design.

8.15 Undercarriage (Military – Combat/Trainer Aircraft)

Chapter 10 is dedicated to the undercarriage. Military configuration study also requires some iteration to position the empennage and undercarriage with regard to the wing because initially the CG position is not known. Weights are estimated from a provisional positioning, and then the positions are fine-tuned through iterations when the CG is known.

8.16 Worked-Out Example – Configuring Military AJT Class Aircraft

As can be appreciated, from the discussions given in a marginal way in the preceding sections, combat aircraft design exercises at undergraduate level are not recommended. This kind of advanced work on combat military designs can be carried only when the basics are well mastered, that is aim of this book. The authors believe that without the considerations described here, a modern combat aircraft design exercise in the classroom would prove no better than an advanced military trainer aircraft with close support capabilities to familiarise the student with a typical mission profile and associated design consideration. Therefore, this introductory book starts the military design exercise with trainer aircraft as an alternative to frontline combat aircraft design.

This simpler introductory design exercise offers sufficient design training towards the readers' understanding of military aircraft combat aircraft design. To quote examples, readers are requested to study aircraft such as the BAe Hawk, MB345 (Italy), MIG AT (Russia), L159 (Czech), YAK 130 (Russia), EAD Mako and the Korean KT50 . All these aircraft have versions for lead-in training to the operational level, as well as a version with light combat capabilities.

In this book, a methodology on military aircraft design considerations is given with worked-out examples of an AJT with a CAS aircraft. Military trainer aircraft, along with a CAS role variant, is a class of military aircraft, which will give some idea of the terminologies and considerations associated with military aircraft design. Therefore, the term *military aircraft* in this book deals with military trainer aircraft with at least one variant in the combat role.

The worked-out AJT example is restricted to the turbofan-powered Hawk class trainer with a single-seat CAS variant. This single-seat CAS variant is obtained by taking out one pilot along with associated instruments and equipment to reduce weight by about 200 kg. However, the CAS version has more advanced avionics, including forward-looking radar making the MTOM heavier than the fully loaded AJT mission load. The AJT has two MTOM, one at Normal Training Configuration (NTC) with no armament; its mass is represented by NTCM (Normal Training Configuration Mass) and the other, with full training weapon load using all hard points represented by MTOM. CAS has only one MTOM with a full combat armament load. Training weapons are different from combat weapons: the former are low-cost practice weapons. The concept of TTOM is introduced in Section 7.10.5.

Military designs require some early decisions on how to configure the aircraft. In general, details can come with the AST of the RFP from the MoD. In this case, a tandem seating arrangement (instructor's seat raised) with a high wing is desired by the Air Force (manufacturers may be consulted). A high wing is considered to be superior for aerodynamics and accessibility. The following are the AJT specifications used in this book.

It is important that a proven reliable engine from a reputed manufacturer should be chosen. There is only one engine in the market that will give this range of thrust; the RR-Turbomeca Adour 861 with 0.75 BPR gas turbine and its variant Adour 951 for the CAS version. The Honeywell ATF120 could compete but this study takes the established, proven engine installed in BAe Hawk, constantly being upgraded to stay abreast with technology.

AJT Specifications/Requirements – as per RFP/AST

Basic mission: Combat training in jet aircraft up to operational conversion.
Payload: two 80 kg pilots and 1500 kg of practise armament.
High visibility raised rear seat of tandem arrangement.
Training mission: Sortie duration of a maximum of 75 min plus 30 min reserve. Two consecutive sorties without refuelling is desired.
Engine: one turbofan with low BPR – no afterburning.
Manoeuvre limits: +7 to −3.5 g for the training mission without armament at NTCM.
Manoeuvre limits: +6 to −3 g for the armament training mission at MTOM.
Maximum level speed = 0.85 Mach at 25 000 ft in clean configuration.
Never exceed speed, V_D = 620 kt (0.95 Mach).
All-digital flight deck.

Since the variant design has also to be considered, its details are provided as listed next. (In-flight refuelling training is desired by some air forces, but it is not dealt with in this example).

CAS Specifications/Requirements

Basic mission: CAS.
Payload: one 90 kg pilot and 2200 kg of armament.
Mission Profile: Hi-Lo-Hi (see Chapter 11).
Sortie duration: 100 min plus 15 min reserve.
Engine: one turbofan with low BPR, afterburning as option.

Manoeuvre limits: +8 to −4 g for the TTOM (lower fuel and armament than MTOM).
Manoeuvre limits: +7 to −3.5 g for the MTOM.
Maximum level speed = 0.85 Mach.

AJT Technology Level Adopted:

Rear fuselage housed single engine.
Aerofoil: NACA 64-210 section.
High-lift device: Single-slotted Fowler flap at the wing trailing edge.
From test data, the following maximum lift coefficients are taken as given next.

Flap setting versus C_{Lmax}

Flap/slat deflection(°)	0/0	8/4	20/10	40/20
C_{Lmax}	1.5	2.0	2.2	2.5

Engine

The Adour 861 with 25.5kN thrust (Ref. 6.1) has length of 1.956 m (77 in.), fan face diameter of 0.56 m (22 in.), maximum depth of 1.04 m, maximum width of 0.75 m and dry weight of 603 kg (1330 lb). The uprated version Adour 951 with 29kN meant for the CAS variant.

Flight Deck

Electronic Flight Information System (EFIS) type pilot display.

Material (simplified statement):

Primary structure is all metal. Some secondary structures are composites. Weight saving is taken as 5–10% depending on the aircraft component in consideration.

Trainer aircraft conceptual study requires special attention. Here, a relatively inexperienced student pilot (typically less than 200 hours) is learning to fly in a stretched flight envelope. A trainer aircraft needs to be safe and forgiving with low-wing loading. It has to satisfy two takeoff weights (mass) conditions: at (i) NTC and at (ii) MTOM with practise weapons load. There is a discrete jump in aircraft mass with weapons load affecting aircraft handling qualities.

Every attempt should be made to conceive a design better than Hawk, at least on paper. AJT's speed capability is slightly curtailed to reduce weight – this does not degrade training obligations, but the maximum speed of the CAS (combat) variant would be slightly higher (combat variant).

As indicated earlier, to develop a new military aircraft configuration it is necessary to make scaled sketches to incorporate new concepts to select a suitable one through design reviews in IPPD environment. Once the concept of AJT is firmed up, at this stage, some idea of weights, wing size and thrust level have to extracted from the statistics and the figures will be formally sized subsequently.

8.16.1 Use of Statistics in the Class of Military Trainer Aircraft

Statistics given in Section 7.10.1 relate to the operational combat class. It is better to generate more refined statistics of the military trainer class of aircraft to work with the task in hand. The authors recommend that the readers make such graphs at a better resolution for the class of aircraft under consideration. Statistical data in military trainer are few and extracting information will require some experience. Readers are suggested to get as many aircraft within the class and adopting to better technology levels.

Since a trainer aircraft has to deal with two takeoff masses (at NTC and at MTO), it is convenient first to extract data for the aircraft masses using the payload as the driver, hoping that its wing and engine sizes from the statistics would give satisfactory results compared to the existing kind. These will be subsequently properly sized – this point will be emphasised again and again.

Figure 8.11 gives the statistics of payload (armament load) versus MTOM. Also shown in the graph is the OEM of the AJT. Aircraft used in the statistics are MB339 (Italy), MIG AT (Russia), L159 (Czech), Hawk

(UK, NTC mass = 6100 kg) and YAK 130 (Russia). After MTOM is decided from the payload, statistics in Figure 8.12 (with large scatter) are used to obtain wing reference area and engine size. The large scatter in Figures 8.11 and 8.12 can only give an indication serving the purpose for initial guesses.

Initially, during the concept definition stage the preliminary design relies on the guesstimate from the statistics to start with. (For trainer aircraft, even *radius of action* is meaningless because practise ranges are normally close to base. Training mission time substitutes for range and with a practise weapons load as payload, are the statistical parameters used to obtain trainer aircraft MTOM. The training mission's endurance is nearly the same for all major air forces, in general. The requirement for a sortie is about 60–75 min (can extend to 90 min) with a reserve of 30 minutes. Two consecutive sorties without refuelling are desired.)

Figures 8.11 and 8.12 give the statistics for the advanced training aircraft class of aircraft and are not meant for its combat CAS variant. With the AJT practise armament load of 1500 kg, corresponding to this load, Figure 8.11 indicates a OEM = 3700 kg, the NTCM = 4800 kg without armament and MTOM of = 6500 kg. The values will be updated when AJT mass is computed in Chapter 9. Corresponding to the NTCM = 4800 kg, Figure 8.12 gives the wing reference area as an S_W of 17 m^2. The same graph gives engine size as 24KN (5390 lb); subsequently, these will be formally sized in Chapter 14.

Its CAS variant has armament load is 2200 kg. While Figures 8.11 and 8.12 are meant for the trainer class of aircraft, they are not applicable to the combat class of aircraft, some preliminary information can be extracted by adding incremental mass changes associated. The changes for the CAS are as follows.

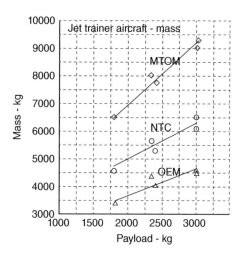

Figure 8.11 Military trainer aircraft mass.

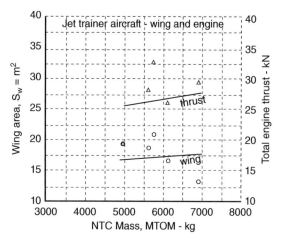

Figure 8.12 Military trainer aircraft. Wing area and engine size.

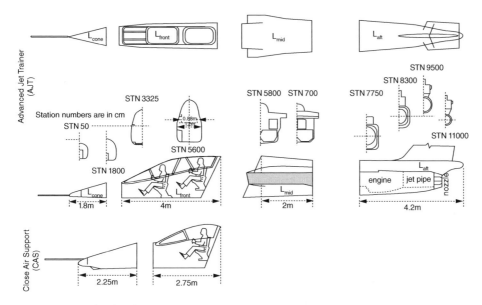

Figure 8.13 AJT fuselage layout.

i. The OEM stays about the same, as the removal of the mass of one crew plus the ejection seat and mass saving from having a slightly shorter fuselage (see Figure 8.13) are taken up as structural reinforcement to support a higher MTOM and manoeuvring load.

ii. Fuel load increment, $\Delta_{Fuel} = 500$ kg, armament load increment, $\Delta_{Payload} = 600$ kg, avionics and systems mass increment, $\Delta_{Systems} = 100$ kg, bigger engine power plant mass increment, $\Delta_{Powerplant} = 160$ kg. This totals mass increment, $\Delta_{MTOM} = 1360$ kg. Therefore, CAS MTOM is expected to be around 7860 kg. The figures would be iterated as more information is generated. The aim is to make new design better than any existing aircraft in the class. This is where the experience of designers counts.

The following subsections systematically develop the preliminary configuration of the AJT.

8.16.2 AJT – Fuselage

The accepted AJT sketch is used to develop the fuselage mouldline section by section. Typically, a rear-mounted military jet aircraft is conveniently split in to four sections as shown in Figure 5.29. The AJT fuselage is split into four sections as shown in Figure 8.13.

Since the military aircraft fuselage houses the engine, intake design is integral to fuselage layout. The choice of intake positioning is given in Section 5.16.1. In this AJT example a high-wing configuration takes the proven type of side mounted intakes. The inlet duct area will be sized in Section 8.16.5. At this stage a statistical value is taken.

Figure 8.13 outlines in detail a tandem seating arrangement for the AJT with the instructor's rear seat raised for better view above the pupil in front. From the sketch, the overall length is 12 m (39.4 ft). Note the varying cross-section of the fuselage housing the turbofan. Provision for internal fuel tanks is made between the two air intakes at each side of the fuselage and in the wing (Figure 8.14). The fuselage is split into front and rear sections to facilitate variant designs. The front fuselage can be replaced by single-seat version with cockpit layout arranged to suit CAS variant. In the example, the front fuselage bears similarity with the Jaguar trainer front fuselage mouldlines.

Intake area at side is $A_{int/2} = 0.117$ m^2 ($\approx$1.26 ft^2) as computed in Section 8.16.5. This is taken into account in developing the AJT fuselage mid-section.

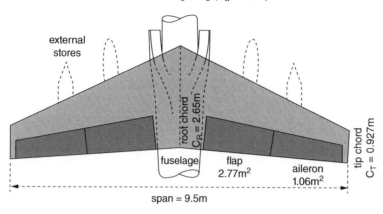

Advance Jet Training Wing ($S_w = 17m^2$)

external stores

root chord $C_R = 2.65$m

fuselage

flap 2.77m²

aileron 1.06m²

tip chord $C_T = 0.927$m

span = 9.5m

Figure 8.14 AJT wing layout.

Step 1: configuring the crew station/flight deck, length L_{front}

A tandem seat with high visibility rear seat (instructor) is raised to have less obstructed forward vision. The high wing is positioned behind the rear pilot at a level that offers good look down view angle. Each seat is 1.6 m including pilot seated. Total front fuselage length = 1.6 m (front pilot) + 1.6 m (rear pilot) + 0.8 m = 4 m (Figure 8.13). Cockpit width for trainer aircraft = 0.88 m allowing the ejection seat to escape the pilot space even under restrained conditions. Fuselage width with structure thickness = 0.88 + (2 × 0.16) = 1.3 m. Below floorboard structure thickness = 0.2 m. A trainer aircraft does not have a protective armour plate.

Step 2: front fuselage closure (configuring the nose cone)

Nose cone (or some kind of ogive shape) = 1.8 m to accommodate small radar and other electronic boxes front of the front fuselage bulkhead. The fuselage reference plane may be positioned at the tip of the nose cone. Establish the fuselage axis.

Step 3: configuring the mid-fuselage

The load-bearing mid-fuselage to which the wing $L_{mid} = 2$ m. It has the main undercarriage storage bay and bifurcated intake ducts converging to the engine face. It may have an attachment arrange with a spine going through the aft end.

Step 4: aft-fuselage closure (configuring the aft end)

Aft fuselage length $L_{aft} = 4.2$ m: there is splitting to reach the engine for its MRO needs. Aft-end shape depends on the type of aircraft configuration chosen. It can have a plane engine exhaust plane as the aft end or may have a closure short in length. Military aircraft have rapid rotation. The fuselage clearance angle, θ, depends on the main-wheel position of the undercarriage relative to the aircraft CG position (see Chapter 10). The typical angle for θ is around 16° to approach high C_{Lmax} at aircraft rotation.

Place the four fuselage subsections along the fuselage axes as referenced along the fuselage reference line. The four sections fuselage adds to total fuselage length = 12 m (39.4 ft). Note that intake is integral to fuselage configuration.

Discussion/(Rationale given in Chapter 5)

Configuring military aircraft fuselage design is different from transport aircraft fuselage design. The military aircraft fuselage houses the engine. Its fuselage has to be developed from the sketche that makes provisions for the intake duct, the undercarriage storage bay in the sketch and then aerodynamically shapes them section by section. With continuously varying cross-sections it becomes difficult to define which diameter to use to establish its fineness ratio. The fuselage length becomes a fall-out of the design. The pilot station deck has a standard fuselage breadth and height, but it may not have a maximum average diameter. In this example, the widest fuselage in the mid-fuselage behind rear pilot with fuselage integrated side mounted intakes. To compare fineness ratio, a uniform dimension is to be used, possible choices are the maximum average diameter D_{ave} where it occurs or an average value for the full length of the fuselage which will be smaller than the maximum average. Combat aircraft have supersonic capability, hence they have a pointed nose cone and its closure is the engine exit plane as the supersonic aircraft is jet propelled. For this example of a high subsonic AJT, the long pitot tube is positioned at the tip of nose cone, but alternatively there can be small pitot tubes at the both sides of front fuselage.

The military fuselage internal structural layout is considerably more complex with integrally milled frames including wing attachment. Integrally milled structures ease maintenance as they suit military operation.

The CAS variant fuselage length is 0.8 m shorter than the AJT version as can be seen in Figure 8.13. There is no credit for fuselage drag and weight reduction is taken from this slightly shorter fuselage length. CAS aircraft avionics are positioned in a way to keep its CG position with respect to the wing unaltered.

8.16.3   AJT – Wing

A high-wing arrangement is the AST requirement – this gives better spanwise lift distribution. It also gives enough under-wing clearance for movement and inspection. Refuelling is done over the wing. Positioning of the wing with respect to the fuselage is about in the middle. Note: a high-wing design does not necessarily need a fairing under the wing to house undercarriage. It can be manufactured in two pieces and attached to the sides of the load-bearing mid-fuselage section.

MTOM of = 6500 kg is taken from the statistics as shown Figure 8.12 corresponding to the armament payload of 1800 kg. The main wing planform shape is that of a low aspect ratio trapezoidal shape. Rigorous aerodynamics optimisation to decide best aspect ratio would prove unrealistic without the structural consideration. Wing span, b, and wing area, S_W, are interrelated $S_W = b^2/AR$. A high rate of roll manoeuvre restricts wing span to minimise the wing root bending moment. In practice, wing span should be decided on in consultation with structural designers. For this reason, instead of taking the aspect ratio from statistics, wing span is considered. The typical aspect ratio in this class of aircraft is within 4.5–6.

Sensitivity studies in Section 14.7 show that within the small variation (Table 14.13), a 0.1 change in aspect ratio would change about 40 kg in weight – this is relatively small amount.

Step 1: Decide the aerofoil section
   For maximum level speed = 0.85 Mach, the aerofoil section chosen is the well-known NACA 64-210 Design $C_L = 0.2$.

Step 2: Establish wing reference (planform) area
   Corresponding to an MTOM of = 6500 kg, Figure 8.13 indicates a wing area, $S_W = 17$ m^2 ($\approx$183 ft^2). This gives a wing loading of 382.35 kg m^{-2} ($\approx$3750 N m^{-2}) ($\approx$78.3 lb ft^{-2}). These preliminary values are meant for *concept definition* that will be finalised by formal sizing (Chapter 16) to *concept refinement*.

Step 3: Establish wing geometric details
   Wing span 9.5 m (31.17 ft) is taken from statistics. This gives the aspect ratio, $AR = b^2/S_W = 9.52/17 = 5.3$, well within the statistics range. AJT is taken as trapezoid with a taper ratio, $\lambda = 0.35$ is closed due to having highest value of the Oswald's efficiency factor, e. Other parameters of interest are twist of 2° (wash out).
   Sweep
      Wing sweep is taken $\Lambda_{1/4} = 20°$. Being high wing and with 20° sweep it will have high roll stability for a military aircraft. With high sweep and lower camber the M_{CR} is raised to 0.8 ($C_{DW} = 0$) and $M_{crit} = 0.85$ ($C_{DW} = 0.0002$) compared to a Bizjet wing. (Boeing definition – see Section 3.13).
   Twist
      The wing twist is taken as −2°, washout.
   Dihedral
      Being of high wing and with 20° sweep it will have high roll stability for a military aircraft. Therefore, an anhedral angle of 2° is used to improve agility by reducing roll stability. It must be understood here that

these form a heuristic approach to design depending on designers' experience and have to be substantiated through CFD and wind-tunnel testing. This is part of the learning process.

Other details

Other details of wing geometry are as follows.

$C_T/C_R = 0.35$ *and* $S_W = b \times (C_T + C_R)/2$

or $17 = 9.5 \times (C_T + C_R)/2$ solving the equations

Root chord, $C_R = 2.65$ m (8.69 ft) and tip chord, $C_T = 0.927$ (3.04 ft).

Using Eq. (3.21), the MAC works out to be 1.928 m (6.325 ft).

Step 4: Establish wing high-lift devices

The adopted technology chose a single-slotted Fowler flap without a LE slat that would be sufficient, saving considerably on costs. Its lift characteristics are given at the beginning of this section.

Step 5: Position wing with respect to fuselage

AJT is configured as a high-wing design. With a rear-mounted engine the CG moves back. At this stage, the mid-chord of the wing $MAC_{1/2}$ is placed behind mid-fuselage, say at 60% of fuselage length from the ZRP. In a conservative estimate, this may start at 58% that will iterate to better accuracy after weights are estimated in Chapter 10. Positioning of the wing relative to the fuselage is an important part of configuring an aircraft. This may not be easy because moving the wing will alter the CG position – an inexperienced engineer could encounter wing chasing.

Step 6: Control surfaces areas and location

Control areas are provisional and are sized in Phase 2. Initially, a company's statistical data of previous experience serve as a good guideline. Aileron, flaps and spoilers are placed behind the wing rear spar, which typically runs straight (or piecewise straight) at about 60–66% of the chord. With a simple trapezoidal wing planform, the rear spar runs straight, which keeps manufacturing costs low and the operation simpler; therefore, it has a lower maintenance cost.

With a third of the wing span exposed, the flap and aileron areas are taken from statistics to be 1.06 m^2 (11.4 ft^2) and the flap area is 2.77 m^2 (29.8 ft^2), as the total of both sides. Single-slotted Fowler action trailing edge flaps and leading-edge slats are chosen. Eventual performance analysis would ascertain whether these assumptions satisfy field performance specifications, if not, the design will have to be iterated with better flap design. Figure 8.14 gives wing configuration obtained from statistics.

Discussion/(Rationale Given in Chapter 4)

In general, the comments made in discussion at the Section 8.11.3 are valid. The AJT has a maximum operating speed of Mach 0.85 with a higher wing sweep of 20°. To cope with high *g* manoeuvres, the aspect ratio is kept lower to AR = 5.3 for the wing, $S_W = 17$ m^2 ($\approx$183 ft^2), resulting wing span, $b = 9.5$ m. Design $C_L = 0.2$ at a training mission speed of 0.75 Mach.

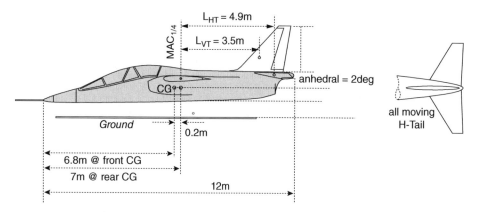

Figure 8.15 AJT tail arm (CG position guessed).

8.16.4   AJT – Empennage

The AJT is configured as a high-wing design. A military aircraft tail arm is shorter than in civil aircraft design and, along with higher rate of manoeuvre, it demands a relatively large empennage.

Step 1: Decide on the aerofoil section
  H-Tail aerofoil NACA 64-210, conventionally mounted on aft fuselage.
  V-Tail aerofoil NACA 64-010, symmetric.

Step 2: Estimate the empennage reference areas from statistics.
  Estimate the H-tail reference area S_{HT} and V-tail reference area S_{VT} from the statistics. From Figure 8.4b (same as Bizjet), corresponding to wing area $S_W = 17$ m², (182.8 ft²), take H-tail reference area $S_{HT} = 4.5$ m² (48.4 ft²) and V-tail reference area $S_{VT} = 3.6$ m² (38.7 ft²). The wing areas obtained from statistics are used to measure the tail arms. These values will be discarded after obtaining the empennage areas using the relation of tail volume coefficients as done in Step 5.

Step 3: Workout empennage planform shape, their positions and determine the tail arms
  Tentatively place H-tail at the extreme end of the fuselage to maximise tail arm and V-tail at the aircraft plane of symmetry, slightly forward so that it is not shielded by the H-tail (keep at least 50% rudder area free). Measure the tail arms $L_{HT} = 4.9$ m (15.74 ft) from the wing $MAC_{1/4\text{-chord}}$ aircraft to the H-tail $MAC_{1/4\text{-chord}}$ and $L_{VT} = 3.5$ m (11.5 ft) measured from the wing $MAC_{1/4\text{-chord}}$ aircraft to the V-tail $MAC_{1/4\text{-chord}}$. These tail arms must satisfy the static stability to be computed about the forward-most and aftward most aircraft CGs (to be established in Chapter 10).

  The wing areas obtained from statistics are used to measure the tail arms. Empennage areas are still preliminary and will be iterated to final size. Figure 8.15 shows the tail arms.

Step 4: Determine tail volume coefficients
  From Table 8.4, take The tail volume coefficients are $C_{HT} = 0.7$ and $C_{VT} = 0.08$. The vertical tail volume coefficient is higher than the civil aircraft example to make sure that sufficient rudder is available for spin recovery.

Step 5: Iterate empennage reference (planform) areas
  H-tail sizing: Eq. (6.1) gives
    $H\text{ - tail reference area}, S_{HT} = (C_{HT}) \times (S_W \times MAC)/L_{HT}$
    Then $S_{HT} = (0.7 \times 17 \times 1.928)/4.9 = 4.7$ m² (50.32 ft²), which is partly buried into fuselage. Discard the earlier value of S_{HT} obtained from statistics.

    This area has to be shared by the elevator and the stabiliser. Normally, the rudder takes 18–25% of the V-tail area; in this case 20% is taken. This gives an elevator area of 0.956 m² (10.3 ft²).

    Finalise the H-tail with other pertinent details: H-tail area = 4.7 m² (50.32 ft²), tail arm = 4.9 m,. With properly computed geometry, find a more accurate L_{HT} and iterate an accurate T-tail geometry.
  V-tail sizing: Eq. (6.2) gives
    $vertical\ tail\ reference\ area\ S_{VT} = (C_{VT}) \times (S_W \times wing\ span)/L_{VT}$
    Hence, $S_{VT} = (0.08 \times 17 \times 9.5)/3.5 = 3.83$ m² (41.1 ft²). Discard the earlier value of S_{VT} obtained from statistics.

    Compare this with what was taken from the statistics in Step 2 – they should be close enough. Accept the values obtained by using the equations. If require, repositions of the H-tail and V-tail relative to the fuselage and the wing considering the aerodynamic, stability, control and structural considerations.

Step 6: Configure the geometric details of the empennage areas
  H-tail
    Typically, H-tail span is 35–40% of wing span. Taking 40% of wing span, the H-tail is taken as 3.8 m (12.5 ft). This gives $AR = 3.25$. $t/c = 10\%$, Make the H-tail sweep, $\Lambda_{1/4} = 25°$, taper ratio, $\lambda = 0.3$. This gives $C_R = 1.9$ m (6.23 ft) and $C_T = 0.57$ m (1.87 ft).
    Using Eq. (3.31), $MAC_{HT} = 1.354$ m (4.44 ft).
  V-tail
    V-tail height = 2 m (Figure 8.15) and V-tail sweep, $\Lambda_{1/4} = 35°$.

Finalise the V-tail design with other pertinent details: V-tail area $= 3.83$ m^2, $\Lambda_{1/4} = 35°$, $t/c = 10\%$, $AR = 1.52$, Span $= 2.135$, $\lambda = 0.26$.

This gives $C_R = 2.2$ m (7.22 ft) and $C_T = 0.572$ m (1.88 ft).

Using Eq. (3.21), $MAC_{VT} = 1.545$ m (5 ft).

Tail arm $= 3.5$ m (11.5 ft), keeping the fin area $= 2.85$ m^2 (30.6 ft^2), the rest being the rudder area.

With properly computed geometry, find a more accurate L_{VT} and iterate more accurately.

Next, settle for the H-tail incidence α_{HT}, if riggable, the value can be fine-tuned after flight tests.

The choices for the empennage aspect ratio, wing sweep and taper ratio are interlinked and follow the same approach as for wing design. The empennage aspect ratio is considerably lower than that of the wing. All these parameters are decided from stability considerations and eventually fine-tuned through CFD analysis and wind-tunnel testing, with the hope that flight test results will not require further tweaking.

Step 7: Establish control surfaces

From statistics take elevator $= 1.5$ m^2 (16.2 ft^2) and rudder $= 0.9$ m^2 (9.7 ft^2).

Initially, the control areas and dimensions of the elevator and the fin are earmarked from statistics and semi-empirical data. At this stage of study, the control surfaces can be postponed until more details are available to accurately size the control areas. In this book, the control surfaces are not sized. Subsequently, in the next design phase, when the finalised aircraft geometry is available, the empennage dimensions are established by formal stability analysis. A worked-out example follows in the next section.

Discussion/(Rationale Given in Chapters 3 and 5)

AJT empennage has the same considerations as in the case for the Bizjet. Like the wing aerofoil choice, the empennage selected aerofoils are as follows.

H-tail aerofoil NACA 64-210, conventionally mounted on aft fuselage.

V-tail aerofoil NACA 64-010, symmetric.

Other empennage geometric parameters are in line with wing design. Empennage sweep is taken to be higher than wing sweep to gain in tail arm lengths.

8.16.5 AJT – Engine/Intake/Nozzle

Engine. The specified engine is the Adour 861 with a 25.5 kN thrust that has a length of 1.956 m (77 in.), fan face diameter of 0.56 m (22 in.), maximum depth of 1.04 m, maximum width of 0.75 m and dry weight of 590 kg (1300 lb). The uprated version of the Adour 951 has a 29 kN mean and dry weight 610 kg (1344 lb) for the CAS variant.

Intake. As indicated earlier, combat aircraft have engines buried inside the fuselage and do not have podded nacelles. It makes the term nacelle redundant; instead the term 'intake' is used. If there is more than one engine, they are kept close coupled within the fuselage to minimise asymmetric thrust in case one engine fails. Figure 8.16 gives a good perspective of where the engine is installed inside the fuselage. Side intakes start just behind the rear pilot so as not to obstruct side view looking below.

Intake area at each side $A_{int/2} = 0.373 \times (D_{fan})^2 = 0.373 \times 0.56^2 = 0.117$ m^2 (≈ 1.26 ft^2).

This area is taken into account when developing the AJT fuselage cross-sections in Section 8.16.2.

Nozzle. Exhaust jet pipes could be longer to suit the engine position with respect to fuselage length. It goes right up to the fuselage end. There could be significant problems with engine exhaust entrainment interfering with the low H-tail. Here, a 'pen-nib' type fuselage profile could save weight by limiting the exhaust pipe length. Military engines do not have large BPR. Mission profiles are throttle dependent during training/operation. Weapons release poses serious considerations for CG shift, aerodynamic asymmetry and store separation problems. These are tackled through careful analysis using CFD and wind-tunnel testing.

Discussion/(Rationale Given in Chapter 5)

Military aircraft have engines embedded in the fuselage, jet propulsion aircraft types naturally at the aft fuselage. They do not have pod and the intake area is integrated with the fuselage either at the sides or as a chin mount under the front fuselage. The intake duct is shaped in a gentle curvature to avoid separation to

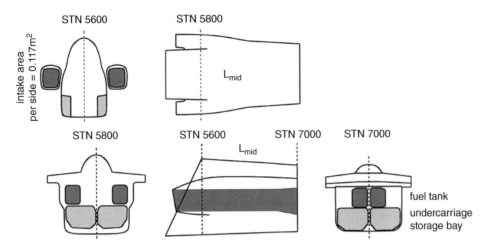

Figure 8.16 AJT intake layout.

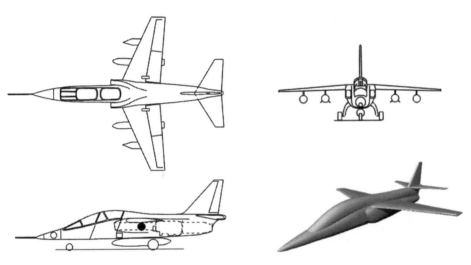

Figure 8.17 AJT three-view drawing and 3D model.

join the circular engine face. The cross-sections need to be developed section by section to contour lines as shown Figure 8.16. The exhaust nozzle duct is generally kept circular and straight.

8.16.6 Undercarriage Positioning

Chapter 10 gives the details of undercarriage (landing gear) design. Undercarriage positioning is CG dependent but at this stage the CG position is not established. Position the undercarriage, estimating the CG position and check rotational tail clearances. Make sure that the aircraft does not tip in any direction for all possible weight distributions.

The main wheels are positioned initially at about 60% of MAC (estimate). This will be revised with iteration as soon as component weights are estimated.

Figure 8.17 gives the three-view diagram of the AJT, showing wing planform and other details.

This will be revised as soon as the aircraft-component weights are estimated and proper CG location established. The next iteration would be after aircraft sizing (Chapter 10). The final iteration is to be carried out after performance estimation (Chapter 14). It is interesting to note how the aircraft is gradually taking shape.

8.16.7 Miscellaneous Considerations – Military Design

CG position: To keep the CG forward, the engine position should be brought as far forward as possible for the layout without creating excessive intake duct curvatures – here is where design experience counts. An engine buried inside the fuselage would require fuselage side intakes with bent ducts joining the engine on the centreline. An intake duct with a gradual bend not exceeding 6° at any point enables the engine position. The bends should be gentle to avoid separation, especially at asymmetrical flight attitudes.

8.16.8 Variant CAS Design

Figure 8.18 also gives the CAS variant of AJT with possible combinations of weapons within a disposable maximum armament load of 2300 kg. The CAS variant is derived from AJT by exchanging the two-seat front fuselage module with a single-seat pilot module. The CG position is unaffected by carefully positioning additional avionics black boxes, especially the forward-looking radar at the nose.

i. Predominantly, engines are buried in the fuselage; hence, there is no wing relief benefit.
ii. A combat role has a wide spectrum of activities, as outlined in Section 4.12. In general, Close support/ground attack aircraft (no afterburning engines) are subsonic with quick turning capabilities, whereas air superiority aircraft have supersonic capabilities with afterburning turbofans.
iii. Modern supersonic combat aircraft configurations show considerable wing-body blending. It becomes relatively difficult to identify the delineation between the fuselage and wing. The manufacturing joint between the wing section and fuselage block is a good place to make the partition, but at the conceptual phase the manufacturing philosophy is unlikely to be finalised. For classroom purposes, one way to decouple the wing and fuselage is to see the blend as a large wing root fairing that allows the fuselage to separate when the fairing part is taken as part of the wing.

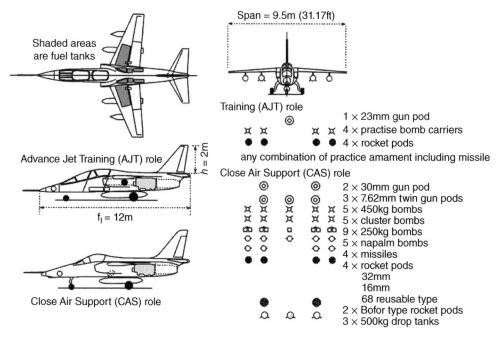

Figure 8.18 AJT and its CAS variant.

Dominantly, all payload is externally mounted except for bigger designs, which could have internal bomb load. Modern military aircraft have an external load that is contoured and flushed with mouldlines. Internally mounted guns are permanent fixtures. Training military aircraft pylons to carry external load are not taken as permanent fixtures.

iv. Consumables (e.g. firing rounds) are internally loaded. In general, for long-range missions, there is more than one crew and some consumables meant for them.

v. Military designs are technology specific and as a result, unlike civil design, military designs exhibit large variations in statistical distribution.

8.16.9 Summary of Worked-Out Military Aircraft Preliminary Details

Chapter 11 sizes the aircraft to final dimensions to freeze the configuration. Thereafter, aircraft and component mass iteration is made.

AJT Requirement Specifications

Payload = 1800 kg	Radius of action = 400 km
HSC Mach = 0.85 @ clean configuration	Initial climb rate = 40 m s^{-1}
Takeoff distance = 1100 m	Initial cruise altitude = 9 km
Landing distance = 1000 m	

Baseline aircraft mass (from statistics – needs to generated from the variant CAS design.)

MTOM = 6500 kg (15 210 lb)	NTCM = 4800 kg (10 800 lb)
OEM = 3700 kg	Fuel mass = 1100 kg (not full tank)
	Full tank capacity = 1400 kg

Baseline External Dimensions
Fuselage (deterministic from capacity) – two-seat tandem arrangement

Length = 12 m (39.4 ft)	
Maximum overall width = 1.8 m	Overall height (depth) = 4.2 m
Cockpit width = 0.88 m	Fineness ratio = 12/1.8 = 6.67

Wing AJT – NACA 64-210.

Aerofoil: NACA 64-210.		
Planform (reference) area = 17 m^2 (183 ft^2)	Span = 9.5 m (31.17 ft)	
Root chord, C_R = 2.65 m (8.69 ft)	Tip chord, C_T = 0.927 m (3.04 ft)	
MAC = 1.928 (6.325 ft)	Taper ratio, λ = 0.35	$\Lambda_{1/4} = 20°$
Dihedral = − 2° (anhedral – high wing)	Twist = 1° (wash out)	
Flap = 2.77 m^2 (29.8 ft^2)	Aileron = 1.06 m^2 (11.4 ft^2)	

H-tail
Aerofoil: NACA 64-210 installed inverted showing a negative camber.

Planform (Reference) area = 4.7 m^2 (51.45 ft^2)	Span = 4.2 m (13.8 ft)
Root chord, C_R = 1.9 m (6.23 ft)	Tip chord, C_T = 0.57 m (1.87 ft)
MAC = 1.354 (4.44 ft)	Taper ratio, λ = 0.3, $\Lambda_{1/4} = 25°$
Aspect ratio = 3.5	Tail arm = 4 m (13.1 ft)
Elevator area = 0.956 m^2 (10.3 ft^2)	

V-tail
Aerofoil: NACA 64-010 symmetrical aerofoil slightly thicker *t/c* with higher sweep.

Planform (Reference) area = 3.83 m² (41.1 ft²)	Span = 2.135 m (7 ft)
Root chord, C_R = 2.2 m (7.22 ft)	Tip chord, C_T = 0.572 m (1.88 ft)
MAC = 1.545 m (5 ft)	Taper ratio, λ = 0.26, $\Lambda_{1/4}$ = 35°
Aspect ratio = 1.52	Tail arm = 4 m (13.1 ft)
Rudder area = 0.98 m² (10.5 ft²)	

Engine

Takeoff static thrust at ISA sea level = 5390 lb.	BPR = 0.75
Dry weight = 603 kg (1330 lb),	
Fan diameter = 0.56 m (22 in.)	Length = 1.956 m (77 in.)
Maximum depth = 1.04 m (3.4 ft)	Maximum width = 0.75 m (2.46 ft)

Nacelle: None as the engine is buried into the fuselage.

CAS Variant (all component dimensions except fuselage length are kept unchanged)
Requirement Specifications

Payload = 2300 kg	Radius of action = 300 nm
HSC Mach = 0.85 @ clean configuration	Initial climb rate = 50 m s⁻¹
Takeoff distance = 1400 m	Initial cruise altitude = 9 km
Landing distance = 1200 m	
CAS aircraft mass (*from statistics*)	
MTOM = 7600 kg	Payload = 2300 kg
OEM = 3700 kg	Fuel mass = 1400 kg (full tank)

Fuselage – single seat

Length = 11.2 m (36.8 ft)	
Maximum overall width = 1.8 m	Overall height (depth) = 4.2 m
Cockpit width = 0.88 m	Fineness ratio = 12/1.8 = 6.67

8.17 Turboprop Trainer Aircraft (TPT)

An example of a TPT aircraft is given here to get some preparation to deal with propeller-powered aircraft design. Unlike the Bizjet and AJT examples, TPT has a front mounted engine with a propeller. This makes the CG come further forward than the other two examples and also requires propeller ground clearance in case of nose wheel collapse. Otherwise, the methodology to configure its geometry follows the same procedure as carried out in the earlier two worked-out examples, depending on whether it is in a civil or military mission role and certified under the appropriate government authorities.

In the past, most of the air forces' pilot training was done in three tiers, starting ab initio pilot training in a small piston engine aircraft at around 1000 kg MTOW class for about 100 hours, then moving to intermediate airmanship training in jet aircraft with some weapons practise before finally moving to advanced training in AJTs. On account of streamlining training syllabi and in an attempt incorporate a more complex approach to combat training, the last four decades saw gradual changes to training syllabi in to two stages dispensing with the piston engine aircraft in the loop. Ab initio pilot training now starts with what was the intermediate level type of aircraft. A TPT presented a good compromise being in the middle of the piston engine type to fast jet propelled trainers. There are not many TPT aircraft that have been designed. It has been noticed that the TPT aircraft capability and size kept growing to COIN capability in a dual role. Some manufacturers offer both the basic version of TPT and the COIN version of TPT. But applications of the COIN version are yet to play a significant role in field action in the international geo-political scenario as the CAS versions of the AJT do a better job.

The authors consider that the two versions of TPT should be kept separate; the trainer version trains and the COIN version is in operation. To retain commonality by maintaining one type of fleet with the COIN version in a training role, as some air forces do, is not cost effective and can become more intimidating to ab initio trainees. It is for this reason that the COIN variant is not dealt with in this book. The trainer version worked out here can be upgraded to COIN version by incorporating a bigger engine in the family, beefing-up structures to carry more load without making any major changes to its external geometry so that the same tools, jigs and fixtures can be used with minor modifications to manufacture at low development cost. Readers may make an attempt to design a COIN variant using the information given here.

The difference between the AJT and the TPT is that the former has an aft-mounted turbofan jet engine, hence its aft fuselage has boat-tail closure like with planar engine exhaust plane while the latter has front mounted turbofan engine with tractor propeller developing the thrust for propulsion. The front mounted engine moves the CG forward, therefore the wing also moves further forward, below the rear pilot seat in the fuselage reference line, compared to an AJT. Due to having a front mounted engine, the nose cone section does not exist splitting the fuselage conveniently into three sections. The pilot section (cockpit) becomes the load-bearing mid-section.

The last one, as the third section, is the aft section with the design methodology similar to that adopted in the case of the Bizjet. Also, the tandem seat arrangement without a rear mount, as in the case of the AJT, makes the rear pilot seat less elevated so as to avoid a hump-like fuselage mouldline that could generate undesirable streamline separation.

The methodology used suggested for a TPT concept definition can be applied to propeller-driven civil aircraft study, of course, to be certified under civil certifying agency.

Therefore, to avoid duplication, only the worked-out example of a military TPT is given next.

TPT Specifications/Requirements

Basic mission: Ab initio to intermediate level military training in a turboprop.
Low wing, tandem seating.
Payload: two 80 kg pilots and 450 kg of practice armament.
Maximum speed = 280 kt (472.7 ft s^{-1}),   Sustained speed = 240 kt (405.2 ft s^{-1}).
Training mission: Sortie duration of a maximum of 75 min plus 30 min reserve.
Two consecutive sorties without refuelling are possible for airmanship training by taking 450 kg of fuel instead
    of armament for which provision for tankage capacity is retained.
Manoeuvre: +7 to −3.5 g for the trainer version at NTCM without practise armament (fully aerobatic).
Manoeuvre: +6 to −3 g for the fully armed version at MTOM.

TPT Technology Level Adopted:

Rear fuselage housed single engine.
Wing aerofoil: NACA 63_2-215.
Design $C_L = 0.2$.
High-lift device: single-slotted Fowler flap at wing trailing edge.
From test data, the following maximum lift coefficients are taken as given next.

Flap setting versus C_{Lmax}

Flap/slat deflection (°)	0/0	8/4	20/10	40/20
C_{Lmax}	1.5	2.0	2.2	2.5

Empennage aerofoil (both H-tail and V-tail) = NACA 0012

Engine:

One turboprop engine.

There are only two types of turboprop available suitable to this proposed study. One is the Pratt and Whitney (Canada) PT6-60A developing uninstalled 1050 SHP (dry mass = 216 kg) and other is the Garrett TPE331-12B developing around uninstalled 985 SHP (dry mass = 185 kg). The lighter Garrett TPE331-12B is chosen. It has length of 100 cm(44 in.), engine height = 68.6 cm (27 in.), inlet area = 0.0484 m^2 (0.52 ft^2) and mass rate flow =7.7 lb s^{-1}.

EFIS type pilot display.

Material (simplified statement):

Primary structure is all metal. Some secondary structures are with composite. Weight saving is taken as 10% as applicable to the component.

8.17.1 Use of Statistics in the Class of Turboprop Trainer Aircraft (TPT)

As the armament mass is small for the intermediate class of military training aircraft, it is better to present the 'useful mass' (= MTOM − OEM) as these values can be obtained from the public domain (Figure 8.19). A new design has to rely on past experiences, even if that means trying out cutting edge new advanced technologies in a scaled down prototype aircraft as technology demonstrator. It is important that readers collect as many aircraft data within the class for any new aircraft design project.

Trainer aircraft has to deal with two takeoff masses (at NTC and at MTO). The NTC is meant for airmanship training and is fully aerobatic without practice weapon load. MTOM is the design maximum load for armament training. It is hoped that its wing and engine sizes from the statistics would give satisfactory results compared to the existing kind. These will be subsequently properly sized; this point is emphasised again and again.

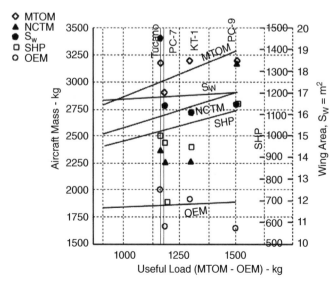

Figure 8.19 TPT statistics – (large scatter).

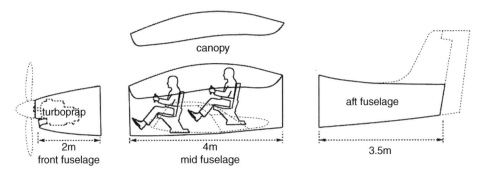

Figure 8.20 TPT aircraft fuselage layout.

It is better to generate more refined statistics of the TPT class of aircraft to work with the task in hand. Unfortunately, there are not many in this class TPT aircraft and the statistics shows some scatter. Given next are the statistics of four TPT class aircraft [2] (RAF Tucano, PC-7, PC-9, KT-1 and PZL Orlik 130 TPTs – Figure 8.19) compared with the proposed TPT worked out in this chapter. PZL Orlik 130 is on the light TPT class. The UTVA Kobac trainer aircraft and the two-seat version TAI Hurkus trainer aircraft data were not sufficiently available. Interestingly, PC-7 and PC-9 OEM are consistently lower than the other two. With only four aircraft data, regression analyses may not show a good rationale. The example worked out in his book leans towards RAF Tucano as the first author is familiarised with the aircraft. It is the modified version of the Brazilian version by Short Brothers Plc and (now Bombardier-Shorts Belfast) and manufactured.

With the proposed TPT specification of 450 kg as practice armament and guessed fuel mass of ≈450 kg gives useful load 900 kg. Figure 8.19 indicates the corresponding masses is expected to be around MTOM = 2800 kg, wing area, $S_W = 16.5$ ft^2 and SHP = 950 as the starting point for mass and CG estimation in Section 10.18. Although the graphs shows the expected OEM = 1800 kg, NTCM = 2400 kg, these may not prove that reliable due to large scatter and are irrelevant (large scatter) at this stage. The mass estimation based on MTOM = 2700 kg will give better values. The fine-tuned weight after proper weight estimation is likely to be different from that indicated by the statistics. The figures would be iterated as more information is generated. The aim is to make new designs better than any existing aircraft in the class. This is where the experience of designers counts.

This a good example on how to make use of a scattered data from few aircraft where the designer has to make decisions based on their experience. In such situation, a newly initiated may lean towards a known design that can be improved and subsequently compared with to establish the merits of the new design.

8.17.2   TPT – Fuselage

Since the military aircraft fuselage houses the engine, intake design is integral to fuselage layout. In this case of front mounted turboprop engine there is a small chin mounted intake (Figure 8.20) of area $A_{int} = 0.0.8$ m^2 (≈0.86 ft^2). The undercarriage is mounted on the low wing with its storage bay. Therefore, the fuselage mould-line has a smooth streamlined contour, hence the cross-sections are not shown. Also the front mounted engine dispenses with the nose cone section of jet aircraft, hence the fuselage is conveniently split into three sections. The tandem seat pilot station (cockpit) becomes the load-bearing mid-fuselage.

Figure 8.20 outlines in detail a tandem seating arrangement for the TPT with the instructor's rear seat raised for better view above the pupil in front. The overall length is 9.5 m (31.16 ft).

Step 1: configuring the crew station/flight deck, length L_{mid}

   The tandem seat pilot station (cockpit) becomes the load-bearing mid-fuselage with length, $L_{mid} = 4$ m (13.12 ft). It has rear seat (instructor) raised to have less obstructed forward vision. Cockpit width for trainer aircraft = 0.88 m allowing ejection seat to eject with pilot under restrained condition. Fuselage width with structure thickness $= 0.88 + (2 \times 0.6) = 1$ m. Below floorboard structure thickness $= 0.2$ m. A trainer aircraft does not have protective armour plate.

Step 2: configuring the front fuselage, length L_{front}

The front fuselage of length, $L_{front} = 2$ m (6.56 ft) houses the turboprop of a bare engine length of 1 m. The chin intake is of short length position in front of the engine intake. The fuselage reference plane may be positioned at the tip of the propeller cone. Establish the fuselage axis.

Step 3: configuring the aft fuselage, length L_{aft}

Aft-fuselage length $L_{aft} = 3.5$ m (11.48 ft). It has a conventionally designed closure to have frames for empennage attachments. Military aircraft have rapid rotation. The fuselage clearance angle, θ, depends on the main-wheel position of the undercarriage relative to the aircraft CG position (see Chapter 7). The typical angle for θ is around 16° to approach high C_{Lmax} at aircraft rotation.

$$\text{Fuselage length} = 4 + 2 + 3.5 = 9.5 \text{ m (31.16 ft)}$$

Place the three fuselage subsections along the fuselage reference line in a seamless manner matching the curvature of both ends.

Discussion/(Rationale given in Chapter 5)

As mentioned in the case of configuring the AJT fuselage, military aircraft fuselage design is different from transport aircraft fuselage design. This example of a front engine mounted TPT fuselage again differs from AJT with rear fuselage mounted jet engine. TPTs have a low wing allowing undercarriage with a wider main-wheel track. The fuselage does not have wheel storage bay as it can be accommodated thicker wing root. In this case, the wheel storage bay has to be housed in fuselage, it is taken in the wing box structure without having to show any bulge in the fuselage. TPT does not have a complicated intake with long curved duct leading to the engine face. A front mounted turboprop engine driving a propeller has a very short intake duct that keeps fuselage mouldlines with smooth contours.

Its fuselage has to be developed from the sketch section by section, keeping the smooth aerodynamic shape. The pilot station deck has a standard fuselage breadth and height. Therefore, even with varying cross-sections, the maximum fuselage is at the rear pilot position and an average fuselage can be determined. The fuselage fineness ratio can be obtained by computing the average fuselage diameter D_{ave}.

8.17.3 TPT – Wing

A low-wing arrangement for the TPT is the AST requirement – this gives better spanwise lift distribution. It also gives enough under-wing clearance for movement and inspection. Refuelling is done over the wing. The undercarriage is mounted on the wing with inward retracted storage bay. The wing planform shape is trapezoidal.

It is to be understood here that these are heuristic approaches to design depending on designers' experience and have to be substantiated through CFD and wind-tunnel testing. This is part of the learning process.

Step 1: Decide the aerofoil selection

For maximum level speed = 280 kt (0.465 Mach) at 20 000 ft, the aerofoil chosen is the well-known NACA 63_2-215. Design $C_L = 0.2$.

Step 2: Establish wing reference (planform) area

The wing area $S_W = 16.5$ m^2 (177.3 ft^2) is taken from the statistics as shown Figure 8.19 corresponding to the NTCM of 2350 kg. This gives a wing loading at MTOM $= 169.7$ kg m^{-2} ($\approx$N m^{-2}) ($\approx$34.74 lb ft^{-2}).

Step 3: Establish wing geometric details

Wing span, b, of 16.36 m (53.5 ft) is taken from statistics. Wing aspect ratio, AR, is interrelated with wing span b and wing area, $S_W = b^2/AR$. This gives a moderate aspect ratio of 6.5, well within the class of aircraft suiting an ab intio training design. In practice, wing span should be decided in consultation with structural designers.

The low speed TPT trapezoid wing has zero sweep and a taper ratio $\lambda = 0.35$ that is closed due to having the highest value of Oswald's efficiency factor, e. Other parameters of interest are a twist of 2° (wash out).

At this low speed, quarter chord sweep $\Lambda_{1/4} = 0°$. For a low wing, at low speed the dihedral angle $= 3°$ to retain some degree of agility to roll for a military trainer aircraft.

Other details

Other details of wing geometry are as follows.

$$C_T/C_R = 0.35 \text{ or } C_T = 0.35 \times C_R$$

and $S_W = b \times (C_T + C_R)/2$
or $16.5 = 10.36 \times (C_T + C_R)/2$ solving the equations
or $(C_T + C_R) = 3.185$
Root chord, $C_R = 2.36$ m (7.74 ft), and tip chord, $C_T = 0.826$ m (2.71 ft).

Using Eq. (3.21), MAC $= 2/3 \times [(2.36 + 0.826) - (2.36 \times 0.826)/(2.36 + 0.826)] = 1.715$ m (5.63 ft).

Adequate wing-body fairing is given to reduce interference drag.

Step 4: Establish wing high-lift devices

The adopted technology chooses a single-slotted Fowler flap without a LE slat that would be sufficient, saving considerably on costs. Their lift characteristics are given at the beginning of this section.

Step 5: Position wing with respect to fuselage

TPT is configured as a low-wing design. At this stage, the quarter chord of wing MAC$_{1/4}$ is placed about at the centre of fuselage. The wing position gets iterated as the aircraft and its components are known (Chapter 10). Positioning of the wing relative to the fuselage is an important part of configuring an aircraft. This may not be easy because moving the wing will alter the CG position – an inexperienced engineer could encounter wing chasing.

Step 6: Control surfaces areas and location

Control areas are provisional and are sized in Phase II. Initially, a company's statistical data of previous experience serve as a good guideline. Aileron, flaps and spoilers are placed behind the wing rear spar, which typically runs straight (or piecewise straight) at about 60–66% of the chord. With a simple trapezoidal wing planform, the rear spar runs straight, which keeps manufacturing costs low and the operation simpler; therefore, it has a lower maintenance cost.

With a third of the wing span exposed, the total flap and aileron areas of both sides are taken from statistics to be 2 m^2 (10.764 ft^2) and 1.5 m^2 (23.68 ft^2), respectively. Eventual performance analysis would ascertain whether these assumptions satisfy field performance specifications, if not, the design will have to be iterated with better flap design. Figure 8.21 gives the TPT wing control surface configuration using statistics.

The wing could be manufactured in two pieces attached to each side of the load-bearing mid-fuselage.

*Discussion/(*Rationale *Given in Chapters 3 and 4)*

TPT maximum operating speed is 280 kt, that is, ≈ 0.46 Mach at 20 000 ft altitude, when compressible effects can be ignored, hence the wing sweep is kept zero to have the best aspect ratio for the wing area,

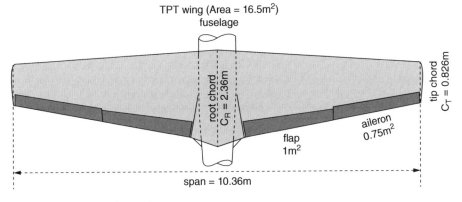

Figure 8.21 TPT aircraft wing layout.

S_W, $=16.5$ m^2 (177.3 ft^2) and a structurally favourable (from statistics) span, b, of 16.36 m (53.5 ft) and taper ratio $\lambda = 0.35$ that gives a high Oswald's efficiency factor, e. The aerofoil section chosen is the well-known NACA 63-212. Design $C_L = 0.5$, which has forgiving stall characterises suits ab initio training.

8.17.4   TPT – Empennage

The TPT is configured as a low-wing design. A military aircraft tail arm is shorter than a civil aircraft design and, along with a higher rate of manoeuvre, it demands a relatively large empennage.

Step 1: Decide the aerofoil section
   H-tail aerofoil NACA 0012: conventionally mounted on aft fuselage.
   V-tail aerofoil NACA 0012: symmetric.

Step 2: Guess empennage reference areas from statistics
   The empennage planform is generally, but not restricted to, a trapezoidal shape. Being a low speed aircraft, make the H-tail sweep $\Lambda_{1/4} = 0°$ and V-tail sweep $\Lambda_{1/4} = 20°$ to gain the tail arm as well extend unshielded area. A strake-like surface could be extended to serve the same aerodynamic gains as for the wing.
   Empennage geometries are still not known. Therefore, guess the H-tail reference area S_{HT} and V-tail reference area S_{VT} from the statistics. From Figure 8.4b, corresponding to wing area $S_W = 16.5$ m^2 (177.3 ft^2), take H-tail reference area $S_{HT} = 3.4$ m^2 (36.54 ft^2) and V-tail reference area $S_{VT} = 2.2$ m^2 ($\approx$23.6 ft^2).

Step 3: Workout empennage planform shape, their positions and determine the tail arms
   First place V-tail at the at symmetric plane at the end of the fuselage taking care of shielding considerations (keep at least 50% rudder area free) as described in Section 6.6.4 when the H-tail is positioned quite ahead of it. The V-tail may turn 1° into the propeller slipstream swirl as yaw compensation. Tentatively place the H-tail and V-tail as shown Figure 8.22 and measure the tail arms $L_{HT} = 5.4$ m (17.7 ft) measured from the wing MAC$_{1/4\text{-chord}}$ aircraft to the H-tail MAC$_{1/4\text{-chord}}$ and $L_{VT} = 5.8$ m (19 ft) measured from the wing MAC$_{1/4\text{-chord}}$ aircraft to the V-tail MAC$_{1/4\text{-chord}}$. Empennage areas are still preliminary and will be iterated to final size. Subsequently, the static stability to be computed about aircraft the forward-most and aft-most aircraft CGs.

Step 4: Determine the tail volume coefficients
   The tail volume coefficients are $C_{HT} = 0.6$ and $C_{VT} = 0.08$. The vertical tail volume coefficient is higher than the civil aircraft example to make sure that sufficient rudder is available for spin recovery.

Step 5: Iterate empennage reference (planform) areas H-tail sizing:
   Eq. (6.1) gives H-tail reference area, $S_{HT} = (C_{HT})(S_W \times MAC)/L_{HT}$.
   Then $S_{HT} = (0.6 \times 16.5 \times 1.715)/5.4 = 4.1$ m^2 (51.45 ft^2), which is partly buried into fuselage. Discard the statistical value of 3.4 m^2 by this new H-tail area, $S_H = 4.1$ m^2. This area has to be shared by the elevator and the stabiliser. No stability credit is taken from the small amount of strake at the root of leading edge, but it acts as strake-like flow modifier to delay separation at a high angle of attack.
   V-tail sizing:
   This type of V-tail configuration requires special consideration in defining the areas. It may be noted that the rudder is positioned past the fuselage to keep rudder area unshielded from H-tail to a large extent. It is a matter of defining the areas to keep this book focused on carrying out related design considerations. The V-tail area S_{VT} using Eq. (3.30) gives the total area as shown shaded in Figure 8.22 plus the extra area of the rudder is kept separate. The total V-tail area is $S_{VT} = 2.2$ m^2 as computed using Eq. (8.21). The shadowed area is used to determine $C_{R\text{-}VT}$, $C_{T\text{-}VT}$ and the height of the V-tail above the fuselage. This rudder arrangement keeps a large portion of it unshielded, which adds to the safety to come out of a spin hands free: an important aspect for ab initio pilot training. No stability credit is taken from the small amount of dorsal fin area, but it acts as strake-like flow modifier to delay separation at high yaw (cross wind capability). This is a special V-tail design consideration.
   Equation (6.2) gives vertical tail reference area $S_{VT} = (C_{VT})(S_W \times wing\ span)/L_{VT}$.

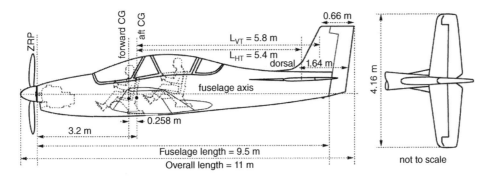

Figure 8.22 TPT aircraft tail arm.

Hence $S_{VT} = (0.08 \times 16.5 \times 9.5)/5.8 = 2.2$ m^2 (23.3 ft^2).
The extra rudder area below $S_{VT} = 0.34$ m^2.
The value can be retained as it came out to be the same as the statistical value.

Accept the values obtained by using the equations. If required, reposition the H-tail and V-tail relative to the fuselage and the wing considering the aerodynamic, stability, control and structural considerations.

Step 6: Configure the geometric details of the empennage areas

H-tail geometry

Finalise the H-tail with other pertinent details. The selected aerofoil gives a $t/c = 12\%$.
Make the H-tail sweep, $\Lambda_{1/4} = 0°$. Typically, H-tail span is 35–40% of wing span. Taking 40% 0f wing span, the H-tail span = 6.54 is taken as 4.16 m (13.62 ft). This gives an AR = 4.0. H-tail root and tip chord can be computed as done for the wing that gives the taper ratio $\lambda = C_T/C_R = 0.5$.
For a trapezoidal planform, $S_{HT} = b \times (C_R + C_T)/2$. This gives $(C_R + C_T) = 2 \times (4.1/4.15) = 1.976$ m.
Leading to: $C_{R\_HT} = 1.32$ m (4.33 ft) and $C_{T\_HT} = 0.66$ m (2.16 ft).
Using Eq. (3.21), $MAC_{HT} = 1.0234$ m (3.36 ft).
Next, settle for the H-tail incidence α_{HT}, if riggable, the value can be fine-tuned after flight tests.

V-tail geometry:

Finalise the V-tail design with other pertinent details: $\Lambda_{1/4} = 20°$, $t/c = 12\%$, height = 1.375 m, $\lambda = 0.4$, fin area = 1.76 m^2 and rudder area = 0.44 m^2. The V-tail root chord and tip chord lengths are given in Figure 8.22 as $C_{R\_VT} = 1.64$ m (5.38 ft) and $C_{T\_VT} = 0.66$ m (2.16 ft).
Using Eq. (3.21), $MAC_{VT} = 1.22$ m (4 ft).

The choices for the empennage aspect ratio, wing sweep and taper ratio are interlinked and follow the same approach as for the wing design. The empennage aspect ratio is considerably lower than that of the wing. All these parameters are decided from stability considerations and eventually fine-tuned through CFD analysis and wind-tunnel testing, with the hope that flight test results will not require further tweaking.

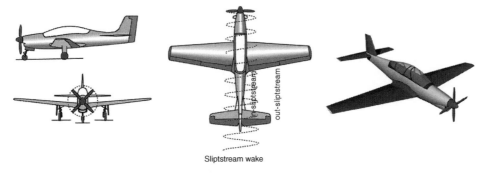

Sliptstream wake

Figure 8.23 Turboprop trainer aircraft (TPT).

Step 7: Establish control surfaces

H-tail geometry:

Normally, the elevator takes 18–25% of the H-tail area; in this case 25% is taken.
Elevator = 1.025 m² (11 ft²).

V-tail geometry:

Normally, the rudder takes 18–25% of the V-tail area; in this case 20% is taken.
This gives a rudder area of 0.44 m² (4.75 ft²).

A TPT has a front mounted engine with a tractor propeller. In case of the inadvertent situation of nose wheel collapse, 7–9 in. tip clearance from ground is required. Undercarriage design is carried out in Chapter 9.

As in the case of the Bizjet (see Step 6 in Section 8.9.3), the TPT control areas and dimensions of the elevator and the fin are earmarked from statistics and semi-empirical data given in the next section.

Step 8: Engine installation and propeller ground clearance

Give the propeller ground clearance to protect against an inadvertent nose wheel collapse. A minimum of 7 in. is required. After finding the matching propeller diameter the extent will be known. For the TPT, about 9 in. is kept at this stage on the progress.

8.17.5 TPT – Intake/Exhaust

The front mounted TPT aircraft have a very short nearly intake duct leading from aircraft intake plane to the engine-face inlet area of 0.048 m² (0.52 ft²). Aircraft intake area is taken as slightly larger than the engine-face inlet area.

Figure 8.23 gives the three-view diagram of the TPT, showing wing planform and other details.

In the following are the TPT details assumed as the concept definition from the statistical data that will be finalised to the formally sized configuration in Chapter 14.

MTOM = 2800 kg (6160 lb)	NTCM = 2350 kg (5170 lb – no armament)	
OEW = 1700 kg (3740 lb)	Practise armament load = 450 kg (≈1000 lb)	
Crew = 400 lb (≈180 kg)	Fuel = 470 kg (1012 lb)	
Typical mid-training Aircraft mass = 2200 kg (4840 lb)		
Maximum speed = 280 kt	Sustained speed = 240 kt (405.2 ft s⁻¹)	

Fuselage

Length = 9.5 m (31.16 ft)	height = 1.6 m (5.25 ft)	
Cockpit width = 0.88 m (35 in.)	Fuselage width = 1.0 m (≈40 in.)	
Fineness ratio = 9.5/1.0 = 9.5		

Wing (wing aerofoil: NACA 63$_2$-215)

Reference area = 16.5 m² (177.3 ft²)	Span = 10.36 m (53.8 ft)	AR = 6.5
Root chord, C_R = 2.36 m (7.74 ft)	Tip chord, C_T = 0.82 m (3.7 ft)	
MAC = 1.715 m (5.62 ft)	Taper ratio, λ = 0.35	$\Lambda_{1/4}$ = 0°
Dihedral = 5° (dihedral – low wing)	Twist = 1° (wash out),	t/c = as specified
Flap = 2.2 m² (23.6 ft²)	Aileron = 1.06 m² (11.4 ft²)	

H-tail (NACA 0012)

Reference area = 4.1 m² (44.2 ft²)	Span = 4.16 m (13.65 ft)	AR = 4.0
Root chord, C_R = 1.32 m (4.33 ft)	Tip chord, C_T = 0.66 m (2.16 ft)	
H-tail MAC = 1.0234 m (3.36 ft)	Taper ratio, λ = 05, $\Lambda_{1/4}$ = 0°	
Elevator area = 1.025 m² (11 ft²)	t/c = 0.12	

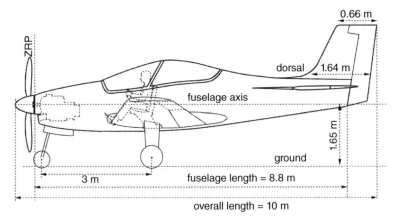

Figure 8.24 Single-seat COIN variant of TPT.

V-tail (NACA 0012)

Reference area = 2.2 m^2 (23.3 ft^2)   Height = 1.375 m (4.52 ft)

Root chord, C_R = 1.64 m (5.38 ft)   Tip chord, C_T = 0.66 m (2.16 ft)

V-tail MAC = 1.0234 m (3.36 ft)   Taper ratio, λ = 04, $\Lambda_{1/4}$ = 20°

Rudder area = 0.44 m^2 (4.74 ft^2) t/c = 0.12

Engine

Takeoff static power at ISA sea level = 975 shp   (Garratt TPT-331-12B)

Dry weight = 185 kg (407 lb)   Engine intake area = 0.048 m^2 (0.52 ft^2)

Length = 1.1 m (43.4 in.)   Height = 0.686 m (27 in.)

COIN variant

A single-seat COIN variant can be derived as suggested in Figure 8.24. The details are given in Section 10.18 but not worked out. The reader may give it a try.

References

1 Abbott, I.H. and von Doenhoff, A.E. (1949). *Theory of Wing Sections, Including a Summary of Airfoil Data*. New York: McGraw Hill.
2 Jane's All the World's Aircraft Manual (yearly publication).
3 Niu, M. (1999). *Airframe Structural Design*. Hong Kong: Commlit Press Ltd.
4 Talay, T.A. (1975). *NASA SP-367; Introduction to the Aerodynamics of Flight*. Langley Research Center.
5 Torenbeek, E. (1982). *Synthesis of Subsonic Airplane Design*. Delft University Press.
6 Nicolai, L.M. and Carichner, G.E. (2010). *Fundamentals of Aircraft and Airship Design*. Reston, VA: AIAA.
7 Roskam, J., *Aircraft Design*, Published by the author as an 8 volume set, 1985–1990.
8 Jenkinson, L.R., Simpson, P., and Rhodes, D. (1999). *Civil Jet Aircraft Design*. Arnold Publishers.
9 Kundu, A.K. (2010). *Aircraft Design*. Cambridge University Press.

9

Undercarriage

9.1   Overview

Since aircraft have to operate from the surface, land or water (or in special situations on snow), in some form their handling on the surface has to be incorporated into the design; wheels on land, floating on water and skidding on snow. There are many operational and design considerations. The mechanical arrangements are a subsystem known as the undercarriage (or the landing gear). Once airborne, the undercarriage becomes an appendage and needs to be dealt with in a manner so as to minimise the penalty associated with it; one way is to retract it to reduce drag and nose-down moment. This is also true for birds as they reposition their legs in flight; some even retract them.

Chapter 8 illustrates how to progress with the concept definition to arrive at a preliminary aircraft configuration of a new aircraft project starting from scratch, to arrive at a tentative aircraft configuration without an undercarriage placed. To progress further, the next task is to lay out the undercarriage position relative to an estimated aircraft centre of gravity (CG) location. Some iteration of wing repositioning may be required to suit the CG location with adequate fuselage clearance angle, γ, and the tipping back angle, β (Figure 5.6). This complete aircraft configuration with undercarriage is still tentative and may have to be iterated again to a more appropriate configuration when the aircraft CG location is accurately estimated in Chapter 10.

This book is not on undercarriage design. This chapter addresses the undercarriage quite extensively but not on its detailed design; rather, it focuses on those aspects related to undercarriage layout (positional geometry and size with respect to aircraft CG) and selection of tyres to develop a feasible aircraft configuration. The chapter begins by introducing the undercarriage, followed by basic definitions, terminologies and information used in the design process and integration with an aircraft. Finally, methodologies for layout of the undercarriage and tyre sizing are presented to complete the aircraft configuration generated thus far. Considerable attention is required to lay out the undercarriage position and to determine tyre size and geometric details so as to avoid hazards during operation. Relevant information on wheel tyres is also presented. Selection of tyres depends not only the load on them but also the airfield strength based on how it is prepared, hence some related information on airfield types is also given. More details on undercarriage design are in the cited references.

The undercarriage is a complex and heavy item and, therefore, expensive to manufacture. It should be made right the first time. Aircraft designers should know the operational basics, leaving the details to those who specialise in the undercarriage as a system that is integrated with an aircraft as a subsystem. Aircraft designers consult with undercarriage specialists during the conceptual stage.

In the past, aircraft manufacturers handled the undercarriage design in a vertically integrated factory setup. Today, its complexity has created specialised organisations (e.g. Messier of France and Dowty, UK) who are dedicated to undercarriage design, thereby making its management and integration more efficient, resulting in better overall designs. However, for smaller aircraft, in the class of club and private use, manufacturers can make their own undercarriages, most of them are of the fixed type. These specialised organisations dedicated to undercarriage design also start in the phases of development studies [1] for a new design in a similar manner to a new aircraft project conducted study, as shown in Chart 2.2.

Conceptual Aircraft Design: An Industrial Approach, First Edition. Ajoy Kumar Kundu, Mark A. Price and David Riordan.
© 2019 John Wiley & Sons Ltd. Published 2019 by John Wiley & Sons Ltd.

The location of the aircraft CG is important in laying out the undercarriage. Initially, the CG position is estimated from statistics and past experience. Once the basics of the undercarriage are explained, Chapter 10 addresses aircraft weight estimation and CG location. An iterative assessment follows to revise the undercarriage positioning due to the differences between the guessed and estimated CG location. The final iteration occurs after the aircraft is sized in Chapter 14.

The undercarriage, as a major component, creates a considerable amount of drag in its extended position during flight. Therefore, its retraction within the aircraft mould lines is necessary to minimise drag. Evolution shows that early designs of a tail-dragging type of undercarriage virtually disappeared and have been replaced by the nose wheel tricycle type. It is interesting that the first nose wheel design undercarriage appeared in 1908 on a Curtiss aircraft. The blowout of tyres during takeoff and landing is dangerous; the *Concorde* crash due to a tyre bursting was an extremely rare event but designers must learn from that situation.

This chapter covers the following topics:

Section 9.2: Introduction to the Undercarriage as a System and its Functions
Section 9.3: Types of Undercarriage
Section 9.4: Undercarriage Layout Relative to the CG, Nomenclature and Definitions
Section 9.5: Undercarriage Retraction and Stowage Issues
Section 9.6: Undercarriage Design Drivers and Considerations
Section 9.7: Undercarriage Performance on the Ground – Turning of an Aircraft
Section 9.8: Types of Wheel Arrangements
Section 9.9: Load on Wheels, Shock Absorber and Deflection
Section 9.10: Runway Pavement Types
Section 9.11: Tyre Nomenclature, Designation and Types
Section 9.12: Tyre Friction with Ground, Rolling and Braking Coefficients
Section 9.13: Undercarriage Layout Methodologies
Section 9.14: Worked-out Examples
Section 9.15: Miscellaneous Considerations
Section 9.16: Worked-Out Examples (Bizjet, AJT and TPT)
Section 9.17: Miscellaneous

Readers will make a comprehensive layout of the nose-wheel-type tricycle undercarriage and position it to fit the aircraft configured in Chapter 8. The first task is to ensure that the layout is safe and satisfies all of its functionality. The wheel and tyre are then sized to complete the layout. This section requires computational work when the aircraft CG position is still unknown. The authors recommend that the readers prepare spreadsheets for repeated calculations because iterations will ensue as the project progresses. Various undercarriage structural design concepts are given in [2] and [3]. Example of a small aircraft undercarriage design is given in [4].

9.2 Introduction

The undercarriage, also known as the landing gear, is an essential aircraft component for the following main functions:

i. Support the aircraft weight at static standing on ground.
ii. Facilitate stable movement/steering during taxi, ground manoeuvre and towing.
iii. For aircraft ground run at relatively high speed for takeoff and landing.
iv. Absorb shock on landing.
v. Braking to reduce speed.

Figure 9.1 Antanov 225 (Mriya) main undercarriage.

Associated design drivers are given next.

i. Ensure the undercarriage layout prevents tipping backwards/forward and side turn-over (Section 9.16).
ii. Ensure gear structural integrity to withstand loads on the gear at all operating conditions.

For these reasons, the authors prefer the term *undercarriage* rather than *landing gear* because the functions encompass more than mere landings. Once an aircraft is airborne, the undercarriage becomes redundant – an appendage that causes drag that can be minimised through retraction.

The main undercarriage of the world's largest aircraft (i.e. Antanov 225) is shown in Figure 9.1. It is a bogey system (see Section 9.3) carrying seven struts (i.e. support shafts with shock absorbers) per side, each carrying two wheels for a total of 32 wheels when the four nose wheels are added ($2 \times 2 \times 7 + 4 = 32$).

The undercarriage stowage bay within the aircraft is compactly sized to the extent that articulation allows. The stowage bay is located in the wing and/or the fuselage, or sometimes in the wing-mounted nacelles, depending on the realistic details of the design considered by aircraft designers at the conceptual stage. It is a challenging task for structural designers to establish a satisfactory design that integrates all the relationships and functionality of the undercarriage with the airframe. The authors recommend keeping the undercarriage layout design as simple as possible for better reliability and maintainability without using too much of the articulation and/or stowage space in an aircraft. More details are given in [1] dedicated to undercarriage design.

Large aircraft are heavy enough to damage a metal runway; therefore, its weight is distributed over many wheels on a bogey system, which itself has articulation for retraction. The undercarriage mass can encompass as much as 7% (typically 4–5%) of the maximum takeoff mass (MTOM) for large aircraft, it can weigh up to 3 tons with a corresponding cost of up to 5% of the aircraft total price and the drag can be 1020% of the total aircraft drag, depending on the size – smaller aircraft have a higher percentage of drag. For small, low-speed aircraft with a low-cost fixed undercarriage without a streamlined shroud, the drag could be as high as nearly a third of the total aircraft drag, incurring about no more than 5% loss of maximum airspeed (drag is proportionate to the square of the velocity). Small aircraft in very light aircraft (VLA)/club trainer class are invariably with fixed undercarriage. They are cheap to manufacture, lighter, safer and easy to maintain; for which a few knots loss of speed can be tolerated.

The undercarriage design should be based on the most critical configuration in the family of derivative aircraft offered. Generally, it is the longest one and therefore the heaviest, requiring the longest strut to clear the aft fuselage at the takeoff rotation. For the smaller version of the family, minor modifications assist in weight savings, yet retain a considerable amount of component commonality that reduces cost. In general, tyres are the same size for all variants.

Other special types of undercarriages are not addressed herein. Today, all 'flying boats' are amphibians with a retractable undercarriage. Undercarriage types are classified in Section 9.3. The Harrier VTOL (Vertical Takeoff and Landing)/STOL (Short Takeoff and Landing) and B52 aircraft have a bicycle-type undercarriage. These are difficult decisions for designers because there are no easier options other than the bicycle type, which requires an outrigger support wheel to prevent the wing from tipping at the sides. Aircraft with skids are intended for application on snow (the skids are mounted on or replace the wheels) or for gliders operating on grass fields. Some 'tail-draggers' get by with using a skid instead of a tail wheel. Special designs use takeoff carts to get airborne; however, landing is another matter.

The approach to undercarriage layout and design is related to tyre technology and airfield/runway construction technology [1]. The topic involves mechanics (kinetics and kinematics) and the role of aerodynamics is applied only when undercarriage extended/fixed. Retraction of undercarriage also retracts the role of aerodynamics.

The authors felt that to keep the undercarriage design course for classroom work and for the freshly inducted readers, the subject of airfield/runway design details (except the types of runways available) are left out. This will avoid entering into complex procedures, for example, pavement classification number (PCN)/aircraft classification number (ACN) methodology and so on, but are supplanted by an easier method; for example, load classification number (LCN)/load classification group (LCG) methodology to select tyres.

9.3 Types of Undercarriage

The undercarriage has A strut attachment point (support point) to the aircraft and it can have more than one strut. Chart 9.1 classifies the various types in an elementary way, as if each support point has one strut with one wheel, with designations similar to a common bicycle. For example, the Airbus 380 aircraft has five support points (i.e. one nose wheel, two fuselage-mounted wheels and two wing-mounted wheels) (see Figure 9.5, later) and many wheels and struts.

A nose wheel-type tricycle undercarriage is, by far, the dominant type, which is the type addressed in this book. The tail wheel type (i.e. fixed undercarriage) causes less drag, which can increase aircraft speed by 2–3%. However, on the ground, the raised nose of tail wheel type impairs forward visibility and is more prone to 'ground looping' (described in Section 9.7.1). Currently, tail wheels are adapted for some light aircraft. Also currently, all transport aircraft have three-point tricycle type undercarriages. This book deals only with the nose wheel-type tricycle undercarriage. The two types are compared Chart 9.2.

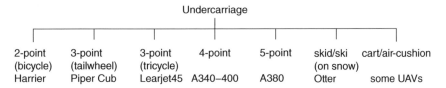

Chart 9.1 Undercarriage types (land based).

Tricycle type undercarriage (Figure 9.2a)	Tail wheel type undercarriage (Figure 9.2b)
1. Good forward vision	1. Raised nose impairs forward vision
2. Easier takeoff and landing	2. Relatively more difficult to land and takeoff
3. Stable ground manoeuvre	3. Prone to ground loop (at landing)
4. Cockpit entry for high wing is easier	4. Cockpit entry for high wing may not be as easy
5. Control stick position held steady in pull at takeoff	5. Control stick need to be pushed and then gently pulled back when the aircraft gains enough speed to attain the nose at a level position
6. Higher undercarriage drag	6. Low tail wheel drag
7. Heavier undercarriage	7. Lighter undercarriage

Chart 9.2 Three-point undercarriage (fixed type).

(a)　　　　　　　　　　　　　　　　　　　　(b)

Figure 9.2 Three-point undercarriages. (a) Three-point tricycle (Beagle Pup) type. (b) Three-point tail wheel (Auster Aiglet) type.

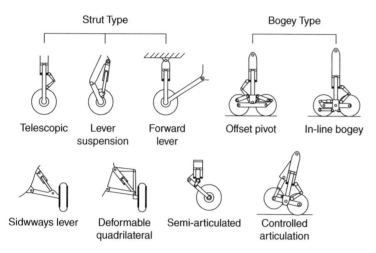

Figure 9.3 Undercarriage strut and bogey types.

Figure 9.2 shows the two types of three-point undercarriages. There is the strut unit together with a wheel called the *gear*: for the tricycle type it has one nose gear and two main gears symmetrically placed about the aircraft centreline. Each gear can have more than one wheel and main gear, and more than one strut. There are many nose and main gears as shown in Figure 9.3.

The simplest form of undercarriage was the earliest rigid axle type not in use any longer. Some form of shock absorber is favoured nowadays. Small aircraft use spring-leaf type undercarriage as a shock absorber. Struts with shock absorbers also are designed in many variations, as shown in Figure 9.3. When one strut has more than one wheel it is seen as a bogey, as shown in the figure. There is a range of bogey designs not included in the figure.

9.4　Undercarriage Description

The undercarriage is seen as a subsystem consisting of a strong support spindle (i.e. strut) with a heavy-duty shock absorber to tackle heavy landings due to a rapid descent, whether inadvertently or on the short runway length of an aircraft carrier ship. Figure 9.4a shows a typical nose gear strut with dual wheel type undercarriage and main gear with dual wheel one strut in bogey arrangement (total four wheels – Figure 9.4b).

The undercarriage has a steering mechanism with *shimmy control* (i.e. control of dynamic instability; wheel oscillation about the support shaft and strut axis). The wheels have heavy-duty brakes that cause the temperature to reach high levels, resulting in a potential fire hazard. Heavy braking requires heavy-duty tyres, which wear out quickly and are frequently replaced with new ones. Most undercarriages are designed to retract; the longer ones have articulated folding kinematics at retraction. The undercarriage retraction mechanism has hydraulic actuation; smaller aircraft may get by with an electrical motor drive.

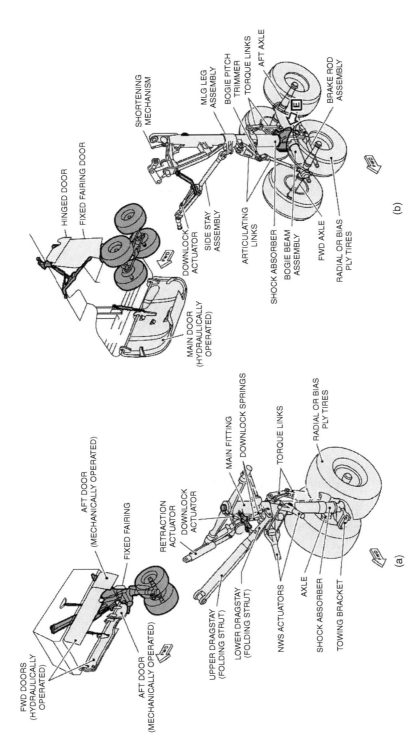

HINGED DOOR

FIXED FAIRING DOOR

SHORTENING MECHANISM

DOWNLOCK ACTUATOR

SIDE STAY ASSEMBLY

ARTICULATING LINKS

SHOCK ABSORBER

BOGIE BEAM ASSEMBLY

FWD AXLE

RADIAL OR BIAS PLY TIRES

MLG LEG ASSEMBLY

BOGIE PITCH TRIMMER

TORQUE LINKS

AFT AXLE

BRAKE ROD ASSEMBLY

MAIN DOOR (HYDRAULICALLY OPERATED)

(b)

AFT DOOR (MECHANICALLY OPERATED)

FIXED FAIRING

RETRACTION ACTUATOR

DOWNLOCK ACTUATOR

MAIN FITTING

DOWNLOCK SPRINGS

TORQUE LINKS

RADIAL OR BIAS PLY TIRES

FWD DOORS (HYDRAULICALLY OPERATED)

AFT DOOR (MECHANICALLY OPERATED)

UPPER DRAGSTAY (FOLDING STRUT)

LOWER DRAGSTAY (FOLDING STRUT)

NWS ACTUATORS

AXLE

SHOCK ABSORBER

TOWING BRACKET

(a)

Figure 9.4 Dual undercarriage strut and tyre. (a) Nose gear. (b) Main gear. Source: Courtesy: Airbus. Airbus. Aircraft Characteristics – Airport and Maintenance Planning, Airbus 330, Airbus SAS, France, January 1993, Rev. January 2014. [5]

9.4.1 Tyre (or Tire)

A tire (in American English and Canadian English) or tyre (in British English, Australian English and others) is a ring-shaped cushion covering that fits around a wheel rim to protect it and to support its weight. It enables better vehicle performance by providing a flexible cushion that absorbs shock of landings while keeping the wheel in close contact with the ground to provide necessary traction and braking during taxi, takeoff and landing. The word itself may be derived from the word 'tie', which refers to the outer steel ring part of a wooden cart wheel that ties the wood segments together.

Unlike automobile, aircraft tyres operate only for short duration (around 90 s for transport jet takeoff) and experience heavy loading conditions especially during touchdown and braking of very heavy aircraft mass. Aircraft tyres may be tube-type or tubeless. This is often used as a means of tyre classification. Tubeless tyres that are made to be used without a tube inserted inside have an inner liner specifically designed to hold air. Tube-type tyres do not contain this inner liner since the tube holds the air from leaking out of the tyre. Tyres that are meant to be used without a tube have the word tubeless on the sidewall. If this designation is absent, the tyre requires a tube. Consult the aircraft manufacturer's maintenance information for any allowable tyre damage and the use of a tube in a tubeless tyre.

9.4.2 Brakes

Early aircraft did not have brakes and relied on their slow speed and ground friction to stop, but accidents kept happening due to being unable to stop when required. As aircraft progressed to operate in higher speed, it became imperative to develop braking system to have better ground control of speed and manoeuvres with their applications. Aircraft turnings are executed with the use of differential brakes along with nose wheel steering. Today all aircraft, big or small, have brakes. There are many kinds to suit the design and economic considerations. Only main gear wheels have brakes.

The dominant brakes types are disc brakes; single disc for small aircraft and multi-disc brakes for larger ones. Most brakes are pedal operated, which are also used for rudder deflection; these two functions are interlinked along with nose wheel steering. Few aircraft designs have lever operated brakes on control sticks which are used along with pedal use.

Small aircraft brakes are mechanically operated. Large aircraft need high application force with hydraulic or pneumatic means. Absorbing high aircraft kinetic energy through the frictional force in braking develops a high amount of heat resulting in an increase in temperature and can become potential fire hazard: this happens.

Wheels, tyres and brakes require regular inspection and maintenance. Brakes are not dealt with in this book.

9.4.3 Wheel Gears

As an aircraft's weight increases, the runway must bear the reaction and retain integrity to keep the vehicle's field performance safe. Heavy commercial transport aircraft are intended to operate from a prepared runway (i.e. Types 2 and 3; see Section 9.12) to stay within the pavement strength; the load per wheel is restricted by distributing the total load over several wheels. Various arrangements for more than one wheel per strut style are shown in Figure 9.5. Aircraft and undercarriage designers must plan for the number of struts, number of wheels per strut and tyre spacing and pressure (which determine the size) to distribute the load. As the aircraft MTOM increases, so does the number of wheels required, as well as considerations for stowing and articulation for retraction.

The fundamental wheel arrangements are single, twin, triple and quadruple on a bogey. Wheel arrangements higher than a quadruple are not seen. The next level is their placement in a dual row as a single tandem, twin tandem (i.e. four wheels), triple tandem (i.e. six wheels) and so forth. Figure 9.5 shows the A380 wheel arrangement model (two main gears, two wing gears and one nose gear). Figure 9.1 shows the wheel bogey of the world's largest aircraft (i.e. the Antanov 225) with twin wheels per strut, for a total of seven struts.

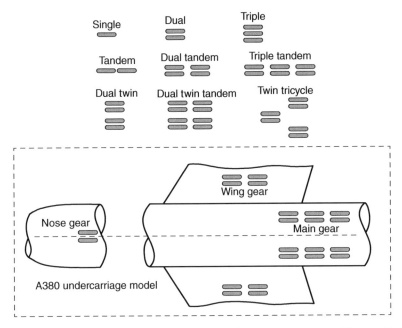

Figure 9.5 Wheel gear arrangement. (Top) Wheel gear arrangement and (bottom) A380 wheel gear arrangement.

9.5 Undercarriage Nomenclature and Definitions

Definitions of the related parameters concerning wheel and strut are shown in Figure 9.6. The position of aircraft CG plays a very important role in laying out undercarriage.

Wheel base: The distance between the front and rear wheel axles in the vertical plane of symmetry.
Wheel tread or *wheel track*: The distance between the main wheels in the lateral plane of the aircraft.

The wheel base and wheel track determine the aircraft turning radius (see Section 9.7.1) on the ground.

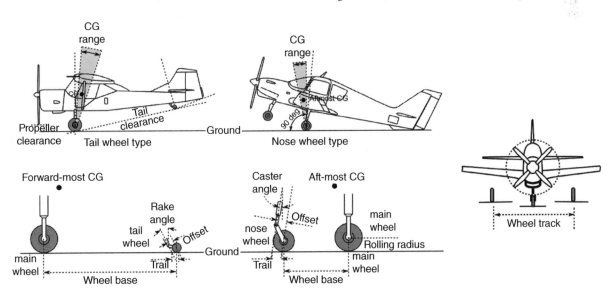

Figure 9.6 Undercarriage layout nomenclature.

9.5.1 Wheel Stability

Figure 9.7 shows the wheel parameters that affect wheel stability.

1. *Caster or rake angle.* Angle between the spindle axis and the vertical line from the ground contact point of the swivel axis.
2. *Caster length.* Perpendicular distance from wheel contact point to ground and spindle axis.
3. *Trail.* Distance from wheel contact point to ground and spindle axis contact point to ground.
4. *Offset.* Perpendicular distance from wheel axis and spindle axis.
5. *Loaded radius.* Distance from wheel axis to ground contact point under static loading.
6. *Rolling radius.* Distance from wheel axis to ground contact point under dynamic loading.

9.5.2 Alignment

Undercarriage wheels are aligned with respect to aircraft, ideally they are kept vertical (camber effects) and straight in the direction of rolling forward path (toe position). As in car wheel alignment, manufactures have set the alignment that can get distorted under impact load and steering load during operational usage, for example, in heavy landing, skidding and so on. Misalignment causes excessive tyre wear at one side and forward movement problems, for example, pulling to one side. Manufacturers give instructions to periodically check wheel alignment. Their maintenance manuals have with procedures on how to check and re-align the wheels. Figure 9.8 gives the two types of alignments.

1. *Camber*: This is the wheel tilt from being vertical. Positive camber is when the top side of the wheel tilts outward and negative camber is when the top side of the wheel tilts inward. Camber affects tyre wear and tear, the ground contact side gets affected more than the other side.

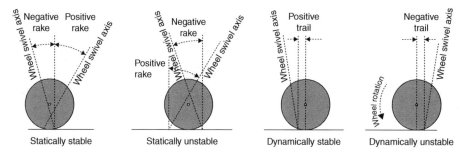

Figure 9.7 Signs for rake angle and trail angle.

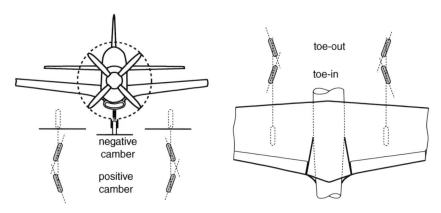

Figure 9.8 Wheel alignment.

2. *Toe position*: This is the wheel tilt in the horizontal plane. Toe-out is when the front side of the wheel tilts outward and toe-in when the front side of the wheel tilts inward. Misaligned toe position causes aircraft get pulled at a side while in forward motion.

Wheel caster as discussed before may be considered as a case for alignment. The caster does not affect tyre wear, it is concerned with steering stability. It affects straight line motion stability. Caster alignment is set by the manufacturer and stays relatively stable, requiring less frequent adjustments.

9.6 Undercarriage Retraction and Stowage

Retraction is required for aircraft operating at more than 150–200 knots. A rapid increase in drag starts building up for speeds of more than 150 knots. There are basically three retraction situations, as shown in Figure 9.9.

1. *No retraction*. The fixed undercarriage is primarily for smaller aircraft or larger aircraft that have a high wing and are operating at low speed (e.g. the Twin Otter and the Shorts 330).
2. *Partial retraction (kneeling position)*. A large wheel bogey with restricted stowage space would have to sacrifice full retraction; however, partial retraction helps considerably to reduce drag.
3. *Full retraction*. Stowage space must be provided for a wheel bogey (i.e. for higher-speed aircraft).

Provision for stowage must be made early in the conceptual design phase. Only the space provision, after consultation with structural and undercarriage designers, is sufficient at this early stage of the project. Typical extended and retracted positions of civil and military types of aircraft are shown in Figure 9.10. The following are areas where the undercarriage can be stowed:

1. *In the wing*. If wing thickness is sufficient, then a maximum of twin wheels can be retracted. Provision for the wing recess is made as early as possible in the design phase. For a thinner wing, if the strut is mounted on the wing, it can go through the wing recess and the wheel to reach the fuselage stowage space (e.g. Learjet45 type; it has a single wheel, the wing thickness does not have sufficient space).
2. *In the fuselage*. This is the dominant pattern for a large undercarriage because the fuselage underbelly could provide generous stowage space. If not, then it can be kept outside encased by a fairing that appears as a bulge (e.g. Antanov225). For fighter aircraft with a very thin wing, the entire undercarriage is mounted on and retracted within the fuselage (e.g. the F104). The coursework advanced jet trainer (AJT) example is a high-wing aircraft (see Figure 9.10) and the undercarriage is stowed in the fuselage.
3. *In an under-the-wing nacelle*. High-wing turboprop aircraft have a long strut; therefore stowing the undercarriage in the nacelle (see Figure 10.18a) slung under the wing reduces the strut length (e.g. the Fokker27 and Saab340).

Once the gear is extended, it must be locked to avoid an inadvertent collapse. A schematic retraction path of an AJT also is shown in Figure 9.10. Retraction kinematics is not addressed in this book. It is assumed that during the conceptual design phase, designers have succeeded in retraction within the stowage space

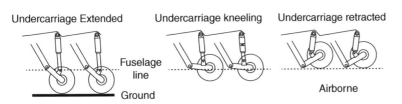

Figure 9.9 Undercarriage retraction and stowage.

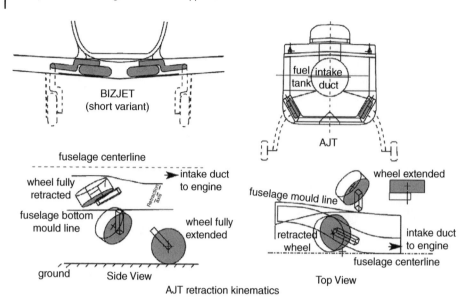

Figure 9.10 Undercarriage stowage space (Bizjet and AJT) and retraction.

provided by aircraft engineers. See [2] and [3] for more details on undercarriage retraction kinematics from aircraft engineers.

9.6.1 Stowage Space Clearances

A tyre expands as the fabric stretches during service. It also expands on account of heat generated during ground operations. It keeps spinning (further enlargement occurs due to centrifugal force of spinning) within the stowage space immediately after retraction. Stowage space within an aircraft should be of the minimum volume occupied by the retracted undercarriage with some clearance to accommodate tyre expansions. Stowage space is dictated by the articulated mechanism for retraction from its unloaded free position. Semi-empirical relations govern the clearance gap to accommodate retraction. As mentioned previously, this book assumes that aircraft designers are in a position to offer proper stowage space with adequate clearances. This book does not discuss stowage space computation. For thin-wing combat aircraft, stowage must be within the tightly packed fuselage, where space is limited.

Unless there is a breakthrough innovation (typically associated with unconventional new designs beyond the scope of this book) on retraction kinematics, the state-of-the-art undercarriage design has been established to maximise compactness. This book addresses articulation in its simplest form. The author recommends using computer aided drawing (CAD) animation to check retraction kinematics and storage space during second-term coursework.

9.7 Undercarriage Design Drivers and Considerations

There are three wheel positions, as shown in Figure 9.11. The application logic for the various types of aircraft is the same. The three positions are as follows:

i. *Normal position*. This is when the aircraft is on the ground and the undercarriage carries the aircraft weight with tyres deflected and the spring compressed.
ii. *Free position*. When an aircraft is airborne, the undercarriage spring is then relieved of aircraft weight and extends to its free position at its maximum length. Stowage space is based on the undercarriage in a free but articulated position.

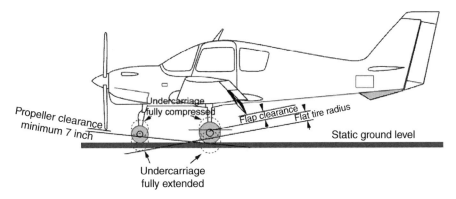

Figure 9.11 Three possible wheel positions.

iii. *Failed/collapsed position.* This is the abnormal case when the spring/oleo collapsed as a result of structural failure, as well as tyres deflated with loss of air pressure. This is the minimum undercarriage length.

The failed position of the aircraft on the ground is the most critical design driver in determining the normal length of the undercarriage strut. The following are design considerations for the failed positions:

i. *Nose wheel failed.* The nose will drop down and the length of the collapsed nose wheel should still prevent the propeller from hitting the ground with adequate clearance.
ii. *Main wheel failed.* There are two scenarios:
   a When one side fails, the wing tilts to one side and it must not touch the ground.
   b If both sides collapse (the most critical situation is when the aircraft rotates for liftoff at the end of the takeoff ground run), it must be ensured that the fully extended flap trailing edges have adequate ground clearance.

Figure 9.12 depicts an important design consideration for fuselage clearance angle, γ, at aircraft rotation for liftoff, when the CG should not go behind the wheel contact point. The fuselage clearance angle, γ, is between 10° and 14° (military trainer up to 15°) and is kept at an angle greater than stall, α_{stall}. The fuselage upsweep angle for clearance is discussed in Section 5.4.1 and it is revised here after the undercarriage layout is completed. The angle β is the angle between the vertical and the line joining the wheel contact point with the ground and the aircraft CG. Ensure that β is greater than γ; otherwise, the CG position will go behind the

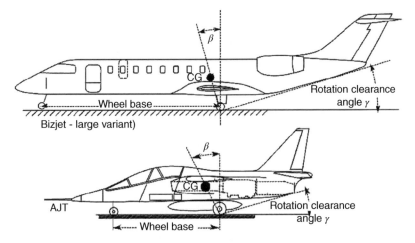

Figure 9.12 Positioning of main wheels and strut length.

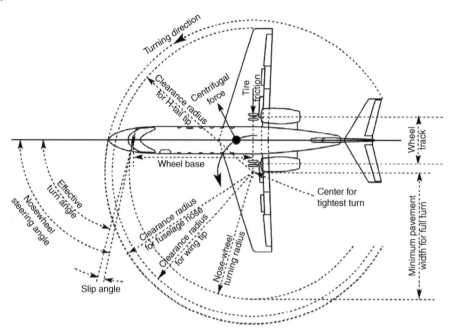

Figure 9.13 Aircraft turn.

wheel contact point. Typical values of angle β are about 15–16°. Both civil and military aircraft types are shown in the figure.

9.7.1 Turning of an Aircraft

Aircraft designers must ensure that an aircraft can turn in the specified radius within the runway width (Figure 9.13). Turning is achieved by steering the nose wheel (i.e. the maximum nose wheel turn is $\approx 78°$) activated by the pilot's foot pedal. There is a slip angle and the effective turn would be approximately 75°.

Pressing the left pedal would steer the nose wheel to the left and vice versa. The tightest turn is achieved when asymmetric braking and thrust (for a multiengine aircraft) are applied. The braked wheel remains nearly stationary. The centre of the turn is slightly away from the braked wheel (see Figure 9.13) and the steered nose wheel guides the turn. The radius of the turn is the distance between the nose wheel and the centre of the turn. Checks must be made to verify that the aircraft nose, outer wing tip and outer H-tail tip are cleared from any obstruction. If the inner wheels were not braked, the turning radius would be higher. Turning is associated with the centrifugal force at the CG and side force at the turning wheels.

Ground Loop. A tail wheel aircraft turning poses a special problem for 'ground looping', particularly when the aircraft is still at speed after landing. If the tail of the aircraft swings out more than necessary in an attempt to keep the aircraft straight using pedal-induced turns, then the centrifugal force of the turn could throw the aircraft rear end outward to the point where the forward-momentum component could move outside the wheel track. This results in instability with an uncontrolled ground loop, which can tilt the aircraft to the point of tipping if the turn-over angle θ (discussed in Section 9.16) is breached.

9.8   Tyre Friction with the Ground: Rolling and Braking Friction Coefficient

Ground movement would experience friction between the tyre and the ground. During the takeoff run, this friction is considered drag that consumes engine power.

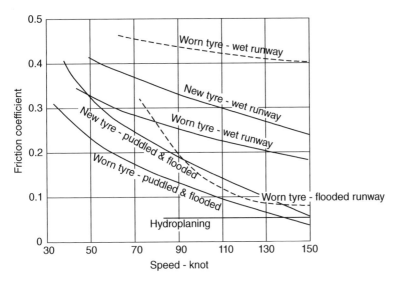

Figure 9.14 Ground friction coefficient.

Table 9.1 Average braking coefficient, μ_b (may interpolate in between).

Aircraft speed – mph	20	40	60	80	100
Dry concrete runway, μ_b	0.85	0.77	0.67	0.57	0.46
Wet concrete runway, μ_b	0.56	0.44	0.35	0.28	0.23
Iced runway	0.1–0.2	0.1–0.2	0.1–0.2	0.1–0.2	0.1–0.2
Tyre load is based on brake coefficient of 0.8					

Figure 9.14 is a representation of the ground-rolling friction coefficient, μ, versus aircraft speed for various types of runways. Conceptual studies use the value for the friction coefficient, μ.

A Type 3 runway (concrete pavement) = 0.02–0.025 (0.025 is recommended for coursework).
A Type 2 runway = 0.025–0.04 (0.03 is recommended for coursework).
A Type 1 runway = 0.04–0.3, depending on the surface type, is as follows:

> hard turf = 0.04
> grass field = 0.04–0.1 (0.05 is recommended for a maintained airfield)
> soft ground = 0.1–0.3 (not addressed in this book)

The braking friction coefficient, μ_b, would be much higher depending on the runway surface condition (e.g. dry, wet, slush, or snow or ice-covered) Table 9.1 shows what the typical value is for μ_b. Locked wheels skid that wear out a tyre to the point of a possible blowout. Most high-performance aircraft that touchdown above 80 knots have an antiskid device when the μ_b value could be as high as 0.7. Slipping wheels are not considered during the conceptual study phase. Tyre-tread selection should be compatible with the runway surface condition (e.g. to avoid hydroplaning). The braking friction coefficient, μ_b = 0.45–0.5, is the average value used in this book.

9.9 Load on Wheels and Shock Absorbers

In its elementary representation, the undercarriage system acts as a *spring-mass* system, shown in Figure 9.15. Shock absorption is accomplished by its main spring and, to a smaller extent, by the tyre pneumatics. Both

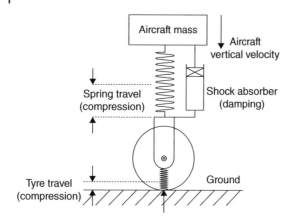

Figure 9.15 Undercarriage as a *spring-mass* system (oleo-pneumatic).

spring and tyre deflect under load. The oleo system acts as a damper; that is, it dissipates kinetic energy of vertical velocity. The strut can act as a spring for the lateral load of the ground friction.

The length of the strut is influenced by the extent that its shock absorber is compressed to the maximum. The minimum strut length is when both tyre and shock absorber collapse simultaneously, yet provide sufficient ground clearance for flaps fully extended (see Figure 9.11). The most critical situation for flap clearance is when the main wheel has collapsed and the nose wheel is at the fully extended position. (In a practical situation, the nose wheel tyre would also remain deflected under load, but the margin of the fully extended position is safer.) The flap trailing edge is at its lowest at aircraft rotation for liftoff. A simultaneous failure of the tyre and shock absorber after decision speed V_1 (see Chapter 15) would force the pilot to continue with the aircraft rotation and liftoff.

Landing impact load is absorbed by undercarriage gear mounted shock absorbers. During landing, as lift is depleting with speed reduction, more aircraft weight is reacting at the ground contact, which increases the spring load of the strut. The energy is stored in the spring. On brake application, the kinetic energy of the aircraft is absorbed by the brake pads, increasing temperature. If the limits are crossed with rapid deceleration, a fire hazard exists.

9.9.1 Load on Wheel Gears

The load on the wheel gears determine the tyre size. *Wheel gear load* is the aircraft weight distributed over the number of wheels. The aircraft CG position could vary, depending on the extent of payload and fuel-load distribution. Therefore, both the forward-most and aft-most CG positions must be considered. (Table 9.4, later, provides an idea of the A380 load.)

As soon as the preliminary undercarriage information is known from the methodology described in this chapter, aircraft weights and the CG can be estimated through the formal procedure described Chapter 10.

Estimating the aft-most CG with the angle $\beta \approx 15$ coinciding with $\approx$40% of the mean aerodynamic chord (MAC) gives a preliminary idea of the main-wheel position relative to the wing. The wing position relative to the fuselage could change when the formal weight and CG estimations are determined after the wing is sized. In that case, the wheel load calculation must be revised. For transport aircraft design, at this stage, the forward-most CG is 20–25% of the MAC ahead of the aft-most CG. For the non-transport category, including combat aircraft design, at this stage the forward-most CG is 15% of the MAC ahead of the aft-most CG. The MTOW rather than the MTOM is used in the computation because the load is a force. (A simplified approach is to divide the main and nose wheel loads as 90% and 10% distribution, which has a reasonable result, but the authors recommend making the formal estimation at the beginning.)

Linear distance is represented by l with associated subscripts; R represents reaction forces. For more than one wheel, the load would then be divided accordingly.

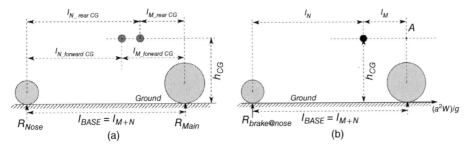

Figure 9.16 Wheel load. (a) Static wheel load. (b) Nose wheel load due to braking at main wheel.

9.9.1.1 Static Condition

See Figure 9.16a. The force balance gives:
  In the static condition

$$MTOW = 2 \times R_{\text{MAIN}} + R_{\text{NOSE}} \tag{9.1}$$

Main wheel

The load distribution between the wheels is shared as follows.
  Taking moment about the nose wheel,

$$l_{\text{BASE}} \times R_{\text{MAIN}} = l_{\text{N}} \times MTOW$$

or

$$R_{\text{MAIN}} = (l_{\text{N}} \times MTOW)/l_{\text{BASE}} \tag{9.2}$$

  Taking moment about the main wheel,

$$l_{\text{BASE}} \times R_{\text{NOSE}} = l_{\text{M}} \times MTOW$$

or

$$R_{\text{NOSE}} = (l_{\text{M}} \times MTOW)/l_{\text{BASE}} \tag{9.3}$$

CG at the aft limit

To compute the maximum main-wheel load at the aft-most CG position, take the moment about the nose wheel. The moment equilibrium equation becomes:

$$l_{\text{BASE}} \times R_{\text{MAIN}} = l_{\text{N\_REAR CG}} \times MTOW$$

or

$$R_{\text{MAIN}} = (l_{\text{N REAR CG}} \times MTOW)/l_{\text{BASE}} \tag{9.4}$$

CG at the forward limit

To compute the maximum nose wheel load at the forward-most CG position, take the moment about the main wheel. The moment equilibrium equation becomes:

$$l_{\text{BASE}} \times R_{\text{NOSE}} = l_{\text{M\_FORWARD CG}} \times MTOW$$

or

$$R_{\text{NOSE}} = (l_{\text{M\_FORWARD CG}} \times MTOW)/l_{\text{BASE}} \tag{9.5}$$

  Normally, the nose wheel has one strut.

Nose wheel load

Nose wheel load has to add the dynamic loading due to main wheel brake application at landing, as shown next.

9.9.1.2 Dynamic Condition – Brake Application – Nose Wheel Load

Additional load comes in the main wheel when the brake is applied on landing or at an aborted takeoff when the moment about ground contact point of the main wheel gives vertical force on the nose wheel through its struts (Figure 9.16b). Typical deceleration, '*a*' on account of brake application is as follows.

Aircraft deceleration (negative sign) aircraft inertia force acting through CG,

$$F_{\text{a/c\_brake}} = -(aW/g) \tag{9.6}$$

Consider only the forces associated with braking (other wheel loads not considered here) as shown in Figure 9.16b. Taking the moment about point, *A*, where the main wheel strut axis meets the CG level, this gives the normal force on the nose wheel due to braking ($F_{\text{N\_brake}}$) and can be computed as follows.

$$\sum_{\text{Moment}} = -(aW/g) \times h_{\text{cg}} + (F_{\text{N\_brake}}) \times l_{\text{base}} = 0.$$

or

$$(F_{\text{N\_brake}}) \times l_{\text{base}} = (aW/g) \times h_{\text{cg}}$$

or

$$(F_{\text{N\_brake}}) = [(aW/g) \times h_{\text{cg}}]/(l_{\text{base}}) \tag{9.7}$$

Therefore, load distribution on wheels due to braking in dynamic conditions is as follows.or

$$R_{\text{MAIN}} = (l_{\text{N}} \times MTOW)/l_{\text{BASE}} \tag{9.8}$$

or

$$R_{\text{NOSE}} = (l_{\text{M}} \times MTOW)/l_{\text{BASE}} + [(aW/g) \times h_{\text{cg}}]/(l_{\text{base}}) \tag{9.9}$$

Civil aircraft have a typical brake friction coefficient, μ, between 0.35 and 0.45, developing deceleration, '*a*', at around 10–12 ft s^{-2}. In the case of military aircraft, brake friction coefficient, μ, is higher, around 0.5 developing deceleration, '*a*' around 12–14 ft s^{-2} for training aircraft and 14–18 ft s^{-2} for combat aircraft. Take the average brake deceleration for classroom examples. Tyre manufacturers recommend to take a value that is 5–7% higher as the built-in margin. The nose wheel typically has one strut (a 225 has two struts).

For more strut on main wheel,

$$L_{\text{M}} = R_{\text{MAIN}}/\text{number of struts} \tag{9.10}$$

Ensure that the load at the nose gear is not too high (i.e. no more than 20% of the MTOW) to avoid a high elevator load to rotate the aircraft for liftoff at takeoff. Also, it must not be too low – that is, not less than 8% of the MTOW; otherwise, there could be steering problems. In the case of nose gear, the critical load is when CG is at its forward limit with brake applied and the main gear critical load is when CG is at its aft limit.

9.10 Energy Absorbed

Both the tyre and the shock absorber absorb the energy to cushion the impact of an aircraft's vertical descent rate on landing in order to maintain structural integrity and avoid the tyre bottoming out (Figure 9.15). Federal Aviation Administration (FAA) safety requirements limit the vertical descent velocity, V_{Vert}, for civil aircraft applications; military specifications limit military applications. Table 9.2 lists limits for various types of aircraft. In turn, V_{Vert} produces a *g*-load at the sudden termination of V_{Vert} at landing – it can be expressed as load factor *n* (see Section 17.5). Equation 17.4 gives $n = (1 + a/g)$ (Table 9.2); it is loosely termed the number of

Table 9.2 Vertical speed.

$V_{\text{Vert}} = < 12$ fps – FAR 23, $n = 3$
$V_{\text{Vert}} = < 12$ fps – FAR 25, $n = 2$
$V_{\text{Vert}} = < 10$ fps – Military transport, $n = 2$
$V_{\text{Vert}} = < 13$ fps – Military trainer, $n =$ maximum 5
$V_{\text{Vert}} = < 17$ fps – Military land-based combat aircraft, $n =$ maximum 6
$V_{\text{Vert}} = < 22$ fps – Military naval (aircraft carrier) -based combat aircraft, $n = 8$

the g-load at the aircraft CG. For undercarriage g-load, it is designated n_l. During landing, n_l takes a positive value.

Take $n_l = (n - 0.67)$ [1]. To keep the safety margin, undercarriage designers keep the load factor below the certification limits.

Table 9.2 gives the values of vertical speeds, V_{Vert}, and the associated certification aircraft load factors 'n' limits (the FAA has a semi-empirical formula for the exact rate of V_{Vert}).

These are extreme values for safety; in practice, 4 fps is a hard landing in a civil aircraft operation. The maximum landing aircraft mass M_L $(= W_L/g)$ is taken as 0.95 MTOM for aircraft with a high wing-loading.

The aircraft kinetic energy from vertical velocity is to be absorbed as given Equation (9.11).

$$E_{\text{ab}} = 0.5 \times M_L \times V_{Vert}^2 \tag{9.11}$$

This is the energy to be absorbed by all the main wheels (number of tyres, m_{tyre}) and struts (number of struts, m_{strut}) at touchdown during landing. The nose wheel touches the ground much later, after the main wheels have already absorbed the impact of landing. Using Equation (9.11), the following is obtained.

$$E_{\text{ab}} = E_{\text{ab\_strut}} + E_{\text{ab\_tyre}} = 0.5 \times M_L \times V_{Vert}^2 \tag{9.12}$$

9.10.1 Energy Absorption by Strut

Let m_{strut} be the number of struts. Assume that a landing is even and all struts have equal deflection of δ_{strut}. Then, using Equation (9.11), energy absorbed by all the struts at aircraft landing weight, W_L $(= M_L g)$. Noting that aircraft is still in motion and has lift that acts as some relief to strut energy absorption and stays as potential energy.

$$E_{\text{ab strut}} = m_{\text{strut}} \times n_l \times W_L \times k_{\text{strut}} \times \delta_{\text{strut}} - (W_L - L) \times m_{\text{strut}} \times \delta_{\text{strut}} \tag{9.13}$$

where k_{strut} is an efficiency factor representing the stiffness of the spring and has values between 0.5 and 0.8 (imperial system), depending on the type of shock absorber used.

In this book, 0.7 is used for modern aircraft and 0.5 is used for small club and homebuilt categories.

9.10.2 Energy Absorption by Tyre

Let m_{tyre} be the number of tyres. Assume that a landing is even and all tyres have an equal deflection of δ_{tyre}. As in the case of the strut, the aircraft is still in motion and has lift that acts as some relief to tyre energy absorption and stays as potential energy. Then, using Equation (9.11), energy absorbed by all the tyres is

$$E_{\text{ab\_tyre}} = m_{\text{tyre}} \times n_l \times W_L \times k_{\text{tyre}} \times \delta_{\text{tyre}} - (W_L - L) \times m_{\text{tyre}} \times \delta_{\text{tyre}} \tag{9.14}$$

where $k_{\text{tyre}} = 0.47$ (imperial system) is an efficiency factor representing the stiffness of all types of tyres.

This is the energy to be absorbed by all the main wheels' tyres (m_t wheels) and struts (m_s struts) at touch down during landing. The nose wheel touches the ground much later when the main wheels have already absorbed the impact of landing as shown in Equation (9.13).

Substitute Equations (9.13) and (9.14) into Equation (9.12) and the following expression is obtained.

$$E_{ab} = 1/2 M_L \times V^2_{Vert} = m_{strut} \times n_l \times W_L \times k_{strut} \times \delta_{strut}$$
$$+ m_{tyre} \times n_l \times W_L \times k_{tyre} \times \delta_{tyre} - n_l \times [(W_L - L) \times (m_{strut} \times \delta_{strut} + m_{tyre} \times \delta_{tyre})] \tag{9.15}$$

Simplifying and substituting the values of k_{strut} and k_{tyre}, the following is obtained.

$$\tfrac{1}{2} \times V^2_{Vert}/g + n_l \times [(1 - L/W_L) \times (m_{strut} \times \delta_{strut} + m_{tyre} \times \delta_{tyre})]$$
$$= (0.7 \times m_{strut} \times \delta_{strut} + 0.47 \times m_t \times \delta_{tyre}) \tag{9.16}$$

All variables except δ_{strut} and δ_{tyre} of Equation (9.16) are known. Tyre deflection δ_{tyre} is dealt with in Section 9.15.2.

9.10.3 Deflection under Load

The total vertical deflection of the main gear strut and tyre during the landing can be computed by using Equation (9.17). Other types of lateral strut deflection during turning and other manoeuvres are not addressed in this book. Total deflection is (see Figure 9.15):

$$\delta = \delta_{strut} + \delta_{tyre} \tag{9.17}$$

Equation (9.25, later) gives (see Section 9.15.2):

$$\delta_{tyre} = (\text{maximum radius at no load}) - (\text{minimum radius under static load}) = D/2 - R_{load},$$

where R_{load} equals the radius of the depressed tyre under load. It can be expressed as a percentage of the maximum radius. For the tyre footprint and inflation pressure, δ_{tyre} can be obtained from tyre data.

Once δ_{tyre} is known, Equation (9.16) computes strut deflection δ_{strut} for a given aircraft configuration at its landing weight, W_L ($= M_L g$).

It is recommended that a cushion be kept in the strut deflection (compression) so that the ends do not hit each other. In general, 1 in. (2.54 cm) is the margin.

9.11 Equivalent Single Wheel Load (ESWL)

Tyre selection is based on the most critical gear load on it (see Section 9.9.1). Tyre selection must match with the runway strength so that the runway surface does not get damaged by gear load subjected on it. Tyre inflation pressure and its foot print on the runway pavement surface contributes to the tyre selection process. Given next are the definitions of terminologies associated with runway pavement surface characteristics required for tyre selection.

9.11.1 Floatation

A heavy aircraft with one wheel per gear may inflict load on runway surface causing the wheel to 'sink' in to it. An airport runway surface must be designed to withstand an aircraft's weight in all operating conditions. Runway pavement loading is known as *floatation*, that is, the aircraft does not sink into the runway surface but sustains operation by floating over it. It is achieved by having a sufficient number of wheels with matching tyres in a cluster to distribute the load on each wheel to have floatation.

Figure 9.5 shows various types of wheel cluster on undercarriage gears. Almost all aircraft nose gear have one strut (A 225 has two struts) but can have many wheels; a dual bogey type has four wheels. The main gears, which carry more than 80% of MTOW, can have many struts with many wheels. It has been found that not all wheels in a cluster per strut carries the same load. This raised the concept of the ESWL. Based on test results and backed by theory, ESWL can be obtained using semi-empirical relations to have an equal load distribution per wheel that, when totalled, will cover the strut load.

Definition: ESWL is the calculated load on a single wheel that will produce the same effect on the runway as the cluster of wheels with similar kinds of tyres, under the same inflation pressure, on a strut will produce. To compute load on a tyre, this is a convenient method that considers ESWL, which takes a conservative approach as all tyres take may not take load in equal distribution.

For more than one wheel per strut, the load per tyre, ESWL calculates what each tyre would produce on the same runway pavement in terms of stress at the same tyre pressure as a single wheel. This is because loads are not shared equally when arranged side by side unlike tandem arrangements. Wheel arrangements would decide ESWL as given here based on statistical means. Readers may consult references for more details on other types of wheel arrangements.

$$\text{Tandem twin wheel, ESWL} = \text{Load per strut}/2 \tag{9.18}$$

$$\text{Side by side dual twin wheel, ESWL} = \text{Load per strut}/(1.5\text{--}1.33)\ (\text{this book assumes } 1.5) \tag{9.19}$$

$$\text{Tandem triple wheel, ESWL} = \text{Load per strut}/3 \tag{9.20}$$

$$\text{Side by side triple wheel, ESWL} = \text{Load per strut}/(1.5\text{--}1.33)\ (\text{this book takes } 1.5) \tag{9.21}$$

$$\text{Twin Tandem (4 wheels), ESWL} = \text{Load per strut}/(3\text{--}2.67) \tag{9.22}$$

Remember that main wheel loads are calculated based on the aft-most CG position and nose wheel loads are based on the forward-most CG position. Take the dynamic load on the nose wheel to be 50% higher than the static load.

9.12 Runway Pavement

Aircraft need to operate from airfield runways to takeoff and land. Runways are therefore prepared in a standard construction method laid down by International Civil Aviation Organization (ICAO)/FAA/MOD to the specified standards for the type planned for. Figure 9.17 gives the types of runways in use.

Among airports, the runway pavement strength varies. A runway is constructed over natural soil of the location and is known as the *subgrade* layer, which differs in characteristics from place to place. The construction is composed of layers (maximum of three for the strongest type) of materials. Layers transfer aircraft load from the top layer to the bottom sub-layer, the load intensity dissipates from maximum at the top layer to low acceptable level. The load-bearing capacity of the top layer has the highest strength and steps down to

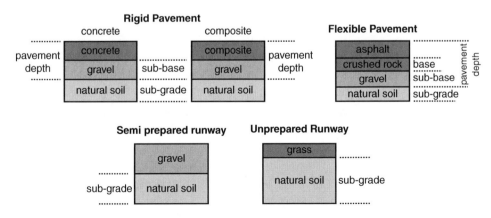

Figure 9.17 Runway classification.

the lowest capacity at the subgrade layer. Cost of the runway construction is proportional to its load-bearing capacity. There are three main types of surfaces [7], (i) Type 1, the unprepared/semi-prepared runway with gravel/grass layer at the top; (ii) Type 2, the *Flexible Pavement* with an asphalt layer at the top and (iii) Type 3, the *Rigid Pavement* with a concrete or composite layer at the top, as shown in Figure 9.17 followed by their descriptions.

1. **Type 1:** *Unprepared Surface.* A grass field or a gravel field, for example, is designated as a Type 1 surface. These are soft runways that are prone to depressions under a heavy load. Low-pressure tyres with a maximum 45–60 lb. per square inch (psi) and a total ESWL load less than 10 000 lb. are the limits of operation on a soft runway. The ground friction is the highest and these airfields are not necessarily long. This type of runway is the least expensive to prepare and they serve remote areas as an additional airfield close to a business centre or as a private airfield. Small utility aircraft can operate from Type 1 airfields.
2. **Type 2:** *Flexible Pavement (Prepared Macadam) Surface.* These are asphalt or tar-topped runways with strength built in by the thick macadam (crushed rocks) filler. These surfaces are less expensive to prepare by using a heavily rolled macadam filler. However, local depressions can cause the surface to undulate, hence they are known as a 'Flexible Pavement'. It requires frequent maintenance with longer downtime. This type of runway can accommodate heavy aircraft such as the B747.
3. **Type 3:** *Rigid Pavement (prepared concrete surface).* This is a rigid concrete runway. These runways are built with pavement-quality concrete (i.e. about a half a metre thick) and are covered by asphalt (e.g. 150 mm thick). All major international airports have Type 3 runways, which can take a load similar to a Type 2 surface and do not have to be as thick. This type is expensive to prepare and maintenance downtime is minimal. Cracks are the typical type of failure that occurs. A Type 3 surface can accommodate heavy aircraft such as the B747 and the A380.

9.13 Airfield/Runway Strength and Aircraft Operating Compatibility

The undercarriage design depends on how the wheels interact with the airfield surface. Aircraft designers must design aircraft to be compatible with designated strength of airfields in order to operate. If the market demand necessitates larger and heavier aircraft, then designers must make the aircraft comply with the pavement strength of existing airfields or the airfield must be reinforced to accept the heavy aircraft. Runway reinforcement depends on new designs; therefore, airport authorities communicate with aircraft manufacturers to remain current with market demand. When the B747s began operating, almost all international airfields needed reinforcement to accept them – some were not operational for several years. There are various methods to determine runway strength, the prevailing ones are briefly discussed next.

9.13.1 California Bearing Ratio (CBR)

The CBR gave the test value of the subgrade soil strength for highway construction. CBR [7] is the ratio of penetration of the subgrade soil (also used for sub-base soil) under test to that of a standard material (crushed lime stone). In the earlier days, these were also used for aerodrome runway construction to determine aircraft tyre pressure. The method was relatively cumbersome and required knowledge of the runway properties in order to make the tyre selection.

9.13.2 Pavement Classification Number (PCN)

As runway construction technology advanced from Type 1 to Type 3, the ICAO proposed in 1977 (implemented in 1983) the PCN assigned to each type of runway (both rigid and flexible types), which uses CBR test values. CBR tests are still carried out to get the subgrade soil strength. PCN takes in to account runway technology that represents the runway strength to choose aircraft tyre and inflation pressure that must be compatible with the type of aircraft.

PCNs correspond to runway properties and their strengths. Transport aircraft operate from a rigid runway and flexible runway; their airport authorities are responsible to assign the PCN of their runways. The runway identification number is in a series starting with the airport's numerical PCN value, followed by '/' with letter 'R' (rigid) or 'F' (flexible) type, followed by '/' with letter indicating subgrade category ('A' for high, 'B' for medium, 'C' for low or 'D' for ultra-low), followed by '/' with letter indicating tyre pressure category ('W' for no limit, 'X' up to 217 psi, 'Y' up to 145 psi, or 'Z' up to 73 psi) and finally followed by '/' with another letter indicating the method of evaluation with letter 'T' for technical study or 'U' from aircraft experience.

9.13.3 Aircraft Classification Number (ACN)

The ACN gives the effect of an aircraft on the runway pavement for a specified standard subgrade soil strength, as defined by ICAO, corresponding to a particular pavement thickness (rigid or flexible type). Aircraft manufacturers supply ACNs for each class of aircraft for various conditions for the operators to choose suitable tyres. The PCN takes in to account runway technology that represents the runway strength to choose aircraft tyre and inflation pressure that must be compatible with the type of aircraft, with the ACN, to be operated on the runway. Therefore, it is possible for a type of aircraft to operate on different airfields as long as it matches with the runway pavement PCN.

9.13.4 ACN/PCN Method

ACN/PCN methodology is a standard approach applicable to larger aircraft under FAR25 certification, now prevalent universally [7]. According to the design manual, the ACN/PCN method is intended only for publication of pavement-strength data in the *Aeronautical Information Publication* (AIP). It is not intended for design or evaluation of pavements, nor does it contemplate the use of a specific method by the airport authority for either the design or evaluation of pavements. The ACN/PCN method is more elaborate and involved. According to the AIP, 'The ACN of an aircraft is numerically defined as two times the derived wheel load, where the derived single wheel load is expressed in thousands of kilograms'. FAA also uses a modified PCN-LCN method with their own rating number, taking into account many other operational factors, which is not dealt with in this book.

9.13.5 Load Classification Number (LCN) and Load Classification Group (LCG)

In 1950, the UK introduced the LCN/LCG method, which was subsequently adopted by the ICAO in 1956. Subsequently, in 1983, ICAO introduced a new classification system, known as the ACN, which represents the aircraft tyre-loading limit and another system that represents the airfield pavement-strength limit, known as the PCN. Both numbers must be the same to operate at an airport without any restrictions.

The ICAO replaced the LCN/LCG method with the ACN-PCN method. However, the simple and effective LCN/LCG method, adopted by ICAO, is still in use in some countries and conversion is needed to use the ACN/PCN method. Given next is a short description of the LCN/LCG method.

A system of classification of the supporting capacity of pavements indicates their ability to support loads without damaging the surface. The LCN is dependent on the gear geometry, tyre pressure and the composition and thickness of the pavement. To simplify further, aircraft manufacturers assign the ESWL of any aircraft expressed in terms of LCN. An airfield is also assigned with an LCN based on airfield strength based on its construction technology adopted. Thus, the ICAO, as an international agency, established ground rules to match aircraft and runway performance requirements.

The ICAO developed the strength classifications of Type 2 and Type 3 runways by designating a LCN that represents the extent of load that a runway can accommodate based on construction characteristics. All Type 2 and Type 3 runways must have a LCN and the aircraft undercarriage design must comply with it. The LCN range of the airfield's type is grouped under the LCG as shown in Table 9.3. For example, an aircraft with the LCN 62 can operate on any airfield with an LCG of I–III. If the LCN of an airfield pavement is larger than the LCN of the aircraft, the aircraft can safely use the pavement. To express the capacity of pavement as a single number, the standard load classification was introduced (shown in Table 9.3). The LCN is airfield-specific and aircraft must comply with it. The method does not differentiate between rigid and flexible runway (Table 9.3).

Table 9.3 Load classification group.

LCN range	LCG	LCN range	LCG	LCN range	LGG
101–120	I	31–50	IV	11–15	VI
76–100	II	16–30	V	10 and below	VII
51–75	III				

The relationship among the LCN, tyre pressure and ESWL is presented in Figure 9.18. This book uses Figure 9.18 to obtain the LCN to select the tyre. The procedure is to first obtain the LCN of the airfield in question. Then, compute the ESWL of the undercarriage. Finally, find the tyre pressure required using manufacturer's tyre chart; this provides a guideline to choose tyre size. Section 9.16.1 outlines the methodology followed by worked-out examples (see the references for more details on other types). Typical examples of aircraft complying with the LCN and the corresponding MTOM and tyre pressures are given in Table 9.4 [7]. The B757, which is twice as heavy as the B737, maintains nearly the same LCN by having more wheels to distribute load per tyre.

The authors were able to locate the Airbus publication for the largest passenger-carrying aircraft, the A380-800F model; pertinent data are listed in Table 9.5. The weight per wheel is distributed relative to the wheel arrangement. The braking deceleration is 10 ft s^{-2}. The horizontal ground load is calculated at a brake coefficient of 0.8. The main landing gears can take as much as 95.5% of the weight.

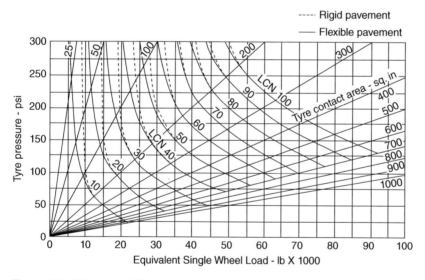

Figure 9.18 ESWL versus LCN.

Table 9.4 Aircraft weight to comply with LCN and corresponding tyre pressure.

Aircraft	MTOM – lb	Tyre pressure – psi	LCN
Fokker F27	45 000	80	19
McDonnell DC-9	65 000	129	39
B737–200	110 000	162	49
B757	210 000	157	50
B707	300 000	180	80

Table 9.5 A380 data.

Maximum Ramp Weight = 592 000 kg (1 305 125 lb).

Maximum Landing Weight = 427 000 kg (941 365 lb).

Zero Fuel Weight = 402 000 kg (886 250 lb).

Tyre size and pressure	*Maximum load per strut*[a]	*CG position*
Nose gear tyre size = 1400 × 530R23 40PR	77 100 kg (169 975 lb)	Forward-most
Nose gear tyre pressure = 11.8 bar (171 psi)	(at 10 ft s^{-2} braking)	(at 36% MAC)
Wing gear tyre size = 56 × 22R24 40PR	112 500 kg (242 025 lb)	Aft-most
Wing gear tyre pressure = 13.6 bar (197 psi)		(at 42.8% MAC)
Body gear tyre size = 56 × 22R24 40PR	168 750 kg (372 025 lb)	Aft-most
Body gear tyre pressure = 13.6 bar (197 psi)		(at 42.9% MAC)

a) Maximum load is at maximum ramp weight and at the limiting CG positions.

9.14 Wheels and Tyres

Aircraft wheels with installed tyres are aircraft components serving the following main functions.

i. Support aircraft weight at static standing on ground.
ii. Facilitate stable movement/steering during taxi and ground manoeuvre and towing.
iii. Aircraft ground run at relatively high speed for takeoff and landing.
iv. Absorb shock on landing.
v. Braking to reduce speed.

Although there are similarities with motor vehicle applications, aircraft wheels and tyres are considerably more complex on account of working under very high loading conditions. Unlike automobile applications, aircraft wheels are not directly driven, but propelled by engine thrust (propeller or reaction type jet). Large transport aircraft and high-performance military aircraft wheels and tyres are highly stressed components operating under sever conditions resulting in corrosion and damage that may contribute to wheel/tyre failures, some leading into catastrophic in nature. Wheel hub failure has occurred.

Wheels and tyres are vendor supplied components complying with specifications laid down by aircraft manufactures. This section introduces design considerations for readers to familiarise themselves with the topic and make matching tyre selections. Given in Refs [9, 12] are the main tyre manufacturers' website details, which give their tyre and wheel data.

In this section, wheels and tyres are discussed in separate subsections. Appendix F gives some supplementary information, tyre glossary and nomenclature in detail, along with tyre data used in this book (taken from Goodyear, but others are as good). Wheel brakes are not dealt with in this book.

9.14.1 Wheels

To save weight, aircraft wheels are made from either aluminium or magnesium alloys and are robust in design requiring little maintenance. Loading on wheels can be severe enough and there have been a few instances of failure. Wheels are design specific for the aircraft and are not interchangeable with other types of aircraft.

Current aircraft wheel constructions mainly consist of two general types: (i) the divided split wheel assembly and (ii) demountable wheel flange assembly. Both of these designs make a wheel easy to assemble.

9.14.1.1 Divided Split Wheel Assembly

These are modern wheels and are made in two parts, each nearly half of the total assembly and are bolted together (Figure 9.19a). An O-ring (tubewell seal) seals the rim surface of the two halves for the use of tubeless

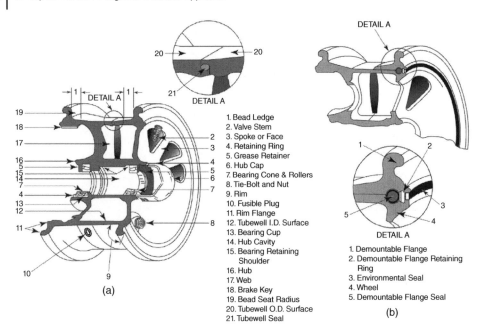

1. Bead Ledge
2. Valve Stem
3. Spoke or Face
4. Retaining Ring
5. Grease Retainer
6. Hub Cap
7. Bearing Cone & Rollers
8. Tie-Bolt and Nut
9. Rim
10. Fusible Plug
11. Rim Flange
12. Tubewell I.D. Surface
13. Bearing Cup
14. Hub Cavity
15. Bearing Retaining Shoulder
16. Hub
17. Web
18. Brake Key
19. Bead Seat Radius
20. Tubewell O.D. Surface
21. Tubewell Seal

1. Demountable Flange
2. Demountable Flange Retaining Ring
3. Environmental Seal
4. Wheel
5. Demountable Flange Seal

Figure 9.19 Aircraft wheel types. Source: Courtesy: http://www.navybmr.com/study. (a) Split wheel assembly. (b) Demountable wheel flange assembly.

tyres, which makes contact with the wheel where its bead tightly sits on the wheel flange ledge taking the load (Figure F1 in Appendix F); in the most severe case is on heavy landing. Heavy braking can generate enough heat to raise the wheel to a high temperature with the possibilities to make a wheel/tyre explode. Thermal plugs work as valves, with a melting point below the temperature limit, are provided to deflate a tyre thus preventing explosion.

9.14.1.2 Demountable Wheel Flange Assembly

The demountable flange wheels are made of one piece that has its outboard side with the flange that can be removed, hence it is called demountable (Figure 9.19b). There is a lock-ring, next to the flange seal, serving as a flange retaining fixture. Tyre change is carried out by removing the flange. These are an older type of main landing gear wheel.

9.14.2 Tyres

The main tyre manufactures in the western world are Bridgestone, Dunlop, Goodyear and Michelin.

Tyre technology developed over time into using modern fabric materials in reinforcing layers encased in rubber that are laid to provide strength. The fundamental materials of modern tyres are synthetic rubber, natural rubber, fabric and wire, along with other compound chemicals. They consist of a tread and a body. The tread provides traction while the body ensures support. Before rubber was invented, the first versions of tyres were simply bands of metal that tied around wooden wheels in order to prevent wear and tear. Today, the vast majority of tyres are pneumatic, comprising a doughnut-shaped body of cords and wires encased in rubber and generally filled with compressed air to form an inflatable cushion. Pneumatic tyres are used on many types of vehicles, such as bicycles, motorcycles, cars, trucks, earthmovers and aircraft. All tyres are identified by the *Aircraft Tyre Serial Number Codes* [9–12].

Tyres of many commercial aircraft are required to be filled with nitrogen or low oxygen air to prevent the internal combustion of the tyre, which may result from overheating brakes producing volatile hydrocarbons from the tyre lining.

9.14.3 Tyre Construction

A ply rating is used to convey the relative strength of an aircraft tyre; high ply ratings give higher strength. There are two kinds of aircraft tyre construction based on direction of ply; for example, 'bias ply' of 'radial ply' as follows. Earlier tyres were of the bias ply type, but these are now only used as a low-speed general aviation tyre. Modern high subsonic aircraft, with high ground speed, require stronger tyres of radial ply type. The 'New Design Tyre' (see Section 9.15.1) are of the bias ply type, cheaper than radial ply, stronger than earlier bias ply types and widely used in current designs of aircraft.

Figure 9.20 gives construction formats of bias ply and radial ply tyres. Although they differ in their characteristics with some fundamental constructions, there are some common layouts as follows. The difference in layout is given separately afterwards.

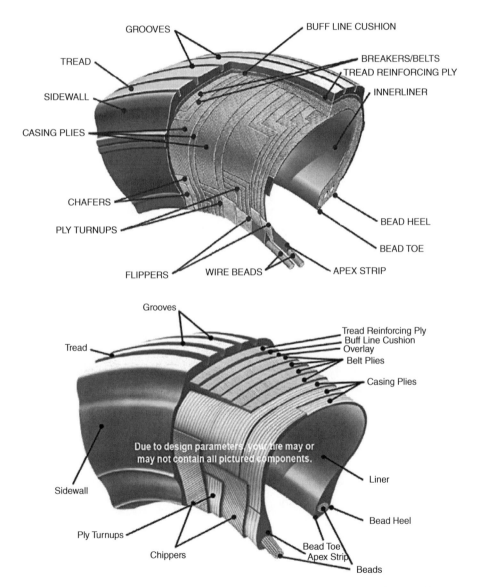

Figure 9.20 Bias ply (top) and radial ply (bottom) tyres. Source: Courtesy: Aircraft tyre and Maintenance, Goodyear June, 2016) https://www.goodyearaviation.com/resources/pdf/aviation_tire_care_june_2016.pdf.

9.14.3.1 Tyre Material

A typical tyre will be a combination of rubber, fabric and steel, each material designed to perform a specific task that helps to achieve the required tyre performance. The tyre casing incorporates fabric of polyester, nylon or rayon cords within the rubber compound casing. These cords add strength to the casing rubber.

Weight is an important factor in tyre components. Heavier tyres can develop excessive heat affecting the rubber compound. Modern polyester compounds have become a dominant material with many favourable qualities. It has high strength, offers good adhesive benefits with rubber and is relatively lighter with good heat dissipation capability. Other competing artificial fabrics, like nylon, rayon and so on, can someway compete with polyester.

9.14.3.2 Bias Ply Aircraft Tyres

These are used for both tube and tubeless types. These are cheaper to manufacture. The earlier types were tubes and used in small aircraft. The *New Design Tyres* are tubeless, widely used in bigger aircraft.

General aviation tyres utilise bias ply construction. Bias ply tyres are constructed with plies of reinforcing materials laid at angles crisscrossing between 30° and 60° (typically at 45°) to the tread centreline or rotation direction of the tyre, embedded in the rubber during the construction process (Figure 9.20). In this manner, the plies have the bias of the fabric from which they are constructed facing the direction of rotation and across the tyre. Hence, they are called bias tyres. The result is flexibility as the sidewall can flex with the fabric plies laid on the bias. This provides balanced strength to the tyre. The '4 ply' in the specification is used to indicate the number of layers, or plies, of reinforcing material, such as nylon, used in the construction of the tyre. In the old days this number was literally the number of layers. However, modern technologies utilised in tyre making, such as Kevlar cords, translate into the use of less actual plies to get the same 'ply rating'. This is better because it saves weight and creates less heat.

Bias aircraft tyres feature a casing that is constructed of alternate layers of rubber-coated ply cords that extend around the beads (see Figure 9.20) and are at alternate angles substantially less than 90° to the centre line of the tread.

9.14.3.3 Radial Ply Aircraft Tyre

These are heavy-duty tyres for both tubeless type, naturally more expensive to manufacture. Radial plies run straight from bead to bead (see Section 9.13.2).

Radial aircraft tyres feature a flexible casing that is constructed of rubber-coated ply cords that extend around the beads and are substantially at 90° to the centreline of the tread. The casing is stabilised by an essentially inextensible circumferential belt (Figure 9.20).

Operational experience has also shown that radial tyres offer a greater overload bearing capacity and withstand under-inflation better. An approximately 10% increase in the footprint area improves the floatation characteristics and reduces hydroplaning on wet runways . In addition, radial tyres do not fail as suddenly as bias tyres do. Some radial tyres currently achieve twice as many landings as conventional bias tyres [9] and [10].

In summary, tyres are divided into two groups, 'radial' and 'bias' tyres. Bias ply tyres are constructed with the carcass plies laid at angles between 30° and 60° to the centre line or rotation direction of the tyre. The succeeding plies are laid with the cord at angles that are opposite to each other. This provides balanced carcass strength. Radial ply tyres have plies at 90°, seen as radial directions to rotation, perpendicular to the sidewall. Radial ply tyres permit higher speed and load operations and are tubeless.

9.14.4 Common Construction Layout Arrangements for Both Bias and Radial Ply Tyres

The tyre casing is the basic element as the frame of the body known as the 'carcass'. At the bottom edge of the frame is a bundle of monofilament steel wires, wound in cable form, could be more than one wire, laid in circular form known as a 'bead bundle', at the bottom of the two edges of the carcass to secure tyre to the wheel. 'Bead heel' and 'bead toe' are the outer and inner edges of the bead bundle to fit secure on wheel flange.

A wedge of rubber, as filler, is affixed to the top of the bead bundle is known as the 'apex strip', which extends up to sidewall areas.

The tyre carcass is constructed in layers, that is, in plies, to form the casing of the tyre. The 'casing plies', also known as a 'carcass ply', are layers of rubber-coated fabric that provide the strength of the tyre. On how the layers are laid as plies makes the difference to the tyre types, bias or radial, as described before. The casing plies are anchored by wrapping them around the wire beads, known as the 'ply turnups'. Above the casing plies are the 'belt plies' made of composite structure of rubber-coated fabric to increase tyre strength tread of the tyre. The chord of the tyre are kept aligned with the circumference of the tyre.

On the top of the carcass is the 'tread re-enforcing casing ply' are one or more layers of fabric that help strengthen (for high-speed operation) and stabilise the rubber 'tread' area of the wheel. This is an important layer acting as the reference layer for re-treading. To cushion the tread above the carcass, a rubber compound cushion layer is laid is known as the 'buff line cushion' between the tread re-enforcing casing ply' and the 'breaker belt' (for bias ply) or 'overlay' ply (for radial ply). This rubber compound cushion layer also serves as adhesion between the top breaker/overlay and the tread re-enforcing casing tread ply and should be thick enough allowing removal for re-treading. Tread on the tyre are circumferential 'grooves', the type that characterises the traction quality of the tyre.

For tubeless tyres, whether with bias or radial ply, have 'ply liner' acting as built-in tube preventing gas to leak into casing plies.

A protective layer of flexible, known as a 'sidewall' is a weather-resistant rubber covering the outer casing ply, extending from tread edge to bead area. Circumferential protrusions that are moulded into both sides of the sidewall of nose wheel tyres, known as 'chines', are meant to deflect water sideways to prevent water ingestion into wing-mounted engines. The 'shoulder' is the area where the tread and the sidewall meet.

9.14.4.1   Difference in Layout Arrangement for both Bias and Radial Ply Tyres

Bias ply

Bias ply tyres have layers of rubberised fabric known as 'flippers' anchor the bead wires. A bias ply tyre can with tube or tubeless. For each case, the following is incorporated.

For tubeless tyre, it has a protective layer of rubber and/or fabric, known as a 'chafer', located between the casing plies and wheel to minimise chafing.

For tube-type tyre, it has a liner, known as a 'ply liner' to prevent tube chafing against the inside ply. A flexible hollow rubber ring, known as a 'ply tube' is inserted inside a pneumatic tyre to hold inflation pressure.

Radial ply

To improve the durability of the tyre in the bead area, there are layers of rubber-coated fabric, known as 'chippers', which are applied at diagonally. Radial ply tyres are tubeless and meant for high-speed operations requiring an 'overlay' layer of reinforcing rubber-coated fabric placed on top of the belts.

Additional comments

Both the 'bias' and 'radial' families have several subgroups according to application and requirements. The group of tyres that will be found on light aircraft (and indeed on most piston powered aircraft) is 'Type III' (see Section 9.15.1).

9.15   Tyre Nomenclature, Classification, Loading and Selection

The pavement-loading (i.e. floatation) limit is one of the drivers for tyre design. This section presents relevant information for preliminary tyre sizing. The FAA regulates tyre standards.

Figure 9.21 gives the nomenclature associated with tyre and wheel rims. The section width (W_G), height (H) and diameter (D), as shown in Figure 9.21. Appendix F gives details of the glossary and nomenclature.

The rim diameter of the hub is designated, d. Under load, the lower half deflects with the radius, R_{load}. The number of wheels and tyre size is related to its load-bearing capacity for inflation pressure and the airfield

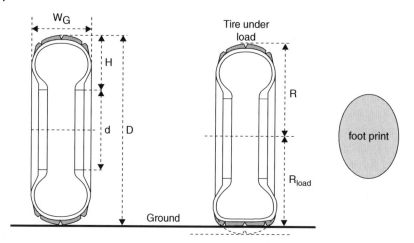

Figure 9.21 Tyre designations.

LCN for an unrestricted operation. For heavy aircraft, the load is distributed over the number of wheel and tyres.

9.15.1 Tyre Type Classification

Tyres are rated based on (i) unloaded inflation pressure, (ii) ply ratings for holding shape under pressure, (iii) maximum static load for the MTOW (i.e. floatation consideration) and (iv) maximum aircraft speed on the ground. Given next is the tyre type classification.

Types I: *Type I* is intended for a fixed undercarriage. It has a one part designation – the nominal diameter.

Type II: This type is becoming obsolete and is no longer produced.

Type III: This type includes *low-pressure* tyres that provide a larger footprint or floatation effect. They have a relatively small rim diameter (d) compared to overall tyre diameter. Speed is limited to less than 160 mph. The tyre two-part designation is expressed by its section width, W_G, and rim diameter, d (Figure 9.21). All dimensions are in inches. For example, a typical small aircraft tyre designation of 6.00–6 means that it has a width of 6.00 in. (in hundredths) and a rim diameter hub of 6 in.

Types IV, V and VI: These types no longer exist.

Type VII: These are high-pressure tyres that are relatively narrower than other types. These are produced both in bias ply and radial tyres. They are widely used in aircraft with pressure levels from 100 to more than 250 psi that operate on Type 2 and Type 3 airfields. Military aircraft tyre pressure can reach as high as 400 psi. Tyre designation is expressed by the overall section diameter (D) and the nominal section width, W_G, with the multiplication sign (×) in between. All dimensions are in inches. For example, 22 × 5.5 has an overall section diameter of 22 in. and a section width of 5.5 in.

New Design Tyre (three-part nomenclature – Type VIII): All newly designed tyres are in this classification. These are produced both as bias and radial ply tyres. These are intended for high-speed aircraft with high tyre inflation pressures. The three-part designation represents (outside diameter, D) × (section width, W_G), (followed by '–' representing bias ply) and (rim diameter, d). Dimensions in FPS are in inches and dimensions in SI are in millimetres but the rim diameter is always in inches. For example, a B747 tyre has the designation, 49 × 19.0–20 (32 ply) and this means that it has an outside diameter of 49 in., a section width of 19 in. and a rim diameter of 20 in. New tyres also have radial types; the three-part designation has an 'R' instead of a hyphen. An example of a radial tyre in SI is 1400 × 530 R 23. There is a special designation that precedes the three-part nomenclature tyres with 'H' indicating that the tyre is designed for a higher percentage deflection.

Basically, there are mainly three types of tyres, for example, (i) Type III, (ii) Type VII and (iii) New Design Tyre (Type VIII) from eight categories, as described herein (see the Michelin, Goodyear, Dunlop and Bridgestone

[9]–[12] data sourcebooks). Type III and Type VII have a two-part designation. The New Design Tyre has a three-part designation. There are small tyres not approved by the FAA that are used in the homebuilt aircraft category.

Several tyre manufacturers are available from which to choose, as in the case of the automobile industry. Tyre manufacturers (e.g. Goodyear, Goodrich, Dunlop and Michelin) publish tyre catalogues, which provide important tyre data (e.g. dimensions and characteristics) in extensive detail. Appendix E lists data from the manufacturers' catalogues needed for the coursework in this book. Aircraft designers have the full range of tyre catalogues and contact tyre manufacturers to stay informed and benefit mutually from new tyre designs.

9.15.2 Tyre Deflection

Under load, a tyre deflects and creates a footprint on the ground. Tyre footprint is approximated to oval shape with major axis is the chord length in ground contact under deflection and minor axis as the tyre width.

Area of the footprint ellipse, $A_{foot} = \pi \times$ (major axis) $\times$ (minor axis)where major axis $= 2 \times \sqrt{(R^2 - R_{load}^2)}$ and minor axis $= W_G$ (see Figure 9.20).

Then,

$$A_{foot} = 2\pi \times \sqrt{(R^2 - R_{load}^2)} \times W_G \tag{9.23}$$

Therefore:

$$\text{Load on tyre} = (\text{footprint} \times \text{tyre pressure}) \tag{9.24}$$

For tyre static deflection:

Tyre static deflection can be expressed as a percentage of the maximum free height (H). Most aircraft tires are designed to operate at 32% deflection, with some at 35%. As a comparison, cars and trucks operate in the 5% to 20% range.

Tyre free height, H, is expressed as (see Figure 9.21)

$$H = (\text{Outside Diameter} - \text{Rim Diameter})/2 = (D - d)/2$$

In this book, take tyre deflection as follows:

$$\delta_{tyre} = 0.3H = 0.3 \times (D - d)/2 \tag{9.25a}$$

For tire static deflection:

$$\delta_{tyre} = (\text{maximum radius at no load}) - (\text{minimum radius under static load}) = D/2 - R_{load} \tag{9.25b}$$

where R_{load} equals the radius of the depressed tyre under load.

As aircraft speed increases, the load also increases on tyres as dynamic loading. During landing impact, the deflection would be higher and would recover sooner, with the tyre acting as a shock absorber. Bottom-out occurs at maximum deflection (i.e. three times the load); therefore, shock absorbers take the impact deflection to prevent a tyre from bottoming out. Typical tyre pressures for the sizes are given in Table 9.6.

Tyre sizing is a complex process and depends on the static and dynamic loads it must sustain. This book addresses tyre sizing for Type 2 and Type 3 runways. One of the largest tyres used by the B747-200F has a main and nose gear tyre size of 49 × 19–20 with an unloaded inflation pressure of 195 psi. Sizes used by existing designs of a class are a good guideline for selecting tyre size.

Use of an unprepared runway (i.e. Type 1) demands a low-pressure tyre; higher pressure tyres are for a metal runway (i.e. Types 2 and 3). The higher the pressure, the smaller is the tyre size. Civil aircraft examples in this book use a Type 3 airfield; military aircraft examples use Type 2 and 3 airfields. Small aircraft use a Type 1 airfield for club usage.

Table 9.6 Typical tyre pressures for a range of aircraft weights.

Weight – lb. (kg)	Pressure – psi (kg cm^{-2})[a)]	Typical tyre size (main wheel)[b)]
<3000 (1360)	≈50 (3.52)	500–5, 600–6
≈5000 (2268)	≈25–50 (1.76– 3.52)	600–6, 700–7, H22 × 8.25–10,
≈10 000 (4990)	≈25–90 (1.76–6.33)	750–6, 850–6, 900–6, 22 × 5,
≈20 000 (9072)	≈45–240 (3.16–16.87)	850–10, 24 × 7.7, 22 × 6.6,
≈50 000 (22 680)	≈60–240 (4.22–16.87)	26 × 6.6, 30 × 7.7, 32 × 7.7, 34 × 9.9
≈100 000 (49 900)	≈ 75–240 (5.27–16.87)	34 × 11, 40 × 12, 15.50 × 20
≈200 000 (90 720)	≈100–240 (7.03–16.87)	44 × 16, 19.00 × 20, 50 × 20
≈300 000 (136 080)	≈110–240 (7.73–16.87)	50 × 20, 20.00 × 20
>500 000 (226 800)	≈150–250 (10.5–16.87)	Not known.

a) Depends on number of wheels.
b) See [9]–[12] for more options. Also consult *Jane's Manual* [8].

9.15.2.1 Inflation

Commercial aircraft tyres are pressurised with nitrogen gas or low oxygen gas for safety reasons. Inert nitrogen gas are produced dry and charged in liquid form in large tanks of mobile 'servicing unit'. Being inert, its anti-filamentary property and being dry there is no moisture content to be free from ice at very low temperatures of high altitude flight contributes to safety. Ambient temperature effects on inflation; A temperate change of 5°F (3°C) produces approximately 1 % (1%) pressure change.

They accept a variety of static and dynamic stresses and must do so dependably in a wide range of operating conditions. As in the case with automobile, aircraft tyres are carefully maintained for safety reasons so as to perform as demanded.

Provided that the wheel load and configuration of the landing gear remain unchanged, the weight and volume of the tyre will decrease with an increase in inflation pressure. From the floatation standpoint, a decrease in the tyre contact area will induce a higher bearing stress on the pavement, thus eliminates certain airports from the aircraft's operational bases. Braking will also become less effective due to a reduction in the frictional force between the tyres and the ground. In addition, the decrease in the size of the tyre, and hence the size of the wheel, could pose a problem if internal brakes are to be fitted inside the wheel rims. The arguments against higher pressure are of such a nature that commercial operators generally prefer the lower pressures in order to maximise tyre life and minimise runway stress. However, too low a pressure can lead to an accident.

9.15.2.2 Aircraft Tyre Pressure and Size

Table 9.6 lists the typical tyre pressures for the range of aircraft weights.

9.15.2.3 Wheel Spacing

Reference [1] gives a formal method to determine dual and tandem arrangement wheel spacing. It is an involved method. In this book, a simplified relation is given taken from the statistic of spacing given in currently operating aircraft. Take the wheel spacing equal to about the tyre height, *H* (see Figure 9.21).

9.16 Configuring Undercarriage Layout and Positioning

Basically, the undercarriage consists of wheels on struts attached to aircraft points. We repeat again that the position of the aircraft CG is the most important consideration when laying out wheel locations relative to an aircraft (see Section 8.2.3). The forward-most aircraft CG position relative to the wheel base and wheel track

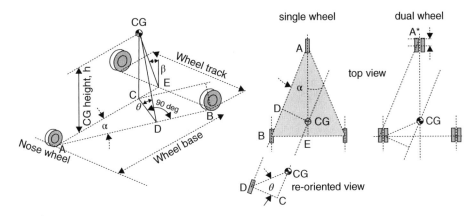

Figure 9.22 Aircraft CG position relative to the undercarriage layout (nose wheel type).

determines the aircraft turn-over characteristics. The geometric parameters in placing wheels relative to the aircraft CG position are shown in Figure 9.21, along with the basic nomenclature of related parameters.

The *turn-over angle*, θ, is the maximum angle for a tilted aircraft with the CG on top of a main wheel; beyond that, the aircraft would turn over on its side. Determination of the angle θ is shown in Figure 9.22. Turn-over tipping is not exactly around the x-axis (i.e. sideways) when a low-wing aircraft could have a wing tip touching the ground before θ is reached. The tipping occurs about the axis joining the nose wheel and main-wheel ground contact point, when the wing leading edge (LE) is likely to hit the ground.

Proper sizing of undercarriage wheel base and while track and their positioning with respect to aircraft CG will establish the aircraft ground handling stability, for example, turn over, tip over and ground looping. From Figure 9.6 it can be seen that for a tail wheel aircraft there will be tip forward if the CG moves ahead of vertical line, and for a nose wheel there will tip backwards if the CG moves behind the vertical line.

Undercarriage layout: to establish the layout geometry with associated angles, Figure 9.22 is used for the tricycle nose wheel type of configuration. It is clear that the CG should fall within the shaded area (middle diagram) at any attitude of the aircraft on ground. Note the considerations of dual wheel undercarriage is slightly different as the point 'A' is not on the nose wheel axle but slightly above it to the point A^* in the last figure. The readers should modify the geometry given here for a single wheel. To simplify, the following symbols are used.

	Linear dimension	Angular dimension
Distance A to $CG = a$	Wheel base $= AE = b$	Angle, $\llcorner BAE = \alpha$
Distance C to $CG = e$	Wheel track $= 2 \times BE = c$	Angle, $\llcorner CD\text{-}CG = \theta$
Distance $CD = f$		
Distance $AC = d$		

From Figure 9.22

$$\alpha = \tan^{-1}(c/2b) = \tan^{-1}[\text{Wheel track}/(2 \times \text{Wheel base})] \tag{9.26}$$

$$f = d\sin\alpha = d\sin\{\tan^{-1}[\text{wheel track}/(2 \times \text{wheel base})]\} \tag{9.27}$$

$$\text{Turn} - \text{over angle}, \theta = \tan^{-1}(e/f), \tag{9.28}$$

The θ limit has to be specified that should not get exceeded to prevent aircraft from turning over.

Turn-over angle, the θ value, depends on the airfield surface and the tendency increases with higher sideways ground friction. For simplification, yet still representative, typical values used in this book are as follows (see the references for more details).

For a paved runway, keep the angle θ less than 60°; for an unprepared field it should be less than 50°. There are aircraft with $\theta = 35$. Most of the aircraft have a θ between 40° and 50°. An aircraft also can tip backwards if its rearmost CG goes behind the main wheel of a tricycle type undercarriage; it can tip forward if its CG is in front of the main wheels of a tail-wheeled aircraft.

Wing position iteration: The wing may have to be moved to have the aft-most CG subtends β at 15°. This may involve some *wing chasing* as the aircraft CG will also move with changing of wing position.

9.16.1 Undercarriage Layout Methodology

After obtaining the necessary information available on the undercarriage as a system, the next step is to systematically lay down the methodology to configure the undercarriage arrangement in order to integrate it with the aircraft conceived in Chapter 8. All aircraft designers benefit from existing designs – this is what is meant by 'experience'; past designs provide a good databank.

First, the undercarriage commonality for variant designs is considered. In general, in civil aircraft design, the baseline aircraft is the middle of the three main sizes (other variants are possible). Therefore, the largest version is more critical to the undercarriage layout for carrying the heaviest load. The methodology described herein should be applied to the largest of the variants and then all other variants should be checked for commonality. At the conceptual design stage, all versions have an identical undercarriage layout except for the wheel base and wheel track. A production version has the scope to shave off metal, making it lighter for smaller variants; this requires only minor changes in the manufacturing setup.

In the following is the step-wise procedure (some iterations will be required to firm up some estimated dimensions to start with) to configure single strut undercarriage layout geometry adopted for the worked-out examples given in Section 9.17. It is to be noted aircraft CG can only be estimated at this stage as the aircraft weights and aircraft CG have yet to be established (Chapter 10).

Step 1: Determine the type of undercarriage: nose wheel (tricycle) or tail wheel. Low-speed smaller aircraft sensitive to weight and drag could have a tail wheel type.

Step 2: If the aircraft operating speed exceeds 175 kt, the undercarriage should be retractable. Decide the retraction kinematics for the storage bay. Keep retraction simple. Retraction articulation of main gear strut installed at fuselage with storage bay in fuselage belly differs from strut mounted to wing with fuselage belly storage bay.

Step 3: For low-wing aircraft, place aircraft CG height at about one-quarter of fuselage height from belly keel line. For high-wing aircraft, place aircraft CG height at about one-third of fuselage height from the belly keel line. This will be accurately determined when aircraft component weights are estimated and CG position is determined in Chapter 19.

Step 4: Place the nose wheel with strut mounted on front fuselage frame (good candidate is the bulkhead) that will give the angle α, no less than 12° and typically within 16°. This will give the undercarriage wheel base. The nose wheel may have to be adjusted to give a wheel base length that gives the aircraft turn-over angle, θ (at the forward CG limit), to a desirable value within the limits.

Step 5: To start with, place the main wheel (single strut) with respect to the fuselage (distance in percentage of fuselage length from the zero reference line placed at the nose) as follows (to be fine-tuned).

	Wing/front mounted engines	Rear mounted engines
Civil aircraft	about 45–50% of the fuselage	about at 50–55% of the fuselage
Military jet trainer aircraft		about at 55–60% of the fuselage
Military turboprop trainer aircraft	about at 40–45% of the fuselage	

Step 6: Guess the main wheel diameter from the tyre manufacturer's chart for 45% of MTOM. Place the wheel in the fuselage belly storage bay as far away from the centre line. Estimate the gear length that will give the fuselage rotational angle, γ, clearance. Together, from Step 5, the CG height, h_{cg}, from the ground can be found. For a wing-mounted strut, this will be the undercarriage wheel track. For a fuselage-mounted strut (mostly high-wing aircraft), the main wheel opens up outside the fuselage to about a quarter of fuselage width (iterate once the preliminary layout is done) – this gives the undercarriage wheel track.

Step 7: Place aft the CG limit just ahead of the angle β, which has to be greater than rotational clearance angle, γ. Readjust wing position so the aft CG limit is around 40% of wing MAC[1] (slightly more for military aircraft trainers).

Step 8: Compute the aircraft turn-over angle θ, with the CG at the forward limit (say at about 20% of wing MAC).

To consider undercarriage maintain component commonalities, there are mainly two options available for variant aircraft designs, as described next.

i. *Low-cost option.* Maintain the same undercarriage for all variants, even when there are performance penalties. In this situation, design the undercarriage for the baseline aircraft in the middle and then use it for other variants. In this option the aircraft rotational clearance angle, γ, changes. By carefully positioning the main gear location, it is possible to keep γ with small changes to an acceptable level. However, to keep commonality, it has to be designed for the heaviest aircraft, hence the smaller ones will have some weight penalty.

ii. *Slightly higher cost option.* Make minor modifications to the undercarriage. In this situation, only the three strut lengths (and shave or add wall thickness while retaining the same external geometry) for each variant are made, yet maintain maximum components and manufacturing-process commonality. This would considerably reduce costs when numerically controlled (NC) machines are used to manufacture. The shock absorber will also require some small modifications. Each variant of aircraft will have its optimised undercarriage without weight penalties. Wheel size can be kept the same. Suitable brake and tyre size options are readily available, as tyres have options for size and brakes are of modular construction. Performance gain could be maximised by this option, offsetting the slightly increased cost of production. One of the biggest gains will be that if the shortest version is light enough then a dual wheel arrangement can be made into single wheel arrangement. This is the desired option and followed in the worked-out example.

In all cases, the nose wheel attachment point remains unchanged. It is possible to keep the wheel track the same as the baseline aircraft that will help to retain retraction and stowage unaffected. However, the wheel base will differ as the wing position moves with respect to the nose gear. It may be possible that such a change will not affect aircraft handling and turn-over problems.

The industry conducts a trade-off study to examine which options offer the maximum cost benefit to operators. This book uses the second option – that is, start with the biggest variant. Iteration is likely to occur after accurate sizing.

9.17 Worked-Out Examples

The worked-out example continues with the aircraft configurations developed in Chapter 8 (both civil and military). The heaviest and the longest in the family is the most critical from the undercarriage design perspective.

---

1 At this stage, the CG limits and CG travel range are estimated using the available statistics. CFD analyses/wind tests are the best way to establish the forward and aft CG limits.

9.17.1 Civil Aircraft: Bizjet

The baseline Bizjet design is in the middle to maintain for undercarriage commonality with each variant having a slightly modified version. For the baseline Bizjet, angle clearances have $\beta = 15$ and $\gamma = 14°$. The undercarriage and tyre sizing for the family of civil aircraft variants are more involved procedures than combat aircraft because the fuselage length and weight changes are relatively high, affecting both the wheel base and the strut load.

The Bizjet undercarriage is of the tricycle type and retractable. CG travel limits are between 15 and 30% of the wing MAC (2.132 m), which is 0.32 m. The wing is its mid MAC at the middle of fuselage. The baseline MTOM is estimated to be 9500 kg (21 000 lb). It will be subsequently sized to a more accurate mass and CG position. The nose wheel load is based on the forward-most CG position.

The main-wheel attachment point is at a reinforced location on the rear-wing spar and is articulated with the ability to fold inward for retraction. It is assumed that there is no problem for stowage space in the fuselage with adequate wheel clearance.

9.17.1.1 Undercarriage Layout and Positioning
Given in Figure 9.23: fuselage length = 15.2 m (50 ft). Wing MAC = 2.132 m, its mid-point is placed at 7.6 m from the zero reference plane (ZRP). CG height from fuselage keel = 0.855 m. Typical tyre diameter = 22 in. (it will be formally selected when the loads are known).

9.17.1.2 Step-by-Step Approach
First, decide the undercarriage wheel base and wheel track geometries that have to be subsequently checked out for turn-over angle. Then guess the vertical CG location, followed by horizontal CG locations for the

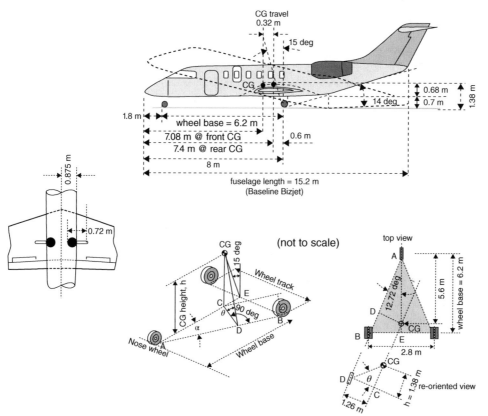

Figure 9.23 Bizjet (large variant) undercarriage positions (CG position estimated).

forward and aft travel limits. Finally, check the turn-over angle. Strictly speaking, undercarriage layout and positioning are not equation based, but based on design judgements. Undercarriage layout and position must simultaneously satisfying several limitation/constraints, for example, angle γ and β limits, CG travel limits and neutral point position with respect to angle, β, and turn-over angle θ associated with angle, α (see Figure 9.23). Some iterations are required (this example is close enough to avoid iteration. If desired, the readers may iterate for some improvement). The step-by-step approach is given next.

Wheelbase

1. Start with the nose gear ground contact point placed at 1.8 m from the ZRP mounted on the front fuselage bulkhead structure. Next, place the main gear ground contact point at 8 m from the ZRP (52.65% of the fuselage length). This gives the wheelbase = (8−1.8) = 6.2 m. Neutral point, NP, of the aircraft is estimated at 50% aircraft fuselage.

Wheel track

2. Fuselage width is 1.75 m (half width = 0.875 m). Consider at this stage tyre radius = 0.25 m and the tyre storage bay is just inside the fuselage and strut tucked into the wing. The main gear location point in the wing would be about 1.4 m from fuselage centre line accommodating a generous distance to size the main gear unloaded extended length. This will allow the main wheel to retract in the fuselage (as shown in Figure 9.10) the edge of the tyre within the fuselage fairings. This main gear extended length of 0.7 m gives an idea for the wheel track as the inward retraction will take at least twice this length. This gives a wheel track of $2 \times 14 = 2.8$ m.

CG positions

Vertical CG location

3. Vertical CG position, $\bar{z}$, is nearly at a constant (very little variation with loading status). Guess the vertical CG location at 0.68 m (38% of fuselage height of 1.78 m) from the bottom fuselage mould-line. Takeoff rotation angle of $\gamma = 14°$ must clear when the main gear is deflected under maximum permissible load. Therefore, at the least, permissible main gear load will be longer and at this stage it is estimated to be 0.7 m measured from ground contact to under the fuselage mouldline. This gives the CG height from ground level = (0.68 + 0.7) = 1.38 m.

Horizontal CG location

4. CG position moves horizontally to some extent depending on the loading status. For Biz-jet it is taken from 15 to 30% MAC, that is, 7.08−0.4 m from ZRP, respectively. This gives CG travel = $(0.15 \times 2.132) = 0.32$ m.

   Rearmost CG position must be ahead of aircraft neutral, which in turn is typically ahead of $\beta = 15°$ limiting angle preventing aircraft tipping backwards. At the CG horizontal line, 15° gives a distance of $1.38 \times \tan 15 = 0.37$ m from the vertical line, that is, 7.63 m from ZRP. Neutral point is estimated at 50% of the fuselage length, at 7.6 m from ZRP, ahead of the β angle line.

   Therefore, aircraft static margin (7.6−7.4) = 0.2 m; that is, 9.3% of MAC.

Checking aircraft turn-over angle (refer to Figure 9.23)

5. Using Equation (9.26), the angle $\alpha = \tan^{-1}(1.4/6.2) = 12.72°$.
   Forward-most CG distance from nose wheel = (7.08−1.8) = 5.28 m.
   Distance CD in Figure 9.23 = $5.6 \times \tan(12.72) = 5.28 \times 0.226 = 1.195 \approx 1.2$ m.
   Therefore, the turn-over angle, $\theta = \tan^{-1}(1.38/1.2) = 49°$deg.
   Baseline Bizjet θ is below 60° for paved runway limit of turn-over angle (unprepared runway limit is 50°). It offers enough of a margin to keep turn-over angle for the variant aircraft (short variant below 47° and long variant slightly above 48°).

9.17.1.3   Wheel load:

Main wheel

At the *aft-most* CG, which gives $l_{N\_REAR} = 6.2 - 0.426 = 5.77$ m (18.94 ft).

Using Equation (9.4), the main-wheel load at the aft-most CG, gives

$R_{MAIN} = (l_{N\_REAR} \times MTOW)/l_{BASE} = (5.77 \times 9500)/6.2 = 8841$ kg (19 450 lb). It is 93% of MTOW.

At the *forward-most* CG, which gives $l_{N\_REAR} = 6.2 - 0.746 = 5.454$ m (17.9 ft).

Using Equation (9.4), the main-wheel load at the forward-most CG, gives

$R_{MAIN} = (l_{N\_REAR} \times MTOW)/l_{BASE} = (5.454 \times 9500)/6.2 = 8357$ kg (18 418 lb). It is 88% of MTOW.

Between the two limits, the two main wheel gears take around 90% of MTOW, a desirable value.

Taking the worst case, the load at each main gear (per strut) is $8841/2 = 4420$ kg.

Taking in to consideration of the growth version, a twin-wheel arrangement is chosen.

Equation (9.18) gives the ESWL $= 4420/1.5 = 2947$ kg (6494 lb).

Nose wheel

Nose wheel and tyre sizing are based on the forward-most CG position; therefore, the aft CG nose wheel load is not computed. The nose wheel load at the forward CG at 7.08 m (23.22 ft) from the ZRP gives $l_{N\_FORWARD} = 8 - 7.08 = 0.92$ m (3 ft). Deceleration force 3 m s^{-2} (9.85 ft s^{-2}).

Equation (9.9) gives

$$R_{NOSE} = (l_N \times MTOW)/l_{BASE} + [(aW/g) \times (h_{cg})]/(l_{BASE})$$
$$= (0.92 \times 9500)/6.2 + [(3 \times 9500)/9.81 \times 1.38]/6.2 = 1410 + 646 = 2056 \text{ kg (4531 lb).}$$

Then single nose wheel ESWL $= 2056$ kg (4531 lb).

9.17.1.4   Tyre selection

Table 9.6 provides several options [9]. From the tyre catalogues, a suitable match for three-part tyre bias ply size with a designation of $22 \times 6.6-10$ (18-ply), an inflation pressure of 260 psi and a maximum wheel load of 10 700 lb. (4855 kg). The maximum speed capability is 200 kt (230 mph – this aircraft class does not exceed 130 kt during approach, touching down at a still slower speed). Although this is about the smallest size for wheel loading, it has some redundancy with a twin wheel. Appendix F gives details on wheels and tyres.

Nose wheel and tyre sizing are based on the forward-most CG position; therefore, the aft CG nose wheel load is not computed. From the tyre catalogues, a suitable match for the nose wheel is with the designation of $17.5 \times 5.75-8$ (14-ply) and an inflation pressure of 220 psi that can take 6050 lb. (2745 kg). The maximum speed capability is 210 mph.

Note that the Bizjet has some margin to keep commonality with the large variant. In a short variant case, it may be possible make a dual wheel main gear a single wheel main gear, with minor down-sizing of the strut and systems.

9.17.1.5   Deflection

Landing is most critical for deflection. Typically, the maximum landing weight is 95% of the MTOM. To calculate deflection on landing (in Table 9.2, FAR25 $V_{Vert} = 12$ fps), the energy to be absorbed is given in Equation (9.7) (it is computed in correlation with FPS to FAR data) as:

$$E_{ab} = 1/2 M_L \times (V_{Vert})^2 = 0.5 \times 0.95 \times 21\,463 \times 144 = 1\,468\,070 \text{ lb ft}^2$$

Simplifying Eq (9.16) and substituting the values of k_{strut} and k_{tire}, total deflection by tyre and strut is as follow:

$$1/2 \times V^2{}_{Vert}/g = n_l \times [(0.7 \times m_{strut} \times \delta_{strut} + 0.47 \times m_{tire} \times \delta_{tire})]$$
$$- [n_l \times \{(1 - L/W_L) \times (m_{strut} \times \delta_{strut} + m_{tire} \times \delta_{tire})\}]$$

Where $n_l = 2$ (Table 9.2)

$$m_{tyre} = 4 \times \text{(number of tyres)}, m_{strut} = 2 \times \text{(number of struts)}$$

At touch down, consider $(L/W_L) = 0.9$

On substitution, the above reduces to as follows:

$$(0.5 \times 144)/64.4 = [(0.7 \times 2 \times \delta_{\text{strut}} + 0.47 \times 4 \times \delta_{\text{tire}})] - [\{0.1 \times (2 \times \delta_{\text{strut}} + 4 \times \delta_{\text{tire}})\}]$$

or

$$1.118 = (1.4 \times \delta_{\text{strut}} + 1.88 \times \delta_{\text{tire}}) - (0.2 \times \delta_{\text{strut}} + 0.4 \times \delta_{\text{tire}})$$

All variables except δ_{strut} and δ_{tire} are known. Tire deflection δ_{tire} can obtained as follows:
Selected Bizjet tyre has designation of $22 \times 6.6\text{-}10$, i.e., $D = 22\text{in}$ and $d = 10\text{in}$
From Eq (9.25a), $\delta_{\text{tyre}} = 0.3 \times (D - d)/2 = 0.3 \times (22 - 10)/2 = 1.8\text{in} = 0.15\text{ft}$
On substation,

$$1.118 = (1.4 \times \delta_{\text{strut}} + 1.88 \times 0.15) - (0.2 \times \delta_{\text{strut}} + 0.4 \times 0.15) = 1.2 \times \delta_{\text{strut}} + 0.222$$

or

$$\delta_{\text{strut}} = (1.118 - 0.222)/1.2 = 0.747\text{ft} = 9.96\text{in} \approx 10\text{in} \ (25.4\text{cm})$$

9.17.1.6 Aircraft LCN

A typical airfield LCN for this aircraft class is low, anywhere from 10 to 20. Wheel loading, LCN and tyre sizing (use Figure 9.18 to compute wheel loading):

Long variant Bizjet aircraft with 14 passengers at a 30 in. pitch
To maintain component commonality, the same undercarriage and tyre size are used. It is left to readers to repeat the calculation as given previously, using the following data:
Long variant aircraft estimated MTOM = 11 000 kg ($\approx$24 244 lb) (to be refined in Chapter 10).
Fuselage length = 17.74 m (56.2 ft):
The CG from the zero reference plane = 8.75 m (28.7 ft).
Fuselage clearance angle may be reduced to $\gamma \approx 14°$, sufficient for rotation yet avoiding major strut modification. $\beta = 15°$.
Wheel base increases by 30 in. (one seat row) to $6.2 + 0.762 \approx 8$ m (at about 55% of the wing MAC).
Wheel track = 2.8 m (no change).
To avoid duplication, it is left to the readers to do the rest. The spring size is changed without affecting strut geometry. It may be possible to retain tyre commonality as the baseline aircraft tyre sufficient growth margin kept in tyre selection.

Short variant Bizjet aircraft with six passengers at a 30 in. pitch
The last task is to check the smallest aircraft size to maintain as much component commonality of the undercarriage and tyre size as possible.
Small variant aircraft MTOM = 7000 kg (15 400 lb) (refined in Chapter 8).
Fuselage length = 13.56 m (44.5 ft).
Wheel base = $6.2\text{-}0.762 \approx 5.5$ m and wheel track = 2.8 m (no change).
To avoid duplication, it is left to the readers to do the rest. The spring size is changed without affecting strut geometry. It may be possible to make the dual wheel a main wheel to adjust the wheel type with spring, strut and system modifications. It would be beneficial to shave off the metal from the strut shank.

9.17.2 Advanced Jet Trainer (Military): AJT

The worked-out example of an AJT aircraft, as developed in Chapter 8 (see Figure 9.14), continues with sizing of the undercarriage and tyres. The combat aircraft tyre sizing for the variant designs is considerably simpler because the affecting geometries are not altered, only the weight changes. This military aircraft example has only two variants, the AJT and CAS versions. The CAS role of the variant aircraft has the same geometric size but is heavier.

Because the affecting geometries (i.e. the wheel base and wheel track) do not change, the cost option logic is less stringent for maintaining undercarriage and tyre component commonality – especially for the worked-out example. In this case, it is designed for the heaviest variant with shaved-off metal for the baseline trainer version. Therefore, only the CAS version design is discussed; it is assumed that the AJT baseline will have lighter struts but the same undercarriage and tyres.

The reference lines are constructed using Figure 9.24. The aircraft centre line is taken conveniently through the centre of the engine exhaust duct. Although at this stage, when the exact CG is not known and placement of the undercarriage is based on a designer's experience, the example of the AJT given here has been sized to avoid repetition. The exercise for readers is to start with their own layout and then iterate to size.

9.17.2.1 Undercarriage Layout and Positioning

AJT and CAS undercarriage are of the tricycle type and retractable. The CAS nose cone and front fuselage is slightly shorter than AJT (Figure 8.14). Use the CAS ZRP the same as for the AJT. Wheel base and wheel track are the same for both the variants. AJT and CAS have angles that are $\beta = 15$ and $\gamma = 15°$. To maintain tyre commonality between both variants, tyres should be selected for the heavier CAS variant.

AJT MTOM = 6500 kg estimated from statistics. A tandem seat arrangement, the aircraft CG travel in this aircraft class would be around 15% (fully loaded) to 30% (aft-most – single pilot, full fuel and no armament) of the wing MAC (1.928 m). With aft heavy rear mounted engine, the wing $MAC_{1/2}$ is placed around at 58% of fuselage length from ZRP.

The fully loaded CAS version has MTOM of 8100 kg. The CAS version of the CG variation is from 20% (fully loaded) to 30% (no armament and full fuel) of MAC.

The nose wheel load is based on the forward-most CG position. An armament payload is placed around the aircraft CG, and the CAS aircraft CG movement is small between a clean configuration and a loaded configuration. The most critical situation for both the main and the nose wheel and the tyre is given: fuselage length = 12 m (39.4 ft) and wing MAC = 1.928 m. Place the main-wheel ground contact point at 62% of the MAC at 7.4 m from ZRP. The aft-most CG position is considered first, placed at 30% of the wing MAC at 7 m (23 ft) from ZRP. CG height from fuselage keel = 0.8 m. Typical tyre diameter = 28 in. (it will be formally selected when the loads are known). CG height from ground is 1.35 m. It will be revised when the CG is known by actual computation.

The methodology is similar to the Bizjet worked out in Section 9.17. Only the heaviest aircraft needs to be considered in this case, as explained previously. The following information is used to determine wheel loading, LCN and tyre sizing (use Figure 9.18 to compute wheel loading):

9.17.2.2 Step-by-step approach – CAS variant

Proceed in the same step-by-step approach given in Section 9.15.1 as in the case of Bizjet. To repeat, undercarriage layout and positioning are not equation based, but based on design judgements. Undercarriage layout and position must simultaneously satisfy several limitation/constraints, for example, angle γ and β limits, CG travel limits and neutral point position with respect to angle, β, and turn-over angle, θ, associated with angle, α (see Figure 9.24).

Wheelbase
1. Start with the nose gear ground contact point placed at 2 m from the ZRP mounted on the front fuselage bulkhead structure. Next, place the main gear ground contact point at 7.4 m from the ZRP (58% of the fuselage length of 12 m). This gives the wheelbase = (7.4–2) = 5.4 m.

Wheel track
2. Fuselage width with intake duct (see Figures 8.13 and 9.24) at is 1.0 m (half width = 0.875 m).

Consider at this stage the tyre radius = 0.28 m and the tyre storage bay is just inside the fuselage and strut in the wing. The main gear location point in the wing would be about 1.3 m from fuselage centre line, accommodating a generous distance to size the main gear unloaded extended length. This will allow the main gear to retract in the fuselage the edge of the tyre within the fuselage fairings. This main gear

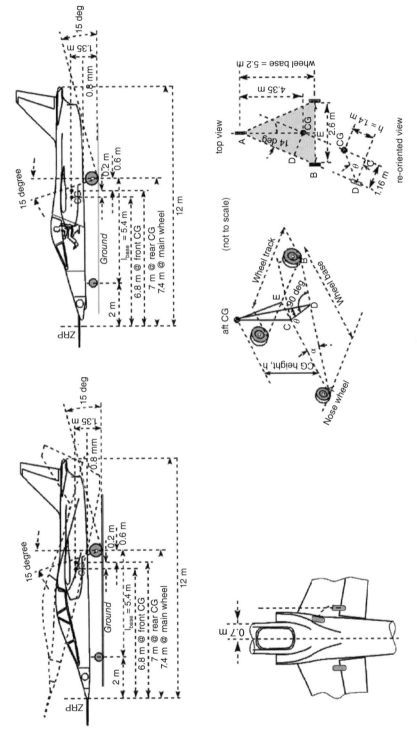

Figure 9.24 AJT undercarriage positions (CG position estimated).

extended length of 0.7 m gives an idea of the wheel track as the inward retraction will require at least twice this length. This gives a wheel track of 2.6 m.

CG positions

Vertical CG location

3. Vertical CG position, $\bar{z}$, is nearly at a constant (very little variation with loading status). Guess vertical CG location at 0.55 m (35% of fuselage height of 1.575 m) from the bottom fuselage mouldline (will be computed in Chapter 10). Takeoff rotation angle of $\gamma = 15°$ must clear at the main gear being deflected under maximum permissible load. Therefore, at the least permissible main gear load will be longer and at this stage is estimated to be 0.8 m measured from ground contact to under the fuselage mouldline. This gives the CG height from ground level $= (0.55 + 0.8) = 1.35$ m.

Horizontal CG location

4. CG position moves horizontally to some extent depending on the loading status. For CAS it is taken from 20 to 30% MAC, that is, 6.8–7 m from ZRP, respectively. This gives CG travel $= 0.2$ m. Rearmost CG position must be at the CG horizontal line, $15°$ gives a distance of $1.35 \times \tan 15 = 0.362$ m from the vertical line that keeps the limiting angle $\beta = 15°$ is behind the aft CG preventing aircraft tipping backwards. Neutral point is estimated at 40% of MAC is at 7.2 from ZRP giving a stability margin of 0.2 m ($\approx$10% of MAC).

Checking aircraft turn-over angle (refer to Figure 9.24)

5. Using Equation (9.26), the angle $\alpha = \tan^{-1}(1.3/5.4) = 13.5°$.
Forward-most CG distance from nose wheel $= (6.8–2) = 4.8$ m.
Distance CD in Figure 9.24 $= 4.8 \times \tan(13.5) = 4.8 \times 0.24 = 1.157 \approx 1.16$ m.
Therefore, the turn-over angle, $\theta = \tan^{-1}(1.35/1.16) = 49.3°$ deg.
AJT the turn-over angle, θ, is below $60°$ for paved runway limit of turn-over angle (unprepared runway limit is $50°$). The CAS variant has the same undercarriage geometric dimension.

9.17.2.3   Wheel load

Main wheel

The main-wheel load is computed at the aft-most CG, which gives $l_{N\_REAR} = 5$ m.
Equation (9.2) gives

$$R_{MAIN} = (l_{N\_REAR} \times MTOW)/l_{BASE}$$

or

$$R_{MAIN} = (5 \times 8100)/5.4 = 7500 \text{ kg} (16\,530 \text{ lb}).$$

The load per strut is 3750 kg (8265 lb).
This gives ESWL = 3750 kg (8265 lb).
Appendix E provides the tyre data from which to choose; there are many options.

Nose wheel

Nose wheel and tyre sizing are based on the forward-most CG position; therefore, the aft CG nose wheel load is not computed. The nose wheel load at the forward CG at 6.8 m (22.3 ft) from the ZRP gives $l_{N\_FORWARD} = 7.4$–$6.8 = 0.6$ m (2 ft). Military aircraft deceleration force 3.5 m s^{-2} (11.5 ft s^{-2}).
Equation (9.9) gives

$$R_{NOSE} = (l_N \times MTOW)/l_{BASE} + [(aW/g) \times (h_{cg})]/(l_{BASE})$$
$$= (0.6 \times 8100)/5.4 + [(3.5 \times 8100)/9.81 \times 1.35]/5.4 = 900 + 722 = 1622 \text{ kg} (3575 \text{ lb}).$$

Then single nose wheel ESWL = 1622 kg (3575 lb).

9.17.2.4 Aircraft LCN

Typical airfield LCN for this class of CAS aircraft is low. From Figure 9.18, the LCN is below 15 for the CAS variant and the AJT is still lower. This means there is good floatation and the aircraft can operate from semi-prepared airfields.

9.17.2.5 Tyre Selection (CAS)

Main wheel. Several options are listed in Table 9.6. From the tyre catalogues, a suitable match for the main wheel is the metric code three-part code (the metric code is used to familiarise readers) with a designation of 435×190–5 (22-ply) and an inflation pressure of 15.5 bar (225 psi) that takes 4030 kg (8880 lb). The maximum speed capability is 375 km/h (233 mph), which is a sufficient margin because this aircraft class does not exceed 240 km/h (130 knots) during an approach. Appendix F gives the details on wheels and tyres.

From the tyre catalogues, a suitable match for the nose wheel is the metric code three-part code with a designation of 450×190–5 (10-ply) and an inflation pressure of 6.2 bar (90 psi) that takes 1734 kg (3822 lb). The maximum speed capability is 370 km/h (230 mph).

The AJT is much lighter with a MTOM of 6500 kg and, therefore, the same tyre at a reduced pressure can be used (or a suitable smaller tyre can be used, if changing the hub is required).

9.17.2.6 Deflection

Deflection is estimated as in the civil aircraft case and therefore is not shown here. The high-wing configuration also shows sufficient clearance. The author suggests that readers undertake the computation.

9.17.2.7 CAS Variant of AJT

This has unchanged geometry. Only the weights are different. Readers are suggested to carry out CAS undercarriage layout and tyre selection.

9.17.3 Turboprop Trainer: TPT

Readers are recommended to carry out the TPT undercarriage layout and tyre selection (Figure 9.25).

Given fuselage length = 9.5 m and wing MAC = 1.715 m. Set ZRP at the forward-most point of the fuselage and not at the tip of propeller spinner as propeller and its spinner have options to choose from many types available.

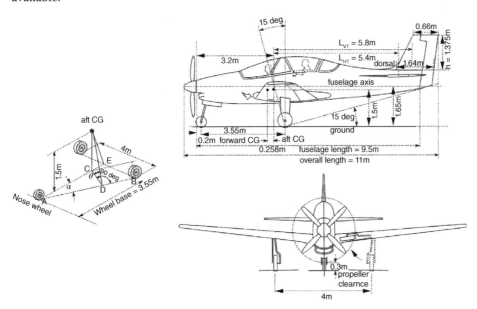

Figure 9.25 TPT undercarriage positions (CG position estimated).

Through trial and error method, the following dimensions are obtained. Being front mounted engine mid wing MAC is placed at 42% of fuselage length, that is, 4 m from ZRP. Place front wheel and main wheel ground contact points at 0.2 and 3.75 m from ZRP. This gives a wheel base = 3.55 m. Having a thick low wing, the undercarriage can be retracted with the wing allowing a wider wheel track of 4 m that gives good ground handling qualities. Propeller ground clearance is 0.3 m.

Forward and aft CG limits are at 15 and 30% MAC, that is, 2.95 and 3.2 m, respectively. CG height from ground = 1.4 m (the Bizjet and AJT examples give some idea). Check the tipping back angle to be $\geq 15°$ and this stays behind the aft CG imit. Check takeoff rotational angle $\approx 15°$.

Checking aircraft turn-over angle (refer to Figure 9.25)

5. Using Equation (9.26), the angle $\alpha = \tan^{-1}(2/3.55) = \tan^{-1}(0.563) = 29.4°$.
   Forward-most CG distance from nose wheel = 3.55–0.25 = 3.3 m.
   Distance $C_D = 3.3 \times \tan(29.4°) = 3.3 \times 0.563 = 1.86$ m.
   Therefore, the turn-over angle, $\theta = \tan^{-1}(1.5/1.86) = \tan^{-1}(0.806) = 39°$.
   AJT the turn-over angle, θ, is below 60° for paved runway limit of turn-over angle (unprepared runway limit is 50°). The CAS variant has the same undercarriage geometric dimension.

The readers may work out the wheel load and tyre size (see Appendix F).

9.18 Discussion and Miscellaneous Considerations

This chapter on undercarriage design is a relatively large, complex and standalone chapter without which an aircraft design cannot be completed. Only the preliminary information, which is needed by aircraft designers to conduct a conceptual study, is presented here. Details of the undercarriage design are implemented by specialists after the go-ahead for a project is obtained. Aircraft designers and undercarriage designers maintain communication to integrate the undercarriage with the aircraft, doing it right the first time.

There is a tendency to minimise undercarriage design work in coursework exercises, possibly because of time constraints. Undercarriage design is an involved procedure. Issues on retraction kinematics, using CAD, may be taken up as separate project as it is not dealt with here.

So far aircraft configuration was based on guesstimates from statistics of past design, for example, MTOM, wing are, engine size, CG travel and location and so on. A good spreadsheet must be prepared from the beginning for the calculations because they are required for subsequent iterations. The next chapter, Chapter 10, refining aircraft component weights will determine MTOM and the CG position more accurately and next iteration of the aircraft configuration with undercarriage installed will bring one step closer to complete conceptual study. Freezing of aircraft configuration to finality will be carried out in a formal manner through aircraft sizing with matched engine to meet the given specification/requirements to start with. This will require aircraft performance analyses for which aircraft drag estimation and engine performance characteristics will be required. These follow Chapter 10, which is devoted to weight and CG estimation.

In summary, the chosen undercarriage should be the tricycle type with retraction. The runway LCN and ESWL decide tyre pressure (the higher level of inflation pressure may be necessary), which in turn decides the number of wheels and struts required. Tyre manufacturers' catalogues list the correct sizes of the tyres.

The methodologies for civil and military aircraft undercarriages and tyre sizing are nearly the same. The differences are in operational requirements. In general, civil aircraft design poses more difficulty in maintaining component commonality.

The cost options for component commonality for variant designs must be decided on early during the conceptual design phase. Trade-off studies on cost versus weight must be conducted.

9.18.1 Undercarriage and Tyre Data

Table 9.7 gives some production aircraft undercarriage and tyre data.

Table 9.7 Some production aircraft undercarriage and tyre data.

Airplane	MTOM lb.	Wheel per/strut	Type	Tyre size	Tyre pressure psi	Turn-radius ft
Cessna 152	2500	1	S	6.00–6		
Beach 58	5500	1	S	6.50–8	56	
Beach 200	12 600	2	T	18 × 5.5	105	
Learjet45	22 000	2	T	22 × 5.75–8	200	
ATR42	41 000	2	T	32 × 8.8R16	126	57
CL600	48 300	2	T	H27 × 8.5–14	175	40
CR200	53 000	2	T	H29 × 9.0–15	162	75
BD700	95 000	2	T	H38 × 12.0–19	200	68
B737–700	140 000	2	T	H40 × 14.5–19	200	68
Airbus 320	170 000	2	T	49 × 19–20		75
B727–200	173 000	2	TT	49 × 17	168	
B707–720	336 000	4	TT	46 × 16	180	
DC8–63	358 000	4	TT	44 × 16	200	
L1011	409 000	4	TT	50 × 20	175	
B747B	775 000	4	DTT	46 × 16	210	159
C130A	124 000	2	ST	56 × 20	65	85
C17	586 000	3	TTT	50 × 21–20	138	90
Hawk	20 000	1	S	650–10	143	
F14	74 300	1	S	37 × 11	245	

Abbreviations: S, single; T, tandem; ST, single tandem; TT, twin tandem; DTT, double twin; TTT, triple twin tandem.

References

 1 Currey, N.D. (1988). *Aircraft Landing Gear Design – Principle and Practice*. AIAA.
 2 Niu, M. (1999). *Airframe Structural Design*. CONMILIT Press.
 3 Roskam, J., *Airplane Design*, Part IV, 2003
 4 Pazmany, L. (1986). *Landing Gear Design for Light Aircraft*. Pazmany Aircraft Corp.
 5 Aircraft Characteristics – Airport and Maintenance Planning, Airbus 330, Airbus SAS, France, January 1993, Revised January 2014.
 6 FAA. (2006). Standard Naming Convention for Aircraft Landing Gear Configurations. (Order 5300.7) 6 October 2006.
 7 Aerodrome Design Manual, Part 3, Pavements, Second Edition, International Civil Aviation Organization, 1983.
 8 Jane's All the World's Aircraft Manual: yearly publications.
 9 Goodyear tyre catalogue. Available online at: www.goodyearaviation.com/img/pdf/datatires.pdf (accessed June 2018).
 10 Aircraft Engineering Data. Michelin. Available online at: https://aircraft.michelin.com/ (accessed June 2018).
 11 Dunlop (2016). Tyre Information. Available online at: www.dunlopaircrafttyres.co.uk/tech_support/ dm1172/DUNLOP_DM1172.pdf (accessed June 2018).
 12 Bridgestone (2018). Tyre Information. Available online at: www.bridgestone.com/products/speciality_tires/ aircraft/products/applications/pdf/tire_specifications.pdf (accessed June 2018).
 13 Aircraft Tire Care and Maintenance Manual (2018). Available online at: https://www.goodyearaviation .com/resources/tirecare.html

10

Aircraft Weight and Centre of Gravity Estimation

10.1 Overview

An aircraft must ascend to heights by defying gravity and sustaining the tiring task of cruise – naturally, this is weight-sensitive. Anyone who has climbed a hill knows about this experience, especially if one has to carry baggage. An inanimate aircraft is no exception; its performance suffers by carrying unnecessary mass (i.e. weight). At the conceptual design stage, aircraft designers have a daunting task of creating a structure not only at a low weight but also at a low cost, without sacrificing safety. Engineers also must be accurate in weight estimation well ahead of manufacture. This chapter presents a formal method to predict an aircraft and its component mass (i.e. weight), which results in locating the centre of gravity (CG) during the conceptual design phase. The aircraft inertia estimation is not within the scope of this book, except finding the area moment of inertia in Chapter 19.

In the past, in the UK and the USA, aircraft mass was expressed in foot pound system (FPS) units in pounds (lb_m) and weight in the same pound terminology (lb_f), a product of acceleration due to gravity and mass. This may create some confusion to those who are starting to use them for the first time. With the use of kg as mass in the System Internationale (SI), the unit for weight is a *newton*, which is calculated as the mass multiplied by gravitational acceleration (9.81 m s^{-2}). This book uses both the FPS and SI systems; sometimes interchangeably.

Aircraft weight depends on the materials used for its manufacture. As stated previously, aircraft conceptual designers must have broad-based knowledge in all aspects of technology; in this case, they must have a sound knowledge of material properties (e.g. strength-to-weight and strength-to-cost ratios). Higher strength-to-weight and strength-to-cost ratios are the desired qualities, but they act in opposition. Higher strength-to-weight ratio material is more expensive, and designers must stay current in materials technology to choose the best compromises. Chapter 19 is devoted to aircraft materials and structures.

In the early days, designers had no choice but to use the best quality wood for aircraft construction material. Today, it is not a viable option for the type of load encountered. Good quality wood has become expensive and also poses an environmental issue. Fortunately, the advent of *duralumin* (i.e. an aluminium alloy) in the 1930s resolved the problem, providing a considerably higher strength-to-weight ratio than wood. Having a mass-produced aluminium alloy also offers a lower material cost-to-strength ratio. Wood is easier to work with, having a low manufacturing infrastructure suitable for homebuilt aircraft, but other civil and military aircraft use predominantly metal alloys and composites. The last two decades have seen a growing use of composite material and more exotic metal alloys offer still better strength-to-weight ratios.

Composites are basically fabric and resin bonded together, generally formed to shape in moulds. The manufacturing process associated with composites is yet to achieve the quality and consistency of metal; hence, at this point, the certifying authorities are compelled to apply reduced values of stress levels to allow for damage tolerance and environmental issues, as well as to keep the factor of safety at 1.5. The manufacturing process also plays a role in deciding the allowable stress level. These considerations can erode the benefits of weight savings. Research on new material, whether metal alloys (e.g. lithium-aluminium and beryllium alloy) or composites (e.g. fabric and resin) or their hybrids, is an area where there is potential to reduce aircraft weight and

Conceptual Aircraft Design: An Industrial Approach, First Edition. Ajoy Kumar Kundu, Mark A. Price and David Riordan.
© 2019 John Wiley & Sons Ltd. Published 2019 by John Wiley & Sons Ltd.

cost. New materials are still relatively expensive, and they are steadily improving in both strength and lower costs.

This chapter covers the following topics:

Section 10.2: Aircraft Mass, Component Mass and CG Position
Section 10.3: Parameters that Act as Drivers for Aircraft Mass
Section 10.4: Aircraft Mass Breakdown Sequence
Section 10.5: Desirable CG Location Relative to Aircraft
Section 10.6: Aircraft Mass Decomposed into Component Groups
Section 10.7: Aircraft Component Mass Estimation Methods
Section 10.8: Civil Aircraft Rapid Mass Estimation Method
Section 10.9: Civil Aircraft Graphical Mass Estimation Method
Section 10.10: Civil Aircraft Semi-Empirical Mass Estimation Method
Section 10.11: Bizjet Example
Section 10.12: Methodology to Establish Aircraft CG with a Bizjet Example
Section 10.13: Military Aircraft Rapid Mass Estimation Method
Section 10.14: Military Aircraft Graphical Method for Mass Estimation
Section 10.15: Military Aircraft Semi-Empirical Mass Estimation Method
Section 10.16: AJT and CAS Examples (Military Aircraft)
Section 10.17: Methodology to Locate Aircraft CG with AJT and CAS Examples
Section 10.18: TPT and COIN Examples

10.1.1 Coursework Content

The coursework task continues linearly with the examples worked out thus far. Readers must estimate aircraft component mass, which gives the aircraft mass and its CG location. This is an important aspect of aircraft design because it determines aircraft performance, stability and control behaviour.

Experience in the industry has shown that weight can only grow. Aircraft performance is extremely sensitive to weight because it must defy gravity. Aerodynamicists want the least weight, whereas stress engineers want the component to be strong so that it will not fail and have the tendency to beef-up a structure. The structure must go through ground tests when revisions may be required. It is easy to omit an item (there are thousands) in weights estimation. Most aeronautical companies have a special division to manage weights by weights-control engineers – a difficult task to perform. The aircraft operators also have *loadmasters* to ensure loading of aircraft for safe operation for the full operational envelope, on ground and in flight.

10.2 Introduction

Because aircraft performance and stability depends on aircraft weight and the CG location, the aircraft weight and its CG position are paramount in configuring an aircraft. The success of a new aircraft design depends considerably on how accurately its weight (mass) is estimated. A pessimistic prediction masks product superiority and an optimistic estimation compromises structural integrity. A pessimistic prediction ends up as a lighter aircraft showing better performance in prototype flight tests and is easy to correct, but an optimistic prediction may come out with a heavier aircraft and may fail to meet the performance requirements when it may prove expensive to lighten the weight by compromising structural integrity.

Once an aircraft is manufactured, the component weights can be easily determined by actual weighing. The aircraft CG then can be accurately determined. However, the problem is in predicting weight and the CG is at the conceptual design stage, before the aircraft is built. When the first prototype is built, the weights engineers have the opportunity to verify the predictions – typically, a four-year wait! Many of the discrepancies result from design changes; therefore, weights engineers must be kept informed in order to revise their estimations. It is a continuous process as long as the product is well supported after the design is completed.

Mass is the product of the solid volume and average density. For an aircraft component (e.g. wing assembled from a multitude of parts and fasteners), it is a laborious process to compute volumes of all those odd-shaped parts. In fact, the difficulty is that the mass prediction of complex components is not easily amenable to theoretical derivations. The typical approach to estimate weights at the conceptual design stage is to use semi-empirical relationships based on theory and statistical data of previously manufactured component masses (a 3D computer aided drawing (CAD) model of parts provides the volume but may not be available in the conceptual design stage).

The mass of each component depends on its load-bearing characteristics, which in turn depend on the operational envelope (i.e. the *V-n* diagram). Each manufacturer has a methodology developed over time from the statistics of their past products combined with the physical laws regarding mass required for the geometry to sustain the load in question. These semi-empirical relations are proprietary information and are not available in the public domain. All manufacturers have developed mass-prediction relationships yielding satisfactory results (e.g. an accuracy of less than ±3% for the type of technology used). The semi-empirical relations of various origins indicate similarity in the physical laws but differ in associated coefficients and indices to suit their application domain (e.g. military or civil, metal, structural and manufacturing philosophy and/or level of desired accuracy). Nowadays, computers are used to predict weight through solid modelling – this is in conjunction with semi-empirical relations. The industry uses more complex forms with involved and intricate manipulations that are not easy to work within a classroom.

The fact is that, no matter how complex academia may propose semi-empirical relations to be to improve accuracy in predicting component mass, it may fall short in supplanting the relationship available in industry based on actual data. Out of necessity, the industry must keep its findings 'commercial in confidence'. At best, industry may interact with academia for mutual benefit. An early publication by Torenbeek [1] with his semi-empirical relations is still widely used in academic circles. Roskam [2] presented three methods (i.e. Torenbeek, Cessna and US DATCOM) that clearly demonstrate the difficulty in predicting mass. Roskam's book presents updated semi-empirical relations corroborated with civil aircraft data showing satisfactory agreement. The equations are not complex – complexity does not serve the purpose of coursework. Readers will have to use industrial formulae when they join a company. This chapter explains the reasons associated with formulating the relationships to ensure that readers understand the semi-empirical relations used in the industry.

The authors recommend the publications by the Society of Allied Weights Engineers (SAWE) (US) as a good source for obtaining semi-empirical relations in the public domain. Some of the relations presented herein are taken from SAWE, Torenbeek, Sechler, Roskam, Niu and Jenkinson ([1–6]). Some of the equations are modified by the authors. It is recommended that readers collect as much component weight data as possible from various manufacturers (both civil and military) to check and modify the correlations and to improvise if necessary. Revision of mass (i.e. weight) data is a continuous process. In each project phase, the weight estimation method is refined to better accuracy. Actual mass is known when components are manufactured, providing an opportunity to assess the mass-prediction methodology. Although strength testing of major aircraft components is a mandatory regulatory requirement before the first flight, structural-fatigue testing continues after many aircraft are already in operation. By the time results are known, it may not prove cost-effective to extensively modify an overdesigned structural member until a major retrofit upgrading is implemented at a later date.

The unavoidable tendency is that aircraft weight grows over time primarily due to modifications (e.g. reinforcements and additions of new components per user requirements). Provision needs to be made at the conceptual stage to manage the weight growth. There are several ways to achieve this, the simplest way is to tweak the engine to produce slightly increased thrust (typically a 1% thrust increase for 4% weight growth – this should prove sufficient in the majority of cases).

The importance of the Six-Sigma approach to make a design right the first time is significant to weights engineers. Many projects have suffered because of prototypes that were heavier than prediction or even experienced component failure in operation resulting in weight growth. The importance of weight prediction should not be underestimated due to not having an analytical approach involving high-level mathematical complexity, as in the case of aerodynamics. Correct weight estimation and its control are vital to aircraft design. One cannot fault stress engineers for their conservatism in ensuring structural integrity – lives depend on it. Weight-control engineers check for discrepancies throughout project development.

Mass-prediction methodology starts with component weight estimation categorised into established groups, as described in Section 10.6. The methodology culminates in overall aircraft weight and locating the CG and its range of variation that can occur in operation. Aircraft inertias are required to assess dynamic behaviour in response to control input but then they are not needed until completion of the conceptual design study – hence, inertia is not addressed in this book. Iteration of the aircraft configuration is required after the CG is located because it is unlikely to coincide with the position guesstimated from statistics.

10.2.1 From 'Concept Definition' to 'Concept Finalisation'

Progress of aircraft conceptual design study up to Chapter 9 led to presenting a preliminary configuration based on confident 'guesstimates' of maximum takeoff mass (MTOM), wing and other geometric parameters, and engine size from the statistics of past designs. From this chapter and onwards, work progresses towards refinement to accurate values as much as is possible, through iterations to update design to a better definition to arrive at the end to *concept definition*, in order to freeze aircraft to, that is, *concept finalisation* in Chapter 14. This completes Phase I of the project and enters into Phase II, the 'Detailed Design Phase'; beyond the scope of this book.

In this chapter, aircraft and component weight estimation are carried out using reliable and well-established semi-empirical formulae, followed by establishing the aircraft CG position. It is expected that the estimated MTOM and CG position will be different from the guessed values. This will necessitate returning to a point to iterate through the subsequent stages to arrive at an improved *concept definition*. The iterative process may require repositioning of wing and undercarriage with respect to the fuselage. The last iteration will be after the aircraft is sized (Chapter 14) with a matched engine that could necessitate altering of wing reference area (like zooming in/out) yet maintain geometric parameters of sweep, taper ratio, dihedral and twist unaffected. Change of wing reference area will also affect empennage reference areas. Geometric changes will also change aircraft drag affecting engine size. All these changes will also change aircraft component weights. Therefore, at least, another iteration is required to fine-tune aircraft configuration to, that is, concept finalisation, to freeze the design. Experience designers can minimise iterative process. This book avoids iterations to save from repetitive work.

Once again, the authors recommend that the readers prepare spreadsheets for repeated calculations because iterations will ensue after the CG is established and the aircraft is sized.

10.3 The Weight Drivers

The factors that drive aircraft weight are listed herein. Aircraft material properties given are typical for comparing relative merits. Material elasticity, E, and density, ρ, provide the strength-to-weight ratio. In the alloys and material categories, there is variation. Chapter 19 deals with aircraft materials and structures.

i. Weight is proportionate to size, indicated by geometry (i.e. length, area and volume).
ii. Weight depends on internal structural-member density – that is, the denser, the heavier.
iii. Weight depends on a specified limit-load factor n (see Chapter 17) for structural integrity.
iv. Fuselage weight depends on pressurisation, engine and undercarriage mounts, doors and so forth.
v. Lifting surface weight depends on the loading, fuel carried, engine and undercarriage mounts and so forth.
vi. Weight depends on the choice of material. There are seven primary types used in aircraft, as follows:
    (a) Aluminium alloy (a wide variety is available – in general, the least expensive)
       typical $E = 11 \times 106$ lb in.$^{-2}$; typical density = 0.1 lb in.$^{-3}$
    (b) Aluminium–lithium alloy (fewer types available – relatively more expensive)
       typical $E = 12 \times 106$ lb in.$^{-2}$; typical density = 0.09 lb in.$^{-3}$

(c) Stainless-steel alloy (hot components around engine – relatively inexpensive)
typical $E = 30 \times 106$ lb in.$^{-2}$; typical density = 0.29 lb in.$^{-3}$

(d) Titanium alloy (hot components around engine – medium-priced but lighter)
typical $E = 16 \times 106$ lb in.$^{-2}$; typical density = 0.16 lb in.$^{-3}$

(e) Composite type varies (e.g. fibreglass, carbon fibre and Kevlar); therefore, there is a wide variety in elasticity and density (price varies from inexpensive to expensive).

(f) Hybrid (metal and composite 'sandwich' – very expensive, e.g. Glare).

(g) Wood (rarely used except for homebuilt aircraft; is not discussed in this book).

The use of composites is increasing, as evidenced in current designs. In this book, the primary load-bearing structures are constructed of metal; secondary structures (e.g. floorboard and flaps) could be made from composites. On the conservative side, it generally is assumed that composites and/or new alloys allow about 10–15% weight saving of the manufacturer's empty mass (MEM) for civil aircraft and about 15–25% of the MEM for military aircraft. Although composites are used in higher percentages, this book remains conservative in approach. All-composite aircraft have been manufactured (mostly small aircraft), although only few in number (except small aircraft). The metal-composite sandwich is used in the Airbus380 and Russia has used aluminium–lithium alloys. In this book, weight change as the consequences of using newer material is addressed by applying factors.

The design drivers for civil aircraft have always been safety and economy. Competition within these constraints kept civil aircraft designs similar to one another.

10.4 Aircraft Mass (Weight) Breakdown

Given next and in Figure 10.1 are definitions of various groups of aircraft mass (weight).

MEM (manufacturer's empty mass)

This is the mass of a finished aircraft as it rolls out of the factory before it is taken to a flight hangar for the first flight.

$$\text{OEM (operator's empty mass)} = \text{MEM} + \text{Crew} + \text{Consumables} \tag{10.1}$$

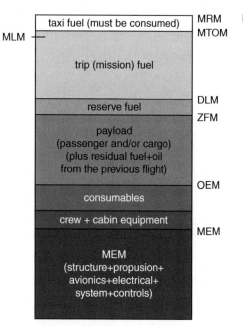

Figure 10.1 Aircraft weights breakdown.

The aircraft is now ready for operation (residual fuel from the previous flight remains).

$$\text{MTOM (maximum takeoff mass)} = \text{OEM} + \text{Payload} + \text{Fuel} \qquad (10.2)$$

The MTOM is the reference mass loaded to the rated maximum design limit. This is also known as the brake release mass (BRM) ready to takeoff. (Not all takeoffs are at MTOM.)

Aircraft are allowed to carry a measured amount of additional fuel for taxiing to the end of the runway, ready for takeoff at the BRM (MTOM). This additional fuel mass would result in the aircraft exceeding the MTOM to the maximum ramp mass (MRM). Taxiing fuel for midsized aircraft would be approximately 100 kg and it must be consumed before the takeoff roll is initiated – the extra fuel for taxiing is not available for the range calculation. On busy airfields, the waiting period in line for takeoff could extend to more than an hour in extreme situations. MRM is also known as the maximum taxi mass (MTM) and is heavier than the MTOM.

$$\text{MRM (maximum ramp mass)} = \text{MTOM} + \text{fuel to taxi to end of runway for takeoff}$$

$$\approx 1.0005 \times \text{MTOM (very large aircraft) to } 1.001 \times \text{MTOM} \qquad (10.3)$$

$$\text{ZFM (zero fuel mass)} = \text{MTOM minus all fuel (non-usable residual fuel remains)} \qquad (10.4)$$

$$\text{DLM (design landing mass)} = \text{MTOM minus trip fuel (reserve fuel remains)} \qquad (10.5)$$

The maximum permissible landing mass is 0.95 MTOM. Fuel dumping is not allowed around built-up areas. Aircraft can land at any mass heavier than design landing mass (DLM), when short hop consecutive sorties are carried out without refuelling.

10.5   Aircraft CG and Neutral Point Positions

Proper distribution of mass (i.e. weight) over the aircraft geometry is key to establishing the CG. It is important for locating the wing, undercarriage, engine and empennage for aircraft stability and control. The convenient method is to first estimate each component weight separately and then position them to satisfy the CG for the maximum takeoff weight (MTOW).

Fuel loads and payloads are variable quantities; hence, the CG position varies relative to overall geometry. Each combination of fuel and payload results in a CG position. The position of CG with respect to aircraft is an important consideration for operations on ground and in flight. There are two extreme CG positions to travel for safe operation. A typical aircraft CG margin that affects aircraft operation is shown in Figure 10.2. On ground, the aircraft should not trip over longitudinally and laterally (see Chapter 9). In flight, the aircraft must remain stable for the full flight envelope. A similar plot can be made for lateral CG movement for the consideration of turn-over issues (not shown here).

Aircraft operators, civil or military, have a team of ground crew who plans and computes the permissible load for the mission sortie. For civil operations, the appropriate containers and packaging are used. For military aircraft, it is the armaments and various types of pods and so on. The team is headed by the 'Loadmaster' who is responsible fuel and cargo/payload loading sequence to ensure safe operation on ground and in flight.

Figure 10.3 shows variations in CG positions for the full range of combinations. Because it resembles the shape of a potato, the CG variation for all loading conditions is sometimes called the 'potato curve'; also embedded in Figure 10.2. Designers must ensure that at no time during loading up to the MTOM does the CG position exceed the loading limits endangering the aircraft to tip over on any side. Loading must be accomplished under supervision. Whereas passengers have free choice in seating, cargo and fuel loading are done in prescribed sequences, with options.

It has been observed that passengers first choose window seats and then, depending on the number of abreast seating, the second choice is made. Figure 10.3a shows the window seating first and the aisle seating last; note the boundaries of front and aft limits. Cargo and fuel loading is accomplished on a schedule with the locus of CG travel in lines. In the figure, the CG of the operator's empty mass (OEM) is at the rear, indicating

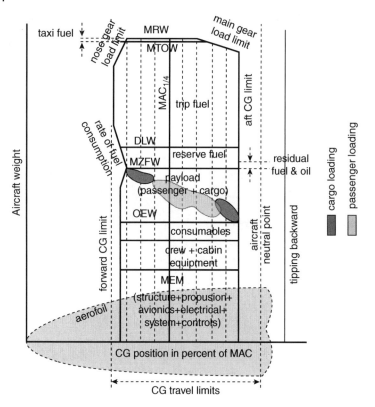

Figure 10.2 Aircraft CG limits.

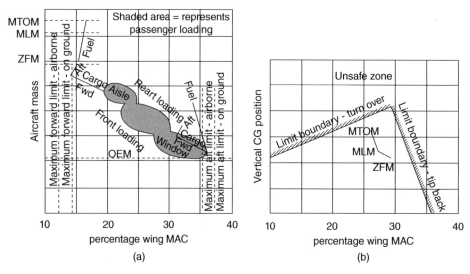

Figure 10.3 Aircraft CG limits travel – range of CG variations. (a) 'Potato' curve showing the horizontal limits (shaded area) and (b) vertical limits.

that the aircraft has aft-mounted engines. For wing-mounted engines, the CG at the OEM moves forward, making the potato curve more erect.

Figure 10.3b represents a typical civil aircraft loading map, which indicates the CG travel to ensure that the aircraft remains in balance within horizontal and vertical limits. Loading starts at the OEM point; if the passengers boarding first opt to sit in the aft end, then the CG can move beyond the airborne aft limit, but it must remain within the ground limit. Therefore, initial forward cargo loading should precede passenger boarding; an early filling of the forward tank fuel is also desirable.

10.5.1 Neutral Point, NP

Aircraft NP is the aerodynamic centre of a weightless aircraft (see Section 3.7). NP is the point where $dM/d\alpha$ is constant, that is, aircraft pitching moment is invariant to attitude changes. In-flight CG has to take into account of the influence wing position relative to fuselage and H-tail size that determine the aircraft aerodynamic centre, a shape dependent aerodynamic parameter. Therefore, the position of NP shifts if external geometric shape changes, for example, between clean and dirty (hi-lift devices and/or undercarriage are in extended position) conditions. Also, the position of CG changes depending on the fuel and payload/cargo loading status; seen as CG travel. The CG position with respect to the aircraft neutral is an important consideration to maintain aircraft stability.

Figure 10.4 shows that the aircraft NP moves backward near ground during field performance (takeoff and landing at dirty configuration) due to the flow field being affected by ground constraints. There is also movement of the CG location depending on the loading (i.e. fuel and/or passengers). It must be ensured that the pre-flight aft-most CG location is forward of the in-flight NP centre by a convenient margin, which should be as low as possible to minimise trim. The distance between aircraft CG and the aircraft NP is known as the *stability margin* of the aircraft. (To repeat that the main-wheel contact point (and strut line) is aft of the aft-most CG, the subtending angle, β, should be greater than the fuselage-rotation angle, γ, as described in Section 9.6.)

10.5.2 Aircraft CG Travel

Aircraft approaches to unstable flight as CG moves close to the NP. CG behind NP makes an aircraft unstable (see Chapter 18). The designers must ensure safe operation for the full operating envelope. (Note that an aircraft can be flown at neutral stability. This requires the pilot to struggle all the time to maintain steady attitude that will tire them out, which is not a safe way to fly. A bicycle ride is a good example where one has to learn to stay in balance.) Advanced military combat aircraft with Fly-by-Wire (FBW) technology can have relaxed static stability to provide quicker responses. That is, the margin between the aft-most CG and the in-flight NP is reduced (it may be even slightly negative), but design considerations relative to the undercarriage position remain unaffected. The following are some points for consideration that determine the extremities of CG travel.

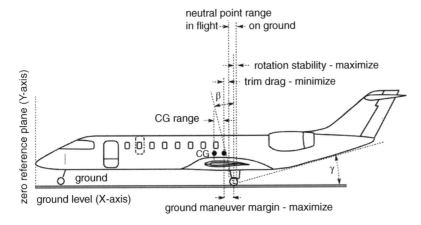

Figure 10.4 Aircraft CG position showing the stability margin.

Forward-Most CG Position

1. If the aircraft is loaded to MTOW or close to it in a manner that makes CG stay forward then load on the nose becomes high and steering problems could show up. The nose gear limit line for high aircraft is shown in Figure 10.2.
2. Aircraft in a dirty configuration with a forward CG position have a high nose down moment requiring adequate elevator power with downward force to keep the aircraft in balance to maintain the desirable attitude. In the case of bad design, trim may run out at the landing configuration with a forward CG when the pilot may not be able to raise the nose up sufficiently to the landing attitude.
3. A larger sized H-tail may overcome this problem at the expense of high stick force. It can now be realised that a compromise is required to size the H-tail.

Aft-Most CG Position

1. The aft-most CG limit is concerned with aircraft safety. As the aircraft CG moves closer to the NP, which must be always behind, the pitch actuation force becomes lighter and the aircraft becomes more responsive with the pilot requiring little force to change pitch attitude. In loose terms, the control stick feels lighter. It suites military pilots, but there is a limit beyond which the pilot will have to struggle with constant attention. A limit has to be given for safe operation, which is the aft limit of the CG.
2. During field performance, aft-most CG at a high load will have a high main gear load (Section 9.7). The main gear limit line for a highly loaded aircraft is shown in Figure 10.2.
3. For pitch stability consideration, a minimum stability margin has to be given suited to the H-tail size. Trim drag is minimum at the aft-most CG position.

CG at MTOM is not necessarily at the forward-most limit. Similarly, the CG of an empty aircraft at OEW (Operator's Empty Weight) is not necessarily at the aft-most position. The CG is always forward of the NP point. The loadmaster's responsibility is to make sure the aircraft CG travel stays within the specified limits for the full operational envelope.

Initially, locations of some of the components (e.g. the wing) were arbitrarily chosen based on designers' past experience, which works well (see Chapter 8). Iterations are required that, in turn, may force any or all of the components to be repositioned. There is flexibility to fine-tune the CG position by moving heavy units (e.g. batteries and fuel storage positions). It is desirable to position the payload around the CG so that any variation will have the least effect on CG movement. Fuel storage should be distributed to ensure the least CG movement; if this is not possible, then an in-flight fuel transfer is necessary to shift weight to maintain the desired CG position (as in the case of supersonic *Concorde*).

10.6 Aircraft Component Groups

The recognised groups of aircraft components are listed in exhaustive detail in the Aircraft Transport Associations (ATAs) publication. This section presents consolidated, generalised groups (for both civil and military aircraft) suitable for studies in the conceptual design phase. Both aircraft classes have similar nomenclature; the differences in military aircraft follow the common listings. Each group includes subgroups of the system at the next level. Care must be taken that items are not duplicated – accurate book-keeping is essential. For example, although the passenger seats are installed in the fuselage, for book-keeping purposes, the fuselage shell and seats are counted separately.

10.6.1 Aircraft Components

$$\text{Structure } (M_{\text{STR}} = M_{\text{FU}} + M_{\text{W}} + M_{\text{HT}} + M_{\text{VT}} M_{\text{N}} + M_{\text{PY}} + M_{\text{UC}} + M_{\text{MISC}}) \tag{10.6}$$

i. Fuselage group (M_{FU})
ii. Wing group (M_{W}) – includes all structural items, for example, flaps, winglets and so on

   iii. Horizontal tail group (M_{HT})
   iv. Vertical tail group (M_{VT})
    v. Nacelle group (M_N and M_{PY} – nacelle and pylon)
   vi. Undercarriage group (M_{UC})
  vii. Miscellaneous (M_{MISC}) – dorsal, ventral fins and so on

Military aircraft have the following to be accounted for:

 viii. Fixed armament (M_{FARM}), for example, internal guns and so on
   ix. Pylon (M_{PYL}) – to carry armament load/drop tank.

This is the basic structure of the aircraft, for example, the fuselage shell (seats are separated under furnishing group) and so on.

$$\text{Power Plant Group } (M_{PP} = M_E + M_{TR} + M_{EC} + M_{FS} + M_{OI}) \tag{10.7}$$

    x. Dry equipped engine (M_E)
   xi. Thrust reverser (M_{TR})
  xii. Engine control system (M_{EC})
 xiii. Fuel system (M_{FS})
 xiv. Engine oil system (M_{OI})

Military aircraft have the following to be accounted for:

  xv. Retarding devise (M_{RD}), for example, brake parachute (M_{RD}).

  These come as a package and are items dedicated to power plant installation. They are mostly bought-out items supplied by specialists.

$$\text{Systems group } (M_{SYS} = M_{ECS} + M_{FC} + M_{HP} + M_{ELEC} + M_{INS} + M_{AV}) \tag{10.8}$$

  xvi. Environmental control system (M_{ECS})
 xvii. Flight control system (M_{FC})
xviii. Hydraulic/pneumatic system (M_{HP}) – sometimes lumped in with other systems
  xix. Electric system (M_{ELEC})
   xx. Instrument system (M_{INS})
  xxi. Avionics system (M_{AV})
 xxii. Oxygen system (M_{OX}).

Military aircraft have the following to be accounted for:

xxiii. Ejection seat system – this is no more treated as furnishing (M_{EJ}).

A wide variety of equipment, these are all vendor-supplied bought-out items.

$$\text{Furnishing } (M_{FUR} = M_{SEAT} + M_{OX} + M_{PN}) \tag{10.9}$$

xxiv. Seat, galleys and other furnishing (M_{SEAT})
 xxv. Paints (M_{PN}).

Note that most of the weight goes to the fuselage, yet this is itemised under a different heading. Paints can be quite heavy. A well-covered B737 with airline livery can use up 75 kg of paint.
  Contingencies (M_{CONT})

xxvi. Contingencies (M_{CONT}) – a margin to allow unspecific weight growth

MEM – the total of these items.
This is the weight of the complete aircraft coming out from the production line for the first time to be airborne. The following items are added to MEM to obtain OEM.

xxvii. Crew – Flight and cabin crews (M_{CREW})
xxviii. Consumables – Food, water and so on. (M_{CON})
(long duration military sortie have consumables for the crew)

OEM – aircraft is now ready for operation.
To make it sortie ready, add the payload and requisite fuel to obtain MRM.

xxix. Fuel (M_{FUEL}) – for the design range, which may not fill all tanks

Civil aircraft specific payload is the passengers and/or cargo is the following:

xxx. Payload (M_{PL} passengers @ 90 kg per passenger – includes baggage)

Military aircraft specific payload is the armament/electronic devices as the following:

xxxi. Armaments (M_{ARM}) – missiles, bombs, firing rounds, drop tanks, electronic pods and so on.

MTOM (Maximum Takeoff Mass) = (MRM – taxi fuel)
That is the aircraft at the edge of runway ready to takeoff.
Civil Aircraft MTOM weight is the sum of weights of all component groups as totalled next.

$$MTOM = \int M(x)\, dx = \sum M_i$$

where subscript 'i' stands for each component group listed here.
For civil aircraft:

$$
\begin{aligned}
MTOM = &(M_{FU}) + (M_W) + (M_{HT}) + (M_{VT}) + (M_N) + (M_{PY}) + (M_{UC}) + (M_{MISC}) \\
&+ (M_E) + (M_{TR}) + (M_{EC}) + (M_{FS}) + (M_{OI}) + (M_{ECS}) + (M_{FC}) + (M_{HP}) \\
&+ (M_{ELEC}) + (M_{INS}) + (M_{AV}) + (M_{SEAT}) + (M_{OX}) + (M_{PN}) + (M_{CONT}) \\
&+ (M_{CREW}) + (M_{CONS}) + M_{PL} + M_{FUEL}
\end{aligned}
\tag{10.10}
$$

For military aircraft:

$$
\begin{aligned}
MTOM = &(M_F) + (M_W) + (M_{HT}) + (M_{VT}) + (M_N) + (M_{PY}) + M_{FARM} + (M_{UC}) + (M_{MISC}) + (M_E) \\
&+ (M_{RD}) + (M_{EC}) + (M_{FS}) + (M_{OI}) + (M_{ECS}) + (M_{FC}) + (M_{HP}) + (M_{ELEC}) \\
&+ (M_{INS}) + (M_{AV}) + (M_{EJ}) + (M_{OX}) + (M_{PN}) + (M_{CONT}) + (M_{CREW}) + (M_{CONS}) \\
&+ (M_{ARM}) + M_{FUEL} + (M_{DT}) + (M_{DT\_FUEL})
\end{aligned}
\tag{10.11}
$$

10.7 Aircraft Component Mass Estimation

Mass estimation at the conceptual design stage must be predicted well in advance of detailed drawings of the parts being prepared. Statistical fitment of data from the past designs is the means to predict component mass at the conceptual design phase. The new designs strive for improvement; therefore, statistical estimation is the starting point. During the conceptual design stage, iterations are necessary when the configuration changes. Typically, there are three ways to make mass (i.e. weight) estimations at the conceptual design stage:

1. *Mass fraction method.* This method relies on the statistical average of mass (Section 10.8). The mass is expressed in terms of percentage (alternatively, as a fraction) of the MTOM. All items should total 100% of the MTOM; this also can be expressed in terms of mass per wing area (i.e. component wing loading). This mass fraction method is accomplished at the cost of considerable approximation. This method is useful in setting up an equation by taking mass fractions as parameters to make analytical studies. This book bypasses the mass fraction method except to tabulate the range of mass fractions in Section 10.8.

2. *Graphical method.* This method consists of plotting component weights of various aircraft already manufactured to fit into a regression curve. The graphical method does not provide fine resolution but it is the fastest. It is difficult to capture the technology level (and types of material) used because there is considerable dispersion. Obtaining details of component mass for statistical analysis from various industries is difficult.

3. *Semi-empirical method.* This method is a considerable improvement in that it uses semi-empirical relations derived from a theoretical foundation and is backed by actual data that have been correlated statistically. The indices and factors in the semi-empirical method can be refined to incorporate the technology level and types of material used. The expressions can be represented graphically, with separate graphs for each class. When grouped together in a generalised manner, they are the graphs in the graphical method described previously.

The three methods are addressed in more detail in the Sections 10.8–10.10. While the first two methods of component mass estimation provide a starting point for Phase I, it will be necessary to refine the estimation of semi-empirical weight estimation equations.

10.7.1 Use of Semi-Empirical Weight Relations versus Use of Weight Fractions

Section 1.11.2 discusses the merit of using the semi-empirical weights formulae. The weight fraction data and related graphs are generated from the statistics of the latest aircraft in operation may not be close enough. Statistical based weight fractions data offer a good check to examine if the results obtained using the semi-empirical weights formulae are in agreement or not. Industries maintain databases of existing aircraft and possible competition aircraft to give some idea of what is expected and serve to initiate the conceptual design study and make comparisons. After that, the typical trend in industry is to use their in-house semi-empirical weights formulae for aircraft components weight estimation with substantiated accuracy within 5% to begin with and gets refined to less than to 2% before the final assembly of the first aircraft is completed. Once manufactured, accurate weights data for components and the whole aircraft can be quickly weighed to obtain the exact value.

10.7.1.1 Weight Fraction Method

Initial weight estimates using weights will, however, offer rather crude values as modern aircraft are increasingly incorporating newer materials for which there may not sufficient data available. Tables 10.1–10.3 give the component weight/mass fractions with respect to MTOM. While the use of weight fractions is reasonable to consider, it will be more convenient to prepare graphs as given in Figure 10.5 for rapid estimation with an error band no less than using the mass fractions. Visual inspection of the graphs allows intelligently deselecting extreme points to narrow the error band to improve results within the class of aircraft. The authors think that the classroom exercises should generate their own data within the category of aircraft closest to their project and plot graphs that will offer the tighter error band with the latest data. Once the preliminary aircraft configuration is established using statistics, it is recommended to use the semi-empirical formulae in the next iteration instead of weight fractions to obtain better accuracy. For this reason, although the weight fraction data are given in Tables 10.1–10.3, these are not used in the worked-out examples.

10.7.1.2 Semi-Empirical Method

Semi-empirical relations are derived from a theoretical foundation and backed by actual data that have been correlated statistically. The work involves adjustment with the best fit curve obtained through regression analyses of the statistics available. The indices and factors in the semi-empirical method can be refined to incorporate the technology level and types of material used. Using semi-empirical formulae bypasses the weight fraction method to get better accuracy.

Table 10.1 Smaller aircraft mass fraction (typically less than around 19 passengers – two-abreast seating).

Group		Small piston aircraft (one piston)		Agricultural aircraft	Small aircraft: two engine (Bizjet utility)	
		One engine	Two engine	One piston	Turboprop	Turbofan
Fuselage	$F_{fu} = M_F/MTOM$	12–15	6–10	6–8	10–11	9–11
Wing	$F_w = M_w/MTOM$	10–14	9–11	14–16	10–12	9–12
H-tail	$F_{ht} = M_{ht}/MTOM$	1.5–2.5	1.8–2.2	1.5–2	1.5–2	1.4–1.8
V-tail	$F_{vt} = M_{vt}/MTOM$	1–1.5	1.4–1.6	1–1.4	1–1.5	0.8–1
Nacelle	$F_n = M_n/MTOM$	1–1.5	1.5–2	1.2–1.5	1.5–1.8	1.4–1.8
Pylon	$F_{py} = M_{py}/MTOM$	0	0	0	0.4–0.5	0.5–0.8
Undercarriage	$F_{uc} = M_{uc}/MTOM$	4–6	4–6	4–5	4–6	3–5
Engine	$F_E = M_E/MTOM$	11–16	18–20	12–15	7–10	7–9
Thrust rev.	$F_{tr} = M_{tr}/MTOM$	0	0	0	0	0
Engine con.	$F_{ec} = M_{ec}/MTOM$	1.5–2.5	2–3	1–2	1.5–2	1.7–2
Fuel sys.	$F_{fs} = M_{fs}/MTOM$	0.7–1.2	1.4–1.8	1–1.4	1–1.2	1.2–1.5
Oil sys.	$F_{os} = M_{os}/MTOM$	0.1–0.3	0.25–0.4	0.1–0.3	0.3–0.5	0.3–0.5
APU		0	0	0	0	0
Flight con. sys	$F_{fc} = M_{fc}/MTOM$	1.5–2	1.4–1.6	1–1.5	1.5–2	1.5–2
Hydr/pneumatic sys	$F_{hp} = M_{hp}/MTOM$	0–0.3	0.3–0.6	0–0.3	0.5–1.5	0.7–1
Electrical	$F_{elc} = M_{elc}/MTOM$	1.5–2.5	2–3	1.5–2	2–4	2–4
Instrument	$F_{ins} = M_{ins}/MTOM$	0.5–1	0.5–1	0.5–1	0.5–1	0.8–1.5
Avionics	$F_{av} = M_{av}/MTOM$	0.2–0.5	0.4–0.6	0.2–0.4	0.3–0.5	0.4–0.6
ECS	$F_{ecs} = M_{ecs}/MTOM$	0–0.3	0.4–0.8	0–0.2	2–3	2–3
Oxygen	$F_{ox} = M_{ox}/MTOM$	0–0.2	0–0.4	0	0.3–0.5	0.3–0.5
Furnishing	$F_{fur} = M_{fur}/MTOM$	2–6	4–6	1–2	6–8	5–8
Misc.	$F_{msc} = M_{msc}/MTOM$	0–0.5	0–0.5	0–0.5	0–0.5	0–0.5
Paint	$F_{pn} = M_{pn}/MTOM$	0.01	0.01	0–0.01	0.01	0.01
Contingency	$F_{con} = M_{con}/MTOM$	1–2	1–2	0–1	1–2	1–2
Manufacturer's Empty Weight (MEW) – %		57–67	60–65	58–62	58–63	55–60
Crew		6–12	6–8	4–6	1–3	1–3
Consumable		0–1	0–1	0	1–2	1–2
OEM – %		65–75	65–70	62–66	60–66	58–64
Payload and fuel are traded						
Payload		12–25	12–20	20–30	15–25	15–20
Fuel		8–14	10–15	8–10	10–20	15–25
MTOM – %		100	100	100	100	100

Lighter/smaller aircraft would show the higher side of the mass fraction.

CAD modelling can give component volume, from which its weight can be determined from the density of the material used. However, keeping account of thousands of components requires considerable attention. Although weight prediction using CAD is improving, as of today, use of semi-empirical formulae is the standard industrial practice. Over the aircraft production span, the aircraft weight keep changing, it always grows with modifications and additional requirements.

Table 10.2 Larger aircraft mass fraction (more than 19 passengers – three-abreast and above seating).

Group		RJ/Mid-size aircraft		Large aircraft	
		Two engines		Turbofan	
		Turboprop	Turbofan	Two engine	Four engine
Fuselage	$F_{fu} = M_F/MTOM$	9–11	10–12	10–12	9–11
Wing	$F_w = M_w/MTOM$	7–9	9–11	12–14	11–12
H-tail	$F_{ht} = M_{ht}/MTOM$	1.2–1.5	1.8–2.2	1–1.2	1–1.2
V-tail	$F_{vt} = M_{vt}/MTOM$	0.6–0.8	0.8–1.2	0.6–0.8	0.7–0.9
Nacelle	$F_n = M_n/MTOM$	2.5–3.5	1.5–2	0.7–0.9	0.8–0.9
Pylon	$F_{py} = M_{py}/MTOM$	0–0.5	0.5–0.7	0.3–0.4	0.4–0.5
Undercarriage	$F_{uc} = M_{uc}/MTOM$	4–5	3.4–4.5	4–6	4–5
Engine	$F_E = M_E/MTOM$	8–10	6–8	5.5–6	5.6–6
Thrust rev.	$F_{tr} = M_{tr}/MTOM$	0	0.4–0.6	0.7–0.9	0.8–1
Engine con.	$F_{ec} = M_{ec}/MTOM$	1.5–2	0.8–1	0.2–0.3	0.2–0.3
Fuel sys.	$F_{fs} = M_{fs}/MTOM$	0.8–1	0.7–0.9	0.5–0.8	0.6–0.8
Oil sys.	$F_{os} = M_{os}/MTOM$	0.2–0.3	0.2–0.3	0.3–0.4	0.3–0.4
APU		0–0.1	0–0.1	0.1	0.1
Flight con. sys	$F_{fc} = M_{fc}/MTOM$	1–1.2	1.4–2	1–2	1–2
Hydr/pneumatic sys	$F_{hp} = M_{hp}/MTOM$	0.4–0.6	0.6–0.8	0.6–1	0.5–1
Electrical	$F_{elc} = M_{elc}/MTOM$	2–4	2–3	0.8–1.2	0.7–1
Instrument	$F_{ins} = M_{ins}/MTOM$	1.5–2	1.4–1.8	0.3–0.4	0.3–0.4
Avionics	$F_{av} = M_{av}/MTOM$	0.8–1	0.9–1.1	0.2–0.3	0.2–0.3
ECS	$F_{ecs} = M_{ecs}/MTOM$	1.2–2.4	1–2	0.6–0.8	0.5–0.8
Oxygen	$F_{ox} = M_{ox}/MTOM$	0.3–0.5	0.3–0.5	0.2–0.3	0.2–0.3
Furnishing	$F_{fur} = M_{fur}/MTOM$	4–6	6–8	4.5–5.5	4.5–5.5
Misc.	$F_{msc} = M_{msc}/MTOM$	0–0.1	0–0.1	0–0.5	0–0.5
Paint	$F_{pn} = M_{pn}/MTOM$	0.01	0.01	0.01	0.01
Contingency	$F_{con} = M_{con}/MTOM$	0.5–1	0.5–1	0.5–1	0.5–1
MEW – %		53–55	52–55	50–54	48–50
Crew		0.3–0.5	0.3–0.5	0.4–0.6	0.4–0.6
Consumable		1.5–2	1.5–2	1–1.5	1–1.5
OEW – %		54–56	53–56	52–55	50–52
Payload and fuel are traded					
Payload		15–18	12–20	18–22	18–20
Fuel		20–28	22–30	20–25	25–32
MTOM – %		100	100	100	100

Lighter aircraft would show the higher side of the mass fraction. Large turbofan aircraft have wing-mounted engines.

10.7.2 Limitations in Use of Semi-Empirical Formulae

The first thing is to understand how semi-empirical weight prediction equations are formulated and practised. Industrial practices are very different from academic studies in aircraft weight prediction. Industrial weights data are the real data by actually weighing the components and sub-components. These real data are the backbones to formulate the weights equations. Using the real weights data, industrial formulated

Table 10.3 Aircraft component weights data.

Aircraft		Weight – lb							
		MTOW	Fuse	Wing	Emp	Nacelle	Eng.	U/C	n
Piston-engined aircraft									
1.	Cessna 182	2 650	400	238	62	34	417	132	5.7
2.	Cessna310A	4 830	319	453	118	129	852	263	5.7
3.	Beech65	7 368	601	570	153	285	1 008	444	6.6
4.	Cessna404	8 400	610	860	181	284	1 000	316	3.75
5.	Herald	37 500	2986	4365	987	830		1625	3.75
6.	Convair240	43 500	4227	3943	922	1213		1530	3.75
Gas-turbine powered aircraft									
7.	Lear25	15 000	1575	1467	361	241	792	584	3.75
8.	Lear45 class[a]	20 000	2300	2056	385	459	1672	779	3.75
9.	Jet star	30 680	3491	2827	879	792	1750	1061	3.75
10.	Fokker27-100	37 500	4122	4408	977	628	2427	1840	3.75
11.	CRJ200 class[a]	51 000	6844	5369	1001			1794	5.75
12.	F28-1000	65 000	7043	7330	1632	834	4495	2759	3.75
13.	Gulf GII (J)	64 800	5944	6372	1965	1239	6570	2011	3.75
14.	MD-9-30	108 000	16 150	11 400	2 780	1430	6410	4170	3.75
15.	B737-200	115 500	12 108	10613	2718	1392	6217	4354	3.75
16.	A320 class[a]	162 000	17 584	17 368	2 855	2580	12 300	6421	3.75
17.	B747-100	710 000	71 850	86 402	11 850	10 031	34 120	31 427	3.75
18.	A380 class[a]	1 190 497	115 205	170 135	24 104		55 200	52 593	3.75

a)   These are not manufacturers' data.

semi-empirical relations and estimation procedures are very elaborate and not suitable in undergraduate classroom usage. They generate different sets of semi-empirical formulae for each type of technology used to maintain high degree of accuracy. Their semi-empirical equations are not mere regression analyses. The regression analyses are backed up by theories and factors to capture the dispersion. It is evolving all the time.

Industries have a separate dedicated group devoted to predicting aircraft weights. They interact with other groups, most importantly the structures group and production planning group, to feed aerodynamic group for aircraft performance analyses. They work full time from the inception to the end of the new aircraft design programme. They have many internal company documents on methodologies, some are over 300 pages. Industrial practices are proprietary information and kept *commercial in confidence*. Some older industrial methods and some technical reports are quoted in some publications but these are not easily available in public domain. It is difficult to find actual weights components weights data of currently manufactured in public domain. Reference [2] is a good source to obtain aircraft weights of older aircraft, mostly they are not in production but it still gives good values for academic usage.

In academies, various authors have formulated their own semi-empirical relations with the data they generate and from some actual data. The work involves analytical derivations taking into account different technologies adopted in designing aircraft structures, suited to technologies adopted, manufacturing philosophies, aircraft performance goals and cost benefits. Due to these complexities, academic methods inherently do not carry high fidelity. In fact, Roskam [2] demonstrated three methods (i.e. Torenbeek, Cessna and US DATCOM) that yield different values, which is typical when using semi-empirical relations. It is suggested that the readers work out using other available methods to examine any discrepancy. This is the real problem associated with different methods used in weight estimations and one has to live with it.

The SAWE was formed to address these issues in [6]. The SAWE [3] consolidated various equations and presented some generic equations with indices that can accept a wide range of user defined values. These served as baseline equations for many users. Also, [4] gives good relations in a simpler form that yield good values compared to many complex ones. The authors used these two sources and presented the semi-empirical relations incorporating some changes in the expressions and adjusted the indices to match with the data in hand.

It is suggested that the instructors collect as many data, as much as possible, of aircraft component weights within the class of aircraft taken up for the study and adjust the available semi-empirical relations for the technology differences, mainly in the choice of material. The state-of-the-art in weight prediction has room for improvement. The advent of solid modelling (i.e. CAD) of components improved the accuracy of the mass-prediction methodology.

Aircraft weight estimates will undergo several iterations. The final one will be after aircraft sizing and engine matching (Chapter 14).

The development of civil aircraft design has remained in the wake of military aircraft evolution. Competition within these constraints kept civil aircraft designs close to one another. As a result, variation in statistics is lower compared to military aircraft designs that are kept in secrecy. It is for this reason civil and military aircraft weight estimation methods are dealt with separately.

CIVIL AIRCRAFT

The following are some general comments with respect to civil aircraft mass estimation.

i. In the case of single engine propeller driven aircraft, the fuselage starts from aft of the engine bulkhead, as the engine nacelle is separately accounted for. These are mostly small aircraft. For wing-mounted nacelles, this is not the case.

ii. A fuselage-mounted undercarriage is shorter and lighter for the same MTOM. Fixed undercarriage mass fraction is lower than the retractable type (typically 10% higher). The extent depends on retraction type.

iii. Three-engine aircraft are not dealt with here. Also fuselage-mounted turboprop powered aircraft are not discussed here. Not many of these kinds of aircraft are manufactured. There is enough information given herein for the readers to adjust mass accordingly for the classes of current operating aircraft.

Turbofan aircraft with a higher speeds would have a longer range compared to turboprop aircraft and, therefore, would have a higher fuel fraction (typically, 2000 nm range will have around 0.26). However, with rigorous analyses using semi-empirical prediction, better accuracy can be achieved that captures the influence of other parameters that remain obscure in mass fraction method.

10.8 Mass Fraction Method – Civil Aircraft

The *mass fraction method* is used to quickly determine the component weight of an aircraft by relating it in terms of a fraction given in the percentage of Mi/MTOM, where the subscript 'i' represents the ith component (see Section 10.6.1 for nomenclature). With a range of variation among aircraft, the tables in this section are not accurate and serve only as an estimate for a starting point of the initial configuration. Roskam [2] provides an exhaustive breakdown of weights for aircraft of relatively older designs. Newer designs show improvements, especially because of the newer materials, better structural design and manufacturing philosophy used.

A range of applicability mass fraction for smaller aircraft is shown in Table 10.1; add another $\pm 10\%$ for extreme designs. A range of applicability mass fraction for larger aircraft is shown in Table 10.2; add another $\pm 10\%$ for extreme designs.

Because mass and weight are interchangeable, differing by the factor g, wing loading can be expressed in either kg m^{-2} or N m^{-2} (multiply 0.204 816 to convert kg m^{-2} to lb ft^{-2}); this chapter uses the former to be consistent with mass estimation. To obtain the component mass per unit wing area (Mi/S_W, kg m^{-2}), the $M_i/MTOM$ is multiplied by the wing loading; that is, $Mi/S_W = (Mi/MTOM) \times (MTOM/S_W)$. Initially, the wing loading is estimated.

Tables 10.1 and 10.2 summarise the component mass fractions, given as a percentage of the MTOM for quick results. The OEM fraction of the MTOM fits well with the graphs (see Figure 7.3). This mass fraction method is not accurate and only provides an estimate of the component mass involved at an early stage of the project. A variance of ±10% is allowed to accommodate the wide range of data. The tables are useful for estimating fuel mass and engine mass, for example, which are required as a starting point for semi-empirical relations.

10.8.1 Mass Fraction Analyses

Using Eqs. (10.1) and (10.2), the MTOM can be written as follows.

$$\text{MTOM} = \text{MEM} + \text{crew (mass)} + \text{consumable (mass)} + \text{fuel (mass)} + \text{payload (mass)}$$

$$= \text{OEM} + \text{fuel mass} + \text{payload mass}$$

In terms of weight with shorter symbols (MTOM = W_0), this relation can be rearranged as follows.

$$W_0 = W_0 \times \left(\frac{W_{OEW}}{W_0}\right) + W_0 \times \left(\frac{W_f}{W_0}\right) + W_0 \times \left(\frac{W_{PL}}{W_0}\right) = \frac{W_{PL}}{1 - \frac{W_{OEW}}{W_0} - \frac{W_f}{W_0}} \tag{10.12}$$

The problem arises on how get accurate OEM and fuel mass fractions. The OEM fraction (W_{OEW}/W_0 – Figure 7.3) and fuel weight fraction (W_f/W_0 – Figure 7.4) are range dependent. There is a wide spread, primarily on account of different design considerations and it becomes difficult to choose the right value.

Equation (15.38) gives the cruise sector range, $R_{\text{cruise}} = \frac{V}{sfc}\left(\frac{L}{D}\right)\ln\left(\frac{W_i}{W_f}\right)$, which gives,

$$\left(\frac{W_{ini}}{W_{final}}\right) = e^{\frac{-(R_{cruise} \times sfc)}{V \times \left(L/D\right)}}, \tag{10.13}$$

where $W_{\text{cruise-fuel}} = (W_{ini} - W_{final})_{\text{cruise}}$.
From which the fuel weight fraction (W_{fuel}/W_0) can be computed as follows.

$$W_{\text{fuel}} = W_{\text{cruise-fuel}} + (W_{\text{Takeoff-fuel}} + W_{\text{climb-fuel}} + W_{\text{descent-fuel}} + W_{\text{landing-fuel}}).$$

Estimations of MTOM using mass fraction analyses have some inherent inadequacies due to not having the fuel consumption during takeoff, climb, descent and landing. These are estimated from statistical values. The fuel fractions for these segments varies to the extent that using a generic factor will be at the loss of fidelity to determine MTOM. The pitfalls of using the statistical values of the mass fractions are highlighted in Section 10.7.1. One has to depend on the tabulated statistical values given in Section 10.8. If such values are picked up from the tabulated values or any empirical relations, then the merits of each design superiority gets masked.

Since the statistics of mass fraction has a wide error band masking the design merits, it is better to avoid using Eq. (10.13) to get the preliminary MTOM of a new aircraft. Estimation of aircraft component mass requires an initial guesstimated MTOW to start with. Therefore, relying on the statistics of MTOM within the class of aircraft for the technology adopted, this book progresses with initial guess of MTOM, in finer resolution. Thereafter, using proven semi-empirical equations that can be made to reflect the merits of the structural efficiency with suitable material selection, aircraft component masses are evaluated. After computing all the aircraft component masses, a more accurate MTOM is obtained to replace the guessed MTOM.

Sections 10.7.2 and 10.10 outline how to use semi-empirical equations to avoid inherent complexities as there are different sets of semi-empirical equations proposed by different authorities. Figure 7.2 shows in a better resolution the relation between payload versus MTOM separated for the range class. Payload (Eq. (10.12)) and range (Eq. (15.38)) dictate the aircraft design. It is suggested that the readers prepare statistical graphs specific to the class of aircraft types that their project embraces.

However, mass fraction analyses using equations such as Eqs. (10.12) and (10.13) offer good insights to make trend analyses, optimisation, rapid sensitivity studies and so on, as the nature of such studies is equation-based, allowing analytical explorations.

10.9   Graphical Method – Civil Aircraft

The graphical method is based on regression analyses of existing designs. To put all the variables affecting weight in graphical form is difficult and may prove impractical because there will be separate trends based on choice of material, manoeuvre loads, fuselage layout (e.g. single or double aisle; single or double deck), type of engine integrated, wing shape, flight control architecture (e.g. FBW is lighter) and so forth. In principle, a graphical representation of these parameters can be accomplished at the expense of simplicity and rapid estimation.

The graphical method is the fastest way to get some idea of component weights as it reads from the given graph within the error band and is no less accurate that the weight fraction method. The problem is to plot such a graph from existing data in a way that a few extreme points are not taken into account. The simplest form, as presented in this section, obtains a preliminary estimate of component and aircraft weight. At the conceptual design stage, when only the technology level to be adopted and the three-view drawing are available to predict weights, the prediction is approximate. However, with some rigorous analyses using semi-empirical formulae, better accuracy can be achieved capturing the influence of various design considerations as listed before.

Not much literature in public domain shows graphical representation. An earlier work (1942; in FPS units) in [7] presents analytical and semi-empirical treatment that culminates in a graphical representation. This was published in the USA before the gas turbine age, when high-speed aircraft were non-existent. Those graphs served the purpose of the day but still give some insight to component weights.

Given in the following are the updated graphs based on the data given in Table 10.3. They are surprisingly representative but give coarse (low accuracy) values good enough to start sizing analysis in Chapter 14. Some sacrifice is made for the simplicity and yet the purpose served. Most of the weight data in the table is taken from Roskam [2] and some is added by the authors. The authors have tried their best to provide readers with data (Table 10.3) but nothing is better than obtained it directly from the manufacturers themselves.

It is interesting to note that all graphs have MTOW as the independent variable. Aircraft component weight depends on MTOW. The heavier the MTOW, the heavier the component weights. That is also reflected in the Chapter 7 graphs. Strictly speaking, wing weight could have been presented as a function of wing reference area, which in turn depends on the sized wing loading (Chapter 16), that is, MTOW.

Figure 10.5 illustrates civil aircraft component weights in FPS units (MTOW instead of MTOM as the independent variable). Aircraft component weight depends on the MTOW; the heavier the MTOW, the heavier the component weights. Strictly speaking, wing weight could have been presented as a function of the wing reference area, which in turn depends on the sized wing loading (i.e. the MTOW).

To use the graph, the MTOW must first be guesstimated from statistics (see Chapter 7). The first graph provides the fuselage, undercarriage and nacelle weights. Piston engine powered aircraft are low-speed aircraft and the fuselage group weight shows their lightness. There are no large piston engine aircraft in comparison to the gas-turbine type. The lower end of the graph represents piston engines; piston engine nacelles can be slightly lighter in weight.

The second graph in Figure 10.5 shows the wing and empennage group weights. The piston- and gas-turbine engine lines are not clearly separated. FBW driven configurations have a smaller wing and empennage (see Chapter 18), as shown in separate lines with lighter weight (i.e. A320 and A380 class). The newer designs have composite structures that contribute to the light weight.

Figure 10.5 shows consistent trends but does not guarantee accuracy equal to semi-empirical relations, which are discussed in the next section.

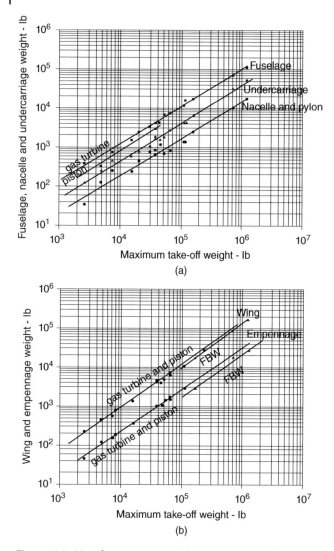

Figure 10.5 Aircraft component weights in pounds – civil aircraft.

10.10   Semi-Empirical Equation Method (Statistical)

Semi-empirical relations are derived from statistical data and tweaked by associated theories to estimate aircraft component mass. It is an involved process to capture the myriad detailed parts. Mass estimation using semi-empirical relations can be inconsistent until a proper one is established. Several forms of semi-empirical weight prediction formulae have been proposed by various analysts, all based on key drivers with refinements as perceived by the proponent. Although all of the propositions have similarity in the basic considerations, their results could differ by as much as 25%. In fact, Roskam [2] describes three methods (i.e. Torenbeek, Cessna and US DATCOM) that yield different values, which is typical when using semi-empirical relations. One of the best ways is to have a known mass data in the aircraft class and then modify the semi-empirical relation for the match; that is, first fine-tune it and then use it for the new design. For a different aircraft class, different fine-tuning is required; the relations provided in this chapter are amenable to modifications.

For coursework, the semi-empirical relations presented in this chapter are from [1] through [6]; some have been modified by the authors and are satisfactory for conventional, all-metal (i.e. aluminium) aircraft. The accuracy depends on how closely aligned is the design. For non-metal and/or exotic metal alloys, adjustments are made depending on the extent of usage.

To demonstrate the effect of the related drivers on mass, their influence is shown as mass increasing by ($\uparrow$) and decreasing by ($\downarrow$) as the magnitude of the driver is increased. For example, $L(\uparrow)$ means that the component weight increases when the length is increased. This is followed by semi-empirical relations to fit statistical data as well as possible.

10.10.1 Fuselage Group (M_F) – Civil Aircraft

A fuselage is essentially a hollow shell to accommodate a payload/crew. The drivers for the fuselage group mass are its length, $L(\uparrow)$, diameter, D_{ave} ($\uparrow$), shell area/volume ($\uparrow$), maximum permissible aircraft velocity, $V(\uparrow)$, pressurisation ($\uparrow$), aircraft load factor and $n(\uparrow)$ and mass increases with engine/undercarriage installation. Maximum permissible aircraft velocity is the dive speed in the *V-n* diagram explained in Chapter 17. For a non-circular fuselage it is the average diameter obtained by taking half of the sum of width and depth of the fuselage. For a rectangular cross-section (invariably unpressurised), it is the average of the fuselage width and height. Length and diameter give the fuselage shell area – the larger the area the greater the weight. Higher velocity and limit load '*n*' would require more material for structural integrity. Installation of engines and/or undercarriage on the fuselage would require additional reinforcement mass. Pressurisation of the cabin increase the fuselage shell hoop stress requiring reinforcement and a rear mounting cargo door is a large addition of mass. (Note again that non-structural items in fuselage, e.g. the furnishing, systems etc., are separately computed.)

Given next are several sets of semi-empirically derived equations by various authors for the transport category of aircraft (the readers may bear in mind the remarks made in the first paragraph that the results may yield different values). Nomenclatures are rewritten in line with what is in this book. The equations are for all-metal (aluminium) aircraft.

By Niu in FPS:

$$W_{Fcivil} = k_1 k_2 \left\{ 2446.4 \left[\begin{array}{l} 0.5(W_{flight\_gross\_weight} + W_{landing\_weight}) \times \left(1 + \dfrac{1.5\Delta P}{4}\right)^{0.5} \times \\ S_{net\_fus\_wetted\_area} \times [0.5\,(W + D)]^{0.5} \times L^{0.6} \times 10^{-4} - 678 \end{array} \right] \right\} \tag{10.14}$$

where

$k_1 = 1.05$ for a fuselage-mounted undercarriage
  $= 1.0$ for a wing-mounted undercarriage
$k_2 = 1.1$ for a fuselage-mounted engine
  $= 1.0$ for a wing-mounted engine
$S_{net\_fus\_wetted\_area}$ = fuselage shell gross area minus cut-outs

Two of Roskam's suggestions are as follows [2]:

1. The General Dynamic method:

$$W_{Fcivil} = 10.43(K_{inlet})^{1.42}(q_D/100)^{0.283}(MTOW/1000)^{0.95}(L/D)^{0.71} \tag{10.15}$$

where

$K_{inlet} = 1.25$ for inlets in or on the fuselage; otherwise, 1.0
q_D = dive dynamic pressure in psf
L = fuselage length
D = fuselage depth

2. The Torenbeek method:

$$W_{Fcivil} = 0.021 K_f \{V_D L_{HT}/(W+D)\}^{0.5} (S_{fus\_gross\_area})^{1.2} \tag{10.16}$$

where

> $K_f = 1.08$ for a pressurised fuselage
> $= 1.07$ for the main undercarriage attached to the fuselage
> $= 1.1$ for a cargo aircraft with a rear door
> V_D = design dive speed in knots equivalent air speed (KEAS)
> $L_{H\,tail}$ = tail arm of the H-tail
> $S_{net\_fus\_wetted\_area}$ = fuselage shell gross area less the cut outs

By Jenkinson (taken from Howe) [8] in SI is given here,

$$M_{Fcivil} = 0.039 \times (2 \times L \times D_{ave} \times V_D{}^{0.5})^{1.5} \tag{10.17}$$

The authors are not in a position to compare which one is better. As mentioned earlier, that depends on the type: weight equations do show some inconsistency. Torenbeek's equation has been in use for a long time. The Eq. (10.17) is the simplest one. Classroom usage may make the choice.

The authors modify Eq. (10.17) as shown in Eq. (10.18). The results appear to have yielded to satisfactory results. This allows capturing more details of the technology level. The authors suggest using Eq. (10.18) for coursework.

$$M_{Fcivil} = c_{fus} \times k_e \times k_p \times k_{uc} \times k_{door} \times (MTOM \times n_{ult})^x \times (2 \times L \times D_{ave} \times V_D{}^{0.5})^y \tag{10.18}$$

where c_{fus} is a generalised constant to fit the regression, as follows:

> $c_{fus} = 0.038$ for small unpressurised aircraft (leaving the engine bulkhead forward)
> $= 0.041$ for a small transport aircraft (≤ 19 passengers)
> $= 0.04$ for 20–100 passengers
> $= 0.039$ for a midsized aircraft
> $= 0.0385$ for a large aircraft
> $= 0.04$ for a double-decked fuselage
> $= 0.037$ for an unpressurised, rectangular-section fuselage

All k-values are 1 unless otherwise specified for the configuration, as follows:

> k_e = for fuselage-mounted engines $= 1.05$–1.07
> k_p = for pressurisation $= 1.08$ up to 40 000-ft operational altitude
> $= 1.09$ above 40 000-ft operational altitude
> $k_{uc} = 1.04$ for a fixed undercarriage on the fuselage
> $= 1.06$ for wheels in the fuselage recess
> $= 1.08$ for a fuselage-mounted undercarriage without a bulge
> $= 1.1$ for a fuselage-mounted undercarriage with a bulge
> $k_{VD} = 1.0$ for low-speed aircraft below Mach 0.3
> $= 1.02$ for aircraft speed $0.3 < \text{Mach} < 0.6$
> $= 1.03$–1.05 for all other high-subsonic aircraft
> $k_{door} = 1.1$ for a rear-loading door

The value of index x depends on the aircraft size: 0 for aircraft with an ultimate load (*nult*) < 5 and between 0.001 and 0.002 for ultimate loads of (*nult*) > 5 (i.e. lower values for heavier aircraft). In general, $x = 0$ for civil aircraft; therefore, $(MTOM \times nult)^x = 1$. The value of index y is very sensitive. Typically, the last term index, y is 1.5, but it can be as low as 1.45. It is best to fine-tune with a known result in the aircraft class and then use it for the new design.

Then for civil aircraft ($n_{ult} < 5$) Eq. (10.18) can be simplified to:

$$M_{Fcivil} = c_{fus} \times k_e \times k_p \times k_{uc} \times k_{door} \times (2 \times L \times D_{ave} \times V_D{}^{0.5})^{1.5} \qquad (10.19)$$

For the club-flying type small aircraft, the fuselage weight with a fixed undercarriage can be written as:

$$M_{Fsmall\_a/c} = 0.038 \times 1.07 \times k_{uc} \times (2 \times L \times D_{ave} \times V_D{}^{0.5})^{1.5} \qquad (10.20)$$

If new materials are used, then the mass changes by the factor of usage. For example, $x\%$ mass is new material that is $y\%$ lighter; the component mass is as follows:

$$M_{Fcivil\_new\text{-}material} = (1 - xy) M_{Fcivil} \qquad (10.21)$$

In a simpler form, if there is some reduction in mass due to lighter material then reduce by that factor. Say there is 10% mass saving then $M_{Fcivil} = 0.9 \times M_{Fcivil\_all\,metal}$.

10.10.2 Wing Group – Civil Aircraft

The wing is a thin, flat, hollow structure. The hollow space is used for fuel storage in sealed wet tanks or in separate tanks fitted in; it also houses control mechanisms, accounted for separately. As an option, the engines can be mounted on the wing. Wing-mounted nacelles are desirable for in-flight wing-load relief; however, for small turbofan aircraft, they may not be possible due to the lack of ground clearances (unless engines are mounted over the wing (Hondajet) or it is a high-wing aircraft (BAe146) – few are manufactured).

The drivers for the wing group mass are its planform reference area, S_W (↑); aspect ratio, AR(↑); quarter-chord wing sweep, $\Lambda_{1/4}$, is (↑); wing-taper ratio, (↑); mean-wing t/c ratio, (↓); maximum permissible aircraft velocity, V(↑); aircraft limit load, n(↑); fuel carried, (↓) and wing-mounted engines, (↓). The AR and wing area give the wing span, b. Because the quarter-chord wing sweep, $\Lambda_{1/4}$ is expressed in the cosine of the angle, it is placed in the denominator, as is the case with the t/c ratio because the increase in the t/c ratio decreases the wing weight due to having better stiffness.

A well-established general analytical wing weight equation published by SAWE [4] is as follows (others are not included):

$$M_W = K \, (M_{dg} N_Z)^{x1} \, S_W{}^{x2} \, AR^{x3} \, (t/c)^{x4} \, (1 + \lambda)^{x5} \, (\cos \Lambda_{1/4})^{x6} \, (B/C)_t{}^{x7} \, S_{CS}{}^{x8} \qquad (10.22)$$

where

    C = wing root chord
    B = width of box beam at wing root
    S_{CS} = wing-mounted control surface reference area
    M_{dg} = MTOM

The authors make some modification to the equation for classroom usage. The term $(M_{dg} N_Z)^{x1}$ in the book nomenclature is $(MTOM \times n_{ult})^{0.48}$. The term $(B/C)_t{}^{x7} \, S_{CS}{}^{x8}$ is replaced by the factor 1.005 and included in the factor K.

The lift load is upward, therefore, mass carried by the wing (e.g. fuel and engines) would relieve the upward bending (like a bow), resulting in stress relief that saves wing weight. Fuel is a variable mass and when it is emptied, the wing does not get the benefit of weight relief; but if aircraft weight is reduced, the fixed mass of the engine offers relief. Rapid methods should be used to obtain engine mass for the first iteration.

Writing the modified equation in terms of this book's notation, Eq. (10.22) is replaced by Eq. (10.23) in SI (the MTOM is estimated):

$$M_W = c_w \times k_{uc} \times k_{sl} \times k_{sp} \times k_{wl} \times k_{re} \times (MTOM \times n_{ult})^{0.48} \times S_W{}^{0.78} \times AR \times (1 + \lambda)^{0.4}$$

$$\times (1 - W_{Fuel\_mass\_in\_wing}/MTOW)^{0.4}/[(\cos \Lambda) \times t/c^{0.4}] \qquad (10.23)$$

where

$c_w = 0.0215$ and flaps are standard fitments to the wing

$k_{uc} = 1.02$ for wing-mounted undercarriage, otherwise 1.0

$k_{sl} = 1.04$ for use of a slat

$k_{sp} = 1.01$ for a spoiler

$k_{wl} = 1.01$ for a winglet (a generalised approach is to have a standard size)

$k_{re} = 1$ for no engine, 0.98 for two engines and 0.95 for four engines (generalised)

If new materials are used, then the mass changes by the factor of usage. For example, x% mass is new material that is y% lighter; the component mass is as follows:

$$M_{Wcivil\_new\text{-}material} = (1 - xy)\, M_{Wcivil} \tag{10.24}$$

In a simpler form, if there is some reduction in mass due to lighter material then reduce by that factor. Say there is 10% mass saving then $M_{Wcivil} = 0.9 \times M_{Wcivil\_all\ metal}$.

10.10.3 Empennage Group – Civil Aircraft

Empennages are also lifting surfaces and use semi-empirical equations similar to those used for the wing. The empennage does not have an engine or undercarriage installation. It may carry fuel, but in this book, fuel is not stored in the empennage. The drivers are the same as those in the wing group mass.

Equation (10.23) is modified to suit empennage mass estimation. Both the horizontal and vertical tail plane mass estimations have a similar form but differ in the values of constants used.

$$M_{EMP\_civil} = k_{emp} \times k_{mat} \times k_{conf}\, (MTOM \times n_{ult})^{0.48} \times S_W^{0.78} \times AR \times (1 + \lambda)^{0.4} / (\cos \Lambda \times t/c^{0.4}) \tag{10.25}$$

If new materials are used, then the mass changes by the factor of usage. For example, x% mass is new material that is y% lighter; the component mass is as follows:

$$M_{EMPcivil\_new\text{-}material} = (1 - xy)\, M_{EMPcivil} \tag{10.26}$$

In a simpler form, if there is some reduction in mass due to a lighter material then reduce by that factor. Say there is 10% mass saving, then $M_{EMPcivil} = 0.9 \times M_{EMPcivil\_all\ metal}$.

Writing the modified equations in terms of the book's notation, Eq. (10.25) is changed to (empennage is now split into a H-tail and V-tail):

Horizontal Tail:

For all tail movement, use $k_{emp} = 0.02$, $k_{conf} = 1.05$, otherwise 1.0.

$$M_{HT} = 0.02 \times k_{conf} \times (MTOM \times n_{ult})^{0.48} \times S_{HT}^{0.78} \times AR_{HT} \times (1 + \lambda_{HT})^{0.4} / (\cos \Lambda_{HT} \times t/c_{HT}^{0.4}) \tag{10.27}$$

Vertical Tail:

For V-tail configurations, use $k_{emp} = 0.0215$ (jet) to 0.025 (prop), $k_{conf} = 1.5$ for T-tail, 1.2 for mid-tail and 1.0 for low tail.

$$M_{VT} = 0.0215 \times k_{conf} \times (MTOM \times n_{ult})^{0.48} \times S_{VT}^{0.78} \times AR_{VT} \times (1 + \lambda_{VT})^{0.4} / (\cos \Lambda_{VT} \times t/c_{VT}^{0.4})$$
$$\tag{10.28}$$

10.10.4 Nacelle Group – Civil Aircraft

The nacelle group can be classified distinctly as a pod that is mounted and interfaced with pylons on the wing or fuselage, or it can be combined. The nacelle size depends on the engine size and type. The nacelle mass semi-empirical relations are as follows.

Jet engine type (includes pylon mass)

The preferred method is to use the factor of the manufacturer supplied dry engine weight for the bypass ratio (BPR) as follows.

$$M_{\text{NAC jet}} = k_{\text{nac}} \times \text{dry engine weight} \tag{10.29}$$

where

$$k_{\text{nac}} = 0.28 \text{ to } 0.3 \text{ for BPR } 4.0$$
$$= 0.30 \text{ to } 0.32 \text{ for BPR } 6.0$$
$$= 0.32 \text{ to } 0.34 \text{ for BPR } 8.0$$

If dry engine weight is not available, then use the factor for the T_{SLS} in kN per engine as follows.

$$M_{\text{NAC jet}} = k_{\text{nac}} \times \textit{thrust} \text{ (kN) per nacelle.} \tag{10.30}$$

where

$$k_{\text{nac}} = 6.2 \text{ for a BPR } 4.0$$
$$= 6.4 \text{ for BPR } 6$$
$$= 6.6 \text{ for BPR } 8$$

Thrust reverser increases the nacelle weight by about 40–50%.

Turboprop Engine Nacelle

Turboprop type pods are slung under the wing or placed above the wing with little pylon, unless it is an aft-fuselage-mounted pusher type (e.g. the Piaggio Avanti). For the same power, turboprop engines are nearly 20% heavier, requiring stronger nacelles; however, they have a small or no pylon.

For a wing-mounted turboprop nacelle:

$$M_{\text{NAC prop}} = 6.5 \times \textit{SHP} \text{ per nacelle} \tag{10.31}$$

For a turboprop nacelle housing an undercarriage:

$$M_{\text{NAC prop uc}} = 8 \times \textit{SHP} \text{ per nacelle} \tag{10.32}$$

For a fuselage-mounted turboprop nacelle with a pylon:

$$M_{\text{NAC prop}} = 7 \times 4 \times \textit{SHP} \text{ per nacelle} \tag{10.33}$$

Piston Engine Nacelle

For fuselage-mounted (single) engine, the nacelle is integral to the fuselage.
For wing-mounted engine, the piston engine nacelle has as follows:

$$M_{\text{NAC\_piston}} = 0.4 \times \textit{HP} \text{ per nacelle} \tag{10.34}$$

For a wing-mounted, piston engine nacelle:

$$M_{\text{NAC\_piston}} = 0.5 \times \textit{HP} \text{ per nacelle} \tag{10.35}$$

If new materials are used, then the mass changes by the factor of usage. For example, x% mass is new material that is y% lighter; the component mass is as follows:

$$M_{\text{NACcivil\_new-material}} = (1 - xy) M_{\text{NACcivil}} \tag{10.36}$$

In a simpler form, if there is some reduction in mass due to lighter material then reduce by that factor. Say there is a 10% mass saving, then $M_{\text{Fcivil}} = 0.9 \times M_{\text{Fcivil\_all metal}}$.

10.10.5 Undercarriage Group – Civil Aircraft

Chapter 9 describes undercarriages and their types in detail. Undercarriage size depends on an aircraft's MTOM. Mass estimation is based on a generalised approach of the undercarriage classes that demonstrate strong statistical relations, as discussed herein.

Tricycle Type (Retractable) – Wing-Mounted (Nose and Main Gear Estimated Together)
For a low-wing mounted undercarriage:

$$M_{UC\ wing} = 0.04 \times \text{MTOM} \tag{10.37}$$

For a mid-wing-mounted undercarriage:

$$M_{UC\ wing} = 0.042 \times \text{MTOM} \tag{10.38}$$

For a high-wing-mounted undercarriage:

$$M_{UC\ wing} = 0.044 \times \text{MTOM} \tag{10.39}$$

Tricycle Type (Retractable) – Fuselage-Mounted (Nose and Main Gear Estimated Together)
These are typically high-wing aircraft. A fuselage-mounted undercarriage usually has shorter struts.

$$M_{UC\ fus} = 0.04 \times \text{MTOM} \tag{10.40}$$

For a fixed undercarriage, the mass is 10–15% lighter; for a tail-dragger, it is 20–25% lighter.

10.10.6 Miscellaneous Group – Civil Aircraft

Carefully examine which structural parts are omitted (e.g. delta fin). Use the mass per unit area for a comparable structure (i.e. a lifting surface or a body of revolution; see Section 10.4). If any item does not fit into the standard groups listed up to Section 10.10.5, then it is included in this group. Typically, this is expressed as:

$$M_{MISC} = 0 \text{ to } 1\% \text{ of the MTOM} \tag{10.41}$$

10.10.7 Power Plant Group – Civil Aircraft

The power plant group consists of the components listed in this section. At the conceptual design stage, they are grouped together to obtain the power plant group mass. It is better to use the engine manufacturer's weight data available in the public domain. However, given here are the semi-empirical relations to obtain the engine mass.

Turbofans

(1) Equipped dry engine mass (M_{E_dry})
(2) Thrust-reverser mass (M_{TR}), if any – mostly installed on bigger engines
(3) Engine control system mass (M_{EC})
(4) Fuel system mass (M_{FS})
(5) Engine oil system mass (M_{OI})

Turboprops

(1) Equipped dry engine mass (M_{E_dry}) – includes reduction in gear mass to drive the propeller
(2) Propeller (M_{PR})
(3) Engine control system mass (M_{EC})
(4) Fuel system mass (M_{FS})
(5) Engine oil system mass (M_{OI})

Piston Engines

(1) Equipped dry engine mass (M_E) – includes reduction gear, if any
(2) Propeller mass (M_{PR})
(3) Engine control system mass (M_{EC})
(4) Fuel system mass (M_{FS})
(5) Engine oil system mass (M_{OI})

In addition, there could be a separate auxiliary power unit (APU) – generally in bigger aircraft – to supply electrical power driven by a gas turbine.

Engine manufacturers supply the equipped dry engine mass (e.g. with fuel pump and generator). The engine thrust-to-weight ratio ($T/M_{\text{dry engine}}$; thrust is measured in Newton) is a measure of dry engine weight in terms of rated thrust (T_{SLS}). Typically, $T/M_{\text{dry engine}}$ varies between 4 and 8 (special-purpose engines can be more than 8). For turboprop engines, the mass is expressed as ($SHP/M_{\text{dry engine}}$); for piston engines, it is ($HP/M_{\text{dry engine}}$).

The remainder of the systems including the thrust reverser (for some turbofans), oil system, engine controls and fuel system are listed here. The total power plant group mass can be expressed semi-empirically (because of the similarity in design, the relationship is fairly accurate). The power plant group mass depends on the size of the engine expressed by the following equations:

For the variant engine mass, only add the increment/decrement of the dry engine mass to the baseline power plant group mass.

Turbofan
Civil aircraft power plant (with no thrust reverser):

$$M_{\text{ENG tf}} = 1.4 \times M_{\text{DRYENG}} \text{ per engine} \tag{10.42}$$

Civil aircraft power plant (with thrust reverser):

$$M_{\text{ENG tf}} = 1.5 \times M_{\text{DRYENG}} \text{ per engine} \tag{10.43}$$

$$M_{\text{ENG\_var}} \text{ per engine} = (1.2 \times M_{\text{DRYENG\_baseline}}) + \Delta M_{\text{ENG\_var}}$$

Turboprop
Civil aircraft power plant:

$$M_{\text{ENG\_tp}} = k_{\text{tp}} \times M_{\text{DRYENG}} \text{ per engine} \tag{10.44}$$

where $1.5 \leq k_{\text{tp}} \leq 1.8$. (due to propeller and gear box)

$$M_{\text{ENG\_var}} \text{ per engine} = (1.2 \times M_{\text{DRYENG\_baseline}}) + \Delta M_{\text{ENG\_var}}$$

Piston Engine
Civil aircraft power plant:

$$M_{\text{ENG ps}} = k_{\text{p}} \times M_{\text{DRYENG}} \text{ per engine} \tag{10.45}$$

where $1.4 \leq k_{\text{p}} \leq 1.5$.

APU (if any)

$$M_{\text{APU}} = 0.001 \text{ to } 0.005 \times M_{\text{DRYENG}} \text{ of an engine} \tag{10.46}$$

10.10.8 Systems Group – Civil Aircraft

The systems group includes flight controls, hydraulics and pneumatics, electrical, instrumentation, avionics, and environmental controls (see Section 10.6.1).

$$M_{\text{SYS}} = 0.1 \text{ to } 0.11 \times MTOW \text{ for large aircraft} > 100 \text{ passengers} \tag{10.47}$$

$$M_{SYS} = 0.12 \text{ to } 0.14 \times MTOW \text{ for smaller transport aircraft of } < 100 \text{ passengers} \qquad (10.48)$$

$$M_{SYS} = 0.05 \text{ to } 0.07 \times MTOW \text{ for unpressurised aircraft} \qquad (10.49)$$

10.10.9 Furnishing Group – Civil Aircraft

This group includes the seats, galleys, furnishings, toilets, oxygen system and paint (see Section 10.6.1). At the conceptual design stage, they are grouped together to obtain the furnishing group.

$$M_{FUR} = 0.07 \text{ to } 0.08 \times MTOW \text{ for large aircraft } > 100 \text{ passengers} \qquad (10.50)$$

$$M_{FUR} = 0.06 \text{ to } 0.07 \times MTOW \text{ for smaller transport aircraft of } < 100 \text{ passengers} \qquad (10.51)$$

$$M_{FUR} = 0.02 \text{ to } 0.025 \times MTOW \text{ for unpressurised aircraft} \qquad (10.52)$$

10.10.10 Contingency and Miscellaneous – Civil Aircraft

A good designer plans for contingencies; that is:

$$M_{CONT} = (0.01 \text{ to } 0.025) \times MTOW \qquad (10.53)$$

Miscellaneous items should also be provided for; that is:

$$M_{MISC} = 0 \text{ to } 1\% \text{ of } MTOW \qquad (10.54)$$

10.10.11 Crew – Civil Aircraft

A civil aircraft crew consists of a flight crew and a cabin crew. Except for very small aircraft, the minimum flight crew is two, with an average of 90 kg per crew member. The minimum number of cabin crew depends on the number of passengers. Operators may employ more than the minimum number, which is listed in Table 10.4.

10.10.12 Payload – Civil Aircraft

A civil aircraft payload is basically the number of passengers at 100 kg per person plus the cargo load. The specification for the total payload capacity is derived from the operator's requirements. The payload for cargo aircraft must be specified from market requirements.

Table 10.4 Minimum cabin crew number for passenger load.

Number of passengers	Minimum number of cabin crew	Number of passengers	Minimum number of cabin crew
≥19	1	200–<250	7
19–<30	2	250–<300	8
21–<50	3	300–<350	9
50–<100	4	350–<400	10
100–<150	5	400–<450	11
150–<200	6	450–<500	12

10.10.13   Fuel – Civil Aircraft

The fuel load is mission specific. For civil aircraft, required fuel is what is needed to meet the design range (i.e. market specification) plus mandatory reserve fuel. It can be determined by the proper performance estimation. At this design stage, statistical data are the only means to estimate fuel load, which is then revised in Chapter 15.

The payload and fuel mass are traded for off-design ranges; that is, a higher payload (if accommodated) for shorter range and vice versa.

10.11   Centre of Gravity Determination

After obtaining the component masses (i.e. weights), it is now time to locate the aircraft CG. A reference-coordinate system is essential for locating the CG position relative to an aircraft. A suggested coordinate system is to use the x-axis along the ground level (or at another suitable level). Typically, the fuselage axis is parallel or nearly parallel to the x-axis. The z-axis is orthogonal to x-axis, passing through the farthest point of the nose cone (i.e. tip), as shown in Figure 10.6. If the x-axis is taken as fuselage axis then the z-axis can be the zero reference plane (ZRP).

The first task is to estimate the CG position for each component group from the statistical data. Figure 10.6 provides generic information for locating the positions. Typical ranges of the CG position relative to the component are given in Table 10.5. At this stage, the extreme forward-most and rearmost CG positions (i.e. x-coordinates) have not been determined and will be done later.

It must be emphasised that the conceptual design phase relies on designers' experience supported by the available statistical data. Aircraft components' CG locates the aircraft CG; therefore, the components must be positioned accordingly. At the conceptual design phase (i.e. not yet manufactured), it is not possible to obtain accurate component weights and their CG locations are yet to evolve. Designers' experience is the way to minimise error. However, errors at this stage do not hinder the progress of the conceptual design, which is revised through iterations for better accuracy. The industry can then confidently predict the accuracy level within ±3 to ±5%, which is sufficient to study the competition before the go-ahead is given.

Immediately after go-ahead is obtained, a considerable amount of budget is released for the Project Definition when major structural details are drawn in CAD to obtain a more accurate component weight and the CG

Figure 10.6 Aircraft component CG locations (see Table 10.5).

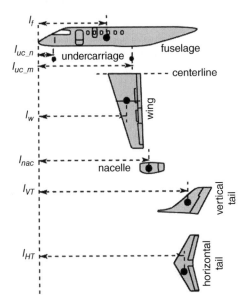

Table 10.5 Typical values of component CG locations – civil aircraft.

	Component	CG and typical % of component characteristic length
1.	Fuselage group	40–45% (military aircraft on higher percentage)
2.	Wing group	No slat – $\approx 30\%$ of MAC
		With slat $\approx 25\%$ MAC
3.	H-tail group	$\approx 30\%$ of $MAC_{\text{H-tail}}$
4.	V-tail group	$\approx 30\%$ of $MAC_{\text{V-tail}}$
5.	Undercarriage group	At wheel centre as positioned (nose and main wheels taken separately)
6.	Nacelle + pylon group	As positioned (wing or fuselage mount) @ $\approx 40\%$ of nacelle length
7.	Miscellaneous	As positioned – use similarity
8.	Power plant group	As positioned (wing or fuselage mount) @ $\approx 50\%$ of engine length
9.	Systems group	As positioned – use similarity (typically 35–40% of fuselage)
10.	Furnishing group	As positioned – use similarity
11.	Contingencies	As positioned (adjustable to ballast)
	MEM	(Need not compute for CG at this moment)
12.	Crew	As positioned
	OEM	Compute
13.	Payload	As positioned (distribute around CG)
14.	Fuel	As positioned (distribute around CG)
	MTOM	Compute
	MRM	Compute

location. Also bought-out items for systems, the undercarriage and power plant are identified and suppliers of these items provide accurate weight and CG data. In fact, during the project definition study, very accurate predictions (less than ± 2 to $\pm 3\%$) can be obtained.

Expressions for $\bar{x}$, $\bar{y}$ and $\bar{z}$ coordinates are given in Eqs. (10.55)–(10.57)

$$\bar{x} = \frac{\sum_{i}^{n}(\text{Component weight} \times \text{distance from nose reference point})}{\text{Aircraft weight}} \tag{10.55}$$

$$\bar{z} = \frac{\sum_{i}^{n}(\text{Component weight} \times \text{distance from ground reference line})}{\text{Aircraft weight}} \tag{10.56}$$

Aircraft weight

$$\bar{y} = 0 \text{ (CG is at the plane of symmetry)} \tag{10.57}$$

If the computation does not indicate the CG within the specified ranges, then move the wing and/or engine(s) to bring CG around the desired percentages of the *mean aerodynamic chord* (MAC) until a satisfactory solution reached. Moving the wing would also move the CG and neutral point – this may require iteration (*wing-chasing*). Fuel tanks can be slightly modified. Batteries are heavy items (or any other heavy items that can be moved) and can be located at a desirable position, that is, fine-tune the CG location.

10.12   Worked-Out Example – Bizjet Aircraft

The semi-empirical relations described in Section 10.10 are now applied to obtain an example of the configuration worked out in Chapters 8 and 9 in the preliminary configuration layout based on guessed MTOM.

In this section, more accurate estimates of component and aircraft mass along with the CG locations of the Bizjet aircraft are established. In case there is an unacceptable difference between the guessed and estimated MTOM, then the preliminary Bizjet aircraft configuration needs to be refined through an iterative process with more accurate data. The iteration process may require the repositioning of aircraft components. From Section 8.9, the following are obtained for the preliminary baseline Bizet aircraft configuration.

MTOM = 9500 kg (initially guessed, and to be refined in this exercise).

Two turbofans (i.e. Honeywell TFE731), each with $T_{SLS} = 17\,235$ N (3800 lbs), BPR 4 and dry weight of 379 kg (836 lbs).

The results from this section are compared with the graphical solutions in Figure 10.3 and in Table 10.3.

10.12.1 Fuselage Group Mass

Consider a 5% weight reduction due to composite usage in non-load-bearing structures (e.g. floorboards).

$$L = 15.24 \text{ m}, D_{ave} = 1.75 \text{ m}, V_D = 380 \text{ KEAS} = 703.76 \text{ km/h} = 195.5 \text{ m s}^{-1} \text{ and}$$

$$c_{fus} = 0.04, k_e = 1.04, k_p = 1.09, k_{uc} = 1.06, \text{ and } k_{mat} = (0.9 + 0.9 \times 0.1) = 0.99$$

Use Eq. (10.19):

$$M_{Fcivil} = c_{fus} \times k_e \times k_p \times k_{uc} \times k_{door} \times (MTOM \times n_{ult})^x \times (2 \times L \times D_{ave} \times V_D^{0.5})^y$$

For civil aircraft having $n_{ult} < 0$, then $(MTOM \times n_{ult})^x = 1$; therefore:

$$M_{Fcivil} = 0.04 \times 1.04 \times 1.09 \times 1.06 \times 1 \times 1 \times (2 \times 15.24 \times 1.75 \times 195.5^{0.5})^{1.5}$$

$$= 0.048 \times (2 \times 15.24 \times 1.75 \times 195.5^{0.5})^{1.5} = 0.048 \times (53.35 \times 13.98)^{1.5}$$

$$= 0.048 \times (745.95)^{1.5} = 0.048 \times 20\,373.3 = 978 \text{ kg}$$

There is a 5% reduction of mass due to use of composites $M_{Fcivil} = 0.95 \times M_{Fcivil\_all \ metal} \approx 930 \text{ kg}$

Checking with Torenbeek's method (Eq. (10.16))

$$W_{Fcivil} = 0.021 K_f \{V_D L_{HT}/(W + D)\}^{0.5} (S_{fus\_gross\_area})^{1.2}$$

$K_f = 1.08, L = 50 \text{ ft}, W = 5.68 \text{ ft}, D = 5.83 \text{ ft}, L_{HT} = 25 \text{ ft}, V_D = 380 \text{ KEAS}, S_{fus\_gross\_area} = 687 \text{ft}^2$

Therefore, $W_{Fcivil} = 0.021 \times 1.08 \times 1.07 \times \{380 \times 25/(5.68 + 5.83)\}^{0.5} \times (687)^{1.2}$

$$= 0.0243 \times (825.4)^{0.5} \times 2537.2 = 0.0243 \times 28.73 \times 2537.2 = 1770 \text{ lb (805 kg)}$$

The higher value of the two (930 kg) is kept. This gives a safer approach to begin with.

10.12.2 Wing Group Mass

Consider that it has a 10% composite secondary structure, that is, $k_{mat} = (0.9 + 0.9 \times 0.1) = 0.99$.

It has no slat making $k_{sl} = 1.0$ and is without a winglet, $k_{wl} = 1$. For the spoiler, $k_{sl} = 1.01$ and for a wing-mounted undercarriage $k_{uc} = 1.02$. Equation (10.23) gives

$$M_W = c_w \times k_{uc} \times k_{sl} \times k_{sp} \times k_{wl} \times k_{re} \times (MTOM \times n_{ult})^{0.48} \times S_W^{0.78} \times AR \times (1 + \lambda)^{0.4}$$

$$\times (1 - W_{Fuel\_mass\_in\_wing}/MTOW)^{0.4}/[(\cos \Lambda) \times t/c^{0.4}]$$

where $c_w = 0.0215$ and flaps are standard fitment to the wing.

$S_W = 30\ m^2$, $n_{ult} = 4.125$, $b = 15\ m$, $AR = 6.75$, $\lambda = 0.375$, Fuel in wing, $M_{WR} = 1140\ kg$, $\Lambda = 14°$ and $t/c = 0.10$. Load factor, $n = 3.8$.

Equation (10.23) becomes:

$$M_W = (0.0215 \times 0.99) \times (1.02 \times 1.01) \times (9500 \times 3.8)^{0.48} \times 29^{0.78} \times 6.75 \times (1 + 0.375)^{0.4}$$

$$\times\ (1 - 1140/9500)^{0.4}/(\cos 14 \times 0.10^{0.4})$$

$$M_W = (0.0213 \times 1.03 \times 154) \times (13.8 \times 6.75 \times 1.136) \times 0.88^{0.4}/(0.97 \times 0.406)$$

$$M_W = 3.38 \times 105.8 \times 0.95/(0.3977) = 339.7/0.3977 = 854\ kg$$

10.12.3 Empennage Group Mass

Use Eq. (10.25).

$$M_{EMPcivil} = k_{emp} \times k_{mat} \times k_{conf} (MTOM \times n_{ult})^{0.48} \times S_W^{0.78} \times AR \times (1 + \lambda)^{0.4}/(\cos \Lambda \times t/c^{0.4})$$

Horizontal Tail:

A conventional all moving tail has $k_{emp} = 0.2$, $k_{conf} = 1.0$. Consider that it has a 20% composite secondary structure, that is, $k_{mat} = (0.8 + 0.9 \times 0.2) = 0.98$.

$MTOM = 9500\ kg$, $n_{ult} = 3.8$, $S_{HT} = 5.5\ m^2$ (exposed), $AR = 3.5$, $\lambda = 0.3$, $\Lambda = 16°$ and $t/c = 0.10$.

Use Eq. (10.27).

$$M_{HT} = (0.02 \times 0.98) \times (9500 \times 3.8)^{0.484} \times 5.5^{0.78} \times 3.5 \times (1 + 0.3)^{0.4}/(\cos 16 \times 0.10^{0.4})$$

$$= 0.0196 \times 160.7 \times 3.8 \times 3.5 \times 1.11/(0.961 \times 0.406) = 45.9/0.39 = 118\ kg$$

Vertical Tail (T-tail):

$k_{conf} = 1.5$. Consider that it has a 20% composite secondary structure, that is, $k_{mat} = (0.8 + 0.9 \times 0.2) = 0.98$.

$MTOM = 9200\ kg$, $n_{ult} = 3.8$, $S_{VT} = 3.5\ m^2$ (exposed), $AR = 2.0$, $\lambda = 0.5$, $\Lambda = 20°$, $t/c = 0.1$ and $V_D = 380\ kt = 703.76\ km/h = 195.5\ m\ s^{-1}$.

Use Eq. (10.28).

$$M_{VT} = (0.0215 \times 0.98 \times 1.5) \times (9500 \times 3.8)^{0.484} \times 3.5^{0.78} \times 2 \times (1 + 0.5)^{0.4}/(\cos 20 \times 0.1^{0.4})$$

$$= 0.0316 \times 160.7 \times 2.66 \times 2 \times 1.176/(0.94 \times 0.406) = 31.8/0.382 = 83\ kg$$

10.12.4 Nacelle Group Mass

The Honeywell TFE731 at dry weight is 379 kg per engine.

Engine thrust = 17 230 N (3800 lb) per engine with a BPR 4.

Use Eq. (10.29): $M_{nac + pylon} = 0.3 \times 379 = 113.7 \approx 114\ kg/nacelle$

Two nacelles = 228 kg.

10.12.5 Undercarriage Group Mass

MTOM = 9500 kg low-wing mount; $M_{U/C}$ *wing* = $0.04 \times 9500 = 380\ kg$.

10.12.6 Miscellaneous Group Mass

Fortunately, there are no miscellaneous structures in the examples considered herein.

10.12.7   Power Plant Group Mass

This is determined from statistics until it is sized in Chapter 14. A typical engine is of the class Allison TFE731–20 turbofan with thrust per engine = 15 570 N to 17 230 N (3500–3800 lbs).

If a manufacturer's dry weight is available, it is better to use it rather than semi-empirical relations. In this case, the manufacturer's dry weight is in the public domain. $M_{\text{DRYENG}} = 379$ kg per engine. This gives:

$$(T/W_{\text{dry engine}}) = 17\,230/(379 \times 9.81) = 4.63$$

The total power plant group mass can be expressed semi-empirically as
$M_{\text{ENG}} = 1.4 \times M_{\text{DRYENG}}$ per engine $= 1.4 \times 379 = 530.5$ kg ≈ 530 kg. For two engines, $M_{\text{ENG}} = 1060$ kg.

10.12.8   Systems Group Mass

$MTOW = 9500$ kg (use Eq. (10.47)), $M_{\text{SYS}} = 0.11 \times 9500 = 1045$ kg.

10.12.9   Furnishing Group Mass

$MTOW = 9500$ kg (use Eq. (10.50)), $M_{\text{FUR}} = 0.065 \times 9500 = 618$ kg.

10.12.10   Contingency and Miscellaneous Group Mass

$MTOW = 9500$ kg (use Eq. (10.53 and 10.54)), $M_{\text{CONT}} = 0.016 \times 9500 = 152$ kg.

10.12.11   Crew Mass

There are two flight crew members and no cabin crew. Therefore, $M_{\text{CREW}} = 2 \times 90 = 180$ kg.

10.12.12   Payload Mass

There are 10 passengers. Therefore, $M_{\text{PL}} = 10 \times 100 + 100 = 1100$ kg.

10.12.13   Fuel Mass

The range requirement is 2000 nm carrying 10 passengers. From statistical data given in Table 10.1, take the highest value of 26% of $MTOW$, $M_{\text{FUEL}} = 0.26 \times 9500 = 2470 \approx 2500$ kg.

10.12.14   Weight Summary

Table 10.7, later, is the weight summary obtained by using the coursework example worked out thus far. The last column provides the estimation as a result of the graphical solution (in bracket – kg).

This computation requires further iterations to be fine-tuned for better accuracy. The CG position is established in the next section, where further iterations will yield a better picture.

Variant Aircraft in the Family

For simplification, linear variations are considered, which should be worked out as a coursework exercise.

Long Variant. Increase payload = 400 kg, fuselage = 300 kg, furnish = 200 kg, fuel = 300 kg and others = 200 kg. Total increase = +1400 kg; $MTOM_{\text{Long}} = 10\,900$ kg (no structural changes).

Table 10.6 Determination of Bizjet CG location.

Item group	Mass (kg)	X (m)	Moment	Z (m)	Moment
Fuselage @ 40%	930	6.08	5655	1.4	1302
Wing @ 30% of MAC	854	7.4	6320	1	854
H-tail	118	14	1652	8	944
V-tail	83	15	1245	3	249
Undercarriage (nose)	110	1.8	198	0.4	44
Undercarriage (main)	270	8	1840	0.5	135
Nacelle/pylon	228	10.5	2415	2	460
Miscellaneous					
Power plant	1060	10.7	11342	2	2120
Systems	1045	6.5	6793	1	1045
Furnishing	618	6	3708	2	1236
Contingencies	152	3	456	1.2	182
MEM	**5470**				
Crew	180	3	540	1.4	252
Consumable	120	4	480	1.5	180
OEM	**5770**				
Payload	1100[a]	6.2 (6.4)	7040	1.1	1210
Fuel	2500	7.8	19500	1	2500
Total moment at MTOM of 9370 ≈ **9400**			69184		12713
		$\bar{x} = 7.36$ m		$\bar{z} = 1.352$ m	
MRM	**9450**				

Bold values highlight the most important parameters.

a)   No frills 10 passengers or a luxury arrangement for eight passengers.

Small Variant. Decrease payload = 500 kg, fuselage = 350 kg, furnish = 250 kg, fuel = 350 kg and others = 250 kg (lightening of the structures). Total decrease = −1700 kg; $MTOM_{\text{Small}} = 7800$ kg.

10.12.15   Bizjet Aircraft Mass and CG Location Example

The x-axis is the fuselage axis and ZRP is the y-axis. Table 10.6 and Eqs. (10.55) and (10.56) are used to locate the CG. SI units are used. Determination of aircraft NP is not done in this book but taken at around 55% of wing MAC.

The CG should lie in the plane of symmetry (there are unsymmetrical aircraft).

The CG locations for the $CG_{\text{Long Variant}}$ and the $CG_{\text{Short Variant}}$ may be worked out by the readers.

10.12.16   First Iteration to Fine-Tune CG Position Relative to an Aircraft and Components

The preliminary aircraft configuration begins in Chapter 8 with a guesstimated MTOM, engine size and CG position. It is unlikely that the computed aircraft mass as worked out in this chapter will match the estimated mass. In fact, the example shows that it is lighter, with a more accurate CG position; therefore, this replaces the estimated values in Chapter 8. The differences from guesstimate and revised in chapter are as follows.

Also note that both mass and CG location are slightly different from preliminary data.

	Estimated at concept definition (Chapter 8)	Revised in Chapter 9 (Figure 9.22) First iteration in Chapter 10
MTOM – kg	9500	9400
MAC leading edge location – m from ZRP	6.77	6.76
Aft CG limit, $\bar{x}$ from ZRP	7.4 m	7.36 m
CG position, $\bar{z}$ from ground – m	1.38 m	1.352 m

The differences are small. From now onwards, the revised values will be used; most importantly the MTOM is now revised to 9400 kg. In principle, the aircraft configuration must be revised at this stage of progress as the first iteration. The worked-out example of the Bizjet shows small differences, hence no iteration is carried out as a simple repositioning of battery, black-boxes and so on will do the job. In that case the angle, β, is kept satisfied. Final sizing is accomplished in Chapter 14, when iteration may be required.

MILITARY AIRCRAFT

10.13   Mass Fraction Method – Military Aircraft

National defence requirements made military designs evolve rapidly, incorporating new technologies at a considerable cost to stay ahead of potential adversary. Whereas miniaturisation of electronic and other equipment reduces onboard mass, increased demand in combat capabilities worked in the opposite direction to add mass. Combat aircraft size kept growing to exceed 35 000 kg for multirole fighter aircraft. Earlier, it was holding on at around 16 000 kg. With improved missile capabilities, aircraft performance demand is now somewhat changed, especially in reaching maximum speed. Combat aircraft takeoff mass varies according to mission requirement. The Typical Takeoff Mass (TTOM) is less than MTOM. It is beneficial to have a small lightweight fighter, but currently with less than 12 000 kg of MTOM, armament capacity and radius of action would suffer in a demanding mission. Light combat aircraft has to be seen to cater mainly for high-subsonic/low supersonic combat missions, unlikely to exceed Mach 1.4. The following points are pertinent to military aircraft component mass estimation methodologies.

  i. Combat aircraft (fighters) have engines buried in the fuselage; hence, there is no wing relief benefit.
 ii. A combat role has a wide spectrum of activities, as outlined in Section 4.12. In general, close support/ground attack aircraft (no afterburning engines) are subsonic with quick turning capabilities, whereas air superiority aircraft have supersonic capabilities with afterburning turbofans.
iii. Modern supersonic combat aircraft configurations show considerable wing-body blending.
 iv. Dominantly, all expendable payload is externally mounted except for bigger designs, which could have internal bomb load. Modern military aircraft have an external load that is contoured and flushed with mouldlines. Internally mounted guns are permanent fixtures, integral to the airframe structure. Training military aircraft pylons to carry external load are not taken as permanent fixtures.
  v. Consumables (e.g. firing rounds) are internally loaded. In general, for long-range missions, there is than one crew and some consumables meant for them.

Military designs are technology specific and, as a result, unlike civil design, military designs exhibit large variations in statistical distribution.

Generic military aircraft component mass is listed in MIL-STD-1374 in exhaustive detail, which is not required in the conceptual design stage. As with civil aircraft design, this section presents a consolidated generalised group for what is required at the conceptual design phase. Note that military aircraft do not have cabin crew and passengers. The payload is weaponry and is mostly carried externally. Section 10.6 gives the military aircraft component nomenclature.

Military aircraft follow the same procedure for mass fraction estimation method as that of civil aircraft but have a different mass fraction (shown in percentage) of Mi/MTOM, where subscript '*i*' represents the *i*th component. All items should total 100% of the MTOM; this also can be expressed in terms of mass per wing area (i.e. component wing loading). Unlike civil aircraft, each generation of military aircraft takes a bigger leap towards advanced technology driven by the requirements of national security more than profit. Reference [9] gives an exhaustive list of weight breakdown for many relatively older military aircraft designs. A new design should show improvements, especially of newer technologies (e.g. new materials, lighter systems). Table 10.7 (combat and trainer aircraft) gives ballpark figures of mass fraction (in percentages) for arriving quickly at

Table 10.7 Military aircraft mass fraction (see Section 10.6.1 for symbols).

Group		Trainer		Combat (all turbofan)	
		Turboprop	Turbofan	Close support fighter	Bomber
Fuselage	$F_{fu} = M_F/MTOM$	10–12		9–11	8–10
Wing	$F_w = M_w/MTOM$	7–9		11–14	7–10
H-Tail	$F_{ht} = M_{ht}/MTOM$	1.4–1.8		1.2–1.6	1–1.5
V-Tail	$F_{vt} = M_{vt}/MTOM$	0.7–0.9		0.8	0.7–1
Intake	$F_{in} = M_{in}/MTOM$	0.8–1.2		1–1.2	1.5–2
Pylon (weapon)	$F_{py} = M_{py}/MTOM$	0.3–0.5		0.2–0.4	0.2–0.3
Undercarriage	$F_{uc} = M_{uc}/MTOM$	3–5		5–7	2.5–4
Fixed armament	$F_{arm} = M_{arm}/MTOM$	1–1.2		1–1.2	1–1.2
Engine	$F_{eng} = M_{eng}/MTOM$	9–11		11–12	8–12
Retarding device	$F_{tr} = M_{tr}/MTOM$	0		0.4–0.6	0.5–0.8
Engine con.	$F_{ec} = M_{ec}/MTOM$	0.5–0.8		1–1.6	0.7–0.9
Fuel sys.	$F_{fs} = M_{fs}/MTOM$	0.5–0.6		2–3	2–3
Oil sys.	$F_{os} = M_{os}/MTOM$	0.2–0.4		0.2–0.4	0.3–0.4
Flight con. sys	$F_{fc} = M_{fc}/MTOM$	2–2.5		2–2.5	1.5–2
Hydr/pneumatic sys	$F_{hp} = M_{hp}/MTOM$	0.5–0.8		0.5–0.8	0.8–1
Electrical	$F_{elc} = M_{elc}/MTOM$	2–2.5		1.5–2	1–1.4
Instrument	$F_{ins} = M_{ins}/MTOM$	0.6–1		0.4–0.5	0.5–0.6
Avionics	$F_{av} = M_{av}/MTOM$	3–4		3–6	4–6
ECS + oxygen	$F_{ecs} = M_{ecs}/MTOM$	0.6–1		0.5–0.8	1–1.2
Seat + escape sys	$F_{fur} = M_{fur}/MTOM$	1.2–1.5		1–1.4	1–1.4
Misc.	$F_{msc} = M_{msc}/MTOM$	0		1	1
Paint	$F_{pn} = M_{pn}/MTOM$	0.01		0.01	0.01
Contingency	$F_{con} = M_{con}/MTOM$	1		1	1
MEW – %		52–56		58–64	49–53
Crew		5–6		0.8–1.2	0.8–1
Consumable		0		0	0
OEW – %		57–60		60–65	50–54
Payload and fuel are traded					
Payload		15–25		15–20	20–25
Fuel		15–20		18–20	20–25
MTOM – %		100		100	100

a starting point for the initial configuration obtained in Chapter 6. This time, a wider variation of ±15% (may exceed in a few designs) may be allowed to accommodate a wide range of variation.

Note that the OEM fraction in the table agrees with the trend of actual aircraft data shown in Figure 10.8. Lighter aircraft would show a higher mass fraction. A fuselage-mounted undercarriage is shorter and lighter for the same MTOM. These tables give some idea of component masses at an early project stage. It is best to use more accurate semi-empirical relations (Section 10.15) to obtain component masses at the conceptual phase. These tables would prove useful to estimate masses (e.g. fuel mass, engine mass etc.) required as a starting point for some semi-empirical relations.

Discussions in Section 10.8.1 on using the mass fraction method mention it is also valid for military aircraft mass estimation.

10.14 Graphical Method to Predict Aircraft Component Weight – Military Aircraft

In addition to the statistical data on military aircraft given in Section 7.10, this section extends statistical data on military aircraft component mass in graphical form, as illustrated in the following figures.

Not much graphical statistical data on military aircraft component mass is available in the public domain. References [9, 10] are a good source of military aircraft component masses. Note that graphs are in the FPS system. As expected, they show wide variation, yet there is a trend through linear regression. Figures 10.7 and 10.8 give the component weights graphs. They show wide dispersion.

10.15 Semi-Empirical Equations Method (Statistical) – Military Aircraft

The semi-empirical relations given in this section are meant for high-subsonic military aircraft mass estimation, for example, light combat aircraft, trainer, close air support (CAS), counter insurgency (COIN) roles

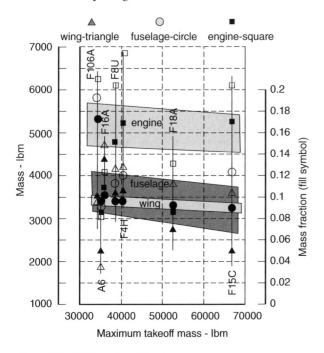

Figure 10.7 Military aircraft fuselage mass.

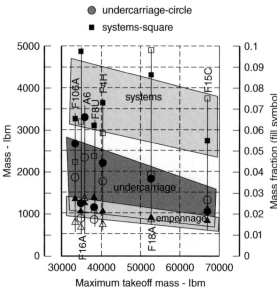

Figure 10.8 Military aircraft wing mass.

and so on. (The reader may refer to [9, 10] for supersonic combat aircraft semi-empirical relations.) The semi-empirical relations are derived from theoretical formulation and have been refined with statistical data. The section is divided into subsections, each with a step-by-step discussion of workflow.

In a similar manner to civil aircraft, military aircraft have their own sets of semi-empirical relations derived from theoretical formulation and refined with statistical data. As in the civil mass estimation, there are several forms of semi-empirical weight prediction formulae proposed by various analysts, all based on the key drivers with refinements as perceived by the proposer. They are similar in nature and yield close enough results. References [1–6, 9, 11] offer more information on weight prediction. The following subsections provide aircraft component weight (mass) semi-empirical formulae, component-by-component.

10.15.1 Military Aircraft Fuselage Group (SI System)

Military aircraft do not have a constant section fuselage and are more densely packed. A combat aircraft fuselage is not pressurised (only cockpit is). In a blended body it is not easy to lineate fuselage line from the wing. The best would be construction specific where the joining line of the wing is the line for fuselage. It is much simpler for the example of the Advanced Jet Trainer (AJT) in hand. In this case take D_{ave} = average of the shape (see Figure 8.14) around the engine. The expression includes fuselage-mounted side/chin intakes. (The AJT does not have a pod – intake weight is taken as integral with fuselage weight.)

$$M_{Fmil} = c_{fus} \times k_{uc} \times k_{mat} \times k_{para} \times k_{intake} \times (MTOM \times n_{ult})^{0.002} \times (L \times D_{ave} \times V_D^{0.5})^{1.52}$$

where $c_{fus} = 0.175$

$k_{uc} = 1.05$ for a fuselage-mounted undercarriage with bulge and 1.03 without bulge.
$k_{mat} = 1.0$ for metal, otherwise make a percentage weight reduction.
$k_{para} = 1.002$ if there is a brake parachute, otherwise 1.
$k_{intake} = 1.005$

$$M_{Fmil} = 0.175 \times 1.0 \times 1.05 \times 1.002 \times 1.005 \times k_{mat} \times (MTOM \times n_{ult})^{0.002} \times (L \times D_{ave} \times V_D^{0.5})^{1.5}$$

$$= 0.185 \times k_{mat} \times (MTOM \times n_{ult})^{0.002} \times (L \times D_{ave} \times V_D^{0.5})^{1.5} \tag{10.58}$$

10.15.2 Military Aircraft Wing Mass (SI System)

The equation is simplified by the authors for classroom usage. The same remarks as made for civil aircraft are also applicable to the military case, except that the values of factors change. Military wings are thinner. In many cases, the wing does not carry fuel (e.g. F104) and does not have winglets.

Writing the modified equation in terms of the book's notation, Eq. (10.21) is changed to:

$$M_{Wmil} = c_w \times k_{uc} \times k_{sl} \times k_{sp} \times k_{wl} \times k_{re} \times (MTOM \times n_{ult})^{0.48} \times S_W{}^{0.78} \times AR \times (1 + \lambda)^{0.4}$$

$$\times (1 - W_{Fuel\_mass\_in\_wing}/MTOW)^{0.4}/[(\cos\Lambda) \times t/c^{0.4}] \tag{10.59}$$

$c_w = 0.0215$ for $AR < 5$ (high subsonic – jet)

$\quad = 0.023$ for $AR > 5$ (low subsonic – propeller)

k_{mat} = effect of material change as in the case of the fuselage

$k_{uc} = 1.002$ for a wing-mounted undercarriage, otherwise 1.0

$k_{sl} = 1.004$ for use of a slat and $k_{sp} = 1.005$ for a spoiler. Flaps are standard.

10.15.3 Aircraft Empennage

The equation is simplified by the authors for classroom usage. The same remarks as made for civil aircraft are also applicable to the military case, except that the values of factors change as follows. Equation (10.25) for civil aircraft gives:

$$M_{EMPmil} = k_{emp} \times k_{mat} \times k_{conf} \times (MTOM \times n_{ult})^{0.48} \times S_W{}^{0.78} \times AR \times (1 + \lambda)^{0.4}/(\cos \Lambda \times t/c^{0.4})$$

k_{mat} = Effect of material change as in the case of fuselage.

The modified equations, in terms of the notations in Eq. (10.25), are as follows.

Horizontal Tail:

For all tail movement, use $k_{emp} = 0.018$, $k_{conf} = 1.05$, otherwise 1.0.

$$M_{HTmil} = 0.018 \times k_{mat} \times k_{conf} \times (MTOM \times n_{ult})^{0.484} \times S_W{}^{0.78} \times AR \times (1 + \lambda)^{0.4}/(\cos \Lambda \times t/c^{0.4}) \tag{10.60}$$

Vertical Tail:

For tail configurations, use $k_{emp} = 0.022$, $k_{conf} = 1.1$ for T-tail, 1.05 for mid-tail and 1.0 for low tail.

$$M_{VTmil} = 0.022 \times k_{mat} \times k_{conf} \times (MTOM \times n_{ult})^{0.484} \times S_W{}^{0.78} \times AR \times (1 + \lambda)^{0.4}/(\cos \Lambda \times t/c^{0.4}) \tag{10.61}$$

10.15.4 Nacelle Mass Example – Military Aircraft

A typical combat aircraft does not have a nacelle and a pod – intake weight is taken to be integral with fuselage weight.

10.15.5 Power Plant Group Mass Example – Military Aircraft

From the engine T_{SLS} the engine dry mass can be obtained from its thrust-to-weight ratio, this typically varies from 5 to 8. It is always better to use engine manufacturer's supplied data that is freely available in the public domain. Otherwise, refer to Table 10.8 to obtain engine dry mass.

For the variant engine mass, only add the increment/decrement of the dry engine mass to the baseline power plant group mass.

The total power plant group mass can be expressed semi-empirically as

Jet engine

$$M_{ENG} \text{ per engine} = 1.2 \times M_{DRYENG} \tag{10.62}$$

$$M_{ENG\_var} \text{ per engine} = (1.2 \times M_{DRYENG\_baseline}) + \Delta M_{ENG\_var}$$

Table 10.8 Typical values of component CG locations – military aircraft.

	Component	Military AJT aircraft
Typical % of component characteristic length		
1.	Fuselage group	45–50%
2.	Wing group	No slat – 30% of MAC
		With slat – 25% MAC
3.	H-tail group	30%
4.	V-tail group	30%
5.	Undercarriage group	At wheel centre
6.	Nacelle + pylon group	Generally not applicable
7.	Miscellaneous	As position – use similarity
8.	Power plant group	70–80%
9.	Systems group	As position – use similarity (typically 35% of fuselage)
10.	Furnishing group	As position – use similarity
11.	Contingencies	As position – use similarity
	MEM	Compute
12.	Crew	As position – use similarity
	OEM	Compute
13.	Payload	As position – use similarity
14.	Fuel	As position – use similarity
	MTOM and MRM	Compute

Turboprop

Civil aircraft power plant:

$$M_{ENG\_tp} = k_{tp} \times M_{DRYENG} \text{ per engine} \tag{10.63}$$

where $1.5 \leq k_{tp} \leq 1.8$. (due to propeller and intake)

$$M_{ENG} \text{ per engine} = (k_{tp} \times M_{DRYENG\_baseline})$$

$$M_{ENG\_var} \text{ per engine} = (k_{tp} \times M_{DRYENG\_baseline}) + \Delta M_{ENG\_var} \text{ (variant engine)}$$

APU (if any)

$$M_{APU} = 0.001 \text{ to } 0.005 \times M_{DRYENG} \text{ of an engine} \tag{10.64}$$

10.15.6 Undercarriage Mass Example – Military Aircraft

Undercarriage size depends on an aircraft's MTOM. A military aircraft undercarriage is heavier in order to withstand hard landings.

Tricycle type (retractable) – wing-mounted (nose and main gear estimated together)

For a low-wing-mounted undercarriage:

$$M_{UC \text{ low wing}} = 0.045 \times MTOM \tag{10.65}$$

For a mid-wing-mounted undercarriage:

$$M_{UC \text{ mid–wing}} = 0.046 \times MTOM \tag{10.66}$$

For a high-wing-mounted have fuselage-mounted undercarriage:

$$M_{\text{UC fus}} = (0.04 \text{ to } 0.45) \times MTOM \tag{10.67}$$

For a fixed undercarriage, the mass is 10–15% lighter; for a tail-dragger, it is 20–25% lighter.

10.15.7 System Mass – Military Aircraft (Higher Mass Fraction for Lighter Aircraft)

System mass comprises of avionics, electrical, plumbing, ECS (oxygen system) and so on.

Take $M_{\text{SYS}} = 0.1 \text{ to } 0.12$ (jet) and $0.12 \text{ to } 0.15$ (propeller) $\times MTOW$ for trainer class aircraft.

Take $M_{\text{SYS}} = 0.16 \text{ to } 0.20 \times MTOW$ for combat class aircraft. $\tag{10.68}$

10.15.8 Aircraft Furnishing (Ejection Seat) – Military Aircraft

$$M_{\text{eject\_seat}} = 90 \text{ kg per crew} \tag{10.69}$$

As such, military aircraft do not have furnishings but the following can be added:

Take $M_{\text{FUR}} = 0$

10.15.9 Miscellaneous Group (M_{MISC}) – Military Aircraft

Carefully examine what structural parts are left out, for example, delta fin and so on. If any item does not fit into the standard groups as listed here – include it in this group. Typically, this be expressed as

$$M_{\text{MISC}} = 0 \text{ to } 1\% \text{ of } MTOM. \tag{10.70}$$

10.15.10 Contingency (M_{CONT}) – Military Aircraft

A good designer keeps provision for contingency, that is,

$$M_{\text{CONT}} = (0.01 \text{ to } 0.25) \times MTOW \tag{10.71}$$

10.15.11 Crew Mass

There are trainer and trainee pilots. Take a mass of 90 kg per crew member.

10.15.12 Fuel (M_{FUEL})

Fuel load is mission specific. This can be determined by proper performance estimation shown in Chapters 11 and 13. At this stage, statistical data is the only means to guesstimate fuel load that will be revised in Chapters 11 and 13.

Military aircraft is mission specific and also has to depend on statistical data at this stage of design until accurate fuel load can be estimated through performance estimation.

10.15.13 Payload (M_{PL})

Military aircraft payload is the armament specified by the user requirements. Internal guns are taken as systems weight.

10.16   CG Determination – Military Aircraft

Table 10.8 gives the CG positions of aircraft components of the classroom examples:
x, y and z are measured from the nose of the aircraft and then convert to MAC_w.

Table 10.8 may be used to determine the CG location. Coordinate origin $x = 0$ at the nose tip, and $z = 0$ at ground level, which is kept horizontal.

Equations (10.54)–(10.56) are valid to compute CG coordinates. See Section 10.12.1 for the classroom Bizjet example to compute $\bar{x}$ and $\bar{z}$. The important 'potato' curve for CG variation for all loading conditions is left out.

The readers should make sure that it has the static margin with full crew. The CG should be at around 18% of MAC when fully loaded and at around 22% when empty, only in between with pilots. CG is always forward of the NP. (Where is it? Take 50% of the MAC.) This is for static stability reasons.

If the computation does not indicate the CG within the specified ranges, then move the wing and/or engine to bring the CG to the desired percentages of the MAC until a satisfactory solution is reached. Fuel tanks can be slightly modified. Batteries are heavy items and can be located at a desirable position to fine-tune the CG location to the desired position.

10.17   Classroom Example of Military AJT/CAS Aircraft Mass Estimation

The user requirement for practice bombs and missiles is 1800 kg, this is a reasonable practise armament load. Therefore, $M_{PL} = 1800$ kg. The fully loaded AJT is estimated to be 6500 kg (Figure 8.12). AJT structure is designed to cater for: $MTOM = 6600$ kg, wing area $S_W = 17$ m^2.

CAS structure will have MEM to withstand the maximum MTOM, but the MEM does not change as is explained next for the incremental mass required to beef up the structure. The structural component commonality can be maintained by beefing up the main load-bearing members of a structure. To the undercarriage is to maintain component commonality without any modification, hence it is sized to CAS MTOM of 7500 kg. CAS armament payload is 2300 kg.

Only the shorter CAS nose cone replaces the longer AJT nose cone as one pilot is removed (Figure 8.14). In summary, both the variants will have same MEM.

AJT manoeuvre limits are as follows.

> Manoeuvre limits: +7 g to −3.5 g for the training mission without armament at 4800 kg
> Manoeuvre limits: +6 g to −3 g for the armament training mission at 6600 kg

CAS manoeuvre limits are as follows.

> Manoeuvre limits: +8 g to −4 g for the TTOM (lower fuel and armament) @ 7000 kg
> Manoeuvre limits: +7 g to −3.5 g for the MTOM @ 7500 kg

The never-exceed speed, $V_D = 620$ kn (0.95 Mach).

10.17.1   AJT Fuselage (Based on CAS Variant)

Consider that it has a 10% weight saving on account of composite usage in secondary structures (10%), that is, $k_{mat} = 0.99$. $MTOM = 6500$ kg. $L = 12.1$ m, $D_{ave} = 0.5 \times (1.8 + 0.9) = 1.35$ m and $V_D = 620$ kt $= 1150$ km/h $= 319.4$ m s^{-1}, $n_{ult} = 6\,g$. Equation (10.58) gives:

$$M_{Fmil} = 0.175 \times k_{mat} \times (MTOM \times n_{ult})^{0.002} \times (L \times D_{ave} \times V_D^{0.5})^{1.5}$$

$$= 0.17325 \times (6500 \times 6)^{0.002} \times (12.1 \times 1.35 \times 319.4^{0.5})^{1.5}$$

$$= 0.17325 \times 1.0213 \times (12.1 \times 1.35 \times 17.87)^{1.5} = 0.177 \times (291.9)^{1.5} = 0.177 \times 4987 = 882 \text{ kg}$$

10.17.2   AJT Wing (Based on CAS Variant)

Consider that it has a 10% weight saving on account of composite usage in secondary structures (10%), that is, $k_{mat} = 0.99$. $MTOM = 6500$ kg. $n_{ult} = 6\,g$, $S_W = 17$ m^2, $AR = 5.3$, $\lambda = 0.3$, $M_{WR} = 828$ kg, $\Lambda = 20°$, $t/c = 0.10$ and $= 1150$ km/h $= 319.4$ m s^{-1}. It has no slat making $k_{sl} = 1$ and has 828 kg fuel in the wing. For the spoiler, $k_{sl} = 1.005$ and the undercarriage is not wing-mounted $k_{uc} = 1$. Equation (10.59) gives:

$$M_W = c_w \times k_{uc} \times k_{sl} \times k_{sp} \times k_{wl} \times k_{re} \times (MTOM \times n_{ult})^{0.48} \times S_W{}^{0.78} \times AR \times (1 + \lambda)^{0.4}$$

$$\times (1 - W_{Fuel\_mass\_in\_wing}/MTOW)^{0.4}/[(\cos\Lambda) \times t/c^{0.4}]$$

$$M_W = 0.0215 \times 0.99 \times 1.005 \times (6500 \times 6)^{0.48} \times 17^{0.78} \times 5.3 \times (1 + 0.3)^{0.4}$$

$$\times (1 - 828/7800)^{0.4}/(\cos 20 \times 0.1^{0.4})$$

$$M_W = (0.02128 \times 159.85) \times (9.115 \times 5.3 \times 1.11) \times 0.8938^{0.4}/(0.94 \times 0.398)$$

$$M_W = 3.4 \times 53.62 \times 0.956/(0.427) = 174.3/0.374 = 466 \text{ kg}$$

10.17.3   AJT Empennage (Based on CAS Variant)

Horizontal Tail: Equation (10.60) gives:

$$M_{HT} = 0.018 \times k_{mat} \times k_{conf} \times (MTOM \times n_{ult})^{0.484} \times S_W{}^{0.78} \times AR \times (1 + \lambda)^{0.4}/(\cos\Lambda \times t/c^{0.4})$$

Consider that it has a 10% weight saving on account of composite usage in secondary structures (10%), that is, $k_{mat} = 0.99$. $MTOM = 6500$ kg.

AJT horizontal tail is all moving, that is, configuration factor $k_{conf} = 1.05$. Note that the exposed area constitutes as the empennage mass as constructed.

$$S_H = 3.91 \text{ (exposed)}, AR = 3.5, \lambda = 0.37, \Lambda = 25°, t/c = 0.10,$$

$$M_{HT} = 0.018 \times 0.99 \times 1.05 \times (6500 \times 6)^{0.48} \times 3.91^{0.78} \times 3.5 \times (1 + 0.37)^{0.4}/(\cos 25 \times 0.1^{0.4})$$

$$= 0.01782 \times 159.85 \times (2.9 \times 3.5 \times 1.134)/(0.906 \times 0.398) = 2.85 \times 11.51/0.36 = 91 \text{ kg}$$

Vertical Tail: Equation (10.61) gives:

$$M_{VT} = 0.022 \times k_{mat} \times k_{conf} \times (MTOM \times n_{ult})^{0.484} \times S_W{}^{0.78} \times AR \times (1 + \lambda)^{0.4}/(\cos\Lambda \times t/c^{0.4})$$

$$S_V = 3.0, AR = 1.52, \lambda = 0.3, \Lambda = 35°, t/c = 0.10,$$

$$M_{VT} = 0.022 \times 0.99 \times (6500 \times 6)^{0.48} \times 3.0^{0.78} \times 1.52 \times (1 + 0.3)^{0.4}/(\cos 35 \times 0.1^{0.4})$$

$$= 0.02178 \times 159.85 \times (2.36 \times 1.52 \times 1.11)/(0.82 \times 0.398) = 3.48 \times 3.892/0.326 = 42 \text{ kg}$$

10.17.4   AJT Nacelle Mass (Based on CAS Variant)

The AJT does not have a pod – the intake weight is taken to be integral to the fuselage weight.

10.17.5 AJT Power Plant Group Mass (Based on AJT Variant)

Suitable engines in the class are the following. Both have BPR = 0.75 and engine diameter = 22 in.
For AJT – Adour 861 producing 26.66KN (6000 lb), dry weight = 590 kg (1300 lb).
For CAS – Adour 871 producing 31.4KN (7060 lb), dry weight = 610 kg (1330 lb).
Taking Adour 871 for AJT (weights will not change with final engine sizing in Chapter 14).
Total power plant group mass can be expressed semi-empirically as (Eq. (10.62) gives):

$$\text{Adour 861 } M_{\text{ENG}} \text{ per engine} = 1.2 \times M_{\text{DRYENG}} = 1.2 \times 603 = 724 \text{ kg.}$$

$$\text{Adour 951 } M_{\text{ENG\_var}} \text{ per engine} = (1.2 \times M_{\text{DRYENG\_baseline}}) + \Delta M_{\text{ENG\_var}} = 724 + 20 = 744 \text{ kg}$$

10.17.6 AJT Undercarriage Mass (Based on CAS Variant)

To maintain commonality, use the heaviest fully loaded CAS mass of 7500 kg. This is the only time CAS maximum mass has been applied to get component mass.
For fuselage-mounted undercarriage use Eq. (10.67).

$$M_{\text{U/C\_fus}} = 0.04 \times 7500 = 300 \text{ kg (nose} = 80 \text{ kg and main} = 220 \text{ kg).}$$

10.17.7 AJT Systems Group Mass (Based on AJT Variant)

Equation (10.65) gives:

$$MTOW = 6500 \text{ kg}, M_{\text{SYS}} = 0.1 \times 6600 = 650 \text{ kg.}$$

10.17.8 AJT Furnishing Group Mass

Ejection seat plus the oxygen system are taken in this group to be 90 kg per crew member.

10.17.9 AJT Contingency Group Mass

Equation (10.71) gives:

$$MTOW = 6500 \text{ kg}, M_{\text{CONT}} = 0.01 \times 6500 = 65 \text{ kg.}$$

10.17.10 AJT Crew Mass

There are trainer and trainee pilots. Therefore, $M_{\text{CREW}} = 2 \times 90 = 180$ kg

10.17.11 AJT Fuel Mass (M_{FUEL})

Fuel load is mission specific. From AJT statistics in Section 6.12.1, $M_{\text{FUEL\_NTC}} = M_{\text{NFC}} - \text{OEW} = 4800 - 3700 = 1100$ kg (tank not full).

It is assumed that practise range is less than 100 miles away and an AJT reaches the range at economic cruise to make two to three passes for armament training at a sortie duration of about 50 minutes. The tankage capacity can hold 1400 kg fuel for ferry flights and, if required, an AJT armament practice sortie can be made longer at a slightly higher MTOM.

10.17.12 AJT Payload Mass (M_{PL})

Military aircraft payload is its armament, which is specified by the user requirements. Internal guns are taken as systems weight. Practice armament (includes gun ammunition) = 1800 kg.

10.18   AJT Mass Estimation and CG Location

Use Table 10.5 and Eqs. (10.40) and (10.41) to locate CG location. SI units (kg and m) are used.

Table 10.9 gives the CG locations of aircraft components of the classroom examples:

The CG location is very close to what was guessed in Section 9.16.2 hence avoided iterations. The angle, β, for NTC and MTOM has also been estimated in Section 9.16.2, within the acceptable range. Readers may compute the CAS CG location. Proper CG positioning can be established after aircraft NP is established when the forward and aft CG limits can be ascertained by fine-tuning component positions. Determination of aircraft NP is not done in this book but taken at around 40% of wing MAC.

The angle, β, for NTC and MTOM has been estimated also in Section 9.16.2, within the acceptable range. This is a satisfactory angle covering maximum fuselage-rotation angle at takeoff.

The external geometry being the same, the CAS variant can be extrapolated from the AJT variant to a fair degree accuracy as reasoned here.

i. The OEM for both the variants stays about the same as the removal of the masses of one crew, plus the ejection seat and mass saving from a slightly shorter fuselage (see Figure 8.14) are taken up as structural reinforcements to support a higher MTOM, manoeuvring load and avionic upgrade. This amounts to 220 kg.

Table 10.9 Typical values of component CG locations – AJT.

Item group	Mass (kg)	X (m)	Moment	Z (m)	Moment
Fuselage	882	6.6	5821	1.3	1164
Wing	466	7.2	3355	1.75	816 822
H-tail	91	11.31	1029	1.9	173
V-tail	42	10.66	448	2.4	100
Undercarriage (nose)	80	2.0	160	0.48	36
Undercarriage (main)	220	7.4	1628	0.5	110
Nacelle + pylon	None				
Miscellaneous	None				
Power plant	724	8.6	6226	1.32	956
Systems	650	5.2	3380	1.2	780 936
Ejection seats	180	4	720	1.32	237
Contingencies	65	4	392	1.32	87
MEM	**3400**				
Crew	180	4	720	1.32	237
OEM	**3580**				
Fuel (wing and fuselage)	1100	7.8	8580	1.65	1815
Total NTC mass	**4680 ≈ 4800**		32 459		6459
CG at NTC mass		$\bar{x} = 6.9$		$\bar{z} = 1.345$ m	
Armament payload	1800	6.6	11 880	1.4	2520
Total MTOM	**6 80 ≈ 6500**		44 339		8979
CG at MTOM		$\bar{x} = 6.82$ m		$\bar{z} = 1.38$ m	

Bold values highlight the most important parameters.

ii. Fuel load increment, $\Delta_{Fuel} = 300$ kg (to full tank capacity, armament load increment, $\Delta_{Payload} = 500$ kg, avionics and systems mass increment, $\Delta_{Systems}$ (small radar) $= 80$ kg, bigger engine power plant mass increment, $\Delta_{Powerplant} = 20$ kg. This totals a mass increment, $\Delta_{MTOM} = 900$ kg. Therefore, CAS MTOM is expected to be around 7500 kg. The figures would be iterated as more information is generated. The aim is to make new design better than any existing aircraft in the class. This is where the experience of designers counts.

The preliminary aircraft configuration begins in Chapter 8 with a guesstimated MTOM, engine size and CG position. It is unlikely that the computed aircraft mass as worked out in this chapter will match the estimated mass. In fact, the example shows that it is lighter, with a more accurate CG position; therefore, this replaces the estimated values in Chapter 8. The differences from guesstimate and revised in chapter are as follows.

	Guessed at concept definition (Chapter 8)	Revised in Chapter 9 (Figure 9.23) First iteration in Chapter 10
MTOM – kg	6500	6500
CG position, $\bar{z}$ from ground – m	1.35 m	1.345 m
CAS variant MTOM – kg	7500	7400

Exact MTOM agreement is, in a way, coincidental, although statistics in Figure 8.12 indicate what is to be expected. The differences in CG locations are small and the revised values are taken as its iteration. In principle, the aircraft configuration must be revised at this stage of progress as the first iteration. Final sizing is accomplished in Chapter 14, when iteration is required. Coursework may require only one iteration cycle of computation.

10.19 Classroom Example of a Turboprop Trainer (TPT) Aircraft and COIN Variant Weight Estimation

The user requirement for practice bombs and missiles is 500 kg, this is a reasonable practise armament load. Therefore, $M_{PL} = 1800$ kg. The fully loaded TPT is estimated to be 2800 kg (Figure 10.9).

Manoeuvre: $+7$ g to -3.5 g for the trainer version at NTCM (Normal Training Configuration Mass) without practise armament (fully aerobatic)

Manoeuvre: $+6$ g to -3 g for fully armed version at MTOM

The following are the TPT details estimated as the concept definition from the statistical data that will be finalised to the formal sized configuration in Chapter 16.

MTOM = 2800 kg (6160 lbm)	NTC-TOM = 2350 kg (5170 lbm)
OEW = 1700 kg (3740 lbm)	
Crew = 180 kg (396 lbm)	Fuel = 500 kg (110 lbm)
Maximum speed = 280 kt	Sustained speed = 240 kt (405.2 ft s^{-1})
MTOM = 2800 kg (6160 lb)	NTCM = 2350 kg (5170 lb – no armament)
OEW = 1700 kg (3740 lb)	Practise armament load = 450 kg ($\approx$1000 lb)
Crew = 400 lb ($\approx$180 kg)	Fuel = 470 kg (1012 lb/kg)
Typical mid-training aircraft mass = 2200 kg (4840 lb)	
Maximum speed = 280 kt	Sustained speed = 240 kt (405.2 ft s^{-1})

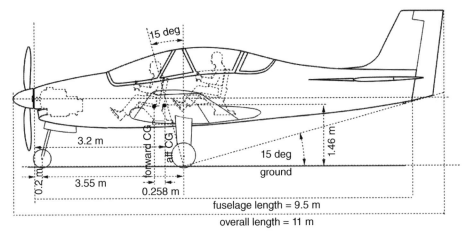

Figure 10.9 Turboprop trainer aircraft (TPT).

Fuselage

Length = 9.5 m (31.16 ft)	height = 1.6 m (5.25 ft)
Cockpit width = 0.88 m (35 in)	Fuselage width = 1.0 m ($\approx$40 in)
Fineness ratio = 9.5/1.0 = 9.5	

Wing (Wing Aerofoil): National Advisory Committee for Aeronautics (NACA) 63_2-215)

Reference area = 16.5 m² (177.3 ft²)	Span = 16.36 m (53.8 ft)	AR = 6.5
Root chord, C_R = 2.36 m (7.74 ft)	Tip chord, C_T = 0.82 m (3.7 ft)	
MAC = 1.715 m (5.62 ft)	Taper ratio, λ = 0.35	$\Lambda_{1/4}$ = 0°
Dihedral = 5° (dihedral – low wing)	Twist = 1° (wash out),	t/c = as specified
Flap = 2.2 m² (23.6 ft²)	Aileron = 1.06 m² (11.4 ft²)	

H-tail (NACA 0012)

Reference area = 4.1 m² (44.2 ft²)	Span = 4.2 m (13.8 ft)	Aspect ratio = 4.0
Root chord, C_R = 1.28 m (4.2 ft)	Tip chord, C_T = 0.67 m (2.2 ft)	
H-tail MAC = 1 m (3.3 ft)	Taper ratio, λ = 05, $\Lambda_{1/4}$ = 0°	
Elevator area = 1.025 m² (11 ft²)	t/c = 0.12	

V-tail (NACA 0012)

Reference area = 2.2 m² (23.3 ft²)	Height = 1.375 m (4.52 ft)
Root chord, C_R = 1.64 m (5.38 ft)	Tip chord, C_T = 0.66 m (2.16 ft)
V-tail MAC = 1.28 m (4.2 ft)	Taper ratio, λ = 04, $\Lambda_{1/4}$ = 20°
Rudder area = 0.44 m² (4.74 t²)	t/c = 0.12

Engine

Takeoff static power at International Standard Atmosphere (ISA) sea level = 975 shp	(Garratt TPT-331-12B)
Dry weight = 182 kg (400 lb),	Engine intake area = 0.048 m² (0.52 ft²)
Length = 1.1 m (43.4 in)	Height = 0.686 m (27 in.)

COIN Variant

A COIN variant can be derived, without affecting the external geometry, by re-engineering the trainer version with 1200 SHP turboprop, increasing armament load to 1000 kg by adding Δ550 kg of combat weapons, increasing fuel capacity to 1000 kg by adding Δ570 kg of fuel that will require adding strength to structure with OEM raised by about Δ400 kg to nearly 3000 kg MTOM. This is merely a suggestion and is not worked out. The readers may give this a try.

10.19.1   TPT Fuselage Example

Consider that it has a 10% weight saving on account of composite usage in secondary structures (10%), that is, $k_{mat} = 0.99$. $MTOM = 2800$ kg. $L = 9.5$ m, $D_{ave} = 0.5 \times (1.8 + 0.9) = 1.35$ m and $V_D = 300$ kt $= 556.5$ mph $= 155$ m s^{-1}, $n_{ult} = 6\ g$. Equation (10.58) gives:

$$M_{Fmil} = 0.175 \ \times k_{mat} \times (MTOM \times n_{ult})^{0.002} \times (L \ \times D_{ave} \ \times V_D{}^{0.5})^{1.5}$$

$$= 0.17325 \times (2800 \times 6)^{0.002} \times (9.5 \times 1.35 \ \times 155^{0.5})^{1.5}$$

$$= 0.17325 \times 1.022 \times (9.5 \times 1.35 \times 12.44)^{1.5} = 0.177 \times (160)^{1.5} = 0.177 \times 2055.8 = 358 \text{ kg}$$

10.19.2   TPT Wing Example

Consider that it has a 10% weight saving on account of composite usage in secondary structures (10%), that is, $k_{mat} = 0.99$. $MTOM = 2800$ kg. $n_{ult} = 6\ g$, $S_W = 16.5$ m^2, $AR = 6.5$, $\lambda = 0.5$, $\Lambda = 0°$, $t/c = 0.15$ and $= 280$ kt $= 144$ m s^{-1}. It has no slat making $k_{sl} = 1.0$, $k_{uc} = 1.002$ and has 470 kg fuel in the wing. For the spoiler, $k_{sl} = 1.005$ and the undercarriage is not wing-mounted $k_{uc} = 1$. Equation (10.59) gives:

$$M_W = c_w \times k_{uc} \times k_{sl} \times k_{sp} \times k_{wl} \times k_{re} \times (MTOM \times n_{ult})^{0.48} \times S_W{}^{0.78} \times AR \times (1 + \lambda)^{0.4}$$

$$\times (1 - W_{Fuel\_mass\_in\_wing}/MTOW)^{0.4}/[(\cos\Lambda) \times t/c^{0.6}]$$

$$M_W = 0.023 \times 0.99 \times 1.002 \times 1.005 \times (2800 \times 6)^{0.48} \times 16.5^{0.78} \times 6.5 \times (1 + 0.5)^{0.4}$$

$$\times (1 - 470/2800)^{0.4}/(\cos 0 \times 0.15^{0.4})$$

$$M_W = (0.023 \times 106.7) \times (8.9 \times 6.5 \times 1.176 \times 0.832^{0.4}/(1.0 \times 0.468)$$

$$M_W = 2.454 \times 57.85 \times 0.929/(0.468) = 132/0.468 = 282 \text{ kg}$$

10.19.3   TPT Empennage Example

Horizontal Tail:
   Equation (10.60) gives:

$$M_{HT} = 0.018 \times k_{mat} \times k_{conf} \times (MTOM \times n_{ult})^{0.484} \times S_W{}^{0.78} \times AR \times (1 + \lambda)^{0.4}/(\cos\Lambda \times t/c^{0.4})$$

Consider that it has a 10% weight saving on account of composite usage in secondary structures (10%), that is, $k_{mat} = 0.99$. $MTOM = 6600$ kg.

AJT horizontal tail is all-moving, that is, configuration factor $k_{conf} = 1.05$. Note that the exposed area constitutes as the empennage mass as constructed.

$$S_H = 4.1, AR = 4, \lambda = 0.5, \Lambda = 5°, t/c = 0.12,$$

$$M_{\text{HT}} = 0.018 \times 0.99 \times 1.05 \times (2800 \times 6)^{0.48} \times 4.1^{0.78} \times 4 \times (1 + 0.5)^{0.4} / (\cos 5 \times 0.12^{0.4})$$

$$= 0.01782 \times 106.7 \times (3 \times 4 \times 1.176) / (0.997 \times 0.428) = 1.9 \times 14.112 / 0.427 = 63 \text{ kg}$$

Vertical Tail: Equation (10.61) gives:

$$M_{\text{VT}} = 0.025 \times k_{\text{mat}} \times k_{\text{conf}} \times (MTOM \times n_{\text{ult}})^{0.484} \times S_W^{0.78} \times AR \times (1 + \lambda)^{0.4} / (\cos \Lambda \times t/c^{0.4})$$

$$S_V = 2.2, AR = 1.8, \lambda = 0.4, \Lambda = 12°, t/c = 0.12,$$

$$M_{\text{VT}} = 0.025 \times 0.99 \times (2800 \times 6)^{0.48} \times 2.2^{0.78} \times 1.8 \times (1 + 0.4)^{0.4} / (\cos 12 \times 0.12^{0.4})$$

$$= 0.02475 \times 106.7 \times (1.85 \times 1.8 \times 1.14) / (0.978 \times 0.428) = 2.64 \times 3.8 / 0.418 = 24 \text{ kg}$$

10.19.4 TPT Nacelle Mass Example

The TPT does not have a pod-intake weight taken to be integral with fuselage weight.

10.19.5 TPT Power Plant Group Mass Example

Use Eq. (10.62):

Engine selected is Garratt TPT-331-12B (include propeller) derated to 1000 SHP.
For TPT – Garratt TPT-331-12B, dry weight = 175 kg (385 lb).
For COIN – Garratt TPT-331-12B at the rated power of 1100 SHP.

$$M_{\text{ENG\_tp}} = k_{\text{tp}} \times M_{\text{DRYENG}} \text{ per engine} = 1.6 \times M_{\text{DRYENG}} \text{ per engine} = 1.6 \times 182 = 327 \text{ kg}$$

$$M_{\text{ENG\_var}} \text{ per engine} = (1.8 \times M_{\text{DRYENG\_baseline}}) + \Delta M_{\text{ENG\_var}}$$

10.19.6 TPT Undercarriage Mass Example (Based on CAS Variant)

To maintain commonality, use the heaviest fully loaded COIN mass of 3100 kg. This is the only time COIN maximum mass has been applied to get component mass.

For a fuselage-mounted undercarriage use Eq. (10.65).

$$M_{\text{U/C\_fus}} = 0.0435 \times 3100 = 135 \text{ kg (nose} = 35 \text{ kg, main} = 100 \text{ kg).}$$

10.19.7 TPT Systems Group Mass Example

Use Eq. (10.68)

$$MTOW = 2800 \text{ kg}, M_{\text{SYS}} = 0.12 \times 2800 = 336 \text{ kg.}$$

10.19.8 TPT Furnishing Group Mass Example

Ejection seat plus oxygen system are taken in this group to be 80 kg per crew member.

10.19.9 TPT Contingency Group Mass Example

Use Eq. (10.71):

$$M_{\text{CONT}} = 0 \text{ for TPT.}$$

10.19.10 TPT Crew Mass Example

There are trainer and trainee pilots. Therefore, $M_{\text{CREW}} = 2 \times 90 = 180$ kg.

10.19.11 TPT Fuel (M_{FUEL})

Fuel load is mission specific. The tankage capacity can hold 1000 kg fuel.

10.19.12 TPT Payload (M_{PL})

Military aircraft payload is its armament, which is specified by the user requirements. Internal guns are taken as systems weight. Practice armament (includes gun ammunition) = 1800 kg.

10.20 Classroom Worked-Out TPT Mass Estimation and CG Location

Use Table 10.5 and Eqs. (10.40) and (10.41) to locate CG location. SI units (kg and m) are used.

Table 10.10 gives the CG locations of aircraft components of the classroom examples:

The CG location is very close to what was estimated in Section 9.16.2, hence iterations were avoided. The angle, β, for NTC and MTOM has also been estimated in Section 9.16.2, within the acceptable range. Readers may compute the CAS CG location. Proper CG positioning can be established after aircraft NP is established when the forward and aft CG limits can be ascertained by fine-tuning component positions. Determination of aircraft NP is not done in the book but taken to be at around 40% of wing MAC.

This is a satisfactory angle covering maximum fuselage-rotation angle at takeoff.

The preliminary aircraft configuration begins in Chapter 8 with a guesstimated MTOM, engine size and CG position. It is unlikely that the computed aircraft mass as worked out in this chapter will match the estimated mass. In fact, the example shows that it is lighter, with a more accurate CG position; therefore, this replaces the estimated values in Chapter 8. The differences from guesstimate and revised in chapter are as follows.

	Guessed at concept definition (Chapter 8)	Revised in Chapter 9 (Figure 9.24) First iteration in Chapter 10
MTOM – kg	2800	2800
CG limit – m from ZRP	3.2	3.23
Aft CG position, $\bar{z}$ from ground – m	1.5	1.52

Exact MTOM agreement is in a way coincidental, although statistics in Figure 8.12 indicate what is to be expected. The differences in CG locations are small and the revised values are taken as its iteration. In principle, the aircraft configuration must be revised at this stage of progress as the first iteration. Final sizing is accomplished in Chapter 14, when iteration is required. Coursework may require only one iteration cycle of computation.

10.20.1 COIN Variant

The role of COIN aircraft is to take airborne military action to encounter insurgent (hence the name) movements against the government, for example, terrorism, paramilitary activities, civil war, guerrilla actions, crime syndicate and so on, within the country or at the borders. These are not exactly in CAS role but carried out at low intensity engagements where heavy armament is not required nor the radius of

Table 10.10 Typical values of component CG locations – TPT.

Item group	Mass (kg)	X (m)	Moment	Z (m)		Moment
Fuselage	358	3.8	1360	1.45		519
Wing	282	3.75	1057	1.2		338
H-tail	63	9.15	576	2.0		126
V-tail	24	9.55	229	2.85		68
Undercarriage (nose)	35	0.22	6.6	0.35		11
Undercarriage (main)	100	3.75	338	0.36		32
Miscellaneous	None					
Power plant	327	0.4	131	1.6		523
Systems	336	1.7	714	1.3		546
Ejection seats	180	3.7	666	1.55		279
Contingencies	0	5.4	140	1.4		36
MEM	**1688 ≈ 1700**					
Crew	180	3.7	666	1.6		288
OEM	**1880 ≈ 1900**					
Fuel (wing and fuselage)	450	3.8	1710	1.8	2.0	810
Total NTC mass	**2350**		7595			3576
CG at NTC mass		$\bar{x} = 3.23$		$\bar{z} = 1.52$		
Armament payload	450	3.6	1800	1.4		7000
Total MTOM	**2800**		9585			4484
CG at MTOM		$\bar{x} = 3.19$		$\bar{z} = 1.495$		

Bold values highlight the most important parameters.

action is far away from conveniently located airbases. The aircraft should be capable of finding, tracking, and attacking targets (observation/attack – US nomenclature OA-X role). The measures are primarily strafing and smaller missiles firing to take vehicles out of action. This type of aircraft need not have very speed but must be capable of high manoeuvre with tight turning capability. A typical high performance COIN armament load is around ±2000 lb. COIN variant of TPT is in many air force inventories, but a COIN variant carrying higher load than 1000 kg would penalise the baseline TPT variant to grow too big for ab initio/intermediate level training to gain proficiency in airmanship, as well as be expensive to operate for the intended role at that level. A true COIN aircraft should be a new design fit for the specific role, possibly be able to keep some components commonality to a considerably lesser extent with the TPT fleet in the air force inventory.

A COIN variant of the TPT example is suggested at the end of Section 8.17.4, as shown in Figure 8.25. Here, the suggestion extends to assist, if the readers would like to refine to a sized COIN variant by taking out the second pilot, ejection seat and flight-deck equipment to the extent of reducing OEM by 250 lb to bring down to a bare-bone OEM of 1650 lb. Add Δ50 lb for COIN avionics, fuel load to 550 lb and armament load of 750 lb to COIN MTOM ≈ 3000 lb, the same as the TPT MTOM. The engine is the same Garratt TPT-331-12B at the rated power of 1100SHP (the TPT engine is derated to 950 SHP). Tail-arms are kept about the same so that the empennage areas retain commonality. Shortening of fuselage length will also shorten the wheel base requiring to shortening the wheel track as well. This can easily be done if the baseline TPT wing structure keeps provision to make minor modifications to move the attachment support members and the wheel well slightly inboard. This is merely a suggestion and is not worked out. The readers may give this a try.

10.21   Summary of Concept Definition

Starting from Chapter 8 with estimated MTOM, S_W and SHP, this chapter culminates in a new aircraft concept definition, with revised MTOM using semi-empirical equations to obtain more accurate weight estimations. This is the first opportunity to revise aircraft configuration if there is large difference between guessed mass and estimated mass. In the example case, as worked out, the table in the previous section shows very little difference, hence no iteration is required. However, the guessed wing reference area, S_W has to be sized and the engine has to be sized to match and satisfy the performance specification to *concept finalisation* that is carried out in Chapter 14. To do the aircraft performance estimation, aircraft drag has to be determined (Chapter 11) and engine performance supplied has to be obtained from the engine manufacturer. The engine and its performance characteristics are extensively dealt with in Chapters 12 and 13. The next three Chapters 11–14 culminate in concept finalisation to freeze the design for management to decide whether to go ahead or not.

The art of guessing is from the designers' past experiences, especially if they have worked on similar aircraft in the past and their current company has the design and test details in that class of aircraft. There is no substitute for it. The designers judge the scatter pattern in the statistics of as many existing aircraft in the class of aircraft to make a decision in estimation based on the new technologies to be adopted (minimise variance, if required to deselect the extreme values) that can slightly deviate from the regression graph. For the newly initiated readers, it is normal to have iterations, possibly more than one: this is a good learning process and should not be neglected.

References

1 Torenbeek, E. (1982). *Synthesis of Subsonic Aircraft Design*. Delft University Press.
2 Roskam, J., Aircraft Design, Volumes 1 to 10, Darcorporation; 2nd edition (December 31, 2003).
3 (1996). *Introduction to Aircraft Weight Engineering*. Los Angeles, USA: Society of Allied Weights Engineers.
4 (2002). *Weight Engineer's Handbook*. Los Angeles, USA: Society of Allied Weights Engineers.
5 Jenkinson, L.R., Simpson, P., and Rhodes, D. (1999). *Civil Jet Aircraft Design*. Arnold Publishing.
6 Niu, M. (1988). *Airframe Structural Design*. Dover Publication CONMILIT Press.
7 Sechler, E.E. and Dunn, L.G. (1963). *Airplane Structural Analysis and Design*. Dover Publications.
8 Howe, D. (2000). *Aircraft Conceptual Design Synthesis*. London: Professional Engineering Ltd.
9 Nicolai, L.M. and Carichner, G. (2010). *Fundamentals of Aircraft Design*. Reston, VA: American Institute of Aeronautics and Astronautics.
10 Janes all the World Aircraft Manual (yearly publication), Jane's Information Group
11 Raymer, D. (2012). *Aircraft Design-a Conceptual Approach*, 5e. Reston, VA: AIAA.

11

Aircraft Drag

11.1 Overview

After the completion of the *concept definition* that concluded in Chapter 10, the next phase is to end the Phase I stage of the new aircraft project with *concept finalisation* through the formal method of aircraft sizing and engine matching exercise in Chapter 14 that will require aircraft performance analyses. This will require aircraft drag polar to be dealt with in this chapter and engine performance data to be dealt with in the next two chapters, Chapters 12 and 13.

An important task for aircraft performance engineers is to make the best possible estimation of all the different types of drag associated with aircraft aerodynamics. Commercial aircraft design is sensitive to the Direct Operating Cost (DOC), which is aircraft drag dependent. Just one count of drag (i.e. $C_D = 0.0001$) could account for several million US dollars in operating cost over the lifespan of a small fleet of midsized aircraft. This becomes increasingly important with the trend in rising fuel costs. Accurate estimation of the different types of drag remains a central theme. (Equally important are other ways to reduce DOC; one of them is reducing manufacturing cost.)

For a century, a massive effort has been made to understand and estimate drag and the work is still continuing. Possibly some of the best work in English language on aircraft drag is compiled by National Advisory Committee for Aeronautics (NACA)/National Aeronautics and Space Administration (NASA), RAE (the Royal Aircraft Establishment), Advisory Group for Aerospace Research and Development (AGARD), Engineering Sciences Data Unit (ESDU), DATCOM (the short name for the *USAF Data Compendium for Stability and Control*) and others [1–11]. These publications indicate that the drag phenomena are still not fully understood [12] and that the way to estimate aircraft drag is by using semi-empirical relations. Computational fluid dynamics (CFD) is gaining ground but it is still some way from supplanting the proven semi-empirical relations. In the case of work on excrescence drag, efforts are lagging.

Two-dimensional surface skin friction drag, elliptically loaded induced drag and wave drag can be accurately estimated – together, they comprise most of the total aircraft drag. The problem arises when estimating drag generated by the 3D effects of the aircraft body, interference effects and excrescence effects. In general, there is a tendency to underestimate aircraft drag. Accurate assessments of aircraft mass, drag and thrust are crucial in the aircraft performance estimation.

This chapter includes the following sections:

Section 11.2: Introduction to Aircraft Drag
Section 11.3: Parasite Drag
Section 11.4: Aircraft Drag Breakdown Structure
Section 11.5: Understanding Drag Polar
Section 11.6: Theoretical Background of Aircraft Drag
Section 11.7: Subsonic Aircraft Drag Estimation Methodology
Section 11.8: Methodology to Estimate Minimum Parasite Drag (C_{Dpmin})
Section 11.9.1: Semi-Empirical Relations to Estimate C_{Dpmin}

Conceptual Aircraft Design: An Industrial Approach, First Edition. Ajoy Kumar Kundu, Mark A. Price and David Riordan.
© 2019 John Wiley & Sons Ltd. Published 2019 by John Wiley & Sons Ltd.

Section 11.10: Excrescence Drag
Section 11.11: Summary of Aircraft Parasite Drag (C_{Dpmin})
Section 11.12: Methodology to Estimate C_{Dp}
Section 11.13: Methodology to Estimate Subsonic Wave Drag
Section 11.14: Summary of Total Aircraft Drag
Section 11.15: Low-Speed Aircraft Drag at Takeoff and Landing
Section 11.16: Drag of Propeller-Driven Aircraft
Section 11.17: Military Aircraft Drag
Section 11.18: Empirical Methodology for Supersonic Drag Estimation
Section 11.19: Bizjet Example – Civil Aircraft
Section 11.20: Turboprop Example – Propeller-Driven Aircraft
Section 11.21: Military Aircraft Example
Section 11.22: Classroom Example – Supersonic Military Aircraft
Section 11.23: Drag Comparison
Section 11.24: Some Concluding Remarks

Coursework content: Readers will carry out aircraft-component drag estimation and obtain the total aircraft drag.

11.2   Introduction

The drag of an aircraft depends on its shape and speed, which are design-dependent, as well as on the properties of air, which are nature-dependent. Drag is a complex phenomenon arising from several sources, such as the viscous effects that result in skin friction and pressure differences as well as the induced flow field of the lifting surfaces and compressibility effects (see Sections 4.9 and 4.11).

The aircraft drag estimate starts with the isolated aircraft components (e.g. wing and fuselage etc.). Each component of the aircraft generates drag, largely dictated by its shape. Total aircraft drag is obtained by summing the drag of all components plus their interference effects when the components are combined. The drag of two isolated bodies increases when they are brought together due to the interference of their flow fields.

The Reynolds Number (*Re*) has a deciding role in determining the associated skin friction coefficient, C_F, over the affected surface and the type, extent, and steadiness of the boundary layer (which affects parasite drag) on it. Boundary-layer separation increases drag and is undesirable; separation should be minimised.

A major difficulty arises in assessing drag of small items attached to an aircraft surface such as instruments (e.g. pitot and vanes), ducts (e.g. cooling), blisters and necessary gaps to accommodate moving surfaces. In addition, there are the unavoidable discrete surface roughness from mismatches (at assembly joints) and imperfections, perceived as defects that result from limitations in the manufacturing processes. Together, from both manufacturing and non-manufacturing origins, they are collectively termed *excrescence drag*.

Currently, accurate total aircraft drag estimation by analytical or CFD methods is not possible. Schmidt of Dornier in the AGARD 256 [10] is categorical about the inability of CFD, analytical methods or even wind-tunnel model-testing to estimate drag. CFD is steadily improving and can predict wing-wave drag (C_{Dw}) accurately but not the total aircraft drag – most of the errors are due to the smaller excrescence effects, interference effects and other parasitic effects. Industrial practices employ semi-empirical relations (with CFD) validated against wind-tunnel and flight tests and are generally proprietary information. Most of the industrial drag data are not available in the public domain. The methodology given in this chapter is a modified and somewhat simplified version of standard industrial practices [1, 3, 7, 8]. The method is validated by comparing its results with the known drag of existing operational aircraft.

The design criterion for today's commercial high-subsonic jet-transport aircraft is that the effects of separation and local shocks are minimised at the Long-Range Cruise (LRC) at the M_{CR} (when compressibility drag is almost equal to zero before the onset of wave drag) condition. At High-Speed Cruise (HSC), a 20-count drag increase is allowed, reaching M_{crit}, due to local shocks (i.e. transonic flow) covering small areas of the aircraft. Modern streamlined shapes maintain low separation at M_{crit}; therefore, such effects are small at HSC. The difference in the Mach number at HSC and LRC for subsonic aircraft is small – on the order of Mach 0.05–Mach 0.075. Aircraft drag characteristics are plotted as *drag polar* (C_D vs C_L).

Strictly speaking drag polar at several speed and altitudes will give better resolution of drag value however, estimation of the drag coefficient at LRC is sufficient because it has a higher C_f, which gives conservative values at HSC when ΔC_{Dw} is added. The LRC condition is by far the longest segment in the mission profile; the industry standard practice uses the LRC drag polar for all parts of the mission profile (e.g. climb and descent). The *Re* at the LRC provides a conservative estimate of drag at the climb and descent segments. At takeoff and landing, the undercarriage and high-lift-device drags must be added.

Supersonic aircraft operate over a wider speed range: the difference between M_{crit} and maximum aircraft speed is on the order of Mach 1.0–Mach 1.2. Therefore, estimation of C_{Dpmin} is required at three speeds: (i) at a speed before the onset of wave drag at M_{CR}, (ii) at M_{crit} and (iii) at maximum speed (say, Mach 2.0).

It is difficult for the industry to absorb drag prediction errors of more than 5% (the goal is to ensure errors of less than 3%) for civil aircraft; overestimating, is better than underestimating. Practitioners are advised to be generous in allocating drag – it is easy to miss a few of the many sources of drag, as shown in the worked-out examples in this chapter. Underestimated drag causes considerable design and management problems; failure to meet customer specifications is expensive for any industry. Subsonic aircraft drag prediction has advanced to the extent that most aeronautical establishments are confident in predicting drag with adequate accuracy. Military aircraft shapes are more complex; therefore, it is possible that predictions will be less accurate.

11.3 Parasite Drag Definition

The components of drag due to viscosity do not contribute to lift. For this reason, it is considered 'parasitic' in nature. For bookkeeping purposes, parasite drag is usually considered separately from other drag sources. The main components of parasite drag are as follows:

i. drag due to skin friction
ii. drag due to the pressure difference between the front and the rear of an object
iii. drag due to the lift-dependent viscous effect and therefore seen as parasitic (to some extent resulting from the non-elliptical nature of lift distribution over the wing); this is a small but significant percentage, of total aircraft drag (at LRC, it is about 2–3%)

All of these components vary (to a small extent) with changes in aircraft incidence (i.e. as C_L changes). The minimum parasite drag, C_{Dpmin}, occurs when shock waves and boundary-layer separation are at a minimum, by design, around the LRC condition. Any change from the minimum condition (C_{Dpmin}) is expressed as ΔC_{Dp}. In summary:

parasite drag (C_{Dp})

$\quad$ = (drag due to skin friction [viscosity] + drag due to pressure difference [viscosity])

$\quad$ = minimum parasite drag(C_{Dpmin}) + incremental parasite drag(ΔC_{Dp})

$\quad$ = $C_{\text{Dpmin}} + \Delta C_{\text{Dp}}$ (both terms have friction and pressure drag) $\hfill$ (11.1)

Oswald's efficiency factor (see Section 4.9.3) is accounted for in the lift-dependent parasite drag, ΔC_{Dp}. The nature of ΔC_{Dp} is specific to a particular aircraft. Numerically, it is small and difficult to estimate.

Parasite drag of a body depends on its form (i.e. shape) and is also known as *form drag*. The form drag of a wing profile is known as *profile drag*. These two terms are not used in this book. In the past, parasite drag in the Foot, Pound, Second (FPS) system was sometimes expressed as the drag force in pound force (lb_f) at 100 ft s^{-1} speed, represented by D100. This practice was useful in its day as a good way to compare drag at a specified speed, but it is not used today.

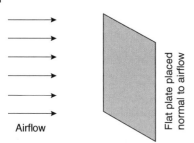

Figure 11.1 Flat plate equivalent of drag.

The current industrial practice using semi-empirical methods to estimate C_{Dpmin} is a time-consuming process. (If computerised, then faster estimation is possible, but the authors recommend relying more on the manual method at this stage.) Parasite drag constitutes half to two-thirds of subsonic aircraft drag. Using the standard semi-empirical methods, the parasite drag units of an aircraft and its components are generally expressed as the drag of the 'equivalent flat plate area' (or 'flat plate drag'), placed normal to airflow as shown in Figure 11.1. These units are in square feet to correlate with literature in the public domain. This is not the same as air flowing parallel to the flat plate and encountering only the skin friction.

The inviscid idealisation of flow is incapable of producing parasite drag because of the lack of skin friction and the presence of full pressure recovery.

11.4 Aircraft Drag Breakdown (Subsonic)

There are many variations and definitions of the book-keeping methods for components of aircraft drag; this book uses the typical US practice [2, 3]. The standard breakdown of aircraft drag is as follows (see Eq. (11.1)):

$$\text{total aircraft drag} = (\text{drag due to skin friction} + \text{drag due to pressure difference})$$

$$+ \text{drag due to lift generation} + \text{drag due to compressibility (wave drag)}$$

$$= \text{parasite drag} \, (C_{\text{Dp}}) + \text{lift} - \text{dependent induced drag} \, (C_{\text{Di}}) + \text{wave drag} \, (C_{\text{Dw}})$$

$$= (\text{minimum parasite drag} \, [C_{\text{Dpmin}}] + \text{incremental parasite drag} \, [\Delta C_{\text{Dp}}])$$

$$+ \text{induced drag} \, (C_{\text{Di}}) + \text{wave drag} \, (C_{\text{Dw}})$$

Eq (4.22) gives $C_{\text{Di}} = C_L{}^2/e\pi AR$
Therefore, the total aircraft drag coefficient is:

$$C_D = C_{\text{Dpmin}} + \Delta C_{\text{Dp}} + C_L{}^2/\pi AR + C_{\text{Dw}} \tag{11.2}$$

The advantage of keeping pure induced drag separate is obvious because it is dependent only on the lifting-surface aspect ratio and is easy to compute. The total aircraft-drag breakdown is shown in Chart 11.1.

It is apparent that the C_D varies with the C_L. When the C_D and the C_L relationship is shown in graphical form, it is known as a *drag polar*, shown in Figure 11.2 (all components of drag are shown in the figure). The C_D versus the $C_L{}^2$ characteristics of Eq. (11.2) are rectilinear, except at high and low C_L values (see Figure 11.16 later), because at a high C_L (i.e. low speed, high angle of attack), there could be additional drag due to separation; at a very low C_L (i.e. high-speed), there could be additional drag due to local shocks. Both effects are nonlinear in nature. Most of the errors in estimating drag result from computing ΔC_{Dp}, three-dimensional effects, interference effects, excrescence effects of the parasite drag and nonlinear range of aircraft drag. Designers should keep $C_{\text{Dw}} = 0$ at LRC and aim to minimise to $\Delta C_{Dp} \approx 0$ (perceived as the design point).

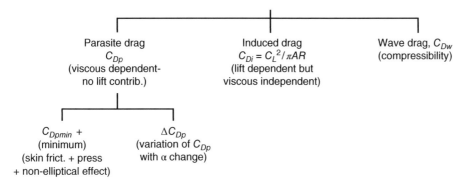

Chart 11.1 Total aircraft drag breakdown.

An aircraft on a long-range mission typically can have a weight change of more than 25% from the initial to the final cruise condition. As the aircraft becomes lighter, its induced drag decreases. Therefore, it is more economical to cruise at a higher altitude to take advantage of having less drag. In practical terms, this is achieved in the *step-climb* technique, or a gradual climb over the cruise range (Chapter 15).

11.4.1 Discussion

Considering air as inviscid makes mathematics simple to quickly obtain aerodynamic properties of a body in airflow, lift, moment and so on, in the linear range of operation. However, it is incapable to predict drag, stall and so on. In the earlier days aircraft designers had to rely on experimental data and semi-empirical relations those matured to sufficient accuracy to send man to the Moon.

With advances made in numerical analyses, today, CFD can approximately solve the exact nonlinear partial differential equations to get comparable results to experimental values (Chapter 23 outlines CFD capabilities). However, unlike getting good results on predicting lift and moment characteristics, CFD capability in predicting drag characteristic accurately and consistently has some way to go. This chapter presents industry standard drag predicting methodology, using well substantiated semi-empirical relationships.

Figure 11.2 plots the Bizjet drag polar and the aircraft components drag values for the Bizjet, as worked out in Section 11.19. It shows that for the Bizjet cruising at M_{crit} (LRC at Mach 0.7), about two-thirds of aircraft drag is viscous dependent, it would exist if air is considered inviscid. There is a small increase in Bizjet drag when flying at M_{div} (HSC at Mach 0.75), on account of a 20 count of wave drag rise ($C_{DW} = 0.002, \approx 6\%$). These percentages of drag breakdown are a good representation of typical transport aircraft drag.

On examining the aircraft-component drag breakdown, it can be seen that just the wing and fuselage (with belly fairing and canopy) together contribute to about 60% of the C_{DPmin}. Adding the nacelle drag, together these three components contribute to three-quarters of the viscous-dependent parasite drag. Aircraft designers concentrates more on these components to get the best configuration as discussed in Chapters 4–6. Nacelle design is dealt in Section 11.9.3.

11.5 Understanding Drag Polar

Aircraft drag polar is one of the most important information to evaluate aircraft capabilities. To evaluate aircraft performance, a good understanding of drag polar is essential. Aircraft drag polar gives the relation between drag and lift at any instant of flight under consideration. If drag polar could conform to expressing it in a simple analytical form, then it would be easier to obtain close-form solutions rather quickly and avoid cumbersome computational effort. Engine performance characteristics are not easy to express accurately in analytical form.

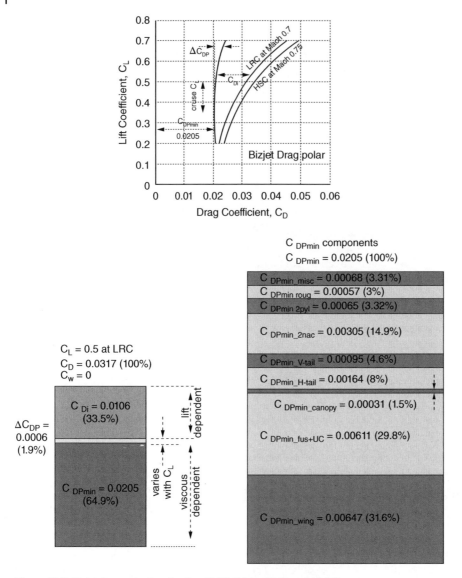

Figure 11.2 Bizjet drag polar (see Section 11.18, Tables 11.13 and 11.16).

11.5.1 Actual Drag Polar

Actual drag polar has the same format as parabolic drag polar but not an exact fit to a parabolic equation, especially at low- and high-speed ranges. The methodology for estimating actual drag using semi-empirical relations is given in this chapter. These semi-empirical relations are based on experimental data from both wind-tunnel and flight tests. These are the best available drag polar for industry standard analyses. Drag polar obtained by this methodology is not easily amenable to representation by *close-form* equations, especially for high-subsonic speed aircraft. To generate graphs that fit a parabolic shape, it may require shaping of graphs with the least loss in accuracy. Considerable insight into aircraft performance can be obtained by manipulating the parabolic drag polar equation. The industrial methods presented in this book give all the information but require laborious computational effort.

11.5.2   Parabolic Drag Polar

As mentioned earlier, *close-form* solutions from the analytical expressions quickly show important aircraft characteristics and prove useful to make a trend analyses. The readers may refer to [1–4] to study analytical treatments. At LRC, when wave drag is simplified to $C_{DW} = 0$, then Eq. (11.2) takes the form as follows

$$C_D = C_{Dpmin} + \Delta C_{Dp} + C_L{}^2/\pi AR \approx C_{Dpmin} + kC_L{}^2 \tag{11.3}$$

where $k = 1/e\pi AR$

The difference between the actual aircraft drag polar $(C_{Dpmin} + \Delta C_{Dp} + C_L{}^2/\pi AR)$ and its corresponding parabolic drag polar $(C_{Dpmin} + kC_L{}^2)$ is in the approximation of the variable lift-dependent parasite drag ΔC_{Dp} to a constant Oswald's efficiency factor 'e' associated with the induced drag C_{Di}. The variable parameter 'e' is approximated to a constant value bringing parabolic drag polar close to actual drag polar around the cruise segment and also gives a good match at climb and descent segments.

Figure 11.3 shows the variation of Oswald's efficiency factor 'e' with Mach number for the Bizjet aircraft at 18 000 lb weight cruising at a 41 000 ft altitude $(C_L \approx 0.512)$. It is evaluated using the following equation and plotted in the figure.

$$k = (C_D - 0.0205)/C_L{}^2$$

where $e = (1/k\pi AR)$

Parabolic drag polar incorporates a value of $k = 0.0447$ $(e \approx 0.95)$ representing average values. At the design C_L (typically, at mid-LRC) it approaches 1. This will cause a discrepancy between parabolic drag polar and actual drag polar values at higher speeds. For flying close to the ground, low speed incorporates drag of a 'dirty configuration' with values accommodating changes in e. For this reason, this book suggests that, for high-subsonic aircraft, one should use actual drag polar for accurate industry standard results, for example, for certification substantiation, preparing the pilot manual and so on, and parabolic drag polar may be used for exploring and establishing aircraft characteristics.

11.5.3   Comparison Between Actual and Parabolic Drag Polar

The parabolic representation of aircraft drag polar offers many advantages. Easy mathematical manipulation gives a quick insight to aircraft characteristics at an early design phase to make improvements, if required. Industry needs drag prediction to be as accurate as possible (in the order of few counts) and, therefore, results

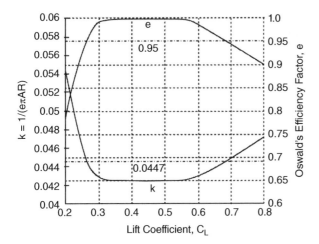

Figure 11.3 Bizjet Oswald's efficiency factor 'e' variation at 41 000 ft altitude (18 000 lb weight, $C_L \approx 0.512$, AR = 7.5).

from approximate analytical expression are not adequate for high-subsonic aircraft, especially when good prediction of drag polar can be obtained through semi-empirical relations substantiated by testing.

This section attempts to present approximated analytical expression for high-subsonic aircraft drag polar. Recall Eq. (11.2), which gives the expression for high-subsonic aircraft drag as

$$C_D = C_{Dpmin} + \Delta C_{Dp} + C_L^2/\pi AR + C_{DW} \tag{11.4}$$

where $(C_{Dpmin} + C_L^2/\pi AR)$ represents parabola in which C_{Dpmin} is the minimum distance of the drag polar from the y-axis representing C_L.

Equation (11.4) is not exactly of parabolic shape, depending on the extent to which it is contributed to by the non-parabolic component of ΔC_{Dp} and the wave drag term C_{DW}. In an incompressible flow regime C_{DW} drops out and ΔC_{Dp} is integrated with the induced drag term with the suitable coefficient 'k' = $1/e\pi AR$ (where e is the Oswald's efficiency factor) to represent the parabola equation as discussed before. A carefully designed aircraft can have $\Delta C_{Dp} \approx 0$ at cruise C_L. The simplification brings Eq. (11.4) into a simpler form as in Eq. (11.5), making it easier to handle.

$$C_D = C_{Dpmin} + kC_L^2 \tag{11.5}$$

This form of representation can be applied to high-speed subsonic aircraft up to *LRC*, that is, up to M_{crit}. When the parabolic part of the C_D is plotted against C_L^2, it is a straight line (Figure 11.16b).

Equation (11.5) can be improved by modifying to more accurate form as shown in Eq. (11.6) (plotted in Figure 11.4a).

$$C_D = C_{Dpmin} + k(C_L - C_{Lm})^2 \tag{11.6}$$

where C_{Dpmin} is at C_{Lm} and not at $C_L = 0$.

In the generalised situation of high-speed subsonic aircraft, C_{Dpmin} is not necessarily at $C_L = 0$. Figure 11.4 typically represents Eqs. (11.4) and (11.5) with drag polar estimated by a semi-empirical method (as done in Table 11.16 later). The various C_L points shown in the graph appear in analytical equations derived in the subsequent sections using the parabolic drag Eq. (11.3).

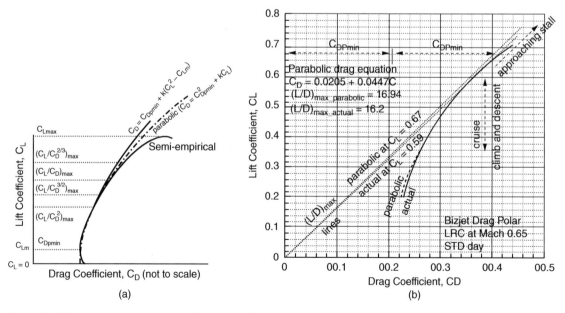

Figure 11.4 Bizjet comparison example. (a) Analytical (parabolic) and semi-empirical drag polar comparison. (b) Drag comparison.

(The most generalised form of aircraft drag polar can be expressed in a polynomial expression as given in Eq. (11.7).

$$C_D = C_{Dpmin} + k_1 C_L + k_2 C_L^2 + k_3 C_L^3 + k_4 C_L^4 + \cdots \cdot \tag{11.7}$$

For high-subsonic aircraft, terms above $k_3 C_L^3$ contribute very little and can be ignored. Even then, the polynomial form is not amenable to an easy close-form solution. The supersonic drag expression can be dealt with using a similar rationale (and its details are not dealt with here).

The expression for parabolic drag polar as given in Eq. (11.5) is

$$C_D = C_{Dpmin} + C_L^2/e\pi\,AR = C_{Dpmin} + kC_L^2 \tag{11.8}$$

where $k = 1/e\pi AR$

Figure 11.4a presents an analytical comparison between actual drag polar with parabolic drag polar. It may be noted that the three graphs are close enough within the operating range below M_{crit}. Figure 11.4b presents the Bizjet example as worked out in Section 11.19. Section 11.4 represents a typical high speed subsonic aircraft operational segments of an LRC (when $C_{DW} = 0$), en-route climb and descent. At HSC, the non-linear effects show up. The Bizjet operates in high subsonic speeds at M_{crit} and, therefore, semi-empirically determined actual drag polar is the preferred one. This is the standard procedure as practiced in industry, which gives credible output.

11.6 Aircraft Drag Formulation

A theoretical overview of drag is provided in this section to show that aircraft geometry is not amenable to the equation for an explicit solution. Even so, CFD is yet to achieve an acceptable result for the full aircraft.

Recall the expression in Eq. (11.2) for the total aircraft drag, C_D, as:

$$C_D = C_{Dparasite} + C_{Di} + C_{Dw} = C_{Dpmin} + \Delta C_{Dp} + C_{Di} + C_{Dw}, \tag{11.9}$$

where $C_{Dparasite} = C_{Dfriction} + C_{Dpressure} = C_{Dpmin} + \Delta C_{Dp}$.

At LRC, when $C_{Dw} \approx 0$, the total aircraft drag coefficient is given by:

$$C_D = C_{Dpmin} + \Delta C_{Dp} + C_{Di} \tag{11.10}$$

The general theory of drag on a 2D body (Figure 11.5a) provides the closed-form Eq. (11.11a). A 2D body has an infinite span. In the diagram, airflow is along the x direction and wake depth is shown in the y direction. The wake is formed due to viscous effects immediately behind the body, where integral operation is applied. Wake behind a body is due to the viscous effect in which there is a loss of velocity (i.e. momentum) and pressure (depletion of energy) as shown in the figure. Measurement and computation across the wake are performed close to the body; otherwise, the downstream viscous effect dissipates the wake profile. Consider an arbitrary control volume (CV) large enough in the y direction where static pressure is equal to free-stream

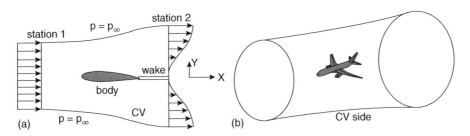

Figure 11.5 Control volume (CV) approach to formulate aircraft drag: (a) 2D body and wake in CV and (b) 3D aircraft in CV.

static pressure (i.e. $p = p_\infty$). The subscript ∞ denotes the free-stream condition. Integration over the y direction on both sides up to the free-stream value gives.

$$D = D_{\text{press}} + D_{\text{skin}} = \int_{-\infty}^{\infty} (p_\infty - p)dy + \int_{-\infty}^{\infty} \rho u(U_\infty - u)dy = \int_{-\infty}^{\infty} [(p_\infty - p) + \rho u(U_\infty - u)]dy \quad (11.11a)$$

An aircraft is a 3D object (Figure 11.5b) with the additional effect of finite wing span that will produce induced drag. In that case, Eq. (11.11a) can be written as

$$D = D_{\text{skin}} + D_{\text{press}} + D_{\text{i}} = \int_{-\infty}^{\infty} \int_{-b/2}^{b/2} [(p_\infty - p) + \rho u(U_\infty - u)]dxdy \quad (11.11b)$$

where b is the span of the wing in the x direction (the axis system has changed).

The finite wing effects on the pressure and velocity distributions result in induced drag D_{i} embedded in the expression on the right-hand side of Eq. (11.11b). Because the aircraft cruise condition (i.e. LRC) is chosen to operate below the onset of wave drag at M_{crit}, the wave drag, D_{w}, is absent; otherwise, it must be added to the expression. Therefore, Eq. (11.11b) can be equated with the aircraft drag expression as given in Eq. (11.9). Finally, Eq. (11.11b) can be expressed in a non-dimensional form, by dividing $\frac{1}{2}\rho_\infty U_\infty^2 S_W$. Therefore

$$\begin{aligned}
C_D &= \frac{1}{S_W} \int_{-\infty}^{\infty} \int_{-b/2}^{b/2} \left[-\frac{(p - p_\infty)}{\frac{1}{2}\rho_\infty U_\infty^2} + \frac{2\rho u}{\rho_\infty U_\infty}\left(1 - \frac{u}{U_\infty}\right) \right] dxdy \\
&= \frac{1}{S_W} \int_{-\infty}^{\infty} \int_{-b/2}^{b/2} \left[-C_p + \frac{2\rho u}{\rho_\infty U_\infty}\left(1 - \frac{u}{U_\infty}\right) \right] dxdy
\end{aligned} \quad (11.12)$$

Unfortunately, the complex 3D geometry of an entire aircraft in Eq. (11.12) is not amenable to easy integration. CFD has discretised the flow field into small domains that are numerically integrated, resulting in some errors. Mathematicians have successfully managed the error level with sophisticated algorithms. The proven industrial-standard, semi-empirical methods are currently the prevailing practice and are backed up by theories and validated by flight tests. CFD assists in the search for improved aerodynamics.

11.7 Aircraft Drag Estimation Methodology (Subsonic)

The semi-empirical formulation of aircraft drag estimation used in this book is a credible method based on [1, 3, 7, 8]. It follows the findings from NACA/NASA, RAE and other research-establishment documents. This chapter provides an outline of the method used. It is clear from Eq. (11.2) that the following four components of aircraft drag are to be estimated:

1. *Minimum parasite drag, C_{Dpmin}* (see Section 11.9).
   Parasite drag is composed of skin friction and pressure differences due to viscous effects that are dependent on the Re. To estimate the minimum parasitic drag, C_{Dpmin}, the first task is to establish geometric parameters such as the characteristic lengths and wetted areas and the Re of the discrete aircraft components.
2. *Incremental parasite drag, ΔC_{Dp}* (see Section 11.9).
   C_{Dp} is a characteristic of a particular aircraft design and includes the lift-dependent parasite drag variation, 3D effects, interference effects and other spurious effects not easily accounted for. There is no theory to estimate ΔC_{Dp}; it is best obtained from wind-tunnel tests or the ΔC_{Dp} of similarly designed aircraft wings and bodies. CFD results are helpful in generating the C_{Dp}-versus-C_L variation.
3. *Induced drag, C_{Di}* (see Section 4.18).
   The pure induced drag, C_{Di}, is computed from the expression

$$C_{\text{Di}} = C_L^2/\pi AR \quad (11.13)$$

4. *Wave drag, C_{Dw}* (see Section 11.13).

The last component of subsonic aircraft drag is the wave drag, C_{Dw}, which accounts for compressibility effects. It depends on the thickness parameter of the body: for lifting surfaces it is the t/c ratio, and for bodies it is the diameter-to-length ratio. CFD can predict wave drag accurately but must be substantiated using wind-tunnel tests. Transport aircraft are designed so that HSC at M_{crit} (e.g. 320 type, ≈ 0.82 Mach) allows a 20-count ($C_{Dw} = 0.002$) drag increase. At LRC, wave drag formation is kept at zero. Compressibility drag at supersonic speed is caused by shock waves.

The methodology presented herein considers fully turbulent flow from the leading edge (LE) of all components. Here, no credit is taken for drag reduction due to possible laminar flow over a portion of the body and lifting surface. This is because it may not always be possible to keep the aircraft surfaces clear of contamination that would trigger turbulent flow. The certifying agencies recommend this conservative approach.

11.8 Minimum Parasite Drag Estimation Methodology

The practised method of computing C_{Dpmin} is first to dissect (i.e. isolate) the aircraft into discrete identifiable components, such as the fuselage, wing, V-tail, H-tail, nacelle and other smaller geometries (e.g. winglets and ventral fins). The wetted area and the Re of each component establish skin friction associated with each component. The 2D flat plate basic means the skin friction coefficient, C_{F_basic}, corresponding to the Re of the component, is determined from Figure 11.24, bottom graph for the flight Mach number. Section 11.19.1 explains the worked-out examples carried out in this book for fully turbulent flow.

The C_Fs arising from the 3D effects (e.g. supervelocity) and wrapping effects of the components are added to the basic flat plate C_{F_basic}. Supervelocity effects result from the 3D nature (i.e. curvature) of aircraft-body geometry where, in the critical areas, the local velocity exceeds the free-stream velocity (hence the term *supervelocity*). The axi-symmetric curvature of a body (e.g. fuselage) is perceived as a wrapping effect when the increased adverse pressure gradient increases the drag. The interference in the flow field is caused by the presence of two bodies in proximity (e.g. the fuselage and wing). The flow field of one body interferes with the flow field of the other body, causing more drag. Interference drag must be accounted for when considering the drag of adjacent bodies or components – it must not be duplicated while estimating the drag of the other body.

The design of an aircraft should be streamlined so that there is little separation over the entire body, thereby minimising parasite drag obtained by taking the total C_F (by adding various C_F, to C_{F_basic}). Hereafter, the total C_F will be known as the C_F. Parasite drag is converted to its flat plate equivalent expressed in '*f*' square feet. In this book, the FPS system is used for comparison with the significant existing data that use the FPS system, although it can be easily converted into the international system of units (SI) system. Being non-dimensional, working in SI will give the same value of drag coefficients. The flat plate equivalent '*f*' is defined as:

$$f_{component} = (A_w \times C_F)_{component}, \tag{11.14}$$

where A_w is the wetted area (unit in ft^2).

The minimum parasite drag C_{Dpmin} of an aircraft is obtained by totalling the contributing fs of all aircraft components with other sundries. Therefore, the minimum parasite drag of the aircraft is obtained by:

$$(C_{Dpmin}) = (f_{component} + sundries)/S_W = (A_w \times C_F)_{component}/S_W \tag{11.15}$$

The stepwise approach to computing C_{Dpmin} is described in the following three subsections.

11.8.1 Geometric Parameters, Reynolds Number and Basic C_F Determination

The Re ($= (\rho_\infty L_{comp} V_\infty)/\mu_\infty$) has the deciding role in determining the skin friction coefficient, C_F, of a component. First, the Re per unit length, speed and altitude are established. Then, the characteristic lengths of each component are determined. The characteristic length L_{comp} of each component is as follows.

Fuselage.   Axial length from the tip of the nose cone to the end of the tail cone (L_{fus})
Wing.   The wing Mean Aerodynamic Chord (MAC)
Empennage.   The MACs of the V-tail and the H-tail
Nacelle.   Axial length from the nacelle highlight plane to the nozzle-exit plane (L_{nac})

Figure 11.24 shows the basic 2D flat plate skin friction coefficient of a fully turbulent flow for local (C_{f_basic}) and average (C_{F_basic}) values. For a partial laminar flow, the C_{F_basic} correction is made using factor f_1, given in Figure 11.25. It has been shown that the compressibility effect increases the boundary layer, thus reducing the local C_F. However, in LRC until the M_{crit} is reached, there is little sensitivity of the C_F change with Mach number variations, therefore, the incompressible C_F line (i.e. the Mach 0 line in Figure 11.24 (bottom graph) is used. At HSC at the M_{crit} and above, the appropriate Mach line is used to account for the compressibility effect. The basic C_F changes with changes in the Re, which depends on speed and altitude of the aircraft. Section 11.2 explains that a subsonic aircraft C_{Dpmin} computed at LRC would cater to the full flight envelope, in this book. For HSC, the wave drag ΔC_{Dw} is to be added.

11.8.2 Computation of Wetted Areas

Computation of the wetted area, Aw, of the aircraft component is shown herein. Skin friction is generated on that part of the surface over which air flows, the so-called wetted area. Wetted area Aw computation has to be accurate as parasite drag is directly proportionate to A_w. A 2% error in area estimation will result in about 1% error in overall subsonic aircraft drag.

Three-dimensional computer aided drawing (CAD) model can give accurate wetted area, A_w. In case CAD data are not available, the component wetted areas can be estimated from manually drawn three-view diagram in a large sheet (say A0 or A1 size) by draftsmen in a different department and then given to aerodynamics group where drag estimation is done. Planimeters prove useful in making accurate area measurements. Normally, three-view 2D drawings have the reference areas of the geometry that can help calibrating the planimeter in use.

11.8.2.1 Lifting Surfaces
These are approximate to the flat surfaces, with the wetted area slightly more than twice the reference area due to some thickness. Care is needed in removing the areas at intersections, such as the wing area buried in the fuselage. A factor k is used to obtain the wetted area of lifting surfaces, as follows:

$A_w = k \times$ (exposed reference area, S_W; the area buried in the body is not included). The factor k may be interpolated linearly for other t/c ratios.
where $k = 2.02$ for $t/c = 0.08\%$

$= 2.04$ for $t/c = 0.12\%$
$= 2.06$ for $t/c = 0.16\%$

11.8.2.2 Fuselage
The fuselage is conveniently divided into sections – typically, for a civil transport aircraft, into a constant cross-section mid-fuselage with varying cross-section front and aft-fuselage closures. The constant cross-section mid-fuselage barrel has a wetted area of A_{wfmid} = perimeter × length. The forward- and aft-closure cones could be sectioned more finely, treating each thin section as a constant section 'slice'. A military aircraft is unlikely to have a constant cross-section barrel and its wetted area must be computed in this way. The wetted areas must be excluded where the wing and empennage join the fuselage or for any other considerations.

11.8.2.3 Nacelle
Only the external surface of the nacelle is considered the wetted area and it is computed in the same way as the fuselage, taking note of the pylon cut-out area. (Internal drag within the intake duct is accounted for as installation effects in engine performance as a loss of thrust.)

11.8.3   Stepwise Approach to Compute Minimum Parasite Drag

The following seven steps are carried out to estimate the minimum parasite drag, C_{Dpmin}:

Step 1. Dissect and isolate aircraft components such as wing, fuselage nacelle and so on. Determine the geometric parameters of the aircraft components such as the characteristic length and wetted areas.

Step 2. Compute *Re* per foot at the LRC condition. Then obtain the component *Re* by multiplying its characteristic length.

Step 3. Determine the basic 2D (Figure 11.24, bottom graph) average skin friction coefficient $C_{\text{F\_basic}}$, corresponding to the Re for each component.

Step 4. Estimate the ΔC_{F} as the increment on account of 3D effects on each component.

Step 5. Estimate the interference drag of two adjacent components. Avoid duplication.

Step 6. Add results obtained in steps 3–5 to get the minimum parasite drag of the component in terms of flat plate equivalent area (ft^2 or m^2); that is, $C_{\text{F}} = C_{\text{F\_2D}} + \sum \Delta C_{\text{F}}$ for the component. $(f)_{\text{comp}} = (A_{\text{w}})_{\text{comp}} \times C_{\text{F}}$, where $(C_{\text{Dpmin}})_{\text{comp}} = (f)_{\text{comp}}/S_{\text{w}}$.

Step 7. Add all the component minimum parasite drags. Then add other drags, for example, trim excrescence drag and so on. Finally, add 3% drag due to surface roughness effects. The aircraft minimum parasite drag is expressed in coefficient form, C_{Dpmin}.

The semi-empirical formulation for each component is given in the following subsections.

11.9   Semi-Empirical Relations to Estimate Aircraft-Component Parasite Drag

Isolated aircraft components are worked on to estimate component parasite drag. The semi-empirical relations given here embed the necessary corrections required for 3D effects. Associated coefficients and indices are derived from actual flight-test data. (Wind-tunnel tests are conducted at a lower *Re* and therefore require correction to represent flight-tested results.) The influence of the related drivers is shown as drag increasing by ↑ and drag decreasing by ↓. For example, an increase of the Re reduces the skin friction coefficient and is shown as *Re* (↓).

11.9.1   Fuselage

The fuselage characteristic length, L_{fus}, is the length from the tip of the nose cone to the end of the tail cone. The wetted area, A_{wf} (↑), and fineness ratio (length/diameter) (↓) of the fuselage are computed. Ensure that cut-outs at the wing and empennage junctions are subtracted. Obtain the *Ref* (↓). The corresponding basic C_{Ff} for the fuselage using (Figure 11.24. bottom graph) is intended for the flat plate at the flight Mach number. Figure 11.24 is accurate and validated over time.

The methodology for the fuselage (denoted by the subscript *f*) is discussed in this section. The Re_{f} is calculated first using the fuselage length as the characteristic length. The semi-empirical formulation is required to correct the 2D skin friction drag for the 3D effects and other influencing parameters, as listed herein. These are incremental values shown by the symbol Δ. There are many incremental effects and it is easy to miss some of them.

 i. 3D effects [1]: 3D effects are on account of surface curvature resulting in change in local flow speed and associated pressure gradients:

   (i) Wrapping:

$$\Delta C_{\text{Ff}} = C_{\text{Ff}} \times [k \times (length/diameter) \times R_e^{-0.2}] \tag{11.16}$$

   where *k* is between 0.022 and 0.025 (take the higher value) and R_e = Reynolds Number of fuselage

   (ii) Supervelocity:

$$\Delta C_{\text{Ff}} = C_{\text{Ff}} \times (diameter/length)^{1.5} \tag{11.17}$$

(iii) Pressure:

$$\Delta C_{Ff} = C_{Ff} \times 7 \times (\text{diameter/length})^3 \tag{11.18}$$

ii. Other effects on the fuselage (increments are given as a percentage of 2D C_{Ff}) are listed herein. The industry has more accurate values of these incremental ΔC_{Ff}. Readers in the industry should not use the values given here – they are intended only for coursework using estimates extracted from industrial data.

(i) *Canopy drag.* There are two types of canopy (Figure 11.6), as follows:

1. *Raised or bubble-type canopy or its variants.* These canopies are mostly associated with military aircraft and smaller aircraft. The canopy drag coefficient $C_{D\pi}$ is based on the frontal cross-sectional area shown in the military-type aircraft in Figure 11.6 (the front view of the raised canopy is shaded). The extent of the raised frontal area contributes to the extent of drag increment and the $C_{D\pi}$ accounts for the effects of canopy rise. $C_{D\pi}$ is then converted to $C_{Ffcanopy} = (A_\pi \times C_{D\pi})/A_{wf}$, where A_{wf} is the fuselage wetted area. The dominant types of a raised or bubble-type canopy and their associated $C_{D\pi}$ are summarised in Table 11.1.

2. *Windshield-type canopy for larger aircraft.* These canopies are typically associated with payload-carrying commercial aircraft from a small Bizjet and larger. Flat panes lower the manufacturing cost but result in a kink at the double curvature nose cone of the fuselage. A curved and smooth transparent windshield avoids the kink that would reduce drag at an additional cost. Smoother types have curved panes with a single or double curvature. Single-curvature panes come in smaller pieces, with a straight side and a curved side. Double curvature panes are the most expensive and considerable attention is required during manufacturing to avoid distortion

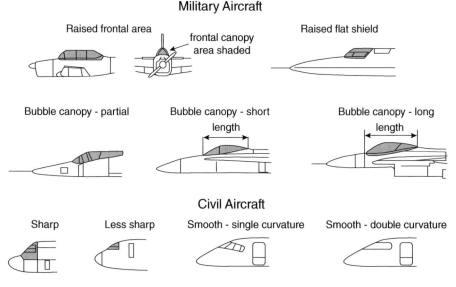

Figure 11.6 Canopy types for drag estimation.

Table 11.1 Typical $C_{D\pi}$ associated with raised or bubble-type canopies.

Raised frontal area (older boxy design – has sharp edges)	$C_{D\pi} = 0.2$
Raised flat shield (reduced sharp edges)	$C_{D\pi} = 0.15$
Bubble canopy (partial)	$C_{D\pi} = 0.12$
Bubble canopy (short)	$C_{D\pi} = 0.08$
Bubble canopy (long)	$C_{D\pi} = 0.06$

Table 11.2 Typical C_{D_π} associated with sharp wind shield type canopies (drag in sq. ft).

2-abreast seating aircraft	0.1 sq. ft	8-abreast seating aircraft	0.4 sq. ft
4-abreast seating aircraft	0.2 sq. ft	10-abreast or more	0.5 sq. ft
6-abreast seating aircraft	0.3 sq. ft		

of vision. The values in square feet in Table 11.2 are used to obtain a sharp-edged windshield-type canopy drag.

Adjust the values for the following variations.	
Kinked windshield (less sharp)	reduce the value by 10%
Smoothed (single-curvature) windshield	reduce the value by 20–30%
Smoothed (double-curvature) windshield	reduce the value by 30–50%

(ii) *Body pressurisation* – fuselage surface waviness, 4–6%

(iii) *Non-optimum* fuselage shape (interpolate the in-between values)

    1. Front fuselage closure ratio, F_{cf} (see Section 5.4.3)

        For $F_{cf} \leq 1.5$: 8%

        For $1.5 \leq F_{cf} \leq 1.75$: 6%

        For $F_{cf} \geq 1.75$: 4%

        For military aircraft type with high nose fineness: 3%

    2. Aft fuselage closure angle, (see Section 5.4.3)

        Less than 10°: 0

        11–12° closure: 1%

        13–14°: 4%

    3. Upsweep closure (see Section 5.4.3) use in conjunction with (4)

        No upsweep: 0

        4° of upsweep: 0–2%

        10° of upsweep: 4–8%

        15° of upsweep: 10–15%

        (interpolate in-between values)

    4. Aft-end cross-sectional shape

        Circular: 0

        Shallow keel: 0–1%

        Deep keel: 1–2%

    5. Rear-mounted door (with fuselage upsweep): 5–10%

(iv) Cabin-pressurisation leakage (if unknown, use higher value): 2–5%

(v) Excrescence (*non-manufacturing* types such as windows)

    1. Windows and doors (use higher values for larger aircraft): 2–4%

    2. Miscellaneous: 1%

(vi) Wing-fuselage belly fairing, if any: 1–5% (use higher value if houses undercarriage)

(vii) Undercarriage fairing – typically for high-wing aircraft (if any fairing): 2–6% (based on fairing protrusion height from fuselage)

iii. The interference drag increment with the wing and empennage is included in the calculation of lifting-surface drag and therefore is not duplicated when computing the fuselage parasite drag. Totalling the C_{Ff} and ΔC_{Ff} from the wetted area A_{wf} of the isolated fuselage, the flat plate equivalent drag, f_{fus} (see Step 6 in Section 11.8.3), is estimated in square feet.

iv. Surface roughness is 2–3%. These effects are from the manufacturing origin, discussed in Section 11.9.5.

Total all the components of parasite drag to obtain C_{Dpmin}, as follows. It should include the excrescence drag increment. Converted into the fuselage contribution to $[C_{Dpmin}]_f$ in terms of aircraft wing area, it becomes:

$$C_{Ff} = (1.02–1.03) \times (\text{Basic } C_{Ff} + \Sigma\Delta C_{Ff})$$ (11.19)

$$f_{fus} = (C_{Ff} + \Delta C_{Ffwrap} + \Delta C_{Ffsupervel} + \Delta C_{Ffpress} + \Delta C_{Ffother} + \Delta C_{Ffrough}) \times A_{wf}$$ (11.20)

$$[C_{Dpmin}]_f = f_{fus}/S_W$$ (11.21)

Because surface roughness drag is the same percentage for all components, it is convenient to total them after evaluating all components. In that case, the term ΔC_{Ff_rough} is dropped from Eq. (11.20).

11.9.2 Wing, Empennage, Pylons and Winglets

The wing, empennage, pylon and winglets are treated as lifting surfaces and use identical methodology to estimate their minimum parasite drag. It is similar to the fuselage methodology except that it does not have the wrapping effect. Here, the interference drag with the joining body (e.g. for the wing, it is the fuselage) is taken into account because it is not included in the fuselage f_{fus}.

The methodology for the wing (denoted by the subscript w) is discussed in this section. The Re_w is calculated first using the wing MAC as the characteristic length. Next, the exposed wing area is computed by subtracting the portion buried in the fuselage and then the wetted area, A_{Ww}, using the k factors for the t/c as in Section 11.8.2.1. Using the Re_w, the basic C_{Fw_basic} is obtained from the graph in Figure 11.24, bottom graph for the flight Mach number. The incremental parasite drag formulae are as follows:

i. 3D effects (taken from [1]):
  i) Supervelocity:

$$\Delta C_{Fw} = C_{Fw} \times K_1 \times (\text{aerofoil thickness/chord ratio})_{ave}$$ (11.22)

  where $K_1 = 1.2–1.5$ for supercritical aerofoil
  $K_1 = 1.6–2$ for conventional aerofoil
  ii) Pressure:

$$\Delta C_{Fw} = C_{Fw} \times 60 \times (\text{aerofoil thickness/chord ratio})_{ave}^4 \times \left(\frac{6}{AR}\right)^{0.125}$$ (11.23)

  where aspect ratio $AR \geq 2$ (modified from [1])
  Last term of this expression includes the effect of non-elliptical lift distribution.
ii. Interference:

$$\Delta C_{Fw} = C_B^2 \times K_2 \times \left\{\frac{0.75 \times (t/c)_{root}^3 - 0.0003}{A_w}\right\}$$ (11.24)

where $K_2 = 0.6$ for high and low wing designs and C_B is the root chord at fuselage intersection. For mid-wing, $K_2 = 1.2$. This is valid for t/c ratio > 0.07. For a t/c ratio below 0.07, take the interference drag

$$\Delta C_{Fw} = 3–5\% \text{ of } C_{Fw}$$ (11.25)

The same relations apply for the V-tail and H-tail.
For pylon interference 10–12%
Interference drag is not accounted for in fuselage drag, it is accounted for in wing drag.
(For the pylon, one interference is at the aircraft side and one is with the nacelle.)
iii. Other effects:
Excrescence (non-manufacturing, e.g. control surface gaps etc.):

Flap gaps	4–5%
Slat gaps	4–5%
Others	4–5%

iv. Surface roughness (to be added later)

The flat plate equivalent of wing drag contribution is (subscript is self-explanatory)

$$f_w = (C_{Fw} + \Delta C_{Ffw\_supervel} + \Delta C_{Fw\_press} + \Delta C_{Fw\_inter} + \Delta C_{Fw\_other} + \Delta C_{Fw\_rough}) \times A_{ww} \qquad (11.26)$$

which can be converted into C_{Dpmin} in terms of aircraft wing area

$$\text{that is, } [C_{Dpmin}]_w = f_w/S_W \qquad (11.27)$$

(Drop the term $\Delta C_{Fwrough}$ of Eq. (11.20) if it is accounted for at the end after computing fs for all components as shown in Eq. (11.28).)

The same procedure is used to compute the parasite drag of empennage, pylons and so on, which are considered as being wing-like lifting surfaces.

$$f_{lifting\_surface} = [(C_{F+} \Delta C_{F\_supervel+} \Delta C_{F\_press} + \Delta C_{F\_inter} + \Delta C_{F\_other} + \Delta C_{F\_rough}) \times A_w]_{lift\ sur} \qquad (11.28)$$

which can be converted into

$$[C_{Dpmin}]_{lifting\_surface} = f_{lifting\_surface}/S_w \qquad (11.29)$$

As before, it is convenient to total $\Delta C_{Ffrough}$ after evaluating all components. In that case, the term $\Delta C_{Ffrough}$ is dropped from Eq. (11.20) and it is accounted for as shown in Eq. (11.21).

11.9.3 Nacelle Drag

The nacelle requires different treatment, with the special consideration of throttle dependent air flowing through as well as over it, like the fuselage. This section provides the definitions and other considerations needed to estimate nacelle parasite drag (see [2, 9, 15, 16]). The nacelle is described in Section 5.12.

The throttle dependent variable of the internal flow passing through the turbofan engine affects the external flow over the nacelle. The dominant changes in the flow field due to throttle dependency are around the nacelle at the lip and aft end. When the flow field around the nacelle is known, the parasite drag estimation method for the nacelle is the same as for the other components but must also consider the throttle dependent effects.

Civil aircraft nacelles are typically pod-mounted. In this book, only the long duct is considered. Military aircraft engines are generally buried in the aircraft shell (i.e. fuselage). A podded nacelle may be thought of as a wrapped-around wing in an axi-symmetric shape like that of a fuselage. The nacelle section shows aerofoil-like sections in Figure 11.7; the important sources of nacelle drag are listed here (a short duct nacelle (see Figure 12.17) is similar except for the fan exhaust coming out at high-speed over the exposed outer surface of the core nozzle, for which its skin friction must be considered):

Nacelles do not have front fuselage like curvature and aft fuselage like upsweep curvatures. Nacelle external flow is throttle dependent and is affected by the internal intake flow when the lip suction effect (Figure 11.7) is considered. Nacelle aft has little separation with the exhaust flow entrainment effect and boat tail (see next) drag takes into account of the pressure drag. Therefore, unlike fuselage drag considerations, the supervelocity and pressure effects are not considered in case of nacelle drag estimation.

Throttle-independent drag (external surface) – at cruise thrust

i. skin friction
ii. wrapping effects of axi-symmetric body excrescence effects (includes non-manufacturing types such as cooling ducts.)

Throttle dependent drag
    inlet drag (front end of the diffuser)
    nacelle base drag (zero for an engine operating at cruise settings and higher)
    boat tail drag (curvature of the nozzle at the aft end of the nacelle)

Definitions and typical considerations for drag estimation associated with the flow field around an isolated long-duct podded nacelle (approximated to circular cross-section) are shown in Figure 11.7. Although there is internal flow through the nacelle, the external geometry of the nacelle may be treated as a fuselage, except that there is a lip section similar to the LE of an aerofoil. The prevailing engine throttle setting is maintained at a rating for LRC or HSC for the mission profile. The *intake drag* and the *base drag/boat tail drag* are explained next.

11.9.3.1   Intake Drag

The intake stream tube flow pattern at cruise is complex that makes intake drag estimation difficult (Figure 11.7). There is spillage during the subcritical operation due to the stream tube being smaller than the cross-sectional area at the nacelle highlight diameter, where external flow turns around the lip creating suction (i.e. thrust). This develops pre-compression, ahead of the intake, when the intake velocity is slower compared to the free-stream velocity expressed in the fraction ($V_{\text{intake}}/V_\infty$). At ($V_{\text{intake}}/V_\infty$) < 0.8. The excess air flow spills over the nacelle lip. The intake lip acts as the LE of a circular aerofoil. The subcritical air flow diffusion ahead of the inlet results in pre-entry drag called *additive drag*. The net effect results in *spillage drag*, as described herein. The spillage drag added to the friction drag at the lip results in the *intake drag*, which is a form of parasite drag. (For the military aircraft intake, see Sections 12.8.2 and 5.16.1.)

$$\text{spillage drag} = \text{additive drag} + \text{lip suction (thrust } sign \text{ changes to } -\text{ve)} \qquad (11.30)$$

intake drag = spillage drag + friction drag at the lip (supervelocity effect).

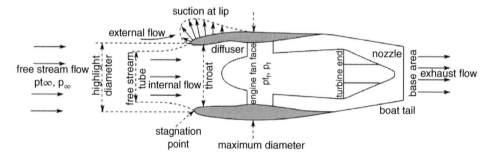

Figure 11.7 Aerodynamic considerations for an isolated long-duct nacelle drag.

Spillage drag =
Additive drag + lip suction (thrust)

Figure 11.8 Throttle dependent spillage drag (subsonic case shown).

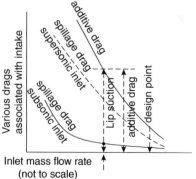

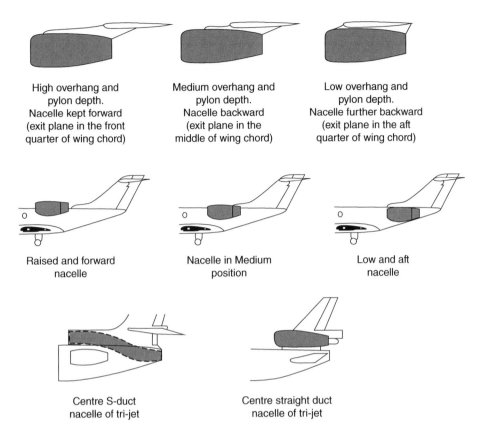

Figure 11.9 Wing and fuselage-mounted nacelle.

Section 11.8 shows intake drag variations with the mass flow rate for both subsonic and supersonic (i.e. sharp LE) intake.

11.9.3.2 Base Drag

The design criteria for the nozzle exit area sizing is such that at LRC, the exit plane static pressure Pe equals the ambient pressure P_∞ (a perfectly expanded nozzle, $P_e = P_\infty$) to eliminate any base drag. At higher throttle settings, when $P_e > P_\infty$, there still is no base drag. At lower settings, for example, idle rating – there is some base drag as a result of the nozzle exit area being larger than what is required.

11.9.3.3 Boat Tail Drag

The long-duct contour for closure (i.e. 'boat tail' shape) at the aft end is shallow enough to avoid separation, especially with the assistance of entrainment effects of the exhaust plume. Hence, the boat tail drag is kept low. At the idle throttle setting, considerable flow separation can occur and the magnitude of boat tail drag would be higher, but it is still small compared to the nacelle drag.

For book-keeping purposes and to avoid conflict with aircraft manufacturers, engine manufacturers generally include internal drag (e.g. ram, diffuser and exhaust-pipe drag) in computing the net thrust of an engine. Therefore, this book only needs to estimate the parasite drag (i.e. external drag) of the nacelle. Intake duct loss is considered engine installation losses expressed as intake-recovery loss. Intake- and exhaust-duct losses are approximately 1–3% in engine thrust at LRC (throttle and altitude dependent). The net thrust of the turbofan, incorporating installation losses, is computed using the engine-manufacturer-supplied programme and data. These manufacturers work in close liaison to develop the internal contour of the nacelle and intake. External nacelle-contour design and airframe integration remain the responsibility as the aircraft manufacturer.

11.9.3.4 Nacelle 3D Effects

The long-duct nacelle characteristic length, L_{nac}, is the length measured from the intake-highlight plane to the exit area plane. The wetted area A_{Wn}, Re and basic C_{Fn} are estimated as for other components. The incremental parasite drag formulae for the nacelle are provided herein. The supervelocity effect around the nacelle lip section is included in the intake drag estimation; hence, it is not computed separately. Similarly, the pressure effect is included in the base/boat tail drag estimation. These two items are addressed this way because of the special consideration of throttle dependency. The following are the relationships used to compute the nacelle drag coefficient C_{Dn}. The stepwise approach to compute this is:

i.

$$\Delta C_{\mathrm{Dn}} \text{ effects (same as fuselage being axi} - \text{symmetric)}: \qquad (11.31)$$

   (i) Wrapping: $\Delta C_{\mathrm{Fn}} = 0.025 \times (diameter/length)^{-1} \times R_e^{-0.2}$

ii. Other incremental effects.
   Drag contributions made by the following effects are given as a percentage of C_{Fn}. These are typical of the generic nacelle design.
   (i) Intake drag at LRC – includes supervelocity effects $\approx$ 40–60% (higher bypass ratio having higher percent)
   (ii) Boat Tail/Base drag (throttle dependent) – includes pressure effects $\approx$ 10–12% (higher value for smaller aircraft)
   (iii) Excrescence (non-manufacturing type, e.g. cooling air intakes etc.) $\approx$ 20–25% (higher value for smaller aircraft)

iii. Interference drag – a podded nacelle in the vicinity of wing or body would have interference drag as follows (per nacelle). For a wing-mounted nacelle, the higher the overhang forward of the wing, the lower the interference drag.
   Typical values of the interference drag by the pylon (each) interacting with the wing or the body) are given in Table 11.3.

iv. Surface roughness (add later $\approx$ 3%)

A long overhang in front of the wing keeps the nacelle free from interference effects, a short overhang has the highest interference. However, there is little variation of inference drag of nacelle mounted on different position at the aft fuselage. Much depends on the proximity of other bodies, for example, wing, empennage and so on. If the nacelle is within one diameter, then interference drag may be increased by another 0.5%. The centre engine is close to fuselage and with V-tail – they have increased interference.

11.9.3.5 Total Nacelle Drag

By adding all the components, the flat plate equivalent of nacelle drag contribution is given by Eq. (11.32).

$$f_{\mathrm{n}} = (C_{\mathrm{Fn}} + \Delta C_{\mathrm{Fn\_wrap}} + \Delta C_{\mathrm{Fn\_intake}} + \Delta C_{\mathrm{fn\_boat\,tail}} + \Delta C_{\mathrm{Fn\_excres}} + \Delta C_{\mathrm{Fn\_rough}}) \times A_{\mathrm{wn}} \qquad (11.32)$$

Table 11.3 Nacelle interference drag (per nacelle).

Wing mounted (Figure 11.9)		Fuselage mounted (Figure 11.9)	
	Interference drag		Interference drag
High (long) overhang	0	Raised	5% of C_{Fn}
Medium overhang	4% of C_{Fn}	Medium	5% of C_{Fn}
Low (short) overhang	7% of C_{Fn}	Low	5% of C_{Fn}
	S-duct		6.5% of C_{Fn}
	Straight duct (centre)		5.8% of C_{Fn}

As before, it is convenient to total $\Delta C_{\text{Ffrough}}$ after evaluating all components. In that case, the term $\Delta C_{\text{F\_rough}}$ is dropped from Eq. (11.32) and it is accounted for as shown in Eq. (11.34). Converted into the nacelle contribution to C_{Dpmin} in terms of aircraft wing area it becomes

$$[C_{\text{Dpmin}}]_n = f_n./S_{\text{W}}. \tag{11.33}$$

In the last three decades, the nacelle drag has been reduced by approximately twice as much as what has been achieved to other aircraft components. This demonstrates the complexity of and unknowns associated with the flow field around nacelles. CFD is important in nacelle design and its integration with aircraft. In this book, nacelle geometry is simplified to the axi-symmetric shape without loss of methodology.

11.9.4 Recovery Factor

Intake duct loss will deplete some flow energy making fan face total pressure, P_{tf}, less than free-stream total pressure, $P_{\text{t}\infty}$. The extent of total loss is expressed as the recovery factor, RF. Thrust loss on account of RF is seen as installation loss and is accounted in installed engine thrust and is dealt in Section 13.2.1.

11.9.5 Excrescence Drag

An aircraft body is not smooth; located all over the body are probes, blisters, bumps, protrusions, surface-protection mats for steps, small ducts (e.g. for cooling) and exhausts (e.g. environmental control and cooling air) – these are unavoidable features. In addition, there are mismatches at subassembly joints – for example, steps, gaps and waviness originating during manufacture and treated as discreet roughness. Pressurisation also causes the fuselage-skin waviness (i.e. areas ballooning up). In this book, excrescence drag is addressed separately as two types:

1. *Manufacturing origin* [16]. This includes aerodynamic mismatches as discreet roughness resulting from tolerance allocation. Aerodynamicists must specify surface smoothness requirements to minimise excrescence drag resulting from the discrete roughness, within the manufacturing-tolerance allocation.

2. *Non-manufacturing origin*. This includes aerials, flap tracks and gaps, cooling ducts and exhausts, bumps, blisters and protrusions.

Excrescence drag due to surface roughness drag is accounted for by using 2–3% of component parasite drag as roughness drag ([1, 7]). As indicated in Step 7 of Section 11.8.3, $f_{\text{comp total}}$ is increased by 3% using a factor of 1.03 after computing all component parasite drags, as follows

$$f_{\text{comp total}} = 1.03(f_f + f_w + f_n + f_p + f_{\text{other comp}}) \tag{11.34}$$

The difficulty in understanding the physics of excrescence drag was summarised by Haines [12] in his review by stating '...one realises that the analysis of some of these early data seems somewhat confused, because three major factors controlling the level of drag were not immediately recognised as being separate effects'. These factors are as follows:

i. how skin friction is affected by the position of the boundary-layer transition
ii. how surface roughness affects skin friction in a fully turbulent flow
iii. how geometric shape (non-planar) affects skin friction

Haines's study showed that a small but significant amount of excrescence drag results from manufacturing origin and was difficult to understand.

11.9.6 Miscellaneous Parasite Drags

In addition to excrescence drag, there are other drag increments such as from the intake drag of the environmental control system (ECS) (e.g. air-conditioning), which is a fixed value depending on the number of passengers) and aerials and trim drag, which are included to obtain the minimum parasite drag of the aircraft.

11.9.6.1 Air-Conditioning Drag

Air-conditioning air is inhaled from the atmosphere through flush intakes that incur drag. It is mixed with hot air bled from a mid-stage of the engine compressor and then purified. Loss of thrust due to engine bleed is accounted for in the engine thrust computation, but the higher pressure of the expunged cabin air causes a small amount of thrust. Table 11.4 shows the air-conditioning drag based on the number of passengers (interpolation is used for the between sizes).

11.9.6.2 Trim Drag

Due to weight changes during cruise, the centre of gravity (CG) could shift, thereby requiring the aircraft to be trimmed in order to relieve the control forces. Change in the trim-surface angle causes a drag increment. The average trim drag during cruise is approximated as shown in Table 11.5, based on the wing reference area (interpolation is used for the between sizes).

11.9.6.3 Aerials

Navigational and communication systems require aerials that extend from an aircraft body, generating parasite drag on the order 0.06–0.1 ft^2, depending on the size and number of aerials installed. For midsized transport aircraft, 0.075 ft^2 is typically used. Therefore:

$$f_{aircraft\_parasite} = f_{comp\_total} + f_{aerial} + f_{air\ cond} + f_{trim} \tag{11.35}$$

11.10 Notes on Excrescence Drag Resulting from Surface Imperfections

Excrescence drag due to surface imperfections is difficult to estimate; therefore, this section provides background on the nature of the difficulty encountered. Capturing all the excrescence effects over the full aircraft in CFD is yet to be accomplished with guaranteed accuracy.

A major difficulty arises in assessing the drag of small items attached to the aircraft surface, such as instruments (e.g. pitot and vanes), ducts (e.g. cooling) and necessary gaps to accommodate moving surfaces. In addition, there is the unavoidable discrete surface roughness from mismatches and imperfections – *aerodynamic*

Table 11.4 Air-conditioning drag.

No. of passengers	Drag – f (ft²)	Thrust – f (ft²)	Net drag – f (ft²)
50	0.1	−0.04	0.06
100	0.2	−0.1	0.1
200	0.5	−0.2	0.4
300	0.8	−0.3	0.5
600	1.6	−0.6	1.0

Table 11.5 Trim drag (approximate).

Wing reference area (ft²)	Trim drag f (ft²)	Wing reference area (ft²)	Trim drag – f (ft²)
200	0.12	2000	0.3
500	0.15	3000	0.5
1000	0.2	4000	0.8

defects – resulting from limitations in the manufacturing processes. Together, all of these drags, from both manufacturing and non-manufacturing origins, are collectively termed *excrescence drag*, which is parasitic in nature [13, 16]. Of particular interest is the excrescence drag resulting from the discrete roughness, within the manufacturing-tolerance allocation, in compliance with the surface smoothness requirements specified by aerodynamicists to minimise drag.

Mismatches at the assembly joints are seen as discrete roughness (i.e. aerodynamic defects) – for example, steps, gaps, fastener flushness and contour deviation – placed normal, parallel or at any angle to the free-stream air flow. These defects generate excrescence drag. In consultation with production engineers, aerodynamicists, that is, specify tolerances to minimise the excrescence drag of the order of 1–3% of the C_{Dpmin}.

The 'defects' are neither at the maximum limits throughout nor uniformly distributed. The excrescence dimension is on the order of less than 0.1 in.; for comparison, the physical dimension of a fuselage is nearly 5000–10 000 times larger. It poses a special problem for estimating excrescence drag; that is, capturing the resulting complex problem in the boundary layer downstream of the mismatch.

The methodology involves first computing excrescence drag on a 2D flat surface without any pressure gradient. On a 3D curved surface with a pressure gradient, the excrescence drag is magnified. The location of a joint of a subassembly on the 3D body is important for determining the magnification factor that will be applied on the 2D flat plate excrescence drag obtained by semi-empirical methods. The body is divided into two zones: Zone 1 (the front side) is in an adverse pressure gradient and Zone 2 is in a favourable pressure gradient ([16] of Chapter 1). Excrescences in Zone 1 are more critical to magnification than in Zone 2. At a LRC flight speed (i.e. at M_{crit} for civil aircraft), shocks are local and subassembly joints should not be placed in Zone 1.

Estimation of aircraft drag uses an average skin friction coefficient C_F (see Figure 11.24, bottom graph, later), whereas excrescence drag estimation uses the local skin friction coefficient C_f (see Figure 11.24, top graph), appropriate to the location of the mismatch. These fundamental differences in drag estimation methods make the estimation of aircraft drag and excrescence drag quite different.

After World War II, efforts continued for the next two decades – especially at the RAE by Gaudet, Winters, Johnson, Pallister and Tillman et al. – using wind-tunnel tests to understand and estimate excrescence drag. Their experiments led to semi-empirical methods subsequently compiled by ESDU as the most authoritative information on the subject. Aircraft and excrescence drag estimation methods still remain state-of-the-art, and efforts to understand the drag phenomena continue.

Surface imperfections inside the nacelle – that is, at the inlet diffuser surface and at the exhaust nozzle – could affect engine performance as loss of thrust. Care must be taken so that the 'defects' do not perturb the engine flow field. The internal nacelle drag is accounted for as an engine installation effect.

11.11 Minimum Parasite Drag

The aircraft C_{Dpmin} can now be obtained from f_{aircraft}. Using Eq. (11.15), the minimum parasite drag of the entire aircraft is $C_{\text{Dpmin}} = (1/S_w) \times f_i$, where f_i is the sum of the total fs of the entire aircraft:

$$C_{\text{Dpmin}} = f_{\text{aircraft}}/S_w \tag{11.36}$$

11.12 ΔC_{Dp} Estimation

Equation (11.2) shows that ΔC_{Dp} is not easy to estimate. ΔC_{Dp} contains the lift-dependent variation of parasite drags due to a change in the pressure distribution with changes in the angle of attack. Although it is a small percentage of the total aircraft drag (it varies from 0 to 10%, depending on the aircraft C_L), it is the most difficult to estimate. There is no proper method available for estimating the ΔC_{Dp} versus C_L relationship; it is

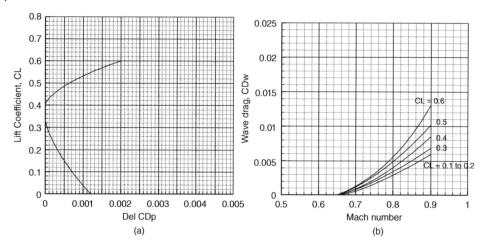

Figure 11.10 Typical ΔC_{Dp} and C_{Dw}. (a) ΔC_{Dp}. (b) C_{Dw}.

design-specific and depends on wing geometry (i.e. planform, sweep, taper ratio, aspect ratio and wing-body incidence) and aerofoil characteristics (i.e. camber and t/c). The values are obtained through wind-tunnel tests and, currently, by CFD.

During cruise, the lift coefficient varies with changes in aircraft weight and/or flight speed. The design-lift coefficient, C_{LD}, is around the mid-cruise weight of the LRC. Let C_{LP} be the lift coefficient when $\Delta C_{Dp} = 0$. The wing should offer C_{LP} at the three-fourths value of the designed C_{LD}. This would permit an aircraft to operate at HSC (at M_{crit}; i.e. at the lower C_L) with almost zero ΔC_{Dp}. Figure 11.10a shows a typical ΔC_{Dp} versus C_L variation. This graph can be used only for coursework in Sections 11.19 and 11.20.

For any other type of aircraft, a separate graph must be generated from wind-tunnel tests and/or CFD analysis. The industry has a large databank to generate such graphs during the conceptual design phase. In general, the semi-empirical method takes a tested wing (with sufficiently close geometrical similarity) ΔC_{Dp} versus C_L relationship and then corrects it for the differences in wing sweep ($\downarrow$), aspect ratio ($\downarrow$), t/c ratio ($\uparrow$), camber and any other specific geometrical differences.

11.13 Subsonic Wave Drag

The thickness parameters of lifting surfaces (e.g. wing) and bodies (e.g. fuselage) have strong influence in drag generation. Thickness gives rise to local super-velocities (higher than free-stream velocities) that increases the local Re altering the skin friction – it does not refer to compressibility. If local velocities are sufficiently high then compressibility effects develop. Semi-empirical formulae to account for the supervelocity effects altering skin friction are given in Section 11.11. This section deals with the compressibility drag, C_{Dw}.

Wave drag is caused by compressibility effects of air as an aircraft approaches high-subsonic speed. Local shocks start to appear on a curved surface as aircraft speed increases. This is in a transonic flow regime, in which a small part of the flow over the body is supersonic while the remainder is subsonic. In some cases, a shock interacting with the boundary layer can cause premature flow separation, thus increasing pressure drag. Initially, it is gradual and then shows a rapid rise as it approaches the speed of sound. The industry practice is to tolerate a 20-count (i.e. $C_D = 0.002$) increase due to compressibility at a speed identified as M_{crit}. Figure 11.10b). At higher speeds, higher wave drag penalties are incurred.

The extent of compressibility drag rise is primarily dependent on the aerofoil design (lifting surfaces – e.g. wing design) and to lesser extent in shaping of the rest of aircraft. In a proper sense, aerofoil design is an iterative process and offers several options to choose from. Once the aircraft configuration is frozen in the conceptual design phase, then the proper C_{Dw} versus Mach graph is obtained and the exact drag rise at LRC can be applied. The difficulty is at the conceptual stage when not much information is available and when approximations are to be made. In this book aircraft configuration is given and its performance is to be estimated. One of the first tasks is to estimate the aircraft drag from the three-view diagram of the given aircraft. Also supplied is its wing characteristics, one of them are the wave drag, C_{Dw}, characteristics. Initially, C_{Dw} has to be assumed from past data and, as the project progresses with more information (through CFD analyses/wind-tunnel tests), it is fine-tuned.

At the conceptual phase of the project, it may prove convenient to keep compressibility drag, $C_{Dw} = 0$ up to LRC. It is permissible, since the drag prediction is constantly updated as more information is available. Some industries use this approach as it offers the advantage to express drag polar in a *close-form* equation of the shape of a parabola ($C_D = C_{Dpmin} + kC_L^2$; there is no C_{Dw} term) with little error in the operating range of cruise, climb and descent. The advantage of having close-form equation is that it can quickly analyse aircraft characteristics to ascertain the LRC Mach number based on, say, best economical cruise or any other criteria. If, however, the proper C_{Dw} versus Mach graph for the aircraft is available (suiting the LRC criteria) then it can be used with the proper C_{Dw} rise when the *close-form* equation gets a little more complicated. Or otherwise a suitable drag rise to account compressibility effect at LRC may be accounted – say, approximately 5–10 counts. This may fall within the average value but at the conceptual stage it is still a guesstimate. The crux is in the accurate book-keeping of drag counts for whatever method is used.

A typical wave drag (C_{Dw}) graph is shown in Figure 11.10b, which can be used for coursework (civil aircraft), described in Section 11.19. Wave drag characteristics are design-specific; each aircraft has its own C_{Dw}, which depends on wing geometry (i.e. planform shape, quarter-chord sweep, taper ratio and aspect ratio) and aerofoil characteristics (i.e. camber and t/c) and to a lesser extent on the shape of the rest of the aircraft. Wind-tunnel testing and *CFD* can predict wave drag accurately but must be verified by flight tests. The industry has a large databank to generate semi-empirically the C_{Dw} graph during the conceptual design phase.

11.14 Total Aircraft Drag

Total aircraft drag is the sum of all drags estimated in Sections 11.8 through 11.12, as follows.

At LRC

$$C_D = C_{Dpmin} + \Delta C_{Dp} + \frac{C_L^2}{\pi AR} \tag{11.37}$$

At HSC (at M_{crit})

$$C_D = C_{Dpmin} + \Delta C_{Dp} + \frac{C_L^2}{\pi AR} + C_{Dw} \tag{11.38}$$

At takeoff and landing, additional drag exists, as explained in the next section.

11.15 Low-Speed Aircraft Drag at Takeoff and Landing

For safety in operation and aircraft structural integrity, aircraft speed at takeoff and landing must be kept as low as possible. At ground proximity, lower speed would provide longer reaction time for the pilot, easing the task of controlling an aircraft at a precise speed. Keeping an aircraft aloft at low speed is achieved by increasing lift through increasing wing camber and area using high-lift devices such as a flap and/or a slat. Deployment of a flap and slat increases drag; the extent depends on the type and degree of deflection. Of course, in this scenario, the undercarriage remains extended, which also would incur a substantial drag increase. At approach to landing, especially for military aircraft, it may require 'washing out' of speed to slow down by using fuselage-mounted speed brakes (in the case of civil aircraft, this is accomplished by wing-mounted spoilers). Extension of all these items is known as a *dirty configuration* of the aircraft, as opposed to a *clean configuration* at cruise. Deployment of these devices is speed-limited in order to maintain structural integrity; that is, a certain speed for each type of device extension should not be exceeded.

After takeoff, typically at a safe altitude of 200 ft, pilots retract the undercarriage, resulting in noticeable acceleration. At about an 800 ft altitude with appropriate speed gain, the pilot retracts the high-lift devices. The aircraft is then in the clean configuration, ready for an en-route climb to cruise altitude; therefore, this is sometimes known as *en-route configuration* or *cruise configuration*.

11.15.1 High-Lift Device Drag

High-lift devices are typically flaps and slats, which can be deployed independently of each other. Some aircraft have flaps but no slats (described in Section 4.15). Flaps and slats conform to the aerofoil shape in the retracted position. The function of a high-lift device is to increase the aerofoil camber when it is deflected relative to the baseline aerofoil. If it extends beyond the wing LE and trailing edge, then the wing area is increased. A camber increase causes an increase in lift for the same angle of attack at the expense of drag increase. Slats are nearly full span, but flaps can be anywhere from part to full span (i.e. flaperon). Typically, flaps are sized up to about two-thirds from the wing root. The flap chord to aerofoil chord (c_c/c) ratio is in the order of 0.2–0.3. The main contribution to drag from high-lift devices is proportional to their projected area normal to free-stream air. The associated parameters affecting drag contributions are as follows:

- type of flap or slat
- extent of flap or slat chord to aerofoil chord (typically, the flap is 20–30% of the wing chord)
- extent of deflection (flap at takeoff is from 7 to 15°; at landing, it is from 25 to 60°)
- gaps between the wing and flap or slat (depends on the construction)
- extent of flap or slat span
- fuselage width fraction of wing span
- wing sweep, t/c, twist and AR

The myriad variables make formulation of semi-empirical relations difficult. References [1, 4, 5] offer different methodologies. It is recommended that practitioners use CFD and test data. Reference [17] gives detailed test results of a double-slotted flap (0.309c) NACA 632-118 aerofoil (Figure 11.11). Both elements of a double-slotted flap move together and the deflection of the last element is the overall deflection. For wing application, this requires an aspect ratio correction, as described in Section 4.9.4.

Figure 11.12 is generated from various sources giving averaged typical values of C_L and C_{D_flap} versus flap deflection. It does not represent any particular aerofoil and is intended only for coursework to be familiar with the order of magnitude involved without loss of overall accuracy. The methodology is approximate; practising engineers should use data generated by tests and CFD.

The simple semi-empirical relation for flap drag given in Eq. (11.39) is generated from flap drag data shown in Figure 11.12. The methodology starts by working on a straight wing (Λ_0) with an aspect ratio of 8, flap-span-to-wing-span ratio (b_f/b) of two-thirds and a fuselage-width-to-wing-span ratio of less than a quarter. Total flap drag on a straight wing (Λ_0) is seen as composed of two-dimensional parasite drag of the flap ($C_{Dp\_flap\_2D}$), change in induced drag due to flap deployment (ΔC_{Di_flap}), and interference generated on deflection ($\Delta C_{Dint\ flap}$). Equation (11.40) is intended for a swept wing. The basic expressions are corrected for other geometries, as given in Eqs. (11.41) and (11.42).

$$\text{Straight wing, } C_{D\_flap\_\Lambda0} = \Delta C_{D\_flap\_2D} + \Delta C_{Di\_flap} + \Delta C_{Dint\_flap} \tag{11.39}$$

$$\text{Swept wing, } C_{D\_flap\_\Lambda\frac{1}{4}} = C_{D\_flap\_\Lambda0} \times \cos\Lambda_{\frac{1}{4}} \tag{11.40}$$

Figure 11.11 NACA 632-118 aerofoil.

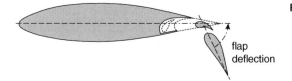

flap
deflection

Figure 11.12 Flap drag.

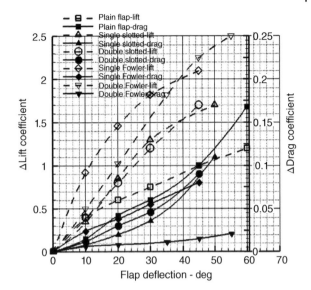

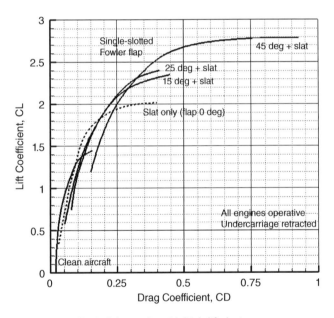

Figure 11.13 Typical drag polar with high-lift devices.

The empirical form of the second term of Eq. (11.39) is given by

$$\Delta C_{Di\_flap} = 0.025 \times (8/AR)^{0.3} \times [(2b)/(3b_f)]^{0.5} \times (\Delta C_L)^2 \tag{11.41}$$

where AR is the wing aspect ratio and (b_f/b) is the flap-to-wing-span ratio.

The empirical form of the third term of Eq. (11.39) is given by

$$\Delta C_{Dint\_flap} = k \times C_{D\_flap\_2D} \tag{11.42}$$

k is 0.1 for single slotted, 0.2 for double slotted, 0.25–0.3 for single Fowler and 0.3–0.4 for double Fowler flap. Lower values may be taken at lower settings.

Figure 11.12 shows the $C_{D\_flap\_2D}$ for various flap types at various deflection angles with the corresponding maximum C_L gain given in Figure 11.12. Aircraft fly well below C_{Lmax}, keeping a safe margin. Increase $C_{Di\,flap}$ by 0.002 if the slats are deployed.

Worked-Out Example

An aircraft has an aspect ratio, $AR = 7.5$, $1/4 = 20°$,

$(b_f/b) = 2/3$, and fuselage-to-wing-span ratio less than 1/4. The flap type is a single-slotted Fowler flap and there is a slat. The aircraft has $C_{Dpmin} = 0.019$. Construct its drag polar.

At 20° deflection:
It is typical for takeoff with $C_L = 2.2$ (approximate) but can be used at landing.
From Figure 11.12, $\Delta C_{D\_flap\_2D} = 0.045$ and $\Delta C_L = 1.46$
From Eq. (11.41), $\Delta C_{Di\_flap} = 0.025 \times (8/7.5)^{0.3} \times [(2/3)/(3/2)]^{0.5} \times (1.46)^2 = 0.025 \times 1.02 \times 2.13 = 0.054$
From Eq. (11.42), $\Delta C_{Dint\_flap} = 0.25 \times 0.045 = 0.01\,125$
$C_{D\_flap\_A0} = 0.045 + 0.054 + 0.01\,125 = 0.11$
With slat on $C_{D\_highlift} = 0.112$.
For the aircraft wing $C_{D\_flap\_A\frac{1}{4}} = C_{D\_flap\_A0} \times \cos\Lambda_0 = 0.112 \times \cos20 = 0.105$
Induced drag $C_{Di} = (C_L^2)/(\pi AR) = (2.2)^2/(3.14 \times 7.5) = 4.48/23.55 = 0.21$
Total aircraft drag, $C_D = 0.019 + 0.105 + 0.21 = 0.334$

At 45° deflection:
It is typical for landing with $C_L = 2.7$ (approximate).
From Figure 11.12, $\Delta C_{D\_flap\_2D} = 0.08$ and $\Delta C_L = 2.1$
From Eq. (11.41), $\Delta C_{Di\_flap} = 0.025 \times (8/7.5)^{0.3} \times [(2/3)/(3/2)]^{0.5} \times (2.1)^2 = 0.025 \times 1.02 \times 4.41 = 0.112$
From Eq. (11.42), $\Delta C_{Dint\_flap} = 0.3 \times 0.08 = 0.024$
$C_{Dp\_flap\_A0} = 0.08 + 0.112 + 0.024 = 0.216$
With slat on $C_{Dp\_highlift} = 0.218$
For the aircraft wing $C_{D\_flap\_A\frac{1}{4}} = C_{D\_flap\_A0} \times \cos\Lambda_0 = 0.218 \times \cos20 = 0.201 \times 0.94 = 0.205$
Induced drag $C_{Di} = (C_L^2)/(\pi AR) = (2.7)^2/(3.14 \times 7.5) = 7.29/23.55 = 0.31$
Total aircraft drag, $C_D = 0.019 + 0.205 + 0.31 = 0.534$.

Drag polar with a high-lift device (Fowler flap) extended is plotted as shown in Figure 11.13 at various deflections. It is cautioned that this graph is intended only for coursework; practising industry-based engineers must use data generated by tests and CFD.

A typical value of C_L/C_D for high-subsonic commercial-transport aircraft at takeoff with flaps deployed is on the order of 10–12; at landing, it is reduced to 6–11.

A more convenient way would that as given in Figure 11.14 and is used for the classroom example (civil aircraft) worked out in Section 11.19.

11.15.2 Dive Brakes and Spoiler Drag

To decrease aircraft speed, whether in combat action or at landing, flat plates – which are attached to the fuselage and shaped to its geometric contour when retracted – are used. They could be placed symmetrically on both sides of the wing or on the upper fuselage (i.e. for military aircraft). The flat plates are deployed during subsonic flight. Use $C_{D\pi\_brake} = 1.2$–2.0 (average 1.6) based on the projected frontal area of the brake to air stream. The force level encountered is high and controlled by the level of deflection. The best position for the dive brake is where the aircraft moment change is the least (i.e. close to the aircraft CG line).

11.15.3 Undercarriage Drag

Undercarriages, fixed or extended (i.e. retractable type), cause considerable drag on smaller, low-speed aircraft. A fixed undercarriage (not streamlined) can cause up to about a third of aircraft parasite drag. When

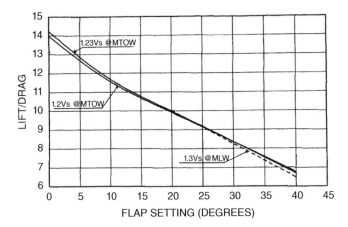

Figure 11.14 Drag polar with single-slotted Fowler flap extended (undercarriage retracted).

Table 11.6 Bare single-wheel drag with side ridge.

Wheel aspect ratio (diameter/width)	3	4	5	6
$C_{D\pi\_wheel}$	0.15	0.25	0.28	0.3

the undercarriage is covered by a streamlined wheel fairing, the drag level can be halved. It is essential for high-speed aircraft to retract the undercarriage as soon as it is safe to do so (like birds). Below a 200 ft altitude from takeoff and landing, an aircraft undercarriage is kept extended.

The drag of an undercarriage wheel is computed based on its frontal area: A_{π_wheel} (product of wheel diameter and width). For twin side-by-side wheels, the gap between them is ignored and the wheel drag is increased by 50% from a single-wheel drag. For the bogey type, the drag also would increase – it is assumed by 10% for each bogey, gradually decreasing to a total maximum 50% increase for a large bogey. Finally, interference effects (e.g. due to undercarriage doors and tubing) would double the total of wheel drag. The drag of struts is computed separately. The bare single-wheel C_{D_wheel} based on the frontal area is in Table 11.6 (wheel aspect ratio $= D/W_b$).

For the smooth side, reduce by half. In terms of an aircraft:

$$C_{Dp\_wheel} = (C_{D\pi\_wheel} \times A_{\pi\_wheel})/S_W.$$

A circular strut has nearly twice the amount of drag compared to a streamlined strut in a fixed undercarriage. For example, the drag coefficient of a circular strut based on its cross-sectional area per unit length is $C_{D\pi\_strut} = 1.0$ as it operates at a low Re during takeoff and landing. For streamlined struts with fairings, it decreases to 0.5–0.6, depending on the type.

Toreenbeck [10] suggests using an empirical formula if details of undercarriage sizes are not known at an early conceptual design phase. This formula is given in the FPS system as follows:

$$C_{D\_UC} = 0.0\,0403 \times (MTOW^{0.785})/S_W \tag{11.43}$$

Understandably, it could result in a slightly higher value (see the following example). If aircraft geometry is available then it is advisable to accept the computed drag values as worked out next.

Worked-Out Example
Continue with the previous example using the largest in the design (i.e. maximum takeoff mass (MTOM) = 24 200 lb. and S_W = 323 ft^2) for the undercarriage size. It has a twin-wheel, single-strut length of 2 ft. (i.e. diameter of 6 in., $A_{\pi\_strul} \approx 0.2$ ft^2) and a main wheel size of a 22 in. diameter and a 9.6 in.

width (i.e. wheel aspect ratio = 3.33, $A_{\pi\_wheel} \approx 1$ ft^2). From Table 11.6, a typical value of $C_{D\pi\_wheel} = 0.18$, based on the frontal area and increased by 50% for the twin wheel (i.e. $C_{D0} = 0.27$). Including the nose wheel (although it is smaller and a single wheel, it is better to be liberal in drag estimation), the total frontal area is about 3 ft^2: (A more accurate approach will be to consider the nose and mail wheels separately.)

$f_{wheel} = 0.27 \times 3 = 0.81$ ft^2 (nose wheel included)
$f_{strut} = 1.0 \times 3 \times 2 \times 0.2 = 1.2$ ft^2 (nose wheel included)

Total $f_{UC} = 2 \times (0.81 + 1.2) = 4.02$ ft2 (100% increase due to interference, doors, tubing, and so on) in terms of $C_{Dpmin\_UC} = 4.02/323 = 0.0124$. Checking the empirical relation in Eq. (11.43), $C_{D\_UC} = 0.00\,403 \times (24\,200^{0.785})/323 = 0.034$, this may be taken if undercarriage geometric details are not available.

11.15.4 One-Engine Inoperative Drag

Mandatory requirements by certifying agencies (e.g. Federal Aviation Administration, FAA and Civil Aeronautics Agency, CAA) specify that multiengine commercial aircraft must be able to climb at a minimum specified gradient with one engine inoperative at a 'dirty' configuration. This immediately safeguards an aircraft in the rare event of an engine failure; and in certain cases, after lift-off. Certifying agencies require backup for mission-critical failures to provide safety regardless of the probability of an event occurring.

Asymmetric drag produced by the loss of an engine would make an aircraft yaw, requiring a rudder to fly straight by compensating for the yawing moment caused by the inoperative engine. Both the failed engine and rudder deflection substantially increase drag, expressed by $C_{Dengine\ out\ +\ rudder}$. Typical values for coursework are in Table 11.7.

11.16 Propeller-Driven Aircraft Drag

Drag estimation of propeller-driven aircraft involves additional considerations. The slipstream of a tractor propeller blows over the nacelle, which blocks the resisting flow. Also, the faster flowing slipstream causes a higher level of skin friction over the downstream bodies. This is accounted for as a loss of thrust, thereby keeping the drag polar unchanged. The following two factors arrest the propeller effects with piston engines (see Chapter 12 for calculating propeller thrust):

1. blockage factor, f_b, for tractor-type propeller: 0.96–0.98 applied to thrust (for the pusher type, there is no blockage; therefore, this factor is not required – i.e. $f_b = 1.0$)
2. a factor, f_h, as an additional profile drag of a nacelle: 0.96–0.98 applied to thrust (this is the slipstream effect applicable to both types of propellers)

Turboprop nacelles have a slightly higher value of f_b than piston-engine types because of a more streamlined shape. For the worked-out example take $f_b \times f_h = 0.98$ as an optimistic design. Typically, it is about 0.96.

Table 11.7 One-engine inoperative drag.

	$\Delta C_{Done\ engine\ out\ +\ ruddert}$
Fuselage-mounted engines	0.003 5
Wing-mounted twin engine	0.004 5
Wing-mounted four engine (outboard failure)	0.005

11.17 Military Aircraft Drag

Military aircraft drag estimation requires additional considerations to account for the weapon system because most are external stores (e.g. missiles, bombs, drop tanks, flares and chaff launchers); few are carried inside the aircraft mouldlines (e.g. guns, ammunition and bombs inside the fuselage bomb bay, if any). Without external carriages, military aircraft are considered at typical configuration (the pylons are not removed – part of a typical configuration). Internal guns without their consumables are considered a typical configuration; with armaments, the aircraft is considered to be in a loaded configuration. In addition, most combat aircraft have a supersonic speed capability, which requires additional supersonic-wave drag.

Rather than drag being due to passenger doors and windows as in a civil aircraft, military aircraft have additional excrescence drag (e.g. gun ports, extra blisters and antennas and pylons) that requires a drag increment. To account for these additional excrescences [3], suggests an increment of the clean flat plate equivalent drag, f, by 28.4%.

Streamlined external store drag is shown in Table 11.8 based on the frontal maximum cross-sectional area. Bombs and missiles flush with the aircraft contour line have minor interference drag and may be ignored at this stage. Pylons and bomb racks create interference, and Eq. (11.24) is used to estimate interference on both sides (i.e. the aircraft and the store). These values are highly simplified at the expense of unspecified inaccuracy; readers should be aware that these simplified values are not far from reality (see [1, 4, 5] for more details).

Military aircraft engines are generally buried into the fuselage and do not have nacelles and associated pylons. Intake represents the air-inhalation duct. Skin friction drag and other associated 3D effects are integral to fuselage drag, but their intakes must accommodate large variations of intake air-mass flow. Military aircraft intakes operate supersonically; their power plants are low bypass turbofans (i.e. on the order of less than 3.0 – earlier designs did not have any bypass). For speed capabilities higher than Mach 1.9, most intakes and exhaust nozzles have an adjustable mechanism to match the flow demand in order to extract the best results. In general, the adjustment aims to keep the V_{intake}/V_{∞} ratio more than 0.8 over operational flight conditions, thereby practically eliminating spillage drag (see Figure 11.8). Supersonic flight is associated with shock-wave drag.

11.18 Supersonic Drag

A well substantiated reference for industrial use is [3], which was prepared by Lockheed as a NASA contract for the National Information Service, published in 1971. A comprehensive method for estimating supersonic drag that is suitable for classroom work is derived from this exercise. The empirical methodology (called the Delta Method) is based on regression analyses of 18 subsonic and supersonic military aircraft (i.e. the T-2B, T37B, KA-3B, A-4F, TA-4F, RA-5C, A-6A, A-7A, F4E, F5A, F8C, F-11F, F100, F101, F104G, F105B, F106A and XB70) and 15 advanced (i.e. supercritical) aerofoils. The empirical approach includes the effects of the following:

Table 11.8 External store drag.

External store	C_{D_π} (based on frontal area)
Drop tanks	0.1–0.2
Bombs (length/diameter < 6)	0.1–0.25
Bombs/missiles (length/diameter > 6)	0.25–0.35

wing geometry (AR, t/c and aerofoil section)
cross-sectional area distribution
CD variation with CL and Mach number

The methodology presented herein follows [3], modified to simplify C_{Dp} estimation resulting in minor discrepancies. The method is limited and may not be suitable to analyse more exotic aircraft configurations. However, this method is a learning tool for understanding the parameters that affect supersonic aircraft drag build-up. Results can be improved when more information is available.

The introduction to this chapter highlights that aircraft with supersonic capabilities require estimation of C_{Dpmin} at three speeds: (i) at a speed before the onset of wave drag; (ii) at M_{crit} and (iii) at maximum speed. The first two speeds follow the same procedure as for the high-subsonic aircraft discussed in Sections 11.7 through 11.15. In the subsonic drag estimation method, the viscous-dependent C_{Dp} varying with the C_L is separated from the wave drag, C_{Dw} (i.e. transonic effects), which also varies with the C_L but is independent of viscosity.

For book-keeping purposes in supersonic flight, such a division between the C_{Dp} and the C_{Dw} is not clear with the C_L variation. In supersonic speed, there is little complex transonic flow over the body, even when the C_L is varied. It is not clear how shock waves affect the induced drag with a change in the angle of attack. For simplicity, however, in the empirical approach presented here, it is assumed that supersonic drag estimation can use the same approach as the subsonic-drag estimation by keeping C_{Dp} and C_{Dw} separate. The C_{Dp} values for the worked-out example are listed in Table 11.24. Here, drag due to shock waves is computed at $C_L = 0$, and C_{Dw} is the additional shock-wave drag due to compressibility varying with $C_L > 0$. The total supersonic-aircraft drag coefficient can then be expressed as follows:

$$C_D = C_{Dpmin} + C_{Dp} + C^2/\pi AR + (C_{DLshock@CL=0}) + C_{Dw} \tag{11.44}$$

It is recommended that in current practice, CFD analysis should be used to obtain the variation of C_{Dp} and C_{Dw} with C_L. Reference [3] was published in 1978 using aircraft data before the advent of CFD. Readers are referred to [1, 4, 5] for other methods. The industry has advanced methodologies, which are naturally more involved.

The aircraft cross-section area distribution should be as smooth as possible, as discussed in Section 5.10.2 (see Figure 5.15). It may not always be possible to use narrowing of the fuselage when appropriate distribution of areas may be carried out.

The stepwise empirical approach to estimate supersonic drag is as follows. The procedure is worked out in detail in Section 11.22.7.

Step 1. Progress in the same manner as for subsonic aircraft to obtain the aircraft-component Re for the cruise flight condition and the incompressible $C_{Fcomponent}$.

Step 2. Increase $C_{Fcomponent}$ in Step 1 by 28.4% as the military aircraft excrescence effect.

Step 3. Compute C_{Dpmin} at the three speeds discussed previously.

Step 4. Compute induced drag using $C_{Di} = C^2/\pi AR$.

Step 5. Obtain C_{Dp} from the CFD and tests or from empirical relations.

Step 6. Plot the fuselage cross-section area versus the length and obtain the maximum area, S_π, and base area, S_b (see e.g. Figure 11.21).

Step 7. Compute the supersonic-wave drag at zero lift for the fuselage and the empennage using graphs; use the parameters obtained in Step 6.

Step 8. Obtain the design C_L and the design Mach number using graphs (see Figures 11.27 and 11.28).

Step 9. Obtain the wave drag, C_{Dw}, for the wing using the graphs.

Step 10. Obtain the wing-fuselage interference drag at supersonic flight using graphs.

Step 11. Total all the drag values to obtain the total aircraft drag and plot as C_D versus C_L.

The worked-out example for the North American RA-5C Vigilante aircraft is a worthwhile coursework exercise. Details of the Vigilante aircraft drag are in [3]. The subsonic drag estimation methodology described in this book differs with what is presented in [3], yet is in agreement with it. The supersonic drag estimation follows the methodology described in [3]. A typical combat aircraft of today is not too different than the Vigilante in configuration details and similar logic can be applied. Exotic shapes (e.g. the F117 Nighthawk) should depend more on information generated from CFD and tests along with the empirical relations. For this reason, the author does not recommend undertaking coursework on exotic aircraft configurations unless the results can be substantiated. Learning with a familiar design that can be substantiated gives confidence to practitioners. Those in the industry are fortunate to have access to more accurate in-house data.

11.19 Coursework Example – Civil Bizjet Aircraft

Figure 11.15a gives a three-view diagram of the Bizjet as the coursework example. Its dissected aircraft components are given in Figure 11.15b.

11.19.1 Geometric and Performance Data

The geometric and performance parameters discussed herein were used in previous chapters. Figure 11.15 illustrates the dissected anatomy of the coursework baseline aircraft.

Aircraft cruise performance for the basic drag polar is computed as follows:

Cruise altitude = 40 000 ft., LRC Mach = 0.65 (630 ft s^{-1}) to 0.7 (678.5 ft s^{-1}) at high altitudes. The Bizjet C_{Dpmin} is evaluated at 0.7 Mach as Bizjets cruises at high altitudes. The 2D flat plate $C_{\text{F\_basic}}$ decreases as Re increases.

Ambient pressure = 391.68 lb ft^{-2}	Ambient temperature = 390 K
Ambient density = 0.000 58 sl ft^{-3}	Ambient viscosity = 2.969 098 47 × 10^{-7} lb. s ft^{-2}

Re/ft. = 1.325×10^6 (use incompressible zero Mach line as explained in Section 11.8.1).
C_L at LRC (Mach 0.7) ≈ 0.5, C_L at HSC (Mach 0.75) ≈ 0.4

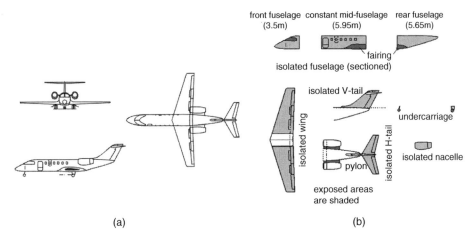

(a)

(b)

Figure 11.15 Bizjet aircraft-component diagram. (a) A three-view diagram of the Bizjet. (b) Dissected aircraft components, isolated.

Fuselage

Fuselage length, $L_f = 15.24$ m (50 ft)   Average diameter = 175.5 cm (70 in.)

Overall width = 175 cm (69 in.)   Overall height (depth) = 178 cm (70 in.)

Average diameter at the constant cross-section barrel $D_f = 1.765$ m (5.74 ft).

Fuselage upsweep angle 10°   Fuselage closure angle 10°

Fineness ratio $= L_f / D_f = 8.633$

Wing

Aerofoil: NACA 65-410 having 10% thickness to chord ratio for design $C_L = 0.4$

Planform reference area $S_W = 30$ m^2 (323 ft^2)   Span = 15 m (49.2 ft),   AR = 7.5

Wing MAC = 2.132 m (7 ft)   Dihedral = 3°   Twist = 2° (wash out) $t/c = 10\%$

Root chord at centreline, $C_R = 2.87$ m (9.4 ft)   Tip chord, $C_T = 1.143$ m (3.75 ft)

Taper ratio $\lambda = 0.4$,   Quarter-chord wing sweep = 14°

H-tail (Tee tail, aerofoil 64-210 – installed with negative camber)

Planform (Reference) area $S_H = 5.88$ m^2 (63.3 ft^2)   Span = 5.15 m (16.7 ft), AR = 4.22

Root chord, $C_R = 1.54$ m (5.04 ft)   Tip chord $C_T = 0.77$ m (2.52 ft)

H-tail MAC = 1.19 m (3.9 ft)   Taper ratio $\lambda = 0.5$, $\Lambda_{1/4} = 15°$

Dihedral = 5°   Elevator = 1.21 m^2 (13 ft^2), $t/c = 10\%$

V-tail (aerofoil 64-010)

Planform (Reference) area $S_V = 4.4$ m^2 (47.3 ft^2)   Height = 2.14 m (7 ft), AR = 2.08

Root chord, $C_R = 2.57$ m (8.43 ft)   Tip chord, $C_T = 1.54$ m (5.05 ft)

V-tail MAC = 2.07 (6.8 ft)   Taper ratio, $\lambda = 0.6$, $\Lambda_{1/4} = 40°$

Rudder = 0.75 m^2 (8 ft^2) $t/c = 10\%$

Nacelle (each – two required)

Length = 2.84 m (9.3 ft)   Maximum diameter = 1.074 m (3.52 ft)

Nacelle fineness ratio = 2.62/1.074 = 2.44

Bare engine (each – two required)

Honeywell TFE731-20 turbofan

Takeoff static thrust at International Standard Atmosphere (ISA) sea level = 3800 lb. (17 235 N) per engine with bypass ratio (BPR) = 4

Dry weight = 379 kg (836 lb)   Fan diameter = 0.716 m (28.2 in.)   Length = 1.547 m (60.9 in.)

11.19.2   Computation of Wetted Areas, *Re* and Basic C_F

The aircraft is dissected in to isolated components as shown in Figure 11.15. The *Re*, wetted area, and basic 2D flat plate C_{F_basic} of each component are worked-out herein. Bizjet C_{Dpmin} is evaluated at 0.7 Mach.

Fuselage

The fuselage is conveniently sectioned into three parts:

1. Front fuselage length, L_{Ff} = 3.5 m with a uniformly varying cross-section
2. Mid-fuselage length L_{Fm} = 5.95 m with an average constant cross-section diameter = 1.765 m
3. Aft-fuselage length L_{Fa} = 5.79 m, with a uniformly varying cross-section

   Wetted area

   Front fuselage, A_{wFf} (no cutout) = 110 ft^2

   Mid-fuselage, A_{wFm} (with two sides of wing cut-outs) = 352 − 2 × 6 = 340 ft^2

   Aft fuselage, A_{wFs} (with empennage cut-outs) = 180 − 10 = 170 ft^2

   Include additional wetted area for the wing-body fairing housing the undercarriage ≈ 50 ft^2

   Total wetted area, A_{wf} = 110 + 340 + 170 + 50 = 670 ft^2

   Fuselage $Re = 50 \times 1.325 \times 10^6 = 6.625 \times 10^7$

   Fineness ratio = L_f /D_f = 8.633

From Figure 11.24, bottom graph (fully turbulent) at LRC, the incompressible basic C_{Ff} = 0.0022.

Wing

Wing exposed reference area = 323 − 50 (area buried in the fuselage) = 273 ft^2

MAC = 2.132 m (7 ft), AR = 7.5

For t/c = 10% of the wing wetted area, A_{ww} = 2.024 × 273 = 552.3 ft^2 (see Section 11.8.2.1)

Root chord, C_R = 2.87 m (9.4 ft)

Tip chord, C_T = 1.143 m (3.75 ft)

Wing Re = $7 \times 1.325 \times 10^6 = 9.275 \times 10^6$

From Figure 11.24, bottom graph at LRC, the incompressible basic C_{Fw} = 0.003

Empennage (same procedure as for the wing)

H-tail

Planform (Reference) area S_H = 5.88 m^2 (63.3 ft^2) AR = 4.22

H-tail MAC = 1.19 m (3.9 ft)   Taper ratio, λ = 0.5

It is a T-tail and it is fully exposed.

for t/c = 10%, the H-tail wetted area, A_{wHT} = 2.024 × 63.3 = 128.12 ft^2

H-tail Re = $3.9 \times 1.325 \times 10^6 = 5.17 \times 10^6$

From Figure 11.24, bottom graph (fully turbulent) at LRC, the incompressible basic C_{F_H-tail} = 0.0032

V-tail

Planform (Reference) area, S_V = 4.4 m^2 (47.3 ft^2) AR = 2.08

Root chord, C_R = 2.57 m (8.43 ft)   Taper ratio, λ = 0.6 $\Lambda_{1/4}$ = 40°

Empennage (same procedure as for the wing)

Exposed reference area $= 47.3 - 7.3$ (approximate area buried in the fuselage) $= 40$ ft^2

For $t/c = 10\%$ the V-tail wetted area, $A_{\text{w\_VT}} = 2.024 \times 40 = 81$ ft^2

$MAC = 2.07$ m (6.8 ft)

V-tail Re $= 6.8 \times 1.325 \times 10^6 = 9.01 \times 10^6$

From Figure 11.24, bottom graph (fully turbulent) at LRC, the incompressible basic $C_{\text{F\_V-tail}} = 0.003$

Nacelle

Length $= 2.84$ m (9.3 ft)   maximum diameter $= 1.074$ m (3.52 ft)

Average diameter ≈ 0.8 m (2.62 ft)   fineness ratio $= 3.55$

Nacelle Re $= 9.3 \times 1.325 \times 10^6 = 1.23 \times 10^7$

Two-nacelle wetted area, $A_{\text{wn}} = (2 \times 3.14 \times 3.55 \times 9.3) - 2 \times 5$ (two pylon cut-outs) $= 197.3$ ft^2

From Figure 11.24, bottom graph (fully turbulent) at LRC, the incompressible basic $C_{\text{F\_nac}} = 0.0029$

Pylon

Each pylon exposed reference area $= 14$ ft^2

Length $= 2.28$ m (7.5 ft), $t/c = 10\%$

Two pylon wetted area $A_{\text{wp}} = 2 \times 2.024 \times 14 = 56.7$ ft^2

Pylon Re $= 7.5 \times 1.325 \times 10^6 = 9.94 \times 10^6$

From Figure 11.24, bottom graph (fully turbulent) at LRC the incompressible basic $C_{\text{Fpylon}} = 0.00\,295$.

Table 11.9 summarises the Bizjet component Re and 2D basic skin friction C_{Fbasic}.

11.19.3 Computation of 3D and Other Effects to Estimate Component C_{Dpmin}

A component-by-component example of estimating C_{Dpmin} is provided in this section. The corrected C_{F} for each component at LRC (i.e. Mach 0.7) is computed in the previous section.

Fuselage

The basic $C_{\text{Ff}} = 0.0022$
3D effects (Eqs. (11.16)–(11.18)).

Table 11.9 Summary of Bizjet component Reynolds Number and 2D basic skin friction C_{Fbasic}.

Parameter	Reference area (ft^2)	Wetted area (ft^2)	Characteristic length (ft)	Reynolds Number	2D $C_{\text{F\_basic}}$
Fuselage	n/a	670	50	6.625×10^7	0.002 2
Wing	323	552.3	7 (MAC_{w})	9.275×10^6	0.003
V-Tail	4.4	81	6.8 (MAC_{VT})	9.01×10^6	0.003
H-Tail	65.3	128.12	3.9 (MAC_{HT})	5.17×10^6	0.003 2
2 × Nacelle	n/a	197.3	9.3	1.23×10^7	0.002 9
2 × Pylon	2 × 14	56.7	7.5	9.94×10^6	0.00 295

Wrapping: $\Delta C_{Ff} = C_{Ff} \times 0.025 \times (\text{length/diameter}) \times Re^{-0.2}$

$$= [0.0022 \times 0.025 \times (8.633)] \times (6.625 \times 107){-0.2}$$
$$= 0.000\,475 \times 0.0273 = 0.000\,013 \ (5.9\% \text{ of basic } C_{Ff})$$

Supervelocity: $\Delta C_{Ff} = C_{Ff} \times (\text{diameter/length})^{1.5} = 0.0022 \times (1/8.633)^{1.5}$

$$= 0.0022 \times (0.1158)^{1.5} = 0.0\,000\,867 \ (3.94\% \text{ of basic } C_{Ff})$$

Pressure: $\Delta C_{Ff} = C_{Ff} \times 7 \times (\text{diameter/length})^3 = 0.0022 \times 7 \times (0.1158)^3$

$$= 0.0154 \times 0.001\,553 = 0.0\,000\,239 \ (1.09\% \text{ of basic } C_{Ff})$$

Other effects on fuselage (see Section 11.9.1)	
Body pressurisation – fuselage surface waviness:	4%
Non-optimum fuselage shape	
(a) nose fineness – for $1.5 \leq F_{cf} \leq 1.75$:	6%
(b) fuselage closure – less than 10°:	0
(c) upsweep closure – 10° upsweep:	6%
(d) aft-end cross-sectional shape – circular:	0%
Cabin-pressurisation leakage (if unknown, use higher value):	4%
Excrescence (non-manufacturing types, e.g. windows)	
(a) windows and doors:	2%
(b) miscellaneous:	1%
Wing-fuselage-belly fairing, if any (higher value if it houses undercarriage):	1%
ECS (see Section 11.8) gives 0.06 ft^2	2.8%
Total other effects $\Delta C_{Ff\_other\ effects}$ increment:	26.8%

Table 11.10 gives the Bizjet fuselage C_{Ff} components.

Add the canopy drag for two-abreast seating $f = 0.1$ ft^2 (see Section 11.9.1).

Table 11.10 Bizjet fuselage ΔC_{Ff} correction (3D and other shape effects).

Item	ΔC_{Ff}	% of $C_{Ffbasic}$
Wrapping	0.000 013	6.9
Supervelocity	0.0 000 867	3.94
Pressure	0.0 000 239	1.09
Body pressurisation		5
Fuselage upsweep of 10°		6
Fuselage closure angle of 9°	0	0
Nose fineness ratio 1.7		4
Aft-end cross-section – circular		
Cabin-pressurisation/leakage		4
Excrescence (windows/doors etc.)		2
Belly fairing		1
Environ. Control System Exhaust (ECS)		2.8
Total ΔC_{Ff}	0.0 001 236	$33.93 \approx 34$

Therefore, the equivalent flat plate area, f, becomes $= C_{Ff} \times A_{wF} + $ canopy drag.

$f_f = 1.34 \times 0.0022 \times 670 + 0.1 = 1.975 + 0.1 = 2.075 \text{ ft}^2$

Surface roughness (to be added later): 3%

Wing

Basic $C_{FW} = 0.003$

3D effects (Eqs. (11.15)–(11.17)):

Supervelocity: $\Delta C_{Fw} = C_{Fw} \times 1.4 \times$ *(aerofoil thickness/chord ratio)*

$$= 0.003 \times 1.4 \times 0.1 = 0.00042 \text{ (14% of basic } C_{Fw})$$

Pressure: $\Delta C_{Fw} = C_{Fw} \times 60 \times$ *(aerofoil thickness/chord ratio)*$^4 \times \left(\dfrac{6}{AR}\right)^{0.125}$

$$= (0.003 \times 60) \times (0.1)^4 \times (6/7.5)^{0.125}$$
$$= 0.18 \times 0.0001 \times 0.9725 = 0.0000175 \text{ (0.58% of basic } C_{Fw})$$

Interference: $\Delta C_{Fw} = C_B{}^2 \times 0.6 \times \left\{ \dfrac{0.75 \times (t/c)^3_{root} - 0.0003}{A_w} \right\}$

$$= 9^2 \times 0.6 \times [\{0.75 \times (0.1)^3 - 0.0003\}/552.3]$$
$$= 81 \times 0.6 \times (0.00075 - 0.0003)/552.3$$
$$= 0.0219/552.3 = 0.00004 \text{ (1.32% of basic } C_{Fw})$$

Other effects:

Excrescence (non-manufacturing, e.g. control surface gaps etc.):

Flap gaps	5%
Others	5%
Total ΔC_{Fw} increment =	25%

Table 11.11 summarises Bizjet wing ΔC_{Fw} correction (3D and other shape effects).

Therefore, in terms of equivalent flat plate area, f, it becomes $= C_{Fw} \times A_{ww}$

$$f_{fus} = 1.26 \times 0.003 \times 552.3 = 2.09 \text{ ft}^2$$

Surface roughness (to be added later) 3%.

Empennage

Since it follows the same procedure as in the case of wing, this is not repeated. The same percentage increment as the case of wing is taken as a classroom exercise. It may be emphasised here that in industry, one must compute systematically as shown in the case of the wing.

Table 11.11 Bizjet wing ΔC_{Fw} correction (3D and other shape effects).

Item	ΔC_{Fw}	% of $C_{Fwbasic}$
Supervelocity	0.00042	14
Pressure	0.0000175	0.58
Interference (wing-body)	0.00004	1.32
Flaps gap		5
Excrescence (others)		5
Total ΔC_{Fw}		26

H-tail:	Wetted area, $A_{\text{wHT}} = 128.12 \text{ ft}^2$	Basic $C_{F\_\text{V-tail}} = 0.0032$
	$f_{\text{HT}} = 1.26 \times 0.0032 \times 128.12 = 0.5166 \text{ ft}^2$	
Surface roughness (to be added later)		3%

V-tail:	Wetted area, $A_{\text{wVT}} = 81 \text{ ft}^2$,	Basic $C_{F\_H\text{-}tail} = 0.003$
It is T-tail configuration having interference from T-tail (add another 1.2%).		
$f_{VT} = 1.26 \times 0.003 \times 81 = 0.306 \text{ ft}^2$		

Nacelle

Fineness ratio = 2.44, Nacelle $Re = 1.23 \times 10^7$.
Wetted area of two nacelles, $A_{\text{wn}} = 197.3 \text{ ft}^2$, Basic $C_{\text{Fnac}} = 0.0029$
3D effects (Eqs. (11.15)–(11.17)):
Wrapping (Eq. (11.10)): $\Delta C_{\text{Fn}} = C_{\text{Fn}} \times 0.025 \times (length/diameter) \times R_e^{-0.2}$

$$= (0.025 \times 0.003 \times 2.44) \times (1.23 \times 10^7)^{-0.2}$$
$$= 0.000\,183 \times 0.0382 = 0.000\,007 \ (0.24\% \text{ of basic } C_{\text{Ff}})$$

Other increments are shown in Table 11.12 for one nacelle.

For two nacelles (shown in wetted area)	$f_n = 1.7224 \times 0.0029 \times 197.3 = 0.986 \text{ ft}^2$
Surface roughness (to be added later)	3%

Pylon

Since it follows the same procedure as in the case of wing, this is not repeated. The same percentage increment as the case of wing is taken as a classroom exercise. It has interference at both sides of pylon.
Each pylon exposed reference area = 14 ft^2.

Length = 2.28 m (7.5 ft), $t/c = 10\%$,	Two pylon wetted area $A_{\text{wp}} = 56.7 \text{ ft}^2$
$7.5 \times 1.325 \times 10^6 = 9.94 \times 10^6$,	Basic $C_{\text{Fpylon}} = 0.00\,295$
For two pylons (shown in wetted area)	$f_{\text{py}} = 1.26 \times 0.00\,295 \times 56.7 = 0.21 \text{ ft}^2$
Surface roughness (to be added later)	3%

Table 11.12 Bizjet nacelle ΔC_{Fn} correction (3D and other shape effects).

Item (one nacelle)	ΔC_{Fn}	% of C_{Fnbasic}
Wrapping (3D effect)	0.000 007	0.24
Excrescence (non-manufacture)		22
Boat Tail (aft end)		10
Base drag (at cruise)	0	0
Intake drag (BPR 4)		40
Total ΔC_{Fn}		72.24

11.19.4 Summary of Parasite Drag

Table 11.13 gives the aircraft parasite drag build-up summary in a tabular form. Surface roughness effect as 3% increase (Eq. (11.28)) in 'f' is added in this table for all surfaces. Wing reference area $S_w = 323$ ft^2, $C_{Dpmin} = f/S_w$. ISA day, 40 000 ft. altitude and Mach 0.7.

11.19.5 ΔC_{Dp} Estimation

Use Figure 11.16 for typical data (requires CFD/testing). The ΔC_{Dp} values used in the Bizjet are given in Table 11.14.

11.19.6 Induced Drag

Formula used $C_{Di} = C_L^2/(3.14 \times 11.5) = C_L^2/23.55$. The ΔC_{Di} values used in Bizjet are given in Table 11.15.

Table 11.13 Bizjet parasite drag build-up summary and C_{Dpmin} estimation.

	Wetted area, A_w ft^2	Basic C_F	ΔC_F	Total C_F	f (ft^2)	C_{Dpmin}
Fuselage + U/C fairing	670	0.002 2	0.000 748	0.002 948	1.975	0.006 11
Canopy					0.1	0.000 31
Wing	552.3	0.003	0.000 78	0.003 784	2.09	0.006 47
V-tail	81	0.003	0.000 786	0.003 786	0.302	0.000 95
H-tail	128.12	0.003 2	0.000 936	0.004 14	0.53	0.001 64
2 × Nacelle	2 × 79	0.002 9	0.002 386	0.005 28	0.986	0.003 05
2 × Pylon	2 × 28	0.002 95	0.000 767	0.003 72	0.21	0.000 65
Rough (3%)	Eq. (11.28)				0.186	0.000 57
Air-condition					0.06	0.000 186
Aerial, lights					0.03	0.000 093
Trim drag					0.13	0.000 4
	Total				6.599 ≈ 6.6	0.020 43 ≈ 0.020 5

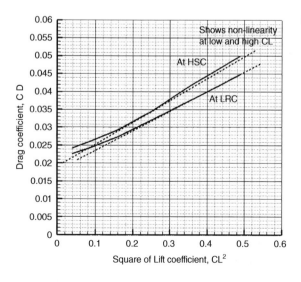

Figure 11.16 Drag polar of Figure 11.2 plotted C_L^2 versus C_D.

Table 11.14 Bizjet ΔC_{Dp} estimation.

C_L	0.1	0.2	0.3	0.4	0.5	0.6	0.7
ΔC_{Dp}	0.0007	0.0003	0.00006	0	0.0006	0.002	0.004

Table 11.15 Bizjet induced drag.

C_L	0.2	0.3	0.4	0.5	0.6	0.7	0.8
C_{Di}	0.0017	0.00382	0.0068	0.0106	0.0153	0.0206	0.0272

Table 11.16 Bizjet total aircraft drag coefficient, C_D.

C_L	0.2	0.3	0.4	0.5	0.6	0.7
C_{Dpmin}			0.0205 from Table 11.13			
ΔC_{Dp} (Table 11.14)	0.0003	0.00006	0	0.0006	0.002	0.003
C_{Di} (Table 11.15)	0.0017	0.00382	0.0068	0.0106	0.0153	0.0206
Total aircraft C_D @ LRC	0.0225	0.02438	0.0273	0.0317	0.0378	0.0441
Wave drag, C_{Dw} (Figure 11.10b)	0.0014	0.0017	0.002	0.0025	0.0032	0.0045
Total aircraft C_D @ HSC	0.0239	0.02608	0.0293	0.0342	0.041	0.0486

11.19.7 Total Aircraft Drag at LRC

Drag polar at LRC can be summed up as shown in Table 11.16.

Drag polar at HSC (Mach 0.75) would require adding wave drag.

This drag polar is plotted in Figure 11.2. $C_L{}^2$ versus C_D is plotted in Figure 11.16. Note the non-linearity at low and high C_L.

11.20 Classroom Example – Subsonic Military Aircraft (Advanced Jet Trainer – AJT)

The classroom example of military aircraft example was conducted on a subsonic AJT/Close Air Support (CAS) type of aircraft of the class BAE Hawk that would follow the same procedure as the civil aircraft drag estimation method. To avoid repetition, only the drag polar and other drag details of the AJT are given in Figure 11.17.

Figure 11.18 gives the three-view diagram of the AJT (of the class of BAE Hawk trainer) and its variant CAS aircraft. Given next is the defence specification for which the design has developed. HSC Mach = 0.85 at a 9 km altitude, LRC Mach = 0.78 at a 9 km altitude

The important external dimensions of the baseline AJT design are as follows.

Fuselage (Baseline)

Length = 12 m (39.4 ft)

Maximum overall width = 1.8 m (5.9 ft)   Overall height (depth) = 4.2 m (13.78 ft)

Cockpit width = 0.88 m (2.89 ft)   Fineness ratio = 12/1.8 = 6.67

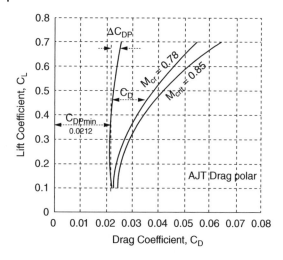

Figure 11.17 AJT drag polar (Section 11.20).

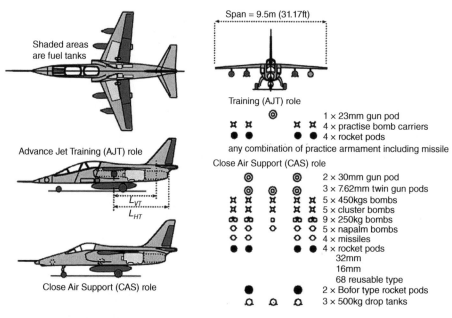

Figure 11.18 AJT and its CAS variant.

Wing

Planform (reference) area = 17 m² (183 ft²)	Span = 9.5 m (31.17 ft), AR = 5.31
Root chord, C_R = 2.65 m (8.69 ft)	Tip chord, C_T = 0.927 m (3.04 ft)
MAC = 1.928 (6.325 ft)	Taper ratio, λ = 0.35, $\Lambda_{1/4}$ = 20°
Dihedral = −2°. (anhedral – high wing)	Twist = 1° (wash out)
Flap = 2.77 m² (29.8 ft²)	Aileron = 1.06 m² (11.4 ft²)

H-tail

Planform (Reference) area $= 4.7$ m^2 (50.32 ft^2)	Span $= 4.2$ m (13.8 ft)
Root chord, $C_R = 1.9$ m (6.23 ft)	Tip chord, $C_T = 0.57$ m (1.87 ft)
MAC $= 1.354$ m (4.44 ft)	Taper ratio, $\lambda = 0.3$, $\Lambda_{1/4} = 25°$
AR $= 3.5$	Tail arm $= 4$ m (13.1 ft)
Elevator area $= 0.956$ m^2 (10.3 ft^2)	$t/c = 10\%$

V-tail

Planform (Reference) area $= 3.83$ m^2 (41.1ft^2)	Span $= 2.135$ m (7 ft)
Root chord, $C_R = 2.2$ m (7.22 ft)	Tip chord $C_T = 0.57$ m (1.87 ft)
MAC $= 1.545$ m (5 ft)	Taper ratio $\lambda = 0.26$, $\Lambda_{1/4} = 35°$
AR $= 1.52$	Tail arm $= 4$ m (13.1 ft)
Rudder area $= 0.98$ m^2 (10.5ft^2)	$t/c = 9\%$

Nacelle: None as the engine is buried into the fuselage.

Engine

Takeoff static thrust at ISA sea level $= 5390$ lb.	BPR $= 0.75$
Fan diameter $= 0.56$ m (22 in.)	Length $= 1.956$ m (77 in.)
Maximum depth $= 1.04$ m (3.4 ft)	Maximum width $= 0.75$ m (2.46 ft)

Baseline aircraft mass (from statistics – needs to generated from the variant CAS design.)

MTOM $= 6500$ kg (15 210 lb)	NTCM $= 4800$ kg (10 800 lb)
Operator's empty mass (OEM) $= 3700$ kg	Fuel mass $= 1100$ kg (maximum capacity 1300 kg)

CAS Variant (all component dimensions except fuselage length are kept unchanged).

The CAS role aircraft is the only variant of the AJT aircraft. Configuration of the CAS aircraft variant is achieved by splitting the AJT front fuselage, then replacing the tandem seat arrangement with a single seat cockpit. The length could be kept the same as the nose cone needs to house more powerful acquisition radar. The front loading of radar and single pilot is placed in a way that the CG location is kept undisturbed. Wing area $= 17$ m^2 (183 ft^2).

Specifications

Payload $= 2500$ kg	Range $=$
$M_{crit} = 0.85$ Mach	$M_{CR} = 0.78$

Fuselage

Length $= 11.2$ m (36.8 ft)	
Maximum overall width $= 1.8$ m	Overall height (depth) $= 4.2$ m
Cockpit width $= 0.88$ m	Fineness ratio $= 11.2/1.8 = 6.22$

Weights

The summary of mass changes is as follows. CAS mass has been derived by removing one pilot, the instrument ejection seat and so on (260 kg) and includes radar and combat avionics (100 kg). There is an increase of 60 kg in engine mass. Internal:

	AJT (kg)	CAS (kg)
OEM	3700	3700
Fuel	1100	1300 (full tank)
Clean aircraft MTOM	4800 (10 582 lb)	5000 (11 023 lb)
Wing loading*, W/S_W	282 kg m^{-2} (57.8 lb ft^{-2})	294 kg m^{-2} (60.23 lb ft^{-2})
Armament mass	1800	2200
MTOM kg (lb)	6500 (14 326 lb)	7500 (16 535 lb)
Wing loading, W/S_W	382.2 kg m^{-2} (78.3 lb ft^{-2})	441.2 kg m^{-2} (90.36 lb ft^{-2})

Armament and fuel could be traded for range. Drop tanks could be used for ferry range.

The drag level of the clean AJT and CAS aircraft may be considered to be about the same. There would be an increase of drag on account of the weapon load. For the CAS aircraft, there is a wide variety, but to give a general perception, the typical drag coefficient increment for armament load is $C_{D\pi} = 0.25$ (includes interference effect) each for five hard points as weapon carrier. Weapon drag is based on maximum cross-section (say 0.8ft^2) area of the weapons.

Parasite drag increment due to armament, $\Delta C_{Dpmin} = (5 \times 0.25 \times 0.8)/183 = 0.0055$, where $S_W = 183$ ft^2.

11.20.1 AJT Details

Drag polar at Mach 0.78 and at Mach 0.85 is tabulated in Table 11.17 and plotted in Figure 11.17. From this, AJT parabolic drag can approximated as $C_D = 0.0212 + 0.07 \, C_L^2$ with $S_W = 183$ ft^2 (gives Oswald's efficiency factor, $e = 0.85$).

11.21 Classroom Example – Turboprop Trainer (TPT)

This an example of the drag estimation for a propeller-driven aircraft. A classroom example of the TPT type aircraft of the class Shorts *Tucano* uses the same procedure as the civil aircraft drag estimation method. However, propeller-driven aircraft drag has to account for the propeller slipstream effects on the wetted surface. Propeller wake has a higher velocity than the aircraft free-stream velocity. This gives a different average skin

Table 11.17 AJT total aircraft drag coefficient, C_D.

C_L	0.1	0.2	0.3	0.4	0.5	0.6
C_{Dpmin}			0.021 2 (readers make work out)			
ΔC_{Dp}	0.000 1	0.000 3	0.000 6	0.001	0.001 4	0.002 6
$C_{Di} = C_L^2/(3.14 \times 5.3)$	0.000 6	0.002 4	0.005 4	0.009 61	0.015	0.021 6
Total aircraft C_D @ 0.78 M	0.021 9	0.023 9	0.027 2	0.031 81	0.037 6	0.045 4
Wave drag, C_{Dw}	0.001 6	0.001 8	0.002	0.002 71	0.003 3	0.005 6
Total aircraft C_D @ 0.85 M	0.023 5	0.025 7	0.029 2	0.034 52	0.040 9	0.051

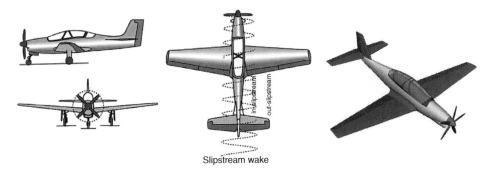

Slipstream wake

Figure 11.19 Turboprop trainer aircraft (TPT).

friction coefficient, C_F. There are also nacelle blockage factors that have to be considered. V-Tail is given 1° offset to counter slipstream rotational effect.

To avoid repetition TPT drag evaluation is done in a short-cut method (based on experience). Some of the flat plate equivalent drags 'fs' are given (using comparable industrial data). It is recommended that the readers should undergo the full methodology. The results should not be much different from the drag polar given here. The three-view diagram of TPT is given in Figure 11.19, showing the slipstream effects. Normal Training Configuration (NTC) is an aircraft is without any weapon load; that is, no external stores.

Given next is the defence specification for which the design has developed. Practise altitude = 25 000 ft. ($\rho = 0.00\,106$ slug ft^{-3} and $\mu = 0.3216 \times 10^{-6}$ lbs ft^{-2}). $C_{DPmin} = 0.023$. The important TPT details are as follows.

MTOM = 2800 kg (6170 lb)	NTC-TOM = 2350 kg (5180 lb)
OEW = 4400 lb (2000 kg)	Typical mid-training aircraft mass = 2200 kg (4850 lb)
Crew = 400 lb (≈180 kg)	Fuel = 1000 lb (≈450 kg)
Maximum speed = 280 kt	Sustained speed = 240 kt (405.2 ft s^{-1})

Fuselage

Length = 30 ft	Overall height = 10 ft
Cockpit width = 35 in	Fineness ratio = $(30 \times 12)/35 = 10.3$

Wing (NACA 63-212)

Reference area = 16.5 m^2 (177.3 ft^2)	Span = 10.36 m (34 ft), AR = 6.5
Root chord, C_R = 2.36 m (7.74 ft)	Tip chord, C_T = 0.82 m (2.7 ft)
MAC = 1.715 m (5.63 ft)	Taper ratio, $\lambda = 0.35$, $\Lambda_{1/4}$ = degrees
Dihedral = 5° (dihedral – low wing)	Twist = 1° (wash out), $t/c = 0.12$
Flap = 2.77 m^2 (29.8 ft^2)	Aileron = 1.06 m^2 (11.4 ft^2)

H-tail (NACA 0012)

Reference area = 4.1 m^2 (44.2 ft^2)	Span = 4.16 m (13.8 ft), AR = 4.0
Root chord, C_R = 1.32 m (4.33 ft)	Tip chord, C_T = 0.66 m (2.17 ft)
H-tail MAC = 1.0234 m (3.36 ft)	Taper ratio, $\lambda = 0.5$, $\Lambda_{1/4} = 25°$
Elevator area = 0.956 m^2 (12 ft^2)	$t/c = 0.12$

V-tail (NACA 0012)

Reference area $= 2.2$ m^2 (23.3 ft^2)	Span $= 1.615$ m (5.3 ft), AR $= 1.52$
Root chord, $C_R = 1.64$ m (5.38 ft)	Tip chord, $C_T = 0.66$ m (2.17 ft)
V-tail MAC $= 1.22$ (4.0 ft)	Taper ratio, $\lambda = 0.4$, $\Lambda_{1/4} = 35°$
Rudder area $= 0.98$ m^2 (8 ft^2)	$t/c = 0.12$

Nacelle:	None as the engine is buried into the fuselage.

11.21.1 TPT Details

ISA day, 25 000 ft. altitude. Wing reference area $S_w = 16.5$ m^2 (177.38 m^2). $C_{Dpmin} = f/S_w$. Cruise $C_L = 0.3$.

Outstream velocity = 405.2 ft s^{-1} ($\approx$ Mach 0.4)

Intstream velocity = 567.28 ft s^{-1} (increase 40%).

Outstream Re/ft. @ 25 000 ft at 405.2 ft s^{-1} = 1.3355×10^6

Intstream Re/ft. @ 25 000 ft and at 567.28 ft s^{-1} = 1.87×10^6

Drag estimation follows the Bizjet methodology incorporating slipstream effects. Note that a TPT does not have nacelles. Intake drag is to be estimated accordingly.

Quick component drag parasite drag estimation: The short-cut method uses Bizjet ΔC_F percentage increases. No credit is taken for partial laminar flow.

Fuselage

The wetted surface of the fuselage is entirely in slipstream; in short, instream.

Fuselage instream Re is 5.6×10^7 and corresponding basic C_{Ff} is 0.0022. Wetted area $A_{wf} = 280$ ft^2.

Unlike Bizjet, the TPT does not have (i) fuselage pressurisation, (ii) windows and doors, (iii) wing-fuselage-belly fairing and (iv) upsweep closure reducing ΔC_{Ff} by 20%. Take (i) wrapping, (ii) supervelocity and (iii) pressure together as 5%.

The basic $C_{Ff} = 0.0022$ (see Table 11.18).

Take the 3D effects, for example, (i) wrapping, (ii) supervelocity and (iii) pressure together as 5% of the basic C_{Ff}.

Other effects on fuselage (see Section 11.9.1)

(a) nose fineness – for $1.5 \leq F_{cf} \leq 1.75 : 6\%$

(b) aft-end cross-sectional shape near circular: 0%

Table 11.18 Summary of TPT component Reynolds Number and 2D basic skin friction C_{Fbasic}.

Parameter	Reference area (ft^2)	Wetted area (ft^2)	Characteristic length (ft)	Reynolds Number	2D C_{F_basic}
Fuselage (instream)	n/a	280	30	5.6×10^7	0.002 2
Wing (in)	200	70	5.5 (MAC_w)[a)]	1.03×10^7	0.002 9
Wing (out)		280	5.5 (kept same)[a)]	9.35×10^6	0.003
V-Tail (in)	22	45	4.2 (MAC_{VT})	9.85×10^6	0.002 98
H-Tail (in)	20	40.6	3.3 (MAC_{HT})[a)]	9.1×10^6	0.003 04
H-Tail (out)	24.2	50	3.3 (MAC_{HT})[a)]	4.4×10^6	0.003 5
Strakes (in)					

V-tail and strakes are instream.

a) Strictly speaking, separate MAC should be taken. The error is small.

Table 11.19 TPT parasite drag build-up summary and C_{Dpmin} estimation.

	Wetted area, A_w ft^2	Basic C_F	ΔC_F	Total C_F	f (ft^2)	C_{Dpmin}
Fuselage + canopy	280	0.002 2	0.000 35	0.002 66	0.81	0.004
Wing in	70	0.002 9	0.000 754	0.003 65	0.256	0.001 28
Wing-out	280	0.003	0.000 78	0.003 78	1.06	0.005 3
V-tail	45	0.002 98	0.000 775	0.003 76	0.17	0.000 85
H-tail-in	40.6	0.003 04	0.000 8	0.003 84	0.156	0.000 78
H-tail-out	50	0.003 5	0.000 91	0.004 41	0.22	0.001 1
Strakes + dorsal					0.05	0.000 25
Engine intake					0.1	0.000 5
Exhaust stub					0.2	0.001
NACA intakes					0.8	0.004
Rough (3%)					0.182	0.000 6
Excrescences					0.106	0.000 53
Aerial, lights					0.05	0.000 4
Trim drag					0.13	0.000 01
Miscellaneous					0.4	0.002
Total					4.5	0.022 6

excrescence (non-manufacturing types): 3%

(c) miscellaneous: 1%

Total ΔC_{Ff} increment: $\approx$15%
Fuselage blockage factors of f_b and f_h are included as engine installation losses reducing thrust by about 3%. Therefore, fuselage blockage factors are not for accounted here.
Add the canopy drag for two-abreast seating $f = 0.1$ ft^2 (see Section 11.9.1).
Therefore, the equivalent flat plate area, f, becomes $= C_{Ff} \times A_{wF}$ + canopy drag.
$f_{fus} = 1.15 \times 0.0022 \times 280 + 0.1 = 0.708 + 0.1 = 0.81$ ft^2.
Surface roughness (to be added later): 3%.
Table 11.19 summarises the TPT fuselage C_{Ff}.

Wing:

Of the wing the wetted A_{wf} surface, approximately 20% is instream and 80% is outstream.
Wetted area $A_{ww} = 320$ ft^2. Instream $= A_{ww\_in} = 70$ ft^2 and outstream $A_{ww\_out} = 250$ ft^2 and
Instream $Re_{in} = 1.03 \times 10^7$ giving basic $C_{FW\_in} = 0.0029$ and outstream $Re_{out} = 1.135 \times 10^7$ giving basic $C_{FW\_out} = 0.003$.
Take the 3D effects, for example, (i) interference, (ii) supervelocity and (iii) pressure together as 16% of the basic C_{Fw}. Other effects on wing:
Excrescence (non-manufacturing, e.g. control surface gaps etc.):

Flap gaps	5%
Others	5%
Total ΔC_{Fw} increment =	26%

Empennage

Since it follows the same procedure as in the case of wing, this is not repeated. The same percentage increment as in the case of wing is taken as a classroom exercise. It may be emphasised here that in industry, one must compute systematically as shown in the case of the wing at 3% surface roughness (to be added later).

H-tail: (of the H-tail the wetted A_{wf} surface, approximately 45% is instream and 55% is outstream). Wetted area $A_{wHT} = 90.6$ ft^2. Instream $= A_{wHT\_in} = 40.6$ ft^2 and outstream $A_{wHT\_out} = 50$ ft^2 Instream $Re_{in} = 9.1 \times 10^6$ giving basic $C_{FW\_in} = 0.00\,304$

$$f_{HT\_in} = 1.26 \times 0.00\,304 \times 40.6 = 0.156 \text{ ft}^2$$

Outstream $Re_{out} = 4.4 \times 10^6$ giving basic $C_{FW\_out} = 0.0035$.

$$f_{HT\_out} = 1.26 \times 0.0035 \times 50 = 0.22 \text{ ft}^2$$

Strakes + dorsal: $f_{strake + dorsal} = 0.05$ ft^2
V-tail: (fully immersed in slipstream (strictly speaking a small portion is outside the slipstream). Wetted area, $A_{wVT} = 45$ ft^2, Basic $C_{F\_H\text{-}tail} = 0.00\,298$

$$f_{VT} = 1.26 \times 0.00\,298 \times 45 = 0.17 \text{ ft}^2$$

Engine intake: There is no nacelle, it is integrated with the fuselage. The TPT has one turboprop engine with low intake mass flow. Only spillage drag and friction drag at the lip are to be estimated. Every aircraft needs to generate a graph of spillage drag versus mass flow for its engine intake like that given in Figure 11.8. TPT intake is sized for takeoff mass flow, which is higher compared with cruise demand. Hence there will be spillage drag (= additive drag + lip suction).

Here, f_{intake} is given as follows (from similar industrial data), $f_{intake} = 0.1$ ft^2

Exhaust stub: The turboprop has a relatively large exhaust stub.	$f_{stub} = 0.2$ ft^2
Excrescences: Excrescences of non-manufacturing origin.	$f_{excr} = 0.108$
Aerial and lights:	$f_{aerial + light} = 0.05$
Trim drag:	$f_{trim} = 0.013$
Miscellaneous:	$f_{misc} = 0.04$ (military aircraft can be dirtier)

Table 11.19 gives the aircraft parasite drag build-up summary in a tabular form. Surface roughness effect as a 3% increase (Eq. (11.28)) in 'f' is added in this table for all surfaces.

Table 11.20 tabulates TPT total aircraft drag coefficient, C_D.

The actual drag polar matches closes with the parabolic drag polar expressed as $C_D = 0.0226 + 0.052\,C_L^2$. The TPT drag polar is plotted in Figure 11.23, later.

Table 11.20 TPT total aircraft drag coefficient, C_D.

C_L	0.1	0.2	0.3	0.4	0.5	0.6
C_{Dpmin}				0.022\,6		
ΔC_{Dp}	0.000\,5	0.000\,2	0.0	0.000\,05	0.000\,4	0.001
$C_{Di} = C_L^2/(3.14 \times 6.5)$	0.000\,5	0.00\,196	0.004\,4	0.007\,84	0.01\,225	0.017\,6
Total aircraft C_D @ 0.4 M	0.023\,6	0.024\,7	0.027	0.030\,5	0.03\,525	0.041\,2

11.22 Classroom Example – Supersonic Military Aircraft

To show proper supersonic drag estimation method, a three-view diagram North American RA-C5 Vigilante aircraft as shown in Figure 11.20, is taken as an example to work out here. Reference [3] gives the Vigilante drag polar to compare this with. Subsonic drag estimation of Vigilante aircraft follows the same procedure as that of civil aircraft example. Therefore, the results of drag at Mach 0.6 (no compressibility) and at Mach 0.9 (at M_{crit}) are worked out in brief and tabulated. The supersonic drag estimation is worked out in detail following the empirical methodology of [3].

11.22.1 Geometric and Performance Data of Vigilante – RA C5 Aircraft

The following pertinent geometric and performance parameters are taken from [3]. The RA-C5 has two crew, engine: two turbo-jets GE J-79-8(N), 75.6 kN, wing span: 16.2 m, length: 22.3 m, height: 5.9 m, wing area: 65.0 m^2, start mass: 27 300 kg, max speed: Mach 2.0, ceiling: 18 300 m, range: 3700 km, armament: nuclear bombs and missiles (only clean a configuration is evaluated).

Fuselage

Fuselage length = 73.25 ft, Average diameter at the maximum cross-section = 9.785 ft (see Figure 11.21).
Fuselage length/diameter = 9.66 (fineness ratio)
Fuselage upsweep angle = 0°, Fuselage closure angle ≈ 0°.

Wing

Planform reference area, S_W = 65.03 m^2 (700 ft^2), Span = 19.2 m (53.14 ft).
AR = 3.73, t/c = 5%, Taper ratio λ = 0.19, Camber = 0.
Wing MAC = 4.63 m (15.19 ft), $\Lambda_{1/4}$ = 39.5° and Λ_{LE} = 43°.
Root chord at centreline = 9.65 m (20 ft) and Tip chord = 1.05 m (3.46 ft).

Empennage

V-tail: S_V = 4.4 m^2 (49.34 ft^2), Span = 3.6 m (11.82 ft). MAC = (8.35 ft), t/c = 4%.
H-tail: S_H = 11.063 m^2 (65.3 ft^2), Span = 11.85 m (32.3 ft). MAC = (9.73 ft), t/c = 4%.
Nacelle/pylon – engine buried in fuselage – no nacelle and pylon
Aircraft cruise performance where the basic drag polar has to be computed (Figure 11.22).
Drag estimated at cruise altitude = 36 152 ft and Mach number = 0.6 (have compressibility drag).
Ambient pressure = 391.68 lb ft^{-2}, Re/ft = 1.381 × 10^6.
Design C_L = 0.365, Design Mach number = 0.896 (M_{crit} is at 0.9), Maximum Mach number = 2.0.

Figure 11.20 North American RA-C5 Vigilante aircraft (very clean wing – no pylon shown).

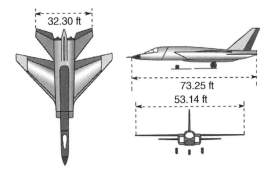

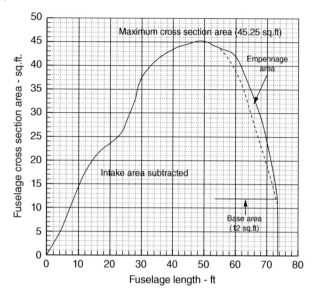

Figure 11.21 Vigilante RA C5 fuselage cross-section area distributions.

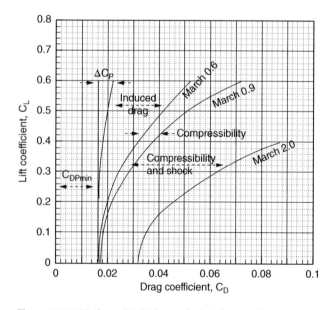

Figure 11.22 Vigilante RA C5 drag polar (not from industry).

11.22.2 Computation of Wetted Areas, *Re* and Basic C_F

The aircraft is first dissected into isolated components to obtain Re, the wetted area and the basic 2D flat plate C_F of each component as listed next. Note that there is no correction factor for C_F at Mach 0.6 (no compressibility drag). The C_F compressibility correction factor (compute this from Figure 11.24b) at Mach 0.9 and Mach 2.0 will be applied later.

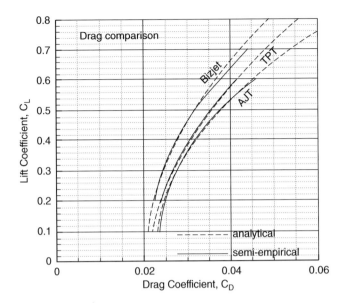

Figure 11.23 Drag comparison.

Fuselage:

Fuselage wetted area $= A_{wf} = 1474 \text{ ft}^2$
Fuselage $Re = 69 \times 1.381 \times 10^6 = 9.53 \times 10^7$ (length trimmed to what is pertinent for Re)
Use Figure 11.24, bottom graph to obtain basic $C_{Ff} = 0.0021$

Wing

Wing wetted area $= A_{ww} = 1144.08 \text{ ft}^2$
Wing $Re = 15.19 \times 1.381 \times 10^6 = 2.1 \times 10^7$
Use Figure 11.24, bottom graph to obtain basic $C_{Fw} = 0.00\,257$

Empennage (same procedure as for wing)

V-tail wetted area $= A_{wVT} = 235.33 \text{ ft}^2$
V-tail $Re = 8.35 \times 1.381 \times 10^6 = 1.2 \times 10^7$
Use Figure 11.24, bottom graph to obtain the basic $C_{F\_V\text{-tail}} = 0.00\,277$
H-tail wetted area $= A_{wHT} = 3811.72 \text{ ft}^2$.
H-tail $Re = 9.73 \times 1.381 \times 10^6 = 1.344 \times 10^7$.
Use Figure 11.24, bottom graph to obtain basic $C_{F\_H\text{-tail}} = 0.002\,705$.

11.22.3 Computation of 3D and Other Effects to Estimate Component C_{Dpmin}

A component-by-component example is given next.

Fuselage:

As before, at Mach 0.6, the basic $C_{Ff} = 0.0021$
3D effects (Eqs. (11.10)–(11.12)):
Wrapping: $\Delta C_{Ff} = C_{Ff} \times 0.025 \times (length/diameter) \times Re^{-0.2}$

$$= 0.025 \times 0.0021 \times (9.66) \times (9.53 \times 10^7)^{-0.2}$$
$$= 0.000\,507 \times 0.0254 = 0.0\,000\,129 \ (0.6\% \text{ of basic } C_{Ff})$$

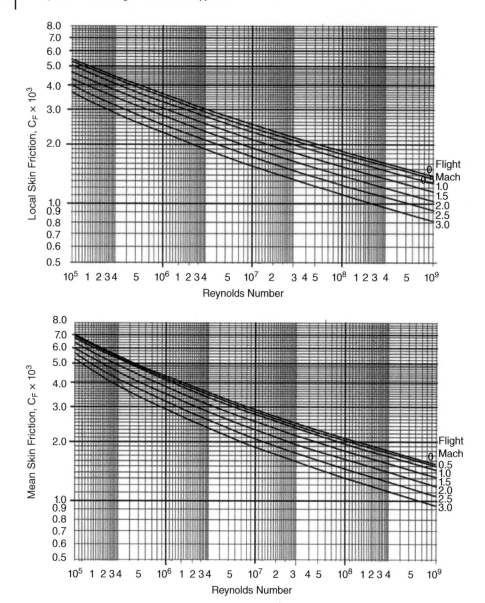

Figure 11.24 Flat pate skin friction coefficients versus Reynolds number (turbulent flow). (*Top*) Local skin friction. (*Bottom*) Mean skin friction.

Supervelocity: $\Delta C_{Ff} = C_{Ff} \times (diameter/length)^{1.5} = 0.0021 \times (1/9.66)^{1.5}$

$$= 0.0021 \times 0.033 = 0.0\,000\,693 \ (3.3\% \text{ of basic } C_{Ff})$$

Pressure: $\Delta C_{Ff} = C_{Ff} \times 7 \times (diameter/length)^3 = 0.0021 \times 7 \times (0.1035)^3$

$$= 0.0147 \times 0.00\,111 = 0.0\,000\,163 \ (0.8\% \text{ of basic } C_{Ff})$$

Other effects on fuselage (this time intake has to be included – see Section 11.9.3.1):

It is to be noted here that [3] suggests to apply a factor of 1.284 to include most of the other effects except for intake. Therefore, unlike the civil aircraft example, it is simplified to the following only:

Intake (little spillage – rest taken in 3D effects): 2%

Table 11.21 Vigilante fuselage ΔC_{Ff} correction (3D and other shape effects).

Item	ΔC_{Ff}	% of $C_{Ffbasic}$
Wrapping	0.000 015	0.6
Supervelocity	0.000 1	3.3
Pressure	0.0 000 274	0.8
Intake (little spillage)		2
Total ΔC_{Ff}	0.00 105	9.7

Total ΔC_{Ff} increment is given in Table 11.21.

Therefore, in terms of equivalent flat plate area, f, it becomes $= C_{Ff} \times A_{wF}$.

$$f = 1.067 \times 0.0021 \times 1474 = 3.3 \text{ ft}^2$$

Add canopy drag, $C_{D\pi} = 0.08$ (approximated from Figure 11.4).

Therefore $c_{anopy} = 0.08 \times 4.5 = 0.4 \text{ ft}^2$. $f_{fus} = 3.3 + 0.36 = 3.66 \text{ ft}^2$

Wing:

From before, at Mach 0.06, the basic $C_F = 0.00 259$.

3D effects (Eqs. (11.15)–(11.17)):

Supervelocity: $\Delta C_{Fw} = C_{Fw} \times 1.4 \times$ (*aerofoil thickness/chord ratio*)

$$= 0.00 257 \times 1.4 \times 0.05 = 0.00 018 \text{ (7\% of basic } C_{Fw})$$

Pressure: $\Delta C_{Fw} = C_{Fw} \times 60 \times$ (*aerofoil thickness/chord ratio*)$^4 \times \left(\dfrac{6}{AR}\right)^{0.125}$

$$= 0.00 257 \times 60 \times (0.05)^4 \times (6/3.73)^{0.125}$$
$$= 0.1542 \times 0.00 000 625 \times 1.06 = 0.00 000 102 \text{ (0.04\% of basic } C_{Fw})$$

Interference: ΔC_{Fw} for thin high wing take 3% of C_{Fw}.

Other effects:

Excrescence (non-manufacturing, e.g. control surface gaps etc.):

Flap/slat gaps	2%
Others (increased later)	0%
Total ΔC_{Fw} increment	12.04%

Table 11.22 summaries Vigilante wing ΔC_{Fw} correction (3D and other shape effects).

Table 11.22 Vigilante wing ΔC_{Fw} correction (3D and other shape effects).

Item	ΔC_{Fw}	% of $C_{Fwbasic}$
Supervelocity	0.000 385	7
Pressure	0.0 000 136	0.04
Interference (wing-body)	0.0 000 328	3
Flap/slat gap		2
Total ΔC_{Fw}		12.04

Therefore, in terms of equivalent flat plate area, f, it becomes $= C_{Fw} \times A_{ww}$

$$f_w = 1.12 \times 0.00\,257 \times 1144.08 = 3.3 \text{ ft}^2$$

Empennage

Since it follows the same procedure as in the case of wing, this is not repeated. The same percentage increment as the case of wing is taken as a classroom exercise. It may be emphasised here that in industry, one must compute systematically as shown in the case of wing.

V-tail: Wetted area, $A_{wVT} = 235.33 \text{ ft}^2$

$$\text{Basic } C_{F\_V\text{-tail}} = 0.00\,277, f_{VT} = 1.12 \times 0.00\,277 \times 235.33 = 0.73 \text{ ft}^2$$

H-tail: Wetted area, $A_{wHT} = 388.72 \text{ ft}^2$

$$\text{Basic } C_{F\_H\text{-tail}} = 0.002\,705, f_{HT} = 1.12 \times 0.002\,705 \times 388.72 = 1.18 \text{ ft}^2$$

11.22.4 Summary of Parasite Drag (ISA Day, 36 182 ft Altitude and Mach 0.9)

Wing reference area $S_w = 700 \text{ ft}^2$, $C_{Dpmin} = f/S_w$.

A Vigilante parasite drag summary is given in Table 11.23.

As was indicated in Section 11.16 [3], offers a correlated factor of 1.284 to include all so-called other effects. Therefore, the final flat plate equivalent drag $f_{aircraft} = 1.284 \times 8.87 = 11.39 \text{ ft}^2$. 28.4% = 11.62 ft² to include military aircraft excrescence.

It gives C_{Dpmin} at Mach 0.6 = 11.39/700 = 0.01 627 ([3] gives 0.1645, close enough).

This is the C_{Dpmin} at flight Mach number before compressibility effects start to show up, that is, it is seen as C_{Dpmin} at incompressible flow. At higher speed there is C_F shift to a lower value. The C_{Dpmin} estimation needs to be repeated with lower C_F at 0.9 Mach and at 2.0 Mach. To avoid repetition to account for compressibility, a factor of 0.97 (ratio of values at Mach 0.9 and Mach 0 in Figure 11.24, bottom graph – a reduction of 3%) is taken at 0.9 Mach. A factor of 0.8 (a reduction of 20%) is taken at 2.0 Mach as shown in Table 11.24. At compressible flow, add the wave drag. At supersonic speed it is contributed to by shock waves.

To stay in line with the methodology presented here, the following values of ΔC_{Dp} have been extracted from [3].

11.22.5 ΔC_{Dp} Estimation

The data for ΔC_{Dp} given in Table 11.24 is extracted from Ref. [3] and is approximate.

Table 11.23 Vigilante parasite drag summary.

Fuselage	3.66 ft²
Wing	3.3 ft²
V-tail	0.73ft²
H-tail	1.18 ft²
Total	8.87 ft²

Table 11.24 Vigilante ΔC_{Dp} estimation.

C_L	0	0.1	0.16	0.2	0.3	0.4	0.5	0.6
ΔC_{Dp}	0.000 8	0.00 015	0	0.000 1	0.000 8	0.00 195	0.003 6	0.006

11.22.6 Induced Drag

Formula used $C_{Di} = C_L^2/(3.14 \times 3.73) = C_L^2/11.71$.

The ΔC_{Di} values used in Vigilante are given in Table 11.25.

11.22.7 Supersonic Drag Estimation

Supersonic flight would have bow shock-wave that is a form of compressibility drag, which is evaluated at zero C_L. Drag increases with change of angle of attack. The difficulty arises in understanding the physics involved with increase in C_L. Clearly, the increase, though lift dependent, has little to do with viscosity unless shock interacts with boundary layer to increase pressure drag. Since the very obligation of design is to avoid such interaction up to a certain C_L, this book treats compressibility drag at supersonic speed as being composed of compressibility drag at zero C_L (that is C_{D_shock}) plus compressibility drag at higher C_L (that is ΔC_{Dw}).

To compute compressibility drag at zero C_L, the following empirical procedure is adopted, taken from [3]. Compressibility drag of an object depends on its thickness parameter, in case of the fuselage it is the fineness ratio and in case of wing it is the thickness to chord ratio. The fuselage (including the empennage) and wing compressibility drags are separately computed and then added along with the interference effects. Extensive use of graphs (Figures 11.25–11.31) is required for the empirical methodology. Compressibility drag values for both at Mach 0.9 and Mach 2.0 are estimated.

Drag estimation at Mach 0.9 follows the same method as worked out in the civil aircraft example and is tabulated in Section 11.19.

Fuselage compressibility drag (includes empennage contribution) at Mach 2.0:
Thickness parameter is fuselage fineness ratio.

Step 1. Plot fuselage cross-section along fuselage length as shown in Figure 11.22 and obtain the maximum cross-section, $S_\pi = 45.25$ ft^2 and the fuselage base $S_b = 12$ ft^2. Find the ratio
Find $(1 + Sb/S\pi) = 1 + 12/45.25 = 1.27$.
Find $S\pi/Sw = 45.25/700 = 0.065$.

Table 11.25 Vigilante induced drag.

C_L	0.2	0.3	0.4	0.5	0.6	0.7
C_{Di}	0.003 42	0.007 68	0.013 7	0.021 4	0.030 7	0.041 8

Figure 11.25 Corrections for laminarisation [3].

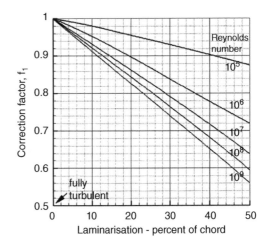

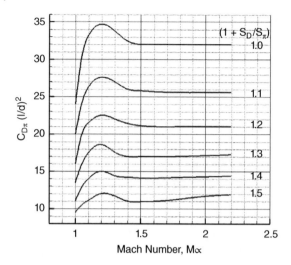

Figure 11.26 Supersonic fuselage compressibility drag [3].

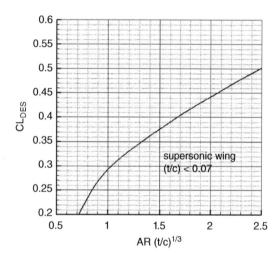

Figure 11.27 Design-lift coefficient [3].

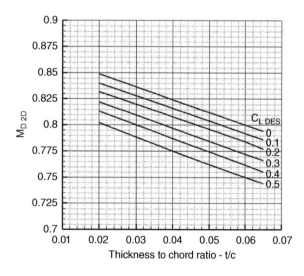

Figure 11.28 Two-dimensional drag divergence Mach number for supersonic aerofoil [3].

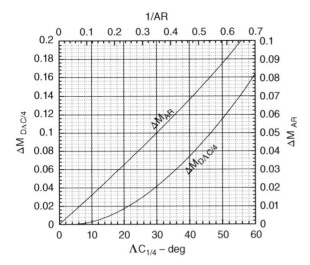

Figure 11.29 Design Mach number [3].

Figure 11.30 Supersonic wing compressibility drag [3].

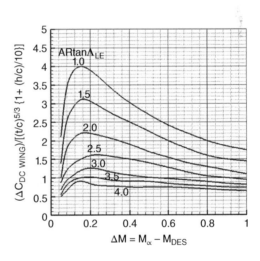

Step 2. Get the fuselage fineness ratio $l/d = 73.3/7.788 = 9.66$ (d is minus intake width).

Obtain $(l/d)^2 = (9.66)^2 = 93.3$.

Step 3. Use Figure 11.26 to get $C_{D\pi}(l/d)^2 = 18.25$ at $M_\infty = 2.0$ for $(1 + S_b/S_\pi) = 1.27$

This gives $C_{D\pi} = 18.25/93.3 = 0.1956$.

Convert this to the fuselage contribution of compressibility drag expressed in terms of wing reference area,

$C_{Dwf} = C_{D\pi} \times (S_\pi/S_w) = 0.1956 \times 0.065 = 0.01271$.

Wing compressibility drag at Mach 2.0:

Step 1. Obtain design C_{L_DES} from Figure 11.27 for supersonic aerofoil for

$AR \times (t/c)^{1/3} = 3.73 \times (0.05)^{1/3} = 1.374$. It gives CL_DES = 0.352.

Test data of CL_DES from [3] gives 0.365, which is close enough. Test data is used.

Step 2. Obtain from Figure 11.28 the two-dimensional design Mach number, $M_{DES\_2D} = 0.784$.

Using Figure 11.29 obtain $\Delta MAR = 0.038$ for $1/AR = 0.268$.

Using Figure 11.29 obtain $\Delta MD\Lambda^{1/4} = 0.067$ for $\Lambda^{1/4} = 37.5°$.

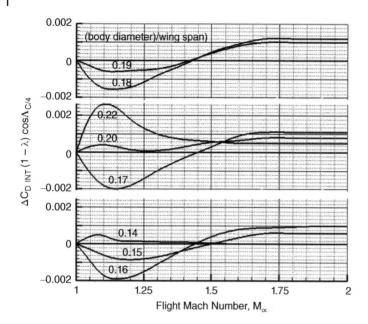

Figure 11.31 Wing-body zero lift interference drag [3].

Step 3. Make the correction to get design Mach as $M_{\text{DES}} = M_{\text{DES\_2D}} + \Delta M_{\text{AR}} + \Delta M_{\text{DA}_{\frac{1}{4}}}$.

Or $M_{\text{DES}} = 0.784 + 0.0.038 + 0.067 = 0.889$.

Then $\Delta M = M_{\infty} - M_{\text{DES}} = 2.0 - -0.889 = 1.111$.
Step 4. Compute $(t/c)^{5/3} \times [1 + (h/c)/10)] = (0.05)^{5/3} \times [1 + (0)/10)] = 0.00679$.
Step 5. Compute $AR \tan \Lambda_{\text{LE}} = 3.73 \times \tan 43 = 3.73 \times 0.9325 = 3.48$.
Step 6. Use Figure 11.30 and the corresponding value of $AR_{\tan \Lambda LE} = 3.48$ in Step 5.

$$[\Delta C_{\text{DC\_WING}}/\{(t/c)^{5/3} \times [1 + (h/c)/10]\}] = 0.675$$

Compute $\Delta C_{\text{DC\_WING}} = 0.675 \times 0.00\,679 = 0.00458$

Finally, interference drag at supersonic speed has to be added to fuselage and wing compressibility drag. The procedure for estimating wing-fuselage interference drag is given next.

Step 1. Compute (fuselage diameter at maximum area/wing span) $= 7.785/53.14 = 0.1465$
Step 2. With a taper ratio, $\lambda = 0.19$, compute $(1 - \lambda)\cos \Lambda_{\frac{1}{4}} = (1 - 0.19) \cos 37.5 = 0.643$.
Step 3. Using Figure 11.31 obtain $\Delta C_{\text{D\_INT}} \times [(1 - \lambda)\cos \Lambda_{\frac{1}{4}}] = 0.00\,048$
Compute $\Delta C_{\text{D\_INT}} = 0.00\,048/0.643 = 0.00\,075$.

Summary of compressibility drag of a Vigilante at zero lift
Compressibility drag at Mach 0.6 (3% reduction on account C_{F} change) and Mach 0.9 (20% reduction on account C_{F} change) has been done as in the case of a civil aircraft and is given in Table 11.26 along with drag at Mach 2.0.

11.22.8 Total Aircraft Drag

Total Vigilante drag at the three Mach numbers is tabulated in Table 11.27. Figure 11.8 gives the Vigilante drag polar at the three aircraft speeds. Figures 11.24–11.31 are re-plotted in this chapter from [3].

Table 11.26 Vigilante supersonic drag summary.

	C_{Dw} at Mach 2.0
Fuselage/empennage contribution	0.01 271
Wing contribution	0.00 458
Wing-fuselage interference (supersonic only)	0.00 075
Total	0.01 804

Table 11.27 Vigilante total aircraft drag coefficient, C_D.

At Mach 0.6

C_L	0.0	0.1	0.2	0.3	0.4	0.5	0.6
C_{Dpmin}				0.01 627			
ΔC_{Dp}	0.000 8	0.00 015	0.000 1	0.000 8	0.00 195	0.003 6	0.006
C_{Di}	0	0.000 854	0.00 342	0.00 768	0.013 7	0.021 4	0.030 7
Aircraft C_D @ Mach 0.6	0.01 707	0.01 727	0.019 8	0.02 475	0.03 192	0.04 127	0.052 3

At Mach 0.9

C_L	0.0	0.1	0.2	0.3	0.4	0.5	0.6
C_{Dpmin}			$0.97 \times 0.01\,627 = 0.01\,582$ (Ref [3] gives 0.01 575)				
ΔC_{Dp}	0.000 8	0.00 015	0.000 1	0.000 8	0.00 195	0.003 6	0.006
C_{Di}	0	0.000 854	0.00 342	0.00 768	0.013 7	0.021 4	0.030 7
Wave drag, C_{Dw}	0.000 3	0.001	0.002	0.003 2	0.005 5	0.01	0.02
Aircraft C_D @ Mach 0.9	0.01 737	0.01 827	0.021 8	0.02 795	0.03 742	0.05 127	0.072 3

C_{Dw} versus C_L is to be taken from CFD/test data. Here it is reduced from Ref. [3]

At Mach 2.0

C_L	0.0	0.1	0.2	0.3	0.4	0.5	0.6
C_{Dpmin}			$0.8 \times 0.01\,627 = 0.013$ (Ref [3] gives 0.01302)				
ΔC_{Dp}	0.000 8	0.00 015	0.000 1	0.000 8	0.00 195	0.003 6	0.006
C_{Di}	0	0.000 854	0.00 342	0.00 768	0.013 7	0.021 4	0.030 7
Shock drag			0.01 804 at zero C_L (C_{D_shock})				
Wave drag, ΔC_{Dw}	0.0	0.003	0.011	0.023	0.041		
Aircraft C_D @ Mach 2.0	0.031 8	0.035	0.045 5	0.062 5	0.087		

ΔC_{Dw} versus C_L is to be taken from CFD/test data.

11.23 Drag Comparison

The following equations are the parabolic drag polar of the aircraft used as classwork examples. Figure 11.23 compares the drag polar values of the Bizjet, the AJT and the TPT.

Bizjet: (Oswald efficiency factor, e, determined in Section 4.9.3.).

Classwork example: Bizjet drag may be approximated as parabolic drag polar as given here.

$$C_D = 0.0205 + 0.0447\, C_L^2 \text{ with } S_W = 323 \text{ ft}^2 \qquad (11.45)$$

(gives Oswald's efficiency factor, $e = 0.95$)

AJT: (Reader's to work out the Oswald efficiency factor, *e*).

Classwork example AJT drag may be approximated as parabolic drag polar as given here.

$$C_D = 0.0212 + 0.068C_L^2 \text{ with } S_W = 183 \text{ ft}^2 \tag{11.46}$$

(gives Oswald's efficiency factor, *e* = 0.85)

Turboprop (TPT): (low-speed aircraft – Oswald efficiency factor, *e*, fits closely with actual polar).

Maximum TPT speed is below 0.5 Mach (incompressible flow). Here parabolic drag is close enough to actual drag polar.

$$C_D = 0.0226 + 0.052C_L^2 \text{ with } S_W = 200 \text{ ft}^2 \tag{11.47}$$

(gives Oswald's efficiency factor, *e* = 0.875)

Figure 11.23 compares actual drag polar with parabolic drag polar for the three classwork examples, that is, the Bizjet, the TPT and the AJT. It may be noted that the three graphs are close enough within the operating range below M_{crit}. Equation (10.5) represents typical high speed subsonic aircraft operational segments of LRC, en-route climb and descent. The TPT flies in incompressible flow region and parabolic drag polar can be used to get good results. At HSC, non-linear effects show up. The Bizjet and AJT operate in high subsonic at M_{crit} and, therefore, semi-empirically determined actual drag polar is the preferred one. This is the standard procedure practised in industry, which gives credible output as well as offering the entire characteristics by close-form solutions.

11.24   Some Concluding Remarks

It is emphasised that, unlike the other chapters, this important chapter has some concluding remarks. Drag estimation is a state-of-the-art and covers a very large territory, as can be seen in this chapter. There is a tendency to underestimate drag primarily on account of missing out some of the multitude of items those need to be considered in the process of estimation.

The object of this chapter is to make readers aware of the sources of drag as well as give a methodology in line with typical industrial practices (without CFD results). Some of the empirical relations are guesstimates based on industrial data available to the authors; those are not in the public domain. The formulation could not possibly cover all aspects of drag estimation methodologies and therefore has to be simplified for classroom usage. For example, the drag of high-lift devices could only give ballpark figures.

Readers are advised to rely on industrial data or generate their own data bank through CFD/tests. The authors would be grateful to receive data and/or substantiated formulation on these that could improve accuracy in future editions (with due acknowledgement).

References

1 Hoerner, S. F. *Fluid Dynamic Drag*, Bricktown, NJ: Published by published by the author. 1965.

2 AGARD LS 67 – Prediction Method for Aircraft Aerodynamic Characteristics, 1974.

3 Feagan, R.C. and Morrision, W.D. (1978). *Delta Method – An Empirical Drag Build-Up Technique*. Lockheed California Company. NASA Contractor Report 151 971.

4 USAF. *USAF Data Compendium for Stability and Control* (DATCOM). https://ipfs.io/ipfs/QmXoypizjW3WknFiJnKLwHCnL72vedxjQkDDP1mXWo6uco/wiki/USAF_Stability_and_Control_DATCOM.html.

5 ESDU. (2012). 90029, An Introduction to Aircraft Excrescence Drag, London.

6 Kundu, A.K., Price, M.A., and Riordan, D. (2016). *Theory and Practice of Aircraft Performance*. John Wiley & Sons, Ltd. ISBN: ISBN: 9871119074175.

7 Kundu, A.K. (2010). *Aircraft Design*. Cambridge University Press. ISBN: ISBN: 978-0-521-88516-4.

8 Kundu, A. K. Subsonic aircraft drag estimation. Shorts Document No: AR/0001, October 1995 (not in public domain).

9 Aerodynamic Drag, *AGARD CP 124. 7 Rue Ancelle, F-92200 Neuilly-sur-Seine, France, April 1973.

10 AGARD AR 256. (1989). Technical status review on drag prediction and analysis from Computational fluid dynamics: State-of-the-art.

11 Torenbeek, E. (1982). *Synthesis of Subsonic Airplane Design.* Springer.

12 Haines, A.B. (1968). Subsonic aircraft drag: an appreciation of present standards. *Aeronautical Journal* 72, No. 687.

13 Young, A.D. and Paterson, J.H. (1981). *AGARD AG264 Aircraft excrescence drag.* 7 Rue Ancelle, F-92200 Neuilly-sur-Seine, France.

14 Seddon, J. and Goldsmith, E. (1965). *Practical Intake Aerodynamic Design.* Reston, VA: AIAA.

15 Farokhi, S. (1991). *Propulsion System Integration. Short Course.* Lawrence, KS: University of Kansas.

16 Kundu, A.K., Watterson, J., MacFadden, R. et al. (2002). Parametric optimisation of manufacturing tolerances at the aircraft surface. *Journal of Aircraft* 39, No 2, AIAA.

---

* The Advisory Group for Aerospace Research and Development (AGARD) was an agency of NATO that existed from 1952 to 1996. AGARD was founded as an Agency of the NATO Military Committee. It was set up in May 1952 with headquarters in Neuilly sur Seine, France.
In a mission statement in the 1982 *History* it published, the purpose involved "bringing together the leading personalities of the NATO nations in the fields of science and technology relating to aerospace"

12

Aircraft Power Plant and Integration

12.1 Overview

The engine may be considered to be the heart of any powered aircraft as a system. For its importance, this largish topic is divided in two chapters; this one gives the fundamentals of the associated theories and installation details when integrated with aircraft. The next Chapter 13 deals purely with engine performances, without which aircraft performance analyses cannot progress; it includes uninstalled and installed thrust/power and fuel flow data of various types of engines. Propeller theory performance and thrust developed by propellers are also included in Chapter 13.

This chapter starts with a brief introduction to engine evolutionary past, followed by classification of the types of engines available and their domain of application, some fundamentals of engine theory, installation details, nacelles and thrust reversers (TRs). Primarily, this chapter deals with gas turbines (both jet and propeller-driven) and to a lesser extent piston engines, which are used only in small general aviation aircraft).

The chapter covers the following:

Section 12.2: Background Information on aircraft Engines and Their Classification
Section 12.3: Definitions Required in this Chapter
Section 12.4: Introduces Types of Aircraft Engines to be Used in this Book
Section 12.5: Gas Turbine Engine Cycles
Section 12.6: Theories Involved to Analyse Engine Performance
Section 12.7: Considerations Required for Engine Integration with Aircraft
Section 12.8: Discusses Topics on Nacelle/Intake
Section 12.9: Discusses Nozzle and Thrust Reversers
Section 12.10: Propeller and Associated Definitions
Section 12.11: Propeller Theories and Charts

Classwork content. This is an important chapter to understand the fundamental theories involved with engine performance. Readers should go through this thoroughly to understand the installation aspects of engine on aircraft.

12.2 Background

Gliders were flying long before the Wright brothers flew but they could not install an engine, even when automobile piston engines were available – these were too heavy. Gustav Weisskopf (Whitehead) made his own engine. The Wright brothers made their own light gasoline engine with the help of Curtiss. Up until World War II, aircraft were designed around available engines. Aircraft sizing was a problem – it was not optimised for the mission role – it was based on the number and/or the size of the existing type of engines that could be installed.

During the late 1930s, Frank Whittle in the UK (Whittle had a hard time developing his case. He died in England, 1996) and Hans von Ohain in Germany (died in the USA, 1998) were working independently and

Conceptual Aircraft Design: An Industrial Approach, First Edition. Ajoy Kumar Kundu, Mark A. Price and David Riordan.
© 2019 John Wiley & Sons Ltd. Published 2019 by John Wiley & Sons Ltd.

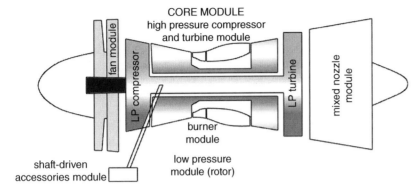

Figure 12.1 Modular concept of gas turbine design (the core module is also known as the gas generator module).

simultaneously on reaction type engines using vane/blade type pre-compression before combustion. Their efforts resulted in today's gas turbines engines. The end of WWII saw gas turbine powered jet aircraft in operation. A good introduction to jet engines is given in [1].

Post-WWII research led to the rapid advancement of gas turbine development to a point when, from a core gas generator module, a family of engines could be made in a modular concept (Figure 12.1) that allows engine designers to offer engines as specified by the aircraft designers. Similar laws in thermodynamic design parameters permit power plants to be scaled (rubberised) to the requisite size around the core gas generator module to meet the demands of the mission requirements. The size and characteristics of engine is determined by matching with the aircraft mission. Now both the aircraft and the engine can be sized to the mission role, thereby improving operational economics. Modular engine design also favours low down times in maintenance.

Potential energy locked in fuel is released through combustion in the form of heat energy. In gas turbine technology, the high energy of the combustion product can be used in two ways: (i) it is converted into increase in kinetic energy (KE) of the exhaust to produce the reactionary thrust (turbojet/turbofan) or (ii) it is further extracted through an additional turbine to drive a propeller (turboprop) to generate thrust.

Initially, the reactionary type engines came as simple straight-through airflow turbojets (Figure 12.5). Subsequently, the turbojet development improved with addition of a fan (long compressor blades visible from outside) in front of the compressor and this type is called a turbofan. Commercial transport aircraft have a higher bypass ratio (BPR) (Section 12.3) requiring a large fan diameter, hence high drag and are they are pod mounted.

Gas turbine components' operating environment demands more complex aerodynamic considerations than aircraft. There are very high stress levels at considerably elevated temperature, yet it has to be made as light as possible, demanding stringent design considerations. Gas turbine parts manufacture is also a difficult task – very tough material has to be machined in a complex 3D shape to a very tight tolerance level (3D printing offers promise to alleviate some of these problems). All these considerations make gas turbine design a very complex technology, possibly second to none. Engine control also involves with very complex microprocessor based management.

Note that gas turbine engines have a wide range of applications from land based large prime movers for power generation and ship usages (civil and military) to weight critical airborne applications. The theories behind all the types have a common base – the hardware design differs, driven by the application requirements and technology level adopted – for example, land-based gas turbines are not weight critical and do not have to stand alone; therefore, they are less constrained. Surface based gas turbines have to run economically for days/months generating a lot of power compared to standalone light aircraft engines running for hours in varying power, altitude, *g*-load and airflow demands. Even the biggest aircraft gas turbine is small compared to the surface-based ones.

Gas turbines design have advanced to incorporate sophisticated microprocessor-based control system with automation, called FADEC (Full Authority Digital Engine Control) working in conjunction with fly-by-wire

(FBW) control of aircraft. The uses of computer-aided drawing (CAD)/computer-aided manufacture (CAM)/computational fluid dynamics (CFD)/finite element method (FEM) are now the standard tools for engine design. Current developments involve having laminar flow in the intake duct, noise and emission reduction.

Liquid-cooled aircraft piston engines over 3000 HP have been built. Except for a few types, they are no longer in production as they are too heavy for the power generated; in their place, gas turbines have taken over. Gas turbine engines have better thrust to weight ratio. Two of the successful piston engines were the WWII types – the Rolls Royce (RR) Merlin and Griffon. They produced 1000–1500HP and weighed around 1500 lb dry. Also, AVGAS (aviation gasoline-petrol) is considerably more expensive than AVTUR (aviation turbine – kerosene). Today, the biggest aircraft piston engine in production is around 500 HP. Of late, diesel fuel piston engines (less than 250 HP) have entered the market for general aviation usage. In the home-built market, MOGAS (automobile gasoline-petrol) powered engines have appeared, approved by certification authorities. Battery powered small engines for very light aircraft are gaining ground (Chapter 24).

Chart 12.1 classifies all types of aircraft engines in current usage. Rocket propulsion is not included in this book.

The application domain of the types dealt with in this book is shown in Figure 12.2 (excluding electric propulsion). High BPR turbofans are meant for high subsonic speeds. At supersonic speeds the bypass ratio comes down to less than three with the benefit of having smaller fan diameter that can be installed inside fuselage. Typically, turboprop powered aircraft speeds are around Mach 0.5 and below (higher speeds exceeding Mach 0.7 has been achieved). Piston engine powered aircraft are at the low speed end.

Turbofans (bypass turbojets) start to compete with turboprops for ranges over 1000 nm on account of time saved as a consequence of higher subsonic flight speed. Fuel cost is not the only consideration in a sector of operation. The combat aircraft power plant uses lower bypass turbofans with smaller fan diameters so that they can be installed within the aircraft fuselage, in earlier days they had straight-through (no bypass) turbojets.

Figure 12.3a gives the thrust to weight ratio of various kind of engines. Figure 12.3b gives the specific fuel consumption (*sfc*) at sea level static takeoff thrust (T_{SLS}) rating in International Standard Atmosphere (ISA) day of various classes of current engines. At cruise, *sfc* would be higher.

Typical levels of *specific thrust* ($F/\dot{m}_a$ – lb (lb s)$^{-1}$) and *sfc* – (lb (h lb)$^{-1}$) of various types of gas turbines are shown Figure 12.4.

Table 12.1 gives various kinds of efficiencies of the different classes of aircraft engines. Table 12.2 gives progress made in the last half a century. It indicates considerable advances made in engine weight savings. Since the 1970s, the engine noise maximum level came into force as a requirement to comply with the certification authorisation. Pollution levels due to noise and emissions are steadily getting lower (Chapter 21).

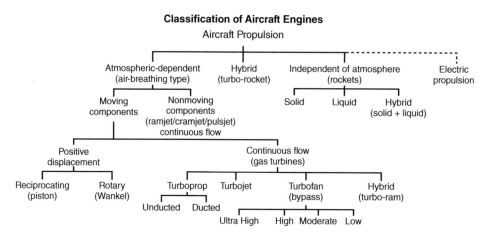

Chart 12.1 Classification of aircraft engines. Source: Reproduced with permission from Cambridge University Press.

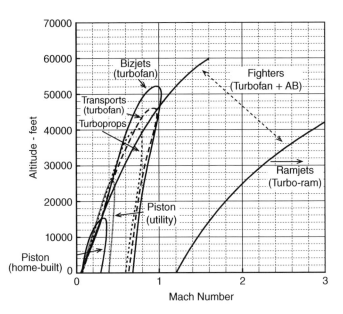

Figure 12.2 Application domains of various types of air-breathing aircraft engines.

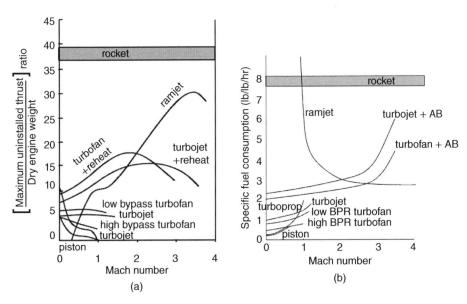

Figure 12.3 Engine performance. (a) Thrust to weight ratio and (b) specific fuel consumption (slightly modified from the diagrams in *Aero Digest*, Vol. 69, No. 1, July 1954).

12.3 Definitions

The following are the definitions of various terminologies used in jet engine performance analysis.

Specific fuel consumption. This is the fuel flow rate required to produce one unit of thrust or *shp*.

$$sfc = (\text{Fuel flow rate})/(\text{Thrust or Power}) \tag{12.1}$$

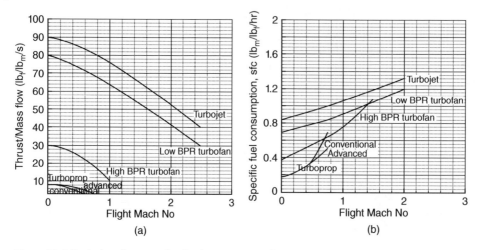

Figure 12.4 Typical performance levels of various gas turbine engines. (a) Specific thrust. (b) Specific fuel consumption.

Table 12.1 Efficiencies of engine types.

	Thermal efficiency	Propulsive efficiency	Overall efficiency
Current types	0.6–0.65	0.75–0.85	0.5–0.55
Propfan	0.52–0.55	0.8–0.85	0.54–0.55
Propfan (ad[a])	0.52–0.55	0.7–0.76	0.46–0.5
High BPR	0.48–0.55	0.62–0.68	0.4–0.42
Low BPR	0.4–0.5	0.55–0.6	0.35–0.38
Turbojet	0.4–0.45	0.45–0.52	0.28–0.32

a) Advanced propfan.
Source. Reproduced with permission from Cambridge University Press.

Table 12.2 Progress in jet engines.

	Thrust/weight ratio
1950s (J69 class)	2.8–3.2
1960s (JT8D, JT3D class)	3.2–3.6
1970s (J79 class)	4.5–7.0
1980s (TF34 class)	6.0–6.5
1990s (F100, F404 class)	7.0–8.0
Current	8.0–10.0

Unit of *sfc* is in lb $(h\,lb)^{-1}$ of thrust produced (in SI units – gm $(s\,N)^{-1}$) – the lower the better. To be more precise, the reaction type engines use '*tsfc*' and propeller-driven ones use '*psfc*', where, '*t*' and '*p*' stand for thrust and power, respectively. For turbofan engines,

$$Bypass\ Ratio, BPR = \frac{secondary\ airmass\ flow\ over\ the\ core\ engine}{primary\ airmass\ flow\ through\ the\ core\ (combustion)} = \dot{m}_s/\dot{m}_p \qquad (12.2)$$

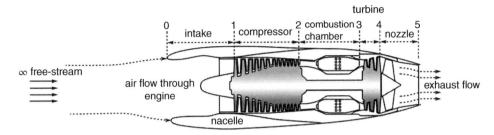

Figure 12.5 Sketch of a simple straight-through turbojet (pod mounted turbojet from [2] – bare engine does not have intake and exhaust nozzle).

Given next are the definitions of various kinds of efficiencies of jet engines. Subscript numbers indicate gas turbine component station numbers as shown in Figure 12.5 (in the figure, subscript 5 represents *e*).

$$\text{Thermal Efficiency}, \eta_t = \frac{\text{mechanical energy produced by the engine}}{\text{Heat energy of}\,(\text{air} + \text{fuel})} = \frac{W_E}{Q}$$

$$\text{or } \eta_t = \frac{V_e^2 - V_\infty^2}{2C_p(T_3 - T_2)} = \left(1 - PR^{\frac{1-\gamma}{\gamma}}\right) \tag{12.3}$$

For a particular aircraft speed, V_∞, the higher the exhaust velocity V_e, the better would be the η_t of the engine. Heat addition at the combustion chamber, $q_{2-3} = C_p(T_3 - T_1) \approx C_p(T_{t3} - T_{t1})$.

$$\text{Propulsive Efficiency}, \eta_p = \frac{\text{useful work done on airplane}}{\text{mechanical energy produced by the engine}}$$

$$\text{or } \eta_p = \frac{W_A}{W_E} = \frac{2\,V_\infty}{V_e + V_\infty} \tag{12.4}$$

For subsonic aircraft, $V_e \gg V_\infty$. Clearly, for a given engine exhaust velocity, V_e, the higher the aircraft speed, the better the propulsion efficiency, η_p, will be. A jet aircraft flying below Mach 0.5 is not desirable – it is better to use propeller-driven aircraft when flying at Mach 0.5 and below.

$$\text{Overall Efficiency}, \eta_o = \frac{\text{useful work done on airplane}}{\text{heat energy of}\,(\text{air} + \text{fuel})} = \frac{W_A}{Q}$$

$$\text{or } \eta_o = \frac{W_A}{Q} = \eta_t \times \eta_p = \frac{(V_e - V_\infty)\,V_\infty}{q_{2-3}} \tag{12.5}$$

It can be shown [2–5] that for non-afterburning (AB) engines, ideally the best overall efficiency, η_o, would be when the engine exhaust velocity, V_e, is twice the aircraft velocity, V_∞. Bypassed turbofans offer such an opportunity at high subsonic aircraft speed.

12.3.1 Recovery Factor, RF

A bare engine on a ground test bed inhales free stream air directly and performs at its best. In contrast, an installed engine in an aircraft can only inhale the airflow diffused through the nacelle/intake duct incurring losses in the pressure head, expressed as recovery loss, defined next.

Intake Pressure Recovery Factor (RF),

$$\text{RF} = (\text{total pressure head at engine fan face})/(\text{free stream air total head}) = p_{t1}/p_{t\infty} \tag{12.6}$$

where subscript '∞' represents free stream condition and subscript '1' represent fan face.

For pod mounted high subsonic jet aircraft at cruise Mach, RF is about 0.98.

For fuselage mounted side intake with engine at aft end, at subsonic Mach, RF is about 0.94–0.96.

For nose mounted pitot type long intake with engine at aft end, at subsonic Mach, RF is about 0.92–0.94.

At supersonic speed, shock loss has to be added (see Section 12.8.2).

12.4 Introduction – Air-Breathing Aircraft Engine Types

This section starts with describing various types of gas turbines followed by introducing the piston engines. Aircraft propulsion depends on the extent of thrust produced by the engine. Chapter 13 presents the available thrust and power from various types of engines. Some statistics of various kinds of aircraft engines are given at the end of this chapter. Gas turbine sizes are progressing in making both larger and smaller engines than the current sizes, that is, expanding the application envelope.

12.4.1 Simple Straight-Through Turbojets

The most elementary form of gas turbine is a simple straight-through turbojet, as shown schematically in Figure 12.5. In this case, the intake airflow goes straight through the whole length of the engine and comes out at a higher velocity and temperature after going through the processes of compression, combustion and expansion. It burns like a stove under pressurised environment. The readers may note the *waisting* of the airflow passage as a result of compressor reducing the volume while the turbine expands. Typically, at long range cruise (LRC) condition, the free stream tube far upstream is narrower in diameter than at the compressor face. As a result, airflow ahead of the intake plane slows down as a pre-compression phase.

Components associated with the thermodynamic processes within the engine are assigned with station numbers as given in Table 12.3.

Overall engine efficiency could be improved if the higher energy of the exhaust gas of a straight through turbojet is extracted through additional turbine that can drive an encased fan in front of a the compressor (when it is called turbofan engine) or a propeller (when it is called turboprop engine).

12.4.2 Turbofan – Bypass Engine (Two Flow – Primary and Secondary)

The energy extraction through the additional turbine lowers the rejected energy at the exhaust resulting in lower exhaust velocity, pressure and temperature. The additional turbine drives a fan in front of q compressor. A large amount of air mass flow through the fan provides thrust. The intake air mass flow is split into two streams (Figure 12.6), the core airflow passes through the combustion chamber of the engine as primary flow and is made to burn while the secondary flow through the fan is bypassed (hence also called bypass engine) around the engine and remains as cold flow. It is for this reason that primary flow is also known as *hot flow* and the secondary bypassed flow as *cold flow*. Figure 12.6b gives a schematic diagram of a turbofan

Table 12.3 Gas turbine station number.

Number	Station	Description
∞	Free stream	Far upstream (if pre-compression is ignored than it is same as 0).
0–1	Intake	A short divergent duct as a diffuser to compress inhaled air mass.
1–2	Compressor	Active compression to increase pressure – temperature rises.
2–3	Burner[a]	Fuel is burned to release the heat energy.
3–4	Turbine	Extracts the power from heat energy to drive the compressor.
4–5	Nozzle[a]	Generally, convergent to increase flow velocity.
e		(station 5 is also known as '*e*' representing the exit plane).

a) Burner is also known as combustion chamber (CC) and the nozzle as exhaust duct.

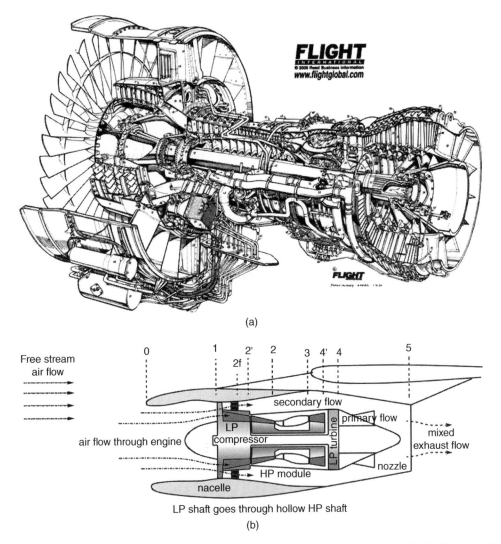

Figure 12.6 Turbofan engine. (a) Pratt and Whitney 2037 Turbofan (Credit: Flightglobal). (b) Schematic diagram of a pod mounted long duct two-shaft turbofan engine. Source: Reproduced with permission from Flightglobal.

engine (top Figure 12.6a is that of bare PW 2037 turbofan). Lower exhaust velocity reduces engine noise. At the design point (LRC), lower exhaust pressure permits the nozzle exit area to be sized to make exit pressure equal to ambient pressure (perfectly expanded nozzle), unlike simple turbojets that can have higher exit pressure.

The readers may note that the component station numbering system follows the same pattern as for the simple straight-through turbojet. The combustion chamber in the middle maintains the same numbers of 2–3. The only difference is the fan exit is with subscript of 'f'. Intermediate stages of compressor and turbine are primed (').

Typically, the BPR – (Eq. (12.2)) for commercial jet aircraft turbofans (high subsonic flight speed of less than 0.98 Mach) is around 4–6 with large fan face diameter and are invariably pod mounted. Of late, turbofans for the new B787 have exceeded a BPR of 8. Airbus 350 and Bombardier C-series turbofans have a BPR = 12. For military application (supersonic flight speed up to Mach 2.5), the BPR is around 1–3. Lower BPR keeps the fan diameter smaller and hence lowers frontal drag so it can be integrated within fuselage.

Multi-spool (shaft) drive shafts offer better efficiency and response characteristics, mostly with two concentric shafts. The shaft driving the low-pressure (LP) section runs inside the hollow shaft high-pressure (HP) section – see the diagram in Figure 12.6. Three shaft turbofans have been designed, but most of the current designs have a twin spool. The recent advent of a geared turbofan offers better fuel efficiency.

Lower fan diameter compared to the propeller permits higher rotational speed and offers scope for having a thinner aerofoil section to extract better aerodynamic benefits. The higher the BPR, the better the fuel economy. Higher BPR demands a larger fan diameter when reduction gears may be required to keep the rpm at a desired level. Ultra-high bypass ratio (UHBPR) turbofans approach the class of ducted fan or ducted propeller or propfan engines. Such engines have been built but their cost versus performance is yet to break into the market arena.

12.4.3 Three Flow Bypass Engine

While the two streams (primary hot core flow and the secondary cold high bypass flow) offer good fuel economy for the high subsonic transport category aircraft operation, they are not is suitable for supersonic military application due to having a large fan face diameter generating high drag. High BPR specific thrust is considerably lower than low bypass ratio of military jet engines. Of current interest for supersonic combat class aircraft engine is pursuing the concept of the adaptive three-stream bypass jet engine. This improves the fuel economy while keeping the front fan face diameter relatively small.

The adaptive three-stream jet engine has the third outer flow as a second layer of bypass over the conventional two stream bypass turbofan. The adaptive three-stream jet engine is a variable cycle engine, because the third airflow stream has a movable internal inlet door that can adjust the second bypass flow during the flight to suit the mission demand. The outer cold third air stream also serves for thermal management of the hot combustion product primary core flow and abates the noise level as well. This arrangement shows improved fuel economy while retaining high specific thrust suiting supersonic military combat aircraft operation. This type of engine is currently in the developmental stage with a technology demonstrator running under test.

12.4.4 Afterburner Engines

The AB is another method of thrust augmentation exclusively meant for supersonic combat category aircraft (the *Concorde* is the only civil aircraft to use an AB). Figure 12.7 gives a cutaway diagram of a modern AB meant for combat aircraft.

The simple straight-through turbojets or low bypass ratio turbofans have a relatively small frontal area to give low drag and have excess air in the exhaust flow. If additional fuel can be burned in the exhaust nozzle beyond the turbine exit plane, then additional thrust can be generated to propel the aircraft at a considerably higher speed and acceleration, thereby also giving the chance to improve propulsive efficiency. However, the reason for using AB arises from the mission demand, for example, at takeoff with high payload, acceleration to engage/disengage during combat/evasion manoeuvres and so on. Mission demand overrides the fact that there is a very high level of energy rejection in the high exhaust velocity. The fuel economy with AB degrades – it takes 80–120% higher fuel burn to gain 30–50% increase in thrust. Nowadays, most supersonic aircraft engines have some BPR when afterburning can be done within in the cooler mixed flow past the turbine section of the primary flow.

12.4.5 Turboprop Engines

Lower speed aircraft could use propellers for thrust generation. Therefore, instead of driving a smaller encased fan (turbofan), a large propeller (known as a turboprop) can be driven by gas turbine to improve efficiency as the exhaust energy can be further extracted to very low exhaust velocity (nearly zero nozzle thrust) (Figure 12.8). There could be some residual jet thrust left at the nozzle exit plane when it needs to be

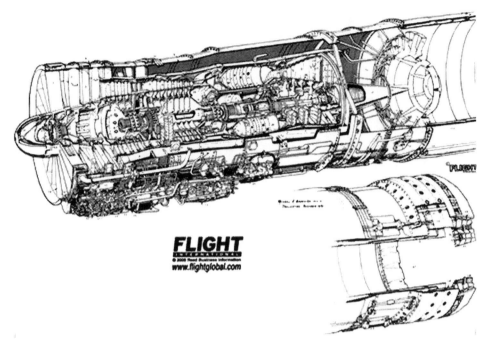

Figure 12.7 Afterburning engine – Volve-Flygmotor RM8 (Credit: Flightglobal). Source: Reproduced with permission from Flightglobal.

added to the propeller thrust. The nozzle thrust is converted into HP and together with the shaft horse power (SHP) generated; it becomes Equivalent Shaft Horse Power (ESHP).

However, a large propeller diameter would limit rotational speed on account of both aerodynamic (transonic blade tips) and structural considerations (centrifugal force). The relative velocity with respect to the propeller blade is the resultant velocity (Figure 12.24) of aircraft speed and the propeller rpm can reach a transonic level when propeller efficiency suffers on account of compressibility effects and possibly trailing edge separation as a result of local shock and boundary layer interaction. Heavy reduction gears are required to bring down propeller rpm to a desirable level. Propeller efficiency drops when aircraft operate at flight speeds above Mach 0.7. For shorter range flights, the turboprop's slower speed does not become time critical to the user and yet it offers better fuel economy. Military transport aircraft are not as time critical as commercial transport and may use a turboprop (e.g. A400). Figure 12.8 gives a schematic diagram of typical turboprop engines. Modern turboprops have up to eight blades (see Section 12.10) allowing reduction of diameter size and operation at a relatively higher rpm. Advanced propeller designs permit aircraft to fly above Mach 0.7.

12.4.6 Piston Engines

Most of the aircraft piston engines are reciprocating types (positive displacement – intermittent combustion), the smaller ones are air-cooled two-stroke cycle – the bigger ones (typically over 200 HP) are liquid-cooled four-stroke cycle. There are a few rotary type positive displacement engines (Wankel) – in principle, they very attractive but they have some sealing problems. Cost wise, the rotary type positive displacement engines are not popular yet on account of these sealing problems. Cost goes down with an increase in the number of production. Figure 12.9 shows aircraft piston engine with its installation components.

To improve high altitude performance (having low air density) supercharging is used. Figure 12.9 shows vane supercharging type for pre-compression. Also aviation fuel (AVGAS – petrol) differs slightly from automobile

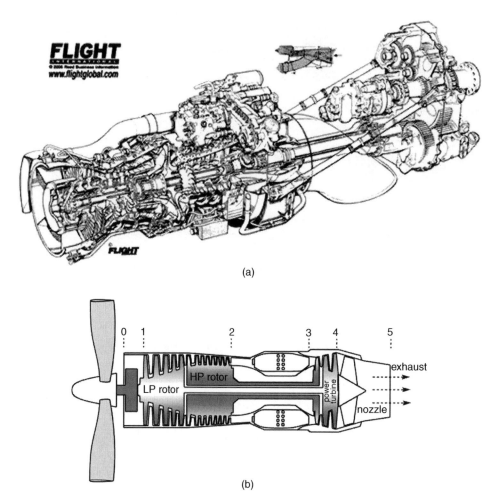

(a)

(b)

Figure 12.8 Aircraft turboprop engine. (a) General Electric CT7 (Credit: Flightglobal). (b) Schematic drawing of a turboprop engine. Source: Reproduced with permission from Flightglobal.

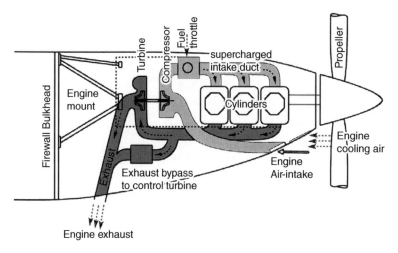

Figure 12.9 Aircraft piston engine and the supercharged scheme.

petrol (MOGAS). Of late, some engines for the home-built category are permitted to use MOGAS. Recently, small diesel engines have been introduced on the market.

Piston engines are the oldest type to power aircraft. Over the life-cycle of an aircraft, gas turbines prove more cost effective for engine sizes over 500 HP. Currently, general aviation aircraft are the main users of piston engines. Recreational small aircraft are invariably powered by piston engines.

12.5 Simplified Representation of a Gas Turbine (Brayton/Joule) Cycle

Figure 12.10a depicts a standard schematic diagram representing a simple straight through turbojet engine as shown in Figure 12.5, with the appropriate station numbers. The thermodynamic cycle associated with gas turbines is known as Joule cycle (also known as Brayton cycle). Figure 12.10b gives the corresponding temperature-entropy diagram of an ideal Joule cycle in which isentropic compression and expansion takes place.

Real engine processes are not isentropic and there are losses involved associated with increased entropy. Comparison of real and ideal cycle is shown in Figure 12.11.

12.6 Formulation/Theory – Isentropic Case (Trend Analysis)

The basic thermodynamic relationships and the related gas turbine equations pertinent to this book are given in Appendix C. These are valid for all types of processes. For more details, see [2–5, 7, 8].

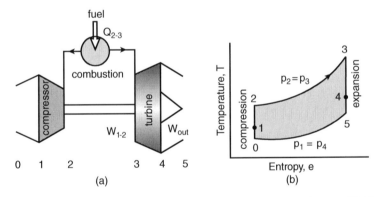

Figure 12.10 Simple straight-through jet. (a) Generic schematic diagram. (b) Ideal Joule cycle.

Figure 12.11 Real and Ideal Joule (Brayton) cycle comparison of a straight-through jet.

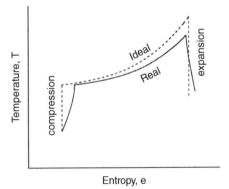

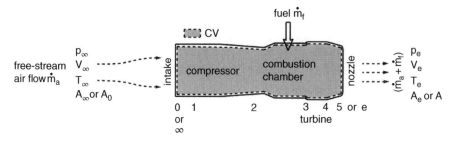

Figure 12.12 Control volume (CV) representation of straight-through turbojet.

12.6.1   Simple Straight-Through Turbojet Engine – Formulation

Consider a control volume (CV – dashed line – note the waist like shape of the bare simple turbojet) representing a straight through axi-symmetric turbojet engine as shown in Figure 12.12. The CV and the component station numbers are as per the sketch and convention shown in Figure 12.6. Gas turbine intake starts with a subscript 0 or ∞ and ends at the nozzle exit plane with subscript 5 or e. Free stream air mass flow rate (MFR), $\dot{m}_a$, is inhaled into the CV at the front face perpendicular to the flow and fuel MFR $\dot{m}_f$ (taken from the onboard fuel tank) is added at the combustion chamber and the product flow rate $(\dot{m}_a + \dot{m}_f)$ is exhausted out of the nozzle plane perpendicular to flow. It is assumed that inlet face static pressure is p_∞, which is fairly accurate. Pre-compression exists but for ideal consideration it has no loss.

No flow crosses the other two lateral boundaries of the CV as it is aligned with the walls of the engine. Force experienced by this CV is the thrust produced by the engine. Consider a cruise condition with aircraft velocity of V_∞. At cruise, the demand for air inhalation is considerably lower than at takeoff. Intake area is sized in between the two demands. At cruise the intake stream tube cross-sectional area is smaller than the intake face area – it is closer to that of exit plane area, A_e (gas exiting at very high velocity). Since this is under ideal conditions, there is no pre-compression loss, station 0 may be considered as having free stream properties with subscript ∞.

From Newton's second Law,

Applied force, F = rate of change of momentum + net pressure force.

Note that the momentum rate is given by the MFR.

Where inlet momentum rate $= \dot{m}_a V_\infty$, Exit momentum rate $= (\dot{m}_a + \dot{m}_f) V_e$

Therefore,

$$\text{the rate of change of momentum} = (\dot{m}_a + \dot{m}_f) V_e - \dot{m}_a V_\infty \tag{12.7}$$

Net pressure force between the intake and exit planes $= (p_e A_e - p_\infty A_\infty)$. The axi-symmetric side pressure at the CV walls cancels out. Typically, at cruise, sufficiently upstream $A_e \approx A_\infty$.

Therefore,

$$F = (\dot{m}_a + \dot{m}_f) V_e - \dot{m}_a V_\infty + A_e (p_e - p_\infty) = \text{net thrust} \tag{12.8}$$

Then

$$(\dot{m}_a + \dot{m}_f) V_e + A_e (p_e - p_\infty) = \text{gross thrust}$$

And $\dot{m}_a V_\infty$ = Ram drag (with a −ve sign, it has to be drag). It is the loss of momentum seen as drag on account of slowing down of the income air as ram effect.

This gives,

net thrust = gross thrust − ram drag.

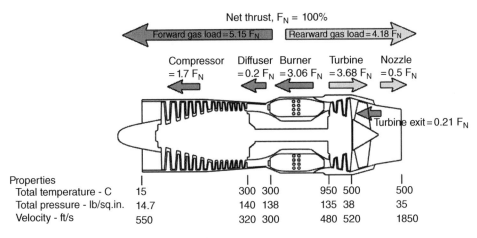

Figure 12.13 Where does the thrust act? [1].

$A_e (p_e - p_\infty)$ = pressure thrust. In general, subsonic commercial transport turbofans have convergent nozzle, its exit area is sized in a way that in cruise, $p_e \approx p_\infty$ (known as a perfectly expanded nozzle). It is different for military engines especially with afterburning, when $p_e > p_\infty$, thus requiring a convergent-divergent nozzle.

For a perfectly expanded nozzle $(p_e - p_\infty)$,

$$\text{net thrust, } F = (\dot{m}_a + \dot{m}_f) V_e - \dot{m}_a V_\infty \qquad (12.9)$$

Further simplification is possible by ignoring the effect of fuel flow, $\dot{m}_f$, since $\dot{m}_a \gg \dot{m}_f$.

Then the thrust for perfectly expanded nozzle is,

$$F = \dot{m}_a V_e - \dot{m}_a V_\infty = \dot{m}_a (V_e - V_\infty) \qquad (12.10)$$

At sea level static takeoff thrust (T_{SLS}) ratings $V_\infty = 0$,
which makes,

$$F = \dot{m}_a V_e + A_e (p_e - p_\infty) \qquad (12.11)$$

The expression indicates that the thrust increase can be achieved by increasing intake air MFR and/or increasing exit velocity.

Equation (12.4) gives propulsive efficiency,

$$\eta_p = \frac{2 V_\infty}{V_e + V_\infty}$$

Clearly, a jet propelled aircraft with low flight speed will have poor propulsive efficiency, η_p. Jet propulsion is favoured for aircraft flight speed above Mach 0.6.

The next question is where does the thrust act? Figure 12.13 [1] shows a typical gas turbine where the thrust is acting. It is all over the engine and the aircraft realises the net thrust transmitted through the engine-mount bolts.

12.6.2   Bypass Turbofan Engines – Formulation

Typically, in this book a long duct nacelle is preferred to obtain better thrust and fuel economy offsetting the weight gain as compared to short duct nacelles (see Figure 12.17). Note that pressure rise across fan (secondary cold flow) is substantially lower than pressure rise of the primary airflow. Also note that the secondary airflow does not have heat addition as in the primary flow. The cooler and lower exit pressure of the fan exit, when mixed with the primary hot flow within the long duct, brings the final pressure lower than the critical pressure $(V_e$ just reaches sonic speed) favouring a perfectly expanded exit nozzle $(p_e = p_\infty)$. Through mixing, there is reduction in jet velocity, which offers a vital benefit of noise reduction to meet the airworthiness requirements. A long duct nacelle exit plane can be sized to expand perfectly.

Primary flow has a subscript designation of p and secondary flow has a subscript designation of s. Therefore, F_p and V_{ep} stand for primary flow thrust and exit velocity, respectively, and F_s and V_{es} stand for bypass-flow thrust and its exit velocity, respectively. Thrust (F) equations of perfectly expanded turbofans are separately computed for primary and secondary flows, and then added to obtain the net thrust, F, of the engine (perfectly expanded nozzle, i.e. $p_e = p_\infty$).

$$F = F_p + F_s = [(\dot{m}_p + \dot{m}_f) \times V_{ep} - \dot{m}_p \times V_\infty] + [\dot{m}_s \times (V_{es} - V_\infty)]$$

Specific thrust in terms of primary flow becomes (f = fuel to air ratio)

$$\text{or } F/\dot{m}_p = [(1 + f) \times V_{ep} + BPR \times V_{es} - V_\infty \times (1 + BPR)] \tag{12.12}$$

If fuel flow is ignored, then

$$F/\dot{m}_p = [V_{ep} - V_\infty] + BPR \times (V_{es} - V_\infty) \tag{12.13}$$

Kinetic energy, $KE = \dot{m}_p[\frac{1}{2}(V_{ep}^2 - V_1^2)] + \dot{m}_s[\frac{1}{2}(V_{sp}^2 - V_\infty^2)]$
or

$$KE/\dot{m}_p = [\frac{1}{2}(V_{ep}^2 - V_1^2)] + BPR \times [\frac{1}{2}(V_{sp}^2 - V_\infty^2)] \tag{12.14}$$

At a given design point, that is, BPR, flight speed V_∞, fuel consumption, and $\dot{m}_p$ are held constant. Then the best specific thrust and KE can be found by varying the fan exit velocity for a given V_{ep}. This is when setting their differentiation with respect to V_{es} equal to zero. (This may be taken as trend analysis for ideal turbofan engines. The real engine analysis is more complex.)

Then by differentiating Eq. (12.13),

$$d(F/\dot{m}_p)/d(V_{es}) = 0 = d(V_{ep})/d(V_{es}) + BPR \tag{12.15}$$

And Eq. (12.14) becomes

$$d(KE/\dot{m}_p)/d(V_{es}) = 0 = V_{ep}\, d(V_{ep})/d(V_{es}) + BPR \times V_{sp} \tag{12.16}$$

Combine Eqs. (12.15) and (12.16),

$$-BPR \times V_{ep} + BPR \times V_{sp} = 0$$

And since $BPR \neq 0$, then the optimum has to be when

$$V_{ep} = V_{sp} \tag{12.17}$$

that is, the best specific thrust is when the primary (hot core flow) exit flow velocity equals the secondary (cold fan flow) exit flow velocity.

Equation (12.4) gave turbojet propulsive efficiency, $\eta_p = \frac{2\,V_\infty}{V_e + V_\infty}$ for a simple turbojet engine, but in the case of a turbofan there are two exit plane velocities: one each for hot core primary flow (V_{ep}) and cold fan secondary flow (V_{es}). Therefore, an equivalent mixed turbofan exit velocity (V_{eq}) could substitute for V_e in Eq. (12.4). Fuel flow rates are small and can be ignored. The equivalent turbofan exit velocity (V_{eq}) is obtained by equating the total thrust (perfectly expanded nozzle) as if it is a turbojet engine with total mass flow ($\dot{m}_p + \dot{m}_s$):
Thus

$$(\dot{m}_p + \dot{m}_s) \times (V_{eq} - V_\infty) = \dot{m}_p \times (V_{ep} - V_\infty) + \dot{m}_s \times (V_{es} - V_\infty)$$

$$\text{or } (1 + BPR) \times (V_{eq} - V_\infty) = (V_{ep} - V_\infty) + BPR \times (V_{es} - V_\infty)$$

$$\text{or } (1 + BPR) \times V_{eq} = (V_{ep} - V_\infty) + BPR \times (V_{es} - V_\infty) + V_\infty \times (1 + BPR) = V_{ep} + BPR \times V_{es}$$

$$\text{or } V_{eq} = [V_{ep} + BPR \times V_{es}]/(1 + BPR) \tag{12.18}$$

Then turbofan propulsive efficiency,

$$\eta_{pf} = \frac{2\,V_\infty}{V_{eq} + V_\infty} \tag{12.19}$$

Large engines could benefit from weight savings by installing short duct turbofans.

12.6.3 Afterburner Engines – Formulation

Figure 12.14 gives a schematic diagram with the station numbers for the afterburning jet engine. To keep numbers consistent with the turbojet numbering system, there is no difference between stations 4 and 5 representing turbine exit condition. Station 5 is the start and station 6 is the end of afterburning. Station 7 is the final exit plane. Figure 12.14 also shows the real cycle of afterburning in a *T-s* diagram.

Afterburning is deployed only in military vehicles (except in the supersonic *Concorde* aircraft) as a temporary thrust augmentation device to meet mission demand at takeoff and/or for fast acceleration/manoeuvre to engage/disengage in combat. Afterburning is applied at full throttle by flicking a fuel switch. The pilot can feel its deployment with sudden increase in '*g*' level in the flight direction. The ground observer would notice a sudden increase in noise level, which can exceed the physical threshold. Afterburning glow is visible at the exit nozzle, in darkness it is depicted as a spectacular plume with supersonic expansion diamonds. In the absence of any downstream rotating machines, the afterburning temperature limit can be raised to around 2000–2200 K at the expense of a large increase in fuel flow (a richer fuel/air ratio than in the core combustion).

The afterburning exit nozzle would invariably run choked (at sonic speed) and would require a convergent-divergent nozzle to make supersonic expansion to increase gain in momentum for thrust augmentation. Typically, to gain a 50% thrust increase, fuel consumption increases by about 80–120%. That is why it is used for short periods, not necessarily in one burst. Interestingly enough, afterburning in bypass engines is an attractive proposition because afterburning inlet temperature is lower. In fact, all modern combat category engines use a low BPR of 1–3 resulting in a smaller frontal area suited to in-fuselage installation.

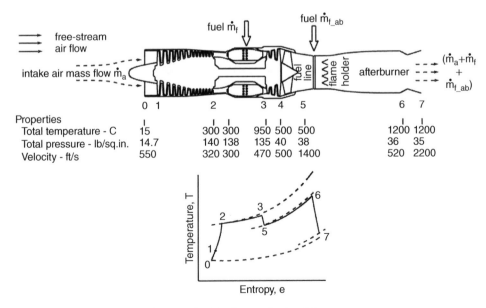

Figure 12.14 Afterburning turbojet and its *T-s* diagram (real cycle).

Losses in afterburning exit nozzles are high – the flame holders and suchlike act as obstructions. It is desirable to diffuse flow speed at the afterburner to around 0.2–0.3 Mach – this results in a small bulge in the jet pipe diameter around that area. The combat aircraft fuselage has to house this.

12.6.4 Turboprop Engines – Formulation

Section 12.4.4 described turboprops. It has considerable similarity with turbojets/turbofans except that the high energy of exhaust jet is utilised to drive a propeller by incorporating additional low-pressure turbine stages as shown in Figure 12.8. Thrust developed by the propellers is the propulsive force for the aircraft. There could be a small amount of residual thrust left at the nozzle exit plane that should be added to the propeller thrust. The relation between thrust power (T_p) and gas turbine shaft power (SHP) is related to the propeller efficiency, η_{prop} as:

$$T_p = SHP \times \eta_{prop} + F \times V_\infty \tag{12.20}$$

ESHP is a convenient way to define the combination of shaft and jet power as follows:

$$ESHP = T_p/\eta_{prop} = SHP + (F \times V_\infty)/\eta_{prop} \tag{12.21}$$

Evidently, aircraft at the static condition will have *ESHP* = *SHP*, as the small thrust at the exit nozzle is not utilised. As speed increases, *ESHP* > *SHP* as long as there is some thrust at the nozzle. Power specific fuel consumption (psfc) and specific power (weight-to-power ratio) are expressed in terms of ESHP.

12.6.4.1 Summary

The formulae give good reasoning on the gas turbine domain of application as shown in Figure 12.2. Turboprops would offer the best economy for design flight speeds at Mach 0.5 and below, suited well to shorter ranges of operation. At higher speeds up to Mach 0.98, turbofans with a high BPR offer better efficiencies (see comments followed after Eq. (12.5)). At supersonic speeds, the BPR is reduced and in most cases an afterburner used. Small aircraft with piston engines up to a size ($\approx \leq 500$ HP) that gas turbine can compete with.

12.7 Engine Integration to Aircraft – Installation Effects

Engine manufacturers supply bare engines to the aircraft manufacturers who install them on to aircraft to integrate. The same type of engines could be used by different aircraft manufacturers who each have their own integration requirements. Engine installation on to aircraft is a specialised technology that aircraft designers must know. Engine integration is carried out by the aircraft manufacturers in consultation with the engine manufacturer. Positioning of the nacelle/intake with respect to aircraft/wing is dealt with in detail in Chapter 8.

A bare engine at the test stand performs differently to an installed engine on an aircraft. Installation effects of engines are those arising from having a nacelle, that is, the losses of intake and exhaust plus off-takes of power (to drive motor/generators etc.) and air bleeds (anti-icing, environmental control etc.) Total loss of thrust at takeoff can be as high as 8–10% of that generated by a bare engine at test bed. At cruise, it can be brought down to less than 5%. Figure 12.15 shows typical off-takes required on account of various installation effects. Aircraft performance engineers use installed engine performance.

Aircraft performance engineers must be given the following data to generate installed engine performance:

 i. Intake and jet pipe losses.
 ii. Compressor air-bleed for the Environmental Control Systems (ECSs), for example, cabin air-conditioning and pressurisation, de-icing/anti-icing and any other purpose.
iii. Power off-takes from engine shaft to drive electric generator, accessories and so on.

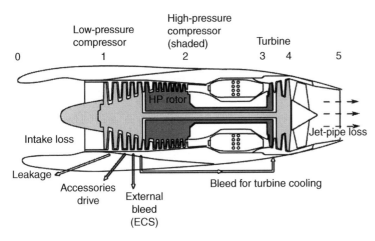

Figure 12.15 Installation effects.

The nacelle is the housing for the engine and interfaces with the aircraft; typically, it would be a pod mount for civil aircraft design. Aircraft with more than one engine would have the pod mounted on the wing and/or on fuselage. It was mentioned earlier that wing mounted nacelles are the best to relieve wing bending under in-flight load. Propeller-driven engine nacelles also have a similar consideration of podded nacelles, modified by the presence of a propeller (Section 12.7.2). Aircraft with one engine are aligned in the plane of aircraft symmetry (engines with propellers can have a small lateral inclination of a degree or two about the aircraft centreline to counter the slipstream and gyroscopic effects from a rotating propeller). Position of the nacelles with respect to the aircraft and their shaping to reduce drag are the important considerations. Military aircraft have engines buried into the fuselage so they do not have nacelles unless designers choose to have pods (some older designs). Military aircraft designers are only to consider intake design as discussed in Section 12.7.3.

12.7.1 Subsonic Civil Aircraft Nacelle and Engine Installation

The nacelle is a multi-functional system comprising of (i) inlet, (ii) exhaust nozzle, (iii) TR, if required and (iv) noise suppression system. The design aim of the nacelle is to minimise associated drag, noise and provide airflow smoothly to the engine at all flight conditions.

As of today, except for *Concorde*, all civil aircraft are subsonic with a maximum speed less than Mach 0.98. All subsonic aircraft use some form of pod mounted nacelle, in a way this has become generic in design. Figure 12.16 shows a turbofan installed in a civil aircraft nacelle pod. The under-wing nacelle is the current best practice but for smaller aircraft, ground clearance issues force a nacelle to be fuselage mounted. An over-wing nacelle like that of VFW614 is a possibility yet to be explored properly (Honda jet aircraft have reintroduced it).

There are two types of podded nacelles. Figure 12.17a shows a long duct nacelle where both the primary and secondary flows mix within it. The mixing increases the thrust and reduces the noise level compared to the short duct nacelles, possibly compensating the weight gain through fuel and cost saving. Figure 12.17b shows short duct nacelle in which the bypassed cold flow does not mix with the hot core flow. The advantage is that there is considerable weight saving by cutting down the length of the outside casing of the nacelle by not extending up to the end. Short duct nacelle length can vary.

The aircraft performance engineers are to substantiate to the certification authorities that the thrust available from the engine after deducting the losses is sufficient to cater for the full flight envelope as specified. In hot and high altitude conditions, it can become critical at aircraft takeoff if (i) the runway is not sufficiently long and/or at high altitude, (ii) if there is an obstruction to clear or (iii) ambient temperature is high. In that case, the aircraft may takeoff with lighter weight. Airworthiness requirements require that the aircraft has to maintain a minimum gradient (Chapter 15) at takeoff with the critical one engine inoperative. Customer requirements could demand more than the minimum requirements.

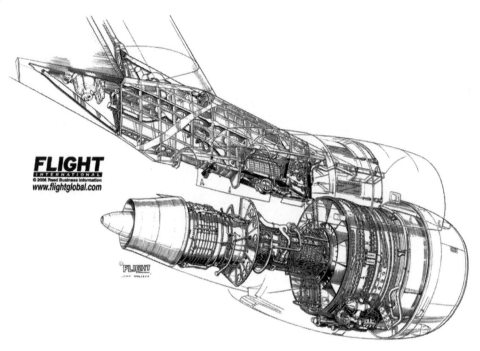

Figure 12.16 Installed turbofan on an aircraft (Credit: Flightglobal). Source: Reproduced with permission from Flightglobal.

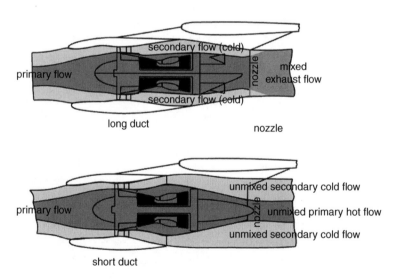

Figure 12.17 Podded nacelle types (courtesy of Bombardier-Aerospace Shorts).

12.7.2 Turboprop Integration to Aircraft

The turboprop nacelle is also a multi-functional system comprising of (i) inlet, (ii) exhaust nozzle and (iii) noise suppression system. Thrust reversing can be achieved by changing the propeller pitch angle sufficiently. There are primarily two types of turboprop nacelles as shown in Figure 12.18. The scoop intake could be above or below (as a chin) the propeller spinner. Interestingly, quite a few turboprop nacelles have an integrated undercarriage mount with storage space in the same nacelle housing as can be seen in Figure 12.18a. The other

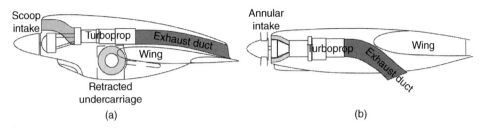

Figure 12.18 Typical wing mounted turboprop installation [9]. (a) Scoop intake. (b) Annular Intake. Source: Reproduced with permission from Flightglobal.

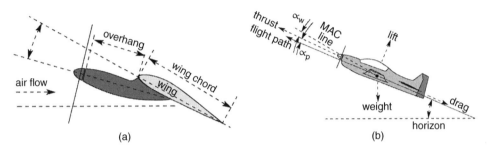

Figure 12.19 Typical turboprop installation parameters. (a) Wing mounted turboprop (note overhang). (b) Fuselage mounted turboprop.

kind is with annular intake as shown in Figure 12.18b. Installation losses are of the same order as discussed in the turbofan installation.

Turboprop nacelle position is dictated by the propeller diameter. The key geometric parameters for turboprop installation are shown in Figure 12.19.

Figure 12.19a shows wing mounted turboprop installation. Overhang should be as far forward the design can take, like that of turbofan overhang to reduce interference drag. For high wing aircraft, the turboprop nacelle is generally under the wing, like the Bombardier Q400 aircraft. For low wing aircraft, nacelles are generally over the wing to get propeller ground clearance. Both types can house the undercarriage. The propeller slipstream assists lift and has a strong effect on static stability, flap deployment aggravates the stability changes. Depending on the extent of wing incidence with respect to the fuselage, there is some angle between wing chord line and the thrust line – typically from 2 to 5°.

A fuselage mounted propeller-driven system arrangement is shown in Figure 12.19b. Note the angle between the thrust line and the wing chord line as it is with wing mounted propeller nacelle. Sometimes, propeller axis is given about a degree down inclination with respect to fuselage axis. This assists longitudinal stability. To counter the propeller slipstream effect, an inclination of a degree or two in the yaw direction can be given. Otherwise, the V-tail can be given such inclination to counter the effect.

Piston engine nacelles on the wing follow the same logic. The older designs had more closely coupled installations.

12.7.3 Combat Aircraft Engine Installation

A combat aircraft has engines integral with the fuselage, mostly buried inside but in some cases, with two engines, they can bulge out to the sides. Therefore, pods do not feature unless they are required for having more than two engines on a large aircraft. Figure 12.20 shows a turbofan installed on to a supersonic combat aircraft.

The external contour of the engine housing is integral to fuselage mouldlines. Early designs had the intake at the front of the aircraft; a pitot type for a subsonic fighter (Sabrejet F86 etc.) and with a movable centre body

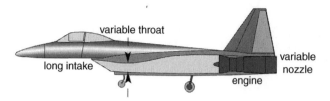

Figure 12.20 Installed engine in a combat aircraft [9].

for a supersonic fighter (MIG 21). The long intake duct snaking inside the fuselage below the pilot seat incurs high losses. Side intakes overtook nose intake designs. A possible choice for side intakes is given in Section 5.12 – primarily, they are side or chin mounted. A plate is kept above the fuselage boundary layer on which the intake is placed. A centre body is required for aircraft speed capability above Mach 1.8, otherwise it can be a pitot intake.

Aircraft performance engineers must be given the following data to generate the installed engine performance.

  i. Intake losses plus losses arising from supersonic shock waves and the duct length.
 ii. Exit nozzle losses. Military aircraft nozzle design is complex.
iii. Additional losses at the intake and nozzle on account of suppression of exhaust temperature for stealth.
 iv. Compressor air-bleed for the ECSs, for example, cabin air-conditioning and pressurisation, de-/anti-icing and any other purpose.
  v. Power off-takes from the engine shaft to drive the electric generator, accessories, for example, pumps and so on.

Military aircraft have excess thrust (with or without afterburner) to cater for hot and high-altitude conditions and operating from short airfields, and they are capable of climbing at steeper angles than civil aircraft. When thrust/weight ratio is more than 1, then the aircraft is able to climb up vertically.

12.8 Intake/Nozzle Design

Since the withdrawal of supersonic *Concorde*, there is currently no supersonic civil aircraft in operation. This section separates subsonic and supersonic intakes as civil and military operational applications, respectively.

12.8.1 Civil Aircraft Subsonic Intake Design

Engine mass flow demand varies a lot as shown in Figure 12.21. To size intake area, the reference cross-section of incoming air mass stream tube is taken at the maximum cruise condition as shown in Figure 12.21a, when it has cross-sectional area almost equal to that of highlight area (i.e. $A_\infty = A_1$). The ratio of MFR in relation with the reference condition (air mass flow at maximum cruise) is a measure of spread the intake would encounter. At the maximum cruise condition, MFR = 1 as a result of $A_\infty = A_1$ (critical operation). At normal cruise conditions the intake air mass demand is lower (MFR < 1 (sub-critical operation), shown in Figure 12.21b). At the maximum at takeoff rating (MFR > 1), intake air mass flow demand is high and the streamline patterns is shown in Figure 12.21c. Variation in intake mass flow demand is quite high.

If the takeoff airflow demand is high enough then a blow-in door can be provided, which closes automatically when the demand drops off. Figure 12.21d shows a typical flow pattern at incidence at high demand, when an automatic blow-in door may be necessary. At idle, the engine is kept running with very little trust generation (MFR « 1). At inoperative conditions, the rotor is kept windmilling to minimise drag. If the rotor seizes due to mechanical failure then there will be considerable drag rise.

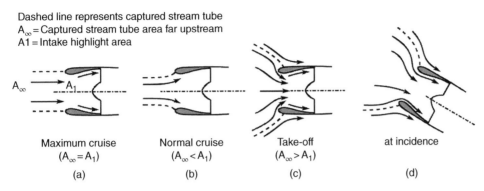

Dashed line represents captured stream tube
A_∞ = Captured stream tube area far upstream
$A1$ = Intake highlight area

Maximum cruise $(A_\infty = A_1)$	Normal cruise $(A_\infty < A_1)$	Take-off $(A_\infty > A_1)$	at incidence
(a)	(b)	(c)	(d)

Figure 12.21 Subsonic intake airflow demand (valid both for civil and military intakes). (a) MFR ratio = 1. (b) MFR ratio < 1. (c) MFR ratio > 1. (d) MFR ratio > 1 (at climb).

Currently, the engine (fan) face Mach number should not exceed Mach 0.5 to avoid degradation due to compressibility effects. At a fan face Mach number above 0.5, the relative velocity at the fan tip region approaches near sonic speed on account of high blade rotational speed.

The purpose of the intake is to serve engine airflow demand as smoothly as it is possible – there should be no flow distortion at the compressor face on account of any separation and/or flow asymmetry. Nacelle intake lip cross-section is designed with the similar logic behind designing the aerofoil leading edge cross-section – to ensure flow does not separate within the flight envelope. More detail on subsonic pod mounted intake design is given in [8–10].

12.8.2 Military Aircraft Supersonic Intake Design

Chart 5.1 described various types of supersonic inlet configurations shown in Section 5.16.1. The two main types are featured at the (i) fuselage nose with a centre body and (ii) fuselage, side mounted. In this section, their aerodynamic considerations are given. Military aircraft have supersonic capabilities and therefore have to manage shock losses (RF, Eq. (12.6) and Figure 12.22) associated with its intake. This can be done by designing supersonic intake oblique shock to develop in stages with a series of deflection angles (typically 5–10°, depending on the number of stages). Normally, three oblique shocks suffice (Figure 12.22) [6, 10]. An increase in number of deflection angles offers diminishing returns to improve RF. The best design is an Oswatitsch curved contour design, which generates infinite weak Mach waves with minimum loss, ideally in an isentropic process; that is, no loss. Oswatitsch intakes are used in hypersonic air-breathing engines.

12.8.2.1 Intake at Fuselage Nose with a Centre Body
Ideally, at design point M_{des}, (at supersonic cruise), the bow shock wave just attaches to the intake lip, which is sharp compared to a subsonic intake lip. Figure 12.23 gives four kinds of flow regime associated with supersonic intake with a fixed centre body. Four types of operating situation arise as follows:

i. At the design flight speed, seen as critical operation, when it is desirable that the oblique shock wave just touches the lip followed by other shocks culminating with a normal shock at the throat beyond which airflow becomes subsonic (Figure 12.23a). The captured free stream tube is same as the highlight area. Intake area is sized to inhale air mass at critical operation.

ii. If the back pressure is lower than the critical operation (on account of throttle action) then more air mass is inhaled and the throat area remains supersonic, pushing back a stronger normal shock. This is known as supercritical operation (Figure 12.23b). The captured free stream tube is still the same as the highlight area and oblique shock position depends on the aircraft speed.

iii. When the back pressure is high, especially when the aircraft is below critical speed, then less air is inhaled, the aircraft is still at supersonic speed, the oblique shock is wider and is followed by a normal shock that

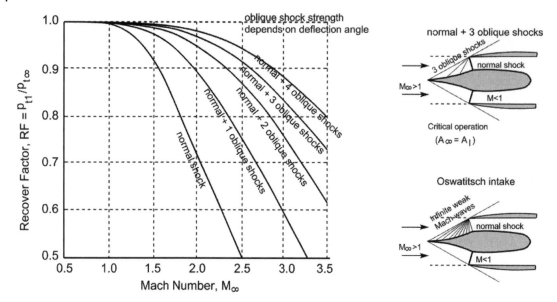

Figure 12.22 Shock Recovery Factor, RF.

pops outside ahead of the nacelle lip and is known as the sub-critical operation (Figure 12.23c). The captured free stream tube is smaller than the highlight area. It is not as efficient as the critical operation but loss is less than supercritical operation.

iv. The last one could happen with a particular combination of aircraft speed (below critical) and air mass inhalation when the normal shock outside keeps oscillating, which starves the engine and it runs erratically, possibly leading to flameout. This is known 'buzz' (Figure 12.23d).

For aircraft operating above Mach 2, a movable centre body keeps the oblique wave at the lip as the shock angle changes with speed change. The simplest centre body is a cone (or half cone for side mounted nacelles). The cone could be stepped to make multiple oblique shocks at reduced intensity; that is, lowering shock loss. A movable centre body (Figure 5.34, MIG21 showing extended and retracted centre body positions) can keep oblique shock at the intake lip and can manage 'buzz' better. Aircraft must avoid buzz occurring and modern designs have a FBW/FADEC to keep clear of it.

12.8.2.2 Intake at the Fuselage Side

Current intake position is at fuselage side or at underbelly, reducing the long intake duct of nose intake by about 25–35%, reducing duct loss and duct weight. For fuselage mounted single engine, a small bend of intake duct is required to have the bifurcated duct to meet before it reaches engine face. For twin engine aircraft, relatively straighter separate ducts offer better design with less duct loss, thus offering better RF. A chin mounted intake escapes duct bifurcation.

Figure 12.24 shows a typical side mounted supersonic intake similar to what an F14 Tomcat fighter aircraft has. It has a rectangular intake seen as a 2D intake, but not exactly. The *Concorde* underwing supersonic intake has similar design considerations.

The supersonic nacelle, as shown in Figure 12.24, is a complex modern design. It dispenses with the older practices of having a centre body by movable doors operated by separately managed actuators. For mass flow demand at low speed operation, especially, at full throttle takeoff rating, the throat opening is kept at maximum. Flap doors are adjusted to reduce throat area with speed increase. At maximum speeds over Mach 2.0, the two stage doors have stepped deflection to develop a total of three weaker oblique shocks, followed by a normal shock, making airflow diffuse to around Mach 0.5 at the engine face. At the top, there is a bleed door

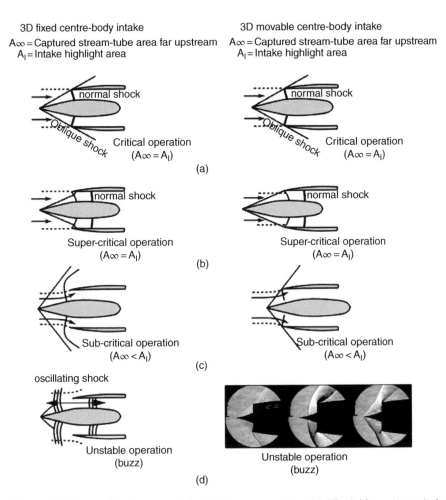

Figure 12.23 Types of ideal supersonic intake demand conditions [6]. (The Schlieren image is that of the buzz phenomenon, the left image has shock inside, then progressing to emerge to its extreme position at the right image and cycles in about 15 Hz [6].)

to serve both as boundary layer bleed as well as a dumping excess intake air mass flow. Normally, these are integrated with FBW/FADEC and activate automatically.

Modern combat aircraft with advance missiles have beyond visual range (BVR) capabilities. In an air-superiority role, rapid manoeuvre at high speed is the combat specification for which a typical maximum speed is of the order of Mach 1.8 ($\approx$F16 variant) when the requirement for a movable centre body/flap doors is not stringent. For side intakes, boundary layer bleed plates serve as the centre body to position oblique shock at the lip (critical design point operation). A Diverterless Supersonic Inlet (DSI – Figure 5.33) eliminates the need for a splitter plate, while compressing the air to slow it down from supersonic to subsonic speeds.

12.9 Exhaust Nozzle and Thrust Reverser (TR)

TRs are not required by the regulatory authorities (FAA: Federal Aviation Administration/CAA: Civil Aviation Authority). These are expensive components, heavy and only applied on ground, yet their impact on aircraft operation is significant on account of having additional safety through better control, reduced time to stop

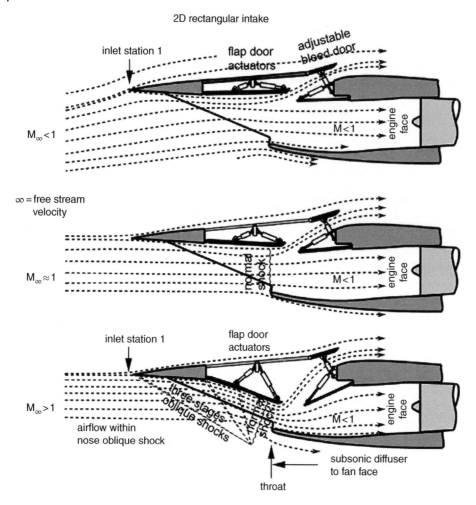

Figure 12.24 Types of ideal supersonic intake demand mechanism [6].

and so on, especially at aborted take-offs and other related emergencies. Airlines want to have TRs even at the cost of increased direct operating cost (DOC).

A general description of TR application in civil aviation is given Section 5.13.1. The TR is part of exhaust nozzle and they are treated together in this section. While this section offers an empirical sizing method for the nozzle, it will not size or design a TR, which is a state of technology in itself. To tackle exhaust nozzles, it is desirable to know about the TR beforehand.

The role of the TR is to retard the aircraft speed by applying thrust in the forward direction, that is, in a reversed application. The rapid retardation by TR application reduces the landing field length. In civil applications it is only applied on the ground. Because of its severity, in-flight application is not permitted. It reduces the wheel brake load hence there is less wear and fewer heating up hazards. The TR is very effective on slippery runways (ice, water etc.) when braking is less effective. A typical example of the benefits of stopping distance when landing on an icy runway with TR application is that it cuts down the field length by less than half. A mid-sized jet transport aircraft would stop at about 4000 ft with a TR, which would be 12 000 ft without it. Without a TR, energy depleted by stopping the aircraft is absorbed by the wheel brake and aerodynamic drag. Application of TR also offers additional intake momentum drag (at full throttle) contributing to energy depletion. The TR is useful to make an aircraft go backwards (e.g. C17) on the ground for parking, alignments and

Figure 12.25 Supersonic nozzle area adjustment and thrust vectoring.

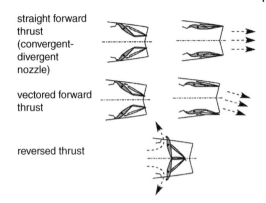

straight forward
thrust
(convergent-
divergent
nozzle)

vectored forward
thrust

reversed thrust

so on – most aircraft with a TR do not use it to make the aircraft go backwards but use a specialised vehicle to push it away.

The TR is integrated on the nacelle and it is the obligation of aircraft manufacturer to design it [12]. Sometimes this is subcontracted to specialist organisations devoted to TR design. Basically, there are two types of civil aircraft TR design, for example, (i) operating on both fan and core flow and (ii) operating on fan flow only. Their choice depends on BPR, nacelle location and customer specification. The next subsection is devoted to introducing the TR.

12.9.1   Civil Aircraft Exhaust Nozzles

Civil aircraft nozzles on which the TR is integrated are conical. Small turbofan aircraft may not need a TR but aircraft of RJ (regional jet) size and above use one. Inclusion of a TR may slightly elongate nozzle length – this will be ignored in this book.

In general, nozzle exit area is sized as a perfectly expanded nozzle ($p_e = p_\infty$) at the LRC condition. At higher engine ratings it has $p_e > p_\infty$. The exit nozzle of a long duct turbofan does not run choked at cruise ratings. At takeoff ratings, the back pressure is high at lower altitude, therefore a long duct turbofan could escape from running choked (low-pressure secondary flow mixes within the exhaust duct). The exhaust nozzle runs in a favourable pressure gradient, hence its shaping means it is relatively simpler to establish its geometrical dimensions. However, this is not simple engineering it for elevated temperature and to suppress noise level.

The nozzle exit plane is at the end of engine. Its length from the turbine exit plane is about 0.8–1.5 fan face diameter. Nozzle exit area diameter may be taken to be coarsely as half to three-quarters of the intake throat diameter in this study.

12.9.2   Military Aircraft TR Application and Exhaust Nozzles

An afterburning military engine TR is integral with the nozzle design and is positioned at the fuselage end. Afterburning engine nozzles always runs choked at the maximum rating and have a variable convergent-divergent (de Laval) nozzle to match with the throttle demands. Of late, all the latest combat aircraft have thrust vectoring capabilities by deflecting the exhaust jet at the desired angles, affecting aircraft manoeuvrability capabilities.

The full extent of TR deployment is shown in Figure 12.25 (the last figure). This has an integral mechanism capable of adjustment for all demands. The diagram shows that the integral mechanism can even provide a mild form of in-flight thrust reversing to spoil thrust (air-braking action) to wash out high speed at its approach to land or in combat to a considerably lower speed.

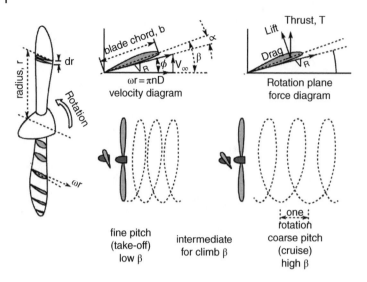

Figure 12.26 Aircraft propeller.

12.10 Propeller

Aircraft flying at speeds less than Mach 0.5 are propeller driven, larger aircraft powered by gas turbines and smaller ones by piston engines. More advanced turboprops have pushed flight speed to over Mach 0.7 (Airbus A400). This book deals with the conventional types of propellers operating at flight speed Mach ≤ 0.7. After a brief introduction to the basics of propeller theory, this section concentrates on the engineering aspects of what is required by the aircraft designers. References [13–17] may be consulted for more details. It is recommended to use certified propellers manufactured by well-known companies.

Propellers are twisted wing like blades that rotate in a plane normal to the aircraft (flight path). They have aerofoil sections that vary from being thickest at the root and to thinnest at the tip chord (Figure 12.26). Thrust generated by propeller is the *lift* produced by the propeller blades in the flight direction. It acts as a propulsive force and is not meant to lift weight unless the thrust line is vectored. In rotation, the tip experiences the highest tangential velocity. The figure shows associated geometries and symbols used in analysis. The three important angles are blade pitch angle, β, angle subtended by the relative velocity, φ, and angle of attack, $\alpha = (\beta - \varphi)$. Also shown is the effect of coarse and fine blade pitch, β. The definition of propeller pitch, p, is given in Section 12.10.1.

The propeller types are shown in Figure 12.27. There can be anywhere from two blades to as many as seven/eight. Smaller aircraft have two to three blades while bigger ones can have anywhere from four to seven/eight. When the propeller is placed in front of the aircraft, it is called a *tractor* and when placed at aft, it is called a *pusher*. The majority of the propellers serve as tractors.

12.10.1 Propeller-Related Definitions

Industry uses propeller charts, those incorporate some special terminologies. Definitions of the necessary terminologies/parameters are given here:

D = Propeller diameter = $2 \times r$
n = Revolution per second (rps)
ω = Angular velocity
N = Number of blades
b = Propeller blade width (varies with radius, r)
P = Propeller power

C_p = Power coefficient (not to confuse with pressure coefficient) = $P/(\rho n^3 D^5)$

T = Propeller thrust

C_{Li} = Integrated design lift coefficient (C_{Ld} = sectional lift coefficient)

C_T = Propeller thrust coefficient = $T/(\rho n^2 D^4)$

β = Blade pitch angle subtended by the blade chord and its rotating plane (also known as geometric pitch)

p = Propeller pitch = no slip distance covered in one rotation = $2\pi r \tan\beta$ (explained previously)

V_R = Relative velocity to the blade element = $\sqrt{(V^2 + \omega^2 r^2)}$ (Blade Mach No: = V_R/a)

φ = Angle subtended by the relative velocity = $\tan^{-1}(V/2\pi nr)$ or $\tan\varphi = V/\pi nD$. It is the pitch angle of the propeller in flight and is not the same as blade pitch, which is independent of aircraft speed.

α = Angle of attack = $(\beta - \varphi)$

J = Advance ratio = $V/(nD) = \pi\tan\varphi$ (a non-dimensional quantity – analogous to α).

AF = Activity factor = $(10^5/16)\int_0^{1.0}(r/R)^3(b/D)d(r/R)$

TAF = Total activity factor = $N \times AF$ (N is number of blades – it gives an indication of power absorbed).

It is found that at around $0.7r$ (tapered propeller) to $0.75r$ (square propeller), the blades give the aerodynamically average value that can be applied uniformly over the entire radius to obtain the propeller performance.

Figure 12.26 shows a two-bladed propeller along with a blade elemental section, dr, at radius r. The propeller has diameter, D. If ω is the angular velocity then the blade element linear velocity at radius r is $\omega r = 2\pi nr = \pi nD$, where n is the number of revolution per unit time. An aircraft with true airspeed of V and with propeller angular velocity of ω, will have blade element moving in a helical path. At any radius, the relative velocity, V_R has angle $\varphi = \tan^{-1}(V/2\pi nr)$. At the tip $\varphi_{tip} = \tan^{-1}(V/\pi nD)$.

However, irrespective of aircraft speed, the inclination of blade angle from the rotating plane can be seen as solid-body screw thread inclination and is known as *pitch angle*, β. The solid-body screw like linear advancement through one rotation is called *propeller pitch, p*. The pitch definition has a problem as unlike mechanical screws, the choice of inclination plane is not standardised. It can be the zero-lift line, which is aerodynamically convenient or the chord line, which is easy to locate or the bottom surface – each plane will give a different pitch. All these planes are interrelated by fixed angles. This book takes the chord line as the reference line for the pitch as shown by blade pitch angle, β, in Figure 12.24. This gives propeller pitch, $p = 2\pi r \tan\beta$.

Since the blade linear velocity ωr varies with radius, its pitch angle needs to be varied as well to make the best use of the blade element aerofoil characteristics. When β is varied in such a way that the pitch is not changed along the radius then the blade has *constant pitch*. This means that β decreases with increase in r (variation in β is about $40°$ from root to tip).

Blade angle of attack,

$$\alpha = (\beta - \varphi) = \tan^{-1}(p/2\pi r) - \tan^{-1}(V/2\pi nr) \tag{12.22}$$

This brings out an analogous non-dimensional parameter, J = advance ratio = $V/(nD) = \pi\tan\varphi$.

Blade pitch should match with aircraft speed, V, to keep blade angle of attack α to produce best lift. To cope with aircraft speed change, it benefits if the blade is rotated (varying the pitch) about its axis through the hub to maintain favourable α at all speeds. It is then called a *variable pitch* propeller. Typically, for pitch variation, the propeller is kept at constant rpm with the help of a *governor*, when it is called a constant speed propeller. Almost all aircraft flying at higher speed will have constant speed variable pitch propeller (when done manually it is β-controlled). The smaller low speed aircraft are with fixed pitch that would run best at one combination of aircraft speed and propeller rpm. If the fixed pitch is meant for cruise then at takeoff (low aircraft speed and high propeller revolution) the propeller would be less efficient. Typically, aircraft designers would like to have fixed pitch propeller matched for the climb, a condition in between cruise and takeoff to minimised the difference between the two extremes. Obviously, for high speed performance it should match the high speed cruise condition. Figure 12.28 shows the benefit of constant speed variable pitch propeller over the speed range.

The β-control can extend to reversing of propeller pitch. A full reverse thrust acts as all the benefits of TR describe in Section 12.9. The pitch control can be made to 'fine-pitch' to produce zero thrust when the aircraft is static. This could assist aircraft to *wash out* speed especially at approach to land.

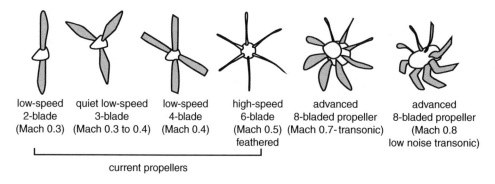

low-speed
2-blade
(Mach 0.3)

quiet low-speed
3-blade
(Mach 0.3 to 0.4)

low-speed
4-blade
(Mach 0.4)

high-speed
6-blade
(Mach 0.5)
feathered

advanced
8-bladed propeller
(Mach 0.7- transonic)

advanced
8-bladed propeller
(Mach 0.8
low noise transonic)

current propellers

Figure 12.27 Multi-bladed aircraft propellers (the low noise transonic propeller is taken from unpublished work by Dr R. Cooper at QUB).

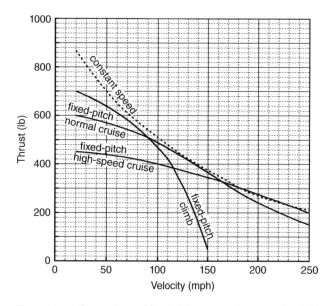

Figure 12.28 Comparison of fixed pitch and constant speed variable pitch propeller ($\approx$200 HP).

When the engine fails, that is the system senses insufficient power, the pilot or the automatic sensing device elects to *feather* the propeller (Figure 12.25). Feathering is changing β to 75–85° (maximum coarse) when propeller slows down to zero rpm – producing net drag/thrust (part of propeller has thrust and the rest drag) to zero.

Windmilling of propeller occurs when the engine has no power and is free to rotate driven by the relative airspeed to the propeller when aircraft is in flight. The β angle is in a fine position.

Installed turboprop thrust is obtained by reducing the thrust by the various loss factors. In addition fuselage factors of fuselage blockage factor f_b and excrescence factor f_h (see Section 11.16) are to be applied further reducing the thrust level.

12.11 Propeller Theory

The fundamentals of propeller performance start with the idealised consideration of momentum theory. Its practical application in industry is based on subsequent 'blade element' theory. Both are presented in this section, followed by the engineering consideration appropriate to the aircraft designers. Industrial practices still use manufacturer supplied propellers and wind tunnel tested generic charts/tables to evaluate performance. There are various forms of propeller charts; the three prevailing ones are (i) the Hamilton Standard [15] (propeller manufacturer) method, (ii) SBAC Standard Method of Performance Estimation [16] and (iii) the NACA (National Advisory Committee for Aeronautics) method [11]. This book uses the Hamilton Standard method used in industry [15]. For designing advanced propellers/propfans to operate at speeds greater than Mach 0.6, CFD is playing an important role to arrive at the best compromise, substantiated by wind tunnel tests. CFD employs more advanced theories, for example, the vortex theory.

12.11.1 Momentum Theory – Actuator Disc

The classical incompressible inviscid momentum theory provides the basis of propeller performance [13]. In the theory, the propeller is represented by a thin *actuator disc* of area, A, placed normal to free stream velocity, V_0. This captures a stream tube within a CV having a front surface sufficiently upstream represented by subscript '0' and at sufficiently downstream by subscript '3' (Figure 12.29). It is assumed that thrust is uniformly distributed over the disc and the tip effects are ignored. Whether the disc is rotating or not is redundant, as flow through it is taken without any rotation. Station numbers just in front and aft of the disc are designated 1 and 2.

The impulse given by the disc (propeller) increases the velocity from the free stream value of V_0, smoothly accelerating, to V_2 behind the disc and continuing to accelerate to V_3 (station 3) until the static pressure equals the ambient pressure p_0. The pressure and velocity distribution along the stream tube is shown in Figure 12.29. There is a jump of static pressure across the disc (from p_1 to p_2) but there is no jump in velocity change.

Newton's law gives that the rate of change of momentum is the applied force; in this case it is the thrust, T. Consider station 2 of the stream tube immediately behind the disc producing the thrust. It has MFR, $\dot{m} = \rho A_{disc} V_2$. The change of velocity is $\Delta V = (V_3 - V_0)$. This is the reactionary thrust experienced at disc through the pressure difference multiplied by its area, A.

Thrust produced by the disc

$$T = \text{pressure across the disc} \times A_{disc} = A_{disc} \times (p_2 - p_1) = \text{rate of change of momentum}$$

$$= \dot{m}\Delta V = \rho A_{disc} \times V_2 \times (V_3 - V_0) \tag{12.23}$$

Equating, Eq. (12.23) can be rewritten as

$$\rho \times (V_3 - V_0) \times V_2 = (p_2 - p_1) \tag{12.24}$$

Figure 12.29 CV showing the stream tube of the actuator disc.

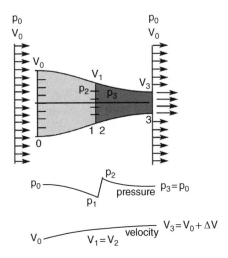

The incompressible flow Bernoulli's equation cannot be applied through the disc imparting energy. Instead, two equations are set up, one for conditions ahead of the disc and the other aft of it. Ambient pressure p_0 is the same everywhere.

Ahead of the disc,

$$p_0 + \tfrac{1}{2}\rho V_0^2 = p_1 + \tfrac{1}{2}\rho V_1^2 \tag{12.25}$$

Aft of the disc,

$$p_0 + \tfrac{1}{2}\rho V_3^2 = p_2 + \tfrac{1}{2}\rho V_2^2 \tag{12.26}$$

Subtracting the front relation from the aft relation,

$$\tfrac{1}{2}\rho(V_3^2 - V_0^2) = (p_2 - p_1) \times \tfrac{1}{2}\rho(V_2^2 - V_1^2) \tag{12.27}$$

Since there is no jump in velocity across the disc, the last term drops out.
Next substitute the value of $(p_2 - p_1)$ from Eq. (12.24) into Eq. (12.27).

$$1/2(V_3^2 - V_0^2) = (V_3 - V_0) \times V_2 \text{ or } (V_3 + V_0) = 2V_2 \tag{12.28}$$

Note that $(V_3 - V_0) = \Delta V$ when subtracted from Eq. (12.28), it gives $2V_3 = 2V_2 + \Delta V$.
Or

$$V_3 = V_2 + \Delta V/2; \text{ this implies that } V_1 = V_0 + \Delta V/2 \tag{12.29}$$

It says that half of the added velocity, $\Delta V/2$, is ahead the disc and the rest, $\Delta V/2$, is added aft of the disc. Using conservation of mass, $A_3 V_3 = A_1 V_1$, Eq. (12.23) becomes

$$T = \rho A_{\text{disc}} V_1 \times (V_3 - V_0) = A_{\text{disc}} (p_2 - p_1)$$

$$\text{or } (p_2 - p_1) = \rho V_1 \times (V_3 - V_0) \tag{12.30}$$

Using Eqs. (12.29) and (12.30), the thrust Eq. (12.23) can rewritten as

$$T = A_{\text{disc}} \rho V_1 \times (V_3 - V_0) = A_{\text{disc}} \rho (V_0 + \Delta V/2) \times \Delta V \tag{12.31}$$

Applying to aircraft, V_0 may be seen as aircraft velocity, V, by dropping the subscript '0'. Then useful work rate (power, P) done on the aircraft is

$$P = TV \tag{12.32}$$

For the ideal flow without the tip effects, the mechanical work produced in the system is the power, P_{ideal}, generated to drive the propeller is force (thrust, T) times velocity, V_1, at the disc.

$$P_{\text{ideal}} = T(V + \Delta V/2) \text{ (this is the maximum possible value in ideal situation)} \tag{12.33}$$

Therefore, ideal efficiency,

$$\eta_i = P/P_{\text{ideal}} = (TV)/[T(V + \Delta V/2)] = 1/[1 + (\Delta V/2V)] \tag{12.34}$$

The real effects will have viscous, propeller tip effects and other installation effects. In other words, to produce the same thrust, the system has to provide more power (in case of piston engine it is seen as Brake Horse Power (BHP) and in case of turboprop as ESHP, where ESHP is the equivalent shaft horse power that converts the residual thrust at the exhaust nozzle to HP, dividing by an empirical factor 2.5. The propulsive efficiency as given in Eq. (12.34) can be written as

$$\eta_p = (TV)/[BHP \text{ or } ESHP] \tag{12.35}$$

This gives,

$$\eta_p/\eta_i = \{(TV)/[BHP \text{ or } SHP]\}/\{1/[1 + (\Delta V/2V)]\}$$

$$= \{(TV)[1 + (\Delta V/2V)]/[BHP \text{ or } SHP]\} \approx 85\text{–}86\% (\text{typically}) \tag{12.36}$$

12.11.2   Blade Element Theory

Practical application of propellers is obtained through blade element theory as given next. Propeller blade cross-sectional profile has the same functions as that of wing aerofoil, that is, to operate at the best L/D.

Figure 12.26 shows that a blade elemental section, dr, at radius r, is valid for any number of blades at any radius, r. As blades are rotating elements, their properties would vary along the radius.

Figure 12.26 gives the velocity diagram showing an aircraft with a flight speed of V with its propeller rotating at n rps, which would make its blade element advance in a helical manner. V_R is the relative velocity to the blade with angle of attack α. Here, β is the propeller pitch angle as defined before. Strictly speaking, each blade rotates in the wake (downwash) of the previous blade, but the current treatment ignores this effect to use propeller charts without appreciable error. Figure 12.26 also gives the force diagram at the blade element in terms of lift, L, and drag, D, that is normal and parallel, respectively, to V_R. Then the thrust, dT, and force, dF (producing torque) on the blade element can be easily obtained, by decomposing lift and drag in the direction of flight and in the plane of propeller rotation, respectively. Integrating over the entire blade length (non-dimensionalised as r/R – an advantage to be applicable to different sizes) would give the thrust, T, and torque producing force, F, by the blade. The root of the hub (with spinner or not) does not produce thrust and integration is normally carried out from 0.2 to the tip, 1.0, in terms of r/R. When multiplied by the number of blades, N, it gives the propeller performance.

Therefore,

$$\text{Propeller thrust, } \boldsymbol{T} = N \times \int_{0.2}^{1.0} dT d(r/R) \tag{12.37}$$

And

$$\text{The force that produces } Torque, \boldsymbol{F} = N \times \int_{0.2}^{1.0} dF d(r/R) \tag{12.38}$$

From the definition, *Advance ratio,* $\boldsymbol{J} = V/(nD)$.

It can be shown that the thrust to power ratio is best when the blade element works at the highest lift to drag ratio (L/D_{max}). It is clear that a fixed pitch blade works best at a particular aircraft speed for the given power rating (rpm) – typically the climb condition is matched for the compromise. It is for this reason, constant speed variable pitch propellers have better performance over a wider aircraft speed range. It is convenient to express thrust and torque in non-dimensional form as given next. From dimensional analysis (note that the denominator does not have the ½):

$$\text{Non} - \text{dimensional thrust, } T_C = \text{Thrust}/(\rho V^2 D^2)$$

$$\text{Thrust coefficient, } \boldsymbol{C_T} = T_C \times J^2 = \text{Thrust} \times [V/(nD)]^2/(\rho V^2 D^2) = \text{Thrust}/(\rho n^2 D^4) \tag{12.39}$$

In the foot–pound system (FPS),

$$C_T = 0.1518 \times \left[\frac{(T/1000)}{\sigma \times (N/1000)^2 \times (D/10)^4} \right] \tag{12.40}$$

where σ = ambient density ratio for altitude performance

$$\text{Non} - \text{dimensional force (for torque), } \boldsymbol{T_F} = F/(\rho V^2 D^2)$$

$$\text{Force coefficient, } \boldsymbol{C_F} = T_F \times J^2 = F \times [V/(nD)]^2/(\rho V^2 D^2) = F/(\rho n^2 D^4) \tag{12.41}$$

$$\text{Torque, } Q = \text{Force} \times \text{Distance} = Fr = C_F \times (\rho n^2 D^4) \times D/2 = (C_F/2) \times (\rho n^2 D^5)$$

$$\text{Torque coefficient, } \boldsymbol{C_Q} = Q/(\rho n^2 D^5) = C_F/2 \tag{12.42}$$

$$\text{Power consumed, } \boldsymbol{P} = 2\pi n \times Q$$

$$\text{Power coefficient, } \boldsymbol{C_P} = P/(\rho n^3 D^5) = 2\pi n \times Q/(\rho n^3 D^5) = 2\pi C_Q = \pi C_F \tag{12.43}$$

In FPS system,

$$C_P = 0.5 \times \left[\frac{(BHP/1000)}{\sigma \times (N/1000)^3 \times (D/10)^5} \right] = \left[\frac{(237.8 \times SHP)}{2000 \times (6/100)^3 \rho n^3 D^5} \right] = \left[\frac{(550 \times SHP)}{\rho n^3 D^5} \right] \quad (12.44)$$

where N stands for r.p.m and n stands for r.p.s. The wider is the blade the higher would be the power absorbed to a point when any further increase would offer diminishing returns in increasing thrust.

A non-dimensional number defined as

Activity Factor, $\boldsymbol{AF} = (10^5/16) \int_0^{1.0} (r/R)^3 (b/D) d(r/R)$ expresses the integrated capacity of the blade element to absorb power. It indicates that increase of blade width outwardly is more effective than at the hub direction.

Total Activity Factor, $\boldsymbol{TAF} = N \times (10^5/16) \int_0^{1.0} (r/R)^3 (b/D) d(r/R)$ expresses the integrated capacity of the total number of blades of a propeller to absorb power.

A piston engine or a gas turbine would drive the propeller. Propulsive efficiency, η_p, can be computed by using Eqs. (12.35), (12.39) and (12.44).

Propulsive efficiency

$$\eta_p = (TV)/[BHP \text{ or } ESHP] = [C_T \times (\rho n^2 D^4) \times V]/[C_P \times (\rho n^3 D^5)]$$
$$= (C_T/C_P) \times [V/(nD)] = (C_T/C_P) \times J \quad (12.45)$$

12.12 Propeller Performance – Use of Charts, Practical Engineering Applications

The theory determines that geometrically similar propellers can be represented in a single non-dimensional chart (propeller graph) from combining the previous non-dimensional parameters such as shown in Figures 12.30 and 12.31 for three bladed propellers and Figures 12.32 and 12.33 for four bladed propellers. A considerable amount of classroom work can be conducted with these graphs. All these graphs and the procedure to estimate propeller performance are taken from [15] courtesy of Hamilton Standard, who kindly permitted use of the graphs in this book. All Hamilton Standard graphs are re-plotted retaining the maximum

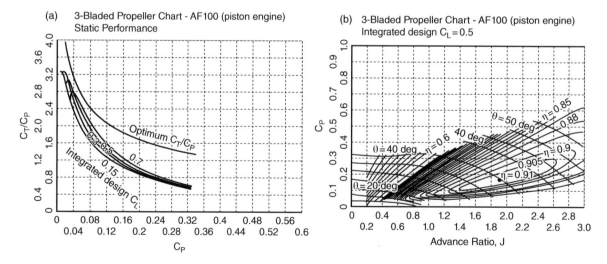

Figure 12.30 Static performance – Three bladed propeller chart – AF100 (for piston engines). (a) Static performance. (b) Propeller performance. (Courtesy of Hamilton Standard – retraced maintaining high fidelity). Source: Reproduced with permission from Cambridge University Press.

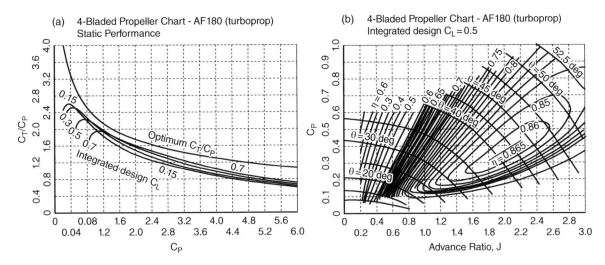

Figure 12.31 Four bladed propeller performance chart, AF180 (for high performance turboprop). (a) Static performance. (b) Propeller performance. (Courtesy of Hamilton Standard – retraced maintaining high fidelity). Source: Reproduced with permission from Cambridge University Press.

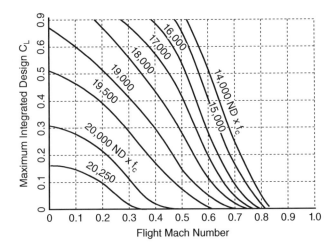

Figure 12.32 Limits of the integrated design C_L to avoid compressibility loss. (Courtesy of Hamilton Standard – retraced maintaining high fidelity). Source: Reproduced with permission from Cambridge University Press.

fidelity. The reference the gives full range of graphs for other types and has charts for propellers with a higher *AF*. The charts use the following relations.

Static computation presents a problem, as when *V* is zero, then $\eta_p = 0$. Different sets of graphs are required to obtain take the values of (C_T/C_P) to compute takeoff thrust as given in Figures 12.30 and 12.31. Finally, Figure 12.32 is meant for selecting the design C_L for the propeller to avoid compressibility loss. Thrust for take-off performance can be obtained from the following equations (FPS): compressibility loss. Thrust for takeoff performance can be obtained from the following equations in FPS:

In flight, thrust:

$$T = (550 \times \text{BHP} \times \eta_p)/V, \text{ where V is in ft s}^{-1}$$

$$= (375 \times \text{BHP} \times \eta_p)/V, \text{ where V is in mph} \tag{12.46}$$

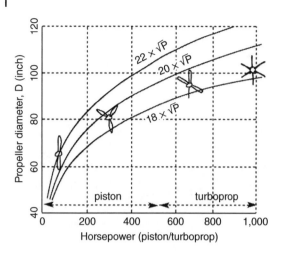

Figure 12.33 Engine power versus propeller diameter (extracted from [13]). Source: Reproduced with permission from Cambridge University Press.

For static performance (takeoff):

$$T_{TO} = [(C_T/C_P) \times (550 \times \text{BHP})]/(nD) \tag{12.47}$$

The book does not deal with propeller design. Aircraft designers are to select a propeller offered by the propeller manufacturer, mostly an off-the-shelf type, unless specially designed ones are used in consultation with aircraft designers. This section gives the considerations necessary for the aircraft designers to select the appropriate propeller to match the size of engine to produce thrust for the full flight envelope.

The readers may note that the propeller charts for the number of blades only use three variables C_p, β and η (subscript p is dropped). It does not specify propeller diameter and rpm. Therefore, similar propellers with same AF and C_{Li} can use the same chart. Aircraft designers will have to choose AF or C_{Li} based on the critical phase of operation. A propeller selection would seek some compromises as optimised performance for full flight envelope is not possible, especially for fixed pitch propellers.

Of late, certification requirements for noise have affected the issues on compromise, especially for the high performance propeller design. High tip Mach number is detrimental to noise and to reduce it, η is compromised by reducing rpm and/or diameter, hence increased J and/or increased number of blades. Increasing the number of blades would increase cost and weight. Propeller curvature suits transonic operation and helps reduce noise.

Equation (12.22) gives the aerodynamic incidence, that is, the blade angle of attack, $\alpha = (\beta - \varphi)$, where φ is determined from aircraft speed and propeller *rpm*, that is, a function of $J = V/nD$. It is desirable to keep α constant along the blade radius to get the best C_{Li}, that is, α is maintained at around 6–8°. The values at $0.7r$ or $0.75r$ being used as the reference point – the propeller chart mentions reference radius.

The combination of the designed propeller rpm is matched with its diameter to keep the operation below experiencing compressibility effects at maximum speed at an altitude. A suitable reduction in gear ratio brings the engine rpm down to the desirable propeller rpm. Figure 12.32 is used to obtain the integrated design C_L for the combination of the propeller rpm and diameter. The factor $ND \times$ (ratio of speed of sound at STD day sea level to the altitude) would establish the integrated design C_L. A spinner at the propeller root is recommended to reduce loss. Typically, majority of production propellers have integrated design C_L within the range of 0.35–0.6. Figure 12.32 gives the limits of integrated design C_L, which should not be exceeded.

The following step-by-step observations/information could prove important to progress propeller performance estimation by using the propeller charts given in Figures 12.30–12.32 (Hamilton Standards).

  i. In this book, the integrated design C_L is taken 0.5. Checking of the integrated design C_{Li} using Figure 12.32 is not used.

  ii. Typical blade activity factor, AF is of the following order:
Low power absorption two to three bladed propellers for home-built flying $= 80 < AF < 90$.

Medium power absorption three to four bladed propellers for piston engines (utility) $= 100 < AF < 120$.
High power absorption four bladed and more propellers for turboprops $= 140 < AF < 200$.

iii. Keep tip Mach number around 0.85 at cruise and make sure that at takeoff rpm it does exceed the value at the second segment climb speed.
iv. Typically, for constant speed variable pitch propeller, β is kept low for takeoff, gradually increasing at climb speed, reaching intermediate values at cruise and a high value at the maximum speed. Automatic blade control mechanism design involves specialised engineering.
v. Propeller diameter in inches can be coarsely determined by an empirical relation $D = K(P)^{0.25}$ where $K = 22$ for two blades, 20 for three blades and 18 for four blades. Power P is the installed power, which is less than the bare engine rating supplied by the engine manufacturer. Figure 12.33 gives the statistics of typical relationship between engine power and propeller diameter. It is a useful graph to deduce the initial size of the propeller empirically.

If n and J are known beforehand, then propeller diameter can be determined using $D = 1056\, V/(NJ)$ in the FPS system.

vi. Maintain at least about 0.5 m (1.6 ft) propeller tip clearance from the ground, for extreme demands this can be slightly reduced. This should take care of nose wheel tyre burst and undercarriage oleo collapse.
vii. At maximum takeoff static power, thrust developed by propellers is about four times the power.

Continue (in FPS) with propeller performance for static takeoff and in-flight cruise separately.

Static performance (Figures 12.30a and 12.31a)

i. Compute the power coefficient, $C_P = (550 \times SHP)/(\rho n^3 D^5)$, where n is in rps. Note ρ for f_c.
ii. From propeller chart find C_T/C_P.
iii. Compute static thrust, $T_S = (C_T/C_P)(33\,000 \times SHP)/ND$, where N is in rpm.

In-flight performance (Figures 12.30b and 12.31b)

i. Compute the advance ratio, $J = V/(nD)$
ii. Compute power coefficient, $C_P = (550 \times SHP)/(\rho n^3 D^5)$, where n is in rps.
iii. From propeller chart find efficiency, η_P.
iv. Compute thrust, T from $\eta_P = (TV)/(550 \times SHP)$, where V is in ft s^{-1}.

If necessary, off-the-shelf propeller blade tips could be slightly chopped off to meet geometrical constraints. Typical penalties are 1% reduction of diameter would affect 0.65% reduction in thrust – for small changes, linear interpolation may be made.

In addition, the fuselage factors blockage factor f_b and excrescence factor f_h (see Section 12.15) should be applied further reducing the thrust level.

12.12.1 Propeller Performance – Blade Numbers $3 \leq N \geq 4$

Using these graphs, linear extrapolation could be made for two and five bladed propellers with similar AF and so on. Reference [15] treats the subject in detail with propeller charts for other AF values.

References

1 Rolls Royce (1996). *The Jet Engine*, 5e. Rolls Royce Plc.
2 Kundu, A.K. (2010). *Aircraft Design*. Cambridge University Press.
3 Saravanamuttoo, H.I.H., Rogers, G.F.C., Cohen, H., and Straznicky, P.V. (2008). *Gas Turbine Theory*. Pearson.
4 Kundu, A.K., Price, M.A., and Riordan, D. (2016). *Theory and Practice of Aircraft Performance*, ISBN: 9871119074175. Wiley.

5 Mattingly, J.D. (1996). *Elements of Gas Turbine Propulsion*. McGraw-Hill.

6 Chima, R.V. (2012). *Analysis of Buzz in a Supersonic Inlet, NASA/TM—2012-217612*. Cleveland, OH: Glenn Research Centre.

7 Oates, G. (1984). *Aerothermodynamics of Gas Turbines and Rocket Propulsion*. AIAA.

8 Swavely, C.E. (1985). *Propulsion System Overview, Pratt and Whitney Short Course*. Pratt and Whitney.

9 Farokhi, S. (1991). *Propulsion System Integration Course*. University of Kansas.

10 Seddon, J. and Goldsmith, E. (1985). *Practical Intake Aerodynamic Design*. AIAA.

11 Gilman, G. (1953). *Propeller Performance Charts for Transport Airplane*. NACA Technical Note 2966, Langley Aeronautical Laboratory, Langley Field, VA.

12 Riordan, D., *Class Lecture Notes at QUB*, 2003.

13 Glauert, H. (1926). *Elements of Aerofoil and Airscerw Theory*. Cambridge University Press.

14 Stinton, D. (1983). *The Design of the Airplane*. BSP Professional Books.

15 Hamilton Standard (1963). *Propeller Performance Mannual*. Hamilton Standard.

16 Society of British Aircraft Constructors (1930). *SBAC Standard Method of Performance Estimation*. London: SBAC.

17 Lan, C.T.E. and Roskam, J. (1981). *Aeroplane Aerodynamics and Performance*. DAR Corporation.

13

Aircraft Power Plant Performance

13.1 Overview

Chapter 12 introduced various types of engines and associated theories. It also briefly discussed the topics associated with the installation effects of an engine to an aircraft. This chapter covers both uninstalled and installed engine performances. The aircraft performance engineers require installed thrust and fuel flow data to estimate aircraft performance. In most industries, aircraft performance engineers generate the installed engine performance using an engine manufacturer supplied computer program amenable to input of the various losses arising from taking engine bleed/off-takes for aircraft systems and intake/nozzle losses.

Uninstalled gas turbine engine performances in a non-dimensional form for some types are given in Sections 13.3–13.5. From the non-dimensional engine data matched engine installed performances are generated. The following types of engine are given in a non-dimensional form (except the piston engine performance). These are validated against industry standard data.

Turbofan engine of BPR ≤ 4 (civil engine of the class $T_{SLS} \approx 3500$ lb)
Turbofan engine of BPR ≥ 4 (civil engine of the class $T_{SLS} \approx 40\,000$ lb)
Medium sized turbofan engine of BPR ≤ 1 (military engine of the class $T_{SLS\_dry} \approx 6000$ lb)
Turboprop engine of the class of 1000 shaft horse power (SHP) at sea level static condition
180 HP Lycoming piston engine

Industry standard engine performance data for classroom usage are not easy to obtain – these are proprietary information. Readers need to be careful in applying engine data – any error in engine data could degrade/upgrade the aircraft performance and corrupt the design. Verification/substantiation of aircraft design is carried out through flight tests of performances. It becomes difficult to locate the source of any discrepancy between predicted and tested performance, whether the discrepancy stems from the aircraft or from the engine or from both. Appropriate engine data may be obtained beyond what is given in this book. It was mentioned earlier that the US contribution to aeronautics is indispensable and their data are generated in units of the foot–pound system (FPS, also known as the Imperial system). Many of the data and worked-out examples are kept in FPS. An extensive list of conversion factors is given in Appendix A.

This chapter includes the following sections:

Section 13.2 Introduction
Section 13.3 Uninstalled Engine Performance
Section 13.4 Installed Engine Performance
Section 13.5 Turboprop Propeller Performance
Section 13.6 Piston Engine
Section 13.7 Engine Performance Grid
Section 13.8 Some Turbofan Data

Conceptual Aircraft Design: An Industrial Approach, First Edition. Ajoy Kumar Kundu, Mark A. Price and David Riordan.
© 2019 John Wiley & Sons Ltd. Published 2019 by John Wiley & Sons Ltd.

Classwork content. This is an important chapter for creating engine performance graphs that will be used for aircraft performance estimation. Readers are to generate thrust and fuel flow levels for the matched engine at various power settings, speed and altitude in standard atmosphere for their projects.

13.2 Introduction

Chapter 12 mentioned that post World War II (WWII) research led to the rapid advancement of gas turbine development to a point when from a core gas generator module, a family of engines can be made in a modular concept (Figure 12.1) that allows engine designers to offer engines as specified by the aircraft designers. Similarity laws in thermodynamic design parameters permit power plants to be scaled (rubberised) to the requisite size around the core gas generator module to meet the demands of the mission requirements.

Gas turbine engine operates at various power demand; they are rated in a band of power settings, that is, in ratings (in this book in terms of engine rpm), maximum at takeoff, then at the next level at climb, then at cruise demand and finally at descent at part throttle to idle power level. In the following, the various power ratings are explained.

13.2.1 Engine Performance Ratings

All power plants have prescribed power ratings as given here. Power settings are decided by the engine rpm and/or by the engine pressure ratio (EPR) at the jet pipe temperature (JPT) that should remain below prescribed levels. Engine life and 'Time Between Overhaul' (TBO) depends how hot and how long it has run for. An engine is identified at its sea level standard day at takeoff at static conditions. For turbofans it is abbreviated to T_{SLS}. The following are the typical power settings (ratings) of engines.

13.2.1.1 Takeoff Rating

This is the highest rating producing the maximum power and is rated at 100% (on static run of turbofans, it is the T_{SLS}). At this rating, the engine runs the hottest and is therefore time limited for longer life and maintenance. For civil engines, this is about 10 minutes and military ones could extend to a little longer. A situation may arise at the one engine inoperative case where the operating engine is subjected to supply more power at the Augmented Power Rating (APR). Not all engines have APR, which exceeds 100% power (by about 5%) for a very short duration (say for about 5 min).

13.2.1.2 Flat-Rated Takeoff Rating

At any engine rating, running at a constant value of revenue passenger miles (RPM), a variation in ambient temperature would make engine thrust vary too. On a cold day, air density increases, which could cause an engine to inhale more air mass flow and therefore produce increased thrust, and vice versa. It is most critical on takeoff rating as the engine is operating at maximum power setting. To protect an engine from structural failure, the fuel control is set in such a manner as to keep thrust level at a set value, a limit known as *flat-rated thrust*, for a limited period (say, around 10 minutes) of operation at takeoff. To ensure 100% thrust is available, on a hotter day the flat rating ambient temperature limit for the flat-rated thrust is set to anywhere from ISA + 15°C to ISA + 20°C depending on the engine design.

A typical flat-rated takeoff setting is shown in Figure 13.1. Thrust at ambient temperatures above the flat-rated limit will drop but will remain constant at the flat-rated value for ambient temperatures colder than the limiting value. In an emergency, for example, if an engine fails at takeoff phase, some turbofans have the capability of the APR extracting more power from the running engine, which can prove helpful just to allow the aircraft to continue with a fly around and back to land. Engine manufacturers like to have the engine inspected if the APR is applied.

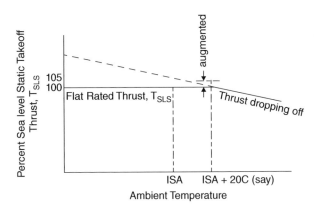

Figure 13.1 Flat-rated thrust at takeoff rating.

13.2.1.3 Maximum Continuous Rating

This is the highest engine rating that can operate continuously at around 90–95% of the maximum power. It is more than that required for climb at a prescribed speed schedule (see Section 15.10.5) for pilot ease and good fuel economy. Operational demand in this rating arises from specific situations, say at an engine failure of multi-engine aircraft, for very fast/steep climb to altitude and so on.

13.2.1.4 Maximum Climb Rating

The climb schedule is carried out at around 85–90% of the maximum power. This is to reduce stress on engine and gain better fuel economy. Typical climb time for a civil aircraft is less than 30–40 min but it can run continuously. It is for this reason that some engines do not include the maximum continuous rating and this is merged with maximum climb rating.

13.2.1.5 Maximum Cruise Rating

This is at around 80–85% of the maximum power matched to the maximum cruise speed capability. Unless there is a need for higher speed, typical cruise is performed at a 70% power rating, which may be called as *cruise rating*. This gives the best fuel economy for the long range cruise (LRC). At holding pattern in the vicinity of airport, engines are run at still lower power barely to maintain altitude and wait till clearance to proceed. The rating depends on the weight of aircraft; at the end of cruise (light weight), a rating about 60–70% would suffice.

13.2.1.6 Idle Rating

This is at around 40–55% of maximum power meant for an engine to run without flameout but produces practically no thrust. This situation arises at descent, at approach or on the ground. During descent, it is found that better economy can be achieved by coming down at part throttle, say at around 60% power rating. This gives a shallower glide slope to cover more distance without consuming much fuel.

Representative engine performances of various types at takeoff rating, maximum climb rating and maximum cruise ratings are given in this chapter for International Standard Atmosphere (ISA) day. Engine manufacturers also supply performance data for non-ISA days; this is more critical for hot and high-altitude conditions where engines would produce considerably less power. To protect engines from heat stress, the fuel control system is tuned to cut-off power generation to a flat-rated value (at ISA day engine rating) up to a hot day that can be 20°C above the ISA day. This book does not deal with non-ISA day performances. In industry, this will be supplied.

13.2.1.7 Recovery Factor, RF

A bare engine on a ground test bed inhales free-stream air directly and performs at its best. In contrast, installed engine in aircraft can only inhale the air flow diffused through the nacelle/intake duct incurring losses in pressure head, expressed as recovery loss as defined next.

Intake Pressure Recovery Factor:

$$RF = \text{(total pressure head at engine fan face)/(free-stream air total head)} = p_{t1}/p_{t\infty} \tag{13.1}$$

where subscript '∞' represents free-stream condition and subscript '1' represent fan face.

For a pod mounted high subsonic jet aircraft at cruise Mach, RF is about 0.98.

For s fuselage mounted side intake with the engine at the aft end, at subsonic Mach, RF is about 0.94–0.96.

For a nose mounted pitot type long intake with the engine at the aft end, at subsonic Mach, RF is about 0.92–0.94.

At supersonic speed, shock loss has to be added (see Section 12.8.2).

Pod mounted nacelles stand isolated on the wing or fuselage by a pylon support, hence these have the least perturbed free-stream air to breadth; with minimal loss of the flow energy yielding to high RF. The RF during cruise speed is 0.98. During takeoff the RF starts low and gradually builds to its high values as aircraft speed increases to cruise speed.

The intake internal geometry depends on the aerodynamic considerations of air inhalation and is discussed in Section 12.8.

13.2.2 Turbofan Engine Parameters

Engine manufacturers supply engine data to aircraft manufacturers who generate the engine data for in-house use. The format of data transfer from engine to aircraft manufacturer has changed considerably with the advent of powerful desktop computers.

In past, engine manufacturers conveniently use tabulated engine data for the full flight envelope of the aircraft under evaluation. The aircraft designers convert these tabulated data into a few non-dimensional graphical plots. This allows expression of engine performance characteristics in fewer graphs for use. Such non-dimensional graphs are generated for each type of turbofan for the particular aircraft incorporating the associated installation losses. Today, instead of supplying data in a tabulated format, the engine manufacturers supply a computer deck for aircraft designers to generated installed engine performance in refined detail for exact off-take demands and installation losses.

The generalised non-dimensional parameters have the fan face diameter, D_{fan} (not the intake diameter) in it, but when expressed for a particular turbofan then D_{fan} can be omitted, as it is of a known constant value. Also the sea level ISA, atmospheric values are brought in to include atmospheric ratios total conditions (subscript 't') at the fan face. These parameters are also used for gas turbine component performance. Overall gas turbine performance can minimise use of some of these parameters.

Non-dimensional thrust, T/δ_t, non-dimensional turbofan air mass flow, $\dfrac{W_a\sqrt{\theta_t}}{\delta_t}$

Non-dimensional RPM, $N/\sqrt{\theta_t}$, non-dimensional turbofan fuel mass flow, $\dfrac{W_f\sqrt{\theta_t}}{\delta_t}$

$\frac{W_a\sqrt{\theta_t}}{\delta_t}$ and $\frac{W_f\sqrt{\theta_t}}{\delta_t}$ are mainly used in gas turbine component performance analyses. Note that T/δ_t is truly not non-dimensional but when multiplied by p_0/D_{fan} it becomes non-dimensional. It is convenient to use T/δ as the non-dimensional parameter where δ is the ambient pressure ratio p/p_0. Today, the engine manufacturer supplied computer deck has made non-dimensionally plotted engine performance data almost redundant.

Turbofan characteristics show that net thrust is a function of engine geometry, pressure ratio, inlet and outlet flow conditions. A given engine shows a net thrust proportional to EPR (i.e. the ratio of pressure at turbine exit to at compressor face). At a particular ambient conditions (temperature and pressure) and aircraft speed, the thrust of a particular turbofan can expressed as a function of engine pressure ratio, EPR (exit nozzle plane total pressure to fan face total pressure). Also, engine RPM can also be considered as an index for thrust and can be related to engine ratings. Although EPR gives more refined thrust values, this book uses RPM as an index for net thrust because it is more convenient.

13.3 Uninstalled Turbofan Engine Performance Data – Civil Aircraft

All thrusts given in this section are bare engine uninstalled thrust, typically as supplied by engine manufacturers [1–3]. In this section, the uninstalled engine performances are given in a non-dimensional form that is scalable for the class of engine. The uninstalled non-dimensional parameter is thrust given in as the fraction of sea level uninstalled static thrust, T/T_{SLS}, in an STD day.

Engine performance is given here for three ratings at STD day, in the format adopted in this book.

i.   Takeoff rating is at the maximum 100% power setting (full throttle) that operates for short duration. The non-dimensional parameter T/T_{SLS} is shown with variation of aircraft speed (Mach number) for three attitudes. At sea level STD day $T/T_{SLS} = 1$.
ii. Maximum Climb Rating is the maximum power setting that can operate continuously at a relatively high rate of fuel consumption. The typical time taken to climb to reach cruise altitude is about 20–30 minutes at 85–95% of the takeoff power setting. The non-dimensional parameter T/T_{SLS} is shown with variation of altitude for three aircraft speed (Mach number).
iii. Maximum Cruise Rating is meant for continuous operation for hours in a better fuel economy typically at 80–85% of the maximum (takeoff) throttle setting. Transport aircraft have two cruise speed, for example, (a) High Speed Cruise (HSC) at M_{crit} allowing 20 count wave drag rise (see Section 11.13) due compressibility effects and (b) LRC to give the best fuel economy at the speed when wave drag is negligibly small so as to be ignored. The non-dimensional parameter T/T_{SLS} is shown with variation of speed (Mach number) for six altitudes.

On integrating with aircraft, there could be a loss of power for the installed engine as discussed in Section 12.7. The extent of loss depends on the mission of the aircraft and technology adopted for engine-airframe integration. To keep it simple, both military and civil installation losses are kept at a similar percentage although the demands of takeoffs are quite different.

At each point of the graph the non-dimensional fraction, T/T_{SLS}, has a value expressed in equation as a factor with symbol 'k' with power rating given in subscript, for example, k_{TO}, k_{cl} and k_{cr}, respectively. These factors are used in the aircraft sizing and engine matching exercises carried out in Chapter 14. Installed k-factors must apply the loss correction as will be shown.

Four types of engines given here cover a wide range of aircraft applications. These are as follows:

i.   Turbofan with a BPR $\approx 4 \pm 1$, for example, in the class of Honeywell TPE731 turbofan, typically used in smaller civil aircraft application, such as Bizjets, small utility aircraft and so on.
ii. Turbofan with a BPR $\approx 6 \pm 1$, for example, in the class of International Aero Engines IAE V2500 turbofan, typically used in larger civil aircraft application, such as early generation 320 aircraft.
iii. Military turbofan with a BPR < 1, for example, in the class of Rolls-Royce/Turbomeca Adour used in the BAe Hawk class of advanced jet trainer (AJT)/close air support (CAS) role.
iv. Turboprop engine, in the class of Garrett TPT 331 used inthe Dornier DO 228 as a utility aircraft, Tucano trainer/COIN (Counter Insurgency) aircraft role and so on.

The latest engines for the Boeing 787 have a BPR ≈ 8 and Airbus 350 BPR ≈ 12. The authors could not obtain a realistic data for the latest class of turbofans. Some spot data points for high bypass ratio (BPR) turbofans are given Section 13.3.3.

The higher the BPR, lesser would be the specific thrust ($T_{SLS}/\dot{m}_a$ – lb (lb s)$^{-1}$). The specific dry engine weight (T_{SLS}/dry engine weight, non-dimensional) increases with higher BPR (Table 13.1).

Table 13.1 Turbofan parameters, BPR and specific thrust.

BPR	$F/\dot{m}_a$ – lb (s lb)$^{-1}$ @ T_{SLS}	T_{SLS}/dry engine weight
Around 4	35–40	4–5
Around 5	32–34	5–6
Around 6	30–34	5–7
Around 8	30–32	6–8

Table 13.2 Turboprop specific horse power for sizes (based on uninstalled SHP).

	$SHP/\dot{m}_a$ (SHP (lb s)$^{-1}$)	SHP_{SLS}/dry engine weight (SHP lb^{-1})
Smaller turboprop ≤1000 SHP	≈0.012 (@ TO)	≈2.2–2.75[a]
Larger turboprops >1000 SHP	≈0.01 (@ TO)	≈2.5–3.0[a]

a) lower factor values for lower SHP.

The higher the SHP, the lower is the specific SHP ($SHP/\dot{m}_a$ – SHP (lb s)$^{-1}$) and the higher the specific dry engine weight (SHP_{SLS}/dry engine weight–lb lb^{-1}). Table 13.2 may be used for these computations.

13.3.1 Performance with BPR ≈ 4 ± 1 (Smaller Engines, e.g. Bizjets)

Uninstalled turbofan (BPR around ≈ 4 ± 1) thrust and fuel flow data in a normalised form are given next.

13.3.1.1 Takeoff Rating

Figure 13.2 gives the uninstalled takeoff thrust in a non-dimensional form for the standard (STD) day for turbofans with BPR 4 ± 1 in the class of the Honeywell TPE731 turbofan. Fuel flow rate remains nearly invariant

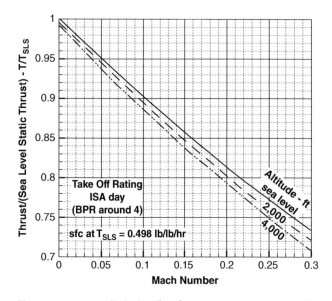

Figure 13.2 Uninstalled takeoff performance (≈BPR 4 ± 1). (Fuel flow rate is nearly constant at takeoff rating).

for the envelope shown in the graph. The specific fuel consumption (sfc) at takeoff rating is taken at the value at the uninstalled T_{SLS} to be 0.478 lb (lb h)$^{-1}$ per engine.

Example: Uninstalled k_{TO}, = 0.905 at Mach 0.1 at sea level. At a 4000 ft altitude it drops down to 0.87. Aircraft sizing exercises use the installed value as worked out in Section 13.4.

13.3.1.2 Maximum Climb Rating

Figure 13.3 gives the uninstalled thrust in a non-dimensional T/T_{SLS} form and sfc flow for the maximum climb rating at a standard day from 5000 up to 50 000 ft altitude for 0.3, 0.5 and 0.7 Mach numbers. Intermediate values may be linearly interpolated. Note that there is a break in thrust generation at about 6000–10 000 ft altitude, depending on Mach number, due to fuel control to keep the EGT within the specified limit.

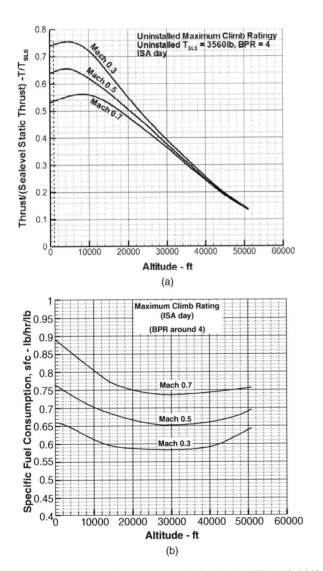

Figure 13.3 Uninstalled maximum climb rating ($\approx$BPR 4 $\pm$ 1). (a) Non-dimensional thrust. (b) Specific fuel consumption.

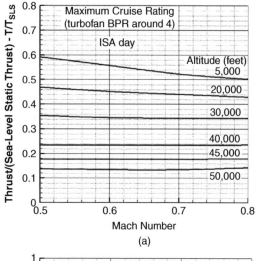

(a)

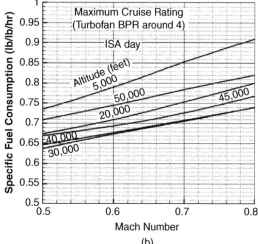

(b)

Figure 13.4 Uninstalled maximum cruise rating ($\approx$BPR 4 ± 1). (a) Non-dimensional thrust. (b) Specific fuel consumption.

Example: Uninstalled $k_{cl}, = T/T_{SLS} = 0.705$ at Mach 0.38 at the 1000 ft altitude. Fuel flow per engine can be computed from Figure 13.3b. Aircraft sizing exercises use the installed value as worked out in Section 13.4.

13.3.1.3   Maximum Cruise Rating

Figure 13.4 gives the uninstalled thrust in a non-dimensional T/T_{SLS} form and *sfc* flow for the maximum cruise rating in standard day form for the standard day from 5000 to 50 000 ft altitude for Mach number variations from 0.5 to 0.8, sufficient for this class of engine-aircraft combination. Intermediate values may be linearly interpolated.

Example: Uninstalled $k_{cr} = T/T_{SLS} = 0.23$ at HSC Mach 0.75 at the 41 000 ft altitude. At HSC, the throttle is cut further back to 80–85% of the maximum throttle setting. At LRC, it is cut further back to 75–80% of the maximum throttle setting. The fuel flow per engine can be computed from Figure 13.4b. Aircraft sizing exercise uses installed value as worked out in Section 13.4.

13.3.1.4   Idle Rating

Uninstalled thrust in a non-dimensional T/T_{SLS} form and *sfc* for idle rating is not given here as it is not used in aircraft sizing exercises. Installed idle rating thrust and fuel flow are given in Figure 13.14, later. Idle ratings produce a small amount of thrust.

Figure 13.5 Uninstalled takeoff performance (≈BPR 6 ± 1).

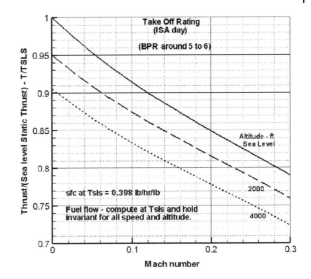

13.3.2 BPR Around ≈6 ± 1 (Larger Engines)

13.3.2.1 Turbofan performance

Bigger engines have higher BPR. Current larger operational turbofans are at $BPR \approx 6 \pm 1$ IAE V2500, which would have non-dimensional engine performance characteristics slightly different to smaller engines as can be compared with Figures 13.2–13.4. Classroom work could follow the same routine as given previously. Appendix D describes a case study of an A320 class aircraft using a IAE V2500 turbofan.

13.3.2.2 Takeoff Rating

Figure 13.5 gives the uninstalled takeoff thrust in a non-dimensional form for the standard day. Typically, at takeoff, the sfc is nearly invariant. The sfc value is given in the graph.

13.3.2.3 Maximum Climb Rating

Figure 13.6 gives the uninstalled maximum climb thrust and fuel flow in a non-dimensional form for the standard day up to 50 000 ft altitude for three Mach numbers. Intermediate values may be linearly interpolated.

13.3.2.4 Maximum Cruise Rating

Figure 13.7 gives the uninstalled maximum cruise thrust and fuel flow in a non-dimensional form for the standard day from 5000 ft to 50 000 ft altitude for Mach numbers varying from 0.5 to 0.8, sufficient for this class of engine-aircraft combination. Intermediate values may be linearly interpolated.

13.3.3 BPR Around ≈10 ± 2 (Larger Engines – Big Jets, Wide Body Aircraft)

The authors could not locate non-dimensional uninstalled engine performance data for these types of high BPR turbofans.

13.3.4 Uninstalled Turbofan Engine Performance Data – Military (BPR < 1)

Military turbofan ratings are slightly different from civil turbofan ratings. Military engines are allowed to run longer at maximum ratings, that is, not only at takeoff but also for fast acceleration in combat. Of course, these still operate within a limited period of time, for example, at takeoff, the extent depends on the design.

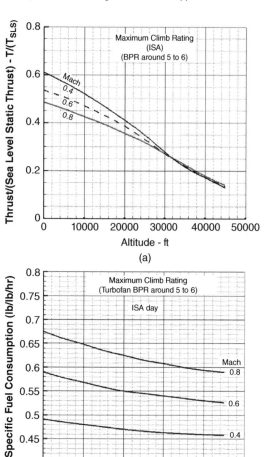

(a)

(b)

Figure 13.6 Uninstalled maximum climb rating ($\approx$BPR 6 ± 1). (a) Non-dimensional thrust. (b) Specific fuel consumption.

For a continuous operation, the engine rpm is throttled down to a maximum of 90% of T_{SLS} as a maximum continuous rating; mostly applied in a demanding situation during combat. For climb, it is further throttled back to a maximum of 85% of T_{SLS} as the maximum climb rating. Typically, maximum continuous rating is close to maximum climb rating. For convenience, some military engine designers have merged the two ratings into one maximum climb or maximum continuous rating.

Military engines operate at considerably varying throttle demands. Here, cruise is less meaningful unless it is a ferry flight. Flight to operation theatre or return to base is not exactly cruising – this period can be executed at lower throttle setting below 75% of T_{SLS}, as the mission demands.

Reheat (afterburning) is added at maximum rating when the throttle is set to a fully forward position. Running at a relatively longer duration at this high power (higher combustion temperature) is at the cost of having a shorter engine life span and shorter TBO. In addition, the hot zone components use more expensive materials to withstand stress at elevated temperatures.

A typical military turbofan engine performance at the maximum rating in the class of Rolls-Royce/Turbomeca Adour suited to the classroom example of an AJT with a derivative in CAS role is given in Figure 13.8. The sfc for this engine at T_{SLS} is 0.75 lb (h lb)$^{-1}$ when operating without reheat and 1.1 lb (h lb)$^{-1}$ when under reheat (afterburner lit). Rated (maximum) air mass flow is 95 lb s^{-1}.

Figure 13.7 Uninstalled maximum cruise rating ($\approx$BPR 6 ± 1).
(a) Non-dimensional thrust. (b) Specific fuel consumption.

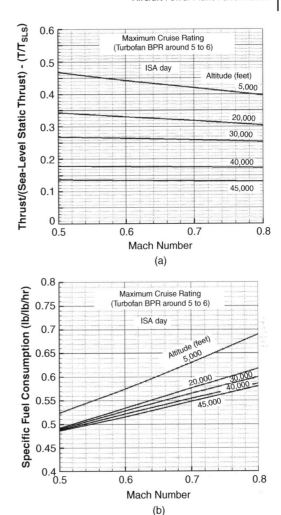

(a)

(b)

Currently, not much information can be supplied on military engines. However, for the classroom example this will prove sufficient as will be shown in subsequent chapters.

13.3.5 Uninstalled Turboprop Engine Performance Data (All Types up to 100 Passenger Class)

Engine performance characteristics vary from type to type. The real engine data is not easy to obtain in the public domain. The graphs in this section are in the class of Garrett TPT 331 representing typical current turboprop engines in the class. The graphs include the small amount of jet thrust available at a 70% rating and higher. The jet power is converted to SHP and the total equivalent shaft horse power (ESHP) is only labelled 'SHP' in the graph. In industry, engine manufacturers supply performance data incorporating exact installation losses for accurate computation.

The sizing exercise provides the required thrust at the specified aircraft performance requirements. Figures 13.9 and 13.10 give the typical turboprop non-dimensional SHP/SHP_{SLS} fraction and sfc (in terms of power specific fuel consumption: psfc) at maximum climb and maximum cruise ratings in a non-dimensional form. Intermediate values may be linearly interpolated from these graphs. Using the propeller performance given in Section 12.12, the SHP at the sea level static condition can be worked out. From this information, engine performance at other ratings can be established for the full flight envelope. Takeoff rating maintains constant power but the thrust changes with speed.

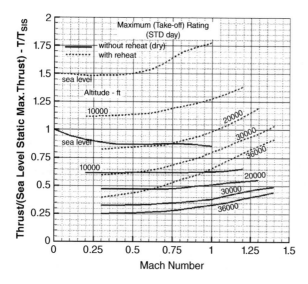

Figure 13.8 Military turbofan engine with and without reheat (BPR = 0.75). Take 90% of maximum rating as the maximum continuous rating.

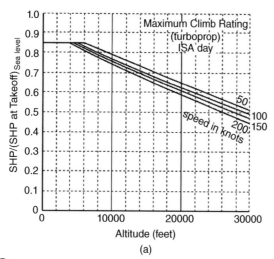

(a)

Figure 13.9 Uninstalled maximum climb rating (turboprop). (a) Shaft horse power. (b) Specific fuel Consumption.

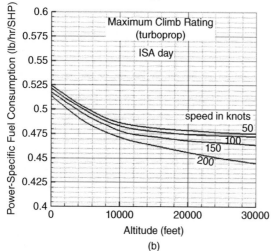

(b)

Figure 13.10 Uninstalled maximum cruise rating (turboprop).
(a) Shaft horse power. (b) Specific fuel flow.

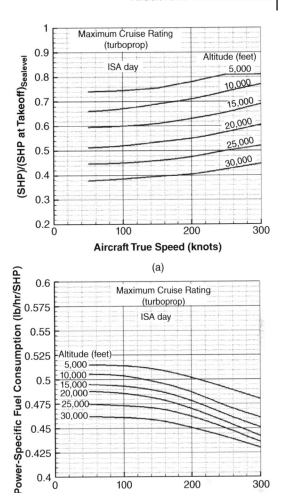

13.3.5.1 Takeoff Rating

The takeoff rating for a turboprop is held constant up to Mach 0.3. In this case, the non-dimensional SHP/SHP_{SLS} fraction is 1.0 and no graph is given. The psfc of the turboprop at takeoff is from 0.475 to 0.6 lb (h shp)$^{-1}$. Section 13.5.2 works out the installed performance. The sizing exercise in Chapter 14 gives the Tucano class military trainer aircraft uninstalled $SHP_{SLS} = 1075$ to get 4000 lb installed thrust.

13.3.5.2 Maximum Climb Rating

Figure 13.9 shows the uninstalled maximum climb SHP and fuel flow (sfc) in a non-dimensional form for the standard day up to a 30 000 ft altitude for four true air speeds from 50 to 200 kt. The break in SHP generation up to a 6000 ft altitude is due to fuel control to keep the EGT low. As in the case of takeoff rating, the SHP remains invariant up to about 5000 ft altitude.

13.3.5.3 Maximum Cruise Rating

Figure 13.10 shows the uninstalled in a non-dimensional maximum cruise SHP/SHP_{SLS} fraction and psfc for the standard day from a 5000 to 30 000 ft altitude for true air speed from 50 to 300 kt.

13.4 Installed Engine Performance Data of Matched Engines to Coursework Aircraft

Section 12.7 describes losses associated with installation effects of engine to aircraft. This section generates the available installed thrust and fuel flow graphs matched for the worked out sized aircraft examples; one Bizjet and one AJT. In addition, a 1075 SHP turboprop engine performance is given for the readers to work out turboprop trainer (TPT) aircraft sizing and an engine matching exercise as well as the TPT performance. The installed engine data is obtained from the non-dimensional data after applying the installation losses and the sized T_{SLS}.

Since the given sfc graphs are based on uninstalled thrust, the fuel flow rates are computed using uninstalled thrust. Therefore, the installed sfc will be higher than the uninstalled sfc. Installation loss at cruise comes down to about half the percentage loss of that at takeoff.

13.4.1 Turbofan Engines (Smaller Engines for Bizjets – BPR ≈ 4)

Figures 13.2–13.4 give the typical uninstalled turbofan thrust in a non-dimensional form in terms of T/T_{SLS} along with corresponding sfc for the Bizjet class of aircraft. Section 14.5 establishes that it requires an installed matched $T_{SLS\_INSTALLED} = 3500$ lb per engine. Table 13.3 summarises the factors associated with installed thrust and fuel flows for the three engine ratings. The data is sufficient for the example taken in this book. Intermediate values may be linearly interpolated. Since all aircraft performance assessments are carried out based on installed engine performance, from now onwards, T_{SLS} stands for installed performance without any subscript.

13.4.1.1 Takeoff Rating (Bizjet) – STD Day

Typically, during the takeoff ground run, the air-conditioning is kept shut-off for a brief period until the undercarriage is retracted after reaching a 35 ft altitude. Taking 4% installation loss at takeoff with air-conditioning shut-off, the uninstalled $T_{SLS\_UNINSTALLED} = 3500/0.96 = 3645$ lb, rounded to 3650 lb per engine for the sized Bizjet. Figure 13.11 gives the installed engine thrust at the takeoff rating with a 4% installation loss. Depending on ambient conditions and how the environmental control system (ECS) is managed, typically installation loss varies from 5 to 7% of the uninstalled sea level static thrust. In that case, the takeoff field length will be slightly longer.

Fuel flow rate is computed from its uninstalled sfc of 0.478 lb $(h\,lb)^{-1}$ at sea level static condition installed sfc $= 0.478/0.96 =$ of 0.498 lb $(h\,lb)^{-1}$. Using installed $T_{SLS} = 3500$ lb per engine, the fuel flow rate is $3500 \times 0.498 = 1743$ lb h^{-1} per engine. Fuel flow is kept nearly constant at takeoff up to the en-route climb segment, when the engine is throttled down to a maximum climb rating computed in the next section.

13.4.1.2 Maximum Climb Rating (Bizjet) – STD Day

Installation loss during climb is taken as 5% of the uninstalled thrust. Using the uninstalled graphs given in Figure 13.3, the installed thrust and fuel flow rates are plotted in Figure 13.12. Estimation of the payload-range would require the full aircraft climb performance up to the cruise altitude. Note that this turbofan has a break

Table 13.3 Summary of installed thrust and fuel flow data per engine at the three ratings – example. All computations are based on $T_{SLS\_UNINSTALLED} = 3650$ lb. per engine.

Rating	Altitude ft	Mach	Loss%	Scaling factor, k	sfc lb $(lb\,h)^{-1}$	Installed thrust lb	Fuel flow lb h^{-1}
Takeoff	0	0	7	0.93	0.498	3500	1772
Max. climb	1000	0.38	6	0.67	0.7	2231	1661
Max. cruise	41000	0.74	4	0.222	0.73	758	578

Figure 13.11 Installed takeoff performance per engine (≈*BPR* 3–4).

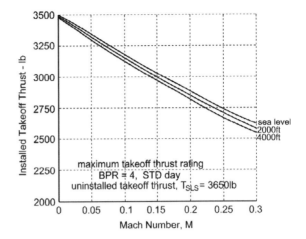

Figure 13.12 Installed maximum climb performance per engine (≈ < BPR 4). (a) Thrust. (b) Fuel flow.

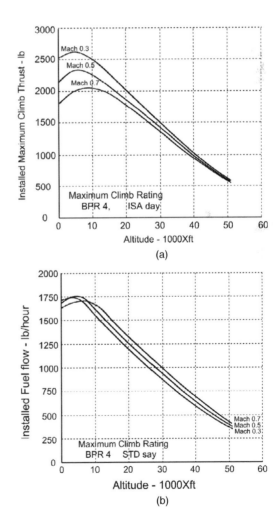

in fuel flow at around 5000 ft–10 000 ft altitude, depending on the flight Mach number, to keep the EGT within the limits resulting in corresponding break in thrust generation.

13.4.1.3 Maximum Cruise Rating (Bizjet) – STD Day

Figure 13.4 gives the uninstalled maximum cruise thrust in a non-dimensional form and sfc from 5000 to 50 000 ft altitude for Mach number variation from 0.5 to 0.8. Take a 4% installation loss at cruise. Based on uninstalled $T_{SLS} = 3560$ lb., the installed thrust and fuel flow at maximum cruise rating is generated from Figure 13.4, as shown in Figure 13.13. Estimation of the payload range would require the full aircraft climb performance up to the cruise altitude.

13.4.1.4 Idle Rating (Bizjet) – STD Day

Figure 13.14 gives the idle thrust and fuel flow at installed idle engine rating for Mach number variation from 0.3 to 0.6 and from sea level to 51 000 ft altitude. The idle rating shows patterns of performance variation quite different from climb and cruise rating. Part throttle rating are not given here but may be interpolated.

13.4.2 Turbofans with BPR Around 6 ± 1 (Larger Engines – Regional Jets and Above)

Bigger engines have a higher BPR. Current big operational turbofans have a BPR around 6 ± 1 (new generation turbofans have achieved a BPR greater than 8). These big engines have performance characteristics

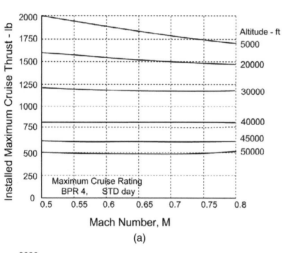

(a)

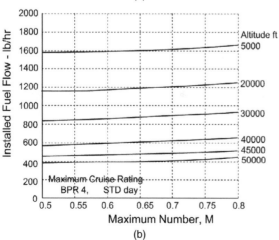

(b)

Figure 13.13 Installed maximum cruise performance per engine ($\approx <$ BPR 4). (a) Thrust. (b) Fuel flow.

Figure 13.14 Installed idle engine rating performance per engine ($\approx < $ BPR 4). (a) Thrust. (b) Fuel flow.

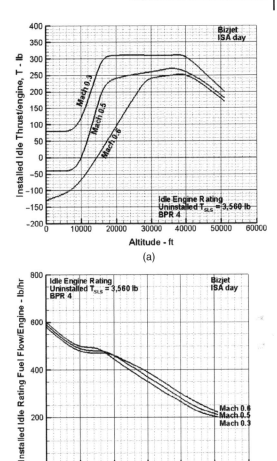

(a)

(b)

slightly different from the smaller ones, specifically the maximum climb rating has no break in thrust with altitude gain. Using Figures 13.10–13.13, the installed thrust and fuel flow rates can be worked out in similar manner.

13.4.3 Military Turbofan (Advanced Jet Trainer/CAS Role – BPR < 1) – STD Day

Military turbofans differ from civil turbofans. Figure 13.8 gives the typical uninstalled thrust in a non-dimensional form in terms of T/T_{SLS} along with corresponding sfc for the AJT/CAS class of aircraft. Section 13.3.4 argues that only one graph at the maximum ratings would prove sufficient for use in this book. In industry, separate graphs are used for each rating.

Installation loss of a military trainer aircraft with single occupancy for airmanship practice sortie can be as low as 3.5% reduction of thrust (combat aircraft can have as high as around 15% loss of thrust). Section 14.7 sizes the AJT that requires installed T_{SLS} of 5800 lb (uninstalled $T_{SLS} = 6000$ lb) for normal training configuration (NTC). Using Figure 13.8, the installed thrust at the maximum (takeoff) rating for the AJT is shown in Figure 13.15. maximum continuous rating (primarily for climb) is at 95% of the maximum takeoff rating. High speed runs are done at 90% of the maximum takeoff rating (combat could demand up to 100% thrust in short bursts). For the CAS role, the thrust values need to be scaled up by 30% with an up-rated engine.

The AJT sfc at T_{SLS} is 0.7 lb (h lb)$^{-1}$ and the fuel flow rate is based on the uninstalled T_{SLS}.

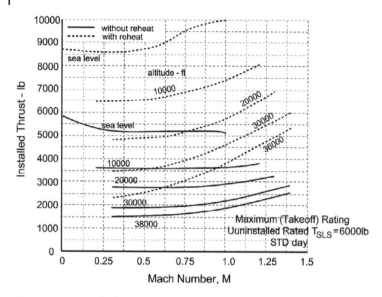

Figure 13.15 Installed maximum rating – military turbofan (BPR = 0.75).

Therefore, fuel flow rate $= 0.7 \times 6000 = 4200\ \text{lb h}^{-1}$.

Unlike civil aircraft, the AJT training profile is throttle-dependent. A training profile operates in varying speeds and altitudes. At NTC (meant for airmanship and navigational training with low level of armament practice), the average throttle setting may be considered to be at about 75–85% of the maximum rating. At 30 000 ft altitude and Mach 0.7 the average operating installed thrust $T = 0.75 \times 1980$ (Figure 13.15) = 1485 lb (uninstalled thrust of 1538 lb) and average sfc is given as $0.7\ \text{lb (h lb)}^{-1}$. It gives the average fuel flow rate $= 0.7 \times 1536 = 1076\ \text{lb h}^{-1}$. Onboard fuel carried is 2425 lb.

13.5 Installed Turboprop Performance Data

The first task is to find a suitable propeller to develop thrust from the turboprop SHP. In a stepwise manner, thrust from a propeller is worked out as a coursework exercise [3, 4]. The method uses the Hamilton Standard charts intended for constant-speed propellers. These charts also can be used for fixed-pitch propellers when the pitch of the propeller should match the best performance at a specific speed: either cruise speed or climb speed. Two forms are shown: (i) from the given thrust, compute the HP (in turboprop case SHP) required and (ii) from the given HP (or SHP), compute the thrust. The starting point is the C_p. Typically, at sea level takeoff rating in a static condition, one SHP produces about 4 lb on an STD day. At the first guesstimate, a factor of 4 is used to obtain SHP to compute the C_p. One iteration may prove sufficient to refine the SHP.

13.5.1 Propeller Performance at STD Day – Worked-Out Example

Problem description. Consider a single, four-bladed, turboprop military trainer aircraft of the Royal Air Force (RAF) class Tucano operating with a constant-speed propeller at N = 2400 rpm giving installed $T_\text{SLS} = 4000$ lbs. The specified aircraft speed is 320 mph at 20 000 ft altitude (i.e. Mach 0.421). For the aircraft speed, the blade activity factor (AF) is taken as 180. Establish its rated SHP at sea level static conditions and thrusts at various speeds and altitudes. All computations are in STD day.

Propeller diameter. The first task is to establish the propeller diameter. From Figure 12.33 for a four-bladed propeller, the diameter is taken to be 98 in. Figure 12.32 establishes the integrated design C_{Li}.

To check, use the empirical relation to determine propeller diameter in Section 12.12. Take $SHP_{SLS} = 4000/4 = 1000$. It gives $D = K(P)^{0.25} = 18 \times (1000)^{0.25} = 101.2$ in., where K for a four-bladed propeller is 18. The empirical relation is close to what is obtained from Figure 12.33. A mean diameter of 100 in. of the two results is accepted.

Aircraft configuration must ensure ground clearance at a collapsed nose-wheel tyre and oleo. A higher number of blades (i.e. higher solidity) could reduce the diameter at the expense of higher cost. For this aircraft class, it is best to retain the permissible largest propeller diameter, keeping the number of blades to four or five. To set an example for the ground clearance, a 2 in. radius is cut off from the tip. Then the final diameter is 96 in. = 8 ft diameter.

Case I Takeoff performance (HP from thrust). This is used to compute the SHP at a sea level takeoff. Guesstimate installed SHP = 4000/4 = 1000 SHP. For a clean nacelle of turboprop an optimistic value of fuselage loss factor ($f_b \times f_h$, see Section 11.16) of 0.98 is taken (typical value).

Taking into account a 5% installation loss at takeoff, the uninstalled $T_{SLS} = 4000/0.95 = 4200$ lb. In addition, apply fuselage loss of 0.98, that is $T_{SLS} = 4200/0.97 = 4300$ lb. It may now be summarised that to obtain 4000 lb installed thrust, the uninstalled rated power is $1000/(0.95 \times 0.98) \approx 1075$ SHP.

Figure 12.32 gives the factor $ND \times f_c = 2400 \times 8 \times 1.0 = 19\,200$. It corresponds to the ratio of the speed of sound at standard day sea level to the altitude, $f_c = 1.0$ and at static condition (Mach 0).

Figure 12.32 gives the integrated design $C_{Li} \approx 0.5$.

Equation (12.44) gives:

$$C_p = (550 \times SHP)/(\rho n^3 D^5)$$

or $C_p = (550 \times 1000)/(0.00238 \times 40^3 \times 8^5) = 550\,000/4\,991\,222 = 0.11$.

Figure 12.30 (four-blade, AF = 180, $C_L = 0.5$) gives $C_T/C_p = 2.4$ corresponding to integrated design $C_{Li} = 0.5$ and $C_p = 0.11$.

Therefore,

$$\text{installed static thrust, } T_{SLS} = (C_T/C_p)(33\,000 \times SHP)/ND = (2.4 \times 550 \times 1000)/(40 \times 8)$$

$$= 1\,320\,000/320 = 4125 \text{ lb.}$$

Therefore, the installed *SHP* is revised to $1000 \times (4000/4125) = 970$ SHP giving installed thrust $T_{SLS} = 4000$ lb. It is close enough to avoid any further iteration. As a safe consideration, it is taken that 4000 lb of thrust is produced by installed 1000 SHP. Corresponding uninstalled SHP is 1075. Readers may carry out an iteration with an installed 970 SHP.

Case II Thrust from SHP (worked out with 1000 SHP installed as the maximum takeoff rating at sea level static conditions on a standard day).

Once the installed SHP_{SLS} is known, the thrust for the takeoff rating can be computed. The turboprop fuel control maintains a constant SHP at takeoff rating keeping it almost invariant. This section computes the thrust available at speeds up to 160 mph to cover lift-off and enter the en-route climb phase. Available thrust is computed at 20, 50, 80, 120 and 160 mph as shown in Table 13.4.

Compute

$$J = V/nD = (1.467 \times V)/(40 \times 8) = 0.00458 \times V, \text{ where } V \text{ is in mph.}$$

power coefficient: $C_p = (550 \times SHP)/(\rho n^3 D^5) = (550 \times 1000)/(0.002\,38 \times 40^3 \times 8^5) = 0.11$.

For integrated design $C_{Li} = 0.5$ and $C_p = 011$. The propeller η_{prop} corresponding to J and C_p is obtained from Figure 12.31 (four-blade, AF = 180, $C_L = 0.5$).

Table 13.4 Propeller installed thrust results.

Velocity, V	20 mph[a]	50 mph	80 mph	120 mph	160 mph
$J = 0.00463 \times$ mph	0.092	0.23	0.37	0.55	0.74
C_p	0.11	0.11	0.11	0.11	0.11
Installed SHP	1000	1000	1000	1000	1000
From Figure 12.29, η_{prop}	0.19	0.4	0.56	0.7	0.77
Installed thrust, T lb	3820	3225	2820	2350	1940

a) Too low a velocity for Figure 13.16. It is close to static conditions and agrees.

13.5.2 Turboprop Performance at STD Day

Based on the methodology shown previously, thrust is obtained from a single, four-bladed (AF = 180, design $C_L = 0.5$), operating with a constant-speed propeller at N = 2400 rpm, which is worked out next at takeoff rating, maximum climb rating and maximum cruise rating.

The installed thrust includes the fuselage factors of fuselage blockage factor f_b and excrescence factor f_h (see Section 11.16).

13.5.2.1 Maximum Takeoff Rating (Turboprop) – STD Day

The psfc of turboprops at takeoff is from 0.475 to 0.6 lb $(\text{h shp})^{-1}$. For an SHP of less than $\approx$1000, use a psfc of 0.6 lb $(\text{h shp})^{-1}$; for more than 1000 use a psfc of approximately 0.48 lb $(\text{h shp})^{-1}$.

Compute thrust: $T = (550\,\eta_{prop} \times 1000)/V = 550\,000 \times \eta_{prop}/V$, where V is in mph (see Table 13.4). Use Eq. (12.46), for the FPS system.

Therefore, fuel flow at SHP_{SLS} is $0.5 \times 1075 = 5313.5$ lb h^{-1}.

From Table 13.2, the intake air mass flow at SHP_{SLS} is $0.011 \times 1075 = 11.83$ lb s^{-1}.

The dry engine weight $= 1075/2.75 = 390$ lb.

At takeoff rating, the engine power is kept nearly constant to a speed when the en-route climb can start at a reduced power setting of maximum climb rating. The sizing exercise in Chapter 14 gives the Tucano class military trainer aircraft with uninstalled $SHP_{SLS} = 1075$, to get a 4000 lb installed thrust. The psfc of the turboprops at takeoff is 0.5 lb $(\text{h shp})^{-1}$ based on uninstalled power (see Section 13.3.5).

Figure 13.16 plots the thrust versus speed at the takeoff rating. In a similar manner, thrust at any speed, altitude and engine rating can be determined from the relevant graphs.

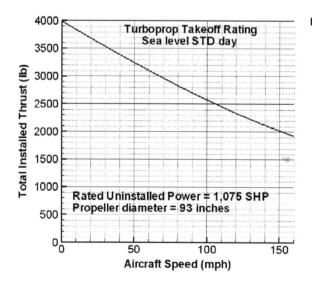

Figure 13.16 Takeoff thrust for the turboprop engine.

Table 13.5 30 000 ft altitude ($\rho = 0.00088$ slug/ft^3, $\sigma = 0.37$) – maximum climb.

True air speed in knots	50	100	150	200
V in mph	57.54	115.08	172.62	230.16
SHP = (factor in Figure 13.9a) $\times$ 1075	543	527	505	482
$J = 0.00533 \times$ ts	0.2665	0.533	0.8	1.066
$C_p = 0.000116 \times$ SHP	0.17	0.165	0.158	0.151
η_{prop} (Figure 12.31),	0.38	0.63	0.77	0.83
Uninstalled thrust, T lb	1345	1064	845	652
Installed thrust – lb	1291	1022	811	626
Sfc – lb (h SHP)$^{-1}$ (Figure 13.9b)	0.39	0.406	0.432	0.463
Fuel flow rate – lb h^{-1}	212	214	218	223

Typically, an engine throttles back from the takeoff rating to the maximum climb rating for an en-route climb. At up to about 4000 ft of altitude, it is kept around 85% of the maximum power and thereafter SHP goes down with altitude.

Therefore, at SHP_{SLS} the fuel flow rate is $0.5 \times 1075 = 537.5$ lb h^{-1}.

Intake air mass flow at SHP_{SLS} is $0.011 \times 1075 = 11.83$ lb s^{-1} (specific power of 0.11 lb (s SHP)$^{-1}$).

13.5.2.2 Maximum Climb Rating – STD Day

Figure 13.9 shows the uninstalled maximum climb SHP and fuel flow (sfc) in a non-dimensional form for the standard day up to a 30 000 ft altitude for four true air speeds from 50 to 200 kt. Intermediate values may be linearly interpolated. The break in SHP generation up to a 6000 ft altitude is due to fuel control to keep the EGT low.

As before, the turboprop class also requires a factor k_{cl} (this varies with speed and altitude) to be applied to the SHP. From Figure 13.9a, a value of 0.85 may be used to obtain the initial climb SHP. In the example, the uninstalled initial climb power is then $0.85 \times 1075 = 914$ SHP. The integrated propeller performance after deducting the installation losses gives the available thrust. Typically, the initial climb starts at a constant European Aviation Safety Agency (EAS) of approximately 200 kt, which is approximately Mach 0.3. At a constant EAS climb, Mach number increases with altitude; when it reaches 0.4, it is held constant. Fuel flow at the initial climb (at 800 ft altitude) is obtained from Figure 13.9b. The example gives $0.522 \times 914 = 477$ lb h^{-1}. With varying values of altitude, climb calculations are performed in small increments of altitude, within which the variation is taken as the mean and is kept constant for the increment.

Figure 13.9 shows the uninstalled maximum climb SHP and fuel flow (sfc) in a non-dimensional form for the standard day up to a 30 000 ft altitude for four true air speeds (TAS) from 50 to 200 kt. Taking 4% as the loss of thrust due to installation effects, the available installed thrust and fuel flow rates at the maximum climb rating are worked out. A sample computational table at 30 000 ft altitude is given in Table 13.5.

Figure 13.17 plots the available installed thrust and fuel flow at the maximum climb rating from sea level to 30 000 ft altitudes. The specified requirement for the initial climb rate for the example in the book is 4500 ft min^{-1}.

13.5.2.3 Maximum Cruise Rating (Turboprop) – STD Day

Figure 13.10 gives the maximum cruise SHP in a non-dimensional form and fuel consumption (sfc) for speeds from TAS of 50–300 kt and altitude up to 30 000 ft. Progressing in a similar way as in the climb performance, the maximum cruise thrust and fuel flow rate can be computed. A sample computation at 30 000 ft altitude is given in Table 13.6. Take 4% loss of thrust due to installation effects.

Figure 13.18 plots the available installed thrust and fuel flow at the maximum cruise rating. The initial maximum cruise speed of the example is 320 mph (270 kt) at 25 000 ft altitude.

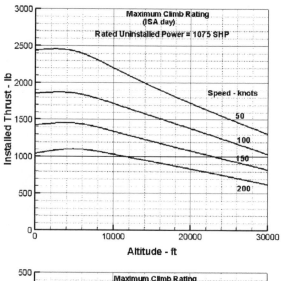

Figure 13.17 Thrust and fuel flow at maximum climb rating.

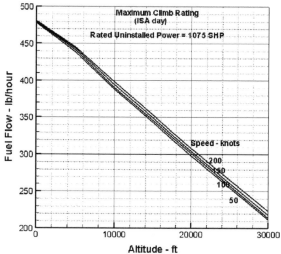

In the example, the design initial maximum cruise speed is given as 300 kt at a 25 000 ft altitude. From Figure 13.10a, the uninstalled power available is $0.525 \times 1075 = 564$ SHP. In Figure 13.10b, the corresponding fuel flow is $0.436 \times 564 = 246$ lb h^{-1}. The integrated propeller performance after deducting the installation losses gives the available installed propeller performance.

13.6 Piston Engine

There is several ways to present piston engine performance. Figure 13.19 gives the Lycoming IO-360 series 180 HP rated piston engine performance graph. It may be noted that the power ratings are given by the

Table 13.6 30 000 ft. altitude ($\rho = 0.00088$ slug/ft^3, $\sigma = 0.37$) – maximum cruise.

True air speed in knots	100	200	250	300
V in mph	115.08	230.16	287.7	345.24
SHP = (factor in Figure 13.10a) $\times$ 1075	414	434.3	457	481
$J = 0.00533 \times$ knots	0.533	1.066	1.3325	1.599
$C_p = 0.000116 \times SHP$	0.13	0.135	0.143	0.146
η_{prop}, from Figure 12.31	0.65	0.83	0.83	0.805
Uninstalled thrust, T lb	877	587	494	420
Installed thrust – lb	833	558	470	400
Sfc – lb (h SHP)$^{-1}$ (Figure 13.10b)	0.461	0.45	0.442	0.43
Fuel flow rate – lb h^{-1}	190.8	195.4	202	206.8

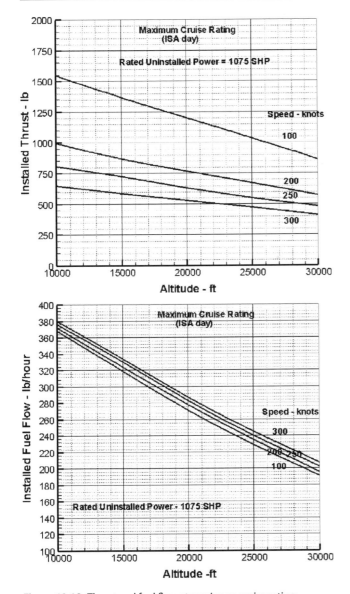

Figure 13.18 Thrust and fuel flow at maximum cruise rating.

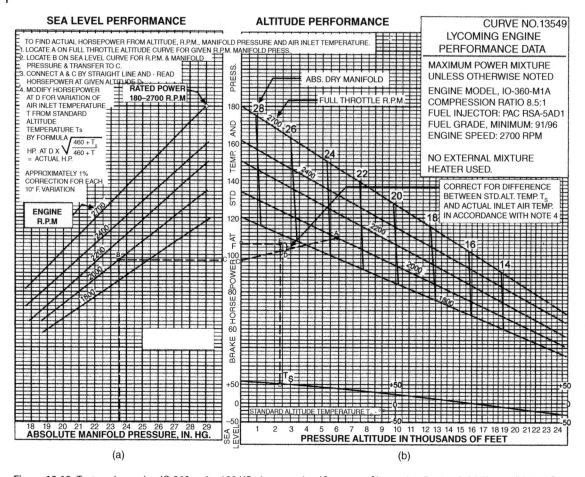

Figure 13.19 Textron-Lycoming IO-360 series 180 HP piston engine (Courtesy of Lycoming Engines). (a) Thrust. (b) Fuel flow. Source: Reproduced with permission of John Wiley & Sons.

rpm. It consists of two graphs to be used together. The first one is at full throttle and part throttle at STD sea level and known as *full throttle sea level* graph and the second one is at full throttle and part throttle at altitudes and known as *full throttle altitude* graph. At each altitude there is a unique value of HP for the combination setting of rpm and manifold pressure. The second graph also gives correction factors for a non-standard day.

Piston engine power is dependent on amount of air mass inhaled, which is indicated by its rpm and manifold pressure, p_{manifold} at a particular ambient condition. Engine rpm is adjusted by throttle control (fuel control). Above a 5000 ft altitude, the air/fuel mixture is controlled by a pilot operated lever to make it leaner. Nowadays, all piston engines are fuel injected, carburettored types are disappearing fast.

A valve controls air mass aspiration and when closed there is no power, that is, $p_{\text{manifold}} = 0$. When the engine is running at full aspiration creating suction and p_{manifold} reads the highest suction values. If there is less propeller load at higher altitudes at the same rpm, then it will generate less power and the valve will be partially closed to inhale less air mass to running at equilibrium. Evidently, at low rpm the aspiration level would be low and there is a limiting p_{manifold} line. Therefore, the variables affecting engine power are rpm, p_{manifold}, altitude and atmospheric temperature for non-standard days. If the engine is supercharged then the graphs will indicate the effect.

The Lycoming IO-360 series power ratings at sea-level STD are as follows:

Takeoff. 180 HP at 2700 rpm consuming 15.6 US gal h^{-1} (this is the normal rated power – 100%)

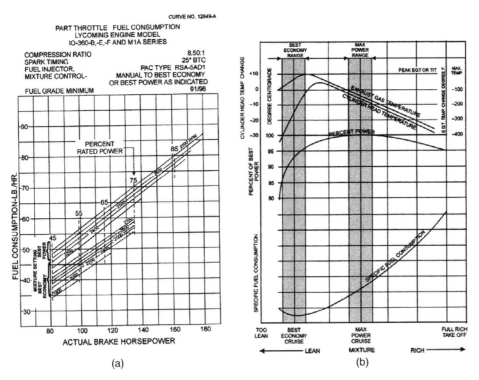

Figure 13.20 Lycoming IO-360 series fuel flow graph Courtesy of Textron-Lycoming Engine. (a) Fuel flow graph. (b) Piston engine operating range. Source: Reproduced with permission of John Wiley & Sons.

Typical climb. at 2500–2650 rpm (80–95% power)
Performance cruise. 135 HP at 2450 rpm consuming 11.0 US gal h^{-1} (75% power)
Economy cruise. 117 HP at 2350 rpm consuming 8.5 US gal h^{-1} (60% power)
Low speed cruise. at 2000–2300 rpm (light weight/holding)
Idle/Descent: below 2000 rpm

Figure 13.19 shows the parameters in graphical form and how to use it. The fuel flow graph is given separately in Figure 13.20.

Use Figure 13.20a is meant for fuel flow. It has two settings, one for *best power* and one for *best economy* – this is adjusted by mixture ratio lever. It can be seen in Figure 13.20b that for higher power and rpm the mixture setting is at best power and for cruise it is at best economy. Evidently, at 2000 rpm of the example is meant for best economy.

Example 13.1 Find power and fuel flow at 2000 rpm and at manifold pressure of 23.75 in Hg at 2400 ft altitude at ISA day. Also find the HP at ISA–20°F day. Take the inlet manifold temperature as the ambient temperature. (The Textron-Lycoming IO-360 series is a normally aspirated fuel injected engine).

Solution:
At ISA day

Step 1: In the first graph locate the point at 2000 rpm and at manifold pressure of 23.75. Draw a horizontal to HP axis giving 96 HP. This is at sea level.

Step 2: In the second graph locate the point 2000 rpm and at manifold pressure of 23.75. It gives 110 HP at 6.400 ft altitude.

Step 3: At 2400 ft altitude, interpolate by joining the two points by straight line and read 102 HP.

Fuel flow is obtained from Figure 13.20. It gives at about $42.5 \, \text{lb h}^{-1} = 7.4 \, \text{US gal h}^{-1}$. The aircraft is at slow speed cruise and possibly of light weight.

At ISA−20°F day

The expression for HP on a non-standard day, $[HP_{act}/HP_{std}] = \sqrt{(T_{std}/T_{act})}$.

Or $HP_{non\_STD} = \sqrt{\{(460 + T_{STD})/(460 + T_{non\_STD})\}} \times HP_{STD}$

At a 2400 ft altitude, $T_{std} = 510°F$, Therefore, $T_{act} = 490°F$

$[HP_{act}/HP_{std}] = \sqrt{(T_{std}/T_{act})} = \sqrt{(510/490)} = 1.02$

It gives,

$HP_{act} = 104 \, \text{HP}$

The expression gives approximately 1% HP variation for 10° R change from standard conditions.

With the old type of analogue gauge, a pilot will not know at what HP the aircraft is operating. Below 5000 ft and with fixed-pitch propeller (typical club/recreational flying) a pilot has only the throttle lever to control power. If he/she wishes to know the power, then he/she will have to record the rpm, manifold pressure, altitude and OAT (Outside Air Temperature) and compute the HP using the method shown. He/she needs to compute it after landing. Some modern electronic flight information system (EFIS) can display the power by processing the engine and air data using onboard computer.

Readers may obtain appropriate engine chart from the manufacturers for other engines. Failing that, this graph may be scaled for classroom usages.

13.7 Engine Performance Grid

Non-dimensional engine performance data as plotted in Figures 13.2–13.10 serve as generalised representations that are converted into actual installed engine performance data as given in Figures 13.11–13.15 for the Bizjet aircraft and Figure 13.15 for AJT aircraft. Since engine performance is affected by atmospheric conditions, thrust changes with altitude as density changes. Therefore, it is convenient to plot engine performance normalised to atmospheric conditions. This is done by using non-dimensional atmospheric property ratios as follows.

Density ratio $= \sigma = $ (density at the altitude ρ)/(density at sea level ρ_0)

Pressure ratio $= \delta = $ (pressure at the altitude p)/(pressure at sea level p_0)

Temperature ratio $= \theta = $ (temperature at the altitude T)/(temperature at sea level T_0)

From thermodynamics,

$\delta = \sigma\theta$

And the ratio of speed of sound

$a/a_0 = \sqrt{(T/T_0)} = \sqrt{\theta}$ or $a = a_0\sqrt{\theta}$

Mach number, $M = V/a$ where V is the aircraft velocity

(note: $\rho_0 = 0.002378 \, \text{slugs/ft}^3, p_0 = 2118.22 \, \text{lb ft}^{-2}, a_0 = 1118.4 \, \text{ft s}^{-1}$ in the imperial system)

In engineering practice, thrust is plotted as T/δ instead of thrust, T, alone. In the past, engineers used F/δ to represent thrust available and T/δ to represent thrust required by the aircraft; another way is to use subscripts, for example, $(T/\delta)_{avail}$ and $(T/\delta)_{reqd}$, respectively. Installed $(T/\delta)_{avail}$ can include intake losses for the particular nacelle for the chosen engine as intake efficiency is known. Drag of aircraft also depends on atmospheric density as well as aircraft weight. In equilibrium flight, thrust equals drag, giving $(T/\delta)_{\text{reqd}}$.

It gives $D = T$. Normalised to atmospheric properties,

$$D/\delta = (T/\delta)_{\text{reqd}}.$$

Note

$$D = 0.5\rho V^2 S_W\, C_D == 0.5(\rho_0\delta)(Ma_0)^2 S_W C_D = 1481 C_D M^2 S_W \delta$$

$$\text{or } D/\delta = 1481 C_D M^2 S_W \tag{13.2}$$

$$\text{or } (T/\delta)_{\text{reqd}} = 1481 C_D M^2 S_W \tag{13.3}$$

Since aircraft drag depends on aircraft weight, $(T/\delta)_{\text{required}}$ can be tied down to aircraft weight as W/δ keeping in line of normalisation. From W/δ the lift coefficient C_L can be determined, which in turn yields induced drag of the aircraft.

$$L = 0.5\rho V^2 S_W\, C_L == 0.5(\rho_0\delta)(Ma_0)^2 S_W C_L = 1481 C_L M^2 S_W \delta$$

$$\text{or } L/\delta = W/\delta = 1481 C_L M^2 S_W \tag{13.4}$$

Finally, by definition, fuel flow is expressed as,

$$FF = sfc \times T$$

$$\text{or } FF/\delta = sfc \times T/\delta \tag{13.5}$$

Chapter 15 dealt with specific range, SR, in detail. It is defined as Nautical Air Miles (nam) covered at the expense of unit fuel consumption.

In still air, Nautical Air Miles (nam) = Nautical Miles (nm) on the ground travelled by the aircraft.

With wind velocity V_W, $nm = nam \pm V_W$, where values are negative for a tail wind and positive for a head wind.

$$\text{Specific range} = SR = \text{nam/lb} = V/(T \times sfc) \tag{13.6}$$

It is convenient to plot using the Mach number, M, instead of velocity, V. Then $M = V/(a_0\sqrt{\theta})$.

Then

$$SR = \text{nam/lb} = M/(a_0\sqrt{\theta})/(T \times sfc)$$

Equation (13.5) can be rewritten as

$$FF/(\delta\sqrt{\theta}) = sfc \times T/(\delta\sqrt{\theta}) \tag{13.7}$$

The two normalised parameters are now $FF/(\delta\sqrt{\theta})$ and $sfc \times T/(\delta\sqrt{\theta})$.

13.7.1 Installed Maximum Climb Rating (TFE731-20 Class Turbofan)

Tables 13.7 and 13.8 show data taken from Figure 13.12 to plot the Bizjet engine performance grid or, simply, the engine grid. This gives available thrust and fuel flow rate of TFE731-20 class turbofan class engine. In this section, only the maximum cruise rating is plotted.

Atmospheric properties

Altitude ft	1500	5000	10 000	20 000	30 000	36 089	40 000	45 000	49 000
δ	0.971	0.832	0.688	0.49	0.297	0.223	0.185	0.146	0.1201
θ		0.9656	0.931	0.8625	0.7937	0.7518	0.7518	0.7518	0.7518
$\sqrt{\theta}$		0.9826	0.965	0.9287	0.891	0.867	0.867	0.867	0.867

Table 13.7 Installed maximum climb thrust: T/engine – lb.

Altitude (ft)	1500	5000	10 000	20 000	30 000	36 089	40 000	45 000
Mach 0.3	2640	2680	2450	1950	1390	1100	910	690
Mach 0.4	2400	2480	2300	1850	1350	1070	880	670
Mach 0.5	2200	2300	2180	1800	1320	1050	860	660
Mach 0.6	1980	2120	2080	1750	1300	1050	860	660
Mach 0.7	1830	1920	2000	1700	1320	1070	870	670

Table 13.8 Installed maximum climb thrust: (T/δ)/engine – lb.

Altitude (ft)	1500	5000	10 000	20 000	30 000	36 089	40 000	45 000
Mach 0.4	2719	3221	3561	3980	4680	4933	4919	4726
Mach 0.4	2472	2980	3343	3776	4546	4798	4757	4589
Mach 0.5	2266	2764	3198	3674	4444	4708	4649	4520
Mach 0.6	2039	2548	3023	3572	4377	4708	4648	4520
Mach 0.7	1885	2308	2907	3470	4377	4798	4702	4590

13.7.2 Maximum Cruise Rating (TFE731-20 Class Turbofan)

Tables 13.9–13.12 show data taken from Figure 13.13 to plot the Bizjet engine grid. This gives available thrust and fuel flow rate of a TFE731-20 class turbofan engine. In this section, only the maximum cruise rating is plotted.

Installed total maximum cruise thrust (T/δ) – lb and fuel flow $FF/(\delta\sqrt{\theta})$ – lb h^{-1} are plotted in Figure 13.21 as the engine grid for the Bizjet application. To obtain constant lines of $FF/(\delta\sqrt{\theta})$, a cross-plot is required using data given in the tables generated previously. These kinds of graphs are useful to evaluate aircraft performance

Table 13.9 Installed maximum cruise thrust: T/engine – lb.

Altitude (ft)	5000	10 000	20 000	30 000	36 089	40 000	45 000	49 000
Mach 0.5	1950	1740	1550	1170	914	770	615	510
Mach 0.6	1840	1650	1500	1155	910	770	615	510
Mach 0.7	1740	1600	1450	1150	910	770	615	510
Mach 0.74	1700	1580	1440	1150	910	770	615	515

Table 13.10 Installed maximum cruise thrust: (T/δ)/engine – lb.

Altitude (ft)	5000	10 000	20 000	30 000	36 089	40 000	45 000	49 000
Mach 0.5	2344	2529	3163	3940	4099	4162	4212	4246
Mach 0.6	2212	2398	3061	3889	4081	4162	4212	4246
Mach 0.7	2092	2326	2959	3872	4081	4162	4212	4246
Mach 0.74	2043.3	2297	2939	3872	4081	4162	4212	4288

Table 13.11 Installed maximum cruise thrust: Corresponding fuel flow: *FF*/engine – lb h^{-1}.

Altitude (ft)	5000	10 000	20 000	30 000	36 089	40 000	45 000	49 000
Mach 0.4	1530	1380	1100	770	620	530	430	370
Mach 0.5	1540	1400	1120	800	640	540	440	380
Mach 0.6	1560	1420	1150	830	660	580	455	390
Mach 0.7	1580	1440	1180	860	690	590	470	400
Mach 0.74	1600	1460	1200	880	710	600	485	410

Table 13.12 Installed maximum cruise thrust: Corresponding fuel flow: *FF*/($\delta \sqrt{\theta}$) per engine – lb h^{-1}.

Altitude (ft)	5000	10 000	20 000	30 000	36 089	40 000	45 000	49 000
Mach 0.4	1871	2079	2417	2910	3206	3304	3397	3552
Mach 0.5	1884	2109	2461	3024	3310	3367	3476	3649
Mach 0.6	1908	2139	2527	3137	3414	3616	3595	3745
Mach 0.7	1932	2169	2593	3250	3568	3678	3714	3842
Mach 0.74	1957	2199	2637	3326	3671	3740	3831	3938

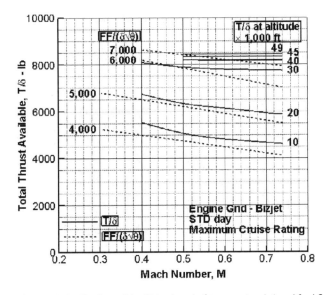

Figure 13.21 Engine grid – Bizjet (total of two engines). (read fuel flow values at the same point of thrust data).

by superimposing on to aircraft grid as done in Figure 13.21. An example of how to use the graph is shown in the next worked 13.2.

Example 13.2   Find the total thrust and fuel flow for the Bizjet flying at 45 000 ft ($\delta = 0.146$ and $\sqrt{\theta} = 0.867$) and Mach 0.5 using the engine grid given in Figure 13.21.

Solution:

From the graph corresponding to Mach 0.5 the (T/δ) at the 45 000 ft altitude line gives:

$(T/\delta) = 8400$ lb or (T/δ) per engine $= 4200$ lb.

It gives total thrust from two engines, $T = 8400 \times 0.146 = 1228.4$ lb, that is, T per engine ≈ 613 lb.

Compare tabulated value that gives (T/δ) per engine $= 4212$ lb and (T) per engine $= 615$ lb.

And the $FF/(\delta\sqrt{\theta})$ is interpolated between the dashed lines at the same point where (T/δ) is read.

In this case, total fuel flow, $FF/(\delta\sqrt{\theta}) = 6980$ lb h^{-1}. Or $FF/(\delta\sqrt{\theta})$ per engine $= 3490$ lb h^{-1}, that is, $FF = 442$ lb h^{-1}.

Compare tabulated value which gives $FF/(\delta\sqrt{\theta})$ per engine $= 3476$ lb h^{-1} and FF per engine $= 440$ lb h^{-1}.
(Tabulated computerised outputs are better than reading from graphs.)

Aircraft performance estimation for non-standard day requires the plots to be further normalised for a non-ISA day condition. It is not progressed further in this book. Readers may refer to aircraft engine text books. Since the 1990s, almost all aircraft manufactures started to use engine manufacturer supplied engine performance programmes (engine deck – computer programs) to obtain engine data at any condition and the use of generalised engine performance graphs gradually thinned out, applied only to where required as discussed in Chapter 15.

13.8   Some Turbofan Data (OPR = Overall Pressure Ratio)

Tables 13.13–13.15 give some currently operating turbofan data.

Table 13.13 Civil engine data – STD day.

Model	Takeoff					Cruise			
	T_{SLS} lb	Fan Diameter, in.	BPR	OPR	Air flow lb s^{-1}	Altitude 1000 ft	Mach	Thrust lb	tsfc lb (lb h)$^{-1}$
CF6-50-C2	52 500	134.1	4.31	30.4	1476	35	0.8	11 555	0.63
CF6-80-C2	52 500	88.4	4.31	28.4	1450	35	0.8	12 000	0.576
GE90-B4	87 400	134	8.4	39.3	3037	35	0.8	17 500	
JT8D-15A	15 500	49.2	1.04	18.6	327	30	0.8	4920	0.779
JT9D-59A	53 000	97	4.9	24.5	1639	35	0.85	11 950	0.666
PW2040	41 700	84.8	8.0	28.6	1210	35	0.85	6500	0.582
PW4052	52 000	97	5.0	28.5	1700				
PW4084	87 500	118.5	13.41	34.4	2550	35	0.83		
CFM56-3	23 500	60	5.0	22.6	655	35	0.85	4890	0.667
CFM56-5C	31 200	72.3	8.6	31.5	1027	35	0.8	6600	0.545
RB211-524B	50 000	85.8	4.5	28.4	1513	35	0.85	11 000	0.643
RB211-535E	40 100	73.9	4.3	25.8	1151	35	0.8	8495	0.607
RB211-882	84 700		8.01	39.0	2640	35	0.83	16 200	0.557
V2528-D5	28 000	63.3	4.7	30.5	825	35	0.8	5773	0.574
ALF502R	6970	41.7	5.7	12.2		35	0.7	2250	0.72
TFE731-20	3500	28.2	3.34	14.4	140	40	0.8	986	0.771
PW300	4750	38.2	4.5	23.0	180	40	0.8	1113	0.675
FJ44	1900	20.9	3.24	12.8	63.3	30	0.7	600	0.75
Olympus 593	38 000			8.3	410	53	2.0	10 030	1.15

Table 13.14 Military engine sea level static data at takeoff – STD day (FPS units as before).

Model	BPR	Weight lb	OPR	Air flow lb s^{-1}	Without afterburner Thrust lb	Without afterburner TSFC lb (lb h)$^{-1}$	With afterburner Thrust lb	With afterburner TSFC lb (lb h)$^{-1}$
P&W F119	0.45	3526	35		23 600		35 400	
P&W F100	0.36	3740	32	254.5	17 800	0.74	29 090	1.94
GE F110	0.77	3950	30.7	270	17 020		29 000	
GE F404	0.27	2320	26	146	12 000	0.84	17 760	1.74
GE F414	0.4	2645	30	170	12 600		22 000	
Snecma-M88	0.3	1980	24	143	11 240	0.78	16 900	1.8

Table 13.15 Turboprop data – STD day.

Model	SHP_{SLS}	Dry weight lb	psfc lb (h SHP)$^{-1}$
RR-250-B17	450	195	
PT6-A	850	328	
TPE-331-12	1100	400	0.5
GE-CT7	1940	805	
AE2100D	4590	1548	

References

1 Kundu, A.K. (2010). *Aircraft Design*. Cambridge University Press.
2 Riordan, D. *Lecture notes to QUB*, 2003.
3 Kundu, A.K., Price, M.A., and Riordan, D. (2016). *Theory and Practice of Aircraft Performance*, ISBN: 9871119074175. Wiley.
4 Generalised Method of Propeller Performance Estimation. Hamilton Standard Report PDB 6101 A, United Aircraft Corporation, June 1963.

14

Aircraft Sizing, Engine Matching and Variant Derivatives

14.1 Overview

Chapter 8 configured a preliminary geometry of a new aircraft based on the laid out specifications. It started with the estimated maximum takeoff mass (MTOM), wing reference area, S_W, and engine size from the available statistics of past designs for the class of aircraft. The fuselage geometry was determined to accommodate what it has to house, for example, in case of civilian aircraft, the number of passengers and in the case of military designs, the engine, fuel volume, equipment and so on. Chapter 9 laid out the undercarriage based on an estimated aircraft aft centre of gravity (CG) position. Empennage size is linked with wing S_W and where it is placed in relation to it. With these preliminary aircraft component geometries, a proper estimate of aircraft MTOM and CG position was evaluated in Chapter 10. At this point, if the estimated MTOM differed (very likely) from the guessed MTOM, the first iteration is required to reconfigure the aircraft primarily affecting the S_W. Together, this presented a 'concept definition' of the new aircraft project at the Phase I stage of conceptual design study.

The next step in the conceptual design study is the size of the aircraft with the proper reference wing area S_W with a matched engine to be more precise than a *guesstimate* taken from statistics.

The aircraft sizing and engine matching routine is a necessary procedure for a new aircraft project starting early at the conceptual design stage. This is to freeze the aircraft configuration to *concept finalisation* to obtain the project go-ahead. The exercise requires aircraft performance analyses for which aircraft drag characteristics should be known along with engine performance details to provide the exact power required. Aircraft drag polar and engine performance capability for the full operational envelope are generated in Chapters 11 and 13, respectively.

The sizing and engine matching exercise are carried out in this chapter to ensure the best configuration to satisfy customer specifications and the mandatory airworthiness requirements. The procedure does not require full aircraft performance analyses. Only the airworthiness requirements and customer specification requirements are checked out in a parametric search as will be shown. Aircraft sizing and engine matching exercise does not guarantee overall aircraft performance in detail but make certain that the new project is capable of meeting the necessary specifications and requirements. After freezing the design to the final configuration through a formal sizing and engine matching exercise, a complete aircraft performance is carried out in Chapter 15. Aircraft performance of the aircraft in extensive details is carried in [1, 2] and may be used as companion books, if required.

It was mentioned earlier that a family of derivative variant aircraft can be offered to cover a wider market at a lower development cost by retaining component commonality. The possible variants for the worked-out examples are offered in Chapter 8. Typically for civil aircraft, the variant designs (at least three in the family) are offered with the baseline aircraft is in the middle plus one bigger growth version and one smaller shrunk version retaining maximum component commonality within all three designs.

Therefore, the objective of the sizing and engine matching exercise is to fine-tune the baseline configuration in the middle is to present a 'satisfactory' configuration offering a family of variant designs. None may be the optimum but together they offer the best fit to satisfy many customers (i.e. operators) and to encompass a wide

Conceptual Aircraft Design: An Industrial Approach, First Edition. Ajoy Kumar Kundu, Mark A. Price and David Riordan.
© 2019 John Wiley & Sons Ltd. Published 2019 by John Wiley & Sons Ltd.

range of payload-range requirements, resulting in increased sales and profitability. To repeat, *optimisation of individual goals through separate design considerations may prove counterproductive and usually prevent the overall (global) optimisation of ownership cost.* Designers' experience is vital to success; it has to be used.

This chapter proposes a formal methodology to reach an appropriate concept definition of the new aircraft project offering variant designs by refining the concept definition based on statistics. At this stage, another iteration will be required if the sized wing reference area, S_W and engine size are different from the initial values taken from statistics. The iteration goes through the full cycle from Chapters 8–11. The newly initiated will invariably face it and must not neglect the iteration.

The two classic important parameters – (i) wing loading (W/S_W) and (ii) thrust loading (T_{SLS}/W) are instrumental in the methodology for aircraft sizing and engine matching. The formal methodology seeks to obtain the sized W/S_W and T_{SLS}/W for the baseline aircraft. These two loadings alone provide sufficient information to conceive aircraft configuration in a preferred aircraft size with a family of variant designs. Empennage size is governed by wing size and location of the CG. This study is possibly the most important aspect in the development of an aircraft, finalising the external geometry.

14.1.1 Summary

Because the preliminary configuration is based on past experience and statistics, an iterative procedure ensues to fine-tune the aircraft converging into the correct size of the wing reference area for a family of variant aircraft designs and matched engines. Wallace [3] provides an excellent presentation on the subject and Loftin [4] provides an aircraft sizing exercise in detail.

This chapter includes the following sections:

Section 14.2: Introduction
Section 14.3: Theory
Section 14.4: Coursework Exercises for Civil Aircraft (Bizjet)
Section 14.5: Sizing Analysis and Variant Designs: Civil Aircraft (Bizjet)
Section 14.6: Coursework Exercises for Military Aircraft (AJT)
Section 14.7: Sizing analysis and Variant Designs: Military Aircraft (AJT)
Section 14.8: Aircraft Sizing Studies and Sensitivity Analyses
Section 14.9: Discussions

Coursework content. This is an important chapter. Readers are to compute the parameters that establish the criteria for aircraft sizing and engine matching. The final size is unlikely to be identical to the preliminary configuration obtained in Chapter 8 using statistics of past designs. It is recommended that the readers make use of spreadsheets to carry out the iterations.

14.2 Introduction

In a systematic manner, the conception of a new aircraft progresses from generating market specifications followed by the preliminary candidate configurations that rely on statistical data of past designs in order to arrive at a baseline design, selected from several candidate configurations. In this chapter, the baseline design of an aircraft is formally sized with a matched engine (or engines) along with the family of variants to finalise the configuration (i.e. external geometry and MTOM. An example from each class of civil (i.e. the Bizjet) and military (i.e. the Advanced Jet Trainer, AJT) aircraft are used to substantiate the methodology. The turboprop trainer (TPT) sizing follows the same routine as done for the AJT and is left to the readers to carry out.

14.2.1 Civil Aircraft

Based on circa 2000 fuel prices, the aircraft cost contributes to the direct operating cost (DOC) three to four times the contribution made by the fuel cost. It is not cost effective for aircraft manufacturers to offer custom-made new designs to each operator with varying payload-range requirements. Therefore, aircraft manufacturers offer aircraft in a family of variant designs. This approach maintains maximum component commonality within the family to reduce development costs and is reflected in aircraft unit-cost savings. In turn, it eases the amortisation of nonrecurring development costs, as sales increase. It is therefore important for the aircraft sizing exercise to ensure that the variant designs are least penalised to maintain commonality of components. This is what is meant by producing satisfying robust designs; these are not necessarily the optimum designs. Sophisticated multi-disciplinary optimisation (MDO) is not easily amenable to ready use. The industries also use parametric search for a satisfying robust design.

To generate a family of variant civil aircraft designs, the tendency is to retain the wing and empennage as almost unchanged while plugging and unplugging the constant fuselage to cope with varying payload capacities. Typically, the baseline aircraft remains as the middle design. The smaller aircraft results in a wing that is larger than necessary, providing better field performances (i.e. takeoff and landing); however, the cruise performance is slightly penalised. Conversely, larger aircraft have smaller wings that improve the cruise performance; the shortfall in takeoff is overcome by providing a higher thrust-to-weight ratio (T_{SLS}/W) and possibly with better high-lift devices, both of which incur additional costs. The baseline aircraft approach speed, V_{app}, initially is kept low enough so that the growth of V_{app} for the larger aircraft is kept within the specifications. Of late, high investment with advanced composite wing-manufacturing method is in a better position to produce adjusted wing sizes for each variant (large aircraft) in a cost effective manner, offering improved economics in the long run. However, for some time to come, metal wing construction is to continue with minimum changes in wing size to maximise component commonality.

Matched engines are also used in a family to meet the variation of thrust (or power) requirements for the aircraft variants. The sized engines are bought-out items supplied by engine manufacturers. Aircraft designers stay in constant communication with engine designers in order to arrive at the family of engines required. A thrust variation of up to $\pm30\%$ from the baseline engine is typically sufficient for larger and smaller aircraft variants from the baseline. Engine designers can produce scalable variants from a proven core gas-generator module of the engine – these scalable projected engines are known loosely as *rubberised* engines. The thrust variation of a rubberised engine does not affect the external dimensions of an engine (typically, the bare engine length and diameter change only around $\pm2\%$). This book uses an unchanged nacelle external dimension for the family variants, although there is some difference in weight for the different engine thrusts. The generic methodology presented in this chapter is the basis for the sizing and matching practice.

14.2.2 Military Aircraft

It follows the same procedure as in the civil aircraft methodology except that the specifications are different. The turn rate capability is an additional parameter that makes a decisive contribution to the design (see Section 14.3).

14.3 Theory

The parameters required for aircraft sizing and engine matching derive from market studies and must satisfy user specifications and the certification agencies' requirements. In general, both civil and military aircraft use similar type of specifications, as given next, as the basic input for aircraft sizing. All performance analyses in this chapter are carried out at an International Standard Atmosphere (ISA) day and all field performances are at sea level.

  i. Payload and range (fuel load): These determine the maximum takeoff weight (MTOW).
 ii. Takeoff field length (TOFL): This determines the engine-power ratings and wing size.
iii. Landing field length (LFL): This determines wing size (baulked landing included).
 iv. Initial maximum cruise speed and altitude capabilities determine wing and engine sizes.
  v. Initial rate of climb establishes wing and engine sizes.

Additional specifications for military aircraft sizing are as follows:

vi. Turn performance *g*-load (in the horizontal plane)
vii. Manoeuvre *g*-load (turning in any plane)
viii. Roll rate (control sizing issue – not dealt with here)

The specifications as requirements for category of aircraft must be satisfied simultaneously. The governing parameters to satisfy TOFL, initial climb, initial cruise, turn rate and landing are wing loading (W/S_W), and thrust loading (T_{SLS}/W).

Normally, thrust sizing for initial climb rates proves sufficient to perform the turning (*g*-load) requirements. It is assumed there is control authority available to execute the manoeuvres. A lower wing aspect ratio (AR) is considered for higher roll rates to reduce the wing-root bending moments as well as to gain in turn rate.

As mentioned previously, an aircraft must simultaneously satisfy the TOFL, initial climb rate, initial maximum cruise speed-altitude capabilities and LFL. Low wing loading (i.e. a larger wing area) is required to sustain low speed at lift-off and touchdown (for a pilot's ease), whereas high wing loading (i.e. a low wing area) is suitable at cruise because high speeds generate the required lift on a smaller wing area. A larger wing area necessary for takeoff/landing results in an excess wing for high-speed cruise. This may require suitable high-lift devices to keep the wing area smaller. The wing area is sized in conjunction with a matched engine for takeoff at maximum takeoff rating, climb at maximum continuous/climb rating, cruise at maximum cruise rating and landing at idle-engine rating. To obtain the minimum wing area for the chosen high-lift device with matched sized engine to satisfy all the requirements is the core of the aircraft sizing and engine matching exercise.

In general, W/S_W varies with time as fuel is consumed and T/W is throttle-dependent. Therefore, the reference design condition of the MTOM and T_{SLS} at sea level per ISA day are used for sizing considerations. This means that the *MTOM*, T_{SLS} and S_W are the only parameters considered for aircraft sizing and engine matching.

In general, wing size variations are associated with changes in all other affecting parameters (e.g. *AR*, λ and wing sweep). However, at this stage, they are kept invariant – that is, the variation in wing size scales the wing span and chord, leaving the general planform unaffected (like zoom in/zoom out).

At this point, readers require knowledge of aircraft performance and the important derivations of the equations used are provided in Chapter 15. Since the aircraft performance analyses has to be carried out for the sized aircraft, this chapter precedes to obtain the sized aircraft first, requiring referring to Chapter 15 for the derivations of the equations used in this chapter. Other proven semi-empirical relations are given in [4]. Although the methodology described herein is the same, the industry practice is more detailed and involved in order to maintain a high degree of accuracy.

Worked-out examples continue with the *Bizjet* (Learjet45 class) for civil aircraft and AJT (BAE Hawk class) for military aircraft. Throughout this chapter, *wing loading* (W/S_W) in the SI system is in $N\,m^{-2}$ (or $kg\,m^{-2}$) to align with the thrust (in Newton) in *thrust loading* (T_{SLS}/W) as a non-dimensional parameter. Imperial units are given.

14.3.1 Sizing for Takeoff Field Length (TOFL) – Two Engines

To keep analyses simple, only a two-engine aircraft is sized and engine matched. This is because the sizing and engine matching exercise requires aircraft drag polar and in this book only the two-engine Bizjet drag polar is worked out. However, three- and four-engine aircraft use exactly the same procedure, equations modified by the number of engines in question.

TOFL is the field length required to clear a 35-ft (10-m) obstacle while maintaining a specified minimum climb gradient, γ, with one engine inoperative and flaps and undercarriage extended (Figure 14.1). The Federal Aviation Regulation (FAR) requirements for a two-engine aircraft minimum second segment climb gradient are given in Section 15.3.3.

For sizing, field length calculations are at the sea level standard day (no wind) and at a zero airfield gradient of paved runway. For further simplification, drag changes are ignored during the transition phase of lift-off to

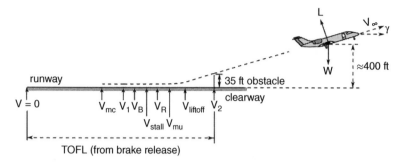

Figure 14.1 Sizing for takeoff.

clear the obstacle (flaring takes about two to four seconds); in other words, the equations applied to *Vlift-off* are extended to V_2.

Chapter 15 addresses takeoff performance in detail. Eq. (15.8) gives

$$\text{TOFL} = \int dS_G = \int V dt = \int V(dV/dV)dt = \int_0^{V_2} (V/a)dV = \sum_{\Delta V_i}^{\Delta V_f} \Delta S_{Gi} \text{ where } dV/dt = a \quad (14.1)$$

where $dV/dt = a$ is the instantaneous acceleration and V is the instantaneous velocity. The aircraft on the ground encounters rolling friction (coefficient $\mu = 0.025$ for a paved, metalled runway). Average acceleration, $\bar{a}$, is taken at $0.7V_2$. FAR requires $V_2 = 1.2V_{\text{stall}}$. This gives $V_2^2 = [2 \times 1.44 \times (W/S_W)]/(\rho C_{\text{Lstall}})$. An aircraft stalls at C_{Lmax}. By substituting the value of V_2, the Eq. (14.1) reduces to:

Its simplified expression for all engines operating is:

$$\text{TOFL} = (1/\bar{a}) \int_0^{V2} V dV = (V_2^2/2\bar{a}) = \frac{1.44 \, W/S}{\rho C_{\text{Lstall}}\bar{a}} \quad (14.2)$$

Writing in terms of wing loading Eq. (14.2) can be written as

$$(W/S_W) = (\text{TOFL} \times \rho \times \bar{a} \times C_{\text{Lstall}})/1.44 \quad (14.3)$$

where average acceleration, $\bar{a} = F/m$ and applied force $F = (T - D) - \mu(W - L)$.

Note that, until lift-off is achieved, $W > L$ and F is the average value at $0.7V_2$.

$$\text{Therefore}, \bar{a} = [(T - D) - \mu(W - L)]g/W = g(T/W)[1 - D/T - \mu W/T + \mu L/T]_{@0.7V2} \quad (14.4)$$

Substituting in Eq. (14.3) it becomes

$$(W/S_W) = (\text{TOFL} \times \rho \times [g(T/W)(1 - D/T - \mu W/T + \mu L/T)] \times C_{\text{Lstall}})/1.44 \quad (14.5)$$

In the foot–pound system (FPS) it can be written as ($\rho = 0.00238$ slugs and $g = 32.2 \, \text{ft s}^{-2}$)

$$(W/S_W) = (\text{TOFL} \times (T/W)(1 - D/T - \mu W/T + \mu L/T) \times C_{\text{Lstall}})/18.85 \quad (14.6a)$$

In the SI system it becomes ($\rho = 1.225 \, \text{kg m}^{-3}$ and $g = 9.81 \, \text{m s}^{-2}$)

$$(W/S_W) = 8.345 \times (\text{TOFL} \times (T/W)(1 - D/T - \mu W/T + \mu L/T) \times C_{\text{Lstall}}) \quad (14.6b)$$

here W/S_W is in N m^{-2} to remain in alignment with the units of thrust in newtons.

Checking of the second-segment climb gradient occurs after aircraft drag estimation, which is explained in Chapter 11. If climb gradient falls short of the requirement, then the T_{SLS} must be increased. In general, the stringent TOFL requirements also likely to satisfy the second segment climb gradient.

14.3.1.1 Civil Aircraft Design: Takeoff

The contribution of the last three terms $(-D/T - \mu W/T + \mu L/T)$ in Eq. (14.4) is minimal and can be omitted at this stage for the sizing calculation. In addition, for the one-engine-inoperative condition after the decision speed V_1, the acceleration slows down, making the TOFL longer than the all engines operative case. Therefore, in the sizing computations to produce the specified TOFL, further simplification is possible by applying a semi-empirical correction factor primarily to compensate for loss of an engine. The correction factors are as follows [2]; all sizing calculations are performed at the MTOW and with T_{SLS}.

For two engines, use a factor of 0.5 (loss of thrust by a half).

Then, Eqs. (14.6a) and (14.6b) in the FPS system reduces to:

$$(W/S_{\text{W}}) = (TOFL \times (T/W) \times C_{\text{Lstall}})/37.5 \tag{14.7a}$$

For the SI system:

$$(W/S_{\text{W}}) = 4.173 \times TOFL \times (T/W) \times C_{\text{Lstall}} \tag{14.7b}$$

For three engines, use a factor of 0.66 (loss of thrust by a third). Then, Eq. (14.6) in the FPS system reduces to:

$$(W/S_{\text{W}}) = (TOFL \times (T/W) \times C_{\text{Lstall}})/28.5 \tag{14.8a}$$

For the SI system:

$$(W/S_{\text{W}}) = 5.5 \times TOFL \times (T/W) \times C_{\text{Lstall}} \tag{14.8b}$$

For four engines, use a factor of 0.75 (loss of thrust by a quarter). Then, Eq. (14.6) in the FPS system reduces to:

$$(W/S_{\text{W}}) = (TOFL \times (T/W) \times C_{\text{Lstall}})/25.1 \tag{14.9a}$$

For the SI system:

$$(W/S_{\text{W}}) = 6.25 \times TOFL \times (T/W) \times C_{\text{Lstall}} \tag{14.9b}$$

14.3.1.2 Military Aircraft Design: Takeoff

Because military aircraft mostly have a single engine, there is no requirement for one engine being inoperative; ejection is the best solution if the aircraft cannot be landed safely. Therefore, Eqs. (14.6a) and (14.6b) can be directly applied (for a multiengine design, the one-engine-inoperative case, generally, uses measures similar to the civil aircraft case). In the FPS system, this can be written as:

$$(W/S_{\text{W}}) = (TOFL \times (T/W)(1 - D/T - \mu W/T + \mu L/T)] \times C_{\text{Lstall}})/18.85 \tag{14.10a}$$

In the SI system, it becomes:

$$(W/S_{\text{W}}) = 8.345 \times (TOFL \times (T/W)(1 - D/T - \mu W/T + \mu L/T)] \times C_{\text{Lstall}}) \tag{14.10b}$$

Military aircraft have a thrust, T_{SLS}/W, that is substantially higher than civil aircraft, which makes $(D/T - \mu W/T + \mu L/T)$ even smaller. Therefore, for a single-engine aircraft, no correction is needed and the simplified equations are as follows:

In the FPS system, this can be written as:

$$(W/S_{\text{W}}) = [TOFL \times (T/W) \times C_{\text{Lstall}}]/18.85 \tag{14.11a}$$

In the SI system, this becomes:

$$(W/S_{\text{W}}) = 8.345 \times (TOFL \times (T/W) \times C_{\text{Lstall}}) \tag{14.11b}$$

14.3.2 Sizing for the Initial Rate of Climb (All Engines Operating)

The initial unaccelerated rate of climb is a user specification and not a FAR requirement; this is when the aircraft is heaviest in climb. In general, the FAR requirement for the one-engine-inoperative gradient provides sufficient margin to give a satisfactory all-engine initial climb rate. However, from the operational perspective, higher rates of climb are in demand when it is sized accordingly. Military aircraft (some with a single engine) requirements stipulate faster climb rates and sizing for the initial climb rate is important. The methodology for aircraft sized to the initial climb rate is described in this section. Sizing exercise for climb is at unaccelerated rate of climb. Figure 14.2 shows a typical climb trajectory.

For a steady-state climb, the Eq. (15.5) gives the expression for rate of climb, $RC = V \times \sin \gamma$.

Steady-state force equilibrium gives $T = D + W \times \sin \gamma$ or $\sin \gamma = (T - D)/W$.

This gives

$$RC = [(\text{T} - \text{D}) \times V]/W = (\text{T}/W - \text{D}/W) \times V \tag{14.12}$$

Equation (14.12) is written as

$$\text{T}/W = RC/\text{V} + (\text{D}/W)$$

$$\text{or } \text{T}/W = RC/\text{V} + [(C_D \times 0.5 \times \rho \times V^2 \times S_W)/W] \tag{14.13}$$

Equation (14.13) is based on climb thrust rating, which is lower than T_{SLS}. Equation (14.13) needs to be written in terms of T_{SLS}. The ratio, $T_{\text{SLS}}/T = k_{\text{cl}}$ varies depending on the engine bypass ratio (BPR).

$$[T_{\text{SLS}}/W]/k_{\text{cl}} = RC/\text{V} + [(C_D \times 0.5 \times \rho \times V^2)S_W/W] \tag{14.14}$$

$$[T_{\text{SLS}}/W] = k_{\text{cl}} \times RC/\text{V} + k_{\text{cl}} \times [(C_D \times 0.5 \times \rho \times V^2)S_W/W] \tag{14.15}$$

The drag polar is now required to compute the relationships given in Equation (14.14).

14.3.3 Sizing to Meet Initial Cruise

There are no FAR or Milspec regulations to meet the initial cruise speed; initial cruise capability is a user requirement. Therefore, both civil and military aircraft sizing for initial cruise use the same equations. At a steady-state level flight, thrust required (aeroplane drag, D) = thrust available (T_a); that is:

$$D = T_a = 0.5\rho V^2 C_D \times S_W \text{ where } T_a \text{ is thrust available} \tag{14.16}$$

Dividing both sides of the equation by the initial cruise weight, $W_{\text{in\_cr}} = k \times MTOW$ due to fuel burned to climb to the initial cruise altitude. The factor k lies between 0.95 and 0.98 (climb performance details are not available during the conceptual design phase), depending on the operating altitude for the class of aircraft, and it can be fine-tuned through iterations: in the coursework exercise, one round of iterations is sufficient. The factor cancels out in the following equation but is required later. Henceforth, in this part of cruise sizing, W represents the $MTOW$, in line with the takeoff sizing:

$$0.5\rho V^2 C_D \times S_W/W = T_a/W \tag{14.17}$$

Figure 14.2 Aircraft climb trajectory.

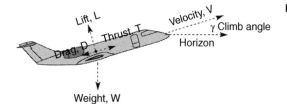

The drag polar gives the C_D value to correspond to the initial cruise C_L (because they are non-dimensional, both the FPS and SI systems provide the same values). Initial cruise:

$$C_L = k \times MTOW/(0.5 \times \rho \times V^2 \times S_W) \tag{14.18}$$

The thrust-to-weight ratio sizing for initial cruise capability is expressed in terms of T_{SLS}. Eq. (14.18) is based on the maximum cruise thrust rating, which is lower than the T_{SLS}. Eq. (14.18) must be written in terms of T_{SLS}. The ratio, $T_{SLS}/T_a = k_{cr}$, varies depending on the engine BPR. The factor k_{cr} is computed from the engine data supplied. Then, Eq. (14.18) can be rewritten as:

$$T_{SLS}/W = k_{cr} \times 0.5\rho V^2 \times C_D/(W/S_W) \tag{14.19}$$

Variation in wing size affects aircraft weight and drag. The question now is: How does the C_D change with changes in W and S_W? (T_a changes do not affect the drag because it is assumed that the physical size of an engine is not affected by small changes in thrust.) The solution method is to work with the wing only – first by scaling the wing for each case and then by estimating the changes in weight and drag and iterating – which is an involved process.

This book simplifies the method by using the same drag polar for all wing-loadings (W/S_W) with little loss of accuracy. As the wing size is scaled up or down (the AR invariant), it changes the parasite drag. The induced drag changes as the aircraft weight increases or decreases. However, the C_D increase with wing growth is divided by a larger wing, which keeps the C_D change minimal.

14.3.4 Sizing for Landing Distance

The most critical case is when an aircraft must land at its maximum landing weight of 0.98 MTOW. In an emergency, an aircraft lands at the same airport for an aborted takeoff procedure, assuming a 2% weight loss due to fuel burn in order to make the return circuit. Pilots prefer to approach as slow as possible for ease of handling at landing.

For the Bizjet class of aircraft, the approach velocity, V_{app} (FAR requirement at $1.3\,V_{stall}$) is less than $125\,kts$ to ensure that it is not constrained by the minimum control speed, V_c. Wing C_{Lstall} is at the landing flap and slat setting. For sizing purposes, an engine is set to the idle rating to produce zero thrust.

At approach: $\sqrt{V_{stall}} = [(0.98W/S_W)/(0.5 \times \rho \times C_{Lstall})]$ (14.20)

At landing $V_{app} = 1.3\,V_{stall}$.

Therefore,

$$V_{app} = 1.3 \times \sqrt{[(0.98W/S_W)/(0.5 \times \rho \times C_{Lstall})]} = 1.793 \times \sqrt{[(W/S_W)/(\rho \times C_{Lstall})]} \tag{14.21}$$

or $(V_{app})^2 \times C_{Lstall} = 3.211 \times (W/S_W)/\rho$

or $(W/S_W) = 0.311 \times \rho \times (V_{app})^2 \times C_{Lstall}$ (14.22)

14.4 Coursework Exercise – Civil Aircraft Design (Bizjet)

Both the FPS and the SI units are used for the worked-out examples. Sizing calculations require the engine data in order to obtain the k factors used. The Bizjet drag polar is provided in Figure 11.2.

14.4.1 Takeoff

Requirements. TOFL $4400\,ft$ ($1341\,m$) to clear a 35-ft height at ISA + sea level. The maximum lift coefficient at takeoff (i.e. flaps down to 20° and no slat) is $C_{Lstall(TO)} = 1.9$ (obtained from testing and computational fluid dynamics (CFD) analysis). Installation losses are taken to 7% that reduces takeoff thrust to $0.93\,T_{SLS}$. Using Eq. (14.7a), the expression reduces to:

Using Eq. (14.7a), the expression reduces to $W/S_W = 4400 \times 1.9 \times (0.93 \times T_{SLS}/W)/37.5 = 207.3 \times (T/W)$.
Using Eq. (14.7b), it becomes $W/S_W = 4.173 \times 1341 \times 1.9 \times (0.93 \times T_{SLS}/W) = 9889.2 \times (T_{SLS}/W)$
The result is computed in Table 14.1.

The industry must also examine other takeoff requirements (e.g. an unprepared runway) and hot and high ambient conditions.

14.4.2 Initial Climb

From the market requirements, an initial climb starts at an 1400-ft altitude ($\rho = 0.00232$ slugs/ft^3) at a speed $V_{EAS} = 250$ kt (Mach 0.38) $= 250 \times 1.68781 = 422$ ft s^{-1} (128.6 m s^{-1}) and the required rate of climb, RC $= 2600$ ft min^{-1} (792.5 m min^{-1}) $= 43.33$ ft s^{-1} (13.2 m s^{-1}). For the class of TFE731 engine data, Figure 13.3 gives the uninstalled $k_{cl} = T/T_{SLS} = 0.705$. Taking 5% installation loss during climb, the installed $k_{cl} = T/T_{SLS} = 0.95 \times 0.705 = 0.67$ that gives $T_{SLS}/T = 1.5$. The undercarriage and high-lift devices are in a retracted position. The result is computed in Table 14.2.

Lift coefficient,

$$C_{Lclimb} = W/(0.5 \times 0.00232 \times 422^2\, S_W) = 0.00484 \times W/S_W$$

Using Eq. (14.14),

$$[T_{SLS}/W]/1.5 = 43.33/422 + (C_D \times 0.5 \times 0.00232 \times 422^2\, S_W)/W)$$

$$T_{SLS}/W = 0.154 + 310 \times C_D \times (S_W/W)$$

14.4.3 Cruise

Requirements. Initial cruise speed must meet the high-speed cruise (HSC) at Mach 0.75 and at 41 000 ft (flying higher than bigger jets in less congested traffic corridors). Using Eq. (14.19), the result is computed in Table 14.3 (for the initial cruise aircraft weight use $k = 0.972$ in.).

In FPS at 41 000 ft:

$$\rho = 0.00056 \text{ slug/ft}^2 \text{ and } V^2 = (0.75 \times 968.076)^2 = 527\,159 \text{ ft}^2 \text{ s}^{-2}$$

In SI at 12 519 m altitude:

$$\rho = 0.289 \text{ kg m}^{-3}, V^2 = (0.75 \times 295.07)^2 = 48\,975 \text{ m}^2 \text{ s}^{-2}.$$

Table 14.1 Bizjet takeoff sizing calculations.

W/S_W (FPS – lb ft^{-2})	40	50	60	70	80
W/S_W (SI – N m^{-2})	1916.9	2395.6	2874.3	3353.7	3832.77
T_{SLS}/W (installed)	0.193	0.24	0.29	0.338	0.386

Table 14.2 Bizjet climb sizing calculations (use Figure 11.2 for the drag polar).

W/S_W (lb ft^{-2})	40	50	60	70	80
W/S_W (N m^{-2})	1916.9	2395.6	2874.3	3353.7	3832.77
C_{Lclimb}	0.194	0.242	0.290	0.339	0.378
C_D (from drag polar)	0.024	0.0246	0.0256	0.0266	0.0282
T_{SLS}/W (installed)	0.34	0.31	0.286	0.272	0.265

Table 14.3 Bizjet cruise sizing.

W/S_W (lb ft^{-2})		40	50	60	70	80
W/S_W (N m^{-2})		1916.9	2395.6	2874.3	3353.7	3832.77
C_L	0.271	0.339	0.4064	0.474	0.542	0.619
C_D (from drag polar)		0.0255	0.0269	0.0295	0.033	0.0368
T/W @ 41 000 ft		0.36	0.305	0.278	0.267	0.26

Equation (14.18) gives the initial cruise:

$$C_L = 0.972\, MTOW/(0.5 \times 0.289 \times 47\,677.5 \times S_W) = 0.0001414\,(W/S_W), \text{where } W/S_W \text{ is in N m}^{-2}.$$

Equation (14.19) gives:

$$T_{SLS}/W = k_{cr} \times 0.5\rho V^2 \times C_D/(W/S_W)$$

For the class of TFE731 engine data, Figure 13.4 gives the uninstalled $k_{cr} = T/T_{SLS} = 0.23$. Taking 4% installation loss during cruise, the installed $k_{cl} = T/T_{SLS} = 0.94 \times 0.23 = 0.222$ that gives $T_{SLS}/T = 4.5$.
In FPS:

$$T_{SLS}/W = 4.5 \times 0.5 \times 0.00056 \times 459\,208.2 \times C_D/(W/S_W) = 565.73 \times C_D/(W/S_W)$$

In SI:

$$T_{SLS}/W = 4.5 \times 0.5 \times 0.289 \times 42\,662.5 \times C_D/(W/S_W) = 27\,741.3 \times C_D/(W/S_W)$$

Using Figure 11.2 for the drag polar, Table 14.3 gives the computed values for Bizjet cruise sizing.

14.4.4 Landing

From the market requirements, $V_{app} = 120$ knots $= 120 \times 1.68781 = 202.5\,\text{ft s}^{-1}$ (61.72 m s^{-1}). Landing $C_{Lmax} = 2.1$ at a 40° flap setting (from testing and CFD analysis). For sizing purposes, the engine is set to the idle rating, producing zero thrust. Using Eq. (14.22) the following is obtained.
In the FPS system, $W/S_W = 0.311 \times 0.002378 \times 2.1 \times (202.5)^2 = 63.8\,\text{lb ft}^{-2}$. In the SI system, $W/S_W = 0.311 \times 1.225 \times 2.1 \times (61.72)^2 = 3052\,\text{N m}^{-2}$. Because the thrust is zero (i.e. idle rating) at landing, the W/S_W remains constant.

14.5 Sizing Analysis – Civil Aircraft (Bizjet)

The four sizing relationships for wing loading, W/S_W, and thrust loading, T_{SLS}/W, meet (i) takeoff, (ii) approach speed for landing, (iii) initial cruise speed and (iv) initial climb rate. These are plotted in Figure 14.3. The circled point in Figure 14.3 is the most suitable for satisfying all four requirements simultaneously. To ensure performance, there is a tendency to use a slightly higher thrust loading T_{SLS}/W; in this case, the choice becomes as follows.
$T_{SLS}/W = 0.34$ at a wing loading of $W/S_W = 64\,\text{lb ft}^{-2}$ (2894 N m^{-2}). This keeps some margin for future weight growth through modifications and some drag growth as shown in Figure 14.3.
So far, the Bizjet aircraft data has been given. This chapter uses the Bizjet size measurements to check the extent of differences in aircraft geometry, engine size and weight. Now is the time for the iterations required to fine-tune the given preliminary configuration. At 20 720 lb (9400 kg) MTOM, the wing planform area is 325 ft^2, close to the original area of 323 ft^2; hence, no iteration is required. Otherwise, it is necessary to revisit the empennage sizing and revise the weight estimates. The T_{SLS} per engine then becomes

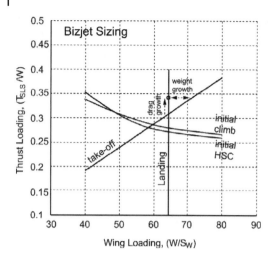

Figure 14.3 Aircraft sizing – civil aircraft.

$0.34 \times 20\,720/2 = 3523\,\text{lbs} \approx 3500\,\text{lb/engine}$. This is very close to the TFE731-20 class of engine; therefore, the engine size and weight remain the same. For this reason, iteration is avoided; otherwise, it must be carried out to fine-tune the mass estimation.

The entire sizing exercise could have been conducted well in advance, even before a configuration was settled – if the chief designer's past experience could guesstimate a close drag polar and MTOM. Statistical data of past designs are useful in guesstimating aircraft close to an existing design. Generating a drag polar requires some experience with extraction from statistical data. Subsequently, the sizing exercise is fine-tuned to better accuracy.

In the industry, more considerations are addressed at this stage – for example, what type of variant design in the basic size can satisfy at least one larger and one smaller capacity (i.e. payload) size. Each design may have to be further varied for more refined variant designs.

14.5.1 Variants in the Family of Aircraft Designs

The family concept of aircraft design is discussed in previous chapters and highlighted again at the beginning of this chapter. Maintaining large component commonality (genes) in a family is a definite way to reduce design and manufacturing costs – in other words, 'design one and get two or more almost free'. This encompasses a much larger market area and, hence, increased sales to generate resources for the manufacturer and nation. The amortisation is distributed over larger numbers, thereby reducing aircraft costs.

Today, all manufacturers produce a family of derivative variants. The Airbus320 series has four variants and more than 3000 have been sold. The Boeing737 family has six variants, offered for nearly four decades, and nearly 6000 have been sold. It is obvious that in three decades, aircraft manufacturers have continuously updated later designs with newer technologies. The latest version of the Boeing737–900 has vastly improved technology compared to the late 1960s 737–100 model. The latest design has a different wing; the resources generated by large sales volumes encourage investing in upgrades – in this case, a significant investment was made in a new wing, advanced cockpit/systems, and better avionics, which has resulted in continuing strong sales in a fiercely competitive market.

The variant concept is market and role driven, keeping pace with technology advancements. Of course, derivatives in the family are not the optimum size (more so in civil aircraft design), but they are a satisfactory size that meets the demands. The unit-cost reduction, as a result of component commonalities, must compromise with the non-optimum situation of a slight increase in fuel burn. Readers are referred to Figure 16.6, which highlights the aircraft unit-cost contribution to DOC as more than three to four times the cost of fuel, depending on payload-range capability. The worked-out examples in the next section offer an idea of three variants in the family of aircraft.

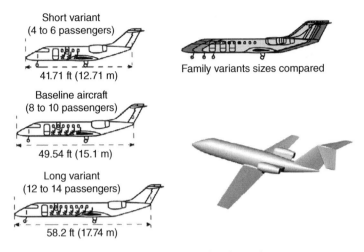

Figure 14.4 Variant designs in the family of civil aircraft.

Table 14.4 Bizjet family variant summary $S_W = 323\ ft^2$.

	Short variant	Baseline	Long variant
MTOM kg (lb)	7800 (17190)	9400 (20720)	10 900 (23580)
W/S_W (lb ft^{-2})	53.22	64.0	73.0[a]
T_{SLS}/W	(0.33)	0.34	(0.35)
T_{SLS}/per engine-lb	2836	3560	4130

a) Long variant requires superior high-lift device for landing.

14.5.2 Bizjet Family

Figure 14.4 shows the final configuration of the family of variants; the baseline aircraft is in the middle. It proposes one smaller (i.e. 4–6 passengers) and one larger (i.e. 14–16 passengers) variant from the baseline design that carries 10–12 passengers by subtracting and adding fuselage plugs from the front and aft of the wing box. The baseline and variant details are worked out in Section 8.9. Table 14.4 gives the summary of the Bizjet variants.

The short variant engine is derated by about 17% resulting in slightly lower engine weight and the long variant engine is up-rated by about 15% resulting in slightly higher engine weight.

14.6 Coursework Exercise – Military Aircraft (AJT)

Military designs follow Milspecs and not FAR for airworthiness requirements. Both FPS and SI units are worked out in the examples. Figure 11.17 gives the AJT drag polar. The military aircraft example of AJT operates in two takeoff weights as (i) at normal training configuration (NTC – clean) at 4800 kg and (ii) at fully loaded for armament training at 6800 kg, that is, a growth of 41.7%. In this example the NTC is more critical to meet the specification of TOFL = 800 m). The readers may work out both the cases. The fully loaded aircraft needs to satisfy the longer field length requirement of 1800 m (<6000 ft), the rest, for example, climb and cruise capabilities, are taken as fallout of the design. After the armament practice run the payload is dropped and the landing weight is the same for both the missions. It may be noted that the AJT should have a close air support (CAS) version. Only initial MTOW and wing area are required for sizing.

14.6.1   Takeoff – Military Aircraft

Requirements. TOFL = 800 m ($\approx$2600 ft) to clear 35 ft (10.7 m) at ISA + sea level at NTC. The maximum lift coefficient at TO (20° flaps down and no slat) is taken as $C_{Lmax\_TO} = C_{Lstall\_TO} = 1.85$. Installation is taken at 3.5%.

Using Eq. (14.11a), the expression becomes (in FPS system)

$$W/S_W = 2600 \times 1.85 \times (0.965 \times T_{SLS}/W)/18.85 = 246.3 \times (T_{SLS}/W)$$

Using Eq. (14.11b), it becomes (in SI system)

$$W/S_w = 8.345 \times 800 \times 1.85 \times (0.965 \times T_{SLS}/W) = 11\,918.3 \times (T_{SLS}/W)$$

The result is computed in Table 14.5 and listed in tabular form.

14.6.2   Initial Climb – Military Aircraft

From market requirement, initial climb speed $V = 350$ kt $= 350 \times 1.68781 = 590.7$ ft s^{-1} and the required rate of climb, RC = 10 000 ft min^{-1} (50 m s^{-1}) = 164 ft s^{-1} (50 m s^{-1}). From the Adour 861 class engine data T_{SLS}/T ratio, $k_{cl} = 1.06$. The result is computed in Table 14.6.

Lift coefficient,

$$C_{Lclimb} = W/(0.5 \times 0.002378 \times 590.7^2 \times S_W) = 0.00241 \times W/S_W$$

Using Eq. (14.15),

$$[T_{SLS}/W]/1.06 = 164/590.7 + [(C_D \times 0.5 \times 0.002378 \times 590.7^2 \times S_W)/W]$$

$$T_{SLS}/W = 0.294 + 440 \times C_D \times (S_W/W)$$

14.6.3   Cruise – Military Aircraft

Market Specification. Initial cruise speed and altitude is 0.75 Mach and 36 000 ft (most of training is takes place below the tropopause), for the initial cruise aircraft weight take, $k = 0.975$ in Eq. (14.14). The result is computed in Table 14.7.

In FPS

$$\text{at 36 000 ft, } \rho = 0.0007 \text{ slug/ft}^2 \text{ and } V^2 = (0.75 \times 968.07)^2 = 726.05^2 = 527\,152.2 \text{ ft}^2 \text{ s}^{-2}$$

Table 14.5 AJT takeoff sizing.

W/S_W (FPS – lb ft^{-2})	40	50	60	70	80	90
W/S_W (SI – N m^{-2})	1916.2	2395.6	2874.3	3353.7	3832.77	4311.5
T_{SLS}/W (uninstalled)	0.163	0.207	0.244	0.284	0.324	0.366

Table 14.6 AJT climb sizing calculations.

W/S (lb ft^{-2})	40	50	60	70	80
W/S (N m^{-2})	1916.2	2395.6	2874.3	3353.7	3832.77
C_{Lclimb}	0.097	0.12	0.145	0.169	0.193
C_D (from Figure 14.16)	0.0222	0.0225	0.0258	0.026	0.0263
T_{SLS}/W (uninstalled)	0.538	0.492	0.483	0.457	0.439

In SI,

$$\text{altitude} = 11\,000 \text{ m}, \rho = 0.364 \text{ kg m}^{-3}, V^2 = (0.75 \times 295.07)^2 = 221.3^2 = 48\,974.8 \text{ m}^2 \text{ s}^{-2}.$$

Equation (14.18) gives initial cruise

$$C_L = 0.975 \times MTOW/(0.5 \times 0.364 \times 48\,974.8 \times S_W)$$

$$= 0.0001094 \times (W/S_W) \text{ where } W/S_W \text{ is in N m}^{-2}.$$

Equation (14.19) gives.

$$T_{SLS}/W = k_{cr} \times 0.5\rho V^2 \times C_D/(W/S_W) \text{ (take factor } 1/k_{cr} = T_{SLS}/T_a = 3.6, \text{ see Figure 13.15)}$$

In FPS

$$T_{SLS}/W = 3.6 \times 0.5 \times 0.0007 \times 527\,152.2 \times C_D/(W/S_W) = 664.2 \times C_D/(W/S_W)$$

In SI

$$T_{SLS}/W = 3.6 \times 0.5 \times 0.364 \times 48\,974.8 \times C_D/(W/S_W) = 32\,088.2 \times C_D/(W/S_W)$$

Once again, make a table (see Table 14.7) and draw a plot.

14.6.4 Landing – Military Aircraft

From the market requirements,

$$V_{app} = 110 \text{ kts} = 110 \times 1.68781 = 185.7 \text{ ft s}^{-1} \text{ (56.6 m s}^{-1}\text{)}$$

Landing $C_{Lstall} = 2.5$ at a 40° double slotted flap setting.
Using Eq. (14.22),
In FPS system,

$$W/S_W = 0.311 \times 0.002378 \times 2.5 \times (185.7)^2 = 63.75 \text{ lb ft}^{-2}$$

In the SI system,

$$W/S_W = 0.311 \times 1.225 \times 2.5 \times (56.6)^2 = 2885 \text{ N m}^{-2}$$

Because at landing the thrust is taken to be zero, the W/S_W remains constant.

14.6.5 Sizing for the Turn Requirement of 4g at Sea Level

In a way, turn sizing is relatively simple and here the parabolic drag polar AJT is used to make use of the equation derived in Chapter 15. The approximated AJT parabolic drag polar is given in Section 11.20.1 as follows.

$$C_D = 0.0212 + 0.07C_L^2 \text{ with } S_W = 183 \text{ ft}^2 \text{ (gives Oswald's efficiency factor, } e = 0.85\text{)}.$$

Table 14.7 AJT cruise sizing.

W/S_W (lb ft^{-2})	50	60	70	80	100
W/S_W (N m^{-2})	2395.6	2874.3	3353.7	3832.8	4791
C_L	0.262	0.314	0.367	0.419	0.524
C_D	0.026	0.0292	0.0315	0.035	0.042
T_{SLS}/W @ 41 000 ft	0.346	0.324	0.30	0.29	0.279

The turn requirement is $n = 4$ at sea level (STD) day, but there is stipulation for speed at which it can be achieved. It gives the opportunity to initiate sizing analyses in many ways different from these methods. The turn sizing is carried out by establishing the relationship between, wing loading, W/S_W, and thrust loading T_{SLS}/W at $n = 4$, the turn speed that has to be establish. Use Eq. (15.58) (derived in the next chapter) and evaluate at three speeds at Mach 0.5 ($V = 558.25 \text{ ft s}^{-1}$), Mach 0.55 ($V = 614.075 \text{ ft s}^{-1}$) and Mach 0.6 ($V = 669.9 \text{ ft s}^{-1}$).

$$n^2 = \left[\frac{0.5\rho V^2}{k\left(W/S_w\right)} \left(\frac{T}{W} - 0.5\rho V^2 \frac{C_{DPmim}}{W/S_W} \right) \right], \text{ on substituting the numeric values, the following is obtained.}$$

$$n^2 = \left[\frac{0.5\rho V^2}{k\left(W/S_w\right)} \left(\frac{T}{W} - 0.5\rho V^2 \frac{C_{DPmim}}{W/S_W} \right) \right].$$

$$\text{or} \quad \left[\frac{T}{W} = \left[\frac{2kn^2 \times \left(W/S_w\right)}{\rho V^2} \right] + \left[\frac{0.5\rho V^2 \times C_{DPmin}}{\left(W/S_w\right)} \right] \right].$$

On substituting the numerical values of $\rho = 0.002378$ slugs/ft^3, $k = 0.07$ and $C_{DPmin} = 0.0212$.

$$\left[\frac{T}{W} \right] = \frac{941 \times \left(W/S_w\right)}{V^2} + \frac{0.0000252 \times V^2}{\left(W/S_w\right)},$$

where T is the installed thrust at maximum continuous rating.

Aircraft sizing thrust loading is expressed in terms of uninstalled sea level static thrust at maximum takeoff rating, T_{SLS_unins}. Maximum continuous rating is taken at 90% of maximum takeoff rating and installation loss is taken as 3.5%. That makes:

installed thrust at maximum continuous rating,

$$T = (0.9 \times 0.965)\, T_{SLS\_unins}$$

or $T_{SLS\_unins} = 1.15\, T$

Tables 14.8–14.10 show the (T/W) values at three (W/S_W) cases of 50 lb ft^{-2}, 60 lb ft^{-2} and 70 lb ft^{-2} for the three speeds at Mach 0.5, 0.55 and 0.6.

Table 14.8 Mach 0.5 ($V = 558.25$ ft s^{-1}), $V^2 = 311\,587.2$ (ft s^{-1})^2.

W/S_W (lb ft^{-2})	50	60	70
V^2	311 587.2	311 587.2	311 587.2
$\frac{941 \times \left(W/S_w\right)}{V^2}$	0.151 001	0.181 201	0.211 401 5
$\frac{0.0000252 \times V^2}{\left(W/S_w\right)}$	0.155 794	0.129 828	0.111 281 1
T/W	0.306 795	0.311 029	0.322 682 6
T_{SLS}/W	0.352 814	0.357 684	0.371 085

Table 14.9 Mach 0.55 ($V = 614.075$ ft s^{-1}), $V^2 = 376\,996$ (ft s^{-1})^2.

W/S_W (lb ft^{-2})	50	60	70
$\frac{941 \times \left(W/S_w\right)}{V^2}$	0.124 802	0.149 763	0.174 723 3
$\frac{0.0000252 \times V^2}{\left(W/S_w\right)}$	0.190 006	0.157 082	0.135 718 6
T/W	0.314 808	0.306 845	0.310 441 9
T_{SLS}/W	0.362 03	0.352 871	0.3570 082

Table 14.10 Mach 0.6 ($V = 669.9 \,\text{ft s}^{-1}$), $V^2 = 448\,632 \,(\text{ft s}^{-1})^2$.

W/S_W (lb ft^{-2})	50	60	70
$\dfrac{941 \times \left(W/S_W \right)}{V^2}$	0.104\,874	0.125\,849	0.146\,824\,1
$\dfrac{0.0000252 \times V^2}{\left(W/S_W \right)}$	0.226\,111	0.188\,425	0.161\,507\,5
T/W	0.330\,985	0.314\,275	0.308\,331\,6
T_{SLS}/W	0.380\,633	0.361\,416	0.354\,581\,4

Figure 14.5 Aircraft sizing – AJT (military).

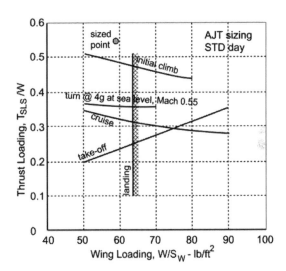

Tables 14.8–14.10 give three lines with close values (T/W). Only the line for Mach 0.55 is plotted in Figure 14.5 as the middle one of the three. With the excess thrust available, AJT can turn at a higher n (see Chapter 15).

14.7   Sizing Analysis – Military Aircraft (AJT)

The methodology for military aircraft is the same as in the case of civil aircraft sizing and engine matching. The five sizing relationships between wing loading, W/S_W, and thrust loading, T_{SLS}/W, would meet (i) takeoff, (ii) approach speed for landing, (iii) initial cruise speed, (iv) initial climb rate and (v) turn rate capability. These are plotted in Figure 14.5.

Military aircraft sizing poses an interesting situation. The variant in combat role, for example, in a CAS role has to carry more armament load externally contributing to drag rise. The overall geometry does not change much except the front fuselage is now redesigned for one pilot saving a weight of about 100 kg (the weight of seat, escape system etc. are replaced by radar, combat avionics). The aircraft still has the same engine tweaked to up-rated thrust level.

Therefore, a conservative sizing of AJT should benefit CAS growth. Figure 14.5 shows the sizing point is at slightly lower wing loading at $W/S_W = 59 \,\text{lb ft}^{-2}$ to benefit CAS performance. Thrust loading is taken as $T_{SLS}/W = 0.55$. The circled point in Figure 14.5 satisfies all requirements simultaneously. If required, iteration can be carried out with a slightly higher value of T_{SLS}/W to meet $n_{max} = 7$ (see Section 15.11.8) as well as benefit the takeoff and initial climb performance of AJT with full practise armament load.

Chapter 10 worked out the mass of the preliminary configuration of AJT aircraft as:

$MTOM = 4800\,\text{kg}\,(10\,582\,\text{lb})$ at NTC, which gives the matched engine thrust $T_{SLS} = 0.55 \times 10,582 \approx 5,830\,\text{lb}$ (25,933 N).

Checking out with the sized wing loading $W/S_W = 59\,\text{lb ft}^{-2}$, the wing area comes out at $185\,\text{ft}^2$ ($17.2\,\text{m}^2$) about 1% error from the preliminary wing area, hence it remains unchanged. The matched engine thrust gives a lower value compared to the statistical estimate of 5860 lb, which is good. Once again, iteration is avoided.

14.7.1 Single Seat Variants in the Family of Aircraft Designs

Military aircraft are no exceptions in offering variant designs depending on their mission role, in addition to the typical 'payload-range' variation. The F16 and F18 have had modifications since they first appeared with an increasing envelope of combat capabilities. The F18 has increased in size. The BAE Hawk100 jet trainer has produced a single seat close support combat derivative, the Hawk200.

The CAS role aircraft is the only variant of the AJT aircraft (Figure 14.6). The details on how it is achieved with associated design changes are described next.

14.7.1.1 Configuration

Configuration of CAS aircraft variant is achieved by splitting the AJT front fuselage, then replacing the tandem seat arrangement with a single seat cockpit. The length could be kept the same as the nose cone needs to house more powerful acquisition radar. The front loading of radar and single pilot is placed in such a way that the CG location is kept undisturbed. Wing area = $17\,\text{m}^2$ ($183\,\text{ft}^2$). The bold values presented highlight the important parameters.

Weights	AJT	CAS
Clean aircraft MTOM	**4800 (10 582 lb)**	**5000 (11 023 lb)**
Wing loading[a], W/S_W	$282\,\text{kg/m}^2$ ($57.8\,\text{lb ft}^{-2}$)	$294\,\text{kg/m}^2$ ($60.23\,\text{lb/ft}^2$).
MTOM kg (lb)	**6500 (14 326 lb)**	**7400 (16 310 lb)**
Wing area, S_W	$17\,\text{m}$ ($183\,\text{ft}^2$)	$17\,\text{m}$ ($183\,\text{ft}^2$)
Wing loading, W/S_W	$382.5\,\text{kg m}^{-2}$ ($78.3\,\text{lb ft}^{-2}$)	$435.3\,\text{kg m}^{-2}$ ($89.1\,\text{lb ft}^{-2}$).

a) Sized wing loading for AJT at NTC came out close to it.

Armament and fuel could be traded for range. Drop tanks could be used for ferry range.

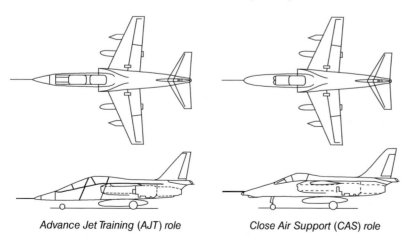

Advance Jet Training (AJT) role   *Close Air Support (CAS) role*

Figure 14.6 Variant designs in the family of military aircraft.

14.7.1.2 Thrust

The CAS variant would require a 30% higher engine thrust variant. This is possible without any change in external dimension but would incur an increase of 60 kg in engine mass.

CAS turbofan (has small bypass) thrust $= 1.3 \times 5300 \approx 6900$ lb (30 700 N).

Thrust loading at MTOM becomes: $T_{SLS}/W = 0.417$ (a satisfactory value).

14.7.1.3 Drag

The drag level of the clean AJT and CAS aircraft may be considered to be about the same. There would be an increase of drag on account of the weapon load. For the CAS aircraft, there is a wide variety, but to give a general perception, the typical drag coefficient increment for armament load is $C_{D_\pi} = 0.25$ (includes interference effect) each for five hard points as weapon carrier. Weapon drag is based on maximum cross section (say 0.8 ft^2) area of the weapons.

Parasite drag increment, $\Delta C_{Dpmin} = (5 \times 0.25 \times 0.8)/183 = 0.0055, where\ S_W = 183\ ft^2$.

14.8 Aircraft Sizing Studies and Sensitivity Analyses

Aircraft sizing studies and sensitivity analyses are not exactly the same thing as outlined next. Together with both sizing and sensitivity study exercises, the aircraft designers are able to find an optimised or satisfying design to satisfy the certification agency and customer simultaneously. The designers and the users will have better insights to make better choices.

14.8.1 Civil Aircraft Sizing Studies

Aircraft sizing exercise freezes aircraft to a unique geometry with little scope to make variation as shown in Figure 14.3 for the Bizjet. In these kinds of studies, aircraft performance requirements from the Federal Aviation Administration (FAA) and specification (customer) are given as: (i) takeoff field length and climb gradients, (ii) initial en-route climb, (iii) initial HSC capability and (iv) LFL and climb gradients at baulked landing are simultaneously satisfied with a matched engine and a unique solution emerges. To search for a compromising to offer a family of variants it depends on objectives, for example, the degree of component components can be retained.

It is a broad-based analysis and could be extensive. Basically, it is an optimisation process to find the best value providing opportunities to make any compromise, if required. The easiest method is to take one variable at a time in a *parametric* search to find on how it is affecting other parameters. Normally, the variations are made in small steps to keep changes in the affecting parameters within the ranges of acceptable errors. In civil aircraft design, the study should continue to examining the objective function, say the DOC. Typically, the results are shown in carpet plots (Figure 14.7) studying up to as many as five variables [5]. This carpet plot shows four variables in one graph with mutual inter-dependency. It shows how aircraft DOC is affected with aircraft drag change, fuel price and aircraft cost for an Airbus 320 class aircraft.

Apart from parametric method of searching there are theoretical methods available. The simpler ones are the Simplex and Gradient methods; those are used in the industry, but the parametric method is most popular.

14.8.1.1 Military Aircraft Sizing Studies

Aircraft sizing exercise freezes aircraft to a unique geometry with little scope to make variations as shown in Figure 14.3 for the Bizjet. The aircraft sizing exercise freezes aircraft to a unique geometry with little scope for variation as shown in Figure 14.5 for the AJT.

In aircraft design sensitivity can be made on determining individual component geometries, for example, for wing examine how its area, AR, thickness to chord ratio, sweep and so on, affects aircraft capabilities as

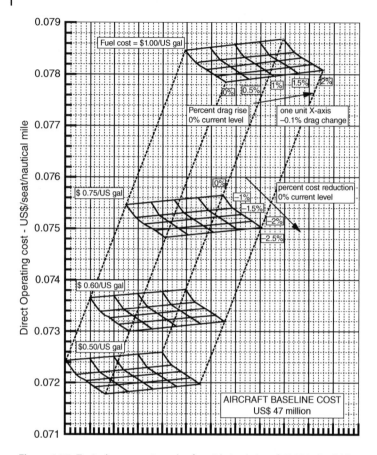

Figure 14.7 Typical *parametric study* of a mid-sized aircraft (A320 class) (shown in a carpet plot showing DOC variations with aircraft drag and costs).

shown in Table 14.11 for the AJT. An example of AJT wing geometry sensitivity study is given next involving what happens with small changes in wing reference area, S_W, aspect ratio, AR, aerofoil thickness to chord ratio, t/c, and wing quarter chord sweep, $\Lambda_{1/4}$.

A more refined analysis could be made with a detailed sensitivity study on various design parameters, such as other geometrical details, materials, structural layout and so on, to address cost versus performance issues to arrive at a 'satisfying' design. This may require some local optimisation with full awareness that the global optimisation is not sacrificed. While a broad-based MDO is the ultimate goal, dealing with large number of parameters in a sophisticated algorithm may not prove easy. It is still researched intensively within academic circles, but industry tends to use MDO in a conservative manner if required in a parametric search, tackling one variable at a time. Industry cannot afford to take risks with any unproven algorithm just because it is elegant bearing promise. Industry takes a more holistic approach to minimise costs without sacrificing safety but may compromise on performance if it pays.

14.9   Discussion

Aircraft sizing exercise can keep open the possibilities for aircraft future growth potential. If possible, even a radically new design can salvage components from older design to maintain parts commonalities.

This culminating section presents some discussion on the classroom examples and their performances options are presented. The sizing exercise gives the simultaneous solution to satisfy the airworthiness and

Table 14.11 AJT sensitivity study.

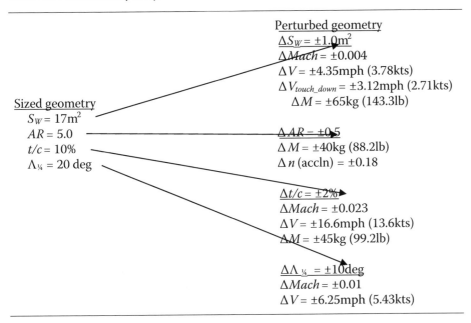

Sized geometry
$S_W = 17\text{m}^2$
$AR = 5.0$
$t/c = 10\%$
$\Lambda_{\frac{1}{4}} = 20$ deg

Perturbed geometry
$\Delta S_W = \pm 1.0\text{m}^2$
$\Delta Mach = \pm 0.004$
$\Delta V = \pm 4.35$mph (3.78kts)
$\Delta V_{touch\_down} = \pm 3.12$mph (2.71kts)
$\Delta M = \pm 65$kg (143.3lb)

$\Delta AR = \pm 0.5$
$\Delta M = \pm 40$kg (88.2lb)
Δn (accln) $= \pm 0.18$

$\Delta t/c = \pm 2\%$
$\Delta Mach = \pm 0.023$
$\Delta V = \pm 16.6$mph (13.6kts)
$\Delta M = \pm 45$kg (99.2lb)

$\Delta \Lambda_{\frac{1}{4}} = \pm 10$deg
$\Delta Mach = \pm 0.01$
$\Delta V = \pm 6.25$mph (5.43kts)

the market requirements. This is an activity in the conceptual design phase when few details about aircraft data are available. The sizing exercise requires specific aircraft performance evaluation in a concise manner using the relevant equations derived thus far. Sizing studies are management issues those are reviewed along with the potential customers to decide to give the go-ahead or not. In sizing and engine matching exercise, the wing loading (W/S_W) and thrust loading (T/W) are the dictating parameters appear in equations for takeoff, second segment climb, en-route climb and maximum speed capability. The first two are FAR requirements and the last two are customer requirements. Information on detailed engine performance is not required during the sizing exercise; projected *rubberised* engine data are used during the sizing and engine matching studies. Payload-range capability of the proposed design needs to be demonstrated as the customer requires. Subsequently, with the details of engine performance and aircraft data, relevant aircraft performance analyses are carried out more accurately to satisfy airworthiness and market requirements.

Once the go-ahead is obtained, then a full blown detailed definition study ensues as Phase II activities with large financial commitments. Refinement continues until accurate data are obtained from flight tests and other refined analyses. Most of the most important aircraft performance equations are in this book in reasonable details supported by worked-out examples.

Sizing in Figure 14.3 (Bizjet) shows the lines of constraints for the various requirements. The sizing point to satisfy all requirements would show different level of margins for each capability. Typically, initial en-route climb rate is the most critical to sizing. Therefore, the takeoff and maximum speed capabilities have considerable margin, which is good as the aircraft can do better than what is required.

From the statistics, experience shows that aircraft mass grows with time. This occurs primarily on account of modifications arising out of mostly from minor design changes and with changing requirements, sometimes even before the first delivery is made. If the new requirements demand a large number of changes then civil aircraft design may appear as a new variant but military aircraft hold on a little longer before a new variant emerges. It is therefore prudent for the designers to keep some margin, especially with some reserve thrust capability, that is, keep the thrust loading, (T/W), slightly higher to begin with. Re-engineering with an up-rated version is expensive.

It can be seen that field performance would require a bigger wing planform area (S_W) than at cruise. It is advisable to keep wing area as small as possible (i.e. high wing loading) by incorporating a superior high-lift capability, which is not only heavy but also expensive. Designers must seek compromise to minimise operating costs. No iteration was needed for the designs worked out in the book.

Section 15.3 derived all the necessary relations to estimate the aircraft cruise segment required to compute aircraft mission payload-range capability and the block time and block fuel required for the mission. In summary, both the close form analytical method for various options for trend analyses using parabolic drag polar and numerical analyses using actual drag polar are analysed and the results are compared. The question now is which method is to be applied when as each one has its own reason of application.

The first thing that needs to be pointed out that actual drag polar is more accurate as it is obtained through proven semi-empirical methods and refined by tests (wind-tunnel and flight). Operators require an accurate manual to maximise city-pair mission planning. To establish the best possibility, for example, to decide the speed and altitude for the sector, is not easy. Here, close form analysis assists to quickly provide answers for the best possibilities. This is then fine-tuned trough flight tests and then incorporated in the operator's manual; nowadays it is digitised and stored at several places, on ground and in the aircraft. In the past flight engineers had to compute prior to any sortie and keep monitored during the sortie flight. By today's standard it is a laborious and time consuming process as today these can be obtained almost instantaneously by the press of buttons.

The sizing point in Figure 14.3 gives a wing loading, $W/S_W = 64$ lb ft^{-2} and a thrust loading, $T/W = 0.34$. It may be noted that there is little margin given for the landing requirement. The maximum landing mass for this design is at 95% of MTOM. A margin for 10% weight growth corresponding drag growth is accommodated in the sizing exercise. Commonality of undercarriage for all variants would start with the design for the heaviest, that is, the growth variant and its bulky components are shaved to lighten them for lower weights. The middle variant is taken as the baseline version. Its undercarriage can be made to accept MTOM growth as a result of OEW (Operating Empty Weight) growth instead of making the wing larger.

It should be borne in mind the recommendation is that civil aircraft should come in a family of variants to cover wider market demand to maximise sale. Truly speaking, none of the three variants are optimised although the baseline is carefully sized in the middle to accept one larger and one smaller variant. Even when the development cost is front loaded, the cost of development of the variant aircraft is low by sharing the component commonality. The low cost is then translated to lowering aircraft price, which can absorb the operating costs of the slightly non-optimised designs.

It is interesting to examine the design philosophy of the Boeing737 family and the Airbus320 family of aircraft variants in the same market arena. Together more than 8000 aircraft have been sold in the world market. This is a no small achievement in engineering. The cost of these aircraft is about $50 million apiece (2005). For airlines with deregulated fare structures, making a profit involves a complex dynamics of design and operation. The cost and operational scenario changes from time to time, for example, rise of fuel cost, terrorist threats and so on.

As early as the 1960s, Boeing saw the potential of keeping component commonality in offering new designs. The B707 was one of the earliest commercial jet transport aircraft carrying passengers. It was followed by a shorter version B720. Strictly speaking the B707 fuselage relied on the KC135 tanker design of the 1950s. From the four-engine B707 came the three-engine B727 and then the two-engine B737, both retaining considerable fuselage commonality. This was one the earliest attempts to utilise the benefits of maintaining component commonality. Subsequently, the B737 started to emerge in different sizes of variants maximising component commonality. The original B737-100 was the baseline design and all other variants that came later, up to the B737-900. It posed certain constraints, especially on the undercarriage length. On the other hand, the A320 serving as the baseline design was in the middle of the family, its growth variant is the A321 and shrunk variants are the A319 and A318. Figure 7.3 gives a good study on how the OEW is affected by the two examples of family of variants. A baseline aircraft starting at the middle of the family would be better optimised and hence in principle would offer a better opportunity to lower production cost of the variants in the family.

Simultaneous failure of two engines of a four-engine aircraft is extremely rare. If it happens after the decision speed and if there is not enough clearway available then it will prove catastrophic. If the climb gradient is not in conflict with the terrain of operation then it is better to takeoff with higher flap settings. If a longer runway is available then lower flap setting could be used. Takeoff speed schedules can slightly exceed the FAR requirements that stipulate the minimum values. There have been cases of all engine failures occurring at cruise on account of volcanic ash in the atmosphere (also in the rare event of fuel starvation). Fortunately, the engines were restarted just before the aircraft would have hit the surface. An all engine failure (A320) by bird strike occurred in 2009 – all survived after the pilot landed on the Hudson River.

14.9.1 The AJT

Military aircraft serves only one customer, the Ministry of Defence (MoD)/United States Department of Defense (DoD) of the nation that designed the aircraft. Front line combat aircraft incorporates the newest technologies at the cutting edge to stay ahead of potential adversaries. Its development cost is high and only few countries can afford to produce advanced designs. International political scenarios indicate a strong demand for combat aircraft even for developing nations who must purchase from abroad. This makes the military aircraft design philosophy different from civil aircraft design. Here, the designers/scientists have a strong voice unlike in civil design where the users dictate terms. Selling combat aircraft to restricted foreign countries is only one way to recover some finance.

Once the combat aircraft performance is well understood over its years of operation, consequent modifications follow capability improvements. Subsequently, a new design replaces the older design – there is a generation gap between the designs. Military modifications for the derivative design are substantial. Derivative designs primarily come from revised combat capabilities with newer types of armament along with all round performance gains. There is also a need for modifications, seen as variants rather than derivatives, to sell to foreign customers. These are quite different to civil aircraft variants, which are simultaneously produced for some time, serving different customers, some operating all the variants.

Advanced trainer aircraft designs have variants serving as combat aircraft in CAS. AJTs are less critical in design philosophy in comparison with front line combat aircraft, but bear some similarity. Typically, AJTs will have one variant in a CAS role produced simultaneously. There is less restriction to export these kinds of aircraft.

The military infrastructure layout influences the aircraft design and here the life cycle cost (LCC) is the prime economic consideration. For military trainer aircraft design, it is better to have a training base close to the armament practice arena, saving time. A dedicated training base may not have as long a runway as major civil runways. These aspects are reflected in the user specifications needed to start a conceptual study. The training mission includes aerobatics and flying with onboard instruments for navigation. Therefore, the training base should be far away from the airline corridors.

The AJT sizing point in Figure 14.7 gives a wing loading, $W/S_\mathrm{W} = 58\,\mathrm{lb\,ft^{-2}}$ and a thrust loading, $T/W = 0.55$. It may be noted that there is a large margin, especially for the landing requirement. The AJT can achieve maximum level speed over Mach 0.88, but this is not demanded as a requirement. Mission weight for the AJT varies substantially; the NTC is at 4800 kg and for armament practice it is loaded to 6600 kg. The margin in the sizing graph covers some increase in loadings (the specification taken in the book is for the NTC only). There is a big demand for higher power for the *CAS* variant. The choice for having an up-rated engine or to have an afterburner depends on the choice of engine and the mission profile.

Competition for military aircraft sales is not as critical as compared to civil aviation sector. The national demand would support the national product for producing a tailor made design with under manageable economics. But the trainer aircraft market can have competition, unfortunately sometimes influenced by other factors that may fail to bring out a national product even if the nation is capable of doing so. The Brazilian design Tucano was re-engineered and underwent extensive modifications by Short Brothers plc of Belfast for the RAF in the UK. The BAE Hawk (UK) underwent made major modifications in the USA for domestic use.

References

1 Kundu, A.K., Price, M.A., and Riordan, D. (2016). *Theory and Practice of Aircraft Performance*. Wiley.
2 Perkins, C.D. and Hage, R.E. (1949). *Airplane Performance, Stability and Control*. Wiley.
3 Wallace, R. E., *Parametric and Optimisation Techniques for Airplane Design Synthesis*, AGARD-LS-56, 1972.
4 Loftin, L. K. Jr, *Subsonic Aircraft: Evolution and Matching of Size to Performance*, NASA RP 1060, 1980.
5 Kundu, A.K. (2010). *Aircraft Design*. Cambridge University Press.

15

Aircraft Performance

15.1 Overview

This is the culminating chapter to assess if the sized aircraft configured, thus far, meets the Federal Aviation Regulation (FAR) requirements and the customer specifications. Classroom work follows in a linear manner from the mock market survey (Chapter 2). The requirements and the specifications dealt with in this chapter comprise of aircraft performance to substantiate the (i) takeoff and (ii) landing field lengths (LFLs), (iii) initial rate of climb, (iv) maximum speed at initial cruise and, especially for civil aircraft design, its (v) payload range plus (vi) the turn rate for the Advanced Jet Trainer (AJT). Chapter 18 computes aircraft Direct Operating Cost (DOC), which should follow the aircraft performance estimation.

Aircraft performance is a subject all on its own. All aeronautical schools offer the subject as a separate course. Therefore, to substantiate the FAR and customer requirements, this chapter deals only with what is required, that is, the related governing equations and examples associated with the items listed above. Note that substantiation of payload-range requires integrated climb and descent performances showing fuel consumed, distance covered and time taken for the flight segments. Readers are recommended to refer to [1] by the authors, as a companion book, dealing with the same aircraft performance in extensive details. The turboprop example is not worked out but there is enough information to compute it in a similar manner.

The rest of the book beyond this chapter has information that aircraft designers should know and apply.

The chapter covers the following

Section 15.2: Introduction
Section 15.3: Takeoff Performance
Section 15.4: Landing Performance
Section 15.5: Climb Performance
Section 15.6: Descent Performance
Section 15.7: Initial Maximum Cruise Speed Capability
Section 15.8: Payload-Range Capability
Section 15.9: In Horizontal Plane (Yaw Plane) – Sustained Coordinated Turn
Section 15.10: Aircraft Performance Substantiation – Bizjet
Section 15.11: Aircraft Performance Substantiation – AJT
Section 15.12: Propeller-Driven Aircraft Performance Substantiation – TPT

15.1.1 Section 15.13: Summarised Discussion of the DesignClasswork Content

At this stage of project design, only the substantiation of requirements/specifications are required for the sized aircraft configurations culminating in 'concept finalisation' as worked out in Chapter 14. Full aircraft performance for the full envelope is carried out after go-ahead is obtained, as done in [1] by the same authors.

The readers will have to compute the following for their projects. This follows that the readers have made engine performance available as shown in Sections 13.4 and 13.5.

Conceptual Aircraft Design: An Industrial Approach, First Edition. Ajoy Kumar Kundu, Mark A. Price and David Riordan.
© 2019 John Wiley & Sons Ltd. Published 2019 by John Wiley & Sons Ltd.

i. Substantiate the takeoff field length (TOFL). This has to demonstrate that the aircraft is capable of meeting the Federal Aviation Administration (FAA) second segment (see Figure 15.3) climb gradient requirement for the most critical case.

ii. Substantiate the initial climb rate. This is not a FAA requirement but a market specification.

iii. Substantiate the initial high speed cruise (HSC) capability. This is also not a FAA requirement but a market specification.

iv. Substantiate the LFL. This has to demonstrate that the aircraft is capable of meeting the FAA requirement for the most critical case.

v. Substantiate the payload-range capability. This is also not a FAA requirement but a market specification. Payload-range computation requires the specific range, integrated climb and descent performance of the aircraft and the readers will have to compute these. Use of spreadsheets is recommended.

For the trainer aircraft, the following are the additional requirements.

vi. Compute turn capability. This is not a Milspec requirement but a MoD (Ministry of Defence) specification.

vii. Compute the training mission profile to substantiate the capability.

If any of the aircraft performances is not met then aircraft has to be configuration iterated (see Chapter 8 up to Chapter 14) until the requirements are met. The spreadsheet method proves convenient for iteration.

Airworthiness regulations differ from country to country. Military certification standards are different from the FAA. In the USA, this is guided by Milspecs. Certification authorities have mandatory requirements to ensure safety at takeoff and landing. For further details on respective certification regulations, the readers may refer to their official websites. Readers are recommended to read references [2–8] for supplementary coverage.

15.2   Introduction

The final outcome of any design is to substantiate aircraft performance for what it is intended to do. In the conceptual design phase, aircraft performance substantiation needs to be carried out mainly on those critical areas as required by the FAR and also customer requirements. Full aircraft performance estimation is done subsequently (beyond the scope of this book). All worked out aircraft performance estimations are done at a standard (STD) day. Non-International Standard Atmosphere (ISA) day performance computations are done in the same way with non-ISA day data.

The sizing exercises in Chapter 14 showed a rapid performance method to generate the relation between wing loading, W/S_W, and thrust loading, T_{SLS}/W, to obtain the sizing point that would simultaneously satisfy the requirements of takeoff/LFLs, initial rate of climb capability and maximum speed at initial cruise. The aircraft sizing point gives the installed maximum sea level takeoff static thrust, T_{SLS} of the matched engines.

Chapter 13 gives generic uninstalled engine performances of *rubberised* engines in a non-dimensional form, from which installed thrust and fuel flow rates at various speeds and altitudes at the power settings of takeoff, maximum climb and maximum cruise ratings by STD day are worked out in Sections 13.4 and 13.5. Using these installed engine performance data, this chapter continues with a more accurate computation of aircraft performance to substantiate the specified takeoff and LFLs, initial rate of climb, maximum speed at initial cruise and payload range. At this point, it may be necessary to revise aircraft configuration if the performance capabilities are not met. If the aircraft performance indicates shortfall (or excess) in meeting the requirements, then the design is iterated for improvement. In classroom work, normally one iteration should prove sufficient, but the readers should not hesitate to fine-tune with iterations; it is a normal industrial practice.

Finally, at the end of the design, the aircraft should be flight tested over the full flight envelope including various safety issues to demonstrate compliance. This concludes *concept finalisation* as the obligation of this book.

15.2.1 Aircraft Speed

Aircraft speed is a vital parameter in computing performance. It is measured from the difference between the total pressure, p_t and static pressure, p_s, expressed as $(p_t - p_s)$. Static pressure is the ambient pressure in which the aircraft is flying and the total pressure, p_t, is aircraft speed dependent. The value of $(p_t - p_s)$ gives the *dynamic head*, which depends on the ambient air density ρ and aircraft velocity (speed). Unlike automobile ground speed measured directly, aircraft ground speed needs to be computed from $(p_t - p_s)$. The pilot reads the gauge reading converted from $(p_t - p_s)$. The following are the various forms of aircraft airspeeds that engineers and pilots use. As can be seen, these require some computations – nowadays onboard computers do it all.

T_{EAS} = True Airspeed (TAS) = Aircraft Ground Speed.

V_i = Gauge reading is what the uninstalled bare instrument supplied by the manufacturer reads. The instrument includes standard adiabatic compressible flow corrections for high subsonic flight at a sea level STD day. But it still requires other corrections as follows:

V_I = Indicated Airspeed (IAS). Manufactured instruments will have some built-in instrumental errors, ΔV_i, (normally very small but an important consideration when close to stall speed). Instrument manufacturers supply the error chart for each instrument. The instrument is calibrated to read correct ground speed at sea level STD day with compressibility corrections. When corrected the instrument reads IAS.

$$V_I = IAS = V_i + \Delta V_i.$$

V_C = Calibrated Airspeed (CAS). Instrument manufacturers calibrate the uninstalled bare instrument for sea level conditions. Once installed on an aircraft, depending where it is installed, the aeroplane flow field will distort instrument readings. Therefore, it will require position error (ΔV_p) corrections by aircraft manufacturers.

$$V_C = CAS = V_I + \Delta V_p. = V_i + \Delta V_i + \Delta V_p$$

V_{EAS} = Equivalent Airspeed (EAS). Air density ρ changes with altitude – it goes down because atmospheric pressure reduces with gain in altitude. Therefore, at the same ground speed (also known as True Air Speed, TAS), the IAS would read lower values at higher altitudes. The mathematical relation between TAS and EAS reflecting the density changes with altitude can be derived as $TAS = EAS/\sqrt{\sigma}$, where σ is the density ratio (ρ/ρ_0) in terms of sea level value, ρ_0. A constant EAS has a dynamic head invariant. For high subsonic flights, it will require an adiabatic compressibility correction (ΔV_c) for the altitude changes.

$$V_{EAS} = EAS = V_C + \Delta V_c. = V_i + \Delta V_i + \Delta V_p + \Delta V_c = TAS\sqrt{\sigma}$$

V_{EAS} is what the pilot in the flight deck reads. In the past, the pilot/flight engineer had to manually correct the instrument and position errors. Today, these corrections are embedded in the aircraft microprocessors off-loading manual effort.

Compressibility corrections for position errors is also there but at this stage such details can be avoided without any loss of conceptual design work undertaken in this book. Supersonic flight will require further adjustments to make the corrections resulting from the shock wave associated with it.

15.2.2 Some Prerequisite Information

The readers will have to find available engine performance data, as done in Chapter 13 in which the installed thrust and fuel flows for the various engines are worked out. The worked-out examples in this chapter uses the following figures.

Turbofan with a BPR (bypass ratio) $\approx 4 \pm 1$, (in the class of Honeywell TPE731 turbofan) – Bizjet application
Figure 13.11 – Installed takeoff thrust and fuel flow per engine
Figure 13.12 – Installed maximum climb thrust and fuel flow per engine
Figure 13.13 – Installed maximum cruise thrust and fuel flow per engine

Figure 13.14 – Installed idle engine thrust and fuel flow per engine

Military Turbofan *with a BPR < 1, (in the class of Rolls-Royce/Turbomeca Adour) – AJT application*

Figure 13.15 – Installed maximum thrust and fuel flow for all the three ratings.

Turboprop engine, in the class of Garrett TPT 331 – TPT application

Figure 13.16 – Installed takeoff thrust and fuel flow per engine

Figure 13.17 – Installed maximum climb thrust and fuel flow per engine

Figure 13.18 – Installed maximum cruise thrust and fuel flow per engine

Using the engine and aircraft data obtained thus far during the conceptual design phase, this chapter checks whether the configured aircraft satisfies the airworthiness (FAR) requirements and the customer specifications in the takeoff/landing, the initial climb rates and the maximum initial cruise speed plus the payload-range capability (civil aircraft).

15.2.3 Cabin Pressurisation

Aircraft payload-range computation requires integrated climb and descent performances. It is topical to introduce the role of cabin pressurisation in assessing aircraft climb and descent performance. Ambient pressure reduces non-linearly with increase in altitude as can be seen in Figure 1.8. At the tropopause it is 472.68 lb ft^{-2} (3.28 lb in^{-2}). For average human physiology, big jets maintain cabin pressure around 8000 ft altitude (ambient pressure of 1571.88 lb ft^{-2} (10.92 lb in^{-2}) resulting in a differential pressure of 7.63 lb in^{-2}). Smaller Bizjets fly at higher altitudes to stay separate from big-jet traffic, as high as 50 000 ft altitude (242.21 lb ft^{-2}, i.e. 1.68 lb in^{-2}) when aircraft structural design considerations are made to maintain the inside cabin pressure at 10 000 ft altitude (1455.33 lb ft^{-2}, i.e. 10.11 lb in^{-2}, a differential pressure of 8.42 lb in^{-2}). For big jets, a maximum of 8.9 lb ft^{-2} (for Bizjets to 9.4 lb in^{-2}) of differential pressure is used in practice. Cabin pressurisation is in use for certain types military aircraft with large fuselage volumes. Combat aircraft have different arrangements for higher altitude provision in the flight deck; pilots wear pressure suits and helmets with masks supplying them with oxygen.

The differential pressure between the outside and inside of the cabin is substantial and it cycles through every sortie flown. Stress and fatigue considerations of fuselage structural design are constrained by weight considerations. Aircraft cabins are built as sealed pressure vessels (very low leakage), kept airtight and provided with complex environment control systems (ECSs), which have pressure control valves to automatically regulate cabin pressure (and air-conditioning). The aircraft crew can select the desired cabin pressure within the design limits. Unless there is a catastrophic failure, the sealed cabin can hold pressure to give enough time for pilots to descend fast enough in case there is a malfunction in pressurisation. If cabin pressure exceeds the limit then automatic safety valves can relieve cabin pressure to bring it down within safe limits.

The classwork example of the Bizjet has cabin pressure selectable from sea level to 10 000 ft altitude, limiting the differential pressure to within 9.4 lb ft^{-2}. During the climb the equivalent Bizjet cabin pressure rate of change is designed to 2600 ft min^{-1}. Bizjet type aircraft engine power is adjusted to have cabin pressurisation at the climb rate to a maximum of about 1800 ft min^{-1} to stay within the ECS system capability, sufficient for the sealed cabin that starts with sea level pressure, and as a result en route climb rate is not required to be restricted. Section 15.10.5 works out the Bizjet climb performance. However, during descent, the cabin needs to gain pressure as it descends to lower altitudes. Average human physical tolerance is taken at 300 ft min^{-1} descent and the ECS system pressurisation capability meets that.

The following equations are derived in this chapter.

   i. TOFL equations
  ii. Climb and descent equations
 iii. Cruise equations
  iv. LFL
   v. Range equations to establish the payload-range capability
  vi. Aircraft turn equations

15.3 Takeoff Performance

At takeoff, the ground run is initiated with an aircraft under maximum thrust (takeoff rating) to accelerate, gaining speed until a suitable safe speed is reached when the pilot initiates rotation of the aircraft by gently pulling back the control stick/wheel (elevator going up) for lift-off. In the case of an inadvertent situation of the critical engine failure, provision is made to cater for safety so that the aircraft can continue with its flight and return immediately to land.

While regression can analyse from the data statistics within the class of aircraft to generate semi-empirical relations that can yield close enough results, they mask the understanding of the physics involved in the takeoff procedure to estimate Balanced Field Length (BFL) (see Figure 15.2) with one engine inoperative. The mandatory certification regulations require complying with specified speed schedules and the determination of the decision speed, V_1, in case of one engine failed, are not captured in the semi-empirical relations. It is for this reason that the authors think it is essential to adopt the formal procedure as practised in industry, except that the treatise is relaxed with simplifications for classroom usage. For example, there is no easy way to determine the C_L and C_D of aircraft in the transient state of operation during a takeoff ground run with speed accelerating but not yet airborne. In this book, average values are taken where necessary. Industries have computer programs to accurately estimate BFL in all ambient conditions that also establish the relation between aircraft weight, airfield altitude and ambient temperature (WAT: weight, altitude and temperature) and their limitations for operational use in all-weather conditions [1]. Takeoff analysis is one of the more involved methods and the readers must perform manual computation for, at least, one set of analysis to appreciate the procedure and assess the labour content, then use a spreadsheet as a labour saving option.

Designers must know the sequence of the takeoff speed schedules stipulated by certification agencies (Figure 15.1). To ensure safety, certification authorities demand a mandatory requirement to cater for taking off with one engine inoperative to clear a 35 ft height representing an obstacle. A one engine inoperative TOFL is computed by considering the BFL when the stopping distance after an engine failure at the decision speed, V_1, is the same as the distance taken to clear the obstacle at maximum takeoff mass (MTOM)

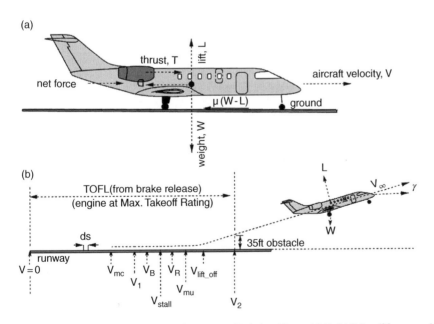

Figure 15.1 Takeoff, first and second segment climb (see Figure 15.3). (a) Takeoff forces and (b) takeoff ground run.

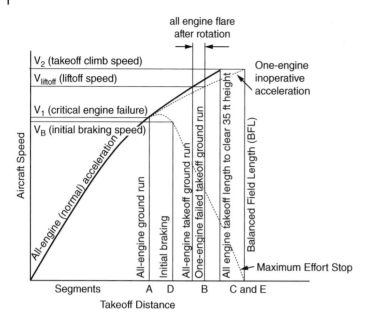

Figure 15.2 Balanced field length consideration.

(Figure 15.1). The speed schedule is related to V_{stall}, a known value for the sized MTOM and type of high-lift device incorporated. These are explained next.

V_1: This is the decision speed. An engine failure below this speed would result in the aircraft not being able to satisfy takeoff within the specified field length, but it will be able to stop. If an engine fails above the V_1 speed, the aircraft should continue with the takeoff operation.

V_{mc}: This is the minimum control speed at which the rudder is effective to control the asymmetry created by one engine failure. This should be lower than the decision speed, V_1, otherwise at the loss of one engine the aircraft cannot be controlled if continuing with takeoff.

V_R: This is the speed when the pilot initiates the action to rotate the aircraft for lift-off. V_R should be more than $1.05 V_{stall}$. Once this is done, reaching V_{LO} and V_2 occurs as a fall-out of the action.

V_{mu}: There is a minimum unstick speed, above which the aircraft can be made to lift off. This should be slightly above V_R. In fact, V_{mu} decides V_R. If the pilot makes an early rotation then V_{mu} may not be sufficient for lift-off and the aircraft will tail drag until it gains sufficient speed for the lift-off.

V_{LO}: This is the speed at which the aircraft lifts off the ground and it is closely associated with V_R. In a case of one engine being inoperative, it should be $V_{LO} \geq 1.05 V_{mu}$.

V_2: This is the takeoff climb speed at 35 ft height, also known as the first segment climb speed. It is also closely associated with V_R. FAR requires that $V_2 = 1.2 V_{stall}$ (at least – it can be higher).

V_B: This is the brake application speed at one engine failure ($V_B > V_1$).

15.3.1   Civil Transport Aircraft Takeoff [FAR (14CFR) 25.103/107/109/149]

To ensure safety, the takeoff procedure must satisfy the airworthiness requirement so that the aircraft will be able to climb to safety or stop safely in a case where one critical engine has failed. FAR (14CFR) 1.1 defines the 'critical engine' whose failure would most adversely affect aircraft takeoff procedure. Evidently, the most outboard engine of a multi-engine aircraft is the critical one. Designing for critical engine failure also covers the safety aspects of the failure of a non-critical engine for an aircraft with more than two engines.

Certification authorities demand mandatory requirements to cater for taking off with one engine inoperative to clear a 35 ft height representing an obstacle and continue with the required climb gradient (Figure 15.3). One

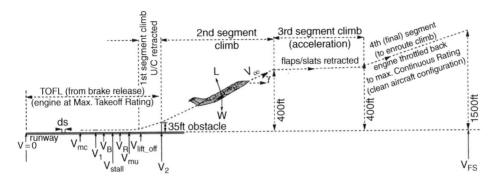

Figure 15.3 Takeoff, first, second, third and fourth segment climb (FAA).

engine inoperative TOFL is computed by considering BFL(Section 15.3.2) when the stopping distance after an engine failure at the decision speed, V_1, is the same as the distance taken to clear the obstacle at MTOM.

FAR (14CFR) 25 requires takeoff substantiation for the both dry and wet runways. This book deals only with the dry runway considerations.

15.3.2 Balanced Field Length (BFL) – Civil Aircraft

The rated TOFL at MTOM is determined by the BFL in the event of an engine failure. Figure 15.2 shows the segments involved in computing BFL. It can be seen that taking off with one engine inoperative (failed) above the decision speed V_1 has three segments to clear 35 ft height. Designers must provide the decision speed V_1 to pilots below which, if an engine fails, then the takeoff has to be aborted for safety reasons. The BFL segments are defined next:

Segment A. The distance covered by all engines operating a ground run until one engine fails at the decision speed V_1.
Segment B. The distance covered by one engine inoperative acceleration from V_1 to V_{LO}.
Segment C. Continue with the flare distance from lift-off speed V_{LO} to clear the 35 ft obstacle height reaching aircraft speed V_2.

In the case of engine failure, past the decision speed V_1, there are two segments (which replace segments B + C), as follows.

Segment D. The distance covered during the reaction time for the pilot to take braking action after an engine fails. (Typically, 3 s is taken as the pilot recognition time and brakes to act, spoiler deployment etc. At engine failure the thrust decay is gradual and within this reaction time before brake application there is a small speed gain shown in the diagram.) Typically, $(V_B \geq V_1)$ but can be taken as $(V_B \approx V_1)$.
Segment E. Distance to stop from V_B to V_0 (maximum brake effort).

$$\text{BFL is established when Segments (B + C) = Segments (D + E)} \tag{15.1}$$

15.3.2.1 Normal All-Engine Operating (Takeoff) – Civil Aircraft
Normal takeoff with all engines ($TOFL_{\text{all\_eng}}$) operating will take considerably lower field length than the one engine failed case ($TOFL_{\text{1eng-faled}}$). To ensure safety, FAR (14CFR) requires a 15% allowance is added to $TOFL_{\text{all\_eng}}$ over the estimated value. Therefore, it must be at least equal to the greater of the two as shown next.

$$TOFL = \text{the highest of BFL and } 1.15 \times TOFL_{\text{all\_engine}} \tag{15.2}$$

15.3.2.2 Unbalanced Field Length (UBFL) – Civil Aircraft

At low aircraft weights, the decision speed V_1 at one engine inoperative can become lower than V_{mc}. Since V_1 should not be below V_{mc} resulting in unbalanced field length (UBFL) as segments (B + C) < Segments (D + E). In this case, the TOFL is taken as segments (A + D + E). In another situation where V_2 < 110% of V_{mc} it makes UBFL when the TOFL is taken as segments (A + B + C). UBFL is not computed in this book as it is not a specification/requirement at the conceptual design phase.

15.3.3 Civil Aircraft Takeoff Segments [FAR (14CFR) 25.107 – Subpart B]

A takeoff process with one engine inoperative (FAR requirement) can be divided into four segments (Figure 15.3) of climb procedure as follows:

first segment[a]	Climb to 35 ft, undercarriage retracting.
second segment[a]	Climb from 35 to 400 ft undercarriage retracted. Engine power is kept at maximum takeoff rating. Drag reduction on account of undercarriage retraction is substantial.
third segment	In this segment the aircraft stays at 400 ft altitude to accelerate to higher speeds (V_2 > 10 kt) to increase stall speed margin. The aircraft is at a clean configuration.
fourth segment	This is the final segment under the takeoff procedure: continues with one engine inoperative and the aircraft must land immediately at a designated airfield. With all engines operating the aircraft continues to accelerate to en route climb at a specified speed schedule. Engine is throttled back to maximum continuous rating. Aircraft speed reaches final segment velocity, V_{FS}, to continue with the en route climb.

a) undercarriage retraction starts as soon as a gradient is established and completes at a height above 35 ft, and can be seen within the first segment by the International Civil Aviation Organization (ICAO) requirement. FAR second segment climb is taken from 35 ft, a known height, as the undercarriage is retracted. The operational takeoff requirements by ICAO are in more detail than the design requirements governed by the FAA.

The capability at the end of the fourth segment is a customer requirement and not a FAR requirement. Tables 15.1 and 15.2 provide the aircraft configuration and power settings at the aircraft takeoff segment FAR requirement.

Table 15.1 Civil aircraft takeoff segment requirements and status (FAR (14CFR) 25.121 Subpart B).

		first segment	second segment	third segment	fourth segment
Altitude – ft		35 to u/c retract	Reach 400 ft	Level at 400 ft	Reach 1500 ft
Speed (reference)		$V_{lift-off}$	V_2	$V_2 + 10$ kt	$V_2 + 10$ kt
Engine rating		Max. Takeoff	Max. Takeoff	Max. Cont.	Max. Cont.
Undercarriage		*Retracting*	Retracted	Retracted	Retracted
Flaps/slats		Extended	Extended	*Retracting*	Retracted
Max. climb gradient (%)	two-engine	0.0	2.4	0	1.2
	three-engine	0.3	2.7	0	1.5
	four-engine	0.5	3.0	0	1.7

The first segment speed schedules are interrelated and expressed in terms of the ratios of V_{stall} as given in Table 15.2. The range of velocity ratios in Table 15.2 are typical and could deviate a little so long as the FAR (marked with * in the table) stipulation is complied with.

Table 15.2 Civil aircraft first segment speed schedule.

	Two-engine	Three-engine	Four-engine
Percentage loss at an engine failure	50	33.3	25
Minimum climb gradient at first segment[a]	0%	0.3%	0.5%
Minimum climb gradient at the second segment[a]	2.4%	2.7%	3%
V_{LO}/V_{stall} (approx..)	1.12–1.14	1.15–1.16	1.17–1.18
V_R/V_{stall} (approx..)	1.1–1.18	1.14–1.18	1.16–1.18
V_{mu}/V_R (approx..)	1.02–1.04	1.02–1.04	1.02–1.04
V_1/V_R (approx.)	0.96–0.98	0.93–0.95	0.9–0.92
V_{mc}/V_1 (approx..)	0.94–0.98	0.94–0.98	0.94–0.98
V_2/V_{stall}	$\geq$1.2	$\geq$1.2	$\geq$1.2

a) Refers to FAR stipulation.

Note that some engines at their takeoff rating can have augmented power rating (APR) that could generate 5% higher thrust than the maximum takeoff thrust for short period of time. These types of engines are not considered in this book.

It may be noted that only V_{stall} is determined from aircraft weight and C_{Lmax} and FAR requires that $V_2 \geq 1.2$ V_{stall}. All other speed schedules are interrelated with regulation requirements. At the conceptual design stage, the ratios are taken from the statistics of past designs and as the project progresses these are refined, initially from wind-tunnel tests and finally from flight tests. Table 15.3 gives the important speed schedule ratios.

Performance engineers must know the sequence of the takeoff speed schedules stipulated by certification agencies as given in Table 15.4.

The higher the thrust loading (T/W), the higher the aircraft acceleration. For small changes, V_R/V_{stall} and V_{LO}/V_{stall} may be linearly decreased with an increase in T/W. The decision speed V_1 with an engine failed is established through iterations as shown in Section 15.10.1.2. In a family of derivate aircraft, the smaller variant can have V_1 approaching values close to V_R.

Military aircraft requirements (MIL-C5011A) are slightly different from the civil requirements. The first segment clearing height is 50 ft instead of 35 ft. Many military aircraft have a single engine where the concept of BFL is not applicable. Military aircraft have to satisfy the Critical Field Length (CFL) as described in Section 15.11.2. The second segment rate of climb is to meet a minimum of 500 ft min^{-1} for multi-engine aircraft.

Table 15.3 Airworthiness regulations at takeoff.

	Civil aircraft design		Military aircraft design
	FAR23	FAR25	MIL-C5011A
Obstacle height at V_2 ft	50	35	50
Takeoff climb speed at 35 ft height V_2	$\geq$1.2 V_{stall}	$\geq$1.2 V_{stall}	$\geq$1.2 V_{stall}
Lift-off velocity, V_{LO}	$\geq$1.1 V_{stall}	$\geq$1.1 V_{stall}	$\geq$1.1 V_{stall}
TOFL (BFL)			
One engine failed	As computed	As computed	As computed
All engines operating	as computed	1.15 × as computed	as computed
Rolling coefficient, μ	not specified	not specified	0.025

Table 15.4 Civil aircraft takeoff speed schedule (FAR requirements).

Speed schedule engine	All-engine operating			One engine inoperative		
	Two-engine	Three-engine	Four-engine	Two-engine	Three-engine	Four-engine
V_2/V_{stall}	≥1.20	≥1.20	≥1.20	≥1.20	≥1.20	≥1.20
V_{LO}/V_{mu} (approx.)	≥1.10	≥1.10	≥1.10	≥1.05	≥1.05	≥1.05
V_R/V_{mc} (approx.)	≥1.05	≥1.05	≥1.05	≥1.05	≥1.05	≥1.05

V_R is selected in such a way (from past experience subsequently finalised by flight tests) that it reaches V_2 speed at 35 ft altitude. Much depends on how the pilot executes rotation; FAR gives the details of requirements between maximum rate and normal rates of rotations. From the FAR requirements it can be seen (Figure 15.1) that $V_2 > V_{LO} > V_{mu} > V_R > V_{stall} > V_1 > V_{mc}$.

15.3.4 Derivation of Takeoff Equations

The derivation of takeoff performance equations is done at sea level STD day, an airfield with no gradient, zero wind conditions and thrust is aligned along the flight path (i.e. no thrust vectoring). The generalised equation for takeoff performance is derived in detail in [1].

15.3.4.1 Acceleration

Let the acceleration of the aircraft be denoted by a. The force balance equation resolved along aircraft velocity vector is given next (Figure 15.1).

Net force on the aircraft, $(W/g)a = \Sigma_{Applied}$ forces along aircraft velocity vector $= \Sigma F_{applied}$

On encountering ground rolling friction, μ. Then at any instant: $(W/g)a = T - D - \mu(W - L)$ (15.3)

where, V and a are instantaneous velocity and acceleration of the aircraft

Rearranging Eq. (15.3),

$$a = (g/W)(T - D - \mu(W - L))$$ (15.4)

In terms of coefficients and rearranging (all-engine operating)

$$a = (g/W)[(T - \mu W) - (D - \mu L)] = (g)\{(T/W - \mu) - (C_D - \mu C_L) \times [(S_W \times q)/W]\}$$ (15.5)

A simplified approach can be taken by taking average acceleration $\bar{a}$ for aircraft at $0.707 V_{LO}$ (average of V_{LO}^2 from zero, that is, at $\sqrt{(V_{LO}^2/2)}$ of the following equation). Then

$$\bar{a} = (g)\left[(\overline{T}/W - \mu) - \left(\frac{0.5\rho S_W C_L V^2}{W}\right)\left(\frac{\overline{C_D}}{\overline{C_L}} - \mu\right)\right]_{0.707V}$$ (15.6)

where S_W is the wing area and $q = 0.5\rho V^2$ is the dynamic head.

The aircraft on the ground encounters rolling friction (coefficient $\mu = 0.025$–0.03 for a paved, metalled runway). At braking friction, $\mu_B = 0.3$–0.5 (for the Bizjet, it is taken as 0.4). Thrust loading (T/W) is obtained from the sizing exercise. Average acceleration, $\bar{a}$, is taken at $0.707 V_{LO}$. FAR requires $V_2 = 1.2 V_{stall}$. This gives $V_2^2 = [2 \times 1.44 \times (W/S_W)]/(\rho C_{Lstall})$. An aircraft stalls at C_{Lmax}.

15.3.4.2 Takeoff Field Length (TOFL) Estimation – Distance Covered from zero to V_2

Figures 15.1 and 15.2 show that aircraft is under all-engine operation up to the decision speed, V_1, and, thereafter, if the decision is made, continues with the takeoff procedure: the aircraft is rotated at V_R to flare out and have lift-off at V_{LO} to reach to V_2 to climb to a 35 ft altitude. It takes about 3–4 s time to reach V_2 from V_R. If aborted at V_1, then the aircraft is braked to stop.

Form Figure 15.1, let dS be the elemental ground run distance covered by the aircraft in infinitesimal time dt. If the instantaneous velocity of the aircraft at that point is V then.

$$dS = Vdt = V(dV/dV)dt = (V/a)dV \text{ where } dV/dt = a \text{ (instantaneous acceleration)} \qquad (15.7)$$

Then the distance, S, covered from the initial velocity $V_i = 0$ to final velocity, V_f, is obtained by integrating Eq. (15.7). Aircraft acceleration and engine thrust varies with aircraft speed. At the conceptual stage of design, taking average values would prove convenient. Taking average acceleration $\bar{a}$ as a constant and using Eq. (15.7), the equation for distance, S, can be written as follows.

$$S = \int dS = \int_0^{V_f} (V/a)dV = \frac{1}{a}\int_0^{V_f} (V)dV = (V_f^2/\bar{a}) \qquad (15.8a)$$

For initial velocity $V_i \neq 0$ the ground run, Eq (15.8) can be re-written in terms of the average condition as follows.

$$S_G = \int dS = \int_{V_i}^{V_f} (V/a)dV = \frac{V_{ave}}{\bar{a}}\int_{V_i}^{V_f} dV = V_{ave} \times (V_f - V_i)/\bar{a} \qquad (15.8b)$$

where using Eq. (15.5), the average acceleration can be written in coefficient form as follows

$\bar{a} = \{g\,[(T/W - \mu) - (C_L Sq/W)(C_D/C_L - \mu)]\}_{ave}$ (in System International (SI) units 'g' drops out).

Equation (15.8) can now be written separately for each segment and then equated for the BFL. The average acceleration $\bar{a}$ is of the segment. A simplified approach can be taken by taking average acceleration $\bar{a}$ for aircraft at $0.707\,\Delta V$ (average of ΔV within the velocity interval, at $\sqrt{(\Delta V^2/2)}$ of the following equation).

$$TOFL = (1/\bar{a})\int_0^{V1} VdV + (1/\bar{a})\int_{V1}^{V_{LO}} VdV + (1/\bar{a})\int_{V_{LO}}^{V2} VdV \text{ (continuing takeoff)} \qquad (15.9)$$

In case it is aborted, the distance, S_{abort}, covered is as follows.

$$S_{abort} = (1/\bar{a})\int_0^{V1} VdV + (1/\bar{a})\int_{V1}^{VB} VdV + (1/\bar{a})\int_{VB}^0 VdV \text{ (braked to stop)} \qquad (15.10)$$

For the BFL,

$$BFL = (1/\bar{a})\int_0^{V1} VdV + (1/\bar{a})\int_{V1}^{V_{LO}} VdV + (1/\bar{a})\int_{V_{LO}}^{V2} VdV$$

$$= (1/\bar{a})\int_0^{V1} VdV + (1/\bar{a})\int_{V1}^{VB} VdV + (1/\bar{a})\int_{VB}^0 VdV$$

The proper choice of the decision speed V_1 will give the BFL. A number of iteration may be required to arrive at the proper V_1, as shown in the classroom example in Section 15.10.1.2.

(In later stages of the project, the computation could be done more accurately in smaller steps of speed increment within which average values of the variables are considered as constants. During takeoff, aircraft C_L and C_D vary with speed changes. In the worked-out examples, average values are taken. Reference [1] describes the takeoff procedure in detail.)

During takeoff, FAR requires $V_2/V_{stall} > 1.2$. The speed gain continues during the rotational (V_R) through aircraft flaring out to V_2. In this phase, typically around 5% change occurs in aircraft velocity through acceleration at the most critical condition; it amounts to about a 5–7 kt speed increase. Therefore, only the proper choice of the decision speed V_1 will give the BFL. A number of iterations may be required to arrive at the proper V_1, as shown in the classroom example in Section 15.7.

Given next are the typical takeoff speeds V_2 (with high-lift device deployed) for some of the larger jet transport currently in operation.

Boeing737: 150 mph	Boeing747: 180 mph	Airbus320: 170 mph
Boeing757: 160 mph	Concorde: 225 mph	Airbus340: 180 mph

15.4   Landing Performance

Computation of LFL uses a similar equation as for computing the TOFL. The difference is that landing encounters deceleration, that is, negative acceleration. Typically, the engine(s) runs idle producing no thrust. Values of friction coefficient, μ, vary at main wheel touch down, then when the nose wheel touches down and when the brakes are applied after nose wheel touchdown (taken here as 2–3 s after touchdown). A considerable amount of heat is generated at full braking and may pose some fire hazard. If brake parachutes are deployed then the drag of the parachute is to be accounted for in the deceleration. With a thrust reverser, the negative thrust needs to be taken into account as the decelerating force. With full flaps extended and with spoilers activated, the aircraft drag is substantially higher than at takeoff.

The landing speed schedules and landing procedure are given in Figure 15.4. Landing performance estimation is carried out at ISA day, sea level, with zero wind and on an airfield without gradient. The landing configuration is with full flaps extended and the aircraft at its landing weight. The approach segment at landing is from 50 ft altitude to touch down. At approach, the FAR requires that the aircraft must have a minimum speed $V_{app} = 1.3\ V_{stall\_land}$. At touchdown, aircraft speed is taken as $V_{TD} = 1.15\ V_{stall\_land}$. Brakes are applied 2 s after touchdown. A typical civil aircraft descent rate at touchdown is anywhere between 12 and 22 ft s^{-1}.

Consider an ideal landing with no floating: it takes about 5 s (glide + flare + nose wheel touchdown) from 50 ft altitude to brake application. An ideal landing does not happen readily and the aircraft floats before touchdown for which the FAA has given a generous allowance by multiplying by a factor of 1.667 for a dry runway (for a wet runway, increased by another 15% not worked out here); that is, increase computed field length by two-thirds more to get the LFL. Generally, this works out to be slightly less than the BFL at MTOM (but not necessarily).

Airworthiness requirements for landing aircraft are given in Table 15.5. All regulations require clearing the threshold height of 50 ft representing an obstacle, as a safety measure. The approach segment starts from a 50 ft height.

Approach velocity, V_{app} is the lowest schedule speed. For an aircraft with fly-by-air, it has relaxed V_{app} to 1.23 V_{stall}, not dealt with in this book.

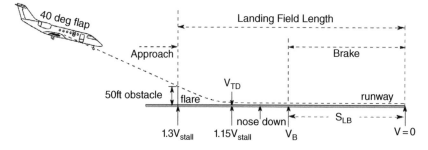

Figure 15.4 Landing.

Table 15.5 Airworthiness regulations at landing.

Design	Civil aircraft design		Military aircraft
Certification	FAR23	FAR25	MIL-C5011A
Approach velocity, V_{app}	$\geq 1.3 V_{stall}$	$\geq 1.3 V_{stall}$	$\geq 1.2 V_{stall}$
Touchdown velocity, V_{TD}	$\geq 1.15 V_{stall}$	$\geq 1.15 V_{stall}$	$\geq 1.1 V_{stall}$
Landing field length, LFL (from 50 ft height)	Distant to stop,	S_G 1.667S_G	Distant to stop, S_G
Braking coefficient, μ_B	Not specified	Not specified	Minimum of 0.30

Military aircraft requirements are slightly different, for example,

$$V_{app} = 1.2V_{stall\_land} \text{ and } V_{TD} = 1.1V_{stall\_land}.$$

The approach has two segments,

1. A steady straight glide path from a 50 ft height and
2. Flaring in a nearly circular arc to level out for touchdown. This would incur higher g.

The distances covered in these two segments depend on how steep the glide path is and how rapid the flaring action. This book will avoid such details of analysis; instead, a simplified approach will be assumed by computing the distance covered during the time taken from a 50 ft height to touch down before the brakes are applied. In this book, this is assumed to be 6 s.

15.4.1 Approach Climb and Landing Climb and Baulked Landing

In the case of a missed approach, the aircraft should be able to climb way above a 50 ft height to make the next attempt to land. The FAR (14CFR) 25.121 subpart B requirements for the approach climb gradient are given in Table 15.6. The aircraft is configured with full flaps, undercarriage retracted and the engine in full takeoff rating.

A discontinued approach takes place above 400 ft with the undercarriage retracted.

In the case of a baulked or called-off landing with all engines operating, the aircraft should be able to climb away from below a 50 ft height to make the next attempt to land. The FAR (14CFR) 25.119 subpart B requirements for the landing climb gradient are 3.2%.

In both these cases, the aircraft is considerably lighter at the end of mission on account of consuming onboard fuel, so there is less reserve.

15.4.2 Derivation of Landing Performance Equations

15.4.2.1 Ground Distance During Glide, S_{glide}

Normally, at a steady approach the aircraft is kept throttled back to idle producing no thrust. Figure 15.4 gives the generalised force diagram. It shows thrust, T, assisted by a component of weight, $W\sin\gamma$, in the descent angle, γ. In equilibrium, lift, L, is balanced by the weight component, $W\cos\gamma$. Equating for equilibrium conditions:

$$L = W\cos\gamma \approx W \text{ (for a small } \gamma)$$

and $D = T + W\sin\gamma$

$$\text{or } \sin\gamma = (D - T)/W = D/W - T/W \approx \gamma \text{ (for a small } \gamma)$$

or $\gamma = C_D/C_L - T/W$

Figure 15.4 gives $S_{glide} = 50/\tan\gamma$. For a small γ, $S_{LG} = 50/\gamma$.

Substituting the relation for γ, it gives

$$S_{glide} = \frac{50}{\left(\frac{C_D}{C_L} - \frac{T}{W}\right)}\text{ft} \tag{15.11}$$

Velocity at touchdown $= 1.15\, V_{stall\_land}.$

Table 15.6 FAA second segment climb gradient at a missed approach.

Number of engines	2	3	4
Second segment climb gradient (%)	2.1	2.4	2.7

However, as mentioned before, instead of using Eq. (15.11) the ground distance during glide, S_{glide} is computed as follows:

$$S_{glide} = (1.3 + 1.15)/2 \, V_{stall\_land} = 6 \times (1.225 \times V_{stall\_land}) \tag{15.12}$$

Ground distance from touchdown to nose wheel on the ground, $S_N = 2 \times 1.15 \, V_{stall\_land}$ (15.13)

Ground distance from V_B to stop, $V = 0$ is formulated next.

To keep the derivation generalised, negative thrust of the thrust reverser is included. However, in the Bizjet example there is no thrust reverser. (Note: the average condition is at $0.707 \, V_{TD}$).

$$S_B = V_{ave} \times (V_B - 0)/\overline{a} \tag{15.14}$$

where

$$\overline{a} = \left(\frac{g}{W}\right)[(-\mu_B W - T_{TR}) - (\overline{C}_D - \mu_B \overline{C}_L)Sq] = (g)[(-\mu_B - T_{TR}/W) - (\overline{D}/W - \mu_B \overline{L}/W)]_{0.707VB} \tag{15.15}$$

Without the thrust reverser, $S_B = (g)[(-\mu_B) - (\overline{D}/W - \mu_B \overline{L}/W)]_{0.707VB}$ (15.16)

LFL,

$$LFL = S_{glide} + S_N + S_B + = 6 \times (1.225 \times V_{stall\_land}) + 2 \times 1.15 V_{stall\_land} + S_B \tag{15.17}$$

15.5 Climb Performance

Climb is possible when available engine thrust is more than the aircraft drag, the excess thrust (thrust minus aircraft drag, $T - D$, is converted into the potential energy of height gain). Figure 15.5 gives the generalised force diagram of an aircraft climb path in the wind axes in a pitch plane (vertical plane, i.e. the plane of symmetry). Here, the thrust vector is aligned to the velocity vector. Aircraft velocity is V climbing at an angle γ with an angle of attack α. To compute integrated climb performance, the climb trajectory is discretised into small steps of altitude (ΔH) within which the variables are considered invariant.

Note that at climb, lift, L (normal to the flight path), is lower than aircraft weight, W (vertically downward). The residual component of the weight is balanced by the excess thrust (see Eq. (15.19)). The combat aircraft or aerobatic aircraft climb angles can be high. Typically, commercial aircraft climb angle is less than 15° ($\cos 15 = 0.96$ and $\sin 15 = 0.23$). Two situations can arise as follows:

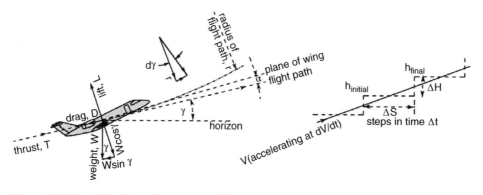

Figure 15.5 Generalised force diagram in pitch (vertical) plane – climb performance.

15.5.1 Derivation of Climb Performance Equations

From Figure 15.5, the force equilibrium gives $(T - D) = mg\sin\gamma + (m)dV/dt$

This gives the gradient, $\sin\gamma = [T - D - (W/g)dV/dt]/W = [(T - D)/W] - [(1/g) \times dV/dt]$

$$(15.18)$$

Write $dV/dt = (dV/dh) \times (dh/dt)$, then;

Rate of climb, $R/C_{accl} = dh/dt = V\,\sin\gamma = V(T - D)/W - (V/g) \times (dV/dh) \times (dh/dt)$ (15.19)

By transposing and collecting dh/dt,

$$R/C_{accl} = dh/dt = \frac{V[(T - D)/W]}{1 + (V/g)(dV/dh)} \tag{15.20a}$$

For un-accelerated rate of climb, Eq. (15.20a) becomes as follows.

$$R/C_{unaccl} = dh/dt = V[(T - D)]/W \tag{15.20b}$$

The rate of climb is a point performance and is valid at any altitude. The term $\frac{V}{g}\left(\frac{dV}{dh}\right)$ is a dimensionless term. It penalises unaccelerated rate (numerator of Eq. (15.20)) of climb depending on how fast the aircraft is accelerating during the climb. Part of the propulsive energy is consumed for speed gain instead of altitude gain. Military aircraft make an accelerated climb in the operational arena when the $\frac{V}{g}\left(\frac{dV}{dh}\right)$ term reduces the rate of climb depending on how fast the aircraft is accelerating. On the other hand, a civil aircraft has no demand for high accelerated climb, instead it makes en route climbs to cruise altitude at a *quasi-steady state* climb by holding aircraft climb speed at a constant EAS/Mach number (very slow rate of speed change as altitude increases). Constant EAS climb makes TAS increase with altitude gain, that is, also the Mach number. A constant indication of speed eases pilot workload. During a *quasi-steady state* climb at constant EAS, the contribution by the $\frac{V}{g}\left(\frac{dV}{dh}\right)$ term is low. The magnitude of the acceleration term reduces with altitude gain and becomes close to zero at the ceiling (defined when $R/C_{accl} = 100\,\text{ft min}^{-1}$). (Remember: $V = V_{EAS}/\sqrt{\sigma}$ and $V_{EAS} = Ma\sqrt{i}$].

15.5.2 Quasi-Steady State Climb

Civil aircraft en route climb is executed at a constant EAS when the acceleration term, $\dot{V} \approx 0$, and $d\gamma/dt = \dot{\gamma} \approx 0$. This makes the $\frac{V}{g}\left(\frac{dV}{dh}\right)$ term in Eq. (15.20) small enough as derived follows.

(The other possibility is to have *quasi-level flight*. This is when the climb angle, γ, is small (typically <10°) but the acceleration term, $\dot{V}$ is not. $\sin\gamma \approx \gamma$ radians, $\cos\gamma \approx 1$. It gives $\gamma^2 << 1$. This is not dealt with in this book.)

15.5.2.1 Constant EAS Climb

The term $\frac{V}{g}\left(\frac{dV}{dh}\right)$ can be worked out in terms of constant EAS as follows.

$$\frac{V}{g}\left(\frac{dV}{dh}\right) = \frac{V_{EAS}\,V_{EAS}}{g\sqrt{\sigma}}\left(\frac{d(1/\sigma)}{dh}\right) = -\frac{V_{EAS}^2}{2g\sigma^2}\left(\frac{d\sigma}{dh}\right) = -\frac{M^2 a^2}{2g\sigma}\left(\frac{d\sigma}{dh}\right) \tag{15.21}$$

Below the troposphere (for air, $\gamma = 1.4$, $R = 287\,\text{J kg K}^{-1}$, $g = 9.81\,\text{m s}^{-2}$)
In SI, Eq. (1.7a) gives in troposphere $T = (288.16 - 0.0065\,h)$ and

$$\sigma = \rho/\rho_0 = (T/T_0)^{[(g/0.0065R)-1]} = (T/T_0)^{[9.81/(0.0065\times287)-1]} = (T/T_0)^{4.255}$$

$$(d\sigma/dh) = 4.225 \times (T/T_0)^{3.255} \times \frac{d\left(\frac{T}{T_0}\right)}{dh} = 4.225 \times (T/T_0)^{3.255} \times \frac{1}{T_0}\left(\frac{dT}{dh}\right) \text{ where } \left(\frac{dT}{dh}\right) = -0.0065$$

Substituting the values $-\dfrac{M^2 a^2}{2g\sigma}\left(\dfrac{d\sigma}{dh}\right)$ of Eq. (15.21)

$$\text{or } \frac{V}{g}\left(\frac{dV}{dh}\right) = -\frac{M^2 a^2}{2g\sigma}\left(\frac{d\sigma}{dh}\right) = -\frac{M^2 a^2}{2g\left(\frac{T}{T_0}\right)^{4.255}} \times 4.225 \times (T/T_0)^{3.255} \times \frac{(-0.0065)}{T_0}$$

$$\text{or } \frac{V}{g}\left(\frac{dV}{dh}\right) = \frac{M^2 a^2}{2g\left(\frac{T}{T_0}\right)} \times 4.255 \times \frac{(0.0065)}{T_0} = \frac{M^2 \gamma R T}{2g\left(\frac{T}{288.07}\right)} \times 4.255 \times \frac{(0.0065)}{288.07}$$

$$\text{or } \frac{V}{g}\left(\frac{dV}{dh}\right) = \frac{M^2 \times 1.4 \times 287}{\left(\frac{2\times9.81}{288.07}\right)} \times 4.255 \times \frac{(0.0065)}{288.07} = \frac{V}{g}\left(\frac{dV}{dh}\right) = 0.566\, M^2 \tag{15.22}$$

Above the tropopause (for air, $\gamma = 1.4$, $R = 287\,\text{J kg K}^{-1}$, $g = 9.81\,\text{m s}^{-2}$, $a = 295.07\,\text{m s}^{-1}$).

From 11 to 20 km the ambient temperature is held constant at 216.65 K. Eq. (1.8g) gives the density relation in this regime as $\rho\,\text{kg m}^{-3} = \rho_1 e^{-\left[\frac{g_0}{RT}\right](h-h_1)}$ where base pressure $\rho_1 = $ is at 11 km.

That gives $\sigma = \rho/\rho_0 = (\rho/\rho_1)(\rho_1/\rho_0) = (\rho/\rho_1)(0.36392/1.225) = 0.297 \times (\rho/\rho_1) = 0.297 \times e^{-\left[\frac{g_0}{RT}\right](h-h_1)}$

or $\sigma = 0.297 \times e^{-\left[\frac{32.2}{287\times216.65}\right](h-11,000)} = 0.297 \times e^{-[0.0001578](h-11,000)} = 0.297 \times e^{(1.7355 - 0.0001578h)}$

Then $(d\sigma/dh) = 0.297 \times e^{(1.7355 - 0.0001578h)}[-0.0001578]$

Then Eq. (15.21) can be written as follows.

$$\frac{V}{g}\left(\frac{dV}{dh}\right) = -\frac{M^2 a^2}{2g\sigma}\left(\frac{d\sigma}{dh}\right) = -\frac{M^2 295.07^2 (-0.0001578)}{2 \times 9.81} = 0.7\, M^2 \tag{15.23}$$

15.5.2.2 Constant Mach Climb

The term $\frac{V}{g}\left(\frac{dV}{dh}\right)$ can be worked out in terms of a constant Mach number climb as follows.

$$\frac{V}{g}\left(\frac{dV}{dh}\right) = \frac{MaM}{g}\left(\frac{da}{dh}\right) \quad (M \text{ being the constant taken out of differentiation.})$$

$$\text{or } \frac{V}{g}\left(\frac{dV}{dh}\right) = \frac{aM^2\sqrt{\gamma R}}{g}\left(\frac{d\sqrt{T}}{dh}\right) = \frac{\gamma R M^2 \sqrt{T}}{g}\left(\frac{d\sqrt{T}}{dh}\right)$$

$$\text{or } \frac{V}{g}\left(\frac{dV}{dh}\right) = \frac{\gamma R M^2 \sqrt{T}}{2g\sqrt{T}}\left(\frac{dT}{dh}\right) \tag{15.24}$$

Below the tropopause (for air, $\gamma = 1.4$, $R = 287\,\text{J kg K}^{-1}$, $g = 9.81\,\text{m s}^{-2}$)

From Eq. (1.7a), atmospheric temperature, T, can be expressed in terms of altitude, h, as follows:

$T = (288 - 0.0065\,h)$ where h is in metres. Substituting the values of γ, R and g, the following is obtained.

Substituting the value in Eq. (15.21), $\dfrac{V}{g}\left(\dfrac{dV}{dh}\right) = \dfrac{1.4 \times 287 \times M^2}{2 \times 9.81}(-0.0065)$

$$\text{or } \frac{V}{g}\left(\frac{dV}{dh}\right) = \frac{401.8 \times M^2}{19.62}(-0.0065) = -0.133\, M^2 \tag{15.25}$$

Above the tropopause (for air, $\gamma = 1.4$, $R = 287\,\text{J kg K}^{-1}$, $g = 9.81\,\text{m s}^{-2}$, $a = 295.07\,\text{m s}^{-1}$)

In the similar manner, the relations above the tropopause can be obtained. Above the tropopause up to 25 km, the atmospheric temperature remains constant at 216.65 K, therefore the speed of sound remains invariant, that is $(da/dh) = 0$ making $\frac{V}{g}\left(\frac{dV}{dh}\right) = 0$.

Table 15.7 The $\frac{V}{g}\left(\frac{dV}{dh}\right)$ value (dimensionless quantity) at constant climb speeds (quasi-steady climb).

	Below tropopause	Above tropopause
At constant EAS	$0.566\,M^2$	$0.7\,M^2$
At constant Mach number	$-0.133\,M^2$	0 (Mach held constant)

Table 15.7 summarises the relation between $\frac{V}{g}\left(\frac{dV}{dh}\right)$ and the constant speed climb schedules as derived before.

To prepare the Pilot Manual, the airspeed needs to be in V_{CAS} and the conversion from V_{EAS} requires incorporating relevant errors associated with the airspeed indicator (see Section 15.2.1). Readers may refer to [1] for details on compressibility correction. Nowadays, ready data is available in the Electronic Flight Information System (EFIS) memory.

With the loss of one engine at the second segment climb, an accelerated climb would penalise the rate of climb. Therefore, second segment climb with one engine inoperative is done at an unaccelerated climb speed, at a speed V_2 with the undercarriage retracted. The unaccelerated climb equation is obtained by dropping the acceleration term in Eq. 15.18, yielding the following equations.

$$T - D = W \sin\gamma \text{ rewriting } \sin\gamma = (T - D)/W$$

$$\text{Unaccelerated rate of climb,} R/C = dh/dt = V \sin\gamma = V \times (T - D)/W \tag{15.26}$$

The en route climb performance parameters vary with altitude. En route climb performance up to cruise altitude is computed at discrete steps of altitude (say in steps of 5000 ft altitude – see Figure 15.5) within which all parameters are considered invariant and taken as average value within the step altitudes. The engineering approach is to compute the integrated distance covered, time taken and fuel consumed to reach the cruise altitude in steps of small altitudes and then summed up. The procedure is shown next.

Infinitesimal time to climb is expressed as $dt = dh/(R/C_{accl})$. The integrated performance within the small steps of altitudes can be written as

$$\Delta t = t_{final} - t_{initial} = (h_{final} - h_{initial})/(R/C_{accl})_{ave} \tag{15.27}$$

$$\text{and } \Delta H = (h_{final} - h_{initial}) \tag{15.28}$$

The distance covered during climb can be expressed as

$$\Delta s = \Delta t \times V_{ave} = \Delta t \times V \cos\gamma \tag{15.29}$$

where V = average aircraft speed within the step altitude
Fuel consumed during climb can be expressed as

$$\Delta fuel = \text{Average fuel flow rate} \times \Delta t \tag{15.30}$$

Military combat aircraft with high (TD) climb is performed in an accelerated climb. The equation for accelerated climb can be derived as follows (Figure 15.5). To keep it simpler, the subscript ∞ is dropped to represent aircraft velocity. This is not dealt with in this book.

Summary
Distance covered during climb, $R_{climb} = \sum\Delta s$ is obtained by summing up the values obtained in the small steps of altitude gain.
Fuel to climb, $Fuel_{climb} = \sum\Delta_{fuel}$ is obtained by summing up the values obtained in the small steps of altitude gain.
Time to climb, $time_{climb} = \sum\Delta t$ is obtained by summing up the values obtained in the small steps of altitude gain.

15.6 Descent Performance

Figure 15.6 shows the aircraft descent force and velocity diagram.

15.6.1 Derivation of Descent Performance Equations

Descent uses the same equations, except that the thrust is less than drag the rate of descent (R/D_{accl}) is opposite to the rate of climb. The rate of descent is expressed as follows:

$$R/D_{accl} = dh/dt = \frac{V[(D-T)/W]}{1+(V/g)(dV/dh)} \tag{15.31}$$

Unlike climb, gravity assists descent, hence it can be performed without any thrust (engine kept at idle rating producing zero thrust). However, passenger comfort and aircraft structural considerations require a controlled descent with maximum rate limited to a certain value depending on the design as explained in Section 15.2.3. Controlled descent is carried out at a part throttle setting. To obtain maximum range, the aircraft should ideally make its descent at the minimum rate. These adjustments will entail varying speed at each altitude. To ease pilot load, descent is made at a constant Mach number and when it has reached the V_{EAS} limit it adopts a constant V_{EAS} descent, in a *quasi-steady state*, as is done for climb. During *quasi-steady state* descent at constant EAS, the contribution by the $\frac{V}{g}\left(\frac{dV}{dh}\right)$ term is low.

Some special situations may be pointed out as follows.

For unaccelerated descent Eq. (15.20b) gives

$$R/D_{unaccl} = dh/dt = \frac{V[(D-T)]}{W} \tag{15.32}$$

At a higher altitude, the prescribed speed schedule for descent is at a constant Mach, hence above the tropopause V_{TAS} is constant and descent is kept in unaccelerated flight.

At zero thrust, Eq. (15.32) becomes

$$R/D_{unaccl} = dh/dt = \frac{VD}{W} \approx \frac{VD}{L} \tag{15.33}$$

It indicates that at constant $V(L/D)$ the $R/C_{descent}$ is the same for all weights.

As in climb, the other parameters of interest during descent are range covered ($R_{descent}$), fuel consumed ($Fuel_{descent}$) and time taken ($time_{descent}$) during descent. There are no FAR requirements for the descent schedule. Descent rate is limited by the cabin pressurisation schedule for passenger comfort. FAR requirements are enforced during approach and landing.

The descent velocity schedule is supplied to cater for the ECS capability (cabin descent rate = 300 ft min^{-1}, actual aircraft descent is higher than that, as explained in Section 15.2.3). This requires a shallow descent gradient, part throttle is required to maintain the gradient with the benefit of distance gained during descent.

Section 15.6 gives the worked example of the Bizjet. Integrated performances for climb to cruise altitude and the descent to sea level are computed and the values for distance covered, time taken and fuel consumed

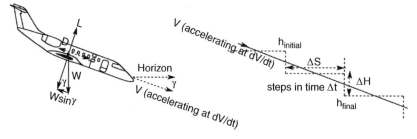

Figure 15.6 Aircraft descent force and velocity diagram.

are estimated to obtain the aircraft payload-range. Reference [1] may be consulted for details of climb and descent performances on this Bizjet example.

15.7 Checking of the Initial Maximum Cruise Speed Capability

At the conceptual design phase (Phase I), the aircraft sizing and engine matching exercise promised the capability to meet the customer requirement of initial maximum cruise speed and needs to be guaranteed through flight test substantiation.

Civil aircraft maximum speed is executed in an HSC in a steady level flight when the available thrust equals the aircraft drag.

The first task is to compute drag for the flight C_L at the maximum cruise speed and then check if the available thrust (at maximum cruise rating of the engine) is sufficient to achieve the required speed. In some cases the available maximum cruise thrust is more than what is required; in that case the engine is adjusted to a slightly lower level. Section 15.10.6 gives the worked-out example.

The long range cruise (LRC) schedule is meant to maximise range and is operated at a lower speed to avoid compressibility drag rise.

15.8 Payload-Range Capability – Derivation of Range Equations

Finally, the civil aircraft must be able to meet the payload-range capability as specified by the market (customer) requirement. The mission range and fuel consumed during the mission are given by the following two equations.

$$\text{Mission range} = R_{\text{climb}} + R_{\text{cruise}} + R_{\text{descent}} \tag{15.34}$$

$$\text{Mission fuel} = Fuel_{\text{climb}} + Fuel_{\text{cruise}} + Fuel_{\text{descent}} \tag{15.35}$$

The method to compute fuel consumption, distance covered, time taken during climb and descent is discussed in Sections 15.5 and 15.6. In this section, the governing equation for cruise range (R_{cruise}), cruise fuel ($Fuel_{\text{cruise}}$) and time taken during cruise are derived.

Let W_i = aircraft initial cruise weight (at the end of climb) and
W_f = aircraft final cruise weight (at the end of cruise)

$$\text{Then, fuel burned during cruise} = Fuel_{\text{cruise}} = W_i - W_f \tag{15.36}$$

At any instant, rate of aircraft weight change, dW = rate of fuel burned (consumed).
In an infinitesimal time dt, the infinitesimal weight change, $dW = sfc \times \text{thrust}\,(T) \times dt$

$$\text{or } dt = dW/(sfc \times T) \tag{15.37}$$

In Eq. (15.37), multiply both the numerator and denominator by weight, W and then equate $T = D$ and $W = L$.

The Eq. (15.37) reduces to

$$dt = \frac{1}{sfc}\left(\frac{W}{T}\right)\left(\frac{dW}{W}\right) = \frac{1}{sfc}\left(\frac{L}{D}\right)\left(\frac{dW}{W}\right) \tag{15.38}$$

$$\text{Elemental range, } ds = V \times dt = \frac{V}{sfc}\left(\frac{L}{D}\right)\left(\frac{dW}{W}\right) \tag{15.39}$$

Therefore, range covered during cruise (R_{cruise}) is the integration of Eq. (15.39) from the initial cruise weight to the final cruise weight. At cruise, V and sfc remain nearly constant. Taking mid-cruise L/D, the change in

L/D can ignored. These can be taken out of the integral sign. Integrating Eq. (15.38) gives the time taken for the R_{cruise}. At cruise $T = D$ and $L = W$.

$$R_{\text{cruise}} = \int ds = \int_{W_f}^{W_i} \frac{V}{sfc}\left(\frac{L}{D}\right)\left(\frac{dW}{W}\right) = \frac{V}{sfc}\left(\frac{L}{D}\right)\ln\left(\frac{W_i}{W_f}\right) \tag{15.40}$$

The value of $\ln(W_i/W_{if}) = k_{1\_range}$ varies from 0.2 to 0.5, longer the range, the higher the value.

Of the Eq. (15.40), the terms W_i and W_f are concerned with fuel consumed during cruise and the term *sfc* stems from the matched engine characteristics. The rest of the terms (*VL/D*) are concerned with aircraft aerodynamics. Aircraft designers aim to increase (*VL/D*) as best as possible to maximise range capability. The aim is not just to maximise (*L/D*) but to maximise (*VL/D*). Expressing in terms of Mach number, it becomes (*ML/D*). To have the best of engine-aircraft gain it is best to maximise (*ML*)/(*sfcD*).

Specific range, *Sp.Rn*, is defined as range cover per unit weight (or mass) of fuel burned.

Using Eq. (15.40),

$$Sp.Rn = R_{\text{cruise}}/\text{cruise fuel} = [k_{1\_range} \times (VL)/(sfcD)]/(W_i - W_f) \tag{15.41}$$

Cruise fuel weight ($W_i - W_f$) can be expressed in terms of maximum takeoff weight (MTOW) and varies from 15 to 40% of MTOW, the longer the range, the higher the value. Let $k_{2\_range} = k_{1\_range}/(0.15{-}0.4)$.

Then Eq. (15.41) reduces to

$$Sp.Rn = [k_{2\_range} \times (VL)/(sfcD)]/MTOW = \left[\left(\frac{\ln(W_i - W_f)}{(W_i - W_f)}\right) \times \frac{(V \times L)}{(sfc \times D)}\right]/(MTOM) \tag{15.42}$$

Equation (15.42) gives a good insight on what can maximise range, that is, a good design to stay ahead in the competition. It says

i. Make aircraft as light (MTOM) as it can be made without sacrificing safety. Material selection and structural efficiency are the keys to it. Integrate with lighter bought-out equipment.
ii. Make superior aerodynamics to lower drag (high *L/D*).
iii. Choose the better aerofoil for good lift keeping the moment low.
iv. Make aircraft to cruise as fast it can within M_{crit} (aerofoil selection).
v. Match the best available engine with the lowest *sfc* (better engine selection).

However, Eq. (15.40) does not address the cost implications. At the end, DOC will dictate the market appeal and the designers will have to compromise performance with cost. This shows the importance of giving due consideration at the conceptual design stage to deciding on the manufacturing philosophy to be adopted to minimise aircraft production cost. These form the essence of good civil aircraft design; easily said but not so easy to achieve as must be experienced by the readers.

Equation (15.40) can be further developed. From the definition of lift coefficient, C_L, the aircraft velocity, V, can be expressed as:

$V = \sqrt{\frac{2W}{\rho S_W C_L}}$, substituting in Eq. (15.40) the cruise range R_{cruise} can be written as

$$R_{\text{cruise}} = \int_{W_f}^{W_i} \frac{1}{sfc}\sqrt{\frac{2W}{\rho S_W C_L}}\left(\frac{L}{D}\right)\left(\frac{dW}{W}\right) = \sqrt{\frac{2}{\rho S_W}}\int_{W_f}^{W_i}\frac{1}{sfc}\left(\frac{\sqrt{C_L}}{C_D}\right)\left(\frac{dW}{\sqrt{W}}\right) \tag{15.43}$$

As mentioned earlier, typically over the cruise range, changes in *sfc*, and *L/D* are small and if the mid-cruise values are taken as averages then they may be treated as constants and are taken out of the integral sign. Then Eq. (15.43) becomes

$$R_{\text{cruise}} = \sqrt{\frac{2C_L}{\rho S_W}}\left(\frac{1}{sfc \times C_D}\right)\int_{W_f}^{W_i}\left(\frac{dW}{\sqrt{W}}\right) = 2\sqrt{\frac{2C_L}{\rho S_W}}\left(\frac{\sqrt{W_i} - \sqrt{W_f}}{sfc \times C_D}\right) \tag{15.44}$$

This equation is known as the Breguet Range formula, originally derived for propeller-driven aircraft, which embedded propeller parameters (jet propulsion was not invented at that time).

LRC is carried out at the best *sfc* and at the maximum value of $\sqrt{C_L/C_D}$ (i.e. L/D) to maximise range. Typically, the best L/D occurs at the mid-cruise condition. For every LRC (say around and over 2500 nm), the aircraft weight difference from initial to final cruise is large. There is a benefit if cruise is carried out at higher altitude when the aircraft becomes lighter. This can be done either in stepped altitudes or by making a gradual shallow climb matching with the gradual lightening of the aircraft.

Sometimes, a mission may demand HSC to save time, in which case Eq. (15.44) is still valid but not operating for the best range. There are many other possibilities on how cruise segments can be executed depending on the sortie requirement as shown in [1].

15.9 In Horizontal Plane (Yaw Plane) – Sustained Coordinated Turn

Turning is a pilot induced manoeuvre. A distinction should be made between a sustained turn in a steady state and an instantaneous turn in an unsteady state. A steady sustained turn is a coordinated turn when the load factor, n, remains constant and thrust is aligned with the velocity vector. Turning has to stay within the permissible load factor limit, n_{max}, (structural consideration) or within the C_{Lstall} (aerodynamic consideration) limit. Operating within these limits, an aircraft is not designed to enter into buffet. Treatment of the instantaneous turn is beyond the scope of this book – refer to [2] for the full derivation.

Forces in wind axes F_W are as shown in Figure 15.7. In a coordinated turn, the roll (bank) angle is φ and turning angle is ψ.

15.9.1 Kinetics of a Coordinated Turn in Steady (Equilibrium) Flight

The equation of motion at an instant of a steady turn has a radius R of turn and aircraft velocity, V, tangential to the flight path. In elemental time Δt, the turning angle is $\Delta \psi$.

Then, the instantaneous angular velocity can be written as $(d\psi/dt)$ in rad s^{-1}.

The instantaneous tangential velocity can be written as $V = R(d/\psi dt)$ in ft s^{-1}.

$$\text{or } (d\psi/dt) = V/R \tag{15.45}$$

The thrust line is taken along the flight path. Turn is in a bank angle of φ with L' as the lift force acting at the aircraft plane of symmetry normal to the instantaneous tangential velocity V (true airspeed).

In the horizontal plane, thrust = drag

In the lateral plane, weight, $W = mg = L' \cos\varphi$

$$\text{or } L' = W/\cos\varphi \tag{15.46}$$

where $L' - L = \Delta L = W/\cos\varphi - W = W(1/\cos\varphi - 1)$

The ratio of the forces, $L'/W = 1/\cos\varphi = n = \sec\varphi$ (aircraft load factor) $\tag{15.47}$

$$\text{This gives, } \sin\varphi = \left(\frac{\sqrt{(n^2 - 1)}}{n} \right) \tag{15.48}$$

where n = aircraft load factor acting in the plane of symmetry of the aircraft along L'.

Perpendicular to a flight path towards the instantaneous centre of turn, the force balance gives: centrifugal force = centripetal force,

$$\text{that is, } mV^2/R = L' \sin\varphi \tag{15.49}$$

Therefore, the centripetal acceleration acting radial (normal to the flight path), $a_n = V^2/R$

And the tangential acceleration (acting along the flight path), $a_t = dV/dt$ $\tag{15.50}$

Combining Eqs. (15.46) and (15.49),

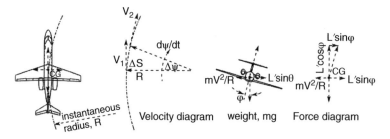

Figure 15.7 Forces on an aircraft coordinated turn in the horizontal plane.

$m = L'\cos\varphi/g = RL'\sin\varphi/V^2$
or $V^2/Rg = \sin\varphi/\cos\varphi = \tan\varphi$
Substituting Eq. (15.45),

$$\tan\varphi = V(d\psi/dt)/g \tag{15.51}$$

$$\text{or angular velocity, } (d\psi/dt) = g\tan\varphi/V \tag{15.52}$$

Eq. (15.49) gives,

$$R = mV^2/L'\sin\varphi \tag{15.53}$$

Using Eq. (15.47) $n = L'/W = L'/mg$, Eq. (15.54) gives the radius of turn

$$R = mV^2/L'\sin\varphi = nV^2/g\sin\varphi = \frac{V^2}{g\sqrt{n^2-1}} \tag{15.54}$$

Load factor, n:

From Eq. (15.53) gives, $R = mV^2/L'\sin\varphi$

$$\text{In equilibrium flight, thrust, } T = \text{drag}, D = 0.5\rho V^2 S_W(C_{DP\min} + kC_L'^2) \tag{15.55}$$

where, C_L' is the lift coefficient while turning under load factor, n.
where $L' = nmg = 0.5\rho V^2 S_W C_L'$

$$\text{or } C_L' = 2nmg/\rho V^2 S_W \tag{15.56}$$

Substituting Eq. (15.56) into Eq. (15.55),
$T = D = 0.5\rho V^2 S_W[C_{DP\min} + k(2nmg/\rho V^2 S_W)^2]$
or $T = 0.5\rho V^2 S_W C_{DP\min} + k(0.5\rho V^2 S_W)(2nW/\rho V^2 S_W)^2]$
or $T = 0.5\rho V^2 S_W C_{DP\min} + k(2n^2 W^2/\rho V^2 S_W)]$

$$\text{or } k(2n^2 W/\rho V^2 S_W)] = T/W - 0.5\rho V^2(S_W/W)C_{DP\min} \tag{15.57}$$

Solve Eq. (15.57) for load factor, n
or $n^2 = \left[\frac{0.5\rho V^2}{k\left(W/S_W\right)}\left(\frac{T}{W} - 0.5\rho V^2\frac{C_{DP\min}}{W/S_W}\right)\right]$

$$\text{Load factor, } n = \left[\frac{0.5\rho V^2}{k\left(W/S_W\right)}\left(\frac{T}{W} - 0.5\rho V^2\frac{C_{DP\min}}{W/S_W}\right)\right]^{\frac{1}{2}} \tag{15.58}$$

15.9.2 Maximum Conditions for a Turn in the Horizontal Plane

Differentiating Eq. (15.58) with respect to V and setting it equal to zero gives the maximum conditions of V_{n_max} as follows.

$$2ndn/dV = 0 = \left[\frac{\rho V}{k\left(W/S_W\right)}\left(\frac{T}{W}\right)\right] - \left[\frac{\rho^2 V^3 C_{DPmin}}{k\left(W/S_W\right)^2}\right] = \left(\frac{\rho V}{k\,W/S_W}\right)\left[\left(\frac{T}{W}\right) - \frac{\rho V^2 C_{DPmin}}{\left(W/S_W\right)}\right]$$

This gives $\dfrac{\rho V_{n\_max}^2 C_{DPmin}}{\left(W/S_W\right)} = \left(\dfrac{T}{W}\right)$

or $(V_{n\_max})^2 = \left(\dfrac{1}{\rho C_{DPmin}}\right)\left(\dfrac{T}{W}\right)\left(\dfrac{W}{S_W}\right)$

or $(V_{n\_max}) = \sqrt{\left(\dfrac{1}{\rho C_{DPmin}}\right)\left(\dfrac{T}{W}\right)\left(\dfrac{W}{S_W}\right)}$ \hfill (15.59)

Substituting in Eq. (15.18), the maximum load factor, n, can be obtained

$$n_{max} = \left[\frac{0.5\rho\left(T/W\right)\left(W/S_W\right)}{k\rho C_{DPmin}\left(W/S_W\right)}\left(\frac{T}{W} - \frac{0.5\rho\left(T/W\right)\left(W/S_W\right)}{\rho C_{DPmin}}\frac{C_{DPmim}}{W/S_W}\right)\right]^{\frac{1}{2}} = \left[\frac{0.5\rho(T/W)^2}{k\rho C_{DPmin}}(1-0.5)\right]^{\frac{1}{2}}$$

or $n_{max} = 0.5\left(\dfrac{T}{W}\right)\left[\dfrac{1}{kC_{DPmin}}\right]^{\frac{1}{2}}$ operating at $(T/W)_{max}$ as T is the only variable \hfill (15.60)

15.10 Aircraft Performance Substantiation – Worked-Out Classroom Examples – Bizjet

As stated earlier that during the Conceptual Design Phase-I, before the 'go-ahead' obtained, only the specified aircraft performance requirements are to be substantiation (see Section 1.8). The specified Bizjet market requirements are given below.

Baseline Version (8 to 10 passengers) - 10 passengers at standard (medium) comfort level.
 Payload: 1,100 kg
 Take-off field length (TOFL) at sea-level STD day $= 4400$ ft
 Landing field length (LFL) at sea-level STD day $= 4400$ ft
 Initial climb rate at sea-level STD day $= 2600$ ft/min
 Initial high speed cruise (HSC) at 41000 ft, STD day $= 0.74$ Mach
 Range: 2,000 miles + reserve
 g-level: 3.8 to -2.

Given below are the sized Bizjet aircraft data (see Section 14.4) for performance substantiation.
 $MTOM = 20\,680$ lb (9400 kg); wing area, $S_W = 323$ ft^2 (30 m^2)
 Bizjet sized wing load, $W/S_W = 64$, $T_{SLS}/W = 0.34$
 At static conditions, $T_{SLS} = 3500$ lb per engine, $T/W_{avg} = 0.308$
 Landing weight $= 15,800$ *lb* (7182 *kg*)
 Dynamic head, $q = \frac{1}{2}\rho V^2 = 0.5 \times 0.002378 V^2 = 0.001189 V^2$
 Equations used: $C_L = \dfrac{2 \times MTOM}{\rho S_W V^2} = \dfrac{20680}{0.001189 \times 323 \times V^2} = \dfrac{53847.6}{V^2}$, $V_{stall} = \sqrt{\dfrac{53847.6}{C_{Lstall}}}$
To make the best use of available data, all computations are done using the FPS system. The results can be subsequently converted to the SI system. Tables 15.8–15.17, present all the data required for Bizjet performance substantiation.

15.10.1   Checking TOFL (Bizjet) – Specification Requirement 4400 ft

During take-off ground run, aircraft accelerates from zero to lift-off speed, V_{lift_off}, causing continuous changes in aircraft aerodynamics characteristic, e.g., lift and drag. An accurate TOFL estimation requires C_L and C_D variation with speed gain. Readers may refer to [1] for the methodology adopted to compute accurate TOFL using a graph showing the varying force parameters during the ground run.

The acceleration term in take-off has a strong contribution to the determination of BFL. Ref [1] shows that TOFL estimation using average acceleration gives result within 1% of the accurate TOFL value for all engine operating case. It is convenient to use average acceleration, $\bar{a}$, term at this stage of study.

Equation 15.6 gives average acceleration, $\bar{a}$, as follows.

$$\bar{a} = (g)\left[(\bar{T}/W - \mu) - \left(\frac{0.5\rho S_W C_L V^2}{W} \right) \left(\frac{\bar{C}_D}{\bar{C}_L} - \mu \right) \right]_{0.707V}$$

The average acceleration, $\bar{a}$, can be applied for the segment with initial velocity, V_i, and final velocity, V_f, i.e., at $0.707V = 0.707 \times (V_f - V_i) + V_i$ of the segment of operation

For the ground run, Eq. 15.8a gives the following.

$$S_G = \int dS = \int_{V_i}^{V_f} (V/a)dV = \frac{V_{ave}}{\bar{a}} \int_{V_i}^{V_f} dV = V_{ave} \times (V_f - V_i)/\bar{a} \tag{15.8b}$$

Table 15.5 may be used for Bizjet aircraft take-off estimation at a sea level ISA day to prepare the spreadsheet for repeated computations. Table 15.8 gives the Bizjet aircraft data generated thus far. To make the best use of the available data all computations are done in the FPS system. The results can be subsequently converted to the SI system.

At V_2 speed, $T@119.25\,kt = 2860$ lb per engine (20° flap)

Rolling friction coefficient on paved runway, $\mu = 0.025$.

Table 15.8 Bizjet performance parameters (takeoff/landing – $W/S_W = 64$ lb ft^{-2}).

Flap setting, deg	0	8	20[a]	Landing[b]
C_{Lmax}	1.55	1.67	1.9	2.2
C_{Dpmin}	0.0205	0.0205	0.0205	0.0205
Rolling friction coefficient, μ	0.03	0.03	0.03	0.03
Braking friction coefficient, μ_B	0.45	0.45	0.45	0.45
V_{stall} @ 20 680 lb ft s^{-1}	186.4	179.5	168.4	136.16[a]
V_{stall}, knots	110.38	106.00	99.38	80.66
V_R, ft s^{-1} (1.11 V_{stall})	205.08	196.91	184.61	149.77
V_R, knots	121.5	116.60	109.32	88.7
V_{LO}, knots (1.15 V_{stall})	126.94	121.90	114.28	92.72
V_{LO} ft s^{-1}	214.38	205.87	193.00	156.58
V_2, knots (1.2V_{stall})	132.46	127.20	119.25	96.75
V_2, ft s^{-1} (1.2V_{stall})	223.7	214.82	201.4	163.40
T/W_{avg} – all-engine	0.308	0.308	0.308	0
T/W_{avg} – single engine	0.154	0.154	0.154	0

a)   Normal takeoff at STD day is carried out with a 20° flap setting. At lower takeoff weight and/or hot and high altitude airfield having longer runway length (4400 ft), the pilot may choose 8° flap that gives a better second segment climb gradient.

b)   Landing at 35–40° flap, engines at idle and V_{stall} at aircraft landing weight of 15 800 lb.

Several decision speeds are worked out to estimate takeoff capabilities. First, the all-engine TOFL is estimated. Next, the BFL is estimated, for which the decision speed V_1 is to be determined in the inadvertent case of one critical engine failure.

15.10.1.1 All-Engine Takeoff – 20° Flap

Table 15.9 gives the values for all-engine TOFL with the Bizjet at 20° flap. Due to extended undercarriage and flap deflection, the aircraft lift to drag ratio degrades to a typical value of approximately 10 for the Bizjet.

$$\bar{a} = 32.2 \times \left[(0.308 - 0.025) - \left(\frac{0.5 \times 0.5 \times 0.002378 \times V_{0.707}^2}{64} \right) (0.1 - 0.025) \right]$$

$$= 32.2 \times (0.283 - 0.0005945 \times V_{0.7}^2/914.2) = 32.2 \times (0.283 - 0.00000065 \times V_{0.707}^2)$$

Flare from V_R to V_2 The aircraft is under manoeuvre load through and in climb, and still accelerates at a lower rate. Aircraft flaring through rotation is a complex physics and is described in detail in [1]. Typically, in this class of aircraft, it takes 3 s. Here the computation is simplified by taking the average velocity from V_R to V_2 to cover field length distance in 3 s.

Taking flare time as 3 s for the Bizjet, the average $V_{LO \, to \, V2} = (192.43 + 202.56)/2 = 197.496 \text{ ft s}^{-1}$

$$S_{Gflare} = 3 \times 197.496 = 592.3 \text{ ft}$$

Computed all-engine operating field length $= 1976 + 592.3 \approx 2568 \text{ ft}$

$$\text{Time taken } t_{all\_eng} \approx \sqrt{[(2 \times 1976)/8.8] + 3} \approx 24 \text{ s}.$$

To ensure safety, the FAR requires an all-engine operating TOFL at a 15% higher margin than the computed value and it may exceed the value obtained for a one engine inoperative BFL value. In that case, the higher value of the two is used.

FAR all-engine operating field length for a Bizjet at 20° flap, $TOFL_{all\_eng} = 1.15 \times 2568 = 2953 \text{ ft}$

$$\text{Corresponding time taken } t_{all\_eng} = \sqrt{[(2 \times 2953)/8.8] + 3} \approx 26 \text{ s}.$$

To compare with the simplified method to compute FL_{all_eng}, take average acceleration at

$$0.707 V_{LO} = 0.707 \times V_{LO} = 0.707 \times 192.91 = 136.387 \text{ ft s}^{-1}. \text{ Then } \bar{a} \text{ ft s}^{-2} = 8.8 \text{ ft s}^{-2}.$$

$$S_{R\_all\_eng} = \frac{V_R^2}{2\bar{a}} = (184.61^2)/(2 \times 8.8) \approx 1936 \text{ ft s}^{-1} \text{ compared to } 1976.3 \text{ ft in Table 15.9. This shows excellent}$$ agreement.

Table 15.9 Segment A – all-engine operating from zero to V_R (see Figure 15.1).

			V_{stall}	V_R
V, kt	80	90	100	110
V, ft s^{-1}	135.04	151.92	168.8	185.68
$0.5\rho S_W$	0.384	0.384	0.384	0.384
$0.707V$	95.47	107.41	119.34	131.28
$\bar{a}_{avg}$	8.91	8.85	8.793	8.73
ΔV	135.04	151.92	168.8	185.68
V_{avg}	67.52	75.96	84.4	92.84
S_G, ft	1024	1304	1620	1976

15.10.1.2 One Engine Inoperative – Balanced Field Takeoff (BFL) Inoperative

Section 15.3.2 gives the BFL takeoff to make sure that the aircraft can stop on the airfield in case an engine fails. It requires a decision speed V_1, below which the aircraft stops and above which the aircraft continues with the takeoff operation with power available to maintain the FAR stipulated minimum climb gradient up to the second segment and then return to land immediately. At first, V_1 is estimated and subsequently establishes the BFL for takeoff. Table 15.9 showing all engine takeoff values can be used up to a speed of V_1.

Segment A – all engines operating up to the decision speed V_1 – BFL run

To determine the decision speed V_1, estimate three speeds, for example, 80, 90 and 100 kt. The all-engine operating case up to these speeds can be taken from Table 15.9.

Segment B – one engine inoperative acceleration from V_1 to lift-off speed, V_R

This phase is a transient one; the aircraft leaves the ground to become airborne. Its lift and drag characteristics are changing fast. With the aircraft in a high drag configuration with one engine inoperative, drag estimation is quite difficult. Therefore, a simplified approach is taken to compute distance covered and time taken from V_1 to lift-off speed, V_{LO}. The simplification gives a reasonable result and conveys the physics involved. Industries using computers and more accurate lift and drag data adopt detailed calculations.

In flight, the lift to drag ratio in such a dirty configuration will be low, in this class in the order of around 9.5. On the ground below rotation speed, the lift is low and so is drag is low; the simplification takes the same value of $(L/D) \approx 9.5$. Therefore, the average value of $(C_D/C_L) \approx 0.105$ is taken. A typical average $C_L \approx 0.8$ is taken, the weight on the wheels is lighter on account of some lift generated on the wing. Because one engine is inoperative there is a loss of power by half $[(T/W)_{avg} = 0.154)]$. The simplified method is used because the difference will be small.

The velocity that would give average acceleration is $V_{0.707} = 0.7 \times (V_R - V_1) + V_1$.
One engine inoperative $\bar{a}$ for the Bizjet after decision V_1 is as follows
20° flap

$$\bar{a} = 32.2 \times \left[(0.154 - 0.025) - \left(\frac{0.5 \times 0.8 \times 0.002378 \times V_{0.707}^2}{64} \right) (0.105 - 0.025) \right]$$

$$= 32.2 \times (0.129 - 0.00095 \times V_{0.707}^2/800) = 32.2 \times (0.129 - 0.000001185 \times V_{0.707}^2)$$

Braking takes place after an engine fails when it has the all-engine acceleration. With the loss of thrust aircraft on account of engine failure the acceleration does not suddenly drop to low levels in a discrete step down, but the aircraft gradually retards to a lower level of acceleration. Therefore, for the interval of computation, the average value has to be taken. This is an important consideration that is often overlooked.
$\bar{a}_{ave} = 0.5 \times (\bar{a}_{all\_engine} + \bar{a}_{one\_engine\_failed})$

Section 15.10.1 gives the ground distance covered, $S_G = V_{ave} \times (V_f - V_i)/\bar{a}$

Table 15.10 computes Segment B, the ground distance covered from V_1 to V_{LO}.

Segment C – Flaring distance with one engine inoperative from V_R to V_2

This is the flaring distance to reach V_2 from V_R. From the statistics, time taken to flare is 2 s. Tables 15.10–15.12 compute the ground distance covered from V_{LO} to V_2 with one engine inoperative for the flap settings. In this segment the aircraft is airborne, hence there is no rolling friction. By taking the average velocity between V_2 and V_R the distance covered during flare is given.

The next step is to compute the stopping distance with the maximum application of brakes.

			V_{stall}	V_R	V_{LO}	V_2
V (kt)	80	90	100	110	114	120
Segment $(B+C)$	2118	1734	1297	781	593	593

Table 15.10 Segment B – the ground distance covered from V_1 to V_{LO} for the 20° flap settings.

			V_{stall}	V_R
V, kt	80	90	100	110
V, ft s^{-1}	135.04	151.92	168.8	185.68
V_{LO}, ft s^{-1}	192.432	192.432	192.432	192.432
ΔV	57.392	40.512	23.632	6.752
V_{avg}	163.74	172.18	180.62	189.06
$0.707V$	175.62	180.56	185.51	190.45
$\bar{A}$	3.416	3.374	3.331	3.286
$\bar{a}_{ave}$	6.162	6.114	6.062	6.006
S_G, ft	1525	1141	704	213

Table 15.11 Segment C – Bizjet one engine ground distance V_R to V_2 (flaring).

V_R at $1.11V_{stall}$ f/s	184.61
V_{LO} at $1.15V_{stall}$ f/s	193.00
V_2 at $1.2V_{stall}$ f/s	201.39
V_{ave} f/s	197.20
S_{Gflair} (3 s), ft	591.6

Table 15.12 Bizjet failure recognition distance (Segment D).

V – knots	80	90	100	114
V ft s^{-1}	135.04	151.92	168.8	192.432
Distance in 3 s at V_1, S_{G_B} – ft	405.12	456	506.4	577.3

Segment D – Engine failed (recognition time)

Table 5.12 computes the ground distance covered from V_1 to V_B.

The distance is covered in 1 s due to pilot recognition time and 2 s for the brakes to act from V_1 to V_B (flap settings are of little consequence).

Table 15.12 shows the Segment D engine failure recognition distance.

Segment E – braking distance from V_B to zero velocity (Flap settings are of little consequence)

Reaction time to apply the brake after the decision speed, V_1 is 3 s. The aircraft continues to accelerate a little in the 3 s but speed returns to V_B – this is ignored. Table 15.13 computes the ground distance covered from V_B to stop.

Aircraft in full brake mode with $\mu_B = 0.4$, all engines shut down and average $C_L = 0.5$,

Using Eq. (15.6), the average acceleration based on $0.707\,V_B$ ($\approx 0.707\,V_1$) reduces to

$$\bar{a} = 32.2 \times \left[(-0.4) - \left(\frac{0.5 \times 0.5 \times 0.002378 \times V_{0.707}^2}{64} \right) (0.1 - 0.4) \right]$$

$$= 32.2 \times [(-0.4) - 0.00000056 \times V_{0.707}^2]$$

Table 15.13 Bizjet stopping distance (Segment E).

V, kn	80	90	100
V, ft s^{-1}	135.04	151.92	168.8
ΔV, ft	135.04	151.92	168.8
V_{avg}, ft	67.52	75.96	84.4
$0.707V$	95.47	107.41	119.34
$\bar{A}$	-12.08	-11.86	-11.62
S_G, ft	755	973	1197
$(D+E)$	1160	1429	1732

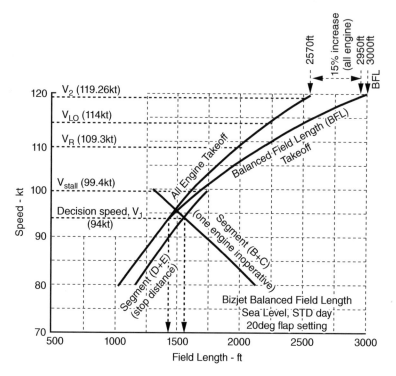

Figure 15.8 Bizjet takeoff.

Select the distance when segments $(B+C)$ equals segment $(D+E)$ as shown in Figure 15.8. With 20° flap, the decision speed V_1 is 94 kt covering a distance of 1570 ft. Then TOFL becomes $1430 + 1570 = 3000$ ft. This is summarised in Table 15.14 and Figure 15.8. Both satisfy the specified TOFL requirement of 4400 ft.

The TOFL requirement is 4400 ft, recommended takeoff procedure uses a 20° flap (Figure 15.9). At lower takeoff weight and/or a longer runway length (4400 ft), the pilot may choose an 8° flap that gives a better second segment climb gradient. With the air-conditioning switched on and at an eight flap setting, the BFL should come close to 4000 ft. The readers may compute for an 8° flap decision speed.

Discussion for takeoff analysis

Increase of flap setting improves the BFL capability at the expense of a loss of climb gradient (this will be shown in the next chapter). With one engine inoperative, the percentage loss of thrust for a two engine aircraft is the highest (50% lost). With one engine failed, the aircraft acceleration suffers and the ground run taken from V_1 to lift-off is higher. Table 15.15 summarises the takeoff performance and the associated speed schedules for

Table 15.14 Bizjet decision speed (Figure 15.8).

	BFL
Flap, degrees	20
Decision speed, V_1, knots	94
Stall speed, V_{stall}, knots	99.3
Rotation speed, V_R, knots	109.3
Lift-off speed, V_{LO}, knots	114
V_2 speed, V_2, knots	119.3
Distance, ft at $(B + C) = (D + E)$, ft	1570
All-engine distance at V_1 speed, ft	1430
BFL, ft	3000
second segment climb gradient (computed in next chapter)	
All-engine distance at V_2 speed, ft	2568 ft
All-engine increase by 15%	2953 ft

Table 15.15 Bizjet takeoff field length summary (Figure 15.8).

Flap setting, degrees	20		8	
	knots	ft s^{-1}	knots	ft s^{-1}
V_{stall} @ 20 680 lb	99.38	167.83	106	179
[b]V_{mc} at 0.94V_1	84.4[b]	142.8	91.2[b]	153.9
V_1 decision speed	90.00	151.92	96.5	163.74
$V_R = 1.05V_{stall}$	109.32	184.61	116.60	196.92
V_{LO} at 1.15V_{stall}	114.28	193.00	121.90	205.8
[a]$V_2 = 1.2\ V_{stall}$ at 35 ft altitude	119.26	201.39	127.2	214.82
V_{mu} at 1.02 V_R (lower than V_{LO})	115.50	188.30	119.00	200.85
BFL, ft	2940		3220	
All-engine takeoff time, s	≈31 s			

a) If required, V_2 can be higher than 1.2 V_{stall}.
b) Aircraft control surfaces (mainly rudder) are sized to get a V_{mc} below V_1 for the pilot to have the speed margin. Here, it arbitrarily chosen to at 75 kt – this needs to be computed (a design task beyond the scope of this book). This is because the lowest V_1 is when the aircraft is lightest (say at the landing weight of 15 800 lb) and at the highest flap deployment (for an inadvertent case, it can be at 40° flap, otherwise at 20° flap). This point is often overlooked in aircraft design coursework. Figure 15.9 gives the Bizjet takeoff speed schedules.

the two flap settings. The ratio of speed schedules can be made to vary for pilot ease, as long as it satisfies FAR requirements. This is an example of the procedure.

Higher flap settings would give more time between decision speed V_1 and the rotation speed V_R. However, it is not a problem if V_1 is close to V_R, so long there is pilot reaction time is available if one engine fails; if not then in a very short time the rotation speed V_R is reached and the aircraft takes off (typically, BFL is considerably less than the available airfield length). Also V_{mu} is close to V_R, hence there is little chance for tail dragging. If the pilot makes an early rotation then V_{mu} may not be sufficient for a lift-off and the aircraft will tail drag until it gains sufficient speed for the lift-off. If the engine fails early enough then the pilot has sufficient time to recognise it and act to abort takeoff.

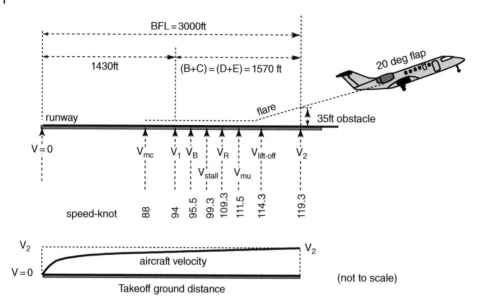

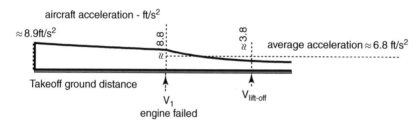

Figure 15.9 Takeoff speed schedule – 20° flap.

With more than two engines, the decision speed V_1 is further away from the rotation speed V_R. The pilot must remain alert as the aircraft speed approaches the decision speed V_1 and must react quickly if an engine fails.

The readers should compute for other weights to prepare graphs for ISA day and ISA + 20 °C.

15.10.2 Checking Landing Field Length (Bizjet) – Specification Requirement 4400 ft

Landing weight of the Bizjet is 15 800 lb (wing loading = 48.92 lb ft^{-2}) and at full flap 40° extended $C_{Lmax} = 2.2$.

$$\text{Therefore, } V_{\text{stall@land}} = \sqrt{\frac{2 \times 15800}{0.002378 \times 323 \times C_L}} = \sqrt{\frac{41141}{2.2}} = 137 \text{ ft s}^{-1}$$

$$V_{\text{appr}} = 1.3 V_{\text{stall@land}} = 178 \text{ ft s}^{-1}$$

$$V_{\text{TD}} = 1.15 V_{\text{stall@land}} = 158 \text{ ft s}^{-1}$$

Average velocity from 50 ft height to touch down = 168 ft s^{-1}.
Distance covered before brake application after 5 s (may differ) from 50 ft height,

$$S_{\text{G\_TD}} = 5 \times 168 = 840 \text{ ft}$$

Aircraft in full brake with $\mu_B = 0.4$, all engines shut down and average $C_L = 0.4$, $C_D/C_L = 0.11$. Section 15.10.1 for average acceleration based on 0.707 $V_{\text{TD}} = 110.6$ ft s^{-1}.
Then $q = 0.5 \times 0.002378 \times 110.6^2 = 14.54$

The deceleration equation becomes, $\bar{a} = (g)\left[(-0.4) - \left(\frac{14.54 \times 0.4}{48.92}\right)(0.11 - 0.4)\right]_{0.707V}$

Deceleration, $\bar{a} = 32.2 \times [(-0.4) + (0.119) \times (0.29)] = -11.8 \text{ ft s}^{-2}$

Distance covered during braking, $S_{G\_0Land} = (158 \times 79)/11.8 = 1058 \text{ ft}$

Landing distance $S_{G\_Land} = 840 + 1058 = 1898 \text{ ft}$

Multiplying by 1.667 (dry runway), the rated LFL $= 1.667 \times 1898 = 3164 \text{ ft}$, within the requirement of 4400 ft. Typically, BFL and LFL are close to one another.

Designers must also check the capability of go-around in case missed approach (approach climb) or baulked landing (landing climb) as described in Section 15.4.1. These cases are not worked out in this book. Generally, airfield is sufficiently long and with robust design ensures that this situation can handled by pilots. Safety can never be compromised.

15.10.3   Checking Takeoff Climb Performance Requirements (Bizjet)

There are three requirements to be substantiated for the climb performance. The three requirements are given here for the two-engine aircraft. The first two are the FAR requirements. Table 5.3 gives aircraft configuration and FAR requirements for the first and second segment climb. The four takeoff climb segments are shown in Figure 15.3.

i.   Check that the FAR first segment climb requirement of positive gradient is maintained.
ii. Check that the FAR second segment climb gradient requirement exceeds 2.4%.
iii. Check the market specification of the initial en route climb rate equals or exceeds 2600 ft min^{-1}. The cabin pressurisation system should cope with the rate of climb. This is a customer specification and not a FAR requirement.

The second segment starts at 35 ft altitude with flaps extended and undercarriage retracted (one engine inoperative) and ends at 400 ft altitude. Aircraft is maintained at V_2 speed to keep the best gradient. A loss of 50% thrust does not favour accelerated climb, which will be low in the case. Engine is at takeoff rating. The available one engine installed thrust is taken from Figure 13.11. The thrust is kept invariant at takeoff rating through first segment and second segment climb. Table 15.16 summarises the first and second segment climbs for both 8° and 20° flap settings.

15.10.4   Checking Initial En Route Rate of Climb – Specification Requirement is 2600 ft min^{-1}

Initial en route climb starts at 1000 ft altitude ($\rho = 0.0023 \text{ slug ft}^{-3}$, $\sigma = 0.9672$), all engines throttled back to maximum climb rating and the aircraft has a clean configuration. The aircraft makes an accelerated climb from V_2 to reach 250 KEAS, which is kept constant in a quasi-steady state climb until it reaches Mach 0.7 at about 32 000 ft altitude, from where the Mach number is held constant in the continued quasi-steady state climb until it reaches the cruise altitude. Fuel consumed during second segment climb is small and taken empirically (from statistics) to be 80 lb (see Table 15.17). Therefore, aircraft weight at the beginning of en route climb, W = 20 600 lb.

At 250 kt (422 ft s^{-1}, Mach 0.35) the aircraft lift coefficient $C_L = M/qS_W = 20\,600/(0.5 \times 0.0023 \times 422^2 \times 323)$ $= 20\,600/66150 = 0.311$.

Clean aircraft drag coefficient from Figure 11.2 at $C_L = 0.311$ gives $C_{Dclean} = 0.0242$.

Clean aircraft drag, $D = 0.0242 \times (0.5 \times 0.0023 \times 422^2 \times 323) = 0.0242 \times 66\,150 = 1600 \text{ lb}$

Available all-engine installed thrust at maximum climb rating from Figure 13.12 at Mach 0.378 is $T = 2 \times 2450 = 4900 \text{ lb}$

From Eq. 15.20a, quasi-steady state rate of climb is given by,

$$R/C_{accl} = \frac{V[(T-D)/W]}{1 + (V/g)(dV/dh)},$$

Table 15.16 First and second segment climb performance. Aircraft flap and undercarriage extended, engine at takeoff rating (one engine inoperative). (Use drag polar in Figures 11.2 and 11.12 to obtain drag.).

	First segment climb		Second segment climb	
	Up to 35 ft altitude $\rho_{ave} = 0.002\,378$ lb ft^{-3} $0.5\rho S_W = 0.3795$		35 ft to 400 ft altitude $\rho_{ave} = 0.002\,35$ lb ft^{-3} $0.5\rho S_W = 0.3747$	
Flap, degress	8	20	8	20
MTOM, kg	20 680	20 680	20 680	20 680
V_{stall}, ft s^{-1} (kt)	179.6 (106.4)	168.4 (99.8)	179.6 (106.4)	168.4 (99.8)
$V_2 =$ ft s^{-1} (Mach)	216.5 (0.194)	202.08 (0.181)	216.5 (0.194)	202.08 (0.181)
$qS_W =$	17 788	15 497.4	17 563	15 301.4
C_L	1.16	1.33	1.177	1.35
C_{Dclean} (Figure 9.2)	0.075	0.1	0.076	0.101
$\Delta C_{D\_one\_eng}$	0.003	0.003	0.003	0.003
ΔC_{Dflap}	0.013	0.032	0.013	0.032
$\Delta C_{D\_u/c}$	0.022	0.022	–	–
C_{D1st_seg}	0.113	0.157	0.092	0.135
Drag, lb	2010	2433	1616	2081
Thrust available, lb	2800	2850	2790	2840
Gradient, %	+ve	+ve	5.4	3.24
FAR	Meets	Meets	Meets	Meets

Table 15.17 Bizjet range (step-climb).

	Aircraft, min weight, lb	Distance, nm	Fuel, lb	Time, min
			(at end of segment)	
Start and taxi out	20 680	0	80[a]	3[a]
Takeoff to 1500 ft	20 600	0	120[a]	5[a]
Climb to 43 000 ft	20 480	170	800	25.5
Initial cruise at 43 000 ft[b]	19 680	670	1460	100.5
Initial cruise at 45 000 ft[c]	18 220	1000	2000	150
Start descent at 45 000 ft[c]	16 220	160	340	30
Approach and land from 1500 ft	15 880	0	80[a]	5[a]
Taxi-in from landing weight (from reserve)[d]	15 800	0	20[a]	3[a]
Stage Total		2000	4800	322 (5.4 h)

a) From operational statistics.
b) Average specific range at 43 000 ft altitude = 0.465 nam lb^{-1} at average weight of $\approx$19 000 lb.
c) Average specific range at 45 000 ft altitude = 0.515 nam lb^{-1} at average weight of $\approx$17 000 lb.
d) Reserve is quite in excess of what is required. In other words, the aircraft can have a little more extra range.

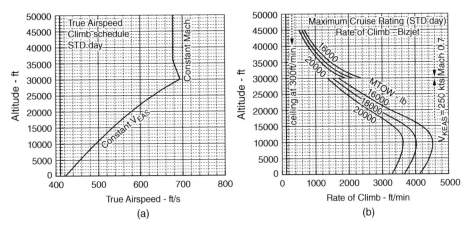

Figure 15.10 Climb point performances. (a) Climb speed schedule and (b) rate of climb.

At quasi-steady state climb, Table 15.7 gives $\frac{V}{g}\left(\frac{dV}{dh}\right) = 0.56\,M^2 = 0.56 \times 0.35^2 = 0.0686$

Hence, $R/C_{\text{quasi-steady}} = \{[422 \times (4900 - 1600)]/20\,680\}/[1 + 0.0686]$

$$= 63.2\,\text{ft s}^{-1} = 3796\,\text{ft min}^{-1}\ (20.7\,\text{m s}^{-1}).$$

The capability satisfies the market requirement of 2600 ft min^{-1} (15.2 m s^{-1}).

(Civil aircraft rate of climb is limited by the cabin pressurisation schedule. The aircraft is limited to 2600 ft min^{-1} at an altitude where the cabin pressurisation rate reaches its maximum capability. Naturally, at a low altitude of 1000 ft, this limit is not applicable.)

15.10.5 Integrated Climb Performance (Bizjet)

Integrated climb performance is not a specification for substantiation – it is used to obtain aircraft payload-range capability, which is a requirement to be substantiated. It is reasoned in Section 15.1 that only the results of the integrated climb performances in graphical form are given as shown in Figures 15.10 and 15.11.

It is convenient to first establish the climb velocity schedule (Figure 15.10a) and the point performances of rate of climb (Figure 15.10b) up to ceiling altitudes and for at least three weights for interpolation. Bizjet carries out quasi-steady state climb at a constant $V_{\text{EAS}} = 250$ kt until it reaches Mach 0.7 and thereafter continues at constant Mach until it reaches the ceiling (rate of climb = 100 ft min^{-1}).

The computations for the integrated climb performance are carried out in steps of approximately 5000 ft altitude (as convenient – at higher altitudes in smaller steps) in which the variables are kept invariant using their mean values (see Figure 15.5). Eqs. (15.27)–(15.29) are used to compute the integrated performances. Figure 15.11 gives the integrated performances of fuel consumed, distance covered and time taken to climb at the desired altitude.

The detailed computations to obtain all the graphs given in this section are shown in the companion book [1]. Classroom instructors may assist the computational work to obtain similar performance graphs for their projects.

15.10.6 Checking Initial High-Speed Cruise (Bizjet) – Specification Requirement of High-Speed Cruise Mach 0.75 at 41 000 ft Altitude

Aircraft at high speed initial cruise is at Mach 0.75 (716.4 ft s^{-1}) at 41 000 ft altitude ($\rho = 0.00055$ slug ft^{-3}). Fuel burned to climb is computed (not shown) to be 700 lb. Aircraft weight at initial cruise is 20 000 lb.

At Mach 0.75 the aircraft lift coefficient, $C_L = MTOM/qS_W = 20\,000/(0.5 \times 0.00055 \times 716.4^2 \times 323) = 20\,000/45627 = 0.438$.

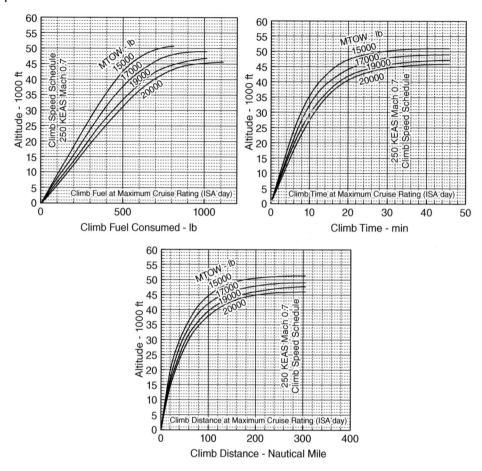

Figure 15.11 Integrated climb performances for the Bizjet (classroom example).

Clean aircraft drag coefficient from Figure 11.2 at $C_L = 0.438$ gives $C_{Dclean} = 0.0305$.

Clean aircraft drag, $D = 0.0305 \times (0.5 \times 0.00055 \times 716.4^2 \times 323) = 0.0305 \times 45\,627 = 1392\,lb$.

Available all-engine installed thrust at the maximum cruise rating at the speed and altitude from Figure 13.13 at Mach 0.75 is $T = 2 \times 750 = 1500\,lb$ (adequate).

The capability satisfies the market requirement of Mach 0.75 at HSC.

15.10.7   Specific Range (Bizjet)

Specific range is a convenient way to present cruise performance. Using Eq. (15.42) the specific range, *Sp.Rn* is computed. It is needed to compute the cruise segment performances (fuel, distance and time). Figure 15.12 gives the specific range for the Bizjet (classroom example).

Cruise segment distance cover is follows.

Cruise range = 2000 − (Distance to climb to 43 000 ft) − (distance to descend from 45 000 ft altitude).

From the *Sp.Rn* values the distance covered is worked out, which in turn gives the time taken for the distance.

15.10.8   Descent Performance (Bizjet) – Limitation Maximum Descent Rate of 1800 ft min^{-1}

It is also reasoned in Section 15.1 that only the results of the integrated descend performance in graphical form is supplied as shown in Figures 15.13 and 15.14. It is convenient to first establish the descent velocity

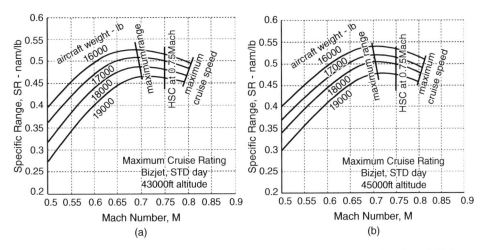

Figure 15.12 Specific range of the Bizjet (classroom example). (a) SR at 41 000 ft altitude and (b) SR at 45 000 ft altitude.

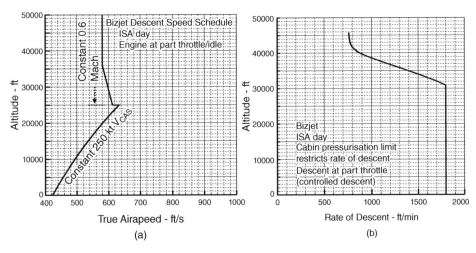

Figure 15.13 Descent point performance. (a) Descend velocity schedule and (b) rate of descent.

schedule (Figure 15.13) and the point performances of the rate of descent (Figure 15.14) up to sea level (valid for all weights – the difference between the weights is ignored). While redoing the calculations by the readers, there could be small differences in the results.

The related governing equations are explained in Section 15.6.1. It also mentions that descent rate is restricted by rate of cabin pressurisation schedule to ensure passenger comfort. Two of the difficulties in computing descent performance are to have the part throttle engine performance and the ECS pressurisation capabilities that dictate the rate of descent, which in turn stipulates the descent velocity schedule. These are not supplied in the book. Classroom instructors are to assist to establish these two graphs. In absence of any information these two may be used. The following simplifications could prove useful.

The first simplification is in obtaining part throttle engine performance as follows.

i. Descent is carried out anywhere at 50–70% of maximum power rating.
ii. Zero thrust at idle rpm at about 40% of the maximum rated power/thrust.

The second simplification is that the descent velocity schedule is supplied to cater for the ECS capability (cabin descent rate = 300 ft min^{-1}, actual aircraft descent is higher than that – Figure 15.13b). This is explained in Section 15.2.3.

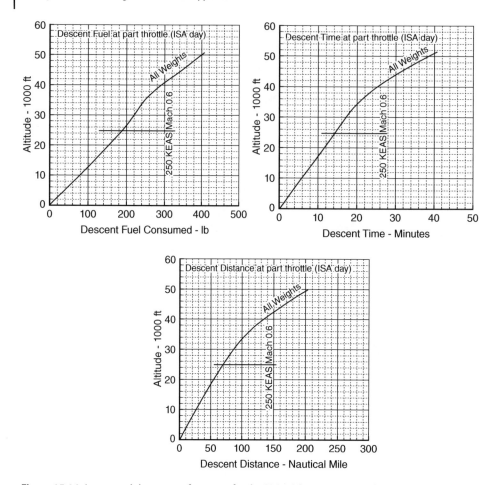

Figure 15.14 Integrated descent performance for the Bizjet (classroom example).

In industry, the exact installed engine performance at each part throttle condition is computed from the engine deck supplied by the manufacturer. Also, the ECS manufacturer supplies the cabin pressurisation capability from which aircraft designers work out the velocity schedule.

The inside cabin pressurisation is restricted to an equivalent rate of 300 ft min^{-1} at sea level to ensure passenger comfort. The aircraft rate of descent is then limited to the pneumatic capability of the ECS. The Bizjet is restricted to a maximum of 1800 ft min^{-1} at any time (for higher performance, at lower altitudes it can be raised up to 2500 ft min^{-1}). Descent speed schedule is to continue at Mach 0.6 from cruise altitude until it reaches the approach height when it changes to constant $V_{EAS} = 250$ kt until the end (for higher performance, it can be raised to Mach 0.7 and $V_{EAS} = 300$ kt). The longest range can be achieved at minimum rate of descent. Evidently, these will require throttle dependent descents to stay within the various limits.

It is convenient to establish first the point performances of velocity schedule (Figure 15.13a) and rate of descent (Figure 15.13b – all weights – variation is small). Descent is performed within the limits of human comfort level. However, in an emergency, rapid descent becomes necessary to compensate for loss of pressure and recover oxygen.

Integrated descent performance is computed in the same way as for climb. The integrated descent performance is computed in steps of approximately 5000 ft altitude (or as convenient) in which the variables are kept invariant. No computational work is shown here.

Figure 15.14 plots the fuel consumed, time taken and distance covered during descent from ceiling altitude to sea level. While redoing the integrated descent performances by the readers, there could be small differences in the results.

15.10.9   Checking out the Payload-Range Capability – Requirement of 2000 nm

A typical transport aircraft mission profile is shown Figure 15.15.

Equations (15.34)–(15.35) give the mission range and fuel consumption expressions as:

$$\text{Mission range} = \text{climb distance } (R_{\text{climb}}) + \text{cruise distance } (R_{\text{cruise}}) + \text{landing distance } (R_{\text{descent}})$$

where $R_{\text{climb}} = \sum \Delta s_{\text{climb}}$ and $R_{\text{descent}} = \sum \Delta s_{\text{descent}}$ computed from steps of heights.

$$\text{mission fuel} = \text{climb fuel } (Fuel_{\text{climb}}) + \text{cruise fuel } (Fuel_{\text{cruise}}) + \text{descent fuel } (Fuel_{\text{descent}}).$$

where $Fuel_{\text{climb}} = \sum \Delta fuel_{\text{climb}}$ and $Fuel_{\text{descent}} = \sum \Delta fuel_{\text{descent}}$ computed from steps of heights.

Minimum reserve fuel is computed for an aircraft holding an 5000 ft altitude at around Mach 0.35–0.4 at about 50% of the maximum rating for 45 min or an 100 nm diversion cruising at 0.5 Mach and 25 000 ft altitude plus 20 min. The amount of reserve fuel has to be decided by the operator suited to the region of operation. The worked-out example takes the first option.

Fuel is consumed during taxiing, takeoff and landing without any range contrition and this fuel is to be added to the mission fuel with the total, known as *block fuel*. The time taken from the start of the engine at the beginning of the mission to the stop of the engine at the end of the mission is known as *block time*, in which a small part of time is not contributing to gain in range. Taken from the statistics of operation, these additional fuel burn and time consumed without contributing to range are given in Table 15.17.

Reserve fuel: At 5000 ft altitude ($\rho = 0.00204$ lb ft^{-3}) and 0.35 Mach (384 ft s^{-1}) it gives $C_L = 0.323$ resulting in $C_D = 0.025$ (Figure 9.2). Equating thrust to drag, $T/\text{engine} = 610$ lb. Corresponding fuel flow rate at this speed altitude is 210 lb/h per engine; holding for 45 min consumes $2 \times 0.75 \times 210 = 315$ lb. It is safer to keep more in reserve if a 100 nm diversion is required, which is not a stipulation if 45 min hold is taken. Therefore, it is decided to carry 600 lb reserve fuel for holding/diversion around landing airfield. The range performance can be improved with a step-climb from 43 000 to 45 000 ft as the aircraft becomes lighter with fuel consumed.

The cruise altitude of the Bizjet starts at 43 000 ft and ends at 45 000 ft (the design range is long to make a step cruise). Taking the average value of cruising at 44 000 ft ($\rho = 0.00048$ slug ft^{-3}), The methods to compute R_{climb} and R_{descent} are discussed in Sections 15.5 and 15.6. Using Figures 15.11 and 15.14, the required values are given as $R_{\text{climb}} = 170$ nm, $Fuel_{\text{climb}} = 800$ lb, $Time_{\text{climb}} = 25$ minutes, $R_{\text{descent}} = 160$ nm and $Fuel_{\text{descent}} = 340$ lb and $Time_{\text{descent}} = 30$ minutes (in part throttle gliding descent). Table 15.17 gives the segment details of aircraft weight, distant covered, fuel burned and time taken. The aircraft is at LRC schedule operating at Mach 0.7.

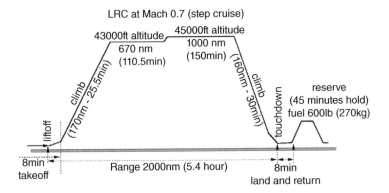

Figure 15.15 Bizjet mission profile.

Table 15.17 gives the following.

$MTOM = 20\,680$ lb (9400 kg)

Landing weight $= 20\,680-5500$ (onboard fuel) $+ 600$ (reserve fuel, not consumed) $= 15\,780$ lb (rounded to 15 800 lb).

Initial and final cruise weights can then be computed from the climb and descent performance data (Figures 15.11 and 15.14, respectively) $= 800$ lb and 340 lb, respectively.

Initial cruise weight $W_{icr} = 20\,680$ (MTOM) $+ 80$ (taxi) $+ 340$ (takeoff) $- 800$ (climb) $= 19\,680$ lb

Final cruise weight $W_{fcr} = 15\,800$ (landing weight) $+ 340$ (descent) $+ 80$ (taxi-in) $= 16\,220$ lb

Available cruise fuel $= 19\,680-16\,220 = 3460$.

Split the fuel load to 1460 lb cruising at 43 000 ft altitude and climb to 45 000 ft to use the 2000 lb fuel.

In operational practice, it is not done in this way. This example is to demonstrate the capability and logic of the operation. It can be seen that the step-climb part has not been computed. It assumes that it is a discrete jump without losing fidelity to get the extent of possible range. Industries meticulously plan stage to make the climb to get the best range. Today, microprocessor-based FBW operation makes the optimised cruise operation that is in a continuous shallow climb. Aircraft Performance Management (APM – see Section 16.6) is integral to flight and fleet management. There are several ways to climb, cruise and descend, treated extensively in Ref. [1].

In case, the specific range graphs are not available, the cruise segment range can also be obtained as shown next. First, compute the thrust required. Then from the *sfc* of the engine the fuel flow rate can be obtained and using the time required for the sector length, fuel required is computed.

From Table 15.17, the mid-cruise weight is $(19\,700 + 16\,240)/2 = 17\,970$ lb.

LRC is at Mach 0.7 (677.7 ft s^{-1}, 401.5 kt). Engine power setting is below maximum cruise rating.

Aircraft lift coefficient, $C_L = 17\,970/(0.5 \times 0.00046 \times 677.7^2 \times 323) = 17\,970/34120 = 0.527$.

From Figure 11.2 the clean aircraft drag coefficient, $C_D = 0.033$.

Aircraft drag, $D = 0.033 \times (0.5 \times 0.00046 \times 677.7^2 \times 323) = 0.033 \times 34\,120 = 1126$ lb.

Thrust required per engine is $1126/2 = 563$ lb. Figure 13.13 is meant for maximum cruise rating to achieve HSC. Therefore, at LRC at Mach 0.7 it is throttled back with reduced fuel flow, *sfc* stays the same. From Figure 13.4, the *sfc* at the speed and altitude is 0.72 lb h^{-1} lb^{-1} Fuel flow for two engines $= 0.72 \times 1126 = 811$ lb h^{-1} at 43 000 ft.

Mission range satisfies the requirement of 2000 nm. Block time for the mission is 5.45 h and block fuel consumed is 4923 lb (2233 kg). On landing, the return taxi-in fuel of 20 lb is taken from the reserve fuel of 600 lb (45 min holding or diversion). Total onboard fuel carried is therefore: $4880 + 600 = 5480 \approx 5500$ lb (2500 kg).

This checks the Bizjet weights breakdown given in Table 15.17.

The payload-range graph is given in Figure 15.16. A summarised discussion of the Bizjet design is given in Section 15.13.1.

The fuel tank has a larger capacity than what is required for the design payload-range. Payload could be traded to increase range until the tanks fill up by trading the payload. Further reduction of payload would make the aircraft lighter, thereby increasing the range.

15.11 Aircraft Performance Substantiation – Military AJT

Military aircraft certification standards are different from civil aircraft certification standards. There is one for customer specifications and the design standards vary from country to country based on their defence requirements. In allied international collaborative projects the customers are only the participating countries, export potential is the fall-out of the project. The certification requirement issues have a similarity in their reasoning but differ in the requirements. Readers are suggested to obtain the regulatory military requirements from their respective Ministry/Department of Defence – these are generally available in public domain. In this book, the substantiation procedure is the same as for the civil aircraft case covered in detail in the previous section.

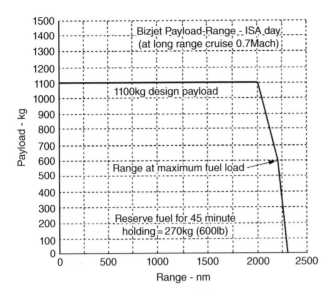

Figure 15.16 Bizjet payload-range capability.

15.11.1.1 AJT data

AJT Requirement Specifications

Payload: 1,800 kg
Take-off field length (TOFL) at sea-level STD day = 3600 ft (1,100 m)
Landing field length (LFL) at sea-level STD day = 3600 ft (1,100 m)
Initial un-accelerated climb rate at sea-level STD day = 10000 ft/min
Initial high speed cruise (HSC) at 30000 ft, STD day = Mach 0.85
Radius of action = 400 km + reserve
g-level: 7 to −3.5.

NTCM[a] = 4800 kg (10 800 lb)	Wing area = 17 m^2 (183 ft^2)
MTOM = 6500 kg (15 210 lb)	Engine data – Figure 13.15

a) Normal Training Configuration Mass.

The computation shown here is a short-cut one to show AJT capability. The readers may undertake the formal approach as was done for the Bizjet.

15.11.1.2 Mission Profile

Fuel load and management depends on the type of mission. Military mission profiles are varied. Figure 15.17 gives a typical normal training profile to gain airmanship and navigational skills in an advanced aircraft.

The training profile segment breakdown is given in Table 15.18.

Block fuel and block time are computed by adding fuel and time consumed in each segment. However, distance covered is meaningless as the aircraft returns to the base. Armament training practise closely follows the combat mission profile. Combat missions depend on target range and the expected adversary defence capability. Two typical missions are shown in Figure 15.18. Air defence would require continuous intelligence information feedback.

The study of a combat mission requires complex analysis by specialists. Defence organisations conduct these studies and are understandably kept confidential in nature. Game theory, twin dome combat simulations and suchlike are some of the tools for such analysis. Actual combat may yet prove quite different as not all is known about the adversary's tactics and capabilities. Detailed study is beyond the scope of this book and possibly of most academies.

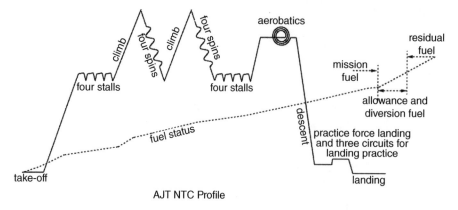

AJT NTC Profile

Figure 15.17 Normal training sortie profile (60 min). A – four turns and four stalls, B and C – two four turn spins, D – aerobatics, E – practise force landing.

Table 15.18 AJT engine ratings – detailed segments (fuel and time computed in Table 15.20).

	Engine rating = % rpm
Taxi and takeoff	60% (idle)
takeoff and climb to 6 km altitude	TO @ 100% then @ 95%
Four turns/stalls	1 min @ 95% + 3 min @ 60%
Climb from 5 to 8 km altitude	95%
Four turn spins	60%
Climb from 5 to 8 km altitude	95%
Four turn spins	60%
Four turns/stalls	1 min @ 95% + 3 min @ 60%
Climb from 5 to 6 km altitude	95%
Aerobatics practise	95%
Descent and practice force landing	2 min @ 95% + 6 min @ 60%
Three circuits for landing practise	Average 80%
Approach, land return taxi	60%
Trainee pilot allowance	95%

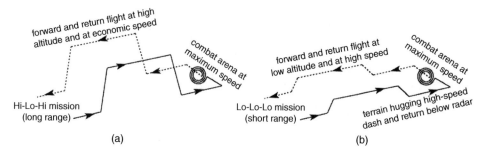

Figure 15.18 Typical combat profile. (a) A hi-lo-hi mission (longer distance) and (b) lo-lo-lo mission (shorter distance).

15.11.2   Checking TOFL (AJT) – Specification Requirement 3600 ft (1100 m)

Being single engine, there is no question regarding BFL. Military aircraft TOFL must satisfy the CFL. The CFL can also have a decision speed V_1, which is determined by whether the pilot can stop within the distance available, or otherwise the pilot needs to eject.

To obtain the CFL for a single engine aircraft, the decision speed is at the critical time just before the pilot initiates rotation at V_R. Then the decision speed V_1 is worked out from the V_R by giving 1 s as the recognition time. Engine failure occurring before this V_1 leads the pilot to stop the aircraft by applying full brakes and deploying any other retarding facilities available, for example a brake-parachute. There should be sufficient runway available for the pilot to stop the aircraft from this V_1. If there is not enough runway available, then a new decision speed V_1, considerably lower than the V_R, has to be worked out in a similar manner as done for the Bizjet. Engine failure occurring after the V_1 means the pilot will have little option other than to eject (if there is not enough runway/clearway available to stop). For multi-engine aircraft, determining the CFL follows the same procedure as the civil aircraft computation but complying with Milspecs.

In this book, the following three (CFL) requirements for the AJT are worked out.

(i)   To meet the takeoff distance of 3600 ft (1100 m) to clear 15 m
(ii) To meet the initial rate of climb (unaccelerated): 10 000 ft min^{-1} (50 m s^{-1})
(iii) To meet the maximum cruise speed of Mach 0.85 at 30 000 ft altitude

MIL-C501A requires a rolling coefficient $\mu = 0.025$ and minimum braking coefficient $\mu_B = 0.3$. Training aircraft will have a good yearly utilisation operated by relatively inexperienced pilots (with about 200 h of flying experience) carrying out a very large number of take-offs and landings on relatively shorter runways. AJT brakes are generally more robust in design giving a brake coefficient, $\mu_B = 0.45$, which is much higher than the minimum Milspec requirement. Figure 15.19 gives the speed schedules of single engine military type aircraft.

Both the Bizjet and the *AJT* have the same class of aerofoil and same types of high-lift devices. Therefore, there is a strong similarity in the wing aerodynamic characteristics. Table 15.19 gives the AJT data pertinent to takeoff performance.

Equation (15.6) gives average acceleration as $\bar{a} = g\,[(T/W - \mu) - (C_L Sq/W)(C_D/C_L - \mu)]$
Using the values given in Table 15.19,

$$\bar{a} = 32.2 \times [(0.55 - 0.025) - (C_L q/58)(0.1 - 0.025)] = 32.2 \times [0.525 - (C_L q/773.33)]$$

15.11.2.1   Takeoff with 8° Flap
Refer to Figure 15.19 showing an AJT takeoff profile.

Distance covered from zero to the decision speed V_1

The decision speed V_1 is established as the speed at 1 s before the rotation speed V_R. Estimate the acceleration of 16 ft s^{-2} (can be computed first and then iterate without estimation).

$$V_1 = 190.6 - 16 = 174.6 \text{ ft s}^{-1}.$$

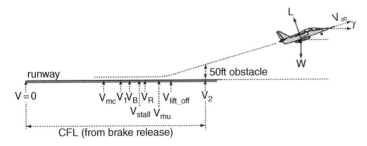

Figure 15.19 AJT takeoff.

Table 15.19 AJT takeoff distance, $S_W = 17\ m^2$ (183 ft^2).

Flap setting – degrees	0	8[a]	20[a]	Landing[a]
C_{Dpmin}	0.0212	0.0212	0.0212	0.0212
C_{Lmax}	1.5	1.65	1.85	2.2
ΔC_{Dflap}	0	0.012	0.03	0.3
$\Delta C_{D\_U/C}$	0.0222	0.022	0.0212	0.021
Rolling friction coefficient, μ	0.025	0.025	0.025	0.025
Braking friction coefficient, μ_B	0.3	0.3	0.3	0.3
V_{stall} @ 10 582 lb – fts^{-1}	180	171.7	162.1	133
V_R – kt (1.688 ft s^{-1}) @ 1.11 V_{stall}	112.9 (190.6)	106.6 (180)	87.5 (147.65)	
V_{LO} – kt (1.688 fts^{-1}) @ 1.12 V_{stall}	114 (192.5)	107.6 (181.6)	88.3 (149)	
V_2 – kt (1.688 ft s^{-1})		122 (206)	115.2 (194.5)	94.5 (159.6)
$(T/W)_{avg}$	0.51	0.51	0.51	0
W/S_W – lb ft^{-2}	58	58	58	46.26
C_D/C_L at ground run		0.1	0.1	0.1

a) Takeoff at 8° and 20° flaps. Landing at 35–40° flap, engines at idle and V_{stall} at aircraft landing weight of 8466 lb.

Aircraft velocity at 0.707 V_1 = 122.22 ft s^{-1}.

$$q = (at\ 0.707 V_1) = 0.5 \times 0.002378 \times 0.5 V_1^{\ 2} = 0.000583 \times V_1^{\ 2} = 8.7$$

Up to V_1, the average $C_L = 0.5$ (still at low incidence)

Then $\bar{a} = 32.2 \times (0.51 - C_L q/773.33) = 32.2 \times (0.51 - 4.35/773.33) = 32.2 \times 0.5004 = 16.24\ ft\ s^{-2}$.

Using Eq. 15.8b, the distance covered until lift-off speed is reached,

$$S_{G\_1} = V_{ave} \times (V_1 - V_0)/\bar{a}\ ft = 87.3 \times (174.6 - 0)/16.24 = 940\ ft$$

However, provision must be made if the engine fails at the decision speed, V_1, when braking has to be applied to stop the aircraft. The designers must make sure that the operating runway length is adequate to stop the aircraft.

If the engine is operating, then the AJT continues with the takeoff procedure when lift-off takes place after rotation is executed at V_R.

Distance covered from zero to lift-off speed V_{LO}

Using the same equation for distance covered up to the decision speed V_1 with the change to lift-off speed, $V_{LO} = 192.5\ ft\ s^{-1}$. Aircraft velocity at 0.707 $V_{LO} = 134.75\ ft\ s^{-1}$.

$$q = (at\ 0.707 V_1) = 0.5 \times 0.002378 \times 0.5 V_{LO}^{\ 2} = 10.586.$$

Up to V_{LO}, the average $C_L = 0.5$ (still at low incidence)
Then $\bar{a} = 32.2 \times (0.51 - C_L q/773.33) = 32.2 \times (0.51 - 4.35/773.33) = 32.2 \times 0.5004 = 16.24\ ft\ s^{-2}$.
Using Eq. (15.8b), the distance covered until lift-off speed is reached,

$$S_{G\_LO} = V_{ave} \times (V_1 - V_0)/\bar{a}\ ft = 96.25 \times (192.5 - 0)/16.24 = 1140\ ft.$$

Distance covered from V_{LO} to V_2

The flaring distance is required to reach V_2 clearing the 50 ft obstacle height from V_{LO}. From statistics; the time taken to flare is 3 s with at 8° flap and at $V_2 = 206\ ft\ s^{-1}$.
In 3 s the aircraft would cover $S_{G\_LO\_V2} = 3 \times 206 = 618\ ft$.

Total takeoff distance

The takeoff length is thus $S_{G\_LO} + S_{G\_LO\_V2} = (1140 + 618) = 1758$ ft, much less than the required TOFL is 3500 ft to cater for the full weight 6800 kg ($\approx$15 000 lb) for armament training. In that case, a higher flap setting of 20° may be required. At a higher flap setting, the aircraft will have a shorter CFL for the same weight.

Stopping distance and the CFL

The next step is to check what is required to stop the aircraft if an engine has failed at V_1. Stopping distance has three segments. CFL is normally longer than TOFL.

1. Distance covered from zero to the decision speed V_1, S_{G_1}. It is computed as before to be 912 ft.
2. Distance covered from V_1 to braking speed V_B, $S_{G\_1\_B}$.
3. Braking distance from V_B ($\approx V_1$) to zero velocity, $S_{G\_B\_0}$.

The other two are estimated here:

Distance covered from V_1 to braking speed V_B

At engine failure, assume 3 s for pilot recognition and taking action. Therefore, the distance covered from V_1 to V_B is $S_{G\_1\_B} = 3 \times 174.6 = 524$ ft

Braking distance from V_B ($\approx V_1$) to zero velocity (flap settings are of little consequence)

The most critical moment of brake failure is at the decision speed V_1 when the aircraft is still on the ground. With full brake coefficient $\mu = 0.45$ and average $C_L = 0.5$, $V_1 = 174.6$ ft s^{-1}
Aircraft velocity at $0.707 V_1 = 122.22$ ft s^{-1}.

$$q = (\text{at } 0.707V_1) = 0.5 \times 0.002378 \times 0.5V_1{}^2 = 0.000583 \times V_1{}^2 = 8.7$$

The equation for average acceleration reduces to

$$\bar{a} = 32.2 \times [(-0.45) - (0.5 \times 8.7)/58)(0.1 - 0.45)] = 32.2 \times [-0.45 + (1.52/58)]$$
$$= 32.2 \times (-0.45 + 0.026) = -15.65.$$

Using Eq. (15.8b), the distance covered till lift-off speed is reached,

$$S_{G\_0} = V_{ave} \times (V_1 - V_0)/\bar{a} \text{ ft} = 87.3 \times (174.6 - 0)/15.65 = 1119 \text{ ft}.$$

Therefore, the minimum runway length (*CFL*) for takeoff should be $= S_{G\_1} + S_{G\_1\_B} + S_{G\_0}$
$TOFL = 940 + 524 + 1119 = 2583$ ft. It is still within the specified requirement of 3600 ft.

The takeoff length is thus 1746 ft, much less than the CFL of 2511 ft computed before. The required TOFL is 3500 ft to cater for the full weight 6800 kg ($\approx$15 000 lb) for armament training. In that case, a higher flap setting of 20° may be required. At a higher flap setting the aircraft will have a shorter CFL for the same weight.

Evidently, the length of the runway available will dictate the decision speed V_1. If the airfield length is much longer then the pilot may have a chance to set down the aircraft if an engine failure occurs immediately after lift-off and it may be possible to stop within the available airfield length, which must have some clearway past the runway end.

Compare with the minimum Milspec requirement of breaking friction, $\mu = 0.3$.

The equation for average acceleration reduces to

$$\bar{a} = 32.2 \times [(-0.3) - (0.5 \times 8.7)/58)(0.1 - 0.3)]$$
$$= 32.2 \times [-0.3 + (0.87/58)] = 32.2 \times (-0.3 + 0.015) = -9.18.$$

Using Eq. 18b, the distance covered from V_B to zero

$$S_{G\_0} = V_{ave} \times (V_1 - V_0)/\bar{a} \text{ ft} = 87.3 \times (174.6 - 0)/9.18 = 1660 \text{ ft}.$$

This value is on the high side.

The minimum runway length for takeoff should be $= S_{G\_1} + S_{G\_1\_B} + S_{G\_0}$

The CFL (critical field length) $= 912 + 524 + 1660 = 3096$ ft is on the high side but within the specification of 3600 ft. It is for this reason that a higher brake coefficient of 0.45 is taken. This is not a problem as nowadays wheels with good brakes have much a higher friction coefficient μ.

A reduction of the decision speed to 140 ft s^{-1} (83 kt) would reduce the $S_{G\_0} = 1068$ ft bringing down the CFL to 2504 ft.

15.11.3 Checking the Second Segment Climb Gradient at 8° Flap

$V_2 = 206$ ft s^{-1} (see Table 15.19) and $W = 10\,580$ lb (4800 kg)

This gives $C_L = (2 \times 10580)/(0.5 \times 0.002378 \times 183 \times 206^2) = 21160/22680 = 0.932$

Clean aircraft drag coefficient from Figure 11.12 gives $C_{Dclean} = 0.098$.

Add $\Delta C_{Dflap} = 0.012$ and $\Delta C_{D\_U/C} = 0.022$ giving $C_D = 0.098 + 0.012 + 0.022 = 0.132$

Therefore drag, $D = 0.132 \times (0.5 \times 0.002378 \times 206^2 \times 183) = 0.134 \times 22680 = 2995$ lb

Available thrust is 5100 lb (from Figure 13.15).

At V_2 speed to clear 50 ft, the climb is at an unaccelerated rate but thereafter it will be at an accelerated rate. From Eq. (15.20b) steady state rate of climb is given by,

$$R/C_{unaccl} = V(T - D)/W,$$

Hence, $R/C_{accl} = \{[206 \times (5100-2995) \times 60]/10\,580\} = (206 \times 2005 \times 60)/(10\,580) = 2388$ ft min^{-1}

The capability satisfies the military requirement of 500 ft min^{-1} (2.54 m s^{-1}). The reader may check for the 20° flap setting.

15.11.4 Checking Landing Field Length (AJT) – Specification Requirement 3600 ft

Keeping a reserve fuel of 440 lb (200 kg), the landing weight of the AJT is 8466 lb (wing loading $= 42.26$ lb ft^{-2}) and at full flap extended $C_{Lmax} = 2.2$.

$$\text{Therefore, } V_{stall@land} = \sqrt{\frac{2 \times 8466}{0.002378 \times 183 \times C_L}} = \sqrt{\frac{38908}{2.2}} = 133 \text{ ft s}^{-1}$$

$V_{appr} = 1.2 V_{stall@land} = 160$ ft s^{-1}
$V_{TD} = 1.1 V_{stall@land} = 146$ ft s^{-1}
Average velocity from 50 ft height to touch down $= 153$ ft s^{-1}.
Distance covered before brake application after five seconds from 50 ft height,
$S_{G\_TD} = 5 \times 153 = 765$ ft.
Aircraft in full brake with $\mu_B = 0.45$, all engines shut down and average $C_L = 0.5$, $C_D/C_L = 0.1$.
Average speed at $0.707 V_{TD} = 107.1$ ft s^{-1}.
Then $q = 0.5 \times 0.002378 \times 107.1^2 = 15.64$

$$\text{Deceleration, } \bar{a} = 32.2 \times [(-0.45) - (C_L q/42.26)(0.1 - 0.45)] = 32.2 \times [-0.45 + (0.15 \times 15.64/42.26)]$$
$$= 32.2 \times [-0.45 + 0.0484] = -12.93 \text{ ft s}^{-2}.$$

Distance covered during braking, $S_{G\_0Land} = (146 \times 73)/12.93 = 824$ ft
Landing distance $S_{G\_Land} = 765 + 824 = 1742$ ft
Multiplying by 1.667, the rated LFL $= 1.667 \times 1742 = 2649 \approx 2650$ ft. It is within the specification of 3600 ft, the margin allowing for higher MTOM $= 15\,210$ lb loaded with armament.

15.11.5 Checking the Initial Climb Performance – Requirement 50 m s^{-1} (10 000 ft min^{-1}) at Normal Training Configuration (NTC)

Military trainers should climb at a much higher rate of climb than civil aircraft. The requirement of 50 m s^{-1} (10 000 ft min^{-1}) at NTC is for unaccelerated climb as a capability for comparison. Unaccelerated rate of climb varies depending on the constant speed (EAS) climb, making comparison difficult. This section shows calculations for both the rates of climb.

This section will check only the en route climb with clean configuration. Unaccelerated climb Eq. (15.20b) is used. MTOM at NTC is 4800 kg (10 582 lb). Wing area $S_W = 17$ m^2 (183 ft^2).

At en route climb, the aircraft has a clean configuration. Under maximum takeoff power, it makes an accelerated climb to 800 ft ($\rho = 0.00232$ slug ft^{-3}, $\sigma = 0.9756$) from the second segment velocity of V_2 to reach a 350 KEAS climb speed schedule to start the en route climb. At en route climb the engine throttle is retarded to maximum climb rating. The quasi-steady state climb schedule holds to 350 KEAS and the aircraft accelerates with altitude gain at a rate of dV/dh until it reaches Mach 0.8 around 25 000 ft, from where the Mach number is held constant until it reaches the cruise altitude. Consider that 100 kg of fuel is consumed to taxi and climb to 800 ft altitude where the aircraft mass is 4700 kg (10 362 lb).

At 350 kt (590.8 ft s^{-1}, Mach 0.49) the aircraft lift coefficient

$$C_L = MTOM/qS_W = 10\,362/(0.5 \times 0.00232 \times 590.8^2 \times 183) = 10\,582/74905 = 0.138.$$

Clean aircraft drag coefficient from Figure 11.17 at $C_L = 0.141$ gives $C_{Dclean} = 0.0225$

$$\text{Clean aircraft drag}, D = 0.0225 \times (0.5 \times 0.002378 \times 590.8^2 \times 183) = 0.0225 \times 74\,905 = 1685 \text{ lb}$$

Available engine installed thrust at maximum continuous rating (95% of maximum thrust given in Figure 13.15) at Mach 0.49 (459.8 ft s^{-1}) is $T = 0.95 \times 5200 = 5016$ lb.

From Eq. (15.20a), the accelerated rate of climb as follows.

$$R/C_{accl} = \frac{V[(T - D)/W]}{1 + (V/g)(dV/dh)}.$$

At quasi-steady state climb, Table 15.5 gives $\dfrac{V}{g}\left(\dfrac{dV}{dh}\right) = 0.56\,M^2 = 0.56 \times 0.49^2 = 0.1345$

On substitution, rate of climb,

$$R/C_{accl} = \{[590.8 \times (5016 - 1685) \times 60]/10362\}/[1 + 0.1345] = 11395.2/1.1345 = 10044.3 \text{ min}^{-1}$$

Therefore, an unaccelerated rate of climb, $R/C = 11\,395$ ft min^{-1}.

Aircraft specification is based on a constant EAS climb rate of 10 000 ft min^{-1}, which is just met. There are many ways to climb that are dealt with in detail in [1]. (Here, the cabin area is small and the pressurised pilot suit allows a high rate of climb.)

15.11.6 Checking the Maximum Speed – Requirement Mach 0.85 at 30 000 ft Altitude at NTC

Aircraft at HSC is at Mach 0.85 (845.5 ft s^{-1}) at 30 000 ft altitude ($\rho = 0.00088$ slug ft^{-3}). Fuel burned to climb is computed (not shown) to be 582 lb. Aircraft weight at the altitude is 10 000 lb.

At Mach 0.85 the aircraft lift coefficient $C_L = MTOM/qS_W = 10\,000/(0.5 \times 0.00088 \times 845.5^2 \times 183) = 10\,000/57561.4 = 0.174$.

Clean aircraft drag coefficient from Figure 11.17 at $C_L = 0.174$ gives $C_{Dclean} = 0.025$ (high speed)

Clean aircraft drag, $D = 0.025 \times (0.5 \times 0.00088 \times 858.5^2 \times 183) = 0.025 \times 57\,561.4 = 1440$ lb.

Available engine installed thrust at maximum cruise rating (90% of the maximum rating) is taken from Figure 13.15 at Mach 0.85 and 30 000 ft altitude as $T = 0.9 \times 2000 = 1800$ lb.

Therefore, the AJT satisfies the customer requirement of Mach 0.85 at HSC.

15.11.7 Compute the Fuel Requirement (AJT)

Other than ferry flight, a military aircraft is not merely dictated by the cruise sector unlike a civil aircraft mission. A short combat time at maximum engine rating, mostly at low altitudes, is a good part of fuel consumed. However, the range to the target area would dictate the fuel required. Long distance ferry flight and the combat arena will require additional fuel carried by drop tanks. Just before combat, the drop tanks (by that time they are empty) could be jettisoned to gain aircraft performance capability. The CAS variant has these types of mission profiles.

A training mission has varied engine demand and it returns to its own base covering no range as can be seen in Figure 15.17. Mission fuel is computed sector by sector of fuel burn as shown next for the classroom example. To compute fuel requirement, climb and specific range graphs for the AJT at NTC are required (Figures 15.20 and 15.21). To compute the varied engine demand of a training mission profile, use *sfc* and Figure 13.15 to establish the fuel flow rate for the throttle settings. The graph is valid for 75% rpm to 100% ratings. Typically, it will have approximately the following values.

At idle (60% rpm) $\approx 8 \, \text{kg} \, \text{min}^{-1}$ At 75% rpm $\approx 11 \, \text{kg} \, \text{min}^{-1}$ At 95% rpm $\approx 16.5 \, \text{kg} \, \text{min}^{-1}$.

Fuel and time consumed for the normal training profile of the AJT is given in Table 15.20.

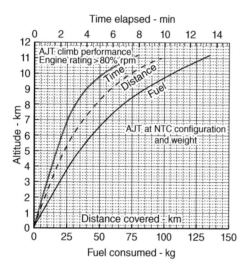

Figure 15.20 AJT climb performance.

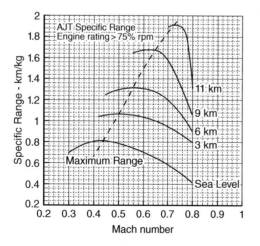

Figure 15.21 AJT specific range.

Table 15.20 AJT mission fuel and time consumed.

	Fuel burned, kg	Time, minutes	Engine rating = % rpm
Taxi and takeoff	60	6	60% (idle)
Takeoff and climb to 6 km altitude	125	5	TO @ 100% then @ 95%
Four turns	50	4	1 min @ 95% + 3 min @ 60%
Four stalls	60	5	1 min @ 95% + 4 min @ 60%
Climb from 5 to 8 km altitude	50	3	95%
Four turn spins	25	3	60%
Climb from 5 to 8 km altitude	50	3	95%
Four turn spins	25	3	60%
Climb from 5 to 6 km altitude	15	1	95%
Aerobatic practise	70	6	95%
Descent and practise force landing	95	8	2 min @ 95% + 6 min @ 60%
Three circuits for landing practise	110	10	Average 80%
Approach, land return taxi	40	4	60%
Trainee pilot allowance	30	2	95%
Total mission fuel	805	59 (≈60)	
Diversion	200		
Residual fuel	105		
Total onboard fuel	1110 (conservative estimate) (internal fuel capacity = 1400 kg)		

15.11.8 Turn Capability – Check n_{max} at the Turn (AJT)

Specified g-level: 7 to −3.5. The specification of $n = 7$ is the maximum permitted g-level (n_{max}) in pitch plane. The specified practice tight turns has $n = 4$ at sea level.

Turning Eq. (15.60) gives $n_{max} = 0.5 \left(\dfrac{T}{W} \right) \left[\dfrac{1}{kC_{DPmin}} \right]^{\frac{1}{2}}$.

To demonstrate the AJT turn capability, let the tight turn maneuver is executed at Mach 0.5 to 0.55 at mid-sortie weight of about 90% of NTC weight at sea level. For the AJT, Figure 14.5 gives the installed thrust loading (T_{SLS}/W) = 0.55. Tight turning is a short duration maneuver executed that can be carried out at full throttle, but in this carried out at maximum continuous rating at $0.9 \times T_{SLS}$. There operating $(T/W) = (0.9 \times 0.55)/(0.9 \times$ NTC weight) ≈ 0.55.

On substitution, becomes it becomes.

$n_{max} = 0.5 \times 0.55 \times \sqrt{[1/(0.07 \times 0.0212)]} = 0.275 \times \sqrt{674} = 7.1$, i.e. have excess capacity to carry out practice tight turns at around 10,000 ft altitude at $n = 4$.

Thrust at 10000 ft turning at a speed Mach 0.5 to 0.55 gives thrust at maximum continuous rating, $T/W = 0.9 \times 0.692 \times (T_{SLS}/W) = 0.623 \times (T_{SLS}/W)$.

On substitution, n_{max} at 10000 ft is as follows.

$n_{max} = 0.5 \times 0.623 \times 0.55 \times \sqrt{[1/(0.07 \times 0.0212)]} = 0.171 \times \sqrt{674} = 4.43$ with a healthy margin over $n = 4$. Tight turning at full throttle permits $n = 4$ at higher altitudes.

15.12 Propeller-Driven Aircraft – TPT (Parabolic Drag Polar)

Turboprop thrust is obtained by extracting turbine power through a propeller and therefore the related equations require additional manipulation incorporating the propeller characteristics to obtain engine power.

This subsection is devoted to deriving the expressions to obtain power required by the turboprops to produce the required thrust. All derivations in this subsection are also applicable to propeller-driven piston engine aircraft.

To develop thrust for equilibrium flight, $(T_P)_{\text{reqd}}$ is the thrust power required from the turboprop engine. It is power $[(ESHP)_{\text{reqd}}]$ and not thrust force. Thrust is obtained by taking into account the propeller efficiency, η_{prop} (propeller characteristics), as given in Eqs. (12.35) and (12.36).

To cruise at C_L at $(L/D)_{\text{max}}$, turboprop aircraft equations use the terms $\frac{C_L^3}{C_D^2}$ and $\left(\frac{C_L}{C_D^2}\right)_{max}$ (not derived here – refer to [1]).

Power required – $(T_P)_{reqd}$

In equilibrium level flight, power required by propeller-driven aircraft is expressed as

$$Drag\,(D) \times aircraft\,velocity\,(V) = DV = (T_P)_{\text{reqd}} \tag{15.61}$$

$$\text{or } (T_P)_{\text{reqd}} = DV = \left(\frac{DW}{L}\right)V = W\left(\frac{D}{L}\right)\sqrt{\frac{2W}{(\rho \times S_W \times C_L)}} = W\sqrt{\frac{2W \times C_D^2}{(\rho \times S_W \times C_L^3)}} \tag{15.62}$$

Figure 15.62 is valid for both turboprop and piston engine applications. In imperial units, the turboprop is measured in shaft power (SHP) and piston engine in horsepower (HP).

Power generated by turboprop engines to produce the propeller thrust is related by Eq. (7.21) incorporating propeller efficiency η_{prop} as

$$(ESHP)_{\text{reqd}} = (T_P)_{\text{reqd}}/\eta_{\text{prop}} \tag{15.63}$$

(*ESHP*: Equivalent Shaft Horse Power.) At a steady level flight $L = W$,

$$(ESHP)_{\text{reqd}} = DV/\eta_{\text{prop}} = (0.5 \times \rho \times V^2 \times S_W \times C_D)V/\eta_{\text{prop}} \tag{15.64}$$

$$= [0.5 \times \rho \times V^2 \times S_W \times (C_{Dpmin} + C_L^2 k)V/\eta_{\text{prop}}$$

$$= 0.5 \times \rho \times V^3 \times S_W \times (C_{Dpmin})/\eta_{\text{prop}} + k \times 0.5 \times \rho \times V^3 \times S_W \times C_L^2/\eta_{\text{prop}}$$

where $k = (1/e\pi AR)$

For a particular aircraft, C_{Dpmin}, k and η_{prop} are constant, which makes this equation

$$= k_1 \times \rho \times V^3 \times S_W \times (C_{Dpmin}) + k_2 \times V^3 \times S_W \times C_L^2 \tag{15.65}$$

where $k_1 = 0.5/\eta_{\text{prop}}$ and $k_2 = (k \times 0.5 \times \rho)/\eta_{\text{prop}}$

15.13   Summarised Discussion of the Design

This is the culminating chapter of the progress made thus far on configuring, sizing and substantiation of classroom examples. It is a good time to review the designs of the Bizjet and the AJT, should they need any revision. Having commonality in design considerations, turboprop aircraft computations are not taken up here. The rest of the book contains information on topics designers must know.

The sizing exercise (Chapter 14) gives the simultaneous solution to satisfy the airworthiness and the market requirements. Wing loading (W/S_W) and thrust loading (T/W) are the dictating parameters appearing in equations for takeoff, second segment climb, en route climb and maximum speed capability. The first two are FAR requirements and the last two are customer requirements. Information on detailed engine performance is not required during the sizing exercise. Substantiation of payload-range estimation, as a customer requirement, is not possible during the sizing exercise as this would require detailed engine performance data.

Subsequently, with the details of engine performance data, relevant aircraft performance analyses are carried out more accurately to satisfy airworthiness and market requirements.

More detailed aircraft performance is estimated during the conceptual design stage but this is beyond the scope of this book. The full aircraft performance is not going to affect aircraft configuration and mass, unless the design review brings out new demands for changes. These are management issues that are reviewed along with the potential customers to decide whether to give the go-ahead or not. Once go-ahead is obtained then a full-blown detailed definition study ensues in the form of Phase II activities with large financial commitments.

The Bizjet sizing in Figure 14.3 and the AJT sizing in Figure 14.5 show the lines of constraints for the various requirements. The AJT can be re-sized to $(T_{SLS}/W) = 0.6$ to meet $n_{max} = 7$. The sizing point to satisfy all requirements would show different level of margins for each capability. Typically, the second segment capability is the most critical to sizing. Therefore, the takeoff and maximum speed capabilities have considerable margin, which is good as the aircraft can do better than what is required.

From the statistics, experience shows that aircraft mass grows with time. This occurs primarily on account of modifications arising out of mostly from minor design changes and with changing requirements, sometimes even before the first delivery is made. If the new requirements demand a large number of changes then civil aircraft designs may appear as a new variants but military aircraft hold on a little longer before a new variant emerges. It is therefore prudent for the designers to keep some margin, especially with some reserve thrust capability, that is, keep the thrust loading, (T/W), slightly higher to begin with. Re-engineering with an up-rated version is expensive.

It can be seen that field performance would require a bigger wing planform area (S_W) than at cruise. It is advisable to keep wing area as small as possible (i.e. high wing loading) by incorporating superior high-lift capability, which is not only heavy but also expensive. Designers must seek a compromise to minimise operating costs (see Chapter 4). This chapter took a more accurate method to substantiate the aircraft performance requirements. No iteration was needed for the designs worked out in this book.

15.13.1 The Bizjet

The sizing point in Figure 14.3 gives a wing loading, $W/S_W = 64\,\text{lb ft}^{-2}$ and a thrust loading, $T/W = 0.34$. It may be noted that there is little margin given for the landing requirement. The maximum landing mass for this design is at 95% of MTOM. If for any reason aircraft OEW (now OEM, Operating Empty Mass) grows then it would be better if the sizing point for the W/S_W is taken a little lower than $64\,\text{lb ft}^{-2}$, say at $62\,\text{lb ft}^{-2}$. A quick iteration would resolve the problem. But this choice is not exercised to keep wing area as small as possible. Instead the aircraft could be allowed to approach to land at a slightly higher speed as LFL is generally about the same length as the TOFL. This is easily achievable as commonality of undercarriages for all variants would start with the design for the heaviest, that is, for the growth variant and then its bulky components are shaved to lighten for the smaller weights. The middle variant is taken as the baseline version. Its undercarriage can be made to accept MTOM growth as a result of OEW growth instead of making the wing larger.

It should be borne in mind that it is recommended that civil aircraft should come in a family of variants to cover wider market demand to maximise sale. Truly speaking, none of the three variants are optimised although the baseline is carefully sized in the middle to accept one larger and one smaller variant. Even when the development cost is front loaded, the cost of development of the variant aircraft is low by sharing the component commonality. The low cost is then translated to lowering aircraft price that can absorb the operating costs of the slightly non-optimised designs.

It is interesting to examine the design philosophy of the Boeing737 family and the Airbus320 family of aircraft variants in the same market arena. Together, more than 10 000 aircraft have been sold in the world market. This is a no small achievement in engineering. The cost of these aircraft is about $50 million apiece (2005). For airlines with deregulated fare structures, making a profit involves complex dynamics of design and operation. The cost and operational scenario changes from time to time, for example, rise of fuel cost, terrorist threat and so on.

As early as the 1960s, Boeing saw the potential for keeping component commonality in offering new designs. The B707 was one of the earliest commercial jet transport aircraft carrying passengers. It followed by a shorter version B720. Strictly speaking, the B707 fuselage relied on the KC135 tanker design of the 1950s. From the four-engine B707 came the three-engine B727 and then the two-engine B737, both retaining considerable fuselage commonality. This was one of the earliest attempts to utilise the benefits of maintaining component commonality. Subsequently, the B737 started to emerge in different sizes of variants maximising component commonality. The original B737-100 was the baseline design and all other variants that came later, up to the B737-900, are larger aircraft. Although B737-100 morphed to a new aircraft as B737-900, retaining commonality in many areas of external geometries, it posed certain constraints, especially on the undercarriage length. On the other hand, the A320 serving as the baseline design was in the middle of the family, its growth variant is the A321 and shrunk variants are the A319 and A318. Figure 7.9 gives a good study on how the OEW is affected by the two examples of family of variants. A baseline aircraft starting at the middle of the family would be better optimised and hence in principle would offer a better opportunity to lower production cost of the variants in the family.

Failure of two engines simultaneously is extremely rare. If it happens after the decision speed and if there is not enough clearway available, then it will prove catastrophic. If the climb gradient is not in conflict with the terrain of operation then it is better to takeoff with higher flap settings. If a longer runway is available then a lower flap setting could be used. Takeoff speed schedules can slightly exceed the FAR requirements that stipulate the minimum values. There have been cases of all-engine failures occurring in cruise on account of volcanic ash in atmosphere (also in the rare event of fuel starvation). Fortunately, the engines were restarted just before the aircraft would have hit the surface. All-engine failure by bird strike also occurred in 2009 – all survived after the pilot ditched in the Hudson River.

15.13.2 The AJT

Military aircraft serves only one customer, the Ministry of Defence (MoD)/Department of Defense (DoD) of the nation that designed the aircraft. Frontline combat aircraft incorporate the newest technologies at the cutting edge to stay ahead of potential adversaries. Development costs are high and only a few countries can afford to produce advanced designs. International political scenarios indicate a strong demand for combat aircraft even for developing nations who must purchase from abroad. This makes military aircraft design philosophy different from civil aircraft design. Here, the designers/scientists have a strong voice unlike in civil design where the users dictate terms. Selling combat aircraft to restricted foreign countries is a way to recover some finance.

Once the combat aircraft performance is well understood over its years of operation, consequent modifications follow capability improvements. Subsequently, a new design replaces the older design – there is a generation gap between them. Military modifications for the derivative design are substantial. Derivative designs primarily come from revised combat capabilities with newer types of armament along with all round performance gains. There is also a need for modifications, seen as variants rather than derivatives, to sell to foreign customers. These are quite different to civil aircraft variants, which are simultaneously produced for some time, serving different customers, some operating all the variants.

Advanced trainer aircraft designs have variants serving as combat aircraft in a close air support (CAS) role. Advanced Jet Trainers (AJTs) are less critical in design philosophy in comparison with front line combat aircraft, but bear some similarity in design considerations. Typically, AJTs will have one variant in the CAS role produced simultaneously. There is less restriction to export these kinds of aircraft.

The military infrastructure layout influences the aircraft design and here the life cycle cost (LCC) is the prime economic consideration. For military trainer aircraft design, it is better to have a training base close to the armament practise arena, saving time. A dedicated training base may not have as long a runway as major civil runways. These aspects are reflected in the user specifications needed to start a conceptual study. The training mission includes aerobatics and flying with onboard instruments for navigation. Therefore, the training base should be far away from the airline corridors.

The AJT sizing point in Figure 14.5 gives a wing loading, $W/S_W = 58$ lb ft^{-2} and a thrust loading, $T/W = 0.55$. It may be noted that there is a large margin, especially for the landing requirement. The AJT can achieve maximum level speed over Mach 0.88, but this is not demanded as a requirement. Mission weight for the AJT varies substantially; the NTC is at 4800 kg and for armament practise it is loaded to 6600 kg. The margin in the sizing graph covers some increase in loadings (the specification taken in the book is for the NTC only). There is a big demand for higher power for the CAS variant. The choice for having an up-rated engine or to have an afterburner depends on the choice of engine and the mission profile.

Competition for military aircraft sales is not as critical compared to the civil aviation sector. The national demand would support the national product for producing a tailor made design with under manageable economics. But the trainer aircraft market can have competition, unfortunately sometimes influenced by other factors that may fail to bring out a national product even if the nation is capable of doing so. The Brazilian design Tucano was re-engineered and underwent massive modifications by Short Brothers plc of Belfast for the Royal Air Force (RAF), UK. The BAE Hawk (UK) underwent made major modifications in the USA for domestic use.

References

1 Kundu, A.K., Price, M.A., and Riordan, D. (2016). *Theory and Practice of Aircraft Performance*. Wiley. ISBN: 9871119074175.

2 Perkins, C.D. and Hege, R.E. (1949). *Airplane Performance, Stability and Control*. Wiley.

3 Anderson J., D. (1999). *Aircraft Performance and Design*. McGraw-Hill.

4 Roskam, J. and Lan, C.T. (1997). *Airplane Aerodynamics and Performance*. Lawrence, USA: DAR Corp.

5 Miele A (1962). *Flight Mechanics, Volume 1*. Addison-Wesley Publishing Company, Inc.

6 Ojha, S.K. (1995). *Flight Performance of Airplane*. AIAA.

7 Talay, T. A. (1975). NASA SP-367. Introduction to the Aerodynamics of Flight. NASA.

8 Airbus Customer Service. (2002). Getting to Grips with Aircraft Performance. Airbus.

16

Aircraft Cost Considerations

16.1   Overview

Civil aviation requires a return on investment with cash flowing back to sustain growth in a self-sustaining manner, with or without some government assistance. Sustainability and growth of civil aviation depend on profitably. In the free market economy, industry and operators face severe competition to survive, forcing them to operate under considerable pressure to manage manufacture and operation in a lean manner. Economic considerations are the main drivers for commercial aircraft operation and to some extent as well as for military aircraft. Military aircraft are driven by defence requirements with the primary objective to meet national defence. Their export potential is a by-product that is restricted to friendly nations with some risk of disclosure of technical confidentiality. There is some difference between the ground rules for costing of military and civil aircraft manufacture and operation. Because of the nature to stay ahead of adversary capabilities, military aircraft designs need to explore newer technologies that are expensive and laborious to ensure safety and effectiveness. Many military projects have had to be abandoned even after prototypes were flown, for example, the TSR2 (UK), Northrop F20 (USA) and so on; their reasons could be different, but the common factor would always be their cost effectiveness. The product must have the appeal to be the best value for money. The readers are encouraged to look into each type of aircraft's project history.

At the conceptual phase of a new aircraft programme, cost information becomes crucial in evolving trade-off studies of cost versus various design parameters to arrive at the best manufacturing cost for the technology adopted. Visibility on costing forces long range planning, helps better understanding of system architecture of the design for trade-off studies to explore alternate designs and to explore the scope for sustainability and eco-friendliness of the product line. The product passes through well-defined stages during its lifetime, for example, conception, design, manufacture, certification, operation, maintenance/modification and finally to disposal at the end of life. Cost information about a past product should be sufficiently comprehensive and available during the conceptual stages of new project. Differential evaluations of product cost and technology, offering reliability and maintainability along with risk analysis, are important considerations in cost management. Cost details also assist preliminary planning for procurement and partnership sourcing through an efficient bid process. The final outcome would be to ensure acquisition of aircraft and its components with an objective to balance the trade-off between cost and performance, which will eventually lead to ensuring affordability and sustainability for the operators over the product's life cycle. Cost analysis stresses the importance in playing a more rigorous role, as an integrated tool embedded in the multidisciplinary systems architecture of aircraft design that helps to arrive at the 'best value', specifically aimed at manufacturing and operational needs.

The scope of cost study will allow those working with highly complex systems architecture of aircraft design to explore cost control beyond current practices and to understand, through trade-off studies, how a diverse range of systems work, allow transfer of best practice and experience of risk management across the operating life of ownership. The concept of this chapter stresses the need for interconnecting cost analysis of different disciplines, at an early stage, exploiting the advantages offered by the advanced digital design and manufacturing processes (Chapter 20). Cost trade-off studies at the conceptual design stage would lead into a 'satisfying' robust design offering least expenditure. Strong multidisciplinary interaction is essential between various design considerations (departments) to attain the overall (global) goal to minimise cost rather than individual (departmental) minimisation. Initially, a proper cost optimisation may not be easily amenable to industrial usage.

This chapter primarily deals with the pertinent aspects of aircraft manufacturing costs and operating costs. While, mainly the commercial aircraft cost considerations are dealt with here, military aircraft cost considerations in the similar and related lines are also briefly covered. This gives a broader perspective of what is meant by economic considerations; the aircraft manufacturing cost that determines the aircraft price, which in turn contributes to the aircraft life-cycle cost (LCC) from 'cradle to grave'. Aircraft operating cost (OC), a component of LCC, starts from the day it is sold to operator.

At the conceptual phase, aircraft manufacturing cost estimation is not easy when not enough information on cost is available. Industry relies on in-house past experiences on cost information to arrive fast at a preliminary but realistic cost; this is meant for the designers. In the subsequent aircraft design phases, as more information become available, the aircraft manufacturing cost estimation is revised to more accurate figures. To understand the complexities involved in manufacturing cost estimation, a rapid cost estimation model is presented in this chapter. This cost model presented here does not reflect practices by any organisation but merely presents a methodology that reflects an industrial line of working out. The framework of this cost modelling is to give some idea of how complex it can be. The worked-out example of the method given in Section 16.3.1 is taken from [1]. The rapid cost model methodology presented herein can be applied to all other aircraft components with their appropriate cost drivers to establish the cost of complete aircraft. Industrial shop-floor data is required to estimate cost. All data are normalised to keep proprietary information 'Commercial in Confidence'.

The core of this rapid cost-modelling methodology is to identify and define the cost drivers/functions of the product, and generate information, which serve as tools for design for manufacturing (DFM) studies in a Concurrent Engineering (also known as Integrated Product and Process Development, IPPD) environment in which designers, production planners, cost estimators and others participate to arrive at the best practice.

Aircraft price contributes to the Aircraft Direct Operating Cost (DOC), which is a part of OC. DOC is the most important parameter that concerns airline operators. The DOC will depend on how many passengers the aircraft carries for what range and the unit is expressed in $/seat-nm (cents per seat nautical miles). There are standard rules, for example, Association of European Airlines (AEA), Aircraft Transport Association (ATA) methods for the DOC comparison (Section 16.5) even when each industry/airline has their own DOC ground rules, which result in different values as compared to what is obtained from the standard methods.

What is to be learnt – This chapter covers the following:

Section 16.2: Introduction to Some Important Aspects of Costing
Section 16.3: Aircraft and Operational Cost
Section 16.4: Rapid Cost Method for Manufacture (from Ref. [2])
Section 16.5: DOC Details and Methods to Compute It
Section 16.6: Aircraft Performance Monitoring (APM)

Classwork content: It is good to have some idea of cost implications in aircraft design and operation. The readers are to estimate the Bizjet DOC. All relevant information to estimate aircraft DOC is given here. Aircraft cost estimation is not carried out in this book, but it offers some idea on the complexities involved in cost estimation. In this book, cost studies do not alter the finalised and substantiated configuration thus far obtained through the worked-out examples.

16.2 Introduction

Aircraft design, construction and operation are very expensive endeavours, something not all nations can afford. Those who can will have to be cost conscious, be it in a totalitarian society or in a free market economy society – the ground rules of accounting may differ but all strive to make the least expensive endeavour for the task envisaged. Success or failure of an aircraft project would depend on its cost effectiveness. Cost consciousness starts from the very conceptual design phase to ensure success in competition. In fact, cost estimation should start prior to conceptual study in a top-down analysis. If funds cannot be managed until the end of the project then beginning it is not viable.

Over the last two decades, the aerospace industry has been increasingly addressing factors such as cost, performance, delivery schedule and quality to satisfy 'customer driven' requirements of affordability by reducing aircraft acquisition cost. Steps to address these factors are to synchronise and integrate design with manufacturing and process-planning as a business strategy that will lower cost of production, while at the same time ensuring reliability and maintainability to lower operational costs. Therefore, there is a need for more rigorous cost assessment at each stage of design in order to cater for the objective of making a more effective value added customer driven product.

There are two kinds of cost to consider; for example, (i) the research, development, design and manufacturing cost (RDDMC – includes testing and production launch cost) and (ii) the operating cost (OC). First the aircraft has to be built before it can be operated for a mission. Research, development, design and test cost occurs once and is termed a non-recurring cost (NRC), but manufacturing cost continues with the production and seen as a recurring cost (RC). Typical RDDMC (i.e. the project cost) of a new civil aircraft project involving a mid-range class of high subsonic aircraft can run into a billion dollars and 4 years' wait until the first delivery can start when return of investment starts to flow back. A new advanced combat aircraft would cost several times more and tax payers bear all. A new very large high subsonic jet transport aircraft project cost (RDDMC) could approach $20 billion.

Operational cost depends on aircraft price, which is known when purchased. It is for this reason aircraft manufacturing cost is analysed first followed by operational cost analysis. It is mentioned here that aircraft cost analysis, as shown here, will not be amenable to be worked out without the instructor's help. It is dependent on industrial data input that is not given here on account of confidentiality. Instructors will have to obtain such data or generate equivalent ones (it is difficult to have realistic ones that can be substantiated) to progress with establishing appropriate indices. There are other methods available to estimate aircraft cost, but their accuracy could be debated without industrial input. The reason behind including aircraft cost estimation in these few pages is to show readers that the relatively simple mathematics involved in cost analysis can prove to be a complex matter. This text should provide some exposure to cost analysis. However, The DOC estimation can easily be carried out if the aircraft price is known (supplied).

Although there is a substantial amount of civil aircraft cost details available in the public domain, the manufactured parts cost are not readily available. Cost estimation models available at one end of the public domain are those of exploratory types researched in academia and at the other end are the generic models by consultants. By plotting a graphical relation between aircraft maximum takeoff weights (MTOWs) and the aircraft prices available in the public domain, some idea of the expected price of a new aircraft can be obtained. These cost models may offer fresh insights but rarely suit industrial practices, for example, the loading of NRC as amortisation, the consequences of learning curves, benefits from better manufacturing philosophy, effects of procurement policies of raw material and finished materials, support cost and so on. This chapter gives a brief outline of the various levels of aircraft cost considerations practised in the free market economy. Based on in-house data, each industry generates specific cost models (with or without external assistance) at a different level of accuracy suited to different departments at various phases of project activities. Estimation of project cost is a laborious task involving large numbers of parameters and databases. A team of cost estimators and accountants devote considerable amounts of time to predicting project cost and subsequently verifying the actual expenditure if it has resulted in being close to prediction. Experience has taught the costing team to use company generated factors to predict estimates. These are not available in the public domain. In a competitive market, cost details are sensitive information and are therefore kept in strict confidence.

A good method to make aircraft cost estimates is to first assess man-hours involved and then use the average man-hour rates (it varies) at the time. Material and bought-out items costs could be obtained (freely) from the suppliers. This book is not in a position to give accurate industrial cost details and it is left to the academic institutes to generate data as required. This chapter merely offers a generic methodology to show a rapid way to predict manufacturing cost, more suited to classroom usage without ignoring what is considered in industry. It is based on a *parametric method*. A normal market situation without any unpredictable trends (global issues) is assumed. At this time, data from the emerging geo-political scenario, national economic infrastructure, increasing fuel prices, emerging technological considerations – for example, on sustainable developments, anti-terrorism design features, passenger health issues and so on – are lean and fluctuating.

Industry relies on in-house past experiences with cost information to arrive a preliminary and fast but realistic cost (say error within 15%, set at a high level architecture of data structure) estimating methodology to help designers to investigate proven new technologies to be adopted to advance the product to a competitive edge and generate specification requirements through parametric studies.

The post *conceptual design study phase* leads into the *project definition phase*, followed by the *detailed design phase* when the manufacturing activities build up to produce the finished aircraft. At later stages of the project when more accurate cost data is available then it allows the use of *analytical cost method* at a lower level of data structure, thereby fine-tuning the cost estimates obtained at the earlier conceptual stages. Finally, when the production is stabilised, accurate manufacturing will be available to compare with the predicted cost and reconcile if there is large discrepancy. The deeper the breakdown of a *parametric method*, the more it converges to being an *analytical method*. The proposed *rapid cost model* based on parametric method is quite different from the analytical cost method. Figure 16.1 shows the levels of cost model architecture to serve various groups at different stages of the programme milestones.

The parametric method is meant for designers while the latter is meant for corporate usage to establish aircraft pricing and have a better understanding of the customer cost goals, constraints and competitive market place requirements. It also serves at the bid stage and other budgetary purposes. The state-of-the-art of cost modelling predicts close enough to the actual incurred cost available after production is stabilised. The success or failure of cost estimation using a parametric method depends on first identifying correct cost drivers and then establishing a good cost relationship with the available 'in-house' data to embed accuracy.

The analytical cost method is time-consuming and has the danger of missing out some of the multitude of details involved. The parametric method is generally meant for designers while the analytical method is meant for corporate usage to establish aircraft pricing and have a better understanding of the customer cost goals, constraints and competitive market requirements. It also serves at the bid stage and other budgetary purposes. The state-of-the-art of cost modelling should predict close enough to the actual cost available after production is stabilised.

Less accurate cost considerations at the conceptual stage, specifically meant for designers, are no less meaningful than what accountants/estimators provide as a management feedback to assess profitability and to run a lean organisation, possibly to the point of anorexic levels. Extending the frontiers of cost saving through IPPD, rather than merely running lean on manpower, would add a new dimension in harnessing human resources by organisations investing in people, where it counts. In fact, the preliminary cost estimates at the higher level

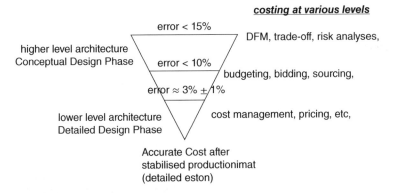

Figure 16.1 Levels of cost modelling.

of architecture flows down to the lower level when more data could be generated as the programme advances through the project milestones. Cost estimations made by different methods should converge within close tolerances, benefiting from in-house past experiences. Cost modelling is industry specific based on individual practices, which can be different from other, but the objective is the same, that is, to obtain cost estimates as accurate as possible.

Ensuring quality while making the product converge on cost (Design for Cost) rather than allowing cost to make the product (Design to Cost) is the essence of cost control. The core of the cost modelling is to identify and define the cost drivers/functions of the product and generate information that serve as tools for Design for Manufacture and Assembly (DFM/A – see Chapter 20) in an IPPD environment. The DFM/A studies lead to design to cost and are a part of the Six-Sigma (Chapter 20) concept to make a product right-first-time that helps to reduce cost. Based on the awareness of customer affordability and requirements, the designing and manufacturing targets costs are established.

Industry will aim to recover all the investment at sales of around 400 aircraft, preferably lower. About 4–6% of the aircraft price is meant to recover the project cost (RDDMC) known as amortisation of investment made. It is for this reason that offering an aircraft in a family concept covers a wider market at considerably lower investment when the cost of amortisation will drop down close to 2–4%. Smaller aircraft will need to break-even at around 200 aircraft sales. In today's practice, civil manufacturers sell pre-production aircraft used for flight testing to recover funds. Military aircraft incorporate unproven new technologies and invest in technology demonstrator aircraft (in a reduced scale) to prove the concept and subsequently substantiate the design by flight testing on pre-production aircraft, some of them could be retained by the manufacturer for future tests.

The general definition of aircraft price is as follows: The aircraft cost includes amortisation of RDDMC but does not include spare and support cost.

Aircraft Price = Aircraft Cost + Profit = Aircraft Acquisition Cost

In this book, the aircraft price and cost are taken synonymously. In fact, the aircraft price is also known as the aircraft acquisition cost. Profit margin is a variable quantity and depends on what the market can bear. This book is not dealing with the aircraft pricing method. In general, profit from a new aircraft sale is rather low. Most of the profits come from sales of spares and maintenance support. Operators will have to depend on the manufacturer as long the aircraft are in operation, say for two to three decades. Manufacturers are in a healthy position for several decades if their products sell in large numbers.

16.3 Aircraft Cost and Operational Cost

For mid-size commercial aircraft operation, the ownership cost contribution to the DOC could be as high as one-third to half of the total DOC (at year 2000 level fuel price). It is for this reason that industry is driven to reduce cost of manufacture and the demand is as high as 25% for cost reduction. Clearly, the design and manufacturing philosophy play a significant role in facilitating manufacturing cost reduction. To understand aircraft cost implications, performance engineers could benefit from a brief overview on the various aspects of costing to appreciate the competition and make route planning accordingly. When fuel price is low then drag could be sacrificed to reduce manufacturing cost; on the other hand, as fuel price increases, the role of drag starts to dominate and aircraft cost could go up. DOC estimation at the initial project phase is continuously revised as a project progresses with more detailed cost information.

Figure 16.2 gives typical high subsonic civil aircraft cost at the 2000 price level in millions of dollars. It reflects the basic (lowest) cost of the aircraft. This graph is generated from a few accurate industrial data that are kept commercial in confidence. Although the trends are similar, the graph needs to be updated with the current values.

In general, exact aircraft cost data are not readily available and the overall accuracy of the graph is not substantiated. The aircraft price varies for each sale depending on the terms, conditions and support involved.

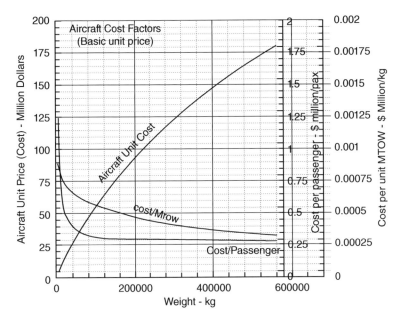

Figure 16.2 Aircraft cost factors.

The values in the figure are crude but offer a feel for the newly initiated on the expected cost of the class of aircraft. Figure 7.2 could be used to obtain the relationship between MTOW versus number of passengers. The basic price of a mid-range 150 passenger class high subsonic turbofan aircraft is taken as US$47 million (2000 level).

Aircraft MTOW reflects the range capability, which varies from type to type. Therefore, strictly speaking, cost factors should be based on MEW (Manufacturer's Empty Weight). The readers should be able to compute MEW from the data given in Chapter 8. In general, bigger aircraft have a longer range (refer to Figure 7.1). Exceptions are when aircraft with low passenger loads have long range missions; for example, the Bombardier Global Express.

Typical cost fractions (with respect to aircraft cost) of various groups of civil aircraft components are given in Table 16.1. This serves as preliminary information for two engine aircraft (four engines slightly higher). It is better to obtain actual data from industry, if possible.

Combat aircraft cost fractions are different. Here, the empty shell structure is smaller but houses very sophisticated avionics black boxes serving the very complex tasks of combat and survivability. Typical cost fractions of various groups of combat aircraft components are given in Table 16.2. In the table, the avionics cost fraction is separated. It serves as preliminary information for two engine aircraft. It is better to obtain actual data from industry, if possible.

In the USA, military aircraft costing uses AMPR (Aeronautical Manufacturer's Planning Report) weight, also known as DCPR (Defence Contractor's Planning Report) weight for the manufacturer to bid. The AMPR weight represents the weight of the empty aircraft shell structural without any bought-out vendor items, for example, engines, undercarriages, avionics packages and so on.

16.3.1 Operating Cost (OC)

The other cost arises during operation after the aircraft is sold – it is seen as an OC. These are the concerns of airline operators. Military OCs have different book-keeping method. Revenue earned from passenger airfare covers the full expenditure of airline operators that covers the aircraft price and support cost. The DOC gives the measure of cost involved with the aircraft mission. Standards for DOC ground rules exist – in the USA it is proposed by the Air Transport Association of America (ATA – 1967) and in Europe by the AEA

Table 16.1 Typical cost fractions of civil aircraft (two engines) at the shop-floor level.

		Cost fraction	Cost fraction
1.	Aircraft empty shell structures[a]		
	Wing shell structure	10–12%	
	Fuselage shell structure	6–8%	
	Empennage shell structure	1–2%	
	Two nacelle shell structure[b]	2–3%	
	Miscellaneous structures	0–1%	
	Sub-total		20–25%
2.	Bought-out vendor items		
	Two turbofan dry bare engines[b]	18–22%	
	Avionics and electrical system	8–10%	
	Mechanical systems[c]	6–10%	
	Miscellaneous[d]	4–6%	
	Subtotal		40–50%
3.	Final assembly to finish (labour intensive)	25–30%	25–30%
	(component sub-assembling, final assembling, equipping/installing, wiring, plumbing, furnishing, finishing, testing etc.)		

a) Individual component subassembly costs fraction.
b) Smaller aircraft engine cost fraction is higher (up to 25%).
c) Includes control linkages, servos, undercarriage etc.
d) Cables, tubing furnishing etc.

(1989 – medium range – [1]) both are comparable. The aircraft designers must be aware of operational needs to make sure that their design serves the expectancies of the operators. In fact, the manufacturers and the operators are in constant dialogue to ensure the current and future products are fine-tuned to the best profitability for all. The big airline operators have permanent representatives located at the place of manufacture to have all-round support and dialogue on all aspects of the product line. Civil aircraft operating cost is of two types as follows.

1. *DOC.* These are the OCs directly involved with the mission flown. Each operator will have their own ground rules depending on the country, pay scales, management policies, fuel price and so on. The standard ground rules are used for comparison of a similar class of product manufactured by different companies. In Europe, the AEA ground rules are accepted as the basis for comparison. This gives a good indication of aircraft capability. A cheaper aircraft may not prove profitable on the long run if its OCs are high.
2. *Indirect operating cost (IOC).* The IOC breakdown in the USA is slightly different from European standards. Given next are the IOC components.

The airline operators have *other costs* those involve training, evaluation, logistic supports, special equipment and ground based resource management that are not directly related to the aircraft design and mission sector operation. These are independent of aircraft type. Together these become the total cost of the operator and are termed LCCs. Unlike DOCs, there is no standard for LCCs proposed by any established associations for commercial aircraft operation. Currently, each organisation has its own ground rules to compute LCC. Together with the DOC they produce Total Operating Cost (TOC). Unlike military aircraft, the impact of 'other costs' of the LCC in commercial aircraft application may be considered to be of a lower order – the DOC covers the bulk of the cost involved. This book is concerned with the DOC only. The breakdown of LCC components is given in Table 16.3.

Table 16.2 Typical cost fractions of combat aircraft (two engines) at the shop-floor level.

		Cost fraction	Cost fraction
1.	Aircraft empty shell structures[a]		
	Wing shell structure	6–7%	
	Fuselage shell structure	4–6%	
	Empennage shell structure	≈1%	
	Two nacelle shell structure[b]	part of fuselage	
	Miscellaneous structures	0–1%	
	Subtotal		12–15%
2.	Bought-out vendor items		
	Two turbofan dry bare engines[b]	25–30%	
	Mechanical systems[c]	5–8%	
	Miscellaneous[d]	1–2%	
	Sub-total		30–40%
3.	Avionics and electrical system	30–35%	30–35%
4.	Final assembly to finish (labour intensive)	12–15%	12–15%
	(component subassembling, final assembling, equipping/installing, wiring, plumbing, furnishing, finishing, testing etc.)		

a) Individual component subassembly costs fraction.
b) Single engine at lower cost fraction.
c) Includes control linkages, servos, undercarriage etc.
d) Cables, tubing furnishing etc.

Table 16.3 Life-cycle cost (civil aircraft).

Aircraft related elements	Passenger related elements	Cargo related elements
property depreciation[a]	handling and insurance	handling and insurance
property amortisation[a]	baggage handling	administration/office
property maintenance[a]	emergencies	sales and support
ground support	administration/office	fees/commissions
administration cost	sales and support	publicity
ground handling/control	publicity	
training	fees/commissions	

a) Ground property (e.g. hanger, equipment etc.).

The military uses LCC (Table 16.4) rather than DOC for the ownership of aircraft in service. In loose terms, this is the cost involved for the entire fleet of the type from cradle to grave including disposal. Military operation has no cash flowing back – there are no paying customers like passengers and cargo handlers. Tax payers bear the full cost of military design and operation. The need for LCC is felt by military operations, which differ considerably from civil operations. Military aircraft OC ground rules are based on total support by the manufacturer for the entire operating life span, which can get extended with fresh contracts. The concept of Design-to-Life-Cycle-Cost (DTLCC) has been suggested but is not yet standardised; this poses problems in providing a credible LCC comparison. It is for this reason that military aircraft operation deals with the LCC,

Table 16.4 Life-cycle cost (military aircraft).

RDDMC	Production	In-service	Disposal
engineering	parts manufacture	operation	storing
ground testing	assembly	maintenance	recycling
technology demonstrator	tooling	ground support	etc.
prototype flight test	deliveries	training	
technical support		post design services	
publication		administration	

although it has various levels of cost breakdown including aircraft related and sortie related costs. Given next is an outline of the breakdown by categorising the elements affecting the military aircraft LCC model.

Of late, the customer driven civil market desires LCC estimation. Academics and researchers are suggesting various types of LCC models. The principles of such approach are directed at cost management and cost control offering advice in assigning responsibilities, effectiveness and on other administrative measures at the conceptual stages in an IPPD environment.

16.4 Rapid Cost Modelling

This section presents a rapid cost-modelling methodology [3] specifically aimed at catering for the classroom needs of DFM/A considerations during the conceptual design phase of commercial transport aircraft. The basic structural philosophy is to cater the DFM/A considerations as early as possible to get a feel for manufacturing cost reduction through trade-off studies. Aircraft component manufacture case studies and operating cost reduction benefits are discussed in Ref. [4].

There are many publications suggesting empirical relations to predict aircraft cost based on various types of aircraft weights, performance capabilities and other details. The empirical relations use coefficients, indices and so on to some degree of success. But without the actual industrial cost details, these empirical relations will find difficult to fine-tune DFM/A gains. A methodology must have inputs based on real data to cater for gains that can be obtained through the application of the fundamentals of modern manufacturing philosophy.

To keep industrial data commercial in confidence, the costing of a new design is estimated in relation to the exact known cost data of an existing (seen as old) design. The actual cost values are kept in s non-dimensionalised form from which the new design is estimated to show design merits through DFM/A to save cost. The methodology is based on a generic turbofan nacelle, typically representing the investigative areas associated with other components of aircraft and makes use of industrial data. Figure 16.3 shows generic nacelle components: (i) nose cowl, (ii) fan cowl – one each side, (iii) core cowl – one each side, with thrust reverser and (iv) aft cowl. The worked-out example given here is that of costing the nose cowl of nacelle only, the existing old design is designated nacelle 'A' as the baseline and the new design is designated nacelle 'B'. The method does not reflect practices by any organisations and does not guarantee accuracy. It is a method and meant to give exposure to the complexities involved in costing for the entire aircraft composed of multitude of components.

The rapid cost model [3] is based on parametric methods in which cost drivers are identified. In the nacelle example 11 drivers are dealt with. From a baseline known cost, the method demonstrates a fast and a relatively accurate prediction along with identifying areas that contribute to cost. A normal market situation without any unpredictable trends (global issues) is assumed for methodology.

The example of the rapid cost methodology concentrates cost modelling of the structural elements of the *nose cowl* of two generic nacelles, A and B, in the same family of aircraft and engines. The methodology uses

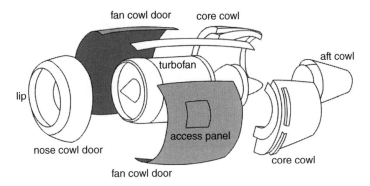

Figure 16.3 Nacelle components.

Table 16.5 Turbofan engine and nacelle data.

	Nacelle A (baseline – existing design)	Nacelle B (new design to estimate)
T_{SLS} – lb	9220	14 000
Engine dry wt. – lb	1625	2 470
Engine face diameter – in.	49	50.6
Nacelle weight – lb	536	860
Nacelle maximum diameter – in.	56	60
Nose cowl length – in.	35	29

indices/factors and it is for this reason two nacelles are used. Nacelle A is an existing product and is taken as the baseline design. All cost data for Nacelle A are known, from which the indices are generated. Nacelle B is of a higher standard of specification and a new design in which the indices are adjusted and then used to predict cost. The two nacelles are compared in Table 16.5. All figures are in the foot-pound system (FPS) obtained from industry. For dissimilar components, a similar methodology can still be applied with extensive data analyses to establish appropriate indices.

While the aerodynamic mouldlines of both the nacelles are similar, their structural design philosophies, hence the subassemblies (tooling concept), differ. Having commonality in the family of designs, the study presents a focused comparative study of the two geometrically 'similar' nose cowls in a complex multidisciplinary interaction affecting cost. The total manufacturing cost of the finished product is taken as the sum of the items given in Table 16.6. The cost of manufacture is not the selling price.

The total manufacturing cost of the finished product is taken as the sum of the items given in Table 16.6. Cost of manufacture is not the selling price, which includes profit.

Table 16.6 Manufacturing cost components.

1. Cost of material (raw and finished product)
2. Cost of parts manufacture
3. Cost of parts assembly to finish the product
4. Cost of support (e.g. rework/concessions/quality)
5. Amortisation of NRCs
6. Miscellaneous cost (other direct costs, contingencies)

Generic nacelles typically represent the investigative areas associated with the design and manufacture of other aircraft components, for example, wing, fuselage and so on of an aircraft. The rapid cost model methodology presented herein can be applied to all other aircraft components with their appropriate cost drivers to establish the cost of complete aircraft. Industrial shop-floor data is required to estimate cost in US dollars. All data are normalised to keep proprietary information commercial in confidence.

16.4.1 Nacelle Cost Drivers

Given next are the 11 specific parameters, in two groups, identified as the design and manufacture sensitive cost drivers for generic nacelles. These cost drivers are applicable to all the four nacelle subassembly components shown in Figure 16.3. Group 1 consists of eight cost drivers, which relate to in-house data within the organisation. The Group 2 cost drivers are not concerned with in-house capability issues, hence these are not within the scope of this treatise. Indices/coefficients obtained during DFM/A study are used.

Group 1 (concern with in-house issues)

i. *Size.* Nacelle size is considered the main parameter in establishing the base cost. Size and weight are correlated. The nacelle cowl size depends on the engine size, that is, primarily the fan diameter (D_F) of the engine, which in turn depends on the thrust (T_{SLS}) ratings as a function of bypass ratio (BPR) and the thermodynamic cycle. Relationship between T_{SLS} and D_{fan} can be expressed as follows:

$$(T_{SLS}) = (K D_{fan}^2)$$ (16.1)

where K = constant of proportionality.

The variants in the family of turbofans are the result of tweaking the baseline design, keeping the core gas generator nearly unchanged. This improves cost effectiveness by keeping component commonality. Hence, the variant fan diameter is marginally affected, the growth variant having better thrust to dry weight ratio (T/W) and vice-versa. As a consequence, the nacelle maximum diameter (D_{max}) and length (L) change to a small extent. The size factor for nacelle, K_{size}, affecting cost is given in semi-empirical form as:

$$K_{size} = \left(\frac{(D_{fan} \times D_{max})_{derivative}}{(D_{fan} \times D_{max})_{baseline}} \right) \times \left(\frac{L_{derivative}}{L_{baseline}} \right) \times \left(\frac{T_{SLS\_derivative}}{T_{SLS\_baseline}} \right)^{0.35}$$ (16.2)

The effect of size on parts fabrication and assembly costs is less pronounced than material cost, unless a very large size calls for drastically different fabrication and assembly philosophies.

ii. *Material.* Parts weight data will give a more accurate cost of material than applying size factor, K_{size}, which may be used when weight details are not available. There are two kinds of material considered based on industrial terminology: (i) raw material and (ii) finished material. The latter is the subcontracted items.

iii. *Geometry.* Double curvature at the nacelle surface requires stretch-formed sheet metal or a complex mould for composites in shaping the mouldlines. Both nacelles are symmetrical to the vertical plane. The nacelle lip cross section is necessarily that of aerofoil section with the crown cut thinner than the keel cut where engine accessories are housed (Figure 16.4). This will not make the outer and inner surfaces concentric. Straight longitudinal and circumferential joints would ease auto-riveting. In short, there are four 'Cs' associated with geometric cost drivers;. Circularity, Concentricity, Cylindricity and Commonality. Nacelles A and B are geometrically similar and therefore will not show difference made by the four C considerations. A geometric cost driver index of 1 is taken for both nacelles as a result of being similar.

iv. *Technical specification.* These standards form the finish and maintainability of the nacelle, for example, the surface smoothness requirements (manufacturing tolerance at the surface), safety issues (fire detection), interchangeability criteria, pollution standards and so on. Figure 16.4 shows the cost versus tolerance relationship taken from [5].

At the wetted surface, Zone 1 (Figure 16.5) is in an adverse pressure gradient requiring tighter tolerances compared to Zone 2 in a favourable pressure gradient. The tighter the tolerance at the wetted surface,

Figure 16.4 Cost versus tolerance.

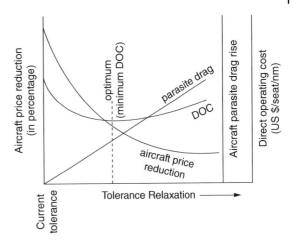

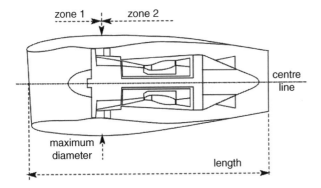

Figure 16.5 Typical nacelle section.

the higher the cost of production on account of increased rework/concessions involved. The technical specifications are similar, so both have an index of 1.

v. *Structural design concept.* Component design concepts contribute to the cost drivers and are taken as NRC amortised over the production run (typically 400 units). The aim is to have a structure with a low part count involving low production man-hours. Comparing with the baseline design of nacelle A, an 'index factor' is associated with the new derivative design. Nacelle B has a more involved design and has an index greater than 1. Manufacturing considerations, discussed in the next paragraph, are integral to the structural design as a part of the DFM/A obligation.

vi. *Manufacturing philosophy.* This is closely linked with the structural design concept as described before. It has two components of cost drivers: (1) the NRC of tool and jig design and (2) the RC during production (parts manufacture and assembly). An expensive tool set-up to cater for rapid learning process and faster assembly time with lower rejection rates (concessions and rework) ensures budgetary provision is front loaded but considerable savings can be obtained. Nacelle B has an NRC index greater than 1 and the RC index is less than 1. Nacelle B is an improvement compared to nacelle A.

vii. *Functionality.* This is concerned with the enhancements required compared to the baseline nacelle design, for example, anti-icing, thrust reversing, treatment of environmental issues of pollution (noise and emission), position of engine accessories, bypass duct type and so on. A factor of 'complexity' may be used to describe the level of sophistication incorporated in the functionality. The nose cowl of both the nacelles has the same functionality due to being in the same family, hence it has a factor of 1: otherwise it needs to be revised. Other components of the nacelle could differ in their functionality.

viii. *Man-hour rates, overheads and so on.* Man-hour rates and overheads are held constant for both the nacelles and therefore the scope of applicability becomes redundant in this study.

Group 2 (does not relate to in-house issues, therefore they are not considered in this paper):

i. *Role.* Basically, This describes the difference between military and commercial aircraft design.
ii. *Scope and condition of supply.* It concerns with the packaging quality of the nacelle supplied to the customer. This is not a design/manufacture issue.
iii. *Programme schedule.* This is an external cost driver not discussed here.

In summary, only four cost drivers in Group A, (i) size, (ii) material, (iii) structural design concept and (iv) manufacturing philosophy, are required in this chapter to establish the cost of component manufacture and their assembly. The other four cost drivers in Group I can be evaluated in a similar manner for nacelles that differ in (i) geometry, (ii) technical specifications, (iii) functionality and (iv) man-hour rates.

16.4.2   Nose Cowl Parts and Subassemblies

The build-work breakdown of the two nacelles from start to finish is grouped in six stages as given in Table 16.7. But the design of their parts design is different. Nacelle A is an existing design and its cost is known; Nacelle B is a later design with lower part count and assembly time, achieved through superior structural and manufacturing considerations through DFM/A studies. The nose cowl consists of pure structures (STR), minor parts (MP – brackets, splices etc.), Engine Built Units (EBU – e.g. anti-icing units, valves etc.) and aircraft general supplies (AGS) (fastener, rivets, nuts, bolts etc.). The EBU costs are separate and are not studied here.

The expensive components are STR and installation of EBU parts. Clearly, these numbers are reduced to almost half, thereby saving cost of the nose cowl of Nacelle B even with a bigger engine. The assembly hours is reduced to nearly half. AGSs are not expensive – their numbers can run high; for example, rivets, nuts, bolts and so on.

16.4.3   Methodology (Nose Cowl Only)

The authors would like to point out that this simple algebraic procedure with elementary mathematics will become a complex work-out. The newly initiated will find it difficult to follow without an instructor's help and will require industrial data to do the classroom work for their project.

The methodology generates the factors/indices from an existing Nacelle A whose cost data is known. Based on the similarity in geometry, these factors/indices are then adjusted with DFM/A considerations and applied to Nacelle B. The conceptual design phase should outline the basis of the manufacturing philosophy under DFM/A consideration relying heavily on experience from Nacelle A. Table 16.8 gives the necessary

Table 16.7 Nose cowl build-work breakdown (abbreviations expanded previously).

		Nacelle A				Nacelle B			
		STR	MP	EBU	AGS	STR	MP	EBU	AGS
1.	Forward bulkhead assembly	4	3	33	482	4	4	1	0
2.	Aft bulkhead assembly	3	0	33	395	3	1	25	644
3.	Primary assembly	1	6	0	393	1	1	10	970
4.	First stage assembly	11	0	105	939	6	0	19	708
5.	Second stage assembly	8	2	78	873	2	2	82	1617
6.	Third stage assembly	0	5	7	1480	0	0	15	95
	Total	28	16	256	4562	16	8	152	4034

Table 16.8 Normalised indices for the eight cost drivers in Group I (* are the main ones).

Cost drivers		Nacelle A (baseline – known)	Nacelle B (to predict)	Remarks
1.	K_{size}	1.0	1.133	
2.	Material (weight ratio)	1.0	1.135	
	(a) Raw material	1.0	See Tables 16.9 and 16.10	B better
	(b) Finished material (subcontracted)	1.0	See Table 16.9	
3.	Geometry	1.0	1	Similar
4.	Technical specifications	1.0	1	Similar
5.	Structural design*	1.0	1.1	NRC
6.	Manufacturing philosophy*			
	(a) Non-recurring (tool/jig design)	1.0	1.2	NRC
	(b) Recurring (manufacture + assembly)	1.0	0.95	Nacelle B better
7.	Functionality	1.0	1	Similar
8.	Man-hour rates	1.0	1	Same

factors/indices for the eight cost drivers (data from industry). The table is followed by expanding the eight cost drivers.

The shop-floor learning characteristic is an important factor in cost consideration. Initially, part fabrication and their assemblies take longer time (actual man-hours) than when it becomes a routine task with a stabilised time frame of standard man-hours, which is initially given as the target time. If actual man-hours do not reach standard man-hours then investigation is required to change the standard man-hours. The faster one learns, the greater the savings, taking fewer man-hours to manufacture. The number of attempts required to reach the standard man-hours varies and the DFM/A study should consider this aspect. In this case, nacelle B has a faster learning curve factor, due to having fewer parts.

i. K_{size}
   The geometric details of the nacelles and engine parameters are given in Table 16.5 to estimate K_{size}.

ii. *Material Cost*
   Material is classified into two categories: (1) raw (sheet metal, bar stock, forging etc.) and (2) finished (e.g. lipskin, engine ring, some welded/cast parts etc. coming from outside as subcontracted items). The weight fractions of both nacelles are given in Table 16.9.
   The unit cost for each type could vary – this depends on the procurement policy (see remarks in the table). The next part of the table gives details of raw material weight fractions. The last column gives the various material costs per unit weight is also normalised with respect to aluminium sheet metal cost. AGS consists of various types of fasteners, for example, blind rivets (more expensive), solid rivets and other types of fasteners. They are classified as raw materials as it is impractical to cost each type separately.

iii. *Cost of Manufacture*
   This is the core of manufacturing cost build-up and takes account of cost drivers of geometry, Technical specifications, manufacturing philosophy, functionality and man-hour rates. For the present study, only the manufacturing philosophy is required to be evaluated as given in the next two subsections of parts fabrication and their assembly.

iv. *Cost of parts fabrication*
   Table 16.10 gives the cost of parts manufacture in a non-dimensional form from the man-hours involved. The actual man-hours taken to manufacture parts for each of the six stages of nacelle A can be obtained from the shop-floor engineering process sheets. Factored indices for nacelle B can be established through DFM/A studies at the conceptual stages of design.

Table 16.9 (Subscript 'T' stands for total weight of nose cowl and A and B stands for each nacelle).

	Nacelle A	Nacelle B		Cost of material per unit weight	
	Weight (W_A/W_{AT})	Weight (W_B/W_{AT})	Weight (W_B/W_{BT})	Nac A	Nac B
Material weight fraction					
All material	1.0	1.135	1.0		
Raw material	0.7136	0.8744	0.77	See below	
Finished material	0.2864	0.2607	0.23	1.0	0.92
Raw material weight fraction (finished material not included)					
Total weight fraction	1.0	1.2253	1.0		
Aluminium alloy sheet	0.2288	0.4778	0.39	1.0	1.0
Aluminium alloy forging	0.1213	0.0	0.0	4.19	4.19
Al alloy honeycomb	0.3104	0.38687	0.3157	2.25	2.25
Titanium alloy	0.2752	0.34254	0.2795	3.5	3.5
Composite	0.0175	0.0	0.0	3.62	2.9[a]
Mech. fasteners (nuts etc.)	0.0366	0.005	0.0041	18.44	18.44
Solid rivets	0.0101	0.0139	0.0113	0.63	0.63

a) No composite in the nose cowl of nacelle B, but is used in core cowls of both the nacelles.

Table 16.10 Man-hours fraction required to fabricate parts (see Table 16.2 for the six stages).

	Nacelle A	Nacelle B
Total man-hours cost for all parts	1.0	1.0878
Learning curve factor	1.0	1.022
Parts in the forward bulkhead assembly – Stage 1 (start):	0.056	0.12
Parts in primary assembly – Stage 2:	0.038	0.003
Parts in aft bulkhead assembly – Stage 3:	0.111	0.130
Parts in first subassembly – Stage 4:	0.623	0.355
Parts second subassembly – Stage 5:	0.073	0.294
Parts in third subassembly – Stage 6 (final):	0.099	0.098

At each stage of parts manufacture, the man-hours are given as a fraction of the total man-hours of all parts. Man-hour rates are kept invariant. The nacelle B learning factor for parts manufacture is about the same as nacelle A, but not for assembly as can be seen in the next section. Nacelle B has fewer parts, saving cost.

$$\text{Manufacturing cost} = \text{rates} \times \text{man-hours} \times \text{learning curve factor}$$

$$\times \text{ size factor} \times \text{manufacturing philosophy}. \tag{16.3}$$

The rates and factors for nacelle A are 1 and details for nacelle B are shown in Table 16.7.

v. *Subassemblies*

Table 16.7 gives nacelles A and B and the nose cowl subassembly details in six stages of processing. Table 16.10 gives the subassembly cost in a non-dimensional form (as a fraction of the total assembly man-hours of all stages). The cost of the pure structure of the nacelle mouldlines is separated from the costs of all other non-structural components – for example, anti-icing ducting, linkages, cables, accessories and so on – that come as a complete EBU fitment for the nacelle ready to accept the turbofan engine. The cost of installing EBUs in the assembly process is taken into account, but not the EBU cost. Assembly cost is expressed as:

$$\text{Assembly cost} = \text{rates} \times \text{man-hours} \times \text{learning curve factor}$$

$$\times \textit{size factor} \times \textit{manufacturing philosophy} \tag{16.4}$$

The rates and factors for nacelle A are 1 and details for nacelle B are shown in Table 16.11. It is here that savings are made through DFM/A.

vi. *Cost of support*

A certain amount of additional cost creeps in when the product fails on inspection to adhere with the desired quality. In that case, rework and/or design concessions are required to salvage the product from being rejected to scrap. These costs are taken as support costs; in general small but hard to determine. A flat rated cost of 5% of the cost of (material + parts manufacture + assembly) is added as a support cost. DFM/A studies attempt to make design and manufacturing considerations to minimise support cost by making the product right-first-time (Six-Sigma concept).

Cost of Amortisation of the NRCs

Table 16.12 gives the two types of NRCs in a non-dimensional form. Amortisation is done over 200 aircraft, that is, distributed over 400 nacelle units.

vii. *Miscellaneous Cost*

These are the unavoidable costs – for example, insurance, packaging and so on – and unforeseen costs kept as contingencies are involved in the supply chain and necessarily charged as manufacturing costs. Normally, these are small amounts and in this study this is taken as 3–5% of the cost of material + parts manufacture + assembly. In industry, the exact amount is available.

16.4.4 Cost Formulae and Results

This section provides the semi-empirical cost formulae to establish the nacelle B cost, for any aircraft component. The input required is relatively simple: (i) geometry with dimensions, (ii) material used, (iii) weights

Table 16.11 Man-hours fraction required to assemble.

	Nacelle A	Nacelle B
Total man-hours required to assemble	1.0	0.7587
Learning curve factor	1.0	0.735
Stage 1 (start): Forward bulkhead assembly	0.1	0.032
Stage 2: Primary assembly	0.116	0.191
Stage 3: Aft bulkhead assembly	0.056	0.141
Stage 4: First subassembly	0.27	0.241
Stage 5: Second subassembly	0.211	0.267
Stage 6: Third subassembly (final)	0.247	0.128

Table 16.12 Non-recurring costs.

	Nacelle A	Nacelle B
Product design cost	1.0	1.1
Methods/tool design cost	1.0	1.1

break down and (iv) a vast array of man-hours required to design, fabricate and assemble up to completion. The factors/indices involved in the design, manufacture and manufacturing considerations are given in Tables 16.8–16.12 obtained through DFM/A studies. The total manufacturing cost of the nacelle is the sum of the individual costs of each of the four nacelle components, as given here.

$$\text{Nacelle cost: } C_N = \sum_{i}^{5} C_i = C_{NC} + C_{FC} + C_{TR} + C_{TC} + C_{EBU} \tag{16.5}$$

where

C_{NC} = Cost of nose cowl
C_{FC} = Cost of fan cowl
C_{TR} = Cost of thrust reverser
C_{TC} = Cost of tail cone assembly
C_{EBU} = Cost of EBU (e.g. anti-icing etc.)

The nose cowl component is the only component studied here. Methodologies for other components follow the same procedure. The cost of each nacelle component is to be individually estimated for each of the six headings in Table 16.7. The nose cowl cost $C_{NoseCowl}$ is the sum of the following six items.

$$C_{NoseCowl} = \sum_{i}^{5} C_i' = C_{Mat} + C_{Fab}' + C_{Asm}' + C_{Sup}' + C_{Amr}' + C_{Misc}' \tag{16.6}$$

where

C_{Mat}' = cost of material
C_{Fab}' = cost of fabrication
C_{Asm}' = cost of assembly
C_{Sup}' = cost of support
C_{Amr}' = cost of amortisation
C_{Misc}' = miscellaneous cost

i. *Nose Cowl Material Cost, C_{Mat}'*

$$C_{Mat}' = \text{cost of material} = \sum_{i}^{n} C_{i\_raw}' + \sum_{i}^{n} C_{i\_finish}' \tag{16.7}$$

where n is the number of different types of materials,
In general, $C_{i\_raw}' = W_i \times u_i \times P_i$,
where W_i = weight of the material,
u_i = standard cost of raw material per unit weight
and P_i = procurement factor (for nose cowl = 1).
Table 16.9 gives seven types of raw material. Parts weight captures the effects of size and therefore the size factor need not be applied here.
$\sum_{i}^{n} C_{i\_finish}'$ is the actual cost and is obtained from the bill of material of the subcontracted items. Therefore, the cost of material C_{Mat}' for the nose cowl in study can be given as:

$$C_{Mat}' = \sum_{i}^{n} C_{i\_raw}' + \sum_{i}^{n} C_{i\_finish}'$$

$$= (W_{al} \times u_{al} \times P_{al})_{sheet} + (W_{al} \times u_{al} \times P_{al})_{forge} + (W_{al} \times u_{al} \times P_{al})_{honeycomb}$$

$$+ W_{ti} \times u_{ti} \times P_{ti} + W_{comp} \times u_{comp} \times P_{comp} + (W_{AGS} \times u_{AGS} \times P_{AGS})_{fasttener}$$

$$+ (W_{AGS} \times u_{AGS} \times P_{AGS})_{rivet} + C_{i\_finish}' \tag{16.8}$$

Using Table 16.9 the following relation can be worked out:

$$\left(\frac{Nacelle\ B\ \cos t\ of\ raw\ material}{Nacelle\ A\ \cos t\ of\ raw\ material} \right) =$$

$$[(0.4778 \times 1)_{sheet} + (0.3868 \times 2.25)_{honeycomb} + (0.342\ 54 \times 3.5)_{ti} + (0.005 \times 18.44)_{fasteners}$$
$$+ (0.0139 \times 0.63)_{rivet}]_{nacelle\_B} / [(0.2288 \times 1)_{sheet} + (0.1213 \times 4.19)_{forge} + (0.3104 \times 2.25)_{honeycomb}$$
$$+ (0.2752 \times 3.5)_{ti} + (0.0175 \times 3.62)_{comp} + (0.0366 \times 18.44)_{fasteners} + (0.0101 \times 0.63)_{rivet}]_{nacelle\_A}$$
$$= 2.6/3.14 = 0.828$$

The finished material consists of a variety of items; these go straight to the assembly line without in-house work involved. Their acquisition policies vary at each negotiation – see Table 16.6.

$$\left(\frac{Nacelle\ B\ \cos t\ of\ finished\ material}{Nacelle\ A\ \cos t\ of\ finished\ material} \right) = 0.92 \times 0.2607/0.2864 = 0.837$$

Table 16.9 shows that 0.77 of nacelle B is composed of raw material and 0.23 is of finished material. Using the proportion

$$\left(\frac{Nacelle\ B\ \cos t\ of\ material}{Nacelle\ A\ \cos t\ of\ material} \right) = 0.77 \times 0.828 + 0.23 \times 0.837 = 0.8302$$

ii. *Nose Cowl Part Fabrication Cost, C'_{Fab}*

Table 16.7 list the number of parts fabricated in each of the six stages of both nacelles. Man-hours required to fabricate each part is a combination of operations, for example, machining, forming, fitting and so on. Nose cowl part fabrication cost is expressed as:

$$Manufacturing\ cost\ (C'_{Fab}) = rates \times man\text{-}hours \times learning\ curve\ factor$$
$$\times size\ factor \times manufacturing\ philosophy. \qquad (16.9)$$

$$or\ C'_{Fab} = (K_{size})^{0.5} \sum_i^m \left[\sum_i^n F_1 F_{2....} F_n \right]_m \times (man\text{-}hour \times rates \times learning\ factor)$$

Table 16.7 illustrates that both the nacelles have six stages ($m = 6$). Each stage has four classes of parts: (1) structure, (ii) minor parts, (iii) AGS and (iv) EBU. The EBU costs are separated from the rest, which are classified into one ($n = 1$).

Though geometrically similar, the DFM/A study made nacelle B with fewer parts to reduce assembly time. While the number of parts is low for nacelle B, part fabrication takes about the same amount of time for both the nacelles. Therefore, the various factors are as follows.

a. Size factor $K_{size} = 1.133$ from Eq. (16.1)
b. Geometry factor $F_1 = 1$ (geometrically similar)
c. Complexity factor, $F_2 = 1$ (functionality issues, same for both the nacelles)
d. Part manufacturing philosophy factor (methods factor), $F_3 = 1.0$
e. Learning curve factor = 1.022 (slightly higher)

Engineering process sheets would give all the information for nacelle A to compute the cost and to non-dimensionalise, all the factors take 1 as baseline value.

Therefore, using Eq. (16.7), parts manufacturing cost of

$$Nacelle\ B = (1.133)^{0.5} \times 1.022 = 1.0878 \times Nacelle\ A.$$

iii. *Nose Cowl Assembly Cost, C'_{Asm}*

Man-hours required to assemble at each stage are given in Table 16.10 in a non-dimensional form. Nose cowl assembly is expressed as:

$$\text{Assembly cost } (C'_{\text{Fab}}) = rates \times man\text{-}hours \times learning\ curve\ factor \times size\ factor$$
$$\times manufacturing\ philosophy \tag{16.10}$$

The rates and factors for nacelle A are 1 and Table 16.2 gives those for Nacelle B.

$$\text{or } C'_{\text{Fab}} = (K_{\text{size}})^{0.25} \sum_i^m \left[\sum_i^n F'_i \right] \times (man\text{-}hours \times rates \times learning\ factor) \tag{16.11}$$

In this case, $m = 6$ stages and n comprises of the following cost driver factors.
a. $F_1' = 0.735$ for the tooling concept (assembly methods – nacelle B takes less time).
b. $F_2' = 1.0$, the complexity factor (functionality – same for both nacelles).
c. $F_3' = 1.0$, aerodynamic smoothness requirements (surface tolerance is the same)
Equation (16.9) reduces to

$$C'_{\text{Fab}} = (K_{\text{size}})^{0.25} \sum_i^6 \left[\sum_i^3 F'_i \right] \times (man\text{-}hours \times rates \times learning\ factor)$$

To simplify, all stages are lumped together to get the nacelle B cost. Taking baseline nacelle A factors/indices to be 1, the nacelle B assembly cost is expressed as follows:

$$\text{Assembly cost of Nacelle B} = (1.133)^{0.25} \times 0.735 = 0.759 \times \text{Nacelle A.}$$

It is interesting to note that a considerable amount of assembly cost can be reduced at the expense of a slightly increased part fabrication cost and, of course, with some increase in tooling cost NRC. We must establish that these factors are the main obligation of DFM/A trade-off studies. Basically, it summarises the man-hours required as compared to the baseline man-hours.

iv. *Nose cowl support cost, C'_{Sup}:*
Note that separate support cost is taken as a percentage of material, fabrication and assembly cost. Support cost arises from rework and concession when build quality is not met (e.g. tolerances). In this paper, it is flat rated as follows:

$$C'_{\text{Sup}} = 0.05 \times (material\ cost + parts\ fabrication\ cost + assembly\ cost) \tag{16.12}$$

v. *Nose Cowl Amortisation cost, C'_{Amr}:*
Cost is amortised over 400 finished products. It is a variant design and hence has a low amortisation cost. This can be included in man-hour rates at all stages or added separately at the end. In this chapter, it is accounted for in man-hour rates and is not separately computed. Typically, because it is produced in twice more in number, the amortisation cost is taken as 2% of (material cost + parts fabrication cost + assembly cost)

$$C'_{\text{Amr}} = \text{cost of amortisation} = (design + methods + tools)\ cost/N \tag{16.13}$$

where, $N = 400$

$$\text{or } C'_{\text{Amr}} = 0.02 \times (material\ cost + parts\ fabrication\ cost + assembly\ cost)$$

vi. *Nose Cowl Miscellaneous Cost, C'_{Misc}*
Miscellaneous costs are the unavoidable ones given next (taken as 3%).

$$C_{\text{Miscp}} = 0.03 \times (material\ cost + parts\ fabrication\ cost + assembly\ cost) \tag{16.14}$$

Total cost of nose cowl of nacelle B
The final cost of the nacelle B nose cowl can now be computed in a non-dimensional form, that is, in relation to nacelle A cost. Taking Eq. (16.4), the following is estimated (cost of amortisation embedded in man-hour rates and costs of support and miscellaneous costs are estimated as 10% of the other costs).

$$C_{\text{NoseCowl}} = \sum_i^5 C_i' = C_{\text{Mat}}' + C_{\text{Fab}}' + C_{\text{Asm}}' + C_{\text{Sup}}' + C_{\text{Misc}}'$$

$$\text{Where } (C_{\text{Sup}}' + C_{\text{Misc}}') = 0.1 \times (C_{\text{Mat}}' + C_{\text{Fab}}' + C_{\text{Asm}}') \tag{16.15}$$

Nacelle B nose cowl cost,

$$C_{\text{NoseCowl\_B}} = 0.8302 C'_{\text{Mat\_NacA}} + 1.0878\, C'_{\text{Fab\_NacA}} + 0.759\, C'_{\text{Asm\_NacA}}$$

$$+ 0.1 \times (0.8302\, C'_{\text{Mat\_NacA}} + 1.0878\, C'_{\text{Fab\_NacA}} + 0.759\, C'_{\text{Asm\_NacA}})$$

$$= 0.9132 C'_{\text{Mat\_NacA}} + 1.1966 C'_{\text{Fab\_NacA}} + 0.8349 C'_{\text{Asm\_NacA}}$$

From the company records, nacelle A cost fractions are as follows:

$$C'_{\text{Mat\_NacA}} / C_{\text{NoseCowl\_A}} = 0.408,\ C'_{\text{Fab\_NacA}} / C_{\text{NoseCowl\_A}} = 0.349\ C'_{\text{Asm\_NacA}} / C_{\text{NoseCowl\_A}} = 0.149$$

Dividing Eq. (16.15) by $C_{\text{NoseCowl\_A}}$, the relative cost of nacelle B is as follows:

$$C_{\text{NoseCowl\_B}} / C_{\text{NoseCowl\_A}} = 0.9132 \times 0.408 + 1.1966 \times 0.349 + 0.8349 \times 0.149 = 0.9146$$

The result shows that, although the two nacelles are geometrically similar, nacelle B with a 13.5% higher thrust engine could be produced at a 8.5% lower cost through DFM considerations in an IPPD environment. Changes in material, structural, tooling and in procurement/sub-contracting policies contribute to cost reduction. A preliminary weight of a new design and procurement policy for the raw material can be established at the conceptual stages (DFM/A studies). The accuracy improves as the project progresses. In the absence of the actual weight, approximations can be made from the geometry. Had it been costed with prevailing empirical relations using weight, size, performance and some manufacturing considerations, then the cost of nacelle B would be higher than A. The prevailing equations do not arrest the subtlety of DFM/A considerations. Section 20.6.1 describes the vast changes that have taken place in the manufacturing technology and that their benefits need to be reflected through a fresh look at formulation.

16.5   Aircraft Direct Operating Cost (DOC)

Each airline generates its own in-house DOC computations. There are variances in the man-hour rates and schedules. Although the ground rules for DOC vary from company to company, the standardisation made by the AEA [1] (1989 – short-medium range) has been accepted as the basis for comparison. The ATA rules are used in the USA.

The NASA report [7] gives American Airlines generated economics in detail. The NASA document proposed an analytical model in 1978 associated with advanced technologies in aircraft design, but it is yet to be accepted as a standard method for comparison. The AEA ground rules appear to have taken into account of all the pertinent points and have become the standard for operational and manufacturing industries in benchmarking/comparison.

Table 16.13 gives the breakdown of DOC components that basically falls under two headings. The ownership cost element depends on aircraft acquisition cost.

The NRC (design and development) of a project and cost of aircraft manufacture contribute to 'ownership cost' while cost of fuel, landing fees and maintenance contribute to 'trip cost elements'. Once the aircraft is purchased, the ownership cost runs even when crews are hired but no flight operation is carried out. Crew cost added to ownership cost become the 'fixed cost' elements. Crew cost added to trip cost is the 'running cost' of the trip (mission sortie). Aircraft price dependent DOC contributions are depreciation, interest, insurance and maintenance (airframe + engine), totalling nearly four times that of the contribution made by fuel cost. Crew

Table 16.13 DOC components.

Fixed cost elements		Trip cost elements
1.	Ownership cost	3. Fuel charges
	(a) Depreciation	4. Maintenance (airframe and engine)
	(b) Interest on loan	5. Navigational and landing charges
	(c) Insurance premium	
2.	Crew salary and cost	

salary/cost and navigational/landing charges are aircraft weight dependent, which is second-order aircraft price dependent but here it is taken based on man-hour rates.

The ownership cost contribution to DOC could be as high as one-third to half of the total DOC (at year 2000 level fuel prices). It is for this reason that the industry was driven to reduce the cost of manufacture and the demand for that was as high as a 25% cost reduction. Clearly, the design philosophy plays a significant role in facilitating manufacturing cost reduction. One of the considerations was to relax some quality issues (Ref. [2] and Chapter 20) sacrificing aerodynamic and structural considerations without sacrificing safety. However, when fuel price rises then considerations for such drivers would be affected. Fuel price had already jumped very high by 2008. Any further increase would require drastic measures on many fronts as return from pure aerodynamic cleanliness at a high investment level may not prove sufficient. These are important considerations during the conceptual design stage. R&D efforts look further, futuristically sometimes, through 'crystal ball gazing'. There is a diminishing return on investment for aerodynamic gains. The fuel price is fluctuating severely and until a stable situation arises, efforts to invest to reduce drag will move slowly. A parallel effort to use a cheaper alternative bio-fuel is underway. Demand for turboprop operation is a reality.

Figure 16.6 shows the DOC components of a mid-size class aircraft. For an average mid-range sector mid-size aircraft, cost contribution to aircraft DOC is three to four times higher than the contribution made by the cost of fuel (2000 level fuel price).

The Passenger Load Factor (LF) is defined as the ratio of occupied seats to the total number of seats available. Typically, an airline would like to break-even the sector DOC by about one-third full, that is, at a LF of 0.33 (33%). Of late, on account of fuel price increase, this figure has gone up. Revenue earned from passengers carried above the break-even LF comes as profit. While some flights could run at 100% LF, the yearly average for a high demand route could be much lower. Passenger accommodation could be in different classes in a fare tier structure, or in one class as decided by the airline. Even in the same class, the airfare can vary depending

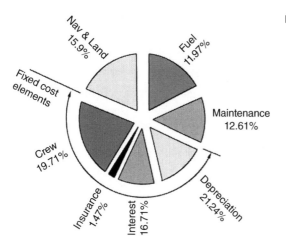

Figure 16.6 DOC breakdown (@ 0.75 cents/US gallon).

Nav & Land 15.9%

Fuel 11.97%

Fixed cost elements

Maintenance 12.61%

Crew 19.71%

Insurance 1.47%

Interest 16.71%

Depreciation 21.24%

on the type of offer made. From airline to airline, the break-even LF varies. With deregulated airfares, the ticket price could vary by the hour, depending on the passenger demand. The standard fare is the ceiling fare of the class and offers better privileges.

16.5.1 Formulation to Estimate DOC

Given next is the DOC formulation, based on the AEA ground rules [2]. The formulae computes component DOC per block hour. To obtain the trip cost, the DOC per block hour is multiplied by the block time. Aircraft performance calculates the block hour and block time for the mission range (Section 13.5.6). The next section works out the DOC values in continuation with the Bizjet example obtained so far.

Normally, DOC is computed for a fleet of aircraft. The AEA suggests a 10-aircraft fleet with 14 years of life span and a residual value of 10% of the total investment made. These values can be changed as will be shown in the next section. Fuel price, insurance rates, salaries and shop-floor man-hour rates vary with time. Engine maintenance cost depends on the type of engine. Here, only the turbofan type is discussed. For other kinds of power plants, the readers are recommended to look at Ref. [2].

Aircraft Price:

$$Total\ Investment = (Aircraft + Engine\ Price) \times (1 + spares\ allowance\ fraction)$$

Readers need to be careful when obtaining the Standard Study Price of the manufacturer. The AEA uses total investment, which includes aircraft delivery price, cost of spares, any change in order and other contractual financial obligations. In the example, the aircraft and engine price is taken as the total investment per aircraft.

$$Outstanding\ capital = Total\ capital\ cost \times (1 - Purchase\ down\ payment\ fraction)$$

Utilisation – per block hour per annum (hours/year)

$$Utilisation,\ U = \left(\frac{3750}{(t + 0.5)} \right) \times t,$$

where t = block time for the mission

Fixed cost elements:

i. Depreciation $= \dfrac{0.9 \times Total\ investment}{14 \times Utilisation}$

ii. Loan interest repayment: $\dfrac{0.053 \times Total\ Capital\ Cost}{Utilisation}$

iii. Insurance premium $= 0.005 \times \left[\dfrac{Aircraft\ Cost}{Utilisation} \right]$

iv. Crew salary and cost

Flight crew – AEA has taken $493 per block hour for a two-crew operation.

Cabin crew – AEA has taken $81 per block hour for each cabin crew.

Trip cost elements:

v. Landing fees $= \left(\dfrac{7.8 \times MTOW\ in\ Tonnes}{t} \right)$

where t = block time for the mission

vi. Navigational charges $= \left(\dfrac{0.5 \times Range\ in\ km}{t} \right) \times \sqrt{\dfrac{MTOW\ in\ Tonnes}{50}}$

where t = block time for the mission

vii. Ground handling charges $= \left(\dfrac{100 \times Payload \ in \ Tonnes}{t} \right)$

where $t =$ block time for the mission

The landing/navigational/ground handling charges are MTOW dependent and ground handling charges are payload dependent. In practice, crew salary is also MTOW dependent but the AEA has kept it invariant.

viii. Airframe maintenance, material and labour

a. Airframe labour $=$

$$\left(0.09 \times W_{airframe} + 6.7 - \frac{350}{(W_{airframe} + 75)} \right) \times \left(\frac{0.8 + 0.68 \times (t - 0.25)}{t} \right) \times R$$

where,

$W_{airframe} =$ the MEW less engine weight in tonnes

$R =$ labour man-hour rate of \$63 per hour at the 1989 level.

$t =$ block time for the mission

b. Airframe material cost

$$\left(\frac{4.2 + 2.2 \times (t - 0.25)}{t} \right) \times C_{airframe}$$

where $C_{airframe} =$ price of aircraft less engine price in millions of dollars.

ix. Engine maintenance, material and labour

c. Engine labour $= 0.21 \times R \times C_1 \times C_3 \times (1 + T)^{0.4}$

where

$R =$ labour man-hour rate of \$63 per hour at 1989 level.

$T =$ sea level static thrust in tonnes

$C_1 = 1.27 - 0.2 \times BPR^{0.2}$ where BPR $=$ bypass ratio

$C_3 = 0.032 \times n_c + K$

where

$n_c =$ Number of compressor stages

$K = 0.50$ for one shaft

$= 0.57$ for two shaft

$= 0.64$ for three shaft

d. Engine material cost $= 2.56 \times (1 + T)^{0.8} \times C_1 \times (C_2 + C_3)$

where

$T =$ sea level static thrust in tonnes

$C_2 = 0.4 \times (OAPR/20)^{1.3} + 0.4$ where, OAPR is the overall pressure ratio

C_1 and C_3 are same as before.

direct engine maintenance cost (labour + material)

$$= Ne \times (\text{engine labour cost + material cost}) \times \frac{(t + 1.3)}{(t - 0.25)}$$

where $Ne =$ number of engines

x. fuel charges $= \dfrac{Block \ fuel \times Fuel \ \cos t}{Block \ time}$

Then, DOC per hour $= (fixed \ charges + trip \ charges)_{\text{per\_hour}}$

And DOC per trip $= t \times (\text{DOC})_{\text{per\_hour}}$

DOC per aircraft mile $= \dfrac{DOC \times 100 \times Block \ time}{Range}$

and DOC per passenger mile per nautical mile $= \dfrac{DOC \times 100 \times Block \ time}{Range \times No. of \ passengers}$

16.5.2 Worked-Out Example of DOC – Bizjet

Based on the formulation given previously, this section works out the DOC of the Bizjet in question. Instead of working on a fleet, only one aircraft is worked-out here. In the following, information (input for DOC calculation) about the Bizjet is given. The man-hour rates are given in [7]. All cost figures are in US dollars and rounded up to the nearest figure.

Table 16.14 gives the aircraft and engine details.

Aircraft price = $7 million
Engine price = $1 million
Total Aircraft Acquisition Cost = $ 8 million (this is taken as total investment per aircraft. Price includes spares and suchlike).

DOC is computed for a single aircraft to obtain the *trip cost* instead of per hour cost. A life span of 14 years is taken with a residual value of 10% of the total investment.

Aircraft price:

Total Investment = $ 8 million

Utilisation – per block hour per annum *(hours/year)*

$$\text{Utilisation}, U = \left(\frac{3750}{(5.38 + 0.5)} \right) \times 5.38 = 637.75 \times 5.38 = 3431 \text{ hours per year}$$

where t = block time for the mission = 5.38 hours.

Fixed cost elements:

i. Depreciation $= \dfrac{0.9 \times 8 \times 10^6}{14 \times 3431} \times 5.38 = \$ 807$ per trip

ii. Loan interest repayment: $= \dfrac{0.053 \times 8 \times 10^6}{3431} \times 5.38 = \$ 665$ per trip

iii. Insurance premium $= 0.005 \times \left[\dfrac{8 \times 10^6}{3431} \times 5.38 \right] = \$63$ per trip

iv. Crew salary and cost
Flight crew $= 493 \times 5.38 = \$ 2652$ per trip for two crew
Cabin crew = 0 as there is no cabin crew.

Table 16.14 Bizjet data for DOC estimation.

Aircraft details	Turbofan details (two engines)	Conversion factors
MTOW – 9400 kg	T_{SLS}/engine – 17.23 kN	1 nm = 1.852 km
OEW[c] – 5800 kg	Dry weight – 379 kg	US gallons = 6.78 lb
MEW – 5519 kg	Bypass ratio – 3.2	1 lb = 0.4535 kg
Payload – 1100 kg[a]	No. of compressor stages – 10[b]	1 ft = 0.3048 m
Range – 2000 nm	Overall compressor ratio – 14	1 kg fuel = 0.3245 Gal
Block time – 5.38 h	No. of shafts = 2	
Block fuel – 2233 kg	Fuel cost = $0.75 per US gallon	

a) 10 passengers.
b) It has one high pressure compressor, four stage low pressure compressors and one fan.
c) OEW = Operator's Empty Weight.

Trip cost elements:

v. Landing fees $= \left(\dfrac{7.8 \times 9.4}{5.38} \times 5.38 \right) = \$\,74$ per trip

vi. Navigational charges $= \left(\dfrac{0.5 \times 2000 \times 1.852}{5.38} \right) \times \sqrt{\dfrac{9.4}{50}} \times 5.38 = \$\,803$ per trip

vii. Ground handling charges $= \left(\dfrac{100 \times 1.1}{5.38} \right) \times 5.38 = \$\,110$ per trip

viii. Airframe maintenance, material and labour

Airframe labour $=$

$$\left(0.09 \times 5.52 \times 1.02 + 6.7 - \dfrac{350}{(5.52 \times 1.02 + 75)} \right) \times \left(\dfrac{0.8 + 0.68 \times (5.38 - 0.25)}{5.38} \right) \times 63$$

$$= (1.853) \times 1.448 \times 63 = \$\,169 \text{ per trip}$$

Airframe material cost $= \left(\dfrac{4.2 + 2.2 \times (5.38 - 0.25)}{5.38} \right) \times 7 \times 5.38 = \$109$ per trip

where $C_{\text{airframe}} = \$7$ million.

Total airframe maintenance (material + labour) $= 910 + 109 = \$\,1019$ per trip

Engine maintenance, material and labour

Engine labour $= 0.21 \times 63 \times 1.018 \times 0.89 \times (1 + 1.72)^{0.4} = \$17.88$ per hour

where $T = 17.23 \text{ KN} = 1.72\,\text{t}$

$C_1 = 1.27 - 0.2 \times 3.20.2 = 1.018$

$C_3 = 0.032 \times 10 + 0.57 = 0.89$

Engine material cost $= 2.56 \times (1 + 1.72)^{0.8} \times 1.018 \times (0.652 + 0.89)$

$= 5.7 \times 1.542 = \$8.79$ per hour

where $C2 = 0.4 \times (14/20)1.3 + 0.4 = 0.652$

C1 and C3 are same as before.

Direct engine maintenance cost (labour + material)

$$= 2 \times (17.88 + 8.79) \times \dfrac{(5.38 + 1.3)}{(5.38 - 0.25)} = \$\,308 \text{ per trip}$$

Fuel charges $= \dfrac{2233 \times 0.3245 \times 0.75}{5.38} \times 5.58 = \$\,544$ per trip

The baseline Bizjet DOC is summarised in Table 16.15.

Then, DOC per hour $= 7045/5.38 = \$1309.5$ per hour

DOC per aircraft nautical mile $= \dfrac{7045}{2000} = \$3.52$ per nautical mile per trip.

DOC per passenger mile per nautical mile $= \dfrac{7045}{2000 \times 10} = \$0.352/\text{nm/passenger}$

See Appendix C, DOC details of a mid-size high subsonic transport aircraft.

Operating Cost of the variants in the family

Large Variant:	*MTOM* = 10 800 kg and 14 passengers.	
	Aircraft price = \$9 million	
	Ownership cost element =	1727
	Crew cost =	3047
	Fuel cost =	625
	Maintenance/operational charges =	2650
	Total operational cost =	8049

Table 16.15 Bizjet summary of DOC per trip (all figures in US Dollars).

Fixed cost elements		
Depreciation	807	
Interest on loan	665	
Insurance premium	63	
Total ownership cost	1535	
Flight crew	2652	
Cabin crew	0	
Total fixed cost elements		4187
Trip cost elements		
Fuel charges	544	
Navigational charges	803	
Landing charges	74	
Ground handling charges	110	
Maintenance (airframe)	1019	
Maintenance (engine)	308	
Total trip cost elements		2858
Total direct operating cost		$7045 per trip

Then, DOC per hour = 8049/5.38 = $1496.4 per hour

$$\text{DOC per aircraft nautical mile} = \frac{8049}{2000} = \$4.025 \text{ per nautical mile per trip.}$$

$$\text{DOC per passenger mile per nautical mile} = \frac{8049}{2000 \times 14} = \$0.2875/\text{nm/passenger}$$

Smaller Variant:	MTOM = 7600 kg and six passengers.	
	Aircraft price = six million	
	Ownership cost element =	1220
	Crew cost =	2201
	Fuel cost =	449
	Maintenance/operational charges =	1910
	Total operational cost =	5780

Then, DOC per hour = 5780/5.38 = $1075 per hour

$$\text{DOC per aircraft nautical mile} = \frac{5,780}{2,000} = \$2.898 \text{ per nautical mile per trip.}$$

$$\text{DOC per passenger mile per nautical mile} = \frac{5796}{2000 \times 6} = \$0.482/\text{nm/passenger}$$

16.6 Aircraft Performance Management

Today's aeronautical industry is involved with the generation and acquisition of large amounts of data in all airline departments, in particular, in airline flight operations, as Aircraft Performance Management information can be used in a number of ways as listed in the following:

i. *Fuel economy*: Reducing fuel burn during operation without sacrificing the mission goal.

ii. *Safety*: Early warning of a potentially fatal failure event.

iii. *Fleet Maintenance*: Reduced mission aborts, fewer grounded aircraft, simplified logistics for fleet deployment.

iv. *Aircraft Maintenance*: Aircraft structure/system maintenance management to cut down cost by avoiding unscheduled inspection, maintenance and troubleshooting by recognising incipient failure stages and taking action in the appropriate time in anticipation.

v. *Operational Maintenance*: Improved flight safety, mission reliability and effectiveness.

vi. *Performance*: Improved aircraft performance and reduced fuel consumption.

vii. *Maintain Status Deck*: Gathering aircraft data in order to determine the actual performance level of each aeroplane (old and new) in the fleet with respect to the manufacturer's book level.

viii. *Flight Tracking (under study for implementation)*: Real time tracking and data streaming on the airwaves

The crucial issue is to identify the type of data needed and for what purpose. From organisation to organisation, the aircraft management terminologies differ based on which aspect is being dealt with in the monitoring process. Some of the terminologies even got patented, signifying their importance. At the core, this means detecting all round anomalies and taking corrective actions in time. Basically, this is a cost saving measure that includes the following, as classified in this book.

i. Aircraft Performance Monitoring (APM)

Commercial aircraft design is sensitive to the DOC, fuel-burn rate during operation (mainly climb and cruise) contributes to nearly a third of the DOC. Just 1% of fuel saved in every mission could account for saving several million US dollars in operating cost over the lifespan of a small fleet of midsized aircraft. These become more important for larger aircraft operations and increasingly important with the increasing fuel costs.

Therefore, it is important to reduce DOC by reducing fuel burn during operation without sacrificing the mission goal. In the past, for pilot ease, the older aircraft operational schedules were set to constant EAS operations for cruise in LRC/HSC. Modern aircraft with microprocessor-based FBW/FADEC control operation allows fine-tuning with synchronised tweaking of the engine to operate at least fuel burn around the prescribed speed schedules. This is seen as a part of the APM. The APM is nothing new but has steadily gained importance facilitated by microprocessor-based aircraft systems architecture to improve all-round performance gains, improved safety and real time flight-tracking during operation.

ii. Engine management (FADEC based) to obtain the lowest fuel consumption for the planned mission.

iii. Aircraft to operate at lowest trim drag (FBW based).

iv. Aircraft Health Monitoring (AHM).

Aircraft structure and system maintenance management should cut down cost by avoiding unscheduled inspection, maintenance and troubleshooting by recognising incipient failure stages and taking action in the appropriate time in anticipation. This is meant for timely and streamlined identification and diagnosis of issues. Performing repairs when the damage is minor increases the aircraft mean time before failure (MTBF) and decreases the mean time to repair (MTTR). It reduces the downtime; where an aircraft that is not utilized is not returning on investment.

The AHM not only monitors the aircraft but also the engines, providing significant overall fuel and emission performance measures. In earlier days, maintenance monitoring was carried out as technology permitted, for example, by oil inspection, vibration checks and so on, but nowadays microprocessor-based technology can have embedded internal sensor read-outs that can offer good insights to aircraft/engine health. These sensors send their data in real-time to the aircraft engine monitoring system in a computer running with specialized software. It detects whether there are anomalies needing corrective actions.

The AHM also facilitates a spare parts flow schedule and stock inventory in a manner that does not block cash. It monitors performance measures for individual airplanes, enabling operators to improve overall average fleet performance.

In military operations, monitoring aircraft structural integrity this an important factor to determine how they fatigue over their lifetime. This generally involves the process of recording aircraft g-levels in three axes (sometimes also includes data from the engine thermal cycling), downloading them into a database with the fatigue model of the aircraft and then calculating the remaining Fatigue Index (FI) of the airframe.

v. The proposed Flight Track Monitoring (AFM, a suggested acronym in this book) – yet to be published. The case of a state-of-the-art aircraft, the MH370 Boeing 777-200, disappearing raises the question of why it can vanish. It brings back the debate about the need for global flight-tracking and data streaming on airwaves that can also be backed up if the emergency locator transmitters failed to emit signals. The aviation authorities are expected to announce new global flight-tracking standards soon. A task force set up by the International Civil Aviation Organization (ICAO) is due to publish global flight-tracking standards in 2016.

An on/off switch is offered to the pilot so they can make a choice whether to activate the option of engaging APM or not.

16.6.1 Methodology

Cash operating costs (COCs), also known as mission trip costs (T_c), consist of the following elements:

i. Fuel cost (F_c – depends on the price at the time).
ii. Crew cost (when airline fleet is in operation).
iii. Maintenance (airframe and engine).
iv. Navigational and landing charges (aircraft weight dependent).

Brief explanation: The faster the aircraft operational speed, the lower the trip time (block time); the saved time can offer a higher level of aircraft utilisation, that is, more sorties can be accommodated per year, earning more revenue. However, an aircraft flying faster or slower than the speed that offers the maximum specific range (SR) will increase cost of fuel burn as fewer nautical miles are covered per unit of fuel burn.

Also, with higher yearly utilisation, the crew cost per flight hour goes down as they are available to make more sorties. When an aircraft is sitting on the ground, crew costs are not attributed to the costs operating that same aircraft. Their salaries are then accounted as fixed cost element (see Table 16.5).

Maintenance cost goes up with faster speed on account of higher utilisation causing more wear and tear.

In summary, trip cost is dependent on block time. Increase in aircraft cruise speed above the speed at maximum SR will decrease the block time, resulting in an increase in fuel cost, an increase in maintenance but a decrease in crew cost. Flying below the speed at a maximum SR will increase the block time resulting in an increase in fuel cost and crew cost, but a decrease in maintenance cost. Figure 16.7 gives a representative graph depicting the relationship as given before. Landing/navigation (aircraft weight depended) costs are not shown, being relatively flat (small percentage) for the aircraft type, although higher utilisation will incur additional costs.

Figure 16.7 Trip cost elements.

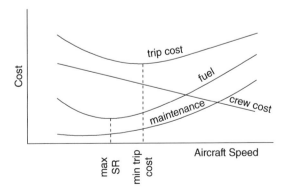

Section 15.6.2 showed that to optimise aircraft performance, it is best to maximise $(ML)/(sfcD)$ for LRC. The APM continually adjusts the aircraft flight parameters to fly around this point. The top graph is the total trip cost that has a minimum at a slightly higher speed than the maximum SR offers. It is closer to maximum $(ML)/(sfcD)$.

This book deals only with that part of the APM involving mission trip cost, primarily in fuel saving as a topic of aircraft performance. Saving fuel consumption in a mission sortie is based on (i) flying aircraft with minimum trim drag that can be easily achieved with FBW aircraft that continuously adjust aircraft centre of gravity (CG) position by fuel transfer to keep trimmed and (ii) continuous thrust adjustment to offer minimum fuel burn with FADEC controlled engines.

It is considered that the aircraft flies at the minimum trim drag and, to keep the explanation simple, only the fuel cost is taken as a significant variable as given in the following expression. The objective of the APM is to minimise F_c by flying at the optimum V.

$(T_c) = f(V)$, that is, trip cost is a function of aircraft velocity, V. By differentiating and setting

it equal to zero, then solving for V gives the optimum.

The relation $(T_c) = f(V)$ is not easily integrated, but the optimum V can be established through parametric studies by lengthy manual computation using the same equations derived in this book. Nowadays, thanks to a FBW/FADEC controlled onboard computer with the aircraft manufacturer's installed software incorporating all the related variables, an aircraft can continuously fly at optimum V to minimise trip cost.

16.6.2 Discussion – The Broader Issues

With the advent of modern microprocessor-based technologies, each industry is stretching the scope of APM, in a broader meaning of the acronym. This can cover a wide territory of aircraft fleet operations to maintain a 'status deck'. It includes a procedure devoted to gathering aircraft data in order to determine the actual performance level of each aeroplane in the fleet with respect to the manufacturer's book level. The performance data given in the Flight Crew Operating Manual reflects this book value. It represents a fleet average of brand new airframe and engines. This level is established in advance of production. Normal scatter of brand new aircraft leads to individual performance above and below the book value.

An APM programme uses information from engine condition monitoring and air data computers to provide fleet operators with fuel burn information. If the fuel burn is higher than expected and the engines are operating as expected then it would imply the engines are pulling an aircraft with a higher than normal aerodynamic drag. This would be caused typically by misrigged/out-of-trim control surfaces or an airframe with poor surface contouring. The APM calculates the SR during the cruise segment. Airlines compare the SR across the fleet to identify poorly performing aircraft and pull them in for overnight maintenance inspections or attaching rigging boards to adjust control surface rest positions, if necessary.

The performance levels are measured in their variations over time. Resulting trends can be made available to the operators' various departments, which perform corrective actions to keep the airline fleet in operation-ready condition.

References

1 *Association of European Airlines Publication*: Short-medium range aircraft – AEA requirements (1989).
2 Kundu, A.K. (2010). *Aircraft Design*. Cambridge University Press.
3 Kundu, A., Curran, R., Crosby, S. et al. (2002). *Rapid Cost Modelling at the Conceptual Stage of Aircraft Design*. Los Angeles, USA: AIAA Aircraft Technology, Integration and Operations Forum.
4 Kundu, A. K., Crosby, S., Curran, R., et. al., (2003). Aircraft component manufacture case studies and operating cost reduction benefit, AIAA Conference ATIO Denver, CO, Nov. 2003.

5 Kundu, A.K. and Watterson, J. (2002). Parametric optimisation of manufacturing tolerances at the aircraft surface. *Journal of Aircraft* 39 (2), AIAA.

6 Apgar, H. (1993). Design to Life-Cycle Cost in aerospace, IAA/AHS/ASEE Aerospace Design Conf. Feb 16–19, 1993, Part G Journal of Aero. Engineering.

7 NASA CR 145190. (1978). A new method for estimating current and future transport aircraft operating cost. March 1978.

Part III

Further Design Considerations

17

Aircraft Load

17.1 Overview

Aircraft performance depends on its structural integrity to withstand the design load level. Aircraft structures must withstand the imposed load during operations; the extent depends on what is expected from the intended mission role. The higher the load, the heavier the structure for its integrity; hence the maximum takeoff weight (MTOW) affecting aircraft performance. Aircraft designers must comply with the mandatory certification regulations to meet the minimum safety standards.

The information provided herein is essential for understanding performance considerations that affect aircraft mass (i.e. weight), hence its performance. The *V-n* diagram depicts the loading condition and the limits of aircraft flight envelope. Only the loads and associated *V-n* diagram in symmetrical flight are discussed herein. Estimation of load is a specialised subject covered in focused courses and textbooks [1, 2]. However, this chapter does outline the key elements of aircraft loads.

This chapter includes the following sections:

Section 17.2: Introduction to Aircraft Load, Buffet and Flutter
Section 17.3: Flight Manoeuvres
Section 17.4: Aircraft Load
Section 17.5: Theory and Definitions (Limit and Ultimate Load)
Section 17.6: Limits (Load Limit and Speed Limit)
Section 17.7: *V-n* Diagram (the Safe Flight Envelope)
Section 17.8: Gust Envelope

Classwork content. The chapter can be skipped if the subject has been learned about in other coursework, otherwise the readers must go through it.

17.2 Introduction

Loads are the external forces applied to an aircraft in its static or dynamic state of existence, in-flight or on-ground. In-flight loads are due to symmetrical flight, unsymmetrical flight or atmospheric gusts from any direction. On-ground loads result from ground handling and field performance (e.g. in static, takeoff and landing). Aircraft designers must be aware of the applied aircraft loads, ensuring that configurations are capable of withstanding them. The subject matter concerns the interaction between aerodynamics and structural dynamics (i.e. deformation occurring under load), a subject that is classified as *aeroelasticity*. Even the simplified assumption of an aircraft as an elastic body requires study beyond the scope of this book. This book addresses rigid aircraft, explained in Section 4.1.

User specifications define the manoeuvre types and speeds that influence aircraft weight, which then dictates aircraft design, affecting aircraft performance. In addition, enough margin must be allocated to cover

Conceptual Aircraft Design: An Industrial Approach, First Edition. Ajoy Kumar Kundu, Mark A. Price and David Riordan.
© 2019 John Wiley & Sons Ltd. Published 2019 by John Wiley & Sons Ltd.

inadvertent excessive load encountered through pilot-induced manoeuvres, or sudden severe atmospheric disturbances (i.e. external input) or a combination of the two scenarios. The limits of these inadvertent situations are derived from historical statistical data and pilots must avoid exceeding the margins. To ensure safety, governmental regulatory agencies have mandatory requirements for structural integrity specified with minimum margins. *Load factor* (not to be confused with the passenger load factor) is a term that expresses structural-strength requirements. The structural regulatory requirements are associated with *V-n* diagrams, which are explained in Section 17.7. In fact, they also require mandatory strength tests to determine ultimate loads, which must be completed before the first flight, with the exceptions of homebuilt and experimental categories of aircraft. In the following sections, buffet and flutter are introduced.

17.2.1 Buffet

The buffet phenomenon occurs at the insipient phase of stall at low-speed 1-*g* level flight. Actually buffet can occur at any speed, whenever flow instability over the wing takes place, for example, during extreme manoeuvres, at higher *g* loads ($n > 1$) when the angle of attack increases to high values. Buffet can also occur at 1-*g* high-speed (called Mach buffet) when transonic flow over-wing (or over any other lifting surface) local shocks interact with the boundary layer to cause unstable separation making the airflow over the wing unsteady; the separation line over the wing keeps fluctuating. In this situation, the H-tail can also get induced to start buffeting. This causes the aircraft to shudder and is a warning to the pilot. The aircraft structure is not necessarily at its maximum loading.

It is clear that every class of aircraft has two stall speeds, hence there are two buffet speeds; one low-speed buffet and the other a Mach buffet at high-speed, the magnitude depends on its weight, altitude and ambient conditions (Figure 17.1). At higher altitudes, because of lower air density, the low-speed buffet goes up while the Mach buffet speed goes down. Low-speed stall buffet is based purely on aircraft aerodynamic design but the high-speed Mach buffet requires the additional requirement of engine capability to attain such speeds in level flying. The excess available thrust between the two limits allows the aircraft to climb. Where the two buffet limit lines merge is the aerodynamic aircraft aerodynamic ceiling height, when the rate of climb reaches zero.

17.2.1.1 Q-Corner – (Coffin Corner)

As the aircraft reaches the aerodynamic ceiling, the buffet margin between the low-speed and high-speed range gradually gets reduced and it becomes very difficult to fly in this small margin. Flying slower will make the aircraft stall and when flying faster, the aircraft starts shuddering. At low-speed stall, the aircraft nose drops and gains speed only to enter into high-speed buffet shudder. If delayed to take recovery action, then control problems may arise and can pose problems in recovery. This zone is loosely termed the 'coffin corner' (Q-corner). Aircraft flight in this zone is carried out but only by test pilots in order to establish the buffet

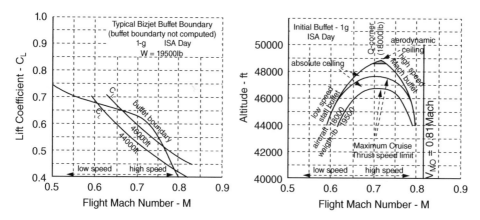

Figure 17.1 Typical buffet boundary.

Table 17.1 Typical permissible *g*-load for civil aircraft.

Type	Ultimate positive *n*	Ultimate negative *n*
FAR25 (14CFR25)		
Transport aircraft below 50 000 lb	3.75	−1 to −2
Transport aircraft above 50 000 lb	$[2.1 + 24\,000/(W + 10\,000)]$ Should not exceed 3.8	−1 to −2
FAR23 (14CFR23)		
Aerobatic category (FAR23 only)	6	−3

boundary to be aware of. Operational flying stays away from the buffet boundary. However, there are exceptions for special missions, for example, the reconnaissance aircraft Lockheed U2 was designed to fly at high altitudes near its ceiling to stay out of missile reach at that time. The Lockheed U2 has a less than 1% margin between low and high speed (within the Q-corner) at either side of the onset of aircraft buffeting. Civil aircraft regulation requires that aircraft should fly 30% below $C_{Lbuffet}$. The buffet '*g*' should not exceed this limit (see Table 17.1). This also gives a margin for the pilot to manoeuvre to a '*g*' level of $n = 1.3$. Although known as the coffin corner, as such, it is not dangerous to fly if pilot can recognise the incipient stage of buffet onset and the 30% margin allows manoeuvring to recover. It is possible the term originated from a different reason than being dangerous to fly – it is better to use the term Q-corner.

The sized engine should have a maximum cruise rating thrust capability to reach the ceiling altitude for the aircraft weight. The absolute aircraft ceiling is actually when rated (normally maximum cruise rating) thrust limited speed equals the stall speed at the aircraft weight, when the aircraft rate of climb is zero: a point could be below high-speed Mach buffet. At a higher thrust rating, say at a maximum continuous rating, the aircraft will have some excess thrust to climb, possibly, limited by high-speed Mach buffet until it reaches the stall buffet at the Q-corner when the aircraft ceases to climb reaching the aerodynamic ceiling for the weight. Any further thrust increase is of no use as the aircraft is already limited by the buffet lines from both sides. Also, note in Figure 17.1 the structural limit line of V_{MO} at 0.81 Mach. The aircraft does not have the capability to reach V_{MO} in level flight but it is possible in a dive. At low altitudes the V_{MO} is given at constant equivalent airspeed (EAS) (in this example 330 kt until it reaches 0.81 Mach at around 27 000 ft altitude) when the structural speed limit is below high-speed Mach buffet limit. At low altitudes, a civil aircraft has no operational role to operate at such a high speed. At cruise altitude, the aircraft is cruise thrust limited (Figure 17.1) and stays below high-speed Mach buffet limit. Military aircraft operate differently but have to fly below their structural limits as well as avoid buffeting.

The buffet boundary for each class of aircraft has to be established for the full flight envelope. At the design stage, simulation analyses can give a preliminary indication of buffet boundary, followed by wind-tunnel tests to serve as guidelines for the pilot to establish the buffet boundary through flight tests for operational usage. Typically, combat aircraft are allowed to tolerate higher levels of buffet compared to transport aircraft for operational reasons. The aircraft design must be free from Mach buffet within the operational envelope.

In the Federal Aviation Administration (FAA) regulations, establishing buffet boundary is a flight test task. The pass/fail assessment is made by the approved test pilot and is solely the result of the pilot's experience and judgement. These data are proprietary information kept commercial in confidence. It is difficult to find examples to give a worked-out example. The readers may contact aircraft manufacturers to get a proper buffet boundary with explanations. An Internet search for the FAA website graphs may give some indications.

17.2.2 Flutter

This is the vibration of the structure that may cause deformation; primarily the wing, but also any other component depending on its stiffness. At transonic speed, the load on the aircraft is high while the shock-boundary layer interaction could result in an unsteady flow causing vibration of the wing, for example. The interaction between aerodynamic forces and structural stiffness is the source of flutter. A weak structure enters into flutter; in fact, if it is too weak, flutter could happen at any speed because the deformation would initiate the unsteady flow. If it is in resonance, then it could be catastrophic – such failures have occurred. Flutter is an aeroelastic phenomenon, while buffet is not. Aircraft flutter is dangerous and must be avoided. At design stage, simulation analyses can give preliminary indication of flutter speed, followed by wind-tunnel tests to improve accuracy followed by flight test substantiation. Being dangerous, flight tests are carried out under utmost caution.

17.3   Flight Manoeuvres

Although throttle-dependent linear acceleration would generate flight load in the direction of the flight path, pilot-induced control manoeuvres could generate the extreme flight loads that may be aggravated by inadvertent atmospheric conditions. Aircraft weight is primarily determined by the air load generated by manoeuvres in the pitch plane. The associated *V-n* diagram described in Section 17.7 provides useful information for aircraft performance analysis. Section 3.6.1 describes the six degrees of freedom for aircraft motion – three linear and three angular. Given here are the three Cartesian coordinate planes of interest.

17.3.1   Pitch-Plane (*x–z*-Plane) Manoeuvre: (Elevator/Canard Induced)

The pitch plane is the symmetrical vertical plane (i.e. *x–z*-plane) in which the elevator/canard-induced motion occurs with angular velocity, q, about the *y*-axis, in addition to the linear velocities in the *x–z*-plane. Changes in the pitch angle due to angular velocity q results in changes in C_L. The most severe aerodynamic loading occurs in this plane.

17.3.2   Roll-Plane (*y–z*-Plane) Manoeuvre: (Aileron Induced)

The aileron-induced motion generates the roll manoeuvre with angular velocity, p, about the *x*-axis, in addition to velocities in the *y–z*-plane. Roll-plane loading is not discussed here.

17.3.3   Yaw-Plane (*y–x*-Plane) Manoeuvre: (Rudder Induced)

The rudder-induced motion generates the yaw (coupled with the roll) manoeuvre with angular velocity, r, about the *z*-axis, in addition to the linear velocities in the *y–x*-plane. Aerodynamic loading of an aircraft due to yaw is also necessary for structural design. It is not discussed here.

17.4   Aircraft Loads

An aircraft is subject to load at any time. The simplest case is an aircraft stationary on the ground experiencing its own weight. Under heavy landing, an aircraft can experience severe loading and there have been cases of structural collapse. Most of these accidents showed failure of the undercarriage, but breaking of the fuselage also has occurred.

17.4.1.1   On-Ground

Loads on the ground are taken up by the undercarriage and then transmitted to the aircraft main structure. Landing gear loads are dependent on the specification of V_{stall}, the maximum allowable sink speed rate at landing and on the maximum takeoff mass (MTOM). The topic is dealt with in greater detail in Chapter 9 concerning undercarriage layout for conceptual study.

17.4.1.2 In Flight

In flight, aircraft loading varies with manoeuvres and/or when gusts are encountered. Early designs resulted in many structural failures in operation. In-flight loading in the pitch plane is the main issue considered in this chapter. The aircraft structure must be strong enough at every point to withstand the pressure field around the aircraft, along with the inertial loads generated by flight manoeuvres. The V-n diagram is the standard way to represent the most severe flight loads that occur in the pitch plane (i.e. the x–z plane).

Civil aircraft designs have conservative limits; there are special considerations for the aerobatic category of aircraft. Military aircraft have higher limits for hard manoeuvres and there is no guarantee that, under threat, a pilot would be able to adhere to the regulations. Survivability requires widening the design limits and strict maintenance routines to ensure structural integrity. Typical human limits are currently taken at $9g$ in sustained manoeuvres and can reach $12g$ for instantaneous loading. Continuous monitoring of the statistical database retrieved from aircraft-mounted black boxes provides feedback to the next generation of aircraft design or at midlife modifications. A g-meter in the flight deck records the g-force and a second needle remains at the maximum g reached in the sortie. If the prescribed limit is exceeded, then the aircraft must be grounded for a major inspection. An important aspect of design is to know what could happen at the extreme points of the flight envelope (i.e. the V-n diagram).

17.5   Theory and Definitions

In steady, level flight, an aircraft is in equilibrium; that is, the lift, L, equals the aircraft weight, W, and the thrust, T, equals drag, D. The wing produces all the lift with a spanwise distribution (see Section 5.4.9). In equation form, for steady level flight:

$$L = W \text{ and } T = D \qquad (17.1)$$

17.5.1   Load Factor, n

Newton's law states that change from an equilibrium state requires an additional applied force, that would associate with some form of acceleration, a. When applied in the pitch plane to increase the angle of incidence, α, initiated by rotation of the aircraft, the additional force appears as an increment in lift, ΔL, resulting in gain in height (Figure 17.2).

From Newton's law,

$$\Delta L = \text{acceleration} \times \text{mass} = a \times W/g \qquad (17.2)$$

The resultant force equilibrium gives

$$L' = L + \Delta L = W + a \times W/g = W(1 + a/g) \qquad (17.3)$$

Load factor, n, is defined as

$$n = (1 + a/g) = L/W + \Delta L/W = 1 + \Delta L/W \qquad (17.4)$$

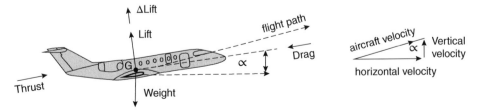

Figure 17.2 Equilibrium flight.

The load factor, *n*, indicates the increase in force contributed by the centrifugal acceleration, *a*. The load factor, $n = 2$, indicates a two-fold increase in weight; that is, a 90-kg person would experience a 180-kg weight. The load factor, *n*, is loosely termed the *g-load*; in this example, it is the 2-*g* load. Unaccelerated normal force on a body is its weight.

A high *g*-load damages the human body, with the human limits of the instantaneous *g*-load higher than for continuous *g*-loads. For a fighter pilot, the limit (i.e. continuous) is taken as 9-*g*; for the civil aerobatic category, it is 6-*g*. Negative *g*-loads are taken as half of the positive *g*-loads. Fighter pilots use pressure suits to control blood flow (i.e. delay blood starvation) to the brain to prevent 'blackouts'. A more inclined pilot seating position reduces the height of the carotid arteries to the brain, providing an additional margin on the *g*-load that causes a blackout.

Because they are associated with pitch-plane manoeuvres, pitch changes are related to changes in the angle of attack, *α*, and the velocity, *V*. Hence, there is variation in C_L, up to its limit of C_{Lmax}, in both the positive and negative sides of the wing incidence to airflow. The relationship is represented in a *V-n* diagram (Figure 17.3). Atmospheric disturbances are natural causes that appear as a gust load from any direction. Aircraft must be designed to withstand this unavoidable situation up to a statistically determined point that would encompass almost all-weather flights, except extremely stormy conditions. Based on the sudden excess in loading that can occur, margins are built in; as explained in the next section.

17.6 Limits – Load and Speeds

Limit load is defined as the maximum load that an aircraft can be subjected to in its life cycle. Under the limit load, any deformation recovers to its original shape and would not affect structural integrity. Structural performance is defined in terms of stiffness and strength. Stiffness is related to flexibility and deformations and has implications for aeroelasticity and flutter. Strength concerns the loads that an aircraft structure is capable of carrying and is addressed within the context of the *V-n* diagram.

To ensure safety, a margin (factor) of 50% increase is enforced through regulations as a *factor of safety* to extend the limit load to the *ultimate load*. A flight load exceeding the limit load but within the ultimate load should not cause structural failure but could affect integrity with permanent deformation. Aircraft are equipped with *g*-meters to monitor the load factor – the *n* for each sortie – and, if exceeded, the airframe must be inspected at prescribed areas and maintained by prescribed schedules that may require replacement of structural components. For example, an aerobatic aircraft with a 6-*g*-limit load will have an ultimate load of 9 *g*. If an in-flight load exceeds 6 *g* (but is below 9 *g*), the aircraft may experience permanent deformation but should not experience structural failure. Above 9 *g*, the aircraft would most likely experience structural failure.

The factor of safety also covers inconsistencies in material properties and manufacturing deviations. However, aerodynamicists and stress engineers should calculate for load and component dimensions such that their errors do not erode the factor of safety. Geometric margins, for example, should be defined such that they add positively to the factor of safety.

> Ultimate load = factor of safety × limit load

For civil aircraft applications, the factor of safety equals 1.5 (FAR 23 (14CAR23) and FAR 25 (14CAR 25), Vol. 3).

17.6.1 Maximum Limit of Load Factor, *n*

This is the required manoeuvre load factor at all speeds up to V_C. (The next section defines speed limits.) Maximum elevator deflection at V_A and pitch rates from V_A to V_D must also be considered. Table 17.2 gives the *g*-limit of various aircraft classes.

Military Aircraft (Military Specification MIL-A-8860, MIL-A-8861 and MIL-A-8870)

Table 17.2 Typical *g*-loads for classes of aircraft.

Club flying	Sports aerobatic	Transport	Fighter	Bomber
+4 to −2	+6 to −3	3.8 to −2	+9 to −4.5	+3 to −1.5

In general, there is factor of safety = 1.5 but this can be modified through negotiation. The *g*-levels taken here are instantaneous limits based on typical human capability.

17.7 *V-n* Diagram

To introduce the *V-n* diagram, the relationship between load factor, *n*, and lift coefficient, C_L, must be understood. The fundament flight operational domain is termed the 'manoeuvre' envelope. The *V-n* envelope also includes the domain termed the 'gust envelope', covering the atmospheric disturbances from statistical weather data that are seen as all-weather conditions, except some extreme situations that must be avoided for operation. An aircraft must be able to perform safely with these two boundaries of speed and load it encounters.

Pitch-plane manoeuvres result in the full spectrum of angles of attack at all speeds within the prescribed boundaries of limit loads. Since C_L varies with changes in attitude and/or aircraft weight, the *V-n* diagram for each altitude and/or weight will be different. Typically, the specification is given at MTOM and sea level at the STD condition. The *V-n* diagram is established out at the specified weight and altitude for detailed analyses. The *V-n* diagram is constructed for the critical altitude, typically at 20 000 ft at its maximum weight at that altitude. In principle, it may be necessary to construct several *V-n* diagrams representing different altitudes.

Depending on the direction of pitch-control input, at any given aircraft speed, positive or negative angles of attack may result. The control input would reach either the C_{Lmax} or the maximum load factor *n*, whichever is the lower of the two. The higher the speed, the greater the load factor, *n*. Compressibility has an effect on the *V-n* diagram. This chapter explains the role of the *V-n* diagram to understand aircraft manoeuvre performance only in the pitch plane.

Figure 17.3 represents a typical *V-n* diagram showing varying speeds within the specified structural load limits. The figure illustrates the variation in load factor with airspeed for manoeuvres. Some points in a *V-n* diagram are of minor interest to configuration studies – for example, at the point *V* = 0 and *n* = 0 (e.g. at the top

Figure 17.3 A typical *V-n* diagram showing load and speed limits (pitch plane).

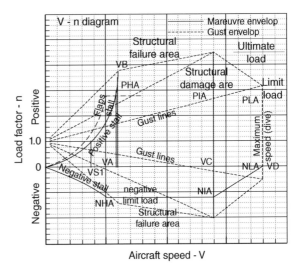

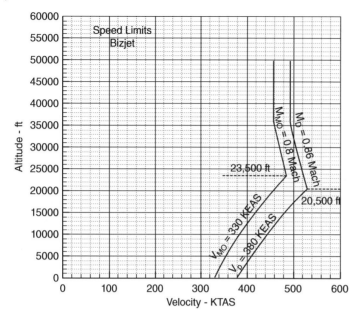

Figure 17.4 Bizjet speed limits.

of the vertical ascent just before the tail slide can occur). Inadvertent situations may take aircraft from within the limit-load boundaries to conditions of ultimate load boundaries. The points of interest are explained in the remainder of this section.

17.7.1 Speed Limits

The *V-n* diagram (Figure 17.4) described in the next section uses various speed limits as defined here [2]. More details of the speed definitions are given in Section 17.8.

V_S – Stalling speed at normal level flight.

V_A – Design manoeuvring speed. FAR (14CFR) specifies, $V_A = V_S \sqrt{n}$. Stalling speed at limit load. In a pitch manoeuvre, an aircraft stalls at a higher speed than v_S. In an accelerated manoeuvre of pitching up, the angle of attack, α, reduces and hence stalls at higher speeds. The tighter the pitch manoeuvre, the higher the stalling speed until it reaches V_A.

V_B – Stalling speed at maximum gust velocity. This is the design speed for maximum gust intensity. The speed should not be less than $V_B = V_S \sqrt{(n_{gust} @ V_C)}$.

V_C – Maximum design cruising (level) speed. Typically, $V_C \geq V_B + 43$ kt. This serves as safety margin (not dealt in this book) above V_B. V_C is the specified operating condition. See Table 17.3 for the relation between V_C and the positive/negative vertical gust velocity at altitudes.

V_D – maximum permissible speed (occurs in dive – sometimes known as placard speed) – structural limitation.

Table 17.3 FAR (14CFR) Part 217.341a Specified gust velocity (in EAS).

Altitude	20 000 ft and below	Altitudes 20 000 ft and above
V_B (rough air gust – maximum)	66 ft s^{-1}	Linear reduction to 38 ft s^{-1} at 50 000 ft
V_C (gust at max design speed)	50 ft s^{-1}	Linear reduction to 25 ft s^{-1} at 50 000 ft
V_D (gust at max dive speed)	25 ft s^{-1}	Linear reduction to 12.5 ft s^{-1} ft s^{-1} at 50 000 ft

V_F – Design flap speeds. (Separate *V-n* diagram. This is not dealt with in this book.)

An aircraft can fly below stall speed if it is in a manoeuvre that compensates loss of lift, or the aircraft attitude is below the maximum angle of attack, α_{max}, for stalling.

Bizjet example

$V_S = 82.6\ K_{EAS}$, $V_C = 328.6\ K_{EAS}$, $V_A = 180\ K_{EAS}$ (see next)
$V_{MO} = 330\ K_{EAS} < 24\,000$ ft altitude, 0.8 Mach $\geq 23\,500$ ft
$V_D = 380\ K_{EAS} < 21\,000$ ft altitude, 0.86 Mach $\geq 20\,500$ ft

17.7.2 Extreme Points of the *V-n* Diagram

Following the corner points of the flight envelope (Figure 17.5) is of interest for the stress and performance engineers. Beefing up of structures would establish an aircraft weight that must be predicted at the conceptual design phase. It also indicates the limitation points of the flight envelope in the pitch plane.

Figure 17.5 shows the various attitudes in pitch plane manoeuvres associated with the *V-n* diagram, each of which are explained next. Note that manoeuvre is a transient situation. The various positions of Figure 17.5 can occur under more than one possibility. Only the attitudes associated by the predominant cases in pitch plane are outlined. Negative *g* occurs when the manoeuvre force is directed in the opposite direction towards the pilot's head, irrespective of his/her orientation with respect to the Earth.

17.7.2.1 Positive Loads

This is when the aircraft (and occupants) experiences force greater than its normal weight. Aircraft stalls at manoeuvres reaching α_{max}. The following lists the various possibilities. In level flight at 1 *g*, aircraft angle of attack, α, increases with deceleration and reaches its maximum value, α_{max}; the aircraft would stall at a speed v_S.

i. *Positive High Angle of Attack (+PHA).* This occurs at pull up manoeuvre raising the aircraft nose in a high pulling *g*-force reaching the limit. The aircraft could stall if pulled harder. At a limit load of '*n*' the aircraft reaches +PHA at aircraft speeds of V_A.

ii. *Positive Intermediate Angle of Attack (+PIA).* This occurs at high-speed level flight when control is actuated to set wing incidence at an angle of attack. Noting that the aircraft has maximum operating speed limit of V_C, when +PIA reaches at the maximum limit load of *n* in manoeuvre – it is in transition or held at an intermediate level of manoeuvre.

iii. *Positive Low Angle of Attack (+PLA).* This occurs when the aircraft gains maximum allowable speed, sometimes in a shallow dive (dive speed, V_D) and then at very small elevator pull (low angle of attack) would hit the maximum limit load of *n*. Some high powered military aircraft can reach V_D at level flight. The higher the speed, the lower the angle of attack, α, would be to reach the limit load – at the highest speed, it would have +PLA.

17.7.2.2 Negative Loads

This occurs when an aircraft (and occupants) experiences forces less than its weight. In an extreme manoeuvre, in bunt (developing −ve *g* in a nose down curved trajectory), the centrifugal force pointing away from the centre of the Earth can cancel weight when the pilot feels weightless during the manoeuvre. The corner points follow the same logic of the positive load description except that the limit load of *n* is in the negative side,

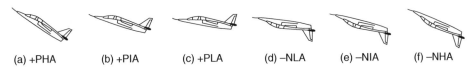

(a) +PHA (b) +PIA (c) +PLA (d) −NLA (e) −NIA (f) −NHA

Figure 17.5 Aircraft angles of attack in pitch plane manoeuvres (see text for the abbreviations).

which is lower due to not being in normal flight regimes. It can occur during aerobatic flight, in combat or in inadvertent situation through atmospheric gusts.

iv. *Negative High Angle of Attack (−NHA).* This is the inverted situation of case (i) explained before. With negative g it has to be in a manoeuvre.

v. *Negative Intermediate Angle of Attack (−NIA).* This is the inverted situation of case (i). The possibility of negative α was mentioned when the elevator is pushed down, called a 'bunting' manoeuvre. Negative α classically occurs at inverted flight at the highest design speed, V_C (coincided with PIA). When it reaches the maximum negative limit load of n, the aircraft takes NIA.

vi. *Negative Low Angle of Attack (−NLA).* This is the inverted situation of case (iii). At V_D, the aircraft should not exceed zero g.

17.7.3 Low-Speed Limit

At low speeds, the maximum load factor is constrained by the aircraft C_{Lmax}. The low-speed limit in a *V-n* diagram is established at the velocity at which the aircraft stalls in an acceleration flight load of n until it reaches the limit-load factor. At higher speeds, the manoeuvre load factor may be restricted to the limit-load factor, as specified by the regulatory agencies. The *V-n* diagram is constructed with V_{EAS} representing the altitude effects.

Let V_{S1} be the stalling speed at $1\,g$ then

$$V_{S1}^{2} = \left(\frac{1}{0.5\rho C_{Lmax}} \right) \left(\frac{W}{S_W} \right)$$

or

$$L = W = (0.5\rho V_{s1}^{2} S_W) C_{Lmax}$$

or

$$W = \frac{\rho C_{Lmax} S_W V_{s1}^{2}}{2} \tag{17.5}$$

This gives,

$$V_{S1} = \sqrt{\frac{2(W/S_W)}{\rho C_{Lmax}}} \tag{17.6}$$

In manoeuvre, let V_{Sn} be the stalling speed at load factor, n, then

$$nW = 0.5\rho V_{sn}^{2} S C_{Lmax} = \frac{\rho C_{Lmax} S_W V_{sn}^{2}}{2} \tag{17.7}$$

or

$$n = \frac{\rho C_{Lmax} S_W V_{sn}^{2}}{2W} = \frac{\rho C_{Lmax} V_{sn}^{2}}{2 \left(W/S_W \right)} \tag{17.8}$$

Use Figure 17.6 to make Mach number correction on C_{Lmax}.
This gives,

$$V_{Sn} = \sqrt{\frac{2n(W/S_W)}{\rho C_{Lmax}}} \tag{17.9}$$

Equating for W in Eqs. (17.6) and (17.7)

$$\frac{\rho C_{Lmax} S_W V_{s1}^{2}}{2} = \frac{\rho C_{Lmax} S_W V_{sn}^{2}}{2n}$$

or

$$V_{S1} = \frac{V_{sn}{}^2}{2n} \text{ that is, } V_{Sn} = V_{S1}\sqrt{n} \qquad (17.10)$$

In terms of Imperial units, Eq. 17.8 becomes,

$$n = \frac{1.688^2 \times \rho_0 C_{Lmax} V_{sn\_KEAS}{}^2}{2\left(W/S_W\right)} = \frac{C_{Lmax} V_{sn\_KEAS}{}^2}{296 \times \left(W/S_W\right)} \qquad (17.11)$$

V_A is the speed at which the positive stall and maximum load factor limits are simultaneously satisfied, that is, $V_A = V_{S1}\sqrt{n_{limit}}$.

The negative side of the boundary can be estimated in a similar fashion. For a cambered aerofoil, C_{Lmax} at a negative angle is less than at a positive angle.

V_C is the design cruise speed. For a transport aircraft, V_C must not be less than $V_B + 43$ kt. The JARs contain more precise definitions plus definitions of several other speeds. In civil aviation, for an aircraft weight less than 50 000 lbs, the maximum manoeuvre load factor is usually +2.17. The appropriate expression to calculate the load factor is given by:

$$n = 2.1 + 24\,000/(W + 10\,000) \text{ up to a maximum of 3.8 } (FAR\,(14CAR)\,217.337b) \qquad (17.12)$$

This is the required manoeuvre load factor at all speeds up to V_C, unless the maximum achievable load factor is limited by stall.

Within the limit load, the negative value of n is -1.0 at speeds up to V_C, decreasing linearly to 0 at V_D (Figure 17.3). Maximum elevator deflection at V_A and pitch rates from V_A to V_D must also be considered.

17.7.4 Manoeuvre Envelope Construction

Computational procedure:

1. Assume $V - K_{EAS}$ and determine the flight Mach number.
2. Apply the Mach number correction (experimental data, Figure 17.6) to revise the value of C_{Lmax}.
3. Compute $q = 0.50.5\rho V_{sn}{}^2$.
4. Estimate n using Eq. 17.8 (Imperial units).

17.7.5 High-Speed Limit

V_D is the maximum design speed. It is limited by the maximum dynamic pressure that the airframe can withstand. At high altitude, V_D may be limited by the onset of high-speed flutter.

Figure 17.6 Effect of Mach number – experimental data [1].

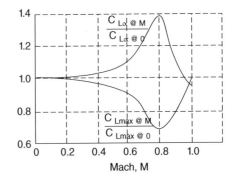

17.8 Gust Envelope

Encountering unpredictable atmospheric disturbance is unavoidable. Weather warnings help but full avoidance is not possible. Gusts can hit aircraft from any angle and the gust envelope can be shown in a separate set of diagrams. The most serious type is the vertical gust (Figure 17.3) affecting load factor n. Vertical gust increases the angle of attack, α, developing ΔL. Regulatory bodies have specified vertical gust rates that must be superimposed on the V-n diagrams to describe limits of operation. It is standard practice to combine manoeuvre and gust envelope in one diagram as shown in Figures 17.2 and 17.5.

The FAR (14CFR) gives a detailed description to establish required gust loads. To keep within the ultimate load, the limits of vertical gust speeds are reduced with increase of aircraft speed. Pilots should fly at a lower speed if high turbulence is encountered. Examining Eq. 17.5, it can be seen that aircraft with low wing loading (W/S_W), flying at high speeds are more affected by gust load. From statistical observations, the regulatory bodies (FAR (14CFR) Part 217.341a) have established maximum gust load at level flight as follows.

Linear interpolation is used to get appropriate velocities between 20 000 ft and 50 000 ft. Construction of V-n diagrams is relatively easy from the aircraft specifications. The corner points of V-n diagrams are specified. Computation to superimpose gust lines is more complex, for which FAR has given some semi-empirical relations.

Flight speed, V_B, is determined by the gust loads. It can be summarised as in Table 17.3.

V_B is the design speed for maximum gust intensity. This definition assumes that the aircraft is in steady level flight at speed V_B when it enters an idealised upward gust of air, which instantaneously increases the aircraft angle of attack and, hence, load factor. The increase in angle of attack must not stall the aircraft, that is, take it beyond the positive and negative stall boundaries.

Except in extreme weather conditions, this gust limit amounts to almost all-weather flying. In gust, the aircraft load may cross the limit load but must not exceed ultimate load as shown in Figure 17.3. If the aircraft crosses the limit load, then appropriate action through inspection is carried out.

Given next is a brief outline to construct a V-n diagram superimposed with gust load. The starting point for the gust envelope is when an aircraft is flying straight and level flight at $n = 1$. This is the starting point in developing the gust envelope. Actually, the gust reaches the aircraft in a velocity gradient and other aircraft characteristics; a factor is used to derive the gust load relationship [1]. A typical V-n diagram with the gust speeds intersecting the lines is illustrated in Figures 17.4 and 17.7. The V-n diagram is plotted with the x-axis in V_{EAS}. Separate V-n diagrams are required for changes in aircraft weight and/or operating altitude.

17.8.1 Gust Load Equations

For a more extensive treatment, the readers may refer to Ref. [1]. The simplest way to consider gust load is to assume that a gust hits sharply in a sudden vertical velocity. Let U_Z be the vertical velocity [1].

Increment of angle of attack $\Delta\alpha$ due to vertical velocity,

$$\Delta\alpha = U_Z/V \tag{17.13}$$

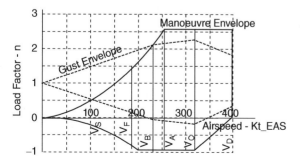

Figure 17.7 Composite *V-n* diagram (typical of the Boeing707 type – courtesy of the Boeing Co. [2]) (A clean configuration at cruise speed and altitude).

If V is in knots then $\Delta\alpha = U_Z/(1.688 \times V)$, U_Z remains in ft s^{-1} and U_{Ze} is EAS in ft s^{-1}.
(Note: U_Z and V are in the same units, typically kept in EAS.)

Considerable simplification is made here to obtain satisfactory estimation. The simplification is made in establishing the value of $C_{L\alpha\_aircraft}$ (in brief $C_{L\alpha\_ac}$). Typically, take $C_{L\alpha\_aircraft}$ to be 10% higher than $C_{L\alpha\_wing}$. In this sub-paragraph, $C_{L\alpha\_aircraft}$ is abbreviated to $C_{L\alpha\_ac}$.

Incremental lift, ΔL, caused by the gust

$$\Delta L = \Delta n W = 0.5 \times \rho \times V^2 \times \Delta C_L S_W = \frac{(\Delta\alpha C_{L\alpha\_aircraft}\rho V^2 S_W)}{2} \tag{17.14a}$$

Noting that $V_{EAS} = \sqrt{\sigma} V_{TAS}$, Eq. 17.14a becomes

$$\Delta L = \Delta n W = \frac{(\Delta\alpha C_{L\alpha\_ac}\rho_0 V_{EAS}^2 S_W)}{2} = \frac{(\Delta\alpha C_{L\alpha\_ac} V_{EAS}^2 S_W)}{2 \times 421.9} \tag{17.14b}$$

When V_{EAS} is in knots, then this equation reduces to

$$\Delta L = \Delta n W = \frac{(\Delta\alpha C_{L\alpha\_ac} \times 1.688^2 \times V_{KEAS}^2 S_W)}{2 \times 421.9} = \frac{(\Delta\alpha C_{L\alpha\_ac} \times V_{KEAS}^2 S_W)}{296} \tag{17.14c}$$

Modification to Eqs. (17.14a–c) is required as a vertical gust does not hit suddenly and sharply – it develops gradually. In addition to the gust gradient, other factors (the Kussner–Wagner effect) affecting the incremental load Δn are aircraft response characteristics (mass and geometry dependent) and lag in lift increments due to gradual increment of angle of attack $\Delta\alpha$ (aeroelastic effect). Incremental load Δn becomes less intense than compared to it encountering a sudden and sharp gust. An empirical factor K_g, known as 'gust alleviation factor' is applied and Eqs. (17.14a–c) becomes as follows.

Using Eq. 17.13 in Eq. 17.14a for $\Delta\alpha$ (velocities in ft s^{-1} true airspeed (TAS), i.e. U_Z in ft s^{-1}).
It becomes

$$\Delta L = \Delta n W = \frac{(\Delta\alpha C_{L\alpha\_ac}\rho V^2 S_W)}{2} = \frac{(U_Z C_{L\alpha\_ac}\rho V S_W)}{2} \tag{17.15a}$$

Eq. (17.14c), in K_{EAS} becomes,

$$\Delta L = \Delta n W = \frac{(U_{Ze} C_{L\alpha\_ac} \times \rho_0 \times 1.688 \times V_{KEAS} S_W)}{2 \times \sqrt{\sigma}}$$
$$= \frac{(U_{Ze} C_{L\alpha\_ac} \times V_{KEAS} S_W)}{421.94 \times 1.185 \times \sqrt{\sigma}} = \frac{(U_{Ze} C_{L\alpha\_ac} \times V_{KEAS} S_W)}{498 \times \sqrt{\sigma}} \tag{17.15b}$$

Use Eq. (17.15a), incremental Δn,

$$\Delta n = \frac{(K_g U_Z C_{L\alpha\_ac}\rho V)}{2\left(W/S_W\right)} \quad \text{where } K_g = (0.88\mu)/(17.3 + \mu) \tag{17.16a}$$

and where $\mu = 2W/(g \times MAC \times \rho \times C_{L\alpha\_ac} \times S_W)$

When V_{EAS} is in knots (V_{KEAS}) then the Eq. (17.15b) reduces to

$$\Delta n = \frac{(K_g U_{Ze} C_{L\alpha\_ac} \times V_{KEAS})}{498\left(W/S_W\right)} \tag{17.16b}$$

The higher the wing aspect ratio, the higher the $C_{L\alpha\_ac}$ and the higher the wing sweep, so the lower the $C_{L\alpha\_ac}$. This means that a straight wing aircraft with a high wing aspect ratio will be more sensitive to loading than a swept wing with low aspect ratio. Military aircraft are more tolerant to gust load compared to turboprop drive civil transport aircraft.

Military aircraft requirements are slightly different from civil designs.

17.8.2 Gust Envelope Construction

Computational procedure

1. Assume $V - K_{EAS}$ and determine the flight Mach number.
2. Apply the Mach number correction (Figure 17.6) to revise the value of C_{Lmax} and $C_{L\alpha}$.
3. Compute $q = 0.50.5\rho V_{sn}^2$.
4. Estimate n using Eq. 17.11 (Imperial units).

Bizjet

Bizjet specification at sea level and 20 680 lb. Manoeuvre limit load $= 3.0$ to -1 and gust load limit $= 4.5$ to -1.17. The n range will increase at lighter weight and higher altitudes.

Maximum design cruising (level) speed V_C at 20 000 ft altitude is the Mach 0.8 ($TAS = 767.3$ ft s^{-1}, $K_{EAS} = 328.6$). Maximum speed at dive is Mach 0.86 (380 K_{EAS}).

Example 17.1   Find the manoeuvre load n and the gust load, n_{gust}, with vertical velocity, $U_{ZE} = 66$ ft s^{-1} on an aircraft at 20 000 lb flying at 200 K_{EAS} (337.6 ft s^{-1}) at a 20 000 ft altitude ($\rho = 0.001\,26$ slug, $\sigma = 0.312$ and $\sqrt{\sigma} = 0.729$), STD day for the following. Maximum operating speed is Mach 0.80 ($V_{MMO} = 380$ K_{EAS}).
$W = 19\,900$ lb, $S_W = 323$ ft^2, ($W/S_W = 61.92$ lb ft^{-2}), $MAC = 7$ ft, $C_{Lmax} = 1.59$.

Solution:
At sea level STD day $V_{EAS} = V_{TAS}$.
$V_S = 82.6\,K_{EAS}$
$V_C = 328.6\,K_{EAS}$
$V_A = 180\,K_{EAS}$ (see next)
$V_D = 380\,K_{EAS}$

To estimate manoeuvre load, n
FAR (14CFR) 217.337b gives $n = 2.1 + 24\,000/(W + 10\,000) = 2.1 + 0.83 = 2.93$.
Take $n = 3.0$ for the additional safety.
Negative manoeuvre load can be estimated in a similar manner.
Use Eq. 17.11,
In Imperial system,

$$n = \frac{C_{Lmax} V_{sn\_KEAS}^2}{296 \times \left(W/S_W\right)},$$

Use Figure 17.6 to correct C_{Lmax} (at low-speed Mach number correction is not applied). To keep simple, the effect of relatively small amount of wing sweep on C_{Lmax} is ignored.
It gives,

$$n = \frac{C_{Lmax} V_{sn\_KEAS}^2}{296 \times \left(W/S_W\right)} = \frac{1.59 \times 200^2}{296 \times 61.92} = 63\,600/18\,328.32 = 3.47$$

Therefore, V_A is at 180 K_{EAS} to the limit of $n = 3.0$ (this is determined by plotting the locus of stall airspeed versus n).
To estimate gust load, n_{gust} Eq. (17.16b).
Incremental gust load Δn

Gust load $n_{gust} = (1 + \Delta n)$

NACA 65–410 given in Appendix F, take the lines for $Re = 9 \times 10^6$, as being close to the flight Reynolds number. It gives a 2D lift-curve slope, $dC_l/d\alpha = a_0 = (1.05 - 0.25)/8 = 0.11/$deg.

For a 3D wing,

$$(dC_L/d\alpha)_w = a = a_0/[1 + (57.3 \, a_0/e\pi AR)] = 0.11/[1 + (57.3 \times 0.11)/0.96 \times 3.14 \times 7.5)]$$

$$= 0.11/[1 + (7.3/22.6)] = 0.11/1.32 = 0.0833 \, (57.296 \times 0.833 = 4.77)$$

is for the wing.

For an aircraft increase by 10%, that is,

$$(dC_L/d\alpha)_a = 1.1 \times 4.77 = 17.25/\text{rad}$$

Mach number correction (Figure 17.6) at 200 K_{EAS}

$$C_{L\alpha\_ac} = 1.04 \times 17.25 = 17.46/\text{rad}$$

From Eq. (17.16a),

$$\mu = 2W/(g \times MAC \times \rho \times C_{L\alpha\_ac} \times S_W)$$

or

$$\mu = (2 \times 20\,000)/(32.2 \times 7 \times 0.001\,26 \times 17.46 \times 323) = 40\,000/500.9 = 79.86$$

and

$$K_g = (0.88\mu)/(17.3 + \mu) = (0.88 \times 79.86)/(17.3 + 79.86) = 70.28/817.16 = 0.825$$

Use Eq. (17.16a)

$$\Delta n = \frac{(K_g U_{Ze} C_{L\alpha\_ac} \times V_{KEAS})}{498 \left(W/S_w \right)} = \frac{(0.825 \times 66 \times 5.46 \times 200)}{498 \times 61.92} = 1.93$$

Gust load $n_{\text{gust}} = (1 + \Delta n) = 1 + 1.93 = 2.93$

The readers may construct the full flight envelope (see the problem assignment in Appendix E).

References

1 Niu, M. (1999). *Airframe Structural Design*. Hong Kong: Commlit Press Ltd.
2 Boeing Document, Jet Aircraft Performance Methods (company usage) n.d..

18

Stability Considerations Affecting Aircraft Design

18.1 Overview

Aircraft performance depends on aircraft stability characteristics. Aircraft stability analyses are subjects of their own but it is necessary to explain some of the pertinent points that affect aircraft performance. Substantiation of aircraft performance by itself would not prove sufficient if the aircraft stability characteristics do not offer satisfactory handling qualities and safety. These characteristics concern aircraft *flying qualities* (National Aeronautics and Space Administration (NASA) has codified the main flying qualities). Many good designs have had to go through considerable modifications to substantiate aircraft performance and handling criteria. The position of the aircraft centre of gravity (CG) plays an important role in inherent aircraft motion characteristics, which can be seen in how an aircraft performs. This chapter gives a qualitative understanding of the geometrical arrangements of aircraft components affecting aircraft performance associated with stability. Only the equations governing static stability are given to explain some of the important roles in aircraft performance.

This book is not on aircraft stability and control analyses. More details on the subject are given in [1–5]. New generation aircraft incorporate artificial stability technologies – for example, the use of *fly-by-wire* (FBW) technology that are seen as *control-configured vehicles* (CCVs) – is not dealt with in this book.

The chapter covers the following:

Section 18.2: Introduces Stability Considerations Affecting Aircraft Design
Section 18.3: Static and Dynamic Stability
Section 18.4: Theory
Section 18.5: Current Statistical Trends in Empennage Sizing Parameters
Section 18.6: Stick Force
Section 18.6: Inherent Aircraft Motions as Characteristics of Design
Section 18.7: Discusses Aircraft Spinning
Section 18.8: Design Consideration for Stability – Civil Aircraft
Section 18.9: Non-linear Effects – Military Aircraft
Section 18.10: Active Control Technology
Section 18.11: Summary of Design Consideration for Stability

Classwork content: There is little classroom work involved in this chapter.

18.2 Introduction

Inherent aircraft performance and stability are a result of CG location, wing and empennage sizing/shaping, fuselage and nacelle sizing/shaping, and their relative locations. This chapter highlights some of the lessons learned on how aircraft components relative to each other affect aircraft performance. Designers have to ensure that there is adequate trim authority (trim should not run out) at any condition.

Conceptual Aircraft Design: An Industrial Approach, First Edition. Ajoy Kumar Kundu, Mark A. Price and David Riordan.
© 2019 John Wiley & Sons Ltd. Published 2019 by John Wiley & Sons Ltd.

Pitching motion of aircraft is in the plane of aircraft symmetry (about the y-axis – elevator actuated) and is uncoupled with any other types of motion. Directional (about the z-axis – rudder actuated) and lateral (about the x-axis – aileron actuated) motions are not in the plane of symmetry. Activating any of the controls, for example, the rudder or ailerons, causes coupled aircraft motion in both the directional and lateral planes.

Flight tests reveal whether the aircraft satisfies the flying qualities and the safety considerations. Practically all projects require some sort of minor tailoring and/or rigging of control surfaces to improve the flying qualities as a consequence of flight tests.

18.3 Static and Dynamic Stability

It is pertinent here to briefly review the terms *static* and *dynamic stability*. Stability analyses examine what happens to an aircraft when it is subjected to forces and moments applied by a pilot and/or induced by external atmospheric disturbances. There are two types of stability, as follows:

i. *Static stability.* This is concerned with the instantaneous tendency of an aircraft when disturbed during equilibrium flight. The aircraft is statically stable if it has restoring moments when disturbed; that is, it shows a tendency to return to the original equilibrium state. However, this does not cover what happens in the due course of time. The recovery motion can overshoot into oscillation, which may not return to the original equilibrium flight.

ii. *Dynamic stability.* This is the time history of an aircraft response after it has been disturbed, which is a more complete picture of aircraft behaviour. A statically stable aircraft may not be dynamically stable, as explained in subsequent discussions. However, it is clear that a statically unstable aircraft also is dynamically unstable and a dynamically stable aircraft is also statically stable. Establishing static stability before dynamic stability is for procedural convenience.

Pitch-plane stability may be compared to a spring-mass system, as shown in Figure 18.1a. The oscillating characteristics are represented by the spring-mass system, with the resistance to the rate of oscillation as the damping force (i.e. proportional to pitch rate, $\dot{q}$) and the spring compression proportional to pitch angle θ. Figure 18.1b shows the various possibilities of the vibration modes. Stiffness is represented by the stability margin, which is the distance between the CG and the neutral point (NP). The higher the force required for deforming, the greater the stiffness. Damping results from the rate of change and is a measure of resistance (i.e. how fast the oscillation fades out); the higher the H-tail area, the more the damping effect. An aircraft only requires adequate stability; making it more stable than what is required poses other handling difficulties in the overall design.

The aircraft motion in 3D space is represented in the three planes of the Cartesian coordinate system (Section 3.6). Aircraft have six degrees of freedom of motion in 3D space. They are decomposed into three

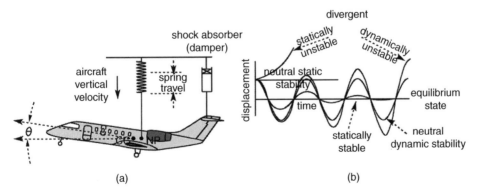

Figure 18.1 Aircraft stability compared with a spring-mass system. (a) Aircraft as spring-mass system. (b) Typical response characteristics.

planes; each exhibits its own stability characteristics, as listed here. The sign conventions associated with the pitch, yaw and roll stabilities need to be learned (they follow the right-handed rule). The brief discussion of the topic here only covers what is necessary in this chapter. The early stages of stability analyses are confined to small perturbations – that is, small changes in all flight parameters.

i. *Longitudinal Stability in the Pitch-Plane.* The pitch-plane is the x–z-plane of aircraft symmetry. The linear velocities are u along the x-axis and w along the z-axis. Angular velocity about the y-axis is $\dot{q}$, known as *pitching* (+ve nose up). Pilot-induced activation of the elevator changes the aircraft pitch. In the plane of symmetry, the aircraft motion is uncoupled; that is, motion is limited only to the pitch-plane.

ii. *Directional Stability in the Yaw-Plane.* The yaw-plane is the x–y-plane and is not in the aircraft plane of symmetry. Directional stability is also known as *weathercock stability* because of the parallel to a weather-cock. The linear velocities are u along the x-axis and v along the y-axis. Angular velocity about the z-axis is $\dot{r}$, known as *yawing* (+ve nose to the left). Yaw can be initiated by the rudder; however, pure yaw by the rudder alone is not possible because yaw is not in the plane of symmetry. Aircraft motion is coupled with motion in the other plane, the y–z-plane.

iii. *Lateral Stability in the Roll-Plane.* The roll-plane is the y–z-plane and also is not in the aircraft plane of symmetry. The linear velocities are v along the y-axis and w along the z-axis. Angular velocity about the x-axis is $\dot{p}$, known as *rolling* (+ve when the right wing drops). Rolling can be initiated by the aileron but a pure roll by the aileron alone is not possible because roll is associated with yaw. To have a pure rolling motion in the plane, the pilot must activate both the yaw and roll controls.

It is convenient now to explain the static and dynamic stability in the pitch-plane using diagrams. The pitching motion of an aircraft is in the plane of symmetry and is uncoupled; that is, motion is limited only to the pitch-plane. The static and dynamic behaviour in the other two planes has similar character-istics, but it is difficult to depict the coupled motion of yaw and roll. These are discussed separately in Sections 18.3.2 and 18.3.4.

Figure 18.2 depicts the stability characteristics of an aircraft in the pitch-plane, which provide the time history of aircraft motion after it is disturbed from an initial equilibrium. It shows that aircraft motion is in an equilibrium level flight – here, motion is invariant with time. Readers may examine what occurs when forces and moments are applied.

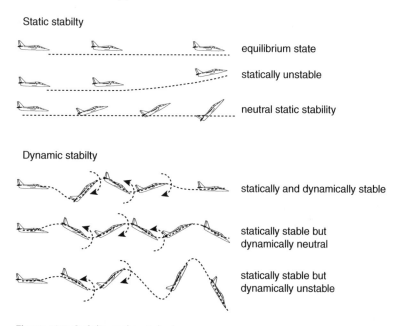

Static stabilty

equilibrium state

statically unstable

neutral static stability

Dynamic stabilty

statically and dynamically stable

statically stable but dynamically neutral

statically stable but dynamically unstable

Figure 18.2 Stability in the pitch-plane.

A statically and dynamically stable aircraft tends to return to its original state even when it oscillates about the original state. An aircraft becomes statically and dynamically unstable if the pitching motion diverges outward – it neither oscillates nor returns to the original state. The third diagram in Figure 18.2 provides an example of neutral static stability – in this case, the aircraft does not have a restoring moment. It remains where it was after the disturbance and requires an applied pilot effort to force it to return to the original state. The tendency of an aircraft to return to the original state is a good indication of what could happen in time: static stability makes it possible but does not guarantee that an aircraft will return to the equilibrium state.

As an example of dynamic stability, Figure 18.2 also shows the time history for when an aircraft returns to its original state after a few oscillations. The time taken to return to its original state is a measure of the aircraft's damping characteristics – the higher the damping, the faster the oscillations fade out. A statically stable aircraft showing a tendency to return to its original state can be dynamically unstable if the oscillation amplitude continues to increase, as shown in the last diagram of Figure 18.2. When the oscillations remain invariant to time, the aircraft is statically stable but dynamically neutral – it requires an application of force to return to the original state.

18.3.1 Longitudinal Stability – Pitch-Plane (Pitch Moment, *M*)

Figure 18.3 depicts the conditions for aircraft longitudinal static stability. In the pitch-plane, by definition, the angle of attack (AOA), α, is positive when an aircraft nose is above the direction of free-stream velocity.

A nose-up pitching moment is considered a positive. Static-stability criteria require that the pitching moment curve exhibit a negative slope, so that an increase in the AOA, α, causes a restoring negative (i.e. nose-down) pitching moment. At equilibrium, the pitching moment is equal to zero ($C_m = 0$) when it is in trimmed condition (α_{trim}). The higher the static margin (see Figure 18.3), the greater the slope of the curve (i.e. the greater is the restoring moment). Using the spring analogy, the stiffness is higher for the response.

The other requirement for static stability is that at a zero AOA, there should be a positive nose-up moment, providing an opportunity for equilibrium at a positive AOA ($+\alpha_{trim}$), typical in any normal flight segment.

18.3.2 Directional Stability – Yaw-Plane (Yaw Moment, *N*)

Directional stability can be compared to longitudinal stability but it occurs in the yaw-plane (i.e. the x–y-plane about the z-axis), as shown in Figure 18.4. By definition, the angle of sideslip, β, is positive when the free-stream

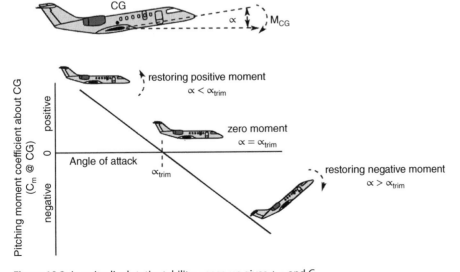

Figure 18.3 Longitudinal static stability – nose-up gives $+\alpha$ and C_m.

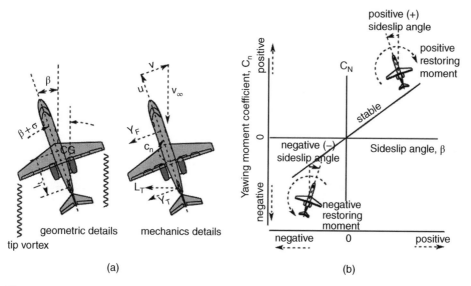

Figure 18.4 Direction stability (see Section 18.4 for more explanation). (a) Side slip angle β. (b) Directional stability criteria.

velocity vector, V, relative to aircraft is from the right (i.e. the aircraft nose is to the left of the velocity). V has component aircraft velocities u along the x-axis and v along the y-axis, subtending the sideslip angle $\beta = \tan^{-1}(v/u)$. The aerodynamics of the directional is dealt with in Section 6.7.3 and shown in Figure 6.11.

The V-tail is subjected to an angle of incidence $(\beta + \sigma)$, where σ is the sidewash angle generated by the wing vortices (like the downwash angle in longitudinal stability). Static-stability criteria require that an increase in the sideslip angle, β, should generate a restoring moment, C_n, that is positive when turning the nose to the right. The moment curve slope of C_n is positive for stability. At zero β, there is no yawing moment (i.e. $C_n = 0$).

Yaw motion invariably couples with roll motion because it is not in the plane of symmetry. In yaw, the windward wing works more to create a lift increase while the lift decreases on the other wing, thereby generating a rolling moment. Therefore, a pure yaw motion is achieved by the use of compensating, opposite ailerons. The use of an aileron is discussed in the next section.

18.3.3 Lateral Stability – Roll-Plane (Roll Moment, L)

Roll stability is more difficult to analyse compared to longitudinal and lateral stabilities. A banked aircraft attitude through pure roll keeps the aircraft motion in the plane of symmetry the plane of symmetry and does not provide any restoring moment. However, roll is always coupled with a yawed motion, as explained previously. As a roll is initiated, the sideslip velocity, v, is triggered by the weight component toward the down-wing side, as shown in Figure 18.5. Then (see the previous section), the sideslip angle is $\beta = \tan^{-1}(v/u)$. The positive angle of roll, φ, is when the right wing drops as shown in the figure (the aircraft is seen from the rear showing the V-tail but no windscreen). A positive roll angle generates a positive sideslip angle, β.

Recovery from a roll is possible as a result of the accompanying yaw (i.e. coupled motion) with the restoring moment contributed by increasing the lift acting on the wing that has dropped. Roll static-stability criteria require that an increase in the roll angle, φ, creates a restoring moment coefficient, C_l (not to be confused with the sectional aerofoil-lift coefficient). The restoring moment has a negative sign.

Having a coupled motion with the sideslip, Figure 18.6 shows that C_l is plotted against the sideslip angle β, not against the roll angle φ because it is β that generates the roll stability. The sign convention for restoring the rolling moment with respect to β must have negative $C_{l\beta}$; that is, with an increase of roll angle φ, the sideslip angle β increases to provide the restoring moment. An increase in β generates a restoring roll moment due to the dihedral. At zero φ, there is no β; hence, the zero rolling moment ($C_l = 0$).

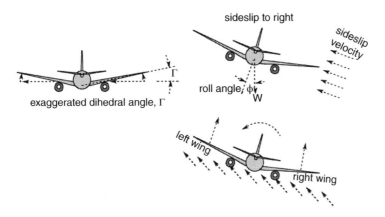

Figure 18.5 Lateral stability – rear view, right wing drop.

Figure 18.6 Lateral stability – fuselage contributions.

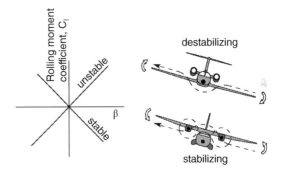

The wing dihedral angle, Γ, is one way to increase roll stability, as shown in Figure 18.5. The dropped wing has an airflow component from below the wing generating lift, while at the other side, the airflow component reduces the AOA (i.e. the lift reduction creates a restoring moment).

The position of the wing relative to the aircraft fuselage has a role in lateral stability, as shown in Figure 18.6. At yaw, the relative airflow about the low wing has a component that reduces the AOA; that is, the reduction of lift and the other side act in opposite ways: a destabilising effect that must be compensated for by the dihedral, as explained previously. Conversely, a high-wing aircraft has an inherent roll stability that acts opposite to a low wing design. If it has too much stability, then the anhedral (−ve dihedral) is required to compensate it. Many high-wing aircraft have an anhedral (e.g. the Harrier and the BAe RJ series).

An interesting situation occurs with a wing sweepback on a high-speed aircraft, as explained in Figure 18.7. At sideslip, the windward wing has an effectively reduced sweep; that is, the normal component of air velocity increases, creating a lift increment, whereas the leeward wing has an effectively increased sweep with a slower normal velocity component, thereby losing lift. This effect generates a rolling moment, which can be quite

Figure 18.7 Effect of wing sweep on roll stability.

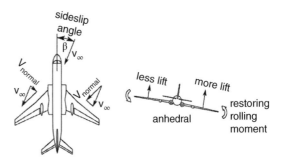

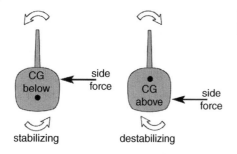

Figure 18.8 Vertical tail contribution to roll viewed from rear (modified from [6]).

powerful for high-swept wings; even for low wing aircraft, it may require some anhedral to reduce the excessive roll stability (i.e. stiffness) – especially for military aircraft, which require a quick response in a roll. Tu-104 in Figure 18.7 is a good example of a low wing military aircraft with a high sweep coupled with an anhedral.

The side force by the fuselage and V-tail contributes to the rolling moment [6], as shown in Figure 18.8. If the V-tail area is large and the fuselage has a relatively smaller side projection, then the aircraft CG is likely to be below the resultant side force, thus increasing the stability. Conversely, if the CG is above the side force, then there is a destabilising effect.

18.3.4   Summary of Forces, Moments and their Sign Conventions

The following information summarises forces and moments.

	Longitudinal Static stability	Directional Stability	Lateral Stability
Angles	Pitch angle α	Side slip angle β	Roll angle, Φ
Positive angle	nose up	nose to left	right wing down
Moment coefficient	C_m	C_n	C_l
Positive moment	Nose up	Nose left	Right wing down

18.4   Theory

Forces and moments affect aircraft motion. In a steady level flight (in equilibrium), the summation of all forces is zero; the same applies to the summation of moments. When not in equilibrium, the resultant forces and moments cause the aircraft to manoeuvre. The following sections provide the related equations for each of the three aircraft planes.

18.4.1   Pitch-Plane

In equilibrium, $\sum$Force = 0, when drag = thrust and lift = weight.

An imbalance in drag and thrust changes the aircraft speed until equilibrium is reached. Drag and thrust act nearly collinearly; if they did not, pitch trim would be required to balance out the small amount of pitching moment it can develop. The same is true with the wing lift and weight, which are rarely collinear and generate a pitching moment. This scenario also must be trimmed, with the resultant lift and weight acting collinearly.

Together in equilibrium, $\sum$moment = 0. Any imbalance results in an aircraft rotating about the y-axis. Figure 18.9 shows the generalised forces and moments (including the canard) that act in the pitch-plane. The forces are shown normal and parallel to the aircraft reference lines (i.e. body axes). Lift and drag are obtained by resolving the forces of Figure 18.9 into perpendicular and parallel directions to the free-stream velocity vector (i.e. aircraft velocity). The forces can be expressed as lift and drag coefficients, dividing by qS_w,

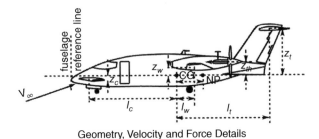

Geometry, Velocity and Force Details

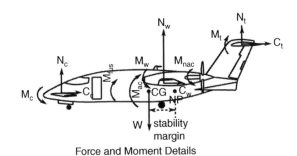

Force and Moment Details

Figure 18.9 Generalised force and moment in pitch-plane.

where q is the dynamic head and S_w is the wing reference area. Subscripts identify the contribution made by the respective components. The arrowhead directions of component moments are arbitrary – they must be assessed properly for the components. With its analysis, Figure 18.9 gives a good idea of where to place aircraft components relative to the aircraft CG and NP. The static margin is the distance between the NP and the CG.

The generalised expression for the moment equation can be written as given in Eq. (18.1). The equation sums up all the moments about the aircraft CG. In the trimmed condition the aircraft moment about the aircraft CG must be zero ($M_{ac\_cg} = 0$).

$$M_{ac\_cg} = (N_c \times l_c + C_c \times z_c + M_c)_{canard} + (N_w \times l_w + C_w \times z_w + M_w)_{wing} + (N_t \times l_{ht} + C_t \times z_t + M_t)_{tail}$$
$$+ M_{fus} + M_{nac} + (thrust \times z_{th} + Nac\_Drag \times z_{th}) + \text{any other item} \tag{18.1}$$

The forces and the moments of each of the components are estimated first. When assembled together as aircraft, each component would be influenced by the flow fields of the others (e.g. flow over horizontal tail is affected by the wing flow). Therefore, a correction factor, η, is applied. This is shown in Eq. (18.2) written in coefficient form, dividing by $qS_w c$, where q is the dynamic head, c as the wing mean aerodynamic chord (MAC) and S_w is the wing reference area. Subscripts identify the contribution made by the respective components. The component moments coefficients are initially computed as isolated bodies and then converted to the reference wing area.

$$C_{mcg} = [C_{Nc} (S_c/S_w)(l_c/c) + C_{Cc} (S_c/S_w)(z_c/c) + C_{mc}(S_c/S_w)]_{for\ canard} +$$
$$[C_{Nw} (l_a/c)\eta_w + C_{Cw} (z_a/c)\,\eta_w + C_{mw}\,\eta_w]_{for\ wing} +$$
$$[C_{Nt} (S_t/S_w)(l_{ht}/c)\eta_t + C_{Ct} (S_t/S_w)(z_t/c)\,\eta_t + C_{mt} (S_t/S_w)\eta_t]_{for\ tail} +$$
$$C_{mfus} + C_{nac} + (thrust \times z_{th} + Nac\_Drag \times z_{th})/qS_w c \tag{18.2}$$

where

i. $\eta\,(= q_i/q_\infty)$ represent the wake effect of lifting surfaces behind another lifting surface producing downwash, q_i is the incident dynamic head and q_∞ is the free-stream dynamic head.

   ii. The vertical distances (z) of each component could be above or below the CG, depending on configuration as described next.

     a. For fuselage-mounted engines, z_{th} is likely to be above the aircraft CG and its thrust would generate nose-down moment. Under-wing slung engines have z_{th} below the CG generating nose-up moment. For most military aircraft, the thrust line is very close to the CG and hence for a preliminary analysis, the z_{th} term can be ignored, that is, no moment is generated with thrust (unless vectored).

     b. Low wing drag below CG (z_a) will have a nose-down moment and vice versa for a high wing. Mid-wing positions have to be noted on which side of CG it lies and could have (z_a) small enough to be ignored for preliminary analysis.

     c. Position of H-tail shows the same effect as for the wing, but would be invariably above the CG. For a low H-tail, z_{th} can be ignored. In general, the drag generated by the H-tail is small and can be ignored. This is also true for a Tee-tail design.

     d. For the same reason, the contribution by the canard vertical distance, z_c can be ignored.

In summary, Eq. (18.2) can be further simplified by comparing the order of magnitude of the contributions of the various terms. At first, the following simplifications are suggested.

   i. The vertical 'z' distance of the canard and the wing from the CG is small. Therefore the terms with z_c and z_a can be dropped.

   ii. The canard and H-tail reference areas are much smaller that the wing reference area and their C ($\approx$drag) component force is less than a tenth of their lifting forces. Therefore, the terms with C_{Cc} (S_c/S_w) and C_{Ct} (S_t/S_w) can be dropped (even for a Tee-tail but it is better to check its overall contribution).

   iii. A high or low wing will have z_a with opposite signs. For a mid-wing, z_a may be small enough to be ignored.

The Eq. (18.2) can be simplified as given here:

$$C_{mcg} = [C_{Nc} (S_c/S_w)(l_c/c) + C_{mc}(S_c/S_w)] + [C_{Nw} (l_a/c)\,\eta_w + C_{mw}\,\eta_w] +$$
$$[C_{Nt} (S_t/S_w)(l_{ht}/c)\,\eta_t + C_{mt} (S_t/S_w)\,\eta_t] + C_{mfus} + C_{nac} + (thrust \times z_{th} + Nac\_Drag \times z_{th})/qS_w c \quad (18.3)$$

The conventional aircraft

The conventional aircraft is without canard. Then the Eq. (18.3) can be further simplified to Eq. (18.4). The conventional aircraft CG is possibly ahead of the wing MAC. In this case, the H-tail needs to have negative lift to trim the moment generated by the wing and body.

$$C_{mcg} = [C_{Nw} (l_a/c) + C_{mw}] + [C_{Nt} (S_t/S_w)(l_{ht}/c)\,\eta_t + C_{mt} (S_t/S_w)\,\eta_t] + C_{mfus} + C_{nac} +$$
$$(thrust \times z_{th} + Nac\_Drag \times z_{th})/qS_w c \quad (18.4)$$

   Normal forces can now be resolved in terms of lift and drag and for small angles of α, the cosine of the angle is taken as 1. The drag components of all the C_N are very small and can be neglected.

   Then the first term, $C_{Nw} (l_a/c) + C_{mw} \approx C_{Lw} (l_a/c) + C_{mw}$

   And the second term, $C_{Nt} (S_t/S_w)(l_t/c)\,\eta_t + C_{mt} (S_t/S_w)\,\eta_t \approx C_{Lt} (S_t/S_w)(l_{ht}/c)\,\eta_t + C_{mt} (S_t/S_w)\,\eta_t$

   Where $C_{mt} (S_t/S_w)\,\eta_t << C_{Lt} (S_t/S_w)(l_{ht}/c)\,\eta_t$, hence the moment contribution by the horizontal tail is represented as $C_{m\_HT} = C_{Lt} (S_H/S_w)(l_t/c)\,\eta_{HT}$.

   Then Eq. (18.4) is rewritten as (note the sign):

$$C_{mcg} = [C_{Lw} (l_a/c) + C_{mw}] + C_{m\_HT} + C_{mfus} + C_{nac} + (thrust \times z_{th} + Nac\_Drag \times z_{th})/qS_w c \quad (18.5)$$

where

$$C_{m\_HT} = -C_{LHT} \times [(l_t/S_{HT})/(S_w c)]\,\eta_{HT} = -C_{HT}\,\eta_{HT} C_{LHT}. \quad (18.6)$$

For conventional aircraft, C_{LHT} has a downward direction to keep the nose up, hence it carries negative sign. Here

$$C_{HT} = \text{horizontal tail volume coefficient} = (l_{ht}/S_{HT})/(S_w c) \quad (18.7)$$

(introduced in Section 6.6.2, now derived here)

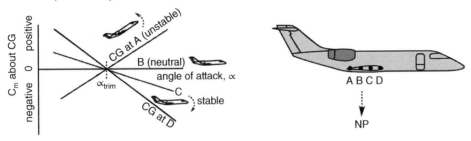

Figure 18.10 Pitch stability.

Then Eq. (18.5) without engines becomes as follows:

$$C_{mcg} = [C_{Lw} (l_a/c) + C_{mw}] - C_{m\_HT} + C_{mfus} \tag{18.8}$$

The second diagram in Figure 18.10 shows the stability effects of different CG positions on a conventional aircraft. The stability margin is the distance between the aircraft CG and the NP (i.e. a point through which the resultant force of the aircraft passes). When the CG is forward of the NP, then the static margin has a positive sign and the aircraft is statically stable. The stability increases as the CG moves farther ahead of the NP.

There is a convenient range from the CG margin in which the aircraft design exhibits the most favourable situation. In Figure 18.10, the position B is where the CG coincides with the NP and shows neutral stability (i.e. at a zero stability margin) – the aircraft can still be flown with the pilot's efforts controlling the aircraft attitude. In fact, an aircraft with relaxed stability can have a small negative margin that requires little force to make rapid manoeuvres – these aircraft invariably have a FBW control architecture (see Section 18.10) in which the aircraft is flown continuously controlled by a computer.

Engine thrust has a powerful effect on stability. If it is placed above and behind the CG, such as in an aft-fuselage-mounted nacelle, it causes an aircraft nose-down pitching moment with thrust application. For an under slung wing nacelle ahead of the CG, the pitching moment is with the aircraft nose up. It is advisable for the thrust line to be as close as possible to the aircraft CG (i.e. a small *ze* to keep the moment small). High-lift devices also affect aircraft pitching moments and it is better that these devices be a small arm's-length from the CG.

In summary, designers must carefully consider where to place components to minimise the pitching moment contribution, which must be balanced by the tail at the expense of some drag affecting aircraft performance – this is unavoidable but can be minimised.

18.4.2 Yaw-Plane

The equation of motion in the yaw-plane can be set up in a similar manner as in the pitch-plane. The weathercock stability of the vertical tail contributes to the restoring moment (also refer to Section 6.7.3).

Figure 18.4 depicts moments in the yaw-plane. In the diagram the aircraft is yawing to the left with positive yaw angle β. This would generate a destabilising moment by the fuselage with the moment ($N_F = Y_F \times l_f$), where Y_F is the resultant side force by the fuselage and l_f is the distance from the CG (see Figure 6.11). Contributions by wing, H-tail and nacelle are small (small projected areas and/or shielded by fuselage projected area). The restoring moment is positive when it tends to turn the nose to the right to realign with the airflow. The weathercock stability of the vertical tail causes the restoring moment ($N_{VT} = Y_T \times l_t$), where Y_T is the resultant side force on the V-tail (for small angles of ($\beta + \sigma$), it can be approximated as the lift generated by the V-tail, L_{VT}) and l_t is the distance of L_T from the CG. Therefore, the total aircraft yaw moment, N (conventional aircraft), is the summation of N_F and N_{VT} as given in Eq. (18.9).

$$N_{ac\_cg} = N_F + N_{VT} \tag{18.9}$$

At equilibrium flight,

$$N_{ac\_cg} = 0 \text{ that is, } N_{VT} = -N_F \tag{18.10}$$

In coefficient form, the fuselage contribution can be written as

$$C_{nf} = -k_n k_{Rl} N_F [(S_f l_f)/(S_w b)]\beta \tag{18.11}$$

where

k_n = an empirical wing-body interference factor
k_{Rl} = an empirical correction factor
S_f = the project side area of the fuselage
l_f = the fuselage length
b = wing span

In coefficient form, the V-tail contribution can be written as (L_{VT} in coefficient form is C_{LVT}).

$$C_{nVT} = [(l_{vt}/S_{VT})/(S_w c)] \eta_{VT} C_{LVT} = C_{VT} \eta_{VT} C_{LVT} \tag{18.12}$$

where,

$$C_{VT} = \text{vertical tail volume coefficient} = (l_{vt}/S_{VT})/(S_w c) \tag{18.13}$$

The Eq. (18.13) in coefficient form becomes:

$$C_{n\_cg} = -k_n k_{Rl} N_F [(S_f l_f)/(S_w b)]\beta + C_{VT} \eta_{VT} C_{LVT} \tag{18.14}$$

18.4.3   Roll-Plane

As explained earlier, the roll stability derives mainly from the following three aircraft features:

i. *Wing dihedral, Γ.*
   Slide slip angle β will increase AOA, α, on the windward wing, $\Delta\alpha = (V \sin \Gamma)/u$ generating $\Delta Lift$.
   For small dihedral and slide slip angle, $\beta = v/u$ that would approximate $\Delta\alpha = \beta\Gamma$.
   The restoring moment is the result of $\Delta Lift$ generated by $\Delta\alpha$. It is quite powerful – normally, for a low wing Γ it is between 1 and 3°, depending on the wing sweep. For straight wing aircraft, the maximum dihedral rarely exceeds 5°.
ii. *Wing position relative to the fuselage* (Figure 18.10).
   Section 18.3.3 explains the contribution to rolling moment caused by different wing positions relative to the fuselage. Semi-empirical methods are used to determine the extent rolling moment contribution.
iii. *Wing sweep at quarter chord, $\Lambda_{1/4}$.*
   The lift produced by a swept wing is a function of the component of velocity V_n, normal, to the $c_{1/4}$ line, that is, in steady rectilinear flight: $V_n = V\cos. \Lambda_{1/4}$
   When the aircraft side slips with angle, β, the component of velocity normal to the $c_{1/4}$ line becomes (small β) $V'_n = V\cos(\Lambda_{1/4} - \beta) = V(\cos\Lambda_{1/4} + \beta\sin\Lambda_{1/4}\beta)$.
   For the leeward wing $V'_{n\_lw} = V\cos(\Lambda_{1/4} + \beta) = V(\cos\Lambda_{1/4} - \beta\sin\Lambda_{1/4}\beta)$.
   Evidently, the windward wing has $V'_n > V_n$ and vice versa, hence, it will offer $\Delta Lift$ as a restoring moment in conjunction with lift decrease on the leeward wing.
   As $\Lambda_{1/4}$ increases, the restoring moment becomes powerful enough so that it has to be compensated for by the use of a wing anhedral.

18.5 Current Statistical Trends for Horizontal and Vertical Tail Coefficients

Figure 7.8 provides statistics for current aircraft (21 civil and 9 military aircraft types), plotted separately for the H-tail and the V-tail. It is advised that readers create separate plots to generate their own aircraft statistics for the particular aircraft class in which they are interested to obtain an appropriate average value. For civil aircraft designs, the typical H-tail area is about a quarter of the wing reference area. The V-tail area varies from 12% of the wing reference area, S_W, for large, long aircraft to 20% for smaller, short aircraft.

Military aircraft require more control authority for greater manoeuvrability and they have shorter tail arms that require larger tail areas. The H-tail area is typically about 30–40% of the wing reference area. The V-tail area varies from 20 to 25% of the wing reference area. Supersonic aircraft have a movable tail for control. If a V-tail is too large, then it is divided in two halves.

Modern aircraft with FBW technology can operate with more relaxed stability margins, especially for military aircraft designs; therefore, they require smaller empennage areas compared to older conventional designs (see Section 18.10).

18.6 Stick Force – Aircraft Control Surfaces and Trim Tabs

Control actuation require pilot effort; the bigger the aircraft, the bigger are the control surfaces, requiring corresponding higher forces that can induce pilot workload stress. This situation shows up from aircraft as small as 2000 lb weight. The design objective is therefore to alleviate pilot stick (pitch and roll) and pedal forces (yaw) to an qualitatively acceptable level stipulated by the certification agencies and accepted by qualitative flight tests.

It is desirable that the empennage is sized and rigged in a manner such that pilot will need to apply the least amount of force to actuate control deflection during the desired flight path; typically, for civil aircraft during the longest sector, that is at cruise. Military designs require special considerations discussed in Section 18.9. Empennage sizing must guarantee to offer adequate authority in the full flight envelope, with special attention to the critical manoeuvre. Figure 18.11 shows wing and empennage control surfaces and associated trim tabs.

Aerodynamic force (in this case the lift force) on movable control surface will give rise to a moment about any line normal to it. At the control hinge axis line, as it does not develop bending, it is associated with a hinge moment. Taking the simple case of an elevator as shown in Figure 18.12, a pilot initiates a moment though the control stick (pull in this case) to counter the control surface hinge moment to keep the surface deflected for the necessary control of aircraft pitch.

Here, only the concept of trimabilty is introduced to establish the incorporation of trim tabs on the control surfaces as shown in Figure 18.12 with a simplified linkage diagram. To relieve the pilot workload, especially when large force is required to control, designers have discovered various methods to achieve that. An aerodynamic trim tab deflects in the opposite direction to the elevator through a fixed linkage. This develops an opposite elevator hinge moment to relieve some pilot (pull) force.

18.6.1 Stick Force

To obtain the gearing ratio, the relation between stick force and elevator deflection, the following are considered. To simplify, ignore the presence of the trim as shown in Figure 18.12. Consider a pilot applying a pull force, P, moving a distance of $\Delta s = (l_s \times \delta_s)$ by rotating control stick of length, l_s, rotating through an angle of δ_s to deflect the elevator up (+ve direction) by δ_e. The aerodynamic force on the elevator makes a hinge moment, H_e, about its axis. Equating the moments for equilibrium in the stick fixed (held by the pilot) condition, the following relation is derived.

$$\sum M = 0 = P \times (l_s \times \delta_s) + H_e \times \delta_e \qquad (18.15)$$

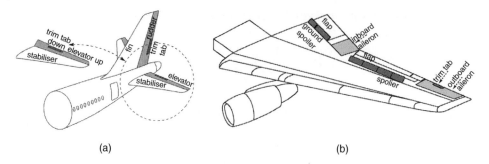

Figure 18.11 Aircraft control surfaces. (a) Empennage control. (b) Wing control.

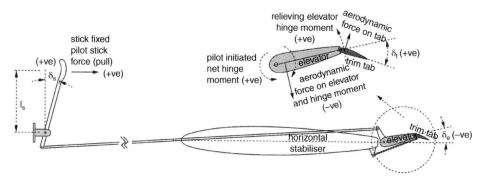

Figure 18.12 H-tail control surfaces (simplified linkage diagram).

or

$$P = GH_e,\tag{18.16}$$

where, $G = -[\delta_e/(l_s \times \delta_s)]$ = gearing ratio.

Expressing in terms of hinge moment coefficient, Eq. (6.2) is written as

$$P = G \times C_{he} \times (\tfrac{1}{2}\rho V^2) S_e c_e,\tag{18.17}$$

where S_e is the elevator reference planform area of the elevator and c_e is the MAC.

Equation (18.16) indicates that pull force, P, increases at the square of speed increase. Also, the bigger the aircraft, the bigger the elevator reference planform area, S_e, and the longer the mean chord length. In fact, a small aircraft the size of a Cessna 152 (TOGW = 1650 lb) will require a pull force that could make a pilot tired on frequent manoeuvres. Here, a movable the tab at the end of elevator can relieve pull force if it can be deflected in the opposite direction to elevator deflection as shown in Figure 18.12. In this case, the aerodynamic force on the tab is upward while the aerodynamic force on the elevator is downward to make aircraft pitch up, say for climbing. The tab linkage arm about the elevator hinge line is considerably longer than the elevator force moment arm about its hinge line. While it can neutralise the elevator moment, the elevator force about aircraft CG is sufficient to make aircraft respond to pilot action. Since the tab has capability to neutralise the elevator moment, it is known as the trim tab.

The analyses procedure adopted for elevator is also applicable to rudder and aileron hinge moment analyses and they also have trim tabs. Certification agencies have stipulated the limit of pilot force required in a harmonised manner for the three axes of controls. There many ways to relieve pilot forces, the advanced aircraft require power assisted controls. Other methods to relieve pilot force are by balancing control surfaces, for example with counter weights, aerodynamically with horn balance and so on. Trim design and control surface balancing are not tackled in this book.

Determining hinge moments is notoriously difficult. Empennage and other movable control design and sizing are topics of their own and is beyond the scope of this book. In the advent of Computational Fluid Dynamics (CFD) as a tool for empennage design, the past practice of using semi-empirical datasheets has gradually receded today. At the early stages of conceptual design phase, the use of statistical data of past empennage sizes in the class is a good starting point. This is adopted in this book. Eventually, qualitative assessments by several test-pilots will tailor the prototype aircraft empennage size. Typically, it requires fine-tuning, then undertaking major changes as CFD analyses followed by wind-tunnel test verifications give closer design suggestions.

18.7 Inherent Aircraft Motions as Characteristics of Design

Once an aircraft is built, its flying qualities (how aircraft performs) are the result of the effects of its mass (i.e. inertia), CG location, static margin, wing geometry, empennage areas and control areas. Flying qualities are based on a pilot's assessment of how an aircraft behaves under applied forces and moments. The level of ease or difficulty in controlling an aircraft is a subjective assessment by a pilot. In a marginal situation, recorded test data may satisfy airworthiness regulations yet may not prove satisfactory to the pilot. Typically, several pilots evaluate aircraft flying qualities to resolve any debatable points.

Practically all modern aircraft incorporate active control technology (FBW) to improve flying qualities. This is a routine design exercise and provides considerable advantage in overcoming any undesirable behaviour, which is automatically and continuously corrected. Described herein are six important flight dynamics of particular design interest. They are based on fixed responses associated with small disturbances, making the rigid-body aircraft motion linear. Military aircraft have additional considerations as a result of non-linear, hard manoeuvres, which are discussed in Section 18.9. The six flight dynamics (seen as aircraft performance) are as follows:

Short-period oscillation	Slow spiral
Phugoid motion (long-period oscillation)	Roll subsidence
Dutch Roll	Spin

18.7.1 Short-Period Oscillation and Phugoid Motion (Long-Period Oscillation)

The diagrams in Figure 18.13 show an exaggerated aircraft flight path (i.e. altitude changes in the pitch-plane). In the pitch-plane, there are two different types of aircraft dynamics that result from the damping experienced when an aircraft has a small perturbation. The two longitudinal modes of motion are as follows:

1. *Short-period oscillation* (SPO) is associated with pitch change (α change) in which the H-tail plane acts as a powerful damper (see Figure 18.1). If a disturbance (e.g. a sharp flick of the elevator and return) causes the aircraft to enter this mode, then recovery is also quick for a stable aircraft. The H-tail acts like an aerodynamic spring that naturally returns to equilibrium. The restoring moment comes from the force imbalance generated by the AOA, α, created by the disturbance. Damping (i.e. resistance to change) comes as a force generated by the tail plane, and the stiffness (i.e. force required) comes from the stability margin. The heavy damping of the H-tail resists changes to make a quick recovery.

The bottom diagram of a short period in Figure 18.13 plots the variation of the AOA, α, with time. All aircraft have a short-period mode and it is not problematic for pilots. A well-designed aircraft oscillatory motion is almost unnoticeable because it damps out in about one cycle. Although aircraft velocity is only slightly affected, the AOA, α, and the vertical height are related. Minimum α occurs at maximum vertical displacement and maximum α occurs at about the original equilibrium height. The damping action offered by the H-tail quickly smooths out the oscillation; that is, one oscillation takes a few seconds (typically, 1–5 s). The exact magnitude of the period depends on the size of the aircraft and its static margin.

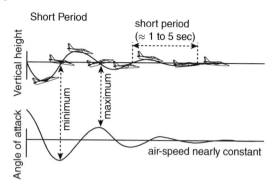

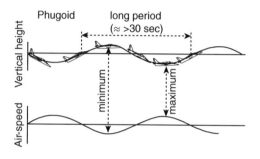

If the H-tail plane area is small, then damping is minimal and the aircraft requires more oscillations to recover.

2. *Phugoid motion* is the slow oscillatory aircraft motion in the pitch-plane, as shown in the bottom diagram in Figure 18.13. It is known as *long-period oscillation* (LPO) – the period can last from 30 s to more than 1 min. Typically, a pilot causes the LPO by a slow up and down movement of the elevator. In this case, the AOA, α, remains almost unchanged while in the oscillatory motion. The aircraft exchanges altitude gain (i.e. increases in potential energy, PE) for decreases in velocity (i.e. decreases in kinetic energy, *KE*). The phugoid motion has a long period, during which time the KE and PE exchange. Because there is practically no change in the AOA, α, the H-tail is insignificant in the spring-mass system. Here, another set of spring-mass is activated but is not shown in schematic form (it results from the aircraft configuration and inertia distribution – typically, it has low damping characteristics). These oscillations can continue for a considerable time and fade out comparatively slowly.

The frequency of a phugoid oscillation is inversely proportional to an aircraft's speed. Its damping also is inversely proportional to the aircraft L/D ratio. A high L/D ratio is a measure of aircraft performance efficiency. Reducing the L/D ratio to increase damping is not preferred; modern designs with a high L/D ratio incorporate automatic active control (e.g. FBW) dampers to minimise a pilot's workload. Conventional designs may have a dedicated automatic damper at a low cost. Automatic active control dampers are essential if the phugoid motion has undamped characteristics. All aircraft have an inherent phugoid motion. In general, the slow motion does not bother a pilot – it is easily controlled by attending to it early. The initial onset, because it is in slow motion, sometimes can escape a pilot's attention (particularly when instrument-flying), which requires corrective action and contributes to pilot fatigue.

18.7.2 Directional/Lateral Modes of Motion

Aircraft motion in the directional (i.e. yaw) and the lateral (i.e. roll) planes is coupled with sideslip and roll; therefore, it is convenient to address the lateral and directional stability together. These modes of motion are relatively complex in nature. Sections 23.143 to 23.181 in FAR 23 (14CFR23), address airworthiness aspects of these modes of motion. Spinning is a post-stall phenomenon and is discussed separately in Section 18.7.3.

Figure 18.14 Spiral mode of motion showing divergence [6].

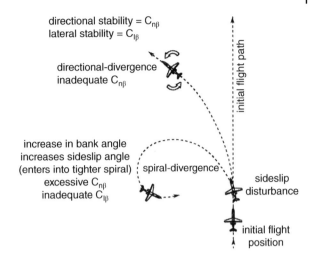

The four typical modes of motion are (i) directional divergence, (ii) spiral, (iii) Dutch roll and (iv) roll subsidence. The limiting situation of directional and lateral stability produces two types of motion. When yaw stability is less than roll stability, the aircraft can enter directional divergence. When roll stability is less than yaw stability, the aircraft can enter spiral divergence. Figure 18.14 shows the two extremes of directional and spiral divergence. The Dutch roll occurs along the straight initial path, as shown in Figure 18.15. The wing acts as a strong damper to the roll motion; its extent depends on the wing aspect ratio. A large V-tail is a strong damper to the yaw motion. It is important to understand the role of damping in stability. When configuring an aircraft, designers need to optimise the relationship between the wing and V-tail geometries. The four modes of motion are as follows:

i. *Directional Divergence.* This results from directional (i.e. yaw) instability. The fuselage is a destabilising body, and if an aircraft does not have a sufficiently large V-tail to provide stability, then sideslip increases accompanied by some roll, with the extent depending on the roll stability. The condition can continue until the aircraft is broadside to the relative wind, as shown in Figure 18.14 [6].

ii. *Spiral.* However, if the aircraft has a large V-tail with a high degree of directional (i.e. yaw) stability but is not very stable laterally (i.e. roll) (e.g. a low wing aircraft with no dihedral or sweep), then the aircraft banks as a result of rolling while side-slipping.

This is a non-oscillatory motion with characteristics that are determined by the balance of directional and lateral stability. In this case, when an aircraft is in a bank and side-slipping, the side force tends to turn the plane into the relative wind. However, the outer wing is travelling faster, generating more lift and the

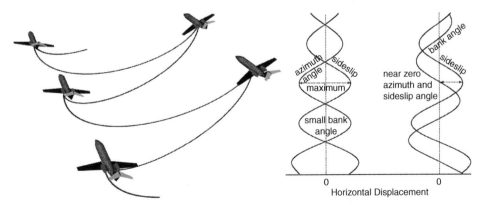

Figure 18.15 Dutch roll motion.

aircraft rolls to a still higher bank angle. If poor lateral stability is available to negate the roll, the bank angle increases. The aircraft continues to turn into the sideslip in an ever-increasing (i.e. tighter) steeper spiral, which is *spiral divergence* (Figure 18.14). Spiral divergence is strongly affected by C_{lr}.

The initiation of a spiral is typically very slow and is known as a *slow spiral*. The time taken to double the amplitude from the initial state is long – 20 s or more. The slow build-up of a spiral mode motion can cause high bank angles before a pilot notices an increase in the *g*-force. If a pilot does not notice the change in horizon, this motion may become dangerous. Night-flying without proper experience in instrument-flying has cost many lives due to spiral divergence. Trained pilots should not experience the spiral mode as dangerous – they would have adequate time to initiate recovery actions. A 747 has a non-oscillatory spiral mode that damps to half amplitude in 95 s under typical conditions; many other aircraft have unstable spiral modes that require occasional pilot input to maintain a proper heading.

iii. *Dutch Roll*. A Dutch roll is a combination of yawing and rolling motions, as shown in Figure 18.15. It can happen at any speed, developing from the use of the stick (i.e. aileron) and rudder, which generates a rolling action when in yaw. If a sideslip disturbance occurs, the aircraft yaws in one direction and, with strong roll stability, then rolls away in a countermotion. The aircraft 'wags its tail' from side to side, so to speak. The term *Dutch roll* derives from the rhythmic motion of Dutch ice skaters swinging their arms and bodies from side to side as they skate over wide frozen areas.

When an aircraft is disturbed in yaw, the V-tail performs a role analogous to the H-tail in SPO; that is, it generates both a restoring moment proportional to the yaw angle and a resisting, damping moment proportional to the rate of yaw. Thus, one component of the Dutch roll is a damped oscillation in yaw. However, lateral stability responds to the yaw angle and the yaw rate by rolling the wings of the aircraft. Hence, the second component of a Dutch roll is an oscillation in a roll. The Dutch roll period is short, in the order of a few seconds. In other words, the main contributors to the Dutch roll are two forms of static stability: the directional stability provided by the V-tail and the lateral stability provided by the effective dihedral and sweep of the wings – both forms offer damping. In response to an initial disturbance in a roll or yaw, the motion consists of a combined lateral-directional oscillation. The rolling and yawing frequencies are equal but slightly out of phase, with the roll motion leading the yawing motion.

Snaking is a pilot term for a Dutch roll, used particularly at approach and landing when a pilot has difficulty aligning with the runway using the rudder and ailerons. Automatic control using yaw dampers is useful in avoiding the snaking/Dutch roll. Today, all modern transport aircraft have some form of yaw damper. The FBW control architecture serves the purpose well.

All aircraft experience Dutch roll mode when the ratio of static directional stability and dihedral effect (i.e. roll stability) lies between the limiting conditions for spiral and directional divergences. A Dutch roll is acceptable as long as the damping is high; otherwise, it becomes undesirable. The characteristics of a Dutch roll and the slow spiral are both determined by the effects of directional and lateral stability; a compromise is usually required. Because the slow-spiral mode can be controlled relatively easily, slow-spiral stability is typically sacrificed to obtain satisfactory Dutch roll characteristics.

High directional stability ($C_{n\beta}$) tends to stabilise the Dutch roll mode but reduces the stability of the slow-spiral mode. Conversely, a large, effective dihedral (rolling moment due to sideslip, $C_{l\beta}$) stabilises the spiral mode but destabilises the Dutch roll motion. Because sweep produces an effective dihedral and because low wing aircraft often have a excessive dihedral to improve ground clearance, Dutch roll motions often are poorly damped on swept-wing aircraft.

iv. *Roll Subsidence*. The fourth lateral mode is also non-oscillatory. A pilot commands the roll rate by application of the aileron. Deflection of the ailerons generates a rolling moment, but the aircraft has a roll inertia and the roll rate builds up. Very quickly, a steady roll rate is achieved when the rolling moment generated by the ailerons is balanced by an equal and opposite moment proportional to the roll rate. When a pilot has achieved the desired bank angle, the ailerons are neutralised and the resisting rolling moment very rapidly damps out the roll rate. The damping effect of the wings is called *roll subsidence*.

18.7.3 Spinning

Spinning of an aircraft is a post-stall phenomenon. An aircraft stall occurs in the longitudinal plane; its nose drops as a result of loss of lift at stall. Unavoidable manufacturing asymmetry in geometry and/or asymmetric load application makes one wing stall before the other. This creates a rolling moment and causes an aircraft to spin around the vertical axis, following a helical trajectory while losing height – even though the elevator has maintained in an up position. The vertical velocity is relatively high (i.e. descent speed in the order of 30–60 m s^{-1}), which maintains adequate rudder authority, whereas the wings have stalled, losing aileron authority. Therefore, recovery from a spin is by the use of the rudder, provided it is not shielded by the H-tail (Section 6.7.6). Spinning is different than spiralling; it occurs in a helical path and not spiral. In a spiral motion, there is a large bank angle; in spinning, there is only a small bank angle. In a spiral, the aircraft velocity is sufficiently high and recovery is primarily achieved by using opposite ailerons. Spin recovery is achieved using the rudder and then the elevator. After straightening the aircraft with the rudder, the elevator authority is required to bring the aircraft nose down in order to gain speed and exit the stall.

There are two types of spin: a steep and a flat-pitch attitude of an aircraft. The type of spin depends on the aircraft inertia distribution. Most general aviation aircraft have a steep spin with the aircraft nose pointing down at a higher speed, making recovery easy – in fact, the best aircraft recover on their own when the controls are released (i.e. hands off). Conversely, the rudder authority in a flat spin may be low. A military aircraft with a wider inertia distribution can enter into a flat spin from which recovery is difficult and, in some cases, impossible. A flat spin for transport aircraft is unacceptable. Records show that the loss of aircraft in a flat spin is primarily from not having sufficient empennage authority in the post-stall wake of the wing.

The prediction of aircraft spinning characteristics is still not accurate. Although theories can establish the governing equations, theoretical calculations are not necessarily reliable because too many variables are involved that require accurate values not easily obtainable. Spin tunnels are used to predict spin characteristics, but the proper modelling on a small scale raises questions about its accuracy. In particular, the initiation of the spin (i.e. the throwing technique of the model into the tunnel) is a questionable art subjected to different techniques. On many occasions, spin-tunnel predictions did not agree with flight tests; there are only a few spin tunnels in the world.

The best method to evaluate aircraft spinning is in the flight test. This is a relatively dangerous task for which adequate safety measures are required. One safe method is to drop a large 'dummy' model from a flying 'mother' aircraft. The model has onboard, real-time instrumentation with remote-control activation. This is an expensive method. Another method is to use a drag chute as a safety measure during the flight test of the piloted aircraft. Spin tests are initiated at a high altitude; if a test pilot finds it difficult to recover, the drag chute is deployed to pull the aircraft out of a spin. The parachute is then jettisoned to resume flying. If a test pilot is under a high g-strain, the drag chute can be deployed by ground command, where the ground crew maintains real-time monitoring of the aircraft during the test. Some types of military aircraft may not recover from a spin once it has been established. If a pilot does not take corrective measures in the incipient stage, then ejection is the routine procedure. FBW technology avoids entering spins because air data recognise the incipient stage and automatic-recovery measures take place.

18.8 Design Considerations for Stability – Civil Aircraft

From the discussions on aircraft behaviour in small disturbances, it is clear that aircraft geometry and mass distribution both play important roles in the design of aircraft with satisfactory flying qualities. The position of CG is obtained by arranging aircraft components relative to each other to suit good in-flight static stability and on-ground stability for all operational envelopes. The full aircraft and its components moments are estimated semi-empirically (DATCOM, RAE sheets etc.) as soon as drawings are available and is followed through during the next phase; the prediction is improved through wind-tunnel tests and CFD analyses. At the conceptual design stage, the control area on wing and empennage (flap, aileron, rudder and elevator) are sized empirically from past experience (and DATCOM/RAE sheets). However, the CG position in relation to the aircraft NP would be tuned subsequently.

Chapter 6 described the aerodynamic design of the major aircraft components. Chapter 14 considered the sizing of the wing and empennage and also established the matched engine size. While statistics of past designs proved vital to configure the empennage, the placement of components with respect to each other was based on the designer's experience, which should form a starting point for the conceptual design phase.

The important points affecting aircraft configurations are reviewed here (with some repetition of what has been discussed earlier).

i. *Fuselage.* The fuselage has a destabilising effect – fuselage lift (though very small) and moment adds to instability – its minimisation is desired. As well as to keep cost down, the fuselage may be kept straight (with least camber). Mass distribution should keep inertia close to the fuselage centreline. A Blended Wing Body (BWB) will require special considerations.

The fuselage length and width are determined from the payload specification. Length to average diameter ratio for baseline aircraft version could be around 10. Its closure angles are important, especially the gradual closure of the aft end that should not have upsweep more than what is necessary, even for a rear-loading door arrangement that cannot escape from having an upsweep. Front closure is blunter and must provide adequate vision polar without too much upper profile curvature.

For a pressurised cabin, the cross-section should be maintained close to circular shape. Vertical elongation of cross-section should be kept to the minimum to accommodate below-floor space requirements. For small aircraft, fuselage depth elongation could be due to placement of the wing-box and for larger ones the container size. Care must be taken for the wing-box not to interfere with the interior cabin space. A generous fairing at wing-body junction and for fuselage-mounted undercarriage bulge is recommended. An unpressurised fuselage may have straight sides (rectangular cross-section) to reduce cost of production. In general, a rectangular fuselage cross-section is used in conjunction with a high wing. Undercarriages for high-wing aircraft will have fuselage bulges.

ii. *Wing.* Typically, an isolated wing has a destabilising effect unless it has a reflex at the trailing edge (tail is integrated into the wing such as in case of all wing aircraft, e.g. delta wing, BWB, etc.). The greater the wing camber, the more the destabilising effect. Optimising aerofoil with a high lift/drag ratio (L/D) and with the least C_{m_wing} is a difficult task not discussed here. Wind-tunnel test and CFD analysis are the ways to seek any compromise. It is assumed here that the aerodynamicists have found a suitable aerofoil with least destabilising moment for the best L/D ratio. The classroom worked-out example has taken a proven aerofoil from the proven National Advisory Committee for Aeronautics (NACA) series.

Sizing of aircraft, as described in Chapter 14, decides the wing reference area. Its structures philosophy settles the aspect ratio (maximising wing aspect ratio is the aim but at the conceptual design stage, it starts with improving upon the past statistics on which the designer can be confident on its structural integrity under load). The wing sweep is obtained from the design maximum cruise speed. It has been found that, in general, a wing taper ratio around 0.4–0.5 is satisfactory. Twist and dihedral at the conceptual stage are based on past experience and data sheets.

Positioning of the wing with respect to the fuselage depends on the mission role but is sometimes swayed by customer preference. A high or low wing position affects stability in opposite ways (Figure 18.8). The wing dihedral is established in conjunction with its sweep and position with respect to the fuselage. Typically, a high wing is given an anhedral and a low wing dihedral, which also helps ground clearance of the wing tips. In extreme design situations a low wing can have an anhedral (Figure 18.7) and a high wing a dihedral. There are case-based 'gull-wing' designs, typically for flying boats. Passenger-carrying aircraft are dominantly low-winged, but there is no reason why they should not have high wings, and there are a few such successful designs. Wing-mounted propeller-driven aircraft favour high wings for ground clearance, but there are low wing propeller-driven aircraft with longer undercarriage struts. Military transports invariably have high wings to facilitate rear loading of bulky items.

iii. *Nacelle.* The stability effects of a nacelle are similar to that of a fuselage. An isolated nacelle is destabilising but, when integrated with the aircraft, its relative position with respect to the aircraft CG would decide its effect on the aircraft, for example, an aft-mounted nacelle increases stability and forward-mounted nacelles on the wing decrease stability. The stability contribution of a nacelle may also be throttle-dependent (engine power effects).

The position of the nacelles on an aircraft is dictated by the aircraft size. The best position is on the wing, thereby providing bending relief in flight. A large forward overhang of a nacelle decreases air flow interference with wing. For small aircraft, the ground clearance mitigates against wing-mounting and for such aircraft nacelles are mounted on the aft fuselage. An over-wing nacelle mount for small aircraft is a possibility, a practice that is yet to gain ground. Even fuselage-mounted nacelles will have to adjust their position with respect to how close the vertical height from the aircraft CG is without having the jet efflux interfering with the empennage in close proximity.

iv. *Fuselage + Wing + Nacelle.* It is good practice to assemble these three components without the empennage to check the total moment in all three planes of reference. The CG position is established with the empennage installed, then it is removed for stability assessment. This helps when designing the empennage, as discussed in the next paragraph. Figure 12.10 shows the typical trends of pitching moments of the isolated components; together they will have destabilising effect, that is, have a positive slope. The aim is to minimise the slope, that is, to have the least destabilising moment.

Equation (18.2) of this chapter gives a good insight into the pitching moment contribution from the geometrical arrangement. It can be seen that minimising the vertical distance of the components from the aircraft CG also minimises their pitching moment contributions.

v. *Empennage.* The empennage configuration is of primary importance in aircraft design. The reference sizes are established through using statistical values of tail volume coefficients but the positioning and shaping of the empennage require considerable study. This now gives another opportunity to check whether the statistical values are adequate. Sweeping of the empennage increases the tail arm and could also enhance appearance; even some very low-speed small aircraft incorporate sweep. Chart 6.1 and Section 6.4 give some of the many possibilities for empennage configuration.

A conventional aircraft H-tail would have a negative camber, the extent depends on the moment produced by the aircraft's tailless configuration as described in the previous paragraphs. For large wing-mounted turbofan aircraft, the best position is to have a low H-tail mounted on the fuselage, which has a robust structure to take the tail load. A Tee-tail on swept vertical tail would increase the tail arm but should be avoided unless it is essential such as when dictated by the aft fuselage-mounted engines. Tee-tail drag is destabilising and would require larger areas if it is in the wing wake at near stalled attitudes. The V-tail would require a heavier structure to support the Tee-tail load. Small turbofan aircraft are constrained with aft-fuselage-mounted engines forcing the H-tail to be raised up anywhere from middle to top of the V-tail. Canard configuration offers a wider choice for aircraft CG location. In general, if an aircraft has all the three surfaces (canard, wing and H-tail) then all of these could provide lift with a positive camber of their sectional characteristics. It is possible that in the future civil aircraft of all sizes may feature canard designs.

Typically, a V-tail has a symmetric aerofoil but for propeller-driven airplanes it may be offset by a degree or two to counter the skewed flow around the fuselage (and also some gyroscopic torque).

This discussion forms the basis of the design of any other type of empennage configuration as outlined in Table 6.1. If the designer chooses a twin-boom fuselage then the empennage design must address the structural considerations of having twin booms (tailless aircraft are less manoeuvrable).

vi. *Undercarriage.* A retracted undercarriage obviously does not contribute to aerodynamic loads but when extended, it would generate a substantial drag creating a nose-down moment. To cope with this, there should be enough elevator nose-up authority at near-stall touch-down attitude, the most critical at the forward-most CG position. Designers must make sure that there is adequate trim authority (trim should not run-out) at this condition.

vii. *Use of any other surface.* It is clear by now how stability considerations affect aircraft configuration. Despite careful design when an aircraft prototype is flight tested it may show unsatisfactory *flying qualities*. Then additional surfaces, for example, a ventral fin, delta fin and so on, may be added to alleviate the problem.

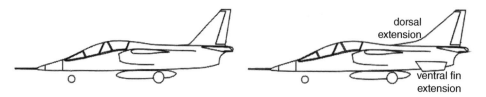

Figure 18.16 Aircraft modifications with additional surfaces – dorsal ventral.

Figure 18.16 gives two examples of such modifications. It is obviously desirable to avoid the need for additional surfaces, which add penalties in both weight and drag.

18.9   Military Aircraft – Non-Linear Effects

Military aircraft often perform extreme manoeuvres involving large disturbances, and hence require non-linear stability analyses. Military aircraft flying qualities are tackled in the MIL-STD-1797A, which supersedes the MIL-STD-8785. In studying the stability of military aircraft, similar design considerations as for civil aircraft (e.g. small disturbances involving linear treatment) are initially used, but additional features associated with large disturbances and involving non-linear treatment must also be considered as indicated next.

  i. Inertial pitch and yaw divergence in roll manoeuvres.
 ii. Aerodynamic yaw departure at high angles of attack.
iii. Wing rock.

These topics are beyond the scope of this book. Considerable data generation is required to initiate studies in these areas. Technology demonstrator aircraft would offer considerable insights to these problems. Even designing a technology demonstrator would require extensive wind-tunnel and CFD analyses at the conceptual stages, as the configuration is still unproven and there are little or no statistical data in use. Wind-tunnel test results may override CFD analyses but in principle they complement each other.

There are yet further problems arising from weapons release simultaneously or asymmetrically causing sudden CG shift that could severely affect aircraft stability. Provision has to be made for quick recovery by fuel transfer in a short time – these required microprocessor-based management incorporated in FBW technology. Stealth of aircraft gives additional constraints to configure aircraft. These give considerable challenges to tackle the stability issues. F117 Nighthawk is a classic example of such a consideration – it is an unstable aircraft and cannot fly without FBW.

A modern two-surface combat aircraft configuration is shown in Figure 18.17. A delta wing design with one large V-tail and a typical swing wing twin-tail configuration are shown.

Supersonic flights would require an all-moving H-tail as shown in the figure. Also, at high AOA it is immersed in its wing trailing vortex system and becomes ineffective. In one or two rare examples the fins (V-tail) are all moving surfaces. In many designs the all moving surfaces are split with some elevator and rudder authority primarily serving as redundancy to protect against failures. A single large V-tail is not desirable for high-performance combat aircraft. It cannot be canted and does not offer stealth. A tall single fin also would generate higher rolling moments in yaw, its stability would depend on the CG position.

The use of twin canted fins (strictly speaking no more a vertical tail) for military aircraft (Figure 5.31) is common nowadays for the following reasons. A twin canted tail is not a Vee-tail as there is a separate horizontal

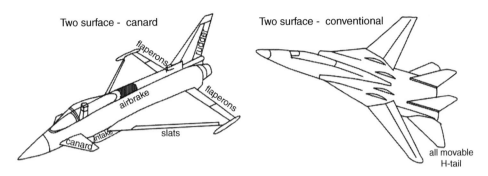

Figure 18.17 Typical modern fighter aircraft.

tail. (A Vee-tail has to combine the work of both pitch and directional control, and the required size becomes very large in order to achieve the required authority.)

 i. It is desirable that a vertical tail (fin) be canted for stealth reasons (deflects radar signals) and hence two such tails are required. A twin fin also halves the size easing structural considerations with very little weight penalty.
 ii. When the aircraft is yawed, an upwind canted fin is less effective than a downwind fin but together they provide the desired authority.
 iii. Twin-tail aircraft do not need to have a separate speed brake. To achieve the braking action, the two rudders are deflected in opposite directions, similar for the spoiler and flaps.

Actuators are designed to cope with the desired rate of control surface movements, which can be from $30°\text{ s}^{-1}$ to as high as $80°\text{ s}^{-1}$. Using FBW technology, the movement of leading-edge slats has a programmed relationship with the AOA. Aircraft roll rate could be as high as $200°\text{ s}^{-1}$.

At supersonic speed the aerodynamic centre moves aft, making the aircraft more stable. At low speed, as the aerodynamic centre moves forward, thrust vectoring in the pitch-plane ($\pm20°$) is helpful. Thrust vectoring is mainly used in low-speed extreme manoeuvres. At high Mach or high q (low altitude) conditions, AOA is low.

High-performance combat aircraft forebody shapes are important. A circular cross-section with high fineness ratio is less stable at high AOA. But with a small fineness ratio it may become acceptable. A vertically elongated cross-section is undesirable, while a horizontally elongated cross-section is better. A good solution is to have a Vee lower section (see F22 – Figure 5.31) for radar deflection and the upper part being horizontally elongated. The chine causes vortex generation at high AOA, which provides additional lift to assist the aircraft to manoeuvre at some unusual attitudes.

The early 1990s demonstrations by MIGs doing the spectacular 'Cobra' manoeuvre was initially seen more as stunt-flying by many experts, and yet today's designs may exceed such a capability, demonstrating combat potential in a twin-dome combat simulator and/or mock combat practise in flight. Fortunately, in most advanced countries the operators (air forces) and designers work together to understand and explore advanced capabilities.

Delta or all wing designs use wing reflex. If the V-tail is eliminated to avoid radar signature then splitting of the ailerons for directional control is necessary (as in the B2 bomber). Such features will invariably require FBW designs.

It is easy to see that advanced military aircraft configuration design is much more complex, and incorporate features not yet used in civil aircraft designs. Very little technical information is available in the public domain on these areas of complexity.

18.10 Active Control Technology (ACT) – Fly-by-Wire (FBW)

It is clear by now that stability considerations have an important part in the design of aircraft configurations. At the conceptual stage, although the related geometrical parameters were taken from statistical data of past designs and subsequently sized, this chapter gives a rationale on their role. It is also shown that to control the inherent aircraft motions, a *feed-back control* system such as a *stability augmentation system* (SAS – e.g. a yaw damper), *control augmentation system* (CAS) and so on, were routinely deployed for some time. In this last section of this chapter, the rationale continues with a brief discussion on how the *feed-back control* system has advanced into the latest technologies such as FBW, Fly-by-Light (FBL) and so on, known collectively as ACT. Today, practically all types of larger aircraft incorporate some form of ACT.

The advantages of FBW have been mentioned in various sections of this book. The concept is nothing new. FBW is basically a feed-back control system based on the use of digital data (Figure 18.18). The figure shows control of one axis and can be done for all the three axes. Earlier SAS and CAS had a mechanical linkage from the pilot to the controls. FBW does not have a direct mechanical linkage (hence the name). It permits the transmission of several digital signal sources through one communications system known as *multiplexing*. A microprocessor is in the loop that continuously processes air data (flight parameters) to keep the aircraft in a desired motion with or without pilot command. Aircraft controls laws (algorithm relating the pilot command to control surface demand and aircraft motion, height, speed – involves equations of motion, aircraft coefficients and stability parameters) are embedded into the computers to keep the aircraft within the permissible flight envelop. Under the command of a human pilot, the computer acts as a subservient flier. The computer continuously monitors the aircraft behaviour and acts accordingly, ensuring safety that the human pilot cannot match.

Figure 18.18 shows a schematic diagram of the FBW feed-back arrangement for pitch control [7]. The flight control computer takes the pilot's steering commands, which are compared with the necessary commands for the aircraft stability to ensure safety and the control surfaces are activated accordingly. The computers are constantly fed with air data, that is, aircraft's speed, altitude, attitude and so on. Built into the computer are the aircraft's limitations, which enable to calculate the optimum control surface movements. Steering commands are no longer linked mechanically from the cockpit to the control surface, but via electric wiring. FBW flight control systems seem the ideal technology to ensure safety and reduce pilot workload.

Because analogue point-to-point wire bundles are inefficient and cumbersome means of interconnecting the sensors, computers, actuators, indicators and other equipment onboard the modern military vehicle, a serial digital multiplex data bus was developed. MIL-STD-1553 (in use since 1983) defines all aspects of the bus (a subsystem of electrical lines for communication – named after electrical bus-bars), therefore, many groups working with the military have chosen to adopt it. The MIL-1553 multiplex data bus provides integrated, centralised system control and a standard interface for all equipment connected to the bus. The bus concept

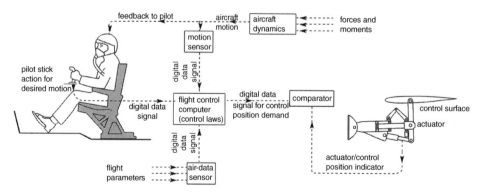

Figure 18.18 Schematic diagram of FBW [7].

provides a means by which all bus traffic is available to be accessed with a single connection for testing and interfacing with the system.

FBW reacts considerably faster than the conventional control system and does not have fatigue problems. One of the strong drivers for incorporating FBW in military aircraft design is the ability to operate at relaxed stability (even extending to slightly unstable condition) used for rapid manoeuvre (increased agility) as a result of very little stiffness in the system. The normal pilot finds it very difficult to control an unstable aircraft without assistance. A computer is needed and a regulator will supply the necessary stability. This system will not generate the natural stability of conventional aircraft, but will automatically trim the aircraft to the desired flight conditions. Progress in FBW systems has depended to a great extent on the progress of onboard computer power.

Evidently, an aircraft flying under relaxed stability using FBW would not have the same requirement in geometrical features in order to provide low stiffness and damping. Hence, stability and control surface sizing would be different from a conventional design. They would be smaller, hence are lighter and with less drag. This is what is meant by a CCV.

Stable designs already have a down-pitching force because of the position of the NP aft of the CG. Any balancing force would need to be generated by a larger downward lift of the horizontal tail. This, again, would decrease the maximum possible lift and increase the trim drag. In an unstable layout (CG moving aft) the elevator's lift is directed upwards to counterbalance the moment. This way, the aircraft's total lift is increased. The aircraft can therefore be designed smaller and lighter and still give the same performance. It shows that there is another benefit from the use of an unstable design: Along with the aircraft's increased agility, there is also a reduction in drag and weight.

The difference between conventional and a CCV design is shown in Figure 18.19 [8] for the longitudinal stability. The wing area and maximum take-off mass (MTOM) of the both designs were kept unchanged and the CCV design yielded in with smaller H-tail area with a larger CG range. The directional stability would exhibit similar gains with a smaller V-tail area, thereby further reducing operating empty mass (OEM) permitting a bigger payload.

The flight control computer takes the pilot's steering commands, which are compared to the commands necessary for aircraft stability to ensure safety and that control surfaces are activated accordingly. Air data are continually fed to the computers (i.e. speed, altitude and attitude). Built into the computer are an aircraft's limitations, which enable it to calculate the optimum control surface movements. Steering commands are no longer linked mechanically from the cockpit to the control surface but rather via electrical wiring. FBW flight control systems seem to be the ideal technology to ensure safety and reduce a pilot's workload.

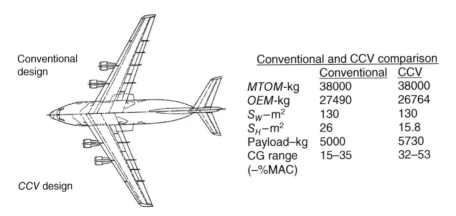

Conventional design

CCV design

Conventional and CCV comparison		
	Conventional	CCV
$MTOM$-kg	38000	38000
OEM-kg	27490	26764
S_W–m^2	130	130
S_H–m^2	26	15.8
Payload–kg	5000	5730
CG range (–%MAC)	15–35	32–53

Figure 18.19 Comparison between a conventional and a CCV design [8].

In summary, FBW offers considerable advantages as listed here:

i. System architecture is simple and flexible, though its design is a complex one.
ii. Consistent handling.
iii. Automatic stabilisation.
iv. Safe manoeuvring up to the envelope limits.
v. Ability to integrate with a wide range of designs, for example, slats, swing wing and so on.
vi. Ability to integrate with engine control through full authority digital electronic control (FADEC), thrust vector and so on.
vii. Use of side stick controller – gives space freedom in cockpit layout and weight saving.
viii. Can incorporate relaxed stability for rapid manoeuvre yet use smaller control surfaces.
ix. Permits complex configuration for stealth that may not be favourable to aerodynamic considerations leading into unstable aircraft like the F117 Nighthawk.
x. Digital data handling permits multiplexing that saves weight.
xi. Overall weight reduction.
xii. Permits standardisation.
xiii. Failure detection.
xiv. Fault isolation.
xv. Built-in tests and monitoring.

FBW has been in use for almost 50 years but its clear advantages were kept secret for military reasons for a considerable time. Early development at the pioneering stages progressed painfully with some mishaps. Nearly two decades later, civil aviation took bold steps and Aircraft Radio Inc. (ARINC) standards emerged to control FBW designs. The Airbus was the first to incorporate full FBW in a major project. The mid-sized A320 twinjet aircraft is the first commercial transport aircraft to incorporate full FBW without manual override. The Habsheim (26 June 1988) and Bangalore (14 February 1990 in front of the first author's residence) disasters did pose questions but today, practically all mid-sized and large transport aircraft incorporate some form of FBW technology.

The FBW system architecture has redundancies built in. During the 1980s, such systems had quadruple-redundant architecture in which each can work independent of the others. Nowadays, with improved reliability, triplex system (with voting and consolidation) dominates design. FBW could be applied to one, two or all three axes of control. The modern systems incorporate all the three axes of control. MIL-STD-1553 specifies that all devices in the system must be connected to a redundant pair of buses. This provides a second path for bus traffic should one of the buses be damaged. Signals are only allowed to appear on one of the two buses at a time. If a message cannot be completed on one bus, the bus controller may switch to the other bus. In some applications more than one bus may be implemented on a given vehicle.

To avoid electromagnetic interference, recently the use of fibre-optics for signalling using light has been developed. Aptly, it is called FBL and is guided by MIL-STD-1773.

In summary, FBW/FBL designs offer weight reduction with smaller wing, empennage, control surfaces, lesser cabling, elimination of mechanical linkages and so on. As a consequence, drag is reduction. In addition, it offers enhanced safety and reliability and improved failure detection.

18.11 Summary of Design Considerations for Stability

18.11.1 Civil Aircraft

Positioning of the wing relative to the fuselage depends on the mission role, but it is sometimes influenced by a customer's preference. A high- or low wing position affects stability in opposite ways (see Figure 18.6). The wing dihedral is established in conjunction with the sweep and position relative to the fuselage. Typically, a high-wing aircraft has an anhedral and a low wing aircraft has a dihedral, which also assist in ground clearance of the wing tips. In extreme design situations, a low wing aircraft can have an anhedral (see Figure 18.7) and a high-wing aircraft can have a dihedral. There are case-based 'gull-wing' designs, which are typically for 'flying boats'. Passenger-carrying aircraft are predominantly low-winged but there is no reason why they should not have high wings; a few successful designs exist. Wing-mounted, propeller-driven aircraft favour a high wing for ground clearance, but there are low wing, propeller-driven aircraft with longer undercarriage struts. Military transport aircraft invariably have a high wing to facilitate the rear loading of bulky items.

A conventional aircraft H-tail has a negative camber, the extent depending on the moment produced by an aircraft's tailless configuration, as described previously. For larger, wing-mounted turbofan aircraft, the best position is a low H-tail mounted on the fuselage, the robust structure of which can accommodate the tail load. A T-tail on a swept V-tail increases the tail arm but should be avoided unless it is essential, such as when dictated by an aft-fuselage-mounted engine. T-tail drag is destabilising and requires a larger area if it is in the wing wake at nearly stalled attitudes. The V-tail requires a heavier structure to support the T-tail load. Smaller turbofan aircraft are constrained with aft-fuselage-mounted engines, which force the H-tail to be raised up from the middle to the top of the V-tail. The canard configuration affords more choices for the aircraft CG location. In general, if an aircraft has all three surfaces (i.e. canard, wing and H-tail), then they can provide lift with a positive camber of their sectional characteristics. It is feasible that future civil aircraft designs of all sizes may feature a canard.

Typically, a V-tail has a symmetric aerofoil but for propeller-driven airplanes, it may be offset by 1 or 2° to counter the skewed flow around the fuselage (as well as gyroscopic torque).

18.11.2 Military Aircraft – Non-Linear Effects

As mentioned in Section 18.9, military aircraft often perform extreme manoeuvres involving large disturbances requiring non-linear stability analyses. The early 1990s saw demonstrations by MIGs doing spectacular 'Cobra' manoeuvres. Aircraft roll rate could be as high as $200° \text{ s}^{-1}$. Military aircraft flying qualities are tackled in MIL-STD-1797A that supersedes the MIL-STD-8785. In studying the stability of military aircraft, similar design considerations to civil aircraft (e.g. small disturbances involving linear treatment) are initially used, but additional features associated with large disturbances and involving non-linear treatment must also be considered as indicated. These topics are beyond the scope of this book.

References

1 Perkins, C.D. and Hege, R.E. (1949). *Airplane Performance, Stability and Control*. New York: Wiley.
2 Etkin, B. (1959). *Dynamics of Flight*. New York: Wiley.
3 Nelson, R.C. (1989). *Flight Stability and Automatic Control*. New York: McGraw Hill.
4 Talay, T.A. (1975). *NASA SP-367; Introduction to the Aerodynamics of Flight*. NASA.
5 McCormick, B.W. (1995). *Aerodynamics, Aeronautics and Flight Mechanics*. Wiley.
6 Centennial of Flight, essay available at: www.centennialofflight.gov/essay/Theories_of_Flight/Stability_II/TH27G16.htm (Retrieved Jun 2018).
7 Kundu, A.K. (2010). *Aircraft Design*. Cambridge University Press. ISBN: 978-0-521-88516-4.
8 Klug, H.G., *Transport Aircraft with Relax/Negative Longitudinal Stability – Results of a Design Study*, AGARD, CP-157, 1975.

19

Materials and Structures

19.1 Overview

Although aerodynamic considerations dominate the *concept finalisation* of a new aircraft, structures and their material selection are an integral part of conceptual design Phase I of the project to generate the aircraft's geometric shape, bearing in mind the need to minimise weight and cost of production without sacrificing structural integrity and operate safely. The underlying workload of structural engineers is stressful, bridging the gap between aerodynamicists and production planners to make the product right the first time (see the Six-Sigma approach to design – Chapter 20).

The essence of this chapter is to consider how the aircraft is to be constructed. That is, what materials should be used and how can they be assembled together to make the aircraft. This is very important because, even at this early stage in the design process, material and manufacturing considerations can significantly affect cost and weight, leading to a reconsideration of the initial concept. The core objective is to select material that gives superior strength-to-weight ratio at an affordable cost. Many decisions at this point can lock in cost and weight that is difficult to recover later in the design process.

This chapter covers:

Section 19.2: Introduction to Structures and Materials for Design
Section 19.3: Function of Structure
Section 19.4: Basic Definitions – Structures
Section 19.5: From Structure to Material
Section 19.6: Basic Definitions – Materials
Section 19.7: Important Considerations for Design
Section 19.8: Materials General Considerations
Section 19.9: Metals
Section 19.10: Wood and Fabric
Section 19.11: Composites
Section 19.12: Structural Considerations
Section 19.13: Rules of Thumb and Concept Checks

It is important to note that this is not a structural design or stress analysis book, so this chapter aims only to give pointers to key aspects, and direct readers to the right places to continue with the detailed design of the structure. Exposing the topics to the readers is an essential obligation.

19.2 Introduction

In the early days of aviation there was really no choice but to use an all wood construction or with fabric cover to wrap over wooden airframe serving as aerodynamic surface. It was effective as there are centuries of experience and knowledge of wood types, their properties and manufacturability. At that time, the available metals

Conceptual Aircraft Design: An Industrial Approach, First Edition. Ajoy Kumar Kundu, Mark A. Price and David Riordan.
© 2019 John Wiley & Sons Ltd. Published 2019 by John Wiley & Sons Ltd.

were relatively heavy and the light ones were soft and corrosive, making wood the preferred choice. But the properties of wood have limitations with it being anisotropic and without enough resistance to impact. Moreover, as a natural material there is significant variability, and quality control is also a major issue. Today, wood is no longer used except in the home-built aircraft category, primarily because wood is the easiest material to work with.

During the 1920s, the combination of progress in engine and aerodynamic technologies allowed aircraft speeds to exceed 200 mph. The corresponding structural loads demanded better materials. Technology changed in the 1930s when Durener Metallwerke, Germany, introduced duralumin, an alloy of aluminium with its higher strength-to-weight ratio, improved anti-corrosion properties and with isotropic properties. It followed rapidly with more variety of alloys with specific manufacturability, damage tolerance, crack propagation and anti-corrosive properties in the form of clad-sheets, rolled bars, ingots and so on. The introduction of metal also brought a new dimension in manufacturing philosophy. This progress in structures, aerodynamics and engine power paved the way for substantial gains in speed, altitude and manoeuvre performance. These improvements were seen, pre-eminently, in the Second World War designs such as the Supermarine Spitfire, the North American P-51, the Focke Wolfe 190, the Mitsubishi Jeero-Sen and so on. The Spitfire was perhaps the finest examples of these advances coming together in a highly innovative machine.

It was not then until the emergence of duralumin, now known as a 2000 series aluminium alloy, that wood had a worthy metallic competitor. Even then, it took the demands of war and mass production to tip the balance in favour of metal. But with materials science continually advancing, aluminium alloys have continued to improve and today's high end alloys have exceptional performance, far beyond those early days. The yield strength of 7075-T6 aluminium alloy is 67% greater than 2011-T6. As mentioned before, wood is no longer used except in home-built aircraft category, and only there primarily because wood is an easy material to work with and is accessible and cheap.

The last three decades have seen the appearance of non-metals, for example, fibreglass/epoxy, Kevlar/epoxy, graphite/epoxy and so on, in progressively increasing usage. These composite materials and are typically laid in layers of fabric and resin. Composites have better strength-to-weight ratios compared to aluminium alloys, but they have anisotropic properties. Since they are laid in moulds during fabrication of parts, difficult curvy 3D shapes can be produced with relative ease. But the quality of composite aircraft components is yet to reach the uniform high standards of metals. The immediate future will see yet more variety of composite material with metal embedded to get the best of both worlds. The Airbus 380, the Airbus 350 and the Boeing 787 are fine examples of how extensively composite material is used. Composite materials and their technology are evolving at a fast rate and the future will see greater variety of composite material with better properties and capabilities at a lower cost. The percentage use of composite material in aircraft will increase further.

The first use of composites was in secondary and tertiary structures where loads are low and failure will not result in catastrophe. But this is no longer the case with composites now extensively used in primary structures. Figure 19.1 gives the composite material usage in the Boeing767 (early 1980s) aircraft [1, 2]. As composite technology progresses, more and more composites are appearing in aircraft, particularly primary load bearing structures.

Table 19.1 compares the extent of increase of composite from an older B747 (1960s) to the relatively newer design of the B777 (1990s). The latest B787 has considerably higher percentage of composite usage at 50%.

Composite materials are becoming incorporated in increasing percentages by weight. There are few small aircraft made of all-composite material but their FAR 23/25 certification procedure is more cumbersome than that of metal construction. The early all-composite aircraft had to struggle to obtain flight-worthy certification due to not having sufficient data to substantiate performance claims. Military certification standards for aircraft structure are different. Their anisotropic properties make composites difficult to analyse and design with, however, they provide the opportunity to tailor the material and create an optimised structure. Recent advances in simulation and design methods makes this achievable. The future is likely to see them used to optimise weight and strength to a much greater extent. A great example of this highly sophisticated structure is the Eurofighter Typhoon, which has a tailored aeroelastic wing. The wing is design such that in flight, under load, it adopts the most aerodynamic shape it can.

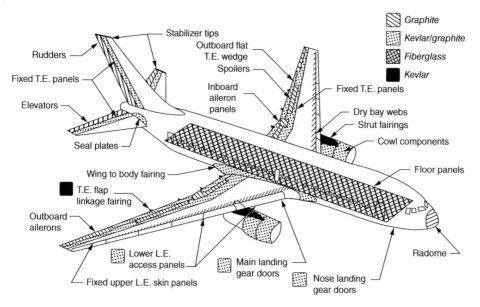

Figure 19.1 Composite material usage in B767. Source: Courtesy of Michael Niu.

Table 19.1 Percentage mass of types of material used in the aircraft structure.

Material	Boeing747	Boeing777	Boeing787
Aluminium alloys	81	70	20
Steel alloys	13	11	10
Titanium alloys	4	7	15
Composite (various types)	1	11	50
The rest	1	1	5

The newer military aircraft designs also use expensive exotic material (e.g. aluminium-lithium alloy, boron alloys etc.) that are yet to prove their cost effectiveness in commercial aircraft. More than half of the Eurofighter's structural mass is made of various types of composite materials. A fifth of the Eurofighter's structural mass is made of Al-Li alloy.

All vehicles that have to defy gravity must necessarily be weight conscious, forcing designers to choose lighter materials, or more precisely, those materials that give a better strength-to-weight ratio. Also implied are the questions of cost of raw material, cost of fabrication and their stability during usage. This section is devoted to making the reader aware that choosing the appropriate material is an involved topic and is an integral part of the study during the conceptual design phase. Aircraft weight and cost are affected by the choice of material and hence the aircraft performance and economy. The success and failure of a new aircraft design depends considerably on the choice of appropriate materials, especially when the number of available varieties is increasing. A choice made at this stage of the design process can embed unavoidable costs for the whole programme that may be impossible to recover at a later stage.

Although not within the scope of this book, it is interesting to note that with the advent and rapid progress made in 3D printing technologies, specific types of materials can be designed and developed to fabricate specific types of components, currently as smaller machine parts rather than larger aircraft structures.

19.3 Function of Structure – Loading

Before considering the definitions, it is worth recapping on what a structure is and the relevant function in aircraft.

The fundamental load pattern is of two types, (i) axial load normal to the surface of a structure member (can be tensile or compressive) and (ii) shear load applied tangentially along the surface of a structure member. Figure 19.2 depicts the loading types on a rectangular cross-section prismatic bar. Note that shear stress comes with its complementary stress [3–6].

Typically, a combination of all types of loads are subjected on a structure composed of many members. Bending of a beam is a good example, as shown later in Figure 19.13. The nature of loading on aircraft components are described in a simplistic manner in the following subsections (for details, see Refs [1, 3–6]).

19.3.1 Wing

The main function of the wing is to provide the lift needed to keep the aircraft airborne. Lift is a pressure force that acts over the surface of the wing, and if the resultant total force is sufficient the aircraft flies aloft. Structurally, the wing acts like a long beam, with the upward lift force being balanced, usually in the middle, by the fuselage and main payload. This is a self-equilibrating system, as there is no ground reaction or supports in that sense. Structurally, therefore, the wing transmits these external forces through the material structure to achieve equilibrium (Figure 19.3).

Of course, depending on the role of the aircraft, and the resulting configuration, the wing can also be required to transmit forces from wing mounted engines, undercarriage, fuel, hydraulic systems and so on. Although this makes for a more complex external loading picture the wing itself still acts like a beam a transmits the forces through the material.

19.3.2 Empennage/Tail

The empennage generates control forces for manoeuvring the aircraft. The individual elements of the empennage can be considered as wings in terms of the forces they transmit and carry within themselves. However, they also therefore transmit such forces through the adjacent structures, for example the fuselage. These forces can be very significant and can cause the structural designers difficulty in ensuring the integrity of the airframe.

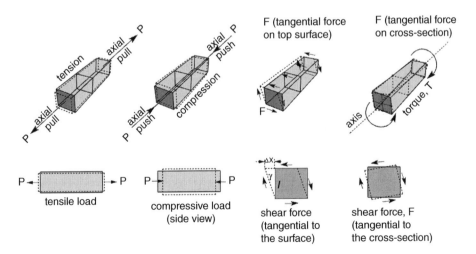

Figure 19.2 Load types on rectangular prismatic bar.

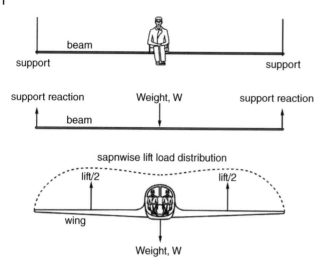

Figure 19.3 Conceptualising the aircraft wing as a beam. Structurally the system is like a playground swing, with the upward force provided by air loads (lift) rather than the ropes on the swing.

19.3.3 Fuselage

In the main category of civil aircraft, the primary function of the fuselage is to provide the space and location for payload and passengers. It must hold them and keep them safe and warm. Conceptually, the weight loading is similar to the wing. In this case the distribution is for the weight of the passengers, assumed to be evenly spread over the fuselage floor (Figure 19.4). Thus far, the fuselage is also like a beam, with a uniformly distributed load over its length, supported by the lift force of the wings, acting usually in the centre of the beam. In addition, the fuselage experiences bending moments from the kinetics of the lifting surfaces.

The fuselage, however, has an additional need where the aircraft flies in the stratosphere and that is to resist pressure. To keep passengers safe, aircraft are pressurised, normally to a minimum pressure equivalent to an altitude of 7000 ft. This additional pressure load provides many challenges for the structural designer, and has been the cause of major catastrophes such as the Comet disasters.

19.3.4 Undercarriage

While the aircraft is on the ground, the entire weight is transmitted through the undercarriage to the ground. In this case they act like stocky legs supporting the great weight of the aircraft. However, when the aircraft

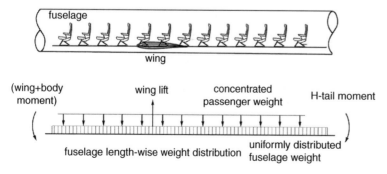

Figure 19.4 Like a beam, the fuselage with the passengers and payload provide the distributed load supported by the wing lift.

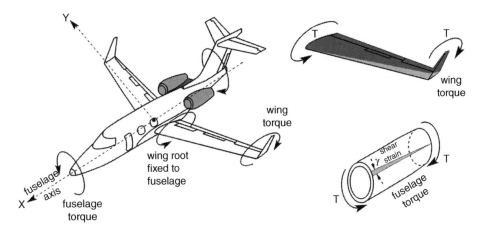

Figure 19.5 Torsional load on an aircraft.

is landing, the shock loads of striking the ground are absorbed through the undercarriage and subsequently transmitted through the wing or fuselage structure to which they are attached. Chapter 9 gives a fuller description of undercarriage design and the corresponding loading scenarios. Undercarriages are therefore not as beam-like as the main airframe elements.

19.3.5   Torsion

In addition to normal load pattern as briefly described before, aircraft components get subjected to torsional load; in simple terms, they get twisted in manoeuvres, asymmetric aerodynamic loadings and so on.

Twisting along the axis of a body occurs as a result of applied torque that produces shear force on the cross-sectional area normal to the axis of the body. Twist causes deformation and becomes significant along a long body, for example, fuselage, wing and so on, as shown in Figure 19.5.

19.4   Basic Definitions – Structures

As with any design problem it is important to have an understanding of the key parameters and definitions relevant to the problem in hand. Such basics are critical as the decisions made at any point in a shared collaborative venture should be equally understood by all team members. For example, the concept of stiffness applies to both material and the overall structure but is different for each. Material stiffness is a fundamental property of the material and so can only be changed by choosing a new material. The stiffness of the structure, however, depends on both the material and its shape. Therefore, structural stiffness can be significantly improved by modifying shape without needing to change the material. The designer needs always then to be clear which aspect is being considered and what the likely outcomes can be. As will be seen in this chapter, a change in structural stiffness can be effected by small or local design changes in a cost-effective manner, but a change in material, say from metal to composite, can significantly impact the entire design and the manufacturing systems. In this section key basic definitions and their relevance to design are given.

19.4.1   Structure

A *structure* is simply the assemblage of material components that transmits and resists the applied loads to keep the aerodynamic shape, keep the passengers safe and comfortable and ensure that the aircraft can fulfil its function over its life.

19.4.2   Load

The *load* is the force exerted on a surface or body. In the case of the whole aircraft, these are normally categorised as air or ground loads. As the aircraft is a dynamic system it encounters many loading scenarios during its operational life, from dynamic loads arising from manoeuvres to impact loads at landing, and inertial loads from releasing stores. In the context of structures, the definitions of various kinds of loads given Chapter 17 is recast here.

19.4.3   Limit Load

The *limit load* for an aircraft is the maximum load it is expected encounter once in its operational life.

19.4.4   Proof Load

The *proof load* is the maximum load the aircraft must withstand without detrimental distortion. Under proof loading conditions, the airframe must not distort permanently more than the specified proof strain (0.1 or 0.2%). This is particularly critical for structures such as flap tracks where any permanent distortion (plastic deformation) would impair functionality. Proof load is defined as:

proof load = limit load × proof factor

The *proof factor* is a factor of safety applied to allow for uncertainty in specifying the limit load. In civil aviation, as regulations are more conservative in defining loads, the proof factor is normally taken as 1, but in military aircraft it is usually taken as 1.125.

19.4.5   Ultimate Load

The *ultimate load* is the maximum load the aircraft can resist without failure. Ultimate load is defined as:

ultimate load = limit load × ultimate factor

The *ultimate factor* is a factor of safety applied to account for uncertainties in a range of aspects, such as the variation of material properties, the degradation of material properties over time, uncertainty in analysis and key assumptions and uncertainty in flight operations (e.g. flying outside the allowable manoeuvre envelope). The ultimate factor is normally taken as 1.5 for all aircraft types.

19.4.6   Structural Stiffness

Structural stiffness is the measure of how much the structure deflects under a load. It is defined as the slope of the load deflection curve. Structural stiffness has several variants referring to the type of load and deflection, for example, bending stiffness or torsional stiffness. The structure itself may be made from elements of many materials. The relation between stiffness, K, load, P, and deflection, δ, is given here.

$$K = (P/\delta) \tag{19.1}$$

19.5   From Structure to Material

These key concepts and definitions thus far relate to the structure as a whole, or at least a large element of structure, such as the wing. At this point, in the design process the designer should at least be aware of these key loads. For example, they should know the total wing load resulting from say a pull-out manoeuvre from a dive, as being an extreme load case. There are a number of structural failure modes that are critical such as wing divergence and control reversal that need to be considered at this level.

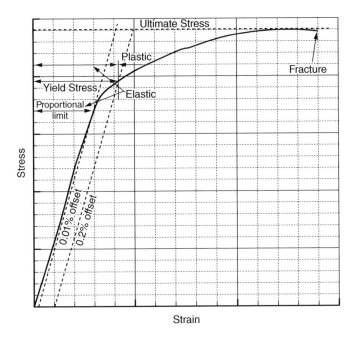

Figure 19.6 Basic stress–strain curve for an aluminium alloy.

However, it should now be becoming clear that in designing the airframe the understanding of loads applied to the aircraft now must be considered in the context of the local structural elements and material itself. Clearly, the ultimate load for the whole wing is dependent on the load carrying capacity of the material, which requires conversion to stress and strain that are the material characteristics used in estimating failure. Loads therefore need to be converted into stresses (f or s) and compared with the allowable proof or ultimate stress of the material to determine whether the design is acceptable. This task has become a major challenge in modern aircraft design. The reason is that different global conditions, for example, pull-out versus landing, will produce different maximum stresses in different local parts of the structure. While the designer cannot at this point know all the details they must be aware of the material characteristics that are important and how these relate to design choices. This is particularly true of stiffness that can affect the configuration choice. Figure 19.6 shows a typical stress–strain curve for an aluminium alloy. A number of key definitions are based around this.

The definitions here are couched in language relevant for the designer to understand in the context of making design decisions. For a fuller understanding and precise definitions it is necessary to refer to specialist structural design books [4–6]. But this simple overview is necessary to understand the dynamics of making decisions in material selection. Many of the properties mentioned bear inverse relationships to one another and trade-offs are necessary, either with the material or the structural shape. These considerations are discussed in the next section.

19.6 Basic Definitions – Materials

Given next are the essential definitions associated with aircraft structural design considerations.

19.6.1 Stress

Section 19.3 mentioned that there are fundamentally two types forces (loads) to be considered in setting up stress equations. The two of loads are given here:

i. Normal force, pull or push, acting perpendicular to surface.
ii. Shear force acting tangential to the surface.

To keep illustration simple, the small amount of dilation on account deformation taking place under the normal force application, as shown in Figure 19.2, is not shown in Figure 19.7 (details can be found in the references given). The stress types are as follows.

The definition of stress is to normalise of load applied on unit area that indicates the intensity of a force on a material (Figure 19.7).

i. Normal stress, σ, is the total applied normal force, P, divided by the area, A, over which it acts.

$$\sigma = (P/A) \tag{19.2}$$

It can be a tensile stress or compressive stress, depending on whether it is pull or push and cane be identified with a subscript.

ii. Shear stress, τ, is the total tangential force applied, F, divided by the area, A, over which it acts.

$$\tau = (F/A) \tag{19.3}$$

The concept of shear stress on thin surface, for example, edges of aircraft metal/composite sheets can be expressed as shear stress per unit length termed as 'shear flow' (not dealt with here).

19.6.2 Strain

Strain is the non-dimensionalised deformation of a material under stress and the two types are as follows.

i. Normal strain, ε, is the ratio of the extension or contraction, x, (or stretch/shrink) of the material under load to its original length, L. It is denoted:

$$\varepsilon = (x/L) \tag{19.4}$$

ii. Shear strain, γ, is the angular deformation under tangential force applied (details can be found in the references given).

$$\gamma = \Delta x/l \tag{19.5}$$

where Δx is the transverse displacement, that is, the angle between the sides of the original shape and deformed shape (Figure 19.7).

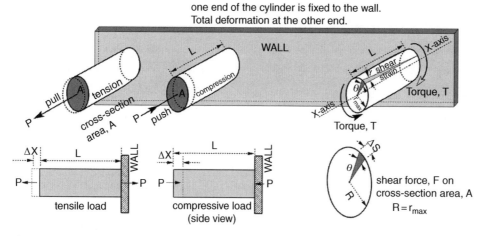

Figure 19.7 Basic stress and strain.

In case of torque, the twist angle, θ, produces circumferential displacement 'dS' on account of the applied torque straining the sold circular by the angle, γ, over the length, L.

Then $dS = \theta R = \gamma L$ where, $R = r_{max}$

or

$$\gamma = (\theta R)/L \tag{19.6}$$

19.6.3   Hooke's Law

The fundamental property of stress–strain relationship within the linear range of application, is shown in Figure 19.6. The stress–strain relationship obeys Hooke's Law, which states that within the linear range the ratio of stress by strain (stress/strain) is constant for both the normal and tangential loads.

The two types of stresses and strains related to stiffness are briefly described next.

i. Under normal load, the material stretches or contracts depending on whether it is pull or push. It is a fundamental property of the material obeying Hooke's Law is called the *Elastic Modulus* (*E*) or the *Young's Modulus*, named in honour of Thomas Young. It is defined as the slope of the stress–strain curve (Figure 19.6).

$$E = \text{Modulus of Elasticity} = (\sigma/\varepsilon) \tag{19.7}$$

For aluminium, $E \approx 69\,\text{GPa}$

ii. Under tangential load, the material deforms as shown in Figures 19.2 and 19.7. Its fundamental property obeying Hooke's Law is called the *Modulus of Rigidity* (*G*). It is defined as follows.

$$G = \text{Modulus of Rigidity} = (\tau/\gamma)$$

For aluminium, $G \approx 25.5\,\text{GPa}$

19.7   Material Properties

The choice material depends on its properties for the role it is required to perform. Given in the following are the different material properties the designers have to consider.

19.7.1   Stiffness

Material stiffness is the measure of how much the material deforms. It is measured from a material's fundamental property of stress–strain relationship within the linear range of application, as shown in Figure 19.6.

19.7.2   Yield Stress

Yield Stress is the stress level in the material that will cause permanent deformation. It is normally defined where the resulting plastic strain is 0.1 or 0.2%. This stress level is often called the 0.1% proof stress.

19.7.3   Fracture Toughness

Fracture Toughness is the stress intensity at which a crack in the material will grow rapidly until failure. It is a measure of the material's ability to resist cracks.

19.7.4 Ductility

Ductility is the ability of a material to deform plastically without fracturing. That is how much the material can be stretched after yielding.

19.7.5 Durability

Durability in the context of aircraft is the ability of the component to perform its function over its life. In aircraft design this is usually in reference to the number of flight hours or load cycles that the component can continue to function over. Cracks exist in materials naturally due to flaws in the material structure and these can grow slowly under loading. As the crack lengthens the local stress increases, eventually reaching the fracture toughness level for the material and resulting in complete failure.

19.8 Considerations with Respect to Design

Armed with the basic definitions, this paragraph discusses about how these material properties are taken into account in the design. When a reader first begins to design real structures or products, the question is usually posed very simply in terms of finding a material that is strong enough to carry the stresses or loads. However, in a real design problem it is more complicated, as the practicalities of making the structure come into play. In the context of an aircraft design a prime driver is the weight of the aircraft, and therefore the material must be used as effectively and efficiently as possible. Significant efficiency can be gained by careful shaping of the material components and the overall structural configuration. This leads directly into considerations of the cost of the structure, which clearly now depends on how material is shaped and assembled as well as the simple cost per unit weight. To determine the cost, therefore, it is necessary to consider the manufacturing process. The challenge in this is that many of these factors have conflicting requirements and what benefits the design in one sense can be detrimental in another. Trade-offs are necessary to achieve a compromise. So real design begins!

There are four main elements then to consider in this: the properties of the material itself, the structural configuration, the manufacturing process and finally the cost.

19.8.1 Material Selection

Material properties are the main drivers to make the choice to select a suitable aircraft component under the nature of the applied load on it. Over the years of development, engineers moved from the natural material, for example, wood to a large variety of man-made alloys and composite materials made from natural materials; each type has different properties to suit specific applications in a selective manner to fabricate aircraft components. Some general materials of interest are given in Table 19.2 (Young's Modulus, E, GigaPascals) and Table 19.3 (density, mega grams m^{-3}), see Ref. [7]. Brief introductory considerations are given in this section on material selection in aircraft design. Readers are suggested to refer to Refs [4, 5, 7–10] for detailed study.

If the maximum stress level required for the design is around 500 MPa, then steel, aluminium alloy, titanium alloy and carbon-fibre reinforced plastics (CFRP) are all possibilities. The strengths of these materials are compared in a bar chart as shown in Figure 19.8.

A quick study of the tables and figure show that, while steel is much stronger than aluminium, it is much heavier. The lightest steel is about 2.5 times denser than the heaviest aluminium alloy. This suggests that the same structure can be made much lighter if aluminium is used. In fact, based on average values the strength-to-weight ratio of these two materials is pretty comparable, meaning that other factors are really needed to help. There is of course a major caveat here in that there is a huge variation in properties for specific material compositions. A further complication in material properties is the inverse relationship between ultimate strength and toughness. As materials are made stronger, they have a tendency to be more brittle and

Table 19.2 Material Young's Modulus, *E* (GPa).

		E (GPa)
Metals		
Ferrous	Cast iron	165–180
	High carbon steels	200–215
	Medium carbon steels	200–216
	Low carbon steels	200–215
	Low alloy steels	201–217
	Stainless steels	189–210
Non-ferrous	Aluminium alloys	68–82
	Copper alloys	112–148
	Lead alloys	12.5–15
	Magnesium alloys	42–47
	Nickel alloys	190–220
	Titanium alloys	90–120
	Zinc alloys	68–95
Ceramics		
Glasses	Borosilicate glass	61–64
	Glass ceramic	64–110
	Silica glass	68–74
	Soda-lime glass	68–72
Porous	Brick	10–50
	Concrete, typical	25–38
	Stone	69–21
Technical	Alumina	215–413
	Aluminium nitride	302–348
	Boron carbide	400–472
	Silicon	140–155
	Silicon carbide	300–460
	Silicon nitride	280–310
	Tungsten carbide	600–720
Composites		
Metal	Aluminium/silicon carbide	81–100
Polymer	CFRP	69–150
	CFRP	15–28
Natural		
	Bamboo	15–20
	Cork	0.013–0.05
	Leather	0.1–0.5
	Wood, typical (longitudinal)	6–20
	Wood, typical (transverse)	0.5–3

(Continued)

Table 19.2 (Continued)

	E (GPa)
Polymers	
Elastomer	
Butile Rubber	0.001–0.002
EVA	0.01–0.04
Isoprene (IR)	0.0014–0.004
Natural rubber (NR)	0.0015–0.0025
Neoprene (CR)	0.0007–0.002
Polyurethane elastomers (elPU)	0.002–0.003
Silicone elastomers	0.005–0.02
Thermoplastic ABS	1.1–2.9
Cellulose polymers (CA)	1.6–2
Ionomer (I)	0.2–0.424
Nylons (PA)	2.62–3.2
Polycarbonate (PC)	2–2.44
PEEK	3.5–4.2
Polyethylene (PE)	0.621–0.896
PET	2.76–4.14
Acrylic (PMMA)	2.24–3.8
Acetal (POM)	2.5–5
Polypropylene (PP)	0.896–1.55
Polystyrene (PS)	2.28–3.34
Polyurethane thermoplastics (tpPU)	1.31–2.07
PVC	2.14–4.14
Teflon (PTFE)	0.4–0.552
Thermoset Epoxies	2.35–3.075
Phenolics	2.76–4.83
Polyester	2.07–4.41
Polymer foams	
Flexible polymer foam (VLD)	0.0003–0.001
Flexible polymer foam (LD)	0.001–0.003
Flexible polymer foam (MD)	0.004–0.012
Rigid polymer foam (LD)	0.023–0.08
Rigid polymer foam (MD)	0.08–0.2
Rigid polymer foam (HD)	0.2–0.48

Note: VLS = very low density, LD = low density, MD = medium density.

have a relatively lower fracture toughness. Thus, if durability is a concern then the material chosen is may have a lower ultimate strength. Therefore, the aircraft designers need to know other material other material properties, for example, their strength, fracture toughness, costs and so on. So, any comparison like these must be made really on the exact material being considered. Many engineers make mistakes by assuming average properties when the material they have to hand is quite different. That said, the average values here are useful to make the point.

Table 19.3 Material density (Mg m^{-3}).

		ρ (Mg m^{-3})
Metals		
Ferrous	Cast irons	7.05–7.25
	High carbon steels	7.8–7.9
	Medium carbon steels	7.8–7.9
	Low carbon steels	7.8–7.9
	Low alloy steels	7.8–7.9
	Stainless steels	7.6–8.1
Non-ferrous	Aluminium alloys	2.5–2.9
	Copper alloys	8.93–8.94
	Lead alloys	10–11.4
	Magnesium alloys	1.74–1.95
	Nickel alloys	8.83–8.95
	Titanium alloys	4.4–4.8
	Zinc alloys	4.95–7
Ceramics		
Glasses	Borosilicate glass	2.2–2.3
	Glass ceramic	2.2–2.8
	Silica glass	2.17–2.22
	Soda-lime glass	2.44–2.49
Porous	Brick	1.9–2.1
	Concrete, typical	2.2–2.6
	Stone	2.5–3
Technical	Alumina	3.5–3.98
	Aluminium nitride	3.26–3.33
	Boron carbide	2.35–2.55
	Silicon	2.3–2.35
	Silicon carbide	3–3.21
	Silicon nitride	3–3.29
	Tungsten carbide	15.3–15.9
Composites		
Metal	Aluminium/silicon carbide	2.66–2.9
Polymer	CFRP	1.5–1.6
	CFRP	1.75–1.97
Natural		
	Bamboo	0.6–0.8
	Cork	0.12–0.24
	Leather	0.81–1.05
	Wood, typical (longitudinal)	0.6–0.8
	Wood, typical (transverse)	0.6–0.8

(Continued)

Table 19.3 (Continued)

			ρ (Mg m^{-3})
Polymers			
Elastomer		Butile rubber	0.9–0.92
		EVA	0.945–0.955
		Isoprene (IR)	0.93–0.94
		Natural rubber (NR)	0.92–0.93
		Neoprene (CR)	1.23–1.25
		Polyurethane elastomers (elPU)	1.02–1.25
		Silicone elastomers	1.3–1.8
Thermoplastic		ABS	1.01–1.21
		Cellulose polymers (CA)	0.98–1.3
		Ionomer (I)	0.93–0.96
		Nylons (PA)	1.12–1.14
		Polycarbonate (PC)	1.14–1.21
		PEEK	1.3–1.32
		Polyethylene (PE)	0.939–0.96
		PET	1.29–1.4
		Acrylic (PMMA)	1.16–1.22
		Acetal (POM)	1.39–1.43
		Polypropylene (PP)	0.89–0.91
		Polystyrene (PS)	1.04–1.05
		Polyurethane thermoplastics (tpPU)	1.12–1.24
		PVC	1.3–1.58
		Teflon (PTFE)	2.14–2.2
Thermoset		Epoxies	1.11–1.4
		Phenolics	1.24–1.32
		Polyester	1.04–1.4
Polymer Foams			
		Flexible polymer foam (VLD)	0.016–0.035
		Flexible polymer foam (LD)	0.038–0.07
		Flexible polymer foam (MD)	0.07–0.115
		Rigid polymer foam (LD)	0.036–0.07
		Rigid polymer foam (MD)	0.078–0.165
		Rigid polymer foam (HD)	0.17–0.47

Because of these contradictory indicators a variety of metrics have been used to aid in material selection. The most common of these are the ratios of strength-to-weight (specific strength) and stiffness to weight (specific stiffness). Other useful measures are modulus to strength and toughness to strength. The best tools for aiding this are the graphs (Figure 19.9) developed by Ashby [8, 9] as part of a materials selection tool suite.

Since cost is not a material property, it is dealt with separately in the next section. For materials with similar properties, cost varies from manufacture to manufacturer. A good example is that of building construction steel with similar specifications, the unit price is considerably lower in the Far East as compared to the prices in the West.

Figure 19.8 Material strength.

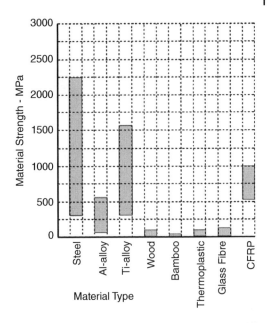

Another important consideration for material selection is the manufacturability of the component. This is dealt with briefly in Section 19.8.1.

19.8.2 Ashby Scatter Plots

Professor Martin Ashby of Cambridge University systematically classified and grouped materials within the class and their associated properties of interest in a series of scatter plots. The applicable ones for this book are given in Figure 19.9. These are (i) Young's Modulus versus density, (ii) Young's Modulus versus strength, (iii) strength versus density and (iv) fracture toughness versus strength and so on, can be found in Refs [8, 9]. In general, engineers like to have material of high strength, high stiffness, high toughness and low weight.

The advantages of Ashby plots are that an overview of material properties can be visualised in few graphs to make a quick comparison of the relative merits to make a decision. These graphs are useful in giving potential material choices and can provide alternatives that were perhaps not originally at the forefront of the designer's mind.

19.8.3 Material Cost Considerations

Of the manufacturing considerations in material choice, the cost of making the structure from the chosen material is critical. Again, an exotic material may provide excellent physical properties but may require new specialist equipment to machine or fabricate. It may be further complicated by being available only from a single supplier or small number of suppliers. A sole supplier can be in a powerful position and significantly affect the final cost of the aircraft.

Another factor, even with more common materials, is the effect of the world markets. Aluminium, for example, is traded in global stock markets and the price can vary widely over time. This is also true for some of the minor additives to make aluminium alloys. Again, it is crucial to know the exact grade of material to have confidence in cost, otherwise the designer must be aware of the risk.

Fundamentally, most companies have estimates for the cost of material for a given weight and use these at design stage. The company will have corporate knowledge of the additional costs of manufacture using materials with which it is familiar. It is where the designer strays into uncharted waters that risk becomes an issue. This was certainly the case for many manufacturers in building the latest generation of composite

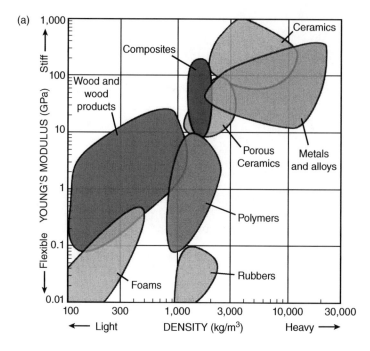

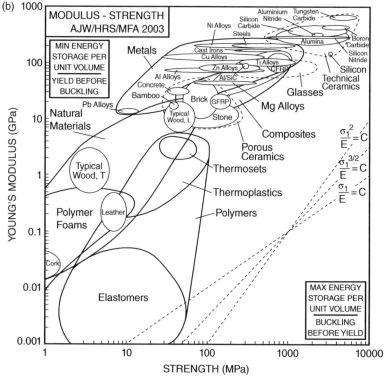

Figure 19.9 Material Properties [8, 9]. (a) Young's Modulus versus density. (b) Young's Modulus versus strength. (c) Strength versus density. (d) Fracture toughness versus strength.

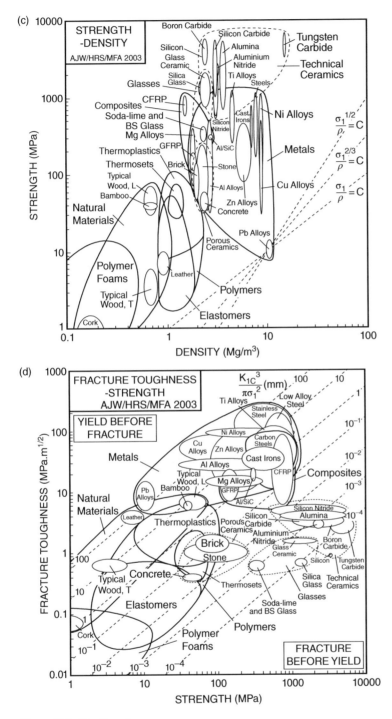

Figure 19.9 (*Continued*)

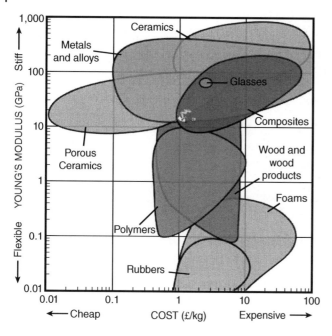

Figure 19.10 Materials cost [8, 9].

aircraft where in many cases a completely new factory was built for the manufacturing programme. These are huge risks and while at the design stage this seems far away, some consideration is always helpful, as the risk can then be taken with some confidence.

Compromise on material properties is required on account of economic factors to remain competitive in the market without compromising aircraft safety but possibly on performance, mainly arising from substitution with a heavier material. An Ashby scatter plot of material cost is given in Figure 19.10.

A more exacting cost of some relevant material, in terms of US$/kg, is given in Figure 19.11 in bar chart form to facilitate relative cost comparison. The comparison allows compromises to be made, if required. It is to note that the material cost varies according to global markets. Therefore, accuracy of the two figures cannot be guaranteed.

19.8.4 Manufacture

While the detailed manufacturing programme is not yet on the horizon for the designer, at this point it is important that some heed is paid to it here. When selecting a material it is very tempting to refer to the kind of diagrams shown here or to use other look up tables available from many accessible resources to garner a range of possible materials. It is a common sight for students, and even experienced engineers when under pressure, to make a decision because a material with the appropriate properties has been identified. However, the many physical characteristics of materials also affect the manufacturing processes. Some, such as titanium and CFRP are very difficult to machine, some are very difficult to bond and some materials when together react and corrode. There are an additional and simple questions for any company: Is the material readily available and is the supply is in sufficient quantity for the project to completion? Are there appropriate manufacturing tools and systems in place to handle the material? These apparently mundane considerations can cause the designer to rethink and perhaps reshape the design so that a more accessible and cost-effective material can be used.

An addition to the older example of Boeing767 given in Figure 19.1, is an example of a current example of the Airbus380 in Figure 19.12. It shows component-wise material selection.

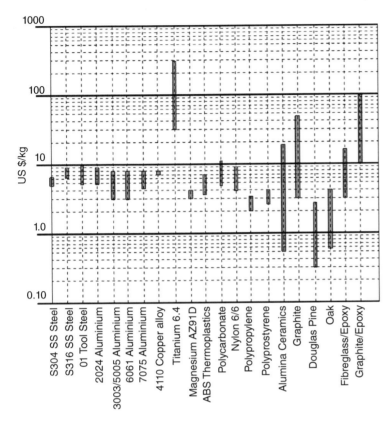

Figure 19.11 Materials cost in bar chart form.

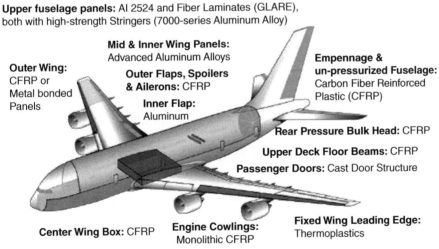

Upper fuselage panels: Al 2524 and Fiber Laminates (GLARE), both with high-strength Stringers (7000-series Aluminum Alloy)

Mid & Inner Wing Panels: Advanced Aluminum Alloys

Outer Wing: CFRP or Metal bonded Panels

Outer Flaps, Spoilers & Ailerons: CFRP

Empennage & un-pressurized Fuselage: Carbon Fiber Reinforced Plastic (CFRP)

Inner Flap: Aluminum

Rear Pressure Bulk Head: CFRP

Upper Deck Floor Beams: CFRP

Passenger Doors: Cast Door Structure

Center Wing Box: CFRP

Engine Cowlings: Monolithic CFRP

Fixed Wing Leading Edge: Thermoplastics

Lower Fuselage Panels: Laser-beam-welded Aluminum Alloys

Figure 19.12 Materials selection – A380. Courtesy Airbus [11].

19.8.5 Integrated Decisions

The preceding sections have indicated four key considerations in material choice. There will always be others in a full design programme, but these four remain dominant. The last and most important consideration is that these aspects are integrated. The decision on the material choice should be based on a consideration of all four aspects: material properties, shape/configuration, manufacture and cost. It is here that the trade-offs begin and choices are exercised such as accepting some weight gain for a cost saving, accepting higher cost for a reduction in weight or shaping the material to keep the stresses low so that a common and well understood material can be used.

19.8.6 Design Exercise

All the information given in Section 19.8 prove useful for the design exercise given here. As an exercise, consider a cantilever beam from above. By using any available resources for material properties and costs, try to enumerate a number of cross-sectional shapes that will keep the cost of each design of the beam within a 10% margin.

As a further exercise, try to design the beam to be half the original weight, but still within 10% cost of the original. If that is too easy, then try half again and so on. Be imaginative with the cross-sections. You can, for example, use hollow cross-sections.

19.9 Structural Configuration

The next consideration for material choice is the shape of the component or configuration of the structure. This is important because in terms of material choice it is the stress level that matters rather than simply the load level. The stress level determines whether failure occurs, and it depends on the geometry of the component as well as the material properties.

Section 19.3 describes the types of loading the main aircraft components have to withstand. The dominant types are bending and torsion dealt with in this section.

19.9.1 Bending

For example, in Figure 19.13 a simple cantilever beam with a rectangular cross-section under bending is shown with one end fixed and one end free. It is a simplistic representation of a wing, the free end is the tip and fixed

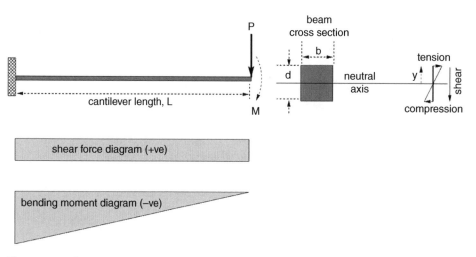

Figure 19.13 Showing cantilever beam and bending stress.

end attached to fuselage. The maximum stress in the cantilever beam like wing spar is at the root joining the fuselage. The beam transverse cross-section has a width of 'b' and depth of '$2d$' and the neutral axis passes through the centroid at middle of the rectangular beam cross-section; 'y' is the distance from the neutral axis.

In the Figure 19.13, without any axial stress, the cantilever beam of length, L, is in pure bending under the vertical load, P, at the free end. This gives the maximum bending moment, M, at the fixed end expressed as.

$$M = (P \times L) \tag{19.8}$$

The shear force is constant along its length. The distribution of shear force and bending moment along the cantilever beam is shown in Figure 19.13.

The loading condition will deflect the beam in to a shallow uniform curvature causing tension above the neutral axis and an equal amount of compression below it. The opposite forces above and below the neutral axis act as resistive forces from pure bending.

The tension makes the upper half stretch and vice versa at the bottom half on account of compression. Therefore, the cross-section has stress distributed in linear variation from '0' at the neutral axis to maximum value at the top and bottom level. The total force, F, equates to zero, top half and bottom have equal and opposite magnitudes. This expressed mathematically as follows.

Let at any elemental area dA of the beam transverse cross-section at a distance 'y' from the neutral axis experiences stress, $\sigma_x = \sigma_{x\_max} \times (y/d)$ with linear variation.

Then the elemental force, dF, on an elemental area dA of the beam transverse cross-section is as follows.

$$dF = [\sigma_{x\_max} \times (y/d)] \times dA,$$

where σ_{x_max} and 'd' have constant values.

Integrating over the cross-section area,

$$F = \int [\sigma_{x\_max} \times (y/d)] \times dA = [(\sigma_{x\_max}/d) \times \int ydA) = 0.$$

Elemental moment, dM on account of elemental force is

$$dM = dF \times y == [\sigma_{x\_max} \times (y/d)] \times dA \times y = [\sigma_{x\_max} \times (y^2/d)] \times dA$$

Integrating moment,

$$M = \int [\sigma_{x\_max} \times (y^2/d)] \times dA = [(\sigma_{x\_max}/d) \times \int y^2\, dA) \tag{19.9}$$

The term $\int y^2 dA$ is the area moment of inertia, I, of about the neutral axis.
Equation (19.9) takes the final form $M = (\sigma_{x\_max}/d) \times I$.or

$$\sigma_{x\_max} = (Md)/I \tag{19.10}$$

The tension makes the upper half to stretch and vice versa at the bottom half on account of compression. Consider the strain due elongation to be ε_x along its axis. The loading condition will deflect the beam in to shallow curvature of radius R. It can be shown [1–3] that for small deflection the following relation holds.

$$\varepsilon_x\, y/R \tag{19.11}$$

The distortion of the transverse cross-section on account of tension and compression is negligibly small and is ignored here. Then applying the elastic property (19.4) in terms of Young's Modulus, E, the following can be written.

$$E = \sigma_x/\varepsilon_x$$

Using Eq. (19.11), at any distance y from the neutral axis, this becomes as follows:

$$\sigma_x/y = E/R = \sigma_{x\_max}/d = M/I \text{ (from Eq.(19.10))} \tag{19.12}$$

where $y = \sigma_x d/\sigma_{x\_max}$.

From Eq. (19.6), Eq. (19.12) can be written as follows

$$\sigma = \frac{My}{I} \tag{19.13}$$

this is known as *elastic flexure formula*

Noting that the rectangular shaped area moment of inertia of about the neutral axis is as follows.

$$I = \int y^2 \, dA = (bd^3/12)$$

Then for rectangular cross-section cantilever beam, Eq. (19.12) becomes:

$$\sigma = \frac{My}{I} = \frac{12PLy}{bd^3} = \tag{19.14}$$

The maximum stress is at $y = d/2$, that makes

$$\sigma_{x\_max} = \frac{6PL}{bd^2} \tag{19.15}$$

The equation indicates that the stress in the beam is inversely proportional to the depth of the beam. The longer the 'd' (beam depth) is, the lesser is the maximum stress, σ_{x_max}. If space is available, then for the beam cross-section area, a deeper beam is preferred. The example given next demonstrates this.

In Figure 19.14, if b_1 is set to be 0.8b, then for the same cross-sectional area, d_1 becomes 1.25d with a net reduction in stress of 20%. In a real design, if such changes are possible then lighter or cheaper materials may be used, or the section is reduced to reduce weight. It is possible to make multiple gains sometimes, by reducing the area and using a lighter material at the same time.

This now provides more freedom in material selection. If the stress is very high, then being able to reduce it may bring other materials into play that can offer benefits in terms of weight or cost or both. For example, reducing the maximum stress from 600 MPa to 300 MPa brings aluminium alloy into competition with titanium, carbon fibre and steel.

Although this example is simple, it is very relevant for the selection of material for an aircraft fuselage or wing. In most design cases a single wing is idealised as a cantilever beam that, although it has a more complicated cross-section, still behaves very much like a beam. It is possible therefore to get very quick estimates of the order of magnitude of stresses in a wing to inform the initial selection of materials.

19.9.2 Torsion

Following on with the idealisation of the wing as a beam it becomes clear that torsional loads arise naturally since the elastic centre and the centre of lift are not coincident (Figure 19.15). Torsion is particularly important

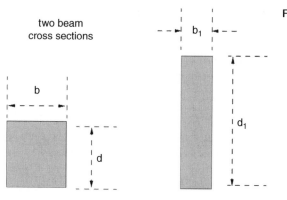

two beam
cross sections

b_1

Figure 19.14 Showing two cross-sections.

b

d

d_1

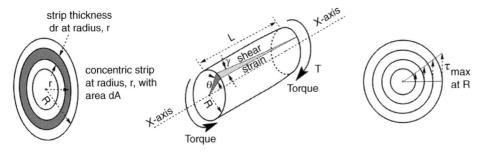

Figure 19.15 Torque on cylindrical solid shaft.

because twist of the wing directly influences the angle of incidence and hence the lift. So, in addition to the stresses the angle of twist is a key response that needs to be known.

When torque is applied to a circular bar, in this case a solid cylindrical shaft, the following assumptions are taken to derive the torque equations.

i. No distortion takes place on the cross-section perpendicular to the cylinder's axis.
ii. Shear strain varies linear from 0 at the centre to γ_{max} at the peripheral surface.
iii. Valid only in the linear range of stress–strain relation obeying Hooke's law, that is, $(\tau/\gamma) = \text{constant} = G$.

This implies that shear stress,

$$\tau = \frac{r}{R} \times \tau_{max} \tag{19.16}$$

linearly varying from 0 at the centre to τ_{max} at the peripheral surface.

Therefore, force, F, on the cross-sectional area, A, is given by

$$F = \int_A \left(\frac{r}{R} \times \tau_{max} \right) \times dA, \tag{19.17}$$

where, dA is the thin concentric cross-sectional strip at radius, r.

Torque,

$$T = \text{force} \times \text{arm}$$

Hence,

$$\text{Torque, } T = \int_A \left(\frac{r}{R} \times \tau_{max} \right) \times r dA = \left(\frac{\tau_{max}}{R} \right) \int_A (r^2) \times dA \tag{19.18}$$

where

$$\int_A (r^2) \times dA = \int_0^R 2\pi r^3 \times dr = \frac{\pi R^4}{2} = \frac{\pi D^4}{32} = I_p, \tag{19.19}$$

the polar moment of inertia.

For a given tube, τ_{max} and $D = 2R$ are constants.

On substituting, the torque Eq. (19.18) can be written as

$$T = \left(\frac{\tau_{max} \times I_p}{R} \right) \tag{19.20}$$

Normally, applied torque, T, can be measured and is known, thus τ_{max} can computed.

$$\tau_{max} = \left(\frac{T \times R}{I_p} \right) \tag{19.21}$$

For a solid circular cylinder,

$$\tau_{\max} = \left(\frac{16 \times T}{\pi D^3} \right) \tag{19.22}$$

Airframe structures are largely thin-walled tubes and hence this basic theory is modified to produce expressions and relationships that are more appropriate for the structural designer. In the case of hollow circular sections:

For *thick tubes*,

$$I_p = \frac{\pi R^4}{2} - \frac{\pi B^4}{2}, \tag{19.23}$$

where B is the internal radius of the tube.

For *thin tubes* of a thickness, t, the stress variation within the thin cross-section section is small and can be approximated by taking the average value at $R_{\text{ave\_thin tube}}$. This gives

$$I_p = 2\pi R_{\text{ave}}^3 \times t. \tag{19.24}$$

As aerofoil shapes are complex this theory has been extended significantly as a general theory of thin-walled tubes, as shown next. Although the treatment is extensive, there are two key equations of use to the wing designer, which can be used to find the skin thickness and the wing twist. Before developing these equations, there are a couple of useful points to note. Firstly, in simple bending, all forces act through the neutral axis, also known as the elastic axis. In symmetric sections, this causes bending about the two main beam axes, but in unsymmetrical sections it also causes twist. The twist is around a point in the section called the shear centre, which is defined as the point in the section through which a shear force will cause no twist. The distance between the shear centre and the elastic centre is a critical design parameter for the structure and should be minimised if possible. Since the centre of lift is typically not through the shear centre of the wing section it means that wings always have a torsional load to deal with. A fuller treatment of these concepts is beyond the scope of this book, but these are useful key points to bear in mind in the structural design of the wing.

The neutral axis passes through the centroid of the area and the equations here are developed assuming that the torque acts along this axis. The underlying assumption in this is that the torsional load is independent from the bending load. This highly idealised assumption is conservative but perfectly suited to this conceptual stage of design to obtain initial estimates of stress and skin thickness.

Figure 19.16 shows a non-circular uniform cross-section tube with a constant thickness, t. It is subjected to torsion, T, about a longitudinal axis passing through its shear centre. The torque, T, produces shear stress, τ, at the tube cross-section, as shown on the elemental circumference length, ds that associates with its complementary stress along its longitudinal direction as shown in the figure.

Then the elemental force dF on the elemental cross-section of length ds can be written as follows.

$$dF = \tau \times t \times ds, \tag{19.25}$$

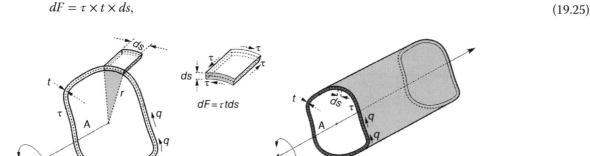

Figure 19.16 Non-circular thin tube.

This is often written in terms of the shear per unit thickness, which is called the shear flow, q, written as:

$$q = \tau \times t \text{ per unit length} \tag{19.26}$$

Force equilibrium gives that force per unit length along the circumference is constant. That means, for the torque loading the shear flow, q, is also constant along the circumference. This is true for non-uniform thickness along the circumference, that is, if thickness, t, reduces then stress, τ, increases and vice versa.

Then torque, T, can be obtained by integrating the whole circumference.

$$T = \oint \tau \times t \times rds = (\tau \times t) \oint rds = q \times \oint rds \tag{19.27}$$

where r is mean radius at ds.

The integral, $\oint rds$ gives twice the cross-sectional area at its mean radius r.

Therefore,

$$T = q \times \oint rds = 2A\tau t = 2Aq \tag{19.28}$$

this is known as the Bredt–Batho formula.

The Bredt–Batho formula is directly useful for estimating the skin thickness needed for a given material choice. The shear stress at any point of the thin tube cross-section (can be non-uniform thickness) can be computed by.

$$\tau = q/t = T/(2At) \tag{19.29}$$

The angle of twist, θ, arising from the applied torque can be obtained by the following expression (not derived).

$$\theta = \frac{Ts}{4A^2 Gt}, \tag{19.30}$$

where G is the modulus of rigidity.

References [4–6] may be consulted for its derivation as well for non-circular cross-section.

19.9.3 Buckling

Buckling is an unstable behaviour, a deformation that can take place on account of axial/in-plane compressive load applied to slender column, thin plate, shell structures and so on. Characteristically, buckling shows large lateral deformation compared to axial deformation. This occurs if there is local misaligned (eccentric) loading and/or imperfections in the material.

Within the elastic limits of the material, such deformation is tolerated as deformation recovers to normality as soon as load is withdrawn. In flight, many aircraft exhibit skin buckling on wing/fuselage surfaces under flight load, loosely described as 'tin canning' or 'wrinkling'. Buckling is not desirable, but if well designed, buckling accommodates flight load variation. Stiffeners/ribs are used to keep buckling deformation within acceptable limits. Modern high performance aircraft have integrally milled skins with low level of buckling.

Only the metal structure buckling characteristics are discussed briefly here. Non-metals also exhibit bucking. Readers may refer to [1, 12] for detailed study. A brief overview is summarised here.

- Structural components under axial/in-plane compression loads are susceptible to buckling – unstable out-of-plane deflection when the applied compression stress reaches a certain level, known as critical buckling stress.
- Depending on the structural geometry and material properties, the compressive stress at which a structure will buckle can be well below that of the material compressive yield stress.
- When defining the allowable compressive design stress, buckling stress must also be considered alongside the material yield stress.

- Buckling on aircraft structures usually takes the form of 'global' buckling (buckling of large or full structural components usually resulting in structural failure) or 'local' buckling (buckling of a small sub-section of a larger component that may not result in structural failure).

The various buckling behaviours applicable to aircraft structures can be considered in two parts: (i) beam/column buckling and (ii) plate buckling, as discussed next at an introductory level. Figure 19.17 shows the basic examples of the two types of buckling. Buckling deformation can be of different patterns.

19.9.3.1 Column/Beam Buckling

Figure 19.17a shows a basic example of buckling of a slender column; a beam like structure supported at one or both ends, such as the stringers on stiffened panels, are susceptible to column buckling. A long, thin column will generally fail due to flexural or Euler elastic buckling and at relatively low buckling loads. The buckling stress of these columns can be predicted by the elastic Euler equation [13, 14].

$$\sigma_{col} = \frac{\pi^2 EI}{A\ L_e^2},$$ (19.31)

where

σ_{Col} = Critical column buckling stress.
E = Material Young's Modulus.
L_e = Column effective length.
I = Column second moment of inertia.
A = Column cross-sectional area.

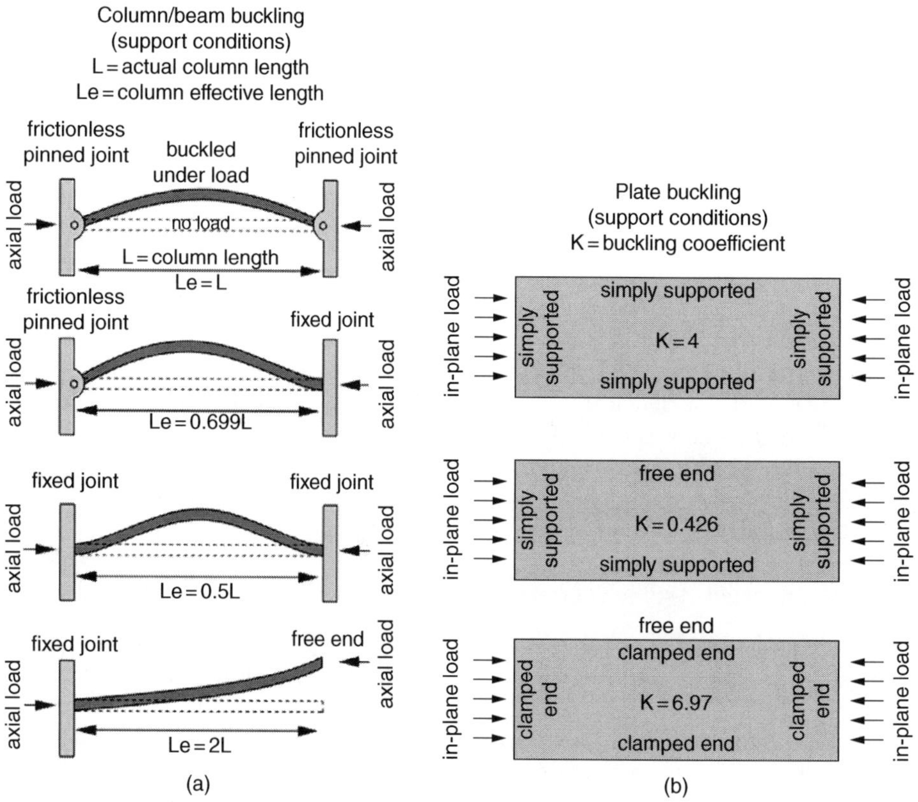

Figure 19.17 Buckling. (a) Column/beam buckling. (b) Plate buckling.

The column buckling stress is dependent on the particular boundary conditions at each end, which can be free, or range from simply supported through to fixed joint (clamped). This is accounted for in the above equation by the use of an 'effective' length parameter, L_e. Figure 19.17a indicates how the column effective length varies with column end boundary conditions.

The boundary conditions on the real structure will vary with joint idealisation and manufacture and assembly methods. More rigid end boundary conditions lead to increased critical buckling stresses. Therefore, for the conservative design of columns constrained at both ends, it should be assumed that the ends are simply supported and that the effective length is equal to the actual column length.

Beams/columns that are shorter and sturdier are more resistant to buckling, but their buckling behaviour can be more complex. In these cases the column failure mechanisms usually involve the interaction of global buckling with localised plastic yielding. Further details can be found in the Secant [14] and Euler–Johnson [1] analysis methods for local-flexural interaction.

It should be noted that buckling of a beam/column typically constitutes global failure of that structural component and the predicted critical column buckling stress should therefore correspond to an ultimate load condition.

19.9.3.2   Plate Buckling

Flat sheets, supported on three or more sides, can also buckle under in-plane compression loads. On an aircraft, these flat sheets usually comprise the wing and fuselage skin components, as well as internal spar or rib webs.

Buckling of these flat sheets between the supporting structures, such as ribs, frames or stringers, is typically considered a 'local' instability. Buckling of the plate sections may not lead to failure of the structure providing the supporting structure remains stable after the local plate sections have buckled. Such structures are described as 'post-buckling' – the total structure remains stable and safe even when local buckling of sub-sections occurs.

The critical buckling stress of a flat uniformly thick plate under uniaxial in-plane compression loading can be found with the following equation;

$$\sigma_{Pl} = \frac{K\pi^2 E}{12(1 - v^2)}\left(\frac{t}{b}\right)^2, \tag{19.32}$$

where

σ_{Pl} = critical plate buckling stress.
E = material Young's Modulus.
v = material Poisson's ratio.
t = thickness of the plate section.
b = width of the compression loaded edges.
K = buckling coefficient.

Similar to column buckling, the critical plate buckling stress is dependent on the support conditions on each side of the plate. The more rigid the support constraint, the more resistant to buckling the plate. Figure 19.17b indicates how the plate buckling varies with the end boundary conditions.

The various levels of constraint/support on each side of the plate is accounted for by the buckling coefficient, K. Various external sources describe the definition of these buckling coefficients in great detail ([12–14]), however, Figure 19.17b describes the three main edge support scenarios and subsequent buckling coefficients which should be used for conservative structural design.

As noted before, a larger stable structure can accommodate local plate buckling of its secondary elements. These 'post-buckled' structures can offer considerable weight savings, however, the local reduction in stiffness and deformation of the flat plate geometry must be taken into consideration. For example, buckling of the external skin on a commercial aircraft skin can have significant impact on aerodynamic performance. Subsequently, local buckling of the wing skin on a commercial aircraft would not typically be allowed below the limit load. However, on the external fuselage skin where the aerodynamic surface is less influential, local plate buckling could be allowed to occur as low as 50% of the limit load.

The methods outlined here model the plate and column buckling stress assuming that the column or plate behaviour is ideal. In reality such perfect conditions are rare and more often than not the column or plate structure has built in imperfections or is loaded out-of-plane caused by general misalignment of the loads. Any geometric imperfection or loading eccentricity will cause the structure to buckle at a critical stress lower than that predicted using these methods.

19.10 Materials – General Considerations

The preceding sections emphasised general principles for consideration when choosing a material for an aircraft design. In this and the following sections, it is worth reviewing some of the key characteristics of materials relevant to aircraft manufacture: metals and alloys, wood, fabric and composite. This is not meant to be an exhaustive guide, but will hopefully provide the reader with some basic knowledge and issues to consider when selecting a material.

Within the general classes of material, there are subclasses of product as alloys, composites and so on that offer appropriate properties to suit the product. In the materials marketplace, there is a huge variety, and it is ever increasing in number with the newer types, which exhibit better properties with every new generation. At the conceptual design phase, engineers must screen and rank the types of materials that would suit their requirements spelling out the limitations and the constraints involved, and if required, change to another type. In the civil domain, designers will typically stay in safe territory with known and common materials that reduces commercial risk. However, military aircraft material often use case-specific material never tried before as they strive to deliver technical advantage.

In general, to select suitable material for a component, the following aspects need taken into consideration:

Property	Production	Operation
Strength	Availability	Erosion, wear and abrasion quality
Stiffness	Fabrication ease	Thermal and electrical characteristics
Toughness	Manufacturability	Plating/galvanic/paint compatibility
Crack propagation	Handling	Compatibility with contact material
Ageing and corrosion		
Tolerance to environment		

As described in the previous section there are usually a number of options available to choose from. It is always a compromise, because many properties work against each other. This makes aircraft weight and cost estimation more complex, and typically engineers will have to identify and compute large number of parts to estimate component weights, and costs, before it is built. It is here 3D modelling in Computer Aided Drawing (CAD) helps as components and assembly weights and costs can be quickly computed and the effect of design changes noted.

The semi-empirical relationship for weight estimation given in Chapter 10 considers an all metal construction. It also indicates how to adjust the prediction if there are some parts made of lighter material. For a rapid method, the OEW may be factored accordingly (only the structural weight is affected – the rest is unchanged).

In most cases designers will look to the most common properties for the initial consideration. As shown in Figure 19.18 initial attention is given to standard properties such as Young's Modulus (material stiffness), ultimate strength and so on. But as per the previous section it is now more common to use ratios of key properties to aid the choice as these give additional insight to the efficacy of the material and structure. In general, for the same Young's Modulus, metals have higher density. But when strength-to-weight ratio (specific strength) is considered then composites overtake metals, that is, engineers can get same strength with lighter components even when the higher factor of safety can eat into the weight savings. Metals show better Young's

Figure 19.18 Comparison of typical aircraft material-stress–strain relationship.

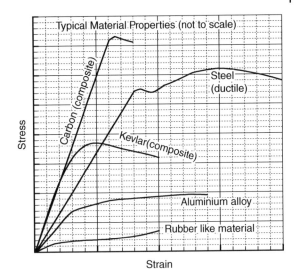

Modulus for the same strength. Metal also shows better fracture toughness for the same Young's Modulus. Another important comparison is the relative cost per unit volume versus Young's Modulus where metal alloys have lower costs. These factors typically make aluminium alloy a much better choice for aircraft than steel, which is counter to the first impression given by typical stress–strain material curves.

Although for design detailed material knowledge is not normally expected, some overview is very useful. The stress–strain curves as per Figures 19.6 and 19.18 should be studied and their general form understood as they reveal additional insights to the material and its usefulness. For example, under load (stress) all materials deform (strain), some more than others but can recover to their original shape when load is removed, provided its application is within a limit. Beyond this load level they will not recover to their original shape. (See references [1, 5–7] for details on stress and strain.) It can be seen then that steel has a pronounced necking and elongation phase before final failure whereas composites just break with a brittle failure. These two characteristics are important in design. In the case of the pronounced stretching evident in aluminium and steel, this means that if either material is used for a component such as a flap track, then the failure stress to be used for design is in fact the yield stress, not the ultimate tensile strength, because the deformation of such a component could cause failure of the whole aircraft, even though it itself has not broken. The designer must understand these issues to make appropriate decisions.

As an example of the level of detail needed, consider the typical stress–strain characteristic of aluminium alloy is shown in Figure 19.18. The figure shows that initially the stress and strain relationship behaves in a linear manner according to Hooke's Law which represents the elastic property of the material. Within the elastic limit, the material strength (how much stress it can bear) and stiffness (how much deformation takes place) are the two main properties to be considered by the designers to make material choice, of course its cost, weight and other properties are also the factors to consider. The maximum point within which linearity holds is the yield point. Past yield point permanent deformation takes place, the material behaves as a plastic and the slope is no longer linear. The highest point in the stress–strain graph is known as its ultimate strength, beyond which the component will keep deforming, ending up in a rupture seen as a catastrophic failure. The linear portion gives.

$$\text{stress/strain} = \text{constant} = \text{Young's Modulus (the slope of the graph)} \tag{19.33}$$

Sometimes raw material is supplied with a small amount of pre-stretching (strain hardening) with permanent deformation set in which the yield point is higher. Typically, some aluminium sheet metals are supplied with 0.2% pre-stretched strain built in (see Figure 19.9a). However, with pre-stretching, the ultimate strength is unchanged. Figure 19.9b compares various types of typical aircraft materials. The steeper the slope indicates higher stiffness, which often has a higher elastic limit. Brittle materials rupture abruptly without much strain build-up, on the other ductile materials exhibit large strain build-up before rupture giving warnings of failure. Rubber like material does not have a linear stress–strain relation.

More generally the pertinent properties associated with materials are as follows (some are shown in the graph). At some point in the design of the aircraft all of these properties are needed.

Brittleness	When sudden rupture takes place under stress application (e.g. glass).
Ductility	This is the opposite of brittleness (e.g. aluminium).
Hardness	This can be seen as a measure of strength.
Resilience	This is a measure of energy stored in elastic manner, i.e. the strain restores when stress is relieved.
Toughness (fracture toughness)	This is a measure of resistance to crack propagation.
Creep resistance	This is a slow deformation with time under load. Here strain can increase without applying much stress.
Wearability	This is a measure of surface degradation mainly when under exposure, such as corrosion.
Fatigue quality	Fatigue sets up with alternate cycling of applied load. A good material should be able to dissipate vibration energy for a number of cycles.
High temperature strength	Ability to hold strength at elevated temperature.

19.11 Metals

As an initial pointer, it is worth noting some of the detailed properties of metals. Again, the reader will need to refer to specialist references for details but this should provide a starting point. The strength and other properties vary from material to material. Table 19.4 gives some important metallic materials used in the aircraft industry (there is a large variation – only typical ones are given).

Table 19.4 Properties of various types of material for comparison (typical values).

Material	Density ρ – lb in^{-3}	Elasticity, Young's Modulus E – 106 lb in^{-2}	Ultimate tensile E/ρ	Specific strength	Relative cost
Aluminium					
Aluminium alloys					
2014-T6 Alclad sheet	0.101	10.7	68	106	Base price
2024-T4 extrusion	0.100	10.7	57	107	Slightly less
7076-T6	0.101	10.3	78	102	Slightly more
7076-T6-extrusion	0.101	10.4	78	103	Slightly more
Steel alloy					
Stainless sheet	0.276	29	177	103	More
Maraging steel	0.283	29	252	94	More
H-11	0.281	30	280	107	More
Titanium alloy					
Lithium-aluminium alloy					
Nickel-alloy	0.3	31	155	103	Much more
Magnesium-alloy	0.064	6.5	40	102	Expensive
Beryllium alloy-rolled	0.067	42.5	65	634	Expensive

It is worth noting the tensile strength of pure aluminium, especially in comparison with wood that is in fact stronger along the grain. Aluminium really only comes into its own when alloyed with key materials such as copper, which significantly increase the strength. The most modern aluminium alloys are some 10 times stronger than pure aluminium.

As design and manufacturing capability has improved along with new variants of metals becoming available, designers are able to capitalise on some special properties and materials manufacturers invest significant effort in developing new alloys with specific capabilities. For example, at the core of a jet engine temperatures become very high, rising above 600°C. Since the properties of metals degrade with increase in temperature, special alloys of steel and titanium have been developed that retain better strength at elevated temperatures. Components experiencing hot temperature have titanium alloys and stainless steel alloys, which come in many varieties. In the quest to find still better materials, nickel, beryllium, magnesium, lithium alloys have appeared. The more the exotic nature of the alloy, the more is the cost. Typically, an aluminium-lithium alloy is three to four times more expensive than duralumin (2005 price level). But aluminium alloys are still the dominant material used in the aircraft industry. There is a variety of aluminium alloys indicating a wide range available for specific usages. Various kinds of aluminium alloys are designated (classified) with numbering system as given in Table 19.5.

In choosing an alloy for use the temper of the material is a critical element to consider. The properties of an alloy significantly change after heat treatment. These are designated with a T after the alloy classification, for example, 2024-T6. There are 10 heat treatment temper codes T1-T10. The most common in aircraft are T6 and T8. T6 is solution heat treated and artificially aged and T8 is solution heat treated then cold worked and artificially aged. Table 19.6 shows the ultimate and yield stress values for grades of alloy and the fatigue strengths. Note that in moving from T3 to T8 increases the ultimate stress only a little, but the yield stress a lot. But even more interesting, the fatigue strength goes down! So, if the structure in question was a fatigue dominated zone then T3 would possibly be a better option, all else being equal. In many modern commercial jet aircraft, 2024 is used as a skin material and 7075 as a stiffener material.

Table 19.5 Aluminium alloys (there are other types without numerical designations).

Series starting with 1xxx	Pure aluminium
Series starting with 2xxx	Aluminium + copper (e.g. 2014-T6 Alclad sheet[a], 2024-T4 extrusion etc.)
Series starting with 3xxx	Aluminium + manganese
Series starting with 4xxx	Aluminium + silicon
Series starting with 5xxx	Aluminium + magnesium
Series starting with 6xxx	Aluminium + magnesium + silicon
Series starting with 7xxx	Aluminium + zinc – high strength, heat treatable, prone to fatigue (e.g. 7076-T6, 7076-T6-extrusion etc.)

a) Both the surfaces of the aluminium alloy sheet are cladded with copper. These have high electrical and thermal conductivity, corrosion resisting, good formability and are not heat treatable.

Table 19.6 Alloy properties – strength.

Alloy	Ultimate	Yield	Fatigue
Aluminium			
2024-T3	440	290	138
2024-T8	455	400	117
7075-T6	572	503	159
Titanium			
Ti-6Al-4V	950	880	240

Although there is a vast range of exotic alloys available they are extremely expensive. For example, titanium alloy Ti-6Al-4V has a tensile strength twice that of 2024 but typically costs about eight times more. This does not at first sight look like a promising value, however, when it comes clear that even above 400°C titanium holds its strength at above 600 MPa the value looks much better.

The data available on material properties under various conditions and tests are vast, and only a materials specialist will really be able to understand the special factors that will make one material win over another. But for most design scenarios, a basic knowledge of the behaviour of materials, and key parameters such as those used here, is sufficient to progress a design.

19.12   Wood and Fabric

Although no longer a prime choice in commercial aircraft manufacture, it is worth some passing notes on wood and fabric as a combination for airframe construction. This is still a valid form and in use in home-built aircraft and, particularly as environmental concerns become more prevalent, wood is now seen as an environmentally sensitive and sustainable material. From a design perspective, one nice aspect of this form of construction is the clarity in which it represents the structural functions of the elements. Wood is used to take the major structural loads with the fabric forming the skin to transfer the air loads through to the structure and carry some of the shear loads in wing torsion. However, such considerations are not foremost in the mind of the designer who simply wishes to design a good aeroplane.

The difficulty with using natural materials such as wood is that they are subject to variation and open to many problems such as diseases. The wood itself therefore needs to be carefully selected and treated to ensure it is of the best quality. Like metals, it must undergo treatments such as drying to ensure its properties are maximised.

There are four main types of wood used: spruce, birch, ash and Douglas fir. Others of course can be used, but these are dominant. Birch and ash are hardwoods and spruce and Douglas fir are softwoods. These are then quite different to work with for shaping and joining. Laminates are also used for specific purposes particularly to build up structure to carry high loads or reduce stress. Modern machine workshop capabilities mean that home construction can be done to a very high standard with accurate and consistent operations. Glues are an essential part of wood construction and again modern adhesives have improved significantly making wooden construction better and significantly improving structural integrity.

In terms of fabric, in the early days of aviation, cotton, linen and even silk were used as the main skin materials to cover the airframe. These were often doped with starch to make them stiff, the most common of which was sago starch, until cellulose dopes were developed such as Emaillite. Modern fabric is almost always synthetic as these are much more durable materials.

Some study of these older forms of construction is useful for any designer as they expose some of the key aspects of aircraft design. For example, in the early days of aviation the fuselage was a simple exposed frame; its function very clearly set to hold the tail in an appropriate position for manoeuvrability, as exemplified by Louis Bleriot's famous aircraft the Bleriot XI. The function was purely structural. Increased understanding of aerodynamics and the significant increase in drag caused by this open framework led to the use of fabric to cover the frame. The function of fabric being an aerodynamic surface reducing the drag caused by the frame. And so the advent of the stressed skin construction that remains dominant today is clear, and its origins in satisfying a functional need equally clear. There are many such insights to be garnered from thinking about older generation aircraft and modern home-built aircraft that are constructed with this method.

19.13   Composite Materials

The preceding section on wood and fabric is interesting in the context of the most modern materials available in aircraft manufacture. These wooden constructions were often laminates, made up from layers of material bonded together to act as a single unit. Essentially, this is what is now commonly referred to as a composite material.

By definition, composite material is the combination of more than one material to bring out product line superior to what was used for fabrication of an item. The concept is as old as adobe buildings made of straw fibre in clay as binder, later reinforced concrete is another example as building material. Nature has abundant examples of composite structures in many kind of plants – for example, flax plan fibres (fabric of Irish linen), cane and so on – that show considerable strength along the fibres. Bakelite, a man-made composite for making small insulating (electrical) components, has been available since the beginning of the twentieth century, but it did not have the physical properties for load bearing structural usage.

As wood became expensive (subsequently with growing concerns on environmental issues), search for synthetic substitute resulted in development of fibreglass reinforced plastics (FRP). Many varieties of FRP appeared exhibiting good strength-to-weight ratio, thermal insulation, less corrosion and so on; most importantly, it can be moulded in complex curvatures in a relatively easy manner compared to metal forming. Its practical usage in aeronautical industries gained ground after World War II. It was not in a position to compete with metal to be used in load bearing structures and there was cost involved, but it showed the promise in the search for superior composite materials to overcome the limitations (see later) of FRPs. Today, it has developed to such an extent so that modern aircraft are made of composite materials, some small aircraft are made of all-composite structures. Composites save weight and are easy to fabricate 3D shapes from. CFRP is the most commonly used composite material.

The fibre element or strand (yarn) of composite materials is made of many filaments. The strands are supported in a matrix resin acting as a binder and this transfers the load to the fibre. The main load bearing material is capable only in the direction of fibre run and hence has unidirectional property. The material is therefore typically anisotropic, unlike isotropic metals that have uniform physical properties in all directions. This book gives only the basics that conceptual aircraft designers should be aware of. For comprehensive coverage on composite materials, the readers may refer to Refs. [2, 10].

19.13.1 Composite Materials, Fibre and Fabric

The fibres are naturally unidirectional with strength along the length. Normally, fibres are woven into fabric, which provides strength in many possible directions. The most common ones are woven orthogonally. The fibres are woven into fabric in such a way that fibres cross each other or in groups, with several possibilities for distributing the fibre properties in different directions as woven. The basis of weaving is to lay vertical (warp) strands and horizontal (weft or fill) stands alternately crossing each other over and under. The woven fabric is then laid in layers sandwiched with resin in between as a bonding compound. In each chosen orientation of fibre pattern to form a composite laminate, the ply number is the number of fabric layers laid.

Some of the most well-known weave patterns are plain, twill, satin, harness weave, basket weave, unidirectional weave and bidirectional weave. In aerospace usage, the plain, twill and satin weave types are dominant, possibly the twin is the most commonly used. These are briefly described in the following and also shown in Figure 19.19. Other types are not dealt with here.

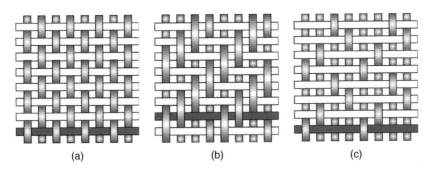

(a)   (b)   (c)

Figure 19.19 Fabric weave type. (a) Plain weave. (b) A 2 × 2 twill weave. (c) Four-harness satin weave.

19.13.1.1   Plain Weave

The simplest composite fabric weave pattern is the 1×1 plain weave when one warp fibre strand goes over one weft strand, then under the next weft strand and so on as shown in (Figure 19.19a). It has good stability (holding strand position, firmness) and satisfactory porosity but is not so amenable to wrap/drape, it tends to wrinkle, that is, it is relatively more difficult to form curvature. This suits for flat surface board-like structures.

19.13.1.2   Twill Weave

This is the next level of weave pattern, more than one strand of equal numbers in both the warp and weft directions alternate like plain weave. It is distinguishable by its herringbone pattern texture. When a bundle of two strands of warp intersect a bundle of two strands of weft, this is known as a 2×2 twill weave (Figure 19.19b). This maintains its stability and is better than plain weave at forming curvature. This type of fabric is widely used in aerospace industry. If the strand count in warp and weft increase in equally numbers, then the fabric is identified by the number of strands; for example, if there is a bundle of four strands intersecting in each direction, it becomes a 4×4 twill weave. Twill is better at forming curvature than plain weave.

19.13.1.3   Satin Weave

This is like twill weave with the difference that the weave pattern warp and weft are not the same. When one weft strand crosses over a bundle of three strands of warp fibre, this is known as a $(3 + 1) = 4$ harness satin weave (Figure 19.19c – also known as Crawford satin). When the one warp strand crosses over a bundle of three strands of weft fibre this is known as a four-shaft satin weave. Because there are is large number of strands in the bundle in close proximity, this gives a shiny texture, hence it is known as satin weave. The strand numbers identify the satin weave, for example, five harness or five shaft satin weave. It is the least stable of the three but has better formability.

There is a large number of possibilities to make twill and satin weave patterns. In addition, there are many other kinds of weave pattern not dealt with here.

19.13.2   Types of Synthetic Composite Material Fibre

There are many types of synthetic composite fibres, for example, fibreglass (E- or S-glass), polyester, carbon, aramid (trade name: Kevlar, Nomex, Conex etc.), boron, graphite and so on, with a wide range of physical properties and cost. Selection of material is based on the suitability for the application at low cost. Currently, carbon fibre is dominantly in use.

19.13.3   Matrix Bond for Composite Material Fibre

Although the composite fibres in themselves are strong, and when woven into fabric sheets they provide strength in different directions, they need to be brought together to really become a super strong material. To achieve this, a matrix material is used to bond fibres and sheets together to make a solid structural unit (Figure 19.20). The matrix material is often a resin of sorts, which when cured forms a solid connection between the fibres and helps them all act as a single unit rather than a loose collection of strands or sheets. The matrix is the support and bonding material that acts as adhesive to bind fabric as reinforcement to transfer and distribute load throughout the structure.

The matrix material is therefore a critical part of the composite material. There are three main types matrix as follows.

   i. Polymer
  ii. Metal
 iii. Ceramic

Composites most often use a polymer matrix, as these resins are more accessible and cost effective. The liquid resin can be more easily poured over the composite fabric and then cured so it sets hard. This type of hard

Figure 19.20 Layers of composite fabric are bonded together with a polymer matrix (e.g. resin) to form a single solid structure.

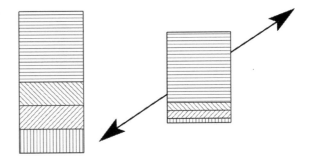

resin is called a *thermoset*. There is a number of types of resin such as acrylic, polyester or polyurethane, but the most common in aircraft manufacture is epoxy resin. In complex structures the resin is drawn through the woven fabric lay-up and cured under pressure in an autoclave.

It is of course also possible to have metal and ceramic matrix materials, but these are far less common and remain a very specialist area of development.

19.13.4 Strength Characteristics of Composite Materials

Since the fibre is the dominant load carrying element within a composite lay-up it is very amenable to designing a material structure that is tailored to specific loading. For example, if most of the end load on a structure acts at 45° then aligning the fibres that way can ensure the material is most effective. This tailoring of the material is one of the most powerful aspects of composites. The challenge, however, is for the analyst to understand fully the many complex loading scenarios and stress distributions in the material to effectively design the material. The governing equations for composite materials are complex and this is no easy task.

The analysis of a composite material to determine failure stresses is an even more complex procedure and even now theories are still being developed that can characterise the failure mechanisms accurately and reliably. Although the fibre is very strong, the resin matrix is far less so and can crack or fail under a much lower load. If such a failure occurs then layers of the composite may separate and no longer act as a unit. This is delamination. There are many other modes of failure, such as fibre pull out, matrix cracking and so on.

As this is such a complicated analysis, most designers use simple approximations to allow some estimation of strength appropriate to this stage in design. Typically, the structure may be treated as a metallic like structure and the stresses compared on a similar basis. This gives an estimate of the potential strength of the structure and hence also its weight. Specialist stress analysts can work through the details at a later stage.

19.13.5 Honeycomb Structural Panels

The preceding sections noted the key materials and general considerations for their use an application in aircraft. When this is brought together with further consideration of the structural configurations and how both geometry (i.e. shape) and material are combined to produce a strong, stiff panel, other 'types' of construction become very useful. One common form arises from the need to increase stiffness and bending strength of panels without having to add much weight. By separating the outer skin layers and filling the interior with a strong shear resistant configuration, a structural strength and stiffness increases significantly. The interior layer is made from hexagonal shells and for that reason these are called 'honeycomb panels'.

Honeycomb panels are extensively used in aircraft industry, mostly of metal fabrication but they can also be made of non-metals, for example, Nomex/Kevlar core with Kevlar/carbon-fibre surface skins layers and so on. Honeycomb panels may be seen as composite structure comprising of top and bottom sheets, sandwiched in between a honeycomb core panel glued with adhesive layers (Figure 19.21). It is named honeycomb as it resembles with columnar hexagonal bee-hive honeycomb. Strength improves with thicker core and skins. A high quality strong adhesive layer is essential, otherwise delamination can occur. Honeycomb panels are orthotropic requiring orientation to suit load application, best applied normally.

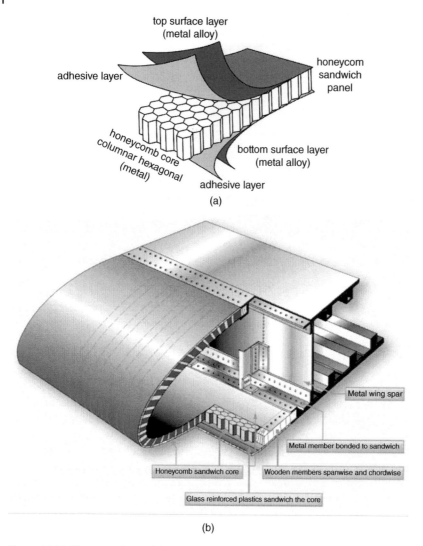

Figure 19.21 Honeycomb panel. (a) Honeycomb panel details. (b) Curved honeycomb panel application on wing. Source: Courtesy: Aircraft Structure (Chapter 1) FAA.

These are generally purpose-built panels to suit applications and are supplied by manufactures in plate form coated with PVDF (*polymerising vinylidenedifluoride*) film to protect from scratches, corrosions and so on, during transportation and in storage. If required, custom-made honeycomb panels are designed and manufactured to suit the structural requirements. With appropriate manufacturing process stress-free curved honeycomb are manufactured. Figure 19.21 shows use of a curved honeycomb panel as the wing leading edge.

The panel cores can be other shapes than hexagonal, for example, corrugated cores or even foam filler but the hexagonal cross-section is most efficient. Small aircraft use a wire-cut foam filler as these are easy and cheap fabricate, yet support loading adequately.

Listed here are some of the advantages and disadvantages of using honeycomb panels.

Advantages:

Honeycomb panels offer the highest strength-to-weight ratio for the materials used. They have excellent rigidity at minimal weight and offer high stiffness.

Honeycomb panels act as insulators for noise vibration propagation: they are used as nacelle acoustic liners.

They are light weight, facilitating easier fabrication processes and save on labour cost. This also reduces the part-count and maintenance costs are relatively low.

Disadvantages:

They are not suitable for concentrated high loads.

Manufacturing quality issues may result into delamination, not easily visible.

If not sealed properly, honeycomb panels are prone to water ingress that is not easily detectable and can cause severe corrosion.

Storage of water filling the hollow cells in the exposed panels can occur and weight gain occurs in a hidden manner. If this happens to the elevator, then flying may prove unsafe.

Applications:

Honeycomb panels suit normal distributed loads. High concentrated load applications should be avoided. The following are some examples of typical applications in aircraft industries.

1. Fuselage floorboards	6. Doors
2. Control surfaces	7. Overhead storage doors
3. Nose cone/radome	8. Galley, furnishing structures
4. Tail cone	9. Cabin interior panelling
5. Nacelle liners/cowlings	10. Support structures

19.14 Structural Configurations

Having considered some of the materials that could be used in the design, the next issue is the overall configuration of the structure: that is, the components and sub-structures when assembled together form the overall airframe. This is important at this point for two reasons. First, with clever design of the structure, much weight can be saved (or lost with poor design). Second, the arrangement of internal members can affect the space available for important things like fuel, control systems, control surface arrangements and so on. In this section, a fresh look is taken to examine what is under the skin of the airframe and highlight key points for consideration. In turn, the constituent components under the skin are considered, for example, fuselage, wings, spars and ribs, and as an example of other major structures, the wing-fuselage junction and the empennage.

19.14.1 Skin

As noted before, in the early days of aviation, most of the structure was an open framework, the wing being the only major element with a surface, being the main lifting surface. As knowledge of aerodynamics, and particularly drag, improved it was recognised that a skin covering the frame enhanced performance significantly. The skin had no structural function, its presence there was to form an impermeable aerodynamic surface.

As structural design methods and philosophies developed, it was recognised that a non-structural skin was a dead weight and its presence could be used more effectively. By using a stressed skin construction, the skin could help improve structural stiffness and strength. But the continuing growth in the size of aircraft to carry many passengers rather than a single adventurous aviator and, subsequently, the advent of stratospheric flight made the need for a structural skin essential, with the loads in a pressurised fuselage being significantly more testing than in early aircraft. This led to the dominance of the semi-monocoque form of airframe that is so prevalent today. In this, the skin both performs its aerodynamic function and carries pressure and some bending loads.

In current designs, functionally, the skin transmits aerodynamic forces to the ribs and stringers in the wing, and similarly to the frames and stiffeners in the fuselage. Additionally, it resists shear from both torsional and bending loads.

19.14.2 Fuselage

Chapter 5 deals with geometric considerations of fuselage design. A typical structural layout of load bearing fuselage components is shown in Figure 19.22. They constitute of circumferential ring-like frames placed to normal and longerons running along the fuselage frame offering structural integrity to withstand imposed load. The skeletal frame is wrapped with a cover, in the figure this is an aluminium alloy skin that also supports tensional load. Figure 19.22b depicts a C-series fuselage frame in a production line showing frames and longerons, as well the floorboard that supports the passenger cabin and the below-floorboard container space (also see Figure 7.16b).

In most modern commercial aircraft the fuselage carries passengers in the stratosphere and it must therefore be pressurised. Of course, for smaller, lighter aircraft, pressurisation is not necessary because the fuselage there mostly has a different purpose, which is to hold the empennage in the right place. In the case of pressurised aircraft, the fuselage is typically circular as this is the most efficient from a stress perspective and is also relatively easy to manufacture. However, in practical terms, a floor is needed to attach seats to and, since this creates an underfloor space, this is used to carry cargo. Consequently, the fuselage is sometimes made slightly elliptical in shape to allow the opportunity to maximise the combination of both passengers and cargo.

Since the fuselage is a pressure vessel it must be sealed. This is done through bulkheads fore and aft (Figure 19.23). The bulkheads are heavy structures. From a manufacturing perspective, it is easier to make them flat stiffened plates, but this is not the most efficient structurally, so they are sometimes made with a slight curvature. This makes them more expensive, but more efficient. Normally, the whole fuselage is pressurised but some aircraft have used a pressurised floor, which is difficult to design and can add weight. Pressure bulkheads cause additional design challenges with the need for cables, ducts and so on to pass through, all of which must be sealed.

19.14.3 Wings

The wing of any aircraft is the critical system to be concerned about. It provides the lift needed for the aircraft to remain aloft and as such has very stringent geometric constraints both in its cross-section and deflected shape under load, both of which are critical to performance. Under the skin of the wing, the structure is conceptually similar to the fuselage. The main structural element in the wing is the spar. The spar carries the bending loads that are in effect reacting the forces to keep the whole aircraft flying (Figure 19.24). The wing

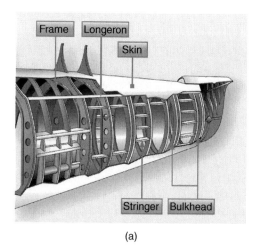
(a)

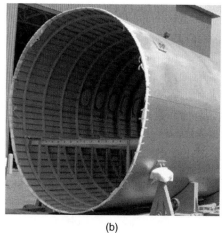
(b)

Figure 19.22 Aircraft fuselage structure. (a) General fuselage structure. (b) C-Series fuselage. Source: Courtesy: Aircraft Structure (Chapter 1) FAA.

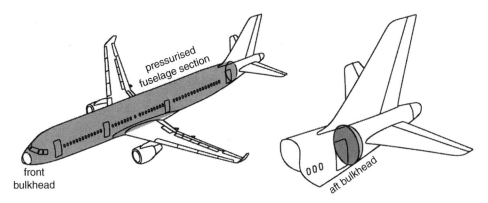

Figure 19.23 Pressurised cabin between two fuselage bulkheads.

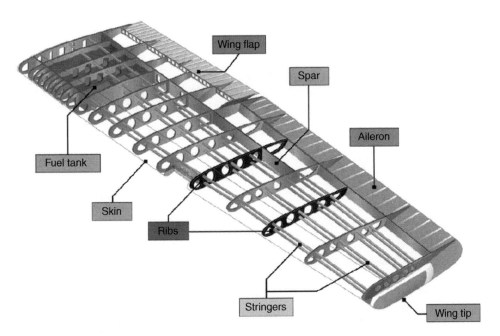

Figure 19.24 Wing structure Source: Courtesy: Aircraft Structure (Chapter 2) FAA.

also has stiffeners, more commonly called *stringers*, in the wing that help the spars with the bending load and help to stabilise the skin to ensure it maintains its aerodynamic shape and does not buckle. It also has ribs periodically spaced along its length that

maintain the aerofoil shape and help to stabilise the skin and the stringers.

Although the primary function of the wing is to provide lift, this is also where control surfaces can be most effectively placed and where fuel can be stored. In many aircraft, the engines and undercarriage are also hung from the wing, adding both complexity and weight to the structure. At this point in the design though, the key thing to consider is the influence of the control surfaces and the need to store fuel. Both these factors directly influence the strength and stiffness of the wing. In particular, the location of front and rear spars is directly connected to these decisions.

Figure 19.25 shows this. If the location of the rear spar is dictated by the chord size of the main flap (c_f), from which it will be attached, then it can be seen that this choice directly impacts the available volume of

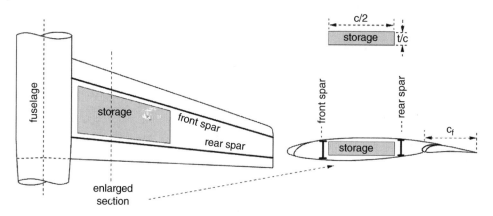

Figure 19.25 Wing planform showing potential fuel storage area.

fuel, if it is stored in the wing centre, and also the stiffness off the wing in that this is now defining the size of the torsion box. Both the torsional stiffness, J, and the shear stress in the skin, τ, are affected directly by the area of the torsion box.

In many cases, the initial sizes from the concept design show that these choices are all easily accommodated for. However, it is very easy to get to a point where a thin wing, needing a larger flap, has an extended range to cover, and suddenly these calculations and corresponding decisions become very important. This simple check is often omitted by students, but it can cause a rethink in the overall concept of the aircraft.

19.14.4 Spars

The primary structural element of the wing, and therefore arguably for the whole aircraft, is the main spar (Figure 19.26). The main spar is essentially a structural beam that carries the wing bending loads. If the spar can be located and designed correctly much of the rest of the design falls easily into place. The spar can be a single entity but wings often have two, one front and one rear, so that together with the skin they form a closed box, which is very strong in torsion. Much of the structural strength of a wing is due to the torsion box. The spar itself is usually a single machined piece, sometime including lightening holes to reduce weight. In the past it was often built up with many rivets and was a heavy component. The spar is idealised as a beam and designed very much as any typically cantilever beam is designed.

Because the wing both bends and twists and, to counter this, the torsion box arrangement (i.e. a box beam) is used, there are usually two spars, one front and one rear. The front one usually takes the greater portion of the bending load and a ratio of 2:3 front to 1:3 rear is a common assumption at this point in the design.

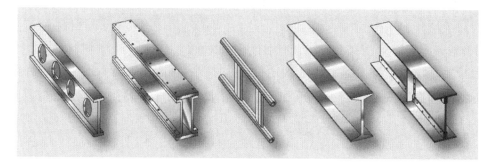

Figure 19.26 Spar types. Source: Courtesy: Aircraft Structure (Chapter 1) FAA.

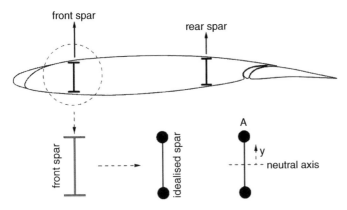

Figure 19.27 Idealised wing spar.

For structural analysis, the spar is usually idealised as a beam with booms and a shear web. The assumption being that the spar caps take all the end load from bending and that the web takes all the shear load. This idealisation is shown in Figure 19.27.

This very simple idealisation is a key concept and allows for a quick estimation of the proposed sizes. It allows for a sanity check on the proposed dimensions of the wing and the material choice. A simple example is shown in Section 19.15 later.

19.14.5 Ribs and Stringers

The main function of the ribs in the wing is to maintain aerodynamic shape of the aerofoil, critical for performance. The ribs act in concert with the skin to distribute the air loads through the structure. They also perform an important stabilising function for the stringers by providing end support for sections and increasing the buckling strength of the structure. Figure 19.28 shows typical wing spars, ribs, stringers and skin.

19.14.6 Fuselage Structural Considerations

Although a circular cross-section is the most desirable from stressing (minimise weight) and manufacture (minimise cost) points of view, the market requirements for the below-cabin floor-space arrangement could result in a cross-section elongated to an oval or elliptical shape. The Boeing 747 with a narrower upper deck width is a unique oval shape in the partial length that it extends. This partial length of the upper deck helps cross-sectional area distribution (see Section 5.10.2) and area ruling. Figure 19.29 shows that wing-box

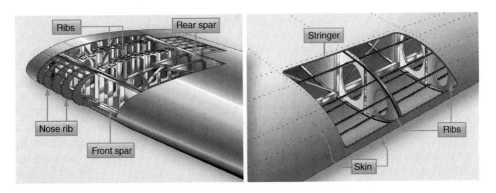

Figure 19.28 Wing Structure. Source: Courtesy: Aircraft Structure (Chapter 1) FAA.

Figure 19.29 Fuselage Structure LD3 container in A300 Source: Wikipedia.

structure passes below the floorboard (near mid-wing aircraft). It shows the underfloor space to accommodate LD3 containers snugly fitted into a circular cross-section fuselage either side of wing box.

A bulkheads are load bearing frames placed transverse to the hull, in this case these are wall-like frames placed transverse to the fuselage shell. Aircraft cruising above 15 000 ft altitude need to be pressurised for passenger safety. All large commercial aircraft have a front pressure bulkhead at the nose cone and an aft pressure bulkheads (or rear pressure bulkhead at the tail cone of the aircraft – Figure 19.23) to seal the fuselage shell from the flight deck in front to the entire usable length of passenger cabin in a leak-proof manner to keep the interior pressurised. Sealing of electrical cables and hydraulic pipes passing through is done as required.

19.14.7   Assembly and Wing Box

Although the preceding sections cover the main elements of the airframe it is worth mentioning two other key elements. The first is the wing carry through box that is prevalent in modern jet aircraft. As noted previously, the torsion box, that is, the section between the front and rear spars, is the key load carrying structure in the aircraft. To maintain structural integrity, this box normally runs through the fuselage and maintains its structural form. The idea being to allow the spar to run interrupted from wing tip to wing tip as any joint will reduce structural strength. There are exceptions to this, but they are all known to be very difficult design challenges. Interestingly, in larger aircraft, the design of the carry through itself is a major design challenge due to the loading intensity at this point.

Figure 4.7 shows typical choices for wing position in relation to fuselage. Figure 19.30 shows wing-box structure passes below the floorboard (near mid-wing aircraft).

Figure 19.30 B787 wing-box structure. Courtesy Boeing.

Where the wing meets the fuselage is also a challenging structural problem for the fuselage. In this case the main fuselage frames here must react the wing shear and torsion loads. Typically, the frames will be very heavy and attached to the wing structure with pins to ensure that only direct axial loads are transferred into the frame itself.

19.14.8 Empennage

Structurally, H- and V-tails are like small wings with smaller planform areas and aspect ratios carrying considerably less load, just what is required for stability and control. They are not. practically required to contribute to aircraft lift, hence there is no requirement for high lift devices. However, the control load has to be sustained by a long and strong fuselage.

The wing joint to the fuselage must ensure the fuselage shell keeps the passenger cabin space to offer maximum passenger space for comfort and movement, hence where and how it is attached requires careful consideration as discussed in Section 4.4.2. This is not the case with the empennage mounted at the tail cone where its narrow volume only houses the control linkages and accessories. Empennage structural layout does not interfere with positioning of control linkages and other devices. While there is similarity with wing structures, the empennage structural design is simpler.

The V-tail is attached to the fuselage at the aircraft symmetric plane; typically, its spars are attached to the near circular fuselage frames. It looks like a banjo and is loosely known as banjo-frame (Figure 19.31a). The rudder is kept free by hinging along the fin trailing end.

Structural considerations of the H-tail attachment to the fuselage are dictated by its position with respect to the fuselage, arising from aerodynamic, stability and control considerations. Aerodynamic loads generated by the empennage require a strong and stiff fuselage structure so that fuselage flexure does not adversely affect empennage control effectiveness. The attachment joint H-tail requires special considerations because part or the whole of the H-tail can be movable (Figure 19.31b). The H-tail attachment could be at the fuselage (low tail) or at the V-tail (T-tail/mid-tail).

A multi-spar empennage arrangement is shown in (Figure 19.31c).

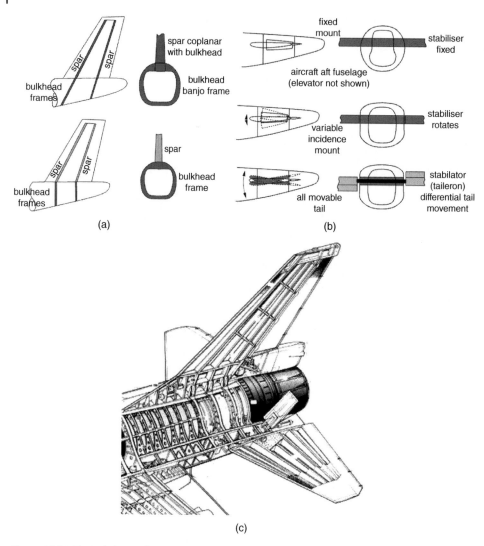

Figure 19.31 Typical choices for empennage joints. (a) Vertical tail. (b) Horizontal tail. (c) Multi-spar empennage. Source: Reproduced with permission from Cambridge University Press.

19.15 Rules of Thumb and Concept Checks

One of the key questions asked at this point in the design process relates to why this is of concern. Having an idea of the overall concept is often considered to be enough and the rest is a detail for later. However, as noted in this chapter, such a philosophy can lead to error and major issues for the design later in the process, sometimes to an extent that cannot be reversed.

In addition to becoming generally informed about the potential materials and structural configurations at this point, it is possible to use these principles and key items of information to carry out a sense check on the major dimensions and assumptions in the design. These are based on sound structural theory and do not require detailed knowledge of the other aspects of performance.

So, key things that can be checked are:

i. What is the maximum bending moment and shear force at the wing root?
ii. Does the structural layout allow for enough fuel?

Figure 19.32 Problem 19.1 loading.

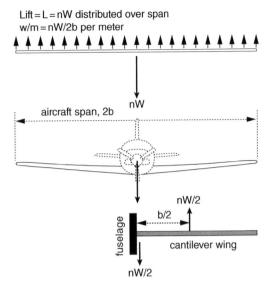

Lift = L = nW distributed over span
w/m = nW/2b per meter

nW

aircraft span, 2b

nW/2

b/2

fuselage

cantilever wing

nW/2

nW/2

idealised free body force diagram (wing semi span)

iii. Does the initial material choice indicate spar caps of a reasonable size?
iv. Is the wing stiff enough?

Exercise 19.1   Examining the example of a light aircraft like a Cessna 172 in practical terms, attempts are made here if the likely size of the internal wing structure will hold the design loads and is something which can be manufactured (Section 4.9 shows how to compute wing load using Schrenk's method).

First, idealise the aircraft in terms of the main wing. Consider that the wing uniform span lift (load) distribution of load 'w' per metre across the wing span of 2b (Figure 19.32). The units for load are N of lb_f for the system adopted. Wing load distribution patter is shown in Figure 4.20.

Next, note the main parameter values coming from the concept design calculations as given here:

Aspect Ratio, $AR = 7.32$	Aircraft weight, $W = 11\,000\,N$
MAC = 1.3 M	Wing area, $S_W = 16.2\,m^2$
Thickness to chord ration, $t/c = 0.195$	Wing fuel storage volume = 212 l
Limit load factor, $n = 3.8$	

Given these, the structure can be idealised ready for analysis. In this case, the loading is approximated as a uniformly distributed load acting over the whole wing surface and the weight of the aircraft acts at the centre of the wing. It is assumed that the limit load factor for this aircraft is +3.8.

To analyse, this structure is considered symmetrical and our analysis model is a cantilever beam representing one wing. Although a distributed load along the beam can be directly analysed, an alternative idealisation is to use the resultant force on that half of the span, which is $W/2$ acting at the mid-point. This will give a slightly different bending moment distribution but will give the same maximum bending moment, and importantly for this point in the design, is suitable for mental arithmetic.

It can now address each of the four questions in turn.

i. What is the maximum bending moment and shear force at the wing root?

The maximum shear at the root of this kind of wing must always be $nW/2$ – from equilibrium.

Maximum shear force at wing root,

$$S_{max} = (nW/2) = \frac{3.8 \times 11000 \times 5.5}{2} = 20.9 \text{ kN}$$

The maximum bending moment from standard formula is:

$$M_{max} = \left(\frac{nW}{2}\right) \times \left(\frac{b}{2}\right) = \left(\frac{3.8 \times 11000W}{2}\right) \times \left(\frac{5.5}{2}\right) = 57.476 \text{ kNm}$$

ii. Does the structural layout allow for enough fuel (see storage dimension in Figure 19.25)?

To do this, an estimate of the volume available in the wing for fuel is computed. If an initial assumption is made that the front spar is at 25% chord and the rear spar is at 75% chord and we have available space out to a quarter of the semi-span, then the volume available (assuming that both wings are used) is:

$$(Volume)_{max} = 2 \times \left(\frac{c}{2}\right) \times \left(\frac{t}{c}\right) \times (0.25b) = 2 \times \left(\frac{1.3}{2}\right) \times (0.195) \times (0.25 \times 5.5) = 0.348 \text{ m}^3$$

This is enough for almost 350 l of fuel. This of the class of Cessna 172 aircraft with a total of 212 l fuel in both the wings.

iii. Does the initial material choice indicate spar caps of a reasonable size (see Figure 19.29)?

To make this estimation based on simple bending theory, use Eq. (19.14). To simplify, use the idealisation from the earlier description where the spars are idealised as booms and a shear web, and will assume that the front spar takes two-thirds of the load.

$$\sigma = \frac{My}{I}$$

where $y = \frac{t}{2 \times c}$ and moment of inertia, $I = 2Ay^2 = 2 \times A \times \left(\frac{1}{4}\right) \times \left(\frac{t}{c}\right)^2 = \left(\frac{A}{2}\right) \times \left(\frac{t}{c}\right)^2$.

From the known maximum bending moment, and assuming then that two-thirds is taken by the front spar, the first task is to obtain an expression for the maximum stress in terms of the key design variables; weight, span and aerofoil thickness.

Therefore,

$$\sigma_{max} = \frac{(M_{max}) \times \left(\frac{t}{2c}\right)}{\left(\frac{A}{2}\right) \times \left(\frac{t}{c}\right)^2} = \frac{(M_{max})}{A \times \left(\frac{t}{c}\right)} = \frac{(n \times W \times b)}{6A \times \left(\frac{t}{c}\right)}$$

By transposing and rearranging this equation, the area of the spar caps can be determined as given here.

$$A = \frac{(n \times W \times b)}{6 \times \sigma_{max} \times \left(\frac{t}{c}\right)}$$

In this example, let's assume the spar will be made from aluminium alloy 2024-T3 with a yield stress of 290 MPa.

$$A = \frac{(3.8 \times 11000 \times 5.5)}{6 \times (290 \times 10^6) \times (0.195)} = 6.78 \times 10^{-4} \text{ m}^2$$

If this was a square block of material, it would be about 26 mm along each side. Assuming sheet material of 6 mm thickness is used to build the spar cap then the section that follows would give an area of $6.8 \times 10^{-4} \text{ m}^2$, which would be adequate.

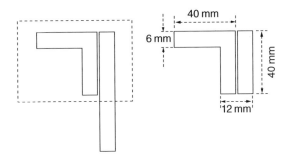

Bearing in mind the maximum thickness of the section is 0.195 m or 195 mm, then this looks like a feasible design, although the spar caps look a little big with respect to the overall size of the wing. With a depth of 195 mm the two caps take up 40% of the depth, which is perhaps a little too much, but something detailed structural sizing can address.

iv. Is the wing stiff enough?

One last check to consider is the overall stiffness of the wing. The quickest check to do at this point is simply to estimate the tip deflection. A reasonably small deflection would be expected in an aircraft this size.

Again, using simple bending theory, for a uniformly distributed load, the equation for tip deflection is:

$$\delta = \frac{w \times b^4}{8EI}$$

In this case $w = nW/2b$, and E for this material is about 70 GN m^{-2}. It must also be remembered that now this is for the whole wing, so the inertia I will be for the two spars. If it can be assumed that they are both the same size, then:

$$\delta = \frac{n \times W \times b^3}{16EI},$$

where $I = A \times \left(\frac{t}{c}\right)^2$.

$$\delta = \frac{3.8 \times 11000 \times 5.5^3}{16 \times (70 \times 10^9) \times (6.78 \times 10^{-4}) \times (0.195)} = 0.241 \text{ m}.$$

At this point, the simple calculations indicate that the size of the structure with this chosen material is in the order of magnitude expected so at this point the design could proceed with this basic concept. However, perhaps there should be some caveats to note for the design team. Although the limit load factor is used as safety factor when the area for the spar booms was calculated, the assumption was that the material was fully stressed. To keep the material within stress limits the size of the spar caps looks a little too big and the deflection of the wing tip a little too much.

So, based on the figures used here the overall sizing looks reasonable and maybe this design can proceed as is. But in most cases, while this can be worked through with a more detailed structural design, it would be worth considering a variation on the wing configuration, either by adjusting the performance, changing the wing planform, changing the material or adding a wing strut to reduce the bending loads. It is worth noting that the real Cessna172 and many aircraft like this have struts to provide additional support for the wing.

The same procedure can be followed for other types of loadings, say torsion and so on. The first task is to compute the maximum stress (normal/shear) and then size the component to withstand the load with safety factor incorporated.

This is the key result here. Some simple calculations have affirmed the sizes of structure for the chosen material are a reasonable starting point, but a question has been raised to the design team that maybe other options either at concept level or detail level should be considered. The team thus has now quantified data on which to base further design decisions and studies. At this early stage of design, this is about as good as can be expected with much further detailed work on all aspects needed to resolve all the uncertainties and finish with a robust aircraft.

19.16   Finite Element Analysis (FEA)/Finite Element Method (FEM)

As modern computational tools have developed, many basic equations have been embedded in computer programmes, but in industry far more sophisticated computational methods are used to obtain more detailed stress and structural loads data. In this introductory chapter, the role of FEA/FEM is only briefly described. In industrial usage, the terms FEA and FEM are synonymous. The FEM is the underlying theory and approach that is used in FEA processes. Industrial application of FEA/FEM is relatively new but has now become the dominant approach for structural design. Although the method was only developed approximately four decades ago, it has now matured to the point where the entire airframe can be analysed in detail.

The FEM has been built upon basic structural theory from continuum mechanics to enable complex geometries and behaviours to be solved. Essentially, the geometry is divided up into small pieces, usually triangles and rectangles or hexahedra and tetrahedra in solids, so that a system of simple equations can be generated to allow the solution to be found in a piecewise or bitesize fashion. The advent of computers has made this a very powerful approach. Sophisticated software applications can do this subdivision (called meshing) very quickly, and automatic solvers can solve the resulting equations in many cases almost instantaneously. In a modern industrial aerospace company it is common to find the internal loads and stresses for linear static loads for the entire structure, and to do this for more than 250 000 load cases. It is a level of detail unimagined by previous generations of engineers, but the underlying principles remain the same.

Finite element systems have also advanced to a point where complex non-linear problems such as buckling, contact and fracture can be analysed and, more recently, multi-disciplinary analysis with aerodynamic loads, thermal loads and electromagnetic loads are all being considered for the one airframe structure.

However, the effort needed to build the computer-aided geometry and ready the models so that this automatic meshing and analysis can take place remains a major effort and, as such, idealisations and simplifications

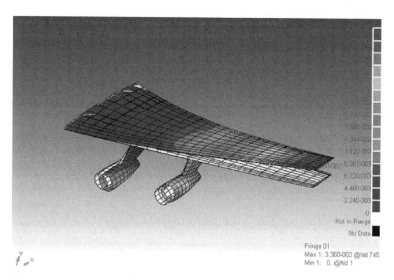

Figure 19.33 Wing FEM analyses – shades indicate stress levels explained in the side bar. Source: www.mscsoftware.com/industry/aerospace.

as used in this book are still very necessary at the conceptual design stage. These very advanced tools are used more extensively in the preliminary stage and completely dominate work in the detailed design stage.

Figure 19.33 is an example. However, in aerospace (and other safety critical industries) structural stress analyses must be validated by physical testing. Such strength tests are mandatory requirements by the certification authorities to allow the first flight.

It can be seen that FEM is a subject on its own. Today, all academic institutes offer the introductory courses at the undergraduate level, as an essential topic of study. However, both designers and structural experts all still carry the fundamentals through and these remain the foundation for supporting key design decisions.

References

1 Michael, C.Y. (1996). *Composite Aircraft Structure*, 3e. Hong Kong: Commilit Press.
2 Michael, C.Y. (1988). *Aircraft Structural Design*. Hong Kong: Commilit Press.
3 Bruhn, E.F. (1965). *Analysis and Design of Flight Vehicle Structures*. Tri-State Offset Company, Cincinnati.
4 Megson, T.H.G. (2016). *Aircraft Structures for Engineering Students*, 6e. Oxford: Butterworth-Heinemann.
5 Benham, P.P. and Crawford, R.J. (1996). *Mechanics of Engineering Materials*, 2e. New Jersey: Prentice Hall.
6 Popov, E.P. (1998). *Mechanics of Materials*, 2e. New Jersey: Prentice Hall.
7 (2003). *Material Data Book*. Cambridge, UK: Cambridge University Engineering Department.
8 Ashby, M.F. (2005). *Materials Selection in Mechanical Design*. New York: Elsevier.
9 Ashby, M. F., Granta Design, *Material and Process Selection* Chart, CES 2009. EDUPACK, Engineering Department, Cambridge, UK, 2005.
10 Baker, A., Dutton, S., and Kelly, D. (2004). *Composite Material for Aircraft Structure*, 2e. Reston, USA: AIAA.
11 Pora, Jérôme. *Composite Materials in the Airbus A380 – From History to Future*, Airbus, Large Aircraft Division 1 Rond Point Maurice Bellonte 31707 Blagnac Cedex, France, Proc. of the 13th International Conference on Composites (ICCM-13), Beijing, China 2001, 1–10.
12 Becker, H. *Handbook of Structural Stability*, NASA Technical Note, NASA-TM-89720, 1957.
13 Bulson, P.S. (1970). *The Stability of Flat Plates*, 1e. London: Chatto & Windus.
14 Timoshenko, S.P. and Gere, J.M. (1961). *Theory of Elastic Stability*, 2e. McGraw-Hill Book Company Inc.

20

Aircraft Manufacturing Considerations

20.1 Overview

Cost analysis and manufacturing technology are subjects on their own requiring specialised sessions in academies. These are not the main topic of this book but are included to make readers aware that the mere classical aeronautical subjects of aerodynamics, structures and propulsion are not sufficient considerations for a successful aircraft design. Cost analysis and manufacturing technology needs to be brought into consideration during the conceptual study integrated with the classical aeronautical subjects. The following terms will be used in this chapter quite extensively. Some of them were referred to earlier.

DBT – Design Built Team. This is a team of hand-picked experienced engineers/specialists drawn from various related disciplines who synthesise designs for Design for Manufacture and Assembly (DFM/A) considerations in multi-disciplinary interactions with the classical subjects.

DFM/A – Design for Manufacture and Assembly (DFM/A). This is an engineering approach with the objective to minimise cost of production without sacrificing design integrity.

IPPD – Integrated Product and Process Development (also known as concurrent engineering). This offers an environment in which DBT uses IPPD to synthesise the trade-off studies in a multi-disciplinary study to arrive at the best value for the product as a global optimum rather than optimising to a particular design study. DFM/A is a part of IPPD.

DFSS – Design for Six-Sigma. The DFSS is an integrated approach to design with the key issues to reduce any scope for mistakes/inefficiencies, that is, to make product 'right-first-time' to prevent waste of company resources. It is a management driven task to extract more from the employees to find new ways to improve upon any routine approach, so that the product is made at the highest quality and lowest cost, and satisfies all of the customer's requirements.

LAM – Lean and Agile Manufacturing. This is a management tool to minimise cost by effective man-management if improvements are to be met in the areas of assembly, system profitability and the working environment.

PLM – Product Life-Cycle Management. This is a business strategy that helps companies share product data, apply common processes and leverage corporate knowledge to develop products from conception to retirement, across the extended enterprise.

MPM – Manufacturing Process Management. This is a management strategy that provides a common environment for manufacturing, pre-planning and cost estimation, as well as detailed production planning, reconciliation (estimate versus actual) analysis and shop-floor work instruction authoring.

PPR – Product, Process and Resource. The hub environment, providing a direct reuse of computer-aided drawing (CAD)-based data as a basis for work instructions.

Commercial aircraft design strategy is steadily evolving, initially being driven by the classical aeronautical subjects, but of late by customer driven design strategies that consider the problems of DFM/A with the objective to minimise cost of production without sacrificing design integrity and laid-down specifications.

Conceptual Aircraft Design: An Industrial Approach, First Edition. Ajoy Kumar Kundu, Mark A. Price and David Riordan.
© 2019 John Wiley & Sons Ltd. Published 2019 by John Wiley & Sons Ltd.

Manufacturing methodologies, jig-less assembly and flyaway tooling concepts facilitate DFM/A. Designing for ease of assembly can be improved in the areas of effectiveness of assembly and product quality.

Chapter 16 stresses the importance of costing in playing a rigorous role as an integrated tool embedded in the multi-disciplinary systems architecture of aircraft design that helps to arrive at a 'best value'. Cost estimation is used to trade-off studies between the classical aeronautical subjects and DFM/A methodology with its guiding principles of part count and man-hour reduction, standardisation of parts and an emphasis on designing for ease of assembly have wider implications for engineers and managers in the manufacturing industry. While specialist groups concentrate on design for their task obligations, be it technology driven, manufacture driven or any other demand, the IPPD environment has to synthesise the trade-off studies to arrive at the best value for the product as a global optimum rather than optimising to a particular design study.

The paradigm of 'better, faster (time) and cheaper to market' has replaced the old mantra of 'higher, faster (speed) and farther' [1]. Aircraft manufacturing companies are meeting the challenges of this new paradigm by assessing how they do things, discarding old methods and working practices for newer 'right-first-time' alternatives; an increase in value of the product will be achieved through improved performance (better), lower cost (cheaper) and in shorter times (faster). The shift of paradigm from the classical aeronautical studies led into the appearance of new considerations of various kinds of 'design for…' terms, more so from the academic circle (Section 20.8). The chapter takes a holistic approach to consolidate aircraft design by consolidating of various 'design for…' considerations. The authors suggest an index of 'Design for Customer' could be introduced as a measure to establish a value for the product.

Digital design and manufacture process (Section 20.10) leads to paperless offices. The advent of digital manufacturing process has greatly facilitated the DFM/A concept. It deals with the role of MPM for the industry. MPM provides a common environment for manufacturing, pre-planning and cost estimation, as well as detailed production planning, reconciliation (estimate versus actual) analysis and shop-floor work instruction authoring. It provides a means to integrate across the full product life-cycle ranging from concept, through field maintenance to retirement (from cradle to grave). Shop-floor execution systems are fed directly from the PPR-Hub environment, providing a direct reuse of CAD-based data as a basis for work instructions. As-built data is captured and available for use within the PPR-Hub for follow-on planning and validation as the product evolves throughout its life cycle.

In a way the automobile manufacturing technology is ahead of aerospace and has implemented digital manufacturing technology quite successfully, advancing into futuristic visions. A successful car design can sell about a million per year lasting for a decade with minor modifications whereas, in peace time, a successful high subsonic commercial transport aircraft is produced at less than 500 per year and is yet to total 10 000 mark in its production tenure. The automobile industry can invest large sums in modern production methods while keeping amortisation cost per car low.

The chapter covers the following:

Section 20.2: Introduces the Role of Manufacturing Consideration
Section 20.3: Describes Design for Manufacture and Assembly (DFM/A)
Section 20.4: Discusses Manufacturing Practices
Section 20.5: Explains the Design for Six-Sigma (DFSS) Concept
Section 20.6: Introduces an Important Aspect of Tolerance Relaxation
Section 20.7: Discusses the Issues of Reliability and Maintainability
Section 20.8: Brings out Various Designs for Consideration – A Holistic Approach
Section 20.9: Proposes an Index of Design for Customer
Section 20.10: Outlines the Concept of Digital Manufacturing

Classwork content: The readers may compute the index of 'Design for Customer'. However this is neither essential nor important as no industry is adopting this system at this stage, it awaits more study. However, the DFM/A considerations could be taken up as second term work. Such studies need not alter the frozen and substantiated configuration so far obtained through the worked-out examples (in industry, DFM/A is carried out in parallel during the conceptual stage). It is good to have some idea of DFM/A implication in aircraft design and operation. However, if it has to be taken up in the second term then it may not prove practical without specialist instructors using realistic data. The chapter only gives a glimpse of the scope of DFM/A during the conceptual study phase.

20.2   Introduction

Today, it is not the operators, who are the only customer, the future trend brings the entire society as a customer of the high-tech aerospace engineering that could make or break any society depending upon how the technology would be used. This is also true for other kinds of technology, for example, nuclear, bio-engineering and so on.

In the past, trade-off studies remained confined to the interaction between aerodynamics, structures and propulsion. Subsequently, during the 1990s, the need for DFM/A considerations in an IPPD environment gained ground. The IPPD process continues to evolve to cater for the customer driven design trend to minimise ownership cost without having to sacrifice integrity, performance, quality, reliability, safety and maintainability. The emerging scenario of the economic downturn demands all-round cost cutting measures to a new high, severely affecting the commercial aircraft industry. In today's economic climate, the roles of reliability, maintainability, recyclability and so on are design and manufacturing process dependent. This chapter introduces an index 'Design for Customer' incorporating value engineering.

The eventual affordability for operators as the 'best buy' (product value) will in turn make the manufacturers thrive. Design considerations should not impose difficulties in their manufacturability. The associated aerodynamic shape and structural design concepts help facilitate easy parts fabrication; their assembly, enhanced interchangability and so on. Selection of bought-out items should be an efficient and cost-effective system leading to better reliability and maintainability during its life span. Recent events have brought in additional constraints of cost effectiveness and environmental issues, these require increased attention. The global sustainable development issues, anti-terrorism design considerations and so on require additional design considerations. The choice of material from the point of view of recycling (disposal) appears as an additional issue when use of composites over metals gains ground.

20.3   Design for Manufacture and Assembly (DFM/A)

The public domain has proliferated with the acronyms like DFM and design for assembly (DFA). It should be noted that these are not a stand-alone concept; there is a relationship between DFM and DFA) to meet the objectives to lower cost of production. In this book, 'fabrication' and 'assembly' are the two components of the 'manufacturing' process and are taken together as DFM/A. Chart 20.1 gives the typical steps of DFM/A application.

DFM/A is concerned with design synthesis of parts fabrication and their assembly, as an integral part of manufacturability. DFM/A analyses competition and risk involved, that is, balancing the trade-off between cost and performance. This will eventually ensure affordability for the operators as the 'best buy'. It is a multi-disciplinary study, searching for aerodynamic mouldlines with surface smoothness requirements (tolerance specification) to minimise performance penalties without imposing difficulties in their manufacturability. Its associated structural design concepts help facilitate easy part fabrication and assembly (low man-hour and low part count, as well as enhanced interchangeability). Selection of bought-out items is made for efficient and cost-effective system integration, leading to better reliability and maintainability during operational life span. Based on the awareness of customer affordability and requirements, the designing and manufacturing targets costs are established. To repeat: it measures the objectives to lower cost of production, improve, reduce manufacturing cycle times and increase the value of the product without sacrificing design integrity, safety and laid specifications.

Chart 20.1 DFM/A steps.

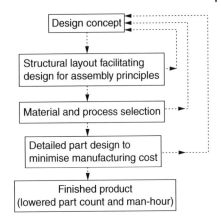

An aircraft is a complex product is made up from a large number of parts. 'Assemblability', as a measure of the relative ease of product assembly, plays a prominent role in productivity. The following are the main goals of DFM/A considerations. These considerations reduce parts count and assembly time.

   i. Improvement of the efficiency of individual parts fabrication
   ii. Improvement of the efficiency of assembly
   iii. Improvement of product quality
   iv. Improvement of the assembly system profitability
   v. Improvement of the working environment within the assembly system
   vi. The product's usefulness in satisfying the need of a customer
   vii. The relative importance of the need being satisfied
   viii. The availability of the product relative to when it is needed
   ix. The best cost of ownership to the customer

20.4   Manufacturing Practices

Depending on the manufacturing philosophy, jigs and fixtures need to be designed for the type of tooling envisaged for part fabrication and assembly. Jigs and fixtures are special holding devices to make fast, accurate fabrication and assembly of parts. Naturally, design of jigs and fixtures starts early during the Phase II of the project along with planning for facility and process layout. These could be expensive requiring additional production launch cost. But there is a payback in saving labour cost when production starts. Investment in the aerospace industry is front loaded.

Accurate dimensioning during fabrication and assembly is important to reduce cost of manufacture and maintenance. The following are deployed to maintain dimensional accuracy.

Definitions. These are not precise definitions but give some idea in the context.

Tools. These are pieces of equipment that cut and shape material in the process of fabrication of parts. These can be hand held or fixed in place. Examples of tools are drills, lathes, hammers, riveters, welders and so on. Tools, jigs and fixtures work in conjunction.

Gauges. These are measuring devices to help accurately locate tools in relation to the fixture in which a work piece is held.

Fixtures. These are special working/clamping devices to facilitate processing, fabrication and assembly. Fixtures are fixed frames designed to hold work pieces in the correct position or to hold several work pieces in the correct position relative to each other. A gauge may be required initially to position a tool for cutting. Fixtures can be large depending on the size of the work piece. These should be very solid and heavy structures to withstand any vibration in the vicinity.

Jigs. These have a similar function to fixtures but in addition they incorporate guides for the tool. Jigs are fixed items like fixtures. Typically, jigs are used for drilling, reaming, welding and so on.

Given next are seven of the many Best Practice Techniques that contribute to DFM/A practices. The basic ideas of the seven use modern manufacturing and tooling philosophies and have moved away from older manual procedures to digital processing (Section 20.10) where the majority of the tasks are performed. Modern methods make extensive use of CAD, Computer-Aided Manufacture (CAM) and CAPP (Computer-Aided Process Planning) to ensure a high standard of accuracy and productivity. Numerically Controlled (NC) machines are part of CAM.

i. *Jig-less assembly.* Designing for ease of assembly should not be restricted exclusively to the task of concept design engineers. Tooling engineers are making a contribution to reducing costs through a jig-less assembly approach to manufacturing. Jig-less assembly is an approach towards reducing the cost and increasing the flexibility of tooling systems for manufacture through the minimisation of product – specific jigs, fixtures and tooling. During the development phase tooling costs are quite high; consequently, savings in this aspect of aircraft manufacture are significant and they also impact the time from concept to market. Jig-less assembly does not mean tool-less assembly, rather, the eradication or at least the reduction of jigs. Simple fixtures may still be needed to hold the parts during particular operations but other methods are being found to correctly locate parts relative to one another. Assembly techniques are simplified by using precision positioned holes in panels and other parts of the structure to 'self-locate' the panels. Here parts serve as jigs. This process, known as determinant assembly uses part-to-part indexing, rather than the conventional part-to-tool systems used in the past.

ii. *Flyaway tooling.* Within the airframe manufacturing industry, it is generally accepted that approximately 10% of the overall manufacturing costs of each airframe can be attributed to the manufacture and maintenance of assembly jigs and fixtures. Traditional 'hard tooling' philosophy requires that the desired quality of the finished structure be built into the tooling. The tooling must therefore be regularly calibrated to ensure build quality. An alternative philosophy, 'flyaway tooling' has been conceived with the purpose of reducing tooling costs and improving build quality. This approach envisages that future airframe components will be designed with integral location features and incorporate positional datums that will transfer into the assembly. This will enable in-process measurement and aid in-service repairs. It may also be possible to design an aerospace structure having sufficient inherent stiffness so that the assembly tooling can be reduced to simple, reusable and re-configurable support structure.

iii. *Gaugeless tooling.* Gaugeless tooling is achieved through using a theodolite system linked through a central processor. Coordinated geometry, obtained directly from CAD, is used to establish the 'hard points' to meet build, interface and interchangability requirements. This is required for the manufacture and periodic inspection of assembly.

iv. *In-line assembly.* Provides a progressive and balanced assembly build sequence, utilising the maximum amount of sub-assemblies in a cellular type environment. This improves interchangeability.

v. *Automatic riveting.* The assembly is at first slave rivetted on the fixture and then moved to the automatic machine. This improves productivity, accuracy, hence quality as human error is minimised. Man-power engaged is also reduced.

vi. *Tolerance relaxation at the wetted surface* [2]. Aircraft surface smoothness requirements are aerodynamically driven with tighter manufacturing tolerances to minimise drag, that is, the tighter the tolerance, the higher the assembly cost. Trade-off studies between surface tolerance and aerodynamic drag rise can reduce manufacturing cost. More details are given in Section 20.6.

vii. *Six-Sigma and supporting Methodologies (see Section 20.5).* An important framework in which DFM/A techniques should be conducted is that of concurrent engineering (IPPD) with its focus on improving the product development process by concentrating on the design stage for the whole life cycle of the product. Management strategy such as DFSS, Lean/Agile Manufacturing and effective man-management also need to be taken into consideration if improvements are to be met in areas of assembly system profitability.

DFM/A should strengthen the team activity at all phases of the design process, thus ensuring that the technical expertise of the participants can be successfully utilised. It is a management tool.

Decisions made during product design have a major impact on cost, defects and cycle time. In fact, about 70% of product cost is locked in during the design process. Design for DFM/A is used to help reduce product complexity through minimisation of parts and fastener count, assembly and manufacturing time and material cost. Additionally, DFM/A application reduces the potential for defects. Robust design, statistical tolerancing and geometric dimensioning and tolerancing actually help reduce defects. A better understanding of 'Design for Manufacture Assembly' (DFM/A), to reduce the cost of production would require detailed study in material selection; different process capabilities and so on are beyond the scope of this book. The DFM/A concept assists the Six-Sigma management strategy.

20.5   Six-Sigma Concept

Robust design has the objective of achieving product designs with very few defects during manufacture, and very few latent defects after the product is delivered to the customer. It focuses on identifying the characteristics of the product that are critical to meeting the product requirements and then seeking the DFSS process capability.

The Design for Six-Sigma (DFSS) is an integrated approach to design with the key issue to reduce any scope for mistakes/inefficiencies, that is, make product 'right-first-time' to prevent a waste of company resources [3]. DFSS is a collection of product design tools and topics used to assist design of products for manufacture by processes operating at Six-Sigma capability. It is a management driven task to extract more from the employees to find new ways to improve on any routine approach, so that the product is manufacturable at the highest quality and lowest cost while satisfying all of the customer's requirements. Six-Sigma helps to expose the 'hidden factory' of waste that robs organisations of profits by a routine approach to issues that go wrong with the product and manufacturing process. The vision of Six-Sigma is as follows.

- Reduce costs and improve margins in a context of declining prices;
- Surpass customer expectations by a margin few competitors can match;
- Improve at a faster rate than the competition and
- Grow a new generation of leaders.

Six-Sigma is a systematic methodology for eliminating defects in products, services and processes while yielding cost and cycle time reductions. By significantly improving process capability it can achieve operational excellence in delivering almost defect free products and services at the lowest possible cost and on time.

For manufactured products, the Six-Sigma Methodology makes use of a variety of managerial, technological and statistical techniques to change the manufacturing processes, the product or both so as to achieve Six-Sigma process capability. The collection of design tools and topics that are used during product design to achieve a Six-Sigma product is called DFSS.

One measure of process capability is in sigma, σ, a statistical measure. For example, when a process is operating at Six-Sigma capability, the long-term yield is 99.99 966% corresponding to 3.4 defects per million opportunities. The demand for Six-Sigma is high guaranteeing robust design. Table 20.1 gives the sigma spread in a statistical histogram of defect levels relative to process capability.

To gain competitive advantage through customer satisfaction, their needs are to be understood. One way to capture customer requirements is through using selected *Quality Function Deployment* (QFD). QFD consists of a series of interlocking matrixes that are used to translate customer requirements into product functional requirements and process characteristics.

However, development and implementation of DFSS is tough. It requires behaviour from employees like: leadership, commitment, professionalism and perseverance so as to overcome the roadblocks of attitudes as heard in phrases, for example 'no time', 'not invented here', 'doesn't apply to us', 'we've been doing it for

Table 20.1 Sigma distribution of defects.

Sigma distribution Process capability	Defects per million
2	308 537
3	66 807
4	6 210
5	233
6	3.4

years', 'I prefer design rules' and 'I refuse to use the tools'. It demands culture change – not an easy process to achieve but possible. One of reasons (mentioned in the Preface) for developing a world changing slower than its potential is due to lack of accountability in higher management who still maintain older attitudes hidden behind seemingly modern views.

20.6 Tolerance Relaxation at the Wetted Surface

One of the best practices to reduce production coat is tolerance relaxation at the wetted aerodynamic surfaces that contribute to parasite drag rise. This section briefly outlines this very important DFM/A consideration of tolerance relaxation. This is an important topic that concerns aerodynamicists and detailed structural designers.

Tolerance relaxation at component manufacture could raise problems of the tolerance chain build-up at the assembly joint. All aspects of tolerance are beyond the scope of this book, here only the tolerance allocation at the aerodynamic surface as aerodynamic smoothness specification is covered [2].

In present day manufacturing philosophy, the following are the main features contributing to excrescence drag.

i. Manufacturing defects (discrete roughness – e.g. steps, gaps, waviness etc.)
ii. Surface contamination with fine particles/dirt adhering on it.
iii. Damage, wear and tear during the life cycle
iv. Fatigue deformation
v. Attachments of small items on the surface (e.g. blisters, antenna, pitot tubes, gaps/holes, cooling air intakes/exhausts etc.).

The first and the last items are the consequences of design considerations. The rest happens during operational usage. The chapter covers only the first item, which gives rise to excrescence drag (parasitic). Non-manufacturing origin of excrescence drag arising from item (v) is treated separately for the estimation of the $C_{Dpmin.}$ To keep excrescence drag within the limit, aerodynamicists specify aircraft smoothness requirements that are translated into tolerance allocations at the sub-assembly joints on the wetted surfaces. If the finish exceeds tolerance limits then it has to be reworked to bring it down within the limit and/or obtain concessions to pass the product to the final line. Tolerance specifications affect aircraft manufacturing cost.

Aircraft wetted surfaces are primarily manufactured from sheet metals/composites. At the sub-assembly joints there will be some mismatches (e.g. steps, gaps, waviness etc.) that need to be kept under tight control by specifying surface smoothness requirements. Mismatches give rise to parasitic drag as an excrescence effect. Aerodynamicists specify aircraft surface smoothness requirements to keep drag rise within a limit. The tighter the tolerance, the more the cost of production. Any tolerance relaxation at the wetted surface could bring down manufacturing cost at the expense of aircraft parasite drag rise, seen as a 'Loss of Quality Function'. It is assumed that the sheet metal/composite at the surface will accommodate a certain degree of tolerance relaxation. In addition, cosmetic appeal is seen as a customer preference. Loss of some cosmetic quality could save cost without unduly penalising parasite drag. However, with increase in fuel price the aerodynamicists need to be more careful in specifying surface smoothness tolerances.

Sources of aircraft surface degeneration: In aircraft application, degeneration of the 'wetted' surface area arises out of surface deviations from the specified level. It has many origins; the important ones are as follows:

i. *Lifting surface* (wing/flaps/empennage etc.):
- Control of leading edge profile and surface panel profiles (aerofoil contour).
- Rivets/fastener flushness for skin joints.
- Component geometry and sub-assembly joint mismatches.
- Fitment of access panels on the surface.

ii. *Bodies of revolution* (fuselage/nacelle etc.):
- Control of nose profile and profile of the rest of the body joined in sections.
- Rivets/fastener flushness for skin joints.
- Component geometry and sub-assembly joint mismatches.
- Fitment of doors, windows and access panels on the surface.

20.6.1   Cost Versus Tolerance Relationship

The relation to establish manufacturing cost, C, at the assembly is given by the sum of all the costs involved, as shown here.

$$\text{manufacturing cost, } C = (\text{basic work time} + \text{rework time}) \times \text{man} - \text{hour cost}$$
$$+ \text{ number of concessions} \times \text{cost of concessions} + \text{non} - \text{recurring costs}$$
$$+ \text{ cost of (support/redeployment/management)} \tag{20.1}$$

Change in tolerance will affect 'rework time', 'number of concession', the 'cost of support' and so on. Tolerance relaxation reduces manufacturing cost as more components and their assemblies become right-first-time but would reach a limit when any further relaxation is of no significant benefit as all the components and their assemblies will require no rework and/or concession for acceptance – it becomes right-first-time. At the limit of relaxation, the cost of manufacturing levels out to what is required for the 'basic' work time and the non-recurring costs. Figure 20.1 illustrates the nature of the cost versus tolerance relationship, a trend common to all features.

The *x*-axis represents the tolerance variation, starting from the existing level to the level when any further relaxation of tolerance has no further benefit in cost reduction. The *y*-axis gives the cost of manufacture. It starts at the existing level of tolerance with current manufacturing cost representing a zero saving. Tolerance relaxation results in cost reduction to the maximum possible level (100%).

Summing up the tolerance relaxation over the entire aircraft, the manufacturing cost can be reduced by few percent while at the same time incurring excrescence (parasitic) drag rise. Aircraft Direct Operating Cost (DOC) will reflect the change in cost reduction and drag rise when a trade-off study could be made. Figure 16.4 shows the trend in the trade-off between cost and tolerance. If the initial tolerance is too tight then relaxation

Figure 20.1 Cost versus tolerance relationship. Manufacturing cost reduces as tolerance is relaxed. Savings = amount reduced from the existing level to a lower level due to tolerance relaxation.

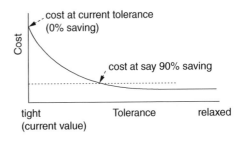

will show reduction up to a point, thereafter DOC would rise as a result of increased fuel burn for the drag rise while aircraft price reduction has levelled out.

Reference [2] has studied the trade-off. It shows that a mid-sized jet aircraft can have an average of about 33% tolerance relaxation; the corresponding net saving in DOC is a reduction of 0.42%. With conservative estimation, given next is the typical aircraft cost reduction through DFM/A studies. Fuel price is taken as US$0.75/US gallon.

i. Approximately, 1.28% DOC saving on account of 2% aircraft cost saving through DFM/A studies involving no drag rise.
ii. Approximately, 0.42% DOC saving on account of 1% aircraft cost saving through tolerance relaxation involving drag rise.

This study demonstrates a total of 1.7% DOC savings, which translates into savings of $530 per sortie for 150 passenger/3000 nm range class of aircraft. With annual utilisation of 500 sorties this totals $265 000 per aircraft. For a fleet of 10 aircraft the savings would amount to $26.5 million in 10 years' time. For smaller aircraft, the percentage savings will be still higher.

This is a good example of how aerodynamics, structures and manufacturing considerations are needed to conceive designs that will result in reduced DOC. Manufacturing cost reduction can be achieved by many other kinds of efforts. This is the aim of the DFSS concept. During the trade-off studies of various design parameters, the benefit of cost estimating activities is to help designers investigate new technologies to be adopted to advance the product to a competitive edge and generate specification requirements (e.g. tolerance allocations etc.), and also analyse the risk involved, balancing the trade-off between cost and performance that will eventually lead to affordability for the operators as the best buy, which, in turn helps the manufacturers thrive.

20.7 Reliability and Maintainability (R&M)

Unreliability is unacceptable. An aircraft as a system must achieve user confidence that it will work as and when required. This entails a multi-disciplinary study for an efficient and cost-effective system integration leading to better reliability and maintainability during the operational life span. In today's economic climate, the roles of reliability, maintainability, recyclability and so on need to be scrutinised for cost control, not just of the in-house product line but also in the supply chain of bought-out items. Even those systems that are perceived as being reliable are only so due to the large amounts of redundancy in the system or vast amounts of corrective maintenance to keep it running. Despite immense efforts to predict and improve the components used in systems, their reliability and maintainability has often remained at the same levels.

The design must guarantee integrity with long times between failures and that it can be repaired in a specific low downtime. The aircraft must have more Time Between Overhaul (TBO) than the competition. It is linked with the reliability of the system as a function of the operational environment and length of operational time. While avionics (and engine) suites come with a well-studied R&M status, many other aircraft components (mainly structures with built-in large amounts of redundancy) have yet to evolve to address maintenance issues at the conceptual design stage. Practically all bought-out items and sub-systems come with reliability figures obtained from rigorous testing. An aircraft as a system maintains a systematic log, recording failures and defects so that they can be followed up with design modifications to make the design more robust, both for those are already built and those to be made.

20.8 The Design Considerations

In the overview of the chapter it was pointed out that public domain literature is flooded with 'design for…' considerations, for example, DFM, DFA, Design for Quality (DFQ), Design for Reliability and Maintenance

(R&M), DFSS, Design for Recycling, Design for Anti-pollution, Design for Life Cycle, Design for Cost (DFC) and so on, heading towards a generic *Design for X*. These considerations led into the appearance of new considerations (16 of them listed in this section), more so from the academic circle. The fresh insight from academia may throw new light but may prove not amenable to industrial implementation. Only recently, the drives for Design for Reliability/Maintenance and DFSS have gone into industrial practices and are still evolving. Industry is yet to tackle decisively the other components of life cycle cost (LCC), for example, training and evaluation, logistic supports, special equipment and so on at the conceptual stages of civil aircraft design, if they can reduce the ownership cost of the operators. Of the various 'design for' considerations, only about a third of them are applicable to DFM/A considerations. A cost model should be robust to support trade-off studies to arrive at the 'best value'.

The new challenge for industry would be to look into all aspects of ownership cost at the conceptual stages of the project status. This is because of performance evaluation based on setting individual goals of cost minimisation at each 'design consideration' may not bring out the global minimum when strong interactions within the multi-disciplines exist. In an IPPD environment of design, the combined effort of various disciplines would provide a better approach to make a product right-first-time at a lower cost. The holistic approach suggests the role of cost modelling as a tool to address all of the considerations simultaneously by facilitating performance versus cost trade-off studies to arrive at the most 'satisfying' product line with the widest coverage of customer. It introduces the term 'Design for Customer' as a measuring index for 'value for money' defined in Section 20.9.

The 16 design considerations appearing as 'design for…' terms can be broadly classified under four categories listed next with brief descriptions. These design considerations require providing the designers with complete information of the product in the conceptual stages in an anticipated manner based on the expertise knowledge and technology level available. The strategy is to make the product to yield specific desired benefits of lowest LCC (in civil application, DOC) in a unified manner, leading into what is termed the 'Design for Customer'.

Category I. Technology Driven Design Considerations

i. *Design for performance.* The classical aircraft design considerations involving aerodynamics, structures, propulsion and systems to minimise fuel consumption. The aeronautical engineers' goal is to make an aircraft light, low drag with a matched engine, with low specific fuel consumption and with bought-out items (engine, avionics, actuators etc.) that offer the best value for money. It has become a matured technology for generic subsonic commercial aircraft design with diminishing return on investment to incorporate advancement.

ii. *Design for Safety.* Crashworthiness, emergency exits and so on are also mature considerations.

iii. *Design for Component Commonality.* The family concept of derivative aircraft design offers considerable benefit from cost reduction by maintaining a large number of component commonality within its variants. Derivative designs can cover a wider market at a much lower unit cost as amortisation of non-recurring cost is distributed over a larger number of units sold. Some of the variant aircraft designs may not be sized to the least fuel burn, but the lower unit cost would offset to a lower DOC. This consideration at the conceptual stage would prove crucial to the success of the product range.

iv. *Design for Reliability and Maintenance (R&M).* Currently, a considerable amount of maintenance resources are planned after the design is made and then are acquired to fit the requirements. This arises from the difficulty of translating feedback (statistical in nature) from the operational arena that could be sufficiently abstract. Design attributes, which could make maintenance difficult demanding additional time and training for highly skilled technicians, need to have more detailed considerations to reduce cost of maintenance. Cost trade-off studies with the attributes of (a) reliability, (b) reparability and (c) fault detection/isolation are the key considerations to be investigated more stringently at the conceptual design stage. Reliability issues are the most important consideration to improve the support environment, in generic terminology to make a *robust* design.

v. *Design for Ecology.* Since the 1970s, environmental issues (anti-pollution) were enforced through government legislation on noise, emission and so on at an additional cost. Use of alternative fuel for sustainability is also a part of this. The growing stringency of existing requirements and additional issues (next paragraph) would only increase the cost of the product. It is approaching a mature technology with diminishing return on investment for improvement.

vi. *Design for Recycling.* Aerospace technology cannot escape from the emphasis on recyclability, an issue gaining strength as can been seen from the topical agenda of 'sustainable development' in the recent UN summit meetings. Design for stripping is an integral part of the Design for Recycling to minimise cost of disassembly. New materials (composites and metals) bring additional considerations for disposal. Cost trade-off studies on LCC versus material selection for recycling could infringe upon marginal gains in weight reduction or cost of fabrication.

vii. *Design for anti-terrorism.* In the offing is the newer demand for 'Design for Anti-Terrorism'. In-flight safety features, for example, protection against terrorist activities are explosion-absorbing airframes, compartmentalising of cabins for isolation and so on, which incur additional cost.

Category II. Manufacture Driven Design Considerations

i. *DFM.* This study is concerned with the appropriate process required for part fabrication. The trade-off study includes cost versus material selection, process selection, use of N-C machines, part commonality and modularity considerations to facilitate assembly and so on. One of the key issues at the conceptual design stage is to have a low part count to reduce assembly time. The lowest part count may not be the cheapest method – compromise may be needed.

ii. *Design for Assembly (DFA).* This is concerned with the lowest man-hours required to assemble parts. Traditional practices in aircraft assemblies consist of numerous components, creating a complex organisational structural in engineering, logistics and management disciplines. This also results in an inefficient use of factory floor space and quality is compromised due to the unnecessary operations and fasteners required to join the mating parts. DFA should minimise manufacturing cost through optimising method engineering with innovative 'Best Practice Techniques' of jig and tool design, be it anything from manual methods to using computerised methods to assemble. Product configurations and detailed design part considerations play important roles in assembly.

iii. *Design for Quality (DFQ).* Adhering to the desired specification requirements is the essence of quality control. One such example is the aerodynamic surface smoothness requirements through surface tolerance specifications at component final assembly. Currently, many of the quality issues are tackled in the post conceptual design stage; these issues need to be advanced to the conceptual design stage.

Category III. Management Driven Design Considerations

i. *DFSS.* This is an integrated approach to design with the key issue to reduce any scope for mistakes/inefficiencies, that is, make product 'right-first-time' to prevent waste of company resources. Section 20.5 has dealt with DFSS. Measure of its success reflects on the final cost of the product, hence an estimation method could indicate at what cost, that is, at what efficiency the Six-Sigma approach is working at.

ii. *Design for Cost/Design to Cost.* This presents the classical question on DTC, DFC or a mixture of both. The tendency of management emphasis on DTC through a 'lean' organisational set-up may prove counterproductive if it carries extreme applications.

Category IV. Operator Driven Design Considerations

i. *Design for Training and Evaluation (T&E).* An area that is currently not under strong consideration at the conceptual design stage. Aircraft DOC estimation does not include the cost of T&E. Design considerations, such as commonality concepts, modular concepts and so on, could reduce T&E costs and therefore need to be addressed at an early stage.

ii. *Design for Logistic Supports.* An operational aspect with second order consideration for civil aircraft design. This is because the existing support system caters for most of the logistic details without infringing on any major changes required in aircraft design unless a special situation arises. Early input from the operators for any design consideration would help in controlling cost.

iii. *Design for Ground Based Resources.* This may also be considered a lower order consideration for civil aircraft design considerations at the conceptual stage, unless special purpose equipment is required (discussed in the next paragraph). In general, ground-based support resources are becoming standardised and can be shared by a large fleet distributing the cost of operation at a lower order of consideration at the conceptual design stage.

iv. *Design for Special Equipment.* This is more meaningful in military applications. If any special purpose equipment has to be introduced for ground-based serviceability, then cost trade-off studies at the conceptual design stage would prove beneficial.

It is important to stress again that separate minimisation of individual costs through the separate design considerations listed here may prove counterproductive and usually prevent the overall minimisation of the ownership cost. In a holistic overview, this chapter introduces the term 'Design for Customer' to unify all the individual considerations at the early stages of design evolution to offer the best value of product satisfying requirements, specifications and integrity to a lower LCC (or DOC). This is a front-loaded investment for eventual savings in LCC (or DOC).

20.9 'Design for Customer' (A Figure of Merit)

Nicolai [3] has presented a meaningful term, 'design for mankind'. This should be the goal for any design and not exclusive to the aerospace industry. The author focuses on specific issues of engineering design and operation by suggesting 'Design for the Customer' as explained in this section.

In a holistic approach, as a tool to address all of these 16 'design for....' considerations simultaneously by trade-off studies, the author suggests an index such as 'Design for Customer' to measure the merits of the product [4]. It applies to establish pricing of the variants in the family of derivatives, but can also include competition aircraft to check the pricing structure. It is to be stated here that the expression for 'Design for Customer' is not substantiated with a large database. It may require fine-tuning to better accuracy. But it conveys the idea that there is a need for such an index to compare merits of any aircraft within the class. It can suggest a pricing policy to arrive at the most 'satisfying' product line offering the best value with the widest sales coverage.

To use the empirical formula, a set of standard parameters is to be established for the baseline aircraft in the class (payload-range). The parameters of family variants and competition aircraft should not differ more than about ±15% to remain within the linear range of variation. The baseline standard parameters of interest are denoted with an asterisk *, for example, the DOC^* in US Dollars per seat/nm, the *unit cost** in million US Dollars and the Delivery time t^* in years (from the placement of order). To evaluate variant designs, it has to be compared with the baseline design. DOC levels out well before it reaches the design range.

The baseline aircraft is designed to have the best 'Lift to Drag' (L/D) ratio at mid-cruise weight (Long Range Cruise conditions). Normally, the L/D characteristic is relatively flat and derivative designs (in the family) that still have an L/D close to the maximum design value of the baseline aircraft. The Breguet Range equation indicates that range is proportional to the square root of the W/S. A shortened variant with lower payload and range will have lower W/S with a de-rated engine. This aircraft has more wing than required. It will have better takeoff performance but slightly degraded range performance. On the other hand, the extended version will have more payload; its weight control may have to be traded for range capability, the takeoff mass invariably increases requiring up-rated engines, especially to make up the takeoff performance on account of a higher W/S (under sized wing).

This formulation is in line for comparison, satisfying the customer's operational requirements for the usefulness of the product in terms of unit cost, operational cost and timing to meet demand. From this definition an increase of value of the product will be achieved through improved performance (better), lower cost (cheaper), and shorter times (faster delivery). It can be seen that DFM/A methodology contributes directly to lower cost, improves quality and reduces manufacturing cycle time, thereby increasing the value of the product.

The higher the value, the better it is for the customer to use a product family that has incorporated a wide range of design considerations to satisfy operational requirements at the best ownership cost and purpose. In absence of standard LCC data, DOC is used. In this context, the following Section 20.9.1 introduces an index 'Design for Customer' [4] that is introduced to compare the value of any other aircraft in the class.

20.9.1 Index for the 'Design for Customer'

The definition relates to the merit of the design by establishing a value index. The suggested definition is as follows:

Design for Customer,

$$K_n = \frac{(DOC^*/DOC) \times (Unit\ Cost^*/Unit\ Cost)}{(t/10t^* + 0.9)}$$

K_n is inversely proportional to DOC and aircraft unit cost, that is, a lower DOC gives a higher K_n. A value greater than 1 is better. Unit cost takes care of engine and aircraft size and the DOC takes care of design merits, passenger number and range capability. Typically, an aircraft with more passengers will have lower a DOC driving the value greater than 1, but it is to be judged with respect to its price and delivery time.

20.9.2 Worked-Out Example

From the worked-out example on the Bizjet, the following are obtained. Since derivative aircraft values are obtained through simplified assumptions, they need to be worked out in better detail. Also note that a linear relationship is taken to work out the example here. This is in order to give a general idea.

The Standard Parameters of the Baseline Aircraft

Unit Cost* in million US Dollars = $8 million, maximum takeoff mass (MTOM) = 9400 kg.
DOC* in US Dollars per seat/nm = $0.352 per nm per passenger (10 passengers).
Delivery time t^* in years (from the placement of order) = 1 year.
Baseline aircraft = 0.8 million $/passenger and $0.000 851/kg maximum takeoff weight (MTOW).

The Parameters of the Extended Variant Aircraft

Unit Cost in million US Dollars = $9 million, MTOM = 10 800 kg.
DOC in US Dollars per seat/nm = $0.2875 per nm per passenger (14 passengers).
Delivery time t in years (from the placement of order) = 1 year.
Large variant aircraft = 0.6428 million $/passenger and $0.000 833/kg MTOW.

$$K_{n\_larger} = \frac{(DOC^*/DOC)}{(UnitCost/UnitCost^*) \times (t/10t^* + 0.9)} = \frac{(0.352/0.2875)}{(9/8) \times (1/10 + 0.9)}$$

$$= 1.224/(1.125 \times 1) = 1.088 \text{ makes it a better value.}$$

The Parameters of the Shortened Variant Aircraft

Unit Cost in million US Dollars = $6 million, MTOM = 7600 kg.
DOC in US Dollars per seat/nm = $0.482 per nm per passenger (six passengers).
Delivery time t in years (from the placement of order) = 1 year.

Small variant aircraft = 1 million \$/passenger and \$ 0.00 079 kg MTOW.

$$K_{n\_smaller} = \frac{(DOC^*/DOC)}{(UnitCost/UnitCost^*) \times (t/10t^* + 0.9)} = \frac{(0.352/0.482)}{(6/8) \times (1/10 + 0.9)}$$

$$= 0.7303/(0.75 \times 1) = 0.974 \text{ evidently it shows a lower value.}$$

In general, the smaller derivative aircraft is penalised as the aircraft is heavier than it should have been if it was a baseline design. The wing area is larger than what is required. The small variant is not an uncompetitive aircraft. There are aircraft in this class with a similar DOC. Of late, several new all composite four-passenger light jet aircraft have emerged, selling for less than \$5 million and have better DOC. These new aircraft are yet to be proven in operational usage. It will be some time before all composite aircraft will overtake the conventional construction. The exampled six-passenger small variant is a robust all metal aircraft with larger cabin volume and high end amenities suiting corporate demand. Together in a mixed fleet of three sizes, the total package offers benefits that are hard to match. When an airline operates a mixed fleet of variants along with the baseline aircraft, then spare parts stock, training and maintenance cost can be shared. Private ownership in this class is growing and there is a place for the two types.

However, some modifications to the small variant can improve the index $K_{n\_smaller}$. If re-engineered with a smaller turbofan (Williams type) and with lighter/cheaper equipment then the price could be brought down to around \$5 million, making it still more attractive but it will lose some component commonality and incur additional development cost. If its tips can be chopped off, at practically no extra cost, then the weight can be brought down to below 7000 kg. The index can reach the value of above 1.

'Design for the Customer' can help manufacturers to decide the aircraft price for the family of variants to give each type a comparative value to the customer. Typically, the smaller variant should be lower priced with low profit. The other aircraft prices are adjusted holding the baseline aircraft price unchanged. That means the price of the bigger aircraft could be raised to compensate the family price structure. The baseline size aims for maximum sales. In the example, the smaller variant is \$6 million extracting a small profit. The baseline aims for the largest sale maximising profit.

In general, route traffic load always increases and as a consequence, the sale of the larger variant increases; therefore, so does the associated profit for the manufacturer making up the relatively less profit from the smaller variant. When a new market emerges for larger traffic load, manufacturers will seize the opportunity for a new baseline design for the new market.

Aircraft price per passenger and unit mass are given next. There is a sharp rise in per passenger price with smaller payload as can be seen in Figure 20.2. In the figure, the dashed lines represent generic data, which is a magnified version of Figure 16.2. The full lines represent the Bizjet family. The small aircraft price is slightly depressed to suit market.

20.9.3 Discussion

This cost model methodology requires simple inputs of (i) geometry with dimensions, (ii) material used and their unit cost/procurement status, (iii) estimated component weights break down and (iv) estimated man-hours to design, fabricate and assemble to completion [5]. Preliminary information of all these could be available at the conceptual stage of design. In fact in industry, the in-house method is simpler using the actual figures directly. While the inputs are simple, to generate proper factors/indices through parametric studies is a complex procedure. In general, weight estimation has matured to a high degree of accuracy, because volume estimation in CAD is very accurate.

Most semi-empirical formulae to get aircraft prices are based on statistics generated from aircraft prices available in the public domain. From the statistics of the available published aircraft price, a linear pattern emerges with some spread band to get some idea of the expected price level of a new aircraft, but this is an inadequate figure for completion analyses as the merits of the new design of get masked. It is for this reason,

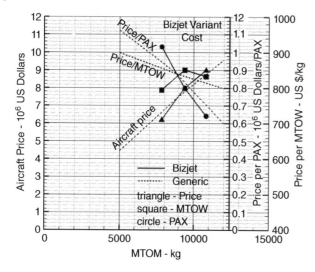

Figure 20.2 Aircraft cost factors of the worked-out Bizjet example.

industries necessarily develop their own cost model based on their past data reflecting company capabilities and challenges faced.

A well-maintained parametric relationship generated from past in-house products or, if possible, obtained from external sources should be able to establish these data. To predict the cost of future designs, technology level comparison is to be made to incorporate relative changes from the baseline design. For any radical departure in design considerations, intelligent modification to the cost model should work for initial estimates, which needs to be refined with a new database, when available. Basically, it summarises the man-hours required to design, fabricate parts and assemble them. The lower the part count, the lower the assembly time should be. The manufacturing philosophy is to optimise the relation between part count and assembly time.

The obligations of DFM study are considerable. Not only does it search for low part count and low manufacturing man-hours, it also has to search for suitable material amenable to lower production cost without sacrificing design integrity while satisfying customer requirements. These may present many conflicting considerations. For example, the use of composites for large components could save weight and cost of manufacture but could increase maintenance cost. Some aircraft component manufacture case studies are given in [6].

In this paper, the DFM study gives an important pointer on the need to search for cheaper procurement of expensive items and on part count reduction/better tooling to lower assembly time. More fundamental, the objective of cost reduction needs to be approached in a holistic manner for the ownership cost of aircraft. DFM studies also play an essential role to trade-off between the front-loaded non-recurring costs in order to reduce the recurring cost of production. A higher investment on tooling with tighter internal tolerances, especially for machined components could reduce cost in the long production run.

The input information improves in an iterative manner as the project advances from the initial conceptual stage to the project definition stage when the design and manufacturing philosophies are firmed up.

Cost models in the public domain are scarce in offering an integrated cost estimation methodology to cater for industrial usage with parameters as considered in this section. The cost implications involved in learning to manufacture, support/procurement policy, miscellaneous cost and so on can make or break an otherwise good design.

A good data bank would lead to a more credible parametric relationship to extract the necessary indices/factors. But it is not easy to obtain a large number of product lines – in aerospace, about 10 different products could be decades of work, by that time the technology change would force the data in to obsolescence. The difficulty in accurate cost estimation should not be underestimated.

20.10   Digital Manufacturing Process Management

Digital Manufacturing Process Management is a newer concept and still evolving. It is software driven. Although industry has already deployed digital manufacture in some areas, its full scope of application is yet to stabilise. Given in this section is a model studied at Queens University Belfast (QUB) [7]. These kinds of studies are carried out in many places proposing many different models. The core fact is that digital manufacture is here to stay and grow to replace/merge with the traditional manufacturing philosophies.

Today, a microprocessor-driven digital manufacturing process is fast overtaking the older methods. In fact, all modern plants have already used it to the extent that it can be advanced. The advantage of microprocessor-based tools is that they are digitally controlled driven by software. These can truly deliver the desired quantum leap in manufacturing assembly techniques for future aircraft. This section briefly outlines the role of MPM and identifies the benefits from PLM through a reconciliation perspective (estimation versus actual) between design and manufacturing engineering disciplines. PLM is a business strategy as a part of MPM as a management strategy. (Note that here the term 'Life Cycle' is used in a generic sense meaning all aspects of a product from concept to retirement.)

Digital manufacturing solutions enable the continuous creation and validation of the manufacturing processes throughout the product's life cycle. It lets manufacturers digitally plan, create, monitor and control production and maintenance processes providing complete coverage of manufacturing processes.

Along with the advent of new processes and techniques, there has also been a greater use of software in the design and engineering of aircraft. CAD, CAM, Computer Assisted Engineering (CAE) and CAPP tools are now used to determine electronically how an aircraft system needs to be built. NC machines are linked with CAM. The new frontier for design focuses on PLM with an emphasis on the manufacturing processes. PLM is a business strategy that helps companies share product data, apply common processes and leverage corporate knowledge to develop products from conception to retirement across the extended enterprise. By including all participants in this process (company departments, business partners, suppliers and operators/customers), PLM enables the entire network to operate as a single entity to conceptualise, design, build and support products.

MPM is segmented into: (i) process detailing and validation, (ii) resource modelling and (iii) process planning simulation. Within each of these segments are several modules, some of which are discussed next.

i. Process Detailing and Validation

This software suite gives engineers the tools to bring the process planning solutions into the application-specific disciplines of manufacturing. It assists engineers in verifying process methodologies with actual product geometry and defining processes within a 3D environment. The human module offers tools that allow users to manipulate accurate standard digital human manikins, referred to as 'workers', and to simulate task activities in the process simulation environment. Thus worker processes can be analysed early in the manufacturing and assembly life cycle. The assembly process simulation module sets a new paradigm for developing manufacturing and maintenance processes. It offers manufacturing engineers and assembly process planners an end-to-end solution by incorporating a single, unified interface for pre-planning, detailed planning, process verification and shop-floor instructions.

ii. Resource Modelling and Simulation.

This software suite gives engineers the tools to develop, create and implement resource, mechanical programming and application routines that are integral with process planning, process detailing and validation solutions. Within this set of solutions, resources such as robots, tooling, fixtures, machinery, automation and ergonomics are defined in a complete manufacturing solution. Digital manufacturing also offers a flexible, object-based, discrete event simulation tool for efficiently modelling, experimenting and analysing facility layout and process flow. Both 2D schematic and 3D physical models are quickly created (rapid prototyping) through push-button interfaces, dialogue boxes and extensive libraries. Real time interaction enables modification of model variables and viewing parameters during runs.

iii. Process Planning and Simulation

The process planning software suite provides an effective process and resource planning support environment. It improves methodically structured planning, early recognition of process risks and reuse of proven processes, traceable changes and decisions and the use of scattered process knowledge. The module is often used during the conceptual design phase, with the process design and alternative manufacturing concepts maturing through all stages up to production. The treatment of the relationships among product, process and manufacturing resource data, including the plant layout, help to avoid planning mistakes and to obtain a precise overview of the needed investment costs, production space and man-power at an early stage.

20.10.1 The Product, Process and Resource (PPR) Hub

Digital manufacturing allows manufacturing engineers and process planners to define, validate, manage and deliver to the shop floor the content needed to manufacture an aircraft. With the combination of the PPR-Hub environment, it is uniquely able to provide savings for an enterprise in the following ways. It reduces risk, time to market and overall cost of manufacturing

- Concurrent engineering design and manufacturing planning and process validation occur before release is final, in the midst of ongoing change.
- Manufacturing. producibility analysis directly influences design thus providing a true 'DFM/A' environment.
- The enterprise is able to capture and reuse manufacturing best practices in a formal way.
- Manufacturing plans are pre-validated in a 3D environment to avoid unexpected problems on the shop floor.
- Unbuildable conditions are found early in the design cycle when the cost of change is minimal.
- Quality targets are met sooner due to reduction (or elimination) of rework and engineering change orders driven from the shop floor.

Through a combination of methodologies, significant reductions can be achieved in time to market, overall cost and effective risk for an aircraft programme. One of the key enablers for reducing the time to market is the ability to support concurrent engineering design and manufacturing planning before release is final. When this is possible, manufacturing producibility analysis directly influences design thus providing a true DFM/A environment. In this scenario, the total cost of the engineering changes that occur over the programme is dramatically lower because these changes are identified and resolved much earlier in the life cycle, before tooling is procured and production starts.

The key technology enabler for the digital manufacturing solution, product/process design and validation is the PPR-Hub and business transformation. This database environment, supporting complex configuration and affectivity rules that are required in the aerospace world provides the infrastructure to allow process and resource planning to occur in the context of the engineering data and a continuously changing environment. In contrast to traditional systems, the PPR-Hub provides the means to explicitly manage product, process and resource objects, and the relationships between them. Since such relationships are explicitly defined and managed within the database, one is able to directly see the impact of changes of one class of object to any other (e.g. 'If a part is changed, which manufacturing plans are affected?').

20.10.2 Integration of CAD/CAM, Manufacturing, Operations and In-Service Domains

One of the major areas for improvement within an aerospace enterprise is in the integration between engineering, manufacturing and operations organisations. The drivers for such integration are to allow the maximum reuse and appropriate consolidation of data and business systems throughout the programme life cycle. The degree to which this is achieved will typically have a large influence on overall costs, both recurring and non-recurring. Traditionally, the engineering and manufacturing (CAD/CAM) domains operate in a self-contained operating unit. As a result, it is difficult if not impossible to actually leverage engineering data downstream in the shop-floor operations area where typically Bills of Material (BOMs) are defined, along with related procurement data and shop-floor work instructions. One major problem with this traditional scenario is that effectively managing engineering changes and reconciling the parallel worlds of CAD/CAM and shop-floor operations is extremely difficult and expensive to do accurately. Digital manufacturing provides a means to integrate these two domains (CAD/CAM and Operations) as well as the in-service domain, through several core technologies and application layers.

An additional benefit of this type of integration is the ability to reduce the number of redundant business systems between the CAD/CAM and Operations organisations. This reduces the recurring and overhead costs. Specifically, the following classes of business systems can be consolidated into the PPR-Hub based solution suite:

- Parts list
- Tool list
- BOM definition
- Bill of resources
- Routings
- Process sheets/work instructions
- Reconciliation analysis – estimate versus actual (e.g. in preparing BOM)
- Simulation-based validation
- Cost estimating
- Production flow analysis
- Human resources

20.10.3   Shop-Floor Interface

One of the key areas where this integrated environment provides value is in the ability to define, evaluate and document various manufacturing alternatives such as alternate routings or resource utilisation, based on evolving conditions in operations. The PPR-based manufacturing database can be used to drive discrete event simulations of the alternatives to determine the impact on material flow, throughput and utilisation under various scheduling and product mix conditions. The process plans, resource allocations and precedence requirements in the PPR-Hub can be further analysed to balance the work across the manufacturing facility and to provide proper utilisation of the workers. This will provide an interface to feed the shop-floor directly from the PPR-Hub database, ensuring optimal reuse of data created in the CAD/CAM and Manufacturing and Planning environments. One of the most significant ways to leverage a PPR-based database is to directly reuse the data as the basis for 3D work instructions on the shop floor. The immediate benefits of this approach are to:

- Eliminate the possibility for a mismatch between shop-floor instructions and engineering data since the instructions are derived directly from the PPR-Hub database, including the related CAD geometry and attributes.
- CAD-based work instructions provide a means to eliminate paper drawings on the shop floor since all required datums, tolerance, notes and related specifications can be embedded within the 3D dataset that also provides all required manufacturing information.
- Intuitive 3D (CAD-based) work instructions, combined with authoritative engineering data and attributes, empower the machinists to perform their job faster and with fewer mistakes. In such cases, the reduction in overall manufacturing flow time and cost can be dramatic.
- Provide a data feedback loop from the shop floor to the manufacturing planning environment to provide visibility on shop-floor-based changes representing as-built product build-up and processing.
- Leverage this data feedback loop and related PPR infrastructure to reconcile and evaluate the differences among the as-designed, as-planned and as-built datasets.

Finally, with respect to the in-service phase of the products' life cycles, this architecture provides a means to also capture and manage the evolving (complex) configuration of the related BOM, and collection of processes performed during ongoing maintenance operations. As appropriate, PPR data defined in one phase of the life cycle can be reused for other phases, thus providing potentially huge savings.

20.10.4 Design for Maintainability and 3D-Based Technical Publication Generation

During the evolution of the product design and manufacturing process planning, the same PPR dataset can be used to validate the maintainability of the product, as well as to develop tech-pub documents containing text, images and movies derived directly from the 3D-based process plans. The core PPR technology supports any number of views of process planning data related to product and resource, and thus provides a way to associatively develop maintenance plans concurrent with manufacturing planning. This ability to concurrently validate the design as well as the maintenance operations for the product is one more example of the significant leverage provided by the PPR data model. This allows the idea of leveraging the results of 3D process planning and analysis directly into web-based technical publications for maintenance operations; for example, (i) analysis of 3D process plans for producibility and (ii) web-based technical publications for maintenance.

3D Enables a New Business Paradigm

- 3D is now leveraged not only for design, but also for manufacturing planning, simulation-based validation, work instruction authoring and delivery to the shop-floor workforce, enabling a true paperless manufacturing process. This is easily extended for 3D maintenance and repair instructions.
- Operational and maintenance scenarios can be simulated using ergonomic analysis early in the design cycle to provide efficiencies later in the life cycle of the product. With a systematic methodology, a true 'Design for Customer' business process can be supported.
- The virtual production mock up eliminates the requirement for prototype parts to prove out mock-ups of production tooling and fixtures, reducing cost and time.
- Tooling orders can be placed much later in the development plan with the latest design revisions incorporated since they will work the first time, eliminating costly change orders to tools and part designs.
- Designs can be modified early in the design cycle to accommodate manual assembly and maintenance tasks, and thus the requirement for special tools can be eliminated.

References

1 Murman, E., Walton, M., and Rebentisch, E. (2000). Challenges in the better, faster, cheaper era of aeronautical design, engineering and manufacturing. *The Aeronautical Journal* 104: 481–489.

2 Kundu, A.K., Watterson, J., MacFadden, R. et. al. (2002). Parametric optimisation of manufacturing tolerances at the aircraft surface. *AIAA Journal of Aircraft* 39: 271–279.

3 Nicolai, L.M. (1975). *Fundamentals of Aircraft Design*. San Jose, CA: METS, Inc.

4 Kundu, A. K. and Curran, R. Design for Customer. ICAS Proceedings, November 2004, Yokohama.

5 Kundu, A.K., Curran, R. and Crosby, S., Rapid Cost Modelling at the Conceptual Stage of Aircraft Design, October 2002, ATIO Conference Paper No: 5401, AIAA, Los Angeles, California, 2002.

6 Kundu, A. K., Crosby, S. and Curran, R., Aircraft component manufacture case studies and operating cost reduction benefit, AIAA Conference ATIO Denver November 2003.

7 Crosby, S. Unpublished document on digital manufacture.

21

Miscellaneous Design Considerations

21.1 Overview

Although the main task of the aircraft configuration is now complete, there are few more topics of interest that require design understanding. Aircraft design is not confined only to the classical aeronautical subjects of aerodynamics (Chapter 3), propulsion (Chapters 12 and 13), materials and structures (Chapter 19). There are many other issues, for example, concerning environment, safety, operational matters, 'end of life' valorisation, some emerging new scenarios and so on. These affect design considerations at the very conceptual stage of the project. Most academic institutions offer separate courses on some of the topics but these can escape attention as the undergraduate curriculum is already packed with the main aeronautical subjects, for example, aerodynamics, propulsion, materials and structures.

This chapter gives a brief overview on the impact made by all round technological advances that need to be considered at the conceptual stages to arrive at a 'satisfying' design. It offers an easy entry to gain some understanding of the specialised topics, some of them may appear out of context during conceptual phase, but these do contribute to design considerations. Detailed study of the topics of this chapter is beyond the scope of this book.

The environmental issues of noise and emission are to comply with the standards specified by the Federal Aviation Administration (FAA) (USA) and the European Aviation Safety Agency (EASA) (Europe) and internationally by the International Civil Aviation Organization (ICAO). The difference between them is small. Aircraft doors (including emergency types) and environment standards, regulated by FAA/ICAO. The FAA is the agency of the United States Department of Transportation responsible for the regulation and oversight, as well as operation, of civil aviation. The authors decided that the role along with the history of FAA should be known as part of aircraft design study, hence they are outlined here. Military aircraft requirements are governed by Milspecs/Defence Standards (DEF).

Flight tests are conducted during the last phase of a new aircraft project, but its scope must be considered during the conceptual design phase, hence they are discussed but not dealt with in detail here. The proximity of ground on a moving aircraft has favourable effects but no credit from this benefit is taken in this book. The gains on account of ground effects during takeoff and landing adds to the safety of the aircraft. A brief discussion on ground effects is given in the chapter. Other topics include aircraft flying in bad weather, flying hazards, end-of-life disposal, Extended Range Twin-Engine Operation (ETOP), flight and human physiology and on some emerging scenarios.

The topics concerning military aircraft design are complex issues and need to be studied separately (specialist books referred to). Earlier chapters indicate the complexity of military aircraft design. These considerations would make advanced military aircraft design a more difficult proposition for the newly initiated undergraduate students.

The chapter covers the following topics:

Section 21.2: Introduces the Topics Considered in this Chapter
Section 21.3: History of FAA – the Role of Regulation

Conceptual Aircraft Design: An Industrial Approach, First Edition. Ajoy Kumar Kundu, Mark A. Price and David Riordan.
© 2019 John Wiley & Sons Ltd. Published 2019 by John Wiley & Sons Ltd.

Section 21.4: Flight Test
Section 21.5: Contribution by the Ground Effect on Takeoff
Section 21.6: Flying in Adverse Environment
Section 21.7: Military Aircraft Flying Hazards
Section 21.8: Aircraft Environmental Issues
Section 21.9: End-of-Life Disposal
Section 21.10: Extended Range Twin-Engine Operation (ETOP)
Section 21.11: Flight and Human Physiology
Section 21.12: Some Emerging Scenario

Classwork content: There is no classroom work in this chapter.

21.2 Introduction

Described next are a few topics chosen to broaden the horizon of readers on topics associated with aircraft design. These few topics are deliberately chosen to broaden the design perspective of the readers. Some of these topics are of relatively recent origin, gradually evolving since the 1970s.

i. *History of FAA (Section 21.3)*
A new comer will find a maze of documents pertaining to airworthiness on the top of what the FAA issues. This is because there is an evolutionary past. The FAA, after its formation, continued with older issues by previous authorities. A brief history should help the newcomer with the structure of Type Certification set-up.

ii. *Flight Testing (Section 21.4)*
Flight tests are carried out by the manufacturers to substantiate the mandatory requirements by FAA and to guarantee customer specifications. It is essential that the aircraft goes through a lengthy series of flight tests at the end of a new aircraft project. Aircraft performance engineers play a significant role in analysing the flight test data to reduce it in a prescribed format for documentation and issuance. The scope of flight test is outlined in Section 21.4.

iii. *Contribution by the ground effect on takeoff (Section 21.5)*
This book does not deal with the ground effects on aircraft performance when flying close to the proximity of ground. Some elementary exposure to the ground effect on takeoff is given in this chapter. Contributions from the ground effect are assessed in detail in the subsequent phases after go-ahead of the new aircraft project is obtained.

iv. *Environment issues (Section 21.6)*
Since the advent of big commercial jets in the 1960s (B707, DC8, Convair990, VC10, etc.) the noise profile started to be a nuisance to residents living in the neighbourhood of airports. Litigation cases started to rise as a result of the damages on property/health. The Federal Aviation Regulations (FAR) stepped in to limit the noise level in a prescribed manner. The aircraft and engine designers strived to reduce/suppress noise at takeoff and landing. Considerable research is continuing to reduce noise arising from supersonic shock waves also known as booms. Currently there are no civil aircraft operating at supersonic speed – Concorde has now been taken out of service after a gloriously long operational period, primarily on account of economic reasons. Supersonic aircraft are not treated in this book; in any case they would fly subsonic at takeoff and landing. References [1–3] may be consulted for more details.
During the 1980s, concerns were raised about climate change in which engine emissions played a contributing role. Once again, the regulatory bodies intervened to set achievable standards to limit pollution caused by the engine exhaust gases.

v. *Flying in adverse environment (Section 21.7)*
This book primarily deals with a zero wind condition with passing mentions of head/tail wind effects. Head/tail wind also influences on field performances at takeoff and landing. With cross-winds there are

control issues and add to drag rise to keep an aircraft aligned to the airfield. While these are not dealt with in this book, industry makes the effort to estimate their effects when preparing a pilot manual. Cruising in head and tailwinds cannot be regarded as flying in adverse environments. This chapter merely outlines those truly adverse environments, for example, foggy weather, icy conditions, turbulent weather, wind shear, jet streams and thunderstorms/lightening. Bird strike events may also be considered adverse environments.

vi. *Fire hazard*

The FAR requires aircraft manufacturers to comply with design considerations such as cabin fire safety, cargo area smoke/fire detection, system, nacelle fire protection, fuel inerting that require approved choices of materials, proper instrumentation, proper installation of fire installation and so on; topics requiring special study and beyond the scope of this book.

Aircraft Safety issues concerning fire hazards on the ground or post-crash evacuation are discussed in Section 22.4. For military aircraft, the extreme measure of ejection is incorporated (Section 22.8).

vii. *Military Aircraft Flying Hazards (Section 21.8)*

In additions to the hazards of flying as described before, military aircraft have hazards in hard manoeuvres to engage or disengage in combat at lower altitudes. Aircraft combat survivability (ACS) is a design consideration to incorporate capability of an aircraft survivability disciplines attempt to maximise the survival of aircraft in wartime. Crew ejection is seen as an integral part of survivability. The last two decades saw stealth technology, as a good survival measure, by minimising aircraft signature. Electronic defence/countermeasures are the other ways to thwart retaliation and increase chances of survivability. These topics are beyond the scope of this book (readers may refer to [4, 5]).

viii. *End-of-life aircraft disposal (Section 21.9)*

In general, civil aircraft operational life is anywhere from 20 to 30 years, depending upon operational demand and on profitability. A few World War II C47s (Dakota) are still flying. There are thousands of aircraft already grounded and in the immediate future thousands more will be grounded forever. Their storage occupies a large land area and disposal does not work in the way that automobiles are disposed of. Old aircraft disposal is a real problem.

ix. *Extended Range Twin-Engine Operation (ETOP) (Section 21.11)*

ETOP allows flying over water up to a distance ensuring that in case of emergency the aircraft can land in 120 minutes under one engine powered up to maximum climb (i.e. continuous) rating.

x. *Some emerging scenarios (Section 21.12)*

The authors would like to bring out the following new topics gaining importance that the next generation of engineers need to consider. The new emerging scenario affecting civil aviation comes from the acts of terrorism. Damage limitation of explosion in the cargo compartment/hold and containing terrorist activities within the cabin are the two areas for design considerations. Damage incurred from runway debris demands a new look to an old problem (the *Concorde* case). In-flight passenger health-care/infection contaminations are becoming as important issues with vastly increased passenger traffic crossing international boundaries.

21.3 History of FAA – the Role of Regulation

Within a decade of the first flight by the Wright brothers, the potential for aircraft application both in civil and military sectors shown an unprecedented promise of growth when the free-market economy rushed in for quick gains through new discoveries to stay ahead of the competition. Proliferation in designs started to show compromises and there were avoidable accidents/incidents. Time and financial constraints forced industries, in certain situations, to rush into production without systematic substantiation of design to guarantee safety.

The history of regulation in the USA began within a decade of the first flight by the Wright brothers. Its intent was to improve safety of the new aviation industry. The FAA can trace its formation to the 1958 Federal Aviation Act (although predecessor agencies dated back to May 1926 in the USA), when the government stepped in. The Air Commerce Act brought in a formal method to maintain a minimum standard within which aircraft could be flown safely. The Act was codified in the *US Department of Commerce Bureau of Air Commerce Aeronautics Bulletin*, No. 7, 'Airworthiness Requirements for Aircraft'. In time, along with rapid progress, it was revised in October 1934 as *Aeronautics Bulletin*, No. 7A in more detail. Special requirements for commercial transport operation were detailed in *Aeronautics Bulletin*, No. 7J.

FAA publications include Airworthiness Directives (AD), Orders and Notices, Advisory Circulars (AC) and the all-important FAR. Aircraft designers and engineers deal with FAA Aircraft Certification Offices (ACO) and the Aircraft Evaluation Group (AEG), whose mission is to assure that aircraft manufacturers offer adequate instructions for continued airworthiness and do so prior to certification.

As aircraft operation in civil sector grew very rapidly, the US Government felt the need for a dedicated department and in 1930 Civil Aeronautics Authority was formed. The new formed set-up made a more authoritative regulation as Civil Air Regulation (CAR) Part 4, replacing the Aeronautics Bulletins issued by earlier set-up. The CAR was in two parts one for larger transport aircraft as Part 4T and one for smaller aircraft as Part 4. The regulations were systematically kept updated and in 1938 new regulations were issued; one for small aircraft as CAR Part 4A and one for large aircraft as CAR Part 4B. What constitutes small and large aircraft is later defined as follows (from these FAR23 and FAR25 evolved):

CAR Part 3 for aircraft MTOW $\leq 12\,500$ lb and
CAR Part 4b for aircraft MTOW $> 12\,500$ lb and
CAR Part 8 for restricted category aircraft.

These regulations came in manual form as Civil Aeronautics Manuals; CAM3 for Part 3, CAM4b for part 4b and CAM8 for Part 8. These contain interpretation of regulations, methods of tests and so on.

World Wars I and II were clear drivers for the explosive growth of aircraft industry, followed by civil aircraft expansion in unprecedented scale. The government needed to further their administrative arm to regulate aircraft design. In 1958 the FAA replaced the Civil Aeronautics Agency (CAA) but maintained the CARs. With progress, in 1965 the CARs were replaced by new regulations issued by FAA, as follows:

FAR 23 replacing CAR Part 3
FAR 25 replacing CAR Part 4b

In the period of 1938–1965 there were other minor amendments. While old designs continued with the CARs, from 1965 new aircraft had to substantiate their designs under the FARs. Once the substantiation formalities are met, the certification authority issues 'Type Certification' for the aircraft type allowing them to operate in the market place. Every modification to the aircraft requires approval from the certification authority.

Like the CAMs for the CARs, outlining detailed explanations and procedures for Type Certification, FAA issued documents as follows:

'Engineering Flight Test Guide' (FAA Order 8110.7) in three categories of aircraft such as (i) Normal Category, (ii) Utility Category, and (iii) Aerobatic Category starting from 1970 and
'Engineering Flight Test Guide' (FAA Order 8110.8) for Transport Category.
Because administrative responsibility required remaining restrictive by not circulating the orders in public domain the above documents were modified and issued as AC as follows. AC are not laws unlike FARs.
Advisory Circular No: 23-8A for 'Flight Test Guide for Certification of Part 3 Airplanes' replacing Engineering Flight Test Guide (FAA Order 8110.7) and.
Advisory Circular No: 25-7 for Transport Category Airplanes replacing Engineering Flight Test Guide (FAA Order 8110.8).

With advent of helicopters, other special purpose aircraft, for example, agriculture spray aircraft, and environment related regulations have their respective FARs. In fact, every country has their certification agencies. Governments in the Western countries have developed and published thorough and systematic

Table 21.1 FAR Categories of Airworthiness standards.

Aircraft types	General aviation	Normal	Transport
Aircraft	FAR Part 23	FAR Part 23	FAR Part 25
Engine	FAR Part 33	FAR Part 33	FAR Part 33
Propeller	FAR Part 35	FAR Part 35	FAR Part 35
Noise	FAR Part 36	FAR Part 36	FAR Part 36
General operations	FAR Part 91	FAR Part 91	FAR Part 91
Agriculture	FAR Part 137		
◄─────► Large commercial transport	Not applicable		FAR Part 121

Table 21.2 Aircraft categories.

Aircraft types	General aviation	Normal	Transport
MTOW – lb	Less than 12 500	Less than 12 500	Over 12 500
No. of engine	Zero or more	More than one	More than one
Type of engine	All types	Propeller only	All types
Flight crew	One	Two	Two
Cabin crew	None	None up to 19 PAX	None up to 19 PAX
Maximum no. of occupants	≤ 9	≤ 23	Unrestricted
Maximum operating altitude	25 000 ft	25 000 ft	Unrestricted

standards to enhance safety. These regulations are available in the public domain (see relevant websites). In civil applications, they are FAR and JAR (the newly formed designation is EASA); both are quite close. The author prefers to work with the established FAR at this point. FAR documentation for certification has branched into many specialist categories as shown in Tables 21.1 and 21.2. Table 21.1 gives the FAR Categories of Airworthiness standards.

FAR criteria for certification include many specialised categories and basic applicability. A partial listing is shown in the Table 21.2 under the heading General Aviation, Normal Category and Transport Category. Aerobatics type aircraft are under General Aviation. (New categories are evolving for the home-built/light aircraft types of small aircraft, not dealt with in this section.)

Code of Federal Regulations (14 CFR)

The Government of United States of America have 50 titles of CFRs published in the Federal Register by the Federal. The FARs are rules prescribed by the FAA governing all aviation activities in the USA under title 14 of the CFRs and title 48 of CFRs is 'Federal Acquisitions Regulations'. To avoid confusion, of late the FAA began to refer to aerospace specific regulations by the term '14 CFR part XX' instead of FAR.

In the UK, the Civil Aviation Authority (CAA), established in 1972, oversees every aspect of aviation in the UK. It also has predecessor agencies dating many years prior to its formation. Recently some regulatory responsibilities in the UK have been passed to the EASA, which became operational in 1972. Based in Cologne, Germany, it is independent under European law with an independent Board of Appeal. The Agency is currently developing close working relationships with its worldwide counterparts.

In military applications, the standards are Milspecs (US) and Defence Standard 970 (earlier AvP 970 – UK); they do differ in places.

21.3.1 The Role of Regulation

Aeroplane design and operation, is reliant on regulatory standards and controls, as well as aeronautical science and the physics of flight. This reliance on regulations is mandatory in order to obtain or retain regulatory agency approval and certification. It is important to maintain the time and experience proven safety standards, represented by mandated regulations, which transcend all other considerations in aviation.

The role of regulation in the design and operation of civil aircraft needs to be acknowledged by all interested parties. Regulation has been founded on empiricism as much as on analytical methods. Many regulations have been written into the law as a result of accidents and/or incidents, many of these being tragic. The regulatory authorities have taken an empirical approach to verify the integrity of complex aircraft systems in order to avoid the uncertainty of sole reliance on analytical solutions, which may prove to be unreliable in actual practice. The air regulations generally pertain to aviation safety including exhaust and noise emissions.

All civil aircraft must hold a Type Certificate, or its equivalent, issued by one or more of the responsible regulatory authorities where the aircraft is to be manufactured and/or entered into operation. The aircraft must also possess an Airworthiness Certificate in its operating venue. The exceptions are aircraft permitted to operate with an experimental certificate in restricted airspace to minimise danger to other aircraft and to those below. These aircraft are generally under development and under the strict scrutiny of the responsible regulatory authority.

Topics of particular concern regarding aircraft performance requirements that must be complied with include the following:

Stalling speed
Takeoff speeds, path, distance, takeoff run and flight path
Accelerate stop distance
Climb with all engines operating and one engine out
Landing and balked landing

These requirements must be met under applicable atmospheric conditions, relative humidity and particular flight conditions, using available propulsive power or thrust, less installation losses and power absorbed by the accessories such as fuel and oil pumps, alternators or generators, fuel control units, environmental control units, tachometers and services, for the appropriate aeroplane configuration.

Special tests required by the responsible regulatory authority will also have their influence, as completion and passage of these tests is mandatory. Testing to FAA, CAA and/or EASA regulations includes simulated bird strikes to engine inlets and windshields, operation under actual lightning conditions, one engine cut out during the most critical flight segment (usually second segment climb for twin-engine powered aircraft), flyover testing of emanated noise levels during takeoff, flyby and landing, cabin pressurisation, engine heating, airframe vibration and flutter, as well as certain operating limitations, to name but a few. The following is a listing of important regulatory standards from the FAA and EASA:

FAR	EASA	
Part 23	CS-23	Normal, utility, acrobatic and commuter category aeroplanes
Part 25	CS-25	Transport category aeroplanes
Part 27	CS-27	Normal category rotorcraft
Part 29	CS-29	Transport category rotorcraft
Part 33	CS-E	Aircraft engines
Part 34	CS-34	Fuel venting and engine exhaust emissions
Part 35	CS-P	Propellers
Part 36	CS-36	Aircraft noise standards

An important advisory publication issued by EASA is AMC-20, 'General Means of Compliance for Airworthiness Products, Parts and Appliances'.

21.4 Flight Test

The proof of aircraft capabilities has to be demonstrated through flight tests – most of them are mandatory requirements. Flight testing is a subject on its own, covering a large task obligation carried out under inter-disciplinary specialisation, for example, aircraft performance, flight dynamics, dealing with wide variety of instrumentations for measurement, avionics, computers and so on. This chapter merely outlines the scope to make aware of the role required by the aircraft performance engineers. Stability and control tests are intertwined with aircraft performance.

Flight test obligations cover the following:

  i. to substantiate requirements by the certification agencies to ensure safety,
 ii. to substantiate customer speciation to ensure guarantee performance,
iii. to check design as needed by the manufacturer (e.g. position errors of pitot-static tube etc.),
 iv. to check modification/repair work carried out on existing designs,
  v. to prepare operational manuals for aircrew and maintenance engineers and
 vi. to prove new technology.

Flight testing is a protracted programme; in most industries it is conducted by a separate department in conjunction with a design bureau, primarily within the aerodynamic departments. The design office has to prepare standards, schedules and check-lists adhering to respective prescribed Flight Test Manuals (FTM) to make a streamlined procedure to stay on time.

Test pilots are specially trained pilots with considerable aircraft engineering background at least up to the level of graduate engineers, some even have Doctoral degrees. Pilot skill to maintain accurate and steady flying is essential. Stability and control response test are carried out under strictest precautions for safety – fatal accidents have occurred in past.

Fight test engineers are also specially trained skilled engineers fully conversant with all kinds of instruments. Calibration of test instruments is a prerequisite and elaborate procedure is followed to ensure credibility of data acquisition. Some instruments have lag-time; pilots and test engineers work together to record steady-state stabilised data. Confidence in raw data is graded from poor to excellent in five classes. Recognising and deleting poor are marginal data improves the test result for usage. The internal cabin volume of a larger transport category aircraft can be utilised to carry on-board recorders and processors to analyses real-time data for the flight test engineers to decide acceptance or to repeat. Bulk of test data are analysed by office-based engineers to prepare reports and manuals.

Typical time frame taken for various designs is given in Table 21.3, which gives typical aircraft flight test details. It is assumed tests continue uninterruptedly without major design revision and that the aircraft is not of a new class (e.g. A380, F35 etc.). Transport aircraft could use their cabin space to carry instruments to record real-time data. Military aircraft uses telemetry to record real-time data.

Aircraft performance figures depend on atmospheric conditions that can categorically be said to be never the same or uniform. It is therefore important to reduce the recorded data into a standard condition such as at zero wind and in an ISA day – this helps comparison to meet requirements in the specified format. The test volume is large and data reduction time using computers is also large. Considerable improvement of data fidelity can be improved if spurious data can be recognised and eliminated and there are formal methods available for its execution. Especially skilled test pilots use precise techniques to minimise bad data.

It is possible to reduce testing time if more aircraft are deployed. Data acquisition and reduction to useable from take a substantial number of man-hours. To ease financial and time constraints meticulous planning of the test schedule is to be prepared in consultation with the certification agencies and customers.

Table 21.3 Typical aircraft flight test details.

Type of aircraft	Typical number of aircraft	Typical time taken (years)	Typical number of sortie
(i) Large commercial transport aircraft	≈4–6	≈2–3	≈2000
(ii) Medium commercial transport aircraft	≈3–5	≈2–3	≈1500
(iii) Business/executive aircraft	≈2–4	≈2	≈1200
(iv) Small aircraft (club flying type)	≈2–3	≈1–2	≈400–800
(v) Military combat aircraft	≈4–6 (2)[a]	≈3–4 (2)[a]	≈2500 (500)[a]
(vi) Military training aircraft	≈2–4 (1)[a]	≈2–3 (1)[a]	≈1500 (200)[a]

a) The additional requirement for armament tests are in brackets.

All flight tests precede with accurate calibration of test equipment on-board and ground based. Possibly, one of the most important aircraft characteristics is to establish its drag polar. Engine manufacturer supplied calibrated engines are required to be installed in the aircraft that are the not the production fitment. The calibrated engines are ground tested and not at altitude. A series of tests are required, sometimes in two aircraft.

Stall and spin tests are essential. Aerodynamic analyses and wind tunnel tests offer some characteristics but a multitude of deviations in variables, one such variable is that limitation in manufacturing technology can mask analyses and in marginal situations the actual tests could prove dangerous. Section 18.7 describes spin tests in more detail.

Stability and control tests to establish aircraft handling qualities also require participation by aircraft performance engineers to establish speed schedules and so on. Establishing flutter speed, high speed at dive and trim run-out tests are critical and could also prove dangerous.

Military aircraft go through considerably more stringent flight tests as they invariably try to incorporate new technologies never tried before. In general, a technology demonstrator aircraft is used as prototype, possibly in a scaled down model, as a proving platform that may undergo many modifications. These can be a lengthy programme until a satisfactory outcome emerges. Details in Table 21.3 do not include these kinds of flight tests.

Both the civil and military aircraft must undertake all weather trials for which aircraft are taken to designated test centres that have arctic weather and hot weather climates, and they are also taken to high altitude stations to prove takeoff capability. Penalties associated with environment control systems (engine air bleeds for cabin requirements, anti-icing protection, other mechanical off-takes and so on all penalise engine power) have to be substantiated to ensure safety within the specified flight envelope.

In summary, the role of aircraft performance engineers in aircraft flight test programmes is substantial. This book covers the fundamental theories that are required for flight test data reduction. (Aircraft performance engineers play some role in aircraft simulator design to supplement the control laws given by stability/control engineers devoted to flight testing.)

21.5 Contribution by the Ground Effect on Takeoff

Ground proximity affects aircraft aerodynamics. Ground as a restricting surface acts as a constraint to the development of wing tip vortices. As a result, there is an increase in C_L and reduction in C_D.

The wing tip vortices hit ground resulting making their vertical velocity to zero. Its mathematical model was proposed as early as the 1920s using Prandtl's lifting-line theory. A mirror image of the wing vortices below the ground plane replaces the ground interacting with wing vortices resulting in zero vertical velocity. The upward velocity of the mirrored vortices increases lift and there is associated reduction in induced drag at the same angle of incidence, α. The closer the wing is to the ground, the higher the effect is and dilutes in a non-linear manner to zero as the gap between wing and ground increases to a height equal to wing span (depending on design, half to twice the wing span). There is also reduction of parasite drag due to some reduction in effective velocity. Text books divide the zones into *in-ground-effect* (IGE) and *out-of-ground-effect* (OGE).

An analysis of ground effect is given in Ref. [6]. It also lists other publications available in the public domain.

However, the accuracy of the theoretical results suffers when flaps are deployed as the flow-field over the wing gets complex. At high flap deflection, experiments show reduction in lift. The IGE has other adverse effects on modern military combat aircraft that use thrust-reverser for short landing; re-circulating reverse flow of engine exhaust gases may lead to possible re-ingestion affecting engine performance. Even vector-thrust for a short takeoff can affect engine performance at this critical moment. Other stability and control problems may also appear.

Chapter 15 dealt with field performance taking no credit for the ground effects. The conservative estimate is accepted by the certification agencies. This section is to make readers aware of aerodynamic effects on aircraft performance and subsequently explore to apply them when required.

Motor gliders with a large wing span and low-wing design staying close to ground benefit from ground effects at takeoff, but at landing tend to float and take time to settle down when lift-dumpers (an appropriate terminology for spoiler or air-brake) are deployed.

Taking the benefits of the ground effect, the Russians successfully designed several versions of 'Ekranoplan' and even operated on the Black Sea. These never surfaced on the Western market but the promise has been felt. Subsequently, Boeing studied a ground effect craft (Pelican) but it was never built. Its potential still exists.

21.6 Aircraft Environmental Issues

Aircraft noise and emission are the two main sources cause environmental concerns. Regulatory authorities had to step to stipulate limits for the aircraft designers to implement. These issues are discussed in the following subsections.

21.6.1 Noise Emissions

Noise is produced by pressure pulses generated from any vibrating source and the pulsating energy is transmitted through air to ear, heard within its audible frequency range (20–20 000 Hz). The intensity and frequency of pulsation are what determine the physical limits of human tolerance. At certain conditions, the acoustic (noise) vibrations can affect aircraft structure. Noise is seen as pollution of environment [1–4].

The intensity of sound energy can be measured by sound pressure level (SPL) and the threshold of hearing value is taken as 20 µPa. The response of the human hearing can be approximated by a logarithmic scale – that is nature. Therefore, the advantage in using a logarithmic scale for noise measurement is to compress the range of SPL extending to well over a million times. The unit of measurement of noise is the 'decibel', shortened to dB, and is based on a logarithmic scale. One 'Bel' is a 10-fold increase in pressure level: that is, 1 Bel $= \log_{10} 10$, 2 Bel $= \log_{10} 100$ and so on. A reading of 0.1 Bel is a decibel or dB, which is antilog$_{10} 0.1 = 1.258$ times increase in sound pressure (intensity) level. A two-fold increase of SPL is $\log_{10} 2 = 0.301$ Bel or 3.01 dB.

Technology required a meaningful scale suited to human hearing. With progress, the units of noise kept changing in line with the technology demand. First to come was the 'A-weighted' scale expressed in dBA suited to being read directly from calibrated instruments (sound meters). Noise is a matter of human reaction to hearing and more than just a mechanical measurement of a physical property. Subsequently, it was felt that human annoyance is a better measure than mere loudness and come the 'Perceived Noise' scale expressed in PNdB, labelled with associated 'Perceived Noise Level' (PNL) as shown in Figure 21.1 from various origins.

Aircraft in motion presented a special situation arising from the duration of noise emanating from an approaching aircraft passing overhead and continues radiating rearwards after passed over. Therefore, for

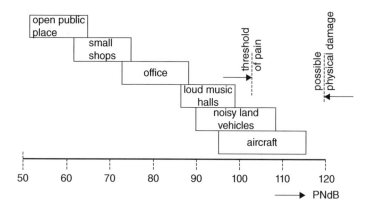

Figure 21.1 Perceived noise level (PNL) expressed in PNdB.

aircraft application it was felt necessary to introduce a time-averaged noise as 'Effective Perceived Noise Level' (EPNL) expressed in EPNdB.

During the 1960s, litigations from damage caused by aircraft noise made the government regulatory bodies to step in to reduce noise and imposed varying limits of EPNdB for various classes of aircraft. Many airports have night-time curfews for noise abatement and control; additional fees are charged for using an airfield between 11 pm and 7 am. Through research and engineering, significant noise reduction has been achieved in spite of engine size increasing to produce several times more thrust.

The US was first to impose noise certification standards for aircraft to operate within the USA. The US airworthiness requirements on noise are governed by FAR Part 35. Aircraft maximum takeoff mass (MTOM) over 12 500 lb must comply with FAR35 noise requirements. The procedure was immediately followed by the international agency governed by ICAO (requirements given in Annex. 16, Vol. I). The difference between the two standards is small. There have been attempts to combine the two into one uniform standard. The readers are recommended to refer to FAR35 and ICAO Annex. 16 for details.

Since the existing big aircraft were the cause of the noise problem, FAR introduced regulations for its abatement in stages, the older aircraft required modification work within a specified date to remain operable. In 1977, The FAA introduced noise level standards in three tiers as follows:

Stage 1	Intended for old aircraft already flying and soon to be phased out, for example the B707, DC8 and so on. These are the noisiest aircraft but least penalised as they are to be grounded soon.
Stage II	Intended for recently manufactured aircraft that have longer lifespans, for example the B737, DC9 and so on. These are also noisy but must be modified to a quieter standard than Stage I. If they are to operate longer, then further modification is necessary to bring the noise level to Stage III standard.
Stage III	Intended for the new designs with the quietest standards.

ICAO standards are given in Annex. 16, Vol. I in ch. 2–10: each chapter addresses different classes of flying vehicle. This book is concerned with chapters 3 and 10 that are basically meant for new aircraft (the first jet aircraft flight after 6 October 1977 and propeller-driven aircraft after 17 November 1988).

To certify aircraft airworthiness, there are three measuring points in the airport vicinity to ensure neighbourhood living is within the specified noise limits. Figure 21.2 gives the distances involved to locate the measuring points. They are as follows:

i. Takeoff reference point −6500 m (3.5 nm) from the brake release (start) point and at an altitude given in Figure 21.2.
ii. Approach reference point −2000 m (1.08 nm) before the touchdown point, which should be within 300 m from the runway threshold line and maintain at least a 3° glide slope with the aircraft at least at an 120 m altitude.

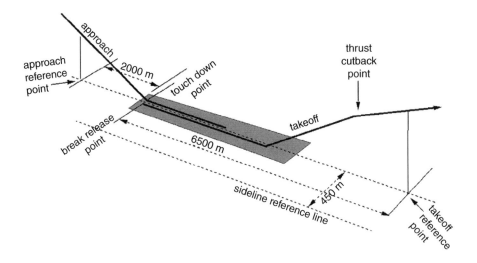

Figure 21.2 Noise measurement points at takeoff and landing.

iii. Side-line reference point −450 m (0.25 nm) from the runway centre-line. At the side-line there are several measuring points along the runway. This is measured at both the sides of the runway.

ICAO, Annex. 16, Vol. I, ch. 3 maximum noise requirements in EPNdB are given in Table 21.4 and is plotted in Figure 21.3.

A typical footprint of the noise profile around runway is shown in Figure 21.4. The engine cut-back area is shown in the figure along with re-established the rated thrust for en route climb. Residential development may avoid the noise footprint areas.

The airframe also produces a significant amount of noise, especially when the aircraft is in a 'dirty' configuration. Figure 21.5 gives the sources of noise emanating from the airframe. The entire wetted surface of the aircraft, more so by the flow interference at the junction of two bodies (e.g. at wing-body junction), produces some degree of noise based on the structure of the turbulent flow causing pressure pulses audible to the human ear. It becomes aggravated when the undercarriage and flaps/slats are deployed creating considerable vortex

Table 21.4 EPNdB limits (Make linear interpolation for in between aircraft masses).

Takeoff (make linear interpolation for in between mass)

No. of engines	2		3		4	
MTOM, kg	≤48 100	≥385 000	≤28 600	≥385 000	≤20 200	≥385 000
EPNdB limit	89	101	89	104	89	106
[a]Cut-back altitude, m		300		260		210

Approach – any number of engines (make linear interpolation for in between mass)

MTOM-kg	≤35 000	≥280 000
EPNdB limit	98	105

Side-line – any number of engines (make linear interpolation for in between mass)

MTOM, kg	≤35 000	≥400 000
EPNdB limit	94	103

a) In certain airports, engine throttle cut-back (lower power setting) are required to reduce noise levels after reaching the altitude as shown in the table. At cut-back, the aircraft should maintain at least a 4% climb gradient. In the event of an engine failure, it should be able to maintain altitude.

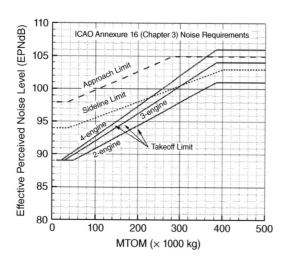

Figure 21.3 ICAO noise requirements.

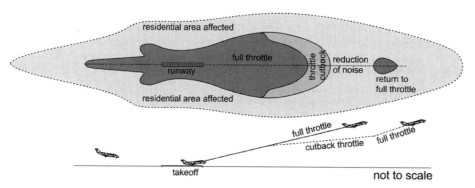

Figure 21.4 Typical noise footprint (≈10 km long) of aircraft showing the engine cut-back profile.

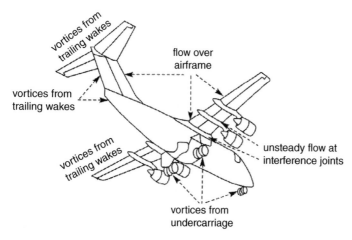

Figure 21.5 Typical sources of noise emanating from an airframe. Source: Reproduced with permission of Cambridge University Press.

flow and unsteady aerodynamics; the frequencies of fluctuations appear as noise. At the conceptual design phase, care must be taken to minimise gaps, provide fillets at the two body junction, make streamlined struts and so on. Noise increases with speed increase. Care is to be taken to eliminate acoustic fatigue in structure and make them damage tolerant; material selection plays an important role (Figure 21.6).

Typical noise levels from various sources are given in Figure 21.7 both at takeoff and at landing. Evidently, engine contribution is the dominant one, which reduces at landing when the engine power is set low and the jet efflux noise reduces substantially. Noise emanating from the airframe is more at landing on account of higher flap/slat settings and the aircraft altitude is lower at the measuring point than at the takeoff measuring point. Since the addition of noise level is done on a logarithmic scale, the total level of noise contribution during takeoff and landing is nearly the same.

The engine and nacelle are the main sources of noise at takeoff when it is running at maximum power. All the gas turbine components, for example, fan blades, compressor blades, combustion chamber walls and turbine blades, generate noise. With the increase of bypass ratio (BPR) the noise level goes down as low exhaust velocity reduces shearing action with the ambient air. The difference in noise between an afterburning turbojet and high BPR turbofan can be as high as 30–EPNdB. Figure 21.7 shows that, at subsonic flight speed, the noise radiation moves ahead of the aircraft.

Propeller-driven aircraft operation has to consider noise emanating (radiation and reflection) from the propellers. Here, the noise reflection pattern depends on the direction of propeller rotation as can be seen in Figure 21.8. The spread of reflected noise also depends on propeller position with respect to the wing and fuselage.

Inside the cabin, the noise comes from the environmental control system (ECS) and needs to be kept at its minimum level. These problems are tackled by the specialists. Details of cabin interior design considerations are taken up in Phase II of the project.

Figure 21.6 Relative noise distributions from various aircraft and engine sources.

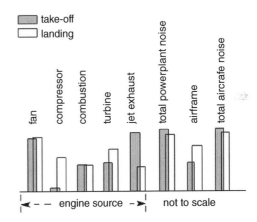

Figure 21.7 In-flight turbofan noise radiation profile.

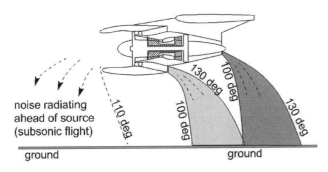

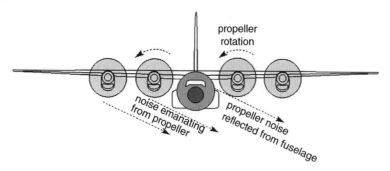

Figure 21.8 Noise considerations for propeller-driven aircraft.

21.6.2 Engine Exhaust Emissions

Currently, the civil aviation sector burns about 12% of the fossil fuel consumed by the worldwide transport industry. It is responsible for about 3% of the annual addition to greenhouse gases and pollutant oxide gases. The environmental debate got hotter on the issues of climate change, depletion of ozone layer and so on, leading into the debate on long term effects of global pollution. Smog is made of nitrogen oxide that affects human pulmonary and respiratory health. The success of the automobile industry in controlling engine emission is clearly evident from the dramatic improvement achieved in many metropolises.

The US Environment Protection Agency (EPA) saw this problem some time ago. The 1980s emphasised the need for government agencies to tackle the engine emission issues. The early 1990s (Kyoto agreement) brought a formal declaration to limit pollution (locally around airports). Currently, there are no regulations for the cruise segment. In the USA, the FAR 34 and for the international body the ICAO (Annex. 16 Vol. II) outline the emission requirements. The EPA has worked closely with both the FAA and the ICAO to make the standards very close to each other. Although military aircraft emission standards are exempt, increasingly they are coming into scrutiny to follow Milspecs standards. Emission is measured by an Emission Index (EI).

Combustion of air (oxygen + nitrogen) and fuel (hydrocarbon plus small amount of sulfur) should ideally produce carbon dioxide (CO_2), water (H_2O), residual oxygen (O_2) and traces of sulfur particles. In practice, the combustion product consists all of these plus an undesirable amount of pollutants such as carbon monoxide (CO – toxic), un-burnt hydrocarbon (UHC), carbon soot (smoke – affect visibility), oxide of sulfur and various oxides of nitrogen (NO_X – affects ozone concentration). The aim of the regulations is to reduce the level of undesirable pollutants by improving combustion technology. Sustenance of air-travel and industry growth would depend on how technology can keep up with the demands for human health preservation.

Lower and slower flying reduces the EI. This is in conflict with market demand of flying higher and faster. Designers will have to make compromises. Reduction of EI is the obligation of engine manufacturer and hence the details of airworthiness requirements on EI are not given here. These can be obtained from the respective FAA and ICAO publications. Aircraft designers will have to depend on engine designers to supply certified engines complying with the regulatory standards. There is no classroom work from this section.

21.7 Flying in Adverse Environments

This section briefly discusses problems associated with flying in icy conditions, foggy weather, turbulent weather, wind shear, jet stream and thunderstorms. Ice, rain, fog and turbulence are adverse environments for aircraft to fly in. These fall in to two groups: (i) an adverse environment with loss of visibility and (ii) an adverse environment due to aerodynamic and stability/control degradation, as follows.

21.7.1 Group 1 – Adverse Environments due to Loss of Visibility

Loss of forward visibility and horizontal reference can affect pilot performance and accidents have happened due to pilots being unable to see or judge flight obstacles and/or aircraft attitude. Lack of visibility is like flying on dark cloudy nights. Fortunately, degraded visibility does not affect aircraft performance, stability and control other than affecting pilot skill. Flight deck instruments prove adequate under skilled hands. There are three main types of environmental conditions that affect pilot visibility, described next.

i. *Atmospheric fog.* There are different kinds of reasons for fog formation and it can appear suddenly. Normally, it occurs at lower altitudes; at higher altitudes the visibility problems arise from flying in cloud. Takeoff and landing become difficult or an obstacle not is seen at any altitude until it is too late. Lack of a horizon reference can make an aircraft enter in to a spiral dive, which is not easy to detect in the initial stages by inexperienced pilots. Commercial aircraft pilots must have instrument-flying training. Forward scanning radar fitted to the aircraft is of considerable assistance. In any case, it is better to wait until the fog is cleared, normally in a relatively short time. In case a pilot confronts foggy conditions on landing, landing should be undertaken under an Instrument Landing System (ILS) supported by the airport control tower. If possible, a pilot should loiter around close by where there is visibility until destination airport fog conditions improve. In the worst case, a pilot could opt for alternative airport provided there is sufficient fuel.

ii. *Flight-deck fog.* This is like what can occur in a motor vehicle. Flight-deck fog has to be removed. The defogging system is like that of a car with an embedded electrical wire in the windscreen. Figure 18.1a shows a generic layout with wipers representing the BAE RJ family in better detail.

iii. *Heavy rain.* Heavy rain can also impair vision. The rain-removal system is also like a car using windscreen wipers. Also, rain-repellent chemicals assist in rain removal. Section 22.7.1 describes the rain-repellent system in better details.

21.7.2 Group 2 – Adverse Environments due to Aerodynamic and Stability/Control Degradation

Flying in adverse environments affecting aircraft performance, stability and control falls into two types as follows.

i. *Aerodynamic degradation.* Ice formation on the aircraft surface affects the skin condition and considerably increases drag. In addition, ice formation on critical areas of lifting surfaces, intake and internal engine components develops into a dangerous situation with loss of lift and power. Ice accretion at the leading edge deforms the required aerofoil contour, disturbing smooth streamline flow causing premature early stall. It is difficult to notice visually clear ice forming at early stages and therefore engineers make provision to remove ice accretion described in better details next. Figure 21.9 shows typical icing envelopes [5].

ii. *Anti-icing, De-icing.* Icing is a natural phenomenon that occurs anywhere in the world depending on weather condition, operating altitude and atmospheric humidity. Ice accumulation on wing/empennage and/or engine intake can have disastrous consequences. Icing increases drag and weight, decreases lift and thrust and even degrade control effectiveness. On the wing/empennage, icing will alter the profile geometry leading to loss of lift. Ice accumulation at the intake will degrade engine performance and can even damage the engine if large chunks are ingested. It is a mandatory requirement to keep the aircraft free from icing degradation. This can be achieved either by anti-icing that never allows ice to form on critical areas, or by de-icing that allows ice build-up to a point and then shed it just before it becomes harmful. De-icing results in blowing chunks of ice away and these should not hit or get ingested by engine.

There are several methods for anti-icing and de-icing. Not all anti-icing, de-icing, defog and rain-removal systems use pneumatics. These are primarily electrical systems but are included in this section to continue with the subject matter. Section 22.7 describes the anti-icing system in better detail.

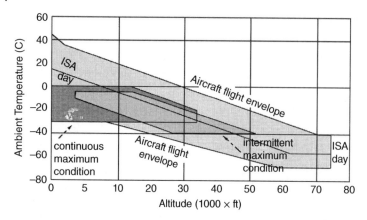

Figure 21.9 Typical icing envelope [3].

i. *Aircraft stability/control degradation.* These are due to influence of weather disturbances, for example, turbulent weather, wind shear, thunderstorms and jet streams are the main causes that affect aircraft stability/control and can degrade to dangerous conditions. These are described separately next. Weather related disturbances primarily concerns with control of aircraft, bearing in mind that there is a sufficient stability margin to control the aircraft satisfactorily in any inadvertent situation. Meteorological offices routinely circulate weather forecasts, which all pilots must study and use to make flight plans. Meteorology is an involved science beyond the scope of this book. Here, only the effects of adverse weather are touched upon. Flight planning must accommodate necessary provisions to encounter bad weather.

ii. *Turbulent weather.* Random unsteady atmospheric wind flow that can change direction is seen as turbulent weather. There are several ways turbulence develops, for example, weather conditions or ground contour/heat distribution developing convective currents. These can come as gusts of wind vibrating the aircraft in motion and sometimes can be severe enough to throw passengers from their positions if they are not secured with restraints such as seat-belts; injuries have happened. Modern transport aircraft are fitted with auto-stabilisation that dampens the shaking of aircraft to a considerable extent but in strong turbulence the aircraft can suffer a heavy, bumpy ride. All passengers must adhere to the cabin safety instructions to stay with seat-belts on for their own safety.

FAA has requirements that transport aircraft must withstand a 66 ft s^{-1} upward gust (see Chapter 17). Typically, a weather report can give the areas under turbulent conditions and good weather radar can detect strong turbulences. Pilots should avoid flying through these conditions by making diversions; normally flight planning accommodates extra fuel for possible diversions, which is usually not a long detour.

A strong downward gust can make an aircraft to lose considerable height. At low altitudes, say at approach, a strong downwind can prove dangerous. The destination airfield offers local weather conditions and pilots remain alert to control their aircraft in any inadvertent conditions.

However, at high altitude around tropopause, turbulence can appear as Clear Air Turbulence (CAT) that is not easy to detect and can be severe. Wind shear and jet stream are associated with CAT.

i. *Jet stream.* Around the tropopause there is a global pattern of periodic tube-like (relatively thin but very wide) wind flow in long stretches from west to east in the mid latitudes of the Northern Hemisphere. When the average wind velocity exceeds 50 kt, it is seen as a jet stream; its core can have a high velocity exceeding 100 kt. This is an unsteady state and its size and strength keep changing. There can be considerable fuel savings in the jet stream when flying eastward and vice-versa in the Southern Hemisphere.

ii. *Wind shear.* An aircraft enters into wind shear when it enters a zone where wind speed changes in magnitude and direction. The aircraft must be able to adjust to these changes; generally, it is controlled through the disturbance in a relatively short time. A downdraft will make an aircraft sink rapidly, which can be observed on the flight deck instruments. Aircraft sinking when flying close to the ground, say at approach to land, can pose a hazardous situation. Pilots need to remain alert at takeoff and landing phases to control the aircraft from hitting ground.

iii. *Thunderstorm/cloud*. Thunderstorms are violent outbursts in the atmosphere and are always associated cumulonimbus cloud (cumulonimbus clouds are nimbus clouds in the form of a rising cumulous cloud). The violent nature can be observed by their swirling pattern moving mainly upward. Their gusts can easily exceed the design limits and can have catastrophic consequences if flown into. Tornados/hurricanes are manifestations of these outbursts. Thunderstorms also produce strong up and down bursts. They must be avoided. Fortunately, thunderstorm areas can be easily located so pilots can detour the area. In low intensity thunderstorms, pilots may fly through after studying the nature but precautionary measures must be taken for both passengers and crew. There are prescribed procedures to fly through permissible thunderstorms.

Lighting strike and hail are associated with thunderstorms. Lightning is the electrical discharge of the static electricity that accumulates. The direction discharge goes towards the maximum potential difference and branches out in different directions; this can be within the cloud or may strike the ground. The vast expanse of thunder clouds develops a very high voltage and current that can be destructive, and this is another reason for pilots to detour the area. Aircraft are equipped to discharge static electricity build-up and the shell insulates the interior. However, they are not designed to fly through severe thunderstorms. Lightning strikes are common and in the majority of cases, in a way, are harmless and the mission continues safely but an extreme strike can create an impact that can cause operational interruptions. Lightning protection is incorporated to reduce lightning strike impact. Most strikes occur between 8000 ft and 16 000 ft altitudes when aircraft are in climb or descent.

Glider pilots seek updraft from thermals to gain altitude. They often take advantage of the upward flow of small cumulous clouds at the early stages that may be seen as patches of cotton floating in interesting shapes. Pilots must be trained to recognise which ones are harmless.

21.7.3 Bird Strikes

Birds can create adverse flying environment. An open span of grass field around the runway as a safety zone is an ideal feeding ground for birds – seeds, grains, inspects, worms and so on abound. Some even make their permanent home there and keep breeding. They can also come from outside in a large flock. Although the birds are used to aircraft noise, they may be spooked into flying in the aircraft flight path during takeoff and landing when their ingestion to the engine is not uncommon. In many instances a single small bird ingestion by the engine may not be felt but experienced pilots can recognise it by observing engine gauges and from the feel. In some occasions, the aircraft continues towards the destination airport and bird strike evidence is found after landing. Aircraft certification authorities have mandatory requirements to demonstrate safety by throwing fresh dead chickens known as the *chicken test*. Most bird strikes occur around an airfield where birds rarely exceed the size of a chicken. However, large bird strike is a different matter.

The recent Hudson River landing of an Airbus320 after both engines failed on account of bird strike (possibly geese) is a well-published case of the kind of accident that can happen. Landing on water has happened many times before but is a glaring example of what can happen with bird strike. At higher altitudes, vulture strikes have occurred many times. There are cases when even older military aircraft with sensitive engines fell due to bird strike from a bird as small as a sparrow.

Birds around an airport can be driven out, at least for a temporary period. Various studies have been made to drive them out with temporary success. Constant surveillance (nowadays with drones), various scaring techniques and environmental planning can keep flight operations continuing. Bird line drones are in use to drive out any wildlife incursion in an airfield area. Over long period of time, these may become an effective measure to some extent. However, stray birds away from an airfield are a random occurrence. Bird strike at high aircraft speed means that the impact force is much higher and can be felt by pilots. Pilots must then land to the nearest available airport for safety reasons; successful landing on the Hudson River is now incorporated as a pilot drill measure if landing on an airfield is not possible. Without a stretch of water and airfield, a pilot has to take action as he/she judges.

21.8 Military Aircraft Flying Hazards

In additions to the hazards of flying as described previously, military aircraft have face additional hazards when flying at low altitudes, especially through terrain as a cover. There are hazards in hard manoeuvres to engage or disengage in combat at lower altitudes. As some measure of safety, military aircraft are fitted with *Ground Proximity Warning Systems*. Of late, modern civil transport aircraft above 30 000 lb MTOW are also fitted with *Ground Proximity Warning Systems*. These instruments sense terrain and ground below to warn pilots in time to react (typically 1 s) according to safety. All types of aircraft, even small recreational aircraft flying through valleys for sightseeing must keep adequate separation distance if a manoeuvre, for example. turning and climb-out, is required. A pilot's understanding of the surroundings, be it terrain or other aircraft, is essential even when a terrain map and *Ground Proximity Warning Systems* are integrated with the aircraft.

Military aircraft jettisoning stores in certain manoeuvres can be hazardous if they hit any part of aircraft. There have been cases when light empty drop-tanks have tumbled at jettison hitting the empennage.

21.8.1 Aircraft Combat Survivability

New combat aircraft capabilities are evaluated in simulated environments in twin platform, the other platform being that of the adversary's capabilities to the extent it can be ascertained in terms of combat scores. These are based on mathematical modelling with measurable parameters, for example, susceptibility (inability to avoid adversary interception), vulnerability (inability to withstand a hit) and killability (inability to both). The general term ACS refers to the capability of aircraft survivability disciplines attempts to maximise the survival of aircraft in wartime. This is in addition to system safety attempts to minimise those conditions known as hazards that can lead to mishaps in environments not arising from combat. These topics are beyond the scope of this book.

Section 22.8 covers some of the systems incorporated as measures for combat survivability.

21.9 End-of-Life Disposal

Like any other machinery, their useful operation ceases on account of wear and tear or being no longer economic to operate when replaced by newer models; the example of a car is a glaring one. Aircraft is no exception on functionality, especially under strict safety/environment regulations. It is estimated that over tens of thousands of civil aircraft will cease to operate in the next two decades, each costing millions of US dollars when purchased new. Already, there are 1000 of disused civil and military aircraft stored in Arizona, California and New Mexico covering 1000 of acres, loosely termed an *aircraft graveyard*. These decommissioned aircraft still carry value as scrap and even some components can be recycled/reused. There are guidelines for storage, offering decommissioning, dissemble, dismantle, recycle/disposal and value extraction. The aircraft designers have to heed these guidelines to facilitate the process of salvaging the aircraft.

Metals sold as scrap can be recycled but more and more composite material is now piling up. Disposal of composite material is difficult as they serve no useful purpose as scrap – attempts are in hand to make them recyclable. Avionics boxes and microprocessors contain toxic material. Fluids in display units are also toxic. It would cost to get rid of the toxic material from the environment. Even incineration plants are specifically designed to keep the efflux clean.

More and more researches are continuing to find suitable materials that are less toxic and can be disposed of cost effectively. This is a matter for material scientists. However, aircraft designers must stay updated with materials technology and choose the correct kinds.

Conceptual aircraft designers should make considerations that will facilitate an easy to dissemble arrangement and use as many recyclable materials and equipment as possible to maximise valorisation and minimise land-fill waste.

21.10 Extended Range Twin-Engine Operation (ETOP)

In past, twin-engine transport aircraft were not allowed to fly over water for safety reasons in case an engine fails losing half the power coupled with any other reason for the other engine not to supply adequate power. Losing an engine increases aircraft drag (see Section 11.15.4). Any twin-engine commercial transport aircraft is restricted to operating within land mass on a route where an emergency airfield is available, in case an emergency demands a landing. Chapter 15 did not compute the ETOP range. In this section, its background is explained and readers may attempt compute the range using the same procedure with the effects of an engine failure and drift down procedure dealt in [6].

However, confidence built up as engines became more reliable when failure at cruise power rating became a rare occurrence. In 1980, the FAA negotiated with operators and manufacturers to allow flying over water up to a distance that ensures that in case of emergency the aircraft can land at a suitable airport in 120 min under one engine powered up to maximum climb (i.e. continuous) rating. The operation is known as an ETOP. With success of ETOPS with a 120 min limit, the FAA subsequently raised the time limit to 180 min (in head-wind less range). This extension of time limit improved the fuel efficiency of operation.

21.11 Flight and Human Physiology

Defying gravity could be dangerous if the laws of the nature are not fully understood ([7–10]). Apart from solving the laws of physics to make aircraft fly and gain altitude there is the additional issue of understanding the nature of human physiology and its limitations. Inadequacy in medical condition of a pilot is an actual threat to survival. Qualifying for medical airworthiness is mandatory for all pilots; its level of stringency differs for the class of aircraft to be operated, combat pilots have the strictest requirements. Even for fit pilots, the influence of drug/alcohol can affect medical airworthiness. Aviation physiology is a subject on its own and is now a mature science for understanding medical human factors, their limitations and pilot response characteristics.

There are two main limitations based on (i) altitude effects and (ii) gravitation loads, which can be different from unit load ($1\,g$) as one experiences in day to day living. These two aspects are briefly described next.

i. *Effects of Altitude.* Pressure and density of atmosphere decreases in non-linear manner (see Section 1.15). Human habitation in Himalayan terrain at around 10 000 ft exists. By far the majority of normal habitation is below 3 000 ft. It has found that humans can sustain unassisted up to 14 000 ft. Therefore, aircraft flying above that altitude must have ECS to keep pressure at an acceptable limit typically around 7000–8000 ft depending on design (see Section 15.2). With adequate pressurisation, oxygen supply for human need is simultaneously met. However, when the inadvertent situation of cabin depressurisation occurs, then an emergency oxygen supply is necessary. The ECS takes engine air flow bleed at about 2–3% degradation of engine performance affecting fuel economy (see Section 12.7). Aircraft make rapid descent to a safe altitude below 10 000 ft.

ii. *Effects of g-load.* In manoeuvre (see Chapter 17), an aircraft experiences g-load. There two kinds of g-load: (a) instantaneous g-load and (b) sustained g-load. Instantaneous g-loads are of short duration lasting a few seconds and humans can tolerate higher loads than sustained g-load, which can last longer (in space travel, initial booster g-load load on astronauts can last in excess of 4 min). Military pilots with g-suits have experienced instantaneous g-load of $12\,g$; extreme aerobatic aircraft are designed to withstand $9\,g$. Sustained g-load for military aircraft is $6\,g$ and for extreme aerobatics aircraft can take $6\,g$. A typical Ferris-wheel/merry-go-round is around 2–$3\,g$.

Aircraft pull outs from a dive/loop in high g-load prevents blood from the heart reaching the brain causing black-outs and a pilot can temporarily lose consciousness, which is of course dangerous. It is for this reason that military aircraft seats are more inclined to reduce the carotid distance, easing blood flow to the brain. Astronauts lay horizontal to withstand sustained g-load. At negative g, a pilot can enter weightlessness with a different kind of reaction, some enjoy this. At extreme high negative g, disorientation of the pilot can occur.

21.11.1 Aircraft Design Considerations for Human Factors

A combat flying mission stretches the pilot to the limits of human capability; in loose terms, an inhuman task. It requires expensive long training of handpicked individuals by rigorous testing of their physical and emotional fitness. High levels of information processing as multitasking of safety critical and mission critical workload is stressful and under physical stress, as described before, can impose considerable fatigue leading to erroneous pilot judgement in executing the correct procedures that can result into catastrophic consequences. Aircraft designers have the obligation to assist pilot cognitive (situation awareness) functions under duress.

Aircraft designers search for inventing appropriate facilities to ease pilot stress, thereby improving operational safety. The examples of autopilot, ILS, microprocessor-based fly-by-wire (FBW)/full authority digital engine control (FADEC), HUD, EFIS, ECS, anti-g pressure suits, voice operated commands, back-to-base system and so on are many examples, most of which have become standard features as the mission demands. Configuring the flight deck to offer a comfortable pilot work environment is an essential design obligation, so are the requirements for passenger cabin configuration.

Military aircraft flight deck configuration and design are considerably more complex than for civil aircraft flight decks. A military flight crew requires a specially designed suit with ECS and pressurisation for high altitude and high-g operation. The flight deck should not bear any light signature during night flying, requiring specially designed 'eyebrow' lights for instrument reading. Ejection seats are standard features for survival design consideration. Vibration and flight deck noise reduction are integral to design.

Some of the automations are old inventions, one such invention is 'autopilot'. Incorporation of all round rapidly growing microprocessor-based systems bear the marks of progress. It will not be long until designers start their conceptual design studies for drone-like fully automated commercial transport aircraft, the cargo versions leading the way.

In the past it has been observed that conversion from older procedures to modern FBW procedures, with multi-functional displays, have caused considerable difficulty with cases of accidents and incidents. Improved training methods and use of flight simulators have practically eliminated such difficulties. Other human factors to consider are the emotional stability under severe workload related stress in emergency, physical incapacity for whatever the reason (apart from being wounded in combat) and so on.

21.11.2 Automation – Unmanned Aircraft Vehicle (UAV)/Unmanned Aircraft System (UAS)

This book is not on UAV/UAS (synonymous terminologies). Readers may refer to Ref. [11] for a detailed treatise on UAV/UAS. However, it may interest readers to examine the consequence of removing human limitations as design constraints, the UAV/UAS offers enhanced scope for performance capability using the technologies dealt with in this book.

The following are the areas affected by the absence of on-board human occupation.

i. The 'g' is level can be relaxed to a considerable higher limited on-board instrument limits, typically 3–4 times higher, if required to still higher values.
ii. Human comfort level of ECS considerations are not required. The system can be designed to extreme weather considerations within the terrestrial flight envelope.
iii. Higher payload capability for not having systems required for humans on-board.
iv. Relaxed airworthiness requirements.

v. UAV/UAS are software driven machines that can be autonomous or ground based pilot controlled, requiring the transmitter-receiver set to communicate.

vi. For long endurance operations, ground-based pilots can avoid fatigue by working in shifts to stay sharp in cognitive function.

vii. On-board pilotless combat aircraft has no human pilot risking their life, at the same time they are vehicles capable of higher performance capabilities. Such vehicles are already in production.

viii. Civil application of a fixed wing bears greater promise but waits infrastructure support and marketability. Copter like drones (UAV/UAS) are already in the early stages of civil applications.

Readers are encouraged to explore possibilities as entrepreneurs to the many areas of possible applications in this early stage of development.

21.12   Some Emerging Scenarios

There are four topics; two arising from terrorist activities, one concerns health issues and one is an old problem arising from aircraft debris on runway. This section is included to keep future designers aware of the kinds of problems one may face.

21.12.1   Counter-Terrorism Design Implementation

Thought is now given to whether there could be ways to counter on-board terrorism. The consideration is topical and yet to be decided on for implementation. It will increase aircraft weight and cost. Some of the areas of design changes are as follows:

i. Make a bullet-proof flight-deck barrier at the cockpit door. Make the cabin compartmentalised to isolate trouble. How effective these are needs to be debated but the aircraft designers must stay ahead with some forethought than afterthought.

ii. Improve the structural integrity of cargo compartment/bay against in-flight explosions. The space below the floorboard needs to be compartmentalised and have a shock absorbing/impact resisting shell structure to retain integrity in the event of a local explosion.

iii. Aircraft flight systems to have automated recovery, homing to the nearest airfield and landing. Military aircraft already have such systems.

21.12.2   Health Issues

A steady annual increase in passenger numbers crossing international boundaries brings in health issues that need to be attended to. This means making space allowance (like cruise ships) to treat/isolate patients. Up until now, these were attended to on an ad hoc basis but manufacturers can increase market appeal with health-care facilities, especially for very large aircraft flying long haul. Cardio-vascular problems, pregnancies, isolated infections and many other kinds of emergency health cases are ever increasing during flights.

21.12.3   Damage From Runway Debris (an Old Problem Needs a New Look)

The catastrophic crash of *Concorde* was a result of runway debris hitting fuel tank making it burst and catch fire. This sends a signal to protect vulnerable areas with stronger impact resistant material. This is a relatively easy task but designers need to examine the point in a new design. This will incur additional weight and cost.

As a preventive measure, regular runway inspection is now a routine procedure for all airfields using well-equipped land vehicles. However, use of drones is gaining ground not only for runway inspection but also readily inaccessible surfaces of aircraft; for example, large aircraft wing/fuselage surfaces using high resolution cameras.

References

1 Smith, M.J.T. (1989). *Aircraft Noise.* Cambridge University Press.

2 Ruijgrok, G.J.J. (2003). *Elements of Aviation Acoustics.* Eburon.

3 ESDU 02020, An Introduction to Aircraft Noise. Available online at https://standards.globalspec.com/std/ 1643052/esdu-02020 (accessed June 2018), ESDU.

4 Gunston, W. (2002). *The Development of Jet and Turbine Aero Engines.* Sparkland, Somerset: Patrick Stephens.

5 Torenbeek, E. (2013). *Advanced Aircraft Design: Conceptual Design, Technology and Optimization of Subsonic Civil Airplanes.* Wiley.

6 Kundu, A.K., Price, M., and Riordan, D. (2016). *Theory and Practice of Aircraft Performance.* Wiley.

7 Garland, D.J., Wise, J.A., and Hopkins, V.J. (1999). *Handbook of Aviation Human Factors.* London: Lawrence Erlbaum Associates Publishers.

8 Dhenin, G. (ed.) (1978). *Aviation Medicine: Health and Clinical Aspects.* London: Tri-Med Books Limited.

9 Society of U.S. Air Force Flight Surgeons. *Flight Surgeons Handbook of Life Support Equipment,* 1 Oct 1987.

10 Shannon, R.H. *Operational aspects of forces on man during escape in the U.S. Air Force,* 1 January 1968–31 December 1970. In AGARD *Linear Acceleration of Impact Type,* June 1971.

11 Grundlach, J. (2014). *Designing Unmanned Aircraft Systems: A Comprehensive Approach,* 2e. Reston, USA: AIAA.

22

Aircraft Systems

22.1 Overview

As mentioned in Chapter 21, conceptual aircraft design is not merely concerned with producing a geometry that is capable of meeting the performance specifications, but also involves early thinking about many issues. This chapter continues with those issues dealing with bought-out items, hardware considerations and selection, human interfaced items, for example, flight deck instruments, that would affect aircraft weight and cost. This chapter gives a brief overview on some of the main aircraft systems that play essential roles in aircraft becoming airworthy and must be known by the designers engaged in conceptual aircraft design to generate aircraft configurations. The hardware system items require space provision that sometimes may affect external aircraft geometry adjustment and in special cases require specific structural components, for example, undercarriage housing, wing-body belly fairings and so on.

Most academic institutions offer separate courses on aircraft systems. In some places these areas could escape attention as the undergraduate curriculum is already packed with the classical aeronautical subjects. While detailed study of these topics is beyond the scope of this book, it is important that the newly initiated must have some feel for what is meant by these topics dealt with in this chapter.

Systems design, including avionics/electrical systems, has a large impact on performance of the propulsion system as well as aircraft mass and the cost frame to remain completive in the market. Designers study the system design implications along with cost benefit analysis for suitability of the solution to the customer. These include the subject of integration of an airframe with systems, for example, installation of equipment and furnishings, routing of harnesses, ducts and pipes and the provision of a high integrity bonding and earthing philosophy. These are studied during the Project Definition Phase (Preliminary Design Phase II), which is beyond the scope of this book.

Aircraft system weight (avionics, electrical, mechanical, hydraulic, pneumatic, actuators etc.) can be 6–10% of maximum takeoff mass (MTOM) (see Tables 10.1, 10.2 and 10.7). The open market offers a large variety to choose from that can affect aircraft cost and weight. The selection of bought-out aircraft systems contributes to the aircraft operational capability for operators to compete in the fare-paying market. Within the class of specifications, typically, bought-out item costs and weights are comparable, hence the manufacturers' integration costs and weights exhibit gradual evolutionary changes. The aircraft external geometry will not be affected by these considerations (unless a radical approach is taken), but there could be weight and cost changes. The semi-empirical weight equations in Chapter 10 are sufficient and can be modified with improved data. In industry, a more accurate weight estimation is carried out to reflect the changes affected by the topics discussed in this chapter.

The topics concerning military aircraft design are complex issues and need to be studied separately (specialist books referred). Earlier chapters indicate the complexity of military aircraft design. These considerations would make advanced military aircraft design a more difficult proposition for the newly initiated undergraduate students.

Conceptual Aircraft Design: An Industrial Approach, First Edition. Ajoy Kumar Kundu, Mark A. Price and David Riordan.
© 2019 John Wiley & Sons Ltd. Published 2019 by John Wiley & Sons Ltd.

The chapter covers the following topics:

Section 22.2: Introduces the Topics Considered in this Chapter
Section 22.3: Environment Issues – How Noise Emission Affects Design
Section 22.4: Safety Issues – Doors, Egress, Escape Systems
Section 22.5: Aircraft Flight Deck (cockpit) and Instruments
Section 22.6: Aircraft Systems, for example, Mechanical, Electrical, Hydraulic, Pneumatic and so on
Section 22.7: Systems Required to Fly in Adverse Environments
Section 22.8: Discusses Military Aircraft Survivability and Stealth Issues

Classwork content: Classroom exercise pertains to checking the Bizjet emergency door compliance with regulatory requirements. Otherwise, there is no other classwork involved. Readers may gather equipment sales brochures (supplied free of charge) from various manufacturers in which dimensions and weights are given. The Internet is a good source.

22.2 Introduction

The following few topics are covered to broaden the perspectives of readers on the system requirement considerations for aircraft design.

i.   Environment issues (noise and engine emission and end of life recycling)
ii. Safety issues (doors, emergency exit, chutes etc.)
iii. Human interface (flight deck description and displays)
iv. System architecture (avionics, electrical, mechanical systems issues etc.)
v.   Military survivability issues (stealth, ejection etc.)
vi. Flying in adverse environments

Environment issues. Figure 21.6 discussed the growing pollution concerns with respect to the climate change requiring the regulatory to set standards to limit the noise and engine exhaust gas emissions levels. Section 22.3 continues dealing with the design considerations associated with these environmental issues.

Safety issues. With increased passenger capacity, in the rare event of fire hazard on ground or post-crash evacuation, the question of quick egress is required by the regulatory bodies to ensure safety. The regulatory requirement stipulates a minimum number of exit doors (main and emergency doors) and slides/chutes that would ensure egress within a specified time. For military aircraft, the extreme measure of ejection is incorporated. Aircraft are equipped with fire detection and protection systems installed inside the aircraft. Section 22.4 deals with design and equipment considerations associated with safety issues.

Human interface. With increased demand of pilot workload, it is important to understand the aircraft flight deck (cockpit) and arrangement of systems. It exposes newly initiated readers to the nature of design features for human interface. It can affect aircraft weight, cost, and the shape of the forward fuselage/canopy area. Section 22.5 describes the flight deck and the associated instruments for both civil and military aircraft.

Systems architecture. Aircraft subsystems are comprised of avionic, electrical, mechanical, hydraulic and pneumatic. The extent of aircraft weight and cost depend on the subsystem design philosophy. Automation and microprocessor-based data management have advanced to cater for wider operational capability without a corresponding increase in pilot workload. This is a fine way to see aircraft as systems and subsystems. Section 22.6 outlines some of the main system architecture to operate an aircraft.

Flying Adverse Environments. Flying through rain, fog and iceconditions are considered as flying in adverse conditions, as discussed in Section 22.7. There are aids to overcome flying in such conditions. Weather radar, Instrument Landing Systems (ILS) and so on are flight instruments to assist pilots operating in adverse environments. The rain-removal system using windshield wipers and rain-repellent chemicals is deployed when flying through heavy rain. Anti- or de-icing systems keep an aircraft from ice accumulation. Lighting strike and hail are associated with thunderstorms. Aircraft are equipped to discharge static electricity build-up and the shell insulates the interior. Section 22.7 outlines some of aircraft systems associated with flying in adverse environments.

Military survivability issues. Military aircraft design for combat survivability has been an active consideration for some time, primarily as a consequence of possible damages occurring during combat. Crew ejection is seen as an integral part of survivability. The last two decades saw stealth technology, as a good survival measure, develop by minimising the aircraft signature. Electronic defence/countermeasures are the other ways to thwart retaliation and increases chances of survivability. Section 22.8 outlines some.

22.3 Environmental Issues (Noise and Engine Emission)

Section 21.61 presented the FAA requirements for take-off and landing noise limits certification. Section 21.6.2 dealt with the engine exhaust gas emissions. Design considerations for noise emission are dealt in this section.

22.3.1 Noise Emissions

The identified main sources of noise generation emanate from the engine, propeller and airframe. Limiting engine and propeller noise is the respective manufacturers' responsibility and beyond the scope of this book. But noise abatement from nacelles and the airframe is aircraft designers' responsibility.

Candidate areas in engine design are the fan, compressor and turbine blade design, gaps in rotating components and also to an extent in combustion chamber design. Power plants are bought-out items for the aircraft manufacturers, who need to make compromises between engine costs versus engine performances in making the choice from what are available in the market. The aircraft and the engine designers are constantly in dialogue to make the best without compromising safety. To reduce noise levels, nacelle designers need to attend the sources of noise as shown in Table 22.1. The aim is to minimise radiated and vibration noise.

Figure 22.1 shows the positions of noise suppression liners placed in various areas and the jet pipe flow mixing arrangements for noise abatement. Exhaust noise suppression is also achieved by using a fluted duct

Table 22.1 Nacelle and turbofan technological challenges to reduce noise.

Nacelle	Fan/compressor and turbine	Burner
Internal liner – intake	i) Improved blade design	i) Efficient burning
i) To absorb fan noise	ii) Blade number optimised	ii) Low vibration
Internal liner – casing/fan duct	iii) Optimise gap around blades	
i) To insulate compressor noise	iv) Minimise support strut vibration	
ii) To absorb burner noise		
Internal liner – jet pipe		
i) To absorb turbine noise		
ii) To absorb burner noise		
iii) To mix hot and cold flows		
iv) To improve exhaust flow mixing		

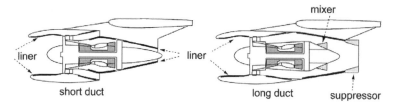

Figure 22.1 Positions of noise liners and suppression/mixing arrangements.

(increase mixing area) at the exit plane. There are many types of liners available on the market and there is scope to improve liner technology. Primarily, there are two types of liners, (i) reactive and (ii) resistive. The reactive ones have honeycomb structures with cells size matched with frequency range, some with different sizes of perforations. The resistive types are noise insulators in layers with screens. Combination of both these types is in use, in which honeycomb perforated layers are sandwiched with insulator screens. Nacelle design is the responsibility of an aircraft manufacturer, even when it is subcontracted to a third party. The engine noise is reflected through the internal surfaces of the nacelle.

At this stage of study, the design considerations for noise reduction would not substantially effect aircraft external configuration other than using proper filleting at two body junctions, making the projected structure streamlined, minimising gaps and so on. The frozen aircraft configurations as obtained in Chapter 8 and then sized in Chapter 14 remain unaffected as the aircraft external geometry is assumed to have taken account of these considerations. Choice of material (nacelle liners, cabin insulators, use of fatigue resistant material etc.) plays an important role and can affect aircraft mass. Engine noise abatement is taken care of by the engine designers.

Computational fluid dynamics (CFD) capabilities in predicting noise has advanced well to make good judgements in improving designs. Substantiation of CFD results would require testing.

The near future (gradually evolving, say, in about two decades from now) will see remarkable improvements in noise abatement in a multi-disciplinary design approach to take benefits from various engineering considerations leading into Blended-Wing-Body shapes. Cambridge University and the Massachusetts Institute of Technology have undertaken feasibility studies on a concept configuration in Figure 22.2 for an Airbus320 class of subsonic commercial jet transport aircraft. They predict to have made the aircraft quieter by 25 dB than the current designs. That would be as quiet to the point of making it a 'silent aircraft'. The shaping of the aircraft is not purely based on noise reduction; it is also driven by all-round aerodynamic considerations, for example, drag reduction, handling qualities and so on. Noise reduction comes from shielding of intake noise by the aircraft body, minimising two body junctions through blending of wing and fuselage and elimination of empennage. Of course, reduction in engine noise itself is a significant part of the exercise. However, to bring the research to a marketable product will take time but the author believes it is sure to come in many sizes; heavy-lift cargo aircraft are good candidates.

Figure 22.2 Concept design of futuristic 'silent aircraft'. [Cambridge-MIT Institute http://silentaircraft.org/sax40]. Reproduced with permission of Cambridge University Press.

As already mentioned, power plants are bought-out items that specify the engine exhaust gas emission level. Aircraft manufacturers are responsible for making their selection on what is best suited for the new aircraft project.

22.4 Safety Issues

This section deals with the regulatory requirement stipulating a minimum number of exit doors (main and emergency doors) and slides/chutes to ensure safety in the unlikely event of any type of hazard. Civil and military requirements differ. Fire detection and protection systems as safety items are installed inside aircraft.

22.4.1 Doors – Emergency Egress

An emergency situation (say from a fire hazard, ditching on water or land etc.) would require fast exit from the cabin to safety. FAA initially imposed a 120 s egress time but, from 1967, changed to a maximum of 90 s. This was possible through the advances made in slide/chute technology. To obtain airworthiness certification, aircraft manufacturers have to demonstrate, by conducting simulated tests, that complete egress is possible within 90 s. The European Aviation Safety Agency (EASA) has similar requirements.

FAR25 Section 25.783 gives the requirements for the main cabin doors and Section 25.807 for the emergency exit doors. There are several kinds of emergency exit doors given in Table 22.2 (all measurements are in inches). All are rectangular in shape with a corner radius. The sizes are minimums and designers can make them bigger. The oversized doors need not be rectangular so long they can inscribe the minimum rectangular size.

All except Type III (inside step up of 20 in. and outside step down of 27 in.) and Type IV (inside step up of 29 in. and outside step down of 36 in.) are from floor level. If Type II is located over the wing then it can have inside step up of 10 in. and outside step down of 17 in. Emergency doors are placed at both the sides of the aircraft and need not be diametrically opposite, but they should be uniformly distributed (no more than 60 ft away), easily accessible for even loading by passengers when it is required. The safety drill by the cabin crew is an important aspect to save lives and all passengers must attend the demonstration, no matter how a frequent flier one is. From type to type, there are differences.

There should be at least one easily accessible external main door. The combination of main doors and emergency doors are at the discretion of the manufacturers who will have to demonstrate evacuation within the stipulated time. Fuselage length will also decide the number of emergency doors as they should not be spaced more than 60 ft away. Table 22.3 gives the minimum number of emergency doors. It is recommended that

Table 22.2 Aircraft door types.

	Position	Minimum height (in.)	Minimum width (in.)	Maximum corner radii (in.)	Passenger number [a]
Type A	Floor level	72	42	7	110
Type B	Floor level	72	32	6	75
Type C	Floor level	48	30	10	55
Type I	Floor level	48	24	8	45
Type II [b]	Floor level	44	20	7	40
Type III	Over wing	36	20	7	35
Type IV	Over wing	26	19	6.3	9

a) The types of doors are related with the minimum number of passengers carried. The higher the number of passengers, the larger is the door size.
b) If Type II is located over the wing then it can have inside step up of 10 in. and outside step down of 17 in.

Table 22.3 Aircraft emergency door types.

Number of passengers	Minimum size emergency door type	Minimum number of emergency doors
1 to 9	Type IV	One in each side of the fuselage
10 to 19	Type III	One in each side of the fuselage[a]
20 to 40	Type II	Two in each side of the fuselage[a]
41 to 110	Type I	Two in each side of the fuselage
>110		

a) one of them could be of one size smaller.

Table 22.4 Door dimensions.

	Step height inside (outside) (in.)	Minimum height (in.)	Minimum width (in.)	Maximum corner radii (in.)
Ventral	–	≥48	≥24	8
Tail cone	24 (27)	42	72	7

more than minimum provision is to be made. Types A, B and C can also be used – they are deployed in larger aircraft.

There could be other types of doors, for example, a door at the tail cone, ventral doors and so on (dimensions given in Table 22.4). Flight crew emergency exit doors are separately provided at the flight deck.

22.4.2 Escape Slide/Chute – Emergency Egress

Where door level is high above the ground, inflatable escape slides/chute are provided as shown in Figure 22.3 When in an emergency situation stairs may not be available (or passengers cannot wait for stairs to arrive), inflatable chutes are used to slide down within the specified time for evacuation. The slides/chutes can serve as rafts with floating attachments.

As aircraft sizes are getting bigger, the technological demand to facilitate quick egress is becoming a challenging task. In March 2006, the Airbus380 could demonstrate that 850 passengers could be evacuated in 80 s (there were some minor injuries). However, a typical A380 passenger load would be less than 650 with a mixed class arrangement. An Airbus has 16 exits but it succeeded in evacuating with only eight doors (half remained closed).

22.4.2.1 Classroom Exercise

Here is a classroom exercise. The configuration developed in Chapter 8 may be checked again. The Bizjet must have the following:

Version	Number of passengers	Emergency door type
Baseline	10	one Type III and one Type IV
Long	14	one Type III and one Type IV
Short	6	one Type IV

Figure 22.3 Inflatable escape slide.

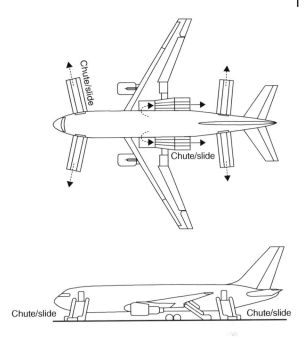

It is better to keep all doors to Type III standards for component commonality that will reduce production cost.

22.5 Aircraft Flight Deck (Cockpit) Layout

Aircraft 'flight deck' is a better term than the older usage of the word 'cockpit', which has its origin in ship design (prevailingly in the sixteenth century) bearing a certain similarity with men working in a confined area under stress as cockerels were made to fight in a pit. 'Crew station' is another term meaning the same as a workplace for the operators of the vehicle of any kind. To standardise terminology, this book uses the word *flight deck*, meant specifically for flying vehicles. The flight deck serves as a man-machine interface by providing (i) outside reference of topography through cabin windows, (ii) on-board instruments to measure flight parameters, (iii) control facilities to operate aircraft safely for the mission role and (iv) to manage aircraft systems, for example, internal environment. Future flight decks with advanced displays could result in visually closed flight decks (TV replacing windows). Front fuselage shapes can be influenced by the flight deck design. Transport aircraft have two pilots sitting side-by-side at a pitch of about 1.2 (smaller aircraft) to 1.4 m (larger aircraft). Understanding the flight deck arrangements will also give some idea of equipment requirements that will give a measure of associated weights involved. It will also give some idea of space requirement and satisfy the adequacy of vision-polar (establishes window size).

Both civil and military pilots have common functions as listed here.

   i. Mission management (planning, checks, takeoff, climb cruise, decent and land)
  ii. Flight path control
 iii. Systems management
 iv. *Communication and
  v. *Navigation
 vi. Routine post flight de-briefing
 vii. *Emergency action when required (some drills differ between civil and military)

The asterisk* refers to this point: civil pilots are assisted by ground control (communication and navigation) while in critical situations, combat pilots will have to manage aircraft on their own – this makes a big difference. Both may require taking emergency action but in the case of a combat pilot, it could be drastic in nature (ejection – see Section 22.8).

In addition, military pilots have intense workload as listed here

  i.   Mission planning (e.g. lo/High combination etc. – see Chapter 15). This is required for mission management (pre-flight briefing may have to be changed if situation demands).
  ii. Target acquisition.
 iii. Weapons management/delivery.
  iv. Defensive measures/manoeuvres.
   v. Counter threats – use of tactics.
  vi. Manage situation when hit.
 vii. In-flight refuelling where applicable and
viii. Detailed post flight briefings in special situations.

The military aircraft flight deck has more stringent design requirements. The civil aircraft flight deck is designed in the wake of military standards and the provision of space is not constrained as much. Aircraft flight deck design has changed dramatically from the early analogue dial displays (four engine aircraft gauges fill the front panel – Figure 22.13, later) to modern microprocessor-based data management in integrated all glass multi-functional displays (MFDs). The MFD is also known as an Electronic Flight Information System (EFIS).

22.5.1   Air Data Instruments and Flight Deck

This section describes the following analogue air data instruments (Figure 22.4) of interest in this book: (i) airspeed indicator (*ASI*), (ii) altimeter, (iii) attitude indicator, (iv) turn-slip indicator, (v) vertical speed indicator (VSI), (vi) Mach number indicator, (vii) heading indicator, (viii) horizontal situation indicator (*H-S* indicator), (ix) angle of attack (*AOA*) indicator and (x) g-meter. These are older types, each displaying a single parameter: as a result there is a clutter of instruments on the flight deck.

All of them use some form of tapping from outside of aircraft. Pitot and static pressure probes and vanes measure most of these parameters. Navigation and communication instruments are beyond the scope of this book.

Figure 22.4 True-air data analogue indicating instruments. *Top row*: Airspeed indicator, altimeter, attitude indicator, turn-slip indicator, vertical speed. *Bottom row*: Mach meter, heading indicator, H-S indicator, angle of attack, *g*-meter. Reproduced with permission of Wiley Publisher.

22.5.2 Altitude Measurement – Altimeter

An altimeter makes use of ambient static pressure taken from a separate probe and not from a pitot-static tube. It records height above ground corresponding to the static pressure. An altimeter consists of an air-tight chamber encasing a differential pressure sensitive hollow vacuumed diaphragm (see Figure 22.5). It is a corrugated bronze aneroid capsule. The chamber is connected to the static pressure probe on aircraft exterior that exerts differential pressure on the hollow diaphragm, which expands or contracts as altitude increases or decreases, respectively. The differential pressure is calibrated to International Standard Atmosphere (ISA) conditions, but needs manual adjustment to conditions as explained next. A knob is provided for the flight crew to make adjustments using the Kollsman window shown in the figure. A radio altimeter (rad-alt) and GPS are used in conjunction to read altitude; these are not discussed here.

At non-standard day the indicated altitude (IA) will be different from true altitude (TA). In a hot day, the TA is higher than the IA and vice versa. The relationship between TA and IA is as follows.

$$\text{True altitude (TA)} = \text{Indicated altitude (IA)} \times \frac{T}{T_{ISA}} \tag{22.1}$$

22.5.3 Airspeed Measuring Instrument – Pitot-Static Tube

A pitot tube is a hollow concentric tube (Figure 22.5) sealed in the front end aligned to airflow in which a stream tube is captured in the inner tube to stagnation converting the kinetic energy of the flow to potential energy of increased total pressure, p_t, as shown in Equations C.9 and C.10.

A pitot-static tube also reads the static pressure, p, of the flow by tapping a hole perpendicular to airflow on the hollow tube. The difference of pressure $\Delta p = (p_t - p)$ is the velocity head, which is calibrated to read

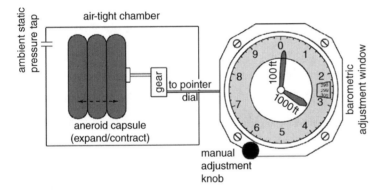

Figure 22.5 Altimeter.

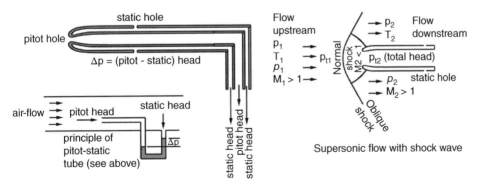

Figure 22.6 Pitot-static tube.

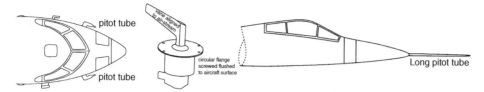

Figure 22.7 Pitot-static probe (centre figure has AOA integrated) on an aircraft. Reproduced with permission of Wiley Publisher.

in indicating gauges. The following subsections are devoted to the theory of calibrating a pitot-static tube reading. Figure 22.6 shows the basic concept of the pitot-static tube to measure airspeed. It consists of two separate concentric hollow tubes sealed from each other. The inner tube is open and aligned to oncoming airflow, which stagnates inside. The outside tube has holes transverse to airflow measuring static pressure as it does not read the normal component of dynamic head of airflow.

The positions and types of pitot-static probe installation on aircraft are shown in Figure 22.7. Probes should be near to the nose of the airplane where the airflow is relatively clean and the boundary layer is thin, minimising the required probe height. Civil aircraft typically have small probes installed, several of them from redundancies and older military installations (the 22.7 shows F104 installation) have a large boom at the nose to suit supersonic flights.

The positions of static holes have to be carefully chosen on fuselage sides where the static pressure matches ambient pressure to have the minimum pressure error. There several holes along the side (Figure 22.8) to take the mean value as pressure distribution over the fuselage changes with altitude changes.

22.5.4 Angle of Attack Probe

AOA is one of the important parameters related to airplane performance and handling. Takeoff and landing take place close to aircraft stall speed, that is, at AOA close to wing stall. Information on AOA also guides the pilot to cruise at close to maximum lift to drag ratio for best fuel economy. Combat manoeuvres can take place at exceptionally high AOAs for which special design features, for example, strake/canard configuration, are incorporated. For the same change in ΔAOA, change in ΔV is higher at high speed than at close to stall (stall warning devices are discussed next).

There is a relation between AOA, lift, aircraft speed and weight. Lift increases with increase in AOA at the same speed. Lift is also weight dependent. Flaps and/or slats increase lift while spoilers/speed brakes spoil lift; all affect aircraft stall characteristics. Aircraft manoeuvres at higher speed can generate high AOA. But when AOA exceeds certain limits then an aircraft stalls, an undesirable situation that can endanger safety. Figure 22.8 shows an AOA probe, which is a vane that aligns with flow velocity. The instrument is calibrated to show the angle at the flight deck for the pilot to notice.

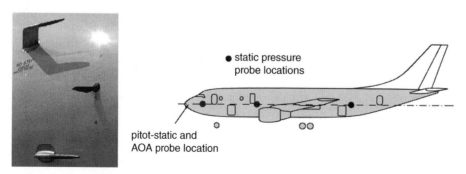

Figure 22.8 Angle of attack probe at the centre in between the pitot-static probes and pilot-static probe location on an aircraft fuselage. Reproduced with permission of Wiley Publisher.

Lack of a stall warning device has caused many accidents. Stalls/spins are most likely to occur in manoeuvring flights when warnings could make a pilot take preventive steps. At low altitudes, especially on approach to land, what AOA an aircraft maintains is critical and its indication keeps the pilot controlling the aircraft safely. AOA indication is air data for safe flying.

The airworthiness requirement requires some form of stall warning device to provide safety in operation. There are many forms of stall warning, for example, audio and/or voice warning sufficient ahead of approaching stall speed, stick shaker or pusher to improve stall avoidance and so on. Fly-by-Wire (FBW) ensures that aircraft have sufficient speed margins driven by built-in microprocessors in line to aircraft control laws. Typically, a pilot is trained to recognise pre-stall buffet characteristics along with the type of stall warning device installed. As speed reduces approaching stall speed, the control stick becomes sloppier.

22.5.5 Vertical Speed Indicator (VSI)

VSI (Figure 22.4) gives the rate of climb or descent or level flight by the aircraft as it performs. Change of altitude records rate of change of ambient (static) pressure that is sensed by a diaphragm inside the instrument and the data is calibrated and indicated. Basically, it records differential static pressure between the instantaneous pressure as it changes and the existing trapped pressure in the instrument. It records best when flight is in vertical plane. It should be maintained to show zero when on the ground.

22.5.6 Temperature Measurement

Total Air Temperature (TAT) and Static Air Temperature (SAT) are part of the air data used to compute aircraft performance. SAT is the atmospheric ambient temperature. TAT (in Equation (C.6) as T_t) is measured by stagnating air velocity, which has ambient temperature SAT (in Equation (C.6 as T). Stagnation is through adiabatic compression converting the kinetic energy of air flow that increases temperature, which makes $TAT > SAT$. Equation C.6 in Appendix C gives the isentropic relation as follows:

$$\left(\frac{T_t}{T}\right)_{isen} = \left(1 + \frac{\gamma - 1}{2}M^2\right)$$

In reality, energy conversion through compression is not perfect and reaches a lower value of $T_r < T_t$. An empirical recovery factor, 'r' is used modify this equation to a realistic value.

$$\left(\frac{T_t}{T}\right) = \left(1 + \frac{\gamma - 1}{2}rM^2\right)$$

Where $r = \frac{(T_r - T)}{(T_t - T)}$ typically with value between 0.75 and 1.0, depending on the design.

Figure 22.9 shows that the TAT probe reads an accurate temperature measurement under all flight conditions, including aircraft ground operations and atmospheric icing conditions. The TAT measurement

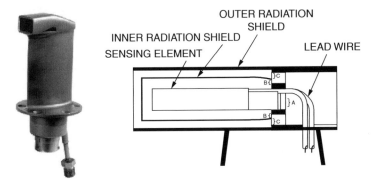

OUTER RADIATION SHIELD

INNER RADIATION SHIELD

SENSING ELEMENT

LEAD WIRE

Figure 22.9 Total and static temperature probe. Reproduced with permission of Wiley Publisher.

is used to determine static temperature, true airspeed computation and so on. The sensor consists of two sensing elements that are insulated and hermetically sealed within the sensor housing. A compensating heater used to de-ice the sensor is integrated into the housing. There are two types of TAT, for example (i) aspirated type and (ii) unaspirated type. Aspirated TAT design includes the addition of an air-to-air ejector, which induces airflow past the sensing element. The increased airflow helps eliminate temperature soaking inaccuracies caused by bright sunshine or hot ramp heat radiation.

22.5.7 Turn/Side Slip Indicator

The turn and side slip indicator is a gyroscope driven instrument as shown next. The dial shows a black ball sealed in a curved liquid filled glass tube with two wires at its centre and a needle or aircraft symbol seen from the aft end as shown in Figure 22.10. The ball measures gravity force and inertial force caused by the turn.

Aircraft manoeuvres may be as follows. A right turn is shown in Figure 22.10.

Straight level flying. Aircraft wings level and ball in centre.

Coordinated turn. In this case when the bank angle is correct then the gravity and inertial forces are equal and the ball stays in the centre but right wing drops to show the turn. This is executed by coordinated application of right rudder and right aileron.

Skid turn. When bank is less then inertia force is greater than gravity force and aircraft skids outward when the ball moves left but right bank angle is shown by the aircraft symbol.

Slip turn. When bank is more than inertia force is less than gravity force and aircraft slips inward when the ball moves right but right bank angle is shown by the aircraft symbol.

22.5.8 Multi-Functional Display (MFD)/Electronic Flight Instrument System (EFIS)

MFD started as a display on a Cathode Ray Tube (CRT) but has now advanced to Liquid Crystal Display (LCD – lighter and clearer) technology. All relevant data for pilot usage, for example, air data, engine data, navigational data and so on, are displayed simultaneously on screen. To avoid clutter, the displays are divided into primary and secondary displays. Separate system displays are accommodated in one or two EFISs – (i) Primary Air data System Display (SD) and (ii) Navigational Display (ND). There are several pages in each type of system. Each display screen can be changed for the specific information. Figure 22.11 shows typical EFIS displays. EFIS/MFD/ND/SD have many pages that can be flipped as desired. Some of the pages are engine pages, cruise pages, flight control pages, fuel pages, electric page, avionics page, oxygen pages, air-bleed pages, air-conditioning page, cabin pressurisation pages, hydraulics pages, undercarriage pages, undercarriage pages, doors pages. Auxiliary Power Unit (APU) pages and so on (a military aircraft has weapons management pages.)

In some designs, the engine display (ED) is shown separately. Figure 22.12 shows the EFIS for a General Aviation piston engine aircraft.

The Primary Flight Display (PFD) consists of air data systems, for example, aircraft speed, altitude, attitude, aircraft reference, ambient conditions and so on. A secondary system consists of ND that gives directional

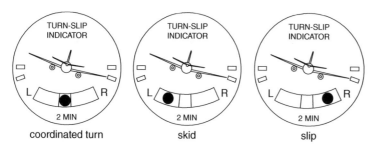

Figure 22.10 Turn and slip indicator showing a right turn.

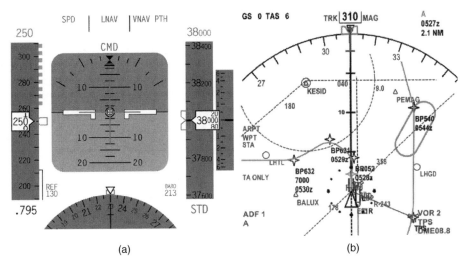

(a)   (b)

Figure 22.11 Multi-functional display (Courtesy of Dynon Avionics). (a) Air data systems display. (b) Navigational display. Reproduced with permission of Dynon Avionics.

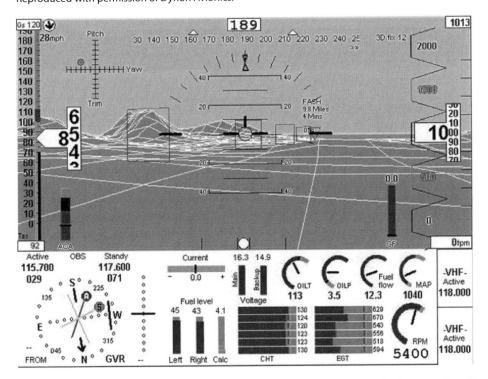

Figure 22.12 Multi-functional display (Courtesy of Dynon Avionics). Reproduced with permission of Dynon Avionics.

bearings (GPS/Inertial system), flight plan, route information, weather information, airport information and so on. For pilot ease, some duplication is done in each type. SD in a separate panel shows the engine data, and all other system data including those required for environment control system (ECS). EFISs have removed the clutter of analogue dials, one for each kind of data. In some designs ED is separately shown. Forward looking weather radar can be with ND or a separate display unit.

Initially, flight decks were supplied also with basic analogue gauges showing air data as redundancies, in case EFIS fails. Currently, with vastly improved reliability in EFISs the old analogue gauges are gradually taken out.

22.5.9 Civil Aircraft Flight Deck

An old style panel with analogue dial gauges is shown in Figure 22.13. With two pilots, some of the displays are duplicated, which also serve as redundancies.

The latest A380 flight deck panel is very different replacing a multitude of gauges with EFISs, which are MFD units. The generic minimum layout of the modern flight deck panel is shown in Figure 22.14. Considerable redundancies are built-in to the display in independent circuits. PFD, NDs are SDs are with several pages displaying a large amount of data.

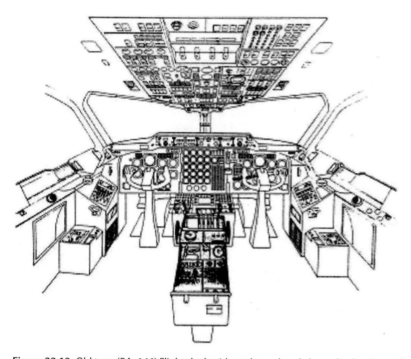

Figure 22.13 Old type (BAe146) Flight deck with analogue head-down display. Reproduced with permission of Wiley Publisher.

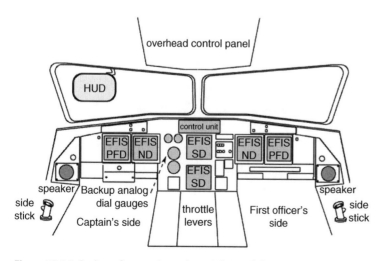

Figure 22.14 Outline of a generic modern civil aircraft flight deck panel with HUD.

22.5.10 Combat Aircraft Flight Deck

Figure 22.15 shows a typical modern flight deck for military aircraft. Note that back-up analogue gauges are provided as well with the MFD type EFIS. On the left-hand side is a hands on Throttle and Stick (HOTAS) (see Section 22.5.12) and the right-hand side has a side stick controller. The figure indicates what kind of data and control the pilot would require. Single pilot work load is exceptionally high when computer assistance is required.

22.5.11 Heads-Up Display (HUD)

Flight deck displays are laid on the instrument panel in front of the pilot who has to look down for flight information, more frequently in critical situations. But when flying close to ground or chasing a target, the pilot should keep their head up looking for external references. This inflicts severe strain on pilots by frequently alternating the head up and down. Engineers have solved the problem to a great extent by projecting a few pieces of important flight information (both primary and navigational data) in bright green light on a transparent glass pane mounted in front of the wind screen as a HUD. Now pilots can get all necessary

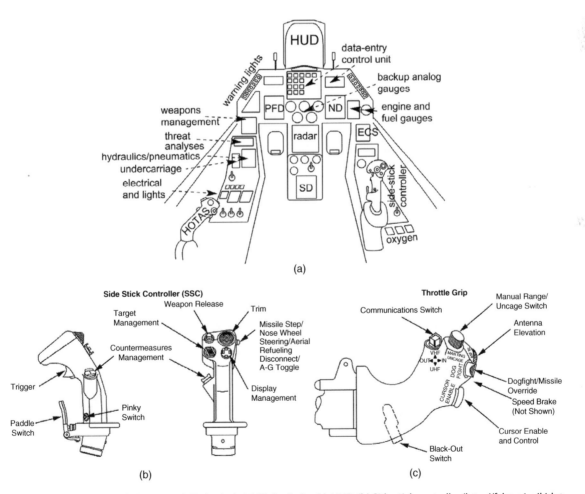

Figure 22.15 Schematic fighter aircraft flight deck. (a) Flight deck with HUD. (b) Side stick controller (http://falcon4.wikidot.com/avionics:hotas). (c) HOTAS (http://falcon4.wikidot.com/avionics:hotas).

Table 22.5 HOTAS control buttons.

On the throttle (left side)		On the control stick (right side)	
Target	Weapons	Trigger	Weapon release
Communications	Antenna	Missile	Sensor select
Radar	In-flight start	Trim	Flight control
Flaps	Dive brakes		

information without having to move their heads, and at the same time he/she can see through the HUD for external reference. Figures 22.14 and 22.15 show the HUD.

Initially, the HUD was installed in combat aircraft but of late the fall-out of the technology has trickled into civil aviation and HUDs are being installed on most of the new medium and big commercial transport aircraft, if operators them.

22.5.12 Hands on Throttle and Stick (HOTAS) and Side Stick Controller

Other examples of providing ease for pilots include incorporating the essentials of weapons management and other requirements embedded on two controls: combat pilots will always keep their hands on the engine throttle control (HOTAS) and the side stick controller. The arrangements of controls buttons on HOTAS (Figure 22.15a) [1] and side stick controller are shown in (Figure 22.15b) [1]. Essential control buttons are ergonomically located (Table 22.5). Most modern aircraft have some form of buttons on flight control stick for communication, trimming and so on.

22.5.13 Helmet Mounted Display (HMD)

Although HUD eased pilots from frequently looking down, its heads-up observation is restricted to forward vision only. Military pilots needed to ease their workload while making visual searches sideways when the HUD is no longer in the line of sight. Engineers came out with a novel device to project flight information on a helmet mounted visor. Now a pilot can turn their head with all relevant information still visible on the transparent visor, through which external reference can be taken.

22.5.14 Voice Operated Control

Voice Operated Control (VOC) through voice recognition has been installed in advanced combat aircraft. It is still in the development stages, All voice commands are also visually displayed. These can prove very effective for pilots operating under severe stress, more so if incapacitated by injury.

All these advancements help a pilot but the systems still require pilot familiarity. Pilots undergo long training and practise sessions to gain familiarity with a clutter of information in a rather claustrophobic presentation. Pilot workload is nearly an inhuman task. They are a special breed of personnel rigorously trained for years to face unknowns in a life or death situation. It is the moral duty of any combat aircraft designer to enhance pilot survivability as best as possible in an integrated manner, embracing all kinds of technologies.

22.6 Aircraft Systems

Figure 2.4 showed the aircraft design process in a systems approach. The definition of 'system' is given in Section 2.4. In that light, an aircraft can be seen as a *system* composed of many *subsystems*. Chart 22.1 gives

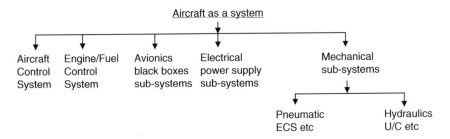

Chart 22.1 Aircraft as a system.

a typical top level subsystem architecture of aircraft as a system. The subsystems can be designed in separate modules and then integrated with the aircraft.

Together, the systems and subsystems mass would be 10–12% of aircraft MTOM. Typically, it amounts to nearly quarter of the Operating Empty Mass (OEM). Practically all the items of aircraft subsystems are bought-out items. A better understanding of the subsystems would improve weight and cost predictions. It is important that good information on subsystem items is required at the conceptual design stage for better weight and cost estimation. Designers are continually assessing cost versus performances of the subsystems to bring out the best value for money.

Mechanical systems are connected by direct linkages, pneumatic and hydraulic means. Bigger undercarriages are actuated by hydraulic means.

22.6.1 Aircraft Control Subsystems

Figure 3.10 in Section 3.6 shows the Cartesian representation of the six degrees of freedom, comprising three linear and three rotational motions. This section offers a brief description of associated control hardware and design considerations (for more details refer to [2] and [3]). Aircraft control system weight is about 1–2% of maximum takeoff weight (MTOW).

Three-axes (pitch, yaw and roll) aircraft control has evolved considerably. Use of trim tabs and aerodynamic/mass balances alleviate hinge moments of the deflecting control surfaces, therefore easing pilot workload. Some of the types in operation are given next.

i. *Wire-pulley type.* This is the basic type. Two wires per axis act as tension cables, moving over low friction pulleys to pull the control surfaces in each direction. Although there are many well designed aircraft using this type of mechanism, it requires frequent maintenance to check tension level and possible fraying of wire strands. If it is under improper tension then the wires can jump out of pulleys making the system inoperable. Other associated problems are dirt getting into, rare occasion of jamming, elastic deformation of support structures leading to loss of tension and so on. Figure 22.16 shows both the wire-pulley (rudder and aileron) and push-pull rod (H-tail) types of control linkages. The push-pull rod is described in the next point (ii).

ii. *Push-pull rod type.* The problems of wire-pulley type are largely overcome by the use of push-pull rods to move the control surfaces. Designers must ensure that the rods do not buckle under compressive load. In general, it is slightly heavier and a little more expensive but worth installing for ease of maintenance. Many aircraft use a combination of wire-pulley and push-pull rod arrangement (Figure 22.16).

iii. *Mechanical control linkage boosted by Power Control Unit (PCU).* As aircraft size increases, the forces required to move the control surfaces increase to a point when pilot work load crosses the specified limit. Power assistance by PCU overcomes the problem. One of the problems of using PCU is that the natural feed-back feel of control forces is obscured. Then artificial feel is incorporated for finer adjustment leading into smoother flight. PCUs can be hydraulic or electric motors driven linearly or rotary actuators (there are many types).

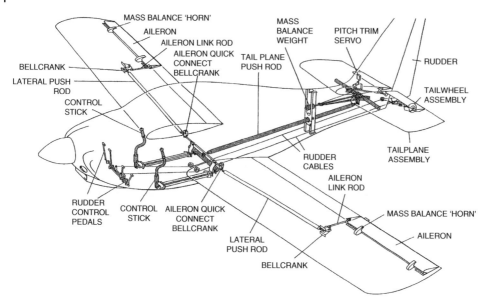

Figure 22.16 Wire-pulley and push-pull rod control system (Courtesy of Europa-Aircraft).

iv. *Electro-mechanical control system.* Large aircraft can save considerable weight by replacing mechanical linkages with electric signals to drive the actuators. Aircraft with FBW use this type of control system (Figure 18.18). Currently, many aircraft routinely use secondary controls (e.g. high-lift devices, spoilers, trim tabs etc.) driven by electrically signalled actuators.

v. *Optically signalled control system.* The latest innovations use optically signalled actuators. Advanced aircraft are already flying with fibreoptic lines to communicate with control systems.

Modern aircraft, especially combat aircraft control systems, have become very sophisticated. A FBW architecture is essential to these very complex systems when aircraft can fly under relaxed stability margins. Enhanced performance requirements and safety issues have increased the design complexities incorporating various kinds of additional control surfaces. Figure 22.17 shows typical subsonic transport aircraft control surfaces.

Figure 22.18 shows the various control surfaces/areas and system retractions required for a three-surface configuration. As can be seen, it has a lot more than what modern civil aircraft have.

Military aircraft control requirements are at higher level on account of the demand for hard manoeuvres and a possible negative stability margin. The F117 is incapable of flying without FBW. Additional controls are canard, intake scheduling and thrust vectoring devices. Fighter aircraft may use stabilators (e.g. F15) in which

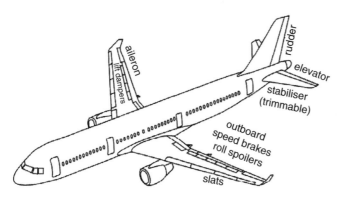

Figure 22.17 Civil aircraft control surfaces.

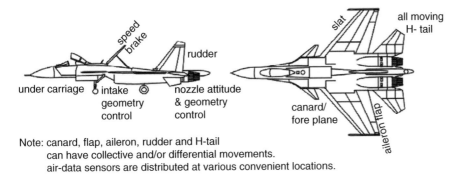

Note: canard, flap, aileron, rudder and H-tail
can have collective and/or differential movements.
air-data sensors are distributed at various convenient locations.

Figure 22.18 Military aircraft control surfaces.

the elevators can move differentially to improve roll capability. Stabilators are used collectively for pitch and differentially for roll control. Also, an aileron and rudder can be interconnecting. There could be automatic control in parallel with the basic system.

22.6.2 Engine and Fuel Control Subsystems

In this section, the engine and fuel control systems are discussed together. Engine and fuel/oil control system must have a fire extinguishing arrangement. A better understanding of engine and fuel/oil control systems can improve weight and cost prediction accuracy. Dry engine weight supplied by engine manufacturer is separately accounted for. The earliest aircraft were piston engine powered. Piston engines use petrol (aviation gasoline: (AVGAS). Diesel powered engines have been introduced recently. Figure 22.19 shows a basic fuel system of a small piston engine powered aircraft.

22.6.2.1 Piston Engine/Fuel Control System (Total System Weight Around 1–1.5% of MTOW)

i. Ignition and starting system.
ii. Throttle to control fuel flow.
iii. Fuel storage (tank) and flow management. This must incorporate fuel refuelling/defuelling and venting arrangements. High-wing small aircraft can have a gravity-fed fuel supply to the engine but most aircraft use fuel pumps. Aerobatic aircraft should be capable of flying inverted, at least for 1 min.
iv. Mixture control to adjust with air density change when altitude changes.
v. Propeller pitch control (see Chapter 12). Small aircraft can have a fixed pitch.
vi. Engine cooling system.
vii. Engine anti-icing system.

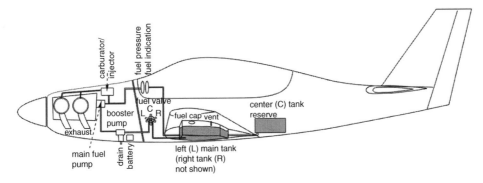

Figure 22.19 Piston engine fuel system.

   viii. Oil system.
   ix. Fire protection system.
   x. Instrumentation and sensor devices.

Small aircraft can get away with storing fuel in the wing. Although a few aircraft have stored fuel in the fuselage, it is better to avoid this form of storage. Fuel in the fuselage can affect larger centre of gravity (CG) shift and, in case of a crash, occupants could be doused by the left over fuel. Fuselage fuel tanks can also come as optional installations to increase range.

In a way, gas turbine engine control at the pilot interface is simpler. It does not require mixture control by the pilots. For turbofans there are no propellers, hence there is no pitch control by pilot. The turbofan engine/fuel control system is listed in the following subsection (see also Figure 22.20).

22.6.2.2 Turbofan Engine/Fuel Control System (Total System Weight Around 1.5–2% of MTOW)

   i. Ignition and starting system.
   ii. Throttle to control fuel flow (thrust adjustment). Big jets have thrust reversers.
   iii. Fuel storage (tank) and flow management. This must incorporate fuel refuelling/de-fuelling and venting arrangements. Some combat aircraft need mid-air refuelling. In an emergency, aircraft should be able to dump (jettison) fuel (an environmental hazard and is discouraged).

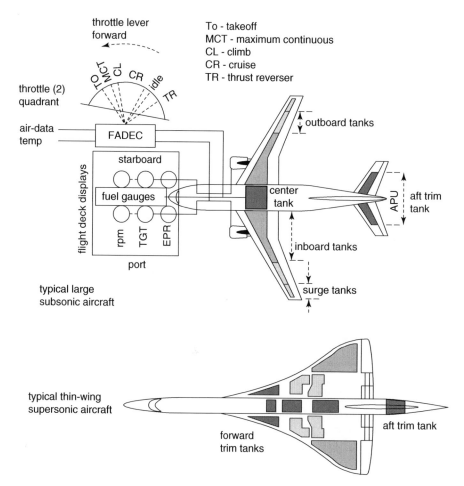

Figure 22.20 Turbofan engine/fuel control system.

iv. Engine cooling system.

v. Engine anti-icing system.

vi. Oil system.

vii. Fire sensing and protecting system.

viii. Built-in tests for fault detection system. There should be flight/ground crew interface.

ix. Instrumentation and sensor devices. Modern military and commercial aircraft engine control is microprocessor-based and known as Full Authority Digital Electronic Control (FADEC). It is linked with FBW using air data for making correct responses as demanded by the pilot. a typical turbofan and fuel/oil control system (Figure 22.20) is dealt with in more detail in the following paragraphs.

22.6.2.3 Fuel Storage and Flow Management

Fuel supply to an engine must be made smoothly and accurately. There must be adequate fuel storage capacity to meet the mission profile with the mandatory excess as a reserve. These requirements are an important part of study during the conceptual design phase. Fuel management is a complex affair. Fuel weight is a large percentage of aircraft MTOW and its consumption from full to empty has the potential for a large movement of CG affecting aircraft stability. It is important that fuel consumption should be managed in a way that will make the least CG shift. In demanding situation, this is achieved through in-flight fuel transfer.

A typical commercial aircraft tank arrangement is shown in Figure 22.20. Storage of fuel is primarily located in the cavity of wing-box that extends from wing root to close to tip (taper wing tip has lower volumetric capacity). Wing fuel storage has an advantage as it is kept close to aircraft CG lowering the range of CG shift. If the volume available in the wing is not adequate, especially for thin winged combat aircraft, then fuselage space can be used for storage. Typically, fuel storage in fuselage can be above or below the floorboards and anywhere forward, rearward and/or at the centre of the wing. Where there is a multitude of tanks then is convenient to collect fuel at a central location before delivering it to the engines. Fuel from tanks at various locations is pumped into a centrally located collector tank following a transfer schedule that would keep CG shift to the minimum. Symmetrical fuel level in wings is also important. Note the compartmentalisations of wing tank. Surge tanks are provided at the wingtips. Internal baffles restrict fuel sloshing. Some long range aircraft have volume available in their stabilisers to balance CG shift by in-flight fuel transfer.

Fuel tank arrangement for thin winged aircraft (supersonic type) is complex as there is not sufficient volume available for the mission. In that case, fuel is provisioned in the fuselage wherever space is available. The example of *Concorde* is shown in Figure 22.20. It carried a substantial amount of fuel and CG shift was minimised by in-flight balancing through fuel transfer from the forward and aft trim tanks. A military aircraft tank arrangement is similar to *Concorde*. There could be as many as 16 tanks, all interconnected to meet the fuel requirements for the mission.

Fuel tanks can be rigid, made of metal or composite material. They can be flexible, made of rubber/neoprene like material. Tanks are installed at the time of component assembly. Flexible tank maintenance requires a change of tanks and can be a laborious task. Most modern aircraft have wet tanks, in which the skin at the joints is treated with sealants. A wet tank system is lighter and more volume efficient but wet tank leakage is a problem and the aircraft requires strict inspection, especially for older aircraft. Sealant technology has improved and wet tanks are favoured.

Heat generated at the stagnant regions of aircraft flying faster than Mach 2.4 can be cooled by re-circulating cold fuel around the hot zones before fed to engine. The pre-heating of fuel also helps combustion.

22.6.3 Emergency Power Supply

Most mid-size and bigger aircraft have an APU installed, performing many functions. It is a small power plant, invariably a turbo-shaft engine, using the same fuel (aviation turbine fuel, AVTUR). When ground facilities are not available then an APU can provide an emergency electric supply, air-conditioning and can start the main aircraft gas-turbines. Interestingly enough, an APU exhaust can reduce aircraft drag, no matter how small it is. A typical example of an aft mounted APU is shown in Figure 22.21 (schematic layout). An APU and

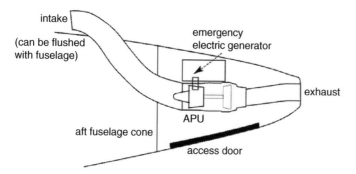

Figure 22.21 Auxiliary power unit (APU).

its installation weight can vary from 100 to 300 kg depending on size. An APU in military aircraft depends on user requirements. APUs can be started using on-board batteries.

The Ram Air Turbine (RAT) is another way to supply emergency power. These are propeller-driven devices mounted on the aircraft surface (at fuselage underbelly) and operate when an aircraft is in motion (Figure 22.22 – schematic layout). RAT is retractable.

22.6.4 Avionics Subsystems

There are a host of avionics black boxes supporting the flight deck and beyond. These black boxes serve navigational, communication, aircraft control and ECSs, record and process all important data to analyse and monitor malfunction and so on.

With increasing features, electrical cable length is long and is relatively heavy. Multiplexing of data transmission considerably reduces cable weight. Of late fibreoptics have been used for data transmission. When used with an FBW system, this is appropriately termed as Fly-by-Light (FBL). FBW is discussed in Chapter 18. This section is meant to make readers aware of design features (hardware) that can help more accurate prediction of weight and cost.

Most avionics boxes have microprocessors. This helps to standardise connections for data flow. The connection of wires is called the bus. The following are the prevailing standards for the bus architecture.

22.6.4.1 Military Application

MIL-STD-1553B: US Military aircraft were first to use data bus architecture, especially to handle large data for FBW and combat operation. In the UK it is covered by Def Stan 00-18.

Figure 22.22 Ram air turbine (RAT).

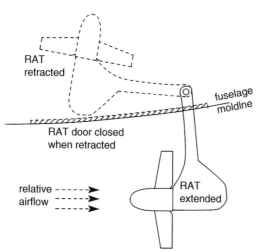

MIL-STD-1773: This is the fibreoptic version of MIL-STD-1553B.

STANAG 3838: This is the NATO standard for bus architecture.

22.6.4.2 Civil Application

ARINC 429 (originated in the 1970s): The success of the military standard was followed by civil standards – it started from scratch in a simplified manner. The A320 was first large transport aircraft to use a full FBW system.

ARINC 629 is a more updated version replacing ARINC 429.

Line Replacement Units (LRUs) are a convenient hardware design to ease installation and maintenance of electrical/avionics transmission and connections following the bus standards. These are constructed in a modular concept as a subassembly and then installed on the aircraft. LRUs are also standardised complying with the bus architecture.

Aircraft communication and navigations equipment form part of the aircraft avionics package. A typical civil aircraft avionics package is listed in Table 22.6.

Typical antenna locations for aircraft communication and navigations are shown in Figure 22.23. These are installed in the symmetrical plane of aircraft. Surveillance aircraft have special large housing for special purpose avionics.

22.6.5 Electrical Subsystems

All aircraft must have some form of electric supply to power aircraft subsystems. Supply of electricity is executed by a combination of generators and batteries. Most modern aircraft require both AC and DC supply. Typical AC voltage is 115 at 400 Hz but there are higher voltage AC supplies. Typically, DC voltage supply is at 28 V. Electric supply control must ensure safety and complying with mandatory requirements.

The following are the systems associated with electrical power.

i. Engine starting and operation. Management of fuel system.
ii. Lighting – both internal and external. Figure 22.24 shows external requirements.

Table 22.6 Aircraft avionics items.

EFIS/MFDs	Computers	Communication	Navigation
System display (SD)	Air data	ATC	DME
Analogue gauges	FBW	VHF	GPS
Radar	FADEC	Television	DME
	Autopilot		VHF

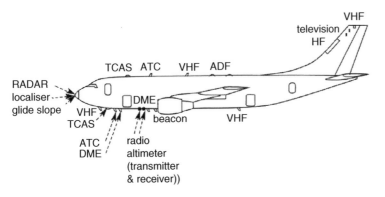

Figure 22.23 Antenna locations.

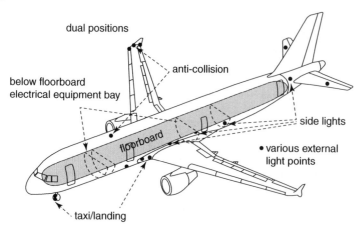

Figure 22.24 Aircraft lighting requirements.

iii. Flight deck instrumentation.
iv. Communication and navigation.
v. Avionics system.
vi. Flight control system using PCU.
vii. Passenger services for civil aircraft.
viii. APU – emergency electrical power generation and supply.
ix. For military aircraft – armament management, electronic defensive/countermeasures and so on.

Typically, electric supply is made at the primary and secondary levels. Engine driven generators supply the primary power. Secondary supply serves before the engine starts and also serves as a standby in emergency situation. Secondary supply comes from batteries, APU or auxiliary system as RAT.

Below the floorboard equipment bay (Figure 22.24) items such as batteries, chargers, power controllers, transformers, inverters and so on are housed.

The weight of the electric system depends on the load requirement. Cable weight is considerable. Avionics systems can be 0.4–4% for civil aircraft and 0.5–5% for military aircraft.

22.6.6 Hydraulic Subsystem

All large aircraft have a hydraulic system. The hydraulic system consists of fluid reservoir, electricity driven pumps, hydraulic lines, valves, pilot interface at the flight deck and so on. These aircraft incorporate hydraulic driven actuators at a higher force level for the following usage.

i. To activate the aircraft control system, for example, elevator, rudder, aileron, high-lift device and so on.
ii. To activate engine thrust reversers.
iii. To activate undercarriage deployment and retraction.
iv. To activate brake application.

Modern civil aircraft hydraulic pressure is 2000 psia (older designs) to 3000 psia (current designs). Military hydraulic oil pressure has reached 8000 psia. Higher pressure lowers system weight but requires stringent design considerations.

Figure 22.25 shows the hydraulic systems scheme of a four engine aircraft. To ensure safety and reliability, at least two independent continuously operating hydraulic systems are laid on separate locations. The port side is identified as a yellow system and the starboard side as a green system. (Airbus has introduced three independent lines – the third one as a blue line). Each line has its own reservoir and functionality. Table 22.7 lists the subsystems activated by a hydraulic system. All systems have gauges, switches, valves, tubing, connectors and so on.

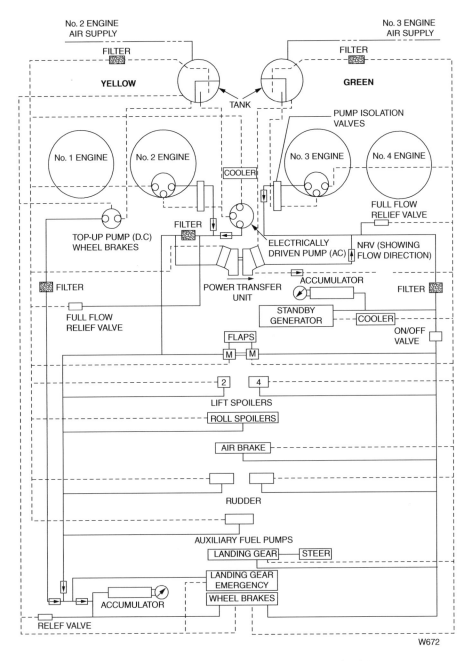

Figure 22.25 Aircraft hydraulics system (Courtesy of BAE Systems – RJ family).

Weight of hydraulic systems depends on the load requirement. Together with pneumatic systems (Section 22.6.7) the hydro-pneumatic system can be 0.4–1% for civil aircraft and 0.5–2% for military aircraft.

22.6.7 Pneumatic System

The pneumatic system comprises the use of high pressure air bled from the engine (gas-turbines) serving (i) the ECS, which consists of cabin pressurisation and air-conditioning, (ii) anti-icing, (iii) de-fogging system

Table 22.7 Hydraulic driven subsystem (BAE RJ family).

Separately in both systems	Yellow system (port side)	Green system (starboard side)
Wheel brake and parking brake	Flap brake	Lift spoilers (centre and outboard)
Power Transfer Unit (PTU)	Roll spoilers	Airbrake
Flap mode	Lift spoilers (inboard)	U/C (Landing gear)
Oil reservoir	U/C emergency lock	Nose gear steering
Accumulator	Rudder control	Wheel brake only
	Standby fuel pumps	Standby AC/DC generator drive
	Internal stair	Engine driven pump (EDP)

and (iv) engine starting. Note that the APU is linked with the pneumatic system. Aircraft oxygen needs are supplied in a separate pneumatic system. The pneumatic system is fitted with pressure reducing shut-off valve (PRSOV) and cross-flow shut-off valve (SOV) to control and isolate air flow as per the scheduled demand.

The other usages of pneumatics are to pressurise hydraulic reservoir, fuel system and water tank, drive accessories, as a medium for rain-repellent and so on. Some thrust reversers are actuated by pneumatics.

22.6.7.1 Environment Control System (ECS) – Cabin Pressurisation/Air-Conditioning

At cruise altitude, atmospheric temperature drops to minus −50°C and below, pressure and density reduces to less than one-fifth and less than one-quarter of sea level values, respectively. Above 14 000 ft altitude, the aircraft interior environment needs to be controlled for crew and passenger comfort and as well equipment protection. Cabin ECS consists of cabin pressurisation and air-conditioning. Smaller unpressurised aircraft flying below 14 000 ft altitude can get away with air-conditioning only, its simplest form uses engine heat mixing with ambient cold air supplied under controlled conditions.

Cabin interior pressure maintained at sea level condition is ideal but it is expensive. Cabin pressurisation is like blowing balloon, fuselage skin bulges out. The large differential between outside and inside pressure requires structural reinforcement making an aircraft heavier and more expensive. It is for this reason aircraft cabin pressure is maintained no higher than 8000 ft, a maximum differential pressure is kept at 8.9 psi. During ascent, the cabin is pressurised gradually. During descent, cabin depressurisation is also done gradually following a prescribed schedule acceptable to the average passenger. Passengers can feel it in the ear as this adjusts with change in pressure difference.

Cabin air-conditioning is an integral part of ECS along with cabin pressurisation. To supply a large passenger load at a uniform pressure and temperature is a specialised design obligation. Engine compressor is bled at some intermediate stage with sufficient pressure and temperature, which carries contamination and is cleaned and moisture removed to an acceptable level. Maintaining proper humidity is also a part of ECS. It is then mixed with cool ambient air. In addition, there is facility for refrigeration facility. There are internal system turbines and compressors driven by the system pressure. The heat exchanger, water extractor, condenser, valves and sensors together form a complex subsystem as shown in Figure 22.26. A generic pattern of supply of air-conditioned air flow in passenger cabin is shown in the figure.

Figure 22.27 depicts the BAE RJ family ECS system.

Avionics black boxes heating up need to be maintained at a level that keeps the equipment from functioning. The equipment bay is below the floorboards. Typically, a separate cooling system is employed to keep equipment cool. Ram air cooling is a convenient and less expensive way to achieve that. Scooping ram air would increase aircraft drag. The cargo compartment also requires some heating.

Advanced military aircraft ECS differs a lot (Figure 22.28). It uses a boot-strap refrigeration system. Of late, a boot-strap refrigeration system has also been deployed in civil applications.

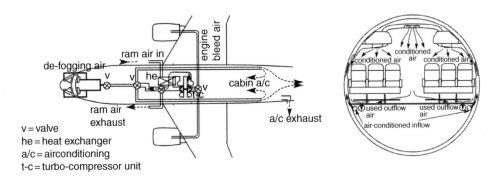

Figure 22.26 Schematic civil aircraft ECS and cabin air flow.

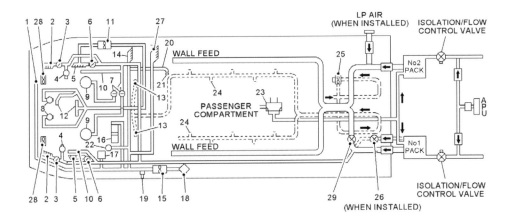

1 PILOTS INSTRUMENT PANEL COOLING
2 FORWARD FLOOR VENTS
3 FORWARD FLOOR AIR SELECTOR VALVE
4 SIDE CONSOLE ADJUSTABLE OUTLET
5 AFT FLOOR VENTS
6 F/DECK AIR SELECTOR FLAP VALVE
7 F/DECK ROOF FLOOD OUTLET SHUT OFF VALVE
8 ADJUSTABLE ROOF OUTLET
9 ROOF FLOOD OUTLET
10 WINDSHIELD PANEL C DE-MIST
11 FLIGHT DECK FAN
12 ROOF INSTRUMENT PANEL COOLING
13 TOILET AND GALLEY AIR OUTLETS
14 1R GALLEY VENTILATION
15 ELECTRICAL EQUIPMENT BAY COOLING FAN
16 AVIONIC EQUIPMENT RACK COOLING

17 PRINTED CIRCUIT BOARD BOX COOLING
18 SMOKE DETECTOR
19 PRESSURE SWITCH
20 CABIN TEMPERATURE SENSOR (CONTROL)
21 CABIN TEMPERATURE SENSOR (INDICATOR)
22 F/DECK TEMPERATURE SENSOR (CONTROL)
23 RECIRCULATION VALVE
24 PASSENGER ADJUSTABLE OUTLETS
25 CABIN FAN
26 RAM AIR ISOLATION VALVE
27 2R GALLEY VENTILATION (WHEN 2R GALLEY INSTALLED)
28 EFIS COOLING FANS (MOD.00950D)
29 FLIGHT DECK COOLING VALVE (MOD.50084A,
     MOD.50084B OR MOD.50084C)

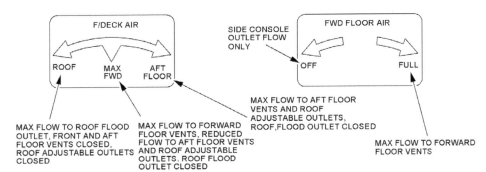

Figure 22.27 BAE RJ family air-conditioning system (Courtesy of BAE Systems).

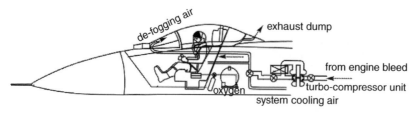

Figure 22.28 Military aircraft ECS.

Figure 22.29 Military oxygen system.

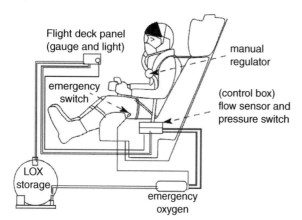

22.6.8 Oxygen System

22.6.8.1 Oxygen Supply

In an inadvertent situation, if there is drop in cabin pressure while the aircraft is still at altitude then oxygen supply for breathing becomes a critical issue. An aircraft system supplies oxygen to the individual passenger through a mask that drops from an overhead panel. Military aircraft have few crew and the oxygen supply is given directly integrated with the pilot suit in a mask as shown in Figure 22.29.

22.7 Flying in Adverse Environments and Passenger Utility

This section deals with aircraft systems required to fly in the following adverse environments.

  i. Anti-icing/de-icing
 ii. De-fogging and rain removal systems
iii. Lightning and sire hazards
 iv. Utility subsystems
  v. Passenger services and utility usages

22.7.1 Anti-Icing/De-Icing Systems

There are several methods for anti-icing and de-icing. Not all anti-icing, de-icing, defog and rain removal systems use pneumatics. These are primarily electrical systems but are included in this section to continue the subject matter. The following are currently used in practice – two of them are not pneumatic but included in this section.

22.7.1.1 Use of Hot Air Blown Through Ducts

This is a pneumatic system and the dominant system for larger civil aircraft. Both anti-icing and de-icing can use a pneumatic system. This is achieved by routing high pressure hot air bled from the mid compressor stage

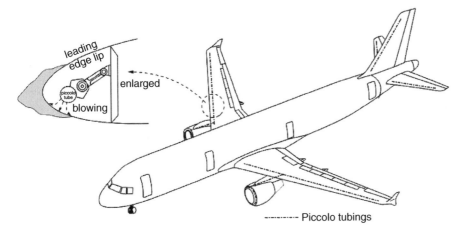

Figure 22.30 Civil aircraft anti-icing subsystem – Piccolo tube and ice formation.

and blown around the critical areas through perforated ducts (known as piccolo tubes, Figure 22.31). Pressure and temperature in the duct is about 25 psi and 200°C. Designers must ensure that damage does not occur on account of overheating. Figure 22.30 shows such a typical system. Figure 22.31 shows a BAE RJ family hot air blown through the ducts in an anti-icing system layout.

22.7.1.2   Use of Electrical Impulses
This is not common but is quite an effective de-icing system. Ice is allowed to accumulate up to a point where vibration generated by electric impulse breaks the ice layer, then it is blown away. It has low power consumption but can be a heavy and expensive system.

22.7.1.3   Use of Chemicals
This is also not a common system, primarily used as anti-icing system. Glycol based anti-freeze is allowed to sweat through small holes in critical areas where the chemical is stored. This is limited to the amount of anti-freeze carried on-board.

Piston engine carburettor and critical instruments must be kept heated for them to keep functioning without icing problems.

22.7.1.4   Use of Boots (Pneumatic/Electric)
Both anti-icing and de-icing can use boots specially designed with an integrated electric heater or passages for hot air flow. Rubber boots are wrapped (capped) around the critical areas (e.g. leading edges of lifting surfaces, propeller leading edges, intake lips etc.) and are heated either by electrical elements or by passing hot air as in the pneumatic case. Electrically heated boots are lighter but can be relatively more expensive. The boot type is used in smaller aircraft. Figure 22.32 shows a typical boot system.

22.7.2   De-Fogging and Rain Removal System

De-fogging and rain removal systems are like a car that uses windscreen wipers and embedded electric wires in the windscreen. There are rain repellent chemicals to assist rain removal. Figure 22.33 shows a generic layout aircraft rain repellent system with wipers.

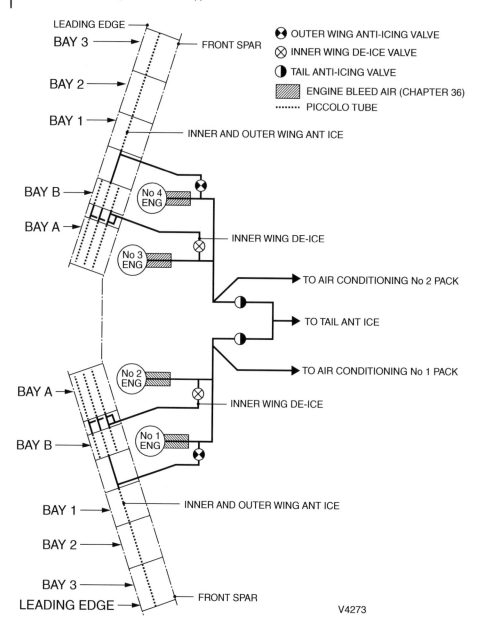

Figure 22.31 BAE RJ family anti-icing system (Courtesy of BAE Systems).

22.7.3 Lightning and Fire Hazards

Lightning strikes are common and in the majority of cases are harmless so a mission continues safely, but extreme strikes can create an impact that can cause operational interruptions. Typical damage causes holes/burn marks on skin. The metal external skin of the aircraft serves as lightning protection. The vulnerable areas (primarily forward and top surfaces) of such strikes have thicker skins. The nature of thunder strike is under continuous investigation. Type of material selection and surface finish and protective features can reduce damage caused by lightning strikes. Composite parts require wire mesh, embedded metal wires, diverter strips and so on to ensure protection from lightning strikes.

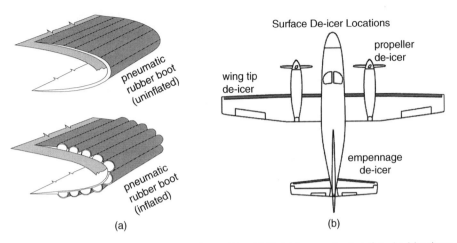

Figure 22.32 Anti-icing subsystem using boots (Goodrich). (a) Pneumatically inflated rubber boot. (b) Boot with electric element to heat.

Figure 22.33 Civil aircraft rain-repellent system.

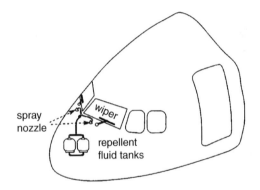

Static electricity builds up during a flight on account of airflow friction over an aircraft surface. The precipitated static electric charge needs to be dissipated through dischargers known as *static wicks* that are installed, typically these are trailing wing tips, on the empennage and so on. Electric charges from lightning are also discharged though these static wicks.

For safe operation, aircraft fire detection systems are installed to detect any occurrence of fire. Installed fire protection systems are then automatically activated to extinguish a fire. There are fire protection systems installed in passenger cabins for cabins to operate manually when required.

22.7.4 Utility Subsystems

Utility systems are composed of water and waste systems as shown in Figure 22.34. Passengers will need water and to relieve themselves. As the number of passengers and duration of flight time increases, the demand for water and waste disposal management also increases. The whole system is self-contained.

Typically, a third of a US gallon of water per passenger (100 US gallons for 300 passengers) is the quantity carried on-board. Both hot and cold water is supplied. Typically, one lavatory per 10 to 15 passengers is provided. Chemicals are used with water for flushing. Waste has to be contained inside the aircraft until it lands, when the system is cleaned and refilled with fresh supplies for the next sortie.

Figure 22.35 represents the RJ family waste water management system in detail.

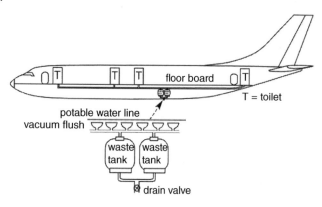

Figure 22.34 Water and waste system.

22.7.5 Passenger Services and Utility Usage

Aircraft need to be prepared for passenger services and utility usage. Several locations are specifically set to prepare an aircraft in a way that they do not interfere with each other and hold up time. Figure 22.36 shows typical utility service points. Access to servicing should not interfere with other activities in and around aircraft. Ice prevention of the water is done by heating the critical areas.

22.7.6 Aircraft Sound Horn

Large aircraft are fitted with pilot operated, siren-like, sound horn by pressing a button (labelled 'GND') located at the instrument panel. This horn is meant for drawing attention of ground crew to initiate communication for whatever may be reason, e.g., breakdown, malfunction, emergency, etc. This horn cannot be used like automobile hooter used as warning in traffic system. In flight, the aircraft horn is inoperable. Aircraft horn is meant for communication and not as traffic warning system.

Aircraft nose wheel strut is fitted with headphone socket, located at a suitable height, for ground crew to communicate with flight-deck pilot/personnel. In case, the ground crew need to initiate dialogue with the flight-deck personnel, a button is also located on the strut to alert flight-deck to engage in communication.

22.8 Military Aircraft Survivability

Being quite different from civil aircraft requirements, military survivability is treated separately in this section. Survivability considerations do not end at what is discussed here, however.

22.8.1 Military Emergency Escape – Egress

During World War II, pilot escape from damaged aircraft was carried out by the pilot climbing out of the flight deck and then jumping out to deploy a parachute to safety. Since then, considerable advancements in escape system technology have been implemented in line with the gain in aircraft speed-altitude capabilities.

Today, military emergency exit takes a drastic measure – just pull the D-ring rip-chord and ejection follows in a sequence of automatic operations. The D-ring is located between the pilot's two legs and at the top of the seat above the headrest to suit the kind of high *g*-load (Figure 22.37). Many designs provide a separate firing handle. A fully equipped ejection seat weighs about 200–400 lb. (90 to 180 kg), depending on the manufacturer and performance capabilities (e.g. at what speed, altitude etc.).

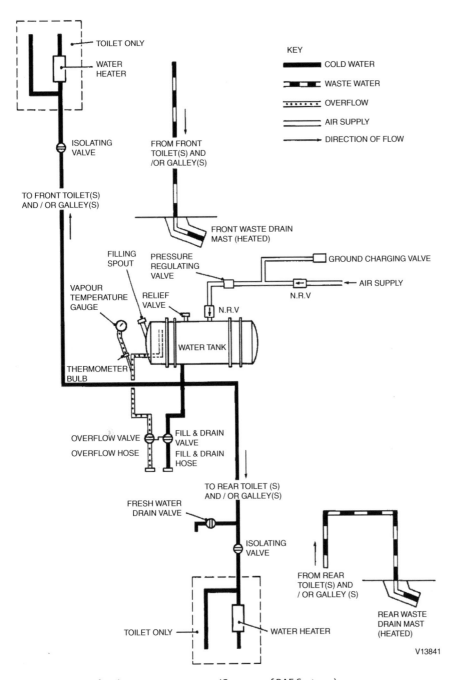

Figure 22.35 RJ family waste water system (Courtesy of BAE Systems).

There are two types of canopy design for egress as follows. The choice depends on the role of the aircraft.

i. Through canopy egress using explosive fracturing an instant before ejection seat egresses. This is done by installing a detonating cord or an explosive charge around or across the canopy (Figure 22.38). Detonating cords are laid on a canopy in a manner that explodes shattering the canopy polycarbonate glass making a sufficient gap for ejection to egress. This prevents possible collision of the pilot with the canopy at a difficult aircraft attitude at the time of ejection.

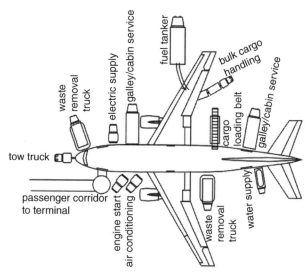

Figure 22.36 Civil aircraft turnaround servicing locations.

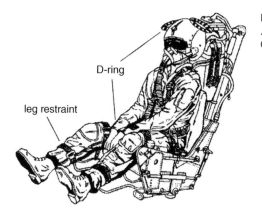

Figure 22.37 Typical military aircraft ejection seat. (http://www.geocities .com/cap17.geo/Ejection.html). Reproduced with permission of Cambridge University Press.

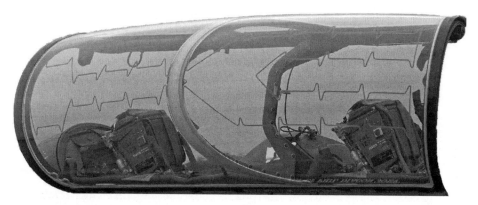

Figure 22.38 Detonation chord laid on canopy glass.

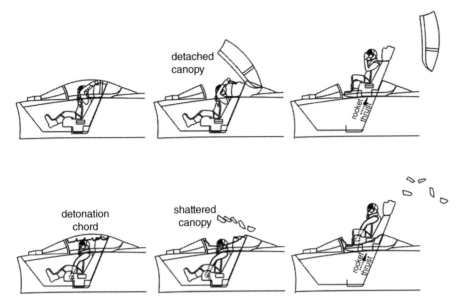

Figure 22.39 Typical ejection sequence from aircraft.

ii. At the pull of D-ring (ejector handle), the whole canopy detaches and punched away just before seat egress takes place.

At the pull of the D-ring, rockets under the seat are fired sending the seat into a ballistic catapult with the pilot strapped in. A typical sequence of ejection from the aircraft is shown in Figures 22.38. After leaving the aircraft, a separation sequence of seat and parachute deployment is shown in Figure 22.39. In the automatic sequence, just before the rockets fire, the canopy is released or made to explode along laid out explosive lining to make hole through which the seat assembly passes through. Hands and legs are kept retrained with straps to avoid injury during escaping through a relatively small canopy opening.

The sustainer-rocket steers the pilot with the seat to a clear space when the seat separates and parachute opens in sequences of operation shown in Figure 22.40. The dynamics of ejection is complex. The ejection can be made at zero altitude (failure at takeoff or landing) – this is known as zero-zero ejection. In case the aircraft is close to ground in inverted flight, the ejection could be successful if there is enough height when the seat could turn around to safe distance. It has to be borne in mind that such escape through an ejection seat is a serious matter and requires pilot drill to avoid injury. Restraints are placed to prevent a pilot's limb flailing and keep the body at safe attitude at the time of passing through the flight deck. Peak *g*-load at ejection can

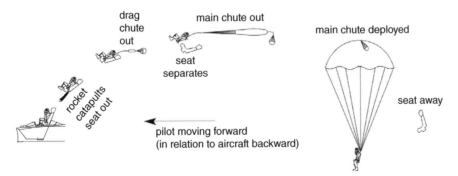

Figure 22.40 Typical ejection sequence showing separation of seat and parachute deployment. Reproduced with permission of Cambridge University Press.

exceed 25 g. A few pilots have suffered injury during ejection but their lives have been saved and they have recovered. Thousands have already been saved without injuries.

There are other kinds of simpler rescue system for basic military trainer type of aircraft. For the home-built category, parachutes can be deployed to bring the entire aircraft down to safety.

There are not many manufacturers in the world who make military ejection seats. The major ones are from the UK (Martin-Baker), the USA (ACES) and Russia (Zvezda). There is not much classroom work for the readers, but they can contact manufacturers for free brochures that can give accurate sizes, weights and some descriptions. These would prove useful for weight and size estimation, and up to date information.

22.8.2   Aircraft Combat Survivability

New combat aircraft capabilities are evaluated in simulated environment in twin platform, the other platform being that of adversary capabilities to the extent it can be ascertained in terms of combat scores. These are based on mathematical modelling with measurable parameters, for example, susceptibility (inability to avoid adversary interception) [4]. Vulnerability (inability to withstand hit) and killability (inability to both). The general term aircraft combat survivability (ACS) refers to the capability of an aircraft survivability disciplines attempt to maximise the survival of aircraft in wartime. This is in addition to system safety attempts to minimise those conditions known as hazards that can lead to a mishap in environments not arising from combat. These topics are beyond the scope of this book (readers may refer to [5]).

22.8.3   Returning to Home Base

An injured/stunned pilot as a result from enemy action could become temporarily unconscious even when the aircraft is still flyable. Combat aircraft could be designed with capabilities to switch to automatic mode for the aircraft to follow pre-programmed sequences to immediately turn back to return to home base so that by the time a pilot regains consciousness, if required, it can land with assistance from ground instructions. Saving one life is worth the investment.

22.8.4   Military Aircraft Stealth Considerations

The most important consideration for military aircraft designers is to maximise the chances for the pilot to survive in a dangerous combat action. The question of survival at combat has many areas to consider. This section will discuss some of the areas that affect aircraft configuration and weight. A brief discussion will be given here on how the aircraft configuration is affected by stealth considerations. Stealth design is an issue of pilot survival.

The parameters affecting combat stealth aircraft design for survival are as follows;

  i. Minimise audio-visual detection – make the airframe as small as possible and eliminate/reduce engine emission (noise and smoke). Very small aircraft may not prove combat effective.
 ii. Minimise radar signature – make the aircraft surface reflect radar beams away and use suitable paint coating to absorb radar emissions.
iii. Minimise heat signature – high temperature of engine exhaust is detectable, especially when an afterburner is used. Incorporation of super-cruise capability makes an aircraft fly at supersonic speed without the use of an afterburner.
 iv. Use of on-board passive systems – these have infra-red search and track, forward looking cameras, night vision aids and so on. This minimises electronic radiation, which can be detected from a distance.
  v. Use of defensive aids – incorporating Beyond Visual Range (BVR) capabilities and other capabilities to detect an enemy without revealing itself.
 vi. Incorporate secure communication – a pilot should be able to receive communications but radio transmission can reveal the aircraft's position. Therefore, the radio system must have a secure system to prevent detection.

vii. Incorporate on-board standalone navigational system – a combat mission may have to fly over unfamiliar territory. Pilots require a terrain map gathered from earlier reconnaissance flights and non-radiative on-board navigational tools to fly as accurately as planned and be certain of the aircraft position at any time and in any place.

The threats from hostile environment are getting more dangerous as technology advances. Pilot survival issues are tackled in the following three main segments of a combat profile.

i. *Before combat* – a surprise entry into the combat zone acts as a preventive measure by minimising the reaction time for hostile retaliatory action, possibly to the point of making it an inefficient attempt. The overriding objective to achieve surprise entry is by making the aircraft approach in stealth. Apart from the electronic counter measures to block signatures and sophisticated standalone navigational capabilities for complex approach route planning, stealth technology affects aircraft configuration. While stealth configuration does not make an aircraft completely invisible, a better term is that it becomes Low Observable (LO); a combat platform that would reduce the warning time.

ii. *During combat* – combat is in an unpredictable scenario and depends on many factors, some unknown, depending on the capabilities of the adversary. At the combat zone, the presence of attack aircraft is known. The aircraft should be capable of extreme manoeuvres. Also, the aircraft should be designed with strong armour plating to protect against penetrative projectiles, especially against small arms firing from the ground.

iii. *After combat* – the scenario is now completely changed. If a mission is successful without getting hit, then there is only one role for the pilot: to escape as effectively as possible – aircraft high speed and altitude capability are now in demand. If hit, then the extent of damage would dictate what action the pilot takes. In case of a catastrophic damage, it will leave no option but to eject. Survival through ejection is discussed in the previous section. If the aircraft is flyable but the pilot is injured to unconsciousness, then the aircraft should be capable of flying back and landing at the home base automatically. This technology is now available and can be a good candidate for commercial aircraft as a counter-terrorist measure. Decision to take what kind of action on account of aircraft damage, or anything in between the two extremes, is for pilots to take appropriate measures and they are trained to tackle such situations.

22.8.5 Low Observable (LO) Aircraft Configuration

A fighter aircraft with LO configuration characteristics for stealth design will have to make compromises to performance (aerodynamic and manoeuvre) affecting weapon carrying capability, thereby limiting combat effectiveness. Aircraft designers will have to make trade-off studies to maximise combat capability. A conventionally designed weapon load will not offer stealth.

In addition, evasive manoeuvres, radar jamming, spraying heat sources and suchlike are other types of measures used to confuse missile attack and so on.

A missile finds its target by homing in through the signals it receives and then gets locked on to it in an attempt to hit the target unless countermeasures can fool the missile system. A stealth design would require suppressing the following parameters given in Table 22.8.

22.8.5.1 Heat Signature

An infra-red seeking homing device is a potent method as long as there is a single clear identifiable target emanating sufficient signal such as from the engine exhaust. The drawback is that the missile can be easily

Table 22.8 Stealth parameters.

1. Heat signature	3. Noise signature
2. Radar signature	4. Visibility

fooled by spraying out many heat flares. Missiles aimed at downward targets or facing the sun could lock on a stationary target elsewhere within capture angle.

Aircraft designers should aim to reduce aircraft heat signature below 350°C by mixing engine exhaust with cool atmospheric air through entrainment at the exhaust. Shielding of the engine exhaust by its wing is an effective method against ground launched heat seeking missiles.

22.8.5.2 Radar Signature

The radar system works by capturing the echoes from the reflected returns of its transmitting radio waves from an object. If the object is such as a moving aircraft, the radar technology adjusts with the Doppler shift phenomena to give an accurate position of the aircraft; its speed altitude and range in real time, which come as alarm and/or homing signal. The fundamental objective of radar stealth is by reducing the radar cross section (RCS) area, that is, reduce the echo strength so that it is noticed much later, thereby reducing the reaction time of the adversary.

RCS area is defined as the projected area of an equivalent perfect reflector with uniform properties in all direction such as from a sphere and which returns the same amount of power per unit solid angle in steradians as the object under consideration.

The intensity of reflected radar beam (echo) depends on the surface on which it is reflecting. The parameters influence reflection are, area, orientation and the nature of the surface. The maximum is when the surface is normal to it and the extent of area capturing the radiation. Figure 22.41 compares echoing from a sphere to a pointed sharp corner of inclined surfaces.

Even a small sphere would offer larger normal (and near normal) surface as compare to a point such as the tip of a nose cone. The inclined surfaces would deflect the reflected beam away. In addition, if the surface is coated with radiation absorbing paints then the echoing would further reduce. Radar absorbing coat is heavy, difficult to maintain and increases cost.

Earlier designs, for example, the F117 Nighthawk had inclined flat plate like surfaces with sharp edges that succeeded in radar signature reduction, evidently not sufficient enough as one was shot down by missile in the Kosovo conflict. The stealth configuration of the time was aerodynamically inefficient – nick named the 'aerodynamicists' nightmare'. The B2 bomber showed improvement with more streamlined shape, engine intake and exhaust over the wing shielding the hot zones against heat signature. The latest F22 Raptor is a fine example of improvement in shaping, incorporating better streamline shape cutting down drag. Figure 22.42 compares configuration and Table 22.9 compares few combat aircraft RCS values.

All modern combat aircraft design will have a specification for maximum RCS area. Therefore, modern combat aircraft design without assessing the RCS area to satisfy requirement is meaningless. Computing RCS area is not difficult but time consuming making it unsuitable for undergraduate classroom work. There are application software packages that can measure RCS area and interactively tailor aerodynamic surfaces with minimum compromise. Currently, there is no such software available in public domain.

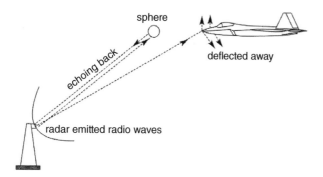

Figure 22.41 Typical comparisons of radar signatures (sphere versus stealth aircraft).

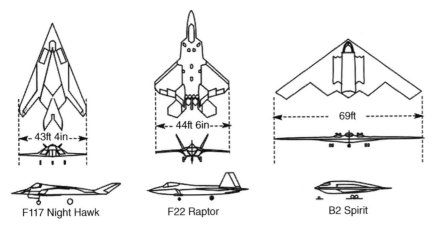

Figure 22.42 Three stealth aircraft configurations.

Table 22.9 RCS values of combat aircraft.

Aircraft type (older designs)	RCS area – m^2 (newer designs)	Aircraft type	RCS area – m^2
F15 eagle	40.5	F117 Nighthawk (1970s)	0.003
B1 bomber	10	B2 Spirit (1980s)	0.0014
SR-71	0.01	F22 Raptor (1990s)	0.0065

The downing of the F117 in Kosovo opened an argument on the extent of stealth affectivity. In addition, stealth features degrades aircraft performance and handling. Advances in missile technology would make stealth technology less effective. But the author thinks that, currently, stealth is still a desirable feature that designers must exploit. The F117 is an older design and stealth technology is also advancing. Aircraft designers have a difficult task to compromise between stealth, performance and cost in a changing environment with newer technologies. The result from future combats will resolve some of the controversies.

References

1 Falconpedia, HOTAS information. Available online at: http://falcon4.wikidot.com/avionics:hotas (accessed June 2018).
2 Moir, I. and Seabridge, A. (2001). Aircraft Systems, Professional Engineering Publishing.
3 Roskam, J. (1989). *Airplane Design*, vol. 3 and 4. USA: DAR Corporation.
4 Advanced Concept Ejection Seat ACES II. Report MDC J4576 Revision D. 1988.
5 Shannon, R.H. (1971). Operational aspects of forces on man during escape in the U.S. Air Force, 1 January 1968–31 December 1970. In: *AGARD Linear Acceleration of Impact Type*.

23

Computational Fluid Dynamics

23.1 Overview

Computational Fluid Dynamics (CFD) is a numerical tool for solving fluid mechanics equations. CFD is a relatively recent development that has grown powerful enough to become an indispensable tool over the last two decades. Originally, it was developed for aeronautical usages, but now pervades all disciplines involving flow phenomena; for example, medical, natural sciences, engineering applications and so on. The built-in codes of the CFD software are algorithms of numerical solutions to fluid mechanics equations. Flow fields that were previously difficult to solve by analytical means, and in some situations impossible, are now accessible by means of CFD.

Today, the aircraft industry uses CFD during the conceptual study phase and continues with detailed analyses in the next stages after the 'go-ahead' is obtained. There are limitations in getting accurate results, but research continues in academic and industrial circles to improve prediction. This chapter aims to make the newly initiated readers aware of the scope of CFD in configuring aircraft geometry. Those who are already exposed to the subject may skip the chapter. This is not a book on CFD, so this chapter does not present a rigorous mathematical approach but an overview.

CFD is a subject on its own requiring considerable knowledge on fluid mechanics, mathematics and computer science. CFD is introduced late in undergraduate studies in the final years when students have covered the prerequisites. Commercial CFD tools are menu driven and one could become proficient soon enough but to interpret the results thus obtained would require considerable experience in the subject matter.

Making an accurate 3D model of the aircraft with complete aircraft drag (CAD) offers considerable reduction in preprocessing time. The CAD software must have compatible format with CFD to transport the drawing models. Together, CAD and CFD provide a CAE (Computer-Aided Engineering) approach to paperless electronic design methods.

There are several good commercial CAD and CFD packages in the market place. Nowadays, all engineering schools have CAD and CFD application software. This book refrains from naming any package.

What is to be learnt. The chapter covers the following:

Section 23.2 Introduces the Concept of CFD
Section 23.3 Introduces the Current Status of CFD
Section 23.4 Presents an Approach Road and Considerations for CFD Analysis
Section 23.5 Presents Some Case Studies
Section 23.6 Suggests the Extent of Classroom Work to be Undertaken
Section 23.7 Gives a Summary of Discussions

Classwork content. There is no classroom work on CFD in the first term. However, it is recommended that CFD studies should be undertaken in the second term after the reader is formally introduced to the subject. It would require appropriate supervision to initiate the task and analyse results. Any CFD work is separated from the scope of what this book can offer. This chapter is only to give the newly initiated a taste of aircraft design work.

23.2 Introduction

Throughout the book, it is shown that the aerodynamic parameters of lift, drag and moments associated with aircraft moving through air are of vital importance. An accurate assessment of these parameters is the goal of the aircraft designers.

Mathematically, lift, drag, and moment of the body can be obtained from integrating the pressure and shear stress field around the aircraft computed from the governing conservation equations (differential or integral forms) of mass, momentum and energy along with the equation of state for the medium of air. Up until the 1970s, wind-tunnel tests were the only way to get the best results to obtain these parameters in various aircraft attitudes representing what can be encountered within the full flight envelope. Semi-empirical formulae generated from vast amounts of test results, backed up by theory, give a good starting point for any conceptual study.

Numerical methods for solving differential equations have been prevailing for some time. Navier–Stokes equations give an accurate representation of the flow field in and around the body under study. But solving these equations for 3D shapes in compressible flow was difficult, if not sometimes impossible. Mathematicians devised methods to discretise differential equations into algebraic form which are solvable even for the difficult non-linear partial differential equations. During the early 1970s, CFD results of simple 2D bodies in inviscid flow were demonstrated to be comparable with wind tunnel test results and analytical solutions.

Industry and national laboratories saw its potential and progressed with in-house research, and in some cases was able to understand complex flow phenomena hitherto unknown. Subsequently, it proliferated into academies and rapid advancement was achieved for the solution techniques. Over time, the methodologies kept improving. The latest techniques discretise the flow field into finite volumes in various sizes (smaller where the fluid properties have steeper variation) matching the wetted surface of the object, which also needs to be divided into cells. The cells do not overlap the adjoining volumes but mesh seamlessly. The mathematical formulation of the small volumes can now be treated algebraically to compute the flux of the conserved properties between the neighbouring cells. Discrete steps of algebraic equations are not calculations of limiting values at a point; therefore, error creeps into the numerical solution. Mathematicians are aware of the problem and struggle with better techniques in making the algorithm minimise the error. This numerical method of solving fluid dynamic problems became computation intensive, requiring computers to tackle a large number of cells; their numbers could run into several millions. The solution technique is thus known as Computational Fluid Dynamics.

The other problem in the 1970s was the inadequacy of the computing power available to tackle the domain comprising of a large number of cells and handle error functions. As computer power increased along with superior algorithms, CFD capability started to become applicable to industry. Today, CFD is a mature method and is well supported by advanced computing power. CFD started in industry and now has finally become an indispensable tool for the industry and research establishments.

One of the difficult areas of CFD simulation lies in turbulence modelling. Over the last two decades, computations of 3D Reynolds-average Navier–Stokes (RANS) equations for complete aircraft configurations became routine. Reference [1] summarises the latest trends in turbulence modelling. A large number of credible application software packages have emerged on the market, some catering for special purpose applications.

As discussed, drag predictions around the design point have improved to become relatively reliable, comparable to WT results (without incurring the cost of testing at high Reynolds number). As discussed, drag predictions around the design point are reliable, comparable to wind-tunnel test results (without incurring the cost of testing at high Reynolds number) stall point, unsteady phenomena, for example, buffet, high angle of attack and separated flows, are not so reliably predicted. CFD still has its limitations. Drag is a viscous dependent phenomenon; inclusion of viscous terms makes the governing equations very complex requiring intensive computational time. Capturing all the elements contributing to drag of a full aircraft is a daunting task – its full representation is yet to achieve credibility in the industrial usages. It is not yet possible to get accurate drag prediction using CFD without manipulating input data based on the designer's experience. However, once set up for the solution, the incremental magnitudes of aerodynamic parameters of a perturbed geometry are well represented in CFD. It is a very useful tool to obtain accurate incremental values of a perturbation geometry from a baseline aircraft configuration with known aerodynamic parameters. It offers a capability to make parametric optimisation to a certain degree (see the next section).

23.3 Current Status

Chapman [2] in his classic review (1979) advocates the indispensability of CFD, as the computers start showing the promise of overtaking experiments as a principal source of detailed information on design. His view is now regarded as an overoptimistic estimate. CFD capabilities can compliment experiments. He assigned the following three main reasons for his conclusion:

i. Experiments cannot represent the real flight envelope (*Re*, temperature etc.) and are limited by flow non-uniformity, wall effects and transient dependent separation.
ii. Very high energy (cost) associated with large tunnels.
iii. CFD is faster and cheaper than experiments to obtain some valuable insight at an initial stage.

He showed the chronology of progress to be in four stages, starting from the late 1960s solving linear potential flow equations, to arriving at a stage when the non-linear Navier–Stokes equations can be tackled.

In his recent paper, Chapman [3] again reviewed the rapid progress achieved in the last decade. With a better understanding of turbulence and with advances in computer technology, both in hardware and software development, researchers have successfully generated aerodynamic results that were impossible to obtain until then.

The latest review by Roache [4] demonstrates that considerable progress has been achieved in CFD, but the promise is still far from being fulfilled in estimating CAD. AGARD report AR256 [5] gives technical status review on drag prediction and analysis from a CFD point of view. In that report, Schmidt categorically stated that 'consistent and accurate prediction of absolute drag for aircraft configuration is currently beyond CFD reach …'. Many researchers were also of the same opinion, stating that the CFD flow modelling was found to be lacking in 'certain respects'. Both agreed that the current state-of-the-art in CFD is still a useful tool at the conceptual stages of design for comparison of shapes and for diagnostic purposes.

An essential route to establish the robustness of any CFD comes through the success of the conceptual model code verification and validation. Roache [4] uses the semantics of 'verification' as solving the equations correctly and 'validation' as solving the correct equations. The process of benchmarking (code to code comparison) results in the selection of the best value for money software, not necessarily the best software on the market.

Experimental results are used to validate and calibrate CFD codes. Various degrees of success have been had in case to case studies. Melnik [6] showed that the current CFD status in aircraft drag prediction of a subsonic jet transport type aircraft wing on a simple circular cross section fuselage had mixed success, as shown in Figure 23.1. Some correlation was achieved after considerable 'tweaking' of the results. The results using these methods are not certifiable because of considerable 'grey' areas.

Currently, industry uses the CFD as a tool for flow field analysis, wherever possible to estimate drag in inviscid flow, for example, induced drag and wave drag, but not used for CAD estimation. In industry, CFD appears as a general purpose tool to simulate flow around objects for qualitative studies and diagnostic purposes. For the design point, design of high-lift system, full RANS simulations are used.

It is difficult to capture large number of 'manufacturing defects' (e.g. steps, gaps, waviness etc., arising from surface smoothness requirements) over the full aircraft. CFD flow field analysis of simple geometries for benchmark work has been carried out, for example on large backward facing steps. One such example is Thangam et al. [7] who describe a detailed study of flow past backward steps to understand turbulence modelling (κ-ε). This type of work does not represent the problems associated with the small geometries of excrescence effects nor can they guarantee accuracy. Another example is Berman's [8] work on a large rearward facing step, but this is not representative of the dimensions of excrescence.

Assessment of excrescence drag using CFD needs a better understanding of the structure of the boundary layer in turbulent flow. While there exists voluminous literature on CFD code generation and qualitative

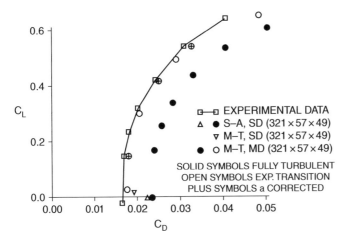

Drag polar for M100 Wing-body Combination; Experimental Data and Simulations Using Code TLNS3D. S–A = Spalart-Allmaras turbulence model, M–T = Menter blended *k*-ω/*k*-ε model. SD = scalar artificial numerical dissipation, MD = matrix dissipation. (From Figure 19 of Melnik et al, 1995).

Figure 23.1 CFD simulation of wing-body drag polar [6]. Reproduced with permission from Cambridge Unversity Press.

assessment of pressure field, no work has been cited about estimating parasite drag of excrescences. As modern CFD software becomes more capable, it may become possible to predict excrescence drag simulating real cases with double curvature in compressible flow with or without shocks or separation.

Reference [9] includes verification of excrescence drag on a flat plate in the absence of a pressure gradient to estimate the excrescence drag on a 2D aerofoil in pressure gradient. The study on an aerofoil [10] may be seen as a precursor to examine the scope for CFD estimates of excrescence drag on the generic 3D aerofoil configuration.

This review provides some earlier CFD studies conducted during the twentieth century. Since then, considerable progress has been made in refining programme algorithms to improve accuracy and rapid improvement at lower cost has made CFD an essential tool for flow analyses in industries and academies.

Based on the Boeing Company's endeavours with CFD, Forrester et al. [11] summarise progress made in the last three decades, as quoted here:

> CFD will continue to see an ever-increasing role in the aircraft development process as long as it continues to add value to the product from the customer's point of view. CFD has improved the quality of aerodynamic design, but has not yet had much effect on the rest of the overall airplane development process. CFD is now becoming more interdisciplinary, helping provide closer ties between aerodynamics, structures, propulsion and flight controls. This will be the key to more concurrent engineering, in which various disciplines will be able to work more in parallel rather than in the sequential manner as is today's practice. The savings due to reduced development flow time can be enormous!

Jameson clearly sums up in [1] in his review of the past and the present of CFD activities then projecting the direction of future work. Figure 23.2 shows the CFD contribution to the Airbus380 aircraft. It corroborates the statement made in [1] that 'CFD has improved the quality of aerodynamic design, but has not yet had much effect on the rest of the overall airplane development process'.

23.4 Approach Road to CFD Analyses

CFD analysis requires preprocessing of the geometric model before computation can start. It comprises of creating an acceptable geometry (2D or 3D) amenable to analysis (say there is no hole for fluid to leak through).

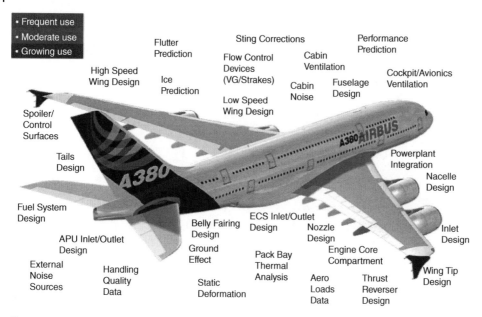

Figure 23.2 CFD contribution to Airbus380 aircraft [1].

A preprocessing package comes with its own CAD to create geometry, specifically suited for a seamless entry to the solver for computation. However, a considerable amount of labour can be saved if the aircraft geometry already created in CAD can be used in CFD. This is possible if care has been taken in creating a geometry that is transportable to CFD preprocessing environments.

The next task is to lay a grid on the geometry for CFD to work on small numbers of cells at a time until the entire domain is tackled. The surface grids should be intelligently laid to capture details where there is large local geometric variation, normally at the junction of two bodies and where there is steep curvature. The next step would be to generate cells in the domain of application, which could be a large flow field space around the body. Evidently, at the far field, the variation in the flow field between the cells is low and therefore could be made larger. The preprocessor is menu driven and gives the options for various kinds of grid generation to select from. Grids must meet the boundary conditions as the physics dictate. Figure 23.3 gives a good example of aircraft geometry (simplified by excluding empennage and nacelle) with structured grids on it and a section of the environment to be analysed. Only one half of the aircraft, because it is symmetrical on the vertical plane, needs to be analysed – the other half is the mirror image.

Figure 23.3b is another example of 3D meshing on a complete aircraft with the nacelle included and in space.

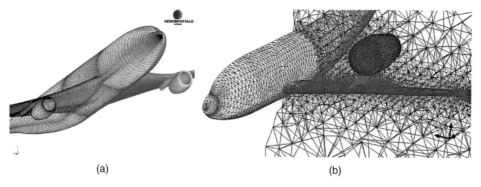

(a)   (b)

Figure 23.3 Wing-fuselage geometry with meshing. (a) On the surface (Courtesy Aerospatiale). (b) On the surface and in the space [12] (denser close to the surface) From Dr. Raymond Devine's PhD work at QUB.

Velocity (*m/s*)
0.000 30.0   60.0   90.0   120.0 142.0

(a)   (b)

Figure 23.4 CFD post-processor visualisation (a) Pressure distribution (Courtesy NASA). (b) Velocity Streamline patterns.

After grid generation in the preprocessor, the model is then introduced into the flow solver, which is also menu driven. The options in the solver are specific and the user must know what to apply. After the solver, a run occurs to compute results (run-time depends on the geometry, type of grid and solver options as well as the computing power).

The results can be examined in many different ways in a post-processor. The important ones on an aircraft body are the C_p distribution, the temperature distribution, streamlines and velocity vectors and so on. The C_p distribution and the temperature distribution are shown in grades of colour representing bands of ranges. Figure 23.4b is a grey scale version of the colour distribution and Figures 23.4a depict the flow field streamlines. The results can also be obtained numerically in tabular form. It is clear now that the readers must have the background and be familiar with the CFD software package. For the newly initiated, it should be conducted under supervision.

Here is a summary of the approach to CFD:

In the Preprocessor (Menu Driven)

Step 1. Create geometry – the 3D geometry of the aircraft
Step 2. Generate grid on the body surface and in the domain of application.
   Match to the boundary conditions.

In the Flow Solver (Menu Driven)

Step 3. Bring the preprocessed geometry and volume mesh into the solver.
   Set boundary and initial conditions. Make appropriate choices for the solver.
   Run the solver.
   Check results and refine (iterate) if necessary (including the grid pattern).

In the Post-Processor (Menu Driven)

Step 4. The result thus obtained from the solver can now be viewed in the solver.
   Select the display format.
Step 5. Analyse results.
Step 6. For a new set up, verify and validate results.

The results can be presented in many ways, for example, C_p distribution, pressure contours, streamlines, velocity patterns, C_L, C_D, L/D or parameters that can be defined by the user. It can depict shock patterns, location, separation and so on, similar to what wind tunnels can do with flow visualisation. These give an insight to the aerodynamics designers need to improve a design for say, best L/D, best aerodynamic moments, compromise shapes to facilitate production and so on.

The results may need tweaking for which iterations would be necessary starting from Step 2 and/or Step 3.

23.5 Some Case Studies

This section includes some elementary examples of some case studies starting with 2D cases as shown in Figure 23.5. The first diagram represents an aerofoil with the grid layout shown.

The domain of analysis is large with anisotropic adaptive grid (Figure 23.5a), which is dense close to the leading edge and trailing edge matching the surface grids, and where there are shocks present. Once the solver is run, the result can be seen in the post-processor showing the Mach number iso-lines (Figure 23.5b). At another run with a different set up, the result is shown in colour spectrum (grey scale version in Figure 23.5c).

CFD analysis of a component should prove relatively easy but flow with internal and external flow through nacelle can be a difficult case. The example is of high-lift device analyses as shown in Figure 23.6a. Nacelle grids for internal and external flow analysis are shown in Figure 23.6b.

While CFD studies around aerofoils exist, flow field analysis around nacelles is rare. Chen et al. [15] present a flow field analysis over a symmetric isolated nacelle using a Euler solver (Figure 23.7). Subsequent studies by Uanishi et al. [16] show confirmation of velocity field obtained by Chen. No work has been located for velocity fields over nacelle using Navier–Stokes solvers.

In a very fine recent analysis [17], it is stated that

> ... the observed scatter in the absolute CFD-based drag estimates is still larger than the desired single drag count error margin that is defined for drag prediction work. Yet, the majority of activities conducted during an aircraft development programme are incremental in nature, that is, testing/computing a number of options and looking for the best relative performance.

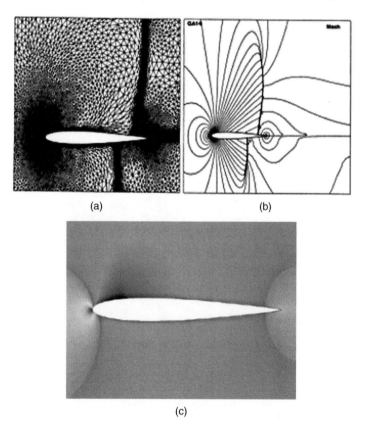

(a) (b)

(c)

Figure 23.5 CFD analysis of a 2-D aerofoil. (a) Preprocessor showing adaptive grid. (b) Post-processor showing Mach isolines. (c) In spectrum. Reproduced with permission from Cambridge University Press.

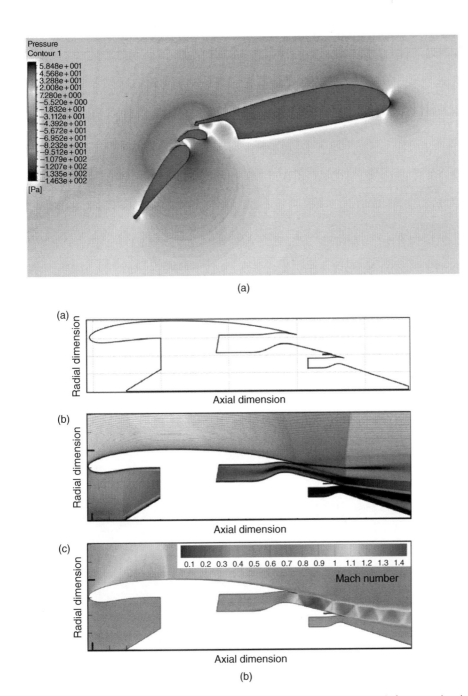

(a)

(b)

Figure 23.6 CFD flow analyses case studies. (a) High-lift device. (b) Nacelle grids for internal and external flow analysis.

23.6   Hierarchy of CFD Simulation Methods

A hierarchy of CFD simulation methods exists in which they are classified according to the physics they are capable of modelling. At the top of the hierarchy are Direct Numerical Simulations (DNSs) simulating time-dependent turbulent flows, capturing the dynamics of the whole temporal and spatial spectrum of eddy

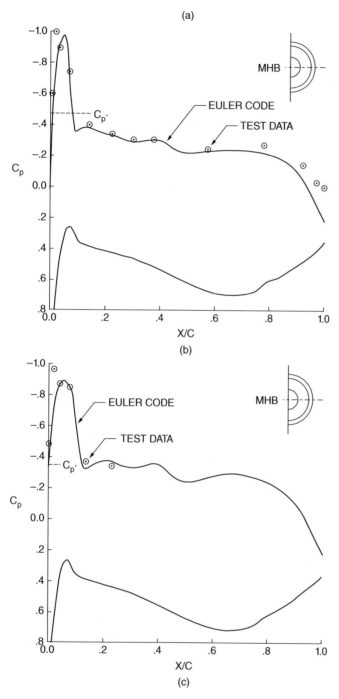

Figure 23.7 CFD results of a nacelle by Chen et al. [16]. (a) Nacelle geometry. Reproduced with permission from Cambridge Unversity Press. Comparisons of C_p at different $M\infty$ values along the maximum half breadth. (b) $M\infty = 0.6$ (c) $M\infty$ 0.84. Reproduced with permission from Cambridge Unversity Press.

sizes. This requires grids and time steps that are finer than the length and time scales at which turbulent energy is dissipated. Low diffusion numerical schemes are necessary. It is very useful for supplementing experimental data and aiding the development of turbulence models, but prohibitively expensive at flight Reynolds numbers. It is not used in the design process. Currently, it is the most sophisticated method.

Large Eddy Simulation (LES) technique. LES takes advantage of the fact that the smallest, dissipative eddies are isotropic. These can be efficiently modelled using simple sub-grid scale models. Meanwhile, the dynamics of the larger eddies, which are anisotropic in nature, is simulated using a grid and time step sufficiently fine to resolve them accurately. The method is, therefore, applicable to flows at relatively large Reynolds numbers, but is still expensive to use as an engineering design tool.

Detached Eddy Simulation (DES) technique. DES can be considered a halfway house between LES and RANS. The method employs a RANS turbulence model (next in the list) for near-wall regions of the flow and a LES-like model away from the wall. The method was first proposed by Spalart et al. in 1997 and is still the subject of research. It may become a standard engineering tool, but at present it is unlikely to be an element of the conceptual and preliminary design toolkits.

Reynolds-averaged Navier–Stokes equations (RANS) technique. The time-dependence of turbulent fluctuations is averaged out to form the RANS equations. This results in the appearance of so-called Reynolds stresses in the equations, and the problem of modelling these (turbulence modelling) arises. There are many turbulence models, but each one falls prey to the fact that turbulence is flow-dependent; consequently, no turbulence model can be generally applicable, and the CFD practitioner must be cognisant of the strengths and failings of the models he/she employs. Nonetheless, RANS does allow relatively inexpensive modelling of complex flows, and when allied to a suitable optimisation method, can be a powerful tool for design synthesis.

Euler equations technique. The Euler equations are obtained when the viscous terms are dropped from the Navier–Stokes equations. This allows fast predictions of pressure distributions and can be usefully employed at the preliminary design stage. Viscous effects can be included by integrating boundary layer methods, and displacing the surface of the aerofoil/wing/aircraft by an amount equal to the local boundary layer displacement thickness.

Full potential flow equations. The full potential flow equations assume that the flow is irrotational. Compressible flows can be modelled, but the 'shocks' that are predicted are isentropic. The method is now quite dated, but can provide very rapid information about pressure distributions, and can, like the Euler equations, be integrated with a boundary layer method.

Panel method. This is simplest of all numerical methods for predicting flow about aircraft and may be seen as a precursor for CFD. The surface of the aircraft is covered with panels, each one a source or sink and some of them, for example, those on lifting surfaces, assigned a bound vortex (with its associated trailing vortex system). The strengths of the sources and bound vortices are initially unknown, but can be determined through the application of the boundary conditions, for example, flow tangency at solid surfaces.

Descending through the hierarchy, the methods provide less and less physical fidelity, but require less and less computational effort. It is conceivable that the panel methods, full potential flow equations, Euler equations and RANS methods could be used in an undergraduate aircraft design project (as a separate task), though not at the conceptual design stage. These methods would give a qualitative pressure distribution pattern to help shape the geometrical details. Whatever method is used, the issue of grid generation must be addressed: more time will be spent on the generation of a suitable mesh than on the prediction of the flow.

A 3D model created in CAD would prove useful at this stage. There should be some planning to prepare the 3D model in CAD in such a way that Boolean operations can build the aircraft model from isolated components, while retaining the isolated components for separate analysis. The wing-fuselage analysis would give tailless pitching moment data, which is useful information to design aircraft horizontal tail and its relative setting with respect to the fuselage to minimise trim drag.

CFD results could be compared with the results obtained through using the semi-empirical relationship, for example, drag (Chapter 9). Generally, it is considered that semi-empirical drag results give good accuracy validated on many aircraft consistent over a long time of usage.

Figure 9.9 presents the wave drag, C_{Dw} for the Mach number. CFD gives a fine opportunity to generate a more accurate viscous independent wave drag versus Mach number. Once the CFD results are available then the data in Figure 9.9 may be replaced and hence a further iteration on the drag polar of the aircraft is obtained. CFD is also a good place to generate ΔC_{Dp} values and can be used for comparison. In general, CFD generated ΔC_{Dp} values should give good values provided the CFD is set up properly.

If the CFD results come within 10% of the results obtained using semi-empirical relations then it may be considered good. Some tweaking of the CFD runs should improve the result – this is where experience helps. Once the CFD is set up to yield good results then it would prove useful to improve and/or modify aircraft configuration through extensive sensitivity studies. The spectrum plots in colour would show hot spots contributing to drag (local shocks, separation etc). These details cannot be seen so easily by other means. Designers can then follow through repairing the hot spots to reduce drag. The opportunities are unique to CFD making it an indispensable tool to optimise configuration for minimum drag.

If there is a large difference between CFD and semi-empirical results then it should be investigated properly.

23.7 Summary of Discussions

CFD simulation is a digital/numerical approach to design incorporated in the virtual design process using computers. The current status is adequate enough to make comparative analysis at low cost and time and therefore must be applied in the early conceptual design phase as soon as a CAD 3D model drawing is available. The development of CFD is not necessarily driven by the aerodynamic considerations by itself but driven by the requirements for having a tool to design a better product for lower cost and less time.

CFD continues to develop with greater computing power at lower costs and less time along with the advances made in the algorithm to tackle solutions offering considerable ease and automation for the users to benefit from. While researchers have achieved a good degree of accuracy in drag prediction for a clean aircraft configuration, its generalised application by engineering users is yet to achieve consistency in the results. Verification and validation of the results from CFD analysis are essential for substantiation and the state-of-the-art is still under scrutiny and continuing to develop. Verification of new CFD software comes before validation and together they involve a protracted process in which research continues.

CFD analyses supplement wind-tunnel tests and they work together to reduce the time and cost frame of a new aircraft project. CFD analyses precedes wind-tunnel tests. The arguments or considerations (pros and cons) of each type of study is briefly described next.

23.7.1 CFD Analyses

Cons. Results are sensitive to the type of software used and how the preprocessor (e.g. grids) is prepared. Errors and discrepancies involved are hard to explain. Analyses of CFD results considerable design experience to correlate with real situation. CFD requires test data (wind tunnel/flight tests) to verify and validate post processed output result. As mentioned before.

Pros. CFD can provide a reasonable understanding of the flow physics quite fast and at low cost. It can select a better configuration from a large number of parametrically varied cases to consider so narrow down the choice, even when as quoted above (Section 23.3) that 'has not yet had much effect on the rest of the overall airplane development process'. Accurate modelling of turbulence is under developmental process. Yet CFD analyses can narrow down the choice to reduce the number of wind tunnel tests required. The number of many perturbed wind tunnel aircraft models required to make parametric optimisation is considerably reduced.

23.7.2 Wind Tunnel Tests

Cons. Expensive and time consuming.

Pros. Test results are realistic, quantifiable as close it can to the actual aircraft capability and errors can be assessed. This assures aircraft capability to plan flight test in a safer manner. CFD helps to minimise the number tests required to get complete information a sought at that phase.

23.7.3 Flight Tests

Flight test culminates the design process in the progressive manner starting with CFD analyses at the conceptual design phase when not much information is available. Flight tests are the final result of the aircraft capability in reality and substantiated for certification.

Industry as a conservative user needs to ensure fidelity of the design. However, CFD is capable of comparing designs to recognise the better ones even if its absolute values remain under scrutiny. This capability brings out the best compromise at an early phase of the project at low cost and time avoiding subsequent costly modifications of aircraft configuration, that is, it gives the opportunity to make the design right-first-time. The design is subsequently tested in a wind tunnel for substantiation. Today, this approach is a matured one requiring little change to the design after the final flight test results.

Industrial effort in CFD is extensive and is not suitable study for an undergraduate course. However, classroom work can follow the industrial approach tackling smaller problems such as those given in Section 23.5.

References

1 Jameson, A. (2015). *Computational Fluid Dynamics – Past, Present and Future*. USA: Stanford University.
2 Chapman, D. (1979). Computational aerodynamics development and outlook. *AIAA Journal*.
3 Chapman, D. (1992). A perspective on Aerospace CFD. Aerospace America.
4 Roache, P.J. (1998). *Verification and Validation in Computational Science and Engineering*. ISBN: 0-913478-08-3.
5 AGARD AR 256. (1989). Technical status review on drag prediction and analysis from computational fluid dynamics: State-of-the-art.
6 Melnik, R. E., Siclari, M. J., and Marconi, F. (1995). An overview of a recent industry effort at CFD Code Validation. AIAA pare 95–2229, 26th AIAA Fluid Dynamic Conference.
7 Thangam, S. and Speziale, C.G. (1992). Turbulent flow past a backward-facing step: a critical evaluation of two-equation method. *AIAA Journal* 30 (5).
8 Berman, H, Anderson, J. D. and Drummond, P. (1982). A numerical solution of the supersonic flow over a rearward-facing step with transverse non-reacting hydrogen injection. AIAA paper No: 82–1002.
9 Kundu, A. K. and Morris, W. H. (1997). Effect of manufacturing tolerances on aircraft aerodynamics and cost. WMC 97 Conference paper No: 711–003.
10 Cook, T. A. (1971). The effects of ridge excrescences and trailing edge control gaps on two-dimensional aerofoil characteristics. ARC R&M No: 3698.
11 Forrester, T., Johnson, E.N., Tinoco, N., and Jong, Y. (2005). *Boeing Commercial Airplanes*, 1115–1151. Seattle: Washington Computers and Fluids 34.
12 Computational Fluid Dynamics is the Future. Website, http://cfd2012.com/wings.html (accessed June 2018).
13 Arafat, A. Website, Compressible aerodynamics of commercial aircraft' simulation project. Available at https://www.simscale.com/forum/t/compressible-aerodynamics-of-commercial-aircraft-simulation-project-by-ali-arafat/1142/2 (accessed June 2018).
14 Goulas, I., Stankowski, T., Otter, J., and MacManus, D. (2016). *Aerodynamic Design of Separate-Jet Exhausts for Future Civil Aero-Engines - Part I: Parametric Geometry Definition and Computational Fluid*

Dynamic Approach. Paper No: GTP-15-1538, Journal of Engineering for Gas Turbines and Power, ASME, March 15, 2016.

15 Chen, H. C., Yu, N.J., Rubbert, P. E. and Jameson, A. (1983). Flow simulation of general nacelle configurations using Euler equations. Conference paper No: AIAA 83–0539.

16 Uanishi, K., Pearson, N. S, Lahnig, T. R. and Leon, R. M. (1991). CFD based 3-D turbofan nacelle design system. AIAA Paper No: A91–16278.

17 Laban, M. (2003). Application of CFD to drag analysis and validation with wind tunnel Data. Nationaal Lucht-en Ruimtevaartlaboratorium. Lecture Series, Von Karman Institute for Fluid Dynamics.

24

Electric Aircraft★

24.1 Overview

The application of an electrically actuated drive system has natural appeal for its efficiency, relative simplicity, reliability, maintainability and cleanliness (eco-friendly environmental issues – emission free). It led to the 'More Electric Aircraft (MEA)' initiative progressing towards all-electric aircraft. Electrically powered aircraft awaited its opportunity, as the energy storage and harvesting technologies kept improving to the point when many innovative aircraft designs have emerged, some heading towards mass production of electric aircraft in the class of 'Light Small Aircraft' (LSA). Future operational prospects for bigger commercial electric aircraft are pursued by big industries.

Being not constrained by fighting against gravity, the automobile industry led the way to make successful production electric cars with increasing sales in the last two decades. Prototype electric aircraft started appearing from about that time, demonstrating the potential though still not viable for mass market on account of high cost, weight and low flight durations – the technology was in its developing stages to make breakthrough for mass market. Since the beginning of the twenty-first century, rapid gains were made in energy storage capacity as well as cost and weight coming down, mainly driven by the automotive industry. The electric aircraft programmes gained impetuous to move forwardly rapidly.

Electric aircraft are invariably propeller driven by electric motors. The energy source to power the motor has options to choose mainly from (i) batteries, (ii) fuel cells, (iii) solar cells and so on. There are other options (not dealt in this book) in their nascent stages of development. The achievement of Solar Impulse 2 circumnavigating the Earth, powered only by solar cells, is proof of the potential, even it means having different a kind of electrical energy harvesting source. In fact, battery powered model aircraft were in existence for nearly half a century, but those designs were not scalable to any usable application as the scaled up batteries types would have been too heavy for a realistic design. The first manned electric powered demonstrator had to wait till the mid-1970s. Today, big and small industries, for example, the Vahana project, (tilt-wing VTOL aircraft – partnership with Airbus), Cessna, Extra (Germany) and so on, are investing to move towards commercial ventures. Urban air mobility (mostly helicopter types, not dealt with here) with flying taxis has now emerged and awaits vehicle certification and infrastructure development to make them operational, possibly within a decade. Electric powered Unmanned Aircraft Vehicles (UAVs)/drones are currently operational, both as winged and as helicopter types.

The authors thought of introducing electric aircraft at this stage as the aerodynamic considerations made so far for conventional aircraft design are still applicable. The additional information included here includes the various types electric drive systems, which are different from the mechanical power plants using fossil/bio-fuels dealt with so far. The several options for the electric drive system, including hybrid types, are briefly discussed. More information on the topic is given in Refs [1–7].

★Partly authored by Dr Danielle Soban, Lecturer, Queens University of Belfast, UK.

Conceptual Aircraft Design: An Industrial Approach, First Edition. Ajoy Kumar Kundu, Mark A. Price and David Riordan.
© 2019 John Wiley & Sons Ltd. Published 2019 by John Wiley & Sons Ltd.

24.2 Introduction

Following the scope of electrically powered aircraft as given in the overview, the first task is to describe the types of electricity storage device and harvesting technologies available in the context of today's design obligations. It is followed by the brief description of the prime mover, that is, the electric motor. Then some generalised theories applicable to both conventional and electrical fixed wing aircraft, followed by specific theories for meant for electric aircraft performance analyses.

The core of the design architecture of a purely electric powered aircraft is built around motor(s) driven by some form of electrical storage, most dominantly by packs of batteries with their controllers to manage the system. The Battery Management System (BMS) controls the electricity usage in an efficient way to get best utilisation. Flight duration of a fully charged battery is still short and does not give enough warning time in case there is inadvertent failure. The well proven hybrid electric propulsion system is a realistic proposition for a safer operation. Hybrid electric propulsion is a coupled system with a fossil fuel powered small engine driving a generator to keep not only the battery charged to extend sortie time but also serving as system redundancies. Airbus two-propeller E-Fan aircraft (Figure 24.1) programme built a prototype battery powered concept demonstrator aircraft with 250 V battery capable of one hour sortie duration. The programme then planned to build a hybrid electric propulsion E-Fan2 as production version with three-hour sortie duration.

At this stage of study, significant progress has been achieved in designing LSA type of small electric aircraft for entrepreneurs to venture into manufacturing as marketable product. However, in spite of rapid progress made, the current status may not be considered as adequate for sensible usage; the sortie duration is not enough. The immediate goal is to achieve sortie duration well over one hour. It is interesting to note that the aircraft lands at the same (almost) weight as it is at the takeoff. There is there is no weight loss on account fossil fuel burn as in case of internal combustion engine powered aircraft.

Electric aircraft offers many attractive propositions as compared to the internal combustion engine powered aircraft, for example, zero emission, substantial noise/vibration reduction. An electric motor has a low part count with better proven reliability and maintainability. The most important advantage is that the power output is independent of ambient conditions, it does not degrade with altitude gain unlike internal combustion engines dependent on air inhalation. The promise of low operating cost is demonstrated. These benefits draw attention for military usage as surveillance UAVs operating stealthily in enemy territories. The civil version UAVs can be used as explore inaccessible terrains. Multi-rotor electrically drones are already showing benefits in delivery services and surveillance.

The electric aircraft and the conventional aircraft are compared in Table 24.1.

The singular obstacle to progress is to have high energy storage density. In spite of having lighter motor for the same power, the battery weight is rather high for the energy stored as compared to fossil/bio fuel. Aircraft aerodynamics is a matured level for LSA class aircraft and so are materials/structures technology. Electrical aircraft certification requires different considerations that are still evolving and may prove cumbersome: that is not an obstacle.

Terminologies Used

The following are the specific parameters used in this chapter related to electric aircraft.

E – Stored energy, Wh. This is the capacity of the storage device

E^* – Mass (gravimetric) *specific energy*, Wh/kg (energy storage per unit mass) $= E/m_{storage}$

P – Power of the storage device, W = energy/time = E/h. This is the rate of energy usage (per hour)

P^* – Mass (gravimetric) *specific power*, W/kg, power per unit of mass

E_V^* – Volumetric specific *energy density* content, Wh/L per litre or Wh/m^3 (energy storage per unit volume)

P_V^* – Volumetric power content, *power density* W/L per litre or W/m^3 (power storage per unit volume)

m_{bat} – Battery mass, kg

P_{bat} – Electric power of the battery, W

$\$^*$ – Cost to energy ratio

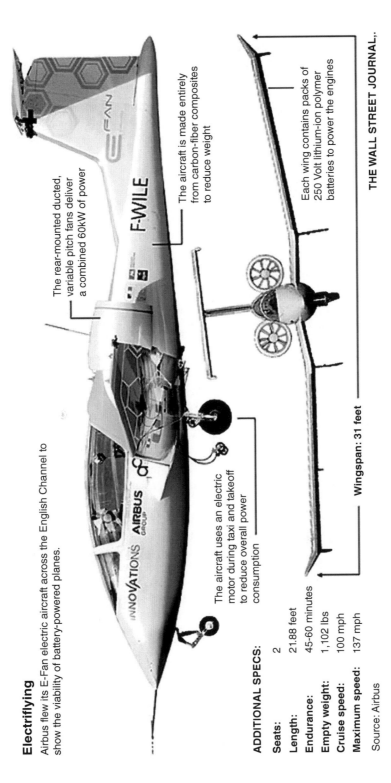

Electrifying

Airbus flew its E-Fan electric aircraft across the English Channel to show the viability of battery-powered planes.

The rear-mounted ducted, variable pitch fans deliver a combined 60kW of power

The aircraft is made entirely from carbon-fiber composites to reduce weight

Each wing contains packs of 250 Volt lithium-ion polymer batteries to power the engines

The aircraft uses an electric motor during taxi and takeoff to reduce overall power consumption

Wingspan: 31 feet

ADDITIONAL SPECS:

Seats:	2
Length:	21.88 feet
Endurance:	45-60 minutes
Empty weight:	1,102 lbs
Cruise speed:	100 mph
Maximum speed:	137 mph

Source: Airbus

THE WALL STREET JOURNAL..

Figure 24.1 Airbus two-propeller E-Fan prototype aircraft (Courtesy: Airbus).

Table 24.1 Battery[a] powered electric aircraft and conventional aircraft are compared.

	Electric aircraft	Conventional aircraft
Prime mover	Electric motor	Internal combustion engine
Prime mover weight	Lighter	Heavier
Energy source	Electricity	Fossil/bio fuel
Energy density	Low, heavy storage	Three to four times higher, lighter storage
Aircraft weight during the mission	Invariant	Reduces with fuel burn
Power variation at constant throttle	Invariant	Reduces with altitude gain
Prime mover part count	Low	High
Maintainability/reliability	Better	Standard
Noise	Very low noise	High
Emission	Zero	Higher
Operation	Limited by battery	In operation

a) Fuel cell powered electric aircraft are same as above except that aircraft weight reduces during the mission.

Energy storage per unit mass, E^*, is the main parameter in designing electric aircraft. It gives aircraft mass and sortie duration, in other words, it acts a measure on the viability of the new electric project. Comparison of E^* of conventional aircraft powered fossil fuel energy with E^* of electric aircraft energy storage gives a good perspective on the differences of system concepts, bearing in mind that unlike conventional aircraft that get lighter with fuel burn, electric aircraft mass does not change. E_V^* gives the volume requirement for the required energy storage. In general, the constraints arising from E^*, is more stringent than E_V^*.

24.3   Energy Storage

The following are the types of electrical energy sources dealt with in this book. Figure 24.2 relates the stored energy to the mass of the electric power source. The Ragone matrices plotted in Figure 24.2a compare the different types.

In order to compare different electric power sources with a battery system, it is necessary to transform their parameters into an equivalent mass specific energy density (mass having equivalent energy) E^* of the complete system. First, we compare the systems as electric power sources, that is, the chain up to the electric connector where the motor controller would be plugged in.

These specific values are shown in Figure 24.2b for various energy storage systems. It can be seen that even the most advanced current battery storage systems fall short of the parameters of Kerosene. While the factor in specific volume is only about 18, the factor in mass specific energy density is in the order of 60.

Typical Energy Content

Fuel	Gravimetric	Volumetric
	MJ/kg	MJ/L
Aviation gasoline jet fuel	43.71	31.00
Kerosene	43.28	35.06
Aviation gasoline jet fuel	46.36	36.86
Aviation gasoline	47.30	33.87

Fuel	Gravimetric	Volumetric
(1 kJ = 0.2778 Wh = 0.000 277 8 kWh)		
(1 MJ = 277.8 Wh = 0.2778 kWh)		
	W/kg	W/l
JP4	12 141	11 395
Lithium-ion	110	300
Lithium-sulfur dioxide	170	190
Lithium-ion polymer	130–1200	300

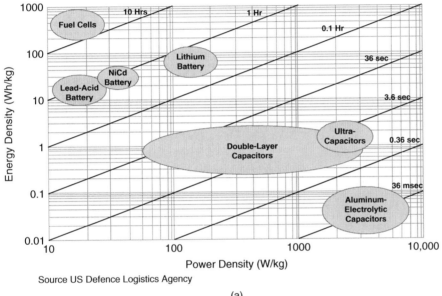

Source US Defence Logistics Agency

(a)

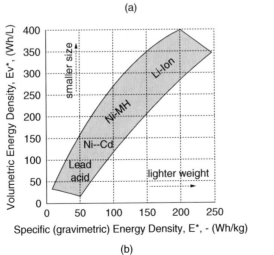

(b)

Figure 24.2 Comparison of different types of systems. (a) The Ragone plot (Courtesy: NASA). (b) Energy density. Reproduced with permission from NASA.

24.3.1 Lithium-Ion Battery

Figure 24.3 shows that lithium-ion offers the best discharge compared to the other kinds mentioned before. Over the time of operation, the voltage decays ±15% from the mean voltage V_{ave}, taken for estimation purposes. The battery is no more useful when the voltage reaches the 'cut-off' point when it can be considered empty, ready to be recharged.

The Boeing787 Dreamliner's aircraft suffered from lithium-ion battery problems at the initial stages of operation. The batteries were getting over heated. causing local fires. The problem was promptly attended to and is now resolved. Lithium-ion batteries are widely used on the consumer market, in defence, aerospace and so on.

Recharging: Lithium-ion batteries can be recharged for more than 10 000 times, but this degrades their capability over time. On colder days, the battery discharges faster when used but stores for longer times when it is not used.

24.3.2 Fuel Cells

In a typical fuel cell, gaseous hydrogen and oxygen are supplied to the anode (−ve) and cathode (+ve) terminals, respectively (Section 24.5.2). Fuel cells convert chemical energy stored in liquid or gaseous form of fuel and oxidants into electrical energy through chemical reactions of positively charge hydrogen ions with some oxidising agent such as an electrolyte. Fuel cells do not require recharging. Their electromotive force is harnessed through the controller as long as fuel flow (FF) continues.

24.3.3 Solar Cell

A photovoltaic (PV) cell converts sunlight photon bombardment directly into electrical energy through a semi-conducting material that has photovoltaic properties. It is the physical property of the PV material and there are many types. The most prevalent material is crystalline silicon (c-Si) is, but the better and more expensive material is copper indium gallium selenide. It produces clean energy without any emission, that is, no pollution.

A PV cell is a small unit, smaller than $200\ cm^2$, producing less than 2 W at its maximum when the Sun's rays are normal to the surface. PV cells have to be stacked in series in a module to yield ≈200 W of power at the maximum. The modules are then arranged in arrays to form a panel to generate sufficient power. Finally, the panels are set in arrays to fit the surface of where it has to be used, in case of aircraft, invariably most of them are on the upper surface of wing and any other suitable place.

Being solely dependent on the Sun, it is dependent on ambient conditions. That means the highest energy harnessing is at noon, developing from a low value in the mornings and gradually reducing to zero at night.

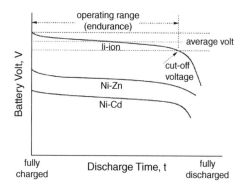
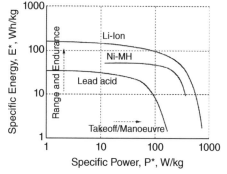

Figure 24.3 Lithium-ion battery characteristics.

Therefore, batteries are required to manage the solar generated electrical energy, storage of any excess production is to be consumed when deficiency arises or it may be used for direct power for a short duration.

24.4 Prime Mover – Motors

The prime mover for aircraft is a brushless DC motor (abbreviated as BLDC) converting electricity power (P_{elec}), mainly from batteries to mechanical power (P_{shaft}) (an AC motor can be used with an inverter). The mechanical power is obtained through the rotating shaft (rpm, ω) developing the torque (τ) required to drive the propeller to generate the thrust required to propel the aircraft. There is a small amount of drive involved in this transmission chain expressed as drive efficiency, η_{drive} that has a high value in the order of 0.95.

$$\eta_{drive} = (P_{shaft}/P_{elec})$$

$$\text{Torque, } \tau = (P_{shaft}/\omega) = (\eta_{drive} \times P_{elec})/\omega \tag{24.1}$$

where P_{elec} = (volt, V) × (current, I).

The efficiency η_{drive} varies. At design load it can have a value as high as 0.95, but at part load it drops off. A controller is required to manage utilisation as efficiently as possible by regulating the battery voltage and amperage.

There are many advantages of having a BLDC motor. This kind of motor design has matured to produce high reliability and good maintainability. A BLDC motor for aircraft propulsion can be designed to have its rated rpm the same as the designed rpm of a matched propeller, thereby eliminating the reduction gear typically associated with internal combustion engines (IC Eng). The following table compares BLDC motors with internal engines considered as the standard.

	BLDC motor	IC Eng
Specific weight (power per unit weight)	>2 × (IC Eng)	Standard
RPM (requirement for a reduction gear)	Requires	Not required
Noise	Silent	Noisy
Emission	Zero emission	Has greenhouse effects
Maintenance cost	Very low	Standard
Reliability	Very high	Standard
Replacement	Fast	Standard
Vibration	Very low	Standard
Operation	Easy[a)]	Standard
Operating cost	Low	Standard

a) Instant restart when required, no mixture control.

However, recharging time for batteries is longer but this is not an energising time, so is not a BLDC motor issue. The BLDC motor specific weight is more than twice that of internal engines, weight saving can be used by heavy battery installation.

24.4.1 Controller

An electric motor requires measured input to provide the required torque. To harmonise the output from electrical storage, a controller is required to manage and distribute electricity accordingly.

24.5 Electric Powered Aircraft Power Train

Electric aircraft are invariably propeller driven by electric motors. Out of several options available, only the power train only options from (i) batteries, (ii) fuel cells and (iii) solar cells are given next. Electric powered aircraft is currently limited to the technology level of power supply (energy) sources. Battery technology has advanced to the point when small recreational class of aircraft are getting ready to enter the market in commercial scale. To move to the next level of larger aircraft for business application, hybrid electric propulsion suits better. Electric powered aircraft have arrived to stay and grow with a potential to compete with conventional fossil fuel powered aircraft. Large aircraft may have to wait more than a decade to make product viable to compete in the market. Electric powered aircraft have many advantages over conventional aircraft.

24.5.1 Battery Powered Aircraft Power Train

A battery cell powered electric aircraft power train is shown in Figure 24.4. The most prevalent type of power aircraft propulsion system is of the lithium type and there several types. Others are nickel-cadmium (NiCad) and nickel-metal hybrid (NiMH). All three types are rechargeable.

A cell stores electrical energy in chemical compounds and is stacked in batteries. The stored energy in the compound can converted to mechanical work such as driving a propeller. Some form of BMS is required to control the energy release from the available energy stored. Its storage capacity is rated by the watt-hour per kilogram (Wh/kg) and the power rating is by watts per kilogram (W/kg). Battery powered system weight stays invariant (see later) from fully charged to fully discharged state. It is emission free and has a low noise level emanating from an aircraft/propeller reselected aerodynamic source.

Current types lithium of batteries do not adequately provide the energy required for practical usage but are progressing towards generating some appeal to make them marketable to a small extent. With the promise shown, these types of batteries are undergoing intense development. There are several types of lithium batteries. The prevailing ones are lithium-ion and lithium-polymer types, similar to those used in the consumer market, giving, comparatively, the highest mass specific power and energy. Cells are stacked in large sizes to give a mass specific energy content of up to of $250 \, \text{Wh} \, \text{kg}^{-1}$. Batteries under development are expected to yield an approximately 10% increase in mass specific energy. Other types in developmental stages include the lithium-sulfur type offering several times more mass-specific energy than the current ones.

24.5.2 Fuel Cell Powered Aircraft Power Train

The fuel cell powered electric aircraft power train is shown in Figure 24.5.

Hydrogen is stored externally in a tank, acts as *energy carrier* and is not a fuel. Oxygen can be harnessed from the atmosphere. Energy release in fuel cells is not a combustion process. In fuel cells, hydrogen and oxygen react to generate electrical energy. The emission is water and operates at a very low noise level, mostly aerodynamic in origin. Naturally, like conventional aircraft, the system weight decreases as the hydrogen fuel is consumed. It is an efficient process but is complex and hence relatively heavy. Unlike batteries, aircraft range depends on the hydrogen storage capacity. Prototype fuel cell powered aircraft is flown. The Fuel Cell Demonstrator project, led by Boeing, converted the Diamond HK-36 Super Dimona motor glider as a technology demonstrator aircraft and it made its successful first flight in early 2008. However, fuel cell technology is heavy and expensive. It awaits considerable development to make it suitable for operational usage.

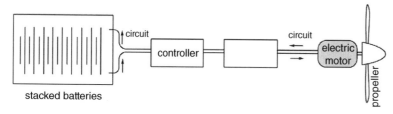

Figure 24.4 Battery powered electric aircraft power train.

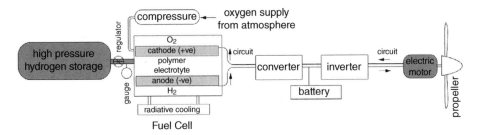

Figure 24.5 Fuel cell powered electric aircraft power train.

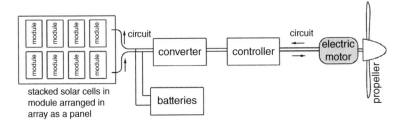

Figure 24.6 Solar cell powered electric aircraft power-train.

24.5.3 Solar Energy (Photovoltaic Cell) Powered Aircraft Power Train

The limiting maximum solar energy available is less than $1000\,\text{W m}^{-2}$ at typical cruise altitude in the stratosphere. In low orbital altitudes, 20–30% more solar energy is available. Figure 24.6 shows a typical scheme for a solar aircraft.

The conversion efficiency from solar energy to electric energy is rather low, current limitation is about 30%. A typical consumer type PV cell has an efficiency $\approx 10\%$ but an aerospace grade PV cell has reached efficiency around $\approx 10\%$.

Evidently, the power output from a single PV cell is very low, therefore it requires a very large area of PV panel coverage. With large wing area and low power output, weight control is a real challenge when designing solar aircraft. Typical solar aircraft wing loading (W/S_W) and thrust loading (T/W) are $\approx 7\,\text{kg m}^{-2}$ and 0.3, respectively. With this kind of W/S_W and T/W the aircraft speed is low, a good speed may be considered to be around 50 km/h.

It is clear that this kind of solar aircraft can only be used in a special mission operating at high altitudes clear of cloud. With readily available solar energy, an aircraft can fly for long distances with well sized and managed batteries, and as a surveillance, communication and navigation platform.

The achievement of the Solar Impulse 2 circumnavigation of the Earth, powered only by solar cells, is a proof of the potential, even it means having different kind of electrical energy harvesting source. Its well managed lithium-polymer battery packs served as backup to fly overnight and getting recharged in the day at cloudless altitudes.

Solar Impulse II Statistics [1]

Crew. 1
Length. 22.4 m (73.5 ft)
Wingspan. 71.9 m (236 ft)
Height. 6.37 m (20.9 ft)
Wing area. 17 248 photovoltaic solar cells cover the top of the wings, fuselage and tailplane for a total area of 269.5 m^2 (rated at 66 kW peak)
Takeoff speed: 36 km h^{-1} (22.4 mph)
Power plant. 4 × electric motors, 4 × 41 kWh lithium-ion batteries (633 kg), providing 13 kW (17.4 hp) each
Propeller diameter. 4 m (13.1 ft)

Maximum speed. 140 km h^{-1} (87 mph)
Cruise speed. 90 km h^{-1} (56 mph) 60 km h^{-1} (37 mph) at night to save power
Service ceiling. 8500 m (27 900 ft) with a maximum altitude of 12 000 m (39 000 ft)

Readers may refer to [2] for more details.

24.6 Hybrid Electric Aircraft (HEA)

The hybrid electric propulsion system is a coupled fossil fuel powered small engine driving a generator to keep a battery charged; not only does it keep batteries charged to extend sortie time but it also serves for system redundancies. The system is well proven in automotive industry and has been in production for more than a decade. HEA followed the car's success. It offers many advantages, for example, it has to have a reliable environment friendly operation at lower cost compared to internal combustion engine powered aircraft. HEA offers the best of both the all-electric aircraft and the piston engine aircraft. Currently, it is being considered for small recreational aircraft usage and a few types have been manufactured. There is good future potential for larger aircraft and development programmes are gaining ground.

Section 24.3 pointed out that the current lithium types of batteries do not adequately provide energy required for practical usage. The prevailing ones are the lithium-ion and lithium-polymer types, similar to those used on the consumer market, giving comparatively highest mass specific power and energy. Cells are stacked to large numbers to give mass specific energy content of up to of 200 Wh kg^{-1}. Therefore, with the typical specific power, it requires large stacking of batteries to get longer sortie endurance. Current market appeal is restricted to small recreational aircraft with sortie duration from 45 to 90 min.

With this energy storage capacity limitation, a hybrid electric propulsion system presents a good compromise for cleaner emissions and lower operating cost with a realistic future to perhaps be installed in larger aircraft for longer endurance to cover a marketable range.

There are three main types of hybrid electric propulsion power train system as shown in Figure 24.7a. The top sketch is the parallel hybrid electric propulsion power train system with an electric motor powered by batteries and the internal combustion engine is coupled through a clutch so the propeller can be driven by either of them or in a shared mode. The size of each type depends on the degree of hybridisation. On the other hand, the middle sketch is the series hybrid electric propulsion power train system where a propeller is powered only by an electric motor and the electrical energy is supplied (i) independently either by the batteries, by the internal combustion engine or by the battery or (ii) together. In this case, an energy storage device is present in the form of capacitors (electric) or a flywheel (mechanical). The last sketch is the hybrid of the hybrid electric system, that is, the series/parallel hybrid electric propulsion power train where the system has the option of operating each system independently of the other, or as a hybrid engine with power sharing to the extent as planned. This is the best of the three but is the heaviest and costliest. With proven reliable components, the parallel hybrid electric propulsion power train system suits small recreational aircraft applications.

Hybrid ratio, HR, is defined as the ratio of electric power supplied to the total power total power supplied as follows.

$$HR = \frac{P_{electric}}{P_{total}}$$

The hybrid system is identified at its design rated value is at the rated maximum,

$$HR_{rated} = \frac{P_{max\_electric}}{P_{max\_total}}$$

The most flexible arrangement is the series/parallel system when the HR can be controlled as required. It is also the most complex, heavy and naturally most costly. Typically, the current usage recreational type aircraft is the parallel hybrid system. Figure 24.7b is the Rotax914 115 HP (87.5 kW) piston engine as a parallel

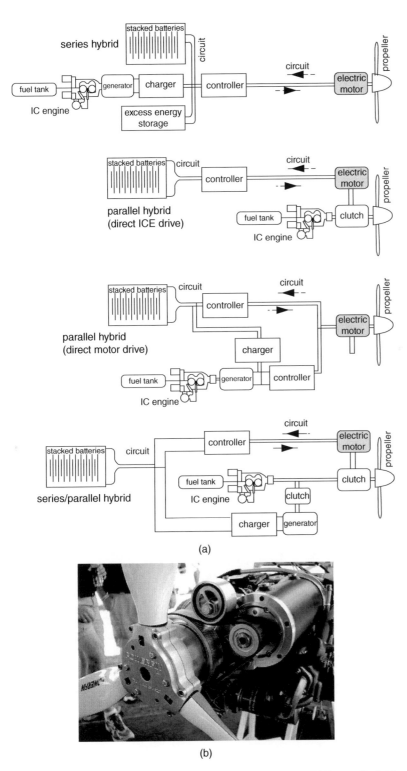

(a)

(b)

Figure 24.7 Hybrid engine powered electric aircraft power train. (a) Types of hybrid propulsion system. (b) Rotax hybrid engine. Reproduced with permission from Rotax.

hybrid engine with a 30 kW brushless DC motor driving a three-bladed 'Powerfin' propeller, appeared in 2009 Oshkosh EAA AirVenture air-show. The Rotax912 hybrid engine, as shown in Figure 24.7b, has $HR_{rated} = 0.255$ with total power available being 155 HP. The electric motor is used to supplement power, especially at takeoff that demands maximum power, thereby the IC engine is sized smaller to make overall economic gains.

24.7 Distributed Electric Propulsion (DEP)

A distributed propulsion system has many advantages, especially for wing-mounted propeller-driven aircraft. It covers more propeller slipstream blown wing surfaces generating higher lift for the same aircraft speed without wing mounted engines. In a way, the four-engine propeller-driven LockheedC130 aircraft can be said to have a distributed propulsion system. However, gas turbines are heavy, expensive and complex systems that override the benefits of distributed internal combustion type distributed engine installations.

On the other hand, electrical motor driven propellers are lighter, simpler, reliable, environmentally friendly, easy to maintain and, as well cost effective, these present a new concept of propulsion system by having a large

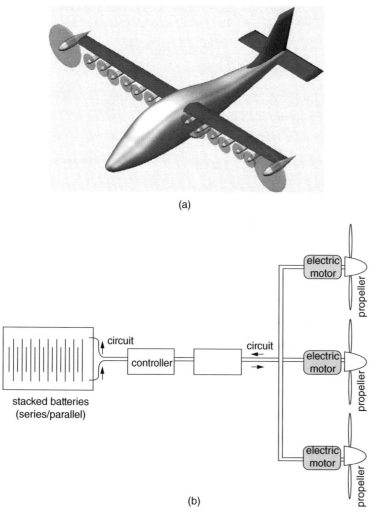

(a)

(b)

Figure 24.8 Distributed electric propulsion (DEP) system. (a) NASA Sceptor X-57 plane (Courtesy: NASA). (b) DEP schematic diagram.

number of electric motor units with the propeller distributed span-wise, as shown in Figure 24.8. The hybrid distributed propulsion system offers good potential at the current state of development.

Each of the individual tractor propeller slipstreams covers the entire wing span. The NASA Sceptor (*Scalable Convergent Electric Propulsion*) X-plane [3] in Figure 24.8a is a good example to study. It has distributed tractor propeller slipstreams that cover the entire wing span and at the tip it is configured with a bigger motor and propeller.

24.8 Electric Aircraft Related Theory/Analyses

The external aerodynamic considerations for conventional aircraft and electrically powered aircraft are essentially the same. The difference lies in the power train. This section presents the associated theories and analyses associated with electric aircraft design considerations in an introductory manner. (Section 24.2 gives some of the symbols used in this section.)

Conversion

1 kW = 1.341 HP.
1 MJ = 0.277 78 kilowatt hour.

24.8.1 Battery and Motor

Battery

There are two types of batteries, the (i) high power and (ii) high energy battery. Primary batteries are not rechargeable but secondary ones are. The smallest unit is a cell and number of these are stacked in series or in parallel to make a battery pack.

Total energy of a battery cannot be consumed as some residual stays unusable. Depth of Discharge (DOD) is expressed as a percentage of the total energy discharged. Battery – 100% DOD means fully discharged, which is to be avoided to extend the number of battery recharging cycles. About 80% DOD is a deep discharge. At this level of DOD a good battery should be able to make more than 1000 cycles of recharging. The current level of battery charge is known as the State of charge (SOC).

Usable battery energy content, E_{bat} = battery mass (m_{bat}) × mass specific energy content (E^*) × DOD × η_{bat}.

$$= \text{power}, P_{bat} \times \text{time}, t_E \tag{24.2}$$

where time, t_E, is the endurance of the battery, that is how long it lasts for the rated DOD.and η_{bat} = inherent characteristics of associated losses that can incur.

Brushless DC motor (BLDC)

Equation (24.1) gives, Torque, t = (P_{shaft}/ω) = $(\eta_{drive} \times P_{elec})/\omega)$, where η_{drive} is drive efficiency and $P_{elec} = (V) \times (I)$.

Typically, loss in electric system is low, simplifying it is considered here, $\eta_{elec} = 0.95$ (all inclusive).

24.8.2 Conventional Aircraft Performance

Electric aircraft are invariably propeller driven. This book deals only with flying at low subsonic flight below 10 000 ft. altitude. Therefore, their parabolic drag polar values are fairly accurate as expressed in Equation (11.4), given again next.

$$C_D = C_{Dpmin} + kC_L^2$$

where $k = 1/(e\pi AR)$.

For equilibrium steady level flight, Aircraft Drag, D = thrust required, T_{req}.
and in steady level flight, lift, L = aircraft weight, W.

Then drag in steady level flight can be expressed in terms of aircraft design characteristic, (L/D) ratio

$$\text{that is,} \ D = D \times (W/W) = D \times (W/L) \tag{24.3}$$

Power required by the aircraft is expressed as

$$\text{Drag } (D) \times \text{ aircraft velocity } (V) = DV = (THP)_{req} \text{ (thrust HP required)} \tag{24.4}$$

(For battery powered aircraft P_{ac} replaces the term THP to relate with the battery power P_{bat})
In coefficient terms, substitute $D = 0.5 \ \rho V^2 S_W C_D = q S_W C_D$ into Equation (24.4)

$$(THP)_{req} = V \times D = (0.5 \times \rho \times V^2 \times S_W \times C_D)V$$

$$= [0.5 \times \rho \times V^2 \times S_W \times (C_{Dpmin} + C_L{}^2 k]V$$

$$= 0.5 \times \rho \times V^3 \times S_W \times \ (C_{Dpmin}) + k \times 0.5 \times \rho \times V^3 \times S_W \times C_L{}^2 \tag{24.5}$$

This is the same equation as Equation (15.46).
FF rate can then be expressed as

$$FF = psfc \times (HP)_{reqd} = psfc \times (T_p)_{reqd}/\eta_{prop} = psfc \times DV/\eta_{prop} \tag{24.6}$$

where '*psfc*' is the power specific fuel consumption for a propeller-driven aircraft.with Imperial units, fuel is consumed in lb. per hour per unit of HP produced, that is, lb./HP/h.

Note that aircraft velocity, $V = \sqrt{\frac{(W/S_W)}{(0.5 \times \rho \times C_L)}}$. In steady level flight (typically in cruise), $L = W$.
In terms of Imperial units, the derivation continues as follows.

$$FF = psfc \times (HP)_{reqd} = (psfc \times (TV)/(550 \times \eta_{prop}) \text{ units in lb/h} \tag{24.7}$$

In equilibrium flight thrust, $T = $ drag, $D = (WC_D/C_L)$.
On substituting in Equation (24.7)

$$FF = \frac{V \times psfc \times C_D \times W}{500 \times \eta_{prop} C_L} = \frac{psfc \times C_D \times W}{500 \times \eta_{prop} \times C_L} \sqrt{\frac{(W/S_W)}{0.5 \times \rho \times C_L}} \tag{24.8}$$

The Breguet range equation derived in Equation (15.42) for jet aircraft is given here.

$$R_{cruise} = \sqrt{\frac{2C_L}{\rho S_W}} \left(\frac{1}{sfc \times C_D} \right) \int_{W_f}^{W_i} \left(\frac{dW}{\sqrt{W}} \right) = 2\sqrt{\frac{2C_L}{\rho S_W}} \left(\frac{\sqrt{W_i} - \sqrt{W_f}}{sfc \times C_D} \right) \tag{24.9}$$

24.8.3 Battery Powered Electric Aircraft Range Equation

Typically, propeller driven battery powered electric aircraft range is short and cruise is carried out at a constant altitude (ρ constant) and constant speed (η_{prop} constant for the propeller). During the sortie, the aircraft weight power does not change and the battery power is unaffected by changes in altitudes but can be affected by the temperature change, which is neglected in this introductory derivations. The range equation is derived in a slightly different approach.

In electric aircraft terminology, using Equation (24.3) and replacing THP with P_{ac}, Equation (24.4) can be written as follows.

$$\text{Aircraft propulsive power delivered by the propeller, } P_{ac} = V \times D = V \times W/(L/D), \tag{24.10}$$

where P_{ac} is the power delivered by the propeller.

The sortie duration is dependent on how long the battery charge lasts, expressed in time 't'.
Equation (24.2) gives the total disposable charge, that is, the energy content $E_{bat} = m_{bat} \times E^* = P_{bat} \times t_E$.

Therefore, the sortie duration, $t = (m_{bat} \times E^*)/P_{bat}$.

Aircraft range, Range, R = velocity × time = $V \times t$.

Taking 10% energy to climb (decent is in idle power), this equation reduces to,

$$\text{Aircraft range, } R = V \times (0.9 m_{bat} \times E^*)/P_{bat} \tag{24.11}$$

The will be some loss in the power train, including that of the propeller producing the thrust, T, to balance the aircraft drag, D.

Define the power train efficiency, η = (propulsive power, P_{ac})/(battery power, P_{bat})

$$\text{or } P_{bat} = P_{ac}/\eta \tag{24.12}$$

Combining Equations (24.2)–(24.6), the following relation is obtained.

$$P_{bat} = V \times W/(\eta L/D) \tag{24.13}$$

Substituting Equation (24.13) into the range Equation (24.11) the following relation is obtained

$$R = V \times (0.9 m_{bat} \times E^*)/P_{bat} = V \times (0.9 m_{bat} \times E^*)/[V \times W/(\eta L/D)]$$

$$\text{Rearranging, } R = [0.9\eta \times E^* \times (m_{bat} \times L/D)]/(W) \tag{24.14}$$

$$\text{or } m_{bat} = \frac{R \times W \times D}{0.9\eta \times E^* \times L} \tag{24.15}$$

Evidently, bigger the battery, the higher the range incurring weight growth.

Case-I: If range, R, is specified then the battery size can be determined.

Case-II: If the battery size ($E^* \times m_{bat}$) is specified then the battery size can be determined.

The wing loading W/S_W, C_L and C_D remain constant.

24.8.4 Fuel Cell Powered Electric Aircraft Range Equation

The Breguet range equation (24.6) is modified for a fuel cell powered propeller driven electric aircraft. Thrust obtained from the propeller is the propulsive force for the aircraft. Retaining the electric aircraft terminologies: the relation between required thrust power (P_{ac}) and available electric aircraft power ($P_{fuel\ cell}$) is related with the propeller efficiency, η_{sys}, which includes propeller efficiency, η_{prop}, which is given next.

$$P_{ac} = P_{fuel\ cell} \times \eta_{sys} \tag{24.16}$$

Let FF_H be the hydrogen FF and *hsfc* be the hydrogen specific fuel consumption, the range Equation (24.8) uses the term V/FF_H and can be written as follows.

$$V/FF_H = 500 \times \eta_{sys}/(hsfc \times D) = 500 \times \eta_{sys}/[hsfc \times (P_{ac})_{reqd}] = \frac{500 \times \eta_{sys} C_L}{hsfc \times C_D W} \tag{24.17}$$

At any instant, rate of aircraft weight change, ($-dW$) = rate of fuel burnt (consumed).

A negative sign indicates aircraft weight reduction; in terms of fuel consumed, the negative sign is removed. In an infinitesimal time dt, the infinitesimal amount of fuel burnt, $dW = FF \times dt$

$$\text{or } dt = dW/FF_H \tag{24.18}$$

$$\text{and elemental range, } ds = V \times dt = V \times dW/FF_H \tag{24.19}$$

Substituting Equation (24.17) in Equation (24.19)

$$ds = \frac{500 \times \eta_{sys} C_L \times dW}{hsfc \times C_D \times W} \text{ft} \tag{24.20}$$

On integrating Equation (24.20) between any two weights, the distance covered can be obtained for propeller-driven aircraft range as follows. Typically, the integration limits are initial cruise weight W_i and final cruise segment weight W_f.

$$\text{Range}, R = \int_{W_i}^{W_f} ds = \int_{W_i}^{W_f} \left(\frac{550 \times \eta_{sys} \times C_L}{hsfc \times C_D} \right) \left(\frac{dW}{W} \right)$$

$$= \int_{W_i}^{W_f} \left(\frac{375 \times \eta_{sys}}{hsfc} \right) \left(\frac{dW}{W \sqrt{\left(\frac{0.5}{\rho} \right) \left(\frac{W}{S_w} \right) \frac{C_D^2}{C_L^3}}} \right) \text{miles} \tag{24.21}$$

Fuel cell powered propeller-driven electric aircraft experience decaying of weight as the stored hydrogen gets consumed as cruise progresses. Then, for the approximation considered before, the Equation (24.20) can be written as

$$\text{Range}, R \text{ in miles} = \int_{W_i}^{W_f} ds = \left(\frac{375 \times \eta_{sys}}{hsfc \sqrt{\left(\frac{0.5}{\rho} \right) \left(\frac{W C_D^2}{S_w C_L^3} \right)}} \right) \int_{W_i}^{W_f} \frac{dW}{W} \tag{24.22}$$

24.9 Electric Powered Aircraft Sizing

Aircraft sizing is an aerodynamic analysis to determine the only two parameters, (i) wing loading (W/S_W) and (ii) thrust loading, (T/W), which will simultaneously satisfy the civil aviation aircraft performance specifications of (i) takeoff field length (TOFL), (ii) initial climb rate, (iii) high speed cruise (HSC) capability at its initial cruise weight and (iv) landing field length (LFL). The sizing analyses do not require consideration of the type of propulsion used, they are assigned for the new aircraft project specifications. After the sizing point of (W/S_W) and (T/W), the designers will have to substantiate that the propulsion system is capable of meeting the thrust requirements for the full flight envelope.

The sizing methodology subsonic jet aircraft presented in Chapter 14 is valid for electric propeller-driven aircraft but requires some modifications to suit the electrical power train. The thrust loading parameter T/W is to be replaced by the power loading parameter, P/W. The relation between thrust (T) and power (P) can be obtained from Equation (24.2).

Battery powered aircraft, for equilibrium flight, $T = D$.

Equation (24.10) gives, $P_{ac} = V \times D = V \times T$.

or $T = P_{ac}/V$ (T in N and P_{ac} in W, $1\,W = 1\,N\,s^{-1}$).

Since both weight and W are symbolised by W, to avoid confusion the weight is expressed as 'mg' instead of W.

Takeoff sizing

Equation (14.6b) gives in system international (SI) system ($\rho = 1.225\,\text{kg m}^{-3}$ and $g = 9.81\,\text{m s}^{-2}$)

$$(mg/S_W) = 8.345 \times (TOFL \times (T/mg)(1 - D/T - \mu \times mg/T + \mu L/T) \times C_{Lstall})$$

Replace T with P_{ac}/V

$$(mg/S_W) = 8.345 \times (TOFL \times (P_{ac}/mg \times V)(1 - DV/P_{ac} - \mu \times mg \times V/P_{ac} + \mu LV/P_{ac}) \times C_{Lstall}) \tag{24.23}$$

Here, (mg/S_W) is in N/m² to remain in alignment with the units of thrust in N.

Initial climb sizing

Equation (14.15) gives, in the SI system

$$[T_{SLS}/mg] = k_{cl} \times RC/V + k_{cl} \times [(C_D \times 0.5 \times \rho \times V^2)S_W/W], \text{ where } k_{cl} = T/T_{SLS}$$

Rewriting in terms of electric aircraft

$$[P_{ac\_SLS}/mg] = k_{cl} \times RC/V + k_{cl} \times [(C_D \times 0.5 \times \rho \times V^2)S_W/mg], \tag{24.24}$$

HSC capability at its initial cruise weight sizing

Equation (14.19b) gives, in the SI system

$$T_{SLS}/mg = k_{cr} \times 0.5\rho V^2 \times C_D/(W/S_W)$$

Rewriting in terms of electric aircraft

$$[P_{ac\_SLS}/mg] = k_{cr} \times 0.5\rho V^2 \times C_D/(mg/S_W) \tag{24.25}$$

LFL sizing

V_{app} for landing is at 1.3 V_{stall} and at zero thrust.
where $V_{app} = 1.793 \times \sqrt{[Wg/S_W)/(\rho \times C_{Lstall})]}$, from Equation (14.21).
Equation (14.22) gives, in the SI system

$$\text{or } (mg/S_W) = 0.311 \times \rho \times (V_{app})^2 \times C_{Lstall} \tag{24.26}$$

Cruise equation to estimate range

Writing in terms of coefficients,

$$P_{ac} = \sqrt{\frac{(W/S_W)}{(0.5 \times \rho \times C_L)}} \times W/(C_L/C_D) = (W/S_W)/(C_L/C_D) \tag{24.27}$$

Dividing both the sides by S_W and collecting the terms, the sizing parameter, power loading,

$$(P_{ac}/S_W) = (W/S_W)/(C_L/C_D) \times \sqrt{\frac{2 \times (W/S_W)}{(\rho \times C_L)}} = \sqrt{\left(\frac{2}{\rho}\right)} \times \left(\frac{W}{S_W}\right)^{3/2} \times \left(\frac{C_D}{C_L^{3/2}}\right) \tag{24.28}$$

Then the sizing parameter of wing loading (W/S_W) can expressed terms of battery power loading $((P_{bat}/S_W)$, as follows.

$$(W/S_W) = \left(\frac{\rho}{2}\right)^{1/3} \times \left(\frac{P_{bat}}{\eta \times S_W}\right)^{2/3} \times \left(\frac{C_L}{C_D^{2/3}}\right) \tag{24.29}$$

where η is the efficiency factor on account of losses in electromechanical power train from battery to propeller. Reference [4] gives a generalised discussion on the sizing of electric powered aircraft.

Example – Converted Electric: Bizjet

Just for a case study, impractical as it may come out, consider the conventional sized Bizjet converted to all-electric battery powered aircraft by replacing the gas turbine power train with an electrical power train.
The sizing point of Bizjet aircraft (Figure 14.3) gives $mg/S_W = 3052$ N m^{-2} and $[T_{SLS}/mg] = 0.34$.
This gives Bizjet maximum takeoff mass (MTOM) = 9400 kg and T_{SLS}/engine = 3500 lb. (15 569 N/eng).
$C_L = 1.9$ at 20° flap and 2.1 at 40° flap.

Table 24.2 Bizjet-electric takeoff sizing.

	Computing and listing in tabular form:				
mg/S_W (SI – N/m^2)	1916.9	2395.6	2874.3	3353.7	3832.77
V – m/s	42.15	47.12	51.62	55.76	59.6
P_{ac}/mg – W/kg	8.17	11.41	15.0	18.9	23.0

Takeoff sizing

Equation (24.22) is reduced to $mg/S_W = 4.173 \times 1341 \times 1.9 \times (0.93 \times P_{ac}/mg \times V) = 9889.2 \times (P_{ac}/mg \times V)$. or P_{ac}/mg – W/kg $= (mgV/S_W)/9889.2$.

Note: in the SI system, V becomes ($\rho = 1.225$ kg m^{-3} and $g = 9.81$ m s^{-2})

$$V = \sqrt{\frac{(mg/S_W)}{(0.5 \times \rho \times C_L)}} = \sqrt{\frac{(mg/S_W)}{(0.5 \times 1.225 \times 1.9)}} = \sqrt{\frac{(mg/S_W)}{1.16375}} = 0.927 \times \sqrt{(mg/S_W)}$$

Figure 14.3 gives the Bizet engine sizing value of $T_{SLS}/mg = 0.34$ at a wing-loading of $W/S_W = 64$lb/ft^2 (2894N/m^2). Converting into battery power, the following can be written (see Table 24.2).

$P_{ac}/mgV = 0.34$, where $V_{stall} = 49.87$m/s or $V = 1.2V_{stall} = 59.84$ m/s

$P_{ac}/mg = 0.34 \times 49.87 = 17$ W/kg

Bizjet $P_{ac} = 9.81 \times 9400 \times 0.017 = 1564$ kW $= 2097$ HP (2 engines)

It is clear that, with the current Li-Ion battery technology level, say at high end of $E^* = 300$ Wh kg^{-1}, it is not feasible to meet a 200 nm range. It will require nearly 10 times more capacity. However, the Bizjet may just be able to do 300 nm if the gas turbine system (1060 kg) and fuel weight (2500 kg), a total of 3560 kg be replaced by battery and motor power train system. [5] gives the Dornier328 turboprop aircraft conversion to a battery powered electric aircraft. It shows that at current battery technology levels ($E^* = 180$ Wh kg^{-1}), the design range reduces to nearly one tenth. With 20% weight reduction and nearly 50% span increase to reduce induced drag, it will require four times higher E^* to get attain two-thirds of the original design range of 2200 km.

Battery powered all-electric aircraft have to wait until battery technology gains ground to make the all-electric battery powered utility type of smaller aircraft marketable to begin with. However, given in Section 24.10.2 is a worked-out example of a small recreational type, with a modest target, which may appeal to market interest.

A better prospect is move in a step-wise fashion with a hybrid type of electric aircraft. The car has established a successful market with a hybrid power train. See Section 24.10 that discusses such a feasibility.

24.10 Discussion

The discussion on electric aircraft is limited to battery powered and hybrid battery powered aircraft as, for some foreseeable future, the photovoltaic solar cell powered aircraft promises for special purpose high altitude platforms are not yet suitable and fuel cell technology is expensive and heavy, lagging behind battery powered aircraft in near future commercial applications. The current limitation of battery specific energy is the main and, strictly speaking the only, constraint; its gravimetric energy density E^* is rather low making an aircraft too heavy for any realistic commercial operation. Battery technology is now well invested in and advancing with encouraging promise for aerospace applications. This technology is now well-established in the automotive industry. Both battery powered all-electric and hybrid cars are in production at progressively increased rates of production.

The battery powered electric aircraft market is struggling, even for recreational small aircraft usage; a sector bearing the best possibility to make commercial gains at this state of technology level. In the near future, short range utility aircraft of less than 500 nm range may enter into the market arena.

Battery powered electric aircraft offer many advantages. Reference [4] compares overall efficiencies of gas turbine propulsion system with electric propulsion system as follows.

	Gas turbine aircraft propulsion		Electric aircraft propulsion (propeller)	
	Turbofan (jet)	Turboprop	Fuel cell	All battery
$\eta_{overall}$	≈ 0.33	≈ 0.39	≈ 0.44	≈ 0.73

Propeller efficiency $\eta_{prop} = 08$ for all systems with a propeller. Gas turbines have high heat rejection in their exhaust gases, losing energy, hence there are lower efficiencies. Losses in electric circuits are low. As soon as the battery technology advances to have E^* over $500\,Wh\,kg^{-1}$, the market opens up for short range utility aircraft and the technology has already achieved E^* around $300\,Wh\,kg^{-1}$. Prediction for batteries with E^* over $1000\,Wh\,kg^{-1}$ has enough confidence to succeed.

Battery powered aircraft will lower the operating cost for lower energy and maintenance costs with better reliability. Other advantages allow a distributed propulsion system that increases aircraft L/D ratio, can easily be made to rotate in opposite directions so the port and starboard sides can produce counterbalance torque and single unit drive installation of contrarotating propellers can be easily achieved. The overall aircraft performance gain is discussed next. Reference [7] covers UAV extensively.

24.10.1 Overall Aircraft Performance with Battery Powered Aircraft

Table 24.1 listed the differences between conventional and battery powered aircraft that affect aircraft performance. The most influential factor is that battery driven motor power is not affected by altitude, that is, its power does not degrade with altitude gain. Also, note that unless it is a hybrid design, aircraft weight also remains invariant as there is no fuel to be consumed to lighten the aircraft as it progresses. Therefore, equations derived in Chapter 15 that reflected weight change need to be modified. For example, the format of range Equation (24.14) is different from the classical Breguet range equation (24.9). Similarly, the climb equation needs to be modified. However, the decent equation is unaffected.

Hybrid aircraft offers the best promise until battery technology replaces IC engines. A rated hybrid ration of $HR_{rated} \approx 0.25$ can be used as a thrust augmenter during takeoff and initial climb, say operating for about 20 min, can get the IC engine (turboprop) size lowered allowing reduced fuel burn, lower weight and have a lower purchase price. A parallel hybrid system with a generator for the IC engine driving only electric motor driven propellers can be arranged in a distributed manner as shown in Figure 24.8. This arrangement increases aircraft L/D ratio that, in turn, can be used for any associated weight penalties. Certification of electric aircraft is evolving.

Given next is a worked-out example of a single seat recreation aircraft to get some idea of electric aircraft design considerations. It basically involves designing the electric power train, the aerodynamic and the structural philosophy remains the same. This book does not deal with electric power train design but only introduces concepts and considerations.

24.10.2 Operating Cost of Battery Powered Aircraft

Unlike conventional fossil fuel powered aircraft, operating cost considerations for battery powered aircraft require different parameters to be taken into account. Battery powered larger commercial aircraft is still not ready, hence operating cost considerations of a small recreational type aircraft is discussed here.

Battery powered aircraft operate at high DOD requiring recharging after each sortie. While the energy cost of battery is fraction of energy cost of aviation gasoline (AVGAS)/aviation turbine (AVTUR) for unit range, aircraft daily utilisation is nearly half of that conventional aircraft on account of long recharging times penalising the 'fixed cost' (see Section 16.4) elements of the sortie. Changing is a relatively time consuming and cumbersome process involving man-hour costs to account for. However, the maintenance cost (about 10% of Direct Operating Cost (DOC) in this class of conventional aircraft) is low. The main attraction is an environmentally friendly clean aircraft with fewer headaches in maintenance.

24.10.3 Battery Powered Unmanned Aircraft Vehicle (UAV)/Unmanned Aircraft System (UAS)

In continuation with Section 21.11.2, fixed winged UAV/UAS are well suited to battery powered aircraft capable of carrying more than three to four times payload for the same power loading at considerably higher speeds. Inability to hover is the only drawback compared with copter-like drones.

24.11 Worked-Out Example

Given here is an example of a recreational battery powered aircraft. Its geometric details, power plant, weight and drag polar are given. The aircraft drag polar and weight estimation were done in an approximate method. The aircraft performance are extracted from comparable aircraft performance. Readers may make proper estimation. The exercise is to find a suitable battery powered power train and examine its feasibility. To get some idea, the example, Equation (24.15) is simplified by considering 100% DOD and the value of power-train efficiency, $\eta = 1$.

Aircraft (AK-E1) Details

Single seat all-composite low wing non-aerobatic aircraft, designated AK-E1, with a tractor propeller
MTOM = 450 kg (990 lb).
Three-bladed propeller diameter = 1.524 m (60 in).
Tapered wing area, $S_W = 9.3\,m^2$ (100 ft²)
Wing span = 8.14 m
Wing taper ratio = 0.5
Flight deck maximum width = 1 m (3.28 ft)
No high lift device
MAC = 1.14 m (3.75 ft)
Aspect ratio = 7.12
Fixed tri-cycle undercarriage
'g' load = +4 to −2
Drag Polar = 0.02 + 0.0526CL2

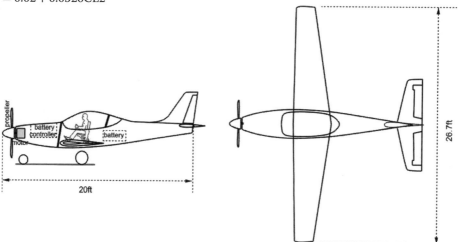

Battery voltage = 300 V, 60 kW (80 HP)

Battery weight = 100 kg (including an additional battery behind the pilot)

Electric motor weight = 40 kg, Continuous power at 2500 rpm

Pilot weight = 80 kg

Aircraft composite structure + systems weight = 230 kg

Power Loading, $P_{ac}/mg = 60/450 = 0.133 \text{ kW kg}^{-1}$ (0.08 HP lb^{-1}) is adequate for the class to give the acceptable performances given next:

Maximum cruise altitude = 10 000 ft. (unpressurised aircraft)

Maximum speed = 100 kt

Sortie duration = 1 h

This gives a range of 400 nm (740 km) – range penalty during climb and descent is ignored

TOFL/LFL = 1000 ft

Solution

Consider the recreational aircraft cruising at 5000 ft ($\rho = 0.002\,04$ slug ft^{-3}) altitude and 100 kt (51.44 m s^{-1}, 168.78 ft s^{-1}).

$$\text{Cruise } C_L = \frac{(990/100)}{0.5 \times 0.00204 \times 168.78^2} = (9.9)/(29.06) = 0.34.$$

$$\text{Cruise } C_D = 0.02 + 0.0526 \times (0.34)^2 = 0.026\,08.$$

$$\text{Cruise } C_L/C_D = 13.$$

Consider a good Li-Ion battery with $E^* = 200$ Wh/$(mg)_{bat}$.

considering 100% DOD and the value of $\eta = 1$.

From Equation (24.15), $m_{bat} = \dfrac{R \times W \times D}{0.9\eta \times E^* \times L} \approx \dfrac{R \times W \times D}{E^* \times L} = \dfrac{740 \times 450}{200 \times 13} = 128$ kg.

With 80% DOD and the value of $\eta = 0.9$, a realistic battery weight becomes $\approx$180 kg.

This will have a cyclic increase in aircraft maximum takeoff weight (MTOW) to reach around 600 kg. Therefore, the endurance of the aircraft has to reduce to around 45 min, which is not bad.

The maximum rated power is 60 kW. For 1 h endurance at average power = 40 kW (no loss is considered at this stage to have a feel on the capability).

Evidently, the aircraft with a 100 kg battery is only be able to fly for about 45 min. If battery E^* can be increased by 40% to 280 kW kg^{-1} then the aircraft can draw market interest. An advanced Li-Ion battery is still very expensive. (Readers may try a hybrid version of this aircraft with, say, half the engine power and double the battery power and capacity.) This book is devoted to conventional aircraft. This example may encourage readers to make an in-depth study of electric aircraft.

References

1 Ross, H. (2009). *Solar Power Aircraft – the True all-Electric Aircraft*. Saville: IBR Aeronautical Consultants, EWADE.

2 Solar impulse. Wikipedia entry: https://en.wikipedia.org/wiki/Solar_Impulse (accessed June 2018).

3 Moore, M.D. *Distributed Electric Propulsion (DEP) Aircraft*. USA: Langley Research Centre, NASA.

4 Hepperle, M. (2012). *Electric Flight – Potential and Limitations*. Braunschweig, Germany: German Aerospace Center.

5 Nam, T., Soban, D., and Mavris, D. (2005). *A Generalised Aircraft Sizing Method and Application to Electric Aircraft*. San Francisco, USA: AIAA.

6 Grundlach, J. (2014). *Designing Unmanned Aircraft Systems: A Comprehensive Approach*, 2e. Reston, USA: AIAA.

7 Isikveren, A.K. (2013). *Future of [More] Electrical Aircraft*. Cape Town, SA: ICAS Biennial Workshop.

Appendix A

Conversions and Important Equations

	Linear			Linear	
1 in.	=	2.54 cm	1 cm	=	0.393 7 in.
1 ft	=	30.48 cm	1 cm	=	0.032 8 ft
1 yd	=	0.914 4 m	1 m	=	1.093 6 yd
1 mile	=	5280 ft	1 m	=	3.280 84 ft
1 mile	=	1.609 3 km	1 nm	=	1.852 km
1 mile	=	0.869 nm	1 km	=	0.621 4 miles
1 nm	=	1.852 km			
1 nm	=	6 076 ft	1 ft	=	0.000 146 nm
1 nm	=	1.150 8 miles	1 km	=	0.54 nm

	Area			Area	
1 sq. in.	=	6.5416 sq. cm	1 sq. cm	=	0.155 sq. in.
1 sq. ft	=	929.03 sq. cm	1 sq. cm	=	0.00108 sq. ft
1 sq. ft	=	0.092903 sq. m	1 sq. m	=	10.764 sq. ft
1 sq. yd	=	0.8361 sq. m	1 sq. m	=	1.196 sq. yd
1 sq. mile	=	2.59 sq. km	1 sq. km	=	0.3861 sq. mile

	Volume			Volume	
1 cu. in.	=	16.387 cu. cm	1 cu. cm	=	0.061 cu. in.
1 cu. ft	=	28 316.85 cu. cm	1 cu. cm	=	0.0000353 cu. ft
1 cu. yd	=	0.764555 cu. m	1 cu. m	=	1.308 cu. yd
1 cu. ft	=	28.317 l	1 l	=	0.0353 cu. ft
1 US gallon	=	3.785 4 l	1 l	=	0.2642 US gallon
1 UK gallon	=	4.546 l	1 l	=	0.22 UK gallon
1 US pint	=	0.0004732 cu. m	1 cu. m	=	2113.376 US pint
1 UK pint	=	0.0005683 cu. m	1 cu. m	=	1 759.754 UK pint
1 quart	=	946.353 cu. cm	1 cu. cm	=	0.001 057 quart

Conceptual Aircraft Design: An Industrial Approach, First Edition. Ajoy Kumar Kundu, Mark A. Price and David Riordan.
© 2019 John Wiley & Sons Ltd. Published 2019 by John Wiley & Sons Ltd.

Density

1 lb./ft³	=	16.1273 g/m³	1 kg/m³	=	0.062 427 96 lb./ft³
1 lb./cu. in.	=	27.68 g/cu. cm	1 g/cu. cm	=	0.03613 lb./cu. in.
1 lb./cu. ft	=	16.0185 kg/cu. m	1 kg/cu. m	=	0.06243 lb./cu. ft

Speed   Speed

1 ft/s	=	1.0973 km/h	1 km/h	=	0.9113 ft/s
1 ft/s	=	0.59232 knots	1 knot (kt)	=	6 077.28 ft/h = 1.6881 mph
1 ft/s	=	0.000189394 mps	1 mph	=	5 280 ft/h = 1.466 7 ft/s
1 ft/min	=	0.00508 m/s	1 m/s	=	196.85 ft/min
1 mph	=	0.447 m/s	1 m/s	=	2.237 mph
1 mph	=	0.869 knots	1 knot	=	1.151 mph
1 knot	=	0.51444 m/s	1 m/s	=	1.944 kt
1 knot	=	1.853 km/h	1 knot	=	1.688 ft/s
1 mph	=	1.4666 ft/s	1 ft/s	=	0.68182 mph

Angle   Angle

1 degree (°)	=	0.01716 rad	1 rad	=	57.296°

Mass   Mass

1 lb	=	0.454 kg	1 kg	=	2.2046 lb

Force   Force

1 lb	=	4.4482 N	1 N	=	0.2248 lb
1 lb	=	0.454 kg	1 kg	=	2.2046 lb
1 oz	=	28.35 g	1 g	=	0.3527 oz
1 oz	=	0.278 N	1 N	=	3.397 oz

Pressure			Pressure		
1 lb./sq. in.	=	6 894.76 Pa	1 Pa	=	0.000145 lb. in.
1 lb./sq. ft	=	44.88 Pa	1 Pa	=	0.02089 lb./sq. ft
1 lb./sq. in.	=	703.07 kg/sq. m	1 kg/sq. m	=	0.0001422 lb./sq. in.
1 lb./sq. ft	=	4.8824 kg/sq. m	1 kg/sq. m	=	0.020482 lb./sq. ft
1 atm	=	1 013.25 millibar	1 millibar	=	0.000987 atm
1 bar	=	14.5 lb./sq. in.	1 atm	=	14.7 lb./sq. in.

Energy			Energy		
1 lb. ft	=	1.356 joule	1 joule	=	0.7376 lb. ft
1 W/h	=	3 600 joule	1 joule	=	0.000278 W/h
1 lb. ft	=	1.356 joule	1 joule	=	0.7376 lb. ft
1 W/h	=	0.001 34 HP/h	1 HP/h	=	745.7 W/h

Power					
1 hp. (550 ft lbf)	=	0.7457 kw	1 kQ	=	1.341022 HP

Thermodynamics

$R_{air} = 287.06\,J/kgK = 53.35\,ft\,lb_f/lb.\,R$

Constants		
AVGAS		
1 US gallon	=	5.75 lb
1 cu. ft	=	43 lb

AVTUR		
1 US gallon (JP4)	=	6.56 lb
1 cu. ft	=	48.6 lb

AVTUR		
1 US gallon (JP5)	=	7.1 lb
1 cu. ft	=	53 lb

Appendix B

International Standard Atmosphere Table Data from Hydrostatic Equations

Altitude (m)	Pressure (N/m^2)	Temperature (K)	Density (kg/m^3)	Viscosity (Ns/m^2)	Sound speed (m/s)	$C_{f\_turbulent}$
0	101327	288.15	1.225	0.00001789	340.3	0.00263
500	95463	284.9	1.16727	0.00001773	338.37	0.00264
1000	89876.7	281.65	1.11164	0.00001757	336.44	0.00266
1500	84558	278.4	1.05807	0.00001741	334.49	0.00267
2000	79497.2	275.15	1.00649	0.00001725	332.53	0.00269
2500	74684.4	271.9	0.95686	0.00001709	330.56	0.00271
3000	70110.4	268.65	0.90912	0.0001693	328.58	0.00273
3500	65765.8	265.4	0.86323	0.00001677	326.59	0.00274
4000	61641.9	262.15	0.81913	0.00001661	324.58	0.00276
4500	57729.9	258.9	0.77678	0.00001644	322.56	0.00278
5000	54021.5	255.65	0.73612	0.00001628	320.53	0.0028
5500	50508.3	252.4	0.69711	0.00001611	318.49	0.00282
6000	47182.5	249.15	0.6597	0.00001594	316.43	0.00284
6500	44036.2	245.9	0.62385	0.00001577	314.36	0.00286
7000	41062.1	242.65	0.5895	0.0000156	312.28	0.00288
7500	38252.7	239.4	0.55663	0.00001543	310.18	0.0029
8000	35601	236.15	0.52517	0.00001526	308.07	0.00292
8500	33100.2	232.9	0.49509	0.00001509	305.94	0.00294
9000	30743.6	229.65	0.46635	0.00001492	303.8	0.00297
9500	28524.7	226.4	0.4389	0.00001474	301.64	0.00299
10000	26437.3	223.15	0.41271	0.00001457	299.47	0.00301
10500	24475.3	219.9	0.38773	0.00001439	297.28	0.00304
11000	**22633**	**216.65**	**0.36392**	**0.00001421**	**295.07**	**0.00306**
11500	20916.8	216.65	0.33633	0.00001421	295.07	0.0031
12000	19331	216.65	0.31983	0.00001421	295.07	0.00314
12500	17865	216.65	0.28726	0.00001421	295.07	0.00318
13000	16511	216.65	0.26548	0.00001421	295.07	0.00323
13500	15259.2	216.65	0.24536	0.00001421	295.07	0.00327
14000	14102.3	216.65	0.22675	0.00001421	295.07	0.00331
14500	13033.2	216.65	0.20956	0.00001421	295.07	0.00336
15000	12045.1	216.65	0.19367	0.00001421	295.07	0.0034

Conceptual Aircraft Design: An Industrial Approach, First Edition. Ajoy Kumar Kundu, Mark A. Price and David Riordan.
© 2019 John Wiley & Sons Ltd. Published 2019 by John Wiley & Sons Ltd.

Altitude (ft)	Pressure (lb./ft^2)	Temperature (R)	Density (slug/ft^3)	Viscosity (10^{-7} lb. s/ft^2)	Sound speed (ft/s)	$C_{f\_turbulent}$
0	2116.22	518.67	0.00238	3.7372	1116.5	0.01449
1000	2040.85	515.1	0.0023	3.7172	1112.6	0.01459
2000	1967.68	511.54	0.00224	3.6971	1108.75	0.0147
3000	1896.64	507.97	0.00217	3.677	1104.88	0.0148
4000	1827.69	504.41	0.00211	3.657	1100.99	0.01491
5000	1760.79	500.84	0.00204	3.637	1097.09	0.1502
6000	1695.89	497.27	0.00198	3.616	1093.178	0.01513
7000	1632.93	493.71	0.00192	3.596	1089.25	0.01525
8000	1571.88	490.14	0.00186	3.575	1085.31	0.01536
9000	1512.7	486.57	0.00181	3.555	1081.35	0.01548
10000	1455.33	483.01	0.00175	3.534	1077.38	0.0156
11000	1399.73	479.44	0.0017	3.513	1073.4	0.01572
12000	1345.87	475.88	0.00164	3.4927	1069.4	0.01585
13000	1293.7	472.31	0.00159	3.4719	1065.39	0.01597
14000	1243.18	468.74	0.00154	3.451	1061.36	0.0161
15000	1194.27	465.18	0.00149	3.43	1057.31	0.01623
16000	1146.92	461.11	0.00144	3.4089	1053.25	0.01637
17000	1101.11	458.05	0.0014	3.388	1049.17	0.0165
18000	1056.8	454.48	0.00135	3.3666	1045.08	0.01664
19000	1013.93	450.91	0.0013	3.3453	1040.97	0.01678
20000	1036.85	447.35	0.00126	3.324	1036.95	0.01693
21000	932.433	443.78	0.00122	3.3025	1032.71	0.01707
22000	893.72	440.21	0.00118	3.281	1028.55	0.01722
23000	856.32	436.65	0.00114	3.26	1024.38	0.01738
24000	820.19	433.08	0.0011	3.238	1020.18	0.01753
25000	785.31	429.52	0.00106	3.216	1015.98	0.01769
26000	751.64	425.95	0.00102	3.1941	1011.75	0.01785
27000	719.15	422.38	0.00099	3.1722	1007.5	0.01802
28000	687.81	418.82	0.00095	3.1502	1003.24	0.01819
29000	657.58	415.25	0.00092	3.128	998.96	0.01836
30000	628.43	411.69	0.00088	3.1059	994.66	0.01854
31000	600.35	408.12	0.00085	3.0837	990.35	0.01872
32000	573.28	404.55	0.00082	3.0614	986.01	0.0189
33000	547.21	400.97	0.00079	3.0389	981.65	0.01909
34000	522.12	397.42	0.00076	3.0164	977.28	0.0193
35000	497.96	393.85	0.00073	2.9938	972.88	0.01948
36089	**472.68**	**389.97**	**0.0007**	**2.969**	**968.08**	**0.0197**

Altitude (ft)	Pressure (lb./ft^2)	Temperature (R)	Density (slug/ft^3)	Viscosity (lb. s/ft^2)	Sound speed (ft/s)	$C_{f\_turbulent}$
37000	452.43	389.97	0.00067	2.969	968.08	0.01999
38000	431.2	389.97	0.00064	2.969	968.08	0.02032
39000	410.97	389.97	0.00061	2.969	968.08	0.02065
40000	391.68	389.97	0.00058	2.969	968.08	0.02099
41000	373.3	389.97	0.00055	2.969	968.08	0.02134
42000	355.78	389.97	0.00053	2.969	968.08	0.02169
43000	339.09	389.97	0.0005	2.969	968.08	0.02205
44000	323.08	389.97	0.00048	2.969	968.08	0.02243
45000	308.01	389.97	0.00046	2.969	968.08	0.02281
46000	299.56	389.97	0.00043	2.969	968.08	0.0232
47000	279.78	389.97	0.00041	2.969	968.08	0.02359
48000	266.65	389.97	0.00039	2.969	968.08	0.024
49000	254.14	389.97	0.00037	2.969	968.08	0.02442
50000	242.21	389.97	0.00036	2.969	968.08	0.02485

Appendix C

Fundamental Equations (See Table of Contents for Symbols and Nomenclature.)

Some elementary yet important equations are listed here. Readers must be able to derive them and appreciate the physics of each term for intelligent application to aircraft design and the performance envelope. The equations are not derived here – readers may refer to any aerodynamic textbook for their derivations.

C.1 Kinetics

Newton's Law: **applied force**, $F = mass \times acceleration = rate\ of\ change\ of\ momentum$.
force $= pressure \times area$ and **work** $= force \times distance$.
Therefore, energy (i.e. rate of work) for the unit mass flow rate m is as follows:

$$\textbf{energy} = force \times (distance/time) = pressure \times area \times velocity = pAV.$$

Steady state conservation equations are as follows:

mass conservation: mass flow rate $m = \rho AV =$ constant $\qquad$ (C.1)

Momentum conservation: $dp = -\rho V dV$ (known as Euler's equation) $\qquad$ (C.2)

With viscous terms, it becomes the Navier–Stokes equation. However, friction forces offered by the aircraft body can be accounted for in the inviscid flow equation as a separate term:

Energy conservation: $C_p T + \frac{1}{2} V^2 =$ constant $\qquad$ (C.3)

From the thermodynamic relation for perfect gas between two stations gives

$$s - s_1 = C_p \ln(T/T_1) - R\ln(p/p_1) = C_v \ln(T/T_1) + R\ln(v/v_1) \qquad (C.4)$$

C.2 Thermodynamics

Using perfect gas laws, entropy change process through shock is an adiabatic process, as follows.

$$(s_2 - s_1)/R = \ln (T_2/T_1)^{\gamma(\gamma-1)} - \ln(p_2/p_1) = \ln\frac{(T_2/T_1)^{\frac{\gamma}{(\gamma-1)}}}{(p_2/p_1)} = \ln\frac{(p_2/p_1)^{\gamma/(\gamma-1)}/(\rho_2/\rho_1)^{\gamma/(\gamma-1)}}{(p_2/p_1)}$$

$$\text{or } (s_2 - s_1)/R = \ln \left[(p_2/p_1)^{1/(\gamma-1)} \left(\frac{\rho_2}{\rho_1} \right)^{-\gamma/(\gamma-1)} \right] = \ln(p_{t1}/p_{t2}) \qquad (C.5)$$

When velocity is stagnated to zero (say in the hole of a pitot tube), then the following equations can be derived for isentropic process. Subscript 't' represents a stagnation property, which is also known as the 'total'

Conceptual Aircraft Design: An Industrial Approach, First Edition. Ajoy Kumar Kundu, Mark A. Price and David Riordan.
© 2019 John Wiley & Sons Ltd. Published 2019 by John Wiley & Sons Ltd.

condition. The equations represent point properties, that is, they are valid at any point of a streamline (γ stands for the ratio of specific heat and M is for Mach number).

$$\frac{Tt}{T} = \left(1 + \frac{\gamma - 1}{2}M^2\right) \tag{C.6}$$

$$\frac{\rho_t}{\rho} = \left(1 + \frac{\gamma - 1}{2}M^2\right)^{\frac{1}{\gamma-1}} \tag{C.7}$$

$$\frac{p_t}{p} = \left(1 + \frac{\gamma - 1}{2}M^2\right)^{\frac{\gamma}{\gamma-1}} \tag{C.8}$$

$$\left(\frac{p_t}{p}\right) = \left(\frac{\rho_t}{\rho}\right)^{\gamma} = \left(\frac{Tt}{T}\right)^{\frac{\gamma}{\gamma-1}} \tag{C.9}$$

The conservation equations yield many other significant equations. In any streamline of a flow process, the conservation laws exchange pressure energy with kinetic energy. In other words, if the velocity at a point is increased, then the pressure at that point falls and vice versa (i.e. Bernoulli's and Euler's equations). The following are a few more important equations. At stagnation, the total pressure, p_t, is given.

Bernoulli's equation for incompressible isentropic flow, $(p_t - p) = \rho V^2/2$ (C.10)

or $p + \rho V^2/2 = \text{constant} = p_t$

Clearly, at any point, if the velocity is increased, then the pressure will fall to maintain conservation. This is the crux of lift generation: the upper surface has lower pressure than the lower surface. For compressible isentropic flow, Euler's (Bernoulli's) equation gives the following:

$$\left[\frac{(p_t - p)}{p} + 1\right]^{\frac{\gamma-1}{\gamma}} - 1 = \frac{(\gamma - 1)V^2}{2a^2} \tag{C.11}$$

There are other important relations using thermodynamic properties, as follows.

From the gas laws (combining Charles Law and Boyle's Law), the equation for state of gas for unit mass is $pv = RT$ where for air

$$R = 287 \, \text{J} \, \text{kg}^{-1} \, \text{K}^{-1} \tag{C.12}$$

$$\gamma = C_p/C_v \text{ and } C_p - C_v = R \text{ that makes } C_p/R = \gamma/(\gamma - 1) \tag{C.13}$$

where $C_p = 1004.68 \, \text{m}^2 \, \text{s}^{-2} \, \text{K}^{-1}$, $C_v = 717.63 \, \text{m}^2 \, \text{s}^{-2} \, \text{K}^{-1}$,
Gas constant, $R = 287.053 \, \text{m}^2 \, \text{s}^{-2} \, \text{K}^{-1}$, Universal Gas constant, $R = 8314.32 \, \text{J} \, (\text{K kmol})^{-1}$
Mean molecular mass of air ($28.96442 \, \text{kg kmol}^{-1}$)

From the energy equation, total temperature, $T_t = T + \dfrac{T(\gamma - 1)V^2}{2RT\gamma} = T + \dfrac{V^2}{2C_p}$ (C.14)

Mach number $= V/a$, where $a = $ speed of sound and $a^2 = \gamma RT = \gamma(p/\rho) = (dp/d\rho)_{\text{isentropic}}$ (C.15)

C.3   Supersonic Aerodynamics

Some supersonic 2D relations are given next and Figure C.1 depicts normal and oblique shocks. Upstream flow properties are represented by subscript '1' and downstream by subscript '2'. The process across a shock is adiabatic, therefore total energy remains invariant; that is, $T_{t1} = T_{t2}$.

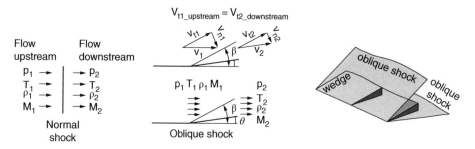

Figure C.1 Shock waves.

C.4 Normal Shock

In this situation, relations between upstream and downstream are unique and do not depend on geometry. If upstream conditions are known, then the downstream properties can be computed from the following relations.

$$\text{Mach number: } M^2_2 = \frac{M_1^2 + \frac{2}{(\gamma-1)}}{\frac{2\gamma}{(\gamma-1)}M_1^2 - 1} \tag{C.16}$$

$$\text{Density ration: } \frac{\rho_2}{\rho_1} = \frac{M_1^2}{\frac{(\gamma-1)}{(\gamma+1)}M_1^2 + \left(\frac{2}{\gamma+1}\right)} = \frac{(\gamma+1)M_1^2}{(\gamma-1)M_1^2 + 2} \tag{C.17}$$

$$\text{Static pressure ratio: } \frac{p_2}{p_1} = \frac{2\gamma}{(\gamma+1)}M_1^2 - \frac{(\gamma-1)}{(\gamma+1)} \tag{C.18}$$

$$\text{Total pressure ratio: } \frac{p_{t2}}{p_{t1}} = \frac{\left[\frac{\left(\frac{\gamma+1}{2}\right)M_1^2}{1+\frac{(\gamma-1)}{2}M_1^2}\right]^{\frac{\gamma}{\gamma-1}}}{\left[\frac{2\gamma}{(\gamma+1)}M_1^2 - \frac{(\gamma-1)}{\gamma+1}\right]^{\frac{1}{\gamma-1}}} = \frac{\left[\frac{(\gamma+1)M_1^2}{2+(\gamma-1)M_1^2}\right]^{\frac{\gamma}{(\gamma-1)}}}{\left[\frac{2\gamma M_1^2-(\gamma-1)}{(\gamma+1)}\right]^{\frac{1}{(\gamma-1)}}} \tag{C.19}$$

$$\text{or } \frac{p_{t2}}{p_1} = \left(\frac{p_{t2}/p_{t1}}{p_1/p_{t1}}\right) = \frac{\left[\frac{(\gamma+1)M_1^2}{2+(\gamma-1)M_1^2}\right]^{\frac{\gamma}{(\gamma-1)}}}{\left[\frac{2\gamma M_1^2-(\gamma-1)}{(\gamma+1)}\right]^{\frac{1}{(\gamma-1)}}}\left(1+\frac{\gamma-1}{2}M^2\right)^{\frac{\gamma-1}{\gamma}}$$

$$= \left[\frac{(\gamma+1)}{2}M_1^2\right]^{\frac{\gamma}{(\gamma-1)}}\left[\frac{2\gamma M_1^2}{(\gamma+1)} - \frac{(\gamma-1)}{(\gamma+1)}\right]^{\frac{1}{(1-\gamma)}} \tag{C.20}$$

$$\text{Total temperature: } T_{t1} = T_{t2} \tag{C.21}$$

$$\text{Total temperature: } \frac{T_2}{T_1} = \frac{\left(\frac{2\gamma}{(\gamma-1)}M_1^2 - 1\right)\left(1 + \frac{(\gamma-1)}{2}M_1^2\right)}{\frac{(\gamma+1)^2}{2(\gamma-1)}M_1^2} \tag{C.22}$$

Once the Mach numbers are known, the static temperatures each side of normal shock can be computed from the isentropic relation given in Equation (C.22).

C.5   Oblique Shock

The physics of 2D supersonic flow past a planar surface and 3D flow past body of revolution differs. In the case of 2D wedge, the flow has two degrees of freedom, one along the flow and the other perpendicular to the flow. But in the case of 3D cone, the flow has three degrees of freedom, the third dimension moving laterally. The difference shows in the physics of flow as the shock strength weaker for the cone flow and also have smaller shock angle for the same subtended object angle at the same free-stream Mach number. This can be checked out in the graphs given next.

A pressure pulse moving through air at speeds higher than the speed of sound will develop as a weak wave pressure cone known as Mach wave. The 2D projection of the Mach wave will have the angle, μ, measured from the centreline. This is a measure of speed and can be computed as follows.

$$\sin \mu = (1/M_1) \text{ or } \mu = \sin^{-1}(1/M_1) \tag{C.23}$$

[example: M_1 at Mach 2 gives $\mu = 30°$]

where μ is a characteristic angle associated with the Mach wave in a limiting sense without any discontinuity in the flow field. In practice, all real objects will have some thickness that will deflect airflow as shown in Figure C.1, in which the object in a wedge shape of angle θ creates a shock wave of angle β, which is always higher than μ; that is, $\beta > \mu$. A shock wave creates a discontinuity across it and is formed by the coalescing of successive Mach waves in the process of flow being deflected. For a given deflection angle, θ, the shock angle, β, can be obtained by the relation given in Equation C.23.

C.6   Supersonic Flow Past a 2D Wedge

Figure C.1, shows the supersonic flow past a wedge shaped object of angle θ creating a shock wave of angle, β. In this situation, relations between upstream and downstream depend on geometry of deflection angle θ and free upstream Mach number M_1 develop an oblique shock angle, β. Higher the upstream Mach number M_1 lowers the oblique shock angle, β. For the same free upstream Mach number M_1, loss through oblique shock is less than normal shock. The relation between free stream Mach number M_1, deflection angle θ, and oblique shock angle, β, can be obtained from the following relation.

$$\tan \beta = 2 \cot \theta \left[\frac{M_1^2 \sin^2 \theta - 1}{M_1^2(\gamma - \cos 2\theta) + 2} \right] \tag{C.24}$$

Geometry relation: $\tan \beta = V_{n1}/V_{t1}$ and $\tan(\beta - \theta) = V_{n2}/V_{t2}$ $\qquad$ (C.25)

From continuity: $V_{n2}/V_{n1} = \dfrac{\rho_2}{\rho_1} = \dfrac{\tan \beta}{\tan(\beta - \theta)}$ $\qquad$ (C.26)

Normal velocities V_{n1} and V_{n2} on oblique shock can be treated as normal shock. In that case, substitute $M_1 \sin \beta$ for M_1 in Eqs. (C.1) to (C.2) obtain relations between properties across oblique shock.

Static pressure ratio: $\dfrac{p_2}{p_1} = \dfrac{2\gamma}{(\gamma + 1)} M_1^2 \sin^2 \beta - \dfrac{(\gamma - 1)}{(\gamma + 1)}$ $\qquad$ (C.27a)

Density ratio: $\dfrac{\rho_2}{\rho_1} = \dfrac{M_1^2 \sin^2 \beta}{\frac{(\gamma-1)}{(\gamma+1)} M_1^2 \sin^2 \beta + \left(\frac{2}{\gamma+1}\right)} = \dfrac{\tan \beta}{\tan(\beta - \theta)}$ $\qquad$ (C.27b)

Static temperature ratio: $\dfrac{T_2}{T_1} = \dfrac{\left(\frac{2\gamma}{(\gamma-1)} M_1^2 \sin^2 \beta - 1\right)\left(1 + \frac{(\gamma-1)}{2} M_1^2 \sin^2 \beta\right)}{\frac{(\gamma+1)^2}{2(\gamma-1)} M_1^2 \sin^2 \beta}$ $\qquad$ (C.27c)

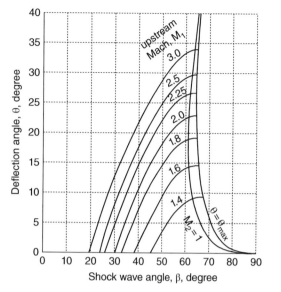
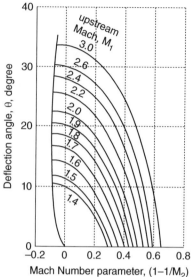

Figure C.2 Shock wave properties.

$$\text{Total temperature}: T_{t1} = T_{t2} \tag{C.28}$$

$$\text{Total pressure ratio}: \frac{p_{t2}}{p_{t1}} = \left[\frac{(\gamma+1)M_1^2\sin^2\beta}{2+(\gamma-1)M_1^2\sin^2\beta}\right]^{\left(\frac{\gamma}{\gamma-1}\right)} \left[\frac{(\gamma+1)}{2\gamma M_1^2\sin^2\beta-(\gamma-1)}\right]^{\frac{1}{(\gamma-1)}} \tag{C.29}$$

$$\text{Downstream Mach number}: M^2_{\;2}\sin^2(\beta-\theta) = \frac{M_1^2\sin^2\beta+\frac{2}{(\gamma-1)}}{\frac{2\gamma}{(\gamma-1)}M_1^2\sin^2\beta-1}$$

$$\text{or } M_2 = \left(\frac{1}{\sin(\beta-\theta)}\right)\sqrt{\frac{1+\left(\frac{\gamma-1}{2}\right)M_1^2\sin^2\beta}{\gamma M_1^2\sin^2\beta-\left(\frac{\gamma-1}{2}\right)}} \tag{C.30}$$

The relation between Mach number M_1, deflection angle θ, and oblique shock angle, β are plotted in Figure C.2. It is recommended readers use the graphs in Figure C.2. These graphs will prove easier to get results to the extent they will be used in this book. Section 4.13 indicates that the complexities involved in supersonic aircraft design are beyond the scope of this book. However, some introductory treatment is carried out to make readers aware of the associated considerations required to design supersonic aircraft.

C.7 Supersonic Flow Past 3D Cone

Supersonic flow physics past 3D bodies differs from 2D planar bodies and therefore Equations (3.23) to (3.28) cannot be used.

Supersonic/hypersonic nose of aircraft and rockets are bodies of revolution resembling an ogive shape. To obtain shock wave properties over an ogive shape is relatively difficult. To simplify, the front portion of an ogive is approximated as a cone and the graphs given in Figure C.3 may be used for initial approximation.

Section 4.12 discusses supersonic wing design.

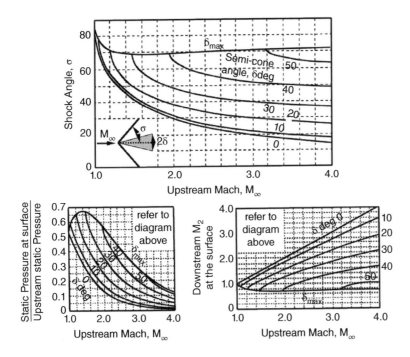

Figure C.3 Properties of shock waves past a 3D cone.

C.8 Incompressible Low Speed Wind Tunnel (Open Circuit)

Refer to the Figure C.4. Let ρ, A and V be the density, cross sectional area and velocity in a settling chamber with subscript 1 and in a test section with subscript 2, respectively. The contraction ratio of the wind tunnel is defined by A_1/A_2.

From conservation of mass flow, $m = \rho_1 A_1 V_1 = \rho_2 A_2 V_2$ (Eq. (C.1))

Being incompressible, $A_1 V_1 = A_2 V_2$

This gives velocity in the test section, $V_2 = (A_1/A_2)/V_1$ (C.31)

Using Bernoulli's theorem (Eq. (C.10)) $p_1 + \rho V_1^2/2 = p_2 + \rho V_2^2/2$ ($\rho_1 = \rho_2 = \rho$, incompressible)

or $V_2^2 = (p_1 - p_2)(\rho/2) + V_1^2/2 = 2(p_1 - p_2)/\rho + [(A_2/A_1)]^2 V_2^2$

Solving $V_2 = \sqrt{\dfrac{2(p_1 - p_2)}{\rho\left[1 - \left(\frac{A_2}{A_1}\right)^2\right]}}$ (C.32)

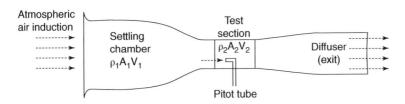

Figure C.4 Open circuit low speed wind tunnel.

Appendix D

Some Case Studies – Aircraft Data

D.1   Airbus320 Class Aircraft

All computations carried out here follow the book's instructions. The results are not from Airbus Industry and they are not responsible for the figures given here. This is used only to substantiate the book methodology with the industry values to gain some confidence. Industry drag data is not available but, at the end, whether payload-range matches with the published data should be checked.

Given:

Long range cruise (LRC) speed and Altitude: Mach 0.75 at 36 089 ft.

Figure D.1 shows a three-view diagram of an Airbus320.

D.1.1   Dimensions (to Scale the Drawing for Detailed Dimensions)

Fuselage: Length = 123.16 ft (scaled measurement differed slightly from the drawings).

Fuselage width = 13.1 ft, Fuselage depth = 13 ft.

Wing reference area (trapezoidal part only) = 1202.5 ft^2, Add the yehudi area = 118.8 ft^2.

Span = 11.85 ft, MAC_{wing} = 11.64 ft, AR = 9.37, $\Lambda_{1/4}$ = 25°, C_R = 16.5 ft, λ = 0.3, t/c_{ave} = 0.12.

H-tail reference area = 330.5 ft^2, $MAC_{H\text{-}tail}$ = 8.63 ft.

V-tail reference area = 235.6 ft^2, $MAC_{V\text{-}tail}$ = 13.02 ft.

Nacelle length = 17.28 ft, Maximum diameter = 6.95 ft.

Pylon = to measure from the drawing.

Reynolds number per foot is given by, $Re_{per\,foot} = (V\rho)/\mu = (aM\rho)/\mu$.

= $[(0.75 \times 968.08)(0.000\,71)]/(0.795\,0 \times 373.718 \times 10^{-9})$ = 1.734×10^6 per foot.

D.2   Drag Computation

D.2.1   Fuselage

Table D.1 gives the basic average 2D flat plate for the fuselage, $C_{Ffbasic}$ = 0.00186. Table D.2 summarises the 3D and other shape effect corrections, ΔC_{Ff}, needed to estimate the total fuselage C_{Ff}.

The total fuselage $C_{Ff} = C_{Ffbasic} + \Delta C_{Ff}$ = 0.00186 + 0.0006875 = 0.002547.

Flat plate equivalent f_f (see Section 9.21.3) = $C_{Ff} \times A_{wf}$ = 0.002547 × 4333 = 11.03 ft^2.

Add canopy drag f_c = 0.3 ft^2 (Ref. 78 and 79).

∴ Total fuselage parasite drag in terms of f_{f+c} = 11.33 ft^2.

Conceptual Aircraft Design: An Industrial Approach, First Edition. Ajoy Kumar Kundu, Mark A. Price and David Riordan.
© 2019 John Wiley & Sons Ltd. Published 2019 by John Wiley & Sons Ltd.

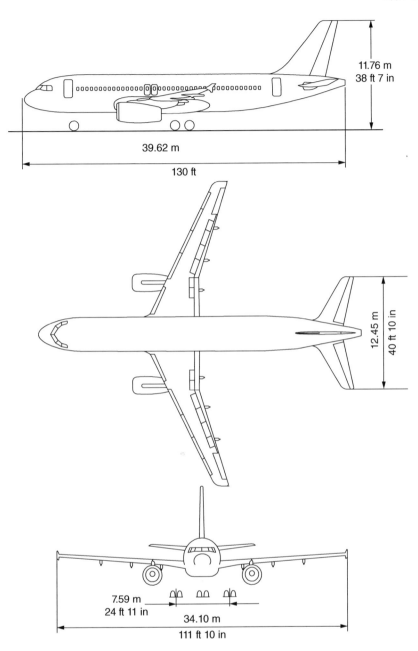

Figure D.1 Airbus 320 three-view with major dimensions (courtesy of Airbus). Reproduced with permission of John Wiley and CUP.

D.2.2   Wing

Table D.1 gives the basic the average 2D flat plate for the wing, $C_{Fwbasic} = 0.00257$ based on the MAC_w.

Important geometric parameters are wing reference area (trapezoidal planform) = 1202.5 ft^2 and the Gross wing planform area (including Yehudi) = 1320.8 ft^2. Table D.3 summarises the 3D and other shape effect corrections needed to estimate the total wing C_{Fw}.

Therefore, the total wing $C_{Fw} = C_{Fwbasic} + \Delta C_{Fw} = 0.00257 + 0.000889 = 0.00345$.

Flat plate equivalent, f_w (Eq. 6.34) $= C_{Fw} \times A_{ww} = 0.00345 \times 2130.94 = 7.35$ ft^2

Table D.1 Reynolds number and 2D basic skin friction C_{Fbasic}. Reproduced with permission of Cambridge University Press.

Parameter	Reference area – ft²	Wetted area ft²	Characteristic length – ft	Reynolds number	2D i_{Fbasic}
Fuselage	n/a	4 333	123.16	2.136×10^8	0.001 86
Wing	1 202.5	2 130.94	11.64 (MAC_w)	2.02×10^7	0.002 55
V-Tail	235	477.05	13.02 (MAC_{VT})	2.26×10^7	0.002 51
H-Tail	330.5	510.34	8.63 (MAC_{HT})	1.5×10^7	0.002 69
2 × Nacelle	n/a	2 × 300	17.28	3×10^7	0.002 38
2 × Pylon	n/a	2 × 58.18	12 (MAC_p)	2.08×10^7	0.002 54

Table D.2 Fuselage ΔC_{Ff} correction (3D and other shape effects). Reproduced with permission of Cambridge University Press.

Item	ΔC_{Ff}	% of $C_{Ffbasic}$
Wrapping	0.000 009 22	0.496
Supervelocity	0.000 1	5.36
Pressure	0.000 016 8	0.9
Fuselage upsweep of 6°	0.000 127	6.8
Fuselage closure angle of 9°	0	0
Nose fineness ratio	0.000163	8.7
Fuselage non-optimum shape	0.0000465	2.5
Cabin pressurisation/leakage	0.000093	5
Passenger windows/doors	0.0001116	6
Belly fairing	0.000039	2.1
Environ. Control Sys Exhaust (ECS)	−0.0000186	−1
Total ΔC_{Ff}	0.0006875	36.9

Table D.3 Wing ΔC_{Fw} correction (3D and other shape effects). Reproduced with permission of Cambridge University Press.

Item	ΔC_{Fw}	% of $C_{Fwbasic}$
Supervelocity	0.000493	19.2
Pressure	0.000032	1.25
Interference (wing-body)	0.000104	4.08
Excrescence (flaps and slats)	0.000257	10
Total ΔC_{Fw}	0.000887	34.53

D.2.3 Vertical Tail

Table D.1 gives the basic average 2D flat plate for the V-tail, $C_{FVTbasic} = 0.002\ 51$ based on the MAC_{VT}. V-tail reference area = 235 ft². Table D.4 summarises the 3D and other shape effect corrections (ΔC_{FVT}) needed to estimate the V-tail C_{FVT}.

Therefore, V-tail $C_{FVT} = C_{FVTbasic} + \Delta C_{FVT} = 0.00251 + 0.000718 = 0.003228$.

Flat plate equivalent f_{VT} (Equation 6.36) = $C_{FVT} \times A_{wVT} = 0.003228 \times 477.05 = 1.54\ \text{ft}^2$.

Table D.4 V-tail ΔC_{FVT} correction (3D and other shape effects). Reproduced with permission of Cambridge University Press.

Item	ΔC_{FVT}	% of $C_{FVTbasic}$
Supervelocity	0.000377	15
Pressure	0.000015	0.6
Interference (V tail-body)	0.0002	8
Excrescence (rudder gap)	0.0001255	5
Total ΔC_{FVT}	0.000718	28.6

D.2.4   Horizotal Tail

Table D.1 gives the basic average 2D flat plate for the H-tail, $C_{FHTbasic} = 0.00269$ based on the MAC_{HT}. The H-tail reference area $S_{HT} = 330.5 \, \text{ft}^2$. Table D.5 summarises the 3D and other shape effect corrections (ΔC_{FHT}) needed to estimate the H-tail C_{FHT}.

Therefore, H-tail $C_{FHT} = C_{FHTbasic} + \Delta C_{FHT} = 0.00269 + 0.000605 = 0.003295$.

Flat plate equivalent f_{HT} (Equation 6.36) $= C_{FHT} \times A_{wHT} = 0.003\,295 \times 510.34 = 1.68 \, \text{ft}^2$.

D.2.5   Nacelle, C_{Fn}

Because the nacelle is a fuselage-like axisymmetric body, the procedure follows the method used for fuselage evaluation, but needs special attention on account of the throttle dependent considerations.

Important geometric parameters are: the nacelle length $= 17.28 \, \text{ft}$. Maximum nacelle diameter $= 6.95 \, \text{ft}$. Average diameter $= 5.5 \, \text{ft}$. Nozzle exit plane diameter $= 3.6 \, \text{ft}$.

Maximum frontal area $= 37.92 \, \text{ft}^2$. Wetted area per nacelle $A_{wn} = 300 \, \text{ft}^2$.

Table D.1 gives the basic average 2D flat plate for the nacelle, $C_{Fnbasic} = 0.002\,38$ based on the nacelle length. Table D.6 summarises the 3D and other shape effect corrections, ΔC_{Fn}, needed to estimate the total nacelle C_{Fn} for one nacelle.

For nacelles, a separate supervelocity effect is not considered as it is taken into account in the throttle-dependent intake drag. Also, pressure drag is accounted for in throttle-dependent base drag.

Thrust reverser drag: The excrescence drag of the thrust reverser is included in Table 7.7 as it does not emanate from manufacturing tolerances. The nacelle is placed well ahead of the wing, hence the nacelle-wing interference drag is minimised and is assumed to be zero.

Therefore, nacelle $C_{Fn} = C_{Fnbasic} + \Delta C_{Fn} = 0.00238 + 0.001777 = 0.00416$.

Flat plate equivalent f_n (Eq. 6.39) $= C_{Fnt} \times A_{wn} = 0.00416 \times 300 = 1.25 \, \text{ft}^2$ per nacelle.

Table D.5 H-tail ΔC_{FHT} correction (3D and other shape effects). Reproduced with permission of Cambridge University Press.

Item	ΔC_{FHT}	% of $C_{FHTbasic}$
Supervelocity	0.0004035	15
Pressure	0.0000101	0.3
Interference (H tail-body)	0.0000567	2.1
Excrescence (elevator gap)	0.0001345	5
Total ΔC_{FHT}	0.000605	22.4

Table D.6 Nacelle ΔC_{Fn} correction (3D and other shape effects). Reproduced with permission of Cambridge University Press.

Item	ΔC_{Fn}	% of $C_{Fnbasic}$
Wrapping (3D effect)	0.0000073	0.31
Excrescence (non-manufacture)	0.0005	20.7
Boat tail (aft end)	0.00027	11.7
Base drag (aft end)	0	0
Intake drag	0.001	41.9
Total ΔC_{Fn}	0.001 777	74.11

D.2.6 Pylon

The pylon is a wing like lifting surface and the procedure is identical to the wing parasite drag estimation. Table D.1 gives the basic average 2D flat plate for the pylon, $C_{Fpbasic} = 0.0025$ based on the MAC_p. The pylon reference area $= 28.8\,\text{ft}^2$ per pylon. Table D.7 summarises the 3D and other shape effect corrections (ΔC_{Fp}) needed to estimate C_{Fp} (one pylon).

Therefore, pylon $C_{Fp} = C_{Fpbasic} + \Delta C_{Fp} = 0.0025 + 0.00058 = 0.00312$.
Flat plate equivalent f_p (Eq. 6.29) $= C_{Fp} \times A_{wp} = 0.182\,\text{ft}^2$ per pylon.

D.2.7 Roughness Effect

The current production standard tolerance allocation offers some excrescence drag. The industry standard takes 3% of the total component parasite drag, which includes the effect of surface degradation in use. The value is $f_{roughness} = 0.744\,\text{ft}^2$ given in the final Table D.8.

D.2.8 Trim Drag

Conventional aircraft will produce some trim drag in cruise and this varies slightly with fuel consumption. For a well-designed aircraft of this class, the trim drag may be taken as $f_{trim} = 0.1\,\text{ft}^2$.

D.2.9 Aerial and Other Protrusions

For this class of aircraft $f_{arial} = 0.005\,\text{ft}^2$.

D.2.10 Air Conditioning

This accounted for in fuselage drag as ECS exhaust. This could provide a small amount of thrust.

Table D.7 Pylon ΔC_{Fp} correction (3D and other shape effects). Reproduced with permission of Cambridge University Press.

Item	ΔC_{Fp}	% of $C_{Fpbasic}$
Supervelocity	0.000 274	10.78
Pressure	0.000 01	0.395
Interference (pylon-wing)	0.000 3	12
Excrescence	0	0
Total ΔC_{Fp}	0.000 584	23

Table D.8 Aircraft parasite drag build-up summary and C_{Dpmin} estimation. Wing reference area $S_w = 1202$ ft^2, $C_{Dpmin} = f/S_w$. ISA day, 36 089 ft altitude and Mach 0.75. Reproduced with permission of Cambridge University Press.

	Wetted area, A_w ft^2	Basic C_F	ΔC_F	Total C_F	f - ft^2	C_{Dpmin}
Fuselage + U/C fairing	4 333	0.001 86	0.000 69	0.002 55	11.03	0.009 18
Canopy					0.3	0.000 25
Wing	2 130.94	0.002 55	0.000 89	0.003 46	7.35	0.006 15
V-tail	477.05	0.002 51	0.000 72	0.003 23	1.54	0.001 28
H-tail	510.34	0.002 69	0.000 61	0.003 3	1.68	0.001 4
2 × Nacelle	2 × 300	0.002 38	0.001 78	0.004 15	2.5	0.002 08
2 × pylon	2 × 58.18	0.002 54	0.000 584	0.003 12	0.362	0.000 3
Rough (3%)					0.744	0.000 62
Aerial					0.005	0.000 004
Trim drag					0.1	0.000 08
	Total				25.611	0.021 3

$C_{Dpmin} = 0.021\ 3$.

D.2.11 Aircraft Parasite Drag Build-Up Summary and C_{Dpmin}

Table D.8 gives the aircraft parasite drag build-up summary in a tabular form.

D.2.12 ΔC_{Dp} Estimation

The ΔC_{Dp} table is constructed, corresponding to the C_L values, as given in Table D.9.

D.2.13 Induced Drag, C_{Di}

Wing Aspect ratio, $AR = \dfrac{span^2}{gross\ wing\ area} = (111.2)^2/1320 = 9.37.$

Induced drag, $C_{Di} = \dfrac{C_L^2}{\pi AR} = 0.034 C_L^2.$
Table D.10 gives the C_{Di} corresponding to each C_L.

Table D.9 ΔC_{Dp} estimation. Reproduced with permission of Cambridge University Press.

C_L	0.2	0.3	0.4	0.5	0.6
ΔC_{Dp}	0.000 44	0	0.000 4	0.001 1	0.001 9

Table D.10 Induced drag. Reproduced with permission of Cambridge University Press.

C_L	0.2	0.3	0.4	0.5	0.6	0.7	0.8
C_{Di}	0.001 36	0.003 06	0.005 44	0.008 5	0.012 24	0.016 7	0.021 8

D.2.14   Total Aircraft Drag

Aircraft drag is given as $C_D = C_{Dpmin} + \Delta C_{Dp} + C_{Di} + [C_{Dw} = 0]$, the total aircraft drag is obtained by adding all the drag components in Table D.11. Note that the low and high values of C_L are beyond the flight envelope. Table D.11 is drawn in Figure D.2.

D.2.15   Engine Rating

Uninstalled sea level static thrust = 25 000 lb per engine.
Installed sea level static thrust = 23 500 lb per engine.

D.2.16   Weights Breakdown (There Could be Some Variation)

Design cruise speed, $V_C = 350$ KEAS.
Design dive speed, $V_D = 403$ KEAS.
Design dive Mach number, $M_D = 0.88$.
Limit load factor = 2.6.
Ultimate load factor = 3.9.
Cabin differential pressure limit = 7.88 psi.

Table D.11 Total aircraft drag coefficient, C_D. Reproduced with permission of Cambridge University Press.

C_L	0.2	0.3	0.4	0.5	0.6
C_{Dpmin}			0.0213 from Table 7.9		
ΔC_{Dp}	0.000 38	0	0.000 4	0.001 1	0.001 9
C_{Di}	0.001 36	0.003 06	0.005 44	0.008 5	0.012 24
Total aircraft C_D	0.023 1	0.024 36	0.027 14	0.030 9	0.035 44

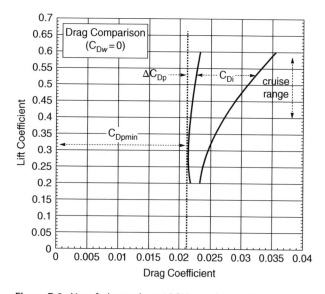

Figure D.2 Aircraft drag polar at LRC. Reproduced with permission of Cambridge University Press.

Component	Weight (lb)	Percentage of MTOW
Wing	14 120	
Flaps + slats	2 435	
Spoilers	380	
Aileron	170	
Winglet	265	
Wing group total	17 370	(above subcomponent weights)
Fuselage group	17 600	(Toernbeek's method)
H-tail group	1845	
V-tail	1010	
Undercarriage group	6425	
Total structure weight	44 250	
Power plant group (two)	15 220	
Control systems group	2280	
Fuel systems group	630	
Hydraulics group	1215	
Electrical systems group	1945	
Avionics systems group	1250	
APU	945	
ECS group	1450	
Furnishing	10 650	
Miscellaneous	4055	
MEW	83 890	
Crew	1520	
Operational items	5660	
OEW	91 070	
Payload (150 × 200)	30 000	
Fuel (see range calculation)	41 240	
MTOW	162 310	

This gives a wing loading = 162 310/1202.5 = 135 lb/ft^2.
And thrust loading = 50 000/162 310 = 0.308.
The aircraft is sized to this specification with better high lift devices.

D.2.17   Payload Range (150 Passengers)

MTOM – 162 000 lb.
On board fuel mass – 40 900 lb.
Payload – 200 × 150 = 30 000 lb.
Long range cruise – Mach 0.75, 36 086 ft (constant condition).
Initial cruise thrust per engine = 4500 lb.
Final cruise thrust per engine = 3800 lb.
Average specific range = 0.09 nm/lb fuel.
An aircraft climbs at 250 KEAS reaching Mach 0.7.
Summary of the mission sector.

Sector	Fuel consumed (lb)	Distance covered (nm)	Time elapsed (min)
Taxi-out	200	0	8
Take-off	300	0	1
Climb	4355	177	30
Cruise	28 400	2560	357
Descent	370	105	20
Approach/land	380	0	3
Taxi-in	135	0	5
Total	34 140	842	424

Diversion fuel calculation:

Diversion distance = 2000 nm, cruising at Mach 0.675 and at 30 000 ft altitude.

Diversion fuel = 2800 lb, contingency fuel (5% of mission fuel) = 1700 lb,

Holding fuel calculation:

Holding time = 30 minutes at Mach 0.35 and at 5000 ft altitude.

Holding fuel = 2600 lb.

Total reserve fuel carried = 2800 + 1700 + 2600 = 7100 lb.

Total onboard fuel carried = 7100 + 34 140 = 41 240 lb.

A consolidated display aircraft performance is given in Figure D.3.

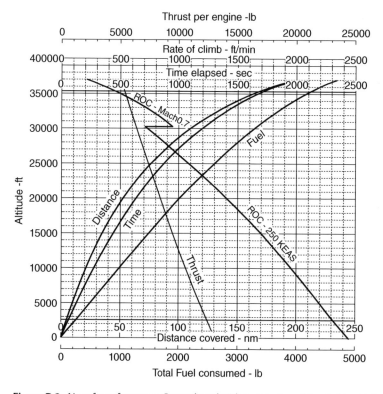

Figure D.3 Aircraft performance. Reproduced with permission of Cambridge University Press.

D.2.18 Cost Calculations (US$ – Year 2000)

Number of passengers = 150.

Yearly utilisation	497 trips per year
Mission (trip) block time	7.05 h
Mission (trip) block distance	2842 nm
Mission (trip) block fuel	34 140 lb (6.68 lb/US gallons)

Fuel cost = 0.6 US$ per US gallon.
Airframe price = 38 million.
Two engines price = 9 million.
Aircraft price = 47 million.
Operating costs per trip – AEA 89 ground rules for medium jet transport aircraft.

Depreciation	6923
Interest	5370
Insurance	473
Flight crew	3482
Cabin crew	2854
Navigation	3194
Landing fees	573
Ground handling	1365
Airframe maintenance	2848
Engine maintenance	1208
Fuel cost	3066 (5 110.8 US gallons)
Total DOC	31 356
DOC/Block hour	4449
DOC/seat	209
DOC/seat/nm	US$ 0.0735/seat/nm

The readers may compare with data available in the public domain.

SUGGESTED EXERCICES FOR THE READERS

Given below are two aircraft:

(1) Model Belfast (B100) - narrow body 100 passenger turbofan aircraft (FAR 25 certification) and
(2) Model AK4 - a four-place utility aircraft (FAR 23 certification).

These aircraft serve as problem assignments for the readers to evaluate their full performances as covered in this book.

[*Solutions for few problems are not supplied*]

D.3   The Belfast (B100) – A Fokker F100 Class Aircraft

The 3-View diagram of B100 is below. It is suggested to enlarge the diagram to A3 size paper and set the scale of the drawing from the dimensions given below.

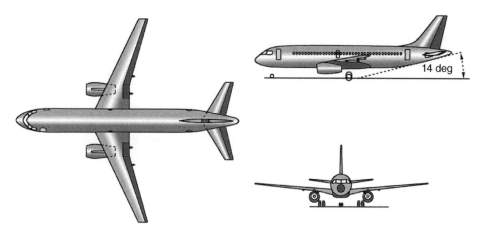

Figure D.4 Belfast 100 (B100).

D.3.1   Customer Specification and Geometric Data

The geometric and performance parameters discussed herein were used in previous chapters. Figure E1 illustrates the dissected anatomy of the given aircraft. The B100 is 5 abreast 110 passenger aircraft.

D.3.1.1   Customer Specification

Payload = 110 passengers	Range = 1500 nm
Take-off distance (*BFL*) = 5200 ft	Landing distance = 5000 ft
Maximum speed = 0.81 Mach	Maximum initial cruise speed = 455 kt (*TAS*)
Initial en-route climb rate = 3000 ft s^{-1}	Initial cruise altitude = 37 000 ft
C_L at *LRC* (Mach 0.77) = 0.5	C_L at *HSC* (Mach 0.82) = 0.43
Maximum dive speed (structural limit) = 0.85 Mach	

D.3.1.2   Family Variants

Short variant B90 = 90PAX	Long variant B130 = 130PAX

D.3.1.3   Fuselage (Circular Constant Section)

Fuselage length, L_f = 33.5 m (110 ft)	Constant diameter = 3.5 m (11.5 ft)
Fuselage upsweep angle 14°	

Measure the fuselage closure angle and fineness ratio from the drawing.

D.3.1.4   Wing

Aerofoil: NACA 65-412 having 12% thickness to chord ratio for design $C_L = 0.45$

Planform reference area, $S_W = 88.26\,\text{m}^2$ ($\approx 950\,\text{ft}^2$)	Span $= 29\,\text{m}$ (95 ft)	Aspect Ratio $= 9.5$
Wing $MAC = 3.52\,\text{m}$ (11.55 ft)	Dihedral $= 5°$, Twist $=-1°$ (wash out)	$t/c = 12\%$
Root Chord at centreline, $C_R = 5.0\,\text{m}$ (16.4 ft)	Tip chord, $C_T = 1.3\,\text{m}$ (4.26 ft)	Taper ration $= 0.26$
Quarter chord wing sweep $= 23.5°$		

D.3.1.5   V-tail (Aerofoil 64-010)

Planform (Reference) area, $S_V = 23\,\text{m}^2$ (248 ft^2)	$AR = 2.08$

D.3.1.6   H-tail (Tee Tail, Aerofoil 64-210 – Installed with Negative Camber)

Planform (Reference) area $S_H = 26.94\,\text{m}^2$ (290 ft^2)	AR $= 4.42$, Dihedral $=5°$

D.3.1.7   Nacelle (Each – 2 Required)

Measure the nacelle length, maximum diameter and nacelle fineness ratio from the drawing.

D.3.2   The B100 Aircraft Weight Summary

D.3.2.1   Weight Summary

$MTOM = 100\,000\,\text{lb}$ (45 455 kg)	$OEW = 59\,500\,\text{lb}$ (27 050 kg)
Wing $= 9800\,\text{lb}$	Installed engine unit $= 10\,500\,\text{lb}$
Fuselage $= 13\,900\,\text{lb}$	Systems $= 7\,900\,\text{lb}$
H-tail $= 1100\,\text{lb}$	Furnishing $= 4\,700\,\text{lb}$
V-tail $= 850\,\text{lb}$	Contingency $= 800\,\text{lb}$
$2 \times$ Pylon $= 1150\,\text{lb}$	Passenger $= 200\,\text{lb/PAX}$
Undercarriage $= 3600\,\text{lb}$	Operator's item $= 5200\,\text{lb}$

D.3.2.2   Engine (Use Figures 13.5 to 13.7 – Uninstalled Values)

Take-off static thrust at *ISA* sea level T_{SLS}/engine $= 17\,000\,\text{lb}$ ($\approx 75\,620\,N$) with $BPR = 5$

sfc @ $T_{SLS} = 0.3452\,\text{lb/lb/h}$	*sfc* @ $T_{max\_cruise} = 0.6272\,\text{lb/lb/h}$

D.3.2.3   Other Pertinent Data

Reserve fuel $= 200\,\text{nm}$ diversion $+ 30$ minutes of holding $+ 5\%$ sector fuel at the specified payload-range.

D.3.2.4 High Lift Devices (Flaps and Slats)

	Clean	Take-off (low setting)	Take-off (mid setting)	Land
C_{LMAX}	1.74	2.6	2.7	3.25

D.3.2.5 B100 Drag

The B100 drag polar equation $C_D = 0.0204 + 0.04C_L^2$

One engine inoperative $\Delta C_D = 0.0023$

Control and asymmetry $\Delta C_D = 0.0015$

Undercarriage $\Delta C_{D\_UC} = 0.0025$

Flap drag ΔC_{D_flap} (flap and slat) = 0.03 (take-off) and 0.068 (land)

D.4 The AK4 (4-Place Utility Aircraft) – Retractable Undercarriage

The 3-View diagram of the AK4 is below. It is suggested to enlarge the diagram to A3 size paper and set the scale of the drawing from the dimensions given below.

D.4.1 Customer Specification and Geometric Data

The geometric and performance parameters of the AK4 is given below.

D.4.1.1 Customer Specification

Payload = 4 occupants	Range = 1000 nm
Take-off distance (*BFL*) = 1000 ft	Landing distance = 800 ft
Cruise altitude = 10 000 ft	Maximum cruise speed = 200 mph
Maximum dive speed (structural limit) = 280 mph	

D.4.1.2 Fuselage

Fuselage length – 24 ft   Maximum width – 50 in.   Maximum height – 50 in.

Fuselage upsweep angle 14°

Measure the fuselage closure angle and fineness ratio from the drawing.

D.4.1.3 Wing

Aerofoil: NACA 63_2-415 having 15% thickness to chord ratio for design $C_L = 0.4$

Planform reference area, $S_W = 140\,\text{ft}^2$ Span = 36 ft   Aspect ratio = 9.257

Wing *MAC* = 3.89 ft   Dihedral = 3°   Twist = −1.5° (wash out)

Root chord at centreline, $C_R = 5$ ft   Tip chord, $C_T = 2.5$ ft   Taper ration = 0.5

Quarter chord wing sweep = 0°

D.4.1.4   V-tail (Aerofoil 64-010)

Planform (reference) area, $S_V = 23\,\text{m}^2\ (248\,\text{ft}^2)$ $AR = 2.08$

D.4.1.5   H-tail (Tee Tail, Aerofoil 64-210 – Installed with Negative Camber)

Planform (reference) area $S_H = 26.94\,\text{m}^2\ (290\,\text{ft}^2)$ AR = 4.42, Dihedral = 5°

D.4.1.6   High Lift Devices (Slotted Flaps)

	Clean	Take-off	Land
C_{LMAX}	1.4	2.0	2.6

AK4 drag polar equation, $C_D = 0.0241 + 0.043C_L{}^2$

$\Delta C_{DPmin\_undercarriage}$	0.008
ΔC_{DPmin_flap}	0.008 (takeoff) 0.01 (land)

D.4.1.7   Engine Details – STD Day Performance (Figure D.5)

Single Lycoming IO-390 piston engine – four cylinder
Sea Level – 210 HP at 2575 rpm, fuel flow 120 lb h^{-1}
Maximum Cruise – 154 HP at 2400 rpm, fuel flow 74 lb h^{-1}
Normal Cruise – 135 HP at 2200 rpm, fuel flow 60 lb h^{-1}
Dry weight – 380 lb, Oil – 8 quarts.
Fuel – Aviation grade 100 octane (AVGAS)

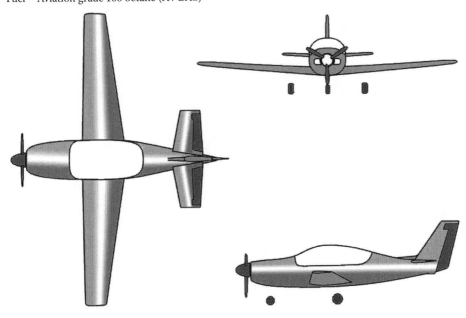

Figure D.5 The AK4.

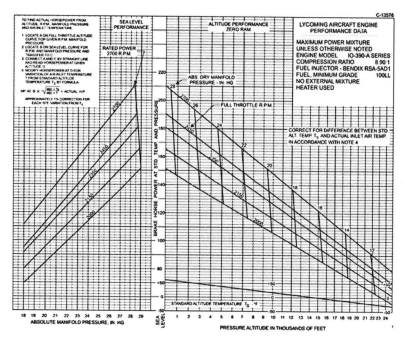

(a) Lycoming IO-390 Horse Power graphs

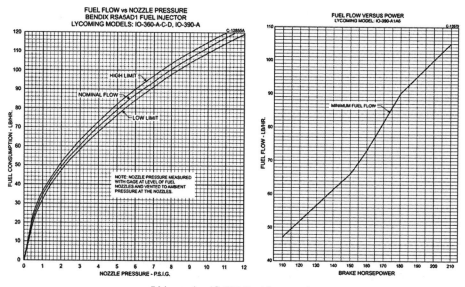

(b) Lycoming IO-390 Fuel flow graphs

Figure D.6 Lycoming IO-390 performance graphs. Source: Courtesy of Textron Lycoming Engine.

D.4.1.8 Engine Data

Lycoming IO-390 performance graphs are given in Figure D.6.

D.4.1.9 Propeller

Three-bladed propeller wit AF = 100, integrated design $C_{Li} = 0.5$, diameter = 5.5 ft
Thrust characteristic given in Figure D.7 below.

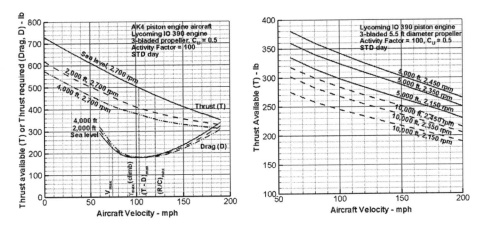

Figure D.7 Propeller performance.

D.4.2   The AK4 Aircraft Its Component Weights (kg)

Wing = 300 lb	Installed engine unit = 450 lb
Fuselage + canopy = 250 lb	Systems = 160 lb
H-tail = 70 lb	Furnishing = 150 lb
V-tail = 50 lb	Contingency = 30 lb
Undercarriage = 140 lb	Total = 1600 lb
Payload (4 × 200 + 100)	900 lb
Fuel + oil	300 lb
MTOM	2800 lb
OEW	1600 lb

Reserve fuel = 200 nm diversion + 30 minutes of holding + 5% sector fuel at the specified payload-range.

Appendix E

Aerofoil Data

Data Courtesy of I. R. Abbott and A. E. Von Doenhoff, *Theory of Wing Sections*.

NACA 4412
(Stations and ordinates given in per cent of aerofoil chord)

Upper surface		Lower surface	
Station	Ordinate	Station	Ordinate
0	0	0	−0
1.25	2.44	1.25	−1.43
2.5	3.39	2.5	−1.95
5.0	4.73	5.0	−2.49
7.5	5.76	7.5	−2.74
10	6.59	10	−2.86
15	7.89	15	−2.88
20	8.80	20	−2.74
25	9.41	25	−2.50
30	9.76	30	−2.26
40	9.80	40	−1.80
50	9.19	50	−1.40
60	8.14	60	−1.00
70	6.69	70	−0.65
80	4.89	80	−0.39
90	2.71	90	−0.22
95	1.47	95	−0.16
100	(0.13)	100	(−0.13)
100	—	100	0

L.E. radius: 1.58
Slope of radius through L.E.: 0.20

Conceptual Aircraft Design: An Industrial Approach, First Edition. Ajoy Kumar Kundu, Mark A. Price and David Riordan.
© 2019 John Wiley & Sons Ltd. Published 2019 by John Wiley & Sons Ltd.

NACA 4415
(Stations and ordinates given in per cent of aerofoil chord)

Upper surface		Lower surface	
Station	Ordinate	Station	Ordinate
0	—	0	0
1.25	3.07	1.25	−1.79
2.5	4.17	2.5	−2.48
5.0	5.74	5.0	−3.27
7.5	6.91	7.5	−3.71
10	7.84	10	−3.98
15	9.27	15	−4.18
20	10.25	20	−4.15
25	10.92	25	−3.98
30	11.25	30	−3.75
40	11.25	40	−3.25
50	10.53	50	−2.72
60	9.30	60	−2.14
70	7.63	70	−1.55
80	5.55	80	−1.03
90	3.08	90	−0.57
95	1.67	95	−0.36
100	(0.16)	100	(−0.16)
100	—	100	0

L.E. radius: 2.48
Slope of radius through L.E.: 0.20

NACA 23012
(Stations and ordinates given in per cent of aerofoil chord)

Upper surface		Lower surface	
Station	Ordinate	Station	Ordinate
0	—	0	0
1.25	2.67	1.25	−1.23
2.5	3.61	2.5	−1.71
5.0	4.91	5.0	−2.26
7.5	5.80	7.5	−2.61
10	6.43	10	−2.92
15	7.19	15	−3.50
20	7.50	20	−3.97
25	7.60	25	−4.28
30	7.55	30	−4.46

Upper surface		Lower surface	
Station	Ordinate	Station	Ordinate
40	7.14	40	−4.48
50	6.41	50	−4.17
60	5.47	60	−3.67
70	4.36	70	−3.00
80	3.08	80	−2.16
90	1.68	90	−1.23
95	0.92	95	−0.70
100	(0.13)	100	(−0.13)
100	—	100	0

L.E. radius: 1.58
Slope of radius through L.E.: 0.305

NACA 65–410
(Stations and ordinates given in per cent of aerofoil chord)

Upper surface		Lower surface	
Station	Ordinate	Station	Ordinate
0	0	0	0
0.372	0.861	0.628	−0.661
0.607	1.061	0.893	−0.781
1.089	1.372	1.411	−0.944
2.318	1.935	2.682	−1.191
4.797	2.800	5.203	−1.536
7.289	3.487	7.711	−1.791
9.788	4.067	10.212	−1.999
14.798	5.006	15.202	−2.314
19.817	5.731	20.183	−2.547
24.843	6.290	25.157	−2.710
29.872	6.702	30.128	−2.814
34.903	6.983	35.097	−2.863
39.936	7.138	40.064	−2.854
44.968	7.153	45.032	−2.772
50.000	7.018	50.000	−2.606
55.029	6.720	54.971	−2.340
60.053	6.288	59.947	−2.004
65.073	5.741	64.927	−1.621
70.085	5.099	69.915	−1.211
75.090	4.372	74.910	−0.792
80.088	3.577	79.912	−0.393

Upper surface		Lower surface	
Station	Ordinate	Station	Ordinate
85.076	2.729	84.924	−0.037
90.057	1.842	89.943	0.0226
95.029	0.937	94.971	0.327
100.000	0	100.000	0

L.E. radius: 0.687
Slope of radius through L.E.: 0168

NACA 64–415
(Stations and ordinates given in per cent of aerofoil chord)

Upper surface		Lower surface	
Station	Ordinate	Station	Ordinate
0	0	0	0
0.299	1.291	0.701	−1.091
0.526	1.579	0.974	−1.299
0.996	2.038	1.504	−1.610
2.207	2.883	2.793	−2.139
4.673	4.121	5.327	−2.857
7.162	5.075	7.838	−3.379
9.662	5.864	10.338	−3.796
14.681	7.122	15.319	−4.430
19.714	8.066	20.286	−4.882
24.756	8.771	25.244	−5.191
29.803	9.260	30.197	−5.372
34.853	9.541	35.147	−5.421
39.904	9.614	40.096	−5.330
44.954	9.414	45.046	−5.034
50.000	9.016	50.000	−4.604
55.040	8.456	54.960	−4.076
60.072	7.762	59.928	−3.478
65.096	6.954	64.904	−2.834
70.111	6.055	69.889	−2.167
75.115	5.084	74.885	−1.504
80.109	4.062	79.891	−0.878
85.092	3.020	84.908	−0.328
90.066	1.982	89.934	−0.086
95.032	0.976	94.968	−0.288
100.00	0	100.000	0

L.E. radius: 1.590
Slope of radius through L.E.: 0168

NACA 64–210
(Stations and ordinates given in per cent of aerofoil chord)

Upper surface		Lower surface	
Station	Ordinate	Station	Ordinate
0	0	0	0
0.431	0.867	0.569	−0.767
0.673	1.056	0.827	−0.916
1.163	1.354	1.337	−1.140
2.401	1.884	2.599	−1.512
4.890	2.656	5.110	−2.024
7.387	3.248	7.613	−2.400
9.887	3.736	10.113	−2.702
14.894	4.514	15.106	−3.168
19.905	5.097	20.095	−3.505
24.919	5.533	25.081	−3.743
29.934	5.836	30.066	−3.892
34.951	6.010	35.049	−3.950
39.968	6.059	40.032	−3.917
44.985	5.898	45.015	−3.748
50.000	5.689	50.000	−3.483
55.014	5.333	54.987	−3.143
60.025	4.891	59.975	−2.749
65.033	4.375	64.967	−2.315
70.038	3.799	69.962	−1.855
75.040	3.176	74.960	−1.386
80.038	2.518	79.962	−0.926
85.033	1.849	84.968	−0.503
90.024	1.188	89.977	−0.154
95.012	0.564	94.988	0.068
100.000	0	100.00	0

L.E. radius: 1.590
Slope of radius through L.E.: 0168

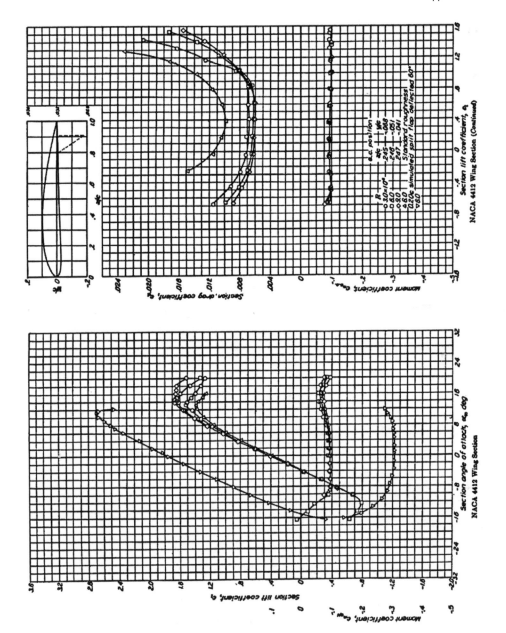

NACA 4412 Wing Section (Continued)

NACA 4412 Wing Section

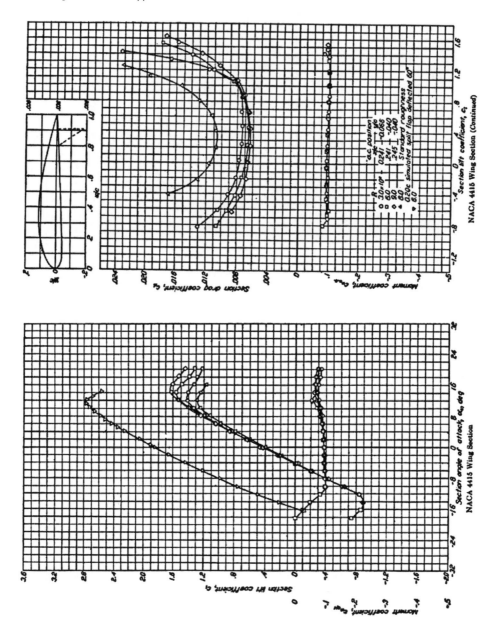

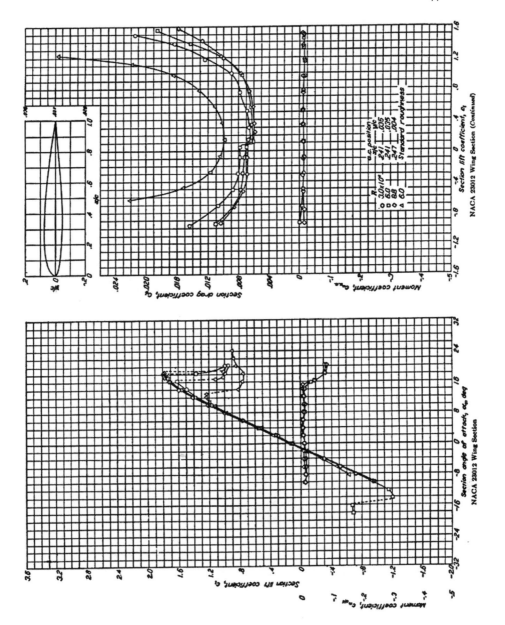

NACA 23012 Wing Section (Continued)

NACA 23012 Wing Section

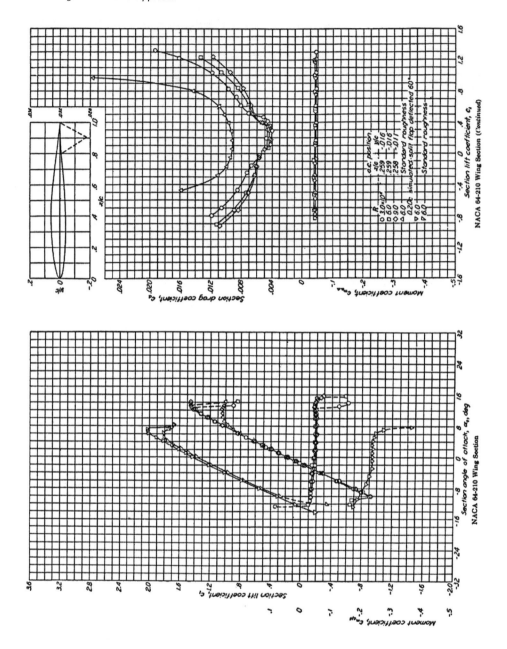

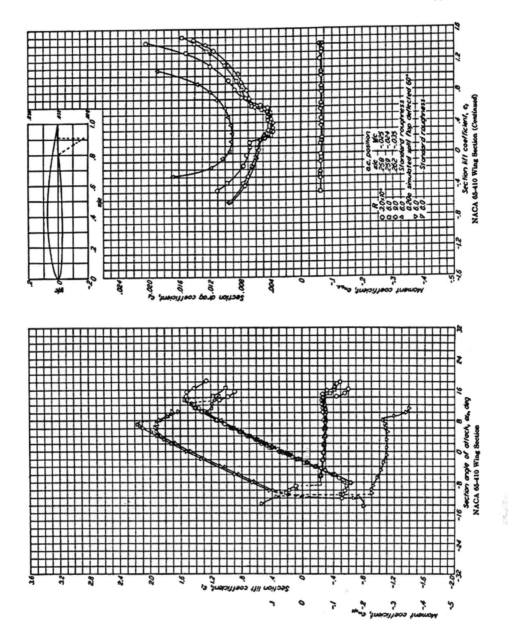

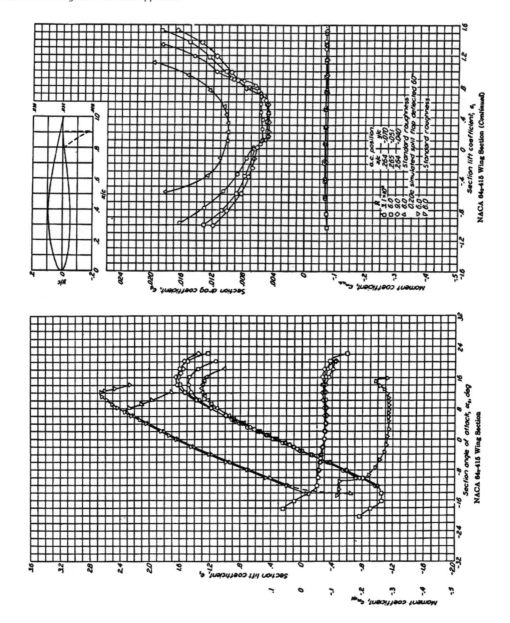

Appendix F

Wheels and Tyres

Typical tyre data/specifications used in this book are given at the end of this Appendix. The following tyre glossary and terminology given here will prove useful. Terminology, glossary and data are extracted from the following web sites. The full range of Goodyear tyre data/specifications is freely downloadable from their web site.

www.goodyearaviation.com/resources/tirecare.html
www.goodyearaviation.com/resources/pdf/tire_specifications_7_2016.pdf
www.aps-aviation.com/wp-content/uploads/goodyear-aircarft-tire-data.pdf
Figure F.1 is drawn by the author.

F.1   Glossary – Bias Tyres

Apex strip. The apex strip is a wedge of rubber affixed to the top of the bead bundle.

Bead heel. The bead heel is the outer bead edge that fits against the wheel flange.

Bead toe. The bead toe is the inner bead edge closest to the tyre centreline.

Breakers. Breakers are reinforcing plies of rubber-coated fabric placed under the buff line cushion to help protect casing plies and to strengthen and stabilise the tread area. They are considered an integral part of the casing construction. The cords of breakers are not substantially aligned with the circumference of the tyre.

Buff line cushion. The buff line cushion is made of rubber compounded to enhance the adhesion between the tread reinforcing ply and the top breaker or casing ply. This rubber layer is of sufficient thickness to allow for the removal of the old tread when the tyre is retreaded.

Casing plies. Plies are layers of rubber-coated fabric (running at alternate angles to one another), which provide the strength of the tyre.

Chafer. A chafer is a protective layer of rubber and/or fabric located between the casing plies and wheel to minimise chafing.

Chines. Also called deflectors, chines are circumferential protrusions that are moulded into the sidewall of some nose tyres that deflect water sideways to help reduce excess water ingestion into the engines. Tyres may have chines on one or both sides, depending on the number of nose tyres on the aircraft.

Flippers. These layers of rubberized fabric help anchor the bead wires to the casing.

Grooves. Circumferential recesses between the tread ribs.

Liner. In tubeless tyres, this inner layer acts as a built-in tube and helps to restrict gas from diffusing into the casing plies. For tube-type tyres the liner helps prevent tube chafing against the inside ply.

Ply turnups. Casing plies are anchored by wrapping them around the wire beads, thus forming the ply turnups.

Sidewall. The sidewall is a protective layer of flexible, weather-resistant rubber covering the outer casing ply, extending from tread edge to bead area.

Tread. The tread is the outer layer of rubber which serves as the only interface between the tyre and the ground. It provides traction for directional control and braking.

Conceptual Aircraft Design: An Industrial Approach, First Edition. Ajoy Kumar Kundu, Mark A. Price and David Riordan.
© 2019 John Wiley & Sons Ltd. Published 2019 by John Wiley & Sons Ltd.

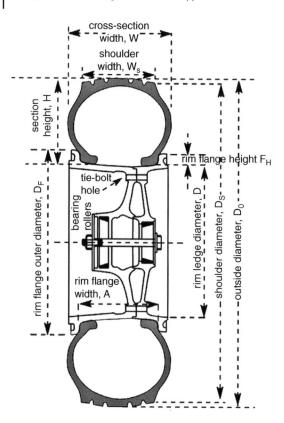

Figure F.1 Wheel and tyre nomenclature.

Tread reinforcing ply. Tread reinforcing plies are one or more layers of fabric that help strengthen and stabilise the tread area for high-speed operation. It also serves as a reference for the buffing process in retreadable tyres.

Tubes. A flexible hollow rubber ring that is inserted inside a pneumatic tyre to hold inflation pressure. Goodyear branded tubes meet SAE standard AS50141.

Wire beads. The beads are hoops of high tensile strength steel wire that anchor the casing plies and provide a firm mounting surface on the wheel.

F.2 Glossary – Radial Bias Tyres

Apex strip. The apex strip is a wedge of rubber affixed to the top of the bead bundle.

Bead heel. The bead heel is the outer bead edge that fits against the wheel flange.

Bead toe. The bead toe is the inner bead edge closest to the tyre centre line.

Belt plies. Belts are a composite structure of rubber-coated fabric that stiffen the tread area for increased landings. The belt plies increase the tyre strength in the tread area. The cords of belts are substantially aligned with the circumference of the tyre.

Buff line cushion. The buff line cushion is made of rubber compounded to enhance the adhesion between the tread reinforcing ply and the overlay. This rubber layer is of sufficient thickness to allow for the removal of the old tread when the tyre is retreaded.

Casing plies. Casing plies are layers of rubber-coated fabric that run radially from bead to bead. The casing plies help provide the strength of the tyre.

Chippers. The chippers are layers of rubber-coated fabric applied at a diagonal angle that improve the durability of the tyre in the bead area.

Chines. Also called deflectors, chines are circumferential protrusions that are moulded into the sidewall of some nose tyres that deflect water sideways to help reduce excess water ingestion into the engines. Tyres may have chines on one or both sides, depending on the number of nose tyres on the aircraft.

Grooves. Circumferential recesses between the tread ribs.

Liner. This inner layer of rubber acts as a built-in tube and helps to restrict gas from diffusing into the casing plies.

Overlay. The overlay is a layer of reinforcing rubber-coated fabric placed on top of the belts to aid in high speed operation.

Ply turnups. Casing plies are anchored by wrapping them around the wire beads, thus forming the ply turnups.

Sidewall. The sidewall is a protective layer of flexible, weather-resistant rubber covering the outer casing ply, extending from tread edge to bead area.

Tread. The tread is the outer layer of rubber which serves as the only interface between the tyre and the ground. It provides traction for directional control and braking.

Tread reinforcing ply. Tread reinforcing plies are one or more layers of rubber-coated fabric that helps strengthen and stabilise the tread area for high-speed operation. This also serves as a reference for the buffing process in retreadable tyres.

Wire beads. The beads are hoops of high tensile strength steel wire that anchor the casing plies and provide a firm mounting surface on the wheel.

F.3 Tyre Terminology

Ply rating. The term 'ply rating' is used to indicate an index to the load rating of the tyre. Years ago when tyres were made from cotton cords, 'ply rating' did indicate the actual number of plies in the carcass. With the development of higher-strength fibres such as nylon, fewer plies are needed to give an equivalent strength. Therefore, the definition of the term 'ply rating' (actual number of cotton plies) has been replaced to mean an index of carcass strength or a load carrying capacity.

Rated load. This is the maximum allowable load that the tyre can carry at the specified rated inflation pressure.

Rated pressure. Rated pressure is the maximum inflation pressure to match the load rating. Aircraft tyre pressures are given for an unloaded tyre; that is, a tyre not on an aeroplane. When the rated load is applied to the tyre, the pressure increases by 4% as a result of a reduction in air volume.

Outside diameter, D_0. This measurement is taken at the circumferential centre line of an inflated tyre.

Section width, W. This measurement is taken at the maximum cross sectional width of an inflated tyre.

Rim diameter, D. This is the nominal diameter of the wheel/rim on which the tyre is mounted.

Section height, H. This measurement can be calculated by using the following formula:

$$H = (\text{Outside Diameter} - \text{Rim Diameter})/2 = (D_0 - D)/2$$

Aspect ratio. Measure of the tyre's cross section shape. This can be calculated by the following formula: Aspect ratio = (Section Height)/(Section Width) = (H/W).

Flange height, FH. This is the height of the wheel rim flange.

Flange diameter, DF. The diameter of the wheel including the flange.

Free height. This measurement can be calculated by using the following formula:

$$\text{Free Height} = (\text{Outside Diameter} - \text{Flange Diameter})/2 = (D_0 - D_F)/2$$

Tyre shoulder: The shoulder of the tyre is where the treaded area meets the sidewall.

Static loaded radius. This is the measurement from the centre of the axle to the runway for a loaded tyre.

Loaded free height. This measurement can be calculated by using the following formula: Loaded Free Height = (Static Loaded Radius − Flange Diameter)/2.

Tyre deflection. A common term used when talking about aircraft tyres is the amount of deflection it sees when rolling under load. The term % Deflection is a calculation made using the following formula: Percent Deflection = (Free Height − Loaded Free Height)/Free Height.

Most aircraft tyres are designed to operate at 32% deflection, with some at 35%. As a comparison, cars and trucks operate in the 5–20% range.

Service load (operational load). Load on the tyre at max aircraft takeoff weight.

Service pressure (operational pressure). Corresponding pressure to provide proper deflection at service load.

Rated speed. Maximum speed to which the tyre is qualified.

F.4   Typical Tyre Data

The full range of tyre data is freely downloadable from the Goodyear web site: www.goodyearaviation.com/resources/pdf/tire_specifications_7_2016.pdf.

Given next are the tyre data used in this book representing typical tyre types currently in use. These are Type III, Type VII and the new type three-part name series also known as Type VIII, both bias and radial ply types in imperial unit and metric unit. From the Goodyear chart, only the pertinent parameters are taken, leaving out inflated tyre dimensions (used for stowage space considerations) and loaded tyre dimensions. Note: some older tyres were designated as either Type III or Type VII. The Type III tyres were mainly used by piston-prop type aircraft with the attribute of low pressure for cushioning and flotation. Type VII tyres were early generation tyres designed for jet aircraft with higher load capacity.

Bizjet worked-out example – Three-part name series, tubeless type tyres:

In the example of a Bizjet tyre 22 × 6.6–10 (18-ply), these have units in inches, 22 is the tyre diameter, 6.6 is the tyre width, the symbol '–' indicates the tyre is of bias ply construction, 10 the rim diameter and has 18 ply (can follow the nomenclature with the designation 18PR).

Main wheel

Size	Rating (ply)	Rated speed (mph)	Rated load (lbs)	Rated inflation (psi)	Max. braking load (lbs)
22 × 6.6–10	18	230 (200 kt)	10 700	260	16 050

Wheel (Rim) size	Width between flanges (in,)	Specified rim diameter (in,)	Flange height (in,)	Minimum ledge width (in,)
22 × 6.6–10	5.5	10	1.0	2.05

Nose wheel

Size	Rating (ply)	Rated speed (mph)	Rated load (lbs)	Rated inflation (psi)	Max. braking load (lbs)
17.5 × 5.75–8	14	210	6050	220	9080

Wheel (Rim) size	Width between flanges (in,)	Specified rim diameter (in,)	Flange height (in,)	Minimum ledge width (in,)
18 × 5.5	4.25	8	0.88	1.4

AJT worked-out example – metric designation (rating in imperial units), tubeless type:

Main wheel

Size	Rating (ply)	Rated speed (mph)	Rated load (lbs)	Rated inflation (psi)	Max. braking load (lb)
450 × 190–5	22	233	8880	225	13 320

Wheel (Rim) size	Width between flanges (in.)	Specified rim diameter (in.)	Flange height (inch)	Minimum ledge width (in.)
450 × 190–5	6.3	5	0.95	2.6

Nose wheel

Size	Rating (ply)	Rated speed (mph)	Rated load (lbs)	Rated inflation (psi)	Max. braking load (lbs)
450 × 190–5	10	230	3822	90	5730

Wheel (Rim) size	Width between flanges (in.)	Specified rim diameter (in.)	Flange height (inch)	Minimum ledge width (in.)
450 × 190–5	6.3	5	0.71	1.38

TPT worked-out example – Type III, tubeless type

Main wheel

Size	Rating (ply)	Rated speed (mph)	Rated load (lbs)	Rated inflation (psi)	Max. braking load (lbs)
6.50–10	8	160	4750	100	6890

Wheel (Rim) size	Width between flanges (in.)	Specified rim diameter (in.)	Flange height (in.)	Minimum ledge width (in.)
6.50–10	4.75	10	0.81	1.10

Nose wheel

Size	Rating (ply)	Rated speed (mph)	Rated load (lbs)	Rated inflation (psi)	Max. braking load (lbs)
500–4.5	6	138 (120 kt)	1650	78	2390

Wheel (Rim) size	Width between flanges (in.)	Specified rim diameter (in.)	Flange height (in.)	Minimum ledge width (in.)
500–4.5	4	4.6	0.65	0.95

Airbus 320 class aircraft (main and nose dual undercarriage) – Three prt name series, tubeless type tyres:

Main wheel

Size	Rating (ply)	Rated speed (mph)	Rated load (lbs)	Rated inflation (psi)	Max. braking load (lbs)
49 × 19–20	32	235	51 900	219	77 800

Wheel (Rim) size	Width between flanges (inch)	Specified rim diameter (inch)	Flange height (inch)	Minimum ledge width (inch)
46 × 16	13.25	20	1.88	3.75

Nose wheel (Radial tyre)

Size	Rating (ply)	Rated speed (mph)	Rated load (lbs)	Rated inflation (psi)	Max. braking load (lbs)
30 × 8.8R15	16	225	14 200	199	21 300

Wheel (Rim) size	Width between flanges (in.)	Specified rim diameter (in.)	Flange height (in.)	Minimum ledge width (in.)
30 × 8.8	7.0	15	1.13	1.65

Small Recreation Aircraft below 3000 lb MTOW (not worked out in this book) – standard tube type tyres

Main wheel

Size	Rating (ply)	Rated speed (mph)	Rated load (lbs)	Rated inflation (psi)	Max. braking load (lbs)
600 × 6	4	120	1150	29	1670

Wheel (Rim) size	Width between flanges (in.)	Specified rim diameter (in.)	Flange height (in.)	Minimum ledge width (in.)
600 × 6	5.0	6	0.75	0.8

Nose wheel

Size	Rating (ply)	Rated speed (mph)	Rated load (lbs)	Rated inflation (psi)	Max. braking load (lbs)
500 × 5	4	120	800	31	1160

Wheel (Rim) size	Width between flanges (in.)	Specified rim diameter (in.)	Flange height (in.)	Minimum ledge width (in.)
500 × 5	3.50	5	0.75	0.8

Index

a

abreast seating
  eight 193, 198, 203, 204, 493
  five 193, 202
  four 193, 195, 201, 327, 493
  nine 193, 204
  seven 193, 196, 198, 203, 204, 327
  six 193, 195, 196, 198, 202, 493
  ten 193, 195, 196, 198, 204, 493
  three 193, 200, 327
  two 193, 195, 196, 198, 200, 204, 327, 340, 440, 493, 515, 525
accelerated stop distance 830
acceleration 10, 11, 36, 37, 87–91, 319, 428, 504, 548, 555, 585, 612, 613, 637, 639–642, 645, 647, 651, 654–658, 660, 671, 673, 718–720, 724, 926
  gravity 36, 37, 428
active control technology (ACT) 262, 743, 752–754
actuator disc 569–570
additive drag 496, 526
advance ratio 567, 571, 575
advanced tactical support (ATF) 293
adverse environment 43, 826–827, 838–842, 848
aerodynamic balancing 248, 260–262
aeroelasticity 328, 333, 358, 359, 715, 720
aerofoil. see also NACA
  camber 86, 100, 103, 117, 122, 167, 263, 357, 504
  centre of pressure 105–109, 125
  characteristics 79, 91–93, 101, 106, 109, 117, 123, 148–150, 502, 503, 567
  chord 107, 115, 139, 163, 504, 948–952
  comparison 105
  five-digit 93–95, 100, 105, 118–120, 122, 124, 164
  four-digit 93–95, 103–105, 118–120, 122, 124, 164
  GAW 99, 118, 120
  mean line 92, 94, 96, 102
  section 97–98, 101, 164

  selection 101, 106, 117, 120–123, 167, 181, 329, 378, 650
  supercritical 97–101, 115, 116, 118, 164, 167, 494, 509
  thickness ratio 91, 115, 120, 162, 494, 516, 531
  thickness-to-chord 115, 123, 162, 181, 182, 272, 328, 329, 332, 344, 494, 516, 531, 628
  Whitcomb 98–100, 115, 116, 208
aerofoil design
  direct method 123–124
  inverse method 123, 124
after-burning gas turbine 350, 362, 372, 461, 548, 549, 555–556, 565, 837
aileron 49, 79, 87, 89, 139, 140, 174–175, 229–231, 236, 244, 256–259, 261, 263, 308, 328, 330, 343, 344, 358, 368, 373, 379, 382, 473, 520, 523, 718, 731, 732, 734, 742, 746, 747, 751, 858, 863, 865, 870, 939
  Frise 261
air data
  computer 710, 752, 869
  instruments 854
air defence 14, 17, 292, 293, 296, 298, 299, 301, 669
air speed
  calibrated (CAS) 633
  equivalent (EAS) 633, 645, 647, 648, 675, 708, 717, 722, 727
  ground 633
  indicated (IAS) 633
  true (TAS) 44, 589, 597, 599, 633, 645, 727, 728, 942
air staff requirement 316
air superiority 10, 14, 292–294, 350, 372, 461, 563
airborne early warning (AEW) 293
airbus xxxvi, lviii, 4, 15, 23, 69, 115, 118, 125, 134, 140, 160, 177, 180, 233, 274–278, 288, 387, 389, 406, 427, 547, 565, 581, 625, 757, 775, 852, 870, 899–901, 932, 933, 964

Conceptual Aircraft Design: An Industrial Approach, First Edition. Ajoy Kumar Kundu, Mark A. Price and David Riordan.
© 2019 John Wiley & Sons Ltd. Published 2019 by John Wiley & Sons Ltd.

aircraft
    centre of gravity   33, 87, 125, 206, 237, 315, 384,
        428–478, 608, 710, 730, lvi
    centre of pressure   125
    civil mission   352
    classification   11–13, 26–27
    classification–military   26–27
    combat   224, 353, 358, 362
    component   liii, lv, lvi, lviii, 35, 52, 67, 84, 88, 117,
        125, 127, 188, 206, 229, 252–253, 268, 272, 292,
        304–310, 318–320, 322–325, 327, 337, 348,
        353–354, 357, 363, 371, 385, 407, 416, 426,
        429–431, 436–446, 455, 457, 461, 463, 464, 468,
        471, 476, 480, 483, 488–500, 510, 511, 608, 683,
        687, 690, 692, 697, 730, 737, 747, 748, 757, 759,
        761, 766, 776, 814, 820, 824
    component groups–mass   436–443
    cost   lvii, lviii, 61, 268, 271, 609, 618, 625, 682–710,
        814, 820, 847
    design choices–civil   291–292
    evolution   10–13, 293, 443
    familiarisation   48–52
    family variants   200–205, 610, 817
    force   89, 242
    load   89–91, 274, 401, 403, 715–729
    mission   128, 162, 181, 192, 270–271, 294, 299, 541,
        628, 667, 687
    moment   33, 264, 267, 506, 737
    motion   86–88, 256, 718, 730–732, 734, 736, 744,
        752
    neutral point (NP)   33, 125, 262, 320, 433–435, 460,
        471, 476, 748
    price   23, 323, 628, 679, 683, 684, 686, 687, 701–703,
        705–707, 814, 819, 941
    sizing   lvii, lx, 10, 34, 36, 62, 65, 269, 272, 294, 316,
        323, 324, 352, 371, 426, 443, 479, 540, 581, 583,
        584, 608–629, 632, 649, 914
    specification   liv, 46, 53, 55–57, 66–71, 181, 315,
        316, 318, 321, 337, 338, 342, 344, 350, 351, 375,
        726
    speed   lxi, 9, 10, 49, 117, 120, 132, 162, 166, 167,
        170, 175, 179, 214, 225, 261, 274, 309, 324, 328,
        387, 397, 412, 413, 448, 481, 502, 503, 506, 536,
        542, 545, 549, 560–562, 564, 567, 571, 574, 580,
        581, 594, 633, 637, 638, 641, 642, 647, 660, 721,
        723, 726, 736, 757, 841, 856, 858, 878, 907, 910
    statistics–civil   271–282
    statistics–military   299–304
    turning   390, 391, 396

    variant   206, 277, 334, 418, 521, 610, 624, 628, 629,
        679, 680
aircraft classification number (ACN)   387, 405
aircraft flight-track monitoring (AFM)   709
aircraft health monitoring (AHM)   708
aircraft lighting   870
aircraft monitoring
    flight-track   709
    health   708
    performance (APM)   xxxv, 24, 668, 708, 710
aircraft performance
    balanced field length   635, 636
    climb   592, 644–647
    constant speed climb   647
    cruise   649–647
    descent   648–649
    grid   602–606
    landing   642–644
    military mission profile   669–670
    monitoring (APM)   xxxv, 24, 668, 708, 710
    payload range   649–651
    takeoff   635–641
aircraft structure. *see* structure
air-to-air refueling   293, 301
airworthiness   liv, lxi, 9, 23–25, 29, 45, 47, 53, 57, 63,
    64, 73–75, 262, 280, 286, 554, 557, 608, 619, 626,
    627, 632, 636, 639, 642, 678, 679, 743, 744, 826,
    828, 829, 831, 834, 838, 843, 844, 851, 857
aisle
    double   197, 199, 203, 274, 445
    single   29, 48, 196, 198, 200, 202, 274, 277, 445
AJT. *see also* worked-out example
    and growth potential   626
    performance   668–677
    undercarriage layout   422, 425
altimeter   44, 45, 854, 855, 869
altitude
    density   43–45
    pressure   37, 43, 44
aluminum alloy   775
amortization   610, 618, 684, 686, 689, 691, 697, 698,
    700, 807, 815
angle
    bank   89, 235, 651, 746, 747, 858
    downwash   145–149, 241, 249, 254, 734
    effective   145–147
    incidence   136, 232
    pitch   558, 566, 567, 571, 718, 731, 736
    roll   734, 736
    yaw   217, 244, 739, 746

angle of attack (AOA) probe   856–857

anhedral   134–136, 181, 182, 230, 232, 235, 239, 263, 265, 291, 328, 329, 331, 357, 367, 368, 373, 520, 735, 736, 740, 748, 754

antenna location   869

anti-icing   328, 335, 356, 360, 556, 560, 693, 694, 697, 698, 832, 839, 865, 867, 871, 874–877

area rule   52, 78, 207–208, 308

Ashby plot   771

aspect ratio (AR)   51, 66, 115, 127, 129, 138, 142, 148, 149, 159, 173, 180, 230, 249, 251, 272, 278, 280, 281, 290, 328, 329, 331, 333, 334, 342, 344–346, 356–359, 367, 368, 370, 373, 374, 378, 379, 381, 449, 473, 482, 494, 502–508, 611, 626, 727, 745, 748, 799, 801, 918, 937, 943, 944, 961

Association of European Airline (AEA)   683, 687–688, 701, 703, 941

atmosphere
   non-standard   43–44
   standard   36–43, 578

automation   19, 321, 351, 541, 821, 844–845, 848, 896

auxiliary power unit (APU)   440, 441, 453, 466, 858, 867, 868, 870, 872, 890, 945

avionics   3, 29, 63, 66, 222, 270, 294, 297, 298, 355, 362, 365, 367, 372, 427, 437, 440, 441, 453, 462, 467, 471, 477, 522, 618, 623, 687–689, 814, 815, 831, 842, 847, 848, 858, 859, 861, 868–870, 872, 885, 939

A-weighted scale   833

axes
   body-fixed   87
   wind   87–89, 644, 651

b

balance
   bevelled trailing edge   258
   control force   260–262
   horn   259, 261, 742
   overhang   258, 261
   trim tab   256, 258, 742, 863

balanced field length (BFL)   635–639, 641, 642, 654–661, 671, 942, 944

banjo frame   333, 799, 800

battery   lxi, 461, 542, 899–902, 904–908, 911–919

beam   137, 449, 759–761, 775–778, 780, 782–783, 796, 797, 801, 802, 882, 884

bending   51, 129, 159, 177, 215, 216, 219, 306, 329, 334, 357, 367, 449, 557, 611, 741, 749, 759, 760, 762, 776–778, 780, 791, 793–797, 800–803

bird strike   629, 680, 827, 830, 841–842

Bizjet. *see* worked-out example

blade element   567, 568, 571–572

Bleriot, Louis   6, 8, 788

bob weight   260–262

Boeing 737 variants, 23, 138, 180, 277, 317

Boeing 787 (F), 180, 581, 757

Boeing Sonic cruiser   167

boundary layer
   bleed   226, 227, 309, 360, 361, 562, 563
   high   81, 110, 217, 309, 502, 563, 716, 888
   laminar   83, 91
   low   83, 110, 114, 115, 205, 563, 716
   turbulent   83, 91, 888

Brayton (Joule) cycle   551

Breguet range equation   817, 912, 913, 917

brittle   766, 785

brushless DC motor (BLDC)   905, 910, 911

buckling   781–784, 797, 804

buffet   651, 716–718, 857

bulkhead   189, 190, 201, 205, 223, 326, 356, 366, 416, 419, 422, 443, 448, 694, 696, 697, 794, 795, 798

BWB. *see* blended wing body

by-pass ratio (BPR)
   high   542, 544, 548, 556, 585, 837
   low   347, 351, 362, 544, 555

by-pass ratio (turbofan)   542, 581, 585, 837

c

cabin
   air flow   873
   crew   70, 75, 200–205, 283, 285, 287, 326, 327, 337, 438, 454, 459, 461, 703, 705, 707, 829, 851, 941
   pressurisation   288, 326, 493, 515, 634, 648, 661, 663, 665, 666, 830, 858, 871–872, 934
   seat layout   197–205
   width   198–205, 338

California Bearing Ration (CBR)   403

canard   6, 51, 52, 66, 167, 170, 230–232, 237–240, 257, 258, 306–308, 337, 358, 718, 736, 738, 749, 755, 856, 864, 865

candidate aircraft   lvi, 66, 127, 229, 317, 322

canopy   223, 224, 352–354, 377, 483, 492, 493, 515, 516, 518, 525, 531, 848, 879–881, 932, 947

cargo
   container   203, 204, 207, 288
   sizes   204, 288

Carry Out Counter Insurgency (COIN)   316, 356, 375, 383, 463, 472–477, 581

case studies   lviii, lix, 67–71, 585, 690, 710, 820, 888, 892–893, 915, 932–947

caster
 angle 392
 length 392
cathode ray tube (CRT) 858
Cayley, George 4
ceiling
 absolute 716, 892
 operating 717
 service 71, 72, 908
center of gravity (CG)
 determination 468
 position 52, 206, 216–218, 238, 239, 243, 263, 315,
  320, 327, 330, 331, 337, 343, 348, 356, 357, 361,
  366–368, 372, 378, 379, 385, 395, 398, 399, 403,
  407, 414–416, 418–420, 422–425, 429, 431, 433,
  435, 436, 455, 459–461, 468, 471, 472, 476, 608,
  710, 739, 748–750
 range 753
centre
 aerodynamic 33, 78, 105–109, 125, 145, 146, 150,
  152, 162, 237, 241, 247, 307, 435, 751
 pressure 105–109, 125, 209
centripetal 38, 651
chandelle 74
Chengdu J-10 187
chord
 mean aerodynamic 107, 139–144, 209, 237, 320,
  398, 456, 490, 738
 root 51, 134, 139, 140, 143, 170, 306, 329, 334, 342,
  346, 349, 357, 368, 373, 374, 379, 381–383, 449,
  473, 494, 512, 513, 520, 521, 523, 524, 527, 943,
  944
 tip 134, 139, 143, 334, 342, 346, 349, 368, 373, 374,
  379, 381–383, 473, 512, 513, 520, 521, 523, 524,
  527, 566, 943, 944
chronology–fighter aircraft 295
circulation 43, 85, 86, 93, 145, 153, 155, 156
Civil Air Regulation (CAR) 828
civil aircraft configuration 288–291, 304, 323–324
civil *versus* military 26–27, 133–134
clearway (CWY) 629, 671, 673, 680
climb
 accelerated 645, 647, 661, 669, 675
 first segment 636–638, 661, 662
 fourth segment 637, 638
 gradient 171, 611, 612, 625, 629, 632, 636, 638, 639,
  643, 654, 656, 658, 659, 661, 674, 680, 835
 initial climb lvi, 282, 373, 374, 597, 611, 614, 616,
  617, 620, 623, 632, 634, 653, 675, 914, 915, 917
 rate of

schedule 579, 647, 675
 unaccelerated 614, 647, 671, 674, 675
 second segment 637, 638
 third segment 637, 638
close air support (CAS) 52, 292, 294, 296, 316, 463,
  519, 581, 619, 680
cobra (Pugachev) 170, 751, 755
cockpit layout. *see* flight deck
Code of Federal Regulation (CFR) 24, 73–75, 829
coefficient
 drag 52, 91, 96, 481, 482, 487, 489, 492, 498, 505,
  507, 510, 519, 526, 537, 625, 661, 664, 668, 674,
  675, 736, 938
 friction 82, 85, 396–397, 400, 480, 489–491, 501,
  523, 530, 642, 654, 672, 674
 lift 91, 96, 103, 112, 114, 116, 149, 154, 160, 162,
  173, 230, 338, 363, 375, 502, 566, 603, 615, 616,
  620, 650, 652, 661, 663, 668, 675, 721
 moment 91, 103, 105, 151, 209, 211, 213, 245, 248,
  253, 733, 734, 736
 pressure 85, 91, 112, 114, 566
 thrust 141–144, 155, 156, 240, 342, 343, 346, 349,
  357, 366, 368–370, 373, 374, 379–383, 473, 512,
  513, 520, 521, 523, 524, 566, 571, 572, 575, 595,
  625, 943, 944
Coffin corner 716–717
combat manoeuvre 267, 856
component, aircraft liii, lv, lvi, lviii, 35, 52, 67, 84, 88,
  117, 125, 127, 188, 206, 229, 252–253, 268, 272,
  292, 304–310, 318–320, 322–325, 327, 337, 348,
  353–354, 357, 363, 371, 385, 407, 416, 426,
  429–431, 436–446, 455, 457, 461, 463, 464, 468,
  471, 476, 480, 483, 488–500, 510, 511, 608, 683,
  687, 690, 692, 697, 730, 737, 747, 748, 757, 759,
  761, 766, 776, 814, 820, 824
composite
 fibre 789–791
 material 3, 84, 85, 207, 239, 428, 757, 758, 766,
  788–793, 805, 842, 867
 strength characteristics 791
compressibility
 correction 114–115, 163, 528, 633, 647
 effect 100, 113–115, 117, 118, 129, 160–167, 181,
  292, 311, 324, 480, 489, 490, 502, 503, 532, 549,
  561, 574, 581
compressible drag 115, 166
computational fluid dynamics (CFD) 25, 49, 80, 128,
  253, 310, 320, 479, 542, 615, 743, 850, 886–897
computer aided drawing (CAD) liii, lx, 27, 32, 33, 36,
  64, 66, 67, 76, 101, 124, 325, 348, 353, 354, 394,

426, 430, 440, 443, 455, 490, 542, 784, 806, 807, 810, 819, 821–823, 886, 888, 890, 895

conceptual study liii, 33, 46, 56, 57, 60, 80, 291, 322, 363, 397, 426, 629, 680, 684, 718, 806, 808, 886, 887

  phases 1–3 and 4, 33, 58, 60, 62–64, 323, 352

concrete surface 404

Congreve, William 22

conservation equations

  energy 887

  mass 887

  momentum 887

container cargo 203, 204, 207, 288

  standard 288, 289

control

  aerodynamic balancing 258, 260–262

  irreversible 258–259

  reversible 258–259

  trim tabs 259–262, 741–743

control augmented system (CAS) 752

control configured vehicle (CCV) 308

control subsystem

  electromechanical 256, 864

  mechanical control linkage 256, 863

  optical signal 256, 864

  push–pull rod type 256, 863

  wire–pulley type 256, 863

control surface 52, 66, 112, 238, 258–262, 264, 307–308, 310, 324, 330, 358, 379, 449, 494, 516, 525, 531, 710, 741, 742, 751–753, 793

control volume (CV) 80, 487, 552

controller 20, 754, 861, 862, 870, 900, 902, 904, 905

converter 907

cost

  aircraft lvii, lviii, 61, 268, 271, 609, 618, 625, 682–710, 814, 820, 847

  cash operating 23, 709

  direct operating 23, 137, 196, 271, 479, 563, 609, 683, 701–707, 813, 918

  drivers 683, 685, 686, 690, 692–695, 700

  factors 20, 687, 820

  fixed 701–703, 705, 707, 709, 918

  fraction 687–689, 701

  frame

    indirect operating 23, 196, 688

    levels 60

    life cycle 30, 294, 297, 351, 629, 680, 682, 683, 689, 690

manufacturing lvii, lviii, 13, 30, 60, 130, 136, 315, 343, 368, 379, 479, 492, 618, 682–686, 690, 691, 695–699, 702, 810, 812–814, 816

  modelling 30, 683, 685, 686, 690–701, 815

  operating 26

  trip 23, 701–703, 705–707, 709, 710

  unit 61, 193, 317, 610, 618, 695, 815, 817–819

counter insurgency (COIN) 463, 581

cowl 217, 219, 221, 227, 336, 690–701

creep 192, 697, 786, 887

cruise

  constant altitude 912

  constant speed 912

  constant thrust 912

  high speed (HSC) 118, 160, 162, 182, 248, 282, 337, 342, 373, 374, 480, 481, 483, 487, 489, 490, 502, 503, 511, 519, 538, 567, 581, 584, 611, 616, 625, 632, 649, 653, 663–664, 668, 669, 675, 708, 914, 915, 942

  long range (LRC) 122, 160, 163, 220, 240, 337, 342, 344, 480–491, 496–498, 501–503, 511, 513, 515, 519, 538, 546, 547, 565, 579, 581, 584, 649, 651, 667–668, 708, 710, 817, 932, 938, 939, 942

Curtiss 6–8, 385, 540

customer 25, 179, 291, 322, 481, 557, 608, 668, 748, 806, 889, 942

cycle

  ideal 551

  real 55, 551

d

da Vinci, Leonardo 4

d'Arlandes, Francois Laurent 4

de Rozier, Pilatre 4

decibel 833

de-fogging 839, 871, 874, 875

de-icing 43, 556, 839, 849, 874–875

density

  air 44, 45, 549, 578, 633, 716, 865

  material 769

descent

  performance 631, 632, 634, 648–649, 664–668

  rate 400, 642, 648, 664–667

  speed 666, 747

design consideration technology-driven

  management-driven 816

  manufacture-driven 816

  operator-driven 816–817

design for

  assembly 808, 814, 816

design for (*contd.*)
  Component Commonality 815
  Customer 807, 808, 815, 817–820, 824
  Life Cycle 815
  manufacture/assembly (DFM/A) lviii, 14, 29, 683, 686, 690, 692–695, 697–701, 806–811, 814, 815, 818, 822
  quality 814, 816
  recycling 815, 816
  Six Sigma 14, 756, 806, 807, 810–812, 814–816
Design-Build-Team (DBT) 58, 64, 806
detonation chord 880
dihedral 134–137, 177, 181, 182, 230, 232, 235, 239, 245, 263, 291, 311, 328, 329, 331, 333, 334, 343, 345, 349, 356–359, 367, 373, 379, 382, 431, 473, 512, 520, 523, 734, 735, 740, 745, 746, 748, 754, 755, 943–945
direct operating cost (DOC)
  breakdown 701, 702
  fixed-cost element 701–703, 705, 707, 709, 918
  formulation 703–707
  trip-cost element 701–703, 706, 707, 709
directional
  divergence 745, 746
  stability 230, 244, 245, 251, 350, 732–734, 744, 746, 753
display
  air data system (SD) 858, 859
  head down 860
  head up (HUD) 844, 860–862
  helmet mounted (HMD) 862
  multifunctional (MFD) 71, 72, 854, 858–859, 861, 869
  navigational (ND) 858, 859
  primary flight (PFD) 858, 860
disposal liv, 57, 58, 221, 236, 294, 351, 682, 689, 690, 808, 816, 825, 827, 842–843, 877
distributed electric propulsion (DEP) 910–911
dive brake. *see* speed brake
diverterless supersonic intake (DSI) 227, 563
doors
  egress 285, 286, 848, 851–853
  slides/chutes 848, 852–853
  types 851
dorsal fin 229, 264, 265, 310, 349, 350, 380
downwash
  angle 146–149, 241, 249, 254, 734
  wing 233, 240–241
drag
  actual 484, 485, 487, 526, 538, 628

base 496–498, 500, 517, 892, 935, 936
boat tail 219, 495–498, 517, 936
breakdown 482–483
canopy 492, 493, 515, 516, 525, 531, 932
dive brake/spoiler 506
empennage 495
estimation lvi, lx, 34, 70, 141, 187, 426, 480, 488–492, 495, 496, 498, 501, 508–511, 519, 522, 524, 527, 533–536, 538, 612, 656, 936
excrescence 479, 480, 491, 494, 499–501, 509, 812, 888, 889, 935, 936
flap 504, 505, 944
flat-plate equivalent 482, 489, 491, 493, 495, 498, 509, 523, 532, 932–936
form 481
formulation 487–489
fuselage 207, 367, 494, 495, 509, 936
high-lift device 481, 504–506
induced 49, 137, 145, 147–149, 153, 154, 159, 160, 162, 166, 175–179, 181, 182, 207, 235, 277, 280, 290, 291, 329, 344, 350, 479, 482, 483, 485, 486, 488, 504, 506, 510, 518–519, 533, 603, 615, 832, 888, 916, 937
intake 213, 496–499, 517, 524, 935, 936
low speed 503–508
methodology 483, 488–491
minimum parasitic 489–491
nacelle 495–499, 501
one-engine inoperative 508, 656
parabolic 484–487, 522, 526, 537, 538, 621, 628, 677–678, 911
parasite 84, 114, 139, 160, 207, 480–483, 485, 488–501, 504, 506, 518, 522, 524–526, 532, 615, 625, 812, 813, 833, 889, 932, 936, 937
profile 84, 98, 120, 344, 481, 508
propeller 508
pylon 494–495, 514, 517
spillage 496, 509, 526
spoiler 506
subsonic 510, 511, 527
supersonic 70, 160, 223, 487, 509–511, 527, 533–537
3D effects 488, 489, 491, 494, 498, 509, 517, 936
total 280, 350, 386, 479–483, 487, 501, 503, 506, 510, 519, 522, 526, 536–537, 938
trim 106, 238, 436, 499, 500, 518, 525, 526, 708, 710, 753, 895, 936, 937
undercarriage 387, 506–508
wave 99, 100, 113, 115, 118, 160, 162, 166, 186, 193, 207, 208, 324, 325, 342, 344, 479–483, 485, 486,

488–490, 502–503, 509, 510, 519, 522, 532, 537, 581, 888, 896

  wing 141, 494, 495, 738

drag polar

  actual 484–487, 526, 538, 628

  flat plate 482

  parabolic 484–487, 526, 537, 538, 621, 628, 677–678, 911

drift down 843

D-ring 878, 881

driver liv, lix, 6, 9, 23, 26, 32, 47, 54, 55, 57, 132, 135, 136, 222, 317, 363, 386, 394, 411, 431–432, 446, 447, 449, 450, 464, 491, 682, 683, 685, 686, 690, 692–695, 700, 702, 753, 766, 822, 826

ductility 766, 786

Dumas, Santos 6

duralumin 9, 428, 757, 787

Dutch roll 265, 743, 745, 746

dynamic head 82, 91, 209, 242, 633, 640, 653, 737, 856

e

Earth

  flat and stationary 87

  round and rotating 87

E-fan 900, 901

efficiency

  battery 913

  overall 544, 545

  power train 913

  propulsive 544, 545, 548, 553, 572

  thermal 544, 545

egress 224, 285, 286, 327, 848, 851–853, 878–882

ejection 27, 38, 71, 72, 224, 355, 358, 365, 366, 377, 437, 467, 470, 471, 475, 477, 522, 613, 747, 827, 844, 848, 849, 854, 878–883

Ekranoplan

electric aircraft

  analyses 911–914

  sizing 914–916

  theory 911–914

electric propulsion 542, 900, 906, 908, 911, 917

electrical subsystem 869–870

electronic flight instrument system (EFIS) 71, 72, 338, 363, 376, 602, 647, 844, 854, 858–861

electronic warfare (EW) 16, 351

elevon 231, 236

emergency power supply 867–868

emerging exit 189, 285–288, 815, 848, 851, 852, 878

emerging scenario 808, 825, 827, 845

empennage

design 120, 231, 234–237, 240, 247, 254, 262–265, 308, 330–334, 344, 357–360, 743, 749

  position 232–235

  types 234–235

end-of-life disposal liv, 825, 842–843

endurance 19, 72, 364, 845, 908, 911, 919, lxi

energy absorbed

  strut 401

  tyre 401–402

engine

  exhaust 51, 214, 215, 219, 223, 233, 234, 304, 354, 355, 360, 366, 370, 375, 422, 545, 550, 826, 830, 833, 838, 848, 849, 851, 882–884

  grid 603–605

  matched lvi, lx, lxi, 25, 33, 127, 128, 276, 315, 323, 426, 431, 577, 578, 590–594, 608–611, 624, 625, 632, 650, 748, 815

engine and fuel control subsystem

  fuel storage and flow management 865, 867

  piston engine 865–866

  turbofan system 866–867

engine installation 217, 224–228, 290, 309, 382, 497, 501, 525, 556–560

engine performance

  installed 556, 560, 577, 580, 590–594, 602, 632, 666

  military gas turbofan 593–594

  piston engine 598–602

  turbofan (civil) 581–589

  turboprop 594–598

  uninstalled 581–589, 632

engine-performance data

  turbofan (civil) 581–589

  turbofan (military) 585–587

  turboprop 587–589

engine position 17, 64, 216, 217, 310, 335, 336, 361, 370, 372

engine ratings

  flat rated 578–579

  idle 497, 584, 611, 615, 617, 648, 670, 677

  maximum climb 579, 583–584

  maximum continuous 579, 586, 588, 593, 611, 622, 638, 675, 677, 717

  maximum cruise 579

  takeoff 578, 582–583

environment adverse 43, 826–827, 838–842, 848–849, 874–878

environmental

  control system (*see* environmental control system (ECS))

environmental (*contd.*)
  issues lxi, 14, 23, 26, 27, 336, 347, 428, 693, 789,
      808, 816, 825, 826, 833–838, 848–851, 899
environmental control system (ECS) 43, 437, 440,
      441, 462, 467, 499, 515, 559, 590, 634, 648, 665,
      666, 837, 843, 844, 859, 863, 871–874, 890
equation of state 887
equivalent airspeed (EAS) 633, 645, 647, 648, 675,
      708, 717, 722, 727
equivalent single wheel load (ESWL) 402–406, 420,
      424, 426
escape slide 852–853
Euler angle 244
Euler's equation 782, 895, 927
European Aviation Safety Agency (EASA) lv, 24, 73,
      286, 825, 829–831, 851
excrescence 479, 480, 482, 493–495, 498, 500, 501,
      509, 511, 515–517, 525, 526, 531, 532, 568, 575,
      596, 812, 888, 889, 934–936
extended twin operation (ETOP) 10, 334, 825, 827,
      843

f

F22 Raptor (F) 884, 885
F117 Nighthawk (F) 16, 511, 750, 754, 884, 885
fabric 8, 394, 408, 410, 411, 428, 784, 788–791,
      959–961
factor of safety 428, 720, 721, 762, 784
family variants 200–205, 610, 619, 817, 942
fatigue 241, 634, 709, 744, 753, 786, 787, 812, 837,
      844, 845, 850
Federal Aviation Administration (FAA)
Federal Aviation Regulation (FAR) 23–25, 27, 34,
      67–70, 73–75, 118, 179, 206, 332, 340, 401, 420,
      611, 612, 614, 615, 619, 629, 631, 632, 636–641,
      655, 661, 662, 678, 680, 720, 722, 725, 726, 728,
      744, 826, 828–830, 838, 941
fence 160, 175, 176
fighter 7, 8, 17, 20, 47, 52, 128, 135, 145, 240, 258, 292,
      293, 297–299, 301, 304, 307, 309, 311, 319, 357,
      359, 393, 461, 462, 559, 562, 720, 721, 861, 883
  basic manoeuvre 17, 128, 258, 292, 293, 357, 359,
      883
fin
  dorsal 229, 264, 265, 310, 349, 350, 380
  ventral 193, 207, 254, 264–265, 310, 350, 437, 489,
      749
fineness ratio (FR) 192–194, 209, 211, 272, 325, 326,
      335, 336, 340, 341, 348, 349, 354, 366, 367, 373,
      374, 378, 382, 473, 491, 512–515, 517, 519, 521,
      523, 527, 533, 535, 751, 934, 942–944
finite element
  analyses (FEA) 804–805
  method (FEM) lix, 27, 63, 124, 542, 804–805
fixtures 130, 350, 373, 375, 408, 461, 809, 810, 821,
      824
flap 52, 175, 240, 308, 376, 436, 504, 562, 629, 657,
      661, 762, 833, 915
flare 509, 637, 640, 642, 655, 656, 672, 884
flight deck
  layout (civil) 860
  layout (military) 224
  level 363, 853
flight envelope 79, 190, 208, 214, 225, 242, 252, 259,
      260, 262, 309, 331, 361, 363, 433, 490, 557, 561,
      574, 580, 587, 632, 715, 717, 719, 723, 729, 741,
      752, 832, 844, 887, 888, 914, 938
flight levels 72, 86, 88, 140, 155, 238, 240, 261, 614,
      649, 678, 716, 717, 719, 722, 723, 726, 732, 736,
      857, 911, 912
flight test liii, lix, lxi, 17, 24, 25, 32, 34, 53, 60, 63, 67,
      74, 175, 217, 220, 227, 233, 239, 241, 253, 254,
      259, 260, 262, 264–266, 307, 317, 328, 329, 333,
      334, 344, 346, 359, 360, 370, 381, 429, 480, 484,
      488, 491, 503, 577, 627, 628, 632, 639, 640, 649,
      686, 690, 717, 718, 731, 741, 747, 749, 825, 826,
      828, 831–832, 896, 897
flotation 962
flutter 66, 216, 219, 234, 260, 715–718, 720, 725, 830,
      832
flyaway tooling 29, 807, 810
fly-by-wire (FBW) lv, 23, 52, 132, 231, 280, 318, 435,
      541, 730, 752–754, 844, 857
flying hazard 825, 827, 842
foot–pound–second (fps) liv, 34, 38, 40, 42, 48, 401,
      420, 445, 447, 463, 481, 573, 575, 596, 613,
      615–617, 619–621, 653, 654, 691
force 7, 75, 140, 207, 260, 390, 420, 481, 542, 612, 678,
      720, 777, 858, 926
  aircraft 89, 242, 741–747
fracture toughness 765, 766, 768, 771, 772, 785, 786
friction
  braking 396–397, 640, 654, 672
  ground rolling 396–397, 640
  skin 81–83, 85, 92, 104, 178, 207, 479–482,
      488–491, 495, 499, 501, 502, 508, 509, 514, 524,
      530, 934
front-closure angle 192, 207, 340, 493, 515, 524
fuel

cell   899, 902, 904, 906, 907, 913, 914, 916, 917

consumption   14, 166, 269, 444, 554, 555, 581, 597, 603, 649, 667, 708, 710, 815, 867, 936

control system   579, 793, 865–867

fraction   275, 276, 301, 443, 444

load   129, 160, 186, 237, 272, 275–276, 299, 301, 302, 365, 398, 433, 455, 467, 470, 472, 476, 477, 610, 668, 669

full authority digital electronic control (FADEC)   351, 754, 867

full authority digital engine control (FADEC)   29, 256, 294, 541, 844

fuselage

   aft-closure angle   192

   aft-closure length   189, 191, 223

   aft-end closure   51, 185, 188, 205, 207, 304, 325, 326, 340, 342, 356

   axis   128, 188, 189, 192, 222, 223, 327, 355, 366, 378, 455, 460, 559

   closure   189–192, 222, 326, 354, 355, 366, 493, 512, 515, 527, 934, 942, 944

   configuration   192, 206, 309, 319, 366

   convertible   327

   cross-section   49, 200, 201, 206, 207, 274, 798, 888

   design   50, 186, 193, 207, 283, 304, 325–327, 353–356, 366, 378, 794

   fineness ratio   192–194, 209, 211, 378, 533, 535

   front-closure angle   192, 207, 340, 493, 515, 524

   front-closure length   189, 191

   front-end closure   190–191, 205, 354, 485

   height   129, 132, 182, 185, 191, 194, 199–203, 211, 223, 326, 340, 416, 419, 424

   layout (civil)   206, 445

   layout (military)   222–224, 354–356

   length   88, 188, 189, 191–193, 196–198, 200–207, 210–212, 223, 234, 246, 263, 272, 278, 280, 283, 287, 326, 331, 340, 341, 343, 344, 349, 360, 366–368, 370, 374, 378, 382, 416, 418, 419, 421, 422, 425, 426, 447, 473, 477, 491, 512, 513, 521, 523, 527, 533, 740, 748, 851, 932, 942, 944

   military   185, 187, 222, 367

   multiboom   233

   nose cone   188, 223

   pitching moment   208–213, 245

   seating   199, 200

   twin boom   235, 749

   upsweep angle   190, 192, 206, 207, 326, 395, 512, 527, 942, 944

width   185, 197–200, 203, 205, 206, 211, 223, 224, 325, 326, 340, 355, 366, 377, 382, 417, 419, 422, 447, 473, 504, 932

zero-reference plane (ZRP)   188, 222, 327, 368, 418, 419, 422, 425, 426, 455, 460

g

Gantt chart   lxi, 47, 64, 65

gas turbine station number   546

gauges   633, 809, 841, 854, 856, 859–861, 869, 870

geopotential altitude   37, 40, 43

glove   138, 142, 160, 182, 328, 329

golf ball   81, 83, 84

graphical method   439, 445–446, 463

green house effects   45, 905

grid

   aircraft   605

   engine, propeller   594–598

   engine, turbofan   606–607

   engine, turboprop   594–598

   ground attack aircraft   292, 351, 372, 462

ground effect   133, 825, 826, 832–833

ground-effect vehicle (GEV)   19

ground rolling   396–397, 640

gust

   envelope   721, 726–729

   load   43, 158, 720, 726–729

Gustav   5

gyroscopic   239, 331, 557, 750, 755

h

hands on throttle and stick (HOTAS)   860, 862

hard turf   397

hardness   786

hazard   43, 285, 384, 388, 390, 398, 564, 642, 825, 827, 842, 848, 851, 866, 874, 876–877, 882

head down display   860

heading indicator   854

head-up display (HUD)   844, 860–862

health issues   685, 845

helical trajectory   747

helmet-mounted display (HMD)   862

hesitation roll

high speed cruise (HSC)   118, 160, 162, 182, 248, 282, 337, 342, 373, 374, 480, 481, 483, 487, 489, 490, 502, 503, 511, 519, 538, 567, 581, 584, 611, 616, 625, 632, 649, 653, 663–664, 668, 669, 675, 708, 914, 915, 942

high-lift
  device   49, 100, 109–112, 117, 123, 170–175, 177,
      182, 256, 259, 272, 276, 277, 282, 283, 321, 328,
      330, 343, 351, 356, 358, 363, 368, 375, 379, 481,
      503–506, 538, 610, 611, 616, 619, 636, 641, 671,
      739, 799, 864, 870, 892, 893, 918, 939, 944, 945
  evolution   171–172
  mechanism   171–172
hi-lo-hi   362, 670
holding   10, 36, 111, 174, 264, 412, 461, 579, 599, 645,
      667, 668, 790, 809, 819, 940, 943, 947
honeycomb   28, 696, 698, 699, 791–793, 850
Hooke's law   765, 779, 785
horizontal-tail (H-Tail)   49, 50, 103, 106, 107, 134, 208,
      229, 231–232, 255, 280, 304, 308, 331, 344–346,
      437, 450, 458, 465, 469, 474, 737, 738, 753, 800,
      895
horse power (HP)   12, 574, 582, 678, 946
H-tail sizing   208, 243, 307, 369, 380
human physiology   9, 634, 825, 843–845
humidity effects   43, 45
hybrid electric aircraft (HEA)   908–910
hydraulic subsystem   870–871
hydrostatic equations   36–43, 923–925
hypersonic
  aircraft   11, 19
  shock wave   169

i
ice accretion   28, 841
icing envelope   839, 840
induced drag   49, 137, 145, 147–149, 153, 154, 159,
      160, 162, 166, 175–177, 179, 181, 182, 207, 235,
      277, 280, 290, 291, 329, 344, 350, 479, 482, 483,
      485, 486, 488, 504, 506, 510, 518–519, 533, 603,
      615, 832, 888, 916, 937
inline assembly   29, 810
installation   17, 66, 175, 190, 217, 335, 336, 350, 361,
      437, 447, 450, 490, 497, 499, 540, 549, 555–560,
      570, 577, 580, 581, 587, 590, 592, 593, 595,–598,
      615–617, 620, 622, 694, 827, 830, 847, 856, 866,
      868, 869, 905, 917
  engine   211, 224–228, 290, 309, 382, 497, 501, 525,
      556–560, 910
instruments
  air data   84
  flight deck   lviii, 351, 839, 840, 847, 870
intake
  design (civil)   52, 214, 225, 228, 309, 310, 360, 361,
      365, 377, 557, 560–561
  design (military)   52, 225, 309–310, 360, 461,
      561–563
  diffuser   214, 497
  highlight diameter   496
  installation   217, 224–228
  integration, engine   556
  position (nacelle)   125, 187, 214–215, 228, 309–310,
      323, 360, 361, 365, 497, 545, 556, 561, 562, 580
  subsonic   361, 560–561
  supersonic   227, 309, 361, 560–564
  throat   214, 220, 361, 565 (*see also* nacelle)
  turboprop   214
  vortex   228
integrated design C_L 566, 573, 574, 594, 595, 946
integrated product and process development (IPPD)
      13–14, 27, 53, 58, 64, 123, 322, 352, 363, 683,
      690, 701, 806–808, 810
interdiction   14, 166, 292, 293, 296, 299
International Council Of Systems Engineering
      (INCOSE)   53
International Standard Atmosphere (ISA)   13, 34,
      36–39, 41, 43–45, 82, 280, 374, 383, 473, 512,
      518, 521, 524, 532, 542, 578–580, 599–602, 610,
      611, 615, 620, 632, 642, 654, 660, 831, 855,
      923–925, 937
isentropic relations   857, 929

j
Jatho, Karl   5
jet stream   43, 234, 270, 827, 838, 840
Johnson, Clarence   10, 501
joined wing aircraft   17, 18
Joint Aviation Authority (JAA)   24
Joint Unmanned Combat Air System (JUCAS)   310
Joule (Brayton), cycle   551

l
lag   727
laminar   17, 28, 78, 80–85, 93, 95, 96, 98, 118, 123,
      181, 261, 489, 490, 542
landing
  baulked   171, 610, 625, 643, 661
  distance   67–69, 71, 72, 373, 374, 615, 661, 667, 674,
      942, 944
  flare   642, 656, 672
  performance   642–644
Langley, Samuel P.   6
lateral stability   136, 230, 265, 732, 734–736, 745, 746
lazy eight   74
Lead   767, 769

leading edge root extension (LERX) 132, 183, 208, 265, 266

life cycle cost (LCC) 27, 30, 294, 297, 629, 680, 683, 688–690, 815–817

lift
  curve slope 93, 108, 111, 115, 116, 149, 151, 212
  dumper 49, 175, 330, 833

lift to drag ratio 51, 79, 128, 178, 185, 280, 281, 305, 571, 655, 656, 817, 856

lightning 350, 830, 841, 874, 876–877

Lilienthal, Otto 5

limit
  negative 724
  positive 42, 351, 401, 717, 720, 721, 723, 725, 726
  ultimate 715, 716, 720, 722, 762, 783, 785

lip contraction ratio 931

lip suction 495, 496, 526

lithium 297, 906, 908

lithium polymer battery 906

lithium-ion battery 903, 904, 907

load
  aircraft 89–91, 274, 401, 403, 715–729
  classification group (LCG) 387, 405–407
  classification number (LCN) 387, 405–407, 412, 421, 422, 425, 426
  in flight 557, 715, 719, 720
  on ground 715, 718
  negative 723–724
  positive 723
  wheel and gears 398–400

load factor
  maximum limit 720–721
  ultimate 720, 762, 938

lo-lo-lo 670

long period oscillation (LPO) 743–744

long range cruise 122, 160, 220, 240, 337, 342, 480–485, 487, 488, 490, 491, 498, 501, 503, 511, 513, 514, 519, 538, 546, 547, 579, 584, 649, 651, 668, 710, 932, 939

longeron 794

longitudinal stability 230, 237, 244, 559, 732–734, 753

loop 64, 375, 752, 844

low observable (LO) 294, 297, 298, 883–885

low speed 95, 98, 99, 103, 110, 120, 128, 129, 137, 139, 171, 175, 182, 232, 276, 277, 328, 331, 342, 357, 358, 378–380, 386, 390, 393, 409, 416, 445, 448, 482, 485, 503–508, 538, 542, 562, 611, 716, 724–725, 749, 751, 931

Lycoming Engine 601, 946

m

macadam surface 404

Mach
  cone 168, 183
  critical 100–101, 113, 114, 160, 328, 356
  number 98, 100, 113–116, 162–166, 168, 237, 245, 278, 329, 333, 357, 481, 485, 489, 490, 510, 527, 534–536, 561, 574, 575, 581, 583–585, 592, 597, 602, 603, 645, 650, 661, 675, 724, 725, 728, 896, 927–930
  wave angle 117, 167–169

maneuver 677
  turn 677

manoeuvre
  envelop 721, 725, 762
  pitch 722
  roll 367, 718, 750
  yaw 134, 153, 230, 259, 260, 380, 680, 718, 741, 750

manufacturer's empty mass (MEM) 272, 432, 437, 444, 456, 460, 466, 468, 477

manufacturing practices 807, 809–811

manufacturing process (digital) 17, 27–29, 56, 124, 188, 428, 480, 501, 682, 766, 774, 792, 807, 808, 811, 821, 824

maritime patrol 293

market 10, 46, 78, 93, 133, 193, 269–271, 315, 404, 455, 542, 608, 631, 683, 771, 819, 827, 847, 887, 899

mass estimation 377, 438, 439, 444, 446, 450, 452, 618

mass flow ratio 497, 552, 560, 569, 926

mass (weight) fraction
  analyses 444–445
  civil 443–445, 464
  method 34, 36, 438, 443–445, 461–463
  military 461–463

material aircraft 428, 431, 784, 785
  composite 492, 784, 785

matrix(ces)
  bond 790–791
  ceramic 791
  cost 56, 789, 790, 811
  metal 28
  polymer 790, 791
  selection 811

maximum takeoff mass (MTOM) 158, 177, 272–277, 291, 299–301, 303, 318–320, 328, 336, 349, 352, 362–365, 376, 386, 418, 421, 426, 431, 433, 436, 438, 439, 443, 444, 456, 457, 462, 463, 466, 468, 471, 472, 474, 476, 478, 507, 608, 609, 611, 618,

maximum takeoff mass (MTOM) (*contd.*)
628, 635, 637, 675, 679, 718, 753, 818, 834, 847, 915

maximum takeoff weight (MTOW)   lvi, 33, 74, 75, 177, 256, 272, 315, 321, 323, 324, 329, 342, 351, 353, 357, 399, 400, 402, 412, 420, 433, 436, 442, 444, 445, 447, 453, 454, 457, 459, 465, 467, 470, 474, 507, 610, 613–615, 619, 650, 684, 687, 703–705, 715, 818, 819, 828, 829, 863, 865–867, 919, 939, 964

maximum taxi mass (MTM)   433

mean aerodynamic chord (MAC)   107, 139–144, 155, 175, 237, 263, 320, 324, 346, 368, 398, 407, 419, 422, 424, 456, 468, 737, 742

metals   27, 28, 32, 63, 321, 338, 351, 363, 376, 386, 408, 421, 428, 430, 432, 756–757, 761, 767, 769, 781, 784–791, 808, 816, 842, 867, 876

military   6, 47, 93, 137, 185, 231, 239, 267, 316, 393, 430, 481, 541, 581, 610, 632, 682, 717, 741, 757, 817, 827, 848

  survey   3, 23, 47, 54–57, 74, 76

military aircraft

  component   304–310, 319, 353–354, 461, 463

  generation   297–299, 730

minimum parasite drag   207, 481, 482, 488–491, 494, 499, 501

mission   11, 17, 20, 26, 50, 52, 56, 57, 117, 128, 129, 160, 181, 185, 207, 230, 252, 270–272, 292–294, 298–300, 304, 318, 319, 321, 351, 461, 467, 540, 548, 555, 578, 586, 624, 628, 643, 668, 669, 676, 684, 703–705, 708, 841, 844, 853, 854, 867, 876, 902

mission profile

  statistics   181, 270, 292, 360, 667

  transport aircraft   667

modular concept   541, 578, 816, 869

modulus

  elastic (Young's modulus)   765–767, 771, 772, 777, 782–785

  rigidity   765, 781

moment of inertia   33, 428, 777–779, 802

Montgolfier, Joseph and Etienne   4

motion, aircraft   86–88, 256, 718, 730–732, 734, 736, 744, 752

motor   407, 839, 899, 905, 911

moving frame   26, 47

Multhopp moving frame   155, 209

multi-disciplinary analyses (MDA)   27, 58, 60, 64, 804

multi-disciplinary optimisation (MDO)   27, 58–60, 290, 322, 610, 626

multifunctional display (MFD)   71, 72, 282, 557, 844, 854, 858–860, 869

multirole fighter   293, 298, 319, 461

Munk, M. M.   209

n

nacelle

  cost   692–694, 698

  drivers   695

  pod   184–228, 266, 268, 334–336, 360

  position   216–219, 232, 234, 289, 290, 335, 559

narrow body   48, 134, 195, 196, 198–203, 274, 277

National Advisory Committee for Aeronautics (NACA)

  five-digit aerofoil   94–95, 105, 120, 122, 164

  four-digit aerofoil   93–95, 103–105, 118–120, 122, 164

  layout   344

  long-duct   338

  nacelle/intake

    drag   524

    external diameter   373

    fuselage-mounted   749

  overwing   749

  pod clearance   219, 220

  position   94

  short-duct   336

  six-digit aerofoil   95–97, 103–105, 114, 118–120, 122, 164

  underwing   367, 378

  vertical position   245

  wing tip   91, 119, 139, 812

neutral point (NP)   lvi, 78, 125, 243, 320, 419, 422, 424, 433–436, 456, 468, 471, 731, 739, 753

Newton, Isaac   4

Night Hawk, F117   16, 511, 750, 754, 884, 885

noise

  emission   335, 360, 830, 833–838, 848–851

  foot print   835, 836

  perceived   833

  propeller   849

  radiation   837

  source   28

  standards   830, 902

  suppression   557, 558, 849

noise level   220, 548, 555, 557, 565, 826, 834, 835, 837, 906

nose wheel   49, 216, 219, 291, 334, 337, 374, 382, 385–387, 390, 395, 396, 398–401, 403, 411, 415–420, 422, 424–426, 508, 642, 644, 962–964

  failed   395

nozzle   187, 213–214, 220, 225, 360–361, 370, 490, 495–497, 546–549, 552–556, 560–565, 577, 581, 935
  exhaust   187, 220–221, 354, 355, 360, 370, 371, 501, 509, 545, 548, 557, 558, 563–565, 570

o

one engine inoperative   70, 215, 251, 280, 290, 508, 557, 578, 611, 613, 614, 635–638, 640, 647, 656–662
operating cost (OC)   lvii, lxi, 23, 30, 271, 288, 322, 479, 628, 679, 683, 684, 687–690, 706, 708, 900, 905, 908, 917–918, 941
operational empty mass (OEM)   272, 274–276, 301, 319, 363, 364, 373, 374, 376, 377, 432–435, 438, 440, 444, 456, 460, 463, 466, 471, 474, 477, 521, 522, 753, 863
  fraction   177, 272, 274, 301
operator's empty weight (OEW)   272, 382, 436, 441, 462, 470, 472, 523, 628, 679, 680, 784, 939, 943, 947
  fraction   272
Oswald's efficiency factor   122, 139, 147–149, 151, 342, 367, 378, 380, 481, 485, 486, 522, 537, 538, 621
  variation   367, 485, 621
overall pressure ratio (OPR)   606–607, 704
oxygen
  supply   68, 69, 71, 72, 843, 874
  system   437, 454, 467, 470, 475, 874

p

parasite drag   84, 114, 139, 207, 480–483, 485, 488–491, 493–501, 506, 518, 522, 526, 532, 615, 812, 813, 833, 889, 932, 936, 937
pavement   390, 397, 402–405, 414. *see also* runway
  classification number (PCN)   387, 400–405
payload range   liv, lx, 14, 56, 130, 192, 270–271, 291, 299, 300, 311, 315, 323, 324, 329, 338, 590, 592, 609, 610, 618, 624, 627, 628, 631, 632, 634, 649–651, 663, 667–669, 678, 817, 932, 939–940, 943, 947
perceived noise level   833, 834
phases of project   33, 684
photo voltaic (PV)   904, 907, 919
phugoid   743–744
physiology
  flight   9, 634, 825, 843–845
  human   9, 634, 825, 843–845
Piaggio   238, 451

piston engine   12, 13, 67, 117, 215, 235, 290, 292, 297, 374, 375, 442, 445, 451, 453, 508, 540, 542, 546, 549–551, 556, 559, 565, 570, 572, 575, 577, 598–602, 678, 858, 865–866, 875, 908, 945
piston engine–supercharged   550, 599
pitch   6, 52, 87, 89, 110, 111, 189, 195, 197, 198, 200–206, 208, 230, 231, 236, 239–241, 244, 252, 253, 255, 258, 265, 283, 306–309, 336, 347, 348, 358, 566, 567, 571, 575, 594, 644, 718–721, 723, 725, 731, 732, 736, 741–743, 750, 853, 865
pitot   225, 309, 360, 367, 480, 500, 545, 559–560, 580, 812, 854, 855, 926
Pline, Joseph   4
pneumatic system   871–874
position error   633, 831
potato curve   433, 434, 468
Power Control Unit (PCU)   256, 259, 863, 870
power train
  fuel cell   906–907
  solar   907–908
Prandtl–Glauert Rule   114
probe
  angle of attack   856–857
  temperature   857
process planning   684, 821, 822, 824
Product, Process and Resource (PPR) Hub   807, 822, 823
propeller
  activity factor   567, 572, 574, 594
  advance ratio   567, 571, 575
  blade element theory   567, 571–572
  blade-pitch angle   566, 567
  constant pitch   567
  definition   566–568
  design C_L 122, 375, 378, 380, 573, 574, 594–596, 946
  momentum theory   568–570
  performance   567, 568, 571–575, 587, 594–596, 598, 947
  power   324, 353, 374, 566
  power coefficient   566, 575
  theory   540, 566, 568–572
  thrust   508, 549, 556, 566, 571, 678
  thrust coefficient   566
  type   566
  variable pitch   567, 569, 571, 575
Pugachev Cobra   170, 751, 755
push pull   256–259, 863, 864

pylon   49, 50, 71, 72, 136, 215–217, 229, 264, 266, 334–336, 348, 350, 373, 437, 440, 441, 450, 451, 460, 461, 466, 471, 490, 494–495, 498, 509, 514, 517, 518, 527, 580, 932, 934, 936, 937, 943

q

Q-corner   716–717
quasi-level   645
quasi-steady   645–648, 661, 663, 673

r

radar
   cross-section   319, 884, 885
   signature   10, 297, 358, 751, 882–885
radar cross-section signature (RCS)   10, 297, 358, 751, 882–885
Ragone plot   903
rain removal   839, 849, 874, 875
rake angle   392
Ram Air Turbine (RAT)   868, 870
Raptor F22   884, 885
rating
   idle   497, 579, 584, 592, 615, 617, 648
   installed   595
   maximum climb   579, 581, 583–593, 596–598, 603–604, 632, 661, 675, 843
   maximum cruise   579, 581, 584, 585, 587–589, 592, 596–598, 603–606, 611, 632, 649, 664, 668, 675, 717
   maximum takeoff   593, 595, 611, 622, 638
   uninstalled   595
reconnaissance   7, 8, 10, 16, 21, 292, 293, 717, 883
recovery factor   219, 225, 499, 545, 561, 562, 580, 857
regression analyses   269, 272, 274, 280, 299–304, 377, 439, 442, 445, 509
regulations   20, 24, 34, 70, 75, 146, 200, 271, 283, 632, 635, 639, 642, 717, 719, 720, 762, 825, 827–831, 834, 838, 842
reliability and maintainability (R&M)   814, 815
remotely piloted vehicles (RPVs)   lix, 19–21, 293
Request for Proposal (RFP)   55–57, 70, 181, 316, 317, 350, 362
required runway distance   396, 424, 426, 564, 637, 640, 661, 671, 827, 835
research design and development (RD&D)   55
resilience   786
return to base   64, 299, 587
revenue passenger mile (RPM)   14, 905

Reynolds number (Re)   80, 101–105, 141, 151, 179, 254, 480, 489–491, 514, 524, 530, 728, 887, 895, 932, 934
   critical   81
rib   128–130, 148, 328, 333, 356, 359, 781, 783, 795, 797, 959, 961
rigid body equation   743
roll   52, 89, 134, 135, 140, 230, 232, 244, 379, 732, 734, 744–746, 865
rudder   49, 79, 87, 89, 230, 232, 234, 235, 240, 244, 247, 251, 254, 256, 259, 261, 263, 265, 308, 331, 346, 349, 369, 370, 374, 380–383, 473, 508, 512, 521, 524, 659, 731, 732, 747, 750, 751, 799, 858, 863, 865, 870, 872, 935
rudder fixed   244–247, 249, 251
ruddervator. *see* vertical tail (V-tail)
runway
   pavement   402–405
   pavement classification   403
   safety area   841
   types   403–404

s

Schrenk's method   155–159, 801
sea-level static thrust   13, 280, 281, 622, 704, 938
Sears–Haack body   207–208
Seat mile cost   271
Samuel Henson   4
sensitivity study   291, 367, 445, 625–627, 896, lvii, lx
separation   28, 81–84, 103, 109, 115, 170, 176, 188, 190, 192, 205, 207, 214, 215, 225, 261, 301, 325, 326, 334, 360, 361, 370–372, 375, 380, 480–482, 489, 495, 497, 502, 549, 561, 716, 842, 881, 888, 889, 891, 896
service
   galley   286
   location   880
   trolley   286
shaft horse power (SHP)   324, 353, 376, 377, 549, 556, 577, 582, 587–590, 594–597, 599, 678
Shevell's method   165–166
shock
   absorber   386, 388, 397–401, 413, 417
   normal   114, 116, 309, 561–562, 928, 929
   oblique   114, 116, 309, 561–563, 927, 929, 930
shop-floor interface   823–824
short period oscillation (SPO)   743–744, 746
side stick controller (SSC)   754, 861, 862
silent aircraft   850
simple straight-through turbojet   546–548, 551–553

Six Sigma   29, 58, 61, 430, 686, 697, 756, 810–812, 816
sizing
 aircraft   lvii, lx, 10, 34, 36, 62, 65, 269, 272, 294, 316, 323, 324, 352, 371, 426, 443, 479, 540, 581, 583, 584, 608–629, 632, 649, 914
 analyses   276, 445, 617–619, 622–625, 914
sizing theory
 initial cruise   614–615, 620, 621, 914, 915
 initial rate of climb   67, 610, 614, 632
 landing   615
 takeoff   614–616, 620, 914, 916
skin
 average   491, 501, 522
 coefficient   82
 friction   81–83, 85, 92, 104, 178, 207, 479–482, 488–491, 495, 499, 501, 502, 508, 509, 530
 local   82, 83
slat   49, 105, 110, 174, 175, 182, 330, 338, 363, 376, 450, 456, 457, 465, 466, 469, 474, 495, 503, 504, 506, 531, 615, 620, 638, 754, 835, 856, 934, 939, 944
slip indicator   858
slow roll   272
solar cell   899, 904–907, 916
Sonic Cruiser   167, 268, 269
sound pressure level (SPL)   833
Space Ship One   10, 22
spar
 front   802
 multi   137, 799, 800
 rear   175, 330, 343, 368, 379, 795, 798, 802
specific density   900, 902
specific energy   lxi, 906, 916
specific fuel consumption (sfc)   13, 542–544, 556, 583–590, 592–594, 596, 597, 599, 649, 650, 668, 676, 815, 943
specific power   597, 900, 908
specific range (SR)   603, 632, 650, 664, 665, 668, 676, 709, 710, 939
specific strength   784, 786
specific thrust   542, 548, 554
speed
 brake application   636
 decision   398, 613, 629, 635–641, 655–660, 671–674, 680
 equivalent air (EAS)   633, 675, 717
 high   11, 83, 95, 99, 130, 131, 166, 238, 329, 385, 407, 411, 482, 484, 487, 495, 563, 565, 567, 593, 611, 663, 716, 717, 726, 832, 856, 883
 lift off   637, 654, 656, 659, 672, 673

 low   84, 95, 98, 110, 129, 137, 232, 276, 328, 331, 357, 358, 379, 393, 409, 416, 449, 482, 485, 491, 503, 542, 556, 611, 716, 724, 751
 minimum control   615, 636
 minimum unstick   636
 rotation   656, 659, 660, 671
speed brake. *see* dive brake
speed schedule
 climb   647, 663
 descent   666, 747
spillage drag   496, 526
spinning   394, 744–745, 747
spiral   743, 745–747, 839
splitter plate   226, 227, 309, 563
spoiler   49, 52, 171, 174, 175, 217, 256, 259, 261, 267, 308, 330, 343, 368, 379, 450, 457, 465, 469, 474, 503, 506, 637, 642, 751, 833, 856, 864, 872, 939
spring-mass   397, 398, 731, 744
square-cubic law   176–177, 277
stabilator   52, 234, 237, 258, 308, 358, 864–865
stabiliser   49, 50, 219, 230, 231, 237, 240, 241, 255, 265, 308, 337, 346, 358, 369, 380, 867
stability
 augmented system   256
 component   252–253
 design considerations   754–755
 directional   230, 244, 245, 251, 350, 732–734, 744, 746, 753
 dynamic   33, 237, 731–736
 lateral   136, 230, 265, 732, 734–736, 745, 746
 margin   231, 243, 256, 258, 265, 280, 329, 358, 424, 435, 436, 731, 739, 741, 744, 840, 864
 pitch plane   731, 733
 static   33, 107, 207, 237, 249, 251, 252, 264, 265, 278, 334, 344, 360, 369, 380, 435, 468, 559, 730–736, 746
 theory   736–740
stall
 post   120, 123, 128, 253–254, 744–745, 747
 type   109, 857
statistics   33, 47, 148, 182, 229, 268, 315, 385, 430, 521, 546, 608, 635, 741, 819, 907
stealth considerations
 heat signature   883–884
 radar signature   884–885
steel   410, 758, 766, 767, 769, 770, 778, 785–787

stick
  fixed   241–244, 249, 264, 741
  force   lix, 32, 110, 111, 171, 237, 243, 259, 261, 262, 436, 741–743
  free   241, 249
stiffness   262, 264, 328, 333, 356, 359, 401, 449, 717, 718, 720, 731, 733, 736, 743, 753, 761–763, 765, 770, 771, 783–785, 791–793, 795, 796, 803, 810
stowage space   386, 393, 394, 418, 962
strain   264, 762–765, 777, 785–786
strake   48, 51, 52, 120, 132, 208, 229, 265, 306, 307, 350, 354, 357, 380, 524–526, 856
strategic bomber   292
stratosphere   38, 41, 42, 45, 760, 794, 907
stress   15, 178, 449, 634, 682, 723, 763–765, 776–781, 783, 785–788, 791, 794, 803, 804, 817, 853, 862
stringer   782, 783, 793, 795, 797
Stringfellow, John   5
structure
  basic definitions   761
  empennage   799–800
  fuselage   50, 137, 761, 794, 797–799
  loading   759–761
  material   759, 766, 791
  properties   765–766
  wing   137, 159, 180, 477, 795, 797, 799, 801
strut
  energy absorption   401
  length   393, 395, 398, 417
subsonic
  high   15, 85, 101, 114, 115, 118, 120, 123, 130, 134, 136, 139, 162, 164, 167, 170, 171, 175, 179, 180, 182, 186, 193, 207, 268, 270, 273, 276, 290, 332, 344, 367, 409, 463, 484–487, 502, 538, 542, 545, 548, 580, 633, 684
  low   98, 114, 118, 123, 164, 179–181, 186, 214, 465, 911
subsystem   53, 384, 388, 752, 814, 848, 862, 863, 870, 872
supersonic
  aircraft   131–132, 223, 561–563, 867
  drag   70, 160, 223, 487, 509–511, 527, 533–536
  effects   115–117, 208
  wing planform   131, 168, 170
surface imperfection   500–501
Sutherland's equation   40, 41
system   9, 50, 87, 154, 186, 240, 268, 315, 384, 436, 500, 540, 577, 613, 661, 682, 728, 731, 759, 806, 827, 847, 895, 899
  aircraft   29, 53, 54, 186, 577, 821, 847–885

t
tail
  arm   52, 232–235, 237–240, 263, 331, 333, 337, 338, 344, 346, 347, 358, 359, 369, 370, 373, 374, 380, 448, 477, 521, 741, 749, 755
  asymmetric   232, 244, 331, 396, 747, 750
  canard   749
  circular   234, 235
  cruciform   232–234, 254, 450, 465, 799
  high   266
  high tee   232–234
  horizontal (H-tail)   49, 50, 103, 106, 107, 134, 208, 229, 231–232, 255, 280, 304, 308, 331, 333, 344–346, 437, 450, 458, 465, 469, 474, 737, 738, 753, 800, 895
  low   49, 134, 232, 233, 339, 344, 387, 450, 465, 799
  mid   232–234, 254, 450, 465, 799
  multi boom   235
  position   lxii, 325, 354, 356
  single boom   232–235
  statistics   237, 239, 252
  tee-tail (T-tail)
    high   232–234
    low   134, 799
    shielding   240
    vee tail (ruddervator)   234–236, 265, 291, 751
    vertical (V-tail)   15, 49, 132, 219, 229, 272, 330, 440, 449, 559, 734, 799, 932
    volume coefficients   233, 237–240, 263, 278–280, 332–334, 344, 346, 358, 359, 369, 380, 738, 749
    Y-tail   234–236
  twin boom   235
taileron   52, 237, 308, 358
tailets   265
takeoff
  balanced field length (BFL)   656–660
  distance   67–69, 71, 72, 373, 374, 671, 942, 944
  run   396, 830
temperature
  ambient   43, 45, 414, 511, 557, 578, 599, 635, 646, 857
  static   857, 858, 928, 929
  total   927, 928, 930
throttle   27, 89, 166, 259, 319, 360, 370, 495–497, 561, 562, 565, 578, 579, 581, 584, 586, 592, 594, 597, 599, 602, 648, 666, 675, 676, 835, 862, 865, 866, 902, 935
thrust loading   268, 280, 282, 294, 302, 303, 324, 353, 609, 611, 617, 622, 623, 627–629, 632, 639, 640, 677–679, 681, 907, 914, 939

thrust reverser
  effect on stopping distance   564, 635, 637, 656
  efflux pattern   221–222
  military   360, 437, 642, 833, 870
  types   336
thrust vectoring   28, 258, 565, 751, 864
time frame   55, 61–62, 64, 65, 695, 831
Tippu, Sultan   22
titanium   219, 325, 432, 696, 758, 766, 767, 769, 774,
      778, 786–788
tolerance   14, 28, 53, 58, 99, 101, 124, 291, 318, 428,
      499, 501, 634, 686, 692, 693, 700, 757, 784, 808,
      810, 812–814, 816, 820, 823, 833, 935, 936
Torenbeek's method   163–165, 457
torque   239, 263, 331, 571, 749, 755, 761, 765,
      779–781, 905, 911, 917
torsion   233, 328, 333, 356, 359, 761, 776, 778–781,
      796, 798, 799, 803
transformation   88, 822
transonic
  aircraft   28, 167, 208
  effect   114–115, 207–208, 510
  wing   167, 168
trim   230, 238, 280, 333, 358, 435, 436, 499, 526, 708,
      730, 738, 741, 749, 862
  tab   256, 259–262, 264, 330, 741–743, 863, 864
trim drag   106, 238, 436, 499, 500, 518, 525, 526, 708,
      710, 753, 895, 936, 937
tropopause   38, 40, 620, 634, 646–648, 840
troposphere   38, 40, 645
true air speed   44, 89, 567, 589, 596, 597, 599, 633, 645,
      651, 727, 858, 942
T-s diagram   555
tube   214, 217, 367, 390, 411, 779, 780, 855, 959, 964
turbofan engine   68, 69, 375, 546, 577, 586, 604, 691,
      697, 866–867
turbojet engine   551–554
turboprop engine   12, 15, 334, 376–378, 451, 453, 526,
      546, 548–550, 556, 577, 581, 587–589, 597, 634,
      678
turboprop trainer aircraft (TPT)   lv, lx, 71, 123, 316,
      318, 320, 374–383, 416, 425–426, 472–477,
      522–526, 537, 538, 581, 590, 609, 677–678, 963
turbulent   80–85, 91, 489, 490, 513, 514, 530, 827, 835,
      838, 840, 888, 893, 895
turn performance
  coordinated   230, 235, 236, 244, 261, 651, 652, 858
  indicator   854
  rate   610, 611, 623, 631
  skid   858
  slip   858
turning of aircraft   396
tyre
  bias ply   409–412, 420
  braking friction   396–397, 640, 654, 672
  carcass ply   411
  classification   390, 412–413
  construction   409
  deflection   402, 413–414, 962
  energy absorbed   401–402
  inflation   402, 412
  loading   405
  material   410
  nomenclature   411–414, 960
  radial ply   409–412
  selection   402, 404, 407, 420, 421, 425
tyre types classification   412–413

u
undercarriage
  bogey type   388, 402, 507
  collapsed position   395
  data   382, 431, 590
  design driver   394–396
  drag   387, 506–508
  energy absorbed   400–402
  free position   394
  layout   384, 386, 387, 391, 395, 414–419, 422, 425,
      718, lxi
  nomenclature   391–393
  normal position   394
  nose wheel   49, 337, 382, 385–387, 390, 395,
      398–401, 403, 411, 415–420, 422, 424, 425, 508,
      575, 964
  retraction   388, 393–394, 638
  stowage   386, 394
  tail-dragging type   49, 385
  tri-cycle type   337, 385, 387, 388, 416, 418, 422, 426,
      452, 466
  types   321, 351, 386, 387
units and dimension   34
unmanned aircraft
  system (UAS)   20–22, 293, 844–845, 918
  vehicle (UAV)   lix, 15, 20–21, 75, 235, 236, 293, 311,
      844–845, 899, 900, 917, 918
unmanned aircraft vehicle (UAV)   lix, 15, 20–21, 75,
      235, 236, 293, 311, 844–845, 899, 900, 917, 918
utilisation   56, 85, 288, 671, 703, 705, 709, 814, 823,
      900, 905, 918, 942
utility system   877

V

vectors   38, 87, 93, 105, 244, 245, 265, 294, 351, 358, 640, 644, 651, 734, 736, 754, 891

velocity   81, 85, 147, 153, 170, 217, 447, 487, 522, 546, 552, 554, 569, 570, 596, 603, 666, 718, 735, 744, 855, 891, 927, 931

ventral fin   193, 207, 254, 264–265, 310, 350, 437, 489, 749

verified design space   14, 46

vertical speed indicator (VSI)   854, 857

vertical tail (V-tail)   15, 16, 49, 50, 132, 229–233, 235, 236, 239, 240, 244–255, 258, 263, 265, 280, 291, 304, 308, 311, 330–333, 338, 344–347, 349, 350, 358, 359, 369, 370, 374, 376, 380–383, 440, 441, 450, 451, 458, 460, 462, 465, 469, 471, 473, 475, 477, 489, 498, 512, 514, 517, 518, 521, 524–527, 529, 532, 559, 734, 736, 739–741, 745, 746, 749–751, 753, 755, 799, 800, 932, 934–935, 937, 939, 943, 945, 947. *see also* tail, vertical (V-tail)

    shielding   240, 331, 333, 358

viscosity   36, 40–43, 79–83, 146, 481, 510, 533, 923–925

V-n diagram   lvii, 329, 357, 430, 447, 715, 716, 718–726

    speed limits   722–723

voice operated control (VOC)   862

von Braun, W.   22

von Ohain, Hans   540

vortex

    generator   132, 175, 176, 207, 215, 310

    lift   52, 120, 138, 170, 229, 265, 266, 307, 358

W

water and waste system   879

wave drag   99, 100, 113, 115, 118, 160, 162, 166, 186, 193, 207, 208, 324, 325, 342, 344, 479–483, 485, 488–490, 502–503, 510, 519, 522, 532, 537, 581, 888, 896

wearability   786

weave

    plain   789, 790

    satin   789, 790

    twill   789, 790

web chart   lxii, 47

wedge

    2D   929–930

    3D   114, 929

weight. *see also* mass (weight) fraction

    breakdown   276, 432–433, 462

data   430, 445, 452, 692

driver   431–432

estimation graphical   439

mass fraction method   lx, 36, 438, 443–445, 461–463

semi-empirical method   34–35, 439, 440, 502, 628

Weisskopf, Gustav   5, 540

wetted area   177–178, 488–494, 498, 513–514, 517, 518, 524–526, 528–529, 532, 934, 935, 937

wheel

    alignment   392

    arrangement   390, 403, 406, 414, 417

    assembly   407–408

    base   222, 391, 414–419, 421, 422, 426, 477

    camber   219, 392

    caster   393

    demountable wheel flange assembly   407, 408

    divided split wheel assembly   407–408

    foot print   402, 410, 412, 414

    load   lvi, 398–400, 403, 405, 414, 418, 420, 422, 424, 426

    loaded radius   392, 961

    offset   388, 392

    rake   392

    retraction   49, 385–388, 390, 393, 394, 416–419, 424, 426

    rolling radius   392

    stability   392

    stowage   49, 386, 393, 394, 417, 418, 962

    toe-in/out   393

    track   lvi, 133, 134, 391, 396, 414–419, 421, 422, 424, 426, 477

Whitcomb aerofoil   115, 116

White Knight   22

Whitte, Frank   540

wide body   15, 48, 69, 134, 195–200, 203–205, 274, 277, 585

wind shear   43, 827, 838, 840

wind tunnel   117, 124, 148, 165, 167, 169, 172, 175, 217, 220, 253, 255, 262, 320, 324, 328, 329, 334, 335, 344, 360, 368, 370, 378, 381, 480, 484, 488, 489, 491, 501–503, 568, 628, 639, 717, 718, 743, 747, 748, 750, 832, 887, 891, 896, 897, 931

wing

    anhedral   740

    area   91, 110, 137, 138, 141–144, 147, 154–157, 162, 165, 170, 171, 174, 177–180, 211, 239, 249, 252, 268, 276–280, 291, 301–302, 304, 311, 319, 321, 323, 324, 339, 344, 345, 351, 353, 364, 367, 369, 377–380, 449, 468, 490, 494, 495, 499, 504, 521,

527, 608, 611, 619, 624, 628, 640, 653, 675, 679, 737, 753, 801, 819, 907, 918
  box   49, 134, 136, 137, 355, 378, 619, 748, 797–799, 867
  configuration   49, 132, 230, 237, 368, 803
  dihedral   134, 291, 328, 329, 343, 735, 740, 748, 754
  generic   128–132, 357
  group   49, 272, 305–308, 436, 449–450, 456, 466
  high   49, 51, 132, 133, 135, 137, 181, 182, 290, 291, 306, 321, 328, 351, 361, 362, 367, 368, 373, 387, 393, 416, 417, 449, 493, 520, 531, 559, 611, 628, 679, 735, 748, 755
  layout   50, 456, 753, 754, 800, 802
  loading   156, 177, 182, 237, 276–278, 294, 301, 303, 324, 367, 401, 438, 443, 445, 522, 611, 612, 617, 623, 624, 627, 632, 678, 679, 907, 914, 916
  low   lxii, 64, 132, 133, 137, 181, 182, 215, 291, 311, 321, 337, 351, 361, 363, 375, 378, 379, 382, 426, 452, 466, 473, 494, 523, 559, 611, 735, 736, 745, 748, 754, 755, 833, 918
  planform   115, 117, 120, 125, 128, 140, 145, 148, 156, 162, 167, 171, 177, 178, 183, 268, 276, 291, 324, 330, 331, 342, 371, 382, 796, 803
  reference area   137, 138, 140, 162, 209, 232, 240, 280, 304, 328, 356, 357, 364, 431, 445, 524, 532, 608, 609, 626, 737, 741, 748, 932
  root chord   51, 170, 329, 357, 449
  shape   445
  span   101, 122, 123, 134, 139, 142, 143, 145, 153–155, 168, 169, 174, 178, 180, 239, 246, 249, 272, 278, 280, 330, 342–344, 346, 357, 367–369, 378, 380, 449, 488, 504, 527, 536, 611, 740, 801, 833, 911, 918
  stall   109, 139–140, 151, 152, 255, 747, 856
  sweep angle   137, 138
  taper ratio   137, 139, 144, 449, 748, 918
  three-surface   306
  tip chord   134, 139
  twist   66, 139–140, 158, 181, 182, 307, 328, 329, 342–344, 357, 367, 780
  two-surface   306
  variable sweep   132, 166–167
wing design
  planform shape   115, 120, 128–132, 145, 148, 162, 178, 183, 263, 278, 296, 305, 342, 356, 357, 367, 378
  position   263, 291, 311, 324, 329, 330, 333, 335, 343, 351, 353, 357, 379, 398, 416, 417, 435, 740, 798
wing downwash   233, 240–241
wing planform area

  cranked   130
  elliptical   128, 145–148, 153, 154, 159
  generic   128–132, 357
  rectangular   138, 151, 179, 180
  tapered   128, 179, 332
wing reference   66, 137, 138, 140, 162, 209, 232, 240, 280, 304, 328, 356, 357, 364, 431, 445, 524, 532, 608, 609, 626, 737, 741, 748, 932
  area   66, 137, 138, 140, 162, 209, 232, 240, 280, 304, 328, 356, 357, 364, 431, 445, 524, 532, 608, 609, 626, 737, 741, 748, 932
winglets   48, 49, 175, 176, 182, 291, 350, 436–437, 450, 457, 465, 489, 494–495, 939
wire pulley   256–258, 863, 864
wood   390, 428, 432, 756, 757, 766–767, 769, 784, 787–789
worked-out example
  AJT
    baseline   71, 72, 422, 519
    CG location   471–472
    discussion   629
    drag evaluation   523
    fuselage layout   365
    growth potential   197, 626
    performance   123, 581, 593, 631, 680
    sizing   629, 679, 681
    variant (CAS)   52, 373–374
    wing   366–368, 373, 469, 626
  Bizjet
    baseline aircraft   lx, 345, 347, 349
    CG location   457, 460
    design for customer   818
    discussion   679–680
    DOC estimation   705
    drag evaluation   523
    empennage layout   344–348
    finalized configuration   249, 367, 370, 640, 683
    fuselage layout   330, 457, 515
    nacelle design   345
    performance   653, 654
    sizing analysis   679
    variants   619
    weight and CG analysis   65, 668
    weight data   459–460
    wing layout   344, 457–458
  propeller performance   594, 597
  supersonic drag evaluation   509–511, 533–536
  turboprop performance   596–598
Wright brothers (Orville and Wilber)   4–6, 230, 540, 827

y

yaw   87, 89, 214, 215, 225, 230–232, 235, 244, 245,
       259–261, 263, 265, 309, 333, 358, 380, 508, 559,
       718, 732, 739, 741, 744–746, 750
Yeager, Chuck   10
Yehudi   138, 142, 160, 182, 329, 932, 933
yield point   785
Young's Modulus   765–767, 771, 772, 777, 782–785

z

zero fuel weight   407
zero reference plane   188, 222, 327, 330, 343, 368,
       418–422, 424–426, 455, 460, 461, 476
zero wind   640, 642, 826, 831

Printed and bound by CPI Group (UK) Ltd, Croydon, CR0 4YY

17/04/2025

14658850-0001